INTRODUCTION TO
GENETIC ANALYSIS

About the Authors

Anthony J. F. Griffiths is a professor of botany, emeritus, at the University of British Columbia. His research focuses on the developmental genetics of fungi using the model fungus *Neurospora crassa*. He has served as president of the Genetics Society of Canada and secretary-general of the International Genetics Federation.

Susan R. Wessler is Distinguished Professor of Genetics in the Department of Botany and Plant Sciences at the University of California, Riverside. Her research focuses on plant transposable elements and their contribution to gene and genome evolution. Dr. Wessler was elected to the National Academy of Sciences in 1998. As a Howard Hughes Medical Institute Professor, she developed and teaches a series of dynamic genome courses in which undergraduates can experience the excitement of scientific discovery.

Sean B. Carroll is an Investigator at the Howard Hughes Medical Institute and Professor of Molecular Biology and Genetics at the University of Wisconsin-Madison. Dr. Carroll is a leader in the field of evolutionary developmental biology and was elected to the National Academy of Sciences in 2007. He is also the author of *Endless Forms Most Beautiful: The Making of the Fittest*, and *Remarkable Creatures*, a finalist for the National Book Award, nonfiction, in 2009.

John Doebley is a professor of genetics at the University of Wisconsin-Madison. He studies the genetics of crop domestication using the methods of population and quantitative genetics. He was elected to the National Academy of Sciences in 2003 and served as the president of the American Genetic Association in 2005. He teaches general genetics and evolutionary genetics at the University of Wisconsin.

INTRODUCTION TO
GENETIC ANALYSIS

TENTH EDITION

Anthony J. F. Griffiths
University of British Columbia

Susan R. Wessler
University of California, Riverside

Sean B. Carroll
Howard Hughes Medical Institute
University of Wisconsin-Madison

John Doebley
University of Wisconsin-Madison

■**=** W. H. Freeman and Company • New York

Executive Editor	Susan Winslow
Associate Director of Marketing	Debbie Clare
Senior Development Editor	Susan Moran
Editorial Assistants	Brittany Murphy, Heidi Bamatter
Supplements Editor	Anna Bristow
Media Editor	Aaron Gass
Publisher	Kate Ahr Parker
Art Director	Diana Blume
Senior Project Editor	Mary Louise Byrd
Photo Editor	Bianca Moscatelli
Illustrations	Dragonfly Media Group
Illustration Coordinator	Janice Donnola
Text Design	Cambraia Fernandes
Layout	Marsha Cohen
Production Coordinator	Paul Rohloff
Composition	Prepare Inc.
Printing and Binding	Quad Graphics

Library of Congress Control Number: 2010936856

Hardcover: ISBN-13: 978-1-4292-2943-8
              ISBN-10: 1-4292-2943-8

Loose Leaf: ISBN-13: 978-1-4292-7277-5
              ISBN-10: 1-4292-7277-5

Printed in the United States of America

Fourth printing

W. H. Freeman and Company
41 Madison Avenue
New York, NY 10010
Houndmills, Basingstoke RG21 6XS, England

www.whfreeman.com

Contents in Brief

Contents

Preface

S ince its first edition in 1974, *Introduction to Genetic Analysis* has emphasized the power and incisiveness of the genetic approach in biological research and its applications. Over its many editions, the text has continuously expanded its coverage as the power of traditional genetic analysis has been extended with the introduction of recombinant DNA technology and then genomics. In the tenth edition, we incorporate the practice of modern genetics into new chapters on population genetics and the inheritance of complex traits.

John Doebley Joins the Author Team

We are delighted to welcome a new coauthor, John Doebley, to the author team. Dr. Doebley is a professor of genetics at the University of Wisconsin-Madison, where he teaches the genetics course with Sean Carroll. Dr. Doebley is an active research worker and teacher in the field of population and evolutionary genetics. His research group is trying to understand the genetic basis of the evolution of new morphological traits in plants.

Completely rewritten and updated chapters on population genetics, quantitative genetics, and evolutionary genetics

In the last several years, the field of genomics has advanced through the completion of full genomic sequences for many species and the development of a variety of genomic-scale analytical methods. These advances in genomics have revolutionized many areas of genetics, especially population and quantitative genetics. The tenth edition of *Introduction to Genetic Analysis* includes completely revised chapters on population and quantitative genetics, written by John Doebley, that integrate classical theory with cutting-edge genomic tools. These chapters examine the application of modern population and quantitative genetics to help the student understand (1) how genetic variation is patterned in human populations, (2) how human populations have adapted to different regions of the world, (3) how DNA forensics is used in criminal trials, (4) how inbreeding is managed in zoo populations, and (5) how the genes contributing to the risk of common diseases can be identified.

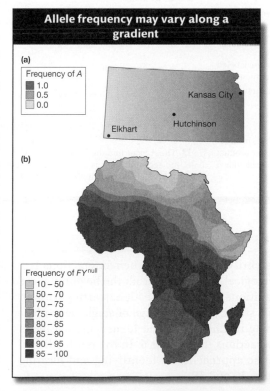

Figure 18-11 (a) Allele frequency variation across Kansas for a hypothetical species of wild sunflower. (b) Frequency variation for the FY^{null} allele of the Duffy blood group locus in Africa. [*From P. C. Sabeti et al., Science 312, 2006, 1614–1620.*]

In addition, the final chapter on the evolution of genes and traits has been heavily revised by Sean Carroll to reflect recent advances in understanding how adaptations arise. New topics include the classic story of the evolution of malarial resistance in humans, multistep evolutionary pathways, and the loss of characters through adaptive changes in regulatory sequences (illustrated by pelvic reduction in fish populations). Together, the chapter's many lucid examples provide a firm empirical foundation for the theory of evolution by natural selection.

Increased focus on visual learning and working with data

A new set of problems titled **"Working with the Figures,"** included at the back of each chapter, asks students incisive questions about figures in the chapter. We have found that students often underappreciate the wealth of information and insight provided by text figures. The new questions encourage students to spend more time thinking about the figures as a way of deepening their understanding of key concepts and analytical methods.

PROBLEMS

Most of the problems are also available for review/grading through the GENETICSPORTAL www.yourgeneticsportal.com

WORKING WITH THE FIGURES

1. In Figure 4-3, would there be any noncrossover meiotic products in the meiosis illustrated? If so, what colors would they be in the color convention used?

2. In Figure 4-6, why does the diagram not show meioses in which two crossovers occur between the same two chromatids (such as the two inner ones)?

3. In Figure 4-8, some meiotic products are labeled parental. Which parent is being referred to in this terminology?

4. In Figure 4-9, why is only locus A shown in a constant position?

5. In Figure 4-10, what is the mean frequency of crossovers per meiosis in the region A–B? The region B–C?

6. In Figure 4-11, is it true to say that from such a cross the product $v\ cv^+$ can have two different origins?

7. In Figure 4-14, in the bottom row four colors are labeled SCO. Why are they not all the same size (frequency)?

8. Using the conventions of Figure 4-15, draw parents and progeny classes from a cross

$$P\ M''''/p\ M' \times p\ M'/p\ M''''$$

9. In Figure 4-17, draw the arrangements of alleles in an octad from a similar meiosis in which the upper product of the first division segregated in an upside-down manner at the second division.

10. In Figure 4-19, what would be the RF between A/a and B/b in a cross in which purely by chance all meioses had four-strand double crossovers in that region?

11. **a.** In Figure 4-21, let GC = A and AT = a, then draw the fungal octad that would result from the final structure (5).

    b. (Challenging) Insert some closely linked flanking markers into the diagram, say P/p to the left and Q/q to the right (assume either cis or trans arrangements). Assume neither of these loci show non-Mendelian segregation. Then draw the final octad based on the structure in part 5.

BASIC PROBLEMS

12. A plant of genotype

$$\frac{A \qquad B}{a \qquad b}$$

is testcrossed with

$$\frac{a \qquad b}{a \qquad b}$$

If the two loci are 10 m.u. apart, what proportion of progeny will be AB/ab?

13. The A locus and the D locus are so tightly linked that no recombination is ever observed between them. If Ad/Ad is crossed with aD/aD and the F_1 is intercrossed,

---

New coverage of modern genetic analysis

One of our goals is to show how identifying genes and their interactions is a powerful tool for understanding biological properties. From the beginning, the student follows the process of a traditional genetic dissection, starting with an overview in Chapter 1, followed by a step-by-step coverage of single-gene identification in Chapter 2, gene mapping in Chapter 4, and identifying pathways and networks by studying gene interactions in Chapter 6. In the tenth edition, we add coverage of the new genomic approaches to identifying and locating genes, which are explored in Chapters 10 and 19.

- A reconceptualized Chapter 1 now provides an overview of how modern genetics works and of the kinds of powerful insights genetics provides that have revolutionized not only biology but many aspects of human society.

- Molecular markers, which are essential to gene identification, are introduced in Chapter 4, in a heavily revised section that introduces the common types of molecular markers and describes how they are detected and mapped. This early introduction of molecular markers helps students understand them as genetic elements that can be mapped just like genes.

- A new section on fine-mapping in Chapter 10 ("Gene Isolation and Manipulation") introduces the genome-based method for gene identification.

- Chapter 19 ("The Inheritance of Complex Traits") discusses using quantitative trait loci (QTL) mapping to locate QTL in the genome and fine-mapping to identify single genes.

- Whole-genome association mapping, used to scan the genome for QTL, is discussed in Chapter 19.

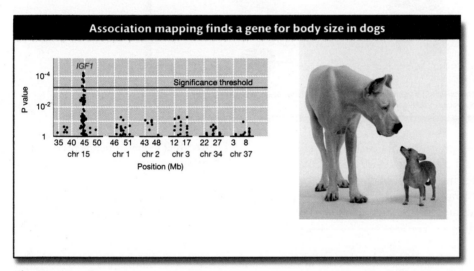

Figure 19-18 Results from an association-mapping experiment for body size in dogs. Each dot in the plot represents the P value for a test of association between an SNP and body size. Dots above the "threshold line" show evidence for a statistically significant association. [*Photo: Tetra Images/Corbis.*]

Focus on key advances in genetics

We have enhanced coverage of several cutting-edge topics in the tenth edition.

Functional RNAs: Coverage of functional RNAs is now woven into the text in multiple chapters:

- Chapter 8 ("RNA: Transcription and Processing") introduces functional RNAs, including new mention of piwi-interacting RNAs and ncRNAs and new discussion of the discovery of miRNAs and their processing in the cell.

- Chapter 12 ("Regulation of Gene Expression in Eukaryotes") ends with a new section on the role of miRNAs in post-transcriptional gene regulation.

- Chapter 15 ("The Dynamic Genome: Transposable Elements") explores the role of the RNAi silencing pathway in preventing the spread of transposable elements and the ability of some transposons, such as MITEs, to evade silencing.

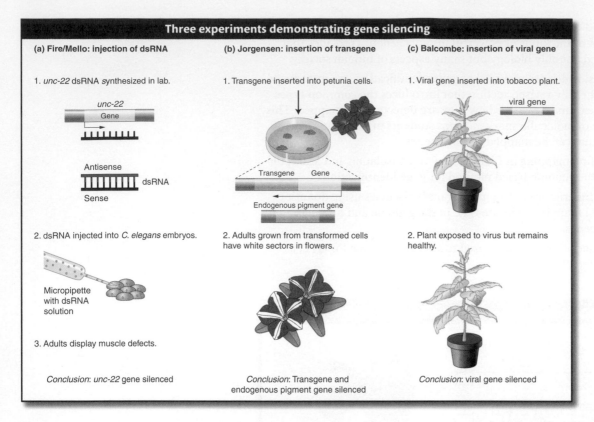

Figure 8-21 Three experiments reveal key features of gene silencing.

Modern Techniques: The techniques chapter appears earlier in the book, as Chapter 10 in the tenth edition. This chapter introduces students to "retail" techniques commonly used in labs, while the first section of the genomics chapter (Chapter 14) focuses on the "wholesale" techniques used for massive genome sequencing projects. Both chapters have been updated to include modern methods used for solving genetics problems, including

- Using PCR in the construction of recombinant DNA molecules and clones
- Finding genes through fine-mapping
- Pyrosequencing
- Next-generation whole-genome sequencing

Comparative Genomics: In revised Chapter 14 ("Genomes and Genomics"), we have enhanced coverage of how comparative genomics informs genetic analysis and reveals crucial differences between organisms.

- A new section, "Phylogenetic Inference," looks at using phylogenies to determine which genomic elements have been gained or lost during evolution.
- A new section, "Comparative Genomics of Humans," examines copy number variations and how they may be adaptive in some populations.
- A **new** discussion of the comparative genomics of color vision contrasts mice and humans.

Enduring Features

Coverage of Model Organisms

The tenth edition retains the enhanced coverage of model systems in formats that are practical and flexible for both students and instructors.

- Chapter 1 introduces some key genetic model organisms and highlights some of the successes achieved through their use.

- Model Organism boxes presented in context where appropriate provide additional information about the organism in nature and its use experimentally.

- A Brief Guide to Model Organisms, at the back of the book, provides quick access to essential, practical information about the uses of specific model organisms in research studies.

- An Index to Model Organisms, on the endpapers at the back of the book, provides chapter-by-chapter page references to discussions of specific organisms in the text, enabling instructors and students to easily find and assemble comparative information across organisms.

Problem Sets

No matter how clear the exposition, deep understanding requires the student to personally engage with the material. Hence our efforts to encourage student problem solving. Building on its focus on genetic analysis, the tenth edition provides students with opportunities to practice problem-solving skills—both in the text and online through the following features:

- **Versatile Problem Sets.** Problems span the full range of degrees of difficulty. They are categorized according to level of difficulty—basic or challenging.

- **NEW Working with the Figures.** A new set of problems included at the back of each chapter asks students pointed questions about figures in the chapter. These questions encourage students to think about the figures and help them to assess their understanding of key concepts.

- **Solved Problems.** Found at the end of each chapter, these worked examples illustrate how geneticists apply principles to experimental data.

- **Unpacking the Problems.** A genetics problem draws on a complex matrix of concepts and information. "Unpacking the Problem" helps students learn to approach problem solving strategically, one step at a time, concept on concept.

- **NEW GENETICS PORTAL** Multiple-choice versions of the end-of-chapter problems are available on our online GeneticsPortal for quick gradable quizzing and easily gradable homework assignments. The **Unpacking the Problem tutorials** from the text have been converted to in-depth online tutorials and expanded to help students learn to solve problems and think like a geneticist.

How Genetics Is Practiced Today

A feature called "What Geneticists Are Doing Today" suggests how genetic techniques are being used today to answer specific biological questions such as, "What is the link between telomere shortening and aging?" or, "How can we find missing components in a specific biological pathway?"

Increased Coverage of Modern Experiments

Building on the text's traditional focus on classical experiments, the molecular chapters present the evidence and reasoning that led to some of the more recent advances. These include the discovery of RNAi and advances in our understanding of eukaryotic gene regulation.

Media and Supplements

GENETICS PORTAL

The *GeneticsPortal* is a dynamic, fully integrated learning environment that brings together all the teaching and learning resources in one place: it features the fully interactive eBook, end-of-chapter practice problems now assignable as homework, animations, and tutorials to help students with difficult-to-visualize concepts.

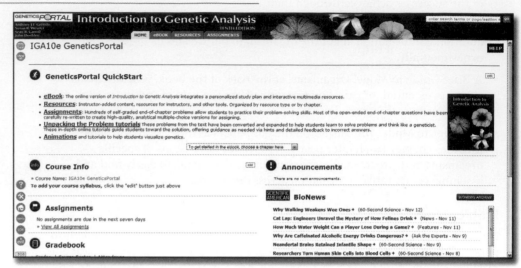

This learning system also includes easy-to-use, powerful assessment tracking and grading tools, a personalized calendar, an announcement center, and communication tools all in one place to help you manage your course. Some examples:

- **Hundreds of self-graded end-of-chapter problems** allow students to practice their problem-solving skills. Most of the open-ended end-of-chapter questions have been carefully rewritten to create high-quality, analytical multiple-choice versions for assigning.

- **Animations** help students visualize genetics.

- **Unpacking the Problem tutorials** from the text have been converted and expanded to help students learn to solve problems and think like a geneticist. These in-depth online tutorials guide students toward the solution, offering guidance as needed via hints and detailed feedback.

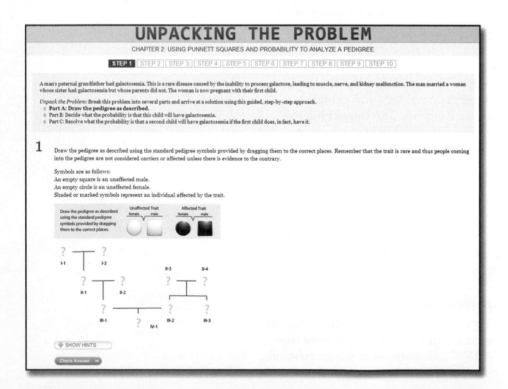

TEACHING RESOURCES FOR INSTRUCTORS

Electronic teaching resources are available from three sources: the GeneticsPortal, the Instructor's Resource DVD, and the Instructor's Resource Web site.

GENETICS PORTAL at http://courses.bfwpub.com/iga10e

Includes all the electronic resources listed below for teachers. Contact your W. H. Freeman sales representative to learn how to log on as an instructor.

Instructor's Resource DVD

(ISBN: 1-4292-7000-4)

The IRDVD contains all text images in PowerPoint and JPEG formats, 45 FLASH Animations, Clicker Questions, Layered PowerPoints, and Assessment Bank.

Password-Protected Instructor's Resource Web Site at www.whfreeman.com/iga10e

Includes all of the electronic resources for teachers listed below except the e-book. Contact your W. H. Freeman sales representative to learn how to log on as an instructor.

Electronic Resources

eBook

(ISBN: 1-4292-3257-9)

The eBook fully integrates the text and its interactive media in a format that features a variety of helpful study tools (full-text, Google-style searching; note taking; bookmarking; highlighting; and more). Available as a stand-alone item or on the GeneticsPortal.

Clicker Questions

Jump-start discussions, illuminate important points, and promote better conceptual understanding during lectures.

Layered PowerPoints

Illuminate challenging topics for students by deconstructing intricate genetic concepts, sequences, and processes step-by-step in a visual format.

All Images from the Text

More than 500 illustrations can be downloaded as JPEGs and PowerPoints. Use high-resolution images with enlarged labels to project clearly for lecture hall presentations. Additionally, these JPEG and PowerPoint files are available without labels for easy customization in PowerPoint.

45 Step-Through and Continuous Play FLASH Animations

These animations were authored by Anthony Griffiths in conjunction with BioStudio Visual Communications. The complete list of animations appears on page xxi.

Assessment Bank

This resource brings together a wide selection of genetics problems for use in testing, homework assignments, or in-class activities. Searchable by topic and provided in MS Word format, the assessment bank offers a high level of flexibility.

Student Solutions Manual

(ISBN: 1-4292-3255-2)

The Student Solutions Manual contains complete worked-out solutions to all the problems in the textbook, including the "Unpacking the Problem" exercises. Available on the GeneticsPortal and the Instructor's Web site as easy-to-print Word files.

Print Resources

Overhead Transparency Set (ISBN: 1-4292-7566-9)
The full-color overhead transparency set contains 150 key illustrations from the text with enlarged labels that project more clearly for lecture hall presentation.

Understanding Genetics: Strategies for Teachers and Learners in Universities and High Schools
(ISBN: 0-7167-5216-6)
Written by Anthony Griffiths and Jolie-Mayer Smith, this collection of articles focuses on problem solving and describes methods for helping students improve their ability to process and integrate new information.

RESOURCES FOR STUDENTS

GENETICS PORTAL at http://courses.bfwpub.com/iga10e

GeneticsPortal 6-month Access Card (ISBN: 1-4292-5320-7)
    The GeneticsPortal contains the following resources for students:

- *Self-Graded End-of-Chapter Problems:* To allow students to practice their problem-solving skills, most of the open-ended end-of-chapter questions have been carefully rewritten to create high-quality, analytical multiple-choice versions for assigning.

- *Online Practice Tests:* Students can test their understanding and receive immediate feedback by answering online questions that cover the core concepts in each chapter. Questions are page referenced to the text for easy review of the material.

- *Step-Through and Continuous Play FLASH Animations:* These animations were authored by Anthony Griffiths in conjunction with BioStudio Visual Communications. The complete list of animations appears on the facing page.

- *Interactive "Unpacking the Problem":* An exercise from the problem set for many chapters is available online in interactive form. As with the text version, each Web-based "Unpacking the Problem" uses a series of questions to step students through the thought processes needed to solve a problem. The online version offers immediate feedback to students as they work through the problems as well as convenient tracking and grading functions. Authored by Craig Berezowsky, University of British Columbia.

Book Companion Site at http://www.whfreeman.com/iga10e
(ISBN: 1-4292-6999-5)
The free book companion site offers the 45 FLASH animations listed above and the **Practice Tests**. Students can test their understanding and receive immediate feedback by answering online questions in **Practice Tests** that cover the core concepts in each chapter.

Student Solutions Manual (ISBN: 1-4292-3255-2)
The Solutions Manual contains complete worked-out solutions to all the problems in the textbook, including the "Unpacking the Problem" exercises. Used in conjunction with the text, this manual is one of the best ways to develop a fuller appreciation of genetic principles.

Other genomic and bioinformatic resources for students:
Text Appendix A, *Genetic Nomenclature,* lists model organisms and their nomenclature.
Text Appendix B, *Bioinformatic Resources for Genetics and Genomics,* builds on the theme of introducing students to the latest genetic research tools by providing students with some valuable starting points for exploring the rapidly expanding universe of online resources for genetics and genomics.

Animations

Forty-five animations developed by Anthony Griffiths are fully integrated with the content and figures in the text chapters. These animations are available on the GeneticsPortal and the Book Companion site.

CHAPTER 1
Three-Dimensional Structure of Nuclear Chromosomes (Figure 1-10)
RFLP Analysis and Gene Mapping

CHAPTER 2
Mitosis (Chapter Appendix 2-1)
Meiosis (Chapter Appendix 2-2)

CHAPTER 3
Meiotic Recombination Between Unlinked Genes by Independent Assortment
  (Figures 3-8 and 3-13)

CHAPTER 4
Meiotic Recombination Between Linked Genes by Crossing Over
  (Figure 4-7)
A Mechanism of Crossing Over: A Heteroduplex Model (Figure 4-21)
A Mechanism of Crossing Over: Genetic Consequences of the Heteroduplex
  Model

CHAPTER 5
Bacterial Conjugation and Mapping by Recombination (Figures 5-11 and 5-17)

CHAPTER 6
Interactions Between Alleles at the Molecular Level, *RR:* Wild-Type
Interactions Between Alleles at the Molecular Level, *rr:* Homozygous
Recessive, Null Mutation
Interactions Between Alleles at the Molecular Level, *r'r':* Homozygous Recessive,
  Leaky Mutation
Interactions Between Alleles at the Molecular Level, *Rr:* Heterozygous,
  Complete Dominance

CHAPTER 7
DNA Replication: The Nucleotide Polymerization Process (Figure 7-15)
DNA Replication: Coordination of Leading and Lagging Strand Synthesis
  (Figure 7-20)
DNA Replication: Replication of a Chromosome (Figure 7-24)

CHAPTER 8
Transcription (Figure 8-4)

CHAPTER 9
Translation: Peptide-Bond Formation (Figure 9-2)
Translation: The Three Steps of Translation (Figure 9-16)
Nonsense Suppression at the Molecular Level: The *rod^{ns}* Nonsense Mutation
  (Figure 9-18)
Nonsense Suppression at the Molecular Level: The tRNA Nonsense Suppressor
  (Figure 9-18)
Nonsense Suppression at the Molecular Level: Nonsense Suppression of the *rod^{ns}*
  Allele (Figure 9-18)

CHAPTER 10
Polymerase Chain Reaction (Figure 10-3)
Finding Specific Cloned Genes by Functional Complementation: Functional
  Complementation of the Gal⁻ Yeast Strain and Recovery of the Wild-Type
  GAL gene

Acknowledgments

We extend our thanks and gratitude to our colleagues who reviewed this edition
and whose insights and advice were most helpful:

Joshua Akey, *University of Washington*
Jonathan Arnold, *University of Georgia*
Nicanor Austriaco, *Providence College*
Paul Babitzke, *Penn State University*
Miriam Barlow, *University of California, Merced*
Isabelle Barrette-Ng, *University of Calgary*

Craig Berezowsky, *University of British Columbia*
David A. Bird, *Mount Royal University*
Clifton P. Bishop, *West Virginia University*
Kerry Bloom, *University of North Carolina, Chapel Hill*
Jay Brewster, *Pepperdine University*
Randy Brewton, *University of Tennessee*

Mirjana M. Brockett, *Georgia Institute of Technology*
Judy Brusslan, *California State University, Long Beach*
Michael A. Buratovich, *Spring Arbor University*
Soochin Cho, *Creighton University*
Matthew H. Collier, *Wittenberg University*
Erin J. Cram, *Northeastern University*
Kenneth A. Curr, *California State University, East Bay*
Ann Marie Davison, *Kwantlen Polytechnic University*
Kim Dej, *McMaster University*
Michael Deyholos, *University of Alberta*
Christine M. Fleet, *Emory & Henry College*
Kimberly Gallagher, *University of Pennsylvania*
Michael A. Gilchrist, *University of Tennessee, Knoxville*
Jamie Lyman Gingerich, *University of Wisconsin, Eau-Claire*
Paul Goldstein, *University of Texas, El Paso*
Julie Goodliffe, *University of North Carolina, Charlotte*
Thomas A. Grigliatti, *University of British Columbia*
Bruce Haggard, *Hendrix College*
Jody L. Hall, *Brown University*
Mike Harrington, *University of Alberta*
Donna Hazelwood, *Dakota State University*
Deborah Hettinger, *Texas Lutheran University*
Brian A. Hyatt, *Bethel University*
Glenn H. Kageyama, *California State Polytechnic University, Pomona*
Pamela Kalas, *University of British Columbia*
Kathleen Karrer, *Marquette University*
Elena L. Keeling, *California Polytechnic State University*
Michele C. Kieke, *Concordia University, St. Paul*
Dubear Kroening, *University of Wisconsin, Fox Valley*
James A. Langeland, *Kalamazoo College*
Janine LeBlanc-Straceski, *Merrimack College*
Brenda G. Leicht, *University of Iowa*
Steven W. L'Hernault, *Emory University*
Stefan Maas, *Lehigh University*
Jeffrey Marcus, *Western Kentucky University*
Michael Martin, *John Carroll University*

Andrew G. McCubbin, *Washington State University*
Debra M. McDonough, *University of New England*
R. A. McGowan, *Memorial University of Newfoundland, St. John's*
Thomas M. McGuire, *Penn State University, Abington*
Leilani M. Miller, *Santa Clara University*
Erin R. Morris, *Baker University*
Rebecca J. Mroczek-Williamson, *University of Arkansas, Fort Smith*
Todd C. Nickle, *Mount Royal University*
Thomas R. Peavy, *California State University, Sacramento*
Michael Perlin, *University of Louisville*
Lynn A. Petrullo, *College of New Rochelle*
David K. Peyton, *Morehead State University*
Jeffrey L. Reinking, *SUNY, New Paltz*
Turk Rhen, *University of North Dakota*
Inder Saxena, *University of Texas at Austin*
Daniel Schoen, *McGill University*
David Scott, *South Carolina State University*
Rebecca L. Seipelt, *Middle Tennessee State University*
Bin Shuai, *Wichita State University*
Elaine A. Sia, *University of Rochester*
Loren C. Skow, *Texas A&M University*
Christopher Somers, *University of Regina*
Marc Spingola, *University of Missouri, St. Louis*
Michael Stock, *Grant MacEwan College, City Centre Campus*
Jared L. Strasburg, *Washington University*
Aram D. Stump, *Adelphi University*
Dan Szymanski, *Purdue University*
Frans E. Tax, *University of Arizona*
Justin Thackeray, *Clark University*
Laura G. Vallier, *Hofstra University*
Jacob Varkey, *Humboldt State University*
Michael K. Watters, *Valparaiso University*
Marta L. Wayne, *University of Florida*
Darla J. Wise, *Concord University*

Sean Carroll would like to thank Leanne Olds for help with the artwork for Chapters 11, 12, 13, 14, and 20. John Doebley would like to thank his University of Wisconsin colleagues Bill Engels, Carter Denniston, and Jim Crow, who shaped his approach to teaching genetics.

The authors also thank the team at W. H. Freeman for their hard work and patience. In particular we thank our developmental editor, Susan Moran; executive editor Susan Winslow; senior project editor Mary Louise Byrd; and copy editor Karen Taschek. We also thank Paul Rohloff, production coordinator; Diana Blume, art director; Marsha Cohen, who designed the page layouts; Bill Page and Janice Donnola, illustration coordinators; Bianca Moscatelli, photo editor; Aaron Gass, media editor; Anna Bristow and Brittany Murphy, supplements editors; and Brittany Murphy and Heidi Bamatter, editorial assistants. Finally, we especially appreciate the marketing and sales efforts of Debbie Clare, associate director of marketing, and the entire sales force.

The Genetics Revolution in the Life Sciences

1

Genetic variation in the color of corn kernels. Each kernel represents a separate individual with a distinct genetic makeup. The photograph symbolizes the history of humanity's interest in heredity. Humans were breeding corn thousands of years before the advent of the modern discipline of genetics. Extending this heritage, corn today is one of the main research organisms in classical and molecular genetics. [*William Sheridan, University of North Dakota; photograph by Travis Amos.*]

Our planet Earth is teeming with life (Figure 1-1), and this living world has been a subject of great curiosity and investigation since the dawn of civilization. However, in the last 60 years a revolution has taken place in our understanding of the living world, and the foundation for this revolution has been the discoveries of genetic research. Today, most of the main questions of biology have been answered through genetics, largely through an understanding of molecular and cellular mechanisms centered on *DNA*. The molecule **deoxyribonucleic acid (DNA)** is the central topic of interest to geneticists, but it has also become a kind of logo for all of the life sciences. The unfolding of our understanding of the nature of DNA and how it operates has not only provided

Figure 1-1 The richness, complexity, and beauty of life have inspired the questions that drive research in biology.

Life on Earth

basic answers to the core questions of all areas of biology, but has also led to spectacular applications in many areas of human endeavor such as medicine and agriculture.

In this chapter, we will provide an overview of the genetics revolution and how it has come to pass. In doing so, we will see how the old mists have parted, leaving us with a clear view of the central processes of life at the cell, organism, and population levels.

First, we need to define genetics. Broadly, *genetics is the study of all aspects of genes.* In turn, **genes** are defined as the fundamental units of biological information. They can be thought of as the words in the language of the living process. Much was known about genes before the discovery of DNA, but now we know that in virtually all cases, genes are composed of DNA. Thus, the discovery of DNA led biological science into a realm called **molecular genetics.** By and large, molecular genetics deals with genes one or a few at a time. However, more recent technological innovations have led to **genomics,** the study of complete gene sets (called **genomes**). Hence, information can be analyzed not only at the level of the "words," but at the more complex level of life's "sentences" and "grammar." Today, the term *genetics* embraces both molecular genetics and genomics.

1.1 The Nature of Biological Information

Life on Earth is represented by all the organisms currently living on the planet. One of the most fascinating properties of life is that it regenerates itself every generation from single cells such as zygotes (fertilized eggs). This regeneration has been going on since the origin of life, and every organism on Earth today, from the smallest such as bacteria to the largest such as whales, is a result of millions of cycles of regeneration. This simple observation has led biologists down the ages to wonder what kind of information is inside these single cells that gives them the ability to rebuild a complex adult organism. The word *information* literally means "that which is necessary to give form." Hence, the question was, *"What constitutes biological information?"* Since the early twentieth century, scientists reasoned that in animals and plants, the information must lie within chromosomes, the worm-shaped, densely staining bodies found within

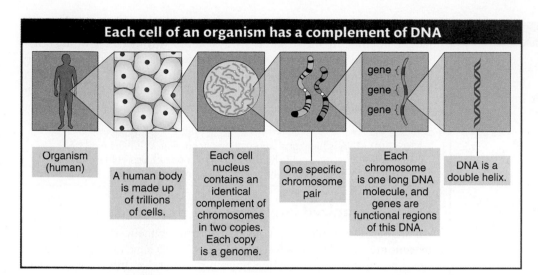

Each cell of an organism has a complement of DNA

Organism (human)

A human body is made up of trillions of cells.

Each cell nucleus contains an identical complement of chromosomes in two copies. Each copy is a genome.

One specific chromosome pair

Each chromosome is one long DNA molecule, and genes are functional regions of this DNA.

gene {
gene {
gene {

DNA is a double helix.

Figure 1-2 Successive enlargements bring the DNA of an organism into focus.

the nuclei of cells (Figure 1-2). Chromosomes were considered likely information carriers because they are passed intact from generation to generation through precisely orchestrated nuclear divisions called *meiosis* and *mitosis*.

In the 1940s, several lines of research showed that the element that carries biological information within the chromosomes is the molecule DNA. Eventually, the detailed molecular structure of DNA was elucidated by James Watson and Francis Crick in the 1950s. They inferred from this structure that DNA contains information written in a **genetic code.** DNA is a linear series of four molecular building blocks called **nucleotides.** The specific *sequence* of nucleotides constitutes the language of the code. DNA, as part of the chromosome, is passed intact from one generation to the next, so all cells in each generation contain the same set of DNA with the same information-containing nucleotide sequence. Hence, one of the big secrets of life had been answered: the architectural blueprint for life is DNA. This discovery was to be a key step in the genetics revolution. In order to see how DNA plays its role, we need to understand its structure and its arrangement in cells.

Before embarking on our broad view of the current state of genetics, it is worth mentioning that most of the items covered in this chapter will be revisited for detailed treatment in subsequent chapters; the treatment here is descriptive rather than analytical, and the goal is to provide a general overview of the subject.

The molecular structure of DNA

A molecule of DNA is made up of two long molecular strands of nucleotides wound around each other in a double helix (Figure 1-3). There are four different kinds of nucleotides in DNA: each nucleotide has a deoxyribose sugar, a phosphate group, and a nitrogenous base. The sugars and phosphates are identical in each nucleotide, but there are four different bases: **adenine (A), thymine (T), guanine (G),** and **cytosine (C).** In each strand, the sugars and phosphate groups form a chain rather like the sides of a ladder. The bases face the center, and each base is hydrogen bonded

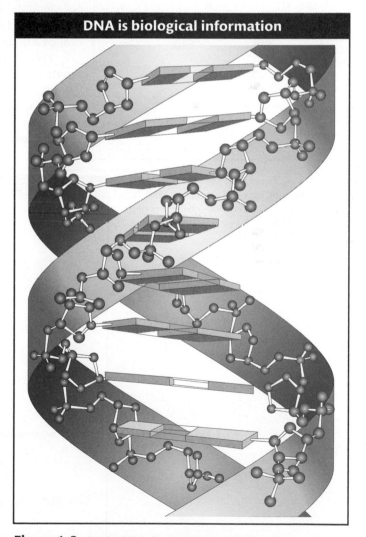

DNA is biological information

Figure 1-3 The double helical structure of DNA, showing the sugar-phosphate backbones in blue and paired bases in brown.

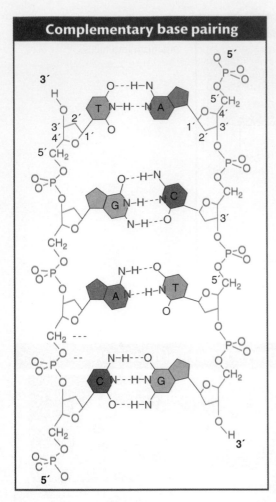

Figure 1-4 A flattened representation of DNA showing how A always pairs with T and G with C. Each row of dots between the bases represents a hydrogen bond.

to the base facing it in the opposite strand to constitute the "rungs" of the ladder: adenine in one strand is always paired with thymine in the other, whereas guanine is always paired with cytosine. This bonding specificity is based on complementarity of shape and charge. It is the sequence of A, T, G, and C on one strand that represents the coded information carried by the DNA molecule (Figure 1-4).

> **Message** DNA is biological information encoded as a sequence of nucleotides. DNA is a double helix of two nucleotide chains held together by complementary pairing of A with T and of G with C.

DNA is organized into genes and chromosomes

An organism's complete set of genetic information, encoded in its DNA, is its **genome.** In eukaryotes (organisms whose cells have nuclei), the bulk of the genome is found in the nuclei, each of which has the same DNA content. The nuclear DNA is divided into physically separate pieces, each a long double helix. An individual chromosome (Figure 1-5) contains just one of these double helices in a highly coiled condition. The set of chromosomes in organisms from the same species has a characteristic number of chromosomes and appearance. An example is seen in Figure 1-6, which shows the chromosomes of a cell from one species of a small Indian deer called a muntjac.

This illustration reveals some interesting general features of chromosomes. The lower part shows the chromosomes from one nucleus, spread out as a result of breaking the nuclear membrane. The chromosomes have been stained by special fluorescent molecular probes called chromosome paints. In this preparation, the probes were designed so that each type of chromosome is painted the same color. This staining reveals that the total of

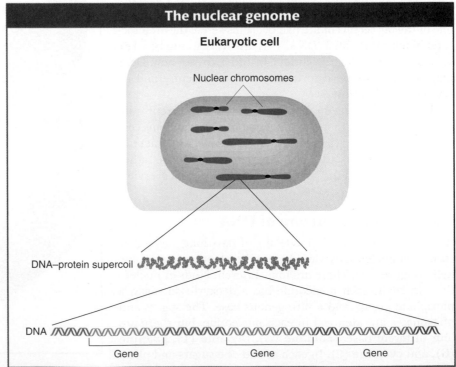

Figure 1-5 The nuclear genome is composed of a species-specific number of chromosomes. One chromosomal region has been expanded to show the arrangement of genes.

six chromosomes is actually two sets of three—a pair of red, a pair of green, and a pair of violet chromosomes. The presence of these pairs points to an important feature of the nuclear genetic material of most animals and plants. Namely, these organisms are **diploid,** meaning that their nuclei contain two complete copies of the genome and so two identical chromosome sets. The number of chromosomes in the basic genomic set is called the **haploid number** (designated n), which for the muntjac is 3. Hence, for this muntjac deer, the diploid state is designated $2n = 6$. Human beings also are diploid, but we have two copies of 23 distinct chromosomes, so in our case $n = 23$ and $2n = 46$. Many eukaryotes such as fungi are **haploid;** that is, their nuclei contain just one chromosome set. For example, the bread mold *Neurospora* is haploid, and $n = 7$. In a diploid, the two members of a chromosome pair are called **homologous chromosomes** or sometimes just **homologs.** The DNA sequences of the members of a homologous pair are virtually the same, even though minor variation in the nucleotide sequence is often present.

A diploid genome visualized

Figure 1-6 The nuclear genome in cells of a female Indian muntjac, a type of small deer ($2n = 6$). The six visible chromosomes are from a cell caught in the process of nuclear division. The three pairs of chromosomes have been stained with chromosome-specific DNA probes, each tagged with a different fluorescent dye (chromosome paint). A nucleus derived from another cell is at the stage between divisions. [*Photograph provided by Fengtang Yang and Malcolm Ferguson-Smith of Cambridge University. Appeared as the cover of Chromosome Research vol. 6, no. 3, April 1998.*]

Each chromosomal DNA molecule contains many functional regions called *genes.* Thus, genes are just segments along one continuous DNA molecule. Genes are the primary carriers of information in the genome, and much of genetics focuses on them. However, there is considerable variation among species in the number and sizes of genes and in the general chromosomal "landscape" (Figure 1-7). For eukaryotes, the number of genes ranges from about 6000 in the yeast *Saccharomyces cerevisiae* to approximately 20,500 in *Homo sapiens* to 32,000 in maize. The sizes of the regions between the genes are also variable between species.

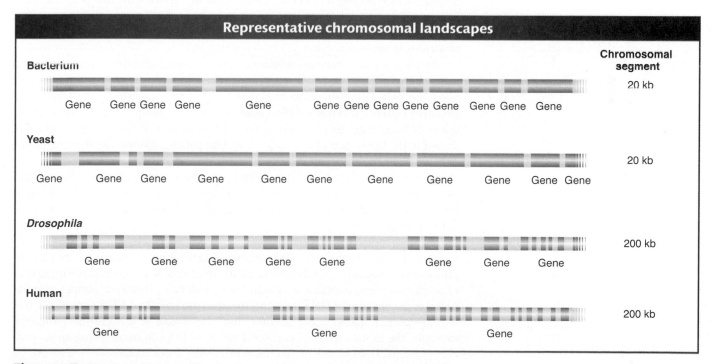

Figure 1-7 The genomes of four different species have very different gene topographies. Light green, introns; dark green, exons; white, regions between the coding sequences (including regulatory regions plus "spacer" DNA). The top two and bottom two illustrations are at different scales.

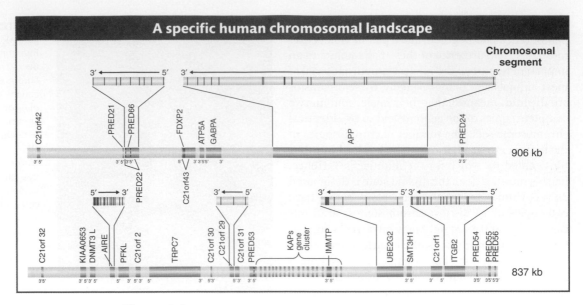

Figure 1-8 Transcribed regions of genes (green) in two segments of chromosome 21, based on the complete sequence for this chromosome. (Two genes, FDXP2 and IMMTP, are in orange to distinguish them from neighboring genes.) Some genes are expanded to show the coding regions ("exons"; black bars) and introns (light green). Vertical labels are gene names (some of known and others of unknown function). The 5′ and 3′ labels show the direction of transcription of the genes. [*After M. Hattori et al., Nature 405, 2000, 311–319.*]

Another surprise emerging from molecular research is that in many species, the functional coding sequence of genes carries internal noncoding inserts called *introns* (see Figure 1-7). The presence of large numbers of introns can make the gene size enormous. A most extreme case is the human gene for the protein dystrophin, which is defective in the disease muscular dystrophy: this gene's introns increase the gene size by a factor of several hundredfold. Two specific segments from the human genome are shown in Figure 1-8 and illustrate the arrangement of introns in some real genes.

Because homologous chromosomes are virtually identical, they carry the same genes in the same relative positions. Thus, in diploids, each gene is present as a **gene pair.** However, notice in Figure 1-6 that, although the nucleus in a body (*somatic*) cell contains pairs of chromosomes, they are not physically paired in the sense of being next to each other. The chromosomes of the ruptured nucleus shown in the lower part of the image reveal no pairing. Notice also that the upper part of the image shows an intact nucleus from another cell, and here again, the chromosomes are clearly not in a paired state; for example, the members of the violet pair are at opposite ends of the nucleus. However, the physical pairing of homologs does take place in the nuclear division known as meiosis, as we will see in Chapter 2.

All DNA molecules in a genome can be separated by size on a gel using a technique called electrophoresis. The number of DNA bands observed after electrophoresis is found to be equal to the haploid chromosome number, confirming that each chromosome contains only one DNA molecule. However, simple calculations on the amount of DNA per cell show that the length of a DNA molecule in a chromosome is always much greater than the length of the chromosome. For example, the human genome is about 1 meter of DNA in total, giving an average chromosomal DNA length of about 4 centimeters. But chromosomes measure on a scale of microns (millionths of a meter). Clearly, the DNA is packaged very efficiently in a chromosome. This packing is achieved by coiling the DNA double

Chromosomal DNA is wrapped around histones

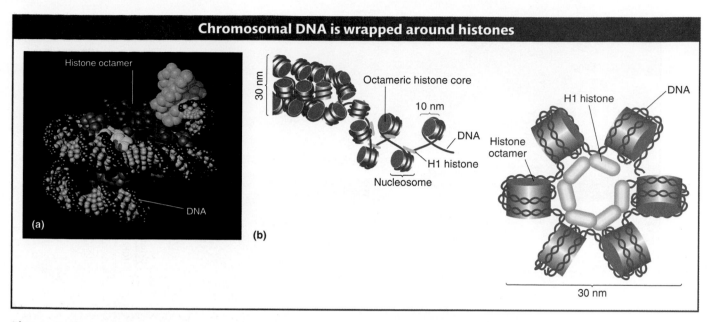

Figure 1-9 (a) A model of a nucleosome shows the DNA wrapped twice around a histone octamer. (b) Side and end views of the coiled chain of nucleosomes, diameter 30 nm, showing histone octomers as purple disks. An additional histone called H1, not part of the octomer, runs down the center of the coil acting as a stabilizer. [(a) Alan Wolffe and Van Moudrianakis; (b) H. Lodish, D. Baltimore, A. Berk, S. L. Zipursky, P. Matsudaira, and J. Darnell, Molecular Cell Biology, 3rd ed. Copyright 1995 by Scientific American Books.]

helix around molecular spools called **nucleosomes** (Figure 1-9). Each nucleosome is composed of eight proteins called **histones.** The DNA–nucleosome chain of a eukaryote is further coiled and folded to the state represented in Figure 1-10. This illustration shows another chromosomal component called the scaffold, which helps organize the three-dimensional structure of a chromosome. The DNA and associated nucleosomes are together called **chromatin,** the stuff of chromosomes. One constricted region of a chromosome called the **centromere** acts as an attachment point to move the chromosome during cell division. The tips of the chromosomes are called **telomeres.** Although telomeres generally have no visible features, they contain specialized DNA sequences needed during chromosome division. Telomeres function like the plastic bands at the ends of shoelaces, preventing the chromosome from fraying.

> **Message** The nuclear genomic DNA of eukaryotes is divided into a discrete number of subunits, each coiled around histone proteins in a chromosome. The main functional regions of the DNA are the genes, which are spaced out along the chromosomal DNA.

Nuclear DNA is not the whole story. In addition to nuclear DNA, a small specialized fraction of eukaryotic genomes is found in mitochondria. Plants also have specialized DNA in their chloroplasts. Together, these DNAs constitute the **extranuclear** genome.

Prokaryotes such as bacteria have no nuclei, so the genome resides unbounded in the cytoplasm. The genome of a prokaryote is generally a single noncoiled chromosome, which in most cases is

Chromosomal condensation by further coiling

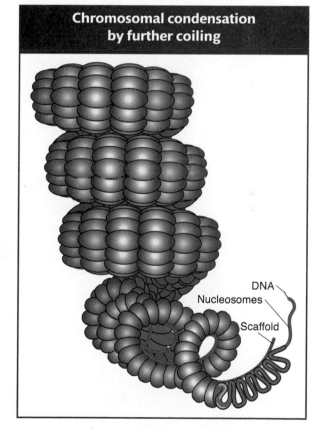

Figure 1-10 The model shows a coiled chromosome in cell division. The loops are so densely packed that only their tips are visible. At one end, the structure is partly uncoiled to show the components.

GENETICS*PORTAL* **ANIMATED ART: 3-D chromosome structure**

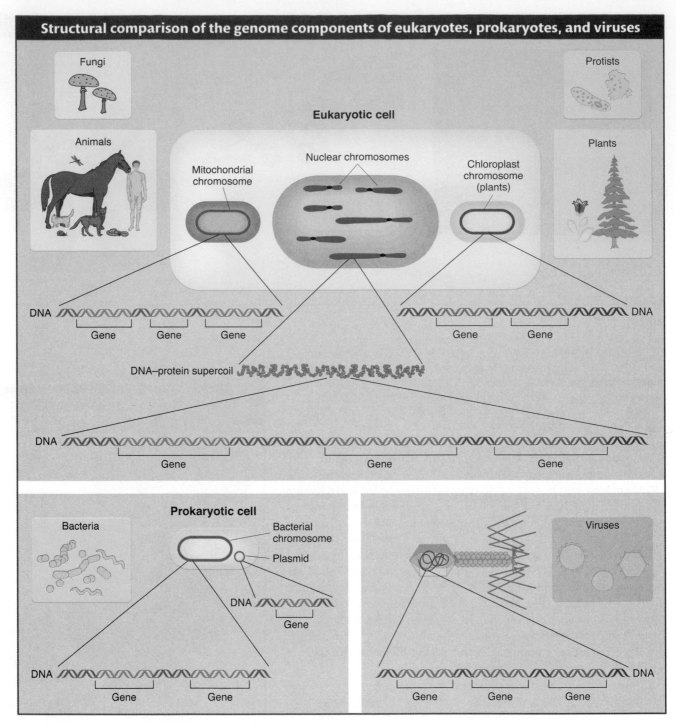

Figure 1-11 Eukaryotes, prokaryotes, and viruses all contain chromosomes on which reside the genes, but there are some differences in the genomes. For example, prokaryotic chromosomes are circular, whereas viral and the nuclear eukaryotic chromosomes are linear. Two eukaryotic organelles—the mitochondria and chloroplasts—contain separate, circular chromosomes.

circular. Prokaryotes often have small circular chromosomes called plasmids in addition to the main chromosome. The genomes of viruses are much smaller and usually linear.

A general representation of genomes is shown in Figure 1-11.

We all know, from trying to build *anything,* that a plan or blueprint is needed. Thus, knowing that the blueprint of life is based on DNA and having an understanding of DNA structure and how DNA is organized in cells was a giant step forward not only for genetics, but for all biology.

1.2 How Information Becomes Biological Form

Once scientists knew the nature of the biological information molecule, the obvious question was, *How* is information contained in the DNA molecule converted into "form," the stuff that we see when looking at an organism? An organism's form is its physical essence, including its size, shape, color, smell, behavior, and so on. The main elements of form in organisms are proteins: when you look at a living organism, you are looking either at proteins or material that has been made by proteins. Proteins can be classified into three basic types: structural, enzymatic, and regulatory. As their name suggests, *structural proteins* contribute to outward physical structure such as hair, nails, and muscle and also to structural elements within the cell such as the cytoskeleton. *Enzymatic proteins* catalyze the reactions going on within cells, reactions that make all the main types of molecules, including proteins themselves, nucleic acids, carbohydrates, and fats. *Regulatory proteins* act to turn on or turn off gene activity at the appropriate time and place. Hence, the main task of the living system is to convert the information of the DNA of genes into proteins.

Molecular geneticists worked out the basic mechanism of conversion soon after the discovery of DNA. The remarkable discovery was that not only is DNA the information storage system for virtually all organisms, but in addition the genetic coding language is virtually the same in all organisms and so is the mechanism whereby DNA is converted into proteins. This remarkable uniformity in the informational system is a result of all organisms sharing a common evolutionary ancestor.

Transcription

In the first stage of the protein-synthesis process, the DNA of a gene is copied to make another linear molecule called **ribonucleic acid (RNA).** The copying process is called **transcription.** RNA is also composed of nucleotides, but the sugar is ribose, and the base uracil replaces the base thymine. Whereas DNA is a double-stranded helix, RNA is single stranded. Nevertheless, the nucleotide sequence of one strand of the DNA double helix is copied precisely onto the nucleotide sequence in RNA, except that uracil appears wherever thymine would appear in the original DNA. In most eukaryotes the initial transcript is modified by excising the introns. The final form of gene transcripts destined for protein synthesis is called **messenger RNA (mRNA).** The word *messenger* is used to convey the idea that this molecule is the vehicle that conveys the information from a gene to the protein-generating machinery. Each transcribed region is flanked by one or more regions that determine when the transcription of that gene will take place and in which cells.

The overall transcriptional unit composed of an mRNA-producing region plus its flanking regulatory elements is the unit we have called a gene. It is in this sense that the gene is the basic functional unit of the genome: a gene is in fact a unit of transcription. The production of eukaryotic mRNA is diagrammed at the top of Figure 1-12.

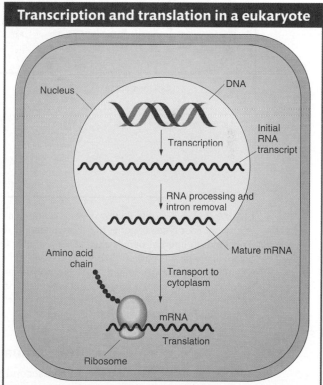

Transcription and translation in a eukaryote

Figure 1-12 In a eukaryotic cell, mRNA is transcribed from DNA in the nucleus and then transported to the cytoplasm for translation into a polypeptide chain.

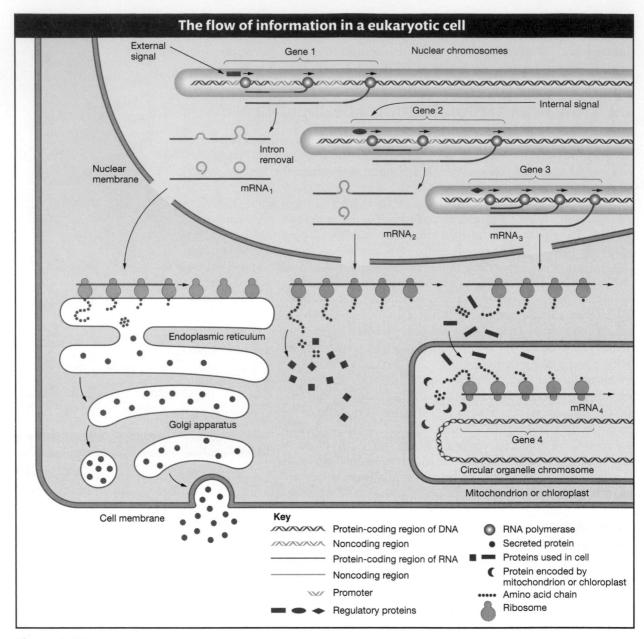

Figure 1-13 Simplified view of gene action in a eukaryotic cell. The basic flow of genetic information is from DNA to RNA to protein. Four types of genes are shown. Gene 1 responds to external regulatory signals and makes a protein for export; gene 2 responds to internal signals and makes a protein for use in the cytoplasm; gene 3 makes a protein to be transported into an organelle; gene 4 is part of the organelle DNA and makes a protein for use inside its own organelle. The promoter is the region. where transcription is initiated, and RNA polymerase is the transcriptional enzyme. Most eukaryotic genes contain introns, regions (generally noncoding) that are cut out in the preparation of functional messenger RNA. Note that many organelle genes have introns and that an RNA-synthesizing enzyme is needed for organelle mRNA synthesis. These details have been omitted from the diagram or the organelle for clarity.

Translation

In the second stage of the protein-synthesis process, each mRNA is **translated** into one specific protein. Hence, the sequence

$$DNA \rightarrow RNA \rightarrow protein$$

has become one of the operational mantras of biology. Indeed, it is one of the greatest insights into biology ever deduced and has been the foundation for much of biological research in the past half century. Like all rules, there are exceptions, and in some situations RNA can be *reverse transcribed* into DNA. For example, reverse transcription is used to maintain the telomeres that form the chromosome tips.

The details of **translation** are complex and intricate, as we will see in Chapter 9, but at the basic level it is quite simple (bottom of Figure 1-12). Every protein has a three-dimensional structure, but essentially it is a long chain of amino acids called a **polypeptide.** There are 20 main amino acids in cells, and it is the various combinations of these 20 that give each protein its specific shape and function. The chain of amino acids is folded or coiled to give the right shape for function.

How does the nucleotide sequence in mRNA become translated into an amino acid sequence in protein? Groups of three nucleotides, called **codons,** constitute the three-letter "words" of the genetic coding language. Every combination of three stands for one of the 20 specific amino acids. The codons in mRNA are "read" consecutively starting at one end by the translational machine, called the ribosome. Hence, a specific linear sequence of nucleotides is converted into a linear sequence of amino acids constituting a specific protein. The translational system involves a number of cellular components.

> **Message** Molecular genetics has shown that biological form is generated by translating the codon sequence of mRNA into the amino acid sequence of protein.

Some RNA molecules are never translated into protein but nevertheless play an important role in themselves. The existence of this general class of **functional RNAs** has been known for some time. Early examples were **ribosomal RNA (rRNA),** part of ribosomes, and **transfer RNAs (tRNAs),** whose role is to carry amino acids to the translational system. Recent research has revealed that there are many more types of functional RNAs that are essential for proper cell function.

The transcription and translational mechanisms outlined are just the bare bones of the complex process of how an undifferentiated zygote becomes a complex organism with many different operating systems. Clearly, the cells that are producing hair in the skin must be acting very differently from those producing insulin in the pancreas. How is this *differentiation* achieved? It is known that each of the trillions of cells of a multicellular organism has the same full complement of DNA, so logically, *different sets of genes must be active in cells of different types.* Indeed, it can be shown that most mRNA molecules are synthesized at specific developmental stages and not others. Gene transcription is controlled by regulatory proteins, and these regulatory proteins in turn are made by other genes in response to specific signals that may come from either outside or inside the cell. Some of the main elements of transcription and translation are illustrated in Figure 1-13, which outlines transcription and translation in a eukaryotic organism.

How does life replicate itself?

Another perennial conundrum of traditional biology was, How can life perpetuate itself through time? People have babies, dogs have puppies, and maple trees have maple seedlings. How can this constancy of form through the ages be achieved? Again the answer lies in DNA, which constitutes the basis of descent through time of both cells and organisms.

The structure of DNA lends itself to replication. Although quite a complex process in its detail, the idea, originally proposed by Watson and Crick, is simple: the two strands of DNA separate and newly synthesized nucleotides are deposited on the old strands, each paired with its appropriate partner, A with T and G with C (Figure 1-14). The nucleotides in the new array are then ligated

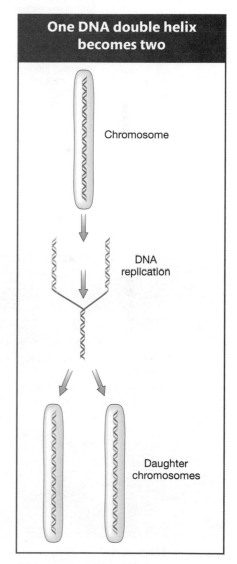

One DNA double helix becomes two

Chromosome

DNA replication

Daughter chromosomes

Figure 1-14 When new cells are made, DNA replication enables a chromosome to become daughter chromosomes and pass into new cells.

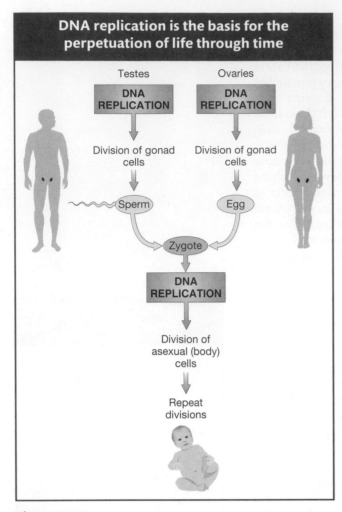

Figure 1-15

Figure 1-16 A nonfunctional version of a skin-pigment gene results in lack of pigment. In this case, both members of the gene pair are mutated. [*Yves Gellie/Gamma-Rapho/Getty Images.*]

(joined) while held in place by the old strand. Thus, two DNA molecules arise, each carrying one of the old separated strands plus a newly synthesized strand. This DNA replication process takes place every time somatic cells divide and also when sex cells (gametes) are formed (Figure 1-15). Hence, it is the way that life perpetuates its blueprint through time, both in producing new generations and in regenerating a new living organism from a single progenitor cell such as a fertilized egg.

> **Message** The perpetuation of life through time is based on high-fidelity replication of a genome's DNA.

Change at the DNA level

Species have established characteristics that define them, so we can (for example) always tell a porpoise from a whale. However, there is a great deal of variation within a species. Much of this is thought to be neutral variation in that it has no demonstrable effect on survival, but it does allow individuals to be distinguished. For example, individual killer whales are easily distinguished by the shape of the dorsal fin and the size and shape of the white body markings.

The basis for variation has long been a topic of great curiosity to humans, particularly as it relates to human variation. Geneticists took a huge step toward understanding variation when they discovered that the DNA in a genome can be changed. Their discovery of the mechanisms of change in DNA has provided a key piece of insight into the basis of variation, knowledge that is now reaping rewards in medicine and many other areas of research.

Comparisons of DNA from different individuals show that the differences are often caused by a minor difference in a gene's DNA sequence. A change to the DNA sequence is called a **mutation.** Mutations occur naturally either as a result of chemical mistakes in DNA processing in the cell or by exposure to environmental agents such as high-energy radiation or reactive chemicals. As random changes to the molecular machine, most mutations are detrimental, but some have no effect or are even advantageous. If mutations arise in the germ cells, such as the egg or sperm, then the mutation can be passed on to progeny and contribute to variation between individuals. A striking example of the potential effect of a mutation in a single gene is seen in the human condition albinism (Figure 1-16).

Mutations can lead to serious conditions. They are the cause of human diseases that are passed on from one generation to the next, known as hereditary diseases. For example, Tay-Sachs disease (affecting nerves) and muscular dystrophy (affecting muscles) are caused by mutations in single genes that radically alter or knock out that gene's function. Such mutations originate in the gonads and so are passed on in egg or sperm. Figure 1-17 shows some examples. Mutations in cells that are not germ cells do not have the same consequences: often such mutations simply kill a cell, which has no impact at all on the organism's func-

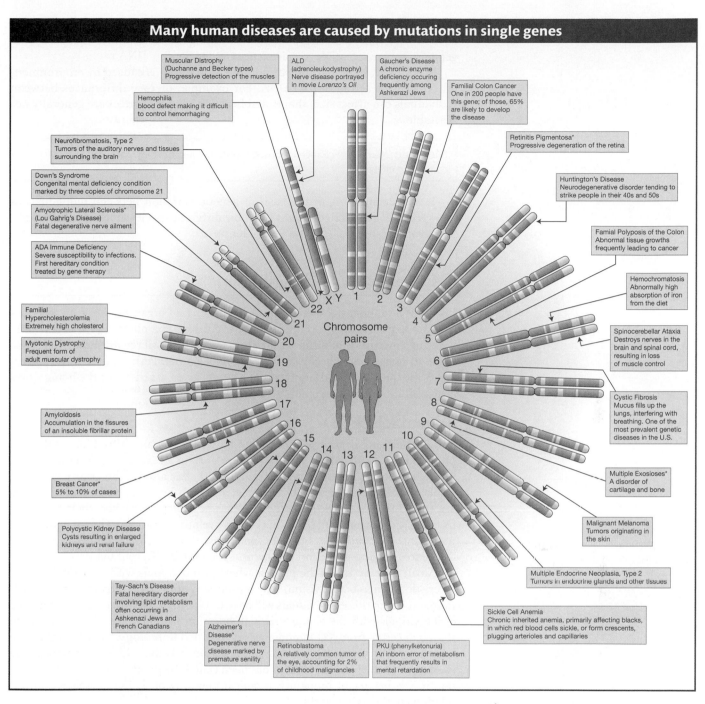

Many human diseases are caused by mutations in single genes

Muscular Distrophy (Duchanne and Becker types) Progressive detection of the muscles

ALD (adrenoleukodystrophy) Nerve disease portrayed in movie *Lorenzo's Oil*

Gaucher's Disease A chronic enzyme deficiency occuring frequently among Ashkerazi Jews

Familial Colon Cancer One in 200 people have this gene; of those, 65% are likely to develop the disease

Hemophilia blood defect making it difficult to control hemorrhaging

Retinitis Pigmentosa* Progressive degeneration of the retina

Neurofibromatosis, Type 2 Tumors of the auditory nerves and tissues surrounding the brain

Huntington's Disease Neurodegenerative disorder tending to strike people in their 40s and 50s

Down's Syndrome Congenital mental deficiency condition marked by three copies of chromosome 21

Famial Polyposis of the Colon Abnormal tissue growths frequently leading to cancer

Amyotrophic Lateral Sclerosis* (Lou Gahrig's Disease) Fatal degenerative nerve ailment

ADA Immune Deficiency Severe susceptibility to infections. First hereditary condition treated by gene therapy

Hemochromatosis Abnormally high absorption of iron from the diet

Familial Hypercholesterolemia Extremely high cholesterol

Spinocerebellar Ataxia Destroys nerves in the brain and spinal cord, resulting in loss of muscle control

Myotonic Dystrophy Frequent form of adult muscular dystrophy

Cystic Fibrosis Mucus fills up the lungs, interfering with breathing. One of the most prevalent genetic diseases in the U.S.

Amyloidosis Accumulation in the fissures of an insoluble fibrillar protein

Breast Cancer* 5% to 10% of cases

Multiple Exosioses* A disorder of cartilage and bone

Polycystic Kidney Disease Cysts resulting in enlarged kidneys and renal failure

Malignant Melanoma Tumors originating in the skin

Tay-Sach's Disease Fatal hereditary disorder involving lipid metabolism often occurring in Ashkenazi Jews and French Canadians

Multiple Endocrine Neoplasia, Type 2 Tumors in endocrine glands and other tissues

Alzheimer's Disease* Degenerative nerve disease marked by premature senility

Retinoblastoma A relatively common tumor of the eye, accounting for 2% of childhood malignancies

PKU (phenylketonuria) An inborn error of metabolism that frequently results in mental retardation

Sickle Cell Anemia Chronic inherited anemia, primarily affecting blacks, in which red blood cells sickle, or form crescents, plugging arterioles and capillaries

X Y 1 2 3 4 5 6 7 8 9 10 11 12 13 14 15 16 17 18 19 20 21 22

Chromosome pairs

Figure 1-17 The positions of the genes mutated in some single-gene diseases, shown on the 23 pairs of chromosomes in a human being. Each chromosome has a characteristic banding pattern. X and Y are the sex chromosomes (XX in women and XY in men). * = one form of the disease. [*Time.*]

tion. In other cases, they can affect the regulatory proteins that control cell division and a growth called a *cancer* (*tumor*) results at that spot.

Recent research has revealed another kind of heritable change of function that is not based on mutations in DNA. One example is chemical modification of certain histones. The role of histones was once thought to be limited to coiling the DNA for chromosome packing, but now it appears they can also carry out a regulatory function by restricting the access of regulatory proteins to the genes, thereby silencing them. Certain environmentally induced chemical changes in histones are self-perpetuating, and the altered gene function they cause can also be handed down to descendants. Such nongenetic changes are called **epigenetic.** Their existence shows that environmental exposure can affect the function of genes, often in a negative way. Current research is aimed at delineating the "epigenome," that part of the genome that is susceptible to epigenetic modification.

In addition, some natural variation in individuals is caused by environmental effects not acting on the DNA. For example, dietary differences between individuals can affect size, shape, and function. These changes are generally not heritable.

> **Message** Hereditary change is caused mostly by mutations in DNA, but also by epigenetic effects.

The understanding of the basis of genetic variation between individuals of one species also provided insight into how different species arise, in other words, how evolution occurs, which we consider next.

1.3 Genetics and Evolution

In addition to the insights it has provided in cell and organismal biology, genetics is now a key component in the study of evolution. The planet Earth is currently home to a multitude of different life-forms, and the fossil record shows that it was home to many more species in the past, now extinct. Perhaps one of the biggest and most controversial questions that has ever been asked about the living world is how all these forms (including humans) arose.

Natural selection

In the nineteenth century, Englishmen Charles Darwin and Alfred Russel Wallace proposed an explanation for the *natural* origin of species. Both men were impressed not only by the vast diversity of life, but also by the clear patterns of similarity between species. For example, although humans, birds, and porpoises are very different species occupying different ecological niches, their forelimbs have the same number of bones in the same relative positions. The interpretation was that these *similarities* between species are due to common ancestry and that *differences* are due to the force of *natural selection* in different habitats.

Natural selection is the process whereby individuals with a particular characteristic (such as better vision) may reproduce better than others in a given environment. Since these individuals will have more offspring, the relative abundance of individuals with the characteristic in question will increase. Similarity due to shared ancestry from a common ancestor is called **homology.** This all-embracing notion of natural selection acting on variation became widely accepted as the **theory of evolution.** This theory has been called the greatest intellectual revolution in the history of humanity, a radical new way of seeing ourselves and our relations to the living world.

Genetics has made a large contribution to the theory of evolution. Wallace and Darwin had no idea what could be the cause of the variation for natural selection to act on, but genetic research has shown that it is change in the DNA that generates variation, which then acts as the raw material for evolution. DNA change can be simple mutation within a gene or larger-scale changes involving whole chromosomes or genomes.

The field of population genetics has provided a complete mathematical underpinning for population change leading to evolution. Furthermore, the study of large-scale chromosomal changes at the genomic level has revealed specific mechanisms for evolution. In the above ways, genetics has provided substantial support for this greatest insight.

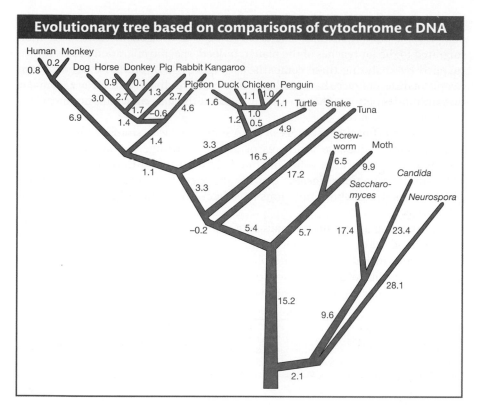

Evolutionary tree based on comparisons of cytochrome c DNA

Figure 1-18 A tree based on the DNA sequence of the cytochrome c gene. The numbers estimate the nucleotide substitutions that have occurred along the lineages in the gene coding for this protein; the distances are proportional to nucleotide differences between organisms. (All differences are small compared to the size of the gene.) [© 1994 Encyclopedia Britannica.]

Constructing evolutionary lineages

An evolutionary tree is a treelike branched diagram that shows the descent of various modern and fossil species through intermediate ancestral forms over time. DNA sequence is a powerful tool for constructing evolutionary trees. Differences in DNA sequences are quantified, and species with similar sequences are placed closer together in the tree of relatedness. Such DNA trees can be used to test patterns of evolutionary relationships previously proposed exclusively on physical homology. They can also reveal new and unexpected taxonomic groupings. DNA homology is often striking; for example, the DNA and amino acid sequence of the gene for the electron-transport-protein cytochrome c is homologous across the range of organisms on the planet, spanning bacteria, fungi, worms, insects, mammals, and so on (Figure 1-18). This type of revelation, together with the demonstration of great homology of the biochemical processes at work in cells, has led to a new awareness that humans are indeed "cousins" to all the life-forms on Earth.

Genetics has provided crucial input on human evolution. The chimpanzee genome sequence shows that chimpanzees are our closest living relatives, supporting Darwin's hypothesis that humans evolved from apes. Recently, DNA obtained from bones of the extinct Neanderthal people (Figure 1-19) has been used to generate a nearly complete Neanderthal genome sequence. As expected, this genome is even closer to ours than chimpanzees'. Furthermore, interesting clues about Neanderthals emerge, such as that one important gene for speech is in the same form as it is in humans, whereas the chimpanzee form is different. This observation leads to the fascinating speculation that Neanderthals had the ability of speech.

Neanderthal man

Figure 1-19 The genome of Neanderthals is the closest to modern humans of all those sequenced. [*imagebroker/Alamy.*]

DNA sequences from people around the world have been compared: these comparisons show that most likely, *Homo sapiens* evolved in Africa and then migrated to the far reaches of the planet. Indeed, specific migration routes can be mapped by analyzing these comparisons (Figure 1-20). Studies on genes from people of different races have made the important finding that there are no major discontinuities between the races, telling us that the race concept is not meaningful at the genetic level.

One spin-off of the discovery of DNA homology is to simplify the daunting task of determining the functions of genes in a huge genome such as the human genome. Because genes of similar structure in different species often have similar functions, insight can be obtained from previous research on homologous genes whose functions have been well established in experimental organisms.

In providing a deep understanding of the way DNA works and changes over time, genetics has given us a novel philosophical view of humanity's position in the universe, including our own evolution. DNA trees show that we are merely the end of one line of a complex web of evolutionary branching. We are not in any special position, but survivors like all other existing species.

Figure 1-20 Comparison of sites in mitochondrial DNA (mtDNA) and Y chromosome DNA reveals the routes taken by *Homo sapiens* in colonizing the planet. Different lines of the same color are the results of studies with different regions of DNA. [*The Genographic Project.*]

Message Genetics has made key contributions to understanding evolution, and conversely knowledge of evolutionary homology at the DNA level allows extrapolation from one species' genetic system to another.

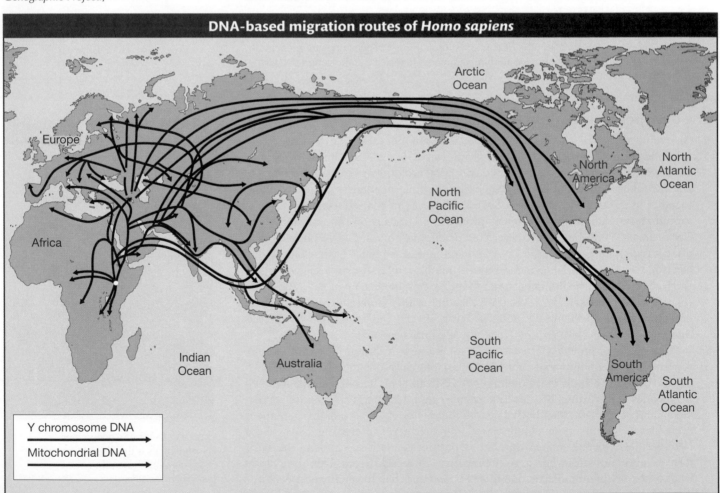

DNA-based migration routes of *Homo sapiens*

1.4 Genetics Has Provided a Powerful New Approach to Biological Research

The genetics revolution has radically influenced the way that biological research is done today. Genetics takes a unique approach to answering biological questions that relies on discovering genes relevant to that question. The investigator starts with a biological function that he or she wants to understand, then looks for mutant genes in which that function has been disrupted. This approach first defines the set of genes underlying the function of interest; then the normal and abnormal functions of those genes can be explored. Seeing how a mutant gene goes wrong can provide great insight about its normal function. Finally, all the genes discovered via such "mutational dissection" can be pieced together to construct the overall system at work in the cell. Every gene identified in this way reveals an important "word" in the genetic program underlying the function, while finding a set of genes all affecting the same function reveals the "sentences" that define the program. This type of genetics works in two ways, called forward and reverse genetics.

Forward genetics

The starting point of **forward genetics** is to treat cells of the normal *wild-type* form of the organism with some agent such as X rays or certain chemicals that causes mutations. Then descendants of these cells (usually organisms growing from them) are screened for abnormal manifestation of the function in question. For example, if we are interested in the biological function "color" and the wild type is purple, then we might look for mutations producing any other color (blue, red, pink, and so on) or even the absence of color (white). The first question asked is, Are these properties inherited as a single mutated gene? That question can be answered by crossing each presumptive mutant organism to a wild-type organism, then inspecting the ratios of wild-type to mutant progeny in the subsequent generations of descendants. The ratios indicating single-gene inheritance were originally established by the "father of genetics," Gregor Mendel, in the 1860s. A gene discovered in this way can be mapped or isolated, often leading to its DNA sequence.

The next step is to determine the function of each gene that has been identified. Returning to our example, we would ask, How does that gene act to influence flower color? The biochemical properties of each mutant obtained are studied at the molecular level and the defective protein encoded by that gene deduced, an important step in piecing together the overall system of reactions responsible for color. Hence, overall forward genetics can be represented by the sequence

$$\text{Mutation} \rightarrow \text{gene discovery} \rightarrow \text{DNA sequence and function}$$

The relatively new field of genomics has facilitated this approach: once a gene for a specific property has been mapped in the genomic sequence, then that gene's sequence is known, and if that gene has been studied in experimental organisms, then because of evolutionary homology, it is very likely that a function is already known for it. For example, human genes for proteins that promote transcription have been identified by their homology with the genes of fruit flies and yeast. Many heritable disorders have complex inheritance (heart disease, diabetes, and cleft palate are some examples) involving several genes; genomic analysis has begun identifying these genes too.

Reverse genetics

The **reverse genetics** approach starts with a gene sequence (probably learned from a genome sequence) that has no known function and then attempts to find that function. As in forward genetics, an important step is to obtain mutations in that gene. Several experimental approaches exist that can target mutations to an individual gene. These approaches are generally termed *directed mutagenesis*. One such approach is to completely knock out the gene's function by eliminating the gene and then to look for the effects on the organism's function. Alterations in the function of the mutant gene reveal aspects of the gene's biochemistry when fulfilling its normal role. (The technique works well for genes that are found as only one copy. It has been discovered from genomics that some genes are present in more than one copy, and in such cases it is possible to knock them all out.) Reverse genetics can be summarized by the sequence

$$\text{Gene (DNA sequence)} \rightarrow \text{mutation} \rightarrow \text{function}$$

Message Both forward and reverse genetics work by analyzing mutations and their effects; by showing how a gene goes wrong, we deduce its normal function.

Manipulating DNA

Like all scientists, geneticists make a large proportion of their inferences by manipulating the system and observing the effects. Hence, there is always a need to manipulate the genome in specific ways. Genomes contain billions of nucleotide pairs and are too big to handle as one unit, and so most DNA manipulation is performed by focusing on parts of the genome, often single genes. Forty years ago, this was impossible, but it is now routine in research and in applications such as medicine and agriculture. How is it possible to get one's hands on a small segment of DNA? The basic approach is called **DNA cloning,** which means taking a DNA fragment and replicating it many times over until there are many copies so that essentially it can be treated like a reagent in a test tube. The process of replicating a DNA sequence is called "amplifying," in the same way as a guitar amplifier multiplies the volume of sound.

Fragments of the genome are obtained by cutting the DNA in some way; for example, by vigorous agitation or scissoring it with certain enzymes. Fragments are inserted individually into a small self-replicating chromosome called a vector (carrier). The vectors with their loads are then introduced individually into separate live bacterial cells. The vector replicates as the cell divides, and its inserted fragment is thus automatically replicated too. As each cell divides repeatedly, it becomes a colony, which contains a *clone* (multiple replica) of one DNA insert.

A DNA clone can be used in a number of ways; for example, the DNA can be modified and reintroduced back into the original organism, introduced into a different organism to create a transgenic organism, or sequenced and the sequence assembled with other cloned sequences to produce a genomic sequence. Cloned DNA is used in many syntheses of industrial proteins, such as the enzymes that make sugar from cornstarch, and of medically important proteins, such as human growth hormone.

Detecting specific sequences of DNA, RNA, and protein

Whether studying gene structure or function, geneticists often need to detect a DNA, RNA, or protein specific to one gene of interest. For example, they might attempt to isolate a gene implicated in a human hereditary disease along with its RNA transcript and the associated protein. How can specific molecules be detected among the thousands of types in the cell? One extensively used method for detecting specific macromolecules in a mixture is **probing.** This method makes use of the specificity of intermolecular binding—for example, the binding affinity of an mRNA to the DNA sequence from which it was transcribed. A mixture of macromolecules is exposed to a molecule called a probe that will bind only with the sought-after macromolecule. The probe is labeled in some way, either by a radioactive atom or by a fluorescent compound, so that the binding product can be easily detected.

Probing for a specific DNA   A cloned gene can act as a probe for finding segments of DNA that have the same or a very similar homologous sequence. For example, if a gene from a fungus has been cloned, it might be used to find the same gene in humans. The human gene for homogentisic acid oxidase (HGO), when mutant, causes the blood disease alkaptonuria. The HGO gene was isolated in fungi before the human genome was sequenced, and a clone of the HGO gene from the fungus *Aspergillus* was used as a probe to detect the human genome fragment containing the HGO gene.

The use of a cloned gene as a probe takes us back to the principle of base complementarity. The probe works because its nucleotide sequence is complementary to its target. The experiment must be done with separated DNA strands, because then the bonding sites of the bases are unoccupied. DNA from the organism under study is extracted and cut with one of the many available types of enzymes that can cut DNA at specific target sequences. The target sequences are at the same positions in all the cells used, and so the enzyme cuts the genome into defined populations of segments of specific sizes. The fragments can be separated into groups of fragments of different sizes (fractionated) by using electrophoresis. After fractionation, the separated fragments are "blotted" onto a piece of porous membrane, where they maintain the same relative positions. This procedure is called a **Southern blot.** After having been heated to separate the DNA strands and hold the DNA in position, the membrane is placed in a solution of the probe. The single-stranded probe will find and bind to its complementary DNA sequence. For example,

TAGGTATCG   Probe

ACTAATCCATAGCTTA Genomic fragment

On the blot, this binding concentrates the label in one spot, as shown in Figure 1-21a. Hence, the position of the relevant DNA on the gel is revealed, and that DNA can be extracted if necessary.

Finding sets of genes using DNA microarrays   When full genomes have been sequenced, genome-wide probing akin to Southern analysis can be performed. An array of DNA fragments representing all the genes in a genome can be glued to the surface of a postage-stamp-size glass slide; this is called a *microarray.* Probes are usually complex mixtures made by converting mRNAs from one tissue (such as a cancer) into DNA called *cDNA.* The microarray is bathed in this labeled probe, and spots of label on the glass reveal which genes were being transcribed in the sample tissue, in this example the cancer. Comparison with noncancerous tissue reveals which genes are active (and inactive) in that

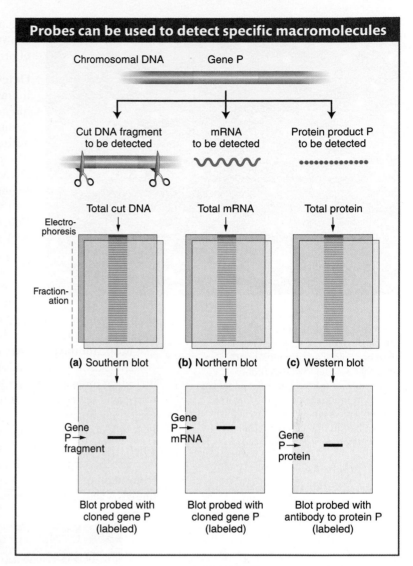

Figure 1-21 A specific gene can be used as a probe to detect that gene or its mRNA in a DNA or RNA mixture, whereas a specific antibody can be used as a probe to detect a specific protein in a mixture of proteins.

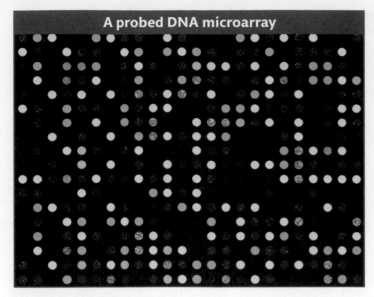

Figure 1-22 Each spot in this microarray is a different DNA sample attached to an inorganic surface. The different colors represent different amounts of labeled cDNA probes (derived from mRNA transcripts) that have bound to the samples in the array. The pattern shows which genes are actively transcribed in that cell type. [*Alfred Pasieka/Photo Researchers.*]

specific type of cancer. An example of such a result is shown in Figure 1-22. This technique also has a wide range of other uses.

Detecting and amplifying sequences using the polymerase chain reaction (PCR)   If a region has been sequenced, it is possible to detect homologs of that region in an unknown sample using the **polymerase chain reaction (PCR).** The method requires inspection of the genome sequence and picking two short single-stranded DNA segments that flank the region in question. These segments can be used as primers to initiate DNA replication across that region. Technical details will be given in Chapter 10, but in brief, the replication process shuttles back and forth across the region, with each synthesized copy acting as the template for a new round of synthesis. The process expands exponentially, giving multiple copies of a DNA segment that is the equivalent of that region (Figure 1-23). The primers will work only if the unknown sample contains a homolog of the target region in question (including of course the primer sequences), and so if any PCR product is obtained, the test acts as a diagnostic for the presence of that DNA in the sample.

The PCR technique has found widespread use throughout the life sciences, including forensics, medicine, and agriculture: it can rapidly detect the presence of specific segments of interest in any type of diagnostics. Target DNA present in very small amounts cannot be detected, but DNA samples can be amplified with PCR, allowing the sequence in question to be identified if it is present. Detecting a particular sequence is often the goal (as in forensics), but the amplified product can also be sequenced and studied further if required. For example, dry tissue from fossil or museum specimens can be subjected to PCR amplification, revealing DNA sequences of animals and plants long extinct.

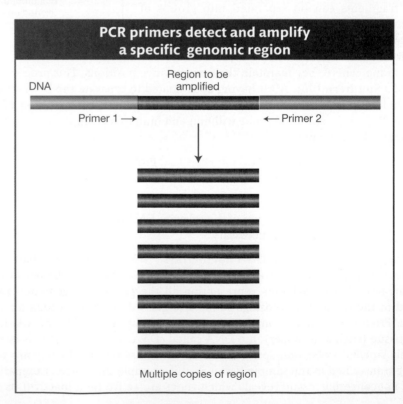

Figure 1-23 Short synthetic DNAs homologous to flanking regions can prime the synthesis of multiple copies of the flanked sequence, providing a large sample of that DNA for analysis.

Probing for a specific RNA It is often necessary to detect an RNA transcript in some particular tissue. For this purpose, a modification of the Southern analysis is useful. Total mRNA is extracted from the tissue, separated into fragments of different sizes using electrophoresis, and blotted onto a membrane (this is called a **Northern blot**). The cloned gene is used as a probe, and its label will highlight the mRNA in question if it is present (see Figure 1-21b).

Probing for a specific protein Probing for proteins is generally performed by using antibodies as probes. An antibody is a protein made by an animal's immune system; it binds with high affinity to a molecule such as a specific protein (which acts as an *antigen*) because the antibody has a specific lock-and-key fit with it. For protein detection, a protein mixture extracted from cells is separated into bands of distinct proteins by electrophoresis and then blotted onto a membrane (this is a **Western blot**). The position of a specific protein of interest on the membrane is revealed by bathing the membrane in a solution of antibody obtained from a rabbit or other host into which the antigen has been previously injected. The position of the protein is revealed by the position of the label that the antibody carries (see Figure 1-21c).

> **Message** Nucleic acids can be used as labeled probes or primers for detecting homologous nucleic acids on gels, on inorganic surfaces, or in solution. Individual proteins can be detected using labeled antibodies.

1.5 Model Organisms Have Been Crucial in the Genetics Revolution

As you read this textbook, you will encounter certain organisms over and over. Organisms such as *Escherichia coli* (a bacterium), *Saccharomyces cerevisiae* (baker's yeast), *Drosophila melanogaster* (fruit fly), and mice have been used repeatedly as the subjects of experiments that have revealed much of what we know about how genetics works. Why does scientific research make use of a relatively small group of organisms?

These species, called **model organisms,** were chosen because they are well suited to study of the biological question under investigation. Part of the suitability of model organisms is biological: that organism should have properties that lend themselves particularly well to that investigation. A suitable model organism also has the benefit of expediency: small organisms that are easy and cheap to maintain and grow quickly are very convenient for research. Because of evolutionary homology, what is learned from a model organism such as the fruit fly can often be applied to other species, even humans.

Some examples of genetic model organisms are as follows (some are illustrated in Figure 1-24). The genomes of all these model organisms have been sequenced.

Because of their small size, billions of bacteria can be used in an experiment. The use of such large numbers permits the detection of extremely rare genetic events. Also, bacteria can be spread on a special solid medium that automatically selects for a specific rare genetic event (such as mutations or new DNA combinations). Thus, the system is said to have a high resolving power between the rare and the wild-type genetic state. Bacteria are also particularly handy because bacteriophages (bacterial viruses) can be used as vectors to carry pieces of DNA from one bacterial cell to another. For these reasons, most of the early discoveries in molecular genetics were made in bacteria. For example, bacteria were the model organisms used in the experiments that revealed the sequence DNA → RNA → protein. Later on, the art of DNA cloning and manipulation was pioneered in bacterial systems.

Some organisms used as models in genetic research

Figure 1-24 Some model organisms. (a) Bacteriophage λ attached to an infected *Escherichia coli* cell; progeny phage particles are maturing inside the cell. (b) *Neurospora* growing on a burnt tree after a forest fire. (c) *Arabidopsis.* (d) *Caenorhabditis elegans.* [*(a) Lee D. Simon/Science Source/Photo Researchers; (b) courtesy of David Jacobson; (c) FloralImages/Alamy; (d) Sinclair Stammers/Photo Researchers.*]

Ascomycete fungi such as baker's yeast (*Saccharomyces cerevisiae*) and the mold *Neurospora crassa* have their products of meiosis enclosed in a small sac, which made them ideal subjects for studies on meiosis and mating. Yeast later became the center of studies of the genes that regulate cell division; many such genes were demonstrated to be important in human cancer.

Arabidopsis thaliana is a miniature flowering plant that can be cultured in large numbers in the greenhouse or laboratory. It has a small genome contained in only five chromosomes. It has been an ideal model for exploring many aspects of plant biology, such as the development of plant parts ranging from roots to flowers in higher plants.

The common fruit fly, *Drosophila melanogaster,* has only four chromosomes in its genome. In the larval stage, these chromosomes have a well-marked pattern of banding that makes it possible to observe large-scale chromosomal alterations, which can then be correlated with genetic changes in morphology and biochemistry. The development of *Drosophila* produces body segments in an anterior-posterior order that exemplifies the basic body plan common to invertebrates and vertebrates, and much has been learned from the fruit fly on this topic.

Mus musculus, the house mouse, has been the model organism for vertebrates, especially for humans. Because of its small size, the mouse has been used in many genetic analyses, including studies of mutation, development, and transgenesis.

> **Message** Most genetic studies are performed on one of a limited number of model organisms, which have features that make them especially suited for scientific study.

1.6 Genetics Changes Society

Many benefits to humanity have resulted from the applications of genetics in medicine, agriculture, and industry. Consider modern agriculture. Most crops and farm animals of today are only distantly related to wild species found in nature since their genomes have been extensively modified by systematic breeding programs. This is also true of garden plants and domestic pets. Although this selection process began centuries ago, traditional and molecular genetics have streamlined it to produce valuable varieties in a much shorter time. Now there is almost no limit to the possible gene combinations that can be produced. Even combinations of genes from different species can be produced by introducing a "foreign" gene into an organism. The foreign gene is called a *transgene,* and the organism into which geneticists have inserted a "foreign" gene is called a *transgenic organism* (Figure 1-25). Crops modified with transgenes for insecticide and herbicide resistance are being widely farmed. Animals have been modified too: for example, certain transgenic goats produce the medically useful anti-blood-clotting protein antithrombin and secrete it in their milk for convenient extraction. Transgenic bacteria are in use industrially for synthesizing important drugs such as human insulin and human growth hormone. Transgenic strains of yeast are used in making the bread we eat and the beer and wine we drink.

In medicine, the results have been just as striking. As we have seen above, many diseases have been identified that are caused by mutations in single genes. Such knowledge allows more rational genetic counseling of families at risk. Perhaps more important, every time a gene causing a disease is identified, it is the beginning of a line of research that will reveal the gene's function and then lead to possible therapy. A good example is in the discovery of the gene for phenylketonuria (PKU), a discovery that led to the alleviation of the disease through special diet.

The ability to modify genomes has inspired the tantalizing hope of correcting genetic disease at the DNA level, a process generally known as *gene therapy.* The development of the technology for transgenesis has made replacing defective genes with normally functioning ones a distinct possibility. Indeed, such gene therapy has been successful in animal models. In humans, more effective methods are needed for delivering the transgenic DNA and ensuring that it will function properly once in the genome. However, there have been some successes. In a recent case, gene therapy partially cured blindness resulting from the disease Lieber's congenital amaurosis, caused by a mutation of a gene active in the retina.

A significant impact of genetics has been in forensics. Each genome, whether human or animal or plant, can be treated to prepare an individualistic *"DNA*

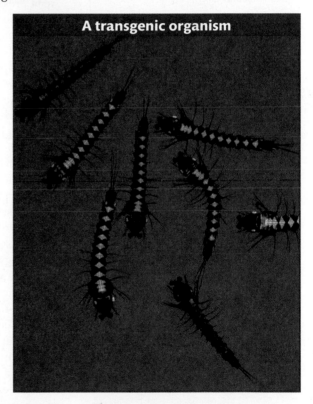

A transgenic organism

Figure 1-25 These transgenic mosquito larvae are expressing a jellyfish gene for green fluorescent protein. The gene is expressed at specific sites in each segment. [*Sinclair Stammers/Photo Researchers.*]

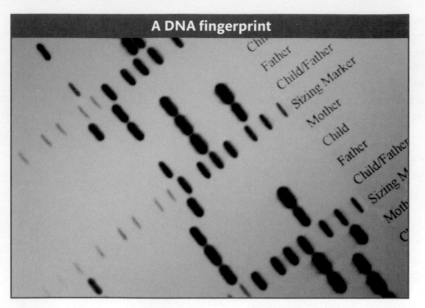

A DNA fingerprint

Figure 1-26 DNA fingerprints from paternity determination investigations. Bands (black stripes) in common show from which parent a child inherited DNA. [*Martin Shields/Alamy.*]

fingerprint" (Figure 1-26). Most DNA fingerprint approaches are based on the observation that certain regions of the genome are found in multiple adjacent copies (*repetitive DNA*), and the number of copies at any chromosomal position is highly individualistic. Comparison of several such sites reveals a personal fingerprint.

DNA fingerprints can be prepared even from minute amounts of bodily fluids (blood, sweat, saliva, semen) by chemically amplifying the DNA using a method called PCR (see above). PCR and DNA fingerprinting have revolutionized the identification of suspects in crimes.

1.7 Genetics and the Future

The genetics revolution has occupied much of the last 100 years, and this chapter so far has shown the great impact of genetics on the life sciences over this time, both on basic research and in applied research areas. In research generally, the continual and rapid advance of genetic technology means that the ability of biologists to genetically dissect all biological functions will undoubtedly improve in ways that we at present can only guess. The pattern has already been established: powerful techniques that once were supreme challenges later become routine and applicable more widely, especially to organisms that are not models and that were thought to be genetically intractable.

Perhaps the biggest challenge will be in the area of development. Although a great deal has been learned from model organisms about how specifically the body plan is laid down and the genes that control this plan, there is still a long journey before a complete and detailed understanding of the building of a living organism is obtained. This information will be directly relevant to human development and medical diagnosis and treatment. No doubt, over the coming decades, genetic disease (and indeed healthy human function) will be much better understood as a result of genetic research.

As population increases put more and more pressure on land and other global resources, society will have to rely more heavily on the most powerful technologies to provide the food, clothing, housing, and health of its members. Inevitably, the power of genetic technologies will be called on. Any powerful new scientific technology leads to ethical dilemmas in its application. Good examples are the nuclear and chemical industries, which although beneficial in many ways have contributed to global pollution and deaths from accidents and exposure to toxins. The question any individual scientist must ask is whether or not his or her discovery will be applicable to the general good. Even discoveries that do not seem directly applicable to societal problems nevertheless contribute to the general pool of knowledge that can be used or abused.

One of the greatest temptations may be in the area of eugenics. Broadly, eugenics has been defined as "improving the quality of human births." When the journey of gene discovery began in the early twentieth century, many human behavioral traits were prematurely attributed to genes. As a result, eugenics movements sprang up in North America and Europe, and laws were passed for the sterilization (and even euthanasia) of people with traits that were considered undesirable in the population. Regrettably, these decisions were based on erroneous genetic understanding, in many cases backed up by societal and political prejudice. However, what might our position be when genetics has advanced to the stage at which we do have a good understanding of complex human genetic conditions?

Gene therapy is just one relevant area: if it is possible for a couple to have a baby free of genetic disease or suffering via gene therapy, why shouldn't the technology be made available and used? Taking the logic further, should parents be allowed to have a baby genetically tailored in specific safe ways—for example, given high intelligence or athletic or musical ability. As more is learned about the basis of our individuality, how will this knowledge affect human freedoms? If a sound genetic basis is found for certain types of antisocial behavior, how will the legal system deal with the responsibilities and rights involved? If we can accurately predict susceptibility to various types of disease, how will this affect our relationships and attitudes toward each other (such as in mate selection), and of course how will health insurance deal with it? One thing is clear: these kinds of societal decisions will need to be made, and ultimately they will depend on a public and a government well educated and knowledgeable in the subject of genetics.

Summary

The impact of genetics on biological research and its applications has been so far reaching that it has been called "the genetics revolution." Genetics is now part of the analytical framework of virtually all areas of biology. It has provided fundamental insights into all the main questions of biology that were previously unanswerable.

One perennial question has been how living systems generate "form" from random components taken in as nutrients. Biological information (that which is needed to generate form) was shown to be encoded in our DNA, the central molecule of life. The information encoded in DNA is the perennial blueprint that is handed down through generations. Form is largely a product of an organism's proteins. The DNA molecule is divided into functional units called genes. Most genes encode a specific protein. The protein is synthesized in two steps: in step 1 (transcription), RNA is transcribed from DNA, and in step 2 (translation), the RNA is "read" to synthesize a protein. The subunits of DNA (nucleotides) are read in groups of three, each corresponding to an amino acid in that gene's protein. DNA structure is ideally suited to enable copies to be made of itself. DNA molecules are replicated every time a cell or an organism reproduces, enabling the information to persist endlessly through time.

Although DNA structure persists through time, it does undergo random change in the process of mutation. Mutation is the source of much variation between individuals of a species. If acted on by natural selection over time, mutation can produce new species in the process of evolution. Genetics has been crucial in showing mechanisms of change relevant to evolution. Even after species diverge during evolution, the DNA sequences of these species continue to show considerable similarity (homology). This DNA homology is convenient for research in that what is learned in one species can be applied to another. DNA homology is used extensively for making evolutionary trees.

The incisiveness of the genetic approach is based on the concept of genetic dissection: a biological function can be picked apart through the use of mutations—each mutation represents another gene in the overall program behind the function in question. Technological advances have allowed individual genes to be isolated, studied, and moved into other species for research and for making "designer organisms." The advent of genomics has expanded genetics to allow complete gene sets (genomes) to be analyzed, further extending the ability to see complete genetic systems at work in normal and disease situations.

Human society has benefited from the genetics revolution. The deep level of understanding that genetics brings about the nature and evolution of life has allowed humans to philosophically see their own and other species in a new way and to design applications in medicine, agriculture, and industry.

The future, with its increasing pressures on resources, will inevitably draw heavily on genetic technology. However, with the likely progress comes a set of ethical dilemmas surrounding the application of the new discoveries, dilemmas regarding human individuality and our treatment of other organisms and the environment. For all these issues, a firm understanding of genetics will be required to make wise decisions.

KEY TERMS

adenine (A) (p. 3)
centromere (p. 7)
chromatin (p. 7)
codon (p. 11)
cytosine (C) (p. 3)

deoxyribonucleic acid (DNA) (p. 1)
diploid (p. 5)
DNA cloning (p. 18)
epigenetic (p. 13)
extranuclear (p. 7)

forward genetics (p. 17)
functional RNA (p. 11)
gene (p. 2)
gene pair (p. 6)
genetic code (p. 3)

genetics (p. 2)
genome (p. 4)
genomics (p. 2)
guanine (G) (p. 3)
haploid (p. 5)
haploid number (p. 5)
histone (p. 7)
homolog (p. 5)
homologous chromosomes (p. 5)
homology (p. 14)
messenger RNA (RNA) (p. 9)

model organism (p. 21)
molecular genetics (p. 2)
mutation (p. 12)
natural selection (p. 14)
Northern blot (p. 21)
nucleosome (p. 7)
nucleotide (p. 3)
polymerase chain reaction (PCR)
  (p. 20)
polypeptide (p. 11)
probing (p. 18)

reverse genetics (p. 17)
ribonucleic acid (RNA) (p. 9)
ribosomal RNA (rRNA) (p. 11)
Southern blot (p. 19)
telomere (p. 7)
theory of evolution (p. 14)
thymine (T) (p. 3)
transcription (p. 9)
transfer RNA (tRNA) (p. 11)
translation (p. 11)
Western blot (p. 21)

PROBLEMS

In each chapter, a set of problems tests the reader's comprehension of the concepts in the chapter and their relation to concepts in previous chapters. Each problem set begins with some problems based on the figures in the chapter, which embody important concepts. These are followed by problems of a more general nature.

Most of the problems are also available for review/grading through the GENETICSPORTAL www.yourgeneticsportal.com.

WORKING WITH THE FIGURES

1. In considering Figure 1-2, if you were to extend the diagram, what would the next two stages of "magnification" beyond DNA be?

2. In considering Figure 1-3,

   a. what do the small blue spheres represent?

   b. what do the brown slabs represent?

   c. do you agree with the analogy that DNA is structured like a ladder?

3. In Figure 1-4, can you tell if the number of hydrogen bonds between adenine and thymine is the same as that between cytosine and guanine? Do you think that a DNA molecule with a high content of A + T would be more stable than one with high content of G + C?

4. From Figure 1-6, can you predict how many chromosomes there would be in a muntjac sperm? How many purple chromosomes would there be in a sperm cell?

5. In examining Figure 1-7, state one major difference between the chromosomal "landscapes" of yeast and *Drosophila*.

6. In Figure 1-8, is it true that the direction of transcription is from right to left as written for all the genes shown in these chromosomal segments?

7. In Figure 1-9, estimate what length of DNA is shown in the right-hand part of the figure.

8. From Figure 1-12, what is the main difference in the locales of transcription and translation?

9. In Figure 1-14, what do the colors blue and gold represent?

10. From Figure 1-17, locate the chromosomal positions of three genes involved in tumor production in the human body.

11. In Figure 1-18, calculate the approximate number of nucleotide differences between humans and dogs in the cytochrome c gene. Repeat for humans and moths. Considering that the gene is several hundred nucleotides long, do these numbers seem large or small to you? Explain.

12. In Figure 1-21, why are colored ladders of bands shown in all three electrophoretic gels? If the molecular labels used in all cases were radioactive, do you think the black bands in the bottom part of the figure would all be radioactive?

BASIC QUESTIONS

13. In this chapter, the statement is made that most of the major questions of biology have been answered through genetics. What are the main questions of biology, and do you agree with the above statement? (State your reasons.)

14. It has been said that the DNA → RNA → protein discovery was the "Rosetta stone" of biology. Do you agree?

15. Who do you think had the greatest impact on biology, Charles Darwin or the research partners James Watson and Francis Crick?

16. How has genetics affected (a) agriculture, (b) medicine, (c) evolution, and (d) modern biological research?

17. Assume for the sake of this question that the human body contains a trillion cells (a low estimate). We know that a human genome contains about 1 meter of DNA. If all the DNA in your body were laid end to end, do you think it could stretch to the Moon and back? Justify your answer with a calculation. (**Note:** The average distance to the Moon is 385,000 kilometers.)

Single-Gene Inheritance

2

The monastery of the father of genetics, Gregor Mendel. A statue of Mendel is visible in the background. Today, this part of the monastery is a museum, and the curators have planted red and white begonias in a grid that graphically represents the type of inheritance patterns obtained by Mendel with peas. [*Anthony Griffiths.*]

W hat kinds of research do biologists do? The central area of research in the biology of all organisms is the attempt to understand the program whereby an organism develops from a fertilized egg into an adult—in other words, what makes an organism the way it is. Usually, this overall goal is broken down into the study of individual biological properties such as the development of plant flower color, or animal locomotion, or nutrient uptake, although biologists also study some general areas such as how a cell works. How do geneticists analyze biological properties? We learned in Chapter 1 that the genetic approach to understanding any biological property is to find the subset of genes in the genome that influence that property, a process sometimes referred to as **gene discovery.** After these genes have been identified, the way in which the genes act to determine the biological property can be elucidated through further research.

There are several different types of analytical approaches to gene discovery, but one of the most widely used relies on the detection of *single-gene inheritance patterns*, the topic of this chapter. Such inheritance patterns may be recog-

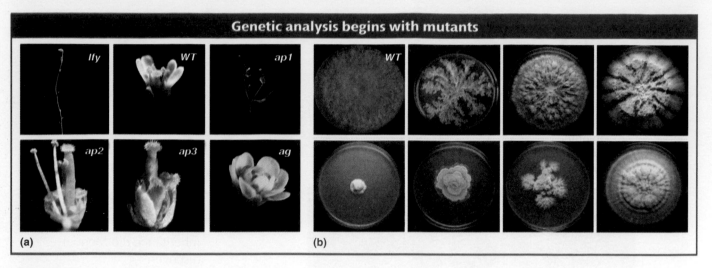

Genetic analysis begins with mutants

(a)   (b)

Figure 2-1 These photographs show the range of mutant phenotypes typical of those obtained in the genetic dissection of biological properties. These cases are from the dissection of floral development in *Arabidopsis thaliana* (a) and hyphal growth in *Neurospora crassa*, a mold (b). WT = wild type [*(a) George Haughn; (b) Anthony Griffiths/Olivera Gavric.*]

nized in the progeny of certain types of controlled matings, which geneticists call **crosses.** The central components in this type of analysis are **mutants,** individual organisms having some altered form of a normal property. The normal form of any property of an organism is called the **wild type,** that which is found "in the wild," or in nature. The genetic modus operandi is to mate an individual showing the property in its wild-type form (for example, a plant with red flowers) to an individual showing a mutated form of the property (for example, a plant with white flowers). The progeny of this cross are interbred and, in their progeny, the ratio of plants with red flowers to those with white flowers will reveal whether a single gene controls that difference in the property under study—in this example, red versus white. By inference, the wild type would be encoded by the wild-type form of the gene and the mutant would be encoded by a form of the same gene in which a mutation event has altered the DNA sequence in some way. Other mutants affecting flower color (perhaps mauve, blotched, striped, and so on) would be analyzed in the same way, resulting overall in a set of defined "flower-color genes." The use of mutants in this way is sometimes called **genetic dissection,** because the biological property in question (flower color in this case) is picked apart to reveal its underlying genetic program, not with a scalpel but with mutants. Each mutant potentially identifies a separate gene affecting that property.

Thus, each gene discovery project begins with a hunt for mutants affecting the biological process that is the focus of the research. The most direct way to obtain mutants is to visually *screen* a very large number of individuals, looking for a chance occurrence of mutants in that population. As examples, some of the results of screening for mutants in two model organisms are shown in Figure 2-1. The illustration shows the effects of mutations on flower development in the plant *Arabidopsis thaliana* and on the development of the mycelium in the mold *Neurospora crassa* (a mycelium is a network of threadlike cells called hyphae). The illustration shows that the development of the properties in question can be altered in a variety of different ways. In the plant, the number or type of floral organs is altered; in the fungus, the growth rate and the number and type of branches are altered in a variety of ways, each of which gives a distinctly abnormal colony morphology. The hope is that each case represents a mutation in a different member of the set of genes responsible for that property. However, genetic changes that are more complex than single-gene changes do occur; furthermore, an abnormal environment also is capable of changing an organism's appearance. Hence, each individual case must be tested to see if it produces descendants in the appropriate ratio that is diagnostic for a mutant caused by the mutation of a *single gene.*

> **Message** The genetic approach to understanding a biological property is to discover the genes that control it. One approach to gene discovery is to isolate mutants and check each one for single-gene inheritance patterns (specific ratios of normal and mutant expression of the property in descendants).

Single-gene inheritance patterns are useful for gene discovery not only in experimental genetics of model organisms, but also in applied genetics. An important example is found in human genetics. Many human disorders, such as cystic fibrosis and Tay-Sachs disease, are inherited as a single mutant gene. After a key gene has been defined in this way, geneticists can zero in on it at the DNA level and try to decipher the basic cellular defect that underlies the disease, possibly leading to new therapies. In agriculture, these same types of single-gene inheritance patterns have led to the discovery of mutations conferring some desirable feature such as disease resistance or better nutrient content. These beneficial mutations have been successfully incorporated into commercial lines of plants and animals.

The rules for single-gene inheritance were originally elucidated in the 1860s by the monk Gregor Mendel, who worked in a monastery in the town of Brno, now part of the Czech Republic. Mendel's analysis is the prototype of the experimental approach to single-gene discovery still used today. Indeed, Mendel was the first to discover any gene! Mendel did not know what genes were, how they influenced biological properties, or how they were inherited at the cellular level. Now we know that genes work through proteins, a topic that we shall return to in later chapters. We also know that single-gene inheritance patterns are produced because genes are parts of chromosomes, and chromosomes are partitioned very precisely down through the generations, as we shall see later in the chapter.

2.1 Single-Gene Inheritance Patterns

Recall that the first step in genetic dissection is to obtain variants that differ in the property under scrutiny. With the assumption that we have acquired a collection of relevant mutants, the next question is whether each of the mutations is inherited as a single gene.

Mendel's pioneering experiments

The first-ever analysis of single-gene inheritance as a pathway to gene discovery was carried out by Gregor Mendel. His is the analysis that we shall follow as an example. Mendel chose the garden pea, *Pisum sativum*, as his research organism. The choice of organism for any biological research is crucial, and Mendel's choice proved to be a good one because peas are easy to grow and breed. Note, however, that Mendel did not embark on a hunt for mutants of peas; instead, he made use of mutants that had been found by others and had been used in horticulture. Moreover, Mendel's work differs from most genetics research undertaken today in that it was not a genetic dissection; he was not interested in the properties of peas themselves, but rather in the way in which the hereditary units that influenced those properties were inherited from generation to generation. Nevertheless, the laws of inheritance deduced by Mendel are exactly those that we use today in modern genetics in identifying single-gene inheritance patterns.

Mendel chose to investigate the inheritance of seven properties of his chosen pea species: pea color, pea shape, pod color, pod shape, flower color, plant height, and position of the flowering shoot. In genetics, the terms **character** and **trait** are used more or less synonymously with **property.** For each of these seven characters, he obtained from his horticultural supplier two lines that showed distinct and contrasting appearances. Today, we would say that, for each character, he studied two contrasting **phenotypes.** A phenotype can be defined as *a form taken*

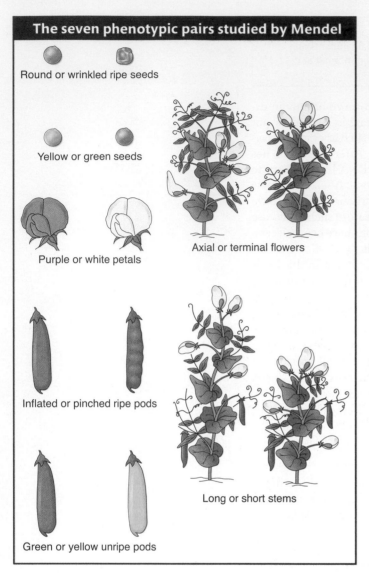

The seven phenotypic pairs studied by Mendel

Round or wrinkled ripe seeds

Yellow or green seeds

Purple or white petals

Axial or terminal flowers

Inflated or pinched ripe pods

Long or short stems

Green or yellow unripe pods

Figure 2-2 For each character, Mendel studied two contrasting phenotypes. [*After S. Singer and H. Hilgard, The Biology of People. Copyright 1978 by W. H. Freeman and Company.*]

by a character. These contrasting phenotypes are illustrated in Figure 2-2. His results were substantially the same for each character, and so we can use one character, pea seed color, as an illustration. All of the lines used by Mendel were **pure lines,** meaning that, for the phenotype in question, all offspring produced by matings within the members of that line were identical. For example, within the yellow-seeded line, all the progeny of any mating were yellow seeded.

Mendel's analysis of pea heredity made extensive use of crosses. To make a cross in plants such as the pea, pollen is simply transferred from the anthers of one plant to the stigmata of another. A special type of mating is a **self,** which is carried out by allowing pollen from a flower to fall on its own stigma. Crossing and selfing are illustrated in Figure 2-3. The first cross made by Mendel mated plants of the yellow-seeded lines with plants of the green-seeded lines. These lines constituted the **parental generation,** abbreviated P. In *Pisum sativum,* the color of the seed (the pea) is determined by its own genetic makeup; hence, the peas resulting from a cross are effectively progeny and can be conveniently classified for phenotype without the need to grow them into plants. The progeny peas from the cross between the different pure lines were found to be all yellow, no matter which parent (yellow or green) was used as male or female. This progeny generation is called the **first filial generation,** or F_1. Hence, the results of these two reciprocal crosses were as follows, where × represents a cross:

female from yellow line × male from green line →
F_1 peas all yellow

female from green line × male from yellow line →
F_1 peas all yellow

The results observed in the descendants of both reciprocal crosses were the same, and so we will treat them as one cross. Mendel grew F_1 peas into plants, and he selfed or intercrossed the resulting F_1 plants to obtain the **second filial generation,** or F_2. The F_2 was composed of 6022 yellow peas and 2001 green peas. Mendel noted that this outcome was very close to a precise mathematical ratio of three-fourths yellow and one-fourth green. Interestingly, the green phenotype, which had disappeared in the F_1, had reappeared in one-fourth of the F_2 individuals, showing that the genetic determinants for green must have been present in the yellow F_1, although unexpressed.

Next Mendel *individually selfed* plants grown from the F_2 seeds. The plants grown from the F_2 green seeds, when selfed, were found to bear only green peas. However, plants grown from the F_2 yellow seeds, when selfed individually, were found to be of two types: one-third of them were pure breeding for yellow seeds, but two-thirds of them gave a progeny ratio of three-fourths yellow seeds and one-fourth green seeds, just as the F_1 plants had.

Another informative cross that Mendel made was a cross of the F_1 with any green-seeded plant; here, the progeny showed the proportions of one-half yellow and one-half green. These two types of matings, the F_1 self and the cross of the F_1 with any green-seeded plant, both gave yellow and green progeny, but in different ratios. These two ratios are represented in Figure 2-4. Notice that the ratios are seen only when the peas in several pods are combined.

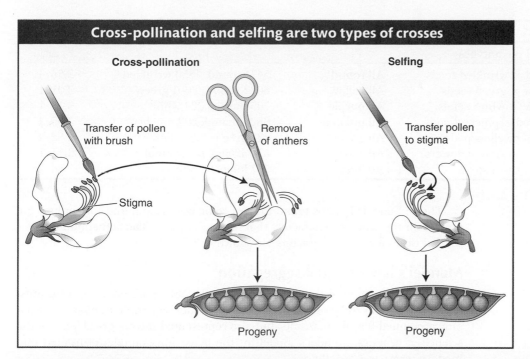

Cross-pollination and selfing are two types of crosses

Cross-pollination

Transfer of pollen with brush

Stigma

Removal of anthers

Progeny

Selfing

Transfer pollen to stigma

Progeny

Figure 2-3 In a cross of a pea plant (*left*), pollen from the anthers of one plant is transferred to the stigma of another. In a self (*right*), pollen is transferred from the anthers to the stigmata of the same plant.

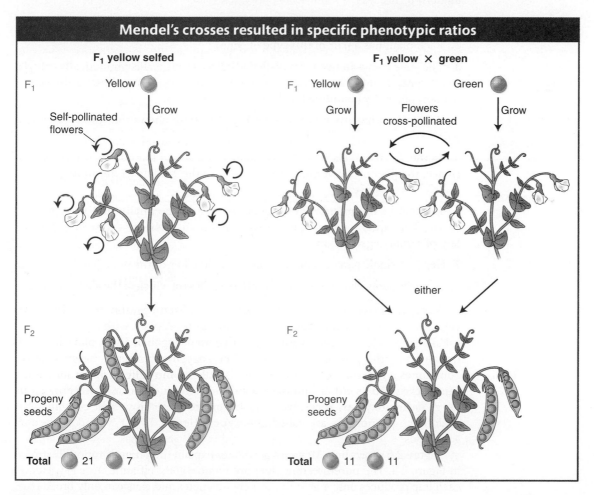

Mendel's crosses resulted in specific phenotypic ratios

F₁ yellow selfed

F_1 Yellow

Grow

Self-pollinated flowers

F_2

Progeny seeds

Total   21   7

F₁ yellow × green

F_1 Yellow   Green

Grow   Flowers cross-pollinated   Grow

or

either

F_2

Progeny seeds

Total   11   11

Figure 2-4 Mendel obtained a 3 : 1 phenotypic ratio in his self-pollination of the F_1 (*left*) and a 1 : 1 phenotypic ratio in his cross of F_1 yellow with green (*right*). Sample sizes are arbitrary.

Table 2-1   Results of All Mendel's Crosses in Which Parents Differed in One Character

Parental phenotypes	F$_1$	F$_2$	F$_2$ ratio
1. round × wrinkled seeds	All round	5474 round; 1850 wrinkled	2.96 : 1
2. yellow × green seeds	All yellow	6022 yellow; 2001 green	3.01 : 1
3. purple × white petals	All purple	705 purple; 224 white	3.15 : 1
4. inflated × pinched pods	All inflated	882 inflated; 299 pinched	2.95 : 1
5. green × yellow pods	All green	428 green; 152 yellow	2.82 : 1
6. axial × terminal flowers	All axial	651 axial; 207 terminal	3.14 : 1
7. long × short stems	All long	787 long; 277 short	2.84 : 1

The 3 : 1 and 1 : 1 ratios found for pea color were also found for comparable crosses for the other six characters that Mendel studied. The actual numbers for the 3 : 1 ratios for those characters are shown in Table 2-1.

Mendel's law of equal segregation

Initially, the meaning of these precise and repeatable mathematical ratios must have been unclear to Mendel, but he was able to devise a brilliant model that not only accounted for all the results, but also represented the historical birth of the science of genetics. Mendel's model for the pea-color example, translated into modern terms, was as follows.

1. A hereditary factor called a **gene** is necessary for producing pea color.

2. Each plant has a pair of this type of gene.

3. The gene comes in two forms called **alleles.** If the gene is phonetically called a "wye" gene, then the two alleles can be represented by Y (standing for the yellow phenotype) and y (standing for the green phenotype).

4. A plant can be either Y/Y, y/y, or Y/y. The slash shows that the alleles are a pair.

5. In the Y/y plant, the Y allele dominates, and so the phenotype will be yellow. Hence, the phenotype of the Y/y plant defines the Y allele as **dominant** and the y allele as **recessive.**

6. In meiosis, the members of a gene pair separate equally into the eggs and sperm. This equal separation has become known as **Mendel's first law** or as the **law of equal segregation.**

7. Hence, a single gamete contains only one member of the gene pair.

8. At fertilization, gametes fuse randomly, regardless of which of the alleles they bear.

Here, we introduce some terminology. A fertilized egg, the first cell that develops into a progeny individual, is called a **zygote.** A plant with a pair of identical alleles is called a **homozygote** (adjective homozygous), and a plant in which the alleles of the pair differ is called a **heterozygote** (adjective heterozygous). Sometimes a heterozygote for one gene is called a **monohybrid.** An individual can be classified as either **homozygous dominant** (such as Y/Y), **heterozygous** (Y/y), or **homozygous recessive** (y/y). In genetics generally, allelic combinations underlying phenotypes are called **genotypes.** Hence, Y/Y, Y/y, and y/y are all genotypes.

Figure 2-5 shows how Mendel's postulates explain the progeny ratios illustrated in Figure 2-4. The pure-breeding lines are homozygous, either Y/Y or y/y. Hence, each line produces only Y gametes or only y gametes and thus can only breed true. When crossed with each other, the Y/Y and the y/y lines produce an F$_1$ generation composed of all heterozygous individuals (Y/y). Because Y is dominant, all F$_1$ individuals are yellow in phenotype. Selfing the F$_1$ individuals can be thought of as

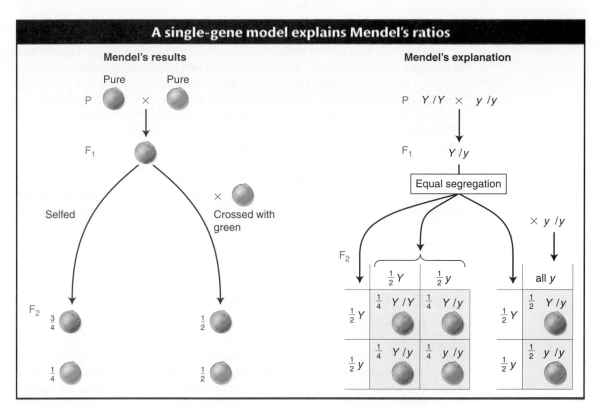

Figure 2-5 Mendel's results (*left*) are explained by a single-gene model (*right*) that postulates the equal segregation of the members of a gene pair into gametes.

a cross of the type $Y/y \times Y/y$. A cross of the type $Y/y \times Y/y$ is sometimes called a **monohybrid cross.** Equal segregation of the Y and y alleles in the heterozygous F_1 results in gametes, both male and female, half of which are Y and half of which are y. Male and female gametes fuse randomly at fertilization, with the results shown in the grid in Figure 2-5. The composition of the F_2 is three-fourths yellow seeds and one-fourth green, a 3:1 ratio. The one-fourth of the F_2 seeds that are green breed true as expected of the genotype y/y. However, the yellow F_2 seeds (totaling three-fourths) are of two genotypes: two-thirds of them are clearly heterozygotes Y/y, and one-third are homozygous dominant Y/Y. Hence, we see that underlying the 3:1 phenotypic ratio in the F_2 is a 1:2:1 genotypic ratio:

$$\left. \begin{array}{l} \tfrac{1}{4}\; Y/Y \;\; \text{yellow} \\[4pt] \tfrac{2}{4}\; Y/y \;\; \text{yellow} \end{array} \right\} \quad \tfrac{3}{4}\; \text{yellow}\; (Y/-)$$

$$\tfrac{1}{4}\; y/y \;\; \text{green}$$

The general depiction of an individual expressing the dominant allele is $Y/-$; the dash represents a slot that can be filled by either another Y or a y. Note that equal segregation is detectable only in the meiosis of a heterozygote. Hence, Y/y produces one-half Y gametes and one-half y gametes. Although equal segregation is taking place in homozygotes too, neither segregation $\tfrac{1}{2}\,Y : \tfrac{1}{2}\,Y$ nor segregation $\tfrac{1}{2}\,y : \tfrac{1}{2}\,y$ is meaningful or detectable at the genetic level.

We can now also explain results of the cross between the plants grown from F_1 yellow seeds (Y/y) and the plants grown from green seeds (y/y). In this case, equal segregation in the yellow heterozygous F_1 gives gametes with a $\tfrac{1}{2}\,Y : \tfrac{1}{2}\,y$ ratio. The y/y parent can make only y gametes, however; so the phenotype of the progeny depends only on which allele they inherit from the Y/y parent. Thus, the $\tfrac{1}{2}\,Y : \tfrac{1}{2}\,y$ *gametic* ratio from the heterozygote is converted into a $\tfrac{1}{2}\,Y/y : \tfrac{1}{2}\,y/y$

genotypic ratio, which corresponds to a 1 : 1 *phenotypic* ratio of yellow-seeded to green-seeded plants. This is illustrated in the right-hand panel of Figure 2-5.

Note that, in defining the allele pairs that underlay his phenotypes, Mendel had identified a gene that radically affects pea color. This identification was not his prime interest, but we can see how finding single-gene inheritance patterns is a process of gene discovery, identifying individual genes that influence a biological property.

> **Message** All 1 : 1, 3 : 1, and 1 : 2 : 1 ratios are diagnostic of single-gene inheritance and are based on equal segregation in a heterozygote.

Mendel's research in the mid-nineteenth century was not noticed by the international scientific community until similar observations were independently published by several other researchers in 1900. Soon research in many species of plants, animals, fungi, and algae showed that Mendel's law of equal segregation was applicable to all eukaryotes and, in all cases, was based on the chromosomal segregations taking place in meiosis, a topic that we turn to in the next section.

2.2 The Chromosomal Basis of Single-Gene Inheritance Patterns

Mendel's view of equal segregation was that the members of a gene pair segregated equally *in gamete formation*. He did not know about the subcellular events that take place when cells divide in the course of gamete formation. Now we understand that gene pairs are located on chromosome pairs and that it is the members of a chromosome pair that actually segregate, carrying the genes with them. The members of a gene pair are segregated as an inevitable consequence.

Single-gene inheritance in diploids

When cells divide, so must the nucleus and its main contents, the chromosomes. To understand gene segregation, we must first understand and contrast the two types of nuclear divisions that take place in eukaryotic cells. When somatic (body) cells divide to increase their number, the accompanying nuclear division is called **mitosis,** a programmed stage of all eukaryotic cell-division cycles (Figure 2-6).

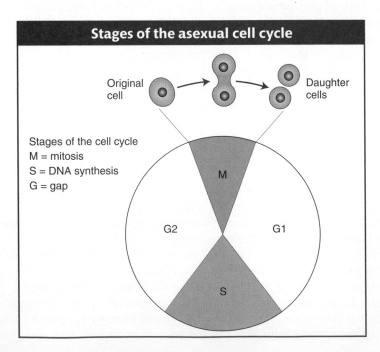

Figure 2-6

Mitosis can take place in diploid or haploid cells. As a result, one progenitor cell becomes two. Hence,

$$\text{either } 2n \longrightarrow 2n + 2n$$
$$\text{or } n \longrightarrow n + n$$

In addition, most eukaryotes have a sexual cycle, and, in these organisms, specialized diploid cells called **meiocytes** are set aside to divide to produce sex cells such as sperm and egg in plants and animals or sexual spores in fungi or algae. Two sequential cell divisions take place, and the two nuclear divisions that accompany them are called **meiosis.** Because there are two divisions, four cells are produced. Meiosis takes place only in diploid cells, and the cells that result (sperm and eggs in animals and plants) are haploid. Hence, the net result of meiosis is

$$2n \longrightarrow n + n + n + n$$

The location of the meiocytes in animal, plant, and fungal life cycles is shown in Figure 2-7.

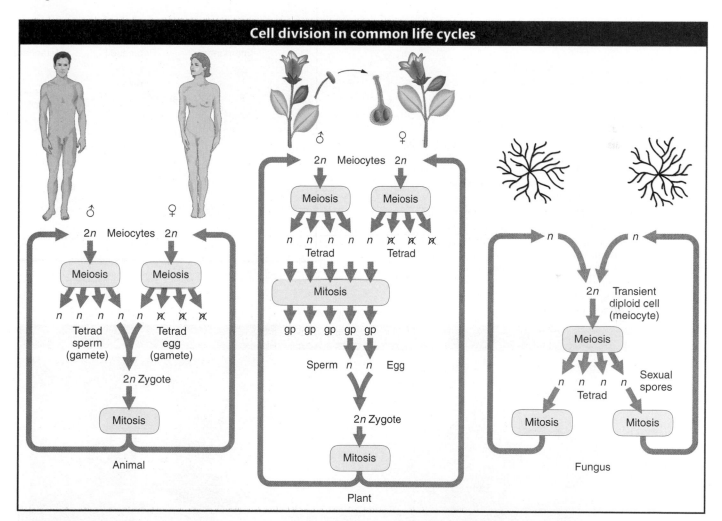

Figure 2-7 The life cycles of humans, plants, and fungi, showing the points at which mitosis and meiosis take place. Note that in the females of humans and many plants, three cells of the meiotic tetrad abort. The abbreviation n indicates a haploid cell, $2n$ a diploid cell; gp stands for gametophyte, the name of the small structure composed of haploid cells that will produce gametes. In many plants such as corn, a nucleus from the male gametophyte fuses with two nuclei from the female gametophyte, giving rise to a triploid ($3n$) cell, which then replicates to form the endosperm, a nutritive tissue that surrounds the embryo (which is derived from the $2n$ zygote).

Figure 2-8 Simplified representation of mitosis and meiosis in diploid cells ($2n$, diploid; n, haploid). (Detailed versions are shown in Appendix 2-1, page 77.)

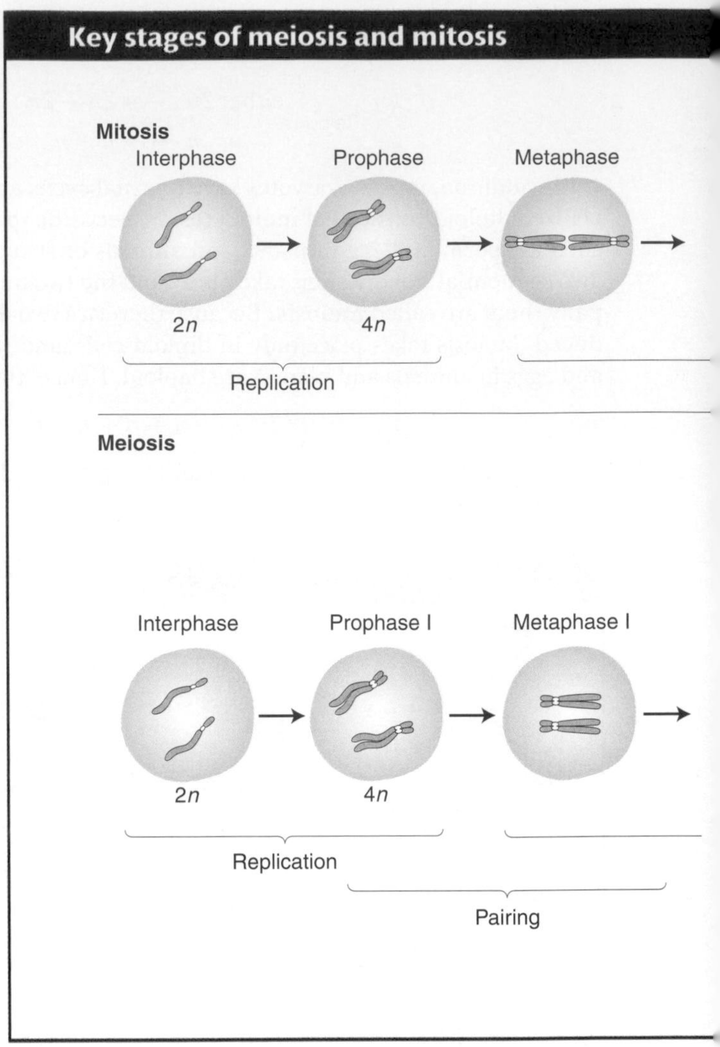

Figure 2-9 (a) In *Hyalophora cecropia*, a silk moth, the normal male chromosome number is 62, giving 31 synaptonemal complexes. In the individual shown here, one chromosome (*center*) is represented three times; such a chromosome is termed *trivalent*. The DNA is arranged in regular loops around the synaptonemal complex. (b) Regular synaptonemal complex in *Lilium tyrinum*. Note (*right*) the two lateral elements of the synaptonemal complex and (*left*) an unpaired chromosome, showing a central core corresponding to one of the lateral elements. [*Courtesy of Peter Moens.*]

Key stages of meiosis and mitosis

Mitosis

Interphase → Prophase → Metaphase →

$2n$ $4n$

Replication

Meiosis

Interphase → Prophase I → Metaphase I →

$2n$ $4n$

Replication

Pairing

Synaptonemal complexes at meiosis

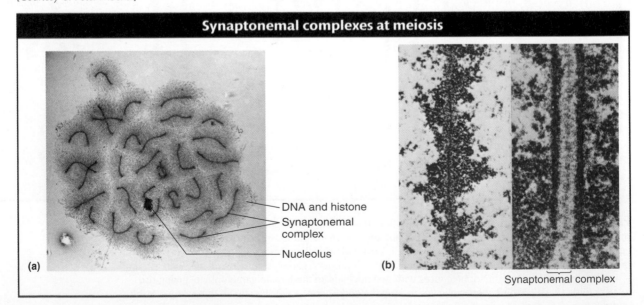

— DNA and histone
— Synaptonemal complex
— Nucleolus

(a)

(b)

Synaptonemal complex

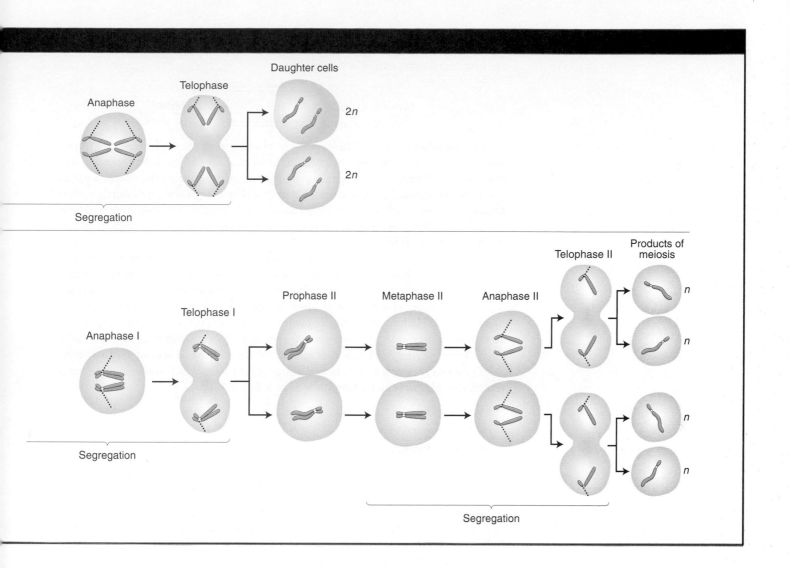

The basic *genetic* features of mitosis and meiosis are summarized in Figure 2-8. To make comparison easier, both processes are shown in a diploid cell. Notice, again, that mitosis takes place in one cell division, and the two resulting "daughter" cells have the same genomic content as that of the "mother" (progenitor) cell. The first key process to note is a premitotic chromosome replication. At the DNA level, this stage is the synthesis, or S, phase (see Figure 2-6), at which the DNA is replicated. The replication produces pairs of identical sister **chromatids,** which become visible at the beginning of mitosis. When a cell divides, each member of a pair of sister chromatids is pulled into a daughter cell, where it assumes the role of a fully fledged chromosome. Hence, each daughter cell has the same chromosomal content as the original cell.

Before meiosis, as in mitosis, chromosome replication takes place to form sister chromatids, which become visible at meiosis. The centromere appears not to divide at this stage, whereas it does in mitosis. Also in contrast with mitosis, the homologous pairs of sister chromatids now unite to form a bundle of four homologous chromatids. This joining of the homologous pairs is called *synapsis*, and it relies on the properties of a macromolecular assemblage called the synaptonemal complex (SC), which runs down the center of the pair (Figure 2-9). Replicate sister chromosomes are together called a **dyad** (from the Greek word for two). The unit comprising the pair of synapsed dyads is called a **bivalent.** The four chromatids that make up a bivalent are called a **tetrad** (Greek for four), to indicate that there are four homologous units in the bundle.

(A parenthetical note. The process of *crossing over* takes place at this tetrad stage. Crossing over changes the combinations of alleles of several different genes but does not directly affect single-gene inheritance patterns; therefore, we will postpone its detailed coverage until Chapter 4. For the present, it is worth noting that, apart from its allele-combining function, crossing over is also known to be a crucial event that must take place in order for proper chromosome segregation in the first meiotic division.)

The bivalents of all chromosomes move to the cell's equator, and, when the cell divides, one dyad moves into each new cell, pulled by spindle fibers attached to the centromeres. In the second cell division of meiosis, the centromeres divide and each member of a dyad (each member of a pair of chromatids) moves into a daughter cell. Hence, although the process starts with the same genomic content as that for mitosis, the two successive segregations result in four haploid cells. Each of the four haploid cells that constitute the four **products of meiosis** contains one member of a tetrad; hence, the group of four cells is sometimes called a tetrad, too. Meiosis can be summarized as follows:

Start: → two homologs

Replication: → two dyads

Pairing: → tetrad

First division: → one dyad to each daughter cell

Second division: → one chromatid to each daughter cell

Research in cell biology has shown that the spindle fibers that pull apart chromosomes are polymers of the molecule tubulin. The pulling apart is caused mainly by a depolymerization and hence shortening of the fibers at the point where they are attached to the chromosomes.

The behavior of chromosomes during meiosis clearly explains Mendel's law of equal segregation. Consider a heterozygote of general type A/a. We can simply follow the preceding summary while considering what happens to the alleles of this gene:

Start: one homolog carries A and one carries a

Replication: one dyad is AA and one is aa

Pairing: tetrad is $A/A/a/a$

First division products: one cell AA, the other cell aa (crossing over can mix these types of products up, but the overall ratio is not changed)

Second division products: four cells, two of type A and two of type a

Hence, the products of meiosis from a heterozygous meiocyte A/a are $\frac{1}{2} A$ and $\frac{1}{2} a$, precisely the equal ratio that is needed to explain Mendel's first law.

Meiosis

Note that, in the present discussion, we have focused on the broad genetic aspects of meiosis, necessary to explain single-gene inheritance. More complete descriptions of the detailed stages of mitosis and meiosis are presented in Appendices 2-1 and 2-2 at the end of this chapter.

Single-gene inheritance in haploids

We have seen that the cellular basis of the law of equal segregation is the segregation of chromosomes in the first division of meiosis. In the discussion so far, the evidence for the equal segregation of alleles in meiocytes of both plants and animals is *indirect*, based on the observation that crosses show the appropriate ratios of progeny expected under equal segregation. Recognize that the gametes in these studies (such as Mendel's) must have come from *many different* meiocytes. However, in some haploid organisms, such as several species of fungi and algae, equal segregation can be observed *directly* within one *individual* meiocyte. The reason is that, in the cycles of these organisms, the four products of a single meiosis are temporarily held together in a type of sac. Baker's yeast, *Saccharomyces cerevisiae*, provides a good example (see the yeast Model Organism box in Chapter 12). In fungi, there are simple forms of sexes called *mating types*. In *S. cerevisiae*, the two mating types are called MATa and MATα, determined by the alleles of one gene. (Note that the symbol for a gene can be more than one letter—in this case, four.) A successful cross must be between strains of opposite mating type—that is, MATa × MATα.

Let's look at a cross that includes a yeast mutant. Normal wild-type yeast colonies are white, but, occasionally, red mutants arise owing to a mutation in a gene in the biochemical pathway that synthesizes adenine. Let's use the red mutant to investigate equal segregation in a single meiocyte. We can call the mutant allele *r* for *red*. What symbol can we use for the normal, or wild-type, allele? In experimental genetics, the wild-type allele for any gene is generally designated by a plus sign, +. This sign is attached as a superscript to the symbol invented for the mutant allele. Hence, the wild-type allele in this example would be designated r^+, but a simple + is often used as shorthand. To see single-gene segregation, the red mutant is crossed with wild type. The cross would have to be between different mating types. For example, if the red mutant happened to have arisen in a MATa strain, the cross would be

$$\text{MAT}\alpha \cdot r^+ \times \text{MATa} \cdot r$$

When two cells of opposite mating type fuse, a diploid cell is formed, and it is this cell that becomes the meiocyte. In the present example (ignoring mating type so as to focus on the red mutation), the diploid meiocyte would be heterozygous r^+/r. Replication and segregation of r^+ and r would give a tetrad of two meiotic products (spores) of genotype r^+ and two of r, all contained within a membranous sac called an **ascus.** Hence,

$$r^+/r \longrightarrow \left. \begin{array}{l} r^+ \\ r^+ \\ r \\ r \end{array} \right\} \text{ tetrad in ascus}$$

The details of the process are shown in Figure 2-10. If the four spores from one ascus are isolated (representing a tetrad

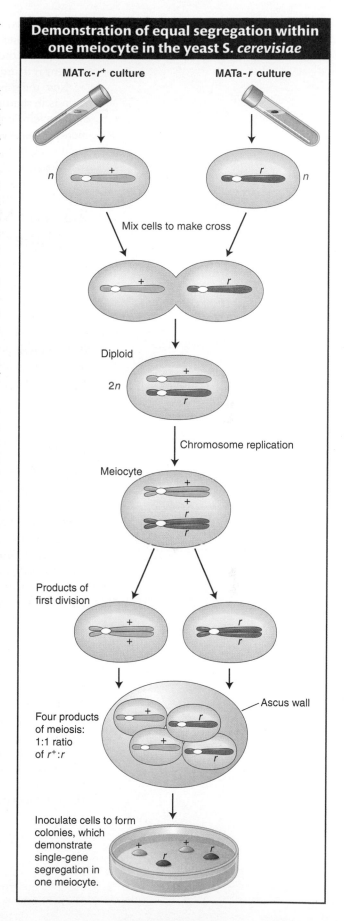

Demonstration of equal segregation within one meiocyte in the yeast *S. cerevisiae*

MATα-r^+ culture   MATa-r culture

Mix cells to make cross

Diploid

$2n$

Chromosome replication

Meiocyte

Products of first division

Four products of meiosis: 1:1 ratio of r^+:r

Ascus wall

Inoculate cells to form colonies, which demonstrate single-gene segregation in one meiocyte.

Figure 2-10 One ascus isolated from the cross + × r leads to two cultures of + and two of r.

of chromatids) and used to generate four yeast cultures, then equal segregation within one meiocyte is revealed directly as two white cultures and two red. If we analyzed the random spores from many meiocytes, we would find about 50 percent red and 50 percent white.

Note the simplicity of haploid genetics: a cross requires the analysis of only one meiosis; in contrast, a diploid cross requires a consideration of meiosis in both the male and the female parent. This simplicity is an important reason for using haploids as model organisms. Another reason is that, in haploids, all alleles are expressed in the phenotype, because there is no masking of recessives by dominant alleles on the other homolog.

2.3 The Molecular Basis of Mendelian Inheritance Patterns

Of course, Mendel had no idea of the molecular nature of the concepts he was working with. In this section, we can begin putting some of Mendel's concepts into a molecular context. Let's begin with alleles. We have used the concept of *alleles* without defining them at the molecular level. What are the *structural differences* between wild-type and mutant alleles at the level of the DNA of a gene? What are the *functional differences* at the protein level? Mutant alleles can be used to study single-gene inheritance without needing to understand their structural or functional nature. However, because a primary reason for embarking on single-gene inheritance is ultimately to investigate a gene's function, we must come to grips with the molecular nature of wild-type and mutant alleles at both the structural and the functional level.

Structural differences between alleles at the molecular level

Mendel proposed that genes come in different forms we now call alleles. What are alleles at the molecular level? When alleles such as *A* and *a* are examined at the DNA level by using modern technology, they are generally found to be identical in most of their sequences and differ only at one or several nucleotides of the thousands of nucleotides that make up the gene. Therefore, we see that the alleles are truly different versions of the same gene. The following diagram represents the DNA of two alleles of one gene; the letter *x* represents a difference in the nucleotide sequence:

| Allele 1 | ███████████████████ |
| Allele 2 | ████ x ██████████████ |

If the nucleotide sequence of an allele changes as the result of a rare chemical "accident," a new allele is created. Such changes are called **mutations:** they can occur anywhere along the nucleotide sequence of a gene. For example, a mutation could be change in the identity of a single nucleotide or the deletion of one or more nucleotides or even the addition of one or more nucleotides.

There are many ways that a gene can be changed by mutation. For one thing, the mutational damage can occur at any one of many different sites. We can represent the situation as follows, where dark blue indicates the normal wild-type DNA sequence and red with the letter *x* represents the altered sequence:

Wild-type allele	*A*	████████████████
Mutant allele	*a'*	████ x ████████████
Mutant allele	*a''*	████████ x ████████
Mutant allele	*a'''*	████████████ x ████

Alleles with mutations are usually recessive because it usually takes only one copy of a wild-type gene to provide normal function. When geneticists use the symbol *A* to represent a wild-type allele, it means one specific sequence of DNA. But when they use the symbol *a* to represent a recessive allele, it is a shorthand that can represent any one of the possible types of damage that can lead to nonfunctional recessive alleles.

Molecular aspects of gene transmission

Replication of alleles during the S phase The first step in transmission of a gene to a subsequent generation of cells or organisms is the formation of sister chromatids., which is a prelude to both mitosis and meiosis. What happens to alleles at the molecular level during the formation of sister chromatids? We know that the primary genomic component of each chromosome is a DNA molecule. This DNA molecule is replicated during the S phase, which precedes both mitosis and meiosis. As we will see in Chapter 7, replication is an accurate process and so all the genetic information is duplicated, whether wild type or mutant. For example, if a mutation is the result of a change in a single nucleotide pair—say, from GC (wild type) to AT (mutant)—then in a heterozygote, replication will be as follows:

$$\text{homolog GC} \longrightarrow \text{replication} \longrightarrow \begin{array}{l} \text{chromatid GC} \\ \text{chromatid GC} \end{array}$$

$$\text{homolog AT} \longrightarrow \text{replication} \longrightarrow \begin{array}{l} \text{chromatid AT} \\ \text{chromatid AT} \end{array}$$

DNA replication before mitosis in a haploid and a diploid are shown in Figure 2-11. This type of illustration serves to remind us that, in our considerations of the mechanisms of inheritance, it is essentially DNA molecules that are being moved around in the dividing cells.

Meiosis and mitosis at the molecular level The replication of DNA during the S phase produces two copies of each

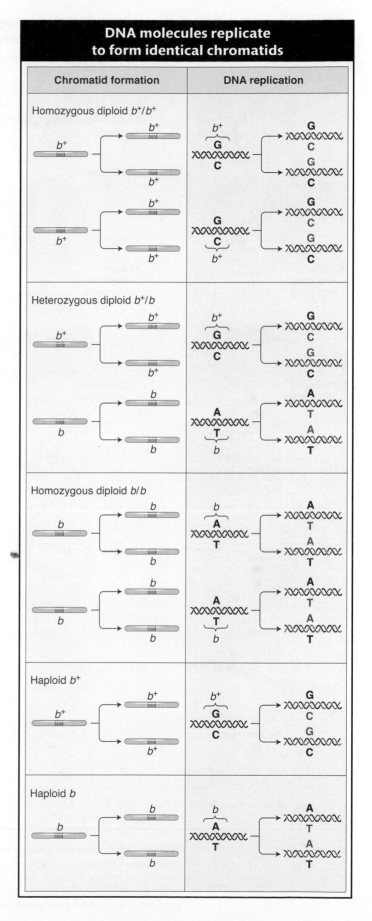

Figure 2-11 Each chromosome divides longitudinally into two chromatids (*left*); at the molecular level (*right*), the single DNA molecule of each chromosome replicates, producing two DNA molecules, one for each chromatid. Also shown are various combinations of a gene with wild-type allele *b*⁺ and mutant form *b*, caused by the change in a single base pair from GC to AT. Notice that, at the DNA level, the two chromatids produced when a chromosome replicates are always identical with each other and with the original chromosome.

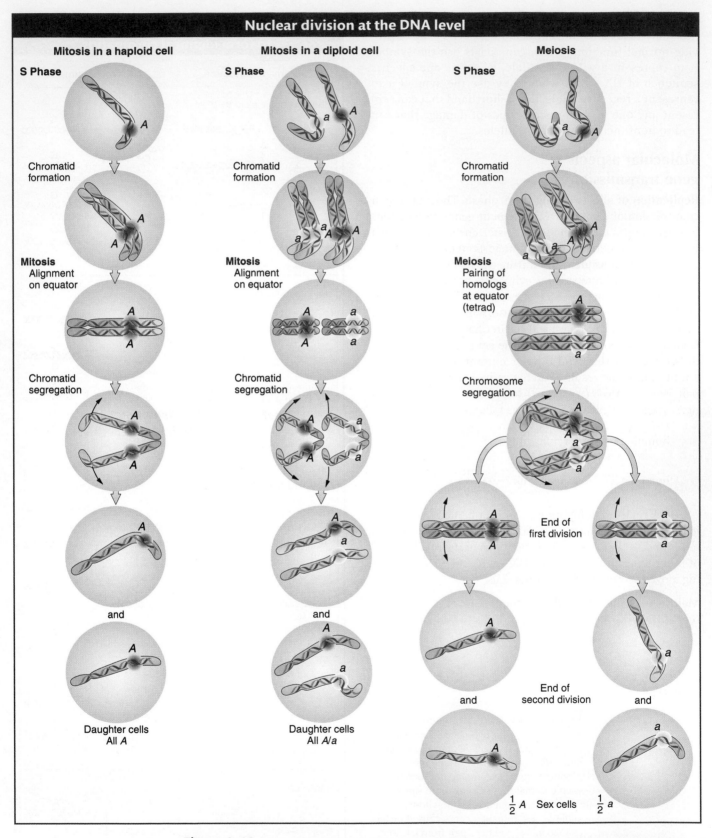

Figure 2-12 DNA and gene transmission in mitosis and meiosis in eukaryotes. The S phase and the main stages of mitosis and meiosis are shown. Mitotic divisions (*left and middle*) conserve the genotype of the original cell. At the right, the two successive meiotic divisions that take place during the sexual stage of the life cycle have the net effect of halving the number of chromosomes. The alleles *A* and *a* of one gene are used to show how genotypes are transmitted in cell division.

allele, *A* and *a*, that can now be segregated into separate cells. Nuclear division visualized at the DNA molecular level is shown in Figure 2-12.

Demonstrating chromosome segregation at the molecular level We have interpreted single-gene phenotypic inheritance patterns in relation to the segregation of chromosomal DNA at meiosis. Is there any way to show DNA segregation directly (as opposed to phenotypic segregation)? The straightforward approach would be to sequence the alleles (say, *A* and *a*) in the parents and the meiotic products: the result would be that one-half of the products would have the *A* DNA sequence and one-half would have the *a* DNA sequence. The same would be true for any DNA sequence that differed in the inherited chromosomes, including those not necessarily inside alleles correlated with known phenotypes such as red and white flowers.

Another way of characterizing and tracking a segment of DNA is to use a *restriction fragment length polymorphism* (*RFLP*). In Chapter 1, we learned that restriction enzymes are bacterial enzymes that cut DNA at specific base sequences in the genome. The target sequences have no biological significance in organisms other than bacteria; they are present purely by chance. Although the target sites are generally found consistently at specific locations, sometimes, on any one chromosome, a specific target site is missing or there is an extra site. If such a site flanks the sequence hybridized by a probe, then a Southern hybridization will reveal an RFLP. Consider this simple example in which one chromosome of one parent contains an extra site not found in the other chromosomes of that type in that cross:

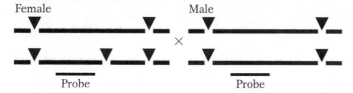

The Southern hybridizations will show two bands in the female and only one in the male. The "heterozygous" fragments will be inherited in exactly the same way as a gene. The results of the preceding cross could be written as follows:

long/short × long/long

and the progeny will be

$\frac{1}{2}$ long/short

$\frac{1}{2}$ long/long

according to the law of equal segregation.

In the preceding example, the RFLP is not associated with any measurable phenotypic difference, and the sites might be within a gene or not. However, sometimes by chance a mutation within a gene's coding sequence that produces a mutant phenotype also introduces a new target site for a restriction enzyme. The new target site provides a convenient molecular tag for the alleles. For example, a recessive mutation in a gene for pigment synthesis might produce an albino mutant allele and, at the same time, a new restriction site. Hence, a probe for the mutant allele detects two fragments in *a* and only one in *A*. The inheritance pattern is shown in Figure 2-13. In this illustration, we see a direct demonstration

Figure 2-13 A recessive mutation that produces allele *a* by chance also introduces a new cutting site for a restriction enzyme. This cutting site allows the inheritance of the mutation to be tracked by using a Southern analysis. The Southern blot detects one DNA fragment in homozygous normal individuals (*A/A*) and two fragments in albino individuals (*a/a*), but it detects three fragments in heterozygous individuals, owing to the presence of the normal and mutant alleles.

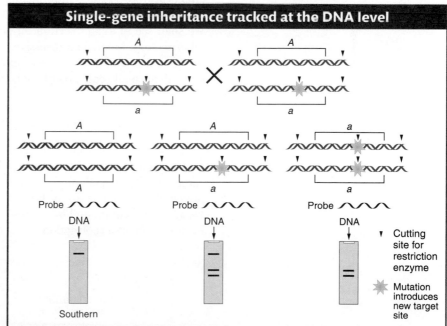

of molecular segregation taking place together with allelic and phenotypic segregation. (Other types of mutations would produce different effects at the level detected by Southern, Northern, and Western analyses.)

A third way of zeroing in on a specific segment of the genome is to use the polymerase chain reaction (PCR; see Chapter 1). As we saw in Chapter 1, parts of the genome contain long stretches of the same sequence of nucleotides repeated over and over again. A segment on one chromosome can contain a different number of repeating sequences from the corresponding segment on a homologous chromosome. Hence, an individual could be heterozygous for a pair of DNA segments, one that is *short* (low number of repeats) on one chromosome and one that is *long* (higher number) on the other chromosome. In this case, the meiotic products will reveal $\frac{1}{2}$ *short* and $\frac{1}{2}$ *long* PCR products, again demonstrating chromosome segregation. These differences in segment length can be detected using PCR. Pairs of PCR primers often span segments that, although generally homologous, contain different numbers of repeating sequences. Hence, PCR amplifications reveal products of different sizes.

When used in the above manner, the direct sequencing, RFLP, and PCR approaches provide what are called **molecular markers.** That is, sequence "alleles," RFLP "alleles," and PCR "alleles" are tags or markers that, although not associated with a specific biological function, can be used to track the inheritance of a segment of a chromosome at some specific position in the same way that the alleles of a gene do.

Message Mendelian inheritance is shown by any segment of DNA on a chromosome: by genes and their alleles and by molecular markers not necessarily associated with any biological function.

Alleles at the molecular level

At the molecular level, the primary phenotype of a gene is the protein it produces. What are the functional differences between proteins that explain the different effects of wild-type and mutant alleles on the properties of an organism?

Let's explore the topic by using the human disease phenylketonuria (PKU). We shall see in a later section on pedigree analysis that the PKU phenotype is inherited as a Mendelian recessive. The disease is caused by a defective allele of the gene that encodes the liver enzyme phenylalanine hydroxylase (PAH). This enzyme normally converts phenylalanine in food into the amino acid tyrosine:

$$\text{phenylalanine} \xrightarrow{\substack{\text{phenylalanine} \\ \text{hydroxylase}}} \text{tyrosine}$$

However, a mutation in the gene encoding this enzyme may alter the amino acid sequence in the vicinity of the enzyme's active site. In this case, the enzyme cannot bind phenylalanine (its substrate) or convert it into tyrosine. Therefore, phenylalanine builds up in the body and is converted instead into phenylpyruvic acid. This compound interferes with the development of the nervous system, leading to mental retardation.

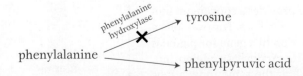

Babies are now routinely tested for this processing deficiency at birth. If the deficiency is detected, phenylalanine can be withheld with the use of a special diet and the development of the disease arrested.

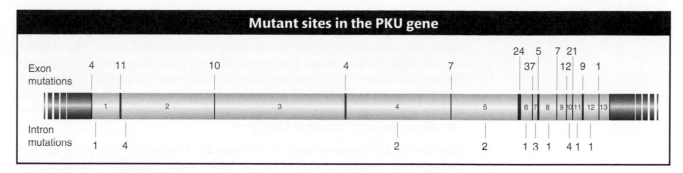

Figure 2-14 Many mutations of the human phenylalanine hydroxylase gene that cause enzyme malfunction are known. The number of mutations in the exons, or protein-encoding regions (black), are listed above the gene. The number of mutations in the intron regions (green, numbered 1 through 13) that alter splicing are listed below the gene. [*After C. R. Scriver, Ann. Rev. Genet. 28, 1994, 141–165.*]

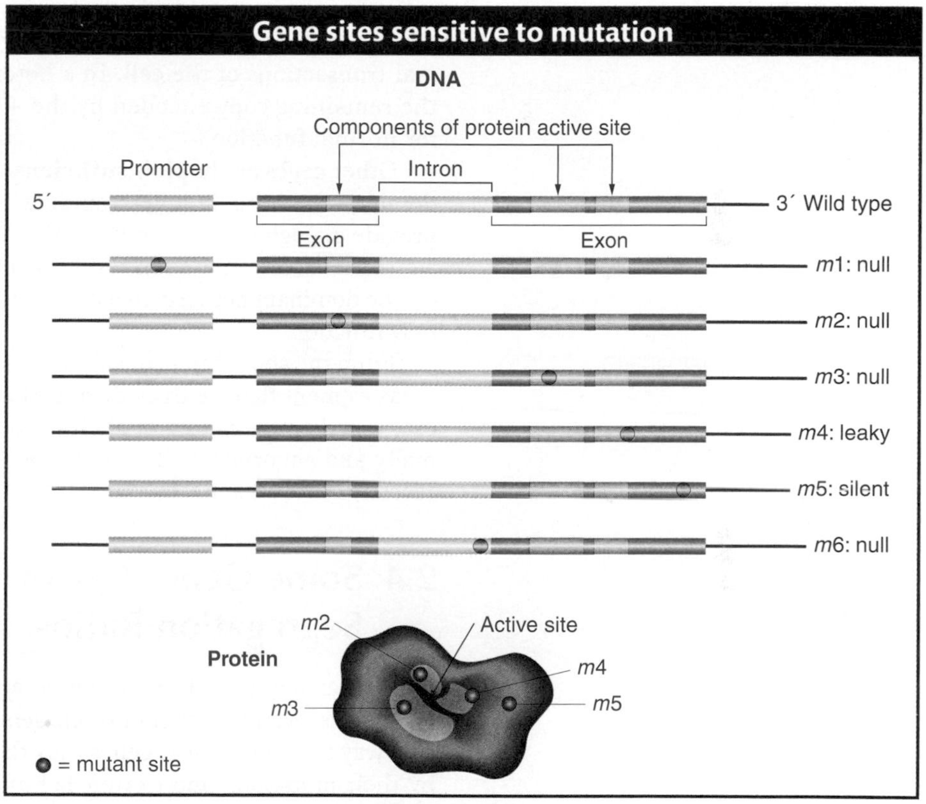

The PAH enzyme is made up of a single type of protein. What changes have occurred in the mutant form of the PKU gene's DNA, and how can such change at the DNA level affect protein function and produce the disease phenotype? Sequencing of the mutant alleles from many PKU patients has revealed a plethora of mutations at different sites along the gene, mainly in the protein-encoding regions, or the exons; the results are summarized in Figure 2-14. They represent a range of DNA changes, but most are small changes affecting only one nucleotide pair among the thousands that constitute the gene. What these alleles have in common is that they encode a defective protein that no longer has normal PAH activity. By changing one or more amino acids, the mutations all inactivate some essential part of the protein encoded by the gene. The effect of the mutation on the function of the gene depends on where within the gene the mutation occurs. An important functional region of the gene is that encoding an enzyme's active site; so this region is very sensitive to mutation. In addition, a minority of mutations are found to be in introns, and these mutations often prevent the normal processing of the primary RNA transcript. Some of the general consequences of mutation at the protein level are shown in Figure 2-15. Many of the mutant alleles are of a type generally called **null alleles:** the proteins encoded by them completely lack PAH function. Other mutant alleles reduce the level of enzyme function; they are sometimes called **leaky mutations,** because some wild-type function seems to "leak" into the mutant phenotype.

Figure 2-15 Mutations in the parts of a gene encoding enzyme active sites lead to enzymes that do not function (null mutations). Mutations elsewhere in the gene may have no effect on enzyme function (silent mutations). Promoters are sites important in transition initiation.

Message Most mutations alter the amino acid sequence of the gene's protein product, resulting in reduced or absent function.

We have been pursuing the idea that finding a set of genes that impinge on the biological property under investigation is an important goal of genetics, because it defines the components of the system. However, finding the *precise* way in which mutant alleles lead to mutant phenotypes is often challenging, requiring not only the identification of the protein products of these genes, but also detailed cellular and physiological studies to measure the effects of the mutations. Furthermore, finding how the set of genes *interacts* is a second level of challenge and a topic that we will pursue later, starting in Chapter 6.

Dominance and recessiveness   With an understanding of how genes function through their protein products, we can better understand dominance and recessiveness. Dominance was defined earlier in this chapter as the phenotype shown by a heterozygote. Hence, formally, it is the *phenotype* that is dominant or recessive, but, in practice, geneticists more often apply the term to alleles. This formal definition has no molecular content, but both dominance and recessiveness can have simple explanations at the molecular level. We introduce the topic here, to be revisited in Chapter 6. Recessiveness is observed in mutations in genes that are functionally **haplosufficient.** Although a diploid cell normally has two wild-type copies of a gene, one copy of a haplosufficient gene provides enough gene product (generally a protein) to carry out the normal transactions of the cell. In a heterozygote (say, $+/m$, where m is a null), the remaining copy encoded by the $+$ allele provides enough protein product for normal function.

Other genes are **haploinsufficient.** In such cases, a null mutant allele will be dominant because, in a heterozygote $(+/M)$, the single wild-type allele cannot provide enough product for normal function.

In some cases, mutation results in a *new function* for the gene. Such mutations can be dominant because, in a heterozygote, the wild-type allele cannot mask this new function.

From the above brief considerations, we see that *phenotype*, the description or measurement that we track during Mendelian inheritance, is an emergent property based on the nature of alleles and the way in which the gene functions normally and abnormally. The same can be said for the descriptions dominant and recessive that we apply to a phenotype.

2.4 Some Genes Discovered by Observing Segregation Ratios

Recall that one general aim of genetic analysis today is to dissect a biological property by discovering the set of single genes that affect it. We learned that an important way to identify these genes is by the phenotypic segregation ratios generated by their mutations—most often 1:1 and 3:1 ratios, both of which are based on equal segregation as defined by Gregor Mendel.

Let's look at some examples that extend the Mendelian approach into a modern experimental setting. Typically, the researcher is confronted by an array of interesting mutant phenotypes that affect the property of interest (such as those depicted in Figure 2-1) and now needs to know whether they are inherited as single mutant alleles. Mutant alleles can be either dominant or recessive, depending on their action; so the question of dominance also needs to be considered in the analysis.

The standard procedure is to cross the mutant with wild type. (If the mutant is sterile, then another approach is needed.) First, we will consider three simple cases that cover most of the possible outcomes:

1. A fertile flower mutant with no pigment in the petals (for example, white petaled in contrast with the normal red)

2. A fertile fruit-fly mutant with short wings

3. A fertile mold mutant that produces excess hyphal branches (hyperbranching)

A gene active in the development of flower color

To begin the process, the white-flowered plant is crossed with the normal wild-type red. All the F_1 plants are red flowered, and, of 500 F_2 plants sampled, 378 are red flowered and 122 are white flowered. If we acknowledge the existence of sampling error, these F_2 numbers are very close to a $\frac{3}{4} : \frac{1}{4}$, or 3 : 1, ratio. Because this ratio indicates single-gene inheritance, we can conclude that the mutant is caused by a recessive alteration in a single gene. According to the general rules of gene nomenclature, the mutant allele for white petals might be called *alb* for *albino* and the wild-type allele would be *alb*⁺ or just +. (The conventions for allele nomenclature vary somewhat among organisms: some of the variations are shown in the appendix on nomenclature.) We conclude that the wild-type allele plays an essential role in producing the colored petals of the plant, a property that is almost certainly necessary for attracting pollinators to the flower. The gene might be implicated in the biochemical synthesis of the pigment or in the part of the signaling system that tells the cells of the flower to start making pigment or in a number of other possibilities that require further investigation. At the purely genetic level, the crosses made would be represented symbolically as

$$
\begin{array}{ll}
P & +/+ \times alb/alb \\
F_1 & \text{all } +/alb \\
F_2 & \frac{1}{4} +/+ \\
 & \frac{1}{2} +/alb \\
 & \frac{1}{4} alb/alb
\end{array}
$$

A gene for wing development

In the fruit-fly example, the cross of the mutant short-winged fly with wild-type long-winged stock yielded 788 progeny, classified as follows:

196 short-winged males

194 short-winged females

197 long-winged males

201 long-winged females

In total, there are 390 short- and 398 long-winged progeny, very close to a 1 : 1 ratio. The ratio is the same within males and females, again within the bounds of sampling error. Hence, from these results, the "short wings" mutant was very likely produced by a dominant mutation. Note that, for a dominant mutation to be expressed, only a single "dose" of mutant allele is necessary; so, in most cases, when the mutant first shows up in the population, it will be in the heterozygous state. (This is not true for a recessive mutation such as that in the preceding plant example, which must be homozygous to be expressed and must have come from the selfing of an unidentified heterozygous plant in the preceding generation.)

When long-winged progeny were interbred, all of their progeny were long winged, as expected of a recessive wild-type allele. When the short-winged progeny were interbred, their progeny showed a ratio of three-fourths short to one-fourth long.

Dominant mutations are represented by uppercase letters or words: in the present example, the mutant allele might be named *SH*, standing for short. Then the crosses would be symbolized as follows:

$$P \qquad +/+ \ \times \ SH/+$$

$$F_1 \qquad \tfrac{1}{2} \ +/+$$

$$\tfrac{1}{2} \ SH/+$$

$$+/+ \ \times \ +/+$$

$$F_2 \qquad \text{all } +/+$$

$$SH/+ \ \times \ SH/+$$

$$F_2 \qquad \tfrac{1}{4} \ SH/SH$$

$$\tfrac{1}{2} \ SH/+$$

$$\tfrac{1}{4} \ +/+$$

This analysis of the fly mutant identifies a gene that is part of a subset of genes that, in wild-type form, are crucial for the normal development of a wing. Such a result is the starting point of further studies that would focus on the precise developmental and cellular ways in which the growth of the wing is arrested, which, once identified, reveal the time of action of the wild-type allele in the course of development.

A gene for hyphal branching

A hyperbranching fungal mutant (such as the buttonlike colony in Figure 2-1) was crossed with a wild-type fungus with normal sparse branching. In a sample of 300 progeny, 152 were wild type and 148 were hyperbranching, very close to a 1:1 ratio. We infer from this single-gene inheritance ratio that the hyperbranching mutation is of a single gene. In haploids, assigning dominance is usually not possible, but, for convenience, we can call the hyperbranching allele *hb* and the wild type *hb⁺* or +. The cross must have been

$$P \qquad\qquad\qquad + \ \times \ hb$$

$$\text{Diploid meiocyte} \qquad +/hb$$

$$F_1 \qquad\qquad\qquad \tfrac{1}{2} \ +$$

$$\tfrac{1}{2} \ hb$$

The mutation and inheritance analysis has uncovered a gene whose wild-type allele is essential for normal control of branching, a key function in fungal dispersal and nutrient acquisition. Now the mutant needs to be investigated to see the location in the normal developmental sequence at which the mutant produces a block. This information will reveal the time and place in the cells at which the normal allele acts.

Sometimes, the severity of a mutant phenotype renders the organism sterile, unable to go through the sexual cycle. How can the single-gene inheritance of sterile mutants be demonstrated? In a diploid organism, a sterile recessive mutant can be propagated as a heterozygote and then the heterozygote can be selfed to produce the expected 25 percent mutants for study. A sterile dominant mutant is a genetic dead end and cannot be propagated sexually, but, in plants and fungi, such a mutant can be easily propagated asexually.

What if a cross between a mutant and a wild type does not produce a 3:1 or a 1:1 ratio as discussed here, but some other ratio? Such a result can be due to the interactions of several genes or to an environmental effect. Some of these possibilities are discussed in Chapter 6.

Forward genetics

In general, the type of approach to gene discovery that we have been following is sometimes called **forward genetics,** an approach to understanding biological function starting with random single-gene mutants and ending with detailed cell and biochemical analysis of them, often including genomic analysis. (We shall see **reverse genetics** at work in later chapters. In brief, it starts with genomic analysis to identify a set of genes as candidates for encoding the biological property of interest, then induces mutants targeted specifically to those genes, and then examines the mutant phenotypes to see if they affect the property under study.)

Message Gene discovery by single-gene inheritance is sometimes called forward genetics. In general, it works in the following sequence:

Choose biological property of interest
↓
Find mutants affecting that property
↓
Check mutants for single-gene inheritance
↓
Identify time and place of action of genes
↓
Zero in on molecular nature of gene by genomic (DNA) analysis

Predicting progeny proportions or parental genotypes by applying the principles of single-gene inheritance

We can summarize the direction of analysis of gene discovery as follows:

Observe phenotypic ratios in progeny →
Deduce genotypes of parents (A/A, A/a, or a/a)

However, the same principle of inheritance (essentially Mendel's law of equal segregation) can also be used to predict phenotypic ratios in the progeny of parents of *known genotypes*. These parents would be from stocks maintained by the researcher. The types and proportions of the progeny of crosses such as $A/A \times A/a$, $A/A \times a/a$, $A/a \times A/a$, and $A/a \times a/a$ can be easily predicted. In summary:

Cross parents of known genotypes → Predict phenotypic ratios in progeny

This type of analysis is used in general breeding to synthesize genotypes for research or for agriculture. It is also useful in predicting likelihoods of various outcomes in human matings in families with histories of single-gene diseases.

After single-gene inheritance has been established, an individual showing the dominant phenotype but of *unknown genotype* can be tested to see if the genotype is homozygous or heterozygous. Such a test can be performed by crossing the individual (of phenotype $A/?$) with a recessive tester strain a/a. If the individual is heterozygous, a $1:1$ ratio will result ($\frac{1}{2} A/a$ and $\frac{1}{2} a/a$); if the individual is homozygous, all progeny will show the dominant phenotype (all A/a). In general, the cross of an individual of unknown heterozygosity (for one gene or more) with a fully recessive parent is called a **testcross,** and the recessive individual is called a **tester.** We will encounter testcrosses many times throughout subsequent chapters; they are very useful in deducing the meiotic events taking place in more complex genotypes such as dihybrids and trihybrids. The use of a fully recessive tester means that meiosis in the tester parent can be ignored because all of its gametes are recessive and do not contribute to the phenotypes of the progeny. An alternative test for heterozygosity (useful if a recessive tester is not available and the organism can be selfed) is simply to self the unknown: if the organism being tested is heterozygous, a 3 : 1 ratio will be found in the progeny. Such tests are useful and common in routine genetic analysis.

> **Message** The principles of inheritance (such as the law of equal segregation) can be applied in two directions: (1) inferring genotypes from phenotypic ratios and (2) predicting phenotypic ratios from parents of known genotypes.

2.5 Sex-Linked Single-Gene Inheritance Patterns

The chromosomes that we have been analyzing so far are autosomes, the "regular" chromosomes that form most of the genomic set. However, many animals and plants have a special pair of chromosomes associated with sex. The sex chromosomes also segregate equally, but the phenotypic ratios seen in progeny are often different from the autosomal ratios.

Sex chromosomes

Most animals and many plants show sexual dimorphism; in other words, individuals are either male or female. In most of these cases, sex is determined by a special pair of **sex chromosomes.** Let's look at humans as an example. Human body cells have 46 chromosomes: 22 homologous pairs of autosomes plus 2 sex chromosomes. Females have a pair of identical sex chromosomes called the **X chromosomes.** Males have a nonidentical pair, consisting of one X and one Y. The **Y chromosome** is considerably shorter than the X. Hence, if we let A represent autosomal chromosomes, we can write

$$\text{females} = 44\text{A} + \text{XX}$$
$$\text{males} = 44\text{A} + \text{XY}$$

At meiosis in females, the two X chromosomes pair and segregate like autosomes, and so each egg receives one X chromosome. Hence, with regard to sex chromosomes, the gametes are of only one type and the female is said to be the **homogametic sex.** At meiosis in males, the X and the Y chromosomes pair over a short region, which ensures that the X and Y separate so that there are two types of sperm, half with an X and the other half with a Y. Therefore, the male is called the **heterogametic sex.**

The inheritance patterns of genes on the sex chromosomes are different from those of autosomal genes. Sex-chromosome inheritance patterns were first investigated in the early 1900s in the laboratory of the great geneticist Thomas Hunt Morgan, using the fruit fly *Drosophila melanogaster* (see the Model Organism box on page 52). This insect has been one of the most important research organisms in genetics; its short, simple life cycle contributes to its usefulness in this regard. Fruit flies have three pairs of autosomes plus a pair of sex chromosomes, again referred to as X and Y. As in mammals, *Drosophila* females have the constitution XX and males are XY. However, the mechanism of sex determination in *Drosophila* differs from that in mammals. In *Drosophila*, the *number of X chromosomes* in relation to the autosomes determines sex: two X's result in a female and one X results in

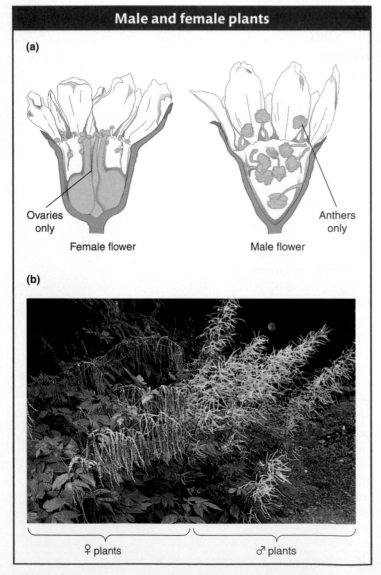

Figure 2-16 Examples of two dioecious plant species are (a) *Osmaronia dioica* and (b) *Aruncus dioicus.* [*(a) Lesley Bohm; (b) Anthony Griffiths.*]

Male and female plants

(a)

Ovaries only

Female flower

Anthers only

Male flower

(b)

♀ plants   ♂ plants

a male. In mammals, the *presence of the Y chromosome* determines maleness and the absence of a Y determines femaleness. However, it is important to note that, despite this somewhat different basis for sex determination, the single-gene inheritance patterns of genes on the sex chromosomes are remarkably similar in *Drosophila* and mammals.

Vascular plants show a variety of sexual arrangements. **Dioecious species** are those showing animal-like sexual dimorphism, with female plants bearing flowers containing only ovaries and male plants bearing flowers containing only anthers (Figure 2-16). Some, but not all, dioecious plants have a nonidentical pair of chromosomes associated with (and almost certainly determining) the sex of the plant. Of the species with nonidentical sex chromosomes, a large proportion have an XY system. For example, the dioecious plant *Melandrium album* has 22 chromosomes per cell: 20 autosomes plus 2 sex chromosomes, with XX females and XY males. Other dioecious plants have no visibly different pair of chromosomes; they may still have sex chromosomes but not visibly distinguishable types.

Sex-linked patterns of inheritance

Cytogeneticists divide the X and Y chromosomes into homologous and differential regions. Again, let's use humans as an example (Figure 2-17). The differential regions, which contain most of the genes, have no counterparts on the other sex chromosome. Hence, in males, the genes in the differential regions are said to be **hemizygous** ("half zygous"). The differential region of the X chromosome contains many hundreds of genes; most of these genes do not take part in sexual function and they influence a great range of human properties. The Y chromosome contains only a few dozen genes. Some of these genes have counterparts on the X chromosome, but most do not. The latter type take part in male sexual function. One of these genes, *SRY*, determines maleness itself. Several other genes are specific for sperm production in males.

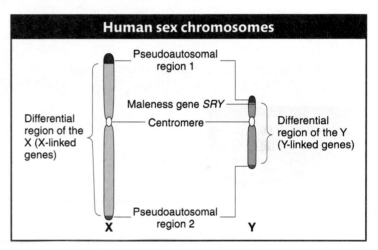

Figure 2-17 Human sex chromosomes contain a differential region and two pairing regions. The regions were located by observing where the chromosomes paired up in meiosis and where they did not.

In general, genes in the differential regions are said to show inheritance patterns called **sex linkage.** Mutant alleles in the differential region of the X chromosome show a single-gene inheritance pattern called **X linkage.** Mutant alleles of the few genes in the differential region of the Y chromosome show **Y linkage.** A gene that is sex-linked can show phenotypic ratios that are different in each sex. In this respect, sex-linked inheritance patterns contrast with the inheritance patterns of genes in the autosomes, which are the same in each sex. If the genomic location of a gene is unknown, a sex-linked inheritance pattern indicates that a gene lies on a sex chromosome.

The human X and Y chromosomes have two short homologous regions, one at each end (see Figure 2-17). In the sense that these regions are homologous, they are autosomal-like, and so they are called **pseudoautosomal regions 1** and **2.** One or both of these regions pairs in meiosis and undergoes crossing over (see Chapter 4 for details of crossing over). For this reason, the X and the Y chromosomes can act as a pair and segregate into equal numbers of sperm.

X-linked inheritance

For our first example of X linkage, we turn to eye color in *Drosophila*. The wild-type eye color of *Drosophila* is dull red, but pure lines with white eyes are available (Figure 2-18). This phenotypic difference is determined by two alleles of a gene located on the differential region of the X chromosome. The

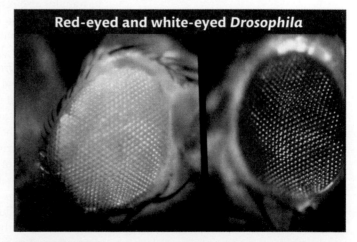

Figure 2-18 The red-eyed fly is wild type, and the white-eyed fly is a mutant. [*Photo Researchers/Getty Images.*]

Model Organism *Drosophila*

Drosophila melanogaster was one of the first model organisms to be used in genetics. It is readily available from ripe fruit, has a short life cycle, and is simple to culture and cross. Sex is determined by X and Y sex chromosomes (XX = female, XY = male), and males and females are easily distinguished. Mutant phenotypes regularly arise in lab populations, and their frequency can be increased by treatment with mutagenic radiation or chemicals. It is a diploid organism, with four pairs of homologous chromosomes ($2n = 8$). In salivary glands and certain other tissues, multiple rounds of DNA replication without chromosomal division result in "giant chromosomes," each with a unique banding pattern that provides geneticists with landmarks for the study of chromosome mapping and rearrangement. There are many species and races of *Drosophila*, which have been important raw material for the study of evolution.

Time flies like an arrow; fruit flies like a banana.
(Groucho Marx)

Drosophila melanogaster, the common fruit fly. [*SLP/Photo Researchers.*]

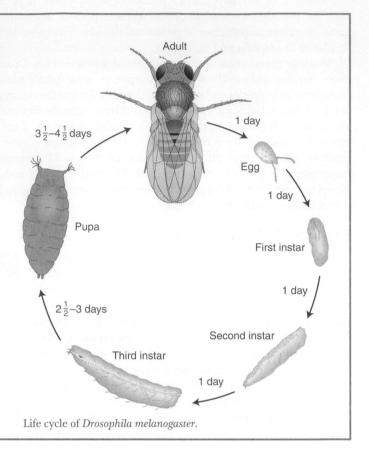

Life cycle of *Drosophila melanogaster*.

mutant allele in the present case is *w* for white eyes (the lowercase letter indicates that the allele is recessive) and the corresponding wild-type allele is w^+. When white-eyed males are crossed with red-eyed females, all the F_1 progeny have red eyes, suggesting that the allele for white eyes is recessive. Crossing these red-eyed F_1 males and females produces a $3:1$ F_2 ratio of red-eyed to white-eyed flies, but *all the white-eyed flies are males.* This inheritance pattern, which shows a clear difference between the sexes, is explained in Figure 2-19. The basis of the inheritance pattern is that all the F_1 flies receive a wild-type allele from their mothers, but the F_1 females also receive a white-eye allele from their fathers. Hence, all F_1 females are heterozygous wild type (w^+/w), and the F_1 males are hemizygous wild type (w^+). The F_1 females pass on the white-eye allele to half their sons, who express it, and to half their daughters, who do not express it, because they must inherit the wild-type allele from their fathers.

The reciprocal cross gives a different result; that is, the cross between white-eyed females and red-eyed males gives an F_1 in which all the females are red eyed but all the males are white eyed. In this case, every female inherited the dominant w^+ allele from the father's X chromosome, whereas every male inherited the recessive *w* allele from its mother. The F_2 consists of one-half red-eyed and one-half white-eyed flies of both sexes. Hence, in sex linkage, we see examples not only of different ratios in different sexes, but also of differences between reciprocal crosses.

Note that *Drosophila* eye color has nothing to do with sex determination, and so we have an illustration of the principle that genes on the sex chromosomes are not necessarily related to sexual function. The same is true in humans: in the discussion of pedigree analysis later in this chapter, we shall see many X-linked genes, yet few could be construed as being connected to sexual function.

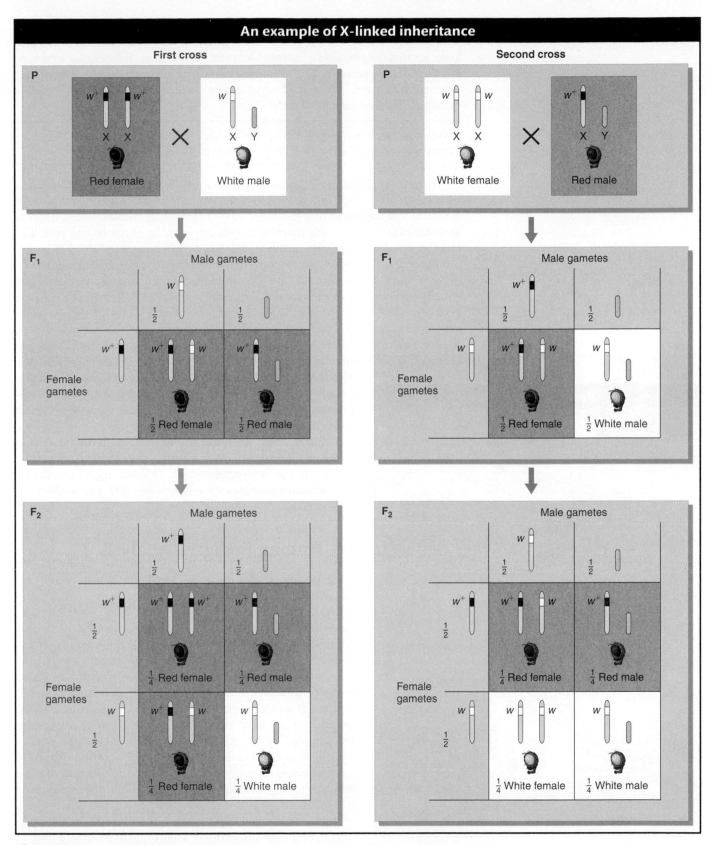

Figure 2-19 Reciprocal crosses between red-eyed (red) and white-eyed (white) *Drosophila* give different results. The alleles are X linked, and the inheritance of the X chromosome explains the phenotypic ratios observed, which are different from those of autosomal genes. (In *Drosophila* and many other experimental systems, a superscript plus sign is used to designate the normal, or wild-type, allele. Here, w^+ encodes red eyes and w encodes white eyes.)

The abnormal allele associated with white eye color in *Drosophila* is recessive, but abnormal alleles of genes on the X chromosome that are dominant also arise, such as the *Drosophila* mutant hairy wing (*Hw*). In such cases, the wild-type allele (*Hw⁺*) is recessive. The dominant abnormal alleles show the inheritance pattern corresponding to that of the wild-type allele for red eyes in the preceding example. The ratios obtained are the same.

> **Message** Sex-linked inheritance regularly shows different phenotypic ratios in the two sexes of progeny, as well as different ratios in reciprocal crosses.

Historically, in the early decades of the twentieth century, the demonstration by Morgan of X-linked inheritance of *white eye* in *Drosophila* was a key piece of evidence that suggested that genes are indeed located on chromosomes, because an inheritance pattern was correlated with one specific chromosome pair. The idea became known as "the chromosome theory of inheritance." At that period in history, it had recently been shown that, in many organisms, sex is determined by an X and a Y chromosome and that, in males, these chromosomes segregate equally at meiosis to regenerate equal numbers of males and females in the next generation. Morgan recognized that the inheritance of alleles of the eye-color gene is exactly parallel to the inheritance of X chromosomes at meiosis; hence, the gene was likely to be on the X chromosome. The inheritance of *white eye* was extended to *Drosophila* lines that had abnormal numbers of sex chromosomes. With the use of this novel situation, it was still possible to predict gene-inheritance patterns from the segregation of the abnormal chromosomes. That these predictions proved correct was a convincing test of the chromosome theory.

Other genetic analyses revealed that, in chickens and moths, sex-linked inheritance could be explained only if the female was the heterogametic sex. In these organisms, the female sex chromosomes were designated ZW and males were designated ZZ.

2.6 Human Pedigree Analysis

Human matings, like those of experimental organisms, provide many examples of single-gene inheritance. However, controlled experimental crosses cannot be made with humans, and so geneticists must resort to scrutinizing medical records in the hope that informative matings have been made (such as monohybrid crosses) that could be used to infer single-gene inheritance. Such a scrutiny of records of matings is called **pedigree analysis.** A member of a family who first comes to the attention of a geneticist is called the **propositus.** Usually, the phenotype of the propositus is exceptional in some way; for example, the propositus might suffer from some type of medical disorder. The investigator then traces the history of the phenotype through the history of the family and draws a family tree, or pedigree, by using the standard symbols given in Figure 2-20.

To see single-gene inheritance, the patterns in the pedigree have to be interpreted according to Mendel's law of equal segregation, but humans usually have few children and so, because of this small progeny sample size, the expected 3:1 and 1:1 ratios are usually not seen unless many similar

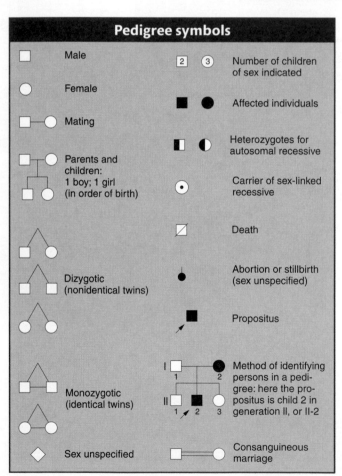

Figure 2-20 A variety of symbols are used in human pedigree analysis. [*After W. F. Bodmer and L. L. Cavalli-Sforza, Genetics, Evolution, and Man. Copyright 1976 by W. H. Freeman and Company.*]

pedigrees are combined. The approach to pedigree analysis also depends on whether one of the contrasting phenotypes is a rare disorder or both phenotypes of a pair are common (in which case they are said to be "morphs" of a polymorphism). Most pedigrees are drawn for medical reasons and therefore concern medical disorders that are almost by definition rare. In this case, we have two phenotypes: the presence and the absence of the disorder. Four patterns of single-gene inheritance are revealed in pedigrees. Let's look, first, at recessive disorders caused by recessive alleles of single autosomal genes.

Autosomal recessive disorders

The affected phenotype of an autosomal recessive disorder is inherited as a recessive allele; hence, the corresponding unaffected phenotype must be inherited as the corresponding dominant allele. For example, the human disease phenylketonuria, discussed earlier, is inherited in a simple Mendelian manner as a recessive phenotype, with PKU determined by the allele p and the normal condition determined by P. Therefore, sufferers of this disease are of genotype p/p, and people who do not have the disease are either P/P or P/p. Note that the term wild type and its allele symbols are not used in human genetics, because wild type is impossible to define.

What patterns in a pedigree would reveal autosomal recessive inheritance? The two key points are that (1) generally the disorder appears in the progeny of unaffected parents and (2) the affected progeny include both males and females. When we know that both male and female progeny are affected, we can infer that we are most likely dealing with simple Mendelian inheritance of a gene on an autosome, rather than a gene on a sex chromosome. The following typical pedigree illustrates the key point that affected children are born to unaffected parents:

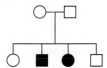

From this pattern, we can deduce a simple monohybrid cross, with the recessive allele responsible for the exceptional phenotype (indicated in black). Both parents must be heterozygotes—say, A/a; both must have an a allele because each contributed an a allele to each affected child, and both must have an A allele because they are phenotypically normal. We can identify the genotypes of the children (in the order shown) as $A/-$, a/a, a/a, and $A/-$. Hence, the pedigree can be rewritten as follows:

$$A/a \longmapsto A/a$$
$$A/- \quad a/a \quad a/a \quad A/-$$

This pedigree does not support the hypothesis of X-linked recessive inheritance, because, under that hypothesis, an affected daughter must have a heterozygous mother (possible) and a hemizygous father, which is clearly impossible because the father would have expressed the phenotype of the disorder.

Notice that, even though Mendelian rules are at work, Mendelian ratios are not necessarily observed in single families, because of small sample size, as predicted earlier. In the preceding example, we observe a 1:1 phenotypic ratio in the progeny of a monohybrid cross. If the couple were to have, say, 20 children, the ratio would be something like 15 unaffected children and 5 with PKU (a 3:1 ratio), but, in a small sample of 4 children, any ratio is possible, and all ratios are commonly found.

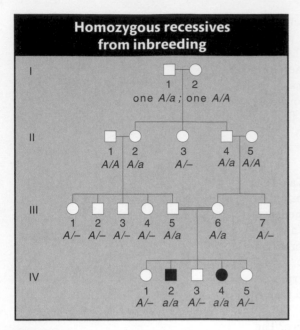

Figure 2-21 Pedigree of a rare recessive phenotype determined by a recessive allele *a*. Gene symbols are normally not included in pedigree charts, but genotypes are inserted here for reference. Persons II-1 and II-5 marry into the family; they are assumed to be normal because the heritable condition under scrutiny is rare. Note also that it is not possible to be certain of the genotype in some persons with normal phenotype; such persons are indicated by *A/−*. Persons III-5 and III-6, who generate the recessives in generation IV, are first cousins. They both obtain their recessive allele from a grandparent, either I-1 or I-2.

The family pedigrees of autosomal recessive disorders tend to look rather bare, with few black symbols. A recessive condition shows up in groups of affected siblings, and the people in earlier and later generations tend not to be affected. To understand why this is so, it is important to have some understanding of the genetic structure of populations underlying such rare conditions. By definition, if the condition is rare, most people do not carry the abnormal allele. Furthermore, most of those people who do carry the abnormal allele are heterozygous for it rather than homozygous. The basic reason that heterozygotes are much more common than recessive homozygotes is that, to be a recessive homozygote, both parents must have the *a* allele, but, to be a heterozygote, only one parent must have it.

The birth of an affected person usually depends on the rare chance union of unrelated heterozygous parents. However, inbreeding (mating between relatives) increases the chance that two heterozygotes will mate. An example of a marriage between cousins is shown in Figure 2-21. Individuals III-5 and III-6 are first cousins and produce two homozygotes for the rare allele. You can see from Figure 2-21 that an ancestor who is a heterozygote may produce many descendants who also are heterozygotes. Hence, two cousins can carry the *same* rare recessive allele inherited from a common ancestor. For two *unrelated* persons to be heterozygous, they would have to inherit the rare allele from *both* their families. Thus, matings between relatives generally run a higher risk of producing recessive disorders than do matings between nonrelatives. For this reason, first-cousin marriages contribute a large proportion of the sufferers of recessive diseases in the population.

What are some other examples of human recessive disorders? Cystic fibrosis is a disease inherited according to Mendelian rules as an autosomal recessive phenotype. Its most important symptom is the secretion of large amounts of mucus into the lungs, resulting in death from a combination of effects but usually precipitated by infection of the respiratory tract. The mucus can be dislodged by mechanical chest thumpers, and pulmonary infection can be prevented by antibiotics; thus, with treatment, cystic fibrosis patients can live to adulthood. The cystic fibrosis gene (and its mutant allele) was one of the first human disease genes to be isolated at the DNA level, in 1989. This line of research eventually revealed that the disorder is caused by a defective protein that transports chloride ions across the cell membrane. The resultant alteration of the salt balance changes the constitution of the lung mucus. This new understanding of gene function in affected and unaffected persons has given hope for more effective treatment.

Human albinism also is inherited in the standard autosomal recessive manner. The mutant allele is of a gene that normally synthesizes the brown or black pigment melanin. The inheritance pattern of this haplosufficient gene is shown together with the cellular defect in Figure 1-8.

> **Message** In human pedigrees, an autosomal recessive disorder is generally revealed by the appearance of the disorder in the male and female progeny of unaffected parents.

Autosomal dominant disorders

What pedigree patterns are expected from autosomal dominant disorders? Here, the normal allele is recessive, and the defective allele is dominant. It may seem paradoxical that a rare disorder can be dominant, but remember that dominance and recessiveness are simply properties of how alleles act in heterozygotes and are not defined in reference to how common they are in the population. A good

Pseudoachondroplasia phenotype

Figure 2-22 The human pseudoachondroplasia phenotype is illustrated here by a family of five sisters and two brothers. The phenotype is determined by a dominant allele, which we can call *D*, that interferes with the growth of long bones during development. This photograph was taken when the family arrived in Israel after the end of World War II. [*UPI/Bettmann News Photos.*]

example of a rare dominant phenotype that shows single-gene inheritance is pseudoachondroplasia, a type of dwarfism (Figure 2-22). In regard to this gene, people with normal stature are genotypically *d/d*, and the dwarf phenotype could in principle be *D/d* or *D/D*. However, the two "doses" of the *D* allele in the *D/D* genotype are believed to produce such a severe effect that this genotype is lethal. If this belief is generally true, all dwarf individuals are heterozygotes.

In pedigree analysis, the main clues for identifying an autosomal dominant disorder with Mendelian inheritance are that the phenotype tends to appear in every generation of the pedigree and that affected fathers or mothers transmit the phenotype to both sons and daughters. Again, the equal representation of both sexes among the affected offspring rules out inheritance through the sex chromosomes. The phenotype appears in every generation because, generally, the abnormal allele carried by a person must have come from a parent in the preceding generation. (Abnormal alleles can also arise de novo by mutation. This possibility must be kept in mind for disorders that interfere with reproduction because, here, the condition is unlikely to have been inherited from an affected parent.) A typical pedigree for a dominant disorder is shown in Figure 2-23. Once again, notice that Mendelian ratios are not necessarily observed in families. As with recessive disorders, persons bearing one copy of the rare *A* allele (*A/a*) are much more common than those bearing two copies (*A/A*); so most affected people are heterozygotes, and virtually all matings that produce progeny with dominant disorders are *A/a* × *a/a*. Therefore, if the progeny of such matings are totaled, a 1:1 ratio is expected of unaffected (*a/a*) to affected (*A/a*) persons.

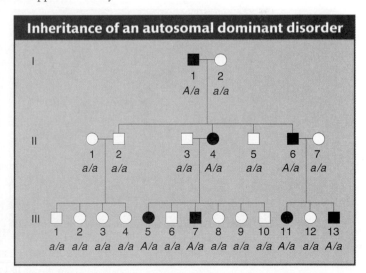

Inheritance of an autosomal dominant disorder

Figure 2-23 Pedigree of a dominant phenotype determined by a dominant allele *A*. In this pedigree, all the genotypes have been deduced.

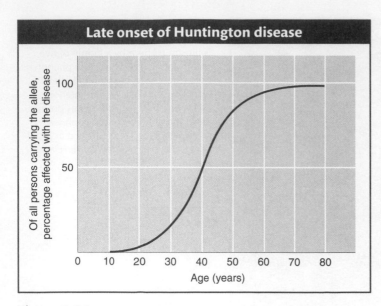

Figure 2-24 The graph shows that people carrying the allele generally do not express the disease until after childbearing age.

Huntington disease is an example of a disease inherited as a dominant phenotype determined by an allele of a single gene. The phenotype is one of neural degeneration, leading to convulsions and premature death. Folk singer Woody Guthrie suffered from Huntington disease. The disease is rather unusual in that it shows late onset, the symptoms generally not appearing until after the person has begun to have children (Figure 2-24). When the disease has been diagnosed in a parent, each child already born knows that he or she has a 50 percent chance of inheriting the allele and the associated disease. This tragic pattern has inspired a great effort to find ways of identifying people who carry the abnormal allele before they experience the onset of the disease. Now there are molecular diagnostics for identifying people who carry the Huntington allele.

Some other rare dominant conditions are polydactyly (extra digits), shown in Figure 2-25, and piebald spotting, shown in Figure 2-26.

Message Pedigrees of Mendelian autosomal dominant disorders show affected males and females in each generation; they also show affected men and women transmitting the condition to equal proportions of their sons and daughters.

Autosomal polymorphisms

In natural populations of organisms, a polymorphism is the coexistence of two or more common phenotypes of a character. The alternative phenotypes of a **polymorphism (morphs)** are often inherited as alleles of a single autosomal gene in the standard Mendelian manner. Among the many human examples are the following dimorphisms: brown versus blue eyes, pigmented versus blond

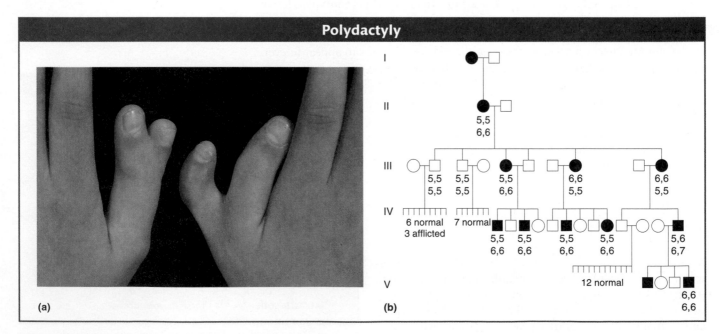

Figure 2-25 Polydactyly is a rare dominant phenotype of the human hands and feet.
(a) Polydactyly, characterized by extra fingers, toes, or both, is determined by an allele *P*. The numbers in the pedigree (b) give the number of fingers in the upper lines and the number of toes in the lower. (Note the variation in expression of the *P* allele.) [(a) Photograph © Biophoto Associates/Science Source.]

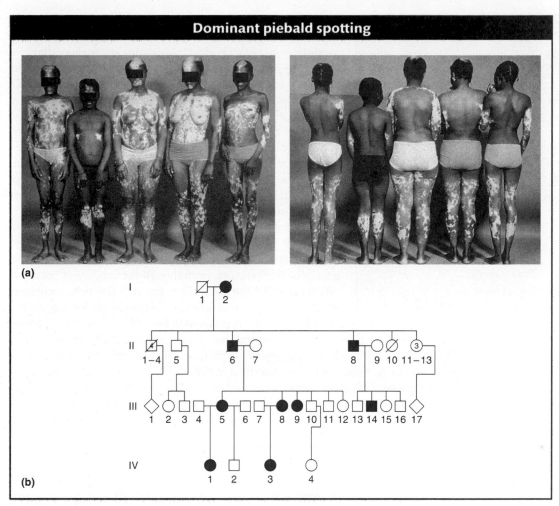

Figure 2-26 Piebald spotting is a rare dominant human phenotype. Although the phenotype is encountered sporadically in all races, the patterns show up best in those with dark skin. (a) The photographs show front and back views of affected persons IV-1, IV-3, III-5, III-8, and III-9 from (b) the family pedigree. Notice the variation in expression of the piebald gene among family members. The patterns are believed to be caused by the dominant allele interfering with the migration of melanocytes (melanin-producing cells) from the dorsal to the ventral surface in the course of development. The white forehead blaze is particularly characteristic and is often accompanied by a white forelock in the hair.

Piebaldism is not a form of albinism; the cells in the light patches have the genetic potential to make melanin, but, because they are not melanocytes, they are not developmentally programmed to do so. In true albinism, the cells lack the potential to make melanin. (Piebaldism is caused by mutations in *c-kit*, a type of gene called a *proto-oncogene*, to be discussed in Chapter 16.) [*(a and b) From I. Winship, K. Young, R. Martell, R. Ramesar, D. Curtis, and P. Beighton, "Piebaldism: An Autonomous Autosomal Dominant Entity," Clin. Genet. 39, 1991, 330.*]

hair, chin dimples versus none, widow's peak versus none, and attached versus free earlobes. In each example, the morph determined by the dominant allele is written first.

The interpretation of pedigrees for polymorphisms is somewhat different from that of rare disorders because, by definition, the morphs are common. Let's look at a pedigree for an interesting human case. A **dimorphism** is the simplest type of polymorphism, with just two morphs. Most human populations are dimorphic for the ability to taste the chemical phenylthiocarbamide (PTC); that is, people can either detect it as a foul, bitter taste or—to the great surprise

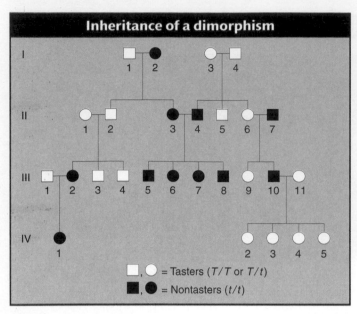

Figure 2-27 Pedigree for the ability to taste the chemical phenylthiocarbamide.

and disbelief of tasters—cannot taste it at all. From the pedigree in Figure 2-27, we can see that two tasters sometimes produce nontaster children, which makes it clear that the allele that confers the ability to taste is dominant and that the allele for nontasting is recessive. Notice in Figure 2-27 that almost all people who marry into this family carry the recessive allele either in heterozygous or in homozygous condition. Such a pedigree thus differs from those of rare recessive disorders, for which the conventional assumption is that all who marry into a family are homozygous normal. Because both PTC alleles are common, it is not surprising that all but one of the family members in this pedigree married persons with at least one copy of the recessive allele.

Polymorphism is an interesting genetic phenomenon. Population geneticists have been surprised at how much polymorphism there is in natural populations of plants and animals generally. Furthermore, even though the genetics of polymorphisms is straightforward, there are very few polymorphisms for which there are satisfactory explanations for the coexistence of the morphs. But polymorphism is rampant at every level of genetic analysis, even at the DNA level; indeed, polymorphisms observed at the DNA level have been invaluable as landmarks to help geneticists find their way around the chromosomes of complex organisms, as will be described in Chapter 4. The population and evolutionary genetics of polymorphisms is considered in Chapters 17 and 19.

> **Message** Populations of plants and animals (including humans) are highly polymorphic. Contrasting morphs are often inherited as alleles of a single gene.

X-linked recessive disorders

Let's look at the pedigrees of disorders caused by rare recessive alleles of genes located on the X chromosome. Such pedigrees typically show the following features:

1. Many more males than females show the rare phenotype under study. The reason is that a female can inherit the genotype only if both her mother *and* her father bear the allele (for example, $X^A X^a \times X^a Y$), whereas a male can inherit the phenotype when *only* the mother carries the allele ($X^A X^a \times X^A Y$). If the recessive allele is very rare, almost all persons showing the phenotype are male.

2. None of the offspring of an affected male show the phenotype, but all his daughters are "carriers," who bear the recessive allele masked in the heterozygous condition. In the next generation, half the sons of these carrier daughters show the phenotype (Figure 2-28).

3. None of the sons of an affected male show the phenotype under study, nor will they pass the condition to their descendants. The reason behind this lack of male-to-male transmission is that a son obtains his Y chromosome from his father; so he cannot normally inherit the father's X chromosome, too. Conversely, male-to-male transmission of a disorder is a useful diagnostic for an autosomally inherited condition.

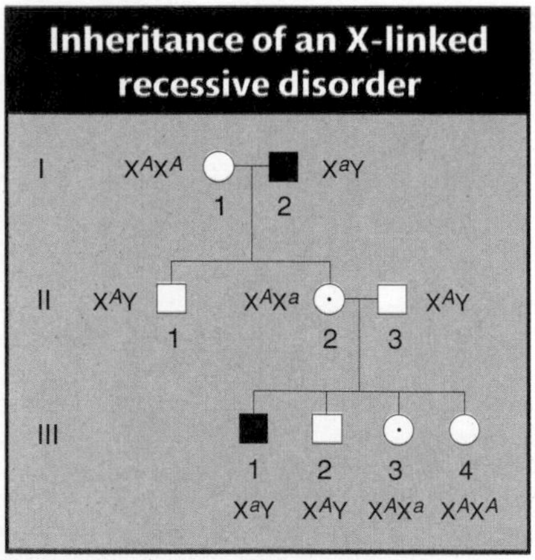

In the pedigree analysis of rare X-linked recessives, a normal female of unknown genotype is assumed to be homozygous unless there is evidence to the contrary.

Perhaps the most familiar example of X-linked recessive inheritance is red–green color blindness. People with this condition are unable to distinguish red from green. The genes for color vision have been characterized at the molecular level. Color vision is based on three different kinds of cone cells in the retina,

Figure 2-28 As is usually the case, expression of the X-linked recessive alleles is only in males. These alleles are carried unexpressed by daughters in the next generation, to be expressed again in sons. Note that III-3 and III-4 cannot be distinguished phenotypically.

each sensitive to red, green, or blue wavelengths. The genetic determinants for the red and green cone cells are on the X chromosome. Red–green color-blind people have a mutation in one of these two genes. As with any X-linked recessive disorder, there are many more males with the phenotype than females.

Another familiar example is *hemophilia*, the failure of blood to clot. Many proteins act in sequence to make blood clot. The most common type of hemophilia is caused by the absence or malfunction of one of these clotting proteins, called *factor VIII*. A well-known pedigree of hemophilia is of the interrelated royal families in Europe (Figure 2-29). The original hemophilia allele in the pedigree

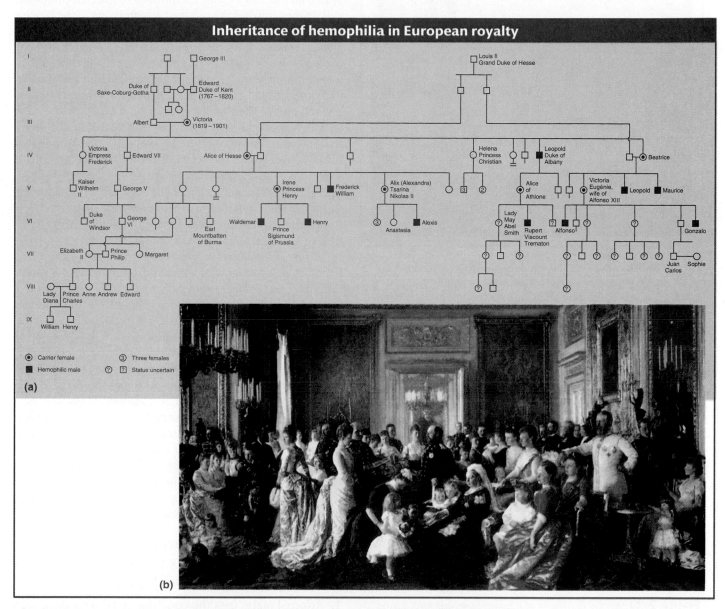

Figure 2-29 A pedigree for the X-linked recessive condition hemophilia in the royal families of Europe. A recessive allele causing hemophilia (failure of blood clotting) arose in the reproductive cells of Queen Victoria or one of her parents through mutation. This hemophilia allele spread into other royal families by intermarriage. (a) This partial pedigree shows affected males and carrier females (heterozygotes). Most spouses marrying into the families have been omitted from the pedigree for simplicity. Can you deduce the likelihood of the present British royal family's harboring the recessive allele? (b) A painting showing Queen Victoria surrounded by her numerous descendants. [*(a) After C. Stern, Principles of Human Genetics, 3rd ed. Copyright 1973 by W. H. Freeman and Company; (b) Lebrecht Music and Arts Photo Library/Alamy.*]

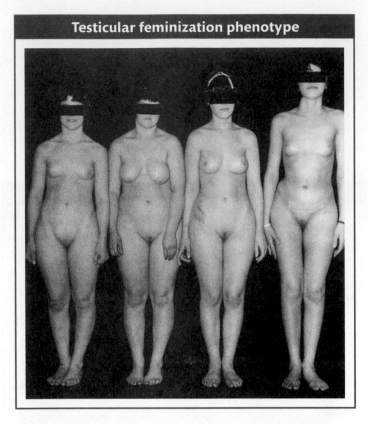

Figure 2-30 The four siblings in this photograph have testicular feminization syndrome (congenital insensitivity to androgens). All four have 44 autosomes plus an X and a Y chromosome, but they have inherited the recessive X-linked allele conferring insensitivity to androgens (male hormones). One of their sisters (not shown), who was genetically XX, was a carrier and bore a child who also showed testicular feminization syndrome. [*Leonard Pinsky, McGill University.*]

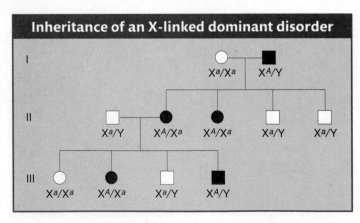

Figure 2-31 All the daughters of a male expressing an X-linked dominant phenotype will show the phenotype. Females heterozygous for an X-linked dominant allele will pass the condition on to half their sons and daughters.

possibly arose spontaneously as a mutation in the reproductive cells of either Queen Victoria's parents or Queen Victoria herself. However, some have proposed that the origin of the allele was a secret lover of Victoria's mother. Alexis, the son of the last czar of Russia, inherited the hemophilia allele ultimately from Queen Victoria, who was the grandmother of his mother, Alexandra. Nowadays, hemophilia can be treated medically, but it was formerly a potentially fatal condition. It is interesting to note that the Jewish Talmud contains rules about exemptions to male circumcision clearly showing that the mode of transmission of the disease through unaffected carrier females was well understood in ancient times. For example, one exemption was for the sons of women whose sisters' sons had bled profusely when they were circumcised. Hence, abnormal bleeding was known to be transmitted through the females of the family but expressed only in their male children.

Duchenne muscular dystrophy is a fatal X-linked recessive disease. The phenotype is a wasting and atrophy of muscles. Generally, the onset is before the age of 6, with confinement to a wheelchair by age 12 and death by age 20. The gene for Duchenne muscular dystrophy encodes the muscle protein dystrophin. This knowledge holds out hope for a better understanding of the physiology of this condition and, ultimately, a therapy.

A rare X-linked recessive phenotype that is interesting from the point of view of sexual differentiation is a condition called *testicular feminization syndrome*, which has a frequency of about 1 in 65,000 male births. People afflicted with this syndrome are chromosomally males, having 44 autosomes plus an X and a Y chromosome, but they develop as females (Figure 2-30). They have female external genitalia, a blind vagina, and no uterus. Testes may be present either in the labia or in the abdomen. Although many such persons marry, they are sterile. The condition is not reversed by treatment with the male hormone androgen, and so it is sometimes called *androgen insensitivity syndrome*. The reason for the insensitivity is that a mutation in the androgen-receptor gene causes the receptor to malfunction, and so the male hormone can have no effect on the target organs that contribute to maleness. In humans, femaleness results when the male-determining system is not functional.

X-linked dominant disorders

The inheritance patterns of X-linked dominant disorders have the following characteristics in pedigrees (Figure 2-31):

1. Affected males pass the condition to all their daughters but to none of their sons.

2. Affected heterozygous females married to unaffected males pass the condition to half their sons and daughters.

This mode of inheritance is not common. One example is hypophosphatemia, a type of vitamin D–resistant rickets. Some forms of hypertrichosis (excess body and facial hair) show X-linked dominant inheritance.

Y-linked inheritance

Only males inherit genes in the differential region of the human Y chromosome, with fathers transmitting the genes to their sons. The gene that plays a primary role in maleness is the **SRY gene,** sometimes called the *testis-determining factor*. Genomic analysis has confirmed that, indeed, the *SRY* gene is in the differential region of the Y chromosome. Hence, maleness itself is Y linked and shows the expected pattern of exclusively male-to-male transmission. Some cases of male sterility have been shown to be caused by deletions of Y-chromosome regions containing sperm-promoting genes. Male sterility is not heritable, but, interestingly, the fathers of these men have normal Y chromosomes, showing that the deletions are new.

There have been no convincing cases of nonsexual phenotypic variants associated with the Y chromosome. Hairy ear rims (Figure 2-32) have been proposed as a possibility, although disputed. The phenotype is extremely rare among the populations of most countries but more common among the populations of India. In some families, hairy ear rims have been shown to be transmitted exclusively from fathers to sons.

> **Message** Inheritance patterns with an unequal representation of phenotypes in males and females can locate the genes concerned to one of the sex chromosomes.

Calculating risks in pedigree analysis

When a disorder with well-documented single-gene inheritance is known to be present in a family, knowledge of transmission patterns can be used to calculate the probability of prospective parents' having a child with the disorder. For example, consider a case in which a newly married husband and wife find out that each had an uncle with Tay-Sachs disease, a severe autosomal recessive disease caused by malfunction of the enzyme hexosaminidase A. The defect leads to the buildup of fatty deposits in nerve cells, causing paralysis followed by an early death. The pedigree is as follows:

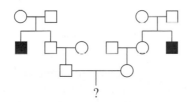

The probability of the couple's first child having Tay-Sachs can be calculated in the following way. Because neither of the couple has the disease, each can only be a normal homozygote or a heterozygote. If both are heterozygotes, then they each stand a chance of passing the recessive allele on to a child, who would then have Tay-Sachs disease. Hence, we must calculate the probability of their both being heterozygotes, and then, if so, the probability of passing the deleterious allele on to a child.

1. The husband's grandparents must have both been heterozygotes (T/t) because they produced a t/t child (the uncle). Therefore, they effectively constituted a monohybrid cross. The husband's father could be T/T or T/t, but we know that the relative probabilities of these genotypes must be 1/4 and 1/2, respectively (the expected progeny ratio in a monohybrid cross is $\frac{1}{4}$ T/T, $\frac{1}{2}$ T/t, and $\frac{1}{4}$ t/t). Therefore, there is a 2/3 probability that the father is a heterozygote (two-thirds

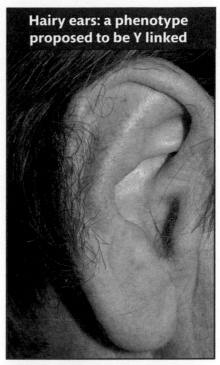

Figure 2-32 Hairy ear rims have been proposed to be caused by an allele of a Y-linked gene. [*Mark Collinson/Alamy.*]

is the proportion of unaffected progeny who are heterozygotes: 1/2 divided by 3/4).

2. The husband's mother is assumed to be T/T, because she married into the family and disease alleles are generally rare. Thus, if the father is T/t, then the mating with the mother was a cross $T/t \times T/T$ and the expected proportions in the progeny (which includes the husband) are $\frac{1}{2}T/T$ and $\frac{1}{2}T/t$.

3. The overall probability of the husband's being a heterozygote must be calculated with the use of a statistical rule called the **product rule,** which states that

> The probability of two independent events both occurring is the product of their individual probabilities.

Because gene transmissions in different generations are independent events, we can calculate that the probability of the husband's being a heterozygote is the probability of his father's being a heterozygote *times* the probability of his father having a heterozygous son, which is $2/3 \times 1/2 = 1/3$.

4. Likewise, the probability of the wife's being heterozygous is also 1/3.

5. If they are both heterozygous (T/t), their mating would be a standard monohybrid cross and so the probability of their having a t/t child is 1/4.

6. Overall, the probability of the couple's having an affected child is the probability of them both being heterozygous and then both transmitting the recessive allele to a child. Again, these events are independent, and so we can calculate the overall probability as $1/3 \times 1/3 \times 1/4 = 1/36$. In other words, there is a 1 in 36 chance of them having a child with Tay-Sachs disease.

In some Jewish communities, the Tay-Sachs allele is not as rare as it is in the general population. In such cases, unaffected people who marry into families with a history of Tay-Sachs cannot be assumed to be T/T. If the frequency of T/t heterozygotes in the community is known, this frequency can be factored into the product-rule calculation. Nowadays, molecular diagnostic tests for Tay-Sachs alleles are available, and the judicious use of these tests has drastically reduced the frequency of the disease in some communities.

Summary

In somatic cell division, the genome is transmitted by mitosis, a nuclear division. In this process, each chromosome replicates into a pair of chromatids and the chromatids are pulled apart to produce two identical daughter cells. (Mitosis can take place in diploid or haploid cells.) At meiosis, which takes place in the sexual cycle in meiocytes, each homolog replicates to form a dyad of chromatids; then, the dyads pair to form a tetrad, which segregates at each of the two cell divisions. The result is four haploid cells, or gametes. Meiosis can take place only in a diploid cell; hence, haploid organisms unite to form a diploid meiocyte.

An easy way to remember the main events of meiosis, by using your fingers to represent chromosomes, is shown in Figure 2-33.

Genetic dissection of a biological property begins with a collection of mutants. Each mutant has to be tested to see if it is inherited as a single-gene change. The procedure fol-

lowed is essentially unchanged from the time of Mendel, who performed the prototypic analysis of this type. The analysis is based on observing specific phenotypic ratios in the progeny of controlled crosses. In a typical case, a cross of $A/A \times a/a$ produces an F_1 that is all A/a. When the F_1 is selfed or intercrossed, a genotypic ratio of $\frac{1}{4}A/A : \frac{1}{2}A/a : \frac{1}{4}a/a$ is produced in the F_2. (At the phenotypic level, this ratio is $\frac{3}{4}A/- : \frac{1}{4}a/a$.) The three single-gene genotypes are homozygous dominant, heterozygous (monohybrid), and homozygous recessive. If an A/a individual is crossed with a/a (a testcross), a 1:1 ratio is produced in the progeny. The 1:1, 3:1, and 1:2:1 ratios stem from the principle of equal segregation, which is that the haploid products of meiosis from A/a will be $\frac{1}{2}A$ and $\frac{1}{2}a$. The cellular basis of the equal segregation of alleles is the segregation of homologous chromosomes at meiosis. Haploid fungi can be used to show equal segregation at the level of a single meiosis (a 1:1 ratio in an ascus).

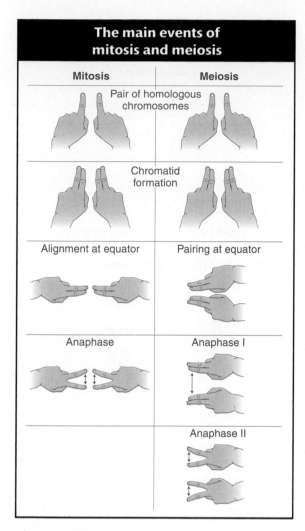

The main events of mitosis and meiosis

Mitosis	Meiosis
Pair of homologous chromosomes	
Chromatid formation	
Alignment at equator	Pairing at equator
Anaphase	Anaphase I
	Anaphase II

Figure 2-33 Using fingers to remember the main events of mitosis and meiosis.

The molecular basis for chromatid production in meiosis is DNA replication. Segregation at meiosis can be observed directly at the molecular (DNA) level if suitable probes are used to detect the morphs of a molecular dimorphism. The molecular force of segregation is the depolymerization and subsequent shortening of microtubules that are attached to the centromeres. Recessive mutations are generally in genes that are haplosufficient, whereas dominant mutations are often due to gene haploinsufficiency.

In many organisms, sex is determined chromosomally, and, typically, XX is female and XY is male. Genes on the X chromosome (X-linked genes) have no counterparts on the Y chromosome and show a single-gene inheritance pattern that differs in the two sexes, often resulting in different ratios in the male and female progeny.

Mendelian single-gene segregation is useful in identifying mutant alleles underlying many human disorders. Analyses of pedigrees can reveal autosomal or X-linked disorders of both dominant and recessive types. The logic of Mendelian genetics has to be used with caution, taking into account that human progeny sizes are small and phenotypic ratios are not necessarily typical of those expected from larger sample sizes. If a known single-gene disorder is present in a pedigree, Mendelian logic can be used to predict the likelihood of children inheriting the disease.

KEY TERMS

allele (p. 32)
ascus (p. 39)
bivalent (p. 38)
character (p. 29)
chromatid (p. 37)
cross (p. 28)
dimorphism (p. 59)
dioecious species (p. 51)
dominant (p. 32)
dyad (p. 37)
first filial generation (F_1) (p. 30)
forward genetics (p. 49)
gene (p. 32)
gene discovery (p. 27)
genetic dissection (p. 28)
genotype (p. 32)

haploid number (p. 00)
haploinsufficient (p. 46)
haplosufficient (p. 46)
hemizygous (p. 51)
heterogametic sex (p. 50)
heterozygote (p. 32)
heterozygous (p. 32)
homogametic sex (p. 50)
homozygote (p. 32)
homozygous dominant (p. 32)
homozygous recessive (p. 32)
law of equal segregation (Mendel's first law) (p. 32)
leaky mutation (p. 45)
meiocyte (p. 35)
meiosis (p. 35)
mitosis (p. 34)

molecular markers (p. 44)
monohybrid (p. 32)
monohybrid cross (p. 33)
morph (p. 58)
mutant (p. 28)
mutation (p. 40)
null allele (p. 45)
parental generation (P) (p. 30)
pedigree analysis (p. 54)
phenotype (p. 29)
polymorphism (p. 58)
product of meiosis (p. 38)
product rule (p. 64)
property (p. 29)
propositus (p. 54)
pseudoautosomal regions 1 and 2 (p. 51)

SOLVED PROBLEMS

This section in each chapter contains a few solved problems that show how to approach the problem sets that follow. The purpose of the problem sets is to challenge your understanding of the genetic principles learned in the chapter. The best way to demonstrate an understanding of a subject is to be able to use that knowledge in a real or simulated situation. Be forewarned that there is no machine-like way of solving these problems. The three main resources at your disposal are the genetic principles just learned, logic, and trial and error.

Here is some general advice before beginning. First, it is absolutely essential to read and understand all of the problem. Most of the problems use data taken from research that somebody actually carried out: ask yourself why the research might have been initiated and what was the probable goal. Find out exactly what facts are provided, what assumptions have to be made, what clues are given in the problem, and what inferences can be made from the available information. Second, be methodical. Staring at the problem rarely helps. Restate the information in the problem in your own way, preferably using a diagrammatic representation or flowchart to help you think out the problem. Good luck.

SOLVED PROBLEM 1. Crosses were made between two pure lines of rabbits that we can call A and B. A male from line A was mated with a female from line B, and the F₁ rabbits were subsequently intercrossed to produce an F₂. Three-fourths of the F₂ animals were discovered to have white subcutaneous fat and one-fourth had yellow subcutaneous fat. Later, the F₁ was examined and was found to have white fat. Several years later, an attempt was made to repeat the experiment by using the same male from line A and the same female from line B. This time, the F₁ and all the F₂ (22 animals) had white fat. The only difference between the original experiment and the repeat that seemed relevant was that, in the original, all the animals were fed fresh vegetables, whereas in the repeat, they were fed commercial rabbit chow. Provide an explanation for the difference and a test of your idea.

Solution

The first time that the experiment was done, the breeders would have been perfectly justified in proposing that a pair of alleles determine white versus yellow body fat because the data clearly resemble Mendel's results in peas. White must be dominant, and so we can represent the white allele as W and the yellow allele as w. The results can then be expressed as follows:

$$P \qquad W/W \times w/w$$

$$F_1 \qquad W/w$$

$$F_2 \qquad \tfrac{1}{4} W/W$$

$$\tfrac{1}{2} W/w$$

$$\tfrac{1}{4} w/w$$

No doubt, if the parental rabbits had been sacrificed, one parent (we cannot tell which) would have been predicted to have white fat and the other yellow. Luckily, the rabbits were not sacrificed, and the same animals were bred again, leading to a very interesting, different result. Often in science, an unexpected observation can lead to a novel principle, and, rather than moving on to something else, it is useful to try to explain the inconsistency. So why did the 3 : 1 ratio disappear? Here are some possible explanations.

First, perhaps the genotypes of the parental animals had changed. This type of spontaneous change affecting the whole animal, or at least its gonads, is very unlikely, because even common experience tells us that organisms tend to be stable to their type.

Second, in the repeat, the sample of 22 F₂ animals did not contain any yellow fat simply by chance ("bad luck"). This explanation, again, seems unlikely, because the sample was quite large, but it is a definite possibility.

A third explanation draws on the principle that genes do not act in a vacuum; they depend on the environment for their effects. Hence, the formula "Genotype + environment = phenotype" is a useful mnemonic. A corollary of this formula is that genes can act differently in different environments; so

genotype 1 + environment 1 = phenotype 1

and

genotype 1 + environment 2 = phenotype 2

In the present problem, the different diets constituted different environments, and so a possible explanation of the results is that the recessive allele w produces yellow fat only when the diet contains fresh vegetables. This explanation is testable. One way to test it is to repeat the experiment again and use vegetables as food, but the parents might be dead by this time. A more convincing way is to breed several of the white-fatted F$_2$ rabbits from the second experiment. According to the original interpretation, some of them should be heterozygous, and, if their progeny are raised on vegetables, yellow fat should appear in Mendelian proportions. For example, if a cross happened to be W/w and w/w, the progeny would be $\frac{1}{2}$ white fat and $\frac{1}{2}$ yellow fat.

If this outcome did not happen and no progeny having yellow fat appeared in any of the matings, we would be forced back to the first or second explanation. The second explanation can be tested by using larger numbers, and if this explanation doesn't work, we are left with the first explanation, which is difficult to test directly.

As you might have guessed, in reality, the diet was the culprit. The specific details illustrate environmental effects beautifully. Fresh vegetables contain yellow substances called xanthophylls, and the dominant allele W gives rabbits the ability to break down these substances to a colorless ("white") form. However, w/w animals lack this ability, and the xanthophylls are deposited in the fat, making it yellow. When no xanthophylls have been ingested, both $W/-$ and w/w animals end up with white fat.

SOLVED PROBLEM 2. Phenylketonuria is a human hereditary disease resulting from the inability of the body to process the chemical phenylalanine, which is contained in the protein that we eat. PKU is manifested in early infancy and, if it remains untreated, generally leads to mental retardation. PKU is caused by a recessive allele with simple Mendelian inheritance.

A couple intends to have children but consult a genetic counselor because the man has a sister with PKU and the woman has a brother with PKU. There are no other known cases in their families. They ask the genetic counselor to determine the probability that their first child will have PKU. What is this probability?

Solution

What can we deduce? If we let the allele causing the PKU phenotype be p and the respective normal allele be P, then the sister and brother of the man and woman, respectively, must have been p/p. To produce these affected persons, all four grandparents must have been heterozygous normal. The pedigree can be summarized as follows:

When these inferences have been made, the problem is reduced to an application of the product rule. The only way in which the man and woman can have a PKU child is if both of them are heterozygotes (it is obvious that they themselves do not have the disease). Both the grandparental matings are simple Mendelian monohybrid crosses expected to produce progeny in the following proportions:

$$\left.\begin{array}{l}\frac{1}{4}\,P/P\\[4pt]\frac{1}{2}\,P/p\end{array}\right\}\quad \text{Normal }\left(\tfrac{3}{4}\right)$$

$$\frac{1}{4}\,p/p\qquad \text{PKU}\left(\tfrac{1}{4}\right)$$

We know that the man and the woman are normal, and so the probability of each being a heterozygote is $2/3$ because, within the $P/-$ class, $2/3$ are P/p and $1/3$ are P/P.

The probability of *both* the man and the woman being heterozygotes is $2/3 \times 2/3 = 4/9$. If both are heterozygous, then one-quarter of their children would have PKU, and so the probability that their first child will have PKU is $1/4$ and the probability of their being heterozygous *and* of their first child's having PKU is $4/9 \times 1/4 = 4/36 = 1/9$, which is the answer.

SOLVED PROBLEM 3. A rare human disease afflicted a family as shown in the accompanying pedigree.

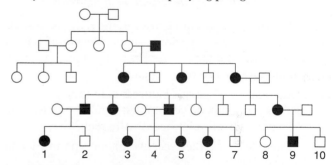

a. Deduce the most likely mode of inheritance.

b. What would be the outcomes of the cousin marriages 1×9, 1×4, 2×3, and 2×8?

Solution

a. The most likely mode of inheritance is X-linked dominant. We assume that the disease phenotype is dominant because, after it has been introduced into the pedigree by the male in generation II, it appears in every generation. We assume that the phenotype is X linked because fathers do not transmit it to their sons. If it were autosomal dominant, father-to-son transmission would be common.

In theory, autosomal recessive could work, but it is improbable. In particular, note the marriages between affected members of the family and unaffected outsiders. If the condition were autosomal recessive, the only way in which these marriages could have affected offspring is if each person marrying into the family were a heterozygote; then the matings would be a/a (affected) $\times A/a$ (unaffected). However, we are told that the disease is rare; in such a case, heterozygotes are highly unlikely to be so common. X-linked recessive inheritance is impossible, because a mating of an affected woman with a normal man could not produce affected daughters. So we can let A represent the disease-causing allele and a represent the normal allele.

b. 1×9: Number 1 must be heterozygous A/a because she must have obtained a from her normal mother. Number 9 must be A/Y. Hence, the cross is $A/a \, ♀ \times A/Y \, ♂$.

Female gametes   Male gametes   Progeny

$$\frac{1}{2}A \Big\langle {\frac{1}{2}A \longrightarrow \frac{1}{4}A/A \; ♀ \atop \frac{1}{2}Y \longrightarrow \frac{1}{4}A/Y \; ♂}$$

$$\frac{1}{2}a \Big\langle {\frac{1}{2}A \longrightarrow \frac{1}{4}A/a \; ♀ \atop \frac{1}{2}Y \longrightarrow \frac{1}{4}a/Y \; ♂}$$

1×4: Must be $A/a \, ♀ \times a/Y \, ♂$.

Female gametes   Male gametes   Progeny

$$\frac{1}{2}A \Big\langle {\frac{1}{2}a \longrightarrow \frac{1}{4}A/a \; ♀ \atop \frac{1}{2}Y \longrightarrow \frac{1}{4}A/Y \; ♂}$$

$$\frac{1}{2}a \Big\langle {\frac{1}{2}a \longrightarrow \frac{1}{4}a/a \; ♀ \atop \frac{1}{2}Y \longrightarrow \frac{1}{4}a/Y \; ♂}$$

2×3: Must be $a/Y \, ♂ \times A/a \, ♀$ (same as 1×4).

2×8: Must be $a/Y \, ♂ \times a/a \, ♀$ (all progeny normal).

PROBLEMS

Most of the problems are also available for review/grading through the GENETICS PORTAL www.yourgeneticsportal.com.

WORKING WITH THE FIGURES

(The first 14 questions require inspection of figures.)

1. In the left-hand part of Figure 2-4, the red arrows show selfing as pollination within single flowers of one F_1 plant. Would the same F_2 results be produced by cross-pollinating two different F_1 plants?

2. In the right-hand part of Figure 2-4, in the plant showing an $11:11$ ratio, do you think it would be possible to find a pod with all yellow peas? All green? Explain.

3. In Table 2-1, state the recessive phenotype in each of the seven cases.

4. Considering Figure 2-8, is the sequence "pairing $\rightarrow$ replication $\rightarrow$ segregation $\rightarrow$ segregation" a good shorthand description of meiosis?

5. Point to all cases of bivalents, dyads, and tetrads in Figure 2-11.

6. In Figure 2-12, assume (as in corn plants) that A encodes an allele that produces starch in pollen and allele a does not. Iodine solution stains starch black. How would you demonstrate Mendel's first law directly with such a system?

7. In the cross diagram on page 43, assume the left-hand individual is selfed. What pattern of radioactive bands would you see in a Southern analysis of the progeny?

8. Considering Figure 2-15, if you had a homozygous double mutant $m3/m3 \; m5/m5$, would you expect it to be mutant in phenotype? (**Note:** This line would have two mutant sites in the same coding sequence.)

9. In which of the stages of the *Drosophila* life cycle (represented in the box on page 52) does meiosis take place?

10. If you assume Figure 2-17 also applies to mice and you irradiate male sperm with X rays (known to inactivate genes), what phenotype would you look for in progeny in order to find cases of individuals with an inactivated *SRY* gene?

11. In Figure 2-19, how does the $3:1$ ratio in the bottom-left-hand grid differ from the $3:1$ ratios obtained by Mendel?

12. In Figure 2-21, assume that the pedigree is for mice, in which any chosen cross can be made. If you bred IV-1 with IV-3, what is the probability that the first baby will show the recessive phenotype?

13. Which part of the pedigree in Figure 2-23 in your opinion best demonstrates Mendel's first law?

14. Could the pedigree in Figure 2-31 be explained as an autosomal dominant disorder? Explain.

BASIC PROBLEMS

15. Make up a sentence including the words *chromosome*, *genes*, and *genome*.

16. Peas (*Pisum sativum*) are diploid and $2n = 14$. In *Neurospora*, the haploid fungus, $n = 7$. If it were possible to fractionate genomic DNA from both species by using pulsed field electrophoresis, how many distinct DNA bands would be visible in each species?

17. The broad bean (*Vicia faba*) is diploid and $2n = 18$. Each haploid chromosome set contains approximately

4 m of DNA. The average size of each chromosome during metaphase of mitosis is 13 μm. What is the average packing ratio of DNA at metaphase? (Packing ratio = length of chromosome/length of DNA molecule therein.) How is this packing achieved?

18. If we call the amount of DNA per genome "*x*," name a situation or situations in diploid organisms in which the amount of DNA per cell is

    a. *x* **b.** 2*x* **c.** 4*x*

19. Name the key function of mitosis.

20. Name two key functions of meiosis.

21. Can you design a different nuclear-division system that would achieve the same outcome as that of meiosis?

22. In a possible future scenario, male fertility drops to zero, but, luckily, scientists develop a way for women to produce babies by virgin birth. Meiocytes are converted directly (without undergoing meiosis) into zygotes, which implant in the usual way. What would be the short- and long-term effects in such a society?

23. In what ways does the second division of meiosis differ from mitosis?

24. Make up mnemonics for remembering the five stages of prophase I of meiosis and the four stages of mitosis.

25. In an attempt to simplify meiosis for the benefit of students, mad scientists develop a way of preventing premeiotic S phase and making do with having just one division, including pairing, crossing over, and segregation. Would this system work, and would the products of such a system differ from those of the present system?

26. Theodor Boveri said, "The nucleus doesn't divide; it is divided." What was he getting at?

27. Francis Galton, a geneticist of the pre-Mendelian era, devised the principle that half of our genetic makeup is derived from each parent, one-quarter from each grandparent, one-eighth from each great-grandparent, and so forth. Was he right? Explain.

28. If children obtain half their genes from one parent and half from the other parent, why aren't siblings identical?

29. State where cells divide mitotically and where they divide meiotically in a fern, a moss, a flowering plant, a pine tree, a mushroom, a frog, a butterfly, and a snail.

30. Human cells normally have 46 chromosomes. For each of the following stages, state the number of nuclear DNA molecules present in a human cell:

    a. Metaphase of mitosis

    b. Metaphase I of meiosis

    c. Telophase of mitosis

    d. Telophase I of meiosis

    e. Telophase II of meiosis

31. Four of the following events are part of both meiosis and mitosis, but only one is meiotic. Which one? (1) Chromatid formation, (2) spindle formation, (3) chromosome condensation, (4) chromosome movement to poles, (5) synapsis.

32. In corn, the allele *f'* causes floury endosperm and the allele *f''* causes flinty endosperm. In the cross *f'/f'* ♀ × *f''/f''* ♂, all the progeny endosperms are floury, but, in the reciprocal cross, all the progeny endosperms are flinty. What is a possible explanation? (Check the legend for Figure 2-7.)

33. What is Mendel's first law?

34. If you had a fruit fly (*Drosophila melanogaster*) that was of phenotype *A*, what test would you make to determine if the fly's genotype was *A/A* or *A/a*?

35. In examining a large sample of yeast colonies on a petri dish, a geneticist finds an abnormal-looking colony that is very small. This small colony was crossed with wild type, and products of meiosis (ascospores) were spread on a plate to produce colonies. In total, there were 188 wild-type (normal-size) colonies and 180 small ones.

    a. What can be deduced from these results regarding the inheritance of the small-colony phenotype? (Invent genetic symbols.)

    b. What would an ascus from this cross look like?

36. Two black guinea pigs were mated and over several years produced 29 black and 9 white offspring. Explain these results, giving the genotypes of parents and progeny.

37. In a fungus with four ascospores, a mutant allele *lys-5* causes the ascospores bearing that allele to be white, whereas the wild type allele *lys-5*+ results in black ascospores. (Ascospores are the spores that constitute the four products of meiosis.) Draw an ascus from each of the following crosses:

    a. *lys-5* × *lys-5*+

    b. *lys-5* × *lys-5*

    c. *lys-5*+ × *lys-5*+

38. For a certain gene in a diploid organism, eight units of protein product are needed for normal function. Each wild-type allele produces five units.

    a. If a mutation creates a null allele, do you think this allele will be recessive or mutant?

    b. What assumptions need to be made to answer part *a*?

39. A *Neurospora* colony at the edge of a plate seemed to be sparse (low density) in comparison with the other colonies on the plate. This colony was thought to be a possible mutant, and so it was removed and crossed with a wild type of the opposite mating type. From this cross, 100 ascospore progeny were obtained. None of the colonies from these ascospores was sparse, all

appearing to be normal. What is the simplest explanation of this result? How would you test your explanation? (**Note:** *Neurospora* is haploid.)

40. From a large-scale screen of many plants of *Collinsia grandiflora*, a plant with three cotyledons was discovered (normally, there are two cotyledons). This plant was crossed with a normal pure-breeding wild-type plant, and 600 seeds from this cross were planted. There were 298 plants with two cotyledons and 302 with three cotyledons. What can be deduced about the inheritance of three cotyledons? Invent gene symbols as part of your explanation.

41. In the plant *Arabidopsis thaliana*, a geneticist is interested in the development of trichomes (small projections). A large screen turns up two mutant plants (A and B) that have no trichomes, and these mutants seem to be potentially useful in studying trichome development. (If they were determined by single-gene mutations, then finding the normal and abnormal functions of these genes would be instructive.) Each plant is crossed with wild type; in both cases, the next generation (F_1) had normal trichomes. When F_1 plants were selfed, the resulting F_2's were as follows:

F_2 from mutant A: 602 normal; 198 no trichomes

F_2 from mutant B: 267 normal; 93 no trichomes

a. What do these results show? Include proposed genotypes of all plants in your answer.

b. Under your explanation to part *a*, is it possible to confidently predict the F_1 from crossing the original mutant A with the original mutant B?

42. You have three dice: one red (R), one green (G), and one blue (B). When all three dice are rolled at the same time, calculate the probability of the following outcomes:

a. 6 (R), 6 (G), 6 (B)

b. 6 (R), 5 (G), 6 (B)

c. 6 (R), 5 (G), 4 (B)

d. No sixes at all

e. A different number on all dice

43. In the pedigree below, the black symbols represent individuals with a very rare blood disease.

If you had no other information to go on, would you think it more likely that the disease was dominant or recessive? Give your reasons.

44. a. The ability to taste the chemical phenylthiocarbamide is an autosomal dominant phenotype, and the inability to taste it is recessive. If a taster woman with a nontaster father marries a taster man who in a previous marriage had a nontaster daughter, what is the probability that their first child will be

(1) A nontaster girl

(2) A taster girl

(3) A taster boy

b. What is the probability that their first two children will be tasters of either sex?

45. John and Martha are contemplating having children, but John's brother has galactosemia (an autosomal recessive disease) and Martha's great-grandmother also had galactosemia. Martha has a sister who has three children, none of whom have galactosemia. What is the probability that John and Martha's first child will have galactosemia?

Unpacking Problem 45

1. Can the problem be restated as a pedigree? If so, write one.

2. Can parts of the problem be restated by using Punnett squares?

3. Can parts of the problem be restated by using branch diagrams?

4. In the pedigree, identify a mating that illustrates Mendel's first law.

5. Define all the scientific terms in the problem, and look up any other terms about which you are uncertain.

6. What assumptions need to be made in answering this problem?

7. Which unmentioned family members must be considered? Why?

8. What statistical rules might be relevant, and in what situations can they be applied? Do such situations exist in this problem?

9. What are two generalities about autosomal recessive diseases in human populations?

10. What is the relevance of the rareness of the phenotype under study in pedigree analysis generally, and what can be inferred in this problem?

11. In this family, whose genotypes are certain and whose are uncertain?

12. In what way is John's side of the pedigree different from Martha's side? How does this difference affect your calculations?

13. Is there any irrelevant information in the problem as stated?

14. In what way is solving this kind of problem similar to solving problems that you have already successfully solved? In what way is it different?

15. Can you make up a short story based on the human dilemma in this problem?

Now try to solve the problem. If you are unable to do so, try to identify the obstacle and write a sentence or two describing your difficulty. Then go back to the expansion questions and see if any of them relate to your difficulty.

46. Holstein cattle are normally black and white. A superb black-and-white bull, Charlie, was purchased by a farmer for $100,000. All the progeny sired by Charlie were normal in appearance. However, certain pairs of his progeny, when interbred, produced red-and-white progeny at a frequency of about 25 percent. Charlie was soon removed from the stud lists of the Holstein breeders. Use symbols to explain precisely why.

47. Suppose that a husband and wife are both heterozygous for a recessive allele for albinism. If they have dizygotic (two-egg) twins, what is the probability that both the twins will have the same phenotype for pigmentation?

48. The plant blue-eyed Mary grows on Vancouver Island and on the lower mainland of British Columbia. The populations are dimorphic for purple blotches on the leaves—some plants have blotches and others don't. Near Nanaimo, one plant in nature had blotched leaves. This plant, which had not yet flowered, was dug up and taken to a laboratory, where it was allowed to self. Seeds were collected and grown into progeny. One randomly selected (but typical) leaf from each of the progeny is shown in the accompanying illustration.

a. Formulate a concise genetic hypothesis to explain these results. Explain all symbols and show all genotypic classes (and the genotype of the original plant).

b. How would you test your hypothesis? Be specific.

49. Can it ever be proved that an animal is not a carrier of a recessive allele (that is, not a heterozygote for a given gene)? Explain.

50. In nature, the plant *Plectritis congesta* is dimorphic for fruit shape; that is, individual plants bear either wingless or winged fruits, as shown in the illustration. Plants were

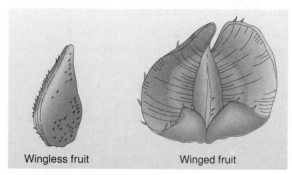

Wingless fruit   Winged fruit

collected from nature before flowering and were crossed or selfed with the following results:

| Pollination | *Number of progeny* | |
	Winged	Wingless
Winged (selfed)	91	1*
Winged (selfed)	90	30
Wingless (selfed)	4*	80
Winged × wingless	161	0
Winged × wingless	29	31
Winged × wingless	46	0
Winged × winged	44	0
Winged × winged	24	0

*Phenotype probably has a nongenetic explanation.

Interpret these results, and derive the mode of inheritance of these fruit-shaped phenotypes. Use symbols. What do you think is the nongenetic explanation for the phenotypes marked by asterisks in the table?

51. The accompanying pedigree is for a rare, but relatively mild, hereditary disorder of the skin.

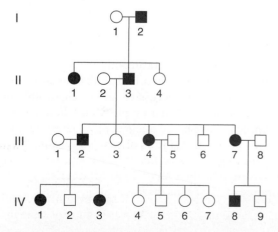

a. How is the disorder inherited? State reasons for your answer.

b. Give genotypes for as many individuals in the pedigree as possible. (Invent your own defined allele symbols.)

c. Consider the four unaffected children of parents III-4 and III-5. In all four-child progenies from parents of these genotypes, what proportion is expected to contain all unaffected children?

52. Four human pedigrees are shown in the accompanying illustration. The black symbols represent an abnormal phenotype inherited in a simple Mendelian manner.

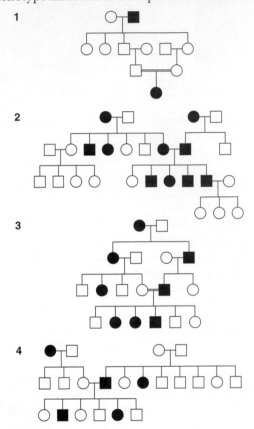

a. For each pedigree, state whether the abnormal condition is dominant or recessive. Try to state the logic behind your answer.

b. For each pedigree, describe the genotypes of as many persons as possible.

53. Tay-Sachs disease (infantile amaurotic idiocy) is a rare human disease in which toxic substances accumulate in nerve cells. The recessive allele responsible for the disease is inherited in a simple Mendelian manner. For unknown reasons, the allele is more common in populations of Ashkenazi Jews of eastern Europe. A woman is planning to marry her first cousin, but the couple discovers that their shared grandfather's sister died in infancy of Tay-Sachs disease.

a. Draw the relevant parts of the pedigree, and show all the genotypes as completely as possible.

b. What is the probability that the cousins' first child will have Tay-Sachs disease, assuming that all people who marry into the family are homozygous normal?

54. The pedigree below was obtained for a rare kidney disease.

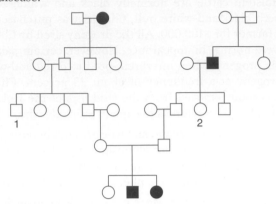

a. Deduce the inheritance of this condition, stating your reasons.

b. If persons 1 and 2 marry, what is the probability that their first child will have the kidney disease?

55. This pedigree is for Huntington disease, a late-onset disorder of the nervous system. The slashes indicate deceased family members.

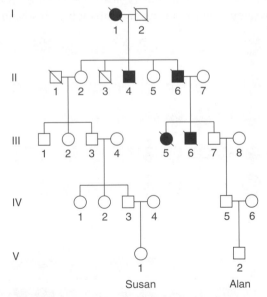

a. Is this pedigree compatible with the mode of inheritance for Huntington disease mentioned in the chapter?

b. Consider two newborn children in the two arms of the pedigree, Susan in the left arm and Alan in the right arm. Study the graph in Figure 2-24 and form an opinion on the likelihood that they will develop Huntington disease. Assume for the sake of the discussion that parents have children at age 25.

56. Consider the accompanying pedigree of a rare autosomal recessive disease, PKU.

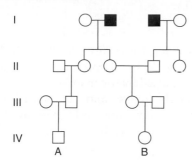

a. List the genotypes of as many of the family members as possible.

b. If persons A and B marry, what is the probability that their first child will have PKU?

c. If their first child is normal, what is the probability that their second child will have PKU?

d. If their first child has the disease, what is the probability that their second child will be unaffected?

(Assume that all people marrying into the pedigree lack the abnormal allele.)

57. A man has attached earlobes, whereas his wife has free earlobes. Their first child, a boy, has attached earlobes.

a. If the phenotypic difference is assumed to be due to two alleles of a single gene, is it possible that the gene is X linked?

b. Is it possible to decide if attached earlobes are dominant or recessive?

58. A rare recessive allele inherited in a Mendelian manner causes the disease cystic fibrosis. A phenotypically normal man whose father had cystic fibrosis marries a phenotypically normal woman from outside the family, and the couple consider having a child.

a. Draw the pedigree as far as described.

b. If the frequency in the population of heterozygotes for cystic fibrosis is 1 in 50, what is the chance that the couple's first child will have cystic fibrosis?

c. If the first child does have cystic fibrosis, what is the probability that the second child will be normal?

59. The allele c causes albinism in mice (C causes mice to be black). The cross $C/c \times c/c$ produces 10 progeny. What is the probability of all of them being black?

60. The recessive allele s causes *Drosophila* to have small wings and the s^1 allele causes normal wings. This gene is known to be X linked. If a small-winged male is crossed with a homozygous wild-type female, what ratio of normal to small-winged flies can be expected

in each sex in the F_1? If F_1 flies are intercrossed, what F_2 progeny ratios are expected? What progeny ratios are predicted if F_1 females are backcrossed with their father?

61. An X-linked dominant allele causes hypophosphatemia in humans. A man with hypophosphatemia marries a normal woman. What proportion of their sons will have hypophosphatemia?

62. Duchenne muscular dystrophy is sex linked and usually affects only males. Victims of the disease become progressively weaker, starting early in life.

a. What is the probability that a woman whose brother has Duchenne's disease will have an affected child?

b. If your mother's brother (your uncle) had Duchenne's disease, what is the probability that you have received the allele?

c. If your father's brother had the disease, what is the probability that you have received the allele?

63. A recently married man and woman discover that each had an uncle with alkaptonuria, otherwise known as "black urine disease," a rare disease caused by an autosomal recessive allele of a single gene. They are about to have their first baby. What is the probability that their child will have alkaptonuria?

64. The accompanying pedigree concerns an inherited dental abnormality, amelogenesis imperfecta.

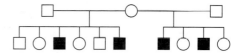

a. What mode of inheritance best accounts for the transmission of this trait?

b. Write the genotypes of all family members according to your hypothesis.

65. A couple who are about to get married learn from studying their family histories that, in *both* their families, their unaffected grandparents had siblings with cystic fibrosis (a rare autosomal recessive disease).

a. If the couple marries and has a child, what is the probability that the child will have cystic fibrosis?

b. If they have four children, what is the chance that the children will have the precise Mendelian ratio of 3:1 for normal:cystic fibrosis?

c. If their first child has cystic fibrosis, what is the probability that their next three children will be normal?

66. A sex-linked recessive allele c produces a red–green color blindness in humans. A normal woman whose father was color blind marries a color-blind man.

a. What genotypes are possible for the mother of the color-blind man?

b. What are the chances that the first child from this marriage will be a color-blind boy?

c. Of the girls produced by these parents, what proportion can be expected to be color blind?

d. Of all the children (sex unspecified) of these parents, what proportion can be expected to have normal color vision?

67. Male house cats are either black or orange; females are black, orange, or calico.

a. If these coat-color phenotypes are governed by a sex-linked gene, how can these observations be explained?

b. Using appropriate symbols, determine the phenotypes expected in the progeny of a cross between an orange female and a black male.

c. Half the females produced by a certain kind of mating are calico, and half are black; half the males are orange, and half are black. What colors are the parental males and females in this kind of mating?

d. Another kind of mating produces progeny in the following proportions: one-fourth orange males, one-fourth orange females, one-fourth black males, and one-fourth calico females. What colors are the parental males and females in this kind of mating?

68. The pedigree below concerns a certain rare disease that is incapacitating but not fatal.

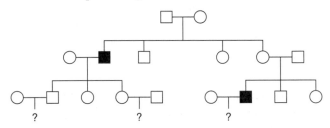

a. Determine the most likely mode of inheritance of this disease.

b. Write the genotype of each family member according to your proposed mode of inheritance.

c. If you were this family's doctor, how would you advise the three couples in the third generation about the likelihood of having an affected child?

69. In corn, the allele s causes sugary endosperm, whereas S causes starchy. What endosperm genotypes result from each of the following crosses?

a. s/s female × S/S male

b. S/S female × s/s male

c. S/s female × S/s male

70. A plant geneticist has two pure lines, one with purple petals and one with blue. She hypothesizes that the phenotypic difference is due to two alleles of one gene. To test this idea, she aims to look for a 3 : 1 ratio in the

F_2. She crosses the lines and finds that all the F_1 progeny are purple. The F_1 plants are selfed and 400 F_2 plants are obtained. Of these F_2 plants, 320 are purple and 80 are blue. Do these results fit her hypothesis well? If not, suggest why.

GENETICS P⊙RTAL **Unpacking the Problem 71.** A man's grandfather has galactosemia, a rare autosomal recessive disease caused by the inability to process galactose, leading to muscle, nerve, and kidney malfunction. The man married a woman whose sister had galactosemia. The woman is now pregnant with their first child.

a. Draw the pedigree as described.

b. What is the probability that this child will have galactosemia?

c. If the first child does have galactosemia, what is the probability that a second child will have it?

CHALLENGING PROBLEMS

72. A geneticist working on peas has a single plant monohybrid Y/y (yellow) and, from a self of this plant, wants to produce a plant of genotype y/y to use as a tester. How many progeny plants need to be grown to be 95% sure of obtaining at least one in the sample?

73. A curious polymorphism in human populations has to do with the ability to curl up the sides of the tongue to make a trough ("tongue rolling"). Some people can do this trick, and others simply cannot. Hence, it is an example of a dimorphism. Its significance is a complete mystery. In one family, a boy was unable to roll his tongue but, to his great chagrin, his sister could. Furthermore, both his parents were rollers, and so were both grandfathers, one paternal uncle, and one paternal aunt. One paternal aunt, one paternal uncle, and one maternal uncle could not roll their tongues.

a. Draw the pedigree for this family, defining your symbols clearly, and deduce the genotypes of as many individual members as possible.

b. The pedigree that you drew is typical of the inheritance of tongue rolling and led geneticists to come up with the inheritance mechanism that no doubt you came up with. However, in a study of 33 pairs of identical twins, both members of 18 pairs could roll, neither member of 8 pairs could roll, and one of the twins in 7 pairs could roll but the other could not. Because identical twins are derived from the splitting of one fertilized egg into two embryos, the members of a pair must be genetically identical. How can the existence of the seven discordant pairs be reconciled with your genetic explanation of the pedigree?

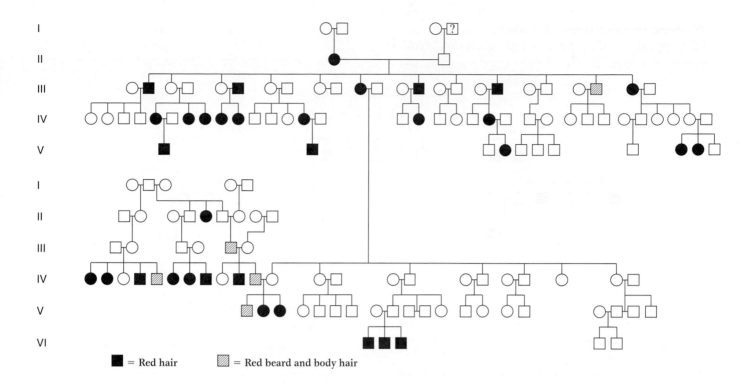

■ = Red hair   ▨ = Red beard and body hair

74. Red hair runs in families, as the pedigree above shows. (Pedigree from W. R. Singleton and B. Ellis, *Journal of Heredity* 55, 1964, 261.)

a. Does the inheritance pattern in this pedigree suggest that red hair could be caused by a dominant or a recessive allele of a gene that is inherited in a simple Mendelian manner?

b. Do you think that the red-hair allele is common or rare in the population as a whole?

75. When many families were tested for the ability to taste the chemical phenylthiocarbamide, the matings were grouped into three types and the progeny were totaled, with the results shown below:

	Children		
Parents	Number of families	Tasters	Non-tasters
Taster × taster	425	929	130
Taster × nontaster	289	483	278
Nontaster × nontaster	86	5	218

With the assumption that PTC tasting is dominant (P) and nontasting is recessive (p), how can the progeny ratios in each of the three types of mating be accounted for?

76. A condition known as icthyosis hystrix gravior appeared in a boy in the early eighteenth century. His skin became very thick and formed loose spines that were sloughed off at intervals. When he grew up, this "porcupine man" married and had six sons, all of whom had this condition, and several daughters, all of whom were normal. For four generations, this condition was passed from father to son. From this evidence, what can you postulate about the location of the gene?

77. The wild-type (W) *Abraxas* moth has large spots on its wings, but the lacticolor (L) form of this species has very small spots. Crosses were made between strains differing in this character, with the following results:

	Parents		Progeny	
Cross	♀	♂	F$_1$	F$_2$
1	L	W	♀ W	♀ $\frac{1}{2}$ L, $\frac{1}{2}$ W
			♂ W	♂ W
2	W	L	♀ L	♀ $\frac{1}{2}$ W, $\frac{1}{2}$ L
			♂ W	♂ $\frac{1}{2}$ W, $\frac{1}{2}$ L

Provide a clear genetic explanation of the results in these two crosses, showing the genotypes of all individual moths.

78. The following pedigree shows the inheritance of a rare human disease. Is the pattern best explained as being caused by an X-linked recessive allele or by an autosomal dominant allele with expression limited to males?

(Pedigree modified from J. F. Crow, *Genetics Notes*, 6th ed. Copyright 1967 by Burgess Publishing Co., Minneapolis.)

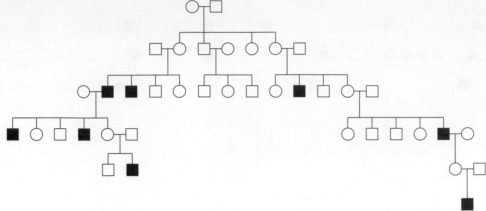

79. A certain type of deafness in humans is inherited as an X-linked recessive trait. A man who suffers from this type of deafness married a normal woman, and they are expecting a child. They find out that they are distantly related. Part of the family tree is shown here.

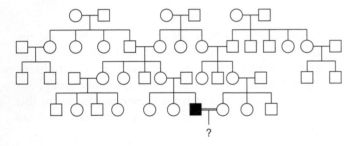

How would you advise the parents about the probability of their child's being a deaf boy, a deaf girl, a normal boy, or a normal girl? Be sure to state any assumptions that you make.

80. The accompanying pedigree shows a very unusual inheritance pattern that actually did exist. All progeny are shown, but the fathers in each mating have been omitted to draw attention to the remarkable pattern.

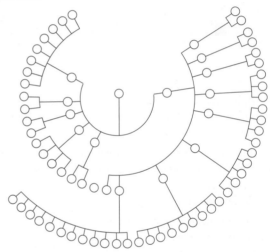

a. Concisely state exactly what is unusual about this pedigree.

b. Can the pattern be explained by Mendelian inheritance?

APPENDIX 2-1   Stages of Mitosis

Mitosis usually takes up only a small proportion of the cell cycle, approximately 5 to 10 percent. The remaining time is the interphase, composed of G1, S, and G2 stages. The DNA is replicated during the S phase, although the duplicated DNA does not become visible until later in mitosis. The chromosomes cannot be seen during interphase (see below), mainly because they are in an extended state and are intertwined with one another like a tangle of yarn.

The photographs below show the stages of mitosis in the nuclei of root-tip cells of the royal lily, *Lilium regale*. In each stage, a photograph is shown at the left and an interpretive drawing at the right.

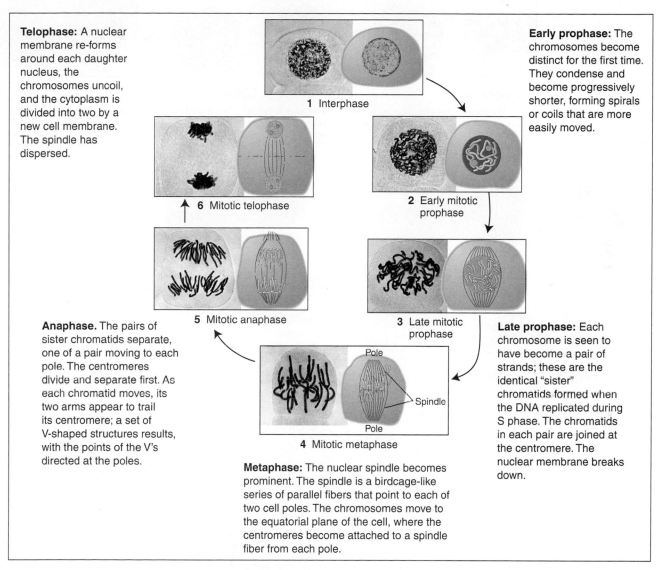

Telophase: A nuclear membrane re-forms around each daughter nucleus, the chromosomes uncoil, and the cytoplasm is divided into two by a new cell membrane. The spindle has dispersed.

6 Mitotic telophase

1 Interphase

Early prophase: The chromosomes become distinct for the first time. They condense and become progressively shorter, forming spirals or coils that are more easily moved.

2 Early mitotic prophase

5 Mitotic anaphase

3 Late mitotic prophase

Anaphase. The pairs of sister chromatids separate, one of a pair moving to each pole. The centromeres divide and separate first. As each chromatid moves, its two arms appear to trail its centromere; a set of V-shaped structures results, with the points of the V's directed at the poles.

Pole

Spindle

Pole

4 Mitotic metaphase

Late prophase: Each chromosome is seen to have become a pair of strands; these are the identical "sister" chromatids formed when the DNA replicated during S phase. The chromatids in each pair are joined at the centromere. The nuclear membrane breaks down.

Metaphase: The nuclear spindle becomes prominent. The spindle is a birdcage-like series of parallel fibers that point to each of two cell poles. The chromosomes move to the equatorial plane of the cell, where the centromeres become attached to a spindle fiber from each pole.

The photographs show mitosis in the nuclei of root-tip cells of *Lilium regale*. [*After J. McLeish and B. Snoad, Looking at Chromosomes. Copyright 1958, St. Martin's, Macmillan.*]

APPENDIX 2-2   Stages of Meiosis

Meiosis consists of two nuclear divisions distinguished as meiosis I and meiosis II, which take place in consecutive cell divisions. Each meiotic division is formally divided into prophase, metaphase, anaphase, and telophase. Of these stages, the most complex and lengthy is prophase I, which is divided into five stages.

The photographs below show the stages of meiosis in the nuclei of root-tip cells of the royal lily, *Lilium regale*. In each stage, a photograph is shown at the left and an interpretive drawing at the right.

1 Leptotene

Prophase I: Leptotene. The chromosomes become visible as long, thin single threads. Chromosomes begin to contract and continue contracting throughout the entire prophase.

2 Zygotene

Prophase I: Zygotene. The threads form pairs as each chromosome progressively aligns, or synapses, along the length of its homologous partner.

3 Pachytene

Prophase I: Pachytene. Chromosomes are thick and fully synapsed. Thus, the number of pairs of homologous chromosomes is equal to the number *n*.

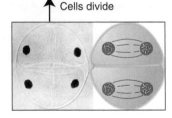

16 Young pollen grains

The tetrad and young pollen grains: In the anthers of a flower, the four products of meiosis develop into pollen grains. In other organisms, the products of meiosis differentiate into other kinds of structures, such as sperm cells in animals.

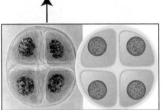

15 The tetrad

 Cells divide

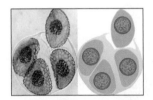

14 Telophase II

Telophase II: The nuclei re-form around the chromosomes at the poles.

Metaphase II: The pairs of sister chromatids arrange themselves on the equatorial plane. Here the chromatids often partly dissociate from each other instead of being closely pressed together as they are in mitosis.

Anaphase II: Centromeres split and sister chromatids are pulled to opposite poles by the spindle fibers.

Prophase II: The haploid number of sister chromatid pairs are now present in the contracted state.

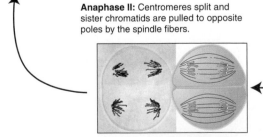

13 Anaphase II

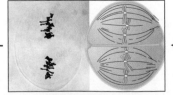

12 Metaphase II

11 Prophase II

The photographs show meiosis sind pollen formation in *Lilium regale*. Note: For simplicity, multiple chiasmata are drawn between only two chromatids; in reality, all four chromatids can take part. [*After J. McLeish and B. Snoad, Looking at Chromosomes. Copyright 1958, St. Martin's, Macmillan.*]

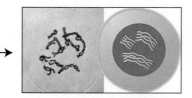

4 Diplotene

Prophase I: Diplotene. Although the DNA has already replicated during the premeiotic S phase, this fact first becomes manifest only in diplotene as each chromosome is seen to have become a pair of sister chromatids. The synapsed structure now consists of a bundle of four homologous chromsomes. The paired homologs separate slightly, and one or more cross-shaped structures called chiasmata (singular, chiasma) appear between nonsister chromatids.

5 Diakinesis

Prophase I: Diakinesis. Further chromosome contraction produces compact units that are very maneuverable.

Metaphase I: The nuclear membrane has disappeared, and each pair of homologs takes up a position in the equatorial plane. At this stage of meiosis; the centromeres do not divide; this lack of division is a major difference from mitosis. The two centromeres of a homologous chromosome pair attach to spindle fibers from opposite poles.

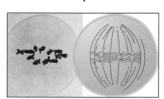

6 Metaphase I

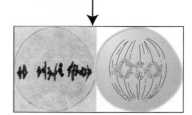

7 Early anaphase I

Anaphase I: The members of each homologous pair move to opposite poles.

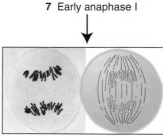

8 Later anaphase I

Telophase I and interphase: The chromosomes elongate and become diffuse, the nuclear membrane re-forms, and the cell divides. After telophase I, there is an interphase, called interkinesis. In many organisms, telophase 1 and interkinesis do not exist or are brief in duration. In any case, there is never DNA synthesis at this time, and the genetic state of the chromosomes does not change.

10 Interphase

Cell divides

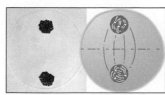

9 Telophase I

Independent Assortment of Genes

3

The Green Revolution in agriculture is fostered by the widespread planting of superior lines of crops (such as rice, shown here) made by combining beneficial genetic traits. [Jorgen Schytte/Peter Arnold.]

This chapter is about the principles at work when two or more cases of single-gene inheritance are analyzed simultaneously. Nowhere have these principles been more important than in plant and animal breeding in agriculture. For example, between the years 1960 and 2000, the world production of food plants doubled. This increase has been dubbed the Green Revolution. What made this Green Revolution possible? In part, the Green Revolution was due to improved agricultural practice, but more important was the development of superior crop genotypes by plant geneticists. These breeders are constantly on the lookout for the chance occurrence of single-gene mutations that significantly increase yield or nutrient value. However, such mutations arise in different lines in different parts of the world. For example, in rice, one of the world's main food crops, the following mutations have been crucial in the Green Revolution:

sd1. This recessive allele results in short stature, making the plant more resistant to "lodging," or falling over, in wind and rain; it also increases the relative amount of the plant's energy that is routed into the seed, the part that we eat.

Figure 3-1 Superior genotypes of crops such as rice have revolutionized agriculture. This photograph shows some of the key genotypes used in rice breeding programs. [*Bloomberg/Getty Images.*]

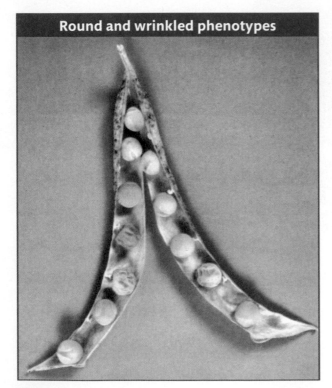

Figure 3-2 Round (*R/R* or *R/r*) and wrinkled (*r/r*) peas are present in a pod of a selfed heterozygous plant (*R/r*). The phenotypic ratio in this pod happens to be precisely the 3 : 1 ratio expected on average in the progeny of this selfing. (Molecular studies have shown that the wrinkled allele used by Mendel is produced by the insertion of a segment of mobile DNA into the gene; see Chapter 14.) [*Madan K. Bhattacharyya.*]

se1. This recessive allele alters the plant's requirement for a specific daylength, enabling it to be grown at different latitudes.

Xa4. This dominant allele confers resistance to the disease bacterial blight.

bph2. This allele confers resistance to brown plant hoppers (a type of insect).

Snb1. This allele confers tolerance to plant submersion after heavy rains.

To make a truly superior genotype, combining such alleles into one line is clearly desirable. To achieve such a combination, mutant lines must be intercrossed two at a time. For instance, a plant geneticist might start by crossing *sd1* and *Xa4*. The F_1 progeny of this cross would carry both mutations but in a heterozygous state. However, most agriculture uses pure lines, which can be efficiently propagated and distributed to farmers. To obtain a pure-breeding doubly mutant *sd1/sd1 · Xa4/Xa4* line, the F_1 would have to be bred further to allow the alleles to "assort" into the desirable combination. Some products of such breeding are shown in Figure 3-1. What principles are relevant here? It depends very much on whether the two genes are on the same chromosome pair or on different chromosome pairs. In the latter case, the chromosome pairs act independently at meiosis, and the alleles of two heterozygous gene pairs are said to show **independent assortment.**

This chapter explains how we can recognize independent assortment and how the principle of independent assortment can be used in strain construction, both in agriculture and in basic genetic research. (Chapter 4 covers the analogous principles applicable to heterozygous gene pairs on the *same* chromosome pair.)

The analytical procedures that pertain to the independent assortment of genes were first developed by the father of genetics, Gregor Mendel. So, again, we turn to his work as a prototypic example.

3.1 Mendel's Law of Independent Assortment

In much of his original work on peas, Mendel analyzed the descendants of pure lines that differed in *two* characters. The following general symbolism is used to represent genotypes that include two genes. If two genes are on different chromosomes, the gene pairs are separated by a semicolon—for example, *A/a;B/b*. If they are on the same chromosome, the alleles on one homolog are written adjacently with no punctuation and are separated from those on the other homolog by a slash—for example, *AB/ab* or *Ab/aB*. An accepted symbolism does not exist for situations in which it is not known whether the genes are on the same chromosome or on different chromosomes. For this situation of unknown position in this book, we will use a dot to separate the genes—for example, *A/a · B/b*. Recall from Chapter 2 that a heterozygote for a *single gene* (such as *A/a*) is sometimes called a monohybrid: accordingly, a *double* heterozygote such as *A/a · B/b* is sometimes called a **dihybrid.** From studying **dihybrid crosses** (*A/a · B/b* × *A/a · B/b*), Mendel came up with his second important principle of heredity.

The pair of characters that he began working with were seed shape and seed color. We have already followed the monohybrid cross for seed color ($Y/y \times Y/y$), which gave a progeny ratio of 3 yellow : 1 green. The seed shape phenotypes (Figure 3-2) were round (determined by allele R) and wrinkled (determined by allele r). The monohybrid cross $R/r \times R/r$ gave a progeny ratio of 3 round : 1 wrinkled as expected (see Table 2-1, page 32). To perform a dihybrid cross, Mendel started with two pure parental lines. One line had wrinkled, yellow seeds. Because Mendel had no concept of the chromosomal location of genes, we must use the dot representation to write the combined genotype initially as $r/r \cdot Y/Y$. The other line had round, green seeds, with genotype $R/R \cdot y/y$. When these two lines were crossed, they must have produced gametes that were $r \cdot Y$ and $R \cdot y$, respectively. Hence, the F$_1$ seeds had to be dihybrid, of genotype $R/r \cdot Y/y$. Mendel discovered that the F$_1$ seeds were round and yellow. This result showed that the dominance of R over r and Y over y was unaffected by the condition of the other gene pair in the $R/r \cdot Y/y$ dihybrid. Next, Mendel selfed the dihybrid F$_1$ to obtain the F$_2$ generation.

The F$_2$ seeds were of four different types in the following proportions:

$\frac{9}{16}$ round, yellow

$\frac{3}{16}$ round, green

$\frac{3}{16}$ wrinkled, yellow

$\frac{1}{16}$ wrinkled, green

a result that is illustrated in Figure 3-3 with the actual numbers obtained by Mendel. This initially unexpected 9 : 3 : 3 : 1 ratio for these two characters seems a lot more complex than the simple 3 : 1 ratios of the monohybrid crosses. Nevertheless, the 9 : 3 : 3 : 1 ratio proved to be a consistent inheritance pattern in peas. As evidence, Mendel also made dihybrid crosses that included several other combinations of characters and found that *all* of the dihybrid F$_1$ individuals produced 9 : 3 : 3 : 1 ratios in the F$_2$. The ratio was another inheritance pattern that required the development of a new idea to explain it.

First, let's check the actual numbers obtained by Mendel in Figure 3-3 to determine if the monohybrid 3 : 1 ratios can still be found in the F$_2$. In regard to seed shape, there are 423 round seeds (315 + 108) and 133 wrinkled seeds (101 + 32). This result is close to a 3 : 1 ratio. Next, in regard to seed color, there are 416 yellow seeds (315 + 101) and 140 green (108 + 32), also very close to a 3 : 1 ratio. The presence of these two 3 : 1 ratios hidden in the 9 : 3 : 3 : 1 ratio was undoubtedly a source of the insight that Mendel needed to explain the 9 : 3 : 3 : 1 ratio, because he realized that it was simply two different 3 : 1 ratios combined at random. One way of visualizing the random combination of these two ratios is with a branch diagram, as follows:

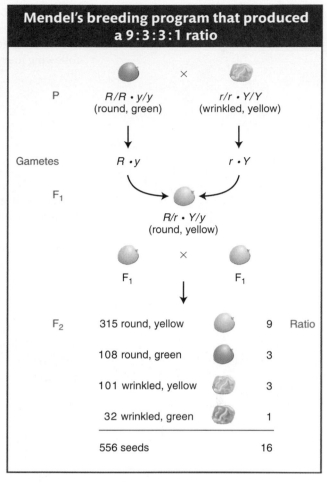

Mendel's breeding program that produced a 9 : 3 : 3 : 1 ratio

P	$R/R \cdot y/y$ (round, green)	$r/r \cdot Y/Y$ (wrinkled, yellow)
Gametes	$R \cdot y$	$r \cdot Y$
F$_1$	$R/r \cdot Y/y$ (round, yellow)	
	F$_1$	F$_1$

F$_2$			Ratio
	315 round, yellow	9	
	108 round, green	3	
	101 wrinkled, yellow	3	
	32 wrinkled, green	1	
	556 seeds	16	

Figure 3-3 Mendel synthesized a dihybrid that, when selfed, produced F$_2$ progeny in the ratio 9 : 3 : 3 : 1.

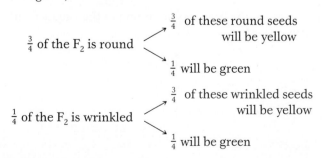

$\frac{3}{4}$ of the F$_2$ is round

$\frac{3}{4}$ of these round seeds will be yellow

$\frac{1}{4}$ will be green

$\frac{1}{4}$ of the F$_2$ is wrinkled

$\frac{3}{4}$ of these wrinkled seeds will be yellow

$\frac{1}{4}$ will be green

The proportions of the four possible outcomes are calculated by using the product rule to multiply along the branches in the diagram. For example, 3/4 of 3/4 is calculated as $3/4 \times 3/4$, which equals 9/16. These multiplications give the following four proportions:

$$\tfrac{3}{4} \times \tfrac{3}{4} = \tfrac{9}{16} \text{ round, yellow}$$

$$\tfrac{3}{4} \times \tfrac{1}{4} = \tfrac{3}{16} \text{ round, green}$$

$$\tfrac{1}{4} \times \tfrac{3}{4} = \tfrac{3}{16} \text{ wrinkled, yellow}$$

$$\tfrac{1}{4} \times \tfrac{1}{4} = \tfrac{1}{16} \text{ wrinkled, green}$$

These proportions constitute the $9:3:3:1$ ratio that we are trying to explain. However, is this exercise not merely number juggling? What could the combination of the two $3:1$ ratios mean biologically? The way that Mendel phrased his explanation does in fact amount to a biological mechanism. In what is now known as **Mendel's second law,** he concluded that *different gene pairs assort independently in gamete formation.* The consequence is that, for two heterozygous gene pairs A/a and B/b, the b allele is just as likely to end up in a gamete with an a allele as with an A allele, and likewise for the B allele. In hindsight, we now know that, for the most part, this "law" applies to genes on different chromosomes. Genes on the same chromosome generally do not assort independently, because they are held together by the chromosome itself. Hence, the modern version of Mendel's second law is stated as in the following Message.

> **Message** Mendel's second law (the principle of independent assortment) states that gene pairs on different chromosome pairs assort independently at meiosis.

We have explained the $9:3:3:1$ phenotypic ratio as two randomly combined $3:1$ phenotypic ratios. But can we also arrive at the $9:3:3:1$ ratio from a consideration of the frequency of gametes, the actual meiotic products? Let us consider the gametes produced by the F_1 dihybrid R/r; Y/y (the semicolon shows that we are now embracing the idea that the genes are on different chromosomes). Again, we will use the branch diagram to get us started because it illustrates independence visually. Combining Mendel's laws of equal segregation and independent assortment, we can predict that

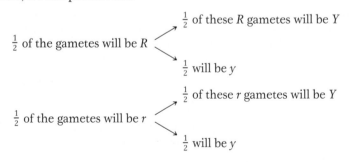

Multiplication along the branches according to the product rule gives us the gamete proportions:

$$\tfrac{1}{4} \ R;Y$$

$$\tfrac{1}{4} \ R;y$$

$$\tfrac{1}{4} \ r;Y$$

$$\tfrac{1}{4} \ r;y$$

These proportions are a direct result of the application of the two Mendelian laws: of segregation and of independence. However, we still have not arrived at the $9:3:3:1$ ratio. The next step is to recognize that, because Mendel did not specify different rules for male and female gamete formation, both the male and the female gametes will show the same proportions just given. The four female gametic types will be fertilized randomly by the four male gametic types to obtain the F_2. The best graphic way of showing the outcomes of the cross is by using a 4×4 grid called a *Punnett square*, which is depicted in Figure 3-4. We have already seen that grids are useful in genetics for providing a visual representation of the data. Their usefulness lies in the fact that their proportions can be drawn according to the genetic proportions or ratios under consideration. In the Punnett square in Figure 3-4, for example, four rows and four columns were drawn to correspond in size to the four genotypes of female gametes and the four of male gametes. We see that there are 16 boxes representing the various gametic fusions and that each box is 1/16th of the total area of the grid. In accord with the product rule, each 1/16th is a result of the fertilization of one egg type at frequency 1/4 by one sperm type also at frequency 1/4, giving the frequency of that fusion as $(1/4)^2$. As the Punnett square shows, the F_2 contains a variety of genotypes, but there are only four phenotypes and their proportions are in the $9:3:3:1$ ratio. So we see that, when we calculate progeny frequencies directly through gamete frequencies, we still arrive at the $9:3:3:1$ ratio. Hence, Mendel's laws explain not only the F_2 phenotypes, but also the genotypes of gametes and progeny that underly the F_2 phenotypic ratio.

Mendel went on to test his principle of independent assortment in a number of ways. The most direct way focused on the $1:1:1:1$ gametic ratio hypothesized to be produced by the F_1 dihybrid $R/r;Y/y$, because this ratio sprang directly from his principle of independent assortment and was the biological basis of the $9:3:3:1$ ratio in the F_2, as shown by the Punnett square. To verify the $1:1:1:1$ gametic ratio, Mendel used a testcross. He testcrossed the F_1 dihybrid with a tester of genotype $r/r;y/y$, which produces only gametes with recessive alleles (genotype $r;y$). He reasoned that, if there were in fact a $1:1:1:1$ ratio of $R;Y$, $R;y$, $r;Y$, and $r;y$ gametes, the progeny proportions of this cross should directly correspond to the gametic proportions produced by the dihybrid; in other words,

$$\tfrac{1}{4}\ R/r;Y/y$$

$$\tfrac{1}{4}\ R/r;y/y$$

$$\tfrac{1}{4}\ r/r;Y/y$$

$$\tfrac{1}{4}\ r/r;y/y$$

These proportions were the result that he obtained, perfectly consistent with his expectations. He obtained similar results for all the other dihybrid crosses that he made, and these tests and other types of tests all showed that he had in fact

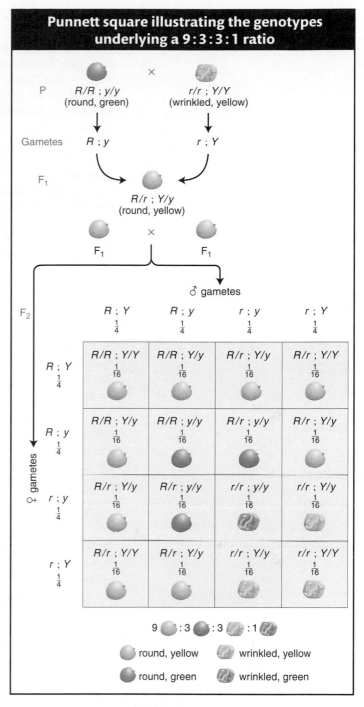

Punnett square illustrating the genotypes underlying a 9:3:3:1 ratio

Figure 3-4 We can use a Punnett square to predict the result of a dihybrid cross. This Punnett square shows the predicted genotypic and phenotypic constitution of the F_2 generation from a dihybrid cross.

devised a robust model to explain the inheritance patterns observed in his various pea crosses.

In the early 1900s, both of Mendel's laws were tested in a wide spectrum of eukaryotic organisms. The results of these tests showed that Mendelian principles were generally applicable. Mendelian ratios (such as $3:1$, $1:1$, $9:3:3:1$, and $1:1:1:1$) were extensively reported, suggesting that equal segregation and independent assortment are fundamental hereditary processes found throughout nature. Mendel's laws are not merely laws about peas, but are laws about the genetics of eukaryotic organisms in general.

As an example of the universal applicability of the principle of independent assortment, we can examine its action in haploids. If the principle of equal segregation is valid across the board, then we should be able to observe its action in haploids, given that haploids undergo meiosis. Indeed, independent assortment can be observed in a cross of the type $A\,;B \times a\,;b$. Fusion of parental cells results in a transient diploid meiocyte that is a dihybrid $A/a\,;B/b$, and the randomly sampled products of meiosis (sexual spores such as ascospores in fungi) will be

$$\tfrac{1}{4}\ A\,;B$$

$$\tfrac{1}{4}\ A\,;b$$

$$\tfrac{1}{4}\ a\,;B$$

$$\tfrac{1}{4}\ a\,;b$$

Hence, we see the same ratio as in the dihybrid testcross in a diploid organism; again, the ratio is a random combination of two monohybrid $1:1$ ratios because of independent assortment.

> **Message** Ratios of $1:1:1:1$ and $9:3:3:1$ are diagnostic of independent assortment in one and two dihybrid meiocytes, respectively.

3.2 Working with Independent Assortment

In this section, we will examine several analytical procedures that are part of everyday genetic research and are all based on the concept of independent assortment. These procedures deal with various aspects of analyzing phenotypic ratios.

Predicting progeny ratios

As stated in Chapter 2, genetics can work in either of two directions: (1) predicting the genotypes of parents by using ratios of progeny or (2) predicting progeny ratios from parents of known genotype. The latter is an important part of genetics concerned with predicting the types of progeny that emerge from a cross and calculating their expected frequencies—in other words, their probabilities. We have already examined two methods for doing so: Punnett squares and branch diagrams. Punnett squares can be used to show hereditary patterns based on one gene pair, two gene pairs, or more. Such grids are good graphic devices for representing progeny, but drawing them is time consuming. Even the 16-compartment Punnett square that we used to analyze a dihybrid cross takes a long time to write out, but, for a trihybrid cross, there are 2^3, or 8, different gamete types, and the Punnett square has 64 compartments. The branch diagram (shown on the facing page) is easier to create and is adaptable for phenotypic, genotypic, or gametic proportions, as illustrated for the dihybrid $A/a\,;B/b$.

Progeny genotypes from a self	Progeny phenotypes from a self	Gametes

$$\frac{1}{4}\,A/A \Longleftrightarrow \begin{array}{l} \frac{1}{4}\,B/B \\ \frac{1}{2}\,B/b \\ \frac{1}{4}\,b/b \end{array}$$

$$\frac{3}{4}\,A/- \nearrow \frac{3}{4}\,B/- \searrow \frac{1}{4}\,b/b$$

$$\frac{1}{2}\,A \nearrow \frac{1}{2}\,B \searrow \frac{1}{2}\,b$$

$$\frac{1}{2}\,A/a \Longleftrightarrow \begin{array}{l} \frac{1}{4}\,B/B \\ \frac{1}{2}\,B/b \\ \frac{1}{4}\,b/b \end{array}$$

$$\frac{1}{4}\,a/a \nearrow \frac{3}{4}\,B/- \searrow \frac{1}{4}\,b/b$$

$$\frac{1}{2}\,a \nearrow \frac{1}{2}\,B \searrow \frac{1}{2}\,b$$

$$\frac{1}{4}\,a/a \Longleftrightarrow \begin{array}{l} \frac{1}{4}\,B/B \\ \frac{1}{2}\,B/b \\ \frac{1}{4}\,b/b \end{array}$$

Note, however, that the "tree" of branches for genotypes is quite unwieldy even in this simple case, which uses two gene pairs, because there are $3^2 = 9$ genotypes. For three gene pairs, there are 3^3, or 27, possible genotypes. To simplify this problem, we can use a statistical approach, which constitutes a third method for calculating the probabilities (expected frequencies) of specific phenotypes or genotypes coming from a cross. The two statistical rules needed are the **product rule** (introduced in Chapter 2) and the **sum rule,** which we will now consider together.

> **Message** The product rule states that the probability of independent events *both* occurring together is the product of their individual probabilities.

The possible outcomes from rolling two dice follow the product rule because the outcome on one die is independent of the other. As an example, let us calculate the probability, p, of rolling a pair of 4's. The probability of a 4 on one die is 1/6 because the die has six sides and only one side carries the number 4. This probability is written as follows:

$$p \text{ (of a 4)} = \tfrac{1}{6}$$

Therefore, with the use of the product rule, the probability of a 4 appearing on both dice is $1/6 \times 1/6 = 1/36$, which is written

$$p \text{ (of two 4's)} = \tfrac{1}{6} \times \tfrac{1}{6} = \tfrac{1}{36}$$

Now for the sum rule:

> **Message** The sum rule states that the probability of *either* of two mutually exclusive events occurring is the sum of their individual probabilities.

(Note that, in the product rule, the focus is on outcomes A *and* B. In the sum rule, the focus is on the outcome A′ *or* A″.)

Dice can also be used to illustrate the sum rule. We have already calculated that the probability of two 4's is 1/36; clearly, with the use of the same type of calculation, the probability of two 5's will be the same, or 1/36. Now we can calculate the probability of either two 4's *or* two 5's. Because these outcomes are mutually exclusive, the sum rule can be used to tell us that the answer is $1/36 + 1/36$, which is 1/18. This probability can be written as follows:

$$p \text{ (two 4's or two 5's)} = \tfrac{1}{36} + \tfrac{1}{36} = \tfrac{1}{18}$$

What proportion of progeny will be of a specific genotype? Now we can turn to a genetic example. Assume that we have two plants of genotypes

$$A/a \, ; b/b \, ; C/c \, ; D/d \, ; E/e$$

and

$$A/a \, ; B/b \, ; C/c \, ; d/d \, ; E/e$$

From a cross between these plants, we want to recover a progeny plant of genotype $a/a \, ; b/b \, ; c/c \, ; d/d \, ; e/e$ (perhaps for the purpose of acting as the tester strain in a testcross). What proportion of the progeny should we expect to be of that genotype? If we assume that all the gene pairs assort independently, then we can do this calculation easily by using the product rule. The five different gene pairs are considered individually, as if five separate crosses, and then the individual probabilities of obtaining each genotype are multiplied together to arrive at the answer:

From $A/a \times A/a$, one-fourth of the progeny will be a/a.

From $b/b \times B/b$, half the progeny will be b/b.

From $C/c \times C/c$, one-fourth of the progeny will be c/c.

From $D/d \times d/d$, half the progeny will be d/d.

From $E/e \times E/e$, one-fourth of the progeny will be e/e.

Therefore, the overall probability (or expected frequency) of obtaining progeny of genotype $a/a \, ; b/b \, ; c/c \, ; d/d \, ; e/e$ will be $1/4 \times 1/2 \times 1/4 \times 1/2 \times 1/4 = 1/256$. This probability calculation can be extended to predict phenotypic frequencies or gametic frequencies. Indeed, there are many other uses for this method in genetic analysis, and we will encounter some in later chapters.

How many progeny do we need to grow? To take the preceding example a step farther, suppose we need to estimate how many progeny plants need to be grown to stand a reasonable chance of obtaining the desired genotype $a/a \, ; b/b \, ; c/c \, ; d/d \, ; e/e$. We first calculate the proportion of progeny that is expected to be of that genotype. As just shown, we learn that we need to examine at least 256 progeny to stand an average chance of obtaining one individual plant of the desired genotype.

The probability of obtaining one "success" (a fully recessive plant) out of 256 has to be considered more carefully. This is the *average* probability of success. Unfortunately, if we isolated and tested 256 progeny, we would very likely have no successes at all, simply from bad luck. From a practical point of view, a more meaningful question would be to ask, What sample size do we need to be 95 percent confident that we will obtain at least one success? (**Note:** This 95 percent confidence value is standard in science.) The simplest way to perform this calculation is to approach it by considering the probability of complete failure—that is, the probability of obtaining no individuals of the desired genotype. In our example, for every individual isolated, the probability of its *not* being the desired type is $1 - (1/256) = 255/256$. Extending this idea to a sample of size n, we see that the probability of no successes in a sample of n is $(255/256)^n$. (This probability is a simple application of the product rule: $255/256$ multiplied by itself n times.)

Hence, the probability of obtaining *at least one success* is the probability of all possible alternative outcomes (this probability is 1) minus the probability of total failure, or $(255/256)^n$. Hence, the probability of at least one success is $1 - (255/256)^n$. To satisfy the 95 percent confidence level, we must put this expression equal to 0.95 (the equivalent of 95 percent).

Therefore,

$$1 - (255/256)^n = 0.95$$

Solving this equation for n gives us a value of 765, the number of progeny needed to virtually guarantee success. Notice how different this number is from the naive expectation of success in 256 progeny. This type of calculation is useful in many applications in genetics and in other situations in which a successful outcome is needed from many trials.

How many distinct genotypes will a cross produce? The rules of probability can be easily used to predict the number of genotypes or phenotypes in the progeny of complex parental strains. For example, in a self of the "tetrahybrid" $A/a;B/b;C/c;D/d$, there will be three genotypes for each gene pair; for example, for the first gene pair, the three genotypes will be A/a, A/A, and a/a. Because there are four gene pairs in total, there will be $3^4 = 81$ different genotypes. In a testcross of such a tetrahybrid, there will be two genotypes for each gene pair (for example, A/a and a/a) and a total of $2^4 = 16$ genotypes in the progeny. Because we are assuming that all the genes are on different chromosomes, all these testcross genotypes will occur at an equal frequency of 1/16.

Using the chi-square test on monohybrid and dihybrid ratios

In genetics generally, a researcher is often confronted with results that are close to an expected ratio but not identical to it. Such ratios can be from monohybrids, dihybrids, or more complex genotypes and with independence or not. But how close to an expected result is close enough? A statistical test is needed to check such ratios against expectations, and the **chi-square test,** or χ^2 test, fulfills this role.

In which experimental situations is the χ^2 test generally applicable? The general situation is one in which observed results are compared with those predicted by a hypothesis. In a simple genetic example, suppose you have bred a plant that you hypothesize on the basis of a preceding analysis to be a heterozygote, A/a. To test this hypothesis, you cross this heterozygote with a tester of genotype a/a and count the numbers of phenotypes with genotypes $A/-$ and a/a in the progeny. Then you must assess whether the numbers that you obtain constitute the expected 1:1 ratio. If there is a close match, then the hypothesis is deemed consistent with the result; whereas if there is a poor match, the hypothesis is rejected. As part of this process, a judgment has to be made about whether the observed numbers are close *enough* to those expected. Very close matches and blatant mismatches generally present no problem, but, inevitably, there are gray areas in which the match is not obvious.

The χ^2 test is simply a way of quantifying the various deviations expected by chance if a hypothesis is true. Take the preceding simple hypothesis predicting a 1:1 ratio, for example. Even if the hypothesis were true, we would not always expect an exact 1:1 ratio. We can model this idea with a barrelful of equal numbers of red and white marbles. If we blindly remove samples of 100 marbles, on the basis of chance we would expect samples to show small deviations such as 52 red:48 white quite commonly and to show larger deviations such as 60 red:40 white less commonly. Even 100 red marbles is a possible outcome, at a very low probability of $(1/2)^{100}$. However, if *any* result is possible at some level of probability even if the hypothesis is true, how can we ever reject a hypothesis? A general scientific convention is that a hypothesis will be rejected as false if there is a probability of less than 5 percent of observing a deviation from expectations at least as large as the one actually observed. The hypothesis might still be true, but we have to make a decision somewhere, and 5 percent is the conventional decision line. The implication is that, although results this far from expectations are expected 5 percent of the time even when the hypothesis is true, we will mistakenly reject the hypothesis in only 5 percent of cases and we are willing to take this chance of error. (This 5 percent is the converse of the 95 percent confidence level used earlier.)

Let's look at some real data. We will test our earlier hypothesis that a plant is a heterozygote. We will let A stand for red petals and a stand for white. Scientists test a hypothesis by making predictions based on the hypothesis. In the present situation, one possibility is to predict the results of a testcross. Assume that we testcross the presumed heterozygote. On the basis of the hypothesis, Mendel's law of equal segregation predicts that we should have 50 percent A/a and 50 percent a/a. Assume that, in reality, we obtain 120 progeny and find that 55 are red and 65 are white. These numbers differ from the precise expectations, which would have been 60 red and 60 white. The result seems a bit far off the expected ratio, which raises uncertainty; so we need to use the χ^2 test. We calculate χ^2 by using the following formula:

$$\chi^2 = \Sigma \, (O - E)^2/E \text{ for all classes}$$

in which E is the expected number in a class, O is the observed number in a class, and Σ means "sum of."

The calculation is most simply performed by using a table:

Class	O	E	$(O - E)^2$	$(O - E)^2/E$
Red	55	60	25	$25/60 = 0.42$
White	65	60	25	$25/60 = 0.42$
				Total $= \chi^2 = 0.84$

Now we must look up this χ^2 value in Table 3-1, which will give us the probability value that we want. The rows in Table 3-1 list different values of *degrees of freedom (df)*. The number of degrees of freedom is the number of independent variables in the data. In the present context, the number of independent variables is simply the number of phenotypic classes minus 1. In this case, df $= 2 = 1 = 1$. So we look only at the 1 df line. We see that our χ^2 value of 0.84 lies somewhere between the columns marked 0.5 and 0.1—in other words, between 50 percent and 10 percent. This probability value is much greater than the cutoff value of 5 percent, and so we accept the observed results as being compatible with the hypothesis.

Some important notes on the application of this test follow:

1. What does the probability value actually mean? It is the probability of observing a deviation from the expected results *at least as large* (not *exactly* this deviation) on the basis of chance if the hypothesis is correct.

2. The fact that our results have "passed" the chi-square test because $p > 0.05$ does not mean that the hypothesis is true; it merely means that the results are compatible with that hypothesis. However, if we had obtained a p value of < 0.05, we would have been forced to reject the hypothesis. Science is all about falsifiable hypotheses, not "truth."

3. We must be careful about the wording of the hypothesis, because tacit assumptions are often buried within it. The present hypothesis is a case in point; if we were to carefully state it, we would have to say that the "individual under

Table 3-1   Critical Values of the χ^2 Distribution

df	0.995	0.975	0.9	0.5	0.1	0.05	0.025	0.01	0.005	df
1	.000	.000	0.016	0.455	2.706	3.841	5.024	6.635	7.879	1
2	0.010	0.051	0.211	1.386	4.605	5.991	7.378	9.210	10.597	2
3	0.072	0.216	0.584	2.366	6.251	7.815	9.348	11.345	12.838	3
4	0.207	0.484	1.064	3.357	7.779	9.488	11.143	13.277	14.860	4
5	0.412	0.831	1.610	4.351	9.236	11.070	12.832	15.086	16.750	5
6	0.676	1.237	2.204	5.348	10.645	12.592	14.449	16.812	18.548	6
7	0.989	1.690	2.833	6.346	12.017	14.067	16.013	18.475	20.278	7
8	1.344	2.180	3.490	7.344	13.362	15.507	17.535	20.090	21.955	8
9	1.735	2.700	4.168	8.343	14.684	16.919	19.023	21.666	23.589	9
10	2.156	3.247	4.865	9.342	15.987	18.307	20.483	23.209	25.188	10
11	2.603	3.816	5.578	10.341	17.275	19.675	21.920	24.725	26.757	11
12	3.074	4.404	6.304	11.340	18.549	21.026	23.337	26.217	28.300	12
13	3.565	5.009	7.042	12.340	19.812	22.362	24.736	27.688	29.819	13
14	4.075	5.629	7.790	13.339	21.064	23.685	26.119	29.141	31.319	14
15	4.601	6.262	8.547	14.339	22.307	24.996	27.488	30.578	32.801	15

test is a heterozygote A/a, these alleles show equal segregation at meiosis, and the A/a and a/a progeny are of equal viability." We will investigate allele effects on viability in Chapter 6, but, for the time being, we must keep them in mind as a possible complication because differences in survival would affect the sizes of the various classes. The problem is that, if we reject a hypothesis that has hidden components, we do not know which of the components we are rejecting. For example, in the present case, if we were forced to reject the hypothesis as a result of the χ^2 test, we would not know if we were rejecting equal segregation or equal viability or both.

4. The outcome of the χ^2 test depends heavily on sample sizes (numbers in the classes). Hence, the test must use *actual numbers*, not proportions or percentages. Additionally, the larger the samples, the more reliable is the test.

Any of the familiar Mendelian ratios considered in this chapter or in Chapter 2 can be tested by using the χ^2 test—for example, $3:1$ (1 df), $1:2:1$ (2 df), $9:3:3:1$ (3 df), and $1:1:1:1$ (3 df). We will return to more applications of the χ^2 test in Chapter 4.

Synthesizing pure lines

Pure lines are among the essential tools of genetics. For one thing, only these fully homozygous lines will express recessive alleles, but the main need for pure lines is in the maintenance of stocks for research. The members of a pure line can be left to interbreed over time and thereby act as a constant source of the genotype for use in experiments. Hence, for most model organisms, there are international stock centers that are repositories of pure lines for use in research. Similar stock centers provide lines of plants and animals for use in agriculture.

Pure lines of plants or animals are made through repeated generations of selfing. (In animals, selfing is accomplished by mating animals of identical genotype.) Selfing a monohybrid plant shows the principle at work. Suppose we start with a population of individuals that are all A/a and allow them to self. We can apply Mendel's first law to predict that, in the next generation, there will $\frac{1}{4}$ A/A, $\frac{1}{2}$ A/a, and $\frac{1}{4}$ a/a. Note that the *heterozygosity* (the proportion of heterozygotes) has halved, from 1 to $\frac{1}{2}$. If we repeat this process of selfing for another generation, all descendants of homozygotes will be homozygous, again, the heterozygotes will halve their proportion to a quarter. The process is shown in the following display:

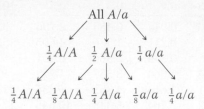

$$\text{All } A/a$$

$$\tfrac{1}{4} A/A \quad \tfrac{1}{2} A/a \quad \tfrac{1}{4} a/a$$

$$\tfrac{1}{4} A/A \quad \tfrac{1}{8} A/A \quad \tfrac{1}{4} A/a \quad \tfrac{1}{8} a/a \quad \tfrac{1}{4} a/a$$

After, say, eight generations of selfing, the proportion of heterozygotes is reduced to $(1/2)^8$, which is 1/256, or about 0.4 percent. Let's look at this process in a slightly different way: we will assume that we start such a program with a genotype that is heterozygous at 256 gene pairs. If we also assume independent assortment, then, after selfing for eight generations, we would end up with an array of genotypes, each having on average only one heterozygous gene (that is, 1/256). In other words, we are well on our way to creating a number of pure lines.

Let us apply this principle to the selection of agricultural lines, the topic with which we began the chapter. We can use as our example the selection of Marquis wheat by Charles Saunders in the early part of the twentieth century. Saunders's goal was to develop a productive wheat line that would have a shorter growing season and hence open up large areas of terrain in northern countries such as Canada and Russia for growing wheat, another of the world's staple foods. He crossed a line having excellent grain quality called Red Fife with a line called Hard Red Calcutta, which, although its yield and quality were poor, matured 20 days earlier than Red Fife. The F_1 produced by the cross was presumably multiply heterozygous for the genes controlling the wheat qualities. From this F_1, Saunders made selfings and selections that eventually led to a pure line that had the combination of favorable properties needed—good-quality grain and early maturation. This line was called Marquis. It was rapidly adopted in many parts of the world.

A similar approach can be applied to the rice lines with which we began the chapter. All the single-gene mutations are crossed in pairs, and then their F_1 plants are selfed or intercrossed with other F_1 plants. As a demonstration, let's consider just four mutations, *1* through *4*. A breeding program might be as follows, in which the mutant alleles and their wild-type counterparts are always listed in the same order:

Figure 3-5 Tomato breeding has resulted in a wide range of lines of different genotypes and phenotypes. [*Mascarucci/Corbis.*]

1/1 ; *+/+* ; *+/+* ; *+/+* × *+/+* ; *2/2* ; *+/+* ; *+/+*   *+/+* ; *+/+* ; *3/3* ; *+/+* × *+/+* ; *+/+* ; *+/+* ; *4/4*

F_1 *1/+* ; *2/+* ; *+/+* ; *+/+* F_1 *+/+* ; *+/+* ; *3/+* ; *4/+*

Self   Self

Select the homozygote *1/1* ; *2/2* ; *+/+* ; *+/+*   Select the homozygote *+/+* ; *+/+* ; *3/3* ; *4/4*

Cross these homozygotes

F_1 *1/+* ; *2/+* ; *3/+* ; *4/+*

Self

Select the homozygote *1/1* ; *2/2* ; *3/3* ; *4/4*

This type of breeding has been applied to many other crop species. The colorful and diverse pure lines of tomatoes used in commerce are shown in Figure 3-5.

Note that, in general when a multiple heterozygote is selfed, a range of different homozygotes is produced. For example, from $A/a;B/b;C/c$, there are two homozygotes for each gene pair (that is, for the first gene, the homozygotes are A/A and a/a), and so there are $2^3 = 8$ different homozygotes possible: $A/A;b/b;C/c$, and $a/a;B/B;c/c$, and so on. Each distinct homozygote can be the start of a new pure line.

Message Repeated selfing leads to an increased proportion of homozygotes, a process that can be used to create pure lines for research or other applications.

Hybrid vigor

We have been considering the synthesis of superior pure lines for research and for agriculture. Pure lines are convenient in that propagation of the genotype from year to year is fairly easy. However, a large proportion of commercial seed that farmers (and gardeners) use is called *hybrid seed*. Curiously, in many cases in which two disparate lines of plants (and animals) are united in an F_1 hybrid (presumed heterozygote), the hybrid shows greater size and vigor than do the two contributing lines (Figure 3-6). This general superiority of multiple heterozygotes is called **hybrid vigor.** The molecular reasons for hybrid vigor are mostly unknown and still hotly debated, but the phenomenon is undeniable and has made large contributions to agriculture. A negative aspect of using hybrids is that, every season, the two parental lines must be

Figure 3-6 (a) Multiple heterozygous hybrid plant on the right next to the two pure lines crossed to make it. (b) Cobs from the same plants. The hybrid cob is in the middle. [(a) and (b) Deana Namuth-Covert, PhD, University of Nebraska.]

grown separately and then intercrossed to make hybrid seed for sale. This process is much more inconvenient than maintaining pure lines, which requires only letting plants self; consequently, hybrid seed is more expensive than seed from pure lines.

From the user's perspective, there is another negative aspect of using hybrids. After a hybrid plant has grown and produced its crop for sale, it is not realistic to keep some of the seeds that it produces and expect this seed to be equally vigorous the next year. The reason is that, when the hybrid undergoes meiosis, independent assortment of the various mixed gene pairs will form many different allelic combinations, and very few of these combinations will be that of the original hybrid. For example, the earlier described tetrahybrid, when selfed, produces 81 different genotypes, of which only a minority will be tetrahybrid. If we assume independent assortment, then, for each gene pair, selfing will produce one-half heterozygotes ($A/a \rightarrow \frac{1}{4} A/A$, $\frac{1}{2} A/a$, and $\frac{1}{4} a/a$). Because there are four gene pairs in this tetrahybrid, the proportion of progeny that will be like the original hybrid $A/a;B/b;C/c;D/d$ will be $(1/2)^4 = 1/16$.

> **Message** Some hybrids between genetically different lines show hybrid vigor. However, gene assortment when the hybrid undergoes meiosis breaks up the favorable allelic combination, and thus few members of the next generation have it.

3.3 The Chromosomal Basis of Independent Assortment

Like equal segregation, the independent assortment of gene pairs on different chromosomes is explained by the behavior of chromosomes during meiosis. Consider a chromosome that we might call number 1; its two homologs could be named 1′ and 1″. If the chromosomes align on the equator, then 1′ might go "north" and 1″ "south" or vice versa. Similarly for a chromosome 2 with homologs 2′ and 2″, 2′ might go north and 2″ south or vice versa. Hence, chromosome 1′ could end up packaged with either chromosome 2′ or 2″, depending on which chromosomes were pulled in the same direction.

Independent assortment is not easy to demonstrate by observing segregating chromosomes under the microscope, because homologs such as 1′ and 1″ do not usually look different, although they might carry minor sequence variation. However, independent assortment can be observed in certain specialized cases. One case was instrumental in the historical development of the chromosome theory.

In 1913, Elinor Carothers found an unusual chromosomal situation in a certain species of grasshopper—a situation that permitted a direct test of whether different chromosome pairs do indeed segregate independently. Studying meioses in the testes of grasshoppers, she found a grasshopper in which one chromosome "pair" had nonidentical members. Such a pair is called a *heteromorphic* pair; presumably, the chromosomes show only partial homology. In addition, the same grasshopper had another chromosome (unrelated to the heteromorphic pair) that had no pairing partner at all. Carothers was able to use these unusual chromosomes as visible cytological markers of the behavior of chromosomes during meiosis. She visually screened many meioses and found that there were two distinct patterns, which are shown in Figure 3-7. In addition, she found that the two patterns were equally frequent. To summarize, if we hold the segregation of the heteromorphic pair constant (brown in the figure), then the unpaired (purple) chromosome can go to either pole equally frequently, half the time with the long form and half the time with the short form. In other words the purple and brown sets were segregating independently. Although these are obviously not typical chromosomes, the results do strongly suggest that different chromosomes assort independently at the first division of meiosis.

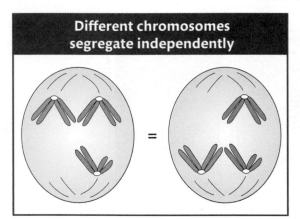

Different chromosomes segregate independently

Figure 3-7 Carothers observed these two equally frequent patterns by which a heteromorphic pair and an unpaired chromosome move into gametes at meiosis.

Independent assortment in diploid organisms

The chromosomal basis of the law of independent assortment is formally diagrammed in Figure 3-8, which illustrates how the separate behavior of two different chromosome pairs gives rise to the 1:1:1:1 Mendelian ratios of gametic types expected from independent assortment. The hypothetical cell has four

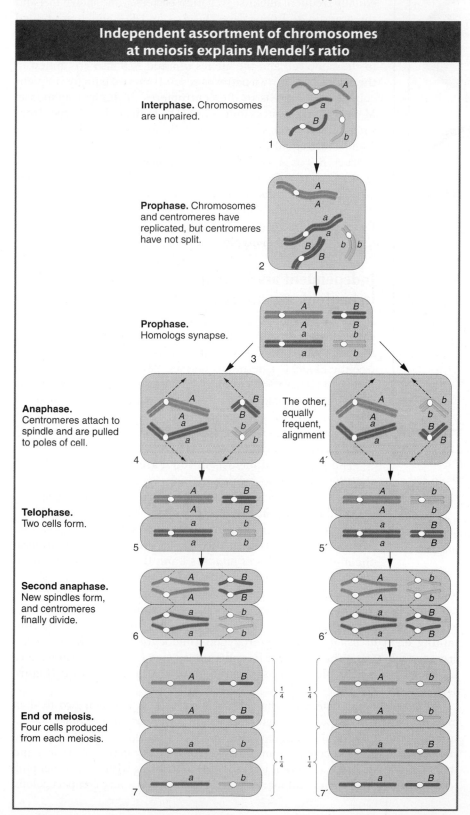

Independent assortment of chromosomes at meiosis explains Mendel's ratio

Interphase. Chromosomes are unpaired.

Prophase. Chromosomes and centromeres have replicated, but centromeres have not split.

Prophase. Homologs synapse.

Anaphase. Centromeres attach to spindle and are pulled to poles of cell.

The other, equally frequent, alignment

Telophase. Two cells form.

Second anaphase. New spindles form, and centromeres finally divide.

End of meiosis. Four cells produced from each meiosis.

Figure 3-8 Meiosis in a diploid cell of genotype *A/a ; B/b*. The diagram shows how the segregation and assortment of different chromosome pairs give rise to the 1:1:1:1 Mendelian gametic ratio.

GENETICS PORTAL **ANIMATED ART: Meiotic recombination between unlinked genes by independent assortment**

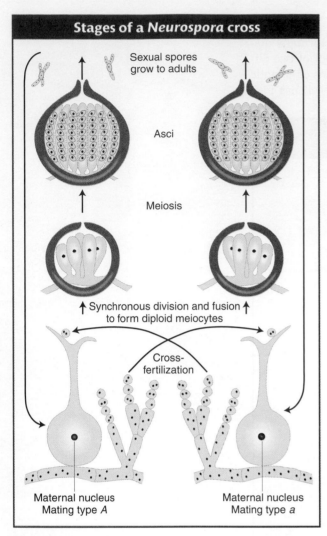

Stages of a *Neurospora* cross

Sexual spores grow to adults

Asci

Meiosis

↑ Synchronous division and fusion ↑
to form diploid meiocytes

Cross-fertilization

Maternal nucleus
Mating type *A*

Maternal nucleus
Mating type *a*

Figure 3-9 The life cycle of *Neurospora crassa,* the orange bread mold. Self-fertilization is not possible in this species: there are two mating types, determined by the alleles *A* and *a* of one gene, and either can act as "female." An asexual spore from the opposite mating type fuses with a receptive hair, and a nucleus from the asexual spore travels down the hair to pair with a female nucleus in the knot of cells. The *A* and *a* pair then undergo synchronous mitoses, finally fusing to form diploid meiocytes.

chromosomes: a pair of homologous long chromosomes and a pair of homologous short ones. The genotype of the meiocytes is A/a ; B/b, and the two allelic pairs, A/a and B/b, are shown on two different chromosome pairs. Parts 4 and 4′ of Figure 3-8 show the key step in independent assortment: there are two equally frequent allelic segregation patterns, one shown in 4 and the other in 4′. In one case, the A/A and B/B alleles are pulled together into one cell, and the a/a and b/b are pulled into the other cell. In the other case, the alleles A/A and b/b are united in the same cell and the alleles a/a and B/B also are united in the same cell. The two patterns result from two equally frequent spindle attachments to the centromeres in the first anaphase. Meiosis then produces four cells of the indicated genotypes from each of these segregation patterns. Because segregation patterns 4 and 4′ are equally common, the meiotic product cells of genotypes A ; B, a ; b, A ; b, and a ; B are produced in equal frequencies. In other words, the frequency of each of the four genotypes is 1/4. This gametic distribution is that postulated by Mendel for a dihybrid, and it is the one that we inserted along one edge of the Punnett square. The random fusion of these gametes results in the $9:3:3:1$ F_2 phenotypic ratio.

Independent assortment in haploid organisms

In the ascomycete fungi, we can actually inspect the products of a single meiocyte to show independent assortment directly. Let's use the filamentous fungus *Neurospora crassa* to illustrate this point (see the Model Organism box on page 98). As we have seen from earlier fungal examples, a cross in *Neurospora* is made by mixing two parental haploid strains of opposite mating type. In a manner similar to that of yeast, mating type is determined by two "alleles" of one gene—in this species, called MAT-A and MAT-a. The way in which a cross is made is shown in Figure 3-9.

The products of meiosis in fungi are sexual spores. Recall that the *ascomycetes* (which include *Neurospora* and *Saccharomyces*) are unique in that, for any given meiocyte, the spores are held together in a membranous sac called an *ascus*. Thus, for these organisms, the products of a single meiosis can be recovered and tested. In the orange bread mold *Neurospora*, the nuclear spindles of meioses I and II do not overlap within the cigar-shaped ascus, and so the four products of a single meiocyte lie in a straight row (Figure 3-10a). Furthermore, for some reason not understood, there is a *postmeiotic mitosis*, which also shows no spindle overlap. Hence, meiosis and the extra mitosis result in a linear ascus containing eight *ascospores*. In a heterozygous meiocyte A/a, if there are no crossovers between the gene and its centromere (see Chapter 4), then there will be two adjacent blocks of ascospores, four of *A* and four of *a* (Figure 3-10b).

Now we can examine a dihybrid. Let's make a cross between two distinct mutants having mutations in different genes on different chromosomes. By assuming that the loci of the mutated genes are both very close to their respective centromeres, we avoid complications due to crossing over between the loci and the centromeres. The first mutant is albino (*a*), contrasting with the normal pink wild type (*a⁺*). The second mutant is biscuit (*b*), which has a very compact colony shaped like a biscuit in contrast with the flat, spreading colony of wild type (b^+). We will assume that the two mutants are of opposite mating type. Hence, the cross is

$$a\,;b^+ \times a^+\,;b$$

Because of random spindle attachment, the following two octad types will be equally frequent:

$a^+\,;b$	$a\,;b$
$a^+\,;b$	$a\,;b$
$a^+\,;b$	$a\,;b$
$a^+\,;b$	$a\,;b$
$a\,;b^+$	$a^+\,;b^+$
$a\,;b^+$	$a^+\,;b^+$
$a\,;b^+$	$a^+\,;b^+$
$a\,;b^+$	$a^+\,;b^+$
50%	**50%**

The equal frequency of these two types is a convincing demonstration of independent assortment occurring in individual meiocytes.

Independent assortment of combinations of autosomal and X-linked genes

The principle of independent assortment is also useful in analyzing genotypes that are heterozygous for both autosomal and X-linked genes. The autosomes and the sex chromosomes are moved independently by spindle fibers attached randomly to their centromeres, just as with two different pairs of autosomes. Some interesting dihybrid ratios are produced. Let's look at an example from *Drosophila*. The cross is between a female with vestigial wings (autosomal recessive, *vg*) and a male with white eyes (X-linked recessive, *w*). Symbolically, the cross is

$$vg/vg\,;+/+\,♀ \;\times\; +/+\,;w/Y\,♂$$

The F$_1$ will be:

Females of genotype $+/vg\,;+/w$

Males of genotype $+/vg\,;+/Y$

These F$_1$ flies must be interbred to obtain an F$_2$. Because the cross is a monohybrid cross for the *autosomal vestigial gene*, both sexes of the F$_2$ will show

Females and males $\frac{3}{4}$ $+/-$ (wild type)

               $\frac{1}{4}$ vg/vg (vestigial)

For the X-linked white eye gene, the ratios will be as follows:

Females $\frac{1}{2}$ $+/+$ and $\frac{1}{2}$ $+/w$ (all wild type)

Males $\frac{1}{2}$ $+/Y$ (wild type) and $\frac{1}{2}$ w/Y (white)

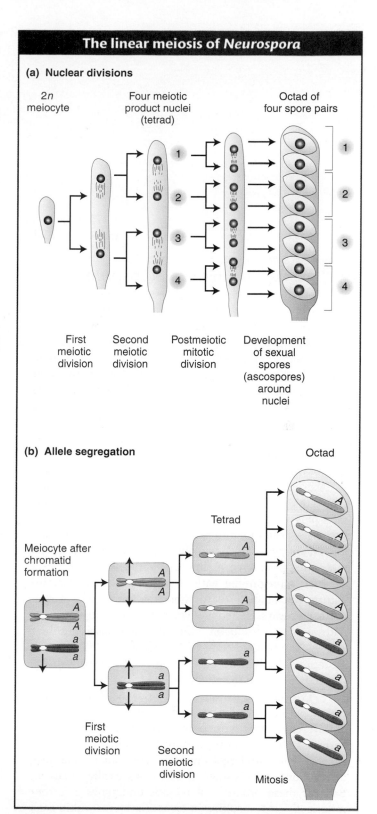

The linear meiosis of *Neurospora*

(a) Nuclear divisions

2*n* meiocyte — Four meiotic product nuclei (tetrad) — Octad of four spore pairs

First meiotic division   Second meiotic division   Postmeiotic mitotic division   Development of sexual spores (ascospores) around nuclei

(b) Allele segregation

Octad

Meiocyte after chromatid formation

Tetrad

First meiotic division   Second meiotic division   Mitosis

Figure 3-10 *Neurospora* is an ideal model system for studying allelic segregation at meiosis. (a) The four products of meiosis (tetrad) undergo mitosis to produce an octad. The products are contained within an ascus. (b) An *A/a* meiocyte undergoes meiosis followed by mitosis, resulting in equal numbers of *A* and *a* products and demonstrating the principle of equal segregation.

If the autosomal and X-linked genes are combined, the F_2 phenotypic ratios will be

Females $\frac{3}{4}$ fully wild type

$\frac{1}{4}$ vestigial

Males $\frac{3}{8}$ fully wild type ($\frac{3}{4} \times \frac{1}{2}$)

$\frac{3}{8}$ white ($\frac{3}{4} \times \frac{1}{2}$)

$\frac{1}{8}$ vestigial ($\frac{1}{4} \times \frac{1}{2}$)

$\frac{1}{8}$ vestigial, white ($\frac{1}{4} \times \frac{1}{2}$)

Hence, we see a progeny ratio that reveals clear elements of both autosomal and X-linked inheritance.

Model Organism *Neurospora*

Neurospora crassa was one of the first eukaryotic microbes to be adopted by geneticists as a model organism. It is a haploid fungus ($n = 7$) found growing on dead vegetation in many parts of the world. When an asexual spore (haploid) germinates, it produces a tubular structure that extends rapidly by tip growth and throws off multiple side branches. The result is a mass of branched threads (called *hyphae*), which constitute a colony. Hyphae have no cross-walls, and so a colony is essentially one cell containing many haploid nuclei. A colony buds off millions of asexual spores, which can disperse and repeat the asexual cycle.

Asexual colonies are easily and inexpensively maintained in the laboratory on a defined medium of inorganic salts plus an energy source such as sugar. (An inert gel such as agar is added to provide a firm surface.) The fact that *Neurospora* can chemically synthesize all its essential molecules from such a simple medium led biochemical geneticists (beginning with George Beadle and Edward Tatum; see Chapter 6) to choose it for studies of synthetic pathways. Geneticists worked out the steps in these pathways by introducing mutations and observing their effects. The haploid state of *Neurospora* is ideal for such mutational analysis because mutant alleles are always expressed directly in the phenotype.

Neurospora has two mating types, MAT-A and MAT-a, which can be regarded as simple "sexes." When colonies of different mating type come into contact, their cell walls and nuclei fuse, resulting in many transient diploid nuclei, each of which undergoes meiosis. The four haploid products of one meiosis stay together in a sac called an *ascus*. Each of these products of meiosis undergoes a further mitotic division, resulting in eight ascospores within each ascus. Ascospores germinate and produce colonies exactly like those produced by asexual spores. Hence, such *ascomycete* fungi are ideal for the study of the segregation and recombination of genes in individual meioses.

The fungus *Neurospora crassa*. (a) Orange colonies of *Neurospora* growing on sugarcane. In nature, *Neurospora* colonies are most often found after fire, which activates dormant ascospores. (Fields of sugarcane are burned to remove foliage before harvesting the cane stalks.) (b) Developing *Neurospora* octads from a cross of wild type to a strain carrying an engineered allele of jellyfish green fluorescent protein fused to histone. The octads show the expected 4:4 Mendelian segregation of fluorescence. In some spores, the nucleus has divided mitotically to form two; eventually, each spore will contain several nuclei. [*(a) Courtesy of David Jacobson; (b) courtesy of Namboori B. Raju.*]

Recombination

The independent assortment of genes at meiosis is one of the main ways by which an organism produces new combinations of alleles. The production of new allele combinations is formally called **recombination.**

There is general agreement that the reason that organisms produce new combinations of alleles is to provide variation as the raw material for natural selection. Recombination is a crucial principle in genetics, partly because of its relevance to evolution but also because of its use in genetic analysis. It is particularly useful for analyzing inheritance patterns of multigene genotypes. In this section, we define recombination in such a way that we would recognize it in experimental results, and we lay out the way in which recombination is analyzed and interpreted.

Recombination is observed in a variety of biological situations, but, for the present, we define it in relation to meiosis.

> **Meiotic recombination** is any meiotic process that generates a haploid product with new combinations of the alleles carried by the haploid genotypes that united to form the meiocyte.

This seemingly wordy definition is actually quite simple; it makes the important point that we detect recombination by comparing the *inputs* into meiosis with the *outputs* (Figure 3-11). The inputs are the two haploid genotypes that combine to form the meiocyte, the diploid cell that undergoes meiosis. For humans, the inputs are the parental egg and sperm. They unite to form a diploid zygote, which divides to yield all the body cells, including the meiocytes that are set aside within the gonads. The output genotypes are the haploid products of meiosis. In humans, these haploid products are a person's own eggs or sperm. Any meiotic product that has a new combination of the alleles provided by the two input genotypes is by definition a **recombinant.**

> **Message** Meiosis generates recombinants, which are haploid meiotic products with new combinations of the alleles carried by the haploid genotypes that united to form the meiocyte.

First, let us look at how recombinants are detected experimentally. The detection of recombinants in organisms with haploid life cycles such as fungi or algae is straightforward. The input and output types in haploid life cycles are the genotypes of individuals rather than gametes and may thus be inferred directly from phenotypes. Figure 3-11 can be viewed as summarizing the simple detection of recombinants in organisms with haploid life cycles. Detecting recombinants in organisms with diploid life cycles is trickier. The input and output types in diploid cycles are gametes. Thus, we must know the genotypes of both input and output gametes to detect recombinants in an organism with a diploid cycle. We cannot detect the genotypes of input or output gametes directly, but we can infer these genotypes by using the appropriate techniques:

- *To know the input gametes,* we use pure-breeding diploid parents because they can produce only one gametic type.

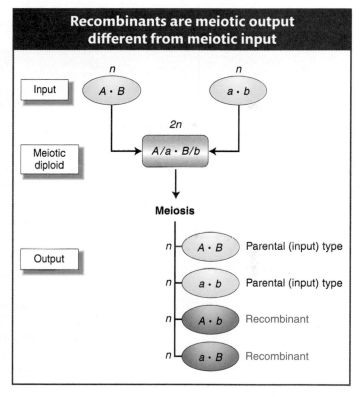

Figure 3-11 Recombinants are those products of meiosis with allele combinations different from those of the haploid cells that formed the meiotic diploid.

Figure 3-12
Recombinant products of a diploid meiosis are most readily detected in a cross of a heterozygote and a recessive tester. Note that Figure 3-11 is repeated as part of this diagram.

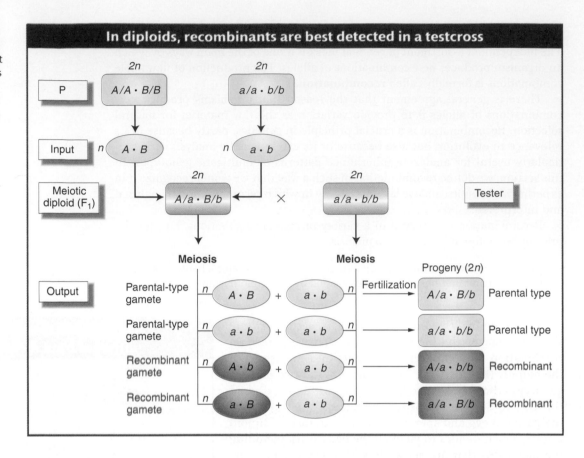

- *To detect recombinant output gametes,* we testcross the diploid individual and observe its progeny (Figure 3-12).

A testcross offspring that arises from a recombinant product of meiosis also is called a *recombinant*. Notice, again, that the testcross allows us to concentrate on *one* meiosis and prevent ambiguity. From a *self* of the F_1 in Figure 3-10, for example, a recombinant $A/A \cdot B/b$ offspring could not be distinguished from $A/A \cdot B/B$ without further crosses.

A central part of recombination analysis is recombinant frequency. One reason for focusing on recombinant frequency is that its numerical value is a convenient test for whether two genes are on different chromosomes. Recombinants are produced by two different cellular processes: the independent assortment of genes on different chromosomes (this chapter) and crossing over between genes on the same chromosome (see Chapter 4). The *proportion* of recombinants is the key idea here because the diagnostic value can tell us whether genes are on different chromosomes. We will deal with independent assortment here.

For genes on separate chromosomes, recombinants are produced by independent assortment, as shown in Figure 3-13. Again, we see the $1:1:1:1$ ratio that we have seen before, but now the progeny of the testcross are classified as either recombinant or resembling the P (parental) input types. Set up in this way, the proportion of recombinants is clearly $\frac{1}{4} + \frac{1}{4} = \frac{1}{2}$, or 50 percent of the total progeny. Hence, we see that independent assortment at meiosis produces a recombinant frequency of 50 percent. If we observe a recombinant frequency of 50 percent in a testcross, we can infer that the two genes under study assort independently. The simplest and most likely interpretation of independent assortment is that the two genes are on separate chromosome pairs. (However, we must note that genes that are very far apart on the *same* chromosome pair can assort virtually independently and produce the same result; see Chapter 4.)

> **Message** A recombinant frequency of 50 percent indicates that the genes are independently assorting and are most likely on different chromosmes.

3.4 Polygenic Inheritance

So far, our analysis in this book has focused on single-gene differences, with the use of sharply contrasting phenotypes such as red versus white petals, smooth versus wrinkled seeds, and long- versus vestigial-winged *Drosophila*. However, a large proportion of variation in natural populations takes the form of *continuous* variation, which is typically found in characters that can take any measurable value between two extremes. Height, weight, and color intensity are examples of such *metric, or quantitative, characters.* Typically, when the metric value of these characters is plotted against frequency in a natural population, the distribution curve is shaped like a bell (Figure 3-14). The bell shape is due to the fact that average values in the middle are the most common, whereas extreme values are rare.

Many cases of continuous variation have a purely environmental basis, little affected by genetics. For example, a population of genetically homozygous plants grown in a plot of ground often show a bell-shaped curve for height, with the smaller plants around the edges of the plot and the larger plants in the middle. This variation can be explained only by environmental factors such as moisture and amount of fertilizer applied. However, many cases of continuous variation do have a genetic basis. Human skin color is an example: all degrees of skin darkness can be observed in populations from different parts of the world, and this variation clearly has a genetic component. In such cases, from several to many alleles interact with a more or less additive effect. The interacting genes underlying hereditary continuous variation are called **polygenes** or **quantitative trait loci (QTLs).** (The term quantitative trait locus needs some definition: *quantitative* is more or less synonymous with continuous; *trait* is more or less synonymous with character or property; *locus,* which literally means place on a chromosome, is more or less synonymous with gene.) The polygenes, or QTLs, for the same trait are distributed throughout the genome; in many cases, they are on different chromosomes and show independent assortment, making them a topic for this chapter. We will show how the inheritance of several heterozygous polygenes (even as few as two) can generate a bell-shaped distribution curve.

Let's consider a simple model that was originally used to explain continuous variation in the degree of redness in wheat seeds. The work was done by Hermann Nilsson-Ehle in the early twentieth century. We will assume two independently assorting gene pairs R_1/r_1 and R_2/r_2. Both R_1

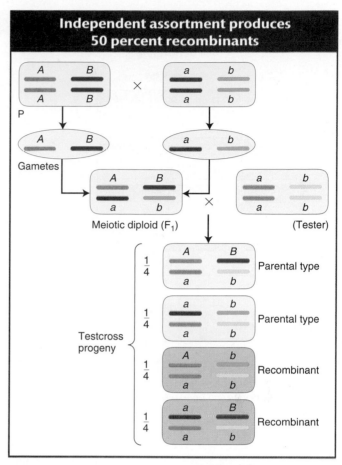

Figure 3-13 This diagram shows two chromosome pairs of a diploid organism with *A* and *a* on one pair and *B* and *b* on the other. Independent assortment produces a recombinant frequency of 50 percent. Note that we could represent the haploid situation by removing the parental (P) cross and the testcross.

GENETICS PORTAL **ANIMATED ART: Meiotic recombination between unlinked genes by independent assortment**

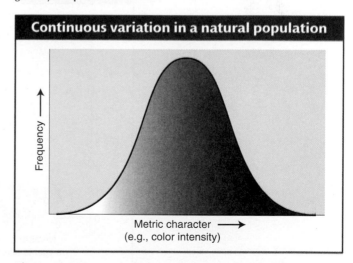

Figure 3-14 In a population, a metric character such as color intensity can take on many values. Hence, the distribution is in the form of a smooth curve, with the most common values representing the high point of the curve. If the curve is symmetrical, it is bell shaped, as shown.

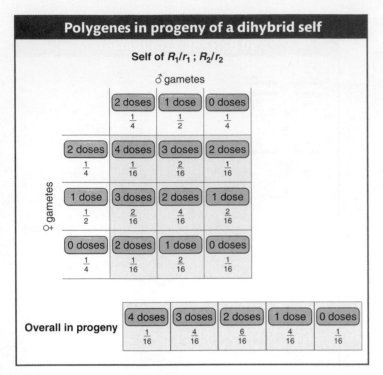

Figure 3-15 The progeny of a dihybrid self for two polygenes can be expressed as numbers of additive allelic "doses."

and R_2 contribute to wheat-seed redness. Each "dose" of an R allele of either gene is additive, meaning that it increases the degree of redness proprotionately. An illustrative cross is a self of a dihybrid $R_1/r_1;R_2/r_2$. Both male and female *gametes* will show the genotypic proportions as follows:

$R_1;R_2$	2 doses of redness
$R_1;r_2$	1 dose of redness
$r_1;R_2$	1 dose of redness
$r_1;r_2$	0 doses of redness

Overall, in this gamete population, one-fourth have two doses, one-half have one dose, and one-fourth have zero doses. The union of male and female gametes both showing this array of R doses is illustrated in Figure 3-15. The number of doses in the progeny ranges from four ($R_1/R_1;R_2/R_2$) down to zero ($r_1/r_2;r_2/r_2$), with all values between.

The proportions in the grid of Figure 3-15 can be drawn as a histogram, as shown in Figure 3-16. The shape of the histogram can be thought of as a scaffold that could be the underlying basis for a bell-shaped distribution curve. When this analysis of redness in wheat seeds was originally done, variation was found within the classes that allegedly represented one polygene "dose" level. Presumably, this variation within a class is the result of environmental differences. Hence, the environment can be seen to contribute in a way that rounds off the sharp shoulders of the histogram bars, resulting in a smooth bell-shaped curve (the red line in the histogram). If the number of polygenes is increased, the histogram more closely approximates a smooth continuous distribution. For example, for a characteristic determined by three polygenes, the histogram is as shown in Figure 3-17.

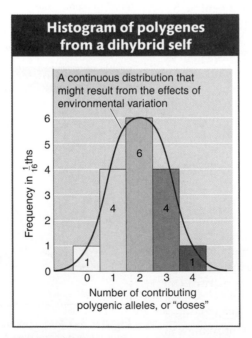

Figure 3-16 The progeny shown in Figure 3-15 can be represented as a frequency histogram of contributing polygenic alleles ("doses").

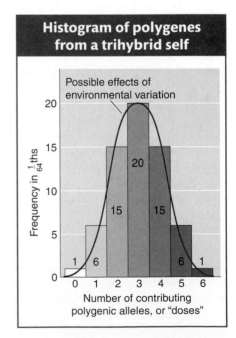

Figure 3-17 The progeny of a polygene trihybrid can be graphed as a frequency histogram of contributing polygenic alleles ("doses").

In our illustration, we used a dihybrid self to show how the histogram is produced. But how is our example relevant to what is going on in natural populations? After all, not all crosses could be of this type. Nevertheless, if the alleles at each gene pair are approximately equal in frequency in the population (for example, R_1 is about as common as r_1), then the dihybrid cross can be said to represent an average cross for a population in which two polygenes are segregating.

Identifying polygenes and understanding how they act and interact are important challenges for geneticists in the twenty-first century. Identifying polygenes will be especially important in medicine. Many common human diseases such as atherosclerosis (hardening of the arteries) and hypertension (high blood pressure) are thought to have a polygenic component. If so, a full understanding of these conditions, which affect large proportions of human populations, requires an understanding of these polygenes, their inheritance, and their function. Today, several molecular approaches can be applied to the job of finding polygenes, and we will consider some in subsequent chapters. Note that polygenes are not considered a special functional class of genes. They are identified as a group only in the sense that they have alleles that contribute to continuous variation.

> **Message** Variation and assortment of polygenes can contribute to continuous variation in a population.

3.5 Organelle Genes: Inheritance Independent of the Nucleus

So far, we have considered only nuclear genes. Although the nucleus contains most of a eukaryotic organism's genes, a distinct and specialized subset of the genome is found in the mitochondria, and, in plants, also in the chloroplasts. These subsets are inherited independently of the nuclear genome, and so they constitute a special case of independent inheritance, sometimes called extranuclear inheritance.

Mitochondria and chloroplasts are specialized organelles located in the cytoplasm. They contain small circular chromosomes that carry a defined subset of the total cell genome. Mitochondrial genes are concerned with the mitochondrion's task of energy production, whereas chloroplast genes are needed for the chloroplast to carry out its function of photosynthesis. However, neither organelle is functionally autonomous, because each relies to a large extent on nuclear genes for its function. Why some of the necessary genes are in the organelles themselves and others are in the nucleus is still something of a mystery, which will not be addressed here.

Another peculiarity of organelle genes is the large number of copies present in a cell. Each organelle is present in many copies per cell, and, furthermore, each organelle contains many copies of its chromosome. Hence, each cell can contain hundreds or thousands of organelle chromosomes. Consider chloroplasts, for example. Any green cell of a plant has many chloroplasts, and each chloroplast contains many identical circular DNA molecules, the so-called chloroplast chromosomes. Hence, the number of chloroplast chromosomes per cell can be in the thousands, and the number can even vary somewhat from cell to cell. The DNA is sometimes seen to be packaged into suborganellar structures called *nucleoids*, which become visible if stained with a DNA-binding dye (Figure 3-18). The DNA is folded within the nucleoid but does not have the type of histone-associated coiling shown by nuclear

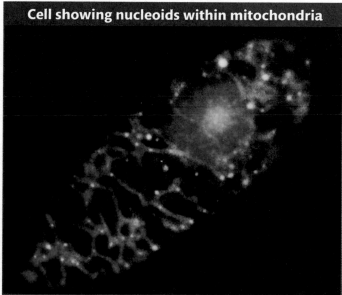

Cell showing nucleoids within mitochondria

Figure 3-18 Fluorescent staining of a cell of *Euglena gracilis*. With the dyes used, the nucleus appears red because of the fluorescence of large amounts of nuclear DNA. The mitochondria fluoresce green, and, within mitochondria, the concentrations of mitochondrial DNA (nucleoids) fluoresce yellow. [*From Y. Huyashi and K. Veda, J. Cell Sci. 93, 1989, 565.*]

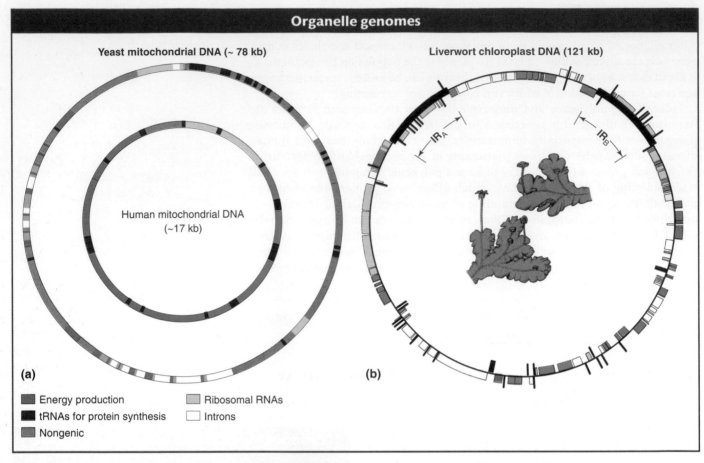

Figure 3-19 DNA maps for mitochondria and chloroplasts. Many of the organelle genes encode proteins that carry out the energy-producing functions of these organelles (green), whereas others (red and orange) function in protein synthesis. (a) Maps of yeast and human mtDNAs. (Note that the human map is not drawn at the same scale as the yeast map.) (b) The 121-kb chloroplast genome of the liverwort *Marchantia polymorpha*. Genes shown inside the map are transcribed clockwise, and those outside are transcribed counterclockwise. IR$_A$ and IR$_B$ indicate inverted repeats. The upper drawing in the center of the map depicts a male *Marchantia* plant; the lower drawing depicts a female. [*From K. Umesono and H. Ozeki, Trends Genet. 3, 1987.*]

chromosomes. The same arrangement is true for the DNA in mitochondria. For the time being, we will assume that all copies of an organelle chromosome within a cell are identical, but we will have to relax this assumption later.

Many organelle chromosomes have now been sequenced. Examples of relative gene size and spacing in **mitochondrial DNA (mtDNA)** and **chloroplast DNA (cpDNA)** are shown in Figure 3-19. Organelle genes are very closely spaced, and, in some organisms, organelle genes can contain introns. Note how most genes concern the chemical reactions taking place within the organelle itself: photosynthesis in chloroplasts and oxidative phosphorylation in mitochondria.

Patterns of inheritance in organelles

Organelle genes show their own special mode of inheritance called **uniparental inheritance:** progeny inherit organelle genes exclusively from one parent but not the other. In most cases, that parent is the mother, a pattern called **maternal inheritance.** Why only the mother? The answer lies in the fact that the organelle chromosomes are located in the cytoplasm and the male and female gametes do not contribute cytoplasm equally to the zygote. In regard to nuclear genes, both parents contribute equally to the zygote. However, the egg contributes the bulk

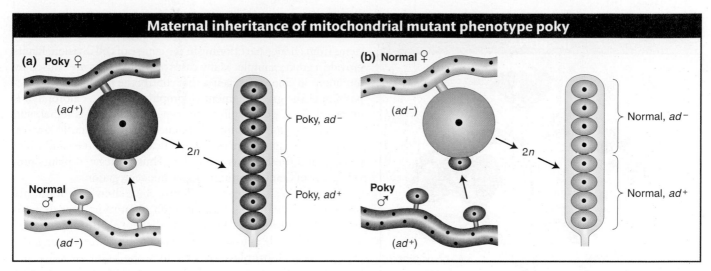

Maternal inheritance of mitochondrial mutant phenotype poky

(a) Poky ♀

(ad⁺)

Normal ♂

(ad⁻)

2n

Poky, ad⁻

Poky, ad⁺

(b) Normal ♀

(ad⁻)

Poky ♂

(ad⁺)

2n

Normal, ad⁻

Normal, ad⁺

of the cytoplasm, whereas the sperm contributes essentially none. Therefore, because organelles reside in the cytoplasm, the female parent contributes the organelles along with the cytoplasm, and essentially none of the organelle DNA in the zygote is from the male parent.

Some phenotypic variants are caused by a mutant allele of an organelle gene, and we can use these mutants to track patterns of organelle inheritance. We will temporarily assume that the mutant allele is present in all copies of the organelle chromosome, a situation that is indeed often found. In a cross, the variant phenotype will be transmitted to progeny if the variant used is the female parent, but not if it is the male parent. Hence, generally, cytoplasmic inheritance shows the following pattern:

mutant female × wild-type male → progeny all mutant

wild-type female × mutant male → progeny all wild type

Indeed, this inheritance pattern is diagnostic of organelle inheritance in cases in which the genomic location of a mutant allele is not known.

Maternal inheritance can be clearly demonstrated in certain mutants of fungi. For example, in the fungus *Neurospora*, a mutant called *poky* has a slow-growth phenotype. *Neurospora* can be crossed in such a way that one parent acts as the maternal parent, contributing the cytoplasm (see Figure 3-9). The results of the following reciprocal crosses suggest that the mutant gene resides in the mitochondria (fungi have no chloroplasts):

poky female × wild-type male → progeny all poky

wild-type female × poky male → progeny all wild type

Sequencing has shown that the poky phenotype is caused by a mutation of a ribosomal RNA gene in mtDNA. Its inheritance is shown diagrammatically in Figure 3-20. The cross includes an allelic difference (*ad* and *ad⁺*) in a nuclear gene in addition to *poky*; notice how the Mendelian inheritance of the nuclear gene is independent of the maternal inheritance of the poky phenotype.

> **Message** Variant phenotypes caused by mutations in cytoplasmic organelle DNA are generally inherited maternally and independent of the Mendelian patterns shown by nuclear genes.

Cytoplasmic segregation

In some cases, cells contain mixtures of mutant and normal organelles. These cells are called *cytohets* or *heteroplasmons*. In these mixtures, a type of **cytoplasmic**

Figure 3-20 Reciprocal crosses of poky and wild-type *Neurospora* produce different results because a different parent contributes the cytoplasm. The female parent contributes most of the cytoplasm of the progeny cells. Brown shading represents cytoplasm with mitochondria containing the *poky* mutation, and green shading represents cytoplasm with wild-type mitochondria. Note that all the progeny in part *a* are poky, whereas all the progeny in part *b* are normal. Hence, both crosses show maternal inheritance. The nuclear gene with the alleles *ad⁺* (black) and *ad⁻* (red) is used to illustrate the segregation of the nuclear genes in the 1 : 1 Mendelian ratio expected for this haploid organism.

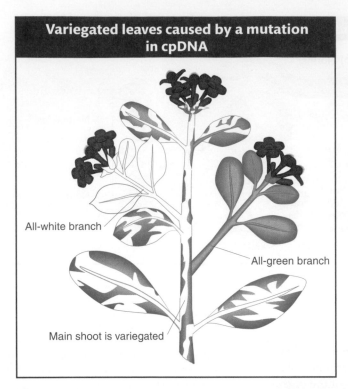

Variegated leaves caused by a mutation in cpDNA

All-white branch

All-green branch

Main shoot is variegated

Figure 3-21 Leaf variegation in *Mirabilis jalapa,* the four-o'clock plant. Flowers can form on any branch (variegated, green, or white), and these flowers can be used in crosses.

segregation can be detected, in which the two types apportion themselves into different daughter cells. The process most likely stems from chance partitioning in the course of cell division. Plants provide a good example. Many cases of white leaves are caused by mutations in chloroplast genes that control the production and deposition of the green pigment chlorophyll. Because chlorophyll is necessary for a plant to live, this type of mutation is lethal, and white-leaved plants cannot be obtained for experimental crosses. However, some plants are variegated, bearing both green and white patches, and these plants are viable. Thus, variegated plants provide a way of demonstrating cytoplasmic segregation.

The four-o'clock plant in Figure 3-21 shows a commonly observed variegated leaf and branch phenotype that demonstrates the inheritance of a mutant allele of a chloroplast gene. The mutant allele causes chloroplasts to be white; in turn, the color of the chloroplasts determines the color of cells and hence the color of the branches composed of those cells. Variegated branches are mosaics of all-green and all-white cells. Flowers can develop on green, white, or variegated branches, and the chloroplast genes of a flower's cells are those of the branch on which it grows. Hence, in a cross (Figure 3-22), the maternal gamete within the flower (the egg cell) determines the color of the leaves and branches of the progeny plant. For

Figure 3-22 The results of the *Mirabilis jalapa* crosses can be explained by autonomous chloroplast inheritance. The large, dark spheres represent nuclei. The smaller bodies represent chloroplasts, either green or white. Each egg cell is assumed to contain many chloroplasts, and each pollen cell is assumed to contain no chloroplasts. The first two crosses exhibit strict maternal inheritance. If, however, the maternal branch is variegated, three types of zygotes can result, depending on whether the egg cell contains only white, only green, or both green and white chloroplasts. In the last case, the resulting zygote can produce both green and white tissue, and so a variegated plant results.

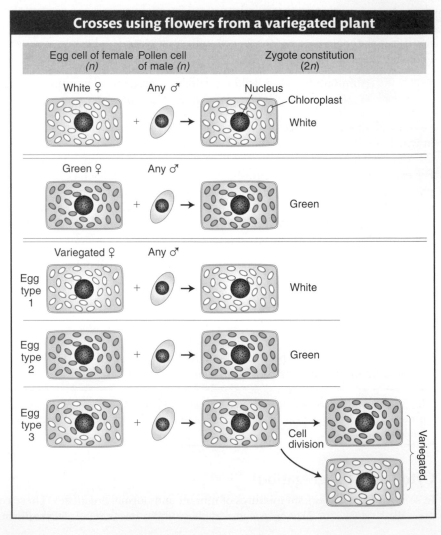

Crosses using flowers from a variegated plant

Egg cell of female *(n)*   Pollen cell of male *(n)*   Zygote constitution *(2n)*

White ♀   Any ♂   Nucleus — Chloroplast   White

Green ♀   Any ♂   Green

Variegated ♀   Any ♂

Egg type 1   White

Egg type 2   Green

Egg type 3   Cell division   Variegated

example, if an egg cell is from a flower on a green branch, all the progeny will be green, regardless of the origin of the pollen. A white branch will have white chloroplasts, and the resulting progeny plants will be white. (Because of lethality, white descendants would not live beyond the seedling stage.)

The variegated zygotes (bottom of Figure 3-22) demonstrate cytoplasmic segregation. These variegated progeny come from eggs that are cytohets. Interestingly, when such a zygote divides, the white and green chloroplasts often segregate; that is, they sort themselves into separate cells, yielding the distinct green and white sectors that cause the variegation in the branches. Here, then, is a direct demonstration of cytoplasmic segregation.

Given that a cell is a population of organelle molecules, how is it ever possible to obtain a "pure" mutant cell, containing only mutant chromosomes? Most likely, pure mutants are created in asexual cells as follows. The variants arise by mutation of a single gene in a single chromosome. Then, in some cases, the mutation-bearing chromosome may by chance increase in frequency in the population within the cell. This process is called *random genetic drift*. A cell that is a cytohet may have, say, 60 percent *A* chromosomes and 40 percent *a* chromosomes. When this cell divides, sometimes all the *A* chromosomes go into one daughter, and all the *a* chromosomes into the other (again, by chance). More often, this partitioning requires several subsequent generations of cell division to be complete (Figure 3-23). Hence, as a result of these chance events, both alleles are expressed in different daughter cells, and this separation will continue through the descendants of these cells. Note that cytoplasmic segregation is not a mitotic process; it does take place in dividing asexual cells, but it is unrelated to mitosis. In chloroplasts, cytoplasmic segregation is a common mechanism for producing variegated (green-and-white) plants, as already mentioned. In fungal mutants such as the *poky* mutant of *Neurospora*, the original mutation in one mtDNA molecule must have accumulated and undergone cytoplasmic segregation to produce the strain expressing the poky symptoms.

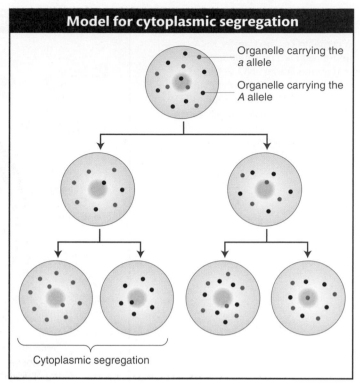

Model for cytoplasmic segregation

Organelle carrying the *a* allele

Organelle carrying the *A* allele

Cytoplasmic segregation

Figure 3-23 By chance, genetically distinct organelles may segregate into separate cells in a number of successive cell divisions. Red and blue dots represent genetically distinguishable organelles, such as mitochondria with and without a mutation.

> **Message** Organelle populations that contain mixtures of two genetically distinct chromosomes often show segregation of the two types into the daughter cells at cell division. This process is called cytoplasmic segregation.

In certain special systems such as in fungi and algae, cytohets that are "dihybrid" have been obtained (say, *AB* in one organelle chromosome and *ab* in another). In such cases, rare crossover-like processes can occur, but such an occurrence must be considered a minor genetic phenomenon.

> **Message** Alleles on organelle chromosomes
> **1.** in sexual crosses are inherited from one parent only (generally the maternal parent) and hence show no segregation ratios of the type nuclear genes do.
> **2.** in asexual cells can show cytoplasmic segregation.
> **3.** in asexual cells can occasionally show processes analogous to crossing over.

Cytoplasmic mutations in humans

Are there cytoplasmic mutations in humans? Some human pedigrees show the transmission of rare disorders only through females and never through males.

Figure 3-24 This map of human mtDNA shows loci of mutations leading to cytopathies. The transfer RNA genes are represented by single-letter amino acid abbreviations: ND = NADH dehydrogenase; COX = cytochrome oxidase; and 12S and 16S refer to ribosomal RNAs. [*After S. DiMauro et al., "Mitochondria in Neuromuscular Disorders," Biochim. Biophys. Acta 1366, 1998, 199–210.*]

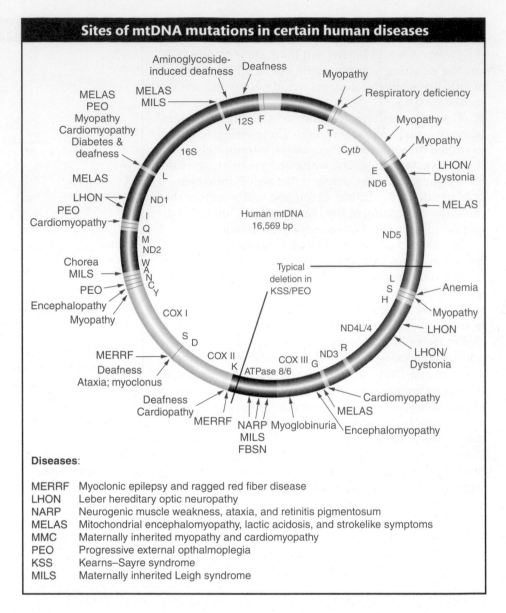

This pattern strongly suggests cytoplasmic inheritance and points to a mutation in mtDNA as the reason for the phenotype. The disease MERRF (myoclonic epilepsy and ragged red fiber) is such a phenotype, resulting from a single base change in mtDNA. It is a disease that affects muscles, but the symptoms also include eye and hearing disorders. Another example is Kearns–Sayre syndrome, a constellation of symptoms affecting the eyes, heart, muscles, and brain that is caused by the loss of part of the mtDNA. In some of these cases, the cells of a sufferer contain mixtures of normal and mutant chromosomes, and the proportions of each passed on to progeny can vary as a result of cytoplasmic segregation. The proportions in one person can also vary in different tissues or over time. The accumulation of certain types of mitochondrial mutations over time has been proposed as a possible cause of aging.

Figure 3-24 shows some of the mutations in human mitochondrial genes that can lead to disease when, by random drift and cytoplasmic segregation, they rise in frequency to such an extent that cell function is impaired. The inheritance of a human mitochondrial disease is shown in Figure 3-25. Note that the condition is always passed to offspring by mothers and never fathers. Occasionally, a mother will produce an unaffected child (not shown), probably owing to cytoplasmic segregation in the gamete-forming tissue.

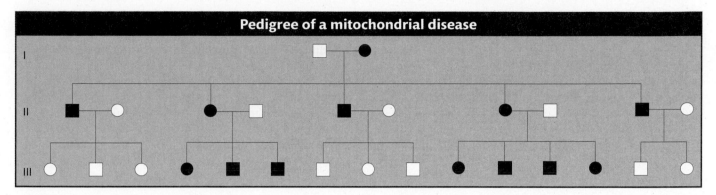

Figure 3-25 This pedigree shows that a human mitochondrial disease is inherited only from the mother.

MtDNA in evolutionary studies

Differences and similarities of homologous mtDNA sequences between species have been used extensively to construct evolutionary trees. Furthermore, it has been possible to introduce some extinct organisms into evolutionary trees using mtDNA sequences obtained from the remains of extinct organisms, such as skins and bones in museums. MtDNA evolves relatively rapidly, so this approach has been most useful in plotting recent evolution such as the evolution of humans and other primates. One key finding is that the "root" of the human mtDNA tree is in Africa, suggesting that *Homo sapiens* originated in Africa and from there dispersed throughout the world.

Summary

Genetic research and plant and animal breeding often necessitate the synthesis of genotypes that are complex combinations of alleles from different genes. Such genes can be on the same chromosome or on different chromosomes; the latter is the main subject of this chapter.

In the simplest case—a dihybrid for which the two gene pairs are on different chromosome pairs—each individual gene pair shows equal segregation at meiosis as predicted by Mendel's first law. Because nuclear spindle fibers attach randomly to centromeres at meiosis, the two gene pairs are partitioned independently into the meiotic products. This principle of independent assortment is called Mendel's second law because Mendel was the first to observe it. From a dihybrid $A/a;B/b$, four genotypes of meiotic products are produced, $A;B$, $A;b$, $a;B$, and $a;b$, all at an equal frequency of 25 percent each. Hence, in a testcross of a dihybrid with a double recessive, the phenotypic proportions of the progeny also are 25 percent (a 1:1:1:1 ratio). If such a dihybrid is selfed, the phenotypic classes in the progeny are $\frac{9}{16}$ $A/-;B/-$, $\frac{3}{16}$ $A/-;$ b/b, $\frac{3}{16}$ $a/a;B/-$, and $\frac{1}{16}$ $a/a;b/b$. The 1:1:1:1 and 9:3:3:1 ratios are both diagnostic of independent assortment.

More complex genotypes composed of independently assorting genes can be treated as extensions of the case for single-gene segregation. Overall genotypic, phenotypic, or gametic ratios are calculated by applying the product rule—that is, by multiplying the proportions relevant to the individual genes. The probability of the occurrence of any of several categories of progeny is calculated by applying the sum rule—that is, by adding their individual probabilities.

In mnemonic form, the product rule deals with "A AND B," whereas the sum rule deals with "A' OR A''." The χ^2 test can be used to test whether the observed proportions of classes in genetic analysis conform to the expectations of a genetic hypothesis, such as a hypothesis of single- or two-gene inheritance. If a probability value of less than 5 percent is calculated, the hypothesis must be rejected.

Sequential generations of selfing increase the proportions of homozygotes, according to the principles of equal segregation and independent assortment (if the genes are on different chromosomes). Hence, selfing is used to create complex pure lines with combinations of desirable mutations.

The independent assortment of chromosomes at meiosis can be observed cytologically by using heteromorphic chromosome pairs (those that show a structural difference). The X and Y chromosomes are one such case, but other, rarer cases can be found and used for this demonstration. The independent assortment of genes at the level of single meiocytes can be observed in the ascomycete fungi, because the asci show the two alternative types of segregations at equal frequencies.

One of the main functions of meiosis is to produce recombinants, new combinations of alleles of the haploid genotypes that united to form the meiocyte. Independent assortment is the main source of recombinants. In a dihybrid testcross showing independent assortment, the recombinant frequency will be 50 percent.

Metric characters such as color intensity show a continuous distribution in a population. Continuous distributions can be based on environmental variation or on variant alleles of multiple genes or on a combination of both. A simple genetic model proposes that the active alleles of several genes (called polygenes) contribute more or less additively to the metric character. In an analysis of the progeny from the self of a multiply heterozygous individual, the histogram showing the proportion of each phenotype approximates a bell-shaped curve typical of continuous variation.

The small subsets of the genome found in mitochondria and chloroplasts are inherited independently of the nuclear genome. Mutants in these organelle genes often show maternal inheritance, along with the cytoplasm, which is the location of these organelles. In genetically mixed cytoplasms (cytohets) the two genotypes (say, wild type and mutant) often sort themselves out into different daughter cells by a poorly understood process called cytoplasmic segregation. Mitochondrial mutation in humans results in diseases that show cytoplasmic segregation in body tissues and maternal inheritance in a mating.

KEY TERMS

chi-square test (p. 89)
chloroplast DNA (cpDNA) (p. 104)
cytoplasmic segregation (p. 105)
dihybrid (p. 82)
dihybrid cross (p. 82)
hybrid vigor (p. 93)
independent assortment (p. 82)

maternal inheritance (p. 104)
meiotic recombination (p. 99)
Mendel's second law (p. 84)
mitochondrial DNA (mtDNA) (p. 104)
polygene (quantitative trait locus) (p. 101)

product rule (p. 87)
quantitative trait locus (QTL) (p. 101)
recombinant (p. 99)
recombination (p. 99)
sum rule (p. 87)
uniparental inheritance (p. 104)

SOLVED PROBLEMS

SOLVED PROBLEM 1. Two *Drosophila* flies that had normal (transparent, long) wings were mated. In the progeny, two new phenotypes appeared: dusky wings (having a semi-opaque appearance) and clipped wings (with squared ends). The progeny were as follows:

Females	Males
179 transparent, long	92 transparent, long
58 transparent, clipped	89 dusky, long
	28 transparent, clipped
	31 dusky, clipped

a. Provide a chromosomal explanation for these results, showing chromosomal genotypes of parents and of all progeny classes under your model.

b. Design a test for your model.

Solution

a. The first step is to state any interesting features of the data. The first striking feature is the appearance of two new phenotypes. We encountered the phenomenon in Chapter 2, where it was explained as recessive alleles masked by their dominant counterparts. So, first, we might suppose that one or both parental flies have recessive alleles of two different genes. This inference is strengthened by the observation that some progeny express only one of the new phenotypes. If the new phenotypes always appeared together, we might suppose that the same recessive allele determines both.

However, the other striking feature of the data, which we cannot explain by using the Mendelian principles from Chapter 2, is the obvious difference between the sexes: although there are approximately equal numbers of males and females, the males fall into four phenotypic classes, but the females constitute only two. This fact should immediately suggest some kind of sex-linked inheritance. When we study the data, we see that the long and clipped phenotypes are segregating in both males and females, but only males have the dusky phenotype. This observation suggests that the inheritance of wing transparency differs from the inheritance of wing shape. First, long and clipped are found in a 3:1 ratio in both males and females. This ratio can be explained if both parents are heterozygous for an autosomal gene; we can represent them as L/l, where L stands for long and l stands for clipped.

Having done this partial analysis, we see that only the inheritance of wing transparency is associated with sex. The most obvious possibility is that the alleles for transparent (D) and dusky (d) are on the X chromosome, because we have seen in Chapter 2 that gene location on this chromosome gives inheritance patterns correlated with sex. If this suggestion is true, then the parental female must be the one sheltering the d allele, because, if the male had the d, he would have been dusky, whereas we were told that he had transparent wings. Therefore, the female parent would be D/d and the male D. Let's see if this suggestion works: if it is true, all female progeny would inherit the D allele from their father, and so all would be transparent winged, as was observed. Half the sons would be D (transparent) and half d (dusky), which also was observed.

So, overall, we can represent the female parent as $D/d\,;L/l$ and the male parent as $D\,;L/l$. Then the progeny would be

Females

$$\frac{1}{2}\,D/D\begin{cases}\xrightarrow{\frac{3}{4}\,L/-}\frac{3}{8}\,D/D\,;L/-\\[4pt]\xrightarrow{\frac{1}{4}\,l/l}\frac{1}{8}\,D/D\,;l/l\end{cases}$$

$$\frac{1}{2}\,D/d\begin{cases}\xrightarrow{\frac{3}{4}\,L/-}\frac{3}{8}\,D/d\,;L/-\\[4pt]\xrightarrow{\frac{1}{4}\,l/l}\frac{1}{8}\,D/d\,;l/l\end{cases}$$

$\frac{3}{4}$ transparent, long

$\frac{1}{4}$ transparent, clipped

Males

$$\frac{1}{2}\,D\begin{cases}\xrightarrow{\frac{3}{4}\,L/-}\frac{3}{8}\,D\,;L/-\quad\text{transparent, long}\\[4pt]\xrightarrow{\frac{1}{4}\,l/l}\frac{1}{8}\,D\,;l/l\quad\text{transparent, clipped}\end{cases}$$

$$\frac{1}{2}\,d\begin{cases}\xrightarrow{\frac{3}{4}\,L/-}\frac{3}{8}\,d\,;L/-\quad\text{dusky, long}\\[4pt]\xrightarrow{\frac{1}{4}\,l/l}\frac{1}{8}\,d\,;l/l\quad\text{dusky, clipped}\end{cases}$$

b. Generally, a good way to test such a model is to make a cross and predict the outcome. But which cross? We have to predict some kind of ratio in the progeny, and so it is important to make a cross from which a unique phenotypic ratio can be expected. Notice that using one of the female progeny as a parent would not serve our needs: we cannot say from observing the phenotype of any one of these females what her genotype is. A female with transparent wings could be D/D or D/d, and one with long wings could be L/L or L/l. It would be good to cross the parental female of the original cross with a dusky, clipped son, because the full genotypes of both are specified under the model that we have created. According to our model, this cross is

$$D/d\,;L/l \times d\,;l/l$$

From this cross, we predict

Females

$$\frac{1}{2}\,D/d\begin{cases}\xrightarrow{\frac{1}{2}\,L/l}\frac{1}{4}\,D/d\,;L/l\\[4pt]\xrightarrow{\frac{1}{2}\,l/l}\frac{1}{4}\,D/d\,;l/l\end{cases}$$

$$\frac{1}{2}\,d/d\begin{cases}\xrightarrow{\frac{1}{2}\,L/l}\frac{1}{4}\,d/d\,;L/l\\[4pt]\xrightarrow{\frac{1}{2}\,l/l}\frac{1}{4}\,d/d\,;l/l\end{cases}$$

Males

$$\frac{1}{2}\,D\begin{cases}\xrightarrow{\frac{1}{2}\,L/l}\frac{1}{4}\,D\,;L/l\\[4pt]\xrightarrow{\frac{1}{2}\,l/l}\frac{1}{4}\,D\,;l/l\end{cases}$$

$$\frac{1}{2}\,d\begin{cases}\xrightarrow{\frac{1}{2}\,L/l}\frac{1}{4}\,d\,;L/l\\[4pt]\xrightarrow{\frac{1}{2}\,l/l}\frac{1}{4}\,d\,;l/l\end{cases}$$

SOLVED PROBLEM 2. Consider three yellow round peas, labeled A, B, and C. Each was grown into a plant and crossed with a plant grown from a green wrinkled pea. Exactly 100 peas issuing from each cross were sorted into phenotypic classes as follows:

A: 51 yellow, round
49 green, round

B: 100 yellow, round

C: 24 yellow, round
26 yellow, wrinkled
25 green, round
25 green, wrinkled

What were the genotypes of A, B, and C? (Use gene symbols of your own choosing; be sure to define each one.)

Solution

Notice that each of the crosses is

yellow, round $\times$ green, wrinkled $\rightarrow$ progeny

Because A, B, and C were all crossed with the same plant, all the differences between the three progeny populations must be attributable to differences in the underlying genotypes of A, B, and C.

You might remember a lot about these analyses from the chapter, which is fine, but let's see how much we can deduce from the data. What about dominance? The key cross for deducing dominance is B. Here, the inheritance pattern is

yellow, round $\times$ green, wrinkled $\rightarrow$ all yellow, round

So yellow and round must be dominant phenotypes because dominance is literally defined by the phenotype of a hybrid. Now we know that the green, wrinkled parent used in each cross must be fully recessive; we have a very convenient situation because it means that each cross is a testcross, which is generally the most informative type of cross.

Turning to the progeny of A, we see a 1:1 ratio for yellow to green. This ratio is a demonstration of Mendel's first law (equal segregation) and shows that, for the character of color, the cross must have been heterozygote $\times$ homozygous recessive. Letting Y represent yellow and y represent green, we have

$$Y/y \times y/y \rightarrow \tfrac{1}{2}\,Y/y\ (\text{yellow}) \rightarrow \tfrac{1}{2}\,y/y\ (\text{green})$$

For the character of shape, because all the progeny are round, the cross must have been homozygous dominant $\times$ homozygous recessive. Letting R represent round and r represent wrinkled, we have

$$R/R \times r/r \rightarrow R/r\ (\text{round})$$

Combining the two characters, we have

$$R/R; Y/y \times y/y; r/r \rightarrow \tfrac{1}{2}\ Y/y; R/r\ \tfrac{1}{2}\ Y/y; R/r$$

Now cross B becomes crystal clear and must have been

$$Y/Y; R/R \times y/y; r/r \rightarrow Y/y; R/r$$

because any heterozygosity in pea B would have given rise to several progeny phenotypes, not just one.

What about C? Here, we see a ratio of 50 yellow : 50 green (1 : 1) and a ratio of 49 round : 51 wrinkled (also 1 : 1). So both genes in pea C must have been heterozygous, and cross C was

$$Y/y; R/r \times y/y; R/r$$

which is a good demonstration of Mendel's second law (independent assortment of different genes).

How would a geneticist have analyzed these crosses? Basically, the same way that we just did but with fewer intervening steps. Possibly something like this: "yellow and round dominant; single-gene segregation in A; B homozygous dominant; independent two-gene segregation in C."

PROBLEMS

Most of the problems are also available for review/grading through the **GENETICSPORTAL** www.yourgeneticsportal.com.

WORKING WITH THE FIGURES

1. Using Table 3-1, answer the following questions:

   a. If χ^2 is calculated to be 17 with 9 df, what is the approximate probability value?

   b. If χ^2 is 17 with 6 df, what is the probability value?

   c. What trend ("rule") do you see in the previous two calculations?

2. Inspect Figure 3-8: which meiotic stage is responsible for generating Mendel's second law?

3. In Figure 3-9,

   a. identify the diploid nuclei

   b. identify which part of the figure illustrates Mendel's first law.

4. Inspect Figure 3-10: what would be the outcome in the octad if on rare occasions a nucleus from the postmeiotic mitotic division of nucleus 2 slipped past a nucleus from the postmeiotic mitotic division of nucleus 3? How could you measure the frequency of such a rare event?

5. In Figure 3-11, if the input genotypes were a · B and A · b, what would be the genotypes colored blue?

6. In Figure 3-13, what are the origins of the chromosomes colored dark blue, light blue, and very light blue?

7. In Figure 3-17, in which bar of the histogram would the genotype $R_1/r_1 \cdot R_2/R_2 \cdot r_3/r_3$ be found?

8. In examining Figure 3-19, what do you think is the main reason for the difference in size of yeast and human mtDNA?

9. In Figure 3-20, what color is used to denote cytoplasm containing wild-type mitochondria?

10. In Figure 3-21, what would be the leaf types of progeny of the apical (top) flower?

11. From the pedigree in Figure 3-25, what principle can you deduce about the inheritance of mitochondrial disease from affected fathers?

BASIC PROBLEMS

12. Assume independent assortment and start with a plant that is dihybrid $A/a; B/b$:

    a. What phenotypic ratio is produced from selfing it?

    b. What genotypic ratio is produced from selfing it?

    c. What phenotypic ratio is produced from test-crossing it?

    d. What genotypic ratio is produced from testcrossing it?

13. Normal mitosis takes place in a diploid cell of genotype $A/a; B/b$. Which of the following genotypes might represent possible daughter cells?

    a. $A; B$

    b. $a; b$

    c. $A; b$

    d. $a; B$

    e. $A/A; B/B$

    f. $A/a; B/b$

    g. $a/a; b/b$

14. In a diploid organism of $2n = 10$, assume that you can label all the centromeres derived from its female parent and all the centromeres derived from its male parent. When this organism produces gametes, how many male- and female-labeled centromere combinations are possible in the gametes?

15. It has been shown that when a thin beam of light is aimed at a nucleus, the amount of light absorbed is proportional to the cell's DNA content. Using this method, the DNA in the nuclei of several different types of cells in a corn plant were compared. The following numbers represent the relative amounts of DNA in these different types of cells:

    0.7, 1.4, 2.1, 2.8, and 4.2

    Which cells could have been used for these measurements? (**Note:** In plants, the endosperm part of the seed is often triploid, $3n$.)

16. Draw a haploid mitosis of the genotype $a^+;b$.

17. In moss, the genes A and B are expressed only in the gametophyte. A sporophyte of genotype $A/a;B/b$ is allowed to produce gametophytes.

  a. What proportion of the gametophytes will be $A;B$?

  b. If fertilization is random, what proportion of sporophytes in the next generation will be $A/a;B/b$?

18. When a cell of genotype $A/a;B/b;C/c$ having all the genes on separate chromosome pairs divides mitotically, what are the genotypes of the daughter cells?

19. In the haploid yeast *Saccharomyces cerevisiae*, the two mating types are known as MATa and MATα. You cross a purple (ad^-) strain of mating type a and a white (ad^+) strain of mating type αα. If ad^- and ad^+ are alleles of one gene, and a and α are alleles of an independently inherited gene on a separate chromosome pair, what progeny do you expect to obtain? In what proportions?

20. In mice, dwarfism is caused by an X-linked recessive allele, and pink coat is caused by an autosomal dominant allele (coats are normally brownish). If a dwarf female from a pure line is crossed with a pink male from a pure line, what will be the phenotypic ratios in the F_1 and F_2 in each sex? (Invent and define your own gene symbols.)

21. Suppose you discover two interesting *rare* cytological abnormalities in the karyotype of a human male. (A karyotype is the total visible chromosome complement.) There is an extra piece (satellite) on *one* of the chromosomes of pair 4, and there is an abnormal pattern of staining on one of the chromosomes of pair 7. With the assumption that all the gametes of this male are equally viable, what proportion of his children will have the same karyotype that he has?

22. Suppose that meiosis occurs in the transient diploid stage of the cycle of a haploid organism of chromosome number n. What is the probability that an individual haploid cell resulting from the meiotic division will have a complete parental set of centromeres (that is, a set all from one parent or all from the other parent)?

23. Pretend that the year is 1868. You are a skilled young lens maker working in Vienna. With your superior new lenses, you have just built a microscope that has better resolution than any others available. In your testing of this microscope, you have been observing the cells in the testes of grasshoppers and have been fascinated by the behavior of strange elongated structures that you have seen within the dividing cells. One day, in the library, you read a recent journal paper by G. Mendel on hypothetical "factors" that he claims explain the results of certain crosses in peas. In a flash of revelation, you are struck by the parallels between your grasshopper studies and Mendel's pea studies, and you resolve to write him a letter. What do you write? (Based on an idea by Ernest Kroeker.)

24. From a presumed testcross $A/a \times a/a$, in which A represents red and a represents white, use the χ^2 test to find out which of the following possible results would fit the expectations:

  a. 120 red, 100 white

  b. 5000 red, 5400 white

  c. 500 red, 540 white

  d. 50 red, 54 white

25. Look at the Punnett square in Figure 3-4.

  a. How many genotypes are there in the 16 squares of the grid?

  b. What is the genotypic ratio underlying the $9:3:3:1$ phenotypic ratio?

  c. Can you devise a simple formula for the calculation of the number of progeny genotypes in dihybrid, trihybrid, and so forth crosses? Repeat for phenotypes.

  d. Mendel predicted that, within all but one of the phenotypic classes in the Punnett square, there should be several different genotypes. In particular, he performed many crosses to identify the underlying genotypes of the round, yellow phenotype. Show two different ways that could be used to identify the various genotypes underlying the round, yellow phenotype. (Remember, all the round, yellow peas look identical.)

26. Assuming independent assortment of all genes, develop formulas that show the number of phenotypic classes and the number of genotypic classes from selfing a plant heterozygous for n gene pairs.

27. **Note:** The first part of this problem was introduced in Chapter 2. The line of logic is extended here.

  In the plant *Arabidopsis thaliana*, a geneticist is interested in the development of trichomes (small projections) on the leaves. A large screen turns up two mutant plants (A and B) that have no trichomes, and these mutants seem to be potentially useful in studying trichome development. (If they are determined by single-gene mutations, then finding the normal and abnormal function of these genes will be instructive.) Each plant was crossed with wild type; in both cases, the next generation (F_1) had normal trichomes. When F_1 plants were selfed, the resulting F_2's were as follows:

  F_2 from mutant A: 602 normal; 198 no trichomes
  F_2 from mutant B: 267 normal; 93 no trichomes

  a. What do these results show? Include proposed genotypes of all plants in your answer.

  b. Assume that the genes are located on separate chromosomes. An F_1 is produced by crossing the original mutant A with the original mutant B. This F_1 is testcrossed: What proportion of testcross progeny will have no trichomes?

28. In dogs, dark coat color is dominant over albino and short hair is dominant over long hair. Assume that these effects are caused by two independently assorting genes, and write the genotypes of the parents in each of the crosses shown here, in which D and A stand for the dark and albino phenotypes, respectively, and S and L stand for the short-hair and long-hair phenotypes.

Parental phenotypes	Number of progeny			
	D, S	D, L	A, S	A, L
a. D, S × D, S	89	31	29	11
b. D, S × D, L	18	19	0	0
c. D, S × A, S	20	0	21	0
d. A, S × A, S	0	0	28	9
e. D, L × D, L	0	32	0	10
f. D, S × D, S	46	16	0	0
g. D, S × D, L	30	31	9	11

Use the symbols *C* and *c* for the dark and albino coat-color alleles and the symbols *S* and *s* for the short-hair and long-hair alleles, respectively. Assume homozygosity unless there is evidence otherwise. (Problem 28 is reprinted by permission of Macmillan Publishing Co., Inc., from M. Strickberger, *Genetics*. Copyright 1968 by Monroe W. Strickberger.)

29. In tomatoes, two alleles of one gene determine the character difference of purple (P) versus green (G) stems, and two alleles of a separate, independent gene determine the character difference of "cut" (C) versus "potato" (Po) leaves. The results for five matings of tomato-plant phenotypes are as follows:

Mating	Parental phenotypes	Number of progeny			
		P, C	P, Po	G, C	G, Po
1	P, C × G, C	321	101	310	107
2	P, C × P, Po	219	207	64	71
3	P, C × G, C	722	231	0	0
4	P, C × G, Po	404	0	387	0
5	P, Po × G, C	70	91	86	77

a. Determine which alleles are dominant.

b. What are the most probable genotypes for the parents in each cross? (Problem 29 is from A. M. Srb, R. D. Owen, and R. S. Edgar, *General Genetics*, 2nd ed. Copyright 1965 by W. H. Freeman and Company.)

30. A mutant allele in mice causes a bent tail. Six pairs of mice were crossed. Their phenotypes and those of their progeny are given in the following table. N is normal phenotype; B is bent phenotype. Deduce the mode of inheritance of this phenotype.

Cross	Parents		Progeny	
	♀	♂	♀	♂
1	N	B	All B	All N
2	B	N	½ B, ½ N	½ B, ½ N
3	B	N	All B	All B
4	N	N	All N	All N
5	B	B	All B	All B
6	B	B	All B	½ B, ½ N

a. Is it recessive or dominant?

b. Is it autosomal or sex-linked?

c. What are the genotypes of all parents and progeny?

31. The normal eye color of *Drosophila* is red, but strains in which all flies have brown eyes are available. Similarly, wings are normally long, but there are strains with short wings. A female from a pure line with brown eyes and short wings is crossed with a male from a normal pure line. The F$_1$ consists of normal females and short-winged males. An F$_2$ is then produced by intercrossing the F$_1$. *Both* sexes of F$_2$ flies show phenotypes as follows:

$\frac{3}{8}$ red eyes, long wings

$\frac{3}{8}$ red eyes, short wings

$\frac{1}{8}$ brown eyes, long wings

$\frac{1}{8}$ brown eyes, short wings

Deduce the inheritance of these phenotypes; use clearly defined genetic symbols of your own invention. State the genotypes of all three generations and the genotypic proportions of the F$_1$ and F$_2$.

 Unpacking Problem 31

Before attempting a solution to this problem, try answering the following questions:

1. What does the word "normal" mean in this problem?

2. The words "line" and "strain" are used in this problem. What do they mean, and are they interchangeable?

3. Draw a simple sketch of the two parental flies showing their eyes, wings, and sexual differences.

4. How many different characters are there in this problem?

5. How many phenotypes are there in this problem, and which phenotypes go with which characters?

6. What is the full phenotype of the F$_1$ females called "normal"?

7. What is the full phenotype of the F$_1$ males called "short winged"?

8. List the F$_2$ phenotypic ratios for each character that you came up with in answer to question 4.

9. What do the F$_2$ phenotypic ratios tell you?

10. What major inheritance pattern distinguishes sex-linked inheritance from autosomal inheritance?

11. Do the F_2 data show such a distinguishing criterion?

12. Do the F_1 data show such a distinguishing criterion?

13. What can you learn about dominance in the F_1? The F_2?

14. What rules about wild-type symbolism can you use in deciding which allelic symbols to invent for these crosses?

15. What does "deduce the inheritance of these phenotypes" mean?

Now try to solve the problem. If you are unable to do so, make a list of questions about the things that you do not understand. Inspect the key concepts at the beginning of the chapter and ask yourself which are relevant to your questions. If this approach doesn't work, inspect the messages of this chapter and ask yourself which might be relevant to your questions.

32. In a natural population of annual plants, a single plant is found that is sickly looking and has yellowish leaves. The plant is dug up and brought back to the laboratory. Photosynthesis rates are found to be very low. Pollen from a normal dark-green-leaved plant is used to fertilize emasculated flowers of the yellowish plant. A hundred seeds result, of which only 60 germinate. All the resulting plants are sickly yellow in appearance.

    a. Propose a genetic explanation for the inheritance pattern.

    b. Suggest a simple test for your model.

    c. Account for the reduced photosynthesis, sickliness, and yellowish appearance.

33. What is the basis for the green-and-white color variegation in the leaves of *Mirabilis*? If the following cross is made,

    variegated ♀ × green ♂

    what progeny types can be predicted? What about the reciprocal cross?

34. In *Neurospora*, the mutant *stp* exhibits erratic stop-and-start growth. The mutant site is known to be in the mtDNA. If an *stp* strain is used as the female parent in a cross with a normal strain acting as the male, what type of progeny can be expected? What about the progeny from the reciprocal cross?

35. Two corn plants are studied. One is resistant (R) and the other is susceptible (S) to a certain pathogenic fungus. The following crosses are made, with the results shown:

    S ♀ × R ♂ → all progeny S
    R ♀ × S ♂ → all progeny R

What can you conclude about the location of the genetic determinants of R and S?

36. A presumed dihybrid in *Drosophila*, B/b ; F/f is testcrossed with b/b ; f/f. (B = black body ; b = brown body; F = forked bristles ; f = unforked bristles.) The results are

Black, forked	230
Black, unforked	210
Brown, forked	240
Brown, unforked	250

Use the χ^2 test to determine if these results fit the results expected from testcrossing the hypothesized dihybrid.

37. Are the following progeny numbers consistent with the results expected from selfing a plant presumed to be a dihybrid of two independently assorting genes, H/h ; R/r? (H = hairy leaves; h = smooth leaves; R = round ovary; r = elongated ovary.) Explain your answer.

hairy, round	178
hairy, elongated	62
smooth, round	56
smooth, elongated	24

38. A dark female moth is crossed with a dark male. All the male progeny are dark, but half the female progeny are light and the rest are dark. Propose an explanation for this pattern of inheritance.

39. In *Neurospora*, a mutant strain called stopper (*stp*) arose spontaneously. Stopper showed erratic "stop and start" growth, compared with the uninterrupted growth of wild-type strains. In crosses, the following results were found:

    ♀ stopper × ♂ wild type → progeny all stopper

    ♀ wild type × ♂ stopper → progeny all wild type

    a. What do these results suggest regarding the location of the stopper mutation in the genome?

    b. According to your model for part *a*, what progeny and proportions are predicted in octads from the following cross, including a mutation *nic3* located on chromosome VI?

    ♀ *stp · nic3* × wild type ♂

40. In polygenic systems, how many phenotypic classes corresponding to number of polygene "doses" are expected in selfs

    a. of strains with four heterozygous polygenes?

    b. of strains with six heterozygous polygenes?

41. In the self of a polygenic trihybrid $R_1/r_1; R_2/r_2; R_3/r_3$, use the product and sum rules to calculate the proportion of progeny with just one polygene "dose."

42. Reciprocal crosses and selfs were performed between the two moss species *Funaria mediterranea* and *F. hygrometrica*. The sporophytes and the leaves of the gametophytes are shown in the accompanying diagram.

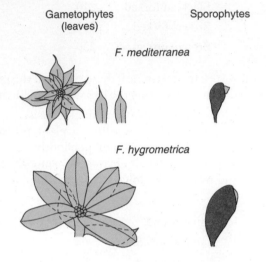

The crosses are written with the female parent first.

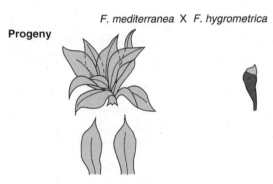

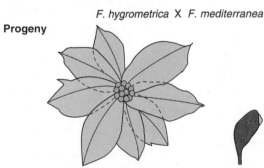

a. Describe the results presented, summarizing the main findings.

b. Propose an explanation of the results.

c. Show how you would test your explanation; be sure to show how it could be distinguished from other explanations.

43. Assume that diploid plant A has a cytoplasm genetically different from that of plant B. To study nuclear-cytoplasmic relations, you wish to obtain a plant with the cytoplasm of plant A and the nuclear genome predominantly of plant B. How would you go about producing such a plant?

44. You are studying a plant with tissue comprising both green and white sectors. You wish to decide whether this phenomenon is due (1) to a chloroplast mutation of the type considered in this chapter or (2) to a dominant nuclear mutation that inhibits chlorophyll production and is present only in certain tissue layers of the plant as a mosaic. Outline the experimental approach that you would use to resolve this problem.

45. Early in the development of a plant, a mutation in cpDNA removes a specific *Bg*III restriction site (*B*) as follows:

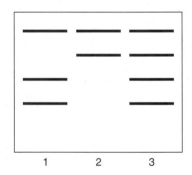

In this species, cpDNA is inherited maternally. Seeds from the plant are grown, and the resulting progeny plants are sampled for cpDNA. The cpDNAs are cut with *Bg*III, and Southern blots are hybridized with the probe P shown. The autoradiograms show three patterns of hybridization:

Explain the production of these three seed types.

CHALLENGING PROBLEMS

46. You have three jars containing marbles, as follows:

jar 1	600 red	and	400 white
jar 2	900 blue	and	100 white
jar 3	10 green	and	990 white

a. If you blindly select one marble from each jar, calculate the probability of obtaining

(1) a red, a blue, and a green.

(2) three whites.

(3) a red, a green, and a white.

(4) a red and two whites.

(5) a color and two whites.

(6) at least one white.

b. In a certain plant, R = red and r = white. You self a red R/r heterozygote with the express purpose of obtaining a white plant for an experiment. What minimum number of seeds do you have to grow to be at least 95 percent certain of obtaining at least one white individual?

c. When a woman is injected with an egg fertilized in vitro, the probability of its implanting successfully is 20 percent. If a woman is injected with five eggs simultaneously, what is the probability that she will become pregnant? (Part *c* is from Margaret Holm.)

47. In tomatoes, red fruit is dominant over yellow, two-loculed fruit is dominant over many-loculed fruit, and tall vine is dominant over dwarf. A breeder has two pure lines: (1) red, two-loculed, dwarf and (2) yellow, many-loculed, tall. From these two lines, he wants to produce a new pure line for trade that is yellow, two-loculed, and tall. How exactly should he go about doing so? Show not only which crosses to make, but also how many progeny should be sampled in each case.

48. We have dealt mainly with only two genes, but the same principles hold for more than two genes. Consider the following cross:

$$A/a\,;B/b\,;C/c\,;D/d\,;E/e \times a/a\,;B/b\,;c/c\,;D/d\,;e/e$$

a. What proportion of progeny will *phenotypically* resemble (1) the first parent, (2) the second parent, (3) either parent, and (4) neither parent?

b. What proportion of progeny will be *genotypically* the same as (1) the first parent, (2) the second parent, (3) either parent, and (4) neither parent?

Assume independent assortment.

49. The accompanying pedigree shows the pattern of transmission of two rare human phenotypes: cataract and pituitary dwarfism. Family members with cataract are shown with a solid left half of the symbol; those with pituitary dwarfism are indicated by a solid right half.

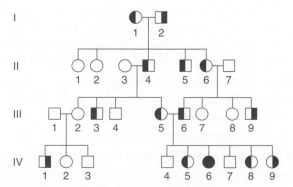

a. What is the most likely mode of inheritance of each of these phenotypes? Explain.

b. List the genotypes of all members in generation III as far as possible.

c. If a hypothetical mating took place between IV-1 and IV-5, what is the probability of the first child's being a dwarf with cataracts? A phenotypically normal child? (Problem 49 is after J. Kuspira and R. Bhambhani, *Compendium of Problems in Genetics.* Copyright 1994 by Wm. C. Brown.)

50. A corn geneticist has three pure lines of genotypes $a/a\,;B/B\,;C/C$, $A/A\,;b/b\,;C/C$, and $A/A\,;B/B\,;c/c$. All the phenotypes determined by *a, b,* and *c* will increase the market value of the corn; so, naturally, he wants to combine them all in one pure line of genotype $a/a\,;b/b\,;c/c$.

a. Outline an effective crossing program that can be used to obtain the $a/a\,;b/b\,;c/c$ pure line.

b. At each stage, state exactly which phenotypes will be selected and give their expected frequencies.

c. Is there more than one way to obtain the desired genotype? Which is the best way?

Assume independent assortment of the three gene pairs. (**Note:** Corn will self- or cross-pollinate easily.)

51. In humans, color vision depends on genes encoding three pigments. The *R* (red pigment) and *G* (green pigment) genes are close together on the X chromosome, whereas the *B* (blue pigment) gene is autosomal. A recessive mutation in any one of these genes can cause color blindness. Suppose that a color-blind man married a woman with normal color vision. The four sons from this marriage were color-blind, and the five daughters were normal. Specify the most likely genotypes of both parents and their children, explaining your reasoning. (A pedigree drawing will probably be helpful.) (Problem 51 is by Rosemary Redfield.)

52. Consider the accompanying pedigree for a rare human muscle disease.

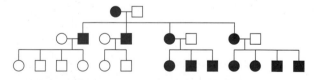

a. What unusual feature distinguishes this pedigree from those studied earlier in this chapter?

b. Where do you think the mutant DNA responsible for this phenotype resides in the cell?

53. The plant *Haplopappus gracilis* has a $2n$ of 4. A diploid cell culture was established and, at premitotic S phase, a radioactive nucleotide was added and was incorporated into newly synthesized DNA. The cells were then removed from the radioactivity, washed, and allowed to proceed through mitosis. Radioactive chromosomes or chromatids

can be detected by placing photographic emulsion on the cells; radioactive chromosomes or chromatids appeared covered with spots of silver from the emulsion. (The chromosomes "take their own photograph.") Draw the chromosomes at prophase and telophase of the first and second mitotic divisions after the radioactive treatment. If they are radioactive, show it in your diagram. If there are several possibilities, show them, too.

54. In the species of Problem 53, you can introduce radioactivity by injection into the anthers at the S phase before meiosis. Draw the four products of meiosis with their chromosomes and show which are radioactive.

55. The DNA double helices of chromosomes can be partly unwound in situ by special treatments. What pattern of radioactivity is expected if such a preparation is bathed in a radioactive probe for

a. a unique gene?

b. dispersed repetitive DNA?

c. ribosomal DNA?

d. telomeric DNA?

e. simple-repeat heterochromatic DNA?

56. If genomic DNA is cut with a restriction enzyme and fractionated by size by electrophoresis, what pattern of Southern hybridization is expected for the probes cited in Problem 55?

57. The plant *Haplopappus gracilis* is diploid and $2n = 4$. There are one long pair and one short pair of chromosomes. The diagrams below (numbered 1 through 12) represent anaphases ("pulling apart" stages) of individual cells in meiosis or mitosis in a plant that is genetically a dihybrid (A/a ; B/b) for genes on different chromosomes. The lines represent chromosomes or chromatids, and the points of the V's represent centromeres. In each case, indicate if the diagram represents a cell in meiosis I, meiosis II, or mitosis. If a diagram shows an impossible situation, say so.

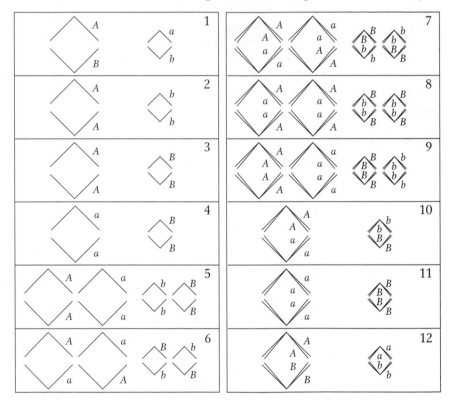

58. The pedigree below shows the recurrence of a rare neurological disease (large black symbols) and spontaneous fetal abortion (small black symbols) in one family. (A slash means that the individual is deceased.) Provide an explanation for this pedigree in regard to the cytoplasmic segregation of defective mitochondria.

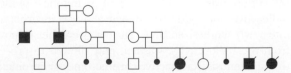

59. A man is brachydactylous (very short fingers; rare autosomal dominant) and his wife is not. Both can taste the chemical phenylthiocarbamide (autosomal dominant; common allele), but their mothers could not.

a. Give the genotypes of the couple.

If the genes assort independently and the couple has four children, what is the probability of

b. all of them being brachydactylous?

c. none being brachydactylous?

d. all of them being tasters?

e. all of them being nontasters?

f. all of them being brachydactylous tasters?

g. none being brachydactylous tasters?

h. at least one being a brachydactylous taster?

60. One form of male sterility in corn is maternally transmitted. Plants of a male-sterile line crossed with normal pollen give male-sterile plants. In addition, some lines of corn are known to carry a dominant nuclear restorer allele (*Rf*) that restores pollen fertility in male-sterile lines.

a. Research shows that the introduction of restorer alleles into male-sterile lines does not alter or affect the maintenance of the cytoplasmic factors for male sterility. What kind of research results would lead to such a conclusion?

b. A male-sterile plant is crossed with pollen from a plant homozygous for *Rf*. What is the genotype of the F_1? The phenotype?

c. The F_1 plants from part *b* are used as females in a testcross with pollen from a normal plant (*rf/rf*). What are the results of this testcross? Give genotypes and phenotypes, and designate the kind of cytoplasm.

d. The restorer allele already described can be called *Rf-1*. Another dominant restorer, *Rf-2*, has been found. *Rf-1* and *Rf-2* are located on different chromosomes. Either or both of the restorer alleles will give pollen fertility. With the use of a male-sterile plant as a tester, what will be the result of a cross in which the male parent is

 (i) heterozygous at both restorer loci?

 (ii) homozygous dominant at one restorer locus and homozygous recessive at the other?

(iii) heterozygous at one restorer locus and homozygous recessive at the other?

(iv) heterozygous at one restorer locus and homozygous dominant at the other?

Mapping Eukaryote Chromosomes by Recombination

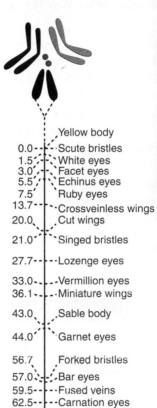

	Yellow body
0.0	Scute bristles
1.5	White eyes
3.0	Facet eyes
5.5	Echinus eyes
7.5	Ruby eyes
13.7	Crossveinless wings
20.0	Cut wings
21.0	Singed bristles
27.7	Lozenge eyes
33.0	Vermillion eyes
36.1	Miniature wings
43.0	Sable body
44.0	Garnet eyes
56.7	Forked bristles
57.0	Bar eyes
59.5	Fused veins
62.5	Carnation eyes
66.0	Bobbed hairs

At the left is a recombination-based map of one of the chromosomes of *Drosophila* (the organism in the image above), showing the loci of genes whose mutations produce known phenotypes. [*Robert Harding Picture Library Ltd/Alamy.*]

KEY QUESTIONS

- What cellular process produces a recombination of linked genes?

- What recombinant frequencies are diagnostic of linkage?

- How can an analysis of recombinant frequencies generate a chromosome map?

- How are recombination maps used in conjunction with physical DNA maps?

OUTLINE

Some of the questions that geneticists want to answer about the genome are, *What genes* are present in the genome? *What functions* do they have? *What positions* do they occupy on the chromosomes? Their pursuit of the third question is broadly called mapping. Mapping is the main focus of this chapter, but all three questions are interrelated, as we will see later in the chapter.

We all have an everyday feeling for the importance of maps in general, and, indeed, we have all used them at some time in our lives to find our way around. Relevant to the focus of this chapter is that, in some situations, *several* maps need to be used simultaneously. A good example in everyday life is in navigating the dense array of streets and buildings of a city such as London, England. A street map that shows the general layout is one necessity. However, the street map is used by tourists and Londoners alike in conjunction with another map, that of the underground railway system. The underground system is so complex and spaghetti-like that, in 1933, an electrical circuit engineer named Harry Beck drew up the streamlined (although distorted) map that has remained to this day an icon of London. The street and underground maps of London are compared in Figure 4-1. Note that the positions of the underground stations and the exact distances between them are of no interest in themselves, except as a way of getting to a destination of interest such as Westminster Abbey. We will see three parallels with the London maps when chromosome maps are used to zero in on individual "destinations," or specific genes. First, several different types of chromosome maps are often necessary and must be used in conjunction; second, maps that contain distortions are still useful; and third, many sites on a chromosome map are charted only because they are useful in trying to zero in on other sites that are the ones of real interest.

Obtaining a map of gene positions on the chromosomes is an endeavor that has occupied thousands of geneticists for the past 80 years or so. Why is it so important? There are several reasons:

1. Gene position is crucial information needed to build *complex genotypes* required for experimental purposes or for commercial applications. For example, in Chapter 6, we will see cases in which special allelic combinations must be put together to explore gene interaction.

2. Knowing the position occupied by a gene provides a way of zeroing in on its *structure and function*. A gene's position can be used to define it at the DNA level. In turn, the DNA sequence of a wild-type gene or its mutant allele is a necessary part of deducing its underlying function.

3. The genes present and their arrangement on chromosomes are often slightly different in related species. For example, the rather long human chromosome number 2 is split into two shorter chromosomes in the great apes. By comparing such differences, geneticists can deduce the *evolutionary genetic mechanisms* through which these genomes diverged. Hence, chromosome maps are useful in interpreting mechanisms of evolution.

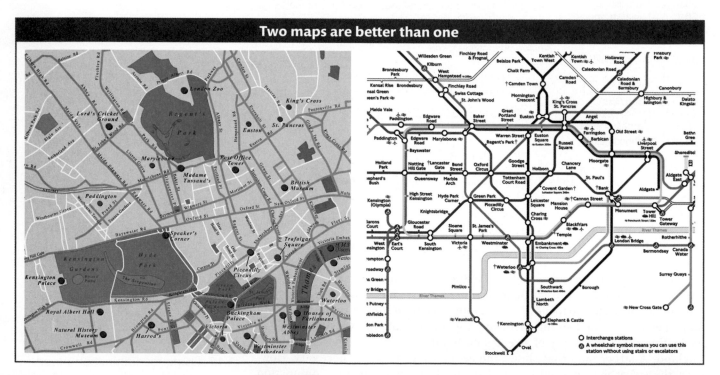

Figure 4-1 These London maps illustrate the principle that, often, several maps are needed to get to a destination of interest. The map of the underground railway ("the Tube") is used to get to a destination of interest such as a street address, shown on the street map. In genetics, two different kinds of genome maps are often useful in locating a gene, leading to an understanding of its structure and function. [*(left)* MAPS.com/CORBIS. *(right)* Transport for London.]

The arrangement of genes on chromosomes is represented diagrammatically as a unidimensional **chromosome map,** showing gene positions, known as **loci** (sing., locus), and the distances between the loci based on some kind of scale. Two basic types of chromosome maps are currently used in genetics; they are assembled by quite different procedures yet are used in a complementary way. *Recombination-based maps,* which are the topic of this chapter, map the loci of genes that have been identified by mutant phenotypes showing single-gene inheritance. *Physical maps* (see Chapter 14) show the genes as segments arranged along the long DNA molecule that constitutes a chromosome. These maps show different views of the genome, but, just like the maps of London, they can be used together to arrive at an understanding of what a gene's function is at the molecular level and how that function influences phenotype.

> **Message** Genetic maps are useful for strain building, for interpreting evolutionary mechanisms, and for discovering a gene's unknown function. Discovering a gene's function is facilitated by integrating information on recombination-based and physical maps.

4.1 Diagnostics of Linkage

Recombination maps of chromosomes are usually assembled two or three genes at a time, with the use of a method called linkage analysis. When geneticists say that two genes are **linked,** they mean that the loci of those genes are on the same chromosome, and, hence, the alleles on any one homolog are physically joined (linked) by the DNA between them. The way in which early geneticists deduced linkage is a useful means of introducing most of the key ideas and procedures in the analysis.

Using recombinant frequency to recognize linkage

In the early 1900s, William Bateson and R. C. Punnett (for whom the Punnett square was named) were studying the inheritance of two genes in sweet peas. In a standard self of a dihybrid F_1, the F_2 did not show the $9:3:3:1$ ratio predicted by the principle of independent assortment. In fact, Bateson and Punnett noted that certain combinations of alleles showed up more often than expected, almost as though they were physically attached in some way. However, they had no explanation for this discovery.

Later, Thomas Hunt Morgan found a similar deviation from Mendel's second law while studying two autosomal genes in *Drosophila*. Morgan proposed linkage as a hypothesis to explain the phenomenon of apparent allele association.

Let's look at some of Morgan's data. One of the genes affected eye color (*pr*, purple, and pr^+, red), and the other gene affected wing length (*vg*, vestigial, and vg^+, normal). The wild-type alleles of both genes are dominant. Morgan performed a cross to obtain dihybrids and then followed with a testcross:

P $\qquad\qquad\qquad pr/pr \cdot vg/vg \times pr^+/pr^+ \cdot vg^+/vg^+$

$\downarrow$

Gametes $\qquad\qquad\quad pr \cdot vg \quad pr^+ \cdot vg^+$

$\downarrow$

F_1 dihybrid $\qquad\qquad pr^+/pr \cdot vg^+/vg$

Testcross:

$pr^+/pr \cdot vg^+/vg ♀ \quad \times \quad pr/pr \cdot vg/vg ♂$

F_1 dihybrid female $\qquad\qquad$ Tester male

Morgan's use of the testcross is important. Because the tester parent contributes gametes carrying only recessive alleles, the phenotypes of the offspring directly reveal the alleles contributed by the gametes of the dihybrid parent, as described in Chapters 2 and 3. Hence, the analyst can concentrate on meiosis in one parent (the dihybrid) and essentially forget about meiosis in the other (the tester). In contrast, from an F_1 *self*, there are *two* sets of meioses to consider in the analysis of progeny: one in the male parent and the other in the female.

Morgan's testcross results were as follows (listed as the gametic classes from the dihybrid):

$$
\begin{array}{lr}
pr^+ \cdot vg^+ & 1339 \\
pr \cdot vg & 1195 \\
pr^+ \cdot vg & 151 \\
pr \cdot vg^+ & \underline{154} \\
& 2839
\end{array}
$$

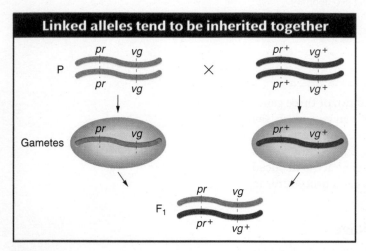

Figure 4-2 Simple inheritance of two genes located on the same chromosome pair. The same genes are present together on a chromosome in both parents and progeny.

Obviously, these numbers deviate drastically from the Mendelian prediction of a 1:1:1:1 ratio expected from independent assortment (approximately 710 in each of the four classes). In Morgan's results, we see that the first two allele combinations are in the great majority, clearly indicating that they are associated, or "linked."

Another useful way of assessing the testcross results is by considering the percentage of recombinants in the progeny. By definition, the recombinants in the present cross are the two types $pr^+ \cdot vg$ and $pr \cdot vg^+$ because they are clearly not the two input genotypes contributed to the F_1 dihybrid by the original homozygous parental flies (more precisely, by their gametes). We see that the two recombinant types are approximately equal in frequency (151 ∼ 154). Their total is 305, which is a frequency of (305/2839) × 100, or 10.7 percent. We can make sense of these data, as Morgan did, by postulating that the genes were linked on the same chromosome, and so the parental allelic combinations are held together in the majority of progeny. In the dihybrid, the allelic conformation must have been as follows:

$$
\frac{pr^+ \qquad vg^+}{pr \qquad vg}
$$

The tendency of linked alleles to be inherited as a package is illustrated in Figure 4-2.

Now let's look at another cross that Morgan made with the use of the same alleles but in a different combination. In this cross, each parent is homozygous for the wild-type allele of one gene and the mutant allele of the other. Again, F_1 females were testcrossed:

P $pr^+/pr^+ \cdot vg/vg \times pr/pr \cdot vg^+/vg^+$

↓

Gametes $pr^+ \cdot vg \qquad pr \cdot vg^+$

↓

F_1 dihybrid $pr^+/pr \cdot vg^+/vg$

Testcross:

$pr^+/pr \cdot vg^+/vg\,♀ \quad \times \quad pr/pr \cdot vg/vg\,♂$

F_1 dihybrid female   Tester male

The following progeny were obtained from the testcross:

$$pr^+ \cdot vg^+ \qquad 157$$
$$pr \cdot vg \qquad 146$$
$$pr^+ \cdot vg \qquad 965$$
$$pr \cdot vg^+ \qquad \underline{1067}$$
$$2335$$

Again, these results are not even close to a 1:1:1:1 Mendelian ratio. Now, however, the recombinant classes are the converse of those in the first analysis, $pr^+ vg^+$ and $pr\ vg$. But notice that their frequency is approximately the same: $(157 + 146)/2335 \times 100 = 12.9$ percent. Again, linkage is suggested, but, in this case, the F_1 dihybrid must have been as follows:

$$\frac{pr^+ \qquad vg}{pr \qquad vg^+}$$

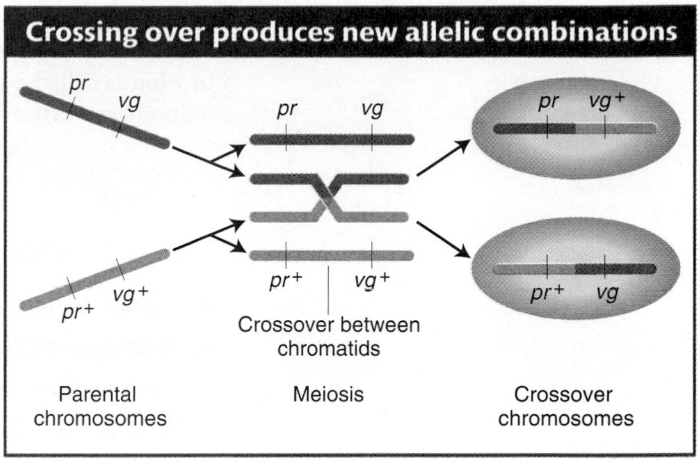

Dihybrid testcross results like those just presented are commonly encountered in genetics. They follow the general pattern:

Two equally frequent nonrecombinant classes totaling
in excess **of 50 percent**

Two equally frequent recombinant classes totaling
less than **50 percent**

> Message When two genes are close together on the same chromosome pair (that is, they are linked), they do not assort independently but produce a recombinant frequency of less than 50 percent. Hence, conversely, a recombinant frequency of less than 50 percent is a diagnostic for linkage.

How crossovers produce recombinants for linked genes

The linkage hypothesis explains why allele combinations from the parental generations remain together: the genes are physically attached by the segment of chromosome between them. But exactly how are *any* recombinants produced when genes are linked? Morgan suggested that, when homologous chromosomes pair at meiosis, the chromosomes occasionally break and exchange parts in a process called **crossing over.** Figure 4-3 illustrates this physical exchange of chromosome segments. The two new combinations are called **crossover products.**

Is there any microscopically observable process that could account for crossing over? At meiosis, when duplicated homologous chromosomes pair with each other—in genetic terms, when the two dyads unite as a bivalent—a cross-shaped structure called a *chiasma* (pl., chiasmata) often forms between two non-sister chromatids. Chiasmata are shown in Figure 4-4. To Morgan, the appearance of the chiasmata visually corroborated the concept of crossing over. (Note that the chiasmata seem to indicate that *chromatids,* not unduplicated chromosomes, participate in a crossover. We will return to this point later.)

> Message For linked genes, recombinants are produced by crossovers. Chiasmata are the visible manifestations of crossovers.

Linkage symbolism and terminology

The work of Morgan showed that linked genes in a dihybrid may be present in one of two basic conformations. In one, the two dominant, or wild-type, alleles are

Crossing over produces new allelic combinations

Parental chromosomes   Meiosis   Crossover chromosomes

Crossover between chromatids

Figure 4-3 The exchange of parts by crossing over may produce gametic chromosomes whose allelic combinations differ from the parental combinations.

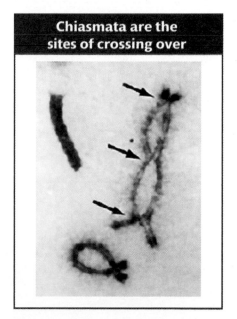

Chiasmata are the sites of crossing over

Figure 4-4 Several chiasmata appear in this photograph taken in the course of meiosis in a male locust. [*G. H. Jones and F. C. H. Franklin, Meiotic Crossing-over: Obligation and Interference, Cell 126:2 (28 July 2006), 246-248.*]

present on the same homolog (as in Figure 4-3); this arrangement is called a **cis conformation** (cis means adjacent). In the other, they are on different homologs, in what is called a **trans conformation** (trans means opposite). The two conformations are written as follows:

Cis AB/ab or $++/ab$

Trans Ab/aB or $+b/a+$

Note the following conventions that pertain to linkage symbolism:

1. Alleles on the same homolog have no punctuation between them.

2. A slash symbolically separates the two homologs.

3. Alleles are always written in the same order on each homolog.

4. As in earlier chapters, genes known to be on different chromosomes (unlinked genes) are shown separated by a semicolon—for example, A/a ; C/c.

5. In this book, genes of *unknown* linkage are shown separated by a dot, $A/a \cdot D/d$.

Evidence that crossing over is a breakage-and-rejoining process

The idea that recombinants are produced by some kind of exchange of material between homologous chromosomes was a compelling one. But experimentation was necessary to test this hypothesis. A first step was to find a case in which the exchange of parts between chromosomes would be visible under the microscope. Several investigators approached this problem in the same way, and one of their analyses follows.

In 1931, Harriet Creighton and Barbara McClintock were studying two genes of corn that they knew were both located on chromosome 9. One affected seed color (*C,* colored; *c,* colorless), and the other affected endosperm composition (*Wx,* waxy; *wx,* starchy). The plant was a dihybrid in cis conformation. However, in one plant, the chromosome 9 carrying the alleles *C* and *Wx* was unusual in that it also carried a large, densely staining element (called a *knob*) on the *C* end and a longer piece of chromosome on the *Wx* end; thus, the heterozygote was

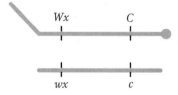

In the progeny of a testcross of this plant, they compared recombinants and parental genotypes. They found that all the recombinants inherited one or the other of the two following chromosomes, depending on their recombinant makeup:

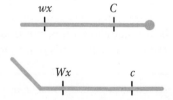

Thus, there was a precise correlation between the *genetic* event of the appearance of recombinants and the *chromosomal* event of crossing over. Consequently, the chiasmata appeared to be the sites of exchange, although what was considered to be the definitive test was not undertaken until 1978.

What can we say about the molecular mechanism of chromosome exchange in a crossover event? The short answer is that a crossover results from the breakage and reunion of DNA. Two parental chromosomes break at the same position, and then each piece joins up with the neighboring piece from the *other* chromosome. In Section 4.8, we will see a model of the molecular processes that allow DNA to break and rejoin in a precise manner such that no genetic material is lost or gained.

> **Message** A crossover is the breakage of two DNA molecules at the same position and their rejoining in two reciprocal recombinant combinations.

Evidence that crossing over takes place at the four-chromatid stage

As already noted, the diagrammatic representation of crossing over in Figure 4-3 shows a crossover taking place at the four-chromatid stage of meiosis; in other words, crossovers are between nonsister *chromatids*. However, it was *theoretically* possible that crossing over took place before replication, at the *two*-chromosome stage. This uncertainty was resolved through the genetic analysis of organisms whose four products of meiosis remain together in groups of four called *tetrads*. These organisms, which we met in Chapters 2 and 3, are fungi and unicellular algae. The products of meiosis of a single tetrad can be isolated, which is equivalent to isolating all four chromatids from a single meiocyte. Tetrad analyses of crosses *in which genes are linked* show many tetrads that contain four different allele combinations. For example, from the cross

$$AB \times ab$$

some (but not all) tetrads contain four genotypes:

$$AB$$
$$Ab$$
$$aB$$
$$ab$$

This result can be explained only if crossovers take place at the four-chromatid stage because, if crossovers took place at the two-chromosome stage, there could only ever be a maximum of two different genotypes in an individual tetrad. This reasoning is illustrated in Figure 4-5.

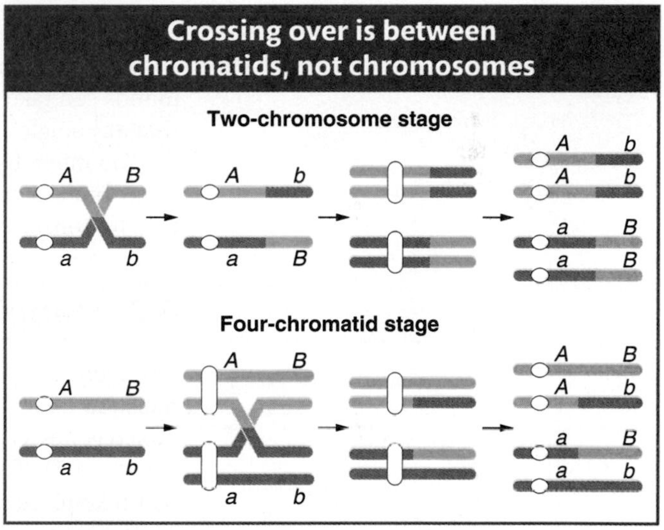

Multiple crossovers can include more than two chromatids

Tetrad analysis can also show two other important features of crossing over. First, in some individual meiocytes, several crossovers can occur along a chromosome pair. Second, in any one meiocyte, these multiple crossovers can exchange material between more than two chromatids. To think about this matter, we need to look at the simplest case: double crossovers. To study double crossovers, we need three linked genes. For example, if the three loci are all linked in a cross such as

$$ABC \times abc$$

many different tetrad types are possible, but some types are informative in the present connection because they can be accounted for only by double crossovers in which more than two chromatids take part. Consider the following tetrad as an example:

$$ABc$$
$$AbC$$
$$aBC$$
$$abc$$

Crossing over is between chromatids, not chromosomes

Two-chromosome stage

Four-chromatid stage

Figure 4-5 Crossing over takes place at the four-chromatid stage. Because more than two different products of a single meiosis can be seen in some tetrads, crossing over cannot take place at the two-strand stage (before DNA replication). The white circle designates the position of the centromere. When sister chromatids are visible, the centromere appears unreplicated.

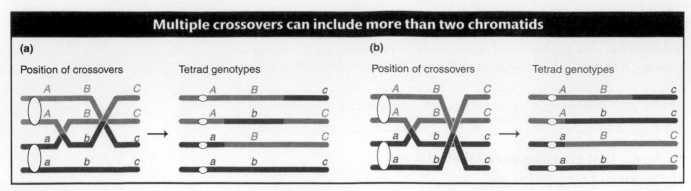

Multiple crossovers can include more than two chromatids

(a)

Position of crossovers   Tetrad genotypes

(b)

Position of crossovers   Tetrad genotypes

Figure 4-6 Double crossovers can include (a) three chromatids or (b) four chromatids.

This tetrad must be explained by two crossovers in which *three* chromatids take part, as shown in Figure 4-6a. Furthermore, the following type of tetrad shows that all *four* chromatids can participate in crossing over in the same meiosis (Figure 4-6b):

$$ABc$$
$$Abc$$
$$aBC$$
$$abC$$

Therefore, for any pair of homologous chromosomes, two, three, or four chromatids can take part in crossing-over events in a single meiocyte. Note, however, that any single crossover is between two chromatids.

You might be wondering about crossovers between *sister* chromatids. They do occur but are rare. They do not produce new allele combinations and so are not usually considered.

4.2 Mapping by Recombinant Frequency

The frequency of recombinants produced by crossing over is the key to chromosome mapping. Fungal tetrad analysis has shown that, for any two specific linked genes, crossovers take place between them in some, but not all, meiocytes (Figure 4-7). The farther apart the genes are, the more likely that a crossover will take place and the higher the proportion of recombinant products will be. Thus, the proportion of recombinants is a clue to the distance separating two gene loci on a chromosome map.

As stated earlier in regard to Morgan's data, the recombinant frequency was significantly less than 50 percent, specifically 10.7 percent. Figure 4-8 shows the general situation for linkage in which recombinants are less than 50 percent. Recombinant frequencies for different linked genes range from 0 to 50 percent, depending on their closeness (see page 132). The farther apart genes are, the more closely their recombinant frequencies approach 50 percent and, in such cases, one cannot decide whether genes are linked or are on different chromosomes. What about recombinant frequencies greater than 50 percent? The answer is that such frequencies are *never* observed, as will be proved later.

Note in Figure 4-7 that a single crossover generates two reciprocal recombinant products, which explains why the reciprocal recombinant classes are generally approximately equal in frequency. The corollary of this point is that the two parental nonrecombinant types also must be equal in frequency, as also observed by Morgan.

Map units

The basic method of mapping genes with the use of recombinant frequencies was worked out by a student of Morgan's. As Morgan studied more and more linked

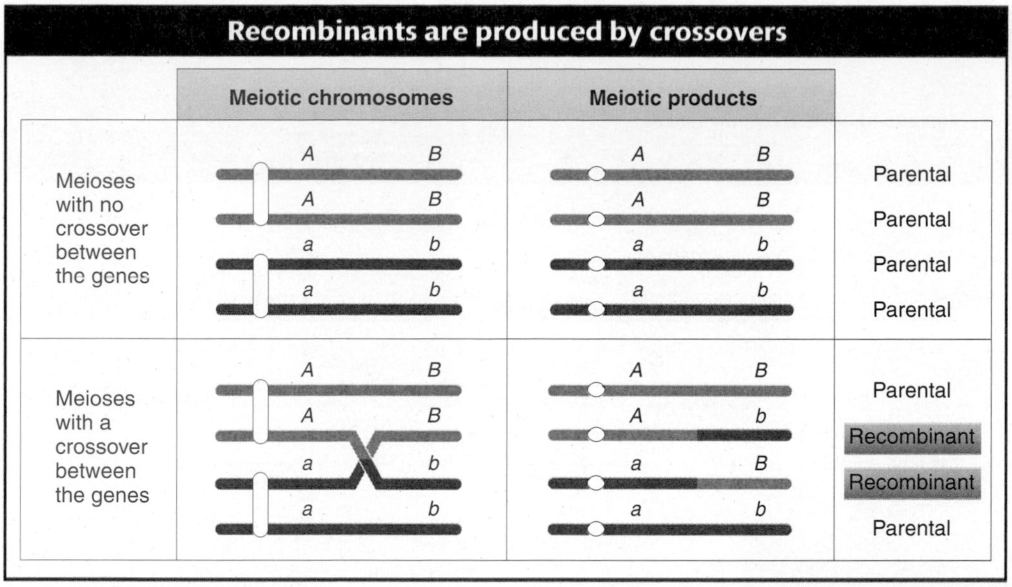

Recombinants are produced by crossovers

	Meiotic chromosomes	Meiotic products	
Meioses with no crossover between the genes	A — B A — B a — b a — b	A — B A — B a — b a — b	Parental Parental Parental Parental
Meioses with a crossover between the genes	A — B A — B a — b a — b	A — B A — b a — B a — b	Parental Recombinant Recombinant Parental

Figure 4-7 Recombinants arise from meioses in which a crossover takes place between nonsister chromatids.

GENETICS*PORTAL* **ANIMATED ART**
Meiotic recombination between linked genes by crossing over

genes, he saw that the proportion of recombinant progeny varied considerably, depending on which linked genes were being studied, and he thought that such variation in recombinant frequency might somehow indicate the actual distances separating genes on the chromosomes. Morgan assigned the quantification of this process to an undergraduate student, Alfred Sturtevant, who also became one of the great geneticists. Morgan asked Sturtevant to try to make some sense of the data on crossing over between different linked genes. In one evening, Sturtevant developed a method for mapping genes that is still used today. In Sturtevant's own words, "In the latter part of 1911, in conversation with Morgan, I suddenly realized that the variations in strength of linkage, already attributed by Morgan to differences in the spatial separation of genes, offered the possibility of determining sequences in the linear dimension of a chromosome. I went home and spent most of the night (to the neglect of my undergraduate homework) in producing the first chromosome map."

As an example of Sturtevant's logic, consider Morgan's testcross results with the *pr* and *vg* genes, from which he calculated a recombinant frequency of 10.7 percent. Sturtevant suggested that we can use this percentage of recombinants as a quantitative index of the linear distance between two genes on a genetic map, or **linkage map,** as it is sometimes called.

The basic idea here is quite simple. Imagine two specific genes positioned a certain fixed distance apart. Now imagine random crossing over along the paired homologs. In some meioses, nonsister chromatids cross over by chance in the chromosomal region between these genes; from these meioses, recombinants are produced. In other meiotic divisions, there are no crossovers between these genes; no recombinants result from these meioses. (See Figure 4-7 for a diagrammatic illustration.) Sturtevant postulated a rough proportionality: the greater the distance between the linked genes, the greater the chance of crossovers in the region between the genes and, hence, the greater the proportion of recombinants that would be produced. Thus, by determining the frequency of recombinants, we can obtain a measure of the map distance between the genes. In fact, Sturtevant defined one **genetic map unit (m.u.)** as that distance between genes for which 1 product of meiosis in 100 is recombinant. For example, the

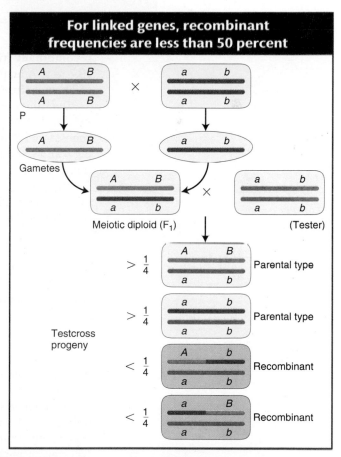

For linked genes, recombinant frequencies are less than 50 percent

Figure 4-8 A testcross reveals that the frequencies of recombinants arising from crossovers between linked genes are less than 50 percent.

Figure 4-9 A chromosome region containing three linked genes. Because map distances are additive, calculation of A–B and A–C distances leaves us with the two possibilities shown for the B–C distance.

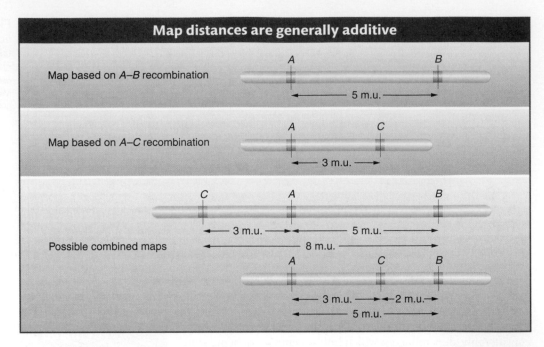

recombinant frequency (RF) of 10.7 percent obtained by Morgan is defined as 10.7 m.u. A map unit is sometimes referred to as a **centimorgan (cM)** in honor of Thomas Hunt Morgan.

Does this method produce a linear map corresponding to chromosome linearity? Sturtevant predicted that, on a linear map, if 5 map units (5 m.u.) separate genes A and B, whereas 3 m.u. separate genes A and C, then the distance separating B and C should be either 8 or 2 m.u. (Figure 4-9). Sturtevant found his prediction to be the case. In other words, his analysis strongly suggested that genes are arranged in some linear order, making map distances additive. (There are some minor but not insignificant exceptions, as we will see later.) Since we now know from molecular analysis that a chromosome is a single DNA molecule with the genes arranged along it, it is no surprise for us today to learn that recombination-based maps are linear since they reflect a linear array of genes.

How is a map represented? As an example, in *Drosophila*, the locus of the eye-color gene and the locus of the wing-length gene are approximately 11 m.u. apart, as mentioned earlier. The relation is usually diagrammed in the following way:

$$pr \qquad\qquad 11.0 \qquad\qquad vg$$

Generally, we refer to the locus of this eye-color gene in shorthand as the "*pr* locus," after the first discovered mutant allele, but we mean the place on the chromosome where *any* allele of this gene will be found, mutant or wild type.

As stated in Chapters 2 and 3, genetic analysis can be applied in two opposite directions. This principle is applicable to recombinant frequencies. In one direction, recombinant frequencies can be used to make maps. In the other direction, when given an established map with genetic distance in map units, we can predict the frequencies of progeny in different classes. For example, the genetic distance between the *pr* and *vg* loci in *Drosophila* is approximately 11 map units. So knowing this value, we know that there will be 11 percent recombinants in the progeny from a testcross of a female dihybrid heterozygote in cis conformation (*pr vg/ pr*$^+$ *vg*$^+$). These recombinants will consist of two reciprocal recombinants of equal frequency: thus, 5.5 percent will be *pr vg*$^+$ and 5.5 percent will be *pr*$^+$ *vg*. We also know that $100 - 11 = 89$ percent will be nonrecombinant in two equal classes, 44.5 percent *pr*$^+$ *vg*$^+$ and 44.5 percent *pr vg*. (Note that the tester contribution *pr vg* was ignored in writing out these genotypes.)

There is a strong implication that the "distance" on a linkage map is a physical distance along a chromosome, and Morgan and Sturtevant certainly intended to imply just that. But we should realize that the linkage map is a *hypothetical* entity constructed from a purely genetic analysis. The linkage map could have been derived without even knowing that chromosomes existed. Furthermore, at this point in our discussion, we cannot say whether the "genetic distances" calculated by means of recombinant frequencies in any way represent actual physical distances on chromosomes. However, physical mapping has shown that genetic distances are, in fact, roughly proportional to recombination-based distances. There are exceptions caused by recombination hotspots, places in the genome where crossing over takes place more frequently than usual. The presence of hot-spots causes proportional expansion of some regions of the map. Recombination blocks, which have the opposite effect, also are known.

A summary of the way in which recombinants from crossing over are used in mapping is shown in Figure 4-10. Crossovers occur more or less randomly along

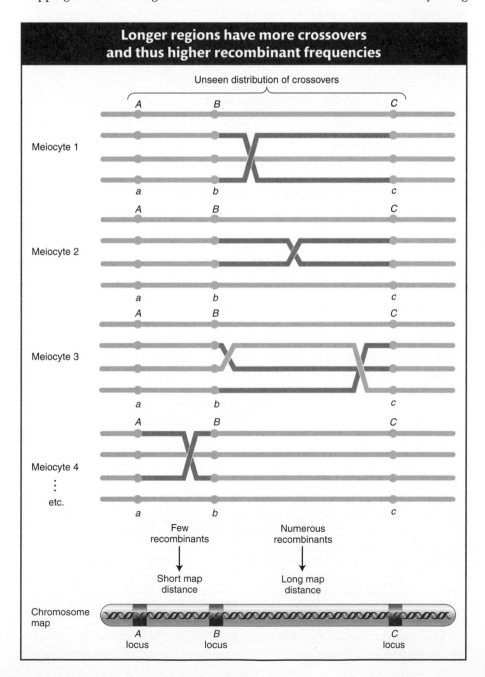

Figure 4-10 Crossovers produce recombinant chromatids whose frequency can be used to map genes on a chromosome. Longer regions produce more crossovers. Brown shows recombinants for that interval.

the chromosome pair. In general, in longer regions, the average number of crossovers is higher and, accordingly, recombinants are more frequently obtained, translating into a longer map distance.

> **Message** Recombination between linked genes can be used to map their distance apart on a chromosome. The unit of mapping (1 m.u.) is defined as a recombinant frequency of 1 percent.

Three-point testcross

So far, we have looked at linkage in crosses of dihybrids (double heterozygotes) with doubly recessive testers. The next level of complexity is a cross of a trihybrid (triple heterozygote) with a triply recessive tester. This kind of cross, called a **three-point testcross** or a **three-factor cross,** is commonly used in linkage analysis. The goal is to deduce whether the three genes are linked and, if they are, to deduce their order and the map distances between them.

Let's look at an example, also from *Drosophila*. In our example, the mutant alleles are v (vermilion eyes), cv (crossveinless, or absence of a crossvein on the wing), and ct (cut, or snipped, wing edges). The analysis is carried out by performing the following crosses:

$$P \qquad v^+/v^+ \cdot cv/cv \cdot ct/ct \ \times \ v/v \cdot cv^+/cv^+ \cdot ct^+/ct^+$$

$$\downarrow$$

$$\text{Gametes} \qquad v^+ \cdot cv \cdot ct \qquad v \cdot cv^+ \cdot ct^+$$

$$F_1 \text{ trihybrid} \qquad v/v^+ \cdot cv/cv^+ \ \cdot \ ct/ct^+$$

Trihybrid females are testcrossed with triple recessive males:

$$v/v^+ \cdot cv/cv^+ \cdot ct/ct^+ \, ♀ \ \times \ v/v \cdot cv/cv \cdot ct/ct \, ♂$$

$$F_1 \text{ trihybrid female} \qquad \text{Tester male}$$

From any trihybrid, only $2 \times 2 \times 2 = 8$ gamete genotypes are possible. They are the genotypes seen in the testcross progeny. The following chart shows the number of each of the eight gametic genotypes in a sample of 1448 progeny flies. The columns alongside show which genotypes are recombinant (R) for the loci taken two at a time. We must be careful in our classification of parental and recombinant types. Note that the parental input genotypes for the triple heterozygotes are $v^+ \cdot cv \cdot ct$ and $v \cdot cv^+ \cdot ct^+$; any combination other than these two constitutes a recombinant.

		Recombinant for loci		
Gametes		v and cv	v and ct	cv and ct
$v \ \cdot \ cv^+ \ \cdot \ ct^+$	580			
$v^+ \cdot \ cv \ \cdot \ ct$	592			
$v \ \cdot \ cv \ \cdot \ ct^+$	45	R		R
$v^+ \cdot \ cv^+ \cdot \ ct$	40	R		R
$v \ \cdot \ cv \ \cdot \ ct$	89	R	R	
$v^+ \cdot \ cv^+ \cdot \ ct^+$	94	R	R	
$v \ \cdot \ cv^+ \cdot \ ct$	3		R	R
$v^+ \cdot \ cv \ \cdot \ ct^+$	5		R	R
	1448	268	191	93

Let's analyze the loci two at a time, starting with the v and cv loci. In other words, we look at just the first two columns under "Gametes" and cover up the third one. Because the parentals for this pair of loci are $v^+ \cdot cv$ and $v \cdot cv^+$, we know that the recombinants are by definition $v \cdot cv$ and $v^+ \cdot cv^+$. There are $45 + 40 + 89 + 94 = 268$ of these recombinants. Of a total of 1448 flies, this number gives an RF of 18.5 percent.

For the v and ct loci, the recombinants are $v \cdot ct$ and $v^+ \cdot ct^+$. There are $89 + 94 + 3 + 5 = 191$ of these recombinants among 1448 flies, and so the RF = 13.2 percent.

For ct and cv, the recombinants are $cv \cdot ct^+$ and $cv^+ \cdot ct$. There are $45 + 40 + 3 + 5 - 93$ of these recombinants among the 1448, and so the RF = 6.4 percent.

Clearly, all the loci are linked, because the RF values are all considerably less than 50 percent. Because the v and cv loci have the largest RF value, they must be farthest apart; therefore, the ct locus must lie between them. A map can be drawn as follows:

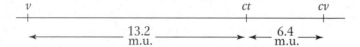

The testcross can be rewritten as follows, now that we know the linkage arrangement:

$$v^+ \, ct \, cv/v \, ct^+ \, cv^+ \quad \times \quad v \, ct \, cv/v \, ct \, cv$$

Note several important points here. First, we have deduced a gene order that is different from that used in our list of the progeny genotypes. Because the point of the exercise was to determine the linkage relation of these genes, the original listing was of necessity arbitrary; the order was simply not known before the data were analyzed. Henceforth, the genes must be written in correct order.

Second, we have definitely established that ct is between v and cv. In the diagram, we have arbitrarily placed v to the left and cv to the right, but the map could equally well be drawn with the placement of these loci inverted.

Third, note that linkage maps merely map the loci in relation to one another, with the use of standard map units. We do not know where the loci are on a chromosome—or even which specific chromosome they are on. In subsequent analyses, as more loci are mapped in relation to these three, the complete chromosome map would become "fleshed out."

> **Message** Three-point (and higher) testcrosses enable geneticists to evaluate linkage between three (or more) genes and to determine gene order, all in one cross.

A final point to note is that the two smaller map distances, 13.2 m.u. and 6.4 m.u., add up to 19.6 m.u., which is greater than 18.5 m.u., the distance calculated for v and cv. Why? The answer to this question lies in the way in which we have treated the two rarest classes of progeny (totaling 8) with respect to the recombination of v and cv. Now that we have the map, we can see that these two rare classes are in fact double recombinants, arising from two crossovers (Figure 4-11).

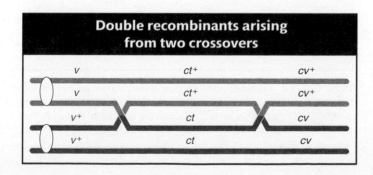

Figure 4-11 Example of a double crossover between two chromatids. Notice that a double crossover produces double recombinant chromatids that have the parental allele combinations at the outer loci. The position of the centromere cannot be determined from the data. It has been added for completeness.

However, when we calculated the RF value for v and cv, we did not count the v ct cv^+ and v^+ ct^+ cv genotypes; after all, with regard to v and cv, they are parental combinations (v cv^+ and v^+ cv). In light of our map, however, we see that this oversight led us to underestimate the distance between the v and the cv loci. Not only should we have counted the two rarest classes, we should have counted each of them *twice* because each represents double recombinants. Hence, we can correct the value by adding the numbers $45 + 40 + 89 + 94 + 3 + 3 + 5 + 5 = 284$. Of the total of 1448, this number is exactly 19.6 percent, which is identical with the sum of the two component values. (In practice, we do not need to do this calculation, because the sum of the two shorter distances gives us the best estimate of the overall distance.)

Deducing gene order by inspection

Now that we have had some experience with the three-point testcross, we can look back at the progeny listing and see that, for trihybrids of linked genes, *gene order* can usually be deduced by inspection, without a recombinant frequency analysis. Typically, for linked genes, we have the eight genotypes at the following frequencies:

two at high frequency

two at intermediate frequency

two at a different intermediate frequency

two rare

Only three gene orders are possible, each with a different gene in the middle position. It is generally true that the double-recombinant classes are the smallest ones, as listed last here. Only one order is compatible with the smallest classes' having been formed by double crossovers, as shown in Figure 4-12; that is, only one order gives double recombinants of genotype v ct cv^+ and v^+ ct^+ cv. A simple

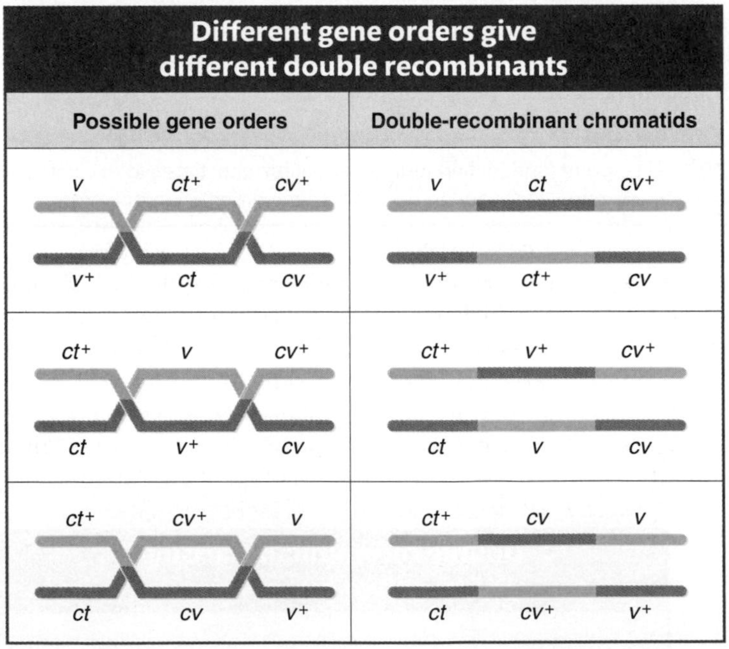

Figure 4-12 The three possible gene orders shown on the left yield the six products of a double crossover shown on the right. Only the first possibility is compatible with the data in the text. Note that only the nonsister chromatids taking part in the double crossover are shown.

rule of thumb for deducing the gene in the middle is that it is the allele pair that has "flipped" position in the double-recombinant classes.

Interference

Knowing the existence of double crossovers permits us to ask questions about their possible interdependence. We can ask, Are the crossovers in adjacent chromosome regions independent events or does a crossover in one region affect the likelihood of there being a crossover in an adjacent region? The answer is that, generally, crossovers inhibit each other somewhat in an interaction called **interference.** Double-recombinant classes can be used to deduce the extent of this interference.

Interference can be measured in the following way. If the crossovers in the two regions are independent, we can use the product rule (see page 87) to predict the frequency of double recombinants: that frequency would equal the product of the recombinant frequencies in the adjacent regions. In the *v-ct-cv* recombination data, the *v-ct* RF value is 0.132 and the *ct-cv* value is 0.064; so, if there is no interference, double recombinants might be expected at the frequency $0.132 \times 0.064 = 0.0084$ (0.84 percent). In the sample of 1448 flies, $0.0084 \times 1448 = 12$ double recombinants are expected. But the data show that only 8 were actually observed. If this deficiency of double recombinants were consistently observed, it would show us that the two regions are not independent and suggest that the distribution of crossovers favors singles at the expense of doubles. In other words, there is some kind of interference: a crossover does reduce the probability of a crossover in an adjacent region.

Interference is quantified by first calculating a term called the **coefficient of coincidence (c.o.c.)**, which is the ratio of observed to expected double recombinants. Interference (I) is defined as $1 - $ c.o.c. Hence,

$$I = 1 - \frac{\text{observed frequency, or number, of double recombinants}}{\text{expected frequency, or number, of double recombinants}}$$

In our example

$$I = 1 - \tfrac{8}{12} = \tfrac{4}{12} = \tfrac{1}{3}, \text{ or 33 percent}$$

In some regions, there are never any observed double recombinants. In these cases, c.o.c. $= 0$, and so $I = 1$ and interference is complete. Interference values anywhere between 0 and 1 are found in different regions and in different organisms.

You may have wondered why we always use heterozygous females for testcrosses in *Drosophila*. The explanation lies in an unusual feature of *Drosophila* males. When, for example, *pr vg/pr$^+$ vg$^+$* males are crossed with *pr vg/pr vg* females, only *pr vg/pr$^+$ vg$^+$* and *pr vg/pr vg* progeny are recovered. This result shows that there is no crossing over in *Drosophila* males. However, this absence of crossing over in one sex is limited to certain species; it is not the case for males of all species (or for the heterogametic sex). In other organisms, there is crossing over in XY males and in WZ females. The reason for the absence of crossing over in *Drosophila* males is that they have an unusual prophase I, with no synaptonemal complexes. Incidentally, there is a recombination difference between human sexes as well. Women show higher recombinant frequencies for the same loci than do men.

With the use of a reiteration of the preceding recombination-based techniques, maps have been produced of thousands of genes for which variant (mutant) phenotypes have been identified. A simple illustrative example from tomato is

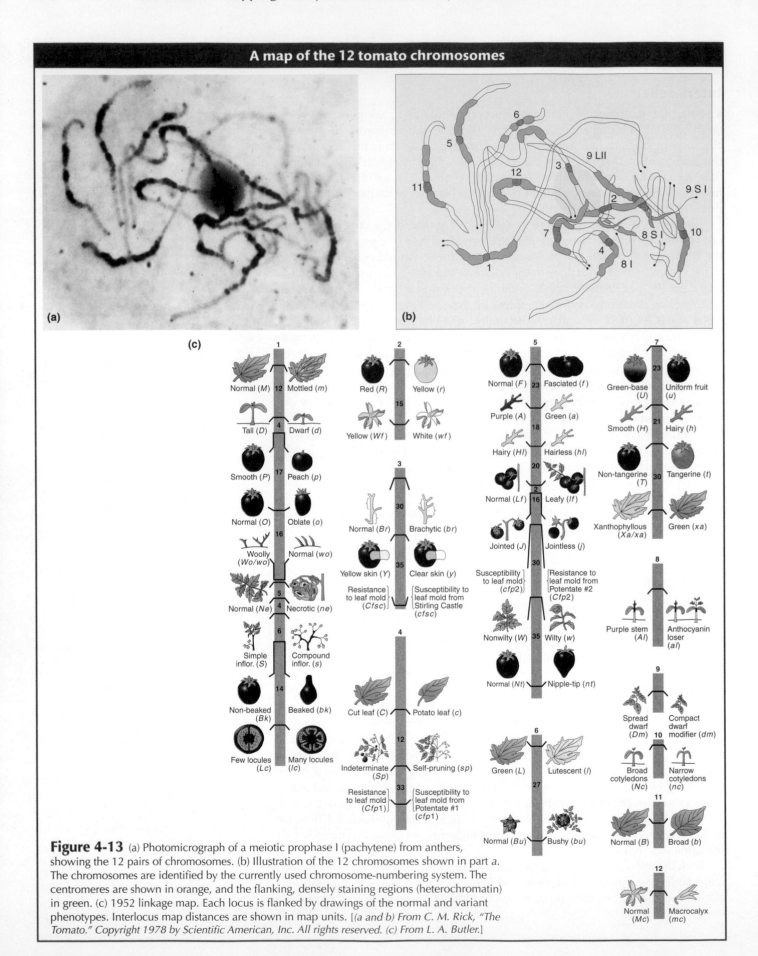

Figure 4-13 (a) Photomicrograph of a meiotic prophase I (pachytene) from anthers, showing the 12 pairs of chromosomes. (b) Illustration of the 12 chromosomes shown in part a. The chromosomes are identified by the currently used chromosome-numbering system. The centromeres are shown in orange, and the flanking, densely staining regions (heterochromatin) in green. (c) 1952 linkage map. Each locus is flanked by drawings of the normal and variant phenotypes. Interlocus map distances are shown in map units. [*(a and b) From C. M. Rick, "The Tomato." Copyright 1978 by Scientific American, Inc. All rights reserved. (c) From L. A. Butler.*]

Figure 4-14

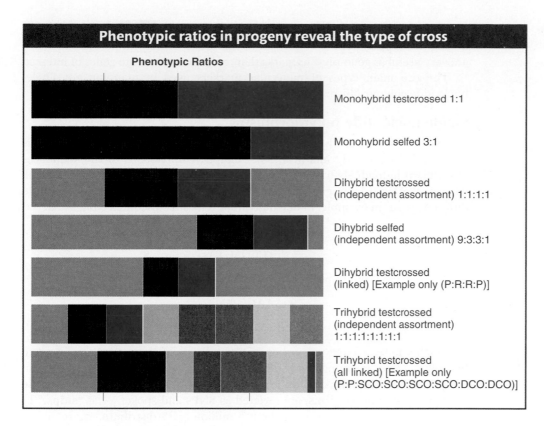

Phenotypic ratios in progeny reveal the type of cross

Phenotypic Ratios

Monohybrid testcrossed 1:1

Monohybrid selfed 3:1

Dihybrid testcrossed
(independent assortment) 1:1:1:1

Dihybrid selfed
(independent assortment) 9:3:3:1

Dihybrid testcrossed
(linked) [Example only (P:R:R:P)]

Trihybrid testcrossed
(independent assortment)
1:1:1:1:1:1:1:1

Trihybrid testcrossed
(all linked) [Example only
(P:P:SCO:SCO:SCO:SCO:DCO:DCO)]

shown in Figure 4-13. The chromosomes are shown as they appear under the microscope, together with chromosome maps based on linkage analysis of various allelic pairs shown with their phenotypes.

Using ratios as diagnostics

The analysis of ratios is one of the pillars of genetics. In the text so far, we have encountered many different ratios whose derivations are spread out over several chapters. Because recognizing ratios and using them in diagnosis of the genetic system under study are part of everyday genetics, let's review the main ratios that we have covered so far. They are shown in Figure 4-14. You can read the ratios from the relative widths of the colored boxes in a row. Figure 4-14 deals with selfs and testcrosses of monohybrids, dihybrids (with independent assortment and linkage), and trihybrids (also with independent assortment and linkage of all genes). One situation not represented is a trihybrid in which only two of the three genes are linked; as an exercise, you might like to deduce the general pattern that would have to be included in such a diagram from this situation. Note that, in regard to linkage, the sizes of the classes depend on map distances. A geneticist deduces unknown genetic states in something like the following way: "a 9 : 3 : 3 : 1 ratio tells me that this ratio was very likely produced by a selfed dihybrid in which the genes are on different chromosomes."

4.3 Mapping with Molecular Markers

So far in this chapter we have mapped gene loci using RF values by counting visible phenotypes produced by the various alleles involved. However, there are also differences in the DNA between two chromosomes that do not produce visibly different phenotypes, either because these DNA differences are not located in genes or they are located in genes but do not alter the product protein. Such sequence differences can be thought of as molecular alleles or **molecular**

markers. Their loci can be mapped by RF values in the same way as alleles producing visible phenotypes. Molecular markers are extremely numerous and hence are very useful as genomic landmarks that can be used to locate genes of interest.

The two main types of molecular markers used in mapping are single-nucleotide polymorphisms and simple-sequence-length polymorphisms.

Single nucleotide polymorphisms

Sequencing has shown that, as expected, the genomic sequences of individuals in a species are mostly identical. For example, comparisons of the sequences of different individuals have revealed that we are about 99.9 percent identical. Almost all of the 0.1 percent difference turns out to be based on single-nucleotide differences. As an example, in one individual, a localized sequence might be

....AA**G**GCTCAT....

....TT**C**CGAGTA....

and, in another, it might be

....AA**A**GCTCAT....

....TT**T**CGAGTA....

Furthermore, a large proportion of these localized sequences are found to be polymorphic, meaning that both molecular "alleles" are quite common in the population. Overall, such differences between individuals are called **single nucleotide polymorphisms**, abbreviated as **SNPs** and spoken of as "**snips.**" In humans, there are thought to be about 3 million SNPs distributed more or less randomly at a frequency of 1 in every 300 to 1000 bases,

Some of these SNPs lie within genes; many do not. In Chapter 2, we saw cases where the change in a single nucleotide pair could produce a new allele, causing a mutant phenotype. The two nucleotide pairs, wild type and mutant, are examples of a SNP. Most SNPs, though, do not produce different phenotypes, either because they do not lie in a gene or because they lie in a gene but both versions of the gene produce the same protein product.

There are two ways to detect an SNP. The first is to sequence a segment of DNA in homologous chromosomes and compare the homologous segments to spot differences. A second way is possible in the case of SNPs located at a restriction enzyme's target site: these SNPs are **restriction fragment length polymorphisms (RFLPs).** In such cases, there will be two RFLP "alleles," or morphs, one of which has the restriction enzyme target and the other of which does not. The restriction enzyme will cut the DNA at the SNP containing the target and ignore the other SNP. The SNPs are then detected as different bands on an electrophoretic gel. RFLP sites can be between or within genes.

Simple sequence length polymorphisms

One of the surprises from molecular genomic analysis is that most genomes contain a great deal of repetitive DNA. Furthermore, there are many types of repetitive DNA. At one end of the spectrum are adjacent multiple repeats of short, simple DNA sequences. The origin of these repeats is not clear, but the feature that makes them useful is that, in different individuals, there are often different numbers of copies. Hence, these repeats are called **simple sequence length polymorphisms (SSLPs).** They are also sometimes called **variable number tandem repeats** or **VNTRs.**

SSLPs commonly have multiple alleles; as many as 15 alleles have been found for an SSLP locus. As a consequence, sometimes 4 alleles (2 from each parent) can be tracked in a pedigree. Two types of SSLPs are useful in mapping and other genome analysis: minisatellite and microsatellite markers. (The word *satellite* in this connection refers to the observation that, when genomic DNA is isolated and fractionated with the use of physical techniques, the repetitive sequences often form a fraction that is physically separate from the rest; that is, it is a satellite fraction in the sense that it is apart from the bulk.)

Minisatellite markers A **minisatellite marker** is based on variation in the number of tandem repeats of a repeating unit from 15 to 100 nucleotides long. In humans, the total length of the unit is from 1 to 5 kb. Minisatellite loci having the same repeating unit but different numbers of repeats are dispersed throughout the genome.

Microsatellite markers A **microsatellite marker** is based on variable numbers of tandem repeats of an even simpler sequence, generally a small number of nucleotides such as a dinucleotide. The most common type is a repeat of CA and its complement GT, as in the following example:

5′ C-A-C-A-C-A-C-A-C-A-C-A-C-A-C-A 3′

3′ G-T-G-T-G-T-G-T-G-T-G-T-G-T-G-T 5′

Detecting simple sequence length polymorphisms

Simple sequence length polymorphisms are detected by taking advantage of the fact that homologous regions bearing different numbers of tandem repeats will be of different lengths. A commonly used procedure for getting at these differences is to use flanking regions as primers in a PCR analysis (see Chapters 1 and 10). PCR replicates the DNA sequences until they are available in enough bulk for further analysis. The different lengths of the amplified PCR products can be detected by the different mobilities of the sequences on an electrophoretic gel. In the case of minisatellites, the patterns produced on the gel are sometimes called **DNA fingerprints.** (These fingerprints are highly individualistic and, hence, have great value in forensics, as detailed in Chapter 18.)

Recombination analysis using molecular markers

When we map the position of a gene whose phenotypes are determined by a single nucleotide difference, we are effectively mapping an SNP. The same technique used to map gene loci can also be used to map SNPs that do not determine a phenotype.

Suppose an individual has a GC base pair at position, say, 5658 on the DNA of one chromosome and an AT at position 5658 on the other chromosome. Such an individual is a molecular heterozygote ("AT/GC") for that DNA position. This fact is useful in mapping because a molecular heterozygote ("AT/GC") can be mapped just like a phenotypic heterozygote A/a. The locus of a molecular heterozygote can be inserted into a chromosomal map by analyzing recombination frequency in exactly the same way as the locus of heterozygous "phenotypic" alleles is inserted. This principle holds even though the variation is usually a silent difference (perhaps not in a gene).

Acting as important "milestones" on the map, molecular markers are useful in orienting the researcher in a quest to find a gene of interest. To understand this point, consider real milestones: they are of little interest in themselves, but are very useful in telling you how close you are to your destination. In a specific genetic example, let's assume that we want to know the map position of a disease gene in mice, perhaps as a way of zeroing in on its DNA sequence. We carry out a number of crosses. In each instance, we cross an individual carrying the disease gene with an individual carrying one of a range of different molecular markers whose map positions are already known. Using PCR, parents and progeny are scored for molecular markers of known map position and then recombination analysis is performed to see if the gene of interest is linked to any of them. The result of these crosses might reveal that the disease gene is 2 map units from one of these markers, which we will call M. The procedure has thus given us an approximate location for the disease gene on the chromosome. The location of the gene for the human disease cystic fibrosis was originally discovered through its linkage to molecular markers known to be located on chromosome 7. This discovery led to the isolation and sequencing of the gene, resulting in the further discovery that it encodes the protein now called *cystic fibrosis transmembrane conductance regulator* (CFTR). The gene for Huntington's disease was also located in this way, leading to the discovery that it encodes the muscle protein *huntingtin*.

The experimental procedure for a hypothetical example might be as follows. Let A and a be the disease-gene alleles and M1 and M2 be alleles of a specific molecular-marker locus. Assume that the cross is $A/a \cdot M1/M2 \times a/a \cdot M1/M1$, a kind of testcross. Progeny would be first scored for the A and a phenotypes, and then DNA would be extracted from each individual and sequenced or otherwise assessed to determine the molecular alleles. Assume that we obtain the following results:

$A/a \cdot$ M1/M1 49 percent	$A/a \cdot$ M2/M1 1 percent
$a/a \cdot$ M2/M1 49 percent	$a/a \cdot$ M1/M1 1 percent

These results tell us that the testcross must have been in the following conformation:

$$A \text{ M1}/a \text{ M2} \times a \text{ M1}/a \text{ M1}$$

and the two progeny genotypes on the right in the list must be recombinants, giving a map distance of 2 map units between the A/a locus and the molecular

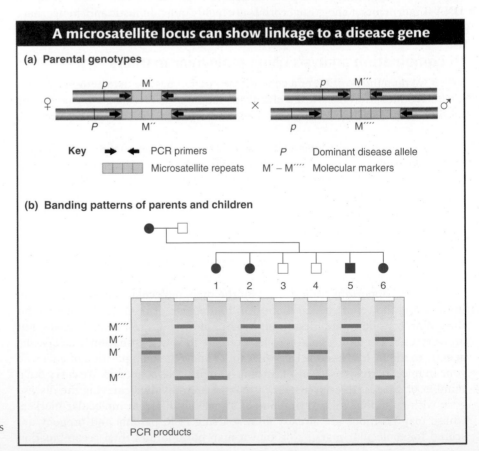

Figure 4-15 A PCR banding pattern is shown for a family with six children, and this pattern is interpreted at the top of the illustration with the use of four different-size microsatellite "alleles," M′ through M′′′′. One of these markers (M′′) is probably linked in cis configuration to the disease allele P. (**Note:** This mating also is not a testcross, yet is informative about linkage.)

locus M1/M2. Hence, we now know the general location of the gene in the genome and can narrow its location down with more finely scaled approaches. In addition, different molecular markers can be mapped to each other, creating a map that can act like a series of stepping-stones on the way to some gene with an interesting phenotype.

Although mapping molecular markers with the use of what are effectively testcrosses is the simplest type of informative analysis, in many analyses (such as those in humans) the molecular markers are not in the form of a testcross. However, because each molecular allele has its own signature, detectable even in heterozygotes, even non-testcrosses are often informative because they enable the detection of recombinants and nonrecombinants. Such an analysis is diagrammed in Figure 4-15.

Figure 4-16 contains some real data showing how molecular markers can flesh out a map of a human chromosome. You can see that the number of mapped molecular markers greatly exceeds the number of mapped genes with mutant phenotypes. Note that SNPs, because of their even higher density, cannot be represented on a whole-chromosome map such as that in Figure 4-16, inasmuch as there would be thousands of them. One centimorgan (1 map unit) of human DNA is a huge segment, estimated as 1 megabase (1 Mb = 1 million base pairs, or 1000 kb). Hence, you can see the need for closely packed molecular markers for a fine-scale analysis that resolves smaller distances. Note that the DNA equivalent of 1 map unit varies a lot between species; for example, in the malarial parasite *Plasmodium falciparium*, 1 map unit = 17 kb.

> **Message** Loci of any DNA heterozygosity can be mapped and used as molecular chromosome markers or milestones.

Figure 4-16 The diagram shows the distribution of all genetic differences that had been mapped to chromosome 1 at the time at which this diagram was drawn. Some markers are genes of known phenotype (their numbers are shaded in green), but most are polymorphic DNA markers (the numbers shaded in mauve and blue represent two different classes of molecular markers). A linkage map displaying a well-spaced-out set of these markers, based on recombinant frequency analyses of the type described in this chapter, is in the center of the illustration. Map distances are shown in centimorgans (cM). At a total length of 356 cM, chromosome 1 is the longest human chromosome. Some markers have also been localized on the chromosome 1 cytogenetic map (right-hand map, called an idiogram), by using techniques described later in this chapter. Having common landmark markers on the different genetic maps permits the locations of other genes and molecular markers to be estimated on each map. Most of the markers shown on the linkage map are molecular, but several genes (highlighted in light green) also are included. [*B. R. Jasney et al., Science, September 30, 1994.*]

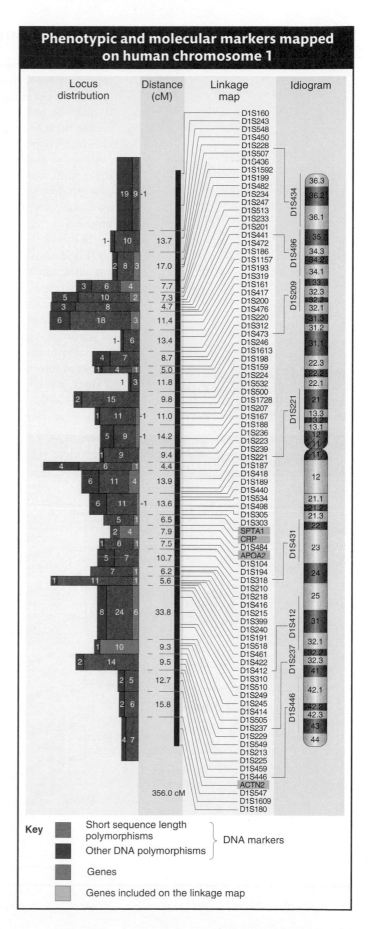

4.4 Centromere Mapping with Linear Tetrads

Centromeres are not genes, but they are regions of DNA on which the orderly reproduction of living organisms absolutely depends and are therefore of great interest in genetics. In most eukaryotes, recombination analysis cannot be used to map the loci of centromeres, because they show no heterozygosity that can enable them to be used as markers. However, in the fungi that produce linear tetrads (see Chapter 3, page 96), centromeres *can* be mapped. We will use the fungus *Neurospora* as an example. Recall that, in fungi such as *Neurospora* (a haploid), the meiotic divisions take place along the long axis of the ascus, and so each meiocyte produces a linear array of eight ascospores, called an **octad.** These eight ascospores constitute the four products of meiosis (a tetrad) plus a postmeiotic mitosis.

In its simplest form, centromere mapping considers a gene locus and asks how far this locus is from its centromere. The method is based on the fact that a different pattern of alleles will appear in a linear tetrad or octad that arises from a meiosis with a crossover between a gene and its centromere. Consider a cross between two individuals, each having a different allele at a locus (say, $A \times a$). Mendel's law of equal segregation dictates that, in an octad, there will always be four ascospores of genotype A and four of a, but how will they be arranged? If there has been no crossover in the region between A/a and the centromere, there will two adjacent blocks of four ascospores in the linear octad (see Figure 3-10, page 97). However, if there has been a crossover in that region, there will be one of four different patterns in the octad, each pattern showing *blocks of two adjacent identical alleles.* Some data from an actual cross of $A \times a$ are shown in the following table.

Figure 4-17 *A* and *a* segregate into separate nuclei at the second meiotic division when there is a crossover between the centromere and the *A* locus.

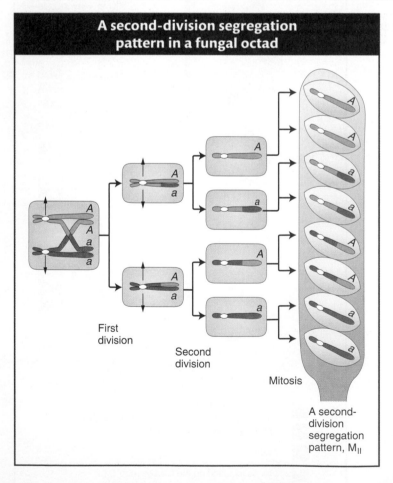

A second-division segregation pattern in a fungal octad

First division

Second division

Mitosis

A second-division segregation pattern, M$_{II}$

		Octads			
A	a	A	a	A	a
A	a	A	a	A	a
A	a	a	A	a	A
A	a	a	A	a	A
a	A	A	a	a	A
a	A	A	a	a	A
a	A	a	A	A	a
a	A	a	A	A	a
126	132	9	11	10	12

Total $= 300$

The first two columns on the left are from meioses with *no* crossover in the region between the A locus and the centromere. The patterns for these columns are called **first-division segregation patterns (M$_I$ patterns)** because the two different alleles segregate into the two daughter nuclei at the first division of meiosis. The other four columns are all from meiocytes *with* a crossover. These patterns are called **second-division segregation patterns (M$_{II}$)** because, as a result of crossing over in the centromere-to-locus region, the A and a alleles are still together in the nuclei at the end of the first division of meiosis (Figure 4-17). There has been no first-division segregation. However, the second meiotic division does segregate the A and a alleles into separate nuclei. Figure 4-17 shows how one of these M$_{II}$ patterns

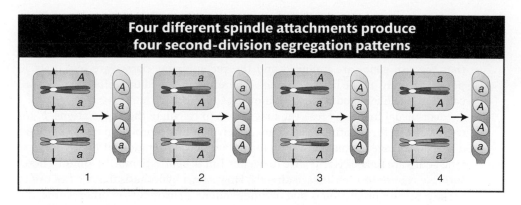

Four different spindle attachments produce four second-division segregation patterns

1   2   3   4

Figure 4-18 In the second meiotic division, the centromeres attach to the spindle at random, producing the four arrangements shown. The four arrangements are equally frequent.

is produced. The other patterns are produced similarly; the difference is that the chromatids move in different directions at the second division (Figure 4-18).

You can see that the frequency of octads with an M_{II} pattern should be proportional to the size of the centromere–A/a region and could be used as a measure of the size of that region. In our example, the M_{II} frequency is $42/300 = 14$ percent. Does this percentage mean that the mating-type locus is 14 map units from the centromere? The answer is no, but this value can be used to calculate the number of map units. The 14 percent value is a percentage of *meioses*, which is not the way that map units are defined. Map units are defined as the percentage of recombinant *chromatids* issuing from meiosis. Because a crossover in any meiosis results in only 50 percent recombinant chromatids (four out of eight; see Figure 4-17), we must divide the 14 percent by 2 to convert the M_{II} frequency (a frequency of *meioses*) into map units (a frequency of recombinant *chromatids*). Hence, this region must be 7 map units in length, and this measurement can be introduced into the map of that chromosome.

4.5 Using the Chi-Square Test for Testing Linkage Analysis

In linkage analysis, the following question often arises: Are these two genes linked? Sometimes the answer is obvious because the recombinant frequency is substantially less than 50 percent, and sometimes not. But, in either situation, the application of an objective statistical test that can support or not support intuition is helpful. The χ^2 test, which we first encountered in Chapter 3, provides a useful way of deciding if two genes are linked. How is the χ^2 test applied to linkage? As discussed earlier in this chapter, we can infer that two genes are linked on the same chromosome if the RF is less than 50 percent. But what about values close to 50 percent?

Let's test a specific set of data for linkage by using χ^2 analysis. Assume that we have crossed pure-breeding parents of genotypes $A/A \cdot B/B$ and $a/a \cdot b/b$ and obtained a dihybrid $A/a \cdot B/b$, which we testcross with $a/a \cdot b/b$. A total of 500 progeny are classified as follows (written as gametes from the dihybrid):

142	$A \cdot B$	parental
133	$a \cdot b$	parental
113	$A \cdot b$	recombinant
112	$a \cdot B$	recombinant

Total: 500

From these data, the recombinant frequency is $225/500 = 45$ percent. On the face of it, this outcome seems as if it is due to linkage because the RF is less than the 50 percent expected from independent assortment. However, it is possible that the two genes are unlinked and the recombinant classes are less than 50 percent merely on the basis of chance. Therefore, we need to perform a χ^2 test to calculate the likelihood of this result owing to chance.

As is usual for the χ^2 test, the first step is to calculate the expectations, E, for each class. In formulating E, we immediately run into a problem: we do not have a precise linkage distance to use to formulate our expectations if the hypothesis "the genes are linked" is true. Are the genes 1 m.u. apart? 10 m.u.? 45 m.u.? Therefore, we cannot test for linkage directly. However, a hypothesis that we *can* use to make a precise prediction is that of independent assortment—in other words, the *absence* of linkage. If the observed results cause us to reject the hypothesis of *no linkage*, then we can infer linkage. This type of hypothesis, called a **null hypothesis,** is generally useful in χ^2 analysis because it provides a precise experimental prediction that can be tested.

How can we calculate gametic E values for the hypothesis that there is no linkage? One way might be to make a simple prediction based on Mendel's first and second laws, as follows:

$$\underline{E \text{ values}}$$

$$0.5\ A \begin{cases} 0.5\ B \longrightarrow 0.25\ A\ ;\ B \\ \\ 0.5\ b \longrightarrow 0.25\ A\ ;\ b \end{cases}$$

$$0.5\ a \begin{cases} 0.5\ B \longrightarrow 0.25\ a\ ;\ B \\ \\ 0.5\ b \longrightarrow 0.25\ a\ ;\ b \end{cases}$$

Hence, we might assert that, if the allele pairs of the dihybrid are assorting independently, there should be a $1:1:1:1$ ratio of gametic types. Therefore, using 1/4 of 500, or 125, as the expected proportion of each gametic class seems reasonable. However, note that the $1:1:1:1$ ratio is expected only if all genotypes are equally viable. It is often the case that genotypes are *not* equally viable, because individuals that carry certain alleles do not survive to adulthood. Therefore, instead of allele ratios of $0.5:0.5$, used earlier, we might see, for example, ratios of $0.6\ A:0.4\ a$ or $0.45\ B:0.55\ b$. We should use these ratios in our predictions of independence.

Let's arrange the observed genotypic classes in a grid to reveal the allele proportions more clearly.

OBSERVED VALUES

		Segregation of A and a		
		A	a	Total
Segregation of	B	142	112	254
B and b	b	113	133	246
	Total	255	245	500

We see that the allele proportions are $255/500$ for A, $245/500$ for a, $254/500$ for B, and $246/500$ for b. Now we calculate the values expected under independent assortment simply by multiplying these allelic proportions. For example, to find the expected number of $A\ B$ genotypes in the sample if the two ratios are combined randomly, we simply multiply the following terms:

Expected, or E, value for $A\,B = (255/500) \times (254/500) \times 500 = 129.54$

With the use of this approach, the entire grid of E values can be completed, as follows:

EXPECTED VALUES

		Segregation of A and a		
		A	a	Total
Segregation of	B	129.54	124.46	254
B and b	b	125.46	120.56	246
	Total	255	245	500

The value of χ^2 is calculated as follows (in which O is the observed value):

Genotype	O	E	$(O - E)^2/E$
AB	142	129.54	1.19
ab	133	120.56	1.29
Ab	113	125.46	1.24
aB	112	124.46	1.25
Total (which equals the χ^2 value) $= 4.97$			

The obtained value of χ^2 (4.97) is used to find a corresponding probability value, p, by using the χ^2 table (see Table 3-1, page 91). First, we need to figure out the degrees of freedom in order to choose the correct row in the table. Generally, in a statistical test, the number of degrees of freedom is the number of nondependent values. Working through the following "thought experiment" will show what this statement means in the present application. In 2×2 grids of data (of the sort used earlier), because the column and row totals are given from the experimental results, specifying any one value within the grid automatically dictates the other three values. Hence, there is only one nondependent value and, therefore, only one degree of freedom. A rule of thumb useful for larger grids is that the number of degrees of freedom is equal to the number of classes represented in the rows minus one multiplied by the number of classes represented in the columns minus one. Applying that rule in the present example gives

$$\mathrm{df} = (2 - 1) \times (2 - 1) = 1$$

Therefore, using Table 3-1, we look along the row corresponding to one degree of freedom until we locate our χ^2 value of 4.97. Not all values of χ^2 are shown in Table 3-1, but 4.97 is close to the value 5.021. Hence, the corresponding probability value is very close to 0.025, or 2.5 percent. This p value is the probability value that we seek, that of obtaining a deviation from expectations this large or larger. Because this probability is less than 5 percent, the hypothesis of independent assortment must be rejected. Thus, having rejected the hypothesis of no linkage, we are left with the inference that indeed the loci are probably linked.

4.6 Accounting for Unseen Multiple Crossovers

In the discussion of the three-point testcross, some parental (nonrecombinant) chromatids resulted from *double* crossovers. These crossovers initially could not be counted in the recombinant frequency, skewing the results. This situation leads to the worrisome notion that *all* map distances based on recombinant frequency might be underestimations of physical distances because undetected multiple crossovers might have occurred, some of whose products would not be recombinant. Several creative mathematical approaches have been designed to get around the multiple-crossover problem. We will look at two methods. First, we examine a method originally worked out by J. B. S. Haldane in the early years of genetics.

A mapping function

The approach worked out by Haldane was to devise a **mapping function,** a formula that relates an observed recombinant-frequency value to a map distance corrected for multiple crossovers. The approach works by relating RF to the mean number of crossovers, m, that must have taken place in that chromosomal segment per meiosis and then deducing what map distance this m value *should* have produced.

To find the relation of RF to m, we must first think about outcomes of the various crossover possibilities. In any chromosomal region, we might expect meioses with 0, 1, 2, 3, 4, or more crossovers. Surprisingly, the only class that is really crucial is the zero class. To see why, consider the following. It is a curious but nonintuitive fact that *any number* of crossovers produces a frequency of 50 percent recombinants *within those meioses.* Figure 4-19 proves this statement for single and double crossovers as examples, but it is true for any number of crossovers. Hence, the true determinant of RF is the relative sizes of the classes with no crossovers (the zero class) compared with the classes with any nonzero number of crossovers.

Now the task is to calculate the size of the zero class. The occurrence of crossovers in a specific chromosomal region is well described by a statistical distribution called the **Poisson distribution.** The Poisson formula in general describes the distribution of "successes" in samples when the average probability of successes is low. An illustrative example is to dip a child's net into a pond of fish: most dips will produce no fish, a smaller proportion will produce one fish, an even smaller proportion two, and so on. This analogy can be directly applied to a chromosomal region, which will have 0, 1, 2, and so forth, crossover "successes" in different meioses. The Poisson formula, given here, will tell us the proportion of the classes with different numbers of crossovers.

$$f_i = (e^{-m}m^i)/i!$$

The terms in the formula have the following meanings:

e = the base of natural logarithms (approximately 2.7)

m = the mean number of successes in a defined sample size

i = the actual number of successes in a sample of that size

f_i = the frequency of samples with i successes in them

$!$ = the factorial symbol (for example, $5! = 5 \times 4 \times 3 \times 2 \times 1$)

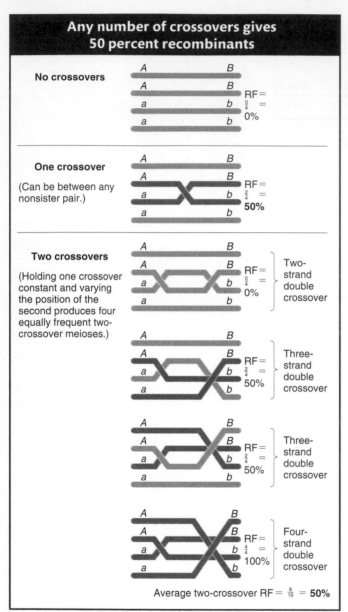

Figure 4-19 Demonstration that the average RF is 50 percent for meioses in which the number of crossovers is not zero. Recombinant chromatids are brown. Two-strand double crossovers produce all parental types; so all the chromatids are orange. Note that all crossovers are between nonsister chromatids. Try the triple crossover class yourself.

The Poisson distribution tells us that the frequency of the $i = 0$ class (the key one) is

$$e^{-m}\frac{m^0}{0!}$$

Because m^0 and $0!$ both equal 1, the formula reduces to e^{-m}.

Now we can write a function that relates RF to m. The frequency of the class with any nonzero number of crossovers will be $1 - e^{-m}$, and, in these meioses, 50 percent (1/2) of the products will be recombinant; so

$$\text{RF} = \tfrac{1}{2}(1 - e^{-m})$$

and this formula is the mapping function that we have been seeking.

Let's look at an example in which RF is converted into a map distance corrected for multiple crossovers. Assume that, in one testcross, we obtain an RF value of 27.5 percent (0.275). Plugging this into the function allows us to solve for m:

$$0.275 = \tfrac{1}{2}(1 - e^{-m})$$

so

$$e^{-m} = 1 - (2 \times 0.275) = 0.45$$

By using a calculator, we can deduce that $m = 0.8$. That is, on average, there are 0.8 crossovers per meiosis in that chromosomal region.

The final step is to convert this measure of crossover frequency to give a "corrected" map distance. All that we have to do to convert into corrected map units is to multiply the calculated average crossover frequency by 50, because, on average, a crossover produces a recombinant frequency of 50 percent. Hence, in the preceding numerical example, the m value of 0.8 can be converted into a corrected recombinant fraction of $0.8 \times 50 = 40$ corrected map units. We see that, indeed, this value is substantially larger than the 27.5 map units that we would have deduced from the observed RF.

Note that the mapping function neatly explains why the maximum RF value for linked genes is 50 percent. As m gets very large, e^{-m} tends to zero and the RF tends to 1/2 or 50 percent.

The Perkins formula

For fungi and other tetrad-producing organisms, there is another way of compensating for multiple crossovers—specifically, double crossovers (the most common type expected). In tetrad analysis of "dihybrids" generally, only three types of tetrads are possible, when classified on the basis of the presence of parental and recombinant genotypes in the products. From a cross $AB \times ab$, they are

Parental ditype (PD)	Tetratype (T)	Nonparental ditype (NPD)
$A \cdot B$	$A \cdot B$	$A \cdot b$
$A \cdot B$	$A \cdot b$	$A \cdot b$
$a \cdot b$	$a \cdot B$	$a \cdot B$
$a \cdot b$	$a \cdot b$	$a \cdot B$

The recombinant genotypes are shown in red. If the genes are linked, a simple approach to mapping their distance apart might be to use the following formula:

$$\text{map distance} = \text{RF} = 100(\text{NPD} + \tfrac{1}{2}\,\text{T})$$

because this formula gives the percentage of all recombinants. However, in the 1960s, David Perkins developed a formula that compensates for the effects of double crossovers. The Perkins formula thus provides a more accurate estimate of map distance:

$$\text{corrected map distance} = 50(\text{T} + 6\,\text{NPD})$$

We will not go through the derivation of this formula other than to say that it is based on the totals of the PD, T, and NPD classes expected from meioses with 0, 1, and 2 crossovers (it assumes that higher numbers are vanishingly rare). Let's look at an example of its use. We assume that, in our hypothetical cross of $A\ B \times a\ b$, the observed frequencies of the tetrad classes are 0.56 PD, 0.41 T, and 0.03 NPD. By using the Perkins formula, we find the corrected map distance between the a and b loci to be

$$50[0.41 + (6 \times 0.03)] = 50(0.59) = 29.5 \text{ m.u.}$$

Let us compare this value with the uncorrected value obtained directly from the RF. By using the same data, we find:

$$\begin{aligned}
\text{uncorrected map distance} &= 100(\tfrac{1}{2}\,\text{T} + \text{NPD}) \\
&= 100(0.205 + 0.03) \\
&= 23.5 \text{ m.u.}
\end{aligned}$$

This distance is 6 m.u. less than the estimate that we obtained by using the Perkins formula because we did not correct for double crossovers.

As an aside, what PD, NPD, and T values are expected when dealing with *unlinked* genes? The sizes of the PD and NPD classes will be equal as a result of independent assortment. The T class can be produced only from a crossover between either of the two loci and their respective centromeres, and, therefore, the size of the T class will depend on the total size of the two regions lying between locus and centromere. However, the formula $\tfrac{1}{2}\,\text{T} + \text{NPD}$ should always yield 0.50, reflecting independent assortment.

> **Message** The inherent tendency of multiple crossovers to lead to an underestimation of map distance can be circumvented by the use of map functions (in any organism) and by the Perkins formula (in tetrad-producing organisms such as fungi).

4.7 Using Recombination-Based Maps in Conjunction with Physical Maps

Recombination maps have been the main topic of this chapter. They show the loci of *genes for which mutant alleles (and their mutant phenotypes) have been found.* The positions of these loci on a map is determined on the basis of the frequency of recombinants at meiosis. The frequency of recombinants is assumed to be proportional to the distance apart of two loci on the chromosome; hence, recombinant frequency becomes the mapping unit. Such recombination-based mapping of genes with known mutant phenotypes has been done for nearly a century. We have seen how sites of molecular heterozygosity (unassociated with mutant phenotypes) also can be incorporated into such recombination maps. Like any heterozygous site, these molecular markers are mapped by recombination and then used to navigate toward a gene of biological interest. We make the perfectly reasonable assumption that a recombination map represents the arrangement of genes on chromosomes, but, as stated earlier, these maps are really hypothetical constructs. In contrast, physical maps are as close to the real genome map as science can get.

The topic of **physical maps** will be examined more closely in Chapter 13, but we can foreshadow it here. A physical map is simply a map of the actual genomic DNA, a very long DNA nucleotide sequence, showing where genes are, how big they are, what is between them, and other landmarks of interest. The units of distance on a physical map are numbers of DNA bases; for convenience the kilobase is the preferred unit. The complete sequence of a DNA molecule is obtained by sequencing large numbers of small genomic fragments and then assembling them into one whole sequence. The sequence is then scanned by a computer, programmed to look for genelike segments recognized by characteristic base sequences including known signal sequences for the initiation and termination of transcription. When the computer's program finds a gene, it compares its sequence with the public database of other sequenced genes for which functions have been discovered in other organisms. In many cases, there is a "hit"; in other words, the sequence closely resembles that of a gene of known function in another species. In such cases, the functions of the two genes also may be similar. The sequence similarity (often close to 100 percent) is explained by the inheritance of the gene from some common ancestor and the general conservation of functional sequences through evolutionary time. Other genes discovered by the computer show no sequence similarity to any gene of known function. Hence, they can be considered "genes in search of a function." In reality, of course, it is the researcher, not the gene, who searches and who must find the function. Sequencing different individual members of a population also can yield sites of molecular heterozygosity, which, just as they do in recombination maps, act as orientation markers on the physical map.

Because physical maps are now available for most of the main genetic model organisms, is there really any need for recombination maps? Could they be considered outmoded? The answer is that both maps are used in conjunction with each other to "triangulate" in determining gene function, a principle illustrated earlier by the London maps. The general approach is illustrated in Figure 4-20, which shows a physical map and a recombination map of the same region of a genome. Both maps contain genes and molecular markers. In the lower part of Figure 4-20, we see a section of a recombination-based map, with positions of genes for which mutant phenotypes have been found and mapped. Not all the genes in that segment are included. For some of these genes, a function may have been discovered on the basis of biochemical or other studies of mutant strains; genes for proteins A and B are examples. The gene in the middle is a "gene of interest" that a researcher has found

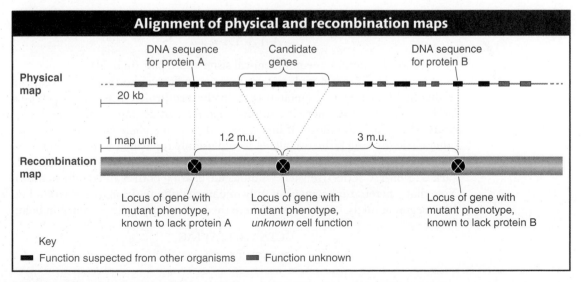

Figure 4-20 Comparison of relative positions on physical and recombination maps can connect phenotype with an unknown gene function.

to affect the aspect of development being studied. To determine its function, the physical map can be useful. The genes in the physical map that are in the general region of the gene of interest on the recombination map become *candidate genes,* any one of which could be the gene of interest. Further studies are needed to narrow the choice to one. If that single case is a gene whose function is known for other organisms, then a function for the gene of interest is suggested. In this way, the phenotype mapped on the recombination map can be tied to a function deduced from the physical map. Molecular markers on both maps (not shown in Figure 4-20) can be aligned to help in the zeroing-in process. Hence, we see that both maps contain elements of function: the physical map shows a gene's possible action at the cellular level, whereas the recombination map contains information related to the effect of the gene at the phenotypic level. At some stage, the two have to be melded to understand the gene's contribution to the development of the organism.

There are several other genetic-mapping techniques, some of which we will encounter in Chapters 5, 16, and 18.

> **Message** The union of recombination and physical maps can ascribe biochemical function to a gene identified by its mutant phenotype.

4.8 The Molecular Mechanism of Crossing Over

In this chapter we have analyzed the genetic consequences of the cytologically visible process of crossing over without worrying about the mechanism of crossing over. However, crossing over is remarkable in itself as a molecular process: how can two large coiled molecules of DNA exchange segments with a precision so exact that no nucleotides are lost or gained?

Studies on fungal octads gave a clue. Although most octads show the expected 4:4 segregation of alleles such as $4A:4a$, some rare octads show aberrant ratios. There are several types, but as an example we will use 5:3 octads (either $5A:3a$ or $5a:3A$). Two things are peculiar about this ratio. First, there is one too many spores of one allele and one too few of the other. Second, there is a *nonidentical sister spore pair.* Normally, postmeiotic replication gives identical sister-spore pairs as follows: the *A A a a* tetrad becomes

A-A A-A a-a a-a

(the hyphens show sister spores). In contrast, an aberrant $5A:3a$ octad must be

*A-A A-A **A-a** a-a*

In other words, there is one nonidentical sister spore pair (in bold).

The observation of a nonidentical sister-spore pair suggests that the DNA of one of the final four meiotic homologs contains **heteroduplex DNA.** Heteroduplex DNA is DNA in which there is a mismatched nucleotide pair in the gene under study. The logic is as follows. If in a cross of $A \times a$, one allele (*A*) is G:C and the other allele (*a*) is A:T, the two alleles would usually replicate faithfully. However, a heteroduplex, which forms only rarely, would be a mismatched nucleotide pair such G:T or A:C (effectively a DNA molecule bearing both *A* and *a* information). Note that a heteroduplex involves only one nucleotide position: the surrounding DNA segment might be as follows, where the heteroduplex site is shown in bold:

GCTAAT**G**TTATTAG

CGATTA**T**AATAATC

At replication to form an octad, a G:T heteroduplex would pull apart and replicate faithfully, with G bonding to C and A bonding to T. The result would be a nonidentical spore pair of G:C (allele *A*) and A:T (allele *a*).

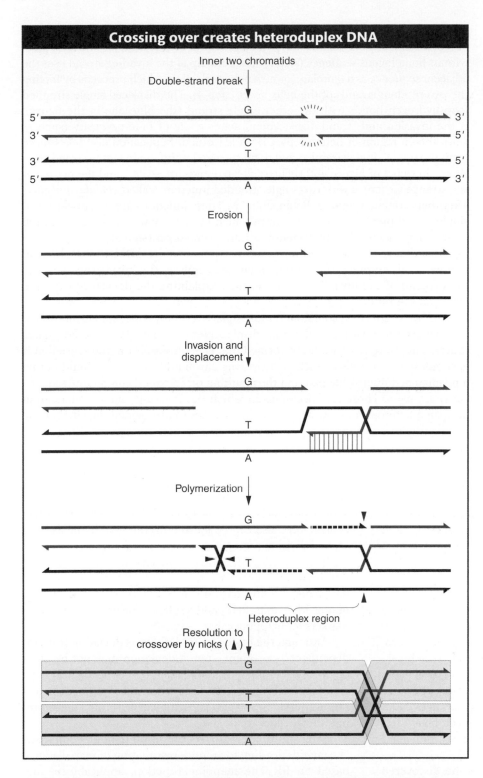

Figure 4-21 A molecular model of crossing over. Only the two chromatids (green and red) participating in the crossover are shown. The 3'-to-5' strand is placed on the inside of both for clarity. The chromatids differ at one site, GC, in one allele (perhaps allele *A*) and AT in the other (perhaps *a*). Only the outcome with mispaired heteroduplex DNA and a crossover are shown. The final crossover products are shaded in yellow and blue.

Nonidentical sister spores (and aberrant octads generally) were found to be statistically correlated with crossing over in the region of the gene concerned, suggesting that crossing over might be based on the formation of heteroduplex DNA.

In the currently accepted model (follow it in Figure 4-21), both the heteroduplex DNA and a crossover are the eventual outcomes of a **double-stranded break** in the DNA of one of the chromatids participating in the crossover. Molecular studies show that broken ends of DNA will promote recombination between different DNAs. In step 1, both chromatids of a pair break in the same location. From the break, DNA is eroded in the 5' strand of each broken end, leaving both

3′ ends single stranded (step 2). One of the single strands "invades" the DNA of the other participating chromatid; that is, it enters the center of the helix and base-pairs with its homologous sequence (step 3). Then the tip of the invading strand uses the adjacent sequence as a template for new polymerization, which proceeds by forcing the two resident strands of the helix apart (step 4). The displaced single-stranded loop hydrogen bonds with the other single strand (the blue one in the figure). If the invasion and strand displacement spans a site of heterozygosity (such as A/a), then a region of heteroduplex DNA is formed. Replication also takes place from the upper single-stranded end to fill the gap left by the invading strand (also shown in step 4 of Figure 4-21). The replicated ends are sealed, and the net result is a strange structure with two single-stranded junctions called *Holliday junctions* after their original proposer, Robin Holliday. These junctions are potential sites of single-strand breakage and reunion; two such events, shown by the darts in the figure, then lead to a complete double-stranded crossover (step 5).

Note that when the invading strand uses the invaded DNA as a replication template, this automatically results in an extra copy of the invaded sequence at the expense of the invading sequence, thus explaining the departure from the expected 4 : 4 ratio.

This same sort of recombination takes place at many different chromosomal sites where the invasion and strand displacement do *not* span a heterozygous mutant site. Here DNA would be formed that is heteroduplex in the sense that it is composed of strands of each participating chromatid, but there would not be a mismatched nucleotide pair and the resulting octad would contain only identical spore pairs. Those rare occasions in which the invasion and polymerization *do* span a heterozygous site are simply lucky cases that provided the clue for the mechanism of crossing over.

Summary

In a dihybrid testcross in *Drosophila,* Thomas Hunt Morgan found a deviation from Mendel's law of independent assortment. He postulated that the two genes were located on the same pair of homologous chromosomes. This relation is called linkage.

Linkage explains why the parental gene combinations stay together but not how the recombinant (nonparental) combinations arise. Morgan postulated that, in meiosis, there may be a physical exchange of chromosome parts by a process now called crossing over. A result of the physical breakage and reunion of chromosome parts, crossing over takes place at the four-chromatid stage of meiosis. Thus, there are two types of meiotic recombination. Recombination by Mendelian independent assortment results in a recombinant frequency of 50 percent. Crossing over results in a recombinant frequency generally less than 50 percent.

As Morgan studied more linked genes, he discovered many different values for recombinant frequency and wondered if these values corresponded to the actual distances between genes on a chromosome. Alfred Sturtevant, a student of Morgan's, developed a method of determining the distance between genes on a linkage map, based on the RF. The easiest way to measure RF is with a testcross of a dihybrid or trihybrid. RF values calculated as percentages can be used as map units to construct a chromosomal map showing the loci of the genes analyzed. In ascomycete fungi, centromeres also can be located on the map by measuring second-division segregation frequencies.

Single nucleotide polymorphisms (SNPs) are single-nucleotide differences in sequences of DNA. Single-sequence length polymorphisms (SSLPs) are differences in the number of repeating units. SNPs and SSLPs can be used as molecular markers for mapping genes.

Although the basic test for linkage is deviation from independent assortment, such a deviation may not be obvious in a testcross, and a statistical test is needed. The χ^2 test, which tells how often observations deviate from expectations purely by chance, is particularly useful in determining whether loci are linked.

Some multiple crossovers can result in nonrecombinant chromatids, leading to an underestimation of map distance based on RF. The mapping function, applicable in any organism, corrects for this tendency. The Perkins formula has the same use in fungal tetrad analysis.

In genetics generally, the recombination-based map of loci conferring mutant phenotypes is used in conjunction with a physical map such as the complete DNA sequence, which shows all the genelike sequences. Knowledge of gene position in both maps enables the melding of cellular function with a gene's effect on phenotype.

The mechanism of crossing over is thought to start with a double-stranded break in one participating chromatid. Erosion leaves the ends single stranded. One single strand invades the double helix of the other participating chroma-tid, leading to the formation of heteroduplex DNA. Gaps are filled by polymerization. The molecular resolution of this structure becomes a full double-stranded crossover at the DNA level.

KEY TERMS

centimorgan (cM) (p. 130)
chromosome map (p. 123)
cis conformation (p. 126)
coefficient of coincidence (c.o.c.) (p. 135)
crossing over (p. 125)
crossover product (p. 125)
DNA fingerprint (p. 139)
double-stranded break (p. 151)
first-division segregation pattern
    (M_I pattern) (p. 142)
genetic map unit (m.u.) (p. 129)
heteroduplex DNA (p. 150)
interference (p. 135)

linkage map (p. 129)
linked genes (p. 123)
locus (p. 123)
mapping function (p. 146)
microsatellite marker (p. 139)
minisatellite marker (p. 139)
molecular marker (p. 137)
null hypothesis (p. 144)
octad (p. 142)
physical map (p. 149)
Poisson distribution (p. 146)
recombinant frequency (RF) (p. 130)
recombination map (p. 123)

restriction fragment length
    polymorphism (RFLP) (p. 138)
second-division segregation pattern
    (M_{II}) (p. 142)
simple sequence length
    polymorphism (SSLP) (p. 138)
single nucleotide polymorphism
    (SNP) (p. 138)
three-factor cross (p. 132)
three-point testcross (p. 132)
trans conformation (p. 126)
variable number tandem repeat
    (VNTR) (p. 138)

SOLVED PROBLEMS

SOLVED PROBLEM 1. A human pedigree shows people affected with the rare nail–patella syndrome (misshapen nails and kneecaps) and gives the ABO blood-group genotype of each person. Both loci concerned are autosomal. Study the pedigree below.

a. Is the nail–patella syndrome a dominant or recessive phenotype? Give reasons to support your answer.

b. Is there evidence of linkage between the nail–patella gene and the gene for ABO blood type, as judged from this pedigree? Why or why not?

c. If there is evidence of linkage, then draw the alleles on the relevant homologs of the grandparents. If there is no evidence of linkage, draw the alleles on two homologous pairs.

d. According to your model, which descendants are recombinants?

e. What is the best estimate of RF?

f. If man III-1 marries a normal woman of blood type O, what is the probability that their first child will be blood type B with nail–patella syndrome?

Solution

a. Nail–patella syndrome is most likely dominant. We are told that it is a rare abnormality, and so the unaffected people marrying into the family are unlikely to carry a presumptive recessive allele for nail–patella syndrome. Let N be the causative allele. Then all people with the syndrome are heterozygotes N/n because all (probably including the grandmother) result from matings with n/n normal people. Notice that the syndrome appears in all three generations—another indication of dominant inheritance.

b. There is evidence of linkage. Notice that most of the affected people—those who carry the N allele—also carry the I^B allele; most likely, these alleles are linked on the same chromosome.

c.
$$\frac{n \qquad i}{n \qquad i} \times \frac{N \qquad I^B}{n \qquad i}$$

(The grandmother must carry both recessive alleles to produce offspring of genotype i/i and n/n.)

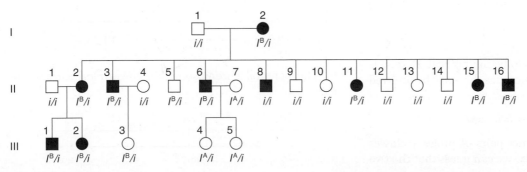

d. Notice that the grandparental mating is equivalent to a testcross; so the recombinants in generation II are

$$\text{II-5}: n\, I^B/n\, i \quad \text{and} \quad \text{II-8}: N\, i/n\, i$$

whereas all others are nonrecombinants, being either $N\, I^B/n\, i$ or $n\, i/n\, i$.

e. Notice that the grandparental cross and the first two crosses in generation II are identical and are testcrosses. Three of the total 16 progeny are recombinant (II-5, II-8, and III-3). The cross of II-6 with II-7 is not a testcross, but the chromosomes donated from II-6 can be deduced to be nonrecombinant. Thus, RF = 3/18, which is 17 percent.

f. (III-1♂) $\dfrac{N \qquad I^B}{n \qquad i} \times \dfrac{n \qquad i}{n \qquad i}$ (normal type O ♀)

↓

Gametes

$$83.0\% \begin{cases} N\, I^B & 41.5\% \quad \longleftarrow \text{nail–patella,} \\ & \qquad\qquad\quad \text{blood type B} \\ n\, i & 41.5\% \end{cases}$$

$$17.0\% \begin{cases} N\, i & 8.5\% \\ n\, I^B & 8.5\% \end{cases}$$

The two parental classes are always equal, and so are the two recombinant classes. Hence, the probability that the first child will have nail–patella syndrome and blood type B is 41.5 percent.

SOLVED PROBLEM 2. The allele *b* gives *Drosophila* flies a black body, and b^+ gives brown, the wild-type phenotype. The allele *wx* of a separate gene gives waxy wings, and wx^+ gives nonwaxy, the wild-type phenotype. The allele *cn* of a third gene gives cinnabar eyes, and cn^+ gives red, the wild-type phenotype. A female heterozygous for these three genes is testcrossed, and 1000 progeny are classified as follows: 5 wild type; 6 black, waxy, cinnabar; 69 waxy, cinnabar; 67 black; 382 cinnabar; 379 black, waxy; 48 waxy; and 44 black, cinnabar. Note that a progeny group may be specified by listing only the mutant phenotypes.

a. Explain these numbers.

b. Draw the alleles in their proper positions on the chromosomes of the triple heterozygote.

c. If appropriate according to your explanation, calculate interference.

Solution

a. A general piece of advice is to be methodical. Here, it is a good idea to write out the genotypes that may be inferred from the phenotypes. The cross is a testcross of type

$$b^+/b \cdot wx^+/wx \cdot cn^+/cn \times b/b \cdot wx/wx \cdot cn/cn$$

Notice that there are distinct pairs of progeny classes in regard to frequency. Already, we can guess that the two largest classes represent parental chromosomes, that the two classes of about 68 represent single crossovers in one region, that the two classes of about 45 represent single crossovers in the other region, and that the two classes of about 5 represent double crossovers. We can write out the progeny as classes derived from the female's gametes, grouped as follows:

$b^+\cdot wx^+\cdot cn$		382
$b \cdot wx \cdot cn^+$		379
$b^+\cdot wx \cdot cn$		69
$b \cdot wx^+\cdot cn^+$		67
$b^+\cdot wx \cdot cn^+$		48
$b \cdot wx^+\cdot cn$		44
$b \cdot wx \cdot cn$		6
$b^+\cdot wx^+\cdot cn^+$		5
		1000

Listing the classes in this way confirms that the pairs of classes are in fact reciprocal genotypes arising from zero, one, or two crossovers.

At first, because we do not know the parents of the triple heterozygous female, it looks as if we cannot apply the definition of recombination in which gametic genotypes are compared with the two parental genotypes that form an individual fly. But, on reflection, the only parental types that make sense in regard to the data presented are $b^+/b^+ \cdot wx^+/wx^+ \cdot cn/cn$ and $b/b \cdot wx/wx \cdot cn^+/cn^+$ because these types are the most common gametic classes.

Now, we can calculate the recombinant frequencies. For *b–wx*,

$$\text{RF} = \frac{69+67+48+44}{1000} = 22.8\%$$

for *b–cn*,

$$\text{RF} = \frac{48+44+6+5}{1000} = 10.3\%$$

and for *wx–cn*,

$$\text{RF} = \frac{69+67+6+5}{1000} = 14.7\%$$

The map is therefore

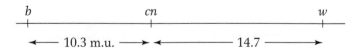

b. The parental chromosomes in the triple heterozygote are

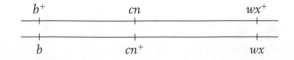

1	2	3	4	5	6	7
$nic^+ \cdot ad$	$nic^+ \cdot ad^+$	$nic^+ \cdot ad^+$	$nic^+ \cdot ad$	$nic^+ \cdot ad$	$nic^+ \cdot ad^+$	$nic^+ \cdot ad^+$
$nic^+ \cdot ad$	$nic^+ \cdot ad^+$	$nic^+ \cdot ad^+$	$nic^+ \cdot ad$	$nic^+ \cdot ad$	$nic^+ \cdot ad^+$	$nic^+ \cdot ad^+$
$nic^+ \cdot ad$	$nic^+ \cdot ad^+$	$nic^+ \cdot ad^+$	$nic \cdot ad$	$nic \cdot ad^+$	$nic \cdot ad$	$nic \cdot ad$
$nic^+ \cdot ad$	$nic^+ \cdot ad^+$	$nic^+ \cdot ad$	$nic \cdot ad$	$nic \cdot ad^+$	$nic \cdot ad$	$nic \cdot ad$
$nic \cdot ad^+$	$nic \cdot ad$	$nic \cdot ad^+$	$nic^+ \cdot ad^+$	$nic^+ \cdot ad$	$nic^+ \cdot ad^+$	$nic^+ \cdot ad$
$nic \cdot ad^+$	$nic \cdot ad$	$nic \cdot ad^+$	$nic^+ \cdot ad^+$	$nic^+ \cdot ad$	$nic^+ \cdot ad^+$	$nic^+ \cdot ad$
$nic \cdot ad^+$	$nic \cdot ad$	$nic \cdot ad$	$nic \cdot ad^+$	$nic \cdot ad^+$	$nic \cdot ad$	$nic \cdot ad^+$
$nic \cdot ad^+$	$nic \cdot ad$	$nic \cdot ad$	$nic \cdot ad^+$	$nic \cdot ad^+$	$nic \cdot ad$	$nic \cdot ad^+$
808	1	90	5	90	1	5

c. The expected number of double recombinants is $0.103 \times 0.147 \times 1000 = 15.141$. The observed number is $6 + 5 = 11$, and so interference can be calculated as

$$I = 1 - (11/15.141) = 1 - 0.726 = 0.274 = 27.4\%$$

SOLVED PROBLEM 3. A cross is made between a haploid strain of *Neurospora* of genotype $nic^+ \ ad$ and another haploid strain of genotype $nic \ ad^+$. From this cross, a total of 1000 linear asci are isolated and categorized as in the table above. Map the ad and nic loci in relation to centromeres and to each other.

Solution

What principles can we draw on to solve this problem? It is a good idea to begin by doing something straightforward, which is to calculate the two locus-to-centromere distances. We do not know if the ad and the nic loci are linked, but we do not need to know. The frequencies of the M_{II} patterns for each locus give the distance from locus to centromere. (We can worry about whether it is the same centromere later.)

Remember that an M_{II} pattern is any pattern that is not two blocks of four. Let's start with the distance between the nic locus and the centromere. All we have to do is add the ascus types 4, 5, 6, and 7, because all of them are M_{II} patterns for the nic locus. The total is $5 + 90 + 1 + 5 = 101$ of 1000, or 10.1 percent. In this chapter, we have seen that, to convert this percentage into map units, we must divide by 2, which gives 5.05 m.u.

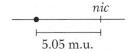

We do the same for the ad locus. Here, the total of the M_{II} patterns is given by types 3, 5, 6, and 7 and is $90 + 90 + 1 + 5 = 186$ of 1000, or 18.6 percent, which is 9.3 m.u.

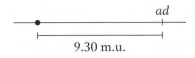

Now we have to put the two together and decide between the following alternatives, all of which are compatible with the preceding locus-to-centromere distances:

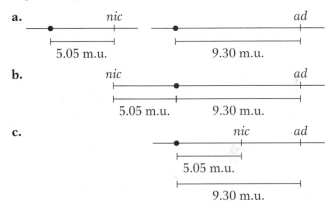

Here, a combination of common sense and simple analysis tells us which alternative is correct. First, an inspection of the asci reveals that the most common single type is the one labeled 1, which contains more than 80 percent of all the asci. This type contains only $nic^+ \cdot ad$ and $nic \cdot ad^+$ genotypes, and they are *parental* genotypes. So we know that recombination is quite low and the loci are certainly linked. This rules out alternative **a**.

Now consider alternative **c**. If this alternative were correct, a crossover between the centromere and the nic locus would generate not only an M_{II} pattern for that locus, but also an M_{II} pattern for the ad locus, because it is farther from the centromere than nic is. The ascus pattern produced by alternative **c** should be

nic^+	ad
nic^+	ad
nic	ad^+
nic	ad^+
nic^+	ad
nic^+	ad
nic	ad^+
nic	ad^+

Remember that the *nic* locus shows M_{II} patterns in asci types 4, 5, 6, and 7 (a total of 101 asci); of them, type 5 is the very one that we are talking about and contains 90 asci. Therefore, alternative **c** appears to be correct because ascus type 5 comprises about 90 percent of the M_{II} asci for the *nic* locus. This relation would not hold if alternative **b** were correct, because crossovers on either side of the centromere would generate the M_{II} patterns for the *nic* and the *ad* loci independently.

Is the map distance from *nic* to *ad* simply $9.30 - 5.05 = 4.25$ m.u.? Close, but not quite. The best way of calculating map distances between loci is always by measuring the recombinant frequency. We could go through the asci and count all the recombinant ascospores, but using the formula $RF = \frac{1}{2} T + NPD$ is simpler. The T asci are classes 3, 4, and 7, and the NPD asci are classes 2 and 6. Hence, $RF = [\frac{1}{2} (100) + 2]/1000 = 5.2$ percent, or 5.2 m.u., and a better map is

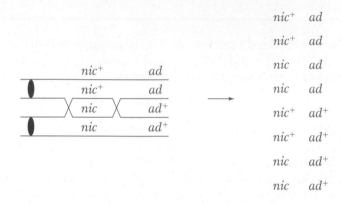

The reason for the underestimation of the *ad*-to-centromere distance calculated from the M_{II} frequency is the occurrence of double crossovers, which can produce an M_I pattern for *ad*, as in ascus type 4:

nic^+	ad
nic^+	ad
nic	ad
nic	ad
nic^+	ad^+
nic^+	ad^+
nic	ad^+
nic	ad^+

PROBLEMS

Most of the problems are also available for review/grading through the ᴳᴱᴺᴱᵀᴵᶜˢ**PORTAL** www.yourgeneticsportal.com

WORKING WITH THE FIGURES

1. In Figure 4-3, would there be any noncrossover meiotic products in the meiosis illustrated? If so, what colors would they be in the color convention used?

2. In Figure 4-6, why does the diagram not show meioses in which two crossovers occur between the same two chromatids (such as the two inner ones)?

3. In Figure 4-8, some meiotic products are labeled parental. Which parent is being referred to in this terminology?

4. In Figure 4-9, why is only locus A shown in a constant position?

5. In Figure 4-10, what is the mean frequency of crossovers per meiosis in the region A–B? The region B–C?

6. In Figure 4-11, is it true to say that from such a cross the product $v\ cv^+$ can have two different origins?

7. In Figure 4-14, in the bottom row four colors are labeled SCO. Why are they not all the same size (frequency)?

8. Using the conventions of Figure 4-15, draw parents and progeny classes from a cross

$$P\ M'''/p\ M' \times p\ M'/p\ M''''$$

9. In Figure 4-17, draw the arrangements of alleles in an octad from a similar meiosis in which the upper product of the first division segregated in an upside-down manner at the second division.

10. In Figure 4-19, what would be the RF between A/a and B/b in a cross in which purely by chance all meioses had four-strand double crossovers in that region?

11. **a.** In Figure 4-21, let GC = A and AT = a, then draw the fungal octad that would result from the final structure (5).

   b. (Challenging) Insert some closely linked flanking markers into the diagram, say P/p to the left and Q/q to the right (assume either cis or trans arrangements). Assume neither of these loci show non-Mendelian segregation. Then draw the final octad based on the structure in part 5.

BASIC PROBLEMS

12. A plant of genotype

$$\frac{A \qquad B}{a \qquad b}$$

is testcrossed with

$$\frac{a \qquad b}{a \qquad b}$$

If the two loci are 10 m.u. apart, what proportion of progeny will be AB/ab?

13. The A locus and the D locus are so tightly linked that no recombination is ever observed between them. If Ad/Ad is crossed with aD/aD and the F_1 is intercrossed,

what phenotypes will be seen in the F$_2$ and in what proportions?

14. The *R* and *S* loci are 35 m.u. apart. If a plant of genotype

$$\frac{R \qquad S}{r \qquad s}$$

is selfed, what progeny phenotypes will be seen and in what proportions?

15. The cross *E/E · F/F × e/e · f/f* is made, and the F$_1$ is then backcrossed with the recessive parent. The progeny genotypes are inferred from the phenotypes. The progeny genotypes, written as the gametic contributions of the heterozygous parent, are in the following proportions:

E · F	$\frac{2}{6}$
E · f	$\frac{1}{6}$
e · F	$\frac{1}{6}$
e · f	$\frac{2}{6}$

Explain these results.

16. A strain of *Neurospora* with the genotype *H · I* is crossed with a strain with the genotype *h · i*. Half the progeny are *H · I*, and the other half are *h · i*. Explain how this outcome is possible.

17. A female animal with genotype *A/a · B/b* is crossed with a double-recessive male (*a/a · b/b*). Their progeny include 442 *A/a · B/b*, 458 *a/a · b/b*, 46 *A/a · b/b*, and 54 *a/a · B/b*. Explain these results.

18. If *A/A · B/B* is crossed with *a/a · b/b* and the F$_1$ is testcrossed, what percentage of the testcross progeny will be *a/a · b/b* if the two genes are **(a)** unlinked; **(b)** completely linked (no crossing over at all); **(c)** 10 m.u. apart; **(d)** 24 m.u. apart?

19. In a haploid organism, the *C* and *D* loci are 8 m.u. apart. From a cross *C d × c D*, give the proportion of each of the following progeny classes: **(a)** *C D*; **(b)** *c d*; **(c)** *C d*; **(d)** all recombinants.

20. A fruit fly of genotype *B R/b r* is testcrossed with *b r/ b r*. In 84 percent of the meioses, there are no chiasmata between the linked genes; in 16 percent of the meioses, there is one chiasma between the genes. What proportion of the progeny will be *B r/b r*?

21. A three-point testcross was made in corn. The results and a recombination analysis are shown in the display below, which is typical of three-point testcrosses (*p* = purple leaves, + = green; *v* = virus-resistant seedlings, + = sensitive; *b* = brown midriff to seed, + = plain). Study the display and answer parts *a* through *c*.

P	$+/+ \cdot +/+ \cdot +/+ \times p/p \cdot v/v \cdot b/b$
Gametes	$+ \cdot + \cdot + \qquad\qquad p \cdot v \cdot b$

a. Determine which genes are linked.

b. Draw a map that shows distances in map units.

c. Calculate interference, if appropriate.

					Recombinant for		
Class	Progeny phenotypes	F$_1$ gametes	Numbers	*p b*	*p–v*	*v b*
1	gre sen pla	$+ \cdot + \cdot +$	3,210			
2	pur res bro	$p \cdot v \cdot b$	3,222			
3	gre res pla	$+ \cdot v \cdot +$	1,024		R	R
4	pur sen bro	$p \cdot + \cdot b$	1,044		R	R
5	pur res pla	$p \cdot v \cdot +$	690	R		R
6	gre sen bro	$+ \cdot + \cdot b$	678	R		R
7	gre res bro	$+ \cdot v \cdot b$	72	R	R	
8	pur sen pla	$p \cdot + \cdot +$	60	R	R	
		Total	10,000	1,500	2,200	3,436

Unpacking Problem 21

1. Sketch cartoon drawings of the P, F$_1$, and tester corn plants, and use arrows to show exactly how you would perform this experiment. Show where seeds are obtained.

2. Why do all the +'s look the same, even for different genes? Why does this not cause confusion?

3. How can a phenotype be purple and brown, for example, at the same time?

4. Is it significant that the genes are written in the order *p-v-b* in the problem?

5. What is a tester and why is it used in this analysis?

6. What does the column marked "Progeny phenotypes" represent? In class 1, for example, state exactly what "gre sen pla" means.

7. What does the line marked "Gametes" represent, and how is it different from the column marked "F$_1$ gametes"?

In what way is comparison of these two types of gametes relevant to recombination?

8. Which meiosis is the main focus of study? Label it on your drawing.

9. Why are the gametes from the tester not shown?

10. Why are there only eight phenotypic classes? Are there any classes missing?

11. What classes (and in what proportions) would be expected if all the genes are on separate chromosomes?

12. To what do the four pairs of class sizes (very big, two intermediates, very small) correspond?

13. What can you tell about gene order simply by inspecting the phenotypic classes and their frequencies?

14. What will be the expected phenotypic class distribution if only two genes are linked?

15. What does the word "point" refer to in a three-point testcross? Does this word usage imply linkage? What would a four-point testcross be like?

16. What is the definition of *recombinant*, and how is it applied here?

17. What do the "Recombinant for" columns mean?

18. Why are there only three "Recombinant for" columns?

19. What do the R's mean, and how are they determined?

20. What do the column totals signify? How are they used?

21. What is the diagnostic test for linkage?

22. What is a map unit? Is it the same as a centimorgan?

23. In a three-point testcross such as this one, why aren't the F_1 and the tester considered to be parental in calculating recombination? (They *are* parents in one sense.)

24. What is the formula for interference? How are the "expected" frequencies calculated in the coefficient-of-coincidence formula?

25. Why does part *c* of the problem say "if appropriate"?

26. How much work is it to obtain such a large progeny size in corn? Which of the three genes would take the most work to score? Approximately how many progeny are represented by one corncob?

22. You have a *Drosophila* line that is homozygous for autosomal recessive alleles *a*, *b*, and *c*, linked in that order. You cross females of this line with males homozygous for the corresponding wild-type alleles. You then cross the F_1 heterozygous males with their heterozygous sisters. You obtain the following F_2 phenotypes (where letters denote recessive phenotypes and pluses denote wild-type phenotypes): $1364 + + +$, $365\ a\ b\ c$, $87\ a\ b +$, $84 + + c$, $47\ a + +$, $44 + b\ c$, $5\ a + c$, and $4 + b +$.

   a. What is the recombinant frequency between *a* and *b*? Between *b* and *c*? (Remember, there is no crossing over in *Drosophila* males.)

   b. What is the coefficient of coincidence?

23. R. A. Emerson crossed two different pure-breeding lines of corn and obtained a phenotypically wild-type F_1 that was heterozygous for three alleles that determine recessive phenotypes: *an* determines anther; *br*, brachytic; and *f*, fine. He testcrossed the F_1 with a tester that was homozygous recessive for the three genes and obtained these progeny phenotypes: 355 anther; 339 brachytic, fine; 88 completely wild type; 55 anther, brachytic, fine; 21 fine; 17 anther, brachytic; 2 brachytic; 2 anther, fine.

   a. What were the genotypes of the parental lines?

   b. Draw a linkage map for the three genes (include map distances).

   c. Calculate the interference value.

24. Chromosome 3 of corn carries three loci (*b* for plant-color booster, *v* for virescent, and *lg* for liguleless). A testcross of triple recessives with F_1 plants heterozygous for the three genes yields progeny having the following genotypes: $305 + v\ lg$, $275\ b + +$, $128\ b + lg$, $112 + v +$, $74 + + lg$, $66\ b\ v +$, $22 + + +$, and $18\ b\ v\ lg$. Give the gene sequence on the chromosome, the map distances between genes, and the coefficient of coincidence.

25. Groodies are useful (but fictional) haploid organisms that are pure genetic tools. A wild-type groody has a fat body, a long tail, and flagella. Mutant lines are known that have thin bodies, are tailless, or do not have flagella. Groodies can mate with one another (although they are so shy that we do not know how) and produce recombinants. A wild-type groody mates with a thin-bodied groody lacking both tail and flagella. The 1000 baby groodies produced are classified as shown in the illustration here. Assign genotypes, and map the three genes. (Problem 25 is from Burton S. Guttman.)

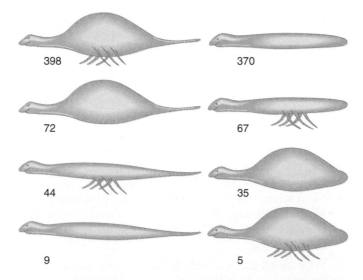

26. In *Drosophila*, the allele dp^+ determines long wings and *dp* determines short ("dumpy") wings. At a separate locus, e^+ determines gray body and *e* determines ebony

body. Both loci are autosomal. The following crosses were made, starting with pure-breeding parents:

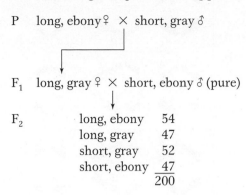

P   long, ebony ♀ × short, gray ♂

F₁   long, gray ♀ × short, ebony ♂ (pure)

F₂   long, ebony   54
     long, gray   47
     short, gray   52
     short, ebony   47
     ──────
     200

Use the χ^2 test to determine if these loci are linked. In doing so, indicate **(a)** the hypothesis, **(b)** calculation of χ^2, **(c)** p value, **(d)** what the p value means, **(e)** your conclusion, **(f)** the inferred chromosomal constitutions of parents, F₁, tester, and progeny.

27. The mother of a family with 10 children has blood type Rh⁺. She also has a very rare condition (elliptocytosis, phenotype E) that causes red blood cells to be oval rather than round in shape but that produces no adverse clinical effects. The father is Rh⁻ (lacks the Rh⁺ antigen) and has normal red blood cells (phenotype e). The children are 1 Rh⁺ e, 4 Rh⁺ E, and 5 Rh⁻ e. Information is available on the mother's parents, who are Rh⁺ E and Rh⁻ e. One of the 10 children (who is Rh⁺ E) marries someone who is Rh⁺ e, and they have an Rh⁺ E child.

a. Draw the pedigree of this whole family.

b. Is the pedigree in agreement with the hypothesis that the Rh^+ allele is dominant and Rh^- is recessive?

c. What is the mechanism of transmission of elliptocytosis?

d. Could the genes governing the E and Rh phenotypes be on the same chromosome? If so, estimate the map distance between them, and comment on your result.

28. From several crosses of the general type $A/A \cdot B/B \times a/a \cdot b/b$ the F₁ individuals of type $A/a \cdot B/b$ were testcrossed with $a/a \cdot b/b$. The results are as follows:

Testcross of F₁ from cross	*Testcross progeny*			
	$A/a \cdot B/b$	$a/a \cdot b/b$	$A/a \cdot b/b$	$a/a \cdot B/b$
1	310	315	287	288
2	36	38	23	23
3	360	380	230	230
4	74	72	50	44

For each set of progeny, use the χ^2 test to decide if there is evidence of linkage.

29. In the two pedigrees diagrammed here, a vertical bar in a symbol stands for steroid sulfatase deficiency, and a horizontal bar stands for ornithine transcarbamylase deficiency.

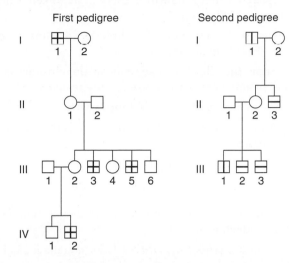

First pedigree   Second pedigree

a. Is there any evidence in these pedigrees that the genes determining the deficiencies are linked?

b. If the genes are linked, is there any evidence in the pedigree of crossing over between them?

c. Draw genotypes of these individuals as far as possible.

30. In the accompanying pedigree, the vertical lines stand for protan color blindness, and the horizontal lines stand for deutan color blindness. These are separate conditions causing different misperceptions of colors; each is determined by a separate gene.

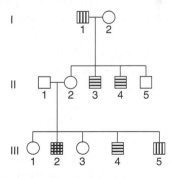

a. Does the pedigree show any evidence that the genes are linked?

b. If there is linkage, does the pedigree show any evidence of crossing over?

Explain your answers to parts *a* and *b* with the aid of the diagram.

c. Can you calculate a value for the recombination between these genes? Is this recombination by independent assortment or by crossing over?

31. In corn, a triple heterozygote was obtained carrying the mutant alleles *s* (shrunken), *w* (white aleurone), and *y* (waxy endosperm), all paired with their normal wild-type alleles. This triple heterozygote was test-crossed, and the progeny contained 116 shrunken, white; 4 fully wild type; 2538 shrunken; 601 shrunken, waxy; 626 white; 2708 white, waxy; 2 shrunken, white, waxy; and 113 waxy.

a. Determine if any of these three loci are linked and, if so, show map distances.

b. Show the allele arrangement on the chromosomes of the triple heterozygote used in the testcross.

c. Calculate interference, if appropriate.

32. a. A mouse cross $A/a \cdot B/b \times a/a \cdot b/b$ is made, and in the progeny there are

$$25\% \ A/a \cdot B/b, \ 25\% \ a/a \cdot b/b,$$
$$25\% \ A/a \cdot b/b, \ 25\% \ a/a \cdot B/b$$

Explain these proportions with the aid of simplified meiosis diagrams.

b. A mouse cross $C/c \cdot D/d \times c/c \cdot d/d$ is made, and in the progeny there are

$$45\% \ C/c \cdot d/d, \ 45\% \ c/c \cdot D/d,$$
$$5\% \ c/c \cdot d/d, \ 5\% \ C/c \cdot D/d$$

Explain these proportions with the aid of simplified meiosis diagrams.

33. In the tiny model plant *Arabidopsis,* the recessive allele *hyg* confers seed resistance to the drug hygromycin, and *her,* a recessive allele of a different gene, confers seed resistance to herbicide. A plant that was homozygous *hyg/hyg · her/her* was crossed with wild type, and the F_1 was selfed. Seeds resulting from the F_1 self were placed on petri dishes containing hygromycin and herbicide.

a. If the two genes are unlinked, what percentage of seeds are expected to grow?

b. In fact, 13 percent of the seeds grew. Does this percentage support the hypothesis of no linkage? Explain. If not, calculate the number of map units between the loci.

c. Under your hypothesis, if the F_1 is testcrossed, what proportion of seeds will grow on the medium containing hygromycin and herbicide?

34. In a diploid organism of genotype A/a ; B/b ; D/d, the allele pairs are all on different chromosome pairs. The two diagrams in the next column purport to show anaphases ("pulling apart" stages) in individual cells. State whether each drawing represents mitosis, meiosis I, or meiosis II or is impossible for this particular genotype.

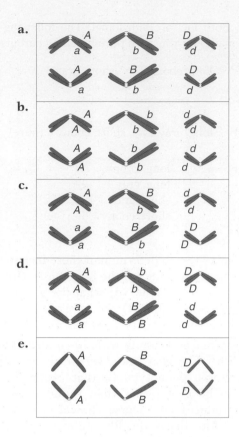

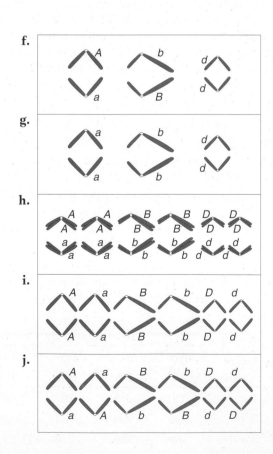

35. The *Neurospora* cross *al-2⁺* × *al-2* is made. A linear tetrad analysis reveals that the second-division segregation frequency is 8 percent.

    a. Draw two examples of second-division segregation patterns in this cross.

    b. What can be calculated by using the 8 percent value?

36. From the fungal cross *arg-6 · al-2* × *arg-6⁺ · al-2⁺*, what will the spore genotypes be in unordered tetrads that are **(a)** parental ditypes? **(b)** tetratypes? **(c)** nonparental ditypes?

37. For a certain chromosomal region, the mean number of crossovers at meiosis is calculated to be two per meiosis. In that region, what proportion of meioses are predicted to have **(a)** no crossovers? **(b)** one crossover? **(c)** two crossovers?

38. A *Neurospora* cross was made between a strain that carried the mating-type allele *A* and the mutant allele *arg-1* and another strain that carried the mating-type allele *a* and the wild-type allele for *arg-1* (+). Four hundred linear octads were isolated, and they fell into the seven classes given in the table below. (For simplicity, they are shown as tetrads.)

    a. Deduce the linkage arrangement of the mating-type locus and the *arg-1* locus. Include the centromere or centromeres on any map that you draw. Label *all* intervals in map units.

    b. Diagram the meiotic divisions that led to class 6. Label clearly.

1	2	3	4	5	6	7
A · arg	*A · +*	*A · arg*	*A · arg*	*A · arg*	*A · +*	*A · +*
A · arg	*A · +*	*A · +*	*a · arg*	*a · +*	*a · arg*	*a · arg*
a · +	*a · arg*	*a · arg*	*A · +*	*A · arg*	*A · +*	*A · arg*
a · +	*a · arg*	*a · +*	*a · +*	*a · +*	*a · arg*	*a · +*
127	125	100	36	2	4	6

Unpacking Problem 38

1. Are fungi generally haploid or diploid?

2. How many ascospores are in the ascus of *Neurospora*? Does your answer match the number presented in this problem? Explain any discrepancy.

3. What is mating type in fungi? How do you think it is determined experimentally?

4. Do the symbols *A* and *a* have anything to do with dominance and recessiveness?

5. What does the symbol *arg-1* mean? How would you test for this genotype?

6. How does the *arg-1* symbol relate to the symbol +?

7. What does the expression *wild type* mean?

8. What does the word *mutant* mean?

9. Does the biological function of the alleles shown have anything to do with the solution of this problem?

10. What does the expression *linear octad analysis* mean?

11. In general, what more can be learned from linear tetrad analysis that cannot be learned from unordered tetrad analysis?

12. How is a cross made in a fungus such as *Neurospora*? Explain how to isolate asci and individual ascospores. How does the term *tetrad* relate to the terms *ascus* and *octad*?

13. Where does meiosis take place in the *Neurospora* life cycle? (Show it on a diagram of the life cycle.)

14. What does Problem 38 have to do with meiosis?

15. Can you write out the genotypes of the two parental strains?

16. Why are only four genotypes shown in each class?

17. Why are there only seven classes? How many ways have you learned for classifying tetrads generally? Which of these classifications can be applied to both linear and unordered tetrads? Can you apply these classifications to the tetrads in this problem? (Classify each class in as many ways as possible.) Can you think of more possibilities in this cross? If so, why are they not shown?

18. Do you think there are several different spore orders within each class? Why would these different spore orders not change the class?

19. Why is the following class not listed?

$$a \cdot +$$
$$a \cdot +$$
$$A \cdot arg$$
$$A \cdot arg$$

20. What does the expression *linkage arrangement* mean?

21. What is a genetic *interval*?

22. Why does the problem state "centromere or centromeres" and not just "centromere"? What is the general method for mapping centromeres in tetrad analysis?

23. What is the total frequency of $A \cdot +$ ascospores? (Did you calculate this frequency by using a formula or by inspection? Is this a recombinant genotype? If so, is it the only recombinant genotype?)

24. The first two classes are the most common and are approximately equal in frequency. What does this information tell you? What is their content of parental and recombinant genotypes?

39. A geneticist studies 11 different pairs of *Neurospora* loci by making crosses of the type $a \cdot b \times a^+ \cdot b^+$ and then analyzing 100 linear asci from each cross. For the convenience of making a table, the geneticist organizes the data as if all 11 pairs of genes had the same designation—*a* and *b*—as shown below. For each cross, map the loci in relation to each other and to centromeres.

Number of asci of type

Cross	$a \cdot b$ $a \cdot b$ $a^+ \cdot b^+$ $a^+ \cdot b^+$	$a \cdot b^+$ $a \cdot b^+$ $a^+ \cdot b$ $a^+ \cdot b$	$a \cdot b$ $a \cdot b^+$ $a^+ \cdot b^+$ $a^+ \cdot b$	$a \cdot b$ $a^+ \cdot b$ $a^+ \cdot b^+$ $a \cdot b^+$	$a \cdot b$ $a^+ \cdot b^+$ $a^+ \cdot b^+$ $a \cdot b$	$a \cdot b^+$ $a^+ \cdot b$ $a^+ \cdot b$ $a \cdot b^+$	$a \cdot b^+$ $a^+ \cdot b$ $a^+ \cdot b^+$ $a \cdot b$
1	34	34	32	0	0	0	0
2	84	1	15	0	0	0	0
3	55	3	40	0	2	0	0
4	71	1	18	1	8	0	1
5	9	6	24	22	8	10	20
6	31	0	1	3	61	0	4
7	95	0	3	2	0	0	0
8	6	7	20	22	12	11	22
9	69	0	10	18	0	1	2
10	16	14	2	60	1	2	5
11	51	49	0	0	0	0	0

40. Three different crosses in *Neurospora* are analyzed on the basis of unordered tetrads. Each cross combines a different pair of linked genes. The results are shown in the following table:

Cross	Parents (%)	Parental ditypes (%)	Tetra-types (%)	Non-parental ditypes (%)
1	$a \cdot b^+ \times a^+ \cdot b$	51	45	4
2	$c \cdot d^+ \times c^+ \cdot d$	64	34	2
3	$e \cdot f^+ \times e^+ \cdot f$	45	50	5

For each cross, calculate:

a. the frequency of recombinants (RF).

b. the uncorrected map distance, based on RF.

c. the corrected map distance, based on tetrad frequencies.

d. the corrected map distance, based on the mapping function.

41. On *Neurospora* chromosome 4, the *leu3* gene is just to the left of the centromere and always segregates at the first division, whereas the *cys2* gene is to the right of the centromere and shows a second-division segregation frequency of 16 percent. In a cross between a *leu3* strain and a *cys2* strain, calculate the predicted frequencies of the following seven classes of linear tetrads where $l = leu3$ and $c = cys2$. (Ignore double and other multiple crossovers.)

(i) $l\,c$ (ii) $l+$ (iii) $l\,c$ (iv) $l\,c$ (v) $l\,c$ (vi) $l+$ (vii) $l+$
  $l\,c$ $l+$ $l+$ $+\,c$ $+\,+$ $+\,c$ $+\,c$
  $+\,+$ $+\,c$ $+\,+$ $+\,+$ $+\,+$ $+\,+$ $+\,+$
  $+\,+$ $+\,c$ $+\,c$ $l+$ $l\,c$ $l+$ $l\,c$

42. A rice breeder obtained a triple heterozygote carrying the three recessive alleles for albino flowers (*al*), brown awns (*b*), and fuzzy leaves (*fu*), all paired with their normal wild-type alleles. This triple heterozygote was testcrossed. The progeny phenotypes were

170	wild type
150	albino, brown, fuzzy
5	brown
3	albino, fuzzy
710	albino
698	brown, fuzzy
42	fuzzy
38	albino, brown

a. Are any of the genes linked? If so, draw a map labeled with map distances. (Don't bother with a correction for multiple crossovers.)

b. The triple heterozygote was originally made by crossing two pure lines. What were their genotypes?

43. In a fungus, a proline mutant (*pro*) was crossed with a histidine mutant (*his*). A nonlinear tetrad analysis gave the following results:

+	+	+	+	+	*his*
+	+	+	*his*	+	*his*
pro	*his*	*pro*	+	*pro*	+
pro	*his*	*pro*	*his*	*pro*	+
	6		82		112

a. Are the genes linked or not?

b. Draw a map (if linked) or two maps (if not linked), showing map distances based on straightforward recombinant frequency where appropriate.

c. If there is linkage, correct the map distances for multiple crossovers (choose one approach only).

44. In the fungus *Neurospora*, a strain that is auxotrophic for thiamine (mutant allele *t*) was crossed with a strain that is auxotrophic for methionine (mutant allele *m*). Linear asci were isolated and classified into the following groups.

Spore pair			Ascus types			
1 and 2	*t* +	*t* +	*t* +	*t* +	*t m*	*t m*
3 and 4	*t* +	*t m*	+ *m*	+ +	*t m*	+ +
5 and 6	+ *m*	+ +	*t* +	*t m*	+ +	*t* +
7 and 8	+ *m*	+ *m*	+ *m*	+ *m*	+ +	+ *m*
Number	260	76	4	54	1	5

a. Determine the linkage relations of these two genes to their centromere(s) and to each other. Specify distances in map units.

b. Draw a diagram to show the origin of the ascus type with only one single representative (second from right).

45. A corn geneticist wants to obtain a corn plant that has the three dominant phenotypes: anthocyanin (A), long tassels (L), and dwarf plant (D). In her collection of pure lines, the only lines that bear these alleles are *AA LL dd* and *aa ll DD*. She also has the fully recessive line *aa ll dd*. She decides to intercross the first two and testcross the resulting hybrid to obtain in the progeny a plant of the desired phenotype (which would have to be *Aa Ll Dd* in this case). She knows that the three genes are linked in the order written and that the distance between the *A/a* and the *L/l* loci is 16 map units and that the distance between the *L/l* and the *D/d* loci is 24 map units.

a. Draw a diagram of the chromosomes of the parents, the hybrid, and the tester.

b. Draw a diagram of the crossover(s) necessary to produce the desired genotype.

c. What percentage of the testcross progeny will be of the phenotype that she needs?

d. What assumptions did you make (if any)?

46. In the model plant *Arabidopsis thaliana* the following alleles were used in a cross:

T = presence of trichomes	*t* = absence of trichomes
D = tall plants	*d* = dwarf plants
W = waxy cuticle	*w* = nonwaxy
A = presence of purple anthocyanin pigment	*a* = absence (white)

The *T/t* and *D/d* loci are linked 26 m.u. apart on chromosome 1, whereas the *W/w* and *A/a* loci are linked 8 m.u. apart on chromosome 2.

A pure-breeding double-homozygous recessive trichomeless nonwaxy plant is crossed with another pure-breeding double-homozygous recessive dwarf white plant.

a. What will be the appearance of the F$_1$?

b. Sketch the chromosomes 1 and 2 of the parents and the F$_1$, showing the arrangement of the alleles.

c. If the F$_1$ is testcrossed, what proportion of the progeny will have all four recessive phenotypes?

47. In corn, the cross *WW ee FF* × *ww EE ff* is made. The three loci are linked as follows:

Assume no interference.

a. If the F$_1$ is testcrossed, what proportion of progeny will be *ww ee ff*?

b. If the F$_1$ is selfed, what proportion of progeny will be *ww ee ff*?

48. The fungal cross + · + × *c* · *m* was made, and *nonlinear (unordered)* tetrads were collected. The results were

+ +	+ +	+ *m*
+ +	+ *m*	+ *m*
c m	*c* +	*c* +
c m	*c m*	*c* +
Total 112	82	6

a. From these results, calculate a simple recombinant frequency.

b. Compare the Haldane mapping function and the Perkins formula in their conversions of the RF value into a "corrected" map distance.

c. In the derivation of the Perkins formula, only the possibility of meioses with zero, one, and two crossovers was considered. Could this limit explain any discrepancy in your calculated values? Explain briefly (no calculation needed).

49. In mice, the following alleles were used in a cross:

W = waltzing gait w = nonwaltzing gait

G = normal gray color g = albino

B = bent tail b = straight tail

A waltzing gray bent-tailed mouse is crossed with a nonwaltzing albino straight-tailed mouse and, over several years, the following progeny totals are obtained:

waltzing	gray	bent	18
waltzing	albino	bent	21
nonwaltzing	gray	straight	19
nonwaltzing	albino	straight	22
waltzing	gray	straight	4
waltzing	albino	straight	5
nonwaltzing	gray	bent	5
nonwaltzing	albino	bent	6
Total			100

a. What were the genotypes of the two parental mice in the cross?

b. Draw the chromosomes of the parents.

c. If you deduced linkage, state the map unit value or values and show how they were obtained.

50. Consider the *Neurospora* cross $+ \; ; + \times f \; ; p$

It is known that the $+/f$ locus is very close to the centromere on chromosome 7—in fact, so close that there are never any second-division segregations. It is also known that the $+/p$ locus is on chromosome 5, at such a distance that there is usually an average of 12 percent second-division segregations. With this information, what will be the proportion of octads that are

a. parental ditypes showing M_I patterns for both loci?

b. nonparental ditypes showing M_I patterns for both loci?

c. tetratypes showing an M_I pattern for $+/f$ and an M_{II} pattern for $+/p$?

d. tetratypes showing an M_{II} pattern for $+/f$ and an M_I pattern for $+/p$?

51. In a haploid fungus, the genes *al-2* and *arg-6* are 30 map units apart on chromosome 1, and the genes *lys-5* and *met-1* are 20 map units apart on chromosome 6. In a cross

$$al\text{-}2 + \; ; + \; met\text{-}1 \times + \; arg\text{-}6 \; ; lys\text{-}5 +$$

what proportion of progeny would be prototrophic $+ + \; ; + +$?

52. The recessive alleles k (kidney-shaped eyes instead of wild-type round), c (cardinal-colored eyes instead of wild-type red), and e (ebony body instead of wild-type gray) identify three genes on chromosome 3 of *Drosophila*. Females with kidney-shaped, cardinal-colored eyes were mated with ebony males. The F_1 was wild type. When F_1 females were testcrossed with kk cc ee males, the following progeny phenotypes were obtained:

k	c	e	3
k	c	+	876
k	+	e	67
k	+	+	49
+	c	e	44
+	c	+	58
+	+	e	899
+	+	+	4
Total			2000

a. Determine the order of the genes and the map distances between them.

b. Draw the chromosomes of the parents and the F_1.

c. Calculate interference and say what you think of its significance.

53. From parents of genotypes $A/A \cdot B/B$ and $a/a \cdot b/b$, a dihybrid was produced. In a testcross of the dihybrid, the following seven progeny were obtained:

$$A/a \cdot B/b, \, a/a \cdot b/b, \, A/a \cdot B/b, \, A/a \cdot b/b,$$
$$a/a \cdot b/b, \, A/a \cdot B/b, \text{ and } a/a \cdot B/b$$

Do these results provide convincing evidence of linkage?

CHALLENGING PROBLEMS

54. Use the Haldane map function to calculate the corrected map distance in cases where the measured RF = 5%, 10%, 20%, 30%, and 40%. Sketch a graph of RF against corrected map distance and use it to answer the question, When should one use a map function?

GENETICSPORTAL **Unpacking the Problem** **55.** An individual heterozygous for four genes, $A/a \cdot B/b \cdot C/c \cdot D/d$, is testcrossed with $a/a \cdot b/b \cdot c/c \cdot d/d$, and 1000 progeny are classified by the gametic contribution of the heterozygous parent as follows:

$a \cdot B \cdot C \cdot D$	42
$A \cdot b \cdot c \cdot d$	43
$A \cdot B \cdot C \cdot d$	140
$a \cdot b \cdot c \cdot D$	145
$a \cdot B \cdot c \cdot D$	6
$A \cdot b \cdot C \cdot d$	9
$A \cdot B \cdot c \cdot d$	305
$a \cdot b \cdot C \cdot D$	310

a. Which genes are linked?

b. If two pure-breeding lines had been crossed to produce the heterozygous individual, what would their genotypes have been?

c. Draw a linkage map of the linked genes, showing the order and the distances in map units.

d. Calculate an interference value, if appropriate.

56. An autosomal allele N in humans causes abnormalities in nails and patellae (kneecaps) called the nail–patella syndrome. Consider marriages in which one partner has the nail–patella syndrome and blood type A and the other partner has normal nails and patellae and blood type O. These marriages produce some children who have both the nail–patella syndrome and blood type A. Assume that unrelated children from this phenotypic group mature, intermarry, and have children. Four phenotypes are observed in the following percentages in this second generation:

nail–patella syndrome, blood type A	66%
normal nails and patellae, blood type O	16%
normal nails and patellae, blood type A	9%
nail–patella syndrome, blood type O	9%

Fully analyze these data, explaining the relative frequencies of the four phenotypes. (See pages 214–215 for the genetic basis of these blood types.)

57. Assume that three pairs of alleles are found in *Drosophila*: x^+ and x, y^+ and y, and z^+ and z. As shown by the symbols, each non-wild-type allele is recessive to its wild-type allele. A cross between females heterozygous at these three loci and wild-type males yields progeny having the following genotypes: 1010 $x^+ \cdot y^+ \cdot z^+$ females, 430 $x \cdot y^+ \cdot z$ males, 441 $x^+ \cdot y \cdot z^+$ males, 39 $x \cdot y \cdot z$ males, 32 $x^+ \cdot y^+ \cdot z$ males, 30 $x \cdot y^+ \cdot z^+$ males, 27 $x \cdot y \cdot z^+$ males, 1 $x^+ \cdot y \cdot z$ male, and 0 $x \cdot y^+ \cdot z^+$ males.

a. On what chromosome of *Drosophila* are the genes carried?

b. Draw the relevant chromosomes in the heterozygous female parent, showing the arrangement of the alleles.

c. Calculate the map distances between the genes and the coefficient of coincidence.

58. From the five sets of data given in the following table, determine the order of genes by inspection—that is, without calculating recombination values. Recessive phenotypes are symbolized by lowercase letters and dominant phenotypes by pluses.

Phenotypes observed in 3-point testcross	Data sets				
	1	2	3	4	5
+ + +	317	1	30	40	305
+ + c	58	4	6	232	0
+ b +	10	31	339	84	28
+ b c	2	77	137	201	107
a + +	0	77	142	194	124
a + c	21	31	291	77	30
a b +	72	4	3	235	1
a b c	203	1	34	46	265

59. From the phenotype data given in the following table for two three-point testcrosses for (1) a, b, and c and (2) b, c, and d, determine the sequence of the four genes a, b, c, and d and the three map distances between them. Recessive phenotypes are symbolized by lowercase letters and dominant phenotypes by pluses.

1		2	
+ + +	669	b c d	8
a b +	139	b + +	441
a + +	3	b + d	90
+ + c	121	+ c d	376
+ b c	2	+ + +	14
a + c	2280	+ + d	153
a b c	653	+ c +	65
+ b +	2215	b c +	141

60. The father of Mr. Spock, first officer of the starship *Enterprise*, came from planet Vulcan; Spock's mother came from Earth. A Vulcan has pointed ears (determined by allele P), adrenals absent (determined by A), and a right-sided heart (determined by R). All these alleles are dominant to normal Earth alleles. The three loci are autosomal, and they are linked as shown in this linkage map:

P ————— A —————————— R
←— 15 m.u. —→ ←—— 20 m.u. ——→

If Mr. Spock marries an Earth woman and there is no (genetic) interference, what proportion of their children will have

a. Vulcan phenotypes for all three characters?

b. Earth phenotypes for all three characters?

c. Vulcan ears and heart but Earth adrenals?

d. Vulcan ears but Earth heart and adrenals?

(Problem 60 is from D. Harrison, *Problems in Genetics.* Addison-Wesley, 1970.)

61. In a certain diploid plant, the three loci *A*, *B*, and *C* are linked as follows:

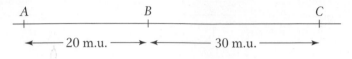

One plant is available to you (call it the parental plant). It has the constitution *A b c/a B C*.

a. With the assumption of no interference, if the plant is selfed, what proportion of the progeny will be of the genotype *a b c/a b c*?

b. Again, with the assumption of no interference, if the parental plant is crossed with the *a b c/a b c* plant, what genotypic classes will be found in the progeny? What will be their frequencies if there are 1000 progeny?

c. Repeat part *b*, this time assuming 20 percent interference between the regions.

62. The following pedigree shows a family with two rare abnormal phenotypes: blue sclerotic (a brittle-bone defect), represented by a black-bordered symbol, and hemophilia, represented by a black center in a symbol. Members represented by completely black symbols have both disorders. The numbers in some symbols are the numbers of those types.

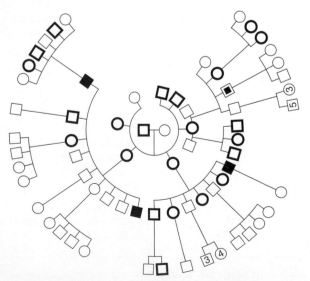

a. What pattern of inheritance is shown by each condition in this pedigree?

b. Provide the genotypes of as many family members as possible.

c. Is there evidence of linkage?

d. Is there evidence of independent assortment?

e. Can any of the members be judged as recombinants (that is, formed from at least one recombinant gamete)?

63. The human genes for color blindness and for hemophilia are both on the X chromosome, and they show a recombinant frequency of about 10 percent. The linkage of a pathological gene to a relatively harmless one can be used for genetic prognosis. Shown here is part of a bigger pedigree. Blackened symbols indicate that the subjects had hemophilia, and crosses indicate color blindness. What information could be given to women III-4 and III-5 about the likelihood of their having sons with hemophilia?

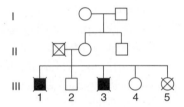

(Problem 63 is adapted from J. F. Crow, *Genetics Notes: An Introduction to Genetics.* Burgess, 1983.)

64. A geneticist mapping the genes *A*, *B*, *C*, *D*, and *E* makes two 3-point testcrosses. The first cross of pure lines is

$$A/A \cdot B/B \cdot C/C \cdot D/D \cdot E/E \times a/a \cdot b/b \cdot C/C \cdot d/d \cdot E/E$$

The geneticist crosses the F_1 with a recessive tester and classifies the progeny by the gametic contribution of the F_1:

$A \cdot B \cdot C \cdot D \cdot E$	316
$a \cdot b \cdot C \cdot d \cdot E$	314
$A \cdot B \cdot C \cdot d \cdot E$	31
$a \cdot b \cdot C \cdot D \cdot E$	39
$A \cdot b \cdot C \cdot d \cdot E$	130
$a \cdot B \cdot C \cdot D \cdot E$	140
$A \cdot b \cdot C \cdot D \cdot E$	17
$a \cdot B \cdot C \cdot d \cdot E$	13
	1000

The second cross of pure lines is $A/A \cdot B/B \cdot C/C \cdot D/D \cdot E/E \times a/a \cdot B/B \cdot c/c \cdot D/D \cdot e/e.$

The geneticist crosses the F$_1$ from this cross with a recessive tester and obtains

$A \cdot B \cdot C \cdot D \cdot E$	243
$a \cdot B \cdot c \cdot D \cdot e$	237
$A \cdot B \cdot c \cdot D \cdot e$	62
$a \cdot B \cdot C \cdot D \cdot E$	58
$A \cdot B \cdot C \cdot D \cdot e$	155
$a \cdot B \cdot c \cdot D \cdot E$	165
$a \cdot B \cdot C \cdot D \cdot e$	46
$A \cdot B \cdot c \cdot D \cdot E$	34
	1000

The geneticist also knows that genes D and E assort independently.

a. Draw a map of these genes, showing distances in map units wherever possible.

b. Is there any evidence of interference?

65. In the plant *Arabidopsis,* the loci for pod length (*L,* long; *l,* short) and fruit hairs (*H,* hairy; *h,* smooth) are linked 16 map units apart on the same chromosome. The following crosses were made:

(i) $L\,H/L\,H \times l\,h/l\,h \rightarrow F_1$

(ii) $L\,h/L\,h \times l\,H/l\,H \rightarrow F_1$

If the F$_1$'s from cross i and cross ii are crossed,

a. what proportion of the progeny are expected to be *l h/l h*?

b. what proportion of the progeny are expected to be *L h/l h*?

66. In corn (*Zea mays*), the genetic map of part of chromosome 4 is as follows, where *w, s,* and *e* represent recessive mutant alleles affecting the color and shape of the pollen:

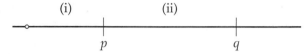

If the following cross is made

$$+ + +/+ + + \times w\,s\,e/w\,s\,e$$

and the F$_1$ is testcrossed with *w s e/w s e,* and if it is assumed that there is no interference on this region of the chromosome, what proportion of progeny will be of genotypes?

a.	+	+	+	**e.**	+	+	*e*
b.	*w*	*s*	*e*	**f.**	*w*	*s*	+
c.	+	*s*	*e*	**g.**	*w*	+	*e*
d.	*w*	+	+	**h.**	+	*s*	+

67. Every Friday night, genetics student Jean Allele, exhausted by her studies, goes to the student union's bowling lane to relax. But, even there, she is haunted by her genetic studies. The rather modest bowling lane has only four bowling balls: two red and two blue. They are bowled at the pins and are then collected and returned down the chute in random order, coming to rest at the end stop. As the evening passes, Jean notices familiar patterns of the four balls as they come to rest at the stop. Compulsively, she counts the different patterns. What patterns did she see, what were their frequencies, and what is the relevance of this matter to genetics?

68. In a tetrad analysis, the linkage arrangement of the *p* and *q* loci is as follows:

(i)　　　　　　　　(ii)

p　　　　　　　q

Assume that
- in region i, there is no crossover in 88 percent of meioses and there is a single crossover in 12 percent of meioses;
- in region ii, there is no crossover in 80 percent of meioses and there is a single crossover in 20 percent of meioses; and
- there is no interference (in other words, the situation in one region does not affect what is going on in the other region).

What proportions of tetrads will be of the following types? **(a)** M_IM_I, PD; **(b)** M_IM_I, NPD; **(c)** M_IM_{II}, T; **(d)** $M_{II}M_I$, T; **(e)** $M_{II}M_{II}$, PD; **(f)** $M_{II}M_{II}$, NPD; **(g)** $M_{II}M_{II}$,T. (**Note:** Here the M pattern written first is the one that pertains to the *p* locus.) **Hint:** The easiest way to do this problem is to start by calculating the frequencies of asci with crossovers in both regions, region i, region ii, and neither region. Then determine what M_I and M_{II} patterns result.

69. For an experiment with haploid yeast, you have two different cultures. Each will grow on minimal medium to which arginine has been added, but neither will grow on minimal medium alone. (Minimal medium is inorganic salts plus sugar.) Using appropriate methods, you induce the two cultures to mate. The diploid cells then divide meiotically and form unordered tetrads. Some of the ascospores will grow on minimal medium. You classify a large number of these tetrads for the phenotypes ARG$^-$ (arginine requiring) and ARG$^+$ (arginine independent) and record the following data:

Segregation of ARG$^-$: ARG$^+$	Frequency (%)
4:0	40
3:1	20
2:2	40

a. Using symbols of your own choosing, assign genotypes to the two parental cultures. For each of the three kinds of segregation, assign genotypes to the segregants.

b. If there is more than one locus governing arginine requirement, are these loci linked?

70. An RFLP analysis of two pure lines $A/A \cdot B/B$ and $a/a \cdot b/b$ showed that the former was homozygous for a long RFLP allele (l) and the latter for a short allele (s). The two were crossed to form an F_1, which was then backcrossed to the second pure line. A thousand progeny were scored as follows:

Aa Bb ss	9
Aa Bb ls	362
aa bb ls	11
aa bb ss	358
Aa bb ss	43
Aa bb ls	93
aa Bb ls	37
aa Bb ss	87

a. What do these results tell us about linkage?

b. Draw a map if appropriate.

c. Incorporate the RFLP fragments into your map.

The Genetics of Bacteria and Their Viruses

5

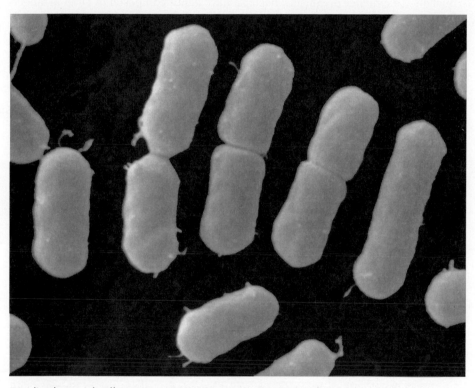

Dividing bacterial cells. [*Custom Medical Stock Photo RM/Getty Images.*]

KEY QUESTIONS

- By what processes do bacteria exchange genes?

- Can these exchange processes be used to map genes producing mutant phenotypes?

- How do phage genomes interact with bacterial genomes?

- How can phage genomes be mapped?

D NA technology is responsible for the rapid advances being made in the genetics of all model organisms. It is also a topic of considerable interest in the public domain. Examples are the highly publicized announcement of the full genome sequences of humans and chimpanzees in recent years and the popularity of DNA-based forensic analysis in television shows and movies (Figure 5-1). These dramatic results, whether in humans, fish, insects, plants, or fungi, are all based on the use of technologies that permit small pieces of DNA to be isolated, carried from cell to cell, and amplified into large samples. The sophisticated systems that permit these manipulations of the DNA of any organism are almost all derived from bacteria and their viruses. Hence, the advance of modern genetics to its present state of understanding was entirely dependent on the development of bacterial genetics, the topic of this chapter.

Even though bacterial genetics has made modern molecular genetics possible, it was never the goal of research in bacterial genetics. Bacteria are biologically

OUTLINE

5.1 Working with microorganisms

5.2 Bacterial conjugation

5.3 Bacterial transformation

5.4 Bacteriophage genetics

5.5 Transduction

5.6 Physical maps and linkage maps compared

The fruits of DNA technology, made possible by bacterial genetics

Figure 5-1 The dramatic results of modern DNA technology, such as sequencing the human genome, were possible only because bacterial genetics led to the invention of efficient DNA manipulation vectors. [*Science, vol. 291, no. 5507 (February 16, 2001), pp. 1145–1434. Image by Ann E. Cutting.*]

important in their own right. They are the most numerous organisms on our planet. They contribute to the recycling of nutrients such as nitrogen, sulfur, and carbon in ecosystems. Some are agents of human, animal, and plant disease. Others live symbiotically inside our mouths and intestines. In addition, many types of bacteria are useful for the industrial synthesis of a wide range of organic products. Hence, the impetus for the genetic dissection of bacteria has been the same as that for multicellular organisms—to understand their biological function.

Bacteria belong to a class of organisms known as **prokaryotes,** which also includes the blue-green algae (now classified as *cyanobacteria*). A key defining feature of prokaryotes is that their DNA is not enclosed in a membrane-bounded nucleus. Like higher organisms, bacteria have genes composed of DNA arranged in a long series on a "chromosome." However, the organization of their genetic material is unique in several respects. The genome of most bacteria is a single molecule of double-stranded DNA in the form of a closed circle. In addition, bacteria in nature often contain extra DNA elements called plasmids. Most plasmids also are DNA circles but are much smaller than the main bacterial genome.

Bacteria can be parasitized by specific **viruses** called **bacteriophages** or, simply, **phages.** Phages and other viruses are very different from the organisms that we have been studying so far. Viruses have some properties in common with organisms; for example, their genetic material can be DNA or RNA, constituting a short "chromosome." However, most biologists regard viruses as nonliving because they cannot reproduce alone. To reproduce, they must parasitize living cells and use the molecular machinery of these cells. Hence, for the study of their genetics, viruses must be propagated in the cells of their host organisms.

When scientists began studying bacteria and phages, they were naturally curious about their hereditary systems. Clearly, bacteria and phages must have hereditary systems because they show a constant appearance and function from one generation to the next (they are true to type). But how do these hereditary systems work? Bacteria, like unicellular eukaryotic organisms, reproduce asexually by cell growth and division, one cell becoming two. This asexual reproduction is quite easy to demonstrate experimentally. However, is there ever a union of different types for the purpose of sexual reproduction? Furthermore, how do the much smaller phages reproduce? Do they ever unite for a sexlike cycle? These questions are pursued in this chapter.

We will see that there are a variety of hereditary processes in bacteria and phages. These processes are interesting because of the basic biology of these forms, but they also act as models—as sources of insight into genetic processes at work in *all* organisms. For a geneticist, the attraction of these forms is that they can be cultured in very large numbers because they are so small. Consequently, it is possible to detect and study very *rare genetic events* that are difficult or impossible to study in eukaryotes.

What hereditary processes are observed in prokaryotes? In asexual cell division, the DNA is replicated but the partitioning of the new copies into daughter cells is accomplished by a mechanism quite different from mitosis. Do mutants arise? Indeed, the process of mutation occurs in asexual cells in much the same way as it does in eukaryotes, thus permitting the genetic dissection of bacterial function.

What about sexual reproduction and recombination? Because the cells and their chromosomes are so small, possible sexlike fusion events are difficult to observe, even with a microscope. Therefore, the general approach to detecting sexlike fusion has been a genetic one based on the detection of recombinants. The logic is that, if different genomes ever do get together in the same cell, they should occasionally produce recombinants. Conversely, if recombinants are detected, with marker A from one parent and B from another, then there must have been some type of "sexual" union. Hence, even though bacteria and phages do not undergo meiosis, the approach to the genetic analysis of these forms is surpris-

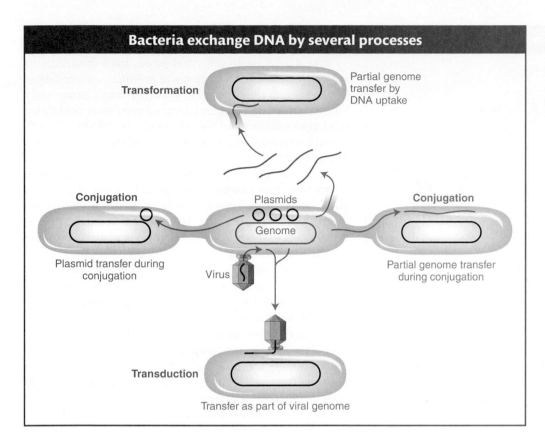

Figure 5-2 Bacterial DNA can be transferred from cell to cell in four ways: conjugation with plasmid transfer, conjugation with partial genome transfer, transformation, and transduction.

ingly similar to that for eukaryotes. The opportunity for genetic recombination in bacteria can arise in several different ways, but, in all cases, two DNA molecules are brought together. However, an important difference from eukaryotes is that, in bacteria, rarely are two complete chromosomes brought together; usually, the union is of one complete chromosome plus a fragment of another. The possibilities are outlined in Figure 5-2.

The first process of gene exchange to be examined in the chapter is *conjugation*, which is the contact and fusion of two different cells. After fusion, one cell, called a donor, sometimes transfers DNA in one direction to the other cell. The transferred DNA may be part or (rarely) all of the bacterial genome. In some cases, one of the extragenomic DNA elements called plasmids, if present, is transferred. Any genomic fragment transferred may recombine with the recipient's chromosome after entry.

A bacterial cell can also take up a piece of DNA from the external environment and incorporate this DNA into its own chromosome, a process called *transformation*. In addition, certain phages can pick up a piece of DNA from one bacterial cell and inject it into another, where it can be incorporated into the chromosome, in a process known as *transduction*.

Phages themselves can undergo recombination when two different genotypes both infect the same bacterial cell (**phage recombination,** not shown in Figure 5-2).

Before we analyze these modes of genetic exchange, let's consider the practical ways of handling bacteria, which are much different from those used in handling multicellular organisms.

5.1 Working with Microorganisms

Bacteria are fast-dividing and take up little space; so they are very convenient to use as genetic model organisms. They can be cultured in a liquid medium or on a

Bacterial colonies, each derived from a single cell

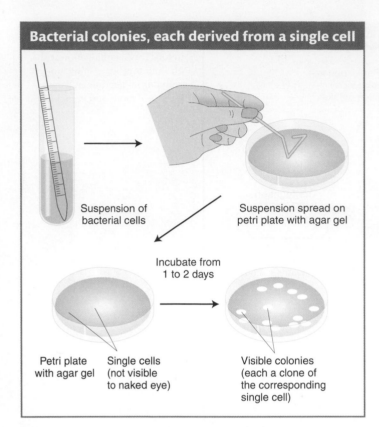

Suspension of bacterial cells

Suspension spread on petri plate with agar gel

Incubate from 1 to 2 days

Petri plate with agar gel

Single cells (not visible to naked eye)

Visible colonies (each a clone of the corresponding single cell)

Figure 5-3 Bacterial phenotypes can be assessed in their colonies. A stock of bacterial cells can be grown in a liquid medium containing nutrients, and then a small number of bacteria from the liquid suspension can be spread on solid agar medium. Each cell will give rise to a colony. All cells in a colony have the same genotype and phenotype.

Distinguishing *lac⁺* and *lac⁻* by using a red dye

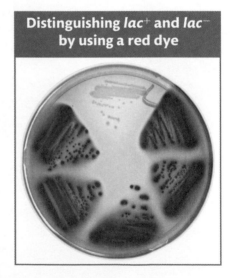

Figure 5-4 Wild-type bacteria able to use lactose as an energy source (*lac⁺*) stain red in the presence of this indicator dye. The unstained cells are mutants unable to use lactose (*lac⁻*). [*Jeffrey H. Miller.*]

solid surface such as an agar gel, as long as basic nutrients are supplied. Each bacterial cell divides asexually from 1 → 2 → 4 → 8 → 16 cells, and so on, until the nutrients are exhausted or until toxic waste products accumulate to levels that halt the population growth. A small amount of a liquid culture can be pipetted onto a petri plate containing solid agar medium and spread evenly on the surface with a sterile spreader, in a process called **plating** (Figure 5-3). The cells divide, but, because they cannot travel far on the surface of the gel, all the cells remain together in a clump. When this mass reaches more than 10^7 cells, it becomes visible to the naked eye as a **colony.** Each distinct colony on the plate has been derived from a single original cell. Members of a colony that have a single genetic ancestor are known as **cell clones.**

Bacterial mutants are quite easy to obtain. Nutritional mutants are a good example. Wild-type bacteria are **prototrophic,** which means that they can grow and divide on **minimal medium**—a substrate containing only inorganic salts, a carbon source for energy, and water. From a prototrophic culture, **auxotrophic** mutants can be obtained: these mutants are cells that will not grow unless the medium contains one or more specific cellular building blocks such as adenine, threonine, or biotin. Another type of useful mutant differs from wild type in the ability to use a specific energy source; for example, the wild type (*lac⁺*) can use lactose and grow, whereas a mutant (*lac⁻*) cannot. Figure 5-4 shows another way of distinguishing *lac⁺* and *lac⁻* colonies by using a dye. In another mutant category, whereas wild types are susceptible to an inhibitor, such as the antibiotic streptomycin, **resistant mutants** can divide and form colonies in the presence of the inhibitor. All these types of mutants allow the geneticist to distinguish different individual strains, thereby providing **genetic markers** (marker alleles) to keep track of genomes and cells in experiments. Table 5-1 summarizes some mutant bacterial phenotypes and their genetic symbols.

The following sections document the discovery of the various processes by which bacterial genomes recombine. The historical methods are interesting in themselves but also serve to introduce the diverse processes of recombination, as well as analytical techniques that are still applicable today.

Table 5-1   Some Genotypic Symbols Used in Bacterial Genetics

Symbol	Character or phenotype associated with symbol
bio⁻	Requires biotin added as a supplement to minimal medium
arg⁻	Requires arginine added as a supplement to minimal medium
met⁻	Requires methionine added as a supplement to minimal medium
lac⁻	Cannot utilize lactose as a carbon source
gal⁻	Cannot utilize galactose as a carbon source
str^r	Resistant to the antibiotic streptomycin
str^s	Sensitive to the antibiotic streptomycin

Note: Minimal medium is the basic synthetic medium for bacterial growth without nutrient supplements.

5.2 Bacterial Conjugation

The earliest studies in bacterial genetics revealed the unexpected process of cell conjugation.

Discovery of conjugation

Do bacteria possess any processes similar to sexual reproduction and recombination? The question was answered by the elegantly simple experimental work of Joshua Lederberg and Edward Tatum, who in 1946 discovered a sexlike process in what became the main model for bacterial genetics, *Escherichia coli* (see the Model Organism box below). They were studying two strains of *E. coli* with different sets of auxotrophic mutations. Strain A$^-$ would grow only if the medium were supplemented with methionine and biotin; strain B$^-$ would grow only if it were supplemented with threonine, leucine, and thiamine. Thus, we can designate the strains as

$$\text{strain A}^-: \quad met^-\ bio^-\ thr^+\ leu^+\ thi^+$$
$$\text{strain B}^-: \quad met^+\ bio^+\ thr^-\ leu^-\ thi^-$$

Model Organism *Escherichia coli*

The seventeenth-century microscopist Antony van Leeuwenhoek was probably the first to see bacterial cells and to recognize their small size: "There are more living in the scum on the teeth in a man's mouth than there are men in the whole kingdom." However, bacteriology did not begin in earnest until the nineteenth century. In the 1940s, Joshua Lederberg and Edward Tatum made the discovery that launched bacteriology into the burgeoning field of genetics: they discovered that, in a certain bacterium, there was a type of sexual cycle including a crossing-over-like process. The organism that they chose for this experiment has become the model not only for prokaryote genetics, but in a sense for all of genetics. The organism was *Escherichia coli*, a bacterium named after its discoverer, the nineteenth-century German bacteriologist Theodore Escherich.

The choice of *E. coli* was fortunate because it has proved to have many features suitable for genetic research, not the least of which is that it is easily obtained, given that it lives in the gut of humans and other animals. In the gut, it is a benign symbiont, but it occasionally causes urinary tract infections and diarrhea.

E. coli has a single circular chromosome 4.6 Mb in length. Of its 4000 intron-free genes, about 35 percent are of unknown function. The sexual cycle is made possible by the action of an extragenomic plasmid called F, which confers a type of "maleness." Other plasmids carry genes whose functions equip the cell for life in specific environments, such as drug-resistance genes. These plasmids have been adapted as gene *vectors*, which are gene carriers that form the basis of the gene transfers at the center of modern genetic engineering.

E. coli is unicellular and grows by simple cell division. Because of its small size (~1 μm in length), *E. coli* can be grown in large numbers and subjected to intensive selection and screening for rare genetic events. *E. coli* research represents the beginning of "black box" reasoning in genetics: through the selection and analysis of mutants, the workings of the genetic machinery could be deduced even though it was too small to be seen. Phenotypes such as colony size, drug resistance, carbon-source utilization, and colored-dye production took the place of the visible phenotypes of eukaryotic genetics.

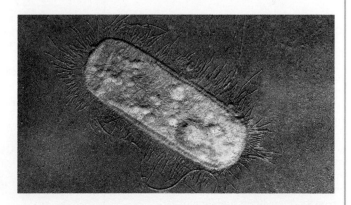

An electron micrograph of an *E. coli* cell showing long flagella, used for locomotion, and fimbriae, proteinaceous hairs that are important in anchoring the cells to animal tissues. (Sex pili are not shown in this micrograph.) [*Biophoto Associates/Science Photo Library.*]

Figure 5-5a displays in simplified form the design of their experiment. Strains A⁻ and B⁻ were mixed together, incubated for a while, and then plated on minimal medium, on which neither auxotroph could grow. A small minority of the cells (1 in 10⁷) was found to grow as prototrophs and, hence, must have been wild type, having regained the ability to grow without added nutrients. Some of the dishes were plated only with strain A⁻ bacteria and some only with strain B⁻ bacteria to act as controls, but no prototrophs arose from these platings. Figure 5-5b illustrates the experiment in more detail. These results suggested that some form of recombination of genes had taken place between the genomes of the two strains to produce the prototrophs.

It could be argued that the cells of the two strains do not really exchange genes but instead leak substances that the other cells can absorb and use for growing. This possibility of "cross-feeding" was ruled out by Bernard Davis in the following way. He constructed a U-shaped tube in which the two arms were separated by a

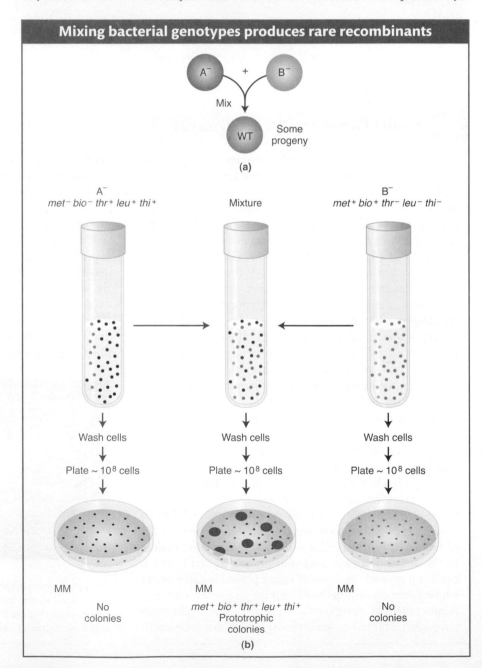

Figure 5-5 With the use of this method, Lederberg and Tatum demonstrated that genetic recombination between bacterial genotypes is possible. (a) The basic concept: two auxotrophic cultures (A⁻ and B⁻) are mixed, yielding prototrophic wild types (WT). (b) Cells of type A⁻ or type B⁻ cannot grow on an unsupplemented (minimal) medium (MM), because A⁻ and B⁻ each carry mutations that cause the inability to synthesize constituents needed for cell growth. When A⁻ and B⁻ are mixed for a few hours and then plated, however, a few colonies appear on the agar plate. These colonies derive from single cells in which genetic material has been exchanged; they are therefore capable of synthesizing all the required constituents of metabolism.

fine filter. The pores of the filter were too small to allow bacteria to pass through but large enough to allow easy passage of any dissolved substances (Figure 5-6). Strain A⁻ was put in one arm, strain B⁻ in the other. After the strains had been incubated for a while, Davis tested the contents of each arm to see if there were any prototrophic cells, but none were found. In other words, *physical contact* between the two strains was needed for wild-type cells to form. It looked as though some kind of genome union had taken place, and genuine recombinants had been produced. The physical union of bacterial cells can be confirmed under an electron microscope and is now called **conjugation** (Figure 5-7).

Discovery of the fertility factor (F)

In 1953, William Hayes discovered that, in the types of "crosses" just described here, the conjugating parents acted *unequally* (later, we will see ways to demonstrate this unequal participation). One parent (and *only* that parent) seemed to transfer some or all of its genome into another cell. Hence, one cell acts as a **donor,** and the other cell acts as a **recipient.** This "cross" is quite different from eukaryotic crosses in which parents contribute nuclear genomes equally.

> **Message** The transfer of genetic material in *E. coli* conjugation is not reciprocal. One cell, the donor, transfers part of its genome to the other cell, which acts as the recipient.

By accident, Hayes discovered a variant of his original donor strain that would not produce recombinants on crossing with the recipient strain. Apparently, the donor-type strain had lost the ability to transfer genetic material and had changed into a recipient-type strain. In working with this "sterile" donor variant, Hayes found that it could regain the ability to act as a donor by association with other donor strains. Indeed, the donor ability was transmitted rapidly and effectively between strains during conjugation. A kind of "infectious transfer" of some factor seemed to be taking place. He suggested that donor ability is itself a hereditary state, imposed by a **fertility factor (F).** Strains that carry F can donate and are designated **F⁺.** Strains that lack F cannot donate and are recipients, designated **F⁻.**

We now know much more about F. It is an example of a small, nonessential circular DNA molecule called a **plasmid** that can replicate in the cytoplasm independent of the host chromosome. Figure 5-8 shows how bacteria can transfer plasmids such as F. The F plasmid directs the synthesis of pili (sing., pilus), projections that initiate contact with a recipient (see Figures 5-7 and 5-8) and draw it closer. The F DNA in the donor cell makes a single-stranded copy of itself in a peculiar mechanism called **rolling circle replication.** The circular plasmid "rolls," and as it turns, it reels out the single-stranded copy like fishing line. This copy passes through a pore into the recipient cell, where the other strand is synthesized, forming a double helix. Hence, a copy of F remains in the donor and another appears in the recipient, as shown in Figure 5-8. Note that the *E. coli* genome is depicted as a single circular chromosome in Figure 5-8. (We will examine the evidence for it later.) Most bacterial genomes are circular, a feature quite

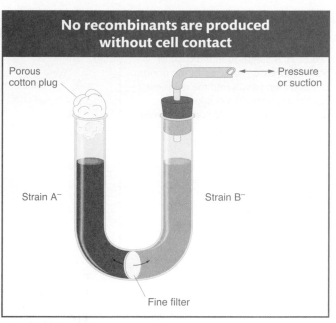

No recombinants are produced without cell contact

Figure 5-6 Auxotrophic bacterial strains A⁻ and B⁻ are grown on either side of a U-shaped tube. Liquid may be passed between the arms by applying pressure or suction, but the bacterial cells cannot pass through the filter. After incubation and plating, no recombinant colonies grow on minimal medium.

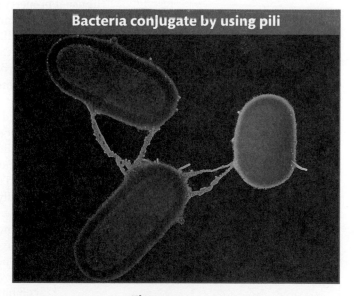

Bacteria conjugate by using pili

Figure 5-7 A donor cell extends one or more projections, or pili, that attach to a recipient cell and pull the two bacteria together. [*Dr. L. Caro / Science Source.*]

Figure 5-8 (a) During conjugation, the pilus pulls two bacteria together. (b) Next, a pilus forms between the two cells. A single-stranded copy of plasmid DNA is produced in the donor cell and then passes into the recipient bacterium, where the single strand, serving as a template, is converted into the double-stranded helix.

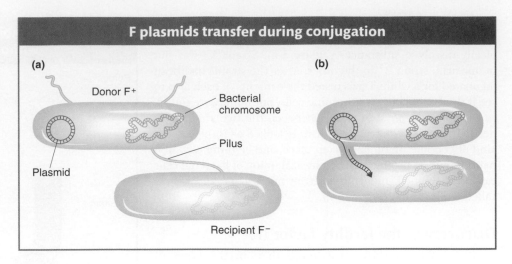

F plasmids transfer during conjugation

(a)

Donor F⁺

Bacterial chromosome

Pilus

Plasmid

Recipient F⁻

(b)

different from eukaryotic nuclear chromosomes. We will see that this feature leads to the many idiosyncrasies of bacterial genetics.

Hfr strains

An important breakthrough came when Luca Cavalli-Sforza discovered a derivative of an F⁺ strain with two unusual properties:

1. On crossing with F⁻ strains, this new strain produced 1000 times as many recombinants as a normal F⁺ strain. Cavalli-Sforza designated this derivative an **Hfr** strain to symbolize its ability to promote a *h*igh *f*requency of *r*ecombination.

2. In Hfr × F⁻ crosses, virtually none of the F⁻ parents were converted into F⁺ or into Hfr. This result is in contrast with F⁺ × F⁻ crosses, in which, as we have seen, infectious transfer of F results in a large proportion of the F⁻ parents being converted into F⁺.

It became apparent that an Hfr strain results from the integration of the F factor into the chromosome, as pictured in Figure 5-9. We can now explain the first unusual property of Hfr strains. During conjugation, the F factor inserted in the chromosome efficiently drives part or all of that chromosome into the F⁻ cell. The chromosomal fragment can then engage in recombination with the recipient chromosome. The rare recombinants observed by Lederberg and Tatum in F⁺ × F⁻ crosses were due to the spontaneous, but rare, formation of Hfr cells in the F⁺ culture. Cavalli-Sforza isolated examples of these rare cells from F⁺ cultures and found that, indeed, they now acted as true Hfr's.

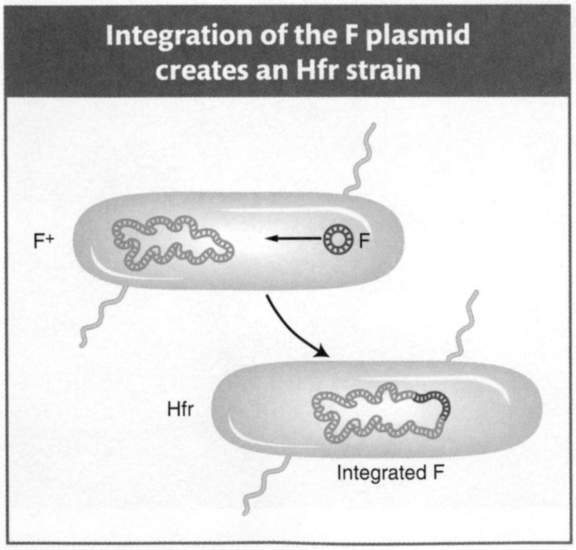

Does an Hfr cell die after donating its chromosomal material to an F⁻ cell? The answer is no. Just like the F plasmid, the Hfr chromosome replicates and transfers a single strand to the F⁻ cell during conjugation. That the transferred DNA is a single strand can be demonstrated visually with the use of special strains and antibodies, as shown in Figure 5-10. The replication of the chromosome ensures a complete chromosome for the donor cell after mating. The transferred strand is converted into a double helix in the recipient cell, and donor genes may become incorporated in the recipient's chromosome through crossovers, creating a recombinant cell (Figure 5-11). If there is no recombination, the transferred fragments of DNA are simply lost in the course of cell division.

Integration of the F plasmid creates an Hfr strain

F⁺

F

Hfr

Integrated F

Figure 5-9 In an F⁺ strain the free F plasmid occasionally integrates into the *E. coli* chromosome, creating an Hfr strain.

Linear transmission of the Hfr genes from a fixed point A clearer view of the behavior of Hfr strains was obtained in 1957, when Elie Wollman and François

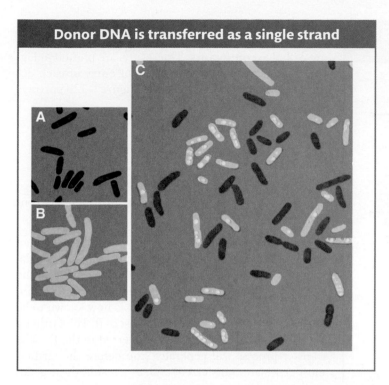

Donor DNA is transferred as a single strand

Figure 5-10 The photographs show a visualization of single-stranded DNA transfer in conjugating *E. coli* cells, with the use of special fluorescent antibodies. Parental Hfr strains (A) are black with red DNA. The red is from the binding of an antibody to a protein normally attached to DNA. The recipient F⁻ cells (B) are green due to the presence of the gene for a jellyfish protein that fluoresces green, and, because they are mutant for a certain gene, their DNA protein does not bind to antibody. When single-stranded DNA enters the recipient, it promotes atypical binding of this protein, which fluoresces yellow in this background. Part C shows Hfr's (unchanged) and exconjugants (cells that have undergone conjugation) with yellow transferred DNA. A few unmated F⁻ cells are visible. [*From M. Kohiyama, S. Hiraga, I. Matic, and M. Radman, "Bacterial Sex: Playing Voyeurs 50 Years Later," Science 301, 2003, p. 803, Fig. 1.*]

Jacob investigated the pattern of transmission of Hfr genes to F⁻ cells during a cross. They crossed

$$\text{Hfr } azi^r \, ton^r \, lac^+ \, gal^+ \, str^s \times \text{F}^- \, azi^s \, ton^s \, luc^- \, gal^- \, str^r$$

(Superscripts "r" and "s" stand for resistant and sensitive, respectively.) At specific times after mixing, they removed samples, which were each put in a kitchen blender for a few seconds to separate the mating cell pairs. This procedure is called **interrupted mating.** The sample was then plated onto a medium containing streptomycin to kill the Hfr donor cells, which bore the sensitivity allele *str*ˢ. The surviving *str*ʳ cells then were tested for the presence of alleles

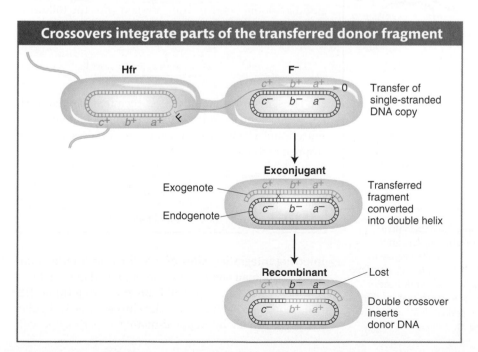

Crossovers integrate parts of the transferred donor fragment

Figure 5-11 After conjugation, crossovers are needed to integrate genes from the donor fragment into the recipient's chromosome and, hence, become a stable part of its genome. GENETICS**PORTAL** ANIMATED ART: **Bacterial conjugation and recombination**

Tracking time of marker entry generates a chromosome map

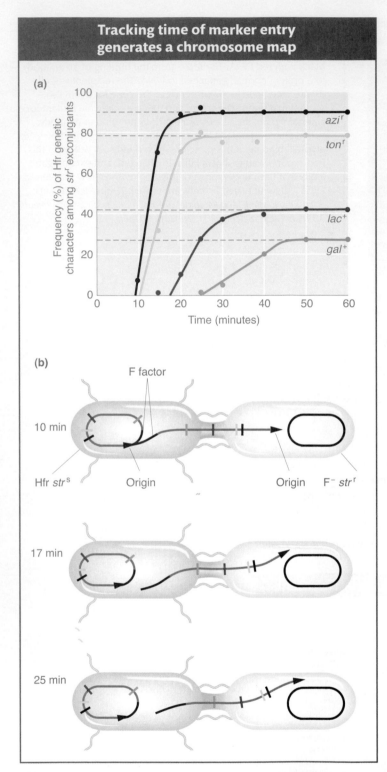

(a)

(b)

F factor

10 min

Hfr *str*^s   Origin   Origin   F⁻ *str*^r

17 min

25 min

Figure 5-12 In this interrupted-mating conjugation experiment, F⁻ streptomycin-resistant cells with mutations in *azi, ton, lac,* and *gal* are incubated for varying times with Hfr cells that are sensitive to streptomycin and carry wild-type alleles for these genes. (a) A plot of the frequency of donor alleles in exconjugants as a function of time after mating. (b) A schematic view of the transfer of markers (shown in different colors) with the passage of time. [(a) After E. L. Wollman, F. Jacob, and W. Hayes, Cold Spring Harbor Symp. Quant. Biol. 21, 1956, 141.]

from the donor genome. Any *str*^r cell bearing a donor allele must have taken part in conjugation; such cells are called **exconjugants.** The results are plotted in Figure 5-12a, showing a time course of entry of each donor allele *azi*^r, *ton*^r, *lac*⁺, and *gal*⁺. Figure 5-12b portrays the transfer of Hfr alleles.

The key elements in these results are

1.   Each donor allele first appears in the F⁻ recipients at a specific time after mating began.

2.   The donor alleles appear in a specific sequence.

3.   Later donor alleles are present in fewer recipient cells.

Putting all these observations together, Wollman and Jacob deduced that, in the conjugating Hfr, single-stranded DNA transfer begins from a fixed point on the donor chromosome, termed the **origin (O),** and continues in a linear fashion. The point O is now known to be the site at which the F plasmid is inserted. The farther a gene is from O, the later it is transferred to the F⁻. The transfer process will generally stop before the farthermost genes are transferred, and, as a result, these genes are included in fewer exconjugants.

How can we explain the second unusual property of Hfr crosses, that F⁻ exconjugants are rarely converted into Hfr or F⁺? When Wollman and Jacob allowed Hfr × F⁻ crosses to continue for as long as 2 hours before disruption, they found that in fact a few of the exconjugants were converted into Hfr. In other words, the part of F that confers donor ability was eventually transmitted but at a very low frequency. The rareness of Hfr exconjugants suggested that the inserted F was transmitted as the *last* element of the linear chromosome. We can summarize the order of transmission with the following generic map, in which the arrow indicates the direction of transfer, beginning with O:

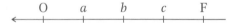

O   *a*   *b*   *c*   F

Thus, almost none of the F⁻ recipients are converted, because the fertility factor is the last element transmitted and usually the transmission process will have stopped before getting that far.

> **Message** The Hfr chromosome, originally circular, unwinds a copy of itself that is transferred to the F⁻ cell in a linear fashion, with the F factor entering last.

Inferring integration sites of F and chromosome circularity   Wollman and Jacob went on to shed more light on how and where the F plasmid integrates to form an Hfr and, in doing so, deduced that the chromosome is circular. They performed interrupted-mating experiments with different, separately derived Hfr strains. Significantly, the order of transmission of the alleles differed from strain to strain, as in the following examples:

Hfr strain	
H	O *thr pro lac pur gal his gly thi* F
1	O *thr thi gly his gal pur lac pro* F
2	O *pro thr thi gly his gal pur lac* F
3	O *pur lac pro thr thi gly his gal* F
AB 312	O *thi thr pro lac pur gal his gly* F

Each line can be considered a map showing the order of alleles on the chromosome. At first glance, there seems to be a random shuffling of genes. However, when some of the Hfr maps are inverted, the relation of the sequences becomes clear.

H (written backward)	F *thi gly his gal pur lac pro thr* O
1	O *thr thi gly his gal pur lac pro* F
2	O *pro thr thi gly his gal pur lac* F
3	O *pur lac pro thr thi gly his gal* F
AB 312 (written backward)	F *gly his gal pur lac pro thr thi* O

The relation of the sequences to one another is explained if each map is the segment of a circle. It was the first indication that bacterial chromosomes are circular. Furthermore, Allan Campbell proposed a startling hypothesis that accounted for the different Hfr maps. He proposed that, if F is a ring, then insertion might be by a simple crossover between F and the bacterial chromosome (Figure 5-13). That

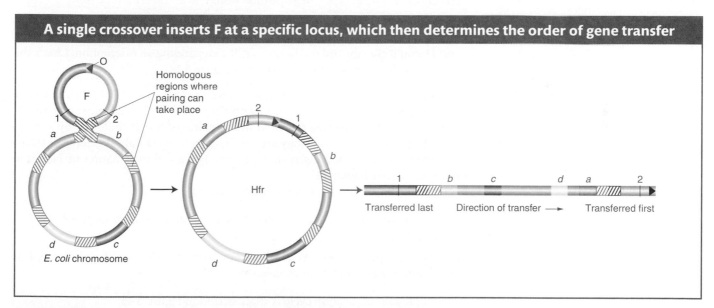

A single crossover inserts F at a specific locus, which then determines the order of gene transfer

Figure 5-13 The insertion of F creates an Hfr cell. Hypothetical markers 1 and 2 are shown on F to depict the direction of insertion. The origin (O) is the mobilization point where insertion into the *E. coli* chromosome occurs; the pairing region is homologous with a region on the *E. coli* chromosome; *a* through *d* are representative genes in the *E. coli* chromosome. Pairing regions (hatched) are identical in plasmid and chromosome. They are derived from mobile elements called *insertion sequences* (see Chapter 14). In this example, the Hfr cell created by the insertion of F would transfer its genes in the order *a, d, c, b*.

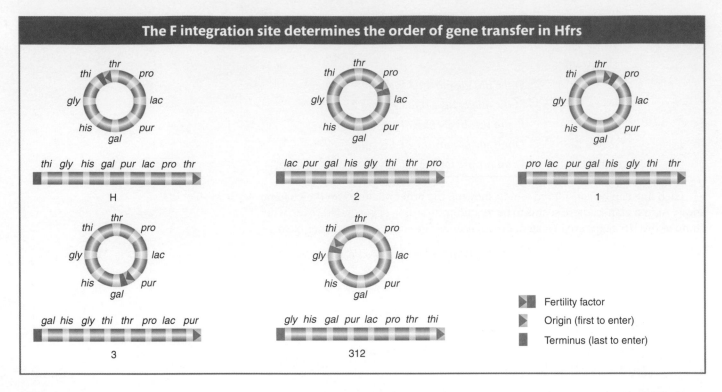

The F integration site determines the order of gene transfer in Hfrs

Figure 5-14 The five *E. coli* Hfr strains shown each have different F plasmid insertion points and orientations. All strains have the same order of genes on the *E. coli* chromosome. The orientation of the F factor determines which gene enters the recipient cell first. The gene closest to the terminus enters last.

being the case, any of the linear Hfr chromosomes could be generated simply by the insertion of F into the ring in the appropriate place and orientation (Figure 5-14).

Several hypotheses—later supported—followed from Campbell's proposal.

1. One end of the integrated F factor would be the origin, where transfer of the Hfr chromosome begins. The **terminus** would be at the other end of F.

2. The orientation in which F is inserted would determine the order of entry of donor alleles. If the circle contains genes *A*, *B*, *C*, and *D*, then insertion between *A* and *D* would give the order *ABCD* or *DCBA*, depending on orientation. Check the different orientations of the insertions in Figure 5-14.

How is it possible for F to integrate at different sites? If F DNA had a region homologous to any of several regions on the bacterial chromosome, any one of them could act as a pairing region at which pairing could be followed by a crossover. These regions of homology are now known to be mainly segments of transposable elements called *insertion sequences*. For a full explanation of insertion sequences, see Chapter 14.

The fertility factor thus exists in two states:

1. The plasmid state: As a free cytoplasmic element, F is easily transferred to F⁻ recipients.

2. The integrated state: As a contiguous part of a circular chromosome, F is transmitted only very late in conjugation.

The *E. coli* conjugation cycle is summarized in Figure 5-15.

Mapping of bacterial chromosomes

Broad-scale chromosome mapping by using time of entry   Wollman and Jacob realized that the construction of linkage maps from the interrupted-mating results would be easy by using as a measure of "distance" the times at which the donor alleles first appear after mating. The units of map distance in this case are minutes. Thus, if b^+ begins to enter the F^- cell 10 minutes after a^+ begins to enter, then a^+ and b^+ are 10 units apart. Like eukaryotic maps based on crossovers, these linkage maps were originally purely genetic constructions. At the time they were originally devised, there was no way of testing their physical basis.

Fine-scale chromosome mapping by using recombinant frequency   For an exconjugant to acquire donor genes as a permanent feature of its genome, the donor fragment must recombine with the recipient chromosome. However, note that time-of-entry mapping is not based on recombinant frequency. Indeed, the units are minutes, not RF. Nevertheless, recombinant frequency can be used for a more fine-scale type of mapping in bacteria, a method to which we now turn.

First, we need to understand some special features of the recombination event in bacteria. Recall that recombination does not take place between two whole genomes, as it does in eukaryotes. In contrast, it takes place between one *complete* genome, from the F^-, called the **endogenote,** and an *incomplete* one, derived from the Hfr donor and called the **exogenote.** The cell at this stage has two copies of one segment of DNA: one copy is part of the endogenote and the other copy is part of the exogenote. Thus, at this stage, the cell is a *partial* diploid, called a **merozygote.** Bacterial genetics is merozygote genetics. A single crossover in a merozygote would break the ring and thus not produce viable recombinants, as

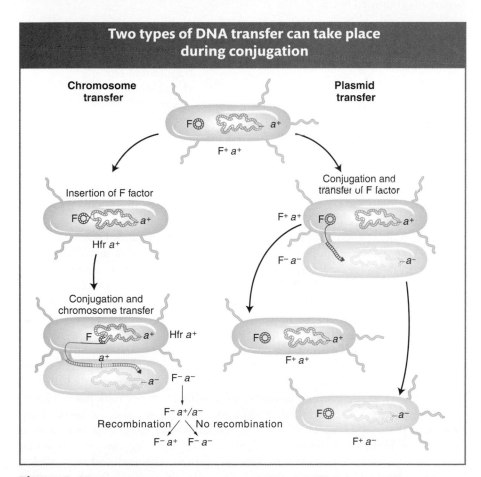

Figure 5-15 Conjugation can take place by partial transfer of a chromosome containing the F factor or by transfer of an F plasmid that remains a separate entity.

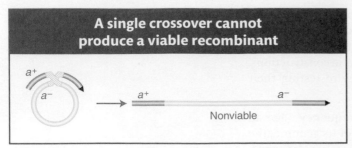

Figure 5-16 A single crossover between exogenote and endogenote in a merozygote would lead to a linear, partly diploid chromosome that would not survive.

shown in Figure 5-16. To keep the circle intact, there must be an even number of crossovers. An even number of crossovers produces a circular, intact chromosome and a fragment. Although such recombination events are represented in a shorthand way as double crossovers, the actual molecular mechanism is somewhat different, more like an invasion of the endogenote by an internal section of the exogenote. The other product of the "double crossover," the fragment, is generally lost in subsequent cell growth. Hence, only one of the reciprocal products of recombination survives. Therefore, another unique feature of bacterial recombination is that we must forget about reciprocal exchange products in most cases.

> **Message** Recombination during conjugation results from a double-crossover-like event, which gives rise to reciprocal recombinants of which only one survives.

With this understanding, we can examine recombination mapping. Suppose that we want to calculate map distances separating three close loci: *met, arg,* and *leu.* To examine the recombination of these genes, we need "trihybrids," exconjugants that have received all three donor markers. Assume that an interrupted-mating experiment has shown that the order is *met, arg, leu,* with *met* transferred first and *leu* last. To obtain a trihybrid, we need the merozygote diagrammed here:

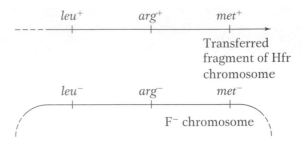

To obtain this merozygote, we must first select stable exconjugants bearing the *last* donor allele, which, in this case, is *leu*+. Why? Because, if we select for the last marker, then we know that such cells at some stage must have contained the earlier markers, too—namely, *arg*+ and *met*+.

The goal now is to count the frequencies of crossovers at different locations. Note that we now have a different situation from the analysis of interrupted conjugation. In mapping by interrupted conjugation, we measure the time of entry of individual loci; to be stably inherited, each marker has to recombine into the recipient chromosome by a double crossover spanning it. However, in the recombinant frequency analysis, we have specifically selected trihybrids as a starting point, and now we have to consider the various possible combinations of the three donor alleles that can be inserted by double crossing over in the various intervals. We know that *leu*+ must have entered and inserted because we selected it, but the *leu*+ recombinants that we select may or may not have incorporated the other donor markers, depending on where the double crossover took place. Hence, the procedure is to first select *leu*+ exconjugants and then isolate and test a large sample of them to see which of the other markers were integrated. Let's look at an example. In the cross Hfr *met*+ *arg*+ *leu*+ *str*s × F− *met*− *arg*− *leu*− *str*r, we would select *leu*+ recombinants and then examine them for the *arg*+ and *met*+ alleles, called the **unselected markers.** Figure 5-17 depicts the types of double-crossover events expected. One crossover must be on the left side of the *leu* marker and the other must be on the right side. Let's assume that the *leu*⁺ exconjugants are of the following types and frequencies:

$$leu^+ \; arg^- \; met^- \quad 4\%$$
$$leu^+ \; arg^+ \; met^- \quad 9\%$$
$$leu^+ \; arg^+ \; met^+ \quad 87\%$$

The double crossovers needed to produce these genotypes are shown in Figure 5-17. The first two types are the key because they require a crossover between *leu* and *arg* in the first case and between *arg* and *met* in the second. Hence, the rela-

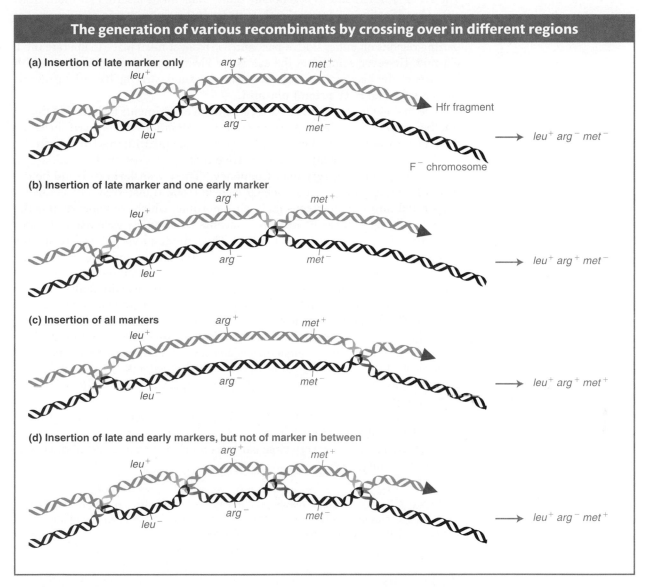

The generation of various recombinants by crossing over in different regions

(a) **Insertion of late marker only**
leu⁺ *arg*⁺ *met*⁺
Hfr fragment
leu⁻ *arg*⁻ *met*⁻
F⁻ chromosome
⟶ *leu*⁺ *arg*⁻ *met*⁻

(b) **Insertion of late marker and one early marker**
leu⁺ *arg*⁺ *met*⁺
leu⁻ *arg*⁻ *met*⁻
⟶ *leu*⁺ *arg*⁺ *met*⁻

(c) **Insertion of all markers**
leu⁺ *arg*⁺ *met*⁺
leu⁻ *arg*⁻ *met*⁻
⟶ *leu*⁺ *arg*⁺ *met*⁺

(d) **Insertion of late and early markers, but not of marker in between**
leu⁺ *arg*⁺ *met*⁺
leu⁻ *arg*⁻ *met*⁻
⟶ *leu*⁺ *arg*⁻ *met*⁺

Figure 5-17 The diagram shows how genes can be mapped by recombination in *E. coli*. In exconjugants, selection is made for merozygotes bearing the *leu*⁺ marker, which is donated late. The early markers (*arg*⁺ and *met*⁺) may or may not be inserted, depending on the site where recombination between the Hfr fragment and the F⁻ chromosome takes place. The frequencies of events diagrammed in parts *a* and *b* are used to obtain the relative sizes of the *leu–arg* and *arg–met* regions. Note that, in each case, only the DNA inserted into the F⁻ chromosome survives; the other fragment is lost.

GENETICS*PORTAL* ANIMATED ART: **Bacterial conjugation and mapping by recombination**

tive frequencies of these types correspond to the sizes of these two regions. We would conclude that the *leu-arg* region is 4 map units and that the *arg-met* is 9 map units.

In a cross such as the one just described, one type of potential recombinants, of genotype *leu⁺ arg⁻ met⁺*, requires four crossovers instead of two (see the bottom of Figure 5-17). These recombinants are rarely recovered, because their frequency is very low compared with that of the other types of recombinants.

F plasmids that carry genomic fragments

The F factor in Hfr strains is generally quite stable in its inserted position. However, occasionally an F factor cleanly exits from the chromosome by a reversal of the recombination process that inserted it in the first place. The two homologous pairing regions on either side re-pair, and a crossover takes place to liberate the F plasmid. However, sometimes the exit is not clean, and the plasmid carries with it a part of the bacterial chromosome. An F plasmid carrying bacterial genomic DNA is called an **F′ (F prime) plasmid.**

The first evidence of this process came from experiments in 1959 by Edward Adelberg and François Jacob. One of their key observations was of an Hfr in which the F factor was integrated near the *lac⁺* locus. Starting with this Hfr *lac⁺* strain, Jacob and Adelberg found an F⁺ derivative that, in crosses, transferred *lac⁺* to F⁻ *lac⁻* recipients at a very high frequency. (These transferrants could be detected by plating on medium lacking lactose.) The transferred *lac⁺* is not incorporated into the recipient's main chromosome, which we know retains the allele *lac⁻* because these F⁺ *lac⁺* exconjugants occasionally gave rise to F⁻ *lac⁻* daughter cells, at a frequency of 1×10^{-3}. Thus, the genotype of these recipients appeared to be F′ *lac⁺*/F⁻ *lac⁻*. In other words, the *lac⁺* exconjugants seemed to carry an F′ plasmid with a piece of the donor chromosome incorporated. The origin of this F′ plasmid is shown in Figure 5-18. Note that the faulty excision occurs because there is another homologous region nearby that pairs with the original. The F′ in our example is called F′ *lac* because the piece of host chromosome that it picked up has the *lac* gene on it. F′ factors have been found carrying many different chromosomal genes and have been named accordingly. For example, F′ factors carrying *gal* or *trp* are called F′ *gal* and F′ *trp,* respectively. Because F *lac⁺*/F⁻ *lac⁻* cells are *lac⁺* in phenotype, we know that *lac⁺* is dominant over *lac⁻*.

Partial diploids made with the use of F′ strains are useful for some aspects of routine bacterial genetics, such as the study of dominance or of allele interaction. Some F′ strains can carry very large parts (as much as one-quarter) of the bacterial chromosome.

> **Message** The DNA of an F′ plasmid is part F factor and part bacterial genome. Like F plasmids, F′ plasmids transfer rapidly. They can be used to establish partial diploids for studies of bacterial dominance and allele interaction.

R plasmids

An alarming property of pathogenic bacteria first came to light through studies in Japanese hospitals in the 1950s. Bacterial dysentery is caused by bacteria of the genus *Shigella.* This bacterium was initially sensitive to a wide array of antibiotics that were used to control the disease. In the Japanese hospitals, however, *Shigella* isolated from patients with dysentery proved to be simultaneously resistant to many of these drugs, including penicillin, tetracycline, sulfanilamide, streptomycin, and chloramphenicol. This resistance to multiple drugs was inherited as a

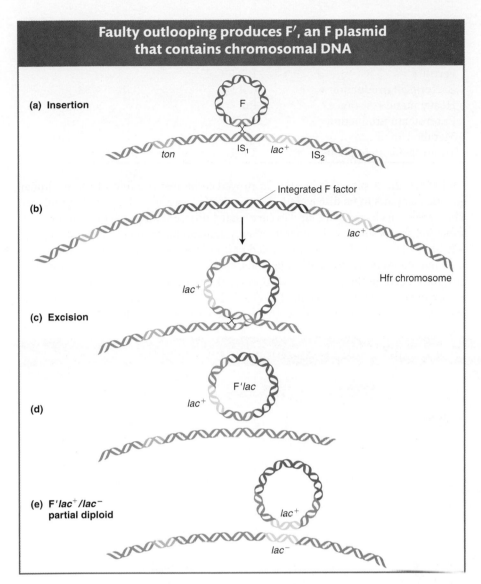

Faulty outlooping produces F′, an F plasmid that contains chromosomal DNA

(a) Insertion

F

ton   IS₁   *lac*⁺   IS₂

(b)

Integrated F factor

lac⁺

Hfr chromosome

(c) Excision

lac⁺

(d)

F′*lac*

lac⁺

(e) F′*lac*⁺/*lac*⁻ partial diploid

lac⁺

lac⁻

Figure 5-18 An F factor can pick up chromosomal DNA as it exits a chromosome. (a) F is inserted in an Hfr strain at a repetitive element identified as IS₁ (insertion sequence 1) between the *ton* and *lac*⁺ alleles. (b) The inserted F factor. (c) Abnormal "outlooping" by crossing over with a different element, IS₂, to include the *lac* locus. (d) The resulting F′ *lac*⁺ particle. (e) F′ *lac*⁺/F⁻ *lac*⁻ partial diploid produced by the transfer of the F′ *lac*⁺ particle to an F⁻ *lac*⁻ recipient. [*From G. S. Stent and R. Calendar, Molecular Genetics, 2nd ed. Copyright 1978 by W. H. Freeman and Company.*]

single genetic package, and it could be transmitted in an infectious manner—not only to other sensitive *Shigella* strains, but also to other related species of bacteria. This talent, which resembles the mobility of the *E. coli* F plasmid, is an extraordinarily useful one for the pathogenic bacterium because resistance can rapidly spread throughout a population. However, its implications for medical science are dire because the bacterial disease suddenly becomes resistant to treatment by a large range of drugs.

From the point of view of the geneticist, however, the mechanism has proved interesting and is useful in genetic engineering. The vectors carrying these multiple resistances proved to be another group of plasmids called **R plasmids.** They are transferred rapidly on cell conjugation, much like the F plasmid in *E. coli*.

Table 5-2   Genetic Determinants Borne by Plasmids

Characteristic	Plasmid examples
Fertility	F, R1, Col
Bacteriocin production	Col E1
Heavy-metal resistance	R6
Enterotoxin production	Ent
Metabolism of camphor	Cam
Tumorigenicity in plants	T1 (in *Agrobacterium tumefaciens*)

In fact, the R plasmids in *Shigella* proved to be just the first of many similar genetic elements to be discovered. All exist in the plasmid state in the cytoplasm. These elements have been found to carry many different kinds of genes in bacteria. Table 5-2 shows some of the characteristics that can be borne by plasmids. Figure 5-19 shows an example of a well-traveled plasmid isolated from the dairy industry.

Engineered derivatives of R plasmids, such as pBR 322 and pUC (see Chapter 20), have become the preferred vectors for the molecular cloning of the DNA of all organisms. The genes on an R plasmid that confer resistance can be used as markers to keep track of the movement of the vectors between cells.

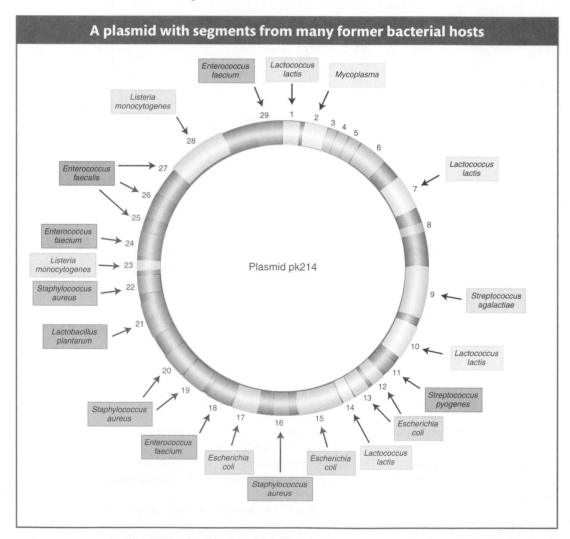

A plasmid with segments from many former bacterial hosts

Plasmid pk214

Figure 5-19 The diagram shows the origins of genes of the *Lactococcus lactis* plasmid pK214. The genes are from many different bacteria. [*Data from Table 1 in V. Perreten, F. Schwarz, L. Cresta, M. Boeglin, G. Dasen, and M. Teuber, Nature 389, 1997, 801–802.*]

On R plasmids, the alleles for antibiotic resistance are often contained within a unit called a *transposon* (Figure 5-20). Transposons are unique segments of DNA that can move around to different sites in the genome, a process called transposition. (The mechanisms for transposition, which occurs in most species studied, will be detailed in Chapter 14.) When a transposon in the genome moves to a new location, it can occasionally embrace between its ends various types of genes, including alleles for drug resistance, and carry them along to their new locations as passengers. Sometimes, a transposon carries a drug-resistance allele to a plasmid, creating an R plasmid. Like F plasmids, many R plasmids are conjugative; in other words, they are effectively transmitted to a recipient cell during conjugation. Even R plasmids that are not conjugative and never leave their own cells can donate their R alleles to a conjugative plasmid by transposition. Hence, through plasmids, antibiotic-resistance alleles can spread rapidly throughout a population of bacteria. Although the spread of R plasmids is an effective strategy for the survival of bacteria, it presents a major problem for medical practice, as mentioned earlier, because bacterial populations rapidly become resistant to any new antibiotic drug that is invented and applied to humans.

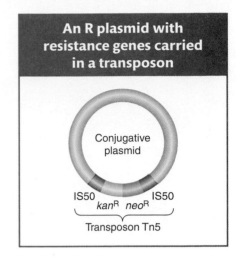

An R plasmid with resistance genes carried in a transposon

Figure 5-20 A transposon such as Tn5 can acquire several drug-resistance genes (in this case, those for resistance to the drugs kanamycin and neomycin) and transmit them rapidly on a plasmid, leading to the infectious transfer of resistance genes as a package. Insertion sequence 50 (IS50) forms the flanks of TN5.

5.3 Bacterial Transformation

Some bacteria can take up fragments of DNA from the external medium, and such uptake constitutes another way in which bacteria can exchange their genes. The source of the DNA can be other cells of the same species or cells of other species. In some cases, the DNA has been released from dead cells; in other cases, the DNA has been secreted from live bacterial cells. The DNA taken up integrates into the recipient's chromosome. If this DNA is of a different genotype from that of the recipient, the genotype of the recipient can become permanently changed, a process aptly termed **transformation.**

The nature of transformation

Transformation was discovered in the bacterium *Streptococcus pneumoniae* in 1928 by Frederick Griffith. Later, in 1944, Oswald T. Avery, Colin M. MacLeod, and Maclyn McCarty demonstrated that the "transforming principle" was DNA. Both results are milestones in the elucidation of the molecular nature of genes. We consider this work in more detail in Chapter 7.

The transforming DNA is incorporated into the bacterial chromosome by a process analogous to the double-recombination events observed in Hfr × F⁻ crosses. Note, however, that, in *conjugation*, DNA is transferred from one living cell to another through close contact, whereas in *transformation*, isolated pieces of external DNA are taken up by a cell through the cell wall and plasma membrane. Figure 5-21 shows one way in which this process can take place.

Transformation has been a handy tool in several areas of bacterial research because the genotype of a strain can be deliberately changed in a very specific way by transforming with an appropriate DNA fragment. For example, transformation is used widely in genetic engineering. More recently, it has been found that even eukaryotic cells can be transformed, by using quite similar procedures, and this technique has been invaluable for modifying eukaryotic cells.

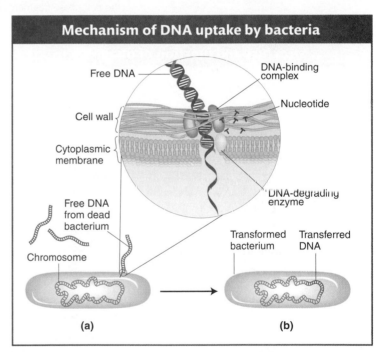

Mechanism of DNA uptake by bacteria

Free DNA — DNA-binding complex — Nucleotide — Cell wall — Cytoplasmic membrane — DNA-degrading enzyme — Free DNA from dead bacterium — Chromosome — Transformed bacterium — Transferred DNA

(a)   (b)

Figure 5-21 A bacterium undergoing transformation (a) picks up free DNA released from a dead bacterial cell. As DNA-binding complexes on the bacterial surface take up the DNA (*inset*), enzymes break down one strand into nucleotides; a derivative of the other strand may integrate into the bacterium's chromosome (b). [*After R. V. Miller, "Bacterial Gene Swapping in Nature." Copyright 1998 by Scientific American, Inc. All rights reserved.*]

Chromosome mapping using transformation

Transformation can be used to measure how closely two genes are linked on a bacterial chromosome. When DNA (the bacterial chromosome) is extracted for transformation experiments, some breakage into smaller pieces is inevitable. If two donor genes are located close together on the chromosome, there is a good chance that sometimes they will be carried on the same piece of transforming DNA. Hence, both will be taken up, causing a **double transformation.** Conversely, if genes are widely separated on the chromosome, they will most likely be carried on separate transforming segments. A genome could possibly take up both segments independently, creating a double transformant, but that outcome is not likely. Hence, in widely separated genes, the frequency of double transformants will equal the product of the single-transformant frequencies. Therefore, testing for close linkage by testing for a departure from the product rule should be possible. In other words, if genes are linked, then the proportion of double transformants will be greater than the product of single-transformant frequencies.

Unfortunately, the situation is made more complex by several factors—the most important being that not all cells in a population of bacteria are competent to be transformed. Nevertheless, at the end of this chapter, you can sharpen your skills in transformation analysis in one of the problems, which assumes that 100 percent of the recipient cells are competent.

> **Message** Bacteria can take up DNA fragments from the surrounding medium. Inside the cell, these fragments can integrate into the chromosome.

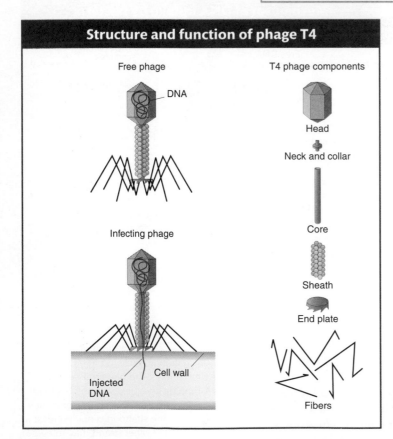

Structure and function of phage T4

Free phage

DNA

Infecting phage

Injected DNA

Cell wall

T4 phage components

Head

Neck and collar

Core

Sheath

End plate

Fibers

Figure 5-22 An infecting phage injects DNA through its core structure into the cell. (*Left*) Bacteriophage T4 is shown as a free phage and then in the process of infecting an *E. coli* cell. (*Right*) The major structural components of T4. [*After R. S. Edgar and R. H. Epstein, "The Genetics of a Bacterial Virus." Copyright 1965 by Scientific American, Inc. All rights reserved.*]

5.4 Bacteriophage Genetics

The word *bacteriophage,* which is a name for bacterial viruses, means "eater of bacteria." These viruses parasitize and kill bacteria. Pioneering work on the genetics of bacteriophages in the middle of the twentieth century formed the foundation of more recent research on tumor-causing viruses and other kinds of animal and plant viruses. In this way, bacterial viruses have provided an important model system.

These viruses can be used in two different types of genetic analysis. First, two distinct phage genotypes can be crossed to measure recombination and hence map the viral genome. Mapping of the viral genome by this method is the topic of this section. Second, bacteriophages can be used as a way of bringing bacterial genes together for linkage and other genetic studies. We will study the use of phages in bacterial studies in Section 5.5. In addition, as we will see in Chapter 20, phages are used in DNA technology as carriers, or vectors, of foreign DNA. Before we can understand phage genetics, we must first examine the infection cycle of phages.

Infection of bacteria by phages

Most bacteria are susceptible to attack by bacteriophages. A phage consists of a nucleic acid "chromosome" (DNA or RNA) surrounded by a coat of protein molecules. Phage types are identified not by species names but by symbols—for example, phage T4, phage λ, and so forth. Figures 5-22 and 5-23 show the structure of

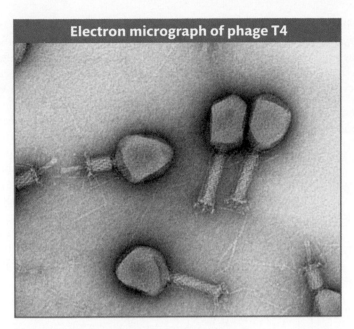

Figure 5-23 Enlargement of the *E. coli* phage T4 reveals details of head, tail, and tail fibers. [*Science Source.*]

phage T4. During infection, a phage attaches to a bacterium and injects its genetic material into the bacterial cytoplasm, as diagrammed in Figure 5-22. An electron micrograph of the process is shown in Figure 5-24. The phage genetic information then takes over the machinery of the bacterial cell by turning off the synthesis of bacterial components and redirecting the bacterial synthetic machinery to make phage components. Newly made phage heads are individually stuffed with replicates of the phage chromosome. Ultimately, many phage descendants are made and are released when the bacterial cell wall breaks open. This breaking-open process is called **lysis.** The population of phage progeny is called the phage **lysate.**

How can we study inheritance in phages when they are so small that they are visible only under the electron microscope? In this case, we cannot produce a visible colony by plating, but we can produce a visible manifestation of a phage by taking advantage of several phage characters.

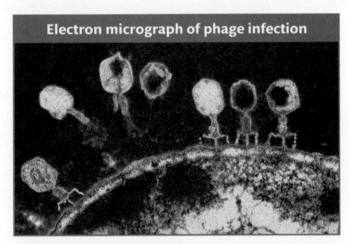

Figure 5-24 Bacteriophages are shown in several stages of the infection process, which includes attachment and DNA injection. [*Dr. L. Caro/Science Photo Library, Photo Researchers.*]

Let's look at the consequences of a phage infecting a single bacterial cell. Figure 5-25 shows the sequence of events in the infectious cycle that leads to the release of progeny phages from the lysed cell. After lysis, the progeny phages infect neighboring bacteria. This cycle is repeated through progressive rounds of infection, and, as these cycles repeat, the number of lysed cells increases exponentially. Within 15 hours after one single phage particle infects a single bacterial cell, the effects are visible to the naked eye as a clear area, or **plaque,** in the opaque lawn of bacteria covering the surface of a plate of solid medium (Figure 5-26). Such plaques can be large or small, fuzzy or sharp, and so forth, depending on the phage genotype. Thus, *plaque morphology* is a phage character that can be analyzed at the genetic level. Another phage phenotype that we can analyze genetically is *host range,* because phages may differ in the spectra of bacterial strains that they can infect and lyse. For example, a specific strain of bacteria might be immune to phage 1 but susceptible to phage 2.

Mapping phage chromosomes by using phage crosses

Two phage genotypes can be crossed in much the same way that we cross organisms. A phage cross can be illustrated by a cross of T2 phages originally studied by Alfred Hershey. The genotypes of the two parental strains in Hershey's cross were $h^- r^+ \times h^+ r^-$. The alleles correspond to the following phenotypes:

h^- : can infect two different *E. coli* strains (which we can call strains 1 and 2)

h^+ : can infect only strain 1

r^- : rapidly lyses cells, thereby producing large plaques

r^+ : slowly lyses cells, producing small plaques

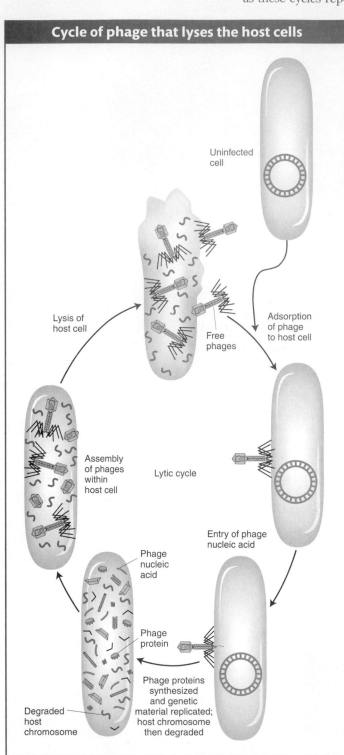

Cycle of phage that lyses the host cells

Uninfected cell

Lysis of host cell

Free phages

Adsorption of phage to host cell

Assembly of phages within host cell

Lytic cycle

Entry of phage nucleic acid

Phage nucleic acid

Phage protein

Phage proteins synthesized and genetic material replicated; host chromosome then degraded

Degraded host chromosome

Figure 5-25 Infection by a single phage redirects the cell's machinery into making progeny phages, which are released at lysis. [*After J. Darnell, H. Lodish, and D. Baltimore, Molecular Cell Biology. Copyright 1986 by W. H. Freeman and Company.*]

A plaque is a clear area in which all bacteria have been lysed by phages

Clear areas, or plaques

Figure 5-26 Through repeated infection and production of progeny phage, a single phage produces a clear area, or plaque, on the opaque lawn of bacterial cells. [*Sue Katz, Rogers State University, Claremore, OK.*]

To make the cross, *E. coli* strain 1 is infected with both parental T2 phage genotypes. This kind of infection is called a **mixed infection** or a **double infection** (Figure 5-27). After an appropriate incubation period, the phage lysate (the progeny phages) is analyzed by spreading it onto a bacterial lawn composed of a mixture of *E. coli* strains 1 and 2. Four plaque types are then distinguishable (Figure 5-28). Large plaques indicate rapid lysis (r^-), and small plaques indicate slow lysis (r^+). Phage plaques with the allele h^- will infect both hosts, forming a clear plaque, whereas phage plaques with the allele h^+ will infect only one host, forming a cloudy plaque. Thus, the four genotypes can be easily classified as parental ($h^- r^+$ and $h^+ r^-$) and recombinant ($h^+ r^+$ and $h^- r^-$), and a recombinant frequency can be calculated as follows:

$$\mathrm{RF} = \frac{(h^+ \, r^+) + (h^- \, r^-)}{\text{total plaques}}$$

If we assume that the recombining phage chromosomes are linear, then single crossovers produce viable reciprocal products. However, phage crosses are subject to some analytical complications. First, several rounds of exchange can take place within the host: a recombinant produced shortly after infection may undergo further recombination in the same cell or in later infection cycles. Second, recombination can take place between genetically similar phages as well as between different types. Thus, if we let P_1 and P_2 refer to general parental genotypes, crosses of $P_1 \times P_1$ and $P_2 \times P_2$ take place in addition to $P_1 \times P_2$. For both these reasons, recombinants from phage crosses are a consequence of a *population* of events rather than defined, single-step exchange events. Nevertheless, *all other things being equal*, the RF calculation does represent a valid index of map distance in phages.

Because astronomically large numbers of phages can be used in phage-recombination analyses, very rare crossover events can be detected. In the 1950s, Seymour Benzer made use of such rare crossover events to map the mutant sites *within* the *rII* gene of phage T4, a gene that controls lysis. For different *rII* mutant alleles arising spontaneously, the mutant site is usually at different positions within the gene. Therefore, when two different *rII* mutants are crossed, a few rare crossovers may take place between the mutant sites, producing wild-type recombinants, as shown here:

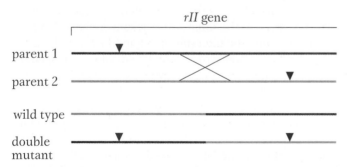

As distance between two mutant sites increases, such a crossover event is more likely. Thus, the frequency of *rII*$^+$ recombinants is a measure of that distance within the gene. (The reciprocal product is a double mutant and indistinguishable from the parentals.)

Benzer used a clever approach to detect the very rare *rII*$^+$ recombinants. He made use of the fact that rII mutants will not infect a strain of *E. coli* called K. Therefore, he made the rII $\times$ rII cross on another strain and then plated the phage lysate on a lawn of strain K. Only *rII*$^+$ recombinants will form plaques on this lawn. This way of finding a rare genetic event (in this case, a recombinant) is a **selective**

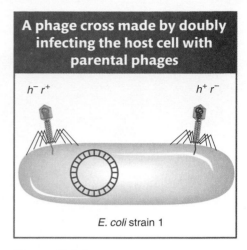

A phage cross made by doubly infecting the host cell with parental phages

$h^- \, r^+$ $\qquad\qquad h^+ \, r^-$

E. coli strain 1

Figure 5-27

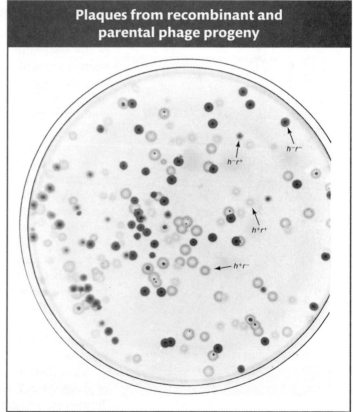

Plaques from recombinant and parental phage progeny

Figure 5-28 These plaque phenotypes were produced by progeny of the cross $h^- \, r^+ \times h^+ \, r^-$. Four plaque phenotypes can be differentiated, representing two parental types and two recombinants. [*From G. S. Stent, Molecular Biology of Bacterial Viruses. Copyright 1963 by W. H. Freeman and Company.*]

system: *only* the desired rare event can produce a certain visible outcome. In contrast, a **screen** is a system in which large numbers of individuals are visually scanned to seek the rare "needle in the haystack."

This same approach can be used to map mutant sites within genes for any organism from which large numbers of cells can be obtained and for which wild-type and mutant phenotypes can be distinguished. However, this sort of intragenic mapping has been largely superseded by the advent of inexpensive chemical methods for DNA sequencing, which identify the positions of mutant sites directly.

> **Message** Recombination between phage chromosomes can be studied by bringing the parental chromosomes together in one host cell through mixed infection. Progeny phages can be examined for both parental and recombinant genotypes.

5.5 Transduction

Some phages are able to pick up bacterial genes and carry them from one bacterial cell to another, a process known as **transduction.** Thus, transduction joins the battery of modes of transfer of genomic material between bacteria—along with Hfr chromosome transfer, F′ plasmid transfer, and transformation.

Discovery of transduction

In 1951, Joshua Lederberg and Norton Zinder were testing for recombination in the bacterium *Salmonella typhimurium* by using the techniques that had been successful with *E. coli.* The researchers used two different strains: one was *phe⁻ trp⁻ tyr⁻*, and the other was *met⁻ his⁻*. We won't worry about the nature of these alleles except to note that all are auxotrophic. When either strain was plated on a minimal medium, no wild-type cells were observed. However, after the two strains were mixed, wild-type prototrophs appeared at a frequency of about 1 in 10^5. Thus far, the situation seems similar to that for recombination in *E. coli.*

However, in this case, the researchers also recovered recombinants from a U-tube experiment, in which conjugation was prevented by a filter separating the two arms. They hypothesized that some agent was carrying genes from one bacterium to another. By varying the size of the pores in the filter, they found that the agent responsible for gene transfer was the same size as a known phage of *Salmonella,* called phage P22. Furthermore, the filterable agent and P22 were identical in sensitivity to antiserum and in immunity to hydrolytic enzymes. Thus, Lederberg and Zinder had discovered a new type of gene transfer, mediated by a virus. They were the first to call this process *transduction.* As a rarity in the lytic cycle, virus particles sometimes pick up bacterial genes and transfer them when they infect another host. Transduction has subsequently been demonstrated in many bacteria.

To understand the process of transduction, we need to distinguish two types of phage cycle. **Virulent phages** are those that immediately lyse and kill the host. **Temperate phages** can remain within the host cell for a period without killing it. Their DNA either integrates into the host chromosome to replicate with it or replicates separately in the cytoplasm, as does a plasmid. A phage integrated into the bacterial genome is called a **prophage.** A bacterium harboring a quiescent phage is described as **lysogenic** and is itself called a **lysogen.** Occasionally, the quiescent phage in a lysogenic bacterium becomes active, replicates itself, and causes the spontaneous lysis of its host cell. A resident temperate phage confers resistance to infection by other phages of that type.

There are two kinds of transduction: generalized and specialized. *Generalized* transducing phages can carry any part of the bacterial chromosome, whereas specialized transducing phages carry only certain *specific* parts.

> **Message** Virulent phages cannot become prophages; they always lyse a cell immediately on entry. Temperate phages can exist within the bacterial cell as prophages, allowing their hosts to survive as lysogenic bacteria; they are also capable of occasional bacterial lysis.

Generalized transduction

By what mechanisms can a phage carry out **generalized transduction?** In 1965, H. Ikeda and J. Tomizawa threw light on this question in some experiments on the *E. coli* phage P1. They found that, when a donor cell is lysed by P1, the bacterial chromosome is broken up into small pieces. Occasionally, the newly forming phage particles mistakenly incorporate a piece of the bacterial DNA into a phage head in place of phage DNA. This event is the origin of the transducing phage.

A phage carrying bacterial DNA can infect another cell. That bacterial DNA can then be incorporated into the recipient cell's chromosome by recombination (Figure 5-29). Because genes on any of the cut-up parts of the host genome

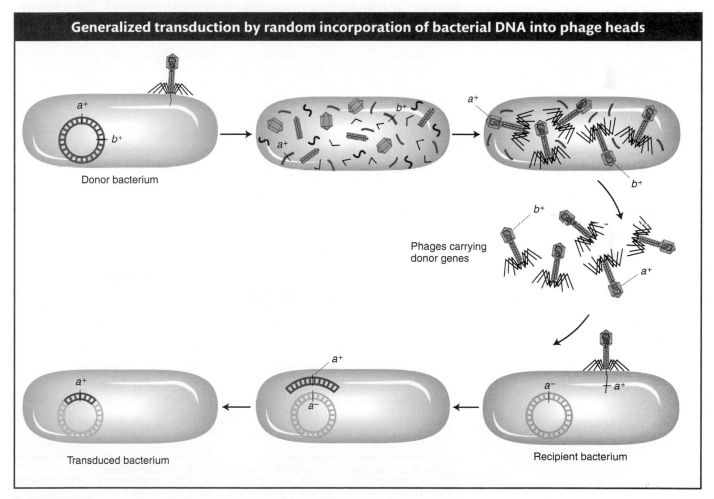

Generalized transduction by random incorporation of bacterial DNA into phage heads

Donor bacterium

Phages carrying donor genes

Recipient bacterium

Transduced bacterium

Figure 5-29 A newly forming phage may pick up DNA from its host cell's chromosome (*top*) and then inject it into a new cell (*bottom right*). The injected DNA may insert into the new host's chromosome by recombination (*bottom left*). In reality, only a very small minority of phage progeny (1 in 10,000) carry donor genes.

Figure 5-30 The diagram shows a genetic map of the *purB*-to-*cysB* region of *E. coli* determined by P1 cotransduction. The numbers given are the averages in percent for cotransduction frequencies obtained in several experiments. The values in parentheses are considered unreliable. [*After J. R. Guest, Mol. Gen. Genet. 105, 1969, 285.*]

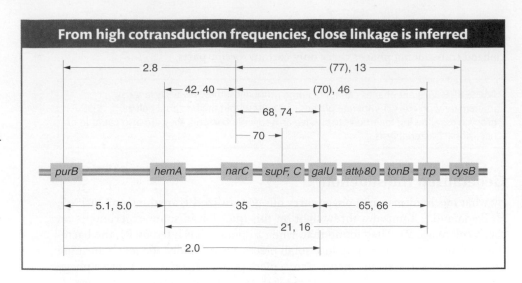

can be transduced, this type of transduction is by necessity of the generalized type.

Phages P1 and P22 both belong to a phage group that shows generalized transduction. P22 DNA inserts into the host chromosome, whereas P1 DNA remains free, like a large plasmid. However, both transduce by faulty head stuffing.

Generalized transduction can be used to obtain bacterial linkage information when genes are close enough that the phage can pick them up and transduce them in a single piece of DNA. For example, suppose that we wanted to find the linkage distance between *met* and *arg* in *E. coli*. We could grow phage P1 on a donor *met*$^+$ *arg*$^+$ strain and then allow P1 phages from lysis of this strain to infect a *met*$^-$ *arg*$^-$ strain. First, one donor allele is selected, say, *met*$^+$. Then the percentage of *met*$^+$ colonies that are also *arg*$^+$ is measured. Strains transduced to both *met*$^+$ and *arg*$^+$ are called **cotransductants.** The *greater* the cotransduction frequency, the *closer* two genetic markers must be (the opposite of most mapping measurements). Linkage values are usually expressed as cotransduction frequencies (Figure 5-30).

By using an extension of this approach, we can estimate the *size* of the piece of host chromosome that a phage can pick up, as in the following type of experiment, which uses P1 phage:

$$\text{donor } leu^+ \, thr^+ \, azi^r \rightarrow \text{recipient } leu^- \, thr^- \, azi^s$$

In this experiment, P1 phage grown on the *leu*$^+$ *thr*$^+$ *azi*^{r} donor strain infect the *leu*$^-$ *thr*$^-$ *azi*^{s} recipient strain. The strategy is to select one or more donor alleles in the recipient and then test these transductants for the presence of the unselected alleles. Results are outlined in Table 5-3. Experiment 1 in Table 5-3 tells us that *leu* is relatively close to *azi* and distant from *thr*, leaving us with two possibilities:

$$\begin{array}{ccc} thr & leu & azi \\ \vert & \vert & \vert \end{array} \quad \text{or} \quad \begin{array}{ccc} thr & azi & leu \\ \vert & \vert & \vert \end{array}$$

Experiment 2 tells us that *leu* is closer to *thr* than *azi* is, and so the map must be

$$\begin{array}{ccc} thr & leu & azi \\ \vert & \vert & \vert \end{array}$$

Table 5-3   Accompanying Markers in Specific P1 Transductions

Experiment	Selected marker	Unselected markers
1	*leu*$^+$	50% are *azi*^{r}; 2% are *thr*$^+$
2	*thr*$^+$	3% are *leu*$^+$; 0% are *ari*^{r}
3	*leu*$^+$ and *thr*$^+$	0% are *azi*^{r}

By selecting for *thr*⁺ and *leu*⁺ together in the transducing phages in experiment 3, we see that the transduced piece of genetic material never includes the *azi* locus because the phage head cannot carry a fragment of DNA that big. P1 can only cotransduce genes less than approximately 1.5 minutes apart on the *E. coli* chromosome map.

Specialized transduction

A generalized transducer, such as phage P22, picks up fragments of broken host DNA at random. How are other phages, which act as specialized transducers, able to carry only certain host genes to recipient cells? The short answer is that a specialized transducer inserts into the bacterial chromosome at one position only. When it exits, a faulty outlooping occurs (similar to the type that produces F′ plasmids). Hence, it can pick up and transduce only genes that are close by.

The prototype of **specialized transduction** was provided by studies undertaken by Joshua and Esther Lederberg on a temperate *E. coli* phage called *lambda* (λ). Phage λ has become the most intensively studied and best-characterized phage.

Behavior of the prophage   Phage λ has unusual effects when cells lysogenic for it are used in crosses. In the cross of an uninfected Hfr with a lysogenic F⁻ recipient [Hfr × F⁻(λ)], lysogenic F⁻ exconjugants with Hfr genes are readily recovered. However, in the reciprocal cross Hfr(λ) × F⁻, the *early* genes from the Hfr chromosome are recovered among the exconjugants, but recombinants for *late* genes are not recovered. Furthermore, lysogenic exconjugants are almost never recovered from this reciprocal cross. What is the explanation? The observations make sense if the λ prophage is behaving as a bacterial gene locus behaves (that is, as part of the bacterial chromosome). Thus, the prophage would enter the F⁻ cell at a specific time corresponding to its position in the chromosome. Earlier genes are recovered because they enter before the prophage. Later genes are not recovered because lysis destroys the recipient cell. In interrupted-mating experiments, the λ prophage does in fact always enter the F⁻ cell at a specific time, closely linked to the *gal* locus.

In an Hfr(λ) × F⁻ cross, the entry of the λ prophage into the cell immediately triggers the prophage into a lytic cycle; this process is called **zygotic induction** (Figure 5-31). However, in the cross of *two* lysogenic cells Hfr(λ) × F⁻(λ), there is no zygotic induction. The presence of any prophage prevents another infecting virus from causing lysis. The prophage produces a cytoplasmic factor that represses the multiplication of the virus. (The phage-directed cytoplasmic repressor nicely

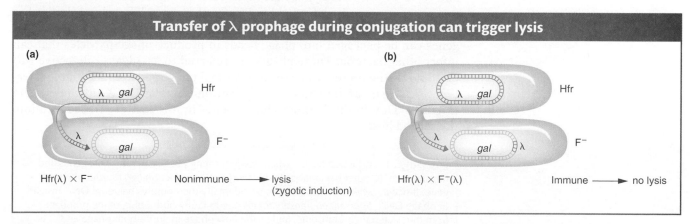

Transfer of λ prophage during conjugation can trigger lysis

(a)

λ   *gal*   Hfr

λ
gal   F⁻

Hfr(λ) × F⁻   Nonimmune ⟶ lysis
                                    (zygotic induction)

(b)

λ   *gal*   Hfr

λ
gal   λ   F⁻

Hfr(λ) × F⁻(λ)   Immune ⟶ no lysis

Figure 5-31 A λ prophage can be transferred to a recipient during conjugation, but the prophage triggers lysis, a process called zygotic induction, only if the recipient has no prophage already—that is, in the case shown in part *a* but not in part *b*.

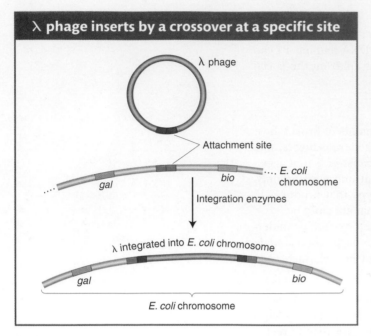

λ phage inserts by a crossover at a specific site

Figure 5-32 Reciprocal recombination takes place between a specific attachment site on the circular λ DNA and a specific region called the attachment site on the *E. coli* chromosome between the *gal* and *bio* genes.

explains the immunity of the lysogenic bacteria, because a phage would immediately encounter a repressor and be inactivated.)

λ insertion The interrupted-mating experiments heretofore described showed that the λ prophage is part of the lysogenic bacterium's chromosome. How is the λ prophage inserted into the bacterial genome? In 1962, Allan Campbell proposed that it inserts by a single crossover between a circular λ phage chromosome and the circular *E. coli* chromosome, as shown in Figure 5-32. The crossover point would be between a specific site in λ, the λ **attachment site,** and an attachment site in the bacterial chromosome located between the genes *gal* and *bio,* because λ integrates at that position in the *E. coli* chromosome.

An attraction of Campbell's proposal is that from it follow predictions that geneticists can test. For example, integration of the prophage into the *E. coli* chromosome should increase the genetic distance between flanking bacterial genes, as can be seen in Figure 5-32 for *gal* and *bio.* In fact, studies show that lysogeny *does* increase time-of-entry or recombination distances between the bacterial genes. This unique location of λ accounts for its specialized transduction.

Mechanism of specialized transduction

As a prophage, λ always inserts between the *gal* region and the *bio* region of the host chromosome (Figure 5-33), and, in transduction experiments, as expected, λ can transduce only the *gal* and *bio* genes.

How does λ carry away neighboring genes? The explanation lies, again, in an imperfect reversal of the Campbell insertion mechanism, like that for F′ formation. The recombination event between specific regions of λ and the bacterial chromosome is catalyzed by a specialized phage-encoded enzyme system that uses the λ attachment site as a substrate. The enzyme system dictates that λ integrates only at a specific point between *gal* and *bio* in the chromosome (see Figure 5-33a). Furthermore, during lysis, the λ prophage normally excises at precisely the correct point to produce a normal circular λ chromosome, as seen in Figure 5-33b(i). Very rarely, excision is abnormal owing to faulty outlooping. In this case, the outlooping phage DNA can pick up a nearby gene and leave behind some phage genes, as seen in Figure 5-33b(ii). The resulting phage genome is defective because of the genes left behind, but it has also gained a bacterial gene, *gal* or *bio.* The abnormal DNA carrying nearby genes can be packaged into phage heads to produce phage particles that can infect other bacteria. These phages are referred to as λdgal (λ-defective *gal*) or λdbio. In the presence of a second, normal phage particle in a double infection, the λdgal can integrate into the chromosome at the λ attachment site (Figure 5-33c). In this manner, the *gal* genes in this case are transduced into the second host.

Message Transduction occurs when newly forming phages acquire host genes and transfer them to other bacterial cells. *Generalized transduction* can transfer any host gene. It occurs when phage packaging accidentally incorporates bacterial DNA instead of phage DNA. *Specialized transduction* is due to faulty outlooping of the prophage from the bacterial chromosome, and so the new phage includes both phage and bacterial genes. The transducing phage can transfer only specific host genes.

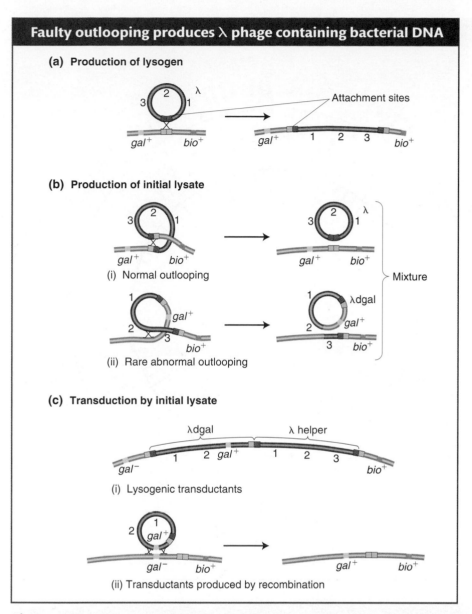

Faulty outlooping produces λ phage containing bacterial DNA

(a) Production of lysogen

Attachment sites

(b) Production of initial lysate

(i) Normal outlooping

(ii) Rare abnormal outlooping

Mixture

(c) Transduction by initial lysate

λdgal   λ helper

(i) Lysogenic transductants

(ii) Transductants produced by recombination

Figure 5-33 The diagram shows how specialized transduction operates in phage λ. (a) A crossover at the specialized attachment site produces a lysogenic bacterium. (b) The lysogenic bacterium can produce a normal λ (i) or, rarely, λdgal (ii), a transducing particle containing the *gal* gene. (c) *gal*⁺ transductants can be produced by either (i) the coincorporation of λdgal and λ (acting as a helper) or (ii) crossovers flanking the *gal* gene, a rare event. The blue double boxes are the bacterial attachment site, the purple double boxes are the λ attachment site, and the pairs of blue and purple boxes are hybrid integration sites, derived partly from *E. coli* and partly from λ.

5.6 Physical Maps and Linkage Maps Compared

Some very detailed chromosomal maps for bacteria have been obtained by combining the mapping techniques of interrupted mating, recombination mapping, transformation, and transduction. Today, new genetic markers are typically mapped first into a segment of about 10 to 15 map minutes by using interrupted mating. Then additional, closely linked markers can be mapped in a more fine-scale analysis with the use of P1 cotransduction or recombination.

Figure 5-34 The 1963 genetic map of *E. coli* genes with mutant phenotypes. Units are minutes, based on interrupted-mating and recombination experiments. Asterisks refer to map positions that are not as precise as the other positions. [*From G. S. Stent, Molecular Biology of Bacterial Viruses. Copyright 1963 by W. H. Freeman and Company.*]

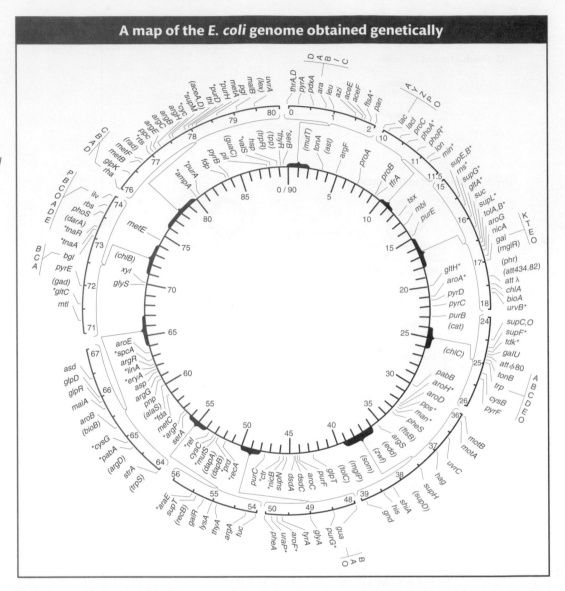

A map of the *E. coli* genome obtained genetically

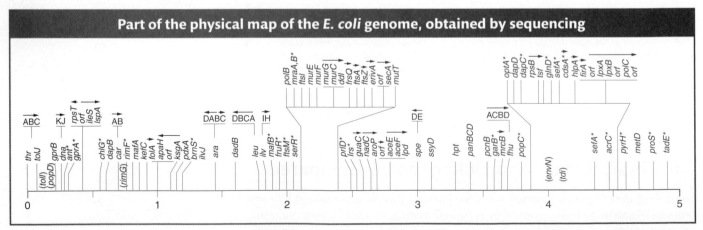

Part of the physical map of the *E. coli* genome, obtained by sequencing

Figure 5-35 A linear scale drawing of a 5-minute section of the 100-minute 1990 *E. coli* linkage map. The parentheses and asterisks indicate markers for which the exact location was unknown at the time of publication. Arrows above genes and groups of genes indicate the direction of transcription. [*From B. J. Bachmann, "Linkage Map of Escherichia coli K-12, Edition 8," Microbiol. Rev. 54, 1990, 130–197.*]

By 1963, the *E. coli* map (Figure 5-34) already detailed the positions of approximately 100 genes. After 27 years of further refinement, the 1990 map depicted the positions of more than 1400 genes. Figure 5-35 shows a 5-minute section of the 1990 map (which is adjusted to a scale of 100 minutes). The complexity of these maps illustrates the power and sophistication of genetic analysis. How well do these maps correspond to physical reality? In 1997, the DNA sequence of the entire *E. coli* genome of 4,632,221 base pairs was completed, allowing us to compare the exact position of genes on the genetic map with the position of the corresponding coding sequence on the linear DNA sequence (the physical map). The full map is represented in Figure 5-36. Figure 5-37 makes a comparison for a segment of both maps. Clearly, the genetic map is a close match to the physical map.

Chapter 4 considered some ways in which the physical map (usually the full genome sequence) can be useful in mapping new mutations. In bacteria, the technique of **insertional mutagenesis** is another way to zero in rapidly on a mutation's position on a known physical map. The technique causes mutations through the random insertion of "foreign" DNA fragments. The inserts inactivate any gene in which they land by interrupting the transcriptional unit. Transposons are particularly useful inserts for this purpose in several model organisms, including bacteria. To map a new mutation, the procedure is as follows. The DNA of a transposon carrying a resistance allele or other selectable marker is introduced by transformation into bacterial recipients that have no active transposons. The transposons insert more or less randomly, and any that land in the middle of a gene cause a mutation. A subset of all mutants obtained will have phenotypes relevant to the bacterial process under study, and these phenotypes become the focus of the analysis.

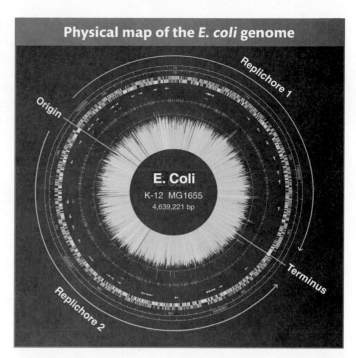

Figure 5-36 This map was obtained from sequencing DNA and plotting gene positions. Key to components from the outside in:

- The DNA replication origin and terminus are marked.
- The two scales are in DNA base pairs and in minutes.
- The orange and yellow histograms show the distribution of genes on the two different DNA strands.
- The arrows represent genes for rRNA (red) and tRNA (green).
- The central "starburst" is a histogram of each gene with lines of length that reflect predicted level of transcription.

[*F. R. Blattner et al., "The Complete Genome Sequence of Escherichia coli K-12," Science 277, 1997, 1453–1462. Image courtesy of Dr. Guy Plunkett III.*]

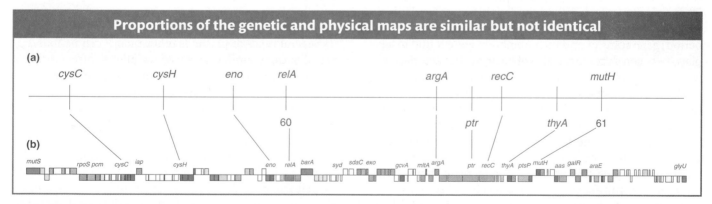

Figure 5-37 An alignment of the genetic and physical maps. (a) Markers on the 1990 genetic map in the region near 60 and 61 minutes. (b) The exact positions of every gene, based on the complete sequence of the *E. coli* genome. (Not every gene is named in this map, for simplicity.) The elongated boxes are genes and putative genes. Each color represents a different type of function. For example, red denotes regulatory functions, and dark blue denotes functions in DNA replication, recombination, and repair. Lines between the maps in parts *a* and *b* connect the same gene in each map. [*F. R. Blattner et al., "The Complete Genome Sequence of Escherichia coli K-12," Science, vol. 277, September 5, 1997, pp. 1453–1462. Image courtesy of Dr. Guy Plunkett III.*]

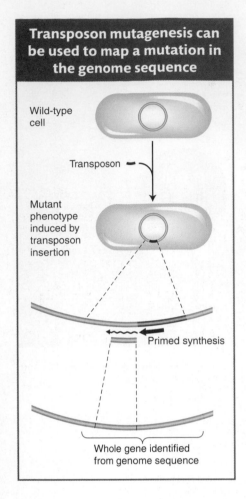

Transposon mutagenesis can be used to map a mutation in the genome sequence

Wild-type cell

Transposon

Mutant phenotype induced by transposon insertion

Primed synthesis

Whole gene identified from genome sequence

The beauty of inserting transposons is that, because their sequence is known, the mutant gene can be located and sequenced. DNA replication primers are created that match the known sequence of the transposon (see Chapter 20). These primers are used to initiate a sequencing analysis that proceeds *outward* from the transposon into the surrounding gene. The short sequence obtained can then be fed into a computer and compared with the complete genome sequence. From this analysis, the position of the gene and its full sequence are obtained. The function of a homolog of this gene might already have been deduced in other organisms. Hence, you can see that this approach (like that introduced in Chapter 4) is another way of uniting mutant phenotype with map position and potential function. Figure 5-38 summarizes the approach.

As an aside in closing, it is interesting that many of the historical experiments revealing the circularity of bacterial and plasmid genomes coincided with the publication and popularization of J. R. R. Tolkien's *The Lord of the Rings*. Consequently, a review of bacterial genetics at that time led off with the following quotation from the trilogy:

One Ring to rule them all, One Ring to find them,
One Ring to bring them all and in the darkness bind them.

Figure 5-38 The insertion of a transposon inserts a mutation into a gene of unknown position and function. The segment next to the transposon is replicated, sequenced, and matched to a segment in the complete genome sequence.

Summary

Advances in bacterial and phage genetics within the past 50 years have provided the foundation for molecular biology and cloning (discussed in later chapters). Early in this period, gene transfer and recombination were found to take place between different strains of bacteria. In bacteria, however, genetic material is passed in only one direction—for example, in *Escherichia coli*, from a donor cell (F⁺ or Hfr) to a recipient cell (F⁻). Donor ability is determined by the presence in the cell of a fertility factor (F), a type of plasmid. On occasion, the F factor present in the free state in F⁺ cells can integrate into the *E. coli* chromosome and form an Hfr cell. When this occurs, a fragment of donor chromosome can transfer into a recipient cell and subsequently recombine with the recipient chromosome. Because the F factor can insert at different places on the host chromosome, early investigators were able to piece the transferred fragments together to show that the *E. coli* chromosome is a single circle, or ring. Interruption of the transfer at different times has provided geneticists with an unconventional method (interrupted mating) for constructing a linkage map of the single chromosome of *E. coli* and other similar bacteria, in which the map unit is a unit of time (minutes).

In an extension of this technique, the frequency of recombinants between markers known to have entered the recipient can provide a finer-scale map distance.

Several types of plasmids other than F can be found. R plasmids carry antibiotic-resistance alleles, often within a mobile element called a transposon. Rapid plasmid spread causes population-wide resistance to medically important drugs. Derivatives of such natural plasmids have become important cloning vectors, useful for gene isolation and study in all organisms.

Genetic traits can also be transferred from one bacterial cell to another in the form of pieces of DNA taken into the cell from the extracellular environment. This process of transformation in bacterial cells was the first demonstration that DNA is the genetic material. For transformation to occur, DNA must be taken into a recipient cell, and recombination must then take place between a recipient chromosome and the incorporated DNA.

Bacteria can be infected by viruses called bacteriophages. In one method of infection, the phage chromosome may enter the bacterial cell and, by using the bacterial metabolic machinery, produce progeny phages that burst the host bacterium. The new phages can then infect other cells. If two phages of different genotypes infect the same host, recombination between their chromosomes can take place.

In another mode of infection, lysogeny, the injected phage lies dormant in the bacterial cell. In many cases, this dormant phage (the prophage) incorporates into the host chromosome and replicates with it. Either spontaneously or under appropriate stimulation, the prophage can leave its dormant state and lyse the bacterial host cell.

A phage can carry bacterial genes from a donor to a recipient. In generalized transduction, random host DNA is incorporated alone into the phage head during lysis. In specialized transduction, faulty excision of the prophage from a unique chromosomal locus results in the inclusion of specific host genes as well as phage DNA in the phage head.

Today, a physical map in the form of the complete genome sequence is available for many bacterial species. With the use of this physical genome map, the map position of a mutation of interest can be precisely located. First, appropriate mutations are produced by the insertion of transposons (insertional mutagenesis). Then the DNA sequence surrounding the inserted transposon is obtained and matched to a sequence in the physical map. This technique provides the locus, the sequence, and possibly the function of the gene of interest.

KEY TERMS

attachment site (p. 196)
auxotroph (p. 172)
bacteriophage (phage) (p. 170)
cell clone (p. 172)
colony (p. 172)
conjugation (p. 175)
cotransductant (p. 194)
donor (p. 175)
double (mixed) infection (p. 191)
double transformation (p. 188)
endogenote (p. 181)
exconjugant (p. 178)
exogenote (p. 181)
F+ (donor) (p. 175)
F− (recipient) (p. 175)
F′ plasmid (p. 184)
fertility factor (F) (p. 175)
generalized transduction (p. 193)

genetic marker (p. 172)
Hfr (high frequency of recombination) (p. 176)
insertional mutagenesis (p. 199)
interrupted mating (p. 177)
lysate (p. 189)
lysis (p. 189)
lysogen (lysogenic bacterium) (p. 192)
merozygote (p. 181)
minimal medium (p. 172)
mixed (double) infection (p. 191)
origin (O) (p. 178)
phage (bacteriophage) (p. 170)
phage recombination (p. 171)
plaque (p. 190)
plasmid (p. 175)
plating (p. 172)
prokaryote (p. 170)

prophage (p. 192)
prototroph (p. 172)
R plasmid (p. 185)
recipient (p. 175)
resistant mutant (p. 172)
rolling circle replication (p. 175)
screen (p. 192)
selective system (p. 191)
specialized transduction (p. 195)
temperate phage (p. 192)
terminus (p. 180)
transduction (p. 192)
transformation (p. 187)
unselected marker (p. 182)
virulent phage (p. 192)
virus (p. 170)
zygotic induction (p. 195)

SOLVED PROBLEMS

SOLVED PROBLEM 1. Suppose that a cell were unable to carry out generalized recombination (*rec*−). How would this cell behave as a recipient in generalized and in specialized transduction? First, compare each type of transduction and then determine the effect of the *rec*− mutation on the inheritance of genes by each process.

Solution

Generalized transduction entails the incorporation of chromosomal fragments into phage heads, which then infect recipient strains. Fragments of the chromosome are incorporated randomly into phage heads, and so any marker on the bacterial host chromosome can be transduced to another strain by generalized transduction. In contrast, specialized transduction entails the integration of the phage at a specific point on the chromosome and the rare incorporation of chromosomal markers near the integration site into the phage genome. Therefore, only those markers that are near the specific integration site of the phage on the host chromosome can be transduced.

Markers are inherited by different routes in generalized and specialized transduction. A generalized transducing phage injects a fragment of the donor chromosome into the recipient. This fragment must be incorporated into the recipient's chromosome by recombination, with the use of the recipient's recombination system. Therefore, a *rec*− recipient will not be able to incorporate fragments of DNA and cannot inherit markers by generalized transduction. On the other hand, the major route for the inheritance of markers by specialized transduction is by integration of the specialized transducing particle into the host chromosome at the specific phage integration site. This integration, which sometimes requires an additional wild-type (helper) phage, is mediated by a phage-specific enzyme system that is independent of the normal recombination enzymes. Therefore, a *rec⁻* recipient can still inherit genetic markers by specialized transduction.

SOLVED PROBLEM 2. In *E. coli*, four Hfr strains donate the following genetic markers, shown in the order donated:

Strain 1:	Q	W	D	M	T
Strain 2:	A	X	P	T	M
Strain 3:	B	N	C	A	X
Strain 4:	B	Q	W	D	M

All these Hfr strains are derived from the same F⁺ strain. What is the order of these markers on the circular chromosome of the original F⁺?

Solution

A two-step approach works well: (1) determine the underlying principle and (2) draw a diagram. Here the principle is clearly that each Hfr strain donates genetic markers from a fixed point on the circular chromosome and that the earliest markers are donated with the highest frequency. Because not all markers are donated by each Hfr, only the early markers must be donated for each Hfr. Each strain allows us to draw the following circles:

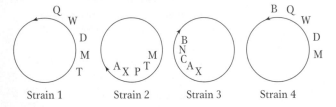

Strain 1	Strain 2	Strain 3	Strain 4

From this information, we can consolidate each circle into one circular linkage map of the order Q, W, D, M, T, P, X, A, C, N, B, Q.

SOLVED PROBLEM 3. In an Hfr × F⁻ cross, *leu⁺* enters as the first marker, but the order of the other markers is unknown. If the Hfr is wild type and the F⁻ is auxotrophic for each marker in question, what is the order of the markers in a cross where *leu⁺* recombinants are selected if 27 percent are *ile⁺*, 13 percent are *mal⁺*, 82 percent are *thr⁺*, and 1 percent are *trp⁺*?

Solution

Recall that spontaneous breakage creates a natural gradient of transfer, which makes it less and less likely for a recipient to receive later and later markers. Because we have selected for the earliest marker in this cross, the frequency of recombinants is a function of the order of entry for each marker. Therefore, we can immediately determine the order of the genetic markers simply by looking at the percentage of recombinants for any marker among the *leu⁺* recombinants. Because the inheritance of *thr⁺* is the highest, *thr⁺* must be the first marker to enter after *leu*. The complete order is *leu, thr, ile, mal, trp*.

SOLVED PROBLEM 4. A cross is made between an Hfr that is *met⁺ thi⁺ pur⁺* and an F⁻ that is *met⁻ thi⁻ pur⁻*. Interrupted-mating studies show that *met⁺* enters the recipient last, and so *met⁺* recombinants are selected on a medium containing supplements that satisfy only the *pur* and *thi* requirements. These recombinants are tested for the presence of the *thi⁺* and *pur⁺* alleles. The following numbers of individuals are found for each genotype:

met⁺ thi⁺ pur⁺	280
met⁺ thi⁺ pur⁻	0
met⁺ thi⁻ pur⁺	6
met⁺ thi⁻ pur⁻	52

a. Why was methionine (Met) left out of the selection medium?

b. What is the gene order?

c. What are the map distances in recombination units?

Solution

a. Methionine was left out of the medium to allow selection for *met⁺* recombinants, because *met⁺* is the last marker to enter the recipient. The selection for *met⁺* ensures that all the loci that we are considering in the cross will have already entered each recombinant that we analyze.

b. Here, a diagram of the possible gene orders is helpful. Because we know that *met* enters the recipient last, there are only two possible gene orders if the first marker enters on the right: *met, thi, pur* or *met, pur, thi*. How can we distinguish between these two orders? Fortunately, one of the four possible classes of recombinants requires two additional crossovers. Each possible order predicts a different class that arises by four crossovers rather than two. For instance, if the order were *met, thi, pur*, then *met⁺ thi⁻ pur⁺* recombinants would be very rare. On the other hand, if the order were *met, pur, thi*, then the four-crossover class would be *met⁺ pur⁻ thi⁺*. From the information given in the table, the *met⁺ pur⁻ thi⁺* class is clearly the four-crossover class and therefore the gene order *met, pur, thi* is correct.

c. Refer to the following diagram:

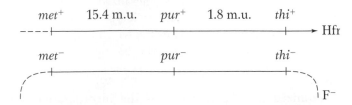

To compute the distance between *met* and *pur*, we compute the percentage of *met⁺ pur⁻ thi⁻*, which is 52/338 = 15.4 m.u. Similarly, the distance between *pur* and *thi* is 6/338 =1.8 m.u.

SOLVED PROBLEM 5. Compare the mechanism of transfer and inheritance of the *lac⁺* genes in crosses with Hfr, F⁺, and F′ *lac⁺* strains. How would an F⁻ cell that cannot undergo normal homologous recombination (*rec*⁻) behave in crosses with each of these three strains? Would the cell be able to inherit the *lac*⁺ gene?

Solution

Each of these three strains donates genes by conjugation. In the Hfr and F⁺ strains, the *lac*⁺ genes on the host chromosome are donated. In the Hfr strain, the F factor is integrated into the chromosome in every cell, and so chromosomal markers can be efficiently donated, particularly if a marker is near the integration site of F and is donated early. The F⁺ cell population contains a small percentage of Hfr cells, in which F is integrated into the chromosome. These cells are responsible for the gene transfer displayed by cultures of F⁺ cells. In the Hfr⁻ and F⁺-mediated gene transfer, inheritance requires the incorporation of a transferred fragment by recombination (recall that two crossovers are needed) into the F⁻ chromosome. Therefore, an F⁻ strain that cannot undergo recombination cannot inherit donor chromosomal markers even though they are transferred by Hfr strains or Hfr cells in F⁺ strains. The fragment cannot be incorporated into the chromosome by recombination. Because these fragments do not possess the ability to replicate within the F⁻ cell, they are rapidly diluted out during cell division.

Unlike Hfr cells, F′ cells transfer genes carried on the F′ factor, a process that does not require chromosome transfer. In this case, the *lac*⁺ genes are linked to the F′ factor and are transferred with it at a high efficiency. In the F⁻ cell, no recombination is required, because the F′ *lac*⁺ strain can replicate and be maintained in the dividing F⁻ cell population. Therefore, the *lac*⁺ genes are inherited even in a *rec*⁻ strain.

PROBLEMS

Most of the problems are also available for review/grading through the ɢᴇɴᴇᴛɪᴄsPⒹRTAL www.yourgeneticsportal.com

WORKING WITH THE FIGURES

1. In Figure 5-2, in which of the four processes shown is a complete bacterial genome transferred from one cell to another?

2. In Figure 5-3, if the concentration of bacterial cells in the original suspension is 200/ml and 0.2 ml is plated onto each of 100 petri dishes, what is the expected average number of colonies per plate?

3. In Figure 5-5,

    a. Why do A⁻ and B⁻ cells not form colonies on the plating medium?

    b. What genetic event do the purple colonies in the middle plate represent?

4. In Figure 5-10c, what do the yellow dots represent?

5. In Figure 5-11, which donor alleles become part of the recombinant genome produced?

6. In Figure 5-12,

    a. Which Hfr gene enters the recipient last? (Which diagram shows it actually entering?)

    b. What is the maximum percentage of cases of transfer of this gene?

    c. Which genes have entered at 25 min? Could they all become part of a stable exconjugant genome?

7. In Figure 5-14, which is the last gene to be transferred into the F⁻ from each of the five Hfr strains?

8. In Figure 5-15, how are each of the following genotypes produced?

    a. F⁺ *a*⁻

    b. F⁻ *a*⁻

    c. F⁻ *a*⁺

    d. F⁺ *a*⁺

9. In Figure 5-17, how many crossovers are required to produce a completely prototrophic exconjugant?

10. In Figure 5-18c, why is the crossover shown occurring in the orange segments of DNA?

11. In Figure 5-19, how many different bacterial species are shown as having contributed DNA to the plasmid pk214?

12. In Figure 5-25, can you point to any phage progeny that could transduce?

13. In Figure 5-28, what are the physical features of the plaques of recombinant phages?

14. In Figure 5-29, do you think that *b*⁺ could be transduced instead of *a*⁺? As well as *a*⁺?

15. In Figure 5-30, which genes show the highest frequencies of cotransduction?

16. In Figure 5-32, what do the half-red, half-blue segments represent?

17. In Figure 5-33, which is the rarest genotype produced in the initial lysate?

18. In Figure 5-38, precisely which gene is eventually identified from the genome sequence?

BASIC PROBLEMS

19. Describe the state of the F factor in an Hfr, F⁺, and F⁻ strain.

20. How does a culture of F⁺ cells transfer markers from the host chromosome to a recipient?

21. With respect to gene transfer and the integration of the transferred gene into the recipient genome, compare

   a. Hfr crosses by conjugation and generalized transduction.

   b. F' derivatives such as F' *lac* and specialized transduction.

22. Why is generalized transduction able to transfer any gene, but specialized transduction is restricted to only a small set?

23. A microbial geneticist isolates a new mutation in *E. coli* and wishes to map its chromosomal location. She uses interrupted-mating experiments with Hfr strains and generalized-transduction experiments with phage P1. Explain why each technique, by itself, is insufficient for accurate mapping.

24. In *E. coli*, four Hfr strains donate the following markers, shown in the order donated:

Strain 1:	M	Z	X	W	C
Strain 2:	L	A	N	C	W
Strain 3:	A	L	B	R	U
Strain 4:	Z	M	U	R	B

   All these Hfr strains are derived from the same F$^+$ strain. What is the order of these markers on the circular chromosome of the original F$^+$?

25. You are given two strains of *E. coli*. The Hfr strain is *arg*$^+$ *ala*$^+$ *glu*$^+$ *pro*$^+$ *leu*$^+$ *T*^{s}; the F$^-$ strain is *arg*$^-$ *ala*$^-$ *glu*$^-$ *pro*$^-$ *leu*$^-$ *T*^{r}. All the markers are nutritional except *T*, which determines sensitivity or resistance to phage T1. The order of entry is as given, with *arg*$^+$ entering the recipient first and *T*^{s} last. You find that the F$^-$ strain dies when exposed to penicillin (*pen*^{s}), but the Hfr strain does not (*pen*^{r}). How would you locate the locus for *pen* on the bacterial chromosome with respect to *arg, ala, glu, pro,* and *leu*? Formulate your answer in logical, well-explained steps and draw explicit diagrams where possible.

GENETICS**PORTAL** **Unpacking the Problem** 26. A cross is made between two *E. coli* strains: Hfr *arg*$^+$ *bio*$^+$ *leu*$^+$ × F$^-$ *arg*$^-$ *bio*$^-$ *leu*$^-$. Interrupted mating studies show that *arg*$^+$ enters the recipient last, and so *arg*$^+$ recombinants are selected on a medium containing *bio* and *leu* only. These recombinants are tested for the presence of *bio*$^+$ and *leu*$^+$. The following numbers of individuals are found for each genotype:

| *arg*$^+$ *bio*$^+$ *leu*$^+$ | 320 | *arg*$^+$ *bio*$^-$ *leu*$^+$ | 0 |
| *arg*$^+$ *bio*$^+$ *leu*$^-$ | 8 | *arg*$^+$ *bio*$^-$ *leu*$^-$ | 48 |

   a. What is the gene order?

b. What are the map distances in recombination percentages?

27. Linkage maps in an Hfr bacterial strain are calculated in units of minutes (the number of minutes between genes indicates the length of time that it takes for the second gene to follow the first in conjugation). In making such maps, microbial geneticists assume that the bacterial chromosome is transferred from Hfr to F$^-$ at a constant rate. Thus, two genes separated by 10 minutes near the origin end are assumed to be the same physical distance apart as two genes separated by 10 minutes near the F$^-$ attachment end. Suggest a critical experiment to test the validity of this assumption.

28. A particular Hfr strain normally transmits the *pro*$^+$ marker as the last one in conjugation. In a cross of this strain with an F$^-$ strain, some *pro*$^+$ recombinants are recovered early in the mating process. When these *pro*$^+$ cells are mixed with F$^-$ cells, the majority of the F$^-$ cells are converted into *pro*$^+$ cells that also carry the F factor. Explain these results.

29. F' strains in *E. coli* are derived from Hfr strains. In some cases, these F' strains show a high rate of integration back into the bacterial chromosome of a second strain. Furthermore, the site of integration is often the site occupied by the sex factor in the original Hfr strain (before production of the F' strains). Explain these results.

30. You have two *E. coli* strains, F$^-$ *str*^{r} *ala*$^-$ and Hfr *str*^{s} *ala*$^+$, in which the F factor is inserted close to *ala*$^+$. Devise a screening test to detect strains carrying F' *ala*$^+$.

31. Five Hfr strains A through E are derived from a single F$^+$ strain of *E. coli*. The following chart shows the entry times of the first five markers into an F$^-$ strain when each is used in an interrupted-conjugation experiment:

A		B		C		D		E	
mal$^+$	(1)	*ade*$^+$	(13)	*pro*$^+$	(3)	*pro*$^+$	(10)	*his*$^+$	(7)
str^s	(11)	*his*$^+$	(28)	*met*$^+$	(29)	*gal*$^+$	(16)	*gal*$^+$	(17)
ser$^+$	(16)	*gal*$^+$	(38)	*xyl*$^+$	(32)	*his*$^+$	(26)	*pro*$^+$	(23)
ade$^+$	(36)	*pro*$^+$	(44)	*mal*$^+$	(37)	*ade*$^+$	(41)	*met*$^+$	(49)
his$^+$	(51)	*met*$^+$	(70)	*str*^{s}	(47)	*ser*$^+$	(61)	*xyl*$^+$	(52)

   a. Draw a map of the F$^+$ strain, indicating the positions of all genes and their distances apart in minutes.

   b. Show the insertion point and orientation of the F plasmid in each Hfr strain.

   c. In the use of each of these Hfr strains, state which allele you would select to obtain the highest proportion of Hfr exconjugants.

32. *Streptococcus pneumoniae* cells of genotype *str*^{s} *mtl*$^-$ are transformed by donor DNA of genotype *str*^{r} *mtl*$^+$ and

(in a separate experiment) by a mixture of two DNAs with genotypes $str^r\ mtl^-$ and $str^s\ mtl^+$. The accompanying table shows the results.

Transforming DNA	*Percentage of cells transformed into*		
	$str^r\ mtl^-$	$str^s\ mtl^+$	$str^r\ mtl^+$
$str^r\ mtl^+$	4.3	0.40	0.17
$str^r\ mtl^- + str^s\ mtl^+$	2.8	0.85	0.0066

a. What does the first row of the table tell you? Why?

b. What does the second row of the table tell you? Why?

33. Recall that, in Chapter 4, we considered the possibility that a crossover event may affect the likelihood of another crossover. In the bacteriophage T4, gene a is 1.0 m.u. from gene b, which is 0.2 m.u. from gene c. The gene order is a, b, c. In a recombination experiment, you recover five double crossovers between a and c from 100,000 progeny viruses. Is it correct to conclude that interference is negative? Explain your answer.

34. You have infected *E. coli* cells with two strains of T4 virus. One strain is minute (m), rapid lysis (r), and turbid (tu); the other is wild type for all three markers. The lytic products of this infection are plated and classified. The resulting 10,342 plaques were distributed among eight genotypes as follows:

$m\ r\ tu$	3467		$m + +$	520
$+ + +$	3729		$+ r\ tu$	474
$m\ r +$	853		$+ r +$	172
$m + tu$	162		$+ + tu$	965

a. Determine the linkage distances between m and r, between r and tu, and between m and tu.

b. What linkage order would you suggest for the three genes?

c. What is the coefficient of coincidence (see Chapter 4) in this cross? What does it signify?

(Problem 34 is reprinted with the permission of Macmillan Publishing Co., Inc., from Monroe W. Strickberger, *Genetics.* Copyright 1968 by Monroe W. Strickberger.)

35. With the use of P22 as a generalized transducing phage grown on a $pur^+\ pro^+\ his^+$ bacterial donor, a recipient strain of genotype $pur^-\ pro^-\ his^-$ is infected and incubated. Afterward, transductants for pur^+, pro^+, and his^+ are selected individually in experiments I, II, and III, respectively.

a. What media are used for these selection experiments?

b. The transductants are examined for the presence of unselected donor markers, with the following results:

I	II	III
$pro^-\ his^-$ 87%	$pur^-\ his^-$ 43%	$pur^-\ pro^-$ 21%
$pro^+\ his^-$ 0%	$pur^+\ his^-$ 0%	$pur^+\ pro^-$ 15%
$pro^-\ his^+$ 10%	$pur^-\ his^+$ 55%	$pur^-\ pro^+$ 60%
$pro^+\ his^+$ 3%	$pur^+\ his^+$ 2%	$pur^+\ pro^+$ 4%

What is the order of the bacterial genes?

c. Which two genes are closest together?

d. On the basis of the order that you proposed in part c, explain the relative proportions of genotypes observed in experiment II.

(Problem 35 is from D. Freifelder, *Molecular Biology and Biochemistry.* Copyright 1978 by W. H. Freeman and Company, New York.)

36. Although most λ-mediated gal^+ transductants are inducible lysogens, a small percentage of these transductants in fact are not lysogens (that is, they contain no integrated λ). Control experiments show that these transductants are not produced by mutation. What is the likely origin of these types?

37. An $ade^+\ arg^+\ cys^+\ his^+\ leu^+\ pro^+$ bacterial strain is known to be lysogenic for a newly discovered phage, but the site of the prophage is not known. The bacterial map is

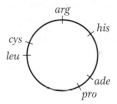

The lysogenic strain is used as a source of the phage, and the phages are added to a bacterial strain of genotype $ade^-\ arg^-\ cys^-\ his^-\ leu^-\ pro^-$. After a short incubation, samples of these bacteria are plated on six different media, with the supplementations indicated in the following table. The table also shows whether colonies were observed on the various media.

	Nutrient supplementation in medium						Presence
Medium	Ade	Arg	Cys	His	Leu	Pro	of colonies
1	−	+	+	+	+	+	N
2	+	−	+	+	+	+	N
3	+	+	−	+	+	+	C
4	+	+	+	−	+	+	N
5	+	+	+	+	−	+	C
6	+	+	+	+	+	−	N

(In this table, a plus sign indicates the presence of a nutrient supplement, a minus sign indicates that a supplement is not present, N indicates no colonies, and C indicates colonies present.)

a. What genetic process is at work here?

b. What is the approximate locus of the prophage?

38. In a generalized-transduction system using P1 phage, the donor is *pur⁺ nad⁺ pdx⁻* and the recipient is *pur⁻ nad⁻ pdx⁺*. The donor allele *pur⁺* is initially selected after transduction, and 50 *pur⁺* transductants are then scored for the other alleles present. Here are the results:

Genotype	Number of colonies
nad⁺ pdx⁺	3
nad⁺ pdx⁻	10
nad⁻ pdx⁺	24
nad⁻ pdx⁻	13
	50

a. What is the cotransduction frequency for *pur* and *nad*?

b. What is the cotransduction frequency for *pur* and *pdx*?

c. Which of the unselected loci is closest to *pur*?

d. Are *nad* and *pdx* on the same side or on opposite sides of *pur*? Explain.

(Draw the exchanges needed to produce the various transformant classes under either order to see which requires the minimum number to produce the results obtained.)

39. In a generalized-transduction experiment, phages are collected from an *E. coli* donor strain of genotype *cys⁺ leu⁺ thr⁺* and used to transduce a recipient of genotype *cys⁻ leu⁻ thr⁻*. Initially, the treated recipient population is plated on a minimal medium supplemented with leucine and threonine. Many colonies are obtained.

a. What are the possible genotypes of these colonies?

b. These colonies are then replica plated onto three different media: (1) minimal plus threonine only, (2) minimal plus leucine only, and (3) minimal. What genotypes could, in theory, grow on these three media?

c. Of the original colonies, 56 percent are observed to grow on medium 1, 5 percent on medium 2, and no colonies on medium 3. What are the actual genotypes of the colonies on media 1, 2, and 3?

d. Draw a map showing the order of the three genes and which of the two outer genes is closer to the middle gene.

40. Deduce the genotypes of the following *E. coli* strains 1 through 4:

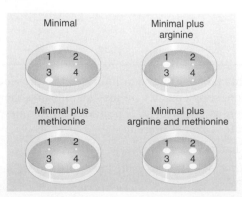

41. In an interrupted-conjugation experiment in *E. coli*, the *pro* gene enters after the *thi* gene. A *pro⁺ thi⁺* Hfr is crossed with a *pro⁻ thi⁻* F⁻ strain, and exconjugants are plated on medium containing thiamine but no proline. A total of 360 colonies are observed, and they are isolated and cultured on fully supplemented medium. These cultures are then tested for their ability to grow on medium containing no proline or thiamine (minimal medium), and 320 of the cultures are found to be able to grow but the remainder cannot.

a. Deduce the genotypes of the two types of cultures.

b. Draw the crossover events required to produce these genotypes.

c. Calculate the distance between the *pro* and *thi* genes in recombination units.

 Unpacking Problem 41

1. What type of organism is *E. coli*?

2. What does a culture of *E. coli* look like?

3. On what sort of substrates does *E. coli* generally grow in its natural habitat?

4. What are the minimal requirements for *E. coli* cells to divide?

5. Define the terms *prototroph* and *auxotroph*.

6. Which cultures in this experiment are prototrophic and which are auxotrophic?

7. Given some strains of unknown genotype regarding thiamine and proline, how would you test their genotypes? Give precise experimental details, including equipment.

8. What kinds of chemicals are proline and thiamine? Does it matter in this experiment?

9. Draw a diagram showing the full set of manipulations performed in the experiment.

10. Why do you think the experiment was done?

11. How was it established that *pro* enters after *thi*? Give precise experimental steps.

12. In what way does an interrupted-mating experiment differ from the experiment described in this problem?

13. What is an exconjugant? How do you think that exconjugants were obtained? (It might include genes not described in this problem.)

14. When the *pro* gene is said to enter after *thi*, does it mean the *pro* allele, the *pro⁺* allele, either, or both?

15. What is "fully supplemented medium" in the context of this question?

16. Some exconjugants did not grow on minimal medium. On what medium would they grow?

17. State the types of crossovers that take part in Hfr × F⁻ recombination. How do these crossovers differ from crossovers in eukaryotes?

18. What is a recombination unit in the context of the present analysis? How does it differ from the map units used in eukaryote genetics?

42. A generalized transduction experiment uses a $metE^+$ $pyrD^+$ strain as donor and $metE^-$ $pyrD^-$ as recipient. $metE^+$ transductants are selected and then tested for the $pyrD^+$ allele. The following numbers were obtained:

$$metE^+ \ pyrD^- \qquad 857$$

$$metE^+ \ pyrD^+ \qquad 1$$

Do these results suggest that these loci are closely linked? What other explanations are there for the lone "double"?

43. An $argC^-$ strain was infected with transducing phage, and the lysate was used to transduce $metF^-$ recipients on medium containing arginine but no methionine. The $metF^+$ transductants were then tested for arginine requirement: most were $argC^+$ but a small percentage were found to be $argC^-$. Draw diagrams to show the likely origin of the $argC^+$ and $argC^-$ strains.

CHALLENGING PROBLEMS

44. Four E. coli strains of genotype $a^+ b^-$ are labeled 1, 2, 3, and 4. Four strains of genotype $a^- b^+$ are labeled 5, 6, 7, and 8. The two genotypes are mixed in all possible combinations and (after incubation) are plated to determine the frequency of $a^+ b^+$ recombinants. The following results are obtained, where M = many recombinants, L = low numbers of recombinants, and 0 = no recombinants:

	1	2	3	4
5	0	M	M	0
6	0	M	M	0
7	L	0	0	M
8	0	L	L	0

On the basis of these results, assign a sex type (either Hfr, F⁺, or F⁻) to each strain.

45. An Hfr strain of genotype $a^+ b^+ c^+ d^- str^s$ is mated with a female strain of genotype $a^- b^- c^- d^+ str^r$. At various times, the culture is shaken vigorously to separate mating pairs. The cells are then plated on agar of the following three types, where nutrient A allows the growth of a^- cells; nutrient B, of b^- cells; nutrient C, of c^- cells; and nutrient D, of d^- cells (a plus indicates the presence of streptomycin or a nutrient, and a minus indicates its absence):

Agar type	Str	A	B	C	D
1	+	+	+	−	+
2	+	−	+	+	+
3	+	+	−	+	+

a. What donor genes are being selected on each type of agar?

b. The following table shows the number of colonies on each type of agar for samples taken at various times after the strains are mixed. Use this information to determine the order of genes *a*, *b*, and *c*.

Time of sampling (minutes)	Number of colonies on agar of type		
	1	2	3
0	0	0	0
5	0	0	0
7.5	100	0	0
10	200	0	0
12.5	300	0	75
15	400	0	150
17.5	400	50	225
20	400	100	250
25	400	100	250

c. From each of the 25-minute plates, 100 colonies are picked and transferred to a petri dish containing agar with all the nutrients except D. The numbers of colonies that grow on this medium are 89 for the sample from agar type 1, 51 for the sample from agar type 2, and 8 for the sample from agar type 3. Using these data, fit gene *d* into the sequence of *a*, *b*, and *c*.

d. At what sampling time would you expect colonies to first appear on agar containing C and streptomycin but no A or B?

(Problem 45 is from D. Freifelder, *Molecular Biology and Biochemistry.* Copyright 1978 by W. H. Freeman and Company.)

46. In the cross Hfr $aro^+ arg^+ ery^r str^s$ × F⁻ $aro^- arg\ ery^\varepsilon$ str^r, the markers are transferred in the order given (with aro^+ entering first), but the first three genes are very close together. Exconjugants are plated on a medium containing Str (streptomycin, to kill Hfr cells), Ery (erythromycin), Arg (arginine), and Aro (aromatic amino acids). The following results are obtained for 300 colonies isolated from these plates and tested for growth on various media: on Ery only, 263 strains grow; on Ery + Arg, 264 strains grow; on Ery + Aro, 290 strains grow; on Ery + Arg + Aro, 300 strains grow.

a. Draw up a list of genotypes, and indicate the number of individuals in each genotype.

b. Calculate the recombination frequencies.

c. Calculate the ratio of the size of the *arg*-to-*aro* region to the size of the *ery*-to-*arg* region.

47. A transformation experiment is performed with a donor strain that is resistant to four drugs: A, B, C, and D. The recipient is sensitive to all four drugs. The treated recipient cell population is divided up and plated on media containing various combinations of the drugs. The following table shows the results.

Drugs added	Number of colonies	Drugs added	Number of colonies
None	10,000	BC	51
A	1156	BD	49
B	1148	CD	786
C	1161	BC	30
D	1139	ABD	42
AB	46	ACD	630
AC	640	BCD	36
AD	942	ABCD	30

a. One of the genes is obviously quite distant from the other three, which appear to be tightly (closely) linked. Which is the distant gene?

b. What is the probable order of the three tightly linked genes?

(Problem 47 is from Franklin Stahl, *The Mechanics of Inheritance*, 2nd ed. Copyright 1969, Prentice Hall, Englewood Cliffs, N.J. Reprinted by permission.)

48. You have two strains of λ that can lysogenize *E. coli*; their linkage maps are as follows:

Strain X   Strain Y

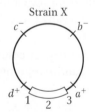
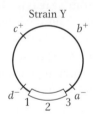

The segment shown at the bottom of the chromosome, designated 1–2–3, is the region responsible for pairing and crossing over with the *E. coli* chromosome. (Keep the markers on all your drawings.)

a. Diagram the way in which λ strain X is inserted into the *E. coli* chromosome (so that the *E. coli* is lysogenized).

b. The bacteria that are lysogenic for strain X can be superinfected by using strain Y. A certain percentage of these superinfected bacteria become "doubly" lysogenic (that is, lysogenic for both strains). Diagram how it will take place. (Don't worry about how double lysogens are detected.)

c. Diagram how the two λ prophages can pair.

d. Crossover products between the two prophages can be recovered. Diagram a crossover event and the consequences.

49. You have three strains of *E. coli*. Strain A is F′ *cys⁺ trp1/cys⁺ trp1* (that is, both the F′ and the chromosome carry *cys⁺* and *trp1*, an allele for tryptophan requirement). Strain B is F⁻ *cys⁻ trp2 Z* (this strain requires cysteine for growth and carries *trp2*, another allele causing a tryptophan require-

ment; strain B is lysogenic for the generalized transducing phage Z). Strain C is F⁻ *cys⁺ trp1* (it is an F⁻ derivative of strain A that has lost the F′). How would you determine whether *trp1* and *trp2* are alleles of the same locus? (Describe the crosses and the results expected.)

50. A generalized transducing phage is used to transduce an *a⁻ b⁻ c⁻ d⁻ e⁻* recipient strain of *E. coli* with an *a⁺ b⁺ c⁺ d⁺ e⁺* donor. The recipient culture is plated on various media with the results shown in the following table. (Note that *a⁻* indicates a requirement for A as a nutrient, and so forth.) What can you conclude about the linkage and order of the genes?

Compounds added to minimal medium	Presence (+) or absence (−) of colonies
CDE	−
BDE	−
BCE	+
BCD	+
ADE	−
ACE	−
ACD	−
ABE	−
ABD	+
ABC	−

51. In 1965, Jon Beckwith and Ethan Signer devised a method of obtaining specialized transducing phages carrying the *lac* region. They knew that the integration site, designated *att80*, for the temperate phage φ80 (a relative of phage λ) was located near *tonB*, a gene that confers resistance to the virulent phage T1:

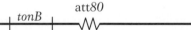

They used an F′ *lac⁺* plasmid that could not replicate at high temperatures in a strain carrying a deletion of the *lac* genes. By forcing the cell to remain *lac⁺* at high temperatures, the researchers could select strains in which the plasmid had integrated into the chromosome, thereby allowing the F′ *lac* to be maintained at high temperatures. By combining this selection with a simultaneous selection for resistance to T1 phage infection, they found that the only survivors were cells in which the F′ *lac* had integrated into the *tonB* locus, as shown here:

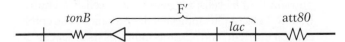

This result placed the *lac* region near the integration site for phage φ80. Describe the subsequent steps that the researchers must have followed to isolate the specialized transducing particles of phage φ80 that carried the *lac* region.

52. Wild-type *E. coli* takes up and concentrates a certain red food dye, making the colonies blood red. Transposon mutagenesis was used, and the cells were plated on food dye. Most colonies were red, but some colonies did not take up dye and appeared white. In one white colony, the DNA surrounding the transposon insert was sequenced, with the use of a DNA replication primer identical with part of the end of the transposon sequence, and the sequence adjacent to the transposon was found to correspond to a gene of unknown function called *atoE,* spanning positions 2.322 through 2.324 Mb on the map (numbered from an arbitrary position zero). Propose a function for *atoE.* What biological process could be investigated in this way and what other types of white colonies might be expected?

Gene Interaction 6

The colors of peppers are determined by the interaction of several genes. An allele *Y* promotes the early elimination of chlorophyll (a green pigment), whereas *y* does not. Allele *R* determines red and *r* determines yellow carotenoid pigments. Alleles *c1* and *c2* of two different genes down-regulate the amounts of carotenoids, causing the lighter shades. Orange is down-regulated red. Brown is green plus red. Pale yellow is down-regulated yellow. [*Anthony Griffiths.*]

The thrust of our presentation in the book so far has been to show how geneticists identify a gene that affects some biological property of interest. We have seen how the approaches of forward genetics can be used to identify individual genes. The researcher begins with a set of mutants, then crosses each mutant with the wild type to see if the mutant shows single-gene inheritance. The cumulative data from such a research program would reveal a set of genes that all have roles in the development of the property under investigation. In some cases, the researcher may be able to identify specific biochemical functions for many of the genes by comparing gene sequences with those of other organisms. The next step, which is a greater challenge, is to deduce how the genes in a set interact to influence phenotype.

How are the gene interactions underlying a property deduced? One molecular approach is to analyze protein interactions directly in vitro by using one protein as "bait" and observing which other cellular proteins attach to it. Proteins that are found to bind to the bait are candidates for interaction in the living cell. Another molecular approach is to analyze mRNA transcripts. The genes that collaborate in some specific developmental process can be defined by the set of RNA transcripts

211

present when that process is going on, a type of analysis now carried out with the use of genome chips (see Chapter 14). Finally, gene interactions and their significance in shaping phenotype can be deduced by *genetic analysis,* which is the focus of this chapter.

Gene interactions can be classified broadly into two categories. The first category consists of interactions between alleles of one locus, broadly speaking variations on dominance. Although this information does not address the range of genes affecting a function, a great deal can be learned of a gene's role by considering allelic interactions. The second category consists of interactions between two or more loci. These interactions reveal the number and types of genes in the overall program underlying a particular biological function.

6.1 Interactions between the Alleles of a Single Gene: Variations on Dominance

There are thousands of different ways to alter the sequence of a gene, each producing a mutant allele, although only some of these mutant alleles will appear in a real population. The known mutant alleles of a gene and its wild-type allele are referred to as **multiple alleles** or an **allelic series.**

One of the tests routinely performed on a new mutant allele is to see if it is dominant or recessive. Basic information about dominance and recessiveness is useful in working with the new mutation and can be a source of insight into the way the gene functions, as we will see in the examples. Dominance is a manifestation of how *the alleles of a single gene* interact in a heterozygote. In any experiment the interacting alleles may be wild and mutant alleles ($+/m$) or two different mutant alleles (m_1/m_2). Several types of dominance have been discovered, each representing a different type of interaction between alleles.

Complete dominance and recessiveness

The simplest type of dominance is **full,** or **complete, dominance.** A fully dominant allele will be expressed when only one copy is present, as in a heterozygote, whereas the alternative allele will be fully recessive. In full dominance, the homozygous dominant cannot be distinguished from the heterozygote; that is, at the phenotypic level, $A/A = A/a$. As mentioned earlier, phenylketonuria (PKU) and many other single-gene human diseases are fully recessive, whereas their wild-type alleles are dominant. Other single-gene diseases such as achondroplasia are fully dominant, whereas, in those cases, the wild-type allele is recessive. How can these dominance relations be interpreted at the cell level?

The disease PKU is a good general model for recessive mutations. Recall that PKU is caused by a defective allele of the gene encoding the enzyme phenylalanine hydroxylase (PAH). In the absence of normal PAH, the phenylalanine entering the body in food is not broken down and hence accumulates. Under such conditions, phenylalanine is converted into phenylpyruvic acid, which is transported to the brain through the bloodstream and there impedes normal development, leading to mental retardation. The reason that the defective allele is recessive is that one "dose" of the wild-type allele P produces enough PAH to break down the phenylalanine entering the body. The PAH gene is said to be *haplosufficient.* Hence, both P/P (two doses) and P/p (one dose) have enough PAH activity to result in the normal cellular chemistry. People with p/p have zero doses of PAH activity. Figure 6-1 illustrates this general notion.

How can we explain fully dominant mutations? There are several molecular mechanisms for dominance. A regularly encountered mechanism is that the wild-type allele of a gene is *haploinsufficient.* In haploinsufficiency, one wild-

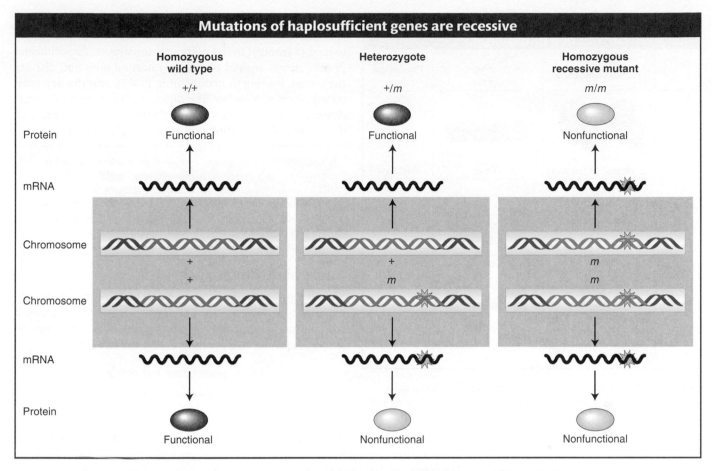

Mutations of haplosufficient genes are recessive

Figure 6-1 In the heterozygote, even though the mutated copy of the gene produces nonfunctional protein, the wild-type copy generates enough functional protein to produce the wild-type phenotype.

type dose is *not* enough to achieve normal levels of function. Assume that 16 units of a gene's product are needed for normal chemistry and that each wild-type allele can make 10 units. Two wild-type alleles will produce 20 units of product, well over the minimum. But consider what happens if one of the mutations is a **null mutation,** which produces a nonfunctional protein. A null mutation in combination with a single wild-type allele would produce 10 + 0 — 10 units, well below the minimum. Hence, the heterozygote (wild type/null) is mutant, and the mutation is, by definition, dominant. In mice, the gene *Tbx1* is haploinsufficient. This gene encodes a transcription-regulating protein (a *transcription factor*) that acts on genes responsible for the development of the pharynx. A knockout of one wild-type allele results in an inadequate concentration of the regulatory protein, which results in defects in the development of the pharyngeal arteries. The same haploinsufficiency is thought to be responsible for DiGeorge syndrome in humans, a condition with cardiovascular and craniofacial abnormalities.

Another important type of dominant mutation is called a **dominant negative.** Polypeptides with this type of mutation act as "spoilers" or "rogues." In some cases, the gene product is a unit of a *homodimeric* protein, a protein composed of two units of the same type. In the heterozygote (+/*M*), the spoiler polypeptide binds to the wild-type polypeptide and distorts it or otherwise interferes with its function. The same type of spoiling can also hinder the functioning of a *heterodimer* composed of polypeptides from different genes. In other cases, the gene product is a monomer, and, in these situations, the mutant binds the substrate, leaving little available on which the wild-type protein can act.

An example of mutations that can act as dominant negatives is found in the gene for collagen protein. Some mutations in this gene give rise to the human phenotype

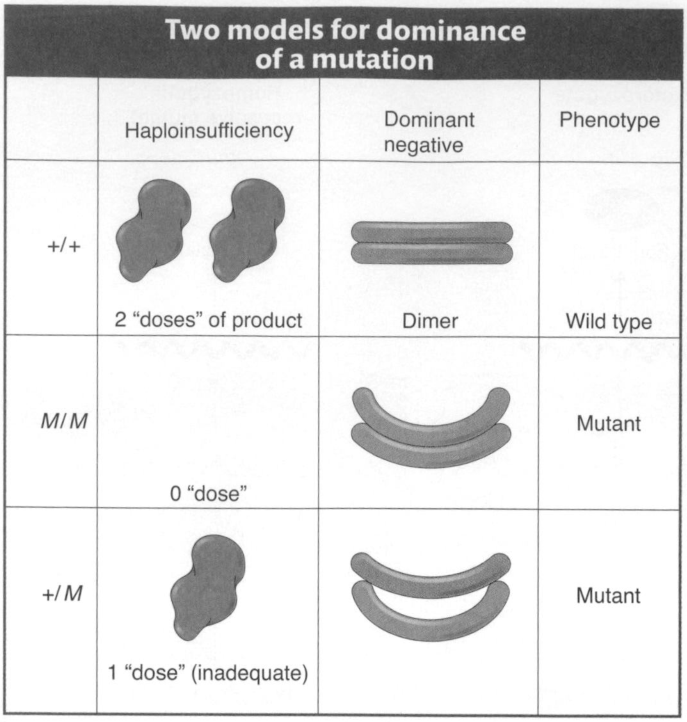

Two models for dominance of a mutation

	Haploinsufficiency	Dominant negative	Phenotype
+/+	2 "doses" of product	Dimer	Wild type
M/M	0 "dose"		Mutant
+/M	1 "dose" (inadequate)		Mutant

Figure 6-2 A mutation may be dominant because (*left*) a single wild-type gene does not produce enough protein product for proper function or (*right*) the mutant allele acts as a dominant negative that produces a "spoiler" protein product.

osteogenesis imperfecta (brittle-bone disease). Collagen is a connective-tissue protein formed of three monomers intertwined (a trimer). In the mutant heterozygote, the abnormal protein wraps around one or two normal ones and distorts the trimer, leading to malfunction. In this way, the defective collagen acts as a spoiler. The difference between haploinsufficiency and the action of a dominant negative as causes of dominance of a mutation is illustrated in Figure 6-2.

Message For most genes, a single copy is adequate for full expression (such genes are haplosufficient), and their null mutations are fully recessive. Harmful mutations of haplosufficient genes are often dominant. Mutations in genes that encode units in homo- or heterodimers can behave as dominant negatives, acting through "spoiler" proteins.

Incomplete dominance

Snapdragons are popular garden plants. When a pure-breeding wild-type four-o'clock line having red petals is crossed with a pure line having white petals, the F_1 has pink petals. If an F_2 is produced by selfing the F_1, the result is

$\frac{1}{4}$ of the plants have red petals

$\frac{1}{2}$ of the plants have pink petals

$\frac{1}{4}$ of the plants have white petals

Figure 6-3 shows these phenotypes. From this $1:2:1$ ratio in the F_2, we can deduce that the inheritance pattern is based on two alleles of a single gene. However, the heterozygotes (the F_1 and half the F_2) are intermediate in phenotype. By inventing allele symbols, we can list the genotypes of the four-o'clocks in this experiment as c^+/c^+ (red), c/c (white), and c^+/c (pink). The occurrence of the intermediate phenotype suggests an **incomplete dominance,** the term used to describe the general case in which the phenotype of a heterozygote is intermediate between those of the two homozygotes, on some quantitative scale of measurement.

How do we explain incomplete dominance at the molecular level? In incomplete dominance, each wild-type allele generally produces a set dose of its protein product. The number of doses of a wild-type allele determines the concentration of a chemical made by the protein, such as pigment. In the snapdragon, two doses produce the most copies of transcript, thus producing the greatest amount of protein and, hence, the greatest amount of pigment, enough to make the flower petals red. One dose produces less pigment, and so the petals are pink. A zero dose produces no pigment.

Codominance

Another variation on the theme of dominance is **codominance,** the expression of both alleles of a heterozygote. A clear example is seen in the human ABO blood groups, where there is codominance of antigen alleles. The ABO blood groups are determined by three alleles of one gene.

Incomplete dominance

Figure 6-3 In snapdragons, a heterozygote is pink, intermediate between the two homozygotes red and white. The pink heterozygote demonstrates incomplete dominance. [*John Kaprielian/Science Source.*]

GENETICS P⊙RTAL ANIMATED ART
Molecular allele interactions

These three alleles interact in several ways to produce the four blood types of the ABO system. The three major alleles are i, I^A, and I^B, but a person can have only two of the three alleles or two copies of one of them. The combinations result in six different genotypes: the three homozygotes and three different types of heterozygotes, as shown in the margin.

In this allelic series, the alleles determine the presence and form of a complex sugar molecule present on the surface of red blood cells. This sugar molecule is an antigen, a cell-surface molecule that can be recognized by the immune system. The alleles I^A and I^B determine two different forms of this cell-surface molecule. However, the allele i results in no cell-surface molecule of this type (it is a null allele). In the genotypes I^A/i and I^B/i, the alleles I^A and I^B are fully dominant over i. However, in the genotype I^A/I^B, each of the alleles produces its own form of the cell-surface molecule, and so the A and B alleles are codominant.

The human disease sickle-cell anemia provides an interesting example of the somewhat arbitrary ways in which we classify dominance. The gene concerned encodes the molecule hemoglobin, which is responsible for transporting oxygen in blood vessels and is the major constituent of red blood cells. There are two main alleles Hb^A and Hb^S, and the three possible genotypes have different phenotypes, as follows:

Genotype	Blood type
I^A/I^A, I^A/i	A
I^B/I^B, I^B/i	B
I^A/I^B	AB
i/i	O

Hb^A/Hb^A: normal; red blood cells never sickle

Hb^S/Hb^S: severe, often fatal anemia; abnormal hemoglobin causes red blood cells to have sickle shape

HB^A/Hb^S: no anemia; red blood cells sickle only under low oxygen concentrations

Figure 6-4 shows an electron micrograph of blood cells including some sickled cells. In regard to the presence or absence of anemia, the Hb^A allele is dominant. In the heterozygote, a single Hb^A allele produces enough functioning hemoglobin to prevent anemia. In regard to blood-cell shape, however, there is incomplete dominance, as shown by the fact that, in the heterozygote, many of the cells have a slight sickle shape. Finally, in regard to hemoglobin itself, there is codominance. The alleles Hb^A and Hb^S encode two different forms of hemoglobin that differ by a single amino acid, and both forms are synthesized in the heterozygote. The A and S forms of hemoglobin can be separated by electrophoresis, because it happens that they have different charges (Figure 6-5). We see that homozygous normal people have one type of hemoglobin (A) and anemics have another (type S), which moves more slowly in the electric field. The heterozygotes have both types, A and S. In other words, there is codominance at the molecular level. The fascinating population genetics of the Hb^A and Hb^S alleles will be considered in Chapter 20.

Sickle-cell anemia illustrates the arbitrariness of the terms *dominance, incomplete dominance,* and *codominance.* The type of dominance inferred depends on the phenotypic level at which the assay is made—organismal, cellular, or molecular. Indeed, the same caution can be applied to many of the categories that scientists use to classify structures and processes; these categories are devised by humans for the convenience of analysis.

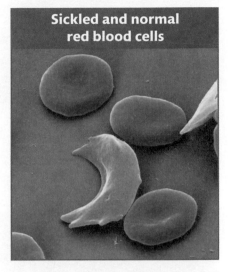

Sickled and normal red blood cells

Figure 6-4 The sickle-shaped cell is caused by a single mutation in the gene for hemoglobin. [*Meckes/Ottawa/Photo Researchers.*]

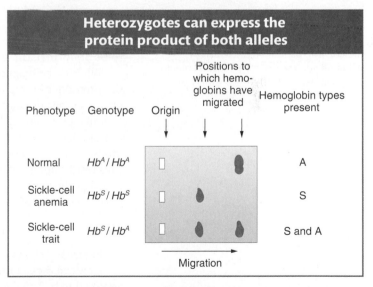

Heterozygotes can express the protein product of both alleles

Figure 6-5 The electrophoresis of normal and mutant hemoglobins. Shown are results produced by hemoglobin from a person with sickle-cell trait (a heterozygote), a person with sickle-cell anemia, and a normal person. The smudges show the positions to which the hemoglobins migrate on the starch gel.

Message The type of dominance is determined by the molecular functions of the alleles of a gene and by the investigative level of analysis.

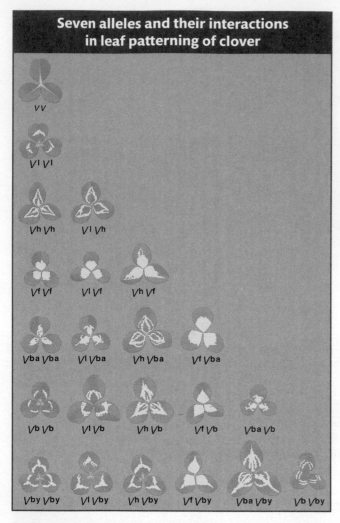

Seven alleles and their interactions in leaf patterning of clover

$v v$

$v^l v^l$

$v^h v^h$ $v^l v^h$

$v^f v^f$ $v^l v^f$ $v^h v^f$

$v^{ba} v^{ba}$ $v^l v^{ba}$ $v^h v^{ba}$ $v^f v^{ba}$

$v^b v^b$ $v^l v^b$ $v^h v^b$ $v^f v^b$ $v^{ba} v^b$

$v^{by} v^{by}$ $v^l v^{by}$ $v^h v^{by}$ $v^f v^{by}$ $v^{ba} v^{by}$ $v^b v^{by}$

Figure 6-6 Multiple alleles determine the chevron pattern on the leaves of white clover. The genotype of each plant is shown below it. There is a variety of dominance interactions. [*After photograph by W. Ellis Davies.*]

The leaves of clover plants show several variations on the dominance theme. Clover is the common name for plants of the genus *Trifolium.* There are many species. Some are native to North America, whereas others grow there as introduced weeds. Much genetic research has been done with white clover, which shows considerable variation among individual plants in the curious V, or chevron, pattern on the leaves. The different chevron forms (and the absence of chevrons) are determined by a series of seven alleles, as seen in Figure 6-6, which shows the many different types of interactions possible for even one allele. In most practical cases many alleles of a gene can be found together in a population, constituting an allelic series. The phenotypes shown by the allelic combinations are many and varied, reflecting the relative nature of dominance: an allele can show dominance with one partner but not with another.

Recessive lethal alleles

An allele that is capable of causing the death of an organism is called a **lethal allele.** In the characterization of a set of newly discovered mutant alleles, a recessive mutation is sometimes found to be lethal. This information is potentially useful in that it shows that the newly discovered gene (of yet unknown function) is essential to the organism's operation. Indeed, with the use of modern DNA technology, a null mutant allele of a gene of interest can now be made intentionally and made homozygous to see if it is lethal and under which environmental conditions. Lethal alleles are also useful in determining the developmental stage at which the gene normally acts. In this case, geneticists look for whether death from a lethal mutant allele occurs early or late in the development of a zygote. The phenotype associated with death can also be informative in regard to gene function; for example, if a certain organ appears to be abnormal, the gene is likely to express through that organ.

What is the diagnostic test for lethality? The test is well illustrated by one of the prototypic examples of a lethal allele, a coat-color allele in mice (see the Model Organism box on page 218). Normal wild-type mice have coats with a rather dark overall pigmentation. A mutation called *yellow* (a lighter coat color) shows a curious inheritance pattern. If any yellow mouse is mated with a homozygous wild-type mouse, a 1:1 ratio of yellow to wild-type mice is always observed in the progeny. This result suggests that a yellow mouse is always heterozygous for the yellow allele and that the yellow allele is dominant over wild type. However, if any two yellow mice are crossed with each other, the result is always as follows:

$$\text{yellow} \times \text{yellow} \rightarrow \tfrac{2}{3} \text{ yellow}, \tfrac{1}{3} \text{ wild type}$$

Figure 6-7 shows a typical litter from a cross between yellow mice.

How can the 2:1 ratio be explained? The results make sense if the yellow allele is assumed to be lethal when homozygous. The yellow allele is known to be of a coat-color gene called *A*. Let's call it A^Y. Hence, the results of crossing two yellow mice are

$$A^Y/A \times A^Y/A$$

Progeny $\tfrac{1}{4} A^Y/A^Y$ lethal

$\tfrac{1}{2} A^Y/A$ yellow

$\tfrac{1}{4} A/A$ wild type

The expected monohybrid ratio of $1:2:1$ would be found among the zygotes, but it is altered to a $2:1$ ratio in the progeny actually seen at birth because zygotes with a lethal A^Y/A^Y genotype do not survive to be counted. This hypothesis is supported by the removal of uteri from pregnant females of the yellow $\times$ yellow cross; one-fourth of the embryos are found to be dead.

The A^Y allele produces effects on two characters: coat color and survival. It is entirely possible, however, that both effects of the A^Y allele result from the same basic cause, which promotes yellowness of coat in a single dose and death in a double dose. In general, the term **pleiotropic** is used for any allele that affects several properties of an organism.

The tailless Manx phenotype in cats (Figure 6-8) also is produced by an allele that is lethal in the homozygous state. A single dose of the Manx allele, M^L, severely interferes with normal spinal development, resulting in the absence of a tail in the M^L/M heterozygote. But in the M^L/M^L homozygote, the double dose of the gene produces such an extreme abnormality in spinal development that the embryo does not survive.

The *yellow* and M^L alleles have their own phenotypes in a heterozygote, but most recessive lethals are silent in the heterozygote. In such a situation, recessive lethality is diagnosed by observing the death of 25 percent of the progeny at some stage of development.

Whether an allele is lethal or not often depends on the environment in which the organism develops. Whereas certain alleles are lethal in virtually any environment, others are viable in one environment but lethal in another. Human hereditary diseases provide some examples. Cystic fibrosis and sickle-cell anemia are diseases that would be lethal without treatment. Furthermore, many of the alleles favored and selected by animal and plant breeders would almost certainly be eliminated in nature as a result of competition with the members of the natural population. The dwarf mutant varieties of grain, which are very high yielding, provide good examples; only careful nurturing by farmers has maintained such alleles for our benefit.

Geneticists commonly encounter situations in which expected phenotypic ratios are consistently skewed in one direction because a mutant allele reduces viability. For example, in the cross $A/a \times a/a$, we predict a progeny ratio of 50 percent A/a and 50 percent a/a, but we might consistently observe a ratio such as 55 percent:45 percent or 60 percent:40 percent. In such a case, the recessive allele is said to be *sublethal* because the lethality is expressed in only some but not all of the homozygous individuals. Thus, lethality may range from 0 to 100 percent, depending on the gene itself, the rest of the genome, and the environment.

We have seen that lethal alleles are useful in diagnosing the time at which a gene acts and the nature of the phenotypic defect that kills. However, maintaining stocks bearing lethal alleles for laboratory use is a challenge. In diploids, recessive lethal alleles can be maintained as heterozygotes. In haploids, heat-sensitive lethal alleles are useful. They are members of a general class of **temperature-sensitive (ts) mutations.** Their phenotype is wild type at the **permissive temperature** (often room temperature) but mutant at some higher **restrictive temperature.** Temperature-sensitive alleles are thought to be caused by mutations that make the protein prone to twist or bend its shape to an inactive conformation at the restrictive temperature. Research stocks

A recessive lethal allele, yellow coat

Figure 6-7 A litter from a cross between two mice heterozygous for the dominant yellow coat-color allele. The allele is lethal in a double dose. Not all progeny are visible. [*Anthony Griffiths.*]

Tailless, a recessive lethal allele in cats

Figure 6-8 A Manx cat. A dominant allele causing taillessness is lethal in the homozygous state. The phenotype of two eye colors is unrelated to taillessness. [*Gerard Lacz/NHPA.*]

Model Organism *Mouse*

The laboratory mouse is descended from the house mouse *Mus musculus*. The pure lines used today as standards are derived from mice bred in past centuries by mouse "fanciers." Among model organisms, it is the one whose genome most closely resembles the human genome. Its diploid chromosome number is 40 (compared with 46 in humans), and the genome is slightly smaller than that of humans (the human genome being 3000 Mb) and contains approximately the same number of genes (current estimate 25,000). Furthermore, all mouse genes seem to have counterparts in humans. A large proportion of genes are arranged in blocks in exactly the same positions as those of humans.

Research on the Mendelian genetics of mice began early in the twentieth century. One of the most important early contributions was the elucidation of the genes that control coat color and pattern. Genetic control of the mouse coat has provided a model for all mammals, including cats, dogs, horses, and cattle. A great deal of work was also done on mutations induced by radiation and chemicals. Mouse genetics has been of great significance in medicine. A large proportion of human genetic diseases have mouse counterparts useful for experimental study (they are called "mouse models"). The mouse has played a particularly important role in the development of our current understanding of the genes underlying cancer.

The mouse genome can be modified by the insertion of specific fragments of DNA into a fertilized egg or into somatic cells. The mice in the photograph have received a jellyfish gene for green fluorescent protein (GFP) that makes them glow green. Gene knockouts and replacements also are possible.

A major limitation of mouse genetics is its cost. Whereas working with a million individuals of *E. coli* or *S. cerevisiae* is a trivial matter, working with a million mice requires a factory-size building. Furthermore, although mice do breed rapidly (compared with humans), they cannot compete with microorganisms for speedy life cycle. Hence, the large-scale selections and screens necessary to detect rare genetic events are not possible.

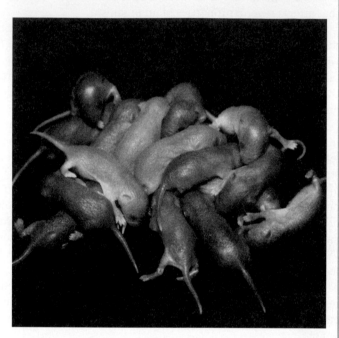

Green-glowing genetically modified mice. The jellyfish gene for green fluorescent protein has been inserted into the chromosomes of the glowing mice. The other mice are normal. [*Eye of Science/Science Source.*]

can be maintained easily under permissive conditions, and the mutant phenotype can be assayed in a subset of individuals by a switch to the restrictive conditions. Temperature-sensitive dominant lethal mutations also are useful. The type of mutation is expressed even when present in a single dose but only when the experimenter switches the organism to the restrictive temperature.

Null alleles for genes identified through genomic sequencing can be made by using a variety of "reverse genetic" procedures that specifically knock out the function of that gene. These will be described in Chapter 13.

> **Message** To see if a gene is essential, a null allele is tested for lethality.

We now turn to the approaches that can be used to detect the interaction between two or more loci.

6.2 Interaction of Genes in Pathways

Genes act by controlling cellular chemistry. Early in the twentieth century, Archibald Garrod, an English physician (Figure 6-9), made the first observation supporting this insight. Garrod noted that several recessive human diseases show defects in what is called metabolism, the general set of chemical reactions taking place in an organism. This observation led to the notion that such genetic diseases are "inborn errors of metabolism." Garrod worked on a disease called alkaptonuria (AKU), or black urine disease. He discovered that the substance responsible for black urine was homogentisic acid, which is present in high amounts and secreted into the urine in AKU patients. He knew that, in unaffected people, homogentisic acid is converted into maleylacetoacetic acid; so he proposed that, in AKU, there is a defect in this conversion. Consequently, homogentisic acid builds up and is excreted. Garrod's observations raised the possibility that the cell's chemical pathways were under the control of a large set of interacting genes. However, the direct demonstration of this control was provided by the later work of Beadle and Tatum on the fungus *Neurospora*.

Discoverer of inborn errors of metabolism

Figure 6-9 British physician Archibald Garrod (1857–1936). [*Science Photo Library.*]

Biosynthetic pathways in *Neurospora*

The landmark study by George Beadle and Edward Tatum in the 1940s not only clarified the role of genes, but also demonstrated the interaction of genes in biochemical pathways. They later received a Nobel Prize for their study, which marks the beginning of all molecular biology. Beadle and Tatum did their work on the haploid fungus *Neurospora*, which we have met in earlier chapters. Their plan was to investigate the genetic control of cellular chemistry. In what has become the standard forward genetic approach, they first irradiated *Neurospora* cells to produce mutations and then tested cultures grown from ascospores for interesting mutant phenotypes relevant to biochemical function. They found numerous mutants that had defective nutrition. Specifically, these mutants were auxotrophic mutants, of the type described for bacteria in Chapter 5. Whereas wild-type *Neurospora* can use its cellular biochemistry to synthesize virtually all its cellular components from the inorganic nutrients and a carbon source in the medium, auxotrophic mutants cannot. In order to grow, such mutants require a nutrient to be supplied (a nutrient that a wild-type fungus is able to synthesize for itself), suggesting that the mutant is defective for some normal synthetic step.

As their first step, Beadle and Tatum confirmed that each mutation that generated a nutrient requirement was inherited as a single-gene mutation because each gave a 1:1 ratio when crossed with a wild type. Letting *aux* represent an auxotrophic mutation,

$$+ \times aux$$
$$\downarrow$$

progeny: $\frac{1}{2} +$ and $\frac{1}{2}$ *aux*

Their second step was to classify the specific nutritional requirement of each auxotroph. Some would grow only if proline was supplied, others methionine, others pyridoxine, others arginine, and so on. Beadle and Tatum decided to focus on arginine auxotrophs. They found that the genes that mutated to give arginine auxotrophs mapped to three different loci on three separate chromosomes. Let's call the genes at the three loci the *arg-1, arg-2,* and *arg-3* genes. A key breakthrough was Beadle and Tatum's discovery that the auxotrophs for each of the three loci

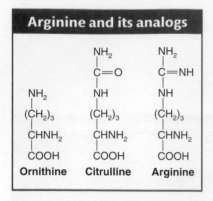

Figure 6-10 The chemical structures of arginine and the structurally related compounds citrulline and ornithine.

differed in their response to the structurally related compounds ornithine and citrulline (Figure 6-10). The *arg-1* mutants grew when supplied with any one of the chemicals ornithine, citrulline, or arginine. The *arg-2* mutants grew when given arginine or citrulline but not ornithine. The *arg-3* mutants grew only when arginine was supplied. These results are summarized in Table 6-1.

Table 6-1   Growth of *arg* Mutants in Response to Supplements

	Supplement		
Mutant	Ornithine	Citrulline	Arginine
arg-1	+	+	+
arg-2	−	+	+
arg-3	−	−	+

Note: A plus sign means growth; a minus sign means no growth.

Cellular enzymes were already known to interconvert such related compounds. On the basis of the properties of the *arg* mutants, Beadle and Tatum and their colleagues proposed a biochemical pathway for such conversions in *Neurospora:*

$$\text{precursor} \xrightarrow{\text{enzyme X}} \text{ornithine} \xrightarrow{\text{enzyme Y}} \text{citrulline} \xrightarrow{\text{enzyme Z}} \text{arginine}$$

This pathway nicely explains the three classes of mutants shown in Table 6-1. Under the model, the *arg-1* mutants have a defective enzyme X, and so they are unable to convert the precursor into ornithine as the first step in producing arginine. However, they have normal enzymes Y and Z, and so the *arg-1* mutants are able to produce arginine if supplied with either ornithine or citrulline. Similarly, the *arg-2* mutants lack enzyme Y, and the *arg-3* mutants lack enzyme Z. Thus, a mutation at a particular gene is assumed to interfere with the production of a single enzyme. The defective enzyme creates a block in some biosynthetic pathway. The block can be circumvented by supplying to the cells any compound that normally comes after the block in the pathway.

We can now diagram a more complete biochemical model:

$$\begin{array}{ccccccc}
& arg\text{-}1^{+} & & arg\text{-}2^{+} & & arg\text{-}3^{+} & \\
& \downarrow & & \downarrow & & \downarrow & \\
\text{precursor} & \xrightarrow{\text{enzyme X}} & \text{ornithine} & \xrightarrow{\text{enzyme Y}} & \text{citrulline} & \xrightarrow{\text{enzyme Z}} & \text{arginine}
\end{array}$$

This brilliant model, which was initially known as the *one-gene–one-enzyme hypothesis,* was the source of the first exciting insight into the functions of genes: genes somehow were responsible for the function of enzymes, and each gene apparently controlled one specific enzyme in a series of interconnected steps in a biochemical pathway. Other researchers obtained similar results for other biosynthetic pathways, and the hypothesis soon achieved general acceptance. All proteins, whether or not they are enzymes, also were found to be encoded by genes, and so the phrase was refined to become **one-gene–one-polypeptide hypothesis.** (Recall that a polypeptide is the simplest type of protein, a single chain of amino acids.) It soon became clear that a gene encodes the *physical structure* of a protein, which in turn dictates its function. Beadle and Tatum's hypothesis became one of the great unifying concepts in biology, because it provided a bridge that brought together the two major research areas of genetics and biochemistry.

(We must add parenthetically that, although the great majority of genes encode proteins, some are now known to encode RNAs that have special functions. All genes are transcribed to make RNA. Protein-encoding genes are transcribed to messenger RNA (mRNA), which is then translated into protein. However, the RNA encoded by a minority of genes is never translated into protein, because the

RNA itself has a unique function. We will call them **functional RNAs.** Some examples are transfer RNAs, ribosomal RNAs, and small cytoplasmic RNAs—more about them in later chapters.)

> **Message** Chemical synthesis in cells is by pathways of sequential steps catalyzed by enzymes. The genes encoding the enzymes of a specific pathway constitute a functionally interacting subset of the genome.

Gene interaction in other types of pathways

The notion that genes interact through pathways is a powerful one that finds application in all organisms. The *Neurospora* arginine pathway is an example of a synthetic pathway, a chain of enzymatic conversions that synthesizes essential nutrients. We can extend the idea again to a human case already introduced, the disease phenylketonuria (PKU), which is caused by an autosomal recessive allele. This disease results from an inability to convert phenylalanine into tyrosine. As a result of the block, phenylalanine accumulates and is spontaneously converted into a toxic compound, phenylpyruvic acid. The PKU gene is part of a synthetic pathway like the *Neurospora* arginine pathway, and part of it is shown in Figure 6-11. The illustration includes several other diseases caused by blockages in steps in this pathway (including alkaptonuria, the disease investigated by Garrod).

Another type of pathway is a *signal-transduction pathway*. This type of pathway is a chain of complex signals from the environment to the genome and from one gene to another. These pathways are crucial to the proper function of an organism. One of the best understood signal-transduction pathways was worked out from a genetic analysis of the mating response in baker's yeast. Recall that two mating types, determined by the alleles MATa and MATα, are necessary for yeast mating to occur. When a cell is in the presence of another cell of opposite mating type, it undergoes a series of changes in shape and behavior preparatory to mating. The mating response is triggered by a signal-transduction pathway requiring the sequential action of a set of genes. This set of genes was discovered through a standard interaction analysis of mutants with aberrant mating response (most were sterile). The steps were pieced together by using the approaches in the next section. The signal that sets things in motion is a mating pheromone (hormone) released by the opposite mating type; the pheromone binds to a membrane receptor, which is coupled to a G protein inside the membrane and activates the protein. The G protein, in turn, sets in motion a series of sequential protein phosphorylations called a kinase cascade. Ultimately, the cascade activates the transcription of a set of mating-specific genes that enable the cell to mate. A mutation at any one of these steps may disrupt the mating process.

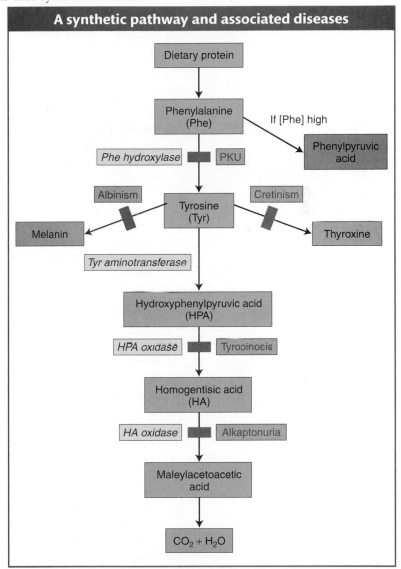

A synthetic pathway and associated diseases

Figure 6-11 A section of the phenylalanine metabolic pathway in humans, including diseases associated with enzyme blockages. The disease PKU is produced when the enzyme phenylalanine hydroxylase malfunctions. Accumulation of phenylalanine results in an increase in phenylpyruvic acid, which interferes with the development of the nervous system. [*After I. M. Lerner and W. J. Libby, Heredity, Evolution, and Society, 2nd ed. Copyright 1976 by W. H. Freeman and Company.*]

Developmental pathways comprise the steps by which a zygote becomes an adult organism. This process involves many genetically controlled steps, including establishment of the anterior-posterior and dorsal-ventral axes, laying down the basic body plan of organs, and tissue differentiation and movement. These steps can require gene regulation and signal transduction. Developmental pathways will be taken up in detail in Chapter 13, but the interaction of genes in these pathways is analyzed in the same way, as we will see next.

6.3 Inferring Gene Interactions

The genetic approach that reveals the interacting genes for a particular biological property is briefly as follows:

Step 1. Obtain many single-gene mutants and test for dominance.

Step 2. Test the mutants for allelism—are they at one or several loci?

Step 3. Combine the mutants in pairs to form **double mutants** to see if the genes interact.

Gene interaction is inferred from the phenotype of the double mutant: if the genes interact, then the phenotype differs from the simple combination of both single-gene mutant phenotypes. If mutant alleles from different genes interact, then we infer that the wild-type genes interact normally as well. In cases in which the two mutants interact, a modified 9:3:3:1 Mendelian ratio will often result.

A procedure that must be carried out before testing interactions is to determine whether each mutation is of a different locus (step 2 above). The mutant screen could have unintentionally favored certain genes. In that case, in a collection of, say, 60 independent mutations, only 20 gene loci might be represented, each with about three mutations. Thus, the set of gene loci needs to be defined, as shown in the next section.

Sorting mutants using the complementation test

How is it possible to decide whether two mutations belong to the same gene? There are several ways. First, each mutant allele could be mapped. Then, if two mutations map to two different chromosomal loci, they are likely of different genes. However, this approach is time consuming on a large set of mutations. A quicker approach often used is the **complementation test.**

In a diploid, the complementation test is performed by intercrossing two individuals that are homozygous for different recessive mutations. The next step is to observe whether the progeny have the wild-type phenotype. If the progeny are wild type, the two recessive mutations must be in *different* genes because the respective wild-type alleles provide wild-type function. In this case, the two mutations are said to have *complemented*. Here, we will name the genes *a1* and *a2,* after their mutant alleles. We can represent the heterozygotes as follows, depending on whether the genes are on the same chromosome or are on different chromosomes:

Different chromosomes:

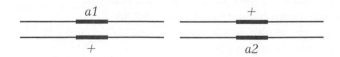

Same chromosome (shown in the trans configuration):

However, if the progeny are *not* wild type, then the recessive mutations must be alleles of the same gene. Because both alleles of the gene are mutants, there is no wild-type allele to help distinguish between two different mutant alleles of a gene whose wild-type allele is a^+. These alleles could have different mutant sites within the same gene, but they would both be nonfunctional. The heterozygote a'/a' would be

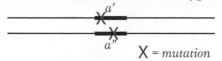

$$\mathsf{X} = mutation$$

At the operational level, complementation is defined as follows.

> **Message** Complementation is the production of a wild-type phenotype when two haploid genomes bearing different recessive mutations are united in the same cell.

Let's illustrate the complementation test with an example from harebell plants (genus *Campanula*). The wild-type flower color of this plant is blue. Let's assume that, from a mutant hunt, we have obtained three white-petaled mutants and that they are available as homozygous pure-breeding strains. They all look the same, and so we do not know a priori whether they are genetically identical. We will call the mutant strains $, £, and ¥ to avoid any symbolism using letters, which might imply dominance. When crossed with wild type, each mutant gives the same results in the F_1 and F_2 as follows:

Harebell plant

Flowers of the harebell plant (*Campanula* species). [*Gregory G. Dimijian/Photo Researchers.*]

$$\text{white } \$ \times \text{blue} \to F_1, \text{ all blue} \to F_2, \tfrac{3}{4} \text{ blue}, \tfrac{1}{4} \text{ white}$$
$$\text{white } £ \times \text{blue} \to F_1, \text{ all blue} \to F_2, \tfrac{3}{4} \text{ blue}, \tfrac{1}{4} \text{ white}$$
$$\text{white } ¥ \times \text{blue} \to F_1, \text{ all blue} \to F_2, \tfrac{3}{4} \text{ blue}, \tfrac{1}{4} \text{ white}$$

In each case, the results show that the mutant condition is determined by the recessive allele of a single gene. However, are they three alleles of one gene, of two genes, or of three genes? Because the mutants are recessive, the question can be answered by the complementation test, which asks if the mutants *complement* one another.

Let us intercross the mutants to test for complementation. Assume that the results of intercrossing mutants $, £, and ¥ are as follows:

$$\text{white } \$ \times \text{white } £ \to F_1, \text{ all white}$$
$$\text{white } \$ \times \text{white } ¥ \to F_1, \text{ all blue}$$
$$\text{white } £ \times \text{white } ¥ \to F_1, \text{ all blue}$$

From this set of results, we can conclude that mutants $ and £ must be caused by alleles of one gene (say, *w1*) because they do not complement, but ¥ must be caused by a mutant allele of another gene (*w2*) because complementation is seen.

How does complementation work at the molecular level? The normal blue color of the harebell flower is caused by a blue pigment called *anthocyanin*. Pigments are chemicals that absorb certain colors of light; in regard to the harebell, the anthocyanin absorbs all wavelengths except blue, which is reflected into the eye of the observer. However, this anthocyanin is made from chemical precursors that are not pigments; that is, they do not absorb light of any specific wavelength and simply reflect back the white light of the sun to the observer, giving a white appearance. The blue pigment is the end product of a series of biochemical conversions of nonpigments. Each step is catalyzed by a specific enzyme encoded by a specific gene. We can explain the results with a pathway as follows:

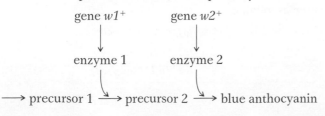

Figure 6-12 Three phenotypically identical white harebell mutants—$, £, and ¥—are intercrossed. Mutations in the same gene (such as $ and £) cannot complement, because the F₁ has one gene with two mutant alleles. The pathway is blocked and the flowers are white. When the mutations are in different genes (such as £ and ¥), there is complementation of the wild-type alleles of each gene in the F₁ heterozygote. Pigment is synthesized and the flowers are blue. (What would you predict to be the result of crossing $ and ¥?)

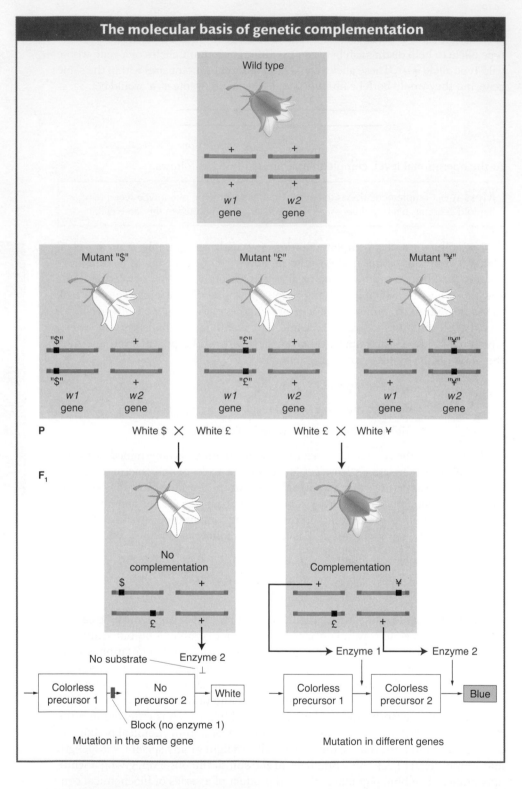

A homozygous mutation in either of the genes will lead to the accumulation of a precursor that will simply make the plant white. Now the mutant designations could be written as follows:

$$\$ \quad w1_\$/w1_\$ \cdot w2^+/w2^+$$
$$£ \quad w1_£/w1_£ \cdot w2^+/w2^+$$
$$¥ \quad w1^+/w1^+ \cdot w2_¥/w2_¥$$

However, in practice, the subscript symbols would be dropped and the genotypes would be written as follows:

$$\$\quad w1/w1 \cdot w2^+/w2^+$$
$$£\quad w1/w1 \cdot w2^+/w2^+$$
$$¥\quad w1^+/w1^+ \cdot w2/w2$$

Hence, an F_1 from $ × £ will be

$$w1/w1 \cdot w2^+/w2^+$$

These F_1 plants will have two defective alleles for *w1* and will therefore be blocked at step 1. Even though enzyme 2 is fully functional, it has no substrate on which to act; so no blue pigment will be produced and the phenotype will be white.

The F_1 plants from the other crosses, however, will have the wild-type alleles for both of the enzymes needed to take the intermediates to the final blue product. Their genotypes will be

$$w1^+/w1 \cdot w2^+/w2$$

Hence, we see that complementation is actually a result of the cooperative interaction of the *wild-type* alleles of the two genes. Figure 6-12 summarizes the interaction of the complementing and noncomplementing white mutants at the genetic and cellular levels.

In a haploid organism, the complementation test cannot be performed by intercrossing. In fungi, an alternative way brings mutant alleles together to test complementation: fusion resulting in a **heterokaryon** (Figure 6-13). Fungal cells fuse readily. When two different strains fuse, the haploid nuclei from the different strains occupy one cell, which is the *heterokaryon* (Greek; different kernels). The nuclei in a heterokaryon do not generally fuse. In one sense, this condition is a "mimic" diploid.

Assume that, in different strains, there are mutations in two different genes conferring the same mutant phenotype—for example, an arginine requirement. We will call these genes *arg-1* and *arg-2*. The genotypes of the two strains can be represented as *arg-1 · arg-2⁺* and *arg-1⁺ · arg-2*. These two strains can be fused to form a heterokaryon with the two nuclei in a shared cytoplasm:

Nucleus 1 is *arg-1 · arg-2⁺*
Nucleus 2 is *arg-1⁺ · arg-2*

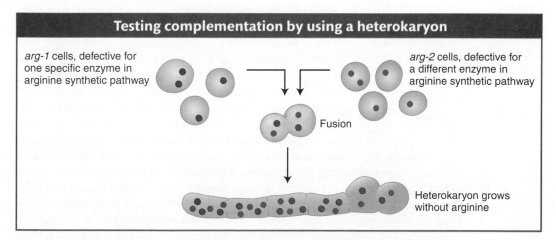

Testing complementation by using a heterokaryon

arg-1 cells, defective for one specific enzyme in arginine synthetic pathway

arg-2 cells, defective for a different enzyme in arginine synthetic pathway

Fusion

Heterokaryon grows without arginine

Figure 6-13 A heterokaryon of *Neurospora* and similar fungi mimics a diploid state. When vegetative cells fuse, haploid nuclei share the same cytoplasm in a heterokaryon. In this example, haploid nuclei with mutations in different genes in the arginine synthetic pathway complement to produce a *Neurospora* that no longer requires arginine.

Because gene products are made in a common cytoplasm, the two wild-type alleles can exert their dominant effect and cooperate to produce a heterokaryon of wild-type phenotype. In other words, the two mutations complement, just as they would in a diploid. If the mutations had been alleles of the same gene, there would have been no complementation.

> **Message** When two independently derived recessive mutant alleles producing similar recessive phenotypes fail to complement, they must be alleles of the same gene.

Independently synthesized and inherited pigments

(a)   (b)   (c)   (d)

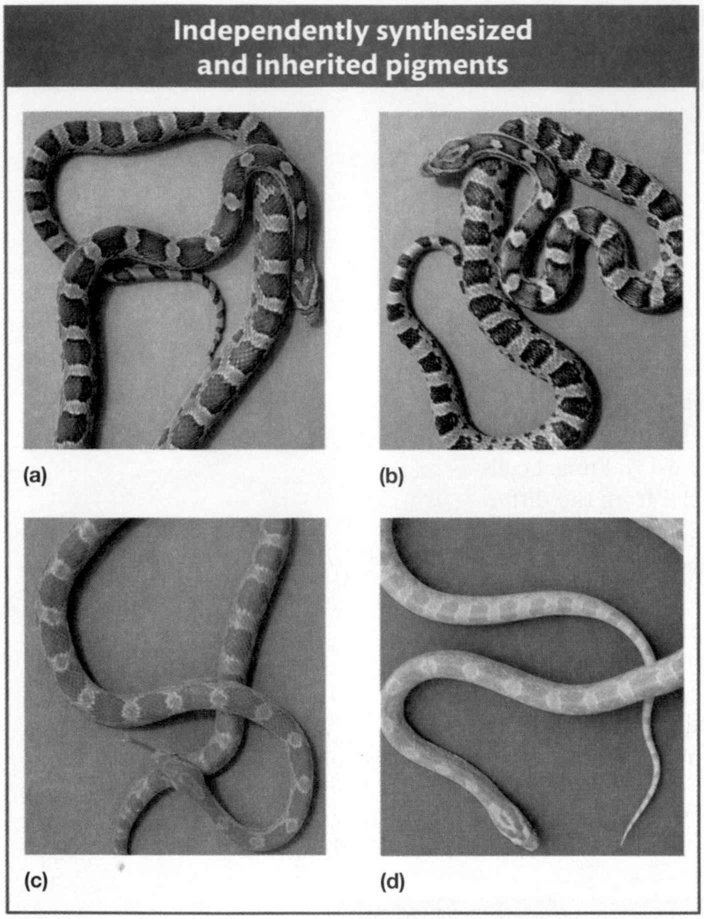

Figure 6-14 In corn snakes, combinations of orange and black pigments determine the four phenotypes shown. (a) A wild-type black-and-orange camouflaged snake synthesizes both black and orange pigments. (b) A black snake does not synthesize orange pigment. (c) An orange snake does not synthesize black pigment. (d) An albino snake synthesizes neither black nor orange pigment. [*Anthony Griffiths.*]

Analyzing double mutants of random mutations

Recall that, to learn whether two genes interact, we need to assess the phenotype of the double mutant to see if it is other than the combination of both single mutations. The double mutant is obtained by intercrossing. The F_1 is obtained as part of the complementation test; so with the assumption that complementation has been observed, suggesting different genes, the F_1 is selfed to obtain an F_2. The F_2 should contain the double mutant. The double mutant may then be identified by looking for Mendelian ratios. For example, if a standard $9:3:3:1$ Mendelian ratio is obtained, the phenotype present in only 1/16 of the progeny represents the double mutant (the "1" in $9:3:3:1$). In cases of gene interaction, however, the phenotype of the double mutant may not be distinct but will match that of one of the single mutants. In this case, a modified Mendelian ratio will result, such as $9:3:4$ or $9:7$.

The standard $9:3:3:1$ Mendelian ratio is the simplest case, expected if there is no gene interaction and if the two mutations under test are on different chromosomes. This $9:3:3:1$ ratio is the null hypothesis: any modified Mendelian ratio representing a departure from this null hypothesis would be informative, as the following examples will show.

The $9:3:3:1$ ratio: no gene interaction As a baseline, let's start with the case in which two mutated genes do not interact, a situation where we expect the $9:3:3:1$ ratio. Let's look at the inheritance of skin coloration in corn snakes. The snake's natural color is a repeating black-and-orange camouflage pattern, as shown in Figure 6-14a. The phenotype is produced by two separate pigments, both of which are under genetic control. One gene determines the orange pigment, and the alleles that we will consider are o^+ (presence of orange pigment) and o (absence of orange pigment). Another gene determines the black pigment, and its alleles are b^+ (presence of black pigment) and b (absence of black pigment). These two genes are unlinked. The natural pattern is produced by the genotype $o^+/-;b^+/-$. A snake that is $o/o;b^+/-$ is black because it lacks the orange pigment (Figure 6-14b), and a snake that is $o^+/-;b/b$ is orange because it lacks the black pigment (Figure 6-14c). The double homozygous recessive $o/o;b/b$ is albino (Figure 6-14d). Notice, however, that the faint pink color of the albino is from yet another pigment, the hemoglobin of the blood that is visible through this snake's skin when the other pigments are absent. The albino snake also clearly shows that there is another element to the skin-pigmentation pattern in addition to pigment: the repeating motif in and around which pigment is deposited.

If a homozygous orange and a homozygous black snake are crossed, the F_1 is wild type (camouflaged), demonstrating complementation:

$$♀ \ o^+/o^+ ; b/b \times ♂ \ o/o ; b^+/b^+$$

(orange)   (black)

↓

$$F_1 \quad o^+/o ; b^+/b$$

(camouflaged)

Here, however, an F_2 shows a standard $9:3:3:1$ ratio:

$$♀ \ o^+/o ; b^+/b \times ♂ \ o^+/o ; b^+/b$$

(camouflaged)   (camouflaged)

↓

$F_2 \quad 9 \ o^+/- ; b^+/-$ (camouflaged)

$3 \ o^+/- ; b/b$ (orange)

$3 \ o/o ; b^+/-$ (black)

$1 \ o/o ; b/b$ (albino)

The $9:3:3:1$ ratio is produced because the two pigment genes act independently at the cellular level.

$$\text{precursor} \xrightarrow{b^+} \text{black pigment}$$
$$\text{precursor} \xrightarrow{o^+} \text{orange pigment} \Bigg\} \text{camouflaged}$$

If the presence of one mutant makes one pathway fail, the other pathway is still active, producing the other pigment color. Only when both mutants are present do both pathways fail, and no pigment of any color is produced.

The $9:7$ ratio: genes in the same pathway The F_2 ratio from the harebell dihybrid cross shows both blue and white plants in a ratio of $9:7$. How can such results be explained? The $9:7$ ratio is clearly a modification of the dihybrid $9:3:3:1$ ratio with the $3:3:1$ combined to make 7; hence, some kind of interaction is inferred. The cross of the two white lines and subsequent generations can be represented as follows:

$$w1/w1 ; w2^+/w2^+ \text{ (white)} \times w1^+/w1^+ ; w2/w2 \text{ (white)}$$

↓

$F_1 \qquad w1^+/w1 ; w2^+/w2 \text{ (blue)}$

$$w1^+/w1 ; w2^+/w2 \times w1^+/w1 ; w2^+/w2$$

↓

$F_2 \qquad 9 \ w1^+/- ; w2^+/- \text{ (blue)} \qquad 9$

$3 \ w1^+/- ; w2/w2 \text{ (white)}$

$3 \ w1/w1 ; w2^+/- \text{ (white)} \Bigg\} \ 7$

$1 \ w1/w1 ; w2/w2 \text{ (white)}$

Clearly, in this case, the only way in which a $9:7$ ratio is possible is if the double mutant has the same phenotypes as the two single mutants. Hence, the modified ratio constitutes a way of identifying the double mutant's phenotype. Furthermore, the identical phenotypes of the single and double mutants suggest that each mutant allele controls a different step in the *same* pathway. The results show that a plant will have white petals if it is homozygous for the recessive mutant allele of *either* gene or *both* genes. To have the blue phenotype, a plant must have at least one copy of the dominant allele of both genes because both are needed to complete the sequential steps in the pathway. No matter which is absent, the same pathway fails, producing the same phenotype. Thus, three of the genotypic classes will produce the same phenotype, and so, overall, only two phenotypes result.

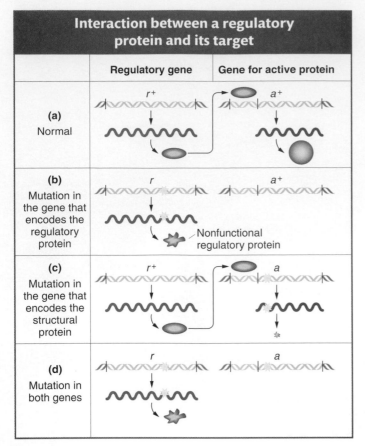

Interaction between a regulatory protein and its target		
	Regulatory gene	**Gene for active protein**
(a) Normal	r^+	a^+
(b) Mutation in the gene that encodes the regulatory protein	r — Nonfunctional regulatory protein	a^+
(c) Mutation in the gene that encodes the structural protein	r^+	a
(d) Mutation in both genes	r	a

Figure 6-15 The r^+ gene encodes a regulatory protein, and the a^+ gene encodes a structural protein. Both must be normal for a functional ("active") structural protein to be synthesized.

The example in harebells entailed different steps in a synthetic pathway. Similar results can come from gene regulation. A regulatory gene often functions by producing a protein that binds to a regulatory site upstream of a target gene, facilitating the transcription of the gene (Figure 6-15). In the absence of the regulatory protein, the target gene would be transcribed at very low levels, inadequate for cellular needs. Let's cross a pure line r/r defective for the regulatory protein to a pure line a/a defective for the target protein. The cross is $r/r;a^+/a^+ \times r^+/r^+;a/a$. The $r^+/r;a^+/a$ dihybrid will show complementation between the mutant genotypes because both r^+ and a^+ are present, permitting normal transcription of the wild-type allele. When selfed, the F_1 dihybrid will also result in a 9:7 phenotypic ratio in the F_2:

Proportion	Genotype	Functional a^+ protein	Ratio
$\frac{9}{16}$	$r^+/-\,;a^+/-$	Yes	9
$\frac{3}{16}$	$r^+/-\,;a/a$	No	
$\frac{3}{16}$	$r/r;a^+/-$	No	7
$\frac{1}{16}$	$r/r;a/a$	No	

Message A 9:7 F_2 ratio suggests interacting genes in the same pathway; absence of either gene function leads to absence of the end product of the pathway.

The 9:3:4 ratio: recessive epistasis A 9:3:4 ratio in the F_2 suggests a type of gene interaction called **epistasis.** This word means "stand upon," referring to the situation in which a double mutant shows the phenotype of one mutation but not the other. The overriding mutation is *epistatic*, whereas the overridden one is *hypostatic*. Epistasis also results from genes being in the same pathway. In a simple synthetic pathway, the epistatic mutation is carried by a gene that is farther upstream (earlier in the pathway) than the gene of the overridden mutation. The mutant phenotype of the upstream gene takes precedence, no matter what is taking place later in the pathway.

Let's look at an example concerning petal-pigment synthesis in the plant blue-eyed Mary (*Collinsia parviflora*). From the blue wild type, we'll start with two pure mutant lines, one with white (w/w) and the other with magenta petals (m/m). The w and m genes are not linked. The F_1 and F_2 are as follows:

$$w/w;m^+/m^+ \text{ (white)} \times w^+/w^+;m/m \text{ (magenta)}$$
$$F_1 \quad w^+/w;m^+/m \text{ (blue)}$$
$$\downarrow$$
$$w^+/w;m^+/m \times w^+/w;m^+/m$$
$$\downarrow$$

F_2		
9 $w^+/-\,;m^+/-$ (blue)		9
3 $w^+/-\,;m/m$ (magenta)		3
3 $w/w;m^+/-$ (white)		4
1 $w/w;m/m$ (white)		

In the F_2, the 9:3:4 phenotypic ratio is diagnostic of recessive epistasis. As in the preceding case, we see, again, that the ratio tells us what the phenotype of the double must be, because the $\frac{4}{16}$ component of the ratio must be a grouping of one single mutant class ($\frac{3}{16}$) plus the double mutant class ($\frac{1}{16}$). Hence, the double

Figure 6-16 Wild-type alleles of two genes (w^+ and m^+) encode enzymes catalyzing successive steps in the synthesis of a blue petal pigment. Homozygous m/m plants produce magenta flowers and homozygous w/w plants produce white flowers. The double mutant w/w; m/m also produces white flowers, indicating that white is epistatic to magenta.

mutant expresses only one of the two mutant phenotypes; so, by definition, white must be epistatic to magenta. (To find the double mutant within the $\frac{4}{16}$ group, white F_2 plants would have to be individually testcrossed.) This interaction is called recessive epistasis because a recessive phenotype (white) overrides the other phenotype. Dominant epistasis will be considered in the next section.

At the cellular level, we can account for the recessive epistasis in *Collinsia* by the following type of pathway (see also Figure 6-16).

$$\text{colorless} \xrightarrow{\text{gene } w^+} \text{magenta} \xrightarrow{\text{gene } m^+} \text{blue}$$

Notice that the epistatic mutation occurs in a step in the pathway leading to blue pigment; this step is upstream of the step that is blocked by the masked mutation.

Another informative case of recessive epistasis is the yellow coat color of some Labrador retriever dogs. Two alleles, B and b, stand for black and brown coats, respectively. The two alleles produce black and brown melanin. The allele e of another gene is epistatic on these alleles, giving a yellow coat (Figure 6-17). Therefore, the genotypes $B/-$; e/e and b/b; e/e both produce a yellow phenotype, whereas $B/-$; $E/-$ and b/b; $E/-$ are black and brown, respectively. This case of epistasis is *not* caused by an upstream block in a pathway leading to dark pigment. Yellow dogs can make black or brown pigment, as can be seen in their noses and lips. The action of the allele e is to prevent the deposition of the pigment in hairs. In this case, the epistatic gene is

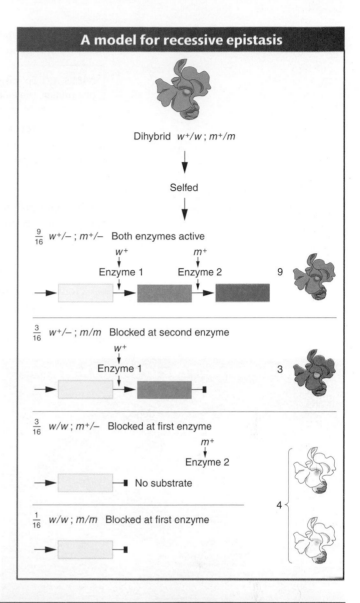

A model for recessive epistasis

Dihybrid w^+/w; m^+/m

Selfed

$\frac{9}{16}$ $w^+/-$; $m^+/-$ Both enzymes active

w^+ m^+

Enzyme 1   Enzyme 2   9

$\frac{3}{16}$ $w^+/-$; m/m Blocked at second enzyme

w^+

Enzyme 1   3

$\frac{3}{16}$ w/w; $m^+/-$ Blocked at first enzyme

m^+

Enzyme 2

No substrate

$\frac{1}{16}$ w/w; m/m Blocked at first enzyme

4

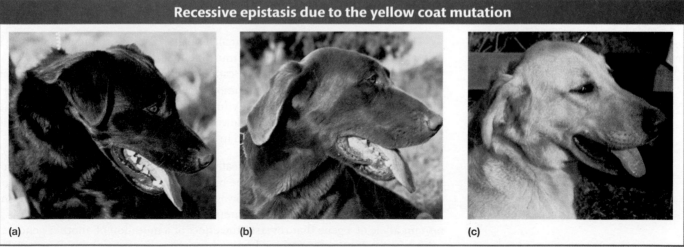

Recessive epistasis due to the yellow coat mutation

(a)   (b)   (c)

Figure 6-17 Three different coat colors in Labrador retrievers. Two alleles B and b of a pigment gene determine (a) black and (b) brown, respectively. At a separate gene, E allows color deposition in the coat, and e/e prevents deposition, resulting in (c) the gold phenotype. Part c illustrates recessive epistasis. [*Anthony Griffiths.*]

developmentally downstream; it represents a kind of developmental target that must be of *E* genotype before pigment can be deposited.

> **Message** Epistasis is inferred when a mutant allele of one gene masks the expression of a mutant allele of another gene and expresses its own phenotype instead.

In fungi, tetrad analysis is useful in identifying a double mutant. For example, an ascus containing half its products as wild type must contain double mutants. Consider the cross

$$a \cdot b^+ \times a^+ \cdot b$$

In some proportion of progeny, the alleles *a* and *b* will segregate together (a nonparental ditype ascus). Such a tetrad will show the following phenotypes:

| wild type | $a^+ \cdot b^+$ | double mutant | $a \cdot b$ |
| wild type | $a^+ \cdot b^+$ | double mutant | $a \cdot b$ |

Hence, the double mutant must be the non-wild-type genotype and can be assessed accordingly. If the phenotype is the *a* phenotype, then *b* is being overridden; if the phenotype is the *b* phenotype, then *a* is being overridden. If both phenotypes are present, then there is no epistasis.

Dominant epistasis due to a white mutation

Figure 6-18 In foxgloves, *D* and *d* cause dark and light pigments, respectively, whereas the epistatic *W* restricts pigment to the throat spots. [*Anthony Griffiths.*]

The 12 : 3 : 1 ratio: dominant epistasis In foxgloves (*Digitalis purpurea*), two genes interact in the pathway that determines petal coloration. The two genes are unlinked. One gene affects the intensity of the red pigment in the petal; allele *d* results in the light red color seen in natural populations of foxgloves, whereas *D* is a mutant allele that produces dark red color (Figure 6-18). The other gene determines in which cells the pigment is synthesized: allele *w* allows synthesis of the pigment throughout the petals as in the wild type, but the mutant allele *W* confines pigment synthesis to the small throat spots. If we self a dihybrid *D/d ; W/w*, then the F_2 ratio is as follows:

$$
\begin{array}{ll}
\left.\begin{array}{l}
9\ D/- ; W/- \text{ (white with spots)} \\
3\ d/d ; W/- \text{ (white with spots)}
\end{array}\right\} & 12 \\
3\ D/- ; w/w \text{ (dark red)} & 3 \\
1\ d/d ; w/w \text{ (light red)} & 1
\end{array}
$$

The ratio tells us that the dominant allele *W* is epistatic, producing the 12 : 3 : 1 ratio. The $\frac{12}{16}$ component of the ratio must include the double mutant class ($\frac{9}{16}$), which is clearly white in phenotype, establishing the epistasis of the dominant allele *W*. The two genes act in a common developmental pathway: *W* prevents the synthesis of red pigment but only in a special class of cells constituting the main area of the petal; synthesis is allowed in the throat spots. When synthesis is allowed, the pigment can be produced in either high or low concentrations.

Suppressors It is not easy to specifically select or screen for epistatic interactions, and cases of epistasis have to be built up by the laborious combination of candidate mutations two at a time. However, for our next type of gene interaction, the experimenter can readily select interesting mutant alleles. A **suppressor** is a mutant allele of a gene that reverses the effect of a mutation of another gene, resulting in a wild-type or near-wild-type phenotype. Suppression implies that the target gene and the suppressor gene normally interact at some functional level in their wild-type states. For example, assume that an allele a^+ produces the normal phenotype, whereas a recessive mutant allele *a* results in abnormality. A recessive mutant allele s at another gene suppresses the effect of a, and so the genotype $a/a \cdot s/s$ will have the wild-type (a^+-like) phenotype. Suppressor alleles sometimes have no effect in the absence of the other mutation; in such a case, the phenotype of $a^+/a^+ \cdot s/s$ would be wild type. In other cases, the suppressor allele produces its own abnormal phenotype.

Screening for suppressors is quite straightforward. Start with a mutant in some process of interest, expose this mutant to mutation-causing agents such as high-energy radiation, and screen the descendants for wild types. In haploids such as fungi, screening is accomplished by simply plating mutagenized cells and look-ing for colonies with wild-type phenotypes. Most wild types arising in this way are merely reversals of the original mutational event and are called **revertants.** However, some will be "pseudorevertants," double mutants in which one of the mutations is a suppressor.

Revertant and suppressed states can be distinguished by appropriate crossing. For example, in yeast, the two results would be distinguished as follows:

$$\text{true revertant } a^+ \times \text{ standard wild-type } a^+$$
$$\downarrow$$

Progeny	all a^+

$$\text{suppressed mutant } a \cdot s \times \text{ standard wild-type } a^+ \cdot s^+$$
$$\downarrow$$

Progeny	$a^+ \cdot s^+$	wild type
	$a^+ \cdot s$	wild type
	$a \cdot s^+$	original mutant
	$a \cdot s$	wild type (suppressed)

The appearance of the original mutant phenotype identifies the parent as a suppressed mutant.

In diploids, suppressors produce various modified F_2 ratios, which are useful in confirming suppression. Let's look at a real-life example from *Drosophila*. The recessive allele pd results in purple eye color when unsuppressed. A recessive allele su has no detectable phenotype itself but suppresses the unlinked recessive allele pd. Hence, $pd/pd; su/su$ is wild type in appearance and has red eyes. The following analysis illustrates the inheritance pattern. A homozygous purple-eyed fly is crossed with a homozygous red-eyed stock carrying the suppressor.

$$pd/pd; su^+/su^+ \text{ (purple)} \times pd^+/pd^+; su/su \text{ (red)}$$
$$\downarrow$$

F_1	all $pd^+/pd; su^+/su$ (red)

Self	$pd^+/pd; su^+/su$ (red) $\times pd^+/pd; su^+/su$ (red)

$$\downarrow$$

F_2			
	9 $pd^+/-; su^+/-$	red	
	3 $pd^+/-; su/su$	red	13
	1 $pd/pd; su/su$	red	
	3 $pd/pd; su^+/-$	purple	3

The overall ratio in the F_2 is 13 red : 3 purple. The $\frac{13}{16}$ component must in-clude the double mutant, which is clearly wild type in phenotype. This ratio is expected from a recessive suppressor that itself has no detectable phenotype. If a recessive suppressor has the *same* phenotype as the target mutation (as is sometimes found), then the F_2 ratio will be 10 : 6 (wild : mutant). This ratio is discussed in the following paragraphs.

A molecular mechanism for suppression

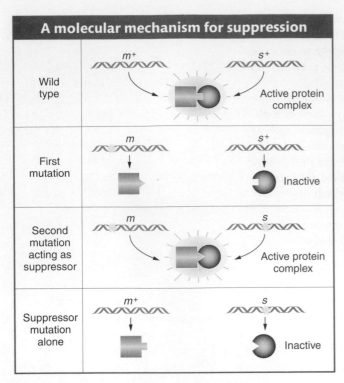

Wild type	m^+ s^+	Active protein complex
First mutation	m s^+	Inactive
Second mutation acting as suppressor	m s	Active protein complex
Suppressor mutation alone	m^+ s	Inactive

Figure 6-19 A first mutation alters the binding site of one protein so that it can no longer bind to a partner. A suppressor mutation in the partner alters the binding site so that both proteins are able to bind once again.

Suppression is sometimes confused with epistasis. However, the key difference is that a suppressor cancels the expression of a mutant allele and restores the corresponding wild-type phenotype. Furthermore, often only two phenotypes segregate (as in the preceding examples) rather than three, as in epistasis.

How do suppressors work at the molecular level? There are many possible mechanisms. A particularly useful type of suppression is based on the physical binding of gene products in the cell—for example, protein–protein binding. Assume that two proteins normally fit together to provide some type of cellular function. When a mutation causes a shape change in one protein, it no longer fits together with the other; hence, the function is lost (Figure 6-19). However, a suppressor mutation that causes a compensatory shape change in the second protein can restore fit and hence normal function. In this figure, if the genotypes were diploids representing an F_2 dihybrid self, then a 10:6 ratio would result; if this were a haploid dihybrid cross (such as $m^+ s^+ \times m\ s$), a 1:3 ratio would result. From suppressor ratios generally, interacting proteins often can be deduced.

Alternatively, in situations in which a mutation causes a block in a metabolic pathway, the suppressor finds some way of bypassing the block—for example, by rerouting into the blocked pathway intermediates similar to those beyond the block. In the following example, the suppressor provides an intermediate B to circumvent the block.

No suppressor

$$A \longrightarrow \cancel{B} \longrightarrow \text{product}$$

With suppressor

$$A \longrightarrow B \longrightarrow \text{product}$$
$$B \nearrow$$

In several organisms, *nonsense suppressors* have been found—mutations in tRNA genes resulting in an anticodon that will bind to a premature stop codon within a mutant coding sequence. Hence, the suppressor allows translation to proceed past the former block and make a complete protein rather than a truncated one. Such suppressor mutations often have little effect on the phenotype other than in suppression.

> **Message** Mutant alleles called suppressors cancel the expression of a mutant allele of another gene, resulting in normal wild-type phenotype.

Modifiers As the name suggests, a **modifier** mutation at a second locus changes the degree of expression of a mutated gene at the first locus. Regulatory genes provide a simple illustration. As in an earlier example, regulatory proteins bind to the sequence of the DNA upstream of the start site for transcription. These proteins regulate the level of transcription. In the discussion of complementation, we considered a null mutation of a regulatory gene that almost completely prevented transcription. However, some regulatory mutations change the level of transcription of the target gene so that either more or less protein is produced. In other words, a mutation in a regulatory protein can down-regulate or up-regulate the transcribed gene. Let's look at an example using a down-regulating regulatory mutation b, affecting a gene A in a fungus such as yeast. We look at the effect of b on a *leaky* mutation of gene A, which is a mutation that produces a protein with some low residual function. We cross a leaky mutation a with the regulatory mutation b:

leaky mutant $a \cdot b^+$ × inefficient regulator $a^+ \cdot b$

Progeny	Phenotype
$a^+ \cdot b^+$	wild type
$a^+ \cdot b$	defective (low transcription)
$a \cdot b^+$	defective (defective protein A)
$a \cdot b$	extremely defective (low transcription of defective protein)

Hence, the action of the modifier is seen in the appearance of two grades of mutant phenotypes *within* the a progeny.

Synthetic lethals In some cases, when two viable single mutants are intercrossed, the resulting double mutants are lethal. In a diploid F_2, this result would be manifested as a $9:3:3$ ratio because the double mutant (which would be the "1" component of the ratio) would be absent. These **synthetic lethals** can be considered a special category of gene interaction. They can point to specific types of interactions of gene products. For instance, genome analysis has revealed that evolution has produced many duplicate systems within the cell. One advantage of these duplicates might be to provide "backups." If there are null mutations in genes in both duplicate systems, then a faulty system will have no backup, and the individual will lack essential function and die. In another instance, a leaky mutation in one step of a pathway may cause the pathway to slow down, but leave enough function for life. However, if double mutants combine, each with a leaky mutation in a different step, the whole pathway grinds to a halt. One version of the latter interaction is two mutations in a protein machine, as shown in Figure 6-20.

In the earlier discussions of modified Mendelian ratios, all the crosses were dihybrid selfs. As an exercise, you might want to calculate the ratios that would be produced in the same systems if testcrosses were made instead of selfs.

A summary of some of the ratios that reveal gene interaction is shown in Table 6-2.

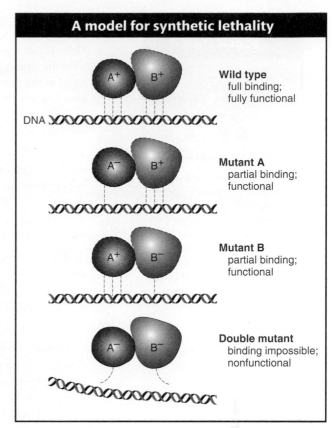

A model for synthetic lethality

DNA

Wild type
full binding; fully functional

Mutant A
partial binding; functional

Mutant B
partial binding; functional

Double mutant
binding impossible; nonfunctional

Figure 6-20 Two interacting proteins perform some essential function on some substrate such as DNA but must first bind to it. Reduced binding of either protein allows some functions to remain, but reduced binding of both is lethal.

Message A range of modified $9:3:3:1$ F_1 ratios can reveal specific types of gene interaction.

Table 6-2 Some Modified F_2 Ratios

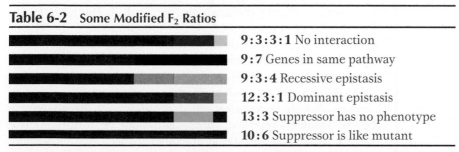

9:3:3:1 No interaction
9:7 Genes in same pathway
9:3:4 Recessive epistasis
12:3:1 Dominant epistasis
13:3 Suppressor has no phenotype
10:6 Suppressor is like mutant

Note: Some of these ratios can be produced with other mechanisms of interaction.

6.4 Penetrance and Expressivity

In the analysis of single-gene inheritance, there is a natural tendency to choose mutants that produce clear Mendelian ratios. In such cases, we can use the phenotype

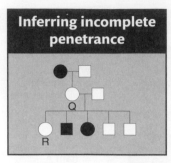

Inferring incomplete penetrance

Figure 6-21 In this human pedigree of a dominant allele that is not fully penetrant, person Q does not display the phenotype but passed the dominant allele to at least two progeny. Because the allele is not fully penetrant, the other progeny (for example, R) may or may not have inherited the dominant allele.

to distinguish mutant and wild-type genotypes with almost 100 percent certainty. In these cases, we say that the mutation is 100 percent *penetrant* into the phenotype. However, many mutations show *incomplete* penetrance: that is, not every individual with the genotype expresses the corresponding phenotype. Thus, **penetrance** is defined as the percentage of *individuals* with a given allele who exhibit the phenotype associated with that allele.

Why would an organism have a particular genotype and yet not express the corresponding phenotype? There are several possible reasons:

1. *The influence of the environment.* Individuals with the same genotype may show a range of phenotypes, depending on the environment. The range of phenotypes for mutant and wild-type individuals may overlap: the phenotype of a mutant individual raised in one set of circumstances may match the phenotype of a wild-type individual raised in a different set of circumstances. Should this matching happen, the mutant cannot be distinguished from the wild type.

2. *The influence of other interacting genes.* Uncharacterized modifiers, epistatic genes, or suppressors in the rest of the genome may act to prevent the expression of the typical phenotype.

3. *The subtlety of the mutant phenotype.* The subtle effects brought about by the absence of a gene function may be difficult to measure in a laboratory situation.

A typical encounter with incomplete penetrance is shown in Figure 6-21. In this human pedigree, we see a normally dominantly inherited phenotype disappearing in the second generation only to reappear in the next.

Another measure for describing the range of phenotypic expression is called **expressivity.** Expressivity measures the degree to which a given allele is expressed at the phenotypic level; that is, expressivity measures the intensity of the phenotype. For example, "brown" animals (genotype *b/b*) from different stocks might show very different intensities of brown pigment from light to dark. As for penetrance, variable expressivity may be due to variation in the allelic constitution of the rest of the genome or to environmental factors. Figure 6-22 illustrates the distinction between penetrance and expressivity. An example of variable expressivity in dogs is found in Figure 6-23.

The phenomena of incomplete penetrance and variable expressivity can make any kind of genetic analysis substantially more difficult, including human pedigree analysis and predictions in genetic counseling. For example, it is often the case that a disease-causing allele is not fully penetrant. Thus, someone could have the allele but not show any signs of the disease. If that is the case, it is difficult to give a clean genetic bill of health to any person in a disease pedigree (for example, person R in Figure 6-21). On the other hand, pedigree analysis can sometimes identify persons who do not express but almost certainly do have a disease genotype (for example, individual Q in Figure 6-21). Similarly, variable expressivity can complicate counseling because persons with low expressivity might be misdiagnosed.

Even though penetrance and expressivity can be quantified, they nevertheless represent "fuzzy" situations because rarely is it possible to identify the specific factors causing variation without substantial extra research.

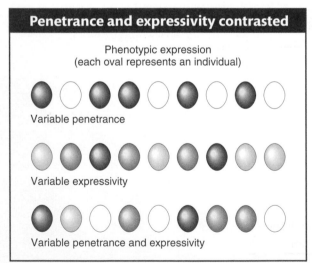

Penetrance and expressivity contrasted

Phenotypic expression
(each oval represents an individual)

Variable penetrance

Variable expressivity

Variable penetrance and expressivity

Figure 6-22 Assume that all the individuals shown have the same pigment allele (*P*) and possess the same potential to produce pigment. Effects from the rest of the genome and the environment may suppress or modify pigment production in any one individual. The color indicates the level of expression.

Message The terms *penetrance* and *expressivity* quantify the modification of gene expression by varying environment and genetic background; they measure, respectively, the percentage of cases in which the gene is expressed and the level of expression.

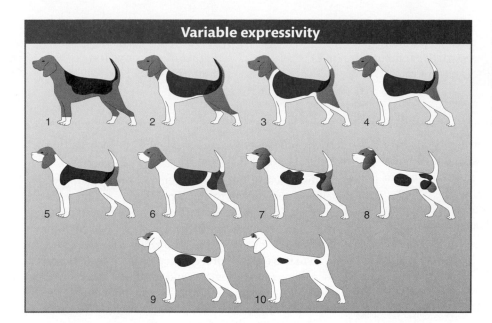

Figure 6-23 Ten grades of piebald spotting in beagles. Each of these dogs has the allele S^P, the allele responsible for piebald spots in dogs. The variation is caused by variation at other loci. [*After Clarence C. Little, The Inheritance of Coat Color in Dogs. Cornell University Press, 1957; and Giorgio Schreiber, J. Hered. 9, 1930, 403.*]

Summary

A gene does not act alone; rather, it acts in concert with many other genes in the genome. In forward genetic analysis, deducing these complex interactions is an important stage of the research. Individual mutations are first tested for their dominance relations, a type of allelic interaction. Recessive mutations are often a result of haplosufficiency of the wild-type allele, whereas dominant mutations are often the result either of haploinsufficiency of the wild type or of the mutant acting as a dominant negative (a rogue polypeptide). Some mutations cause severe effects or even death (lethal mutations). Lethality of a homozygous recessive mutation is a way to assess if a gene is essential in the genome.

The interaction of different genes is a result of their participation in the same or connecting pathways of various kinds—synthetic, signal transduction, or developmental. Genetic dissection of gene interactions begins by the experimenter amassing mutants affecting a character of interest. The complementation test determines whether two distinct recessive mutations are of one gene or of two different genes. The mutant genotypes are brought together in an F_1 individual, and if the phenotype is mutant, then no complementation has occurred and the two alleles must be of the same gene. If the phenotype is wild type, then complementation has occurred, and the alleles must be of different genes.

The interaction of different genes can be detected by testing double mutants, because allele interaction implies interaction of gene products at the functional level. Some key types of interaction are epistasis, suppression, and synthetic lethality. Epistasis is the replacement of a mutant phenotype produced by one mutation with a mutant phenotype produced by mutation of another gene. The observation of epistasis suggests a common developmental or chemical pathway. A suppressor is a mutation of one gene that can restore wild-type phenotype to a mutation at another gene. Suppressors often reveal physically interacting proteins or nucleic acids. Some combinations of viable mutants are lethal, a result known as *synthetic lethality*. Synthetic lethals can reveal a variety of interactions, depending on the nature of the mutations.

The different types of gene interactions produce F_2 dihybrid ratios that are modifications of the standard $9:3:3:1$. For example, recessive epistasis results in a $9:3:4$ ratio.

In more general terms, gene interaction and gene–environment interaction are revealed by variable penetrance (the ability of a genotype to express itself) and expressivity (the quantitative degree of expression of a genotype).

KEY TERMS

allelic series (multiple alleles) (p. 212)
codominance (p. 214)
complementation test (p. 222)
dominant negative mutation (p. 213)

double mutants (p. 222)
epistasis (p. 228)
expressivity (p. 234)
full (complete) dominance (p. 212)
functional RNA (p. 221)

heterokaryon (p. 225)
incomplete dominance (p. 214)
lethal allele (p. 216)
modifier (p. 232)
multiple alleles (allelic series) (p. 212)

null mutation (p. 213)
one-gene–one-polypeptide
    hypothesis (p. 220)
penetrance (p. 234)

permissive temperature (p. 217)
pleiotropic allele (p. 217)
restrictive temperature (p. 217)
revertant (p. 231)

suppressor (p. 230)
synthetic lethal (p. 233)
temperature-sensitive (ts)
    mutations (p. 217)

SOLVED PROBLEMS

SOLVED PROBLEM 1. Most pedigrees show the polydactyly (see Figure 2-25) inherited as a rare autosomal dominant, but the pedigrees of some families do not fully conform to the patterns expected for such inheritance. Such a pedigree is shown here. (The unshaded diamonds stand for the specified number of unaffected persons of unknown sex.)

a. What irregularity does this pedigree show?

b. What genetic phenomenon does this pedigree illustrate?

c. Suggest a specific gene-interaction mechanism that could produce such a pedigree, showing genotypes of pertinent family members.

carry the polydactyly gene inherited from I-1 because they transmit it to their progeny.

c. As discussed in this chapter, environmental suppression of gene expression can cause incomplete penetrance, as can suppression by another gene. To give the requested genetic explanation, we must come up with a genetic hypothesis. What do we need to explain? The key is that I-1 passes the gene on to two types of progeny, represented by II-1, who expresses the gene, and by II-6 and II-10, who do not. (From the pedigree, we cannot tell whether the other children of I-1 have the gene.) Is genetic suppression at work? I-1 does not have a suppressor allele, because he expresses polydactyly.

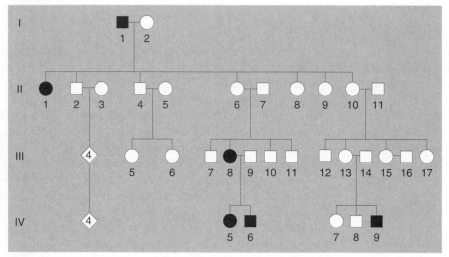

Solution

a. The normal expectation for an autosomal dominant is for each affected individual to have an affected parent, but this expectation is not seen in this pedigree, which constitutes the irregularity. What are some possible explanations?

Could some cases of polydactyly be caused by a different gene, one that is an X-linked dominant gene? This suggestion is not useful, because we still have to explain the absence of the condition in persons II-6 and II-10. Furthermore, postulating recessive inheritance, whether autosomal or sex-linked, requires many people in the pedigree to be heterozygotes, which is inappropriate because polydactyly is a rare condition.

b. Thus, we are left with the conclusion that polydactyly must sometimes be incompletely penetrant. As described in this chapter, some individuals who have the genotype for a particular phenotype do not express it. In this pedigree, II-6 and II-10 seem to belong in this category; they must

So the only person from whom a suppressor could come is I-2. Furthermore, I-2 must be heterozygous for the suppressor gene because at least one of her children does express polydactyly. We have thus formulated the hypothesis that the mating in generation I must have been

$$\text{(I-1) } P/p \cdot s/s \times \text{(I-2) } p/p \cdot S/s$$

where S is the suppressor and P is the allele responsible for polydactyly. From this hypothesis, we predict that the progeny will comprise the following four types if the genes assort:

Genotype	Phenotype	Example
$P/p \cdot S/s$	normal (suppressed)	II-6, II-10
$P/p \cdot s/s$	polydactylous	II-1
$p/p \cdot S/s$	normal	
$p/p \cdot s/s$	normal	

If S is rare, the progeny of II-6 and II-10 are:

Progeny genotype	Example
$P/p \cdot S/s$	III-13
$P/p \cdot s/s$	III-8
$p/p \cdot S/s$	
$p/p \cdot s/s$	

We cannot rule out the possibilities that II-2 and II-4 have the genotype $P/p \cdot S/s$ and that by chance none of their descendants are affected.

SOLVED PROBLEM 2. Beetles of a certain species may have green, blue, or turquoise wing covers. Virgin beetles were selected from a polymorphic laboratory population and mated to determine the inheritance of wing-cover color. The crosses and results were as given in the following table:

Cross	Parents	Progeny
1	blue × green	all blue
2	blue × blue	$\frac{3}{4}$ blue : $\frac{1}{4}$ turquoise
3	green × green	$\frac{3}{4}$ green : $\frac{1}{4}$ turquoise
4	blue × turquoise	$\frac{1}{2}$ blue : $\frac{1}{2}$ turquoise
5	blue × blue	$\frac{1}{4}$ blue : $\frac{1}{4}$ green
6	blue × green	$\frac{1}{2}$ blue : $\frac{1}{2}$ green
7	blue × green	$\frac{1}{2}$ blue : $\frac{1}{4}$ green
		$\frac{1}{4}$ turquoise
8	turquoise × turquoise	all turquoise

a. Deduce the genetic basis of wing-cover color in this species.

b. Write the genotypes of all parents and progeny as completely as possible.

Solution

a. These data seem complex at first, but the inheritance pattern becomes clear if we consider the crosses one at a time. A general principle of solving such problems, as we have seen, is to begin by looking over all the crosses and by grouping the data to bring out the patterns.

One clue that emerges from an overview of the data is that all the ratios are one-gene ratios: there is no evidence of two separate genes taking part at all. How can such variation be explained with a single gene? The answer is that there is variation for the single gene itself—that is, multiple allelism. Perhaps there are three alleles of one gene; let's call the gene w (for wing-cover color) and represent the alleles as w^g, w^b, and w^t. Now we have an additional problem, which is to determine the dominance of these alleles.

Cross 1 tells us something about dominance because all of the progeny of a blue × green cross are blue; hence, blue ap-

pears to be dominant over green. This conclusion is supported by cross 5, because the green determinant must have been present in the parental stock to appear in the progeny. Cross 3 informs us about the turquoise determinants, which must have been present, although unexpressed, in the parental stock because there are turquoise wing covers in the progeny. So green must be dominant over turquoise. Hence, we have formed a model in which the dominance is $w^b > w^g > w^t$. Indeed, the inferred position of the w^t allele at the bottom of the dominance series is supported by the results of cross 7, where turquoise shows up in the progeny of a blue × green cross.

b. Now it is just a matter of deducing the specific genotypes. Notice that the question states that the parents were taken from a polymorphic population, which means that they could be either homozygous or heterozygous. A parent with blue wing covers, for example, might be homozygous (w^b/w^b) or heterozygous (w^b/w^g or w^b/w^t). Here, a little trial and error and common sense are called for, but, by this stage, the question has essentially been answered, and all that remains is to "cross the t's and dot the i's." The following genotypes explain the results. A dash indicates that the genotype may be *either* homozygous or heterozygous in having a second allele farther down the allelic series.

Cross	Parents	Progeny
1	$w^b/w^b \times w^g/-$	w^b/w^g or $w^b/-$
2	$w^b/w^t \times w^b/w^t$	$\frac{3}{4} w^b/- : \frac{1}{4} w^t/w^t$
3	$w^g/w^t \times w^g/w^t$	$\frac{3}{4} w^g/- : \frac{1}{4} w^t/w^t$
4	$w^b/w^t \times w^t/w^t$	$\frac{1}{2} w^b/w^t : \frac{1}{2} w^t/w^t$
5	$w^b/w^g \times w^b/w^g$	$\frac{3}{4} w^b/- : \frac{1}{4} w^g/w^g$
6	$w^b/w^g \times w^g/w^g$	$\frac{1}{2} w^b/w^g : \frac{1}{2} w^g/w^g$
7	$w^b/w^t \times w^g/w^t$	$\frac{1}{2} w^b/- : \frac{1}{4} w^g/w^t : \frac{1}{4} w^t/w^t$
8	$w^t/w^t \times w^t/w^t$	all w^t/w^t

SOLVED PROBLEM 3. The leaves of pineapples can be classified into three types: spiny (S), spiny tip (ST), and piping (nonspiny; P). In crosses between pure strains followed by intercrosses of the F_1, the following results appeared:

Cross	Parental	Phenotypes	
		F_1	F_2
1	ST × S	ST	99 ST : 34 S
2	P × ST	P	120 P : 39 ST
3	P × S	P	95 P : 25 ST : 8 S

a. Assign gene symbols. Explain these results in regard to the genotypes produced and their ratios.

b. Using the model from part *a*, give the phenotypic ratios that you would expect if you crossed (1) the F_1 progeny from piping × spiny with the spiny parental stock and (2) the F_1 progeny of piping × spiny with the F_1 progeny of spiny × spiny tip.

Solution

a. First, let's look at the F_2 ratios. We have clear $3:1$ ratios in crosses 1 and 2, indicating single-gene segregations. Cross 3, however, shows a ratio that is almost certainly a $12:3:1$ ratio. How do we know this ratio? Well, there are simply not that many complex ratios in genetics, and trial and error brings us to the $12:3:1$ quite quickly. In the 128 progeny total, the numbers of $96:24:8$ are expected, but the actual numbers fit these expectations remarkably well.

One of the principles of this chapter is that modified Mendelian ratios reveal gene interactions. Cross 3 gives F_2 numbers appropriate for a modified dihybrid Mendelian ratio, and so it looks as if we are dealing with a two-gene interaction. It seems the most promising place to start; we can return to crosses 1 and 2 and try to fit them in later.

Any dihybrid ratio is based on the phenotypic proportions $9:3:3:1$. Our observed modification groups them as follows:

$$\left.\begin{array}{l} 9\ A/-\ ;B/- \\ 3\ A/-\ ;b/b \end{array}\right\}\ \text{12 piping}$$
$$3\ a/a\ ;B/-\qquad \text{3 spiny tip}$$
$$1\ a/a\ ;b/b\qquad \text{1 spiny}$$

So, without worrying about the name of the type of gene interaction (we are not asked to supply this anyway), we can already define our three pineapple-leaf phenotypes in relation to the proposed allelic pairs A/a and B/b:

$$\text{piping} = A/-\ (B/b\ \text{irrelevant})$$
$$\text{spiny tip} = a/a\ ;B/-$$
$$\text{spiny} = a/a\ ;b/b$$

What about the parents of cross 3? The spiny parent must be $a/a\ ;b/b$, and, because the B gene is needed to produce F_2 spiny-tip leaves, the piping parent must be $A/A\ ;B/B$. (Note that we are *told* that all parents are pure, or homozygous.) The F_1 must therefore be $A/a\ ;B/b$.

Without further thought, we can write out cross 1 as follows:

$$a/a\ ;\ B/B\ \times\ a/a\ ;\ b/b\ \longrightarrow$$
$$a/a\ ;\ B/b\ \Big\langle\begin{array}{l}\nearrow\ \frac{3}{4}\ a/a\ ;B/- \\ \searrow\ \frac{1}{4}\ a/a\ ;\ b/b\end{array}$$

Cross 2 can be partly written out without further thought by using our arbitrary gene symbols:

$$A/A\ ;\ -/-\ \times\ a/a\ ;\ B/B\ \longrightarrow$$
$$A/a\ ;\ B/-\ \Big\langle\begin{array}{l}\nearrow\ \frac{3}{4}\ A/-\ ;\ -/- \\ \searrow\ \frac{1}{4}\ a/a\ ;\ B/-\end{array}$$

We know that the F_2 of cross 2 shows single-gene segregation, and it seems certain now that the A/a allelic pair has a role. But the B allele is needed to produce the spiny-tip phenotype, and so all plants must be homozygous B/B:

$$A/A\ ;\ B/B\ \times\ a/a\ ;\ B/B\ \longrightarrow$$
$$A/a\ ;\ B/B\ \Big\langle\begin{array}{l}\nearrow\ \frac{3}{4}\ A/-\ ;\ B/B \\ \searrow\ \frac{1}{4}\ a/a\ ;\ B/B\end{array}$$

Notice that the two single-gene segregations in crosses 1 and 2 do not show that the genes are *not* interacting. What is shown is that the two-gene interaction is not *revealed* by these crosses—only by cross 3, in which the F_1 is heterozygous for both genes.

b. Now it is simply a matter of using Mendel's laws to predict cross outcomes:

(1)　$A/a\ ;\ B/b\ \times\ a/a\ ;\ b/b\ \longrightarrow \frac{1}{4}\ A/a\ ;\ B/b$

(independent assortment in a standard testcross)

$$\left.\begin{array}{l} \frac{1}{4}\ A/a\ ;\ B/b \\ \frac{1}{4}\ A/a\ ;\ b/b \end{array}\right\}\ \text{piping}$$
$$\frac{1}{4}\ a/a\ ;\ B/b\qquad \text{spiny tip}$$
$$\frac{1}{4}\ a/a\ ;\ b/b\qquad \text{spiny}$$

(2)　$A/a\ ;\ B/b\ \times\ a/a\ ;\ B/b\ \longrightarrow$

$$\frac{1}{2}\ A/a\ \Big\langle\begin{array}{l}\nearrow\ \frac{3}{4}\ B/-\ \longrightarrow\ \frac{3}{8} \\ \searrow\ \frac{1}{4}\ b/b\ \longrightarrow\ \frac{1}{8}\end{array}\Big\}\ \frac{1}{2}\ \text{piping}$$

$$\frac{1}{2}\ a/a\ \Big\langle\begin{array}{l}\nearrow\ \frac{3}{4}\ B/-\ \longrightarrow\ \frac{3}{8}\quad\text{spiny tip} \\ \searrow\ \frac{1}{4}\ b/b\ \longrightarrow\ \frac{1}{8}\quad\text{spiny}\end{array}$$

PROBLEMS

Most of the problems are also available for review/grading through the GENETICSPORTAL www.yourgeneticsportal.com.

WORKING WITH THE FIGURES

1. In Figure 6-1,

   a. what do the yellow stars represent?

   b. explain in your own words why the heterozygote is functionally wild type.

2. In Figure 6-2, explain how the mutant polypeptide acts as a spoiler and what its net effect on phenotype is.

3. In Figure 6-6, assess the allele V^f: is it dominant? recessive? codominant? incompletely dominant?

4. In Figure 6-11,

a. in view of the position of HPA oxidase earlier in the pathway compared to that of HA oxidase, would you expect people with tyrosinosis to show symptoms of alkaptonuria?

b. if a double mutant could be found, would you expect tyrosinosis to be epistatic to alkaptonuria?

5. In Figure 6-12,

a. what do the dollar, pound, and yen symbols represent?

b. why can't the left-hand F_1 heterozygote synthesize blue pigment?

6. In Figure 6-13, explain at the protein level why this heterokaryon can grow on minimal medium.

7. In Figure 6-14, write possible genotypes for each of the four snakes illustrated.

8. In Figure 6-15,

a. which panel represents the double mutant?

b. state the purpose of the regulatory gene.

c. in the situation in panel *b*, would protein from the active protein gene be made?

9. In Figure 6-16, if you selfed 10 different F_2 pink plants, would you expect to find any white-flowered plants among the offspring? Any blue-flowered plants?

10. In Figure 6-19,

a. what do the square/triangular pegs and holes represent?

b. is the suppressor mutation alone wild type in phenotype?

11. In Figure 6-21, propose a specific genetic explanation for individual Q (give a possible genotype, defining the alleles).

BASIC PROBLEMS

12. In humans, the disease galactosemia causes mental retardation at an early age. Lactose (milk sugar) is broken down to galactose plus glucose. Normally, galactose is broken down further by the enzyme galactose-1-phosphate uridyltransferase (GALT). However, in patients with galactosemia, GALT is inactive, leading to a buildup of high levels of galactose, which, in the brain, causes mental retardation. How would you provide a secondary cure for galactosemia? Would you expect this disease phenotype to be dominant or recessive?

13. In humans, PKU (phenylketonuria) is a disease caused by an enzyme inefficiency at step A in the following simplified reaction sequence, and AKU (alkaptonuria) is due to an enzyme inefficiency in one of the steps summarized as step B here:

$$\text{phenylalanine} \xrightarrow{\text{A}} \text{tyrosine} \xrightarrow{\text{B}} CO_2 + H_2O$$

A person with PKU marries a person with AKU. What phenotypes do you expect for their children? All normal, all having PKU only, all having AKU only, all having both PKU and AKU, or some having AKU and some having PKU?

14. In *Drosophila*, the autosomal recessive *bw* causes a dark brown eye, and the unlinked autosomal recessive *st* causes a bright scarlet eye. A homozygote for both genes has a white eye. Thus, we have the following correspondences between genotypes and phenotypes:

$$st^+/st^+ ; bw^+/bw^+ = \text{red eye (wild type)}$$
$$st^+/st^+ ; bw/bw = \text{brown eye}$$
$$st/st ; bw^+/bw^+ = \text{scarlet eye}$$
$$st/st ; bw/bw = \text{white eye}$$

Construct a hypothetical biosynthetic pathway showing how the gene products interact and why the different mutant combinations have different phenotypes.

15. Several mutants are isolated, all of which require compound G for growth. The compounds (A to E) in the biosynthetic pathway to G are known, but their order in the pathway is not known. Each compound is tested for its ability to support the growth of each mutant (1 to 5). In the following table, a plus sign indicates growth and a minus sign indicates no growth.

	Compound tested					
	A	B	C	D	E	G
Mutant 1	−	−	−	+	−	+
2	−	+	−	+	−	+
3	−	−	−	−	−	+
4	−	+	+	+	−	+
5	+	+	+	+	−	+

a. What is the order of compounds A to E in the pathway?

b. At which point in the pathway is each mutant blocked?

c. Would a heterokaryon composed of double mutants 1,3 and 2,4 grow on a minimal medium? Would 1,3 and 3,4? Would 1,2 and 2,4 and 1,4?

16. In a certain plant, the flower petals are normally purple. Two recessive mutations arise in separate plants and are found to be on different chromosomes. Mutation 1 (m_1) gives blue petals when homozygous (m_1/m_1). Mutation 2 (m_2) gives red petals when homozygous (m_2/m_2). Biochemists working on the synthesis of flower pigments in this species have already described the following pathway:

a. Which mutant would you expect to be deficient in enzyme A activity?

b. A plant has the genotype $M_1/m_1 ; M_2/m_2$. What would you expect its phenotype to be?

c. If the plant in part *b* is selfed, what colors of progeny would you expect and in what proportions?

d. Why are these mutants recessive?

17. In sweet peas, the synthesis of purple anthocyanin pigment in the petals is controlled by two genes, *B* and *D*. The pathway is

$$\text{white} \xrightarrow[\text{enzyme}]{\text{gene } B} \text{blue} \xrightarrow[\text{enzyme}]{\text{gene } D} \text{anthocyanin}$$
$$\text{intermediate} \qquad\qquad \text{intermediate} \qquad\qquad \text{(purple)}$$

a. What color petals would you expect in a pure-breeding plant unable to catalyze the first reaction?

b. What color petals would you expect in a pure-breeding plant unable to catalyze the second reaction?

c. If the plants in parts *a* and *b* are crossed, what color petals will the F_1 plants have?

d. What ratio of purple:blue:white plants would you expect in the F_2?

18. If a man of blood-group AB marries a woman of blood-group A whose father was of blood-group O, to what different blood groups can this man and woman expect their children to belong?

19. Most of the feathers of erminette fowl are light colored, with an occasional black one, giving a flecked appearance. A cross of two erminettes produced a total of 48 progeny, consisting of 22 erminettes, 14 blacks, and 12 pure whites. What genetic basis of the erminette pattern is suggested? How would you test your hypotheses?

20. Radishes may be long, round, or oval, and they may be red, white, or purple. You cross a long, white variety with a round, red one and obtain an oval, purple F_1. The F_2 shows nine phenotypic classes as follows: 9 long, red; 15 long, purple; 19 oval, red; 32 oval, purple; 8 long, white; 16 round, purple; 8 round, white; 16 oval, white; and 9 round, red.

a. Provide a genetic explanation of these results. Be sure to define the genotypes and show the constitution of the parents, the F_1, and the F_2.

b. Predict the genotypic and phenotypic proportions in the progeny of a cross between a long, purple radish and an oval, purple one.

21. In the multiple-allele series that determines coat color in rabbits, c^+ encodes agouti, c^{ch} encodes chinchilla (a beige coat color), and c^h encodes Himalayan. Dominance is in the order $c^+ > c^{ch} > c^h$. In a cross of $c^+/c^{ch} \times c^{ch}/c^h$, what proportion of progeny will be chinchilla?

22. Black, sepia, cream, and albino are coat colors of guinea pigs. Individual animals (not necessarily from pure lines) showing these colors were intercrossed; the results are tabulated as follows, where the abbreviations A (albino), B (black), C (cream), and S (sepia) represent the phenotypes:

		Phenotypes of progeny			
Cross	Parental phenotypes	B	S	C	A
1	B × B	22	0	0	7
2	B × A	10	9	0	0
3	C × C	0	0	34	11
4	S × C	0	24	11	12
5	B × A	13	0	12	0
6	B × C	19	20	0	0
7	B × S	18	20	0	0
8	B × S	14	8	6	0
9	S × S	0	26	9	0
10	C × A	0	0	15	17

a. Deduce the inheritance of these coat colors and use gene symbols of your own choosing. Show all parent and progeny genotypes.

b. If the black animals in crosses 7 and 8 are crossed, what progeny proportions can you predict by using your model?

23. In a maternity ward, four babies become accidentally mixed up. The ABO types of the four babies are known to be O, A, B, and AB. The ABO types of the four sets of parents are determined. Indicate which baby belongs to each set of parents: **(a)** AB × O, **(b)** A × O, **(c)** A × AB, **(d)** O × O.

24. Consider two blood polymorphisms that humans have in addition to the ABO system. Two alleles L^M and L^N determine the M, N, and MN blood groups. The dominant allele R of a different gene causes a person to have the Rh+ (rhesus positive) phenotype, whereas the homozygote for r is Rh− (rhesus negative). Two men took a paternity dispute to court, each claiming three children to be his own. The blood groups of the men, the children, and their mother were as follows:

Person	Blood group		
husband	O	M	Rh+
wife's lover	AB	MN	Rh−
wife	A	N	Rh+
child 1	O	MN	Rh+
child 2	A	N	Rh+
child 3	A	MN	Rh−

From this evidence, can the paternity of the children be established?

25. On a fox ranch in Wisconsin, a mutation arose that gave a "platinum" coat color. The platinum color proved very popular with buyers of fox coats, but the breeders could not develop a pure-breeding platinum strain. Every time two platinums were crossed, some normal foxes appeared in the progeny. For example, the repeated matings of the same pair of platinums produced 82 platinum and 38 normal progeny. All other such matings gave similar progeny ratios. State a concise genetic hypothesis that accounts for these results.

26. For a period of several years, Hans Nachtsheim investigated an inherited anomaly of the white blood cells of rabbits. This anomaly, termed the *Pelger anomaly,* is the arrest of the segmentation of the nuclei of certain white cells. This anomaly does not appear to seriously inconvenience the rabbits.

a. When rabbits showing the typical Pelger anomaly were mated with rabbits from a true-breeding normal stock, Nachtsheim counted 217 offspring showing the Pelger anomaly and 237 normal progeny. What appears to be the genetic basis of the Pelger anomaly?

b. When rabbits with the Pelger anomaly were mated with each other, Nachtsheim found 223 normal progeny, 439 showing the Pelger anomaly, and 39 extremely abnormal progeny. These very abnormal progeny not only had defective white blood cells, but also showed severe deformities of the skeletal system; almost all of them died soon after birth. In genetic terms, what do you suppose these extremely defective rabbits represented? Why do you suppose there were only 39 of them?

c. What additional experimental evidence might you collect to support or disprove your answers to part *b*?

d. In Berlin, about 1 human in 1000 shows a Pelger anomaly of white blood cells very similar to that described for rabbits. The anomaly is inherited as a simple dominant, but the homozygous type has not been observed in humans. Can you suggest why, if you are permitted an analogy with the condition in rabbits?

e. Again by analogy with rabbits, what phenotypes and genotypes might be expected among the children of a man and woman who both show the Pelger anomaly?

(Problem 26 is from A. M. Srb, R. D. Owen, and R. S. Edgar, *General Genetics,* 2nd ed. W. H. Freeman and Company, 1965.)

27. Two normal-looking fruit flies were crossed, and, in the progeny, there were 202 females and 98 males.

a. What is unusual about this result?

b. Provide a genetic explanation for this anomaly.

c. Provide a test of your hypothesis.

28. You have been given a virgin *Drosophila* female. You notice that the bristles on her thorax are much shorter than normal. You mate her with a normal male (with long bristles) and obtain the following F$_1$ progeny: $\frac{1}{3}$ short-bristled females, $\frac{1}{3}$ long-bristled females, and $\frac{1}{3}$ long-bristled males. A cross of the F$_1$ long-bristled females with their brothers gives only long-bristled F$_2$. A cross of short-bristled females with their brothers gives $\frac{1}{3}$ short-bristled females, $\frac{1}{3}$ long-bristled females, and $\frac{1}{3}$ long-bristled males. Provide a genetic hypothesis to account for all these results, showing genotypes in every cross.

29. A dominant allele *H* reduces the number of body bristles that *Drosophila* flies have, giving rise to a "hairless" phenotype. In the homozygous condition, *H* is lethal. An independently assorting dominant allele *S* has no effect on bristle number except in the presence of *H*, in which case a single dose of *S* suppresses the hairless phenotype, thus restoring the hairy phenotype. However, *S* also is lethal in the homozygous (*S/S*) condition.

a. What ratio of hairy to hairless flies would you find in the live progeny of a cross between two hairy flies both carrying *H* in the suppressed condition?

b. When the hairless progeny are backcrossed with a parental hairy fly, what phenotypic ratio would you expect to find among their live progeny?

30. After irradiating wild-type cells of *Neurospora* (a haploid fungus), a geneticist finds two leucine-requiring auxotrophic mutants. He combines the two mutants in a heterokaryon and discovers that the heterokaryon is prototrophic.

a. Were the mutations in the two auxotrophs in the *same* gene in the pathway for synthesizing leucine or in two *different* genes in that pathway? Explain.

b. Write the genotype of the two strains according to your model.

c. What progeny and in what proportions would you predict from crossing the two auxotrophic mutants? (Assume independent assortment.)

31. A yeast geneticist irradiates haploid cells of a strain that is an adenine-requiring auxotrophic mutant, caused by mutation of the gene *ade1*. Millions of the irradiated cells are plated on minimal medium, and a small number of cells divide and produce prototrophic colonies. These colonies are crossed individually with a wild-type strain. Two types of results are obtained:

(1) prototroph × wild type : progeny all prototrophic

(2) prototroph × wild type : progeny 75% prototrophic, 25% adenine-requiring auxotrophs

a. Explain the difference between these two types of results.

b. Write the genotypes of the prototrophs in each case.

c. What progeny phenotypes and ratios do you predict from crossing a prototroph of type 2 by the original *ade1* auxotroph?

32. In roses, the synthesis of red pigment is by two steps in a pathway, as follows:

$$\text{colorless intermediate} \xrightarrow{\text{gene } P}$$
$$\text{magenta intermediate} \xrightarrow{\text{gene } Q} \text{red pigment}$$

a. What would the phenotype be of a plant homozygous for a null mutation of gene P?

b. What would the phenotype be of a plant homozygous for a null mutation of gene Q?

c. What would the phenotype be of a plant homozygous for null mutations of genes P and Q?

d. Write the genotypes of the three strains in parts a, b, and c.

e. What F_2 ratio is expected from crossing plants from parts a and b? (Assume independent assortment.)

33. Because snapdragons (*Antirrhinum*) possess the pigment anthocyanin, they have reddish purple petals. Two pure anthocyaninless lines of *Antirrhinum* were developed, one in California and one in Holland. They looked identical in having no red pigment at all, manifested as white (albino) flowers. However, when petals from the two lines were ground up together in buffer in the same test tube, the solution, which appeared colorless at first, gradually turned red.

a. What control experiments should an investigator conduct before proceeding with further analysis?

b. What could account for the production of the red color in the test tube?

c. According to your explanation for part b, what would be the genotypes of the two lines?

d. If the two white lines were crossed, what would you predict the phenotypes of the F_1 and F_2 to be?

34. The frizzle fowl is much admired by poultry fanciers. It gets its name from the unusual way that its feathers curl up, giving the impression that it has been (in the memorable words of animal geneticist F. B. Hutt) "pulled backwards through a knothole." Unfortunately, frizzle fowl do not breed true: when two frizzles are intercrossed, they always produce 50 percent frizzles, 25 percent normal, and 25 percent with peculiar woolly feathers that soon fall out, leaving the birds naked.

a. Give a genetic explanation for these results, showing genotypes of all phenotypes, and provide a statement of how your explanation works.

b. If you wanted to mass-produce frizzle fowl for sale, which types would be best to use as a breeding pair?

35. The petals of the plant *Collinsia parviflora* are normally blue, giving the species its common name, blue-eyed Mary. Two pure-breeding lines were obtained from color variants found in nature; the first line had pink petals, and the second line had white petals. The following crosses were made between pure lines, with the results shown:

Parents	F_1	F_2
blue × white	blue	101 blue, 33 white
blue × pink	blue	192 blue, 63 pink
pink × white	blue	272 blue, 121 white, 89 pink

a. Explain these results genetically. Define the allele symbols that you use, and show the genetic constitution of the parents, the F_1, and the F_2 in each cross.

b. A cross between a certain blue F_2 plant and a certain white F_2 plant gave progeny of which $\frac{3}{8}$ were blue, $\frac{1}{8}$ were pink, and $\frac{1}{2}$ were white. What must the genotypes of these two F_2 plants have been?

 Unpacking Problem 35

1. What is the character being studied?

2. What is the wild-type phenotype?

3. What is a variant?

4. What are the variants in this problem?

5. What does "in nature" mean?

6. In what way would the variants have been found in nature? (Describe the scene.)

7. At which stages in the experiments would seeds be used?

8. Would the way of writing a cross "blue × white," for example, mean the same as "white × blue"? Would you expect similar results? Why or why not?

9. In what way do the first two rows in the table differ from the third row?

10. Which phenotypes are dominant?

11. What is complementation?

12. Where does the blueness come from in the progeny of the pink × white cross?

13. What genetic phenomenon does the production of a blue F_1 from pink and white parents represent?

14. List any ratios that you can see.

15. Are there any monohybrid ratios?

16. Are there any dihybrid ratios?

17. What does observing monohybrid and dihybrid ratios tell you?

18. List four modified Mendelian ratios that you can think of.

19. Are there any modified Mendelian ratios in the problem?

20. What do modified Mendelian ratios indicate generally?

21. What is indicated by the specific modified ratio or ratios in this problem?

22. Draw chromosomes representing the meioses in the parents in the cross blue × white and representing meiosis in the F_1.

23. Repeat step 22 for the cross blue × pink.

36. A woman who owned a purebred albino poodle (an autosomal recessive phenotype) wanted white puppies; so she took the dog to a breeder, who said he would mate the female with an albino stud male, also from a pure stock. When six puppies were born, all of them were black; so the woman sued the breeder, claiming that he replaced the stud male with a black dog, giving her six unwanted puppies. You are called in as an expert witness, and the defense asks you if it is possible to produce black offspring from two pure-breeding recessive albino parents. What testimony do you give?

37. A snapdragon plant that bred true for white petals was crossed with a plant that bred true for purple petals, and all the F_1 had white petals. The F_1 was selfed. Among the F_2, three phenotypes were observed in the following numbers:

white	240
solid purple	61
spotted purple	19
Total	320

a. Propose an explanation for these results, showing genotypes of all generations (make up and explain your symbols).

b. A white F_2 plant was crossed with a solid purple F_2 plant, and the progeny were

white	50%
solid purple	25%
spotted purple	25%

What were the genotypes of the F_2 plants crossed?

38. Most flour beetles are black, but several color variants are known. Crosses of pure-breeding parents produced the following results (see table) in the F_1 generation, and intercrossing the F_1 from each cross gave the ratios shown for the F_2 generation. The phenotypes are abbreviated Bl, black; Br, brown; Y, yellow; and W, white.

Cross	Parents	F_1	F_2
1	Br × Y	Br	3 Br : 1 Y
2	Bl × Br	Bl	3 Bl : 1 Br
3	Bl × Y	Bl	3 Bl : 1 Y
4	W × Y	Bl	9 Bl : 3 Y : 4 W
5	W × Br	Bl	9 Bl : 3 Br : 4 W
6	Bl × W	Bl	9 Bl : 3 Y : 4 W

a. From these results, deduce and explain the inheritance of these colors.

b. Write the genotypes of each of the parents, the F_1, and the F_2 in all crosses.

39. Two albinos marry and have four normal children. How is this possible?

40. Consider the production of flower color in the Japanese morning glory (*Pharbitis nil*). Dominant alleles of either of two separate genes ($A/- \cdot b/b$ or $a/a \cdot B/-$) produce purple petals. $A/- \cdot B/-$ produces blue petals, and $a/a \cdot b/b$ produces scarlet petals. Deduce the genotypes of parents and progeny in the following crosses:

Cross	Parents	Progeny
1	blue × scarlet	$\frac{1}{4}$ blue : $\frac{1}{2}$ purple : $\frac{1}{4}$ scarlet
2	purple × purple	$\frac{1}{4}$ blue : $\frac{1}{2}$ purple : $\frac{1}{4}$ scarlet
3	blue × blue	$\frac{3}{4}$ blue : $\frac{1}{4}$ purple
4	blue × purple	$\frac{3}{8}$ blue : $\frac{4}{8}$ purple : $\frac{1}{8}$ scarlet
5	purple × scarlet	$\frac{1}{2}$ purple : $\frac{1}{2}$ scarlet

41. Corn breeders obtained pure lines whose kernels turn sun red, pink, scarlet, or orange when exposed to sunlight (normal kernels remain yellow in sunlight). Some crosses between these lines produced the following results. The phenotypes are abbreviated O, orange; P, pink; Sc, scarlet; and SR, sun red.

		Phenotypes	
Cross	Parents	F_1	F_2
1	SR × P	all SR	66 SR : 20 P
2	O × SR	all SR	998 SR : 314 O
3	O × P	all O	1300 O : 429 P
4	O × Sc	all Y	182 Y : 80 O : 58 Sc

Analyze the results of each cross, and provide a unifying hypothesis to account for *all* the results. (Explain all symbols that you use.)

42. Many kinds of wild animals have the agouti coloring pattern, in which each hair has a yellow band around it.

a. Black mice and other black animals do not have the yellow band; each of their hairs is all black. This absence of wild agouti pattern is called *nonagouti*. When mice of a true-breeding agouti line are crossed with nonagoutis, the F_1 is all agouti and the F_2 has a 3 : 1 ratio of agoutis to nonagoutis. Diagram this cross, letting A represent the allele responsible for the agouti phenotype and a, nonagouti. Show the phenotypes and genotypes of the parents, their gametes, the F_1, their gametes, and the F_2.

b. Another inherited color deviation in mice substitutes brown for the black color in the wild-type hair. Such brown-agouti mice are called *cinnamons*. When wild-type mice are crossed with cinnamons, all of the F_1 are wild type and the F_2 has a 3 : 1 ratio of wild type to cinnamon. Diagram this cross as in part *a*, letting B

stand for the wild-type black allele and *b* stand for the cinnamon brown allele.

c. When mice of a true-breeding cinnamon line are crossed with mice of a true-breeding nonagouti (black) line, all of the F_1 are wild type. Use a genetic diagram to explain this result.

d. In the F_2 of the cross in part *c*, a fourth color called *chocolate* appears in addition to the parental cinnamon and nonagouti and the wild type of the F_1. Chocolate mice have a solid, rich brown color. What is the genetic constitution of the chocolates?

e. Assuming that the *A/a* and *B/b* allelic pairs assort independently of each other, what do you expect to be the relative frequencies of the four color types in the F_2 described in part *d*? Diagram the cross of parts *c* and *d*, showing phenotypes and genotypes (including gametes).

f. What phenotypes would be observed in what proportions in the progeny of a backcross of F_1 mice from part *c* with the cinnamon parental stock? With the nonagouti (black) parental stock? Diagram these backcrosses.

g. Diagram a testcross for the F_1 of part *c*. What colors would result and in what proportions?

h. Albino (pink-eyed white) mice are homozygous for the recessive member of an allelic pair *C/c*, which assorts independently of the *A/a* and *B/b* pairs. Suppose that you have four different highly inbred (and therefore presumably homozygous) albino lines. You cross each of these lines with a true-breeding wild-type line, and you raise a large F_2 progeny from each cross. What genotypes for the albino lines can you deduce from the following F_2 phenotypes?

Phenotypes of progeny

F_2 of line	Wild type	Black	Cinna-mon	Choco-late	Albino
1	87	0	32	0	39
2	62	0	0	0	18
3	96	30	0	0	41
4	287	86	92	29	164

(Problem 42 is adapted from A. M. Srb, R. D. Owen, and R. S. Edgar, *General Genetics*, 2nd ed. W. H. Freeman and Company, 1965.)

43. An allele *A* that is not lethal when homozygous causes rats to have yellow coats. The allele *R* of a separate gene that assorts independently produces a black coat. Together, *A* and *R* produce a grayish coat, whereas *a* and *r* produce a white coat. A gray male is crossed with a yellow female, and the F_1 is $\frac{3}{8}$ yellow, $\frac{3}{8}$ gray, $\frac{1}{8}$ black, and $\frac{1}{8}$ white. Determine the genotypes of the parents.

44. The genotype *r/r;p/p* gives fowl a single comb, *R/−;P/−* gives a walnut comb, *r/r;P/−* gives a pea comb, and *R/−;p/p* gives a rose comb (see the illustrations). Assume independent assortment.

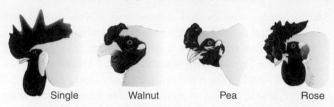
Single   Walnut   Pea   Rose

a. What comb types will appear in the F_1 and in the F_2 and in what proportions if single-combed birds are crossed with birds of a true-breeding walnut strain?

b. What are the genotypes of the parents in a walnut × rose mating from which the progeny are $\frac{3}{8}$ rose, $\frac{3}{8}$ walnut, $\frac{1}{8}$ pea, and $\frac{1}{8}$ single?

c. What are the genotypes of the parents in a walnut × rose mating from which all the progeny are walnut?

d. How many genotypes produce a walnut phenotype? Write them out.

45. The production of eye-color pigment in *Drosophila* requires the dominant allele *A*. The dominant allele *P* of a second independent gene turns the pigment to purple, but its recessive allele leaves it red. A fly producing no pigment has white eyes. Two pure lines were crossed with the following results:

P   red-eyed female × white-eyed male

F_1 purple-eyed females
         red-eyed males
         F_1 × F_1

F_2 both males and females: $\frac{3}{8}$ purple eyed
                    $\frac{3}{8}$ red eyed
                    $\frac{2}{8}$ white eyed

Explain this mode of inheritance and show the genotypes of the parents, the F_1, and the F_2.

46. When true-breeding brown dogs are mated with certain true-breeding white dogs, all the F_1 pups are white. The F_2 progeny from some F_1 × F_1 crosses were 118 white, 32 black, and 10 brown pups. What is the genetic basis for these results?

47. Wild-type strains of the haploid fungus *Neurospora* can make their own tryptophan. An abnormal allele *td* renders the fungus incapable of making its own tryptophan. An individual of genotype *td* grows only when its medium supplies tryptophan. The allele *su* assorts independently of *td;* its only known effect is to suppress the *td* phenotype. Therefore, strains carrying both *td* and *su* do not require tryptophan for growth.

a. If a *td*;*su* strain is crossed with a genotypically wild-type strain, what genotypes are expected in the progeny and in what proportions?

b. What will be the ratio of tryptophan-dependent to tryptophan-independent progeny in the cross of part *a*?

48. Mice of the genotypes $A/A\,;B/B\,;C/C\,;D/D\,;S/S$ and $a/a\,;b/b\,;c/c\,;d/d\,;s/s$ are crossed. The progeny are intercrossed. What phenotypes will be produced in the F_2 and in what proportions? (The allele symbols stand for the following: A = agouti, a = solid (nonagouti); B = black pigment, b = brown; C = pigmented, c = albino; D = nondilution, d = dilution (milky color); S = unspotted, s = pigmented spots on white background.)

49. Consider the genotypes of two lines of chickens: the pure-line mottled Honduran is $i/i\,;D/D\,;M/M\,;W/W$, and the pure-line leghorn is $I/I\,;d/d\,;m/m\,;w/w$, where

I = white feathers, i = colored feathers

D = duplex comb, d = simplex comb

M = bearded, m = beardless

W = white skin, w = yellow skin

These four genes assort independently. Starting with these two pure lines, what is the fastest and most convenient way of generating a pure line that has colored feathers, has a simplex comb, is beardless, and has yellow skin? Make sure that you show

a. the breeding pedigree.

b. the genotype of each animal represented.

c. how many eggs to hatch in each cross, and why this number.

d. why your scheme is the fastest and the most convenient.

50. The following pedigree is for a dominant phenotype governed by an autosomal allele. What does this pedigree suggest about the phenotype, and what can you deduce about the genotype of individual A?

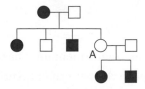

51. Petal coloration in foxgloves is determined by three genes. *M* encodes an enzyme that synthesizes anthocyanin, the purple pigment seen in these petals; *m/m* produces no pigment, resulting in the phenotype albino with yellowish spots. *D* is an enhancer of anthocyanin, resulting in a darker pigment; *d/d* does not enhance. At the third locus, *w/w* allows pigment deposition in petals, but *W* prevents pigment deposition except in the spots and so results in the white, spotted phenotype. Consider the following two crosses:

Cross	Parents	Progeny
1	dark purple × white with yellowish spots	$\frac{1}{2}$ dark purple: $\frac{1}{2}$ light purple
2	white with yellowish spots × light purple	$\frac{1}{2}$ white with purple spots: $\frac{1}{4}$ dark purple: $\frac{1}{4}$ light purple

In each case, give the genotypes of parents and progeny with respect to the three genes.

52. In one species of *Drosophila*, the wings are normally round in shape, but you have obtained two pure lines, one of which has oval wings and the other sickle-shaped wings. Crosses between pure lines reveal the following results:

Parents		F1	
Female	Male	Female	Male
sickle	round	sickle	sickle
round	sickle	sickle	round
sickle	oval	oval	sickle

a. Provide a genetic explanation of these results, defining all allele symbols.

b. If the F_1 oval females from cross 3 are crossed with the F_1 round males from cross 2, what phenotypic proportions are expected for each sex in the progeny?

53. Mice normally have one yellow band on each hair, but variants with two or three bands are known. A female mouse having one band was crossed with a male having three bands. (Neither animal was from a pure line.) The progeny were

Females $\frac{1}{2}$ one band   Males $\frac{1}{2}$ one band

$\frac{1}{2}$ three bands $\frac{1}{2}$ two bands

a. Provide a clear explanation of the inheritance of these phenotypes.

b. In accord with your model, what would be the outcome of a cross between a three-banded daughter and a one-banded son?

54. In minks, wild types have an almost black coat. Breeders have developed many pure lines of color variants for the mink-coat industry. Two such pure lines are platinum (blue gray) and aleutian (steel gray). These lines were used in crosses, with the following results:

Cross	Parents	F_1	F_2
1	wild × platinum	wild	18 wild, 5 platinum
2	wild × aleutian	wild	27 wild, 10 aleutian
3	platinum × aleutian	wild	133 wild
			41 platinum
			46 aleutian
			17 sapphire (new)

a. Devise a genetic explanation of these three crosses. Show genotypes for the parents, the F_1, and the F_2 in the three crosses, and make sure that you show the alleles of each gene that you hypothesize for every mink.

b. Predict the F_1 and F_2 phenotypic ratios from crossing sapphire with platinum and with aleutian pure lines.

55. In *Drosophila*, an autosomal gene determines the shape of the hair, with *B* giving straight and *b* giving bent hairs. On another autosome, there is a gene of which a dominant allele *I* inhibits hair formation so that the fly is hairless (*i* has no known phenotypic effect).

a. If a straight-haired fly from a pure line is crossed with a fly from a pure-breeding hairless line known to be an inhibited bent genotype, what will the genotypes and phenotypes of the F_1 and the F_2 be?

b. What cross would give the ratio 4 hairless : 3 straight : 1 bent?

56. The following pedigree concerns eye phenotypes in *Tribolium* beetles. The solid symbols represent black eyes, the open symbols represent brown eyes, and the cross symbols (X) represent the "eyeless" phenotype, in which eyes are totally absent.

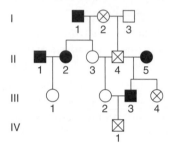

a. From these data, deduce the mode of inheritance of these three phenotypes.

b. Using defined gene symbols, show the genotype of beetle II-3.

57. A plant believed to be heterozygous for a pair of alleles *B/b* (where *B* encodes yellow and *b* encodes bronze) was selfed, and, in the progeny, there were 280 yellow and 120 bronze plants. Do these results support the hypothesis that the plant is *B/b*?

58. A plant thought to be heterozygous for two independently assorting genes (P/p ; Q/q) was selfed, and the progeny were

88 $P/-$; $Q/-$	25 p/p ; $Q/-$
32 $P/-$; q/q	14 p/p ; q/q

Do these results support the hypothesis that the original plant was P/p ; Q/q?

59. A plant of phenotype 1 was selfed, and, in the progeny, there were 100 plants of phenotype 1 and 60 plants of an alternative phenotype 2. Are these numbers compatible with expected ratios of 9 : 7, 13 : 3, and 3 : 1? Formulate a genetic hypothesis on the basis of your calculations.

60. Four homozygous recessive mutant lines of *Drosophila melanogaster* (labeled 1 through 4) showed abnormal leg coordination, which made their walking highly erratic. These lines were intercrossed; the phenotypes of the F_1 flies are shown in the following grid, in which "+" represents wild-type walking and "−" represents abnormal walking:

	1	2	3	4
1	−	+	+	+
2	+	−	−	+
3	+	−	−	+
4	+	+	+	−

a. What type of test does this analysis represent?

b. How many different genes were mutated in creating these four lines?

c. Invent wild-type and mutant symbols and write out full genotypes for all four lines and for the F_1 flies.

d. Do these data tell us which genes are linked? If not, how could linkage be tested?

e. Do these data tell us the total number of genes taking part in leg coordination in this animal?

61. Three independently isolated tryptophan-requiring mutants of haploid yeast are called *trpB*, *trpD*, and *trpE*. Cell suspensions of each are streaked on a plate of nutritional medium supplemented with just enough tryptophan to permit weak growth for a *trp* strain. The streaks are arranged in a triangular pattern so that they do not touch one another. Luxuriant growth is noted at both ends of the *trpE* streak and at one end of the *trpD* streak (see the figure below).

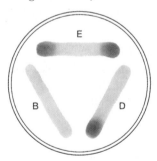

a. Do you think complementation has a role?

b. Briefly explain the pattern of luxuriant growth.

c. In what order in the tryptophan-synthesizing pathway are the enzymatic steps that are defective in mutants *trpB*, *trpD*, and *trpE*?

d. Why was it necessary to add a small amount of tryptophan to the medium to demonstrate such a growth pattern?

CHALLENGING PROBLEMS

62. A pure-breeding strain of squash that produced disk-shaped fruits (see the accompanying illustration) was crossed with a pure-breeding strain having long fruits.

The F$_1$ had disk fruits, but the F$_2$ showed a new phenotype, sphere, and was composed of the following proportions:

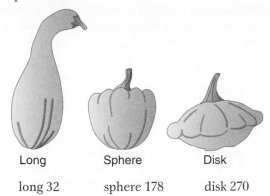

Long	Sphere	Disk
long 32	sphere 178	disk 270

Propose an explanation for these results, and show the genotypes of the P, F$_1$, and F$_2$ generations. (Illustration from P. J. Russell, *Genetics,* 3rd ed. HarperCollins, 1992.)

63. Marfan's syndrome is a disorder of the fibrous connective tissue, characterized by many symptoms, including long, thin digits; eye defects; heart disease; and long limbs. (Flo Hyman, the American volleyball star, suffered from Marfan's syndrome. She died from a ruptured aorta.)

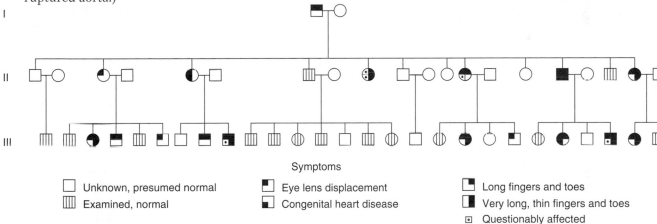

Symptoms

☐ Unknown, presumed normal	◼ Eye lens displacement	◩ Long fingers and toes
▥ Examined, normal	◼ Congenital heart disease	◪ Very long, thin fingers and toes
		⊡ Questionably affected

a. Use the pedigree above to propose a mode of inheritance for Marfan's syndrome.

b. What genetic phenomenon is shown by this pedigree?

c. Speculate on a reason for such a phenomenon.

(Illustration from J. V. Neel and W. J. Schull, *Human Heredity.* University of Chicago Press, 1954.)

64. In corn, three dominant alleles, called *A, C,* and *R,* must be present to produce colored seeds. Genotype $A/-$;$C/-$;$R/-$ is colored; all others are colorless. A colored plant is crossed with three tester plants of known genotype. With tester a/a;c/c;R/R, the colored plant produces 50 percent colored seeds; with a/a;C/C;r/r, it produces 25 percent colored; and with A/A;c/c;r/r, it produces 50 percent colored. What is the genotype of the colored plant?

65. The production of pigment in the outer layer of seeds of corn requires each of the three independently assorting genes *A, C,* and *R* to be represented by at least one dominant allele, as specified in Problem 64. The dominant allele *Pr* of a fourth independently assorting gene is required to convert the biochemical precursor into a purple pigment, and its recessive allele *pr* makes the pigment red. Plants that do not produce pigment have yellow seeds. Consider a cross of a strain of genotype A/A;C/C;R/R;pr/pr with a strain of genotype a/a;c/c;r/r;Pr/Pr.

a. What are the phenotypes of the parents?

b. What will be the phenotype of the F$_1$?

c. What phenotypes, and in what proportions, will appear in the progeny of a selfed F$_1$?

d. What progeny proportions do you predict from the testcross of an F$_1$?

66. The allele *B* gives mice a black coat, and *b* gives a brown one. The genotype *e/e* of another, independently assorting gene prevents the expression of *B* and *b*, making the coat color beige, whereas $E/-$ permits the expression of *B* and *b*. Both genes are autosomal. In the following pedigree, black symbols indicate a black coat, pink symbols indicate brown, and white symbols indicate beige.

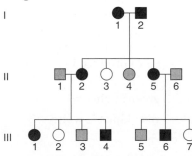

a. What is the name given to the type of gene interaction in this example?

b. What are the genotypes of the individual mice in the pedigree? (If there are alternative possibilities, state them.)

67. A researcher crosses two white-flowered lines of *Antirrhinum* plants as follows and obtains the following results:

pure line 1 × pure line 2

↓

F$_1$ all white

F$_1$ × F$_1$

↓

F$_2$ 131 white

29 red

a. Deduce the inheritance of these phenotypes; use clearly defined gene symbols. Give the genotypes of the parents, F$_1$, and F$_2$.

b. Predict the outcome of crosses of the F$_1$ with each parental line.

68. Assume that two pigments, red and blue, mix to give the normal purple color of petunia petals. Separate biochemical pathways synthesize the two pigments, as shown in the top two rows of the accompanying diagram. "White" refers to compounds that are not pigments. (Total lack of pigment results in a white petal.) Red pigment forms from a yellow intermediate that is normally at a concentration too low to color petals.

pathway I   ···⟶ white$_1$ $\xrightarrow{E}$ blue

pathway II   ···⟶ white$_2$ $\xrightarrow{A}$ yellow $\xrightarrow{B}$ red

⟰
C ⋮

pathway III   ···⟶ white$_3$ $\xrightarrow{D}$ white$_4$

A third pathway, whose compounds do not contribute pigment to petals, normally does not affect the blue and red pathways, but, if one of its intermediates (white$_3$) should build up in concentration, it can be converted into the yellow intermediate of the red pathway.

In the diagram, the letters A through E represent enzymes; their corresponding genes, all of which are unlinked, may be symbolized by the same letters.

Assume that wild-type alleles are dominant and encode enzyme function and that recessive alleles result in a lack of enzyme function. Deduce which combinations of true-breeding parental genotypes could be crossed to produce F$_2$ progeny in the following ratios:

a. 9 purple : 3 green : 4 blue

b. 9 purple : 3 red : 3 blue : 1 white

c. 13 purple : 3 blue

d. 9 purple : 3 red : 3 green : 1 yellow

(**Note:** Blue mixed with yellow makes green; assume that no mutations are lethal.)

69. The flowers of nasturtiums (*Tropaeolum majus*) may be single (S), double (D), or superdouble (Sd). Superdoubles are female sterile; they originated from a double-flowered variety. Crosses between varieties gave the progeny listed in the following table, in which *pure* means "pure breeding."

Cross	Parents	Progeny
1	pure S × pure D	All S
2	cross 1 F$_1$ × cross 1 F$_1$	78 S : 27 D
3	pure D × Sd	112 Sd : 108 D
4	pure S × Sd	8 Sd : 7 S
5	pure D × cross 4 Sd progeny	18 Sd : 19 S
6	pure D × cross 4 S progeny	14 D : 16 S

Using your own genetic symbols, propose an explanation for these results, showing

a. all the genotypes in each of the six rows.

b. the proposed origin of the superdouble.

70. In a certain species of fly, the normal eye color is red (R). Four abnormal phenotypes for eye color were found: two were yellow (Y1 and Y2), one was brown (B), and one was orange (O). A pure line was established for each phenotype, and all possible combinations of the pure lines were crossed. Flies of each F$_1$ were intercrossed to produce an F$_2$. The F$_1$ and the F$_2$ flies are shown within the following square; the pure lines are given at the top and at the left-hand side.

		Y1	Y2	B	O
Y1	F$_1$	all Y	all R	all R	all R
	F$_2$	all Y	9 R	9 R	9 R
			7 Y	4 Y	4 O
				3 B	3 Y
Y2	F$_1$		all Y	all R	all R
	F$_2$		all Y	9 R	9 R
				4 Y	4 Y
				3 B	3 O
B	F$_1$			all B	all R
	F$_2$			all B	9 R
					4 O
					3 B
O	F$_1$				all O
	F$_2$				all O

a. Define your own symbols and list the genotypes of all four pure lines.

b. Show how the F$_1$ phenotypes and the F$_2$ ratios are produced.

c. Show a biochemical pathway that explains the genetic results, indicating which gene controls which enzyme.

71. In common wheat, *Triticum aestivum*, kernel color is determined by multiply duplicated genes, each with an *R* and an *r* allele. Any number of *R* alleles will give red, and a complete lack of *R* alleles will give the white phenotype. In one cross between a red pure line and a white pure line, the F_2 was $\frac{63}{64}$ red and $\frac{1}{64}$ white.

a. How many *R* genes are segregating in this system?

b. Show the genotypes of the parents, the F_1, and the F_2.

c. Different F_2 plants are backcrossed with the white parent. Give examples of genotypes that would give the following progeny ratios in such backcrosses: (1) 1 red : 1 white, (2) 3 red : 1 white, (3) 7 red : 1 white.

d. What is the formula that generally relates the number of segregating genes to the proportion of red individuals in the F_2 in such systems?

72. The following pedigree shows the inheritance of deaf-mutism.

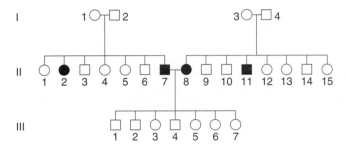

a. Provide an explanation for the inheritance of this rare condition in the two families in generations I and II, showing the genotypes of as many persons as possible; use symbols of your own choosing.

b. Provide an explanation for the production of only normal persons in generation III, making sure that your explanation is compatible with the answer to part *a*.

73. The pedigree below is for blue sclera (bluish thin outer wall of the eye) and brittle bones.

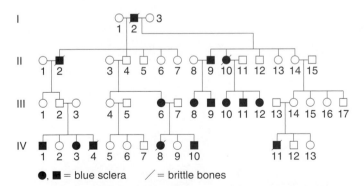

●, ■ = blue sclera   ╱ = brittle bones

a. Are these two abnormalities caused by the same gene or by separate genes? State your reasons clearly.

b. Is the gene (or genes) autosomal or sex-linked?

c. Does the pedigree show any evidence of incomplete penetrance or expressivity? If so, make the best calculations that you can of these measures.

74. Workers of the honeybee line known as *Brown* (nothing to do with color) show what is called "hygienic behavior"; that is, they uncap hive compartments containing dead pupae and then remove the dead pupae. This behavior prevents the spread of infectious bacteria through the colony. Workers of the *Van Scoy* line, however, do not perform these actions, and therefore this line is said to be "nonhygienic." When a queen from the *Brown* line was mated with *Van Scoy* drones, all the F_1 were nonhygienic. When drones from this F_1 inseminated a queen from the *Brown* line, the progeny behaviors were as follows:

$\frac{1}{4}$ hygienic

$\frac{1}{4}$ uncapping but no removing of pupae

$\frac{1}{2}$ nonhygienic

However, when the compartment of dead pupae was uncapped by the beekeeper and the nonhygienic honeybees were examined further, about half the bees were found to remove the dead pupae, but the other half did not.

a. Propose a genetic hypothesis to explain these behavioral patterns.

b. Discuss the data in relation to epistasis, dominance, and environmental interaction.

(**Note:** Workers are sterile, and all bees from one line carry the same alleles.)

75. The normal color of snapdragons is red. Some pure lines showing variations of flower color have been found. When these pure lines were crossed, they gave the following results (see the table):

Cross	Parents	F_1	F_2
1	orange × yellow	orange	3 orange : 1 yellow
2	red × orange	red	3 red : 1 orange
3	red × yellow	red	3 red : 1 yellow
4	red × white	red	3 red : 1 white
5	yellow × white	red	9 red : 3 yellow : 4 white
6	orange × white	red	9 red : 3 orange : 4 white
7	red × white	red	9 red : 3 yellow : 4 white

a. Explain the inheritance of these colors.

b. Write the genotypes of the parents, the F_1, and the F_2.

76. Consider the following F_1 individuals in different species and the F_2 ratios produced by selfing:

F_1	Phenotypic ratio in the F_2		
1 cream	$\frac{12}{16}$ cream	$\frac{3}{16}$ black	$\frac{1}{16}$ gray
2 orange	$\frac{9}{16}$ orange	$\frac{7}{16}$ yellow	
3 black	$\frac{3}{16}$ black	$\frac{3}{16}$ white	
4 solid red	$\frac{9}{16}$ solid red	$\frac{3}{16}$ mottled red	$\frac{4}{16}$ small red dots

If each F_1 were testcrossed, what phenotypic ratios would result in the progeny of the testcross?

77. To understand the genetic basis of locomotion in the diploid nematode *Caenorhabditis elegans*, recessive mutations were obtained, all making the worm "wiggle" ineffectually instead of moving with its usual smooth gliding motion. These mutations presumably affect the nervous or muscle systems. Twelve homozygous mutants were intercrossed, and the F_1 hybrids were examined to see if they wiggled. The results were as follows, where a plus sign means that the F_1 hybrid was wild type (gliding) and "w" means that the hybrid wiggled.

	1	2	3	4	5	6	7	8	9	10	11	12
1	w	+	+	+	w	+	+	+	+	+	+	+
2		w	+	+	+	w	+	w	+	w	+	+
3			w	w	+	+	+	+	+	+	+	+
4				w	+	+	+	+	+	+	+	+
5					w	+	+	+	+	+	+	+
6						w	+	w	+	w	+	+
7							w	+	+	+	w	w
8								w	+	w	+	+
9									w	+	+	+
10										w	+	+
11											w	w
12												w

a. Explain what this experiment was designed to test.

b. Use this reasoning to assign genotypes to all 12 mutants.

c. Explain why the phenotype of the F_1 hybrids between mutants 1 and 2 differed from that of the hybrids between mutants 1 and 5.

78. A geneticist working on a haploid fungus makes a cross between two slow-growing mutants called *mossy* and *spider* (referring to the abnormal appearance of the colonies). Tetrads from the cross are of three types (A, B, C), but two of them contain spores that do not germinate.

Spore	A	B	C
1	wild type	wild type	spider
2	wild type	spider	spider
3	no germination	mossy	mossy
4	no germination	no germination	mossy

Devise a model to explain these genetic results, and propose a molecular basis for your model.

DNA: Structure and Replication 7

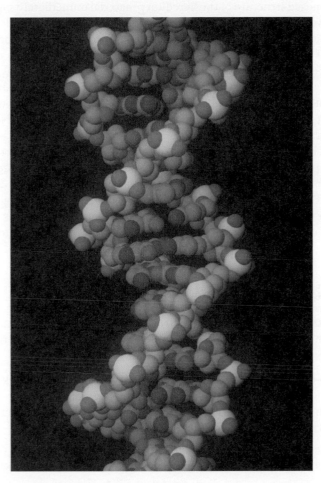

Computer model of DNA. [*Photo Researchers RM/Getty Images.*]

James Watson (an American microbial geneticist) and Francis Crick (an English physicist) solved the structure of DNA in 1953. Their model of the structure of DNA was revolutionary. It proposed a definition for the gene in chemical terms and, in doing so, paved the way for an understanding of gene action and heredity at the molecular level. A measure of the importance of their discovery is that the double-helical structure has become a cultural icon that is seen more and more frequently in paintings, in sculptures, and even in playgrounds (Figure 7-1).

The story begins in the first half of the twentieth century, when the results of several experiments led scientists to conclude that DNA is the genetic material, not some other biological molecule such as a carbohydrate, protein, or lipid.

Figure 7-1 [Neil Grant/Alamy]

DNA is a simple molecule made up of only four different building blocks (the four nucleotide bases). It was thus necessary to understand how this very simple molecule could be the blueprint for the incredible diversity of organisms on Earth.

The model of the double helix proposed by Watson and Crick was built on the results of scientists before them. They relied on earlier discoveries of the chemical composition of DNA and the ratios of its bases. In addition, X-ray diffraction pictures of DNA revealed to the trained eye that DNA is a helix of precise dimensions. Watson and Crick concluded that DNA is a double helix composed of two strands of linked nucleotide bases that wind around each other.

The proposed structure of the hereditary material immediately suggested how it could serve as a blueprint and how this blueprint could be passed down through the generations. First, the information for making an organism is encoded in the sequence of the nucleotide bases composing the two DNA strands of the helix. Second, because of the rules of base complementarity discovered by Watson and Crick, the sequence of one strand dictates the sequence of the other strand. In this way, the genetic information in the DNA sequence can be passed down from one generation to the next by having each of the separated strands of DNA serve as a template for producing new copies of the molecule.

In this chapter, we focus on DNA, its structure, and the production of DNA copies in a process called replication. Precisely how DNA is replicated is still an active area of research more than 50 years after the discovery of the double helix. Our current understanding of the mechanism of replication gives a central role to a protein machine, called the replisome. This complex of proteins coordinates the numerous reactions that are necessary for the rapid and accurate replication of DNA.

7.1 DNA: The Genetic Material

Before we see how Watson and Crick solved the structure of DNA, let's review what was known about genes and DNA at the time that they began their historic collaboration:

1. Genes—the hereditary "factors" described by Mendel—were known to be associated with specific traits, but their physical nature was not understood. Similarly, mutations were known to alter gene function, but precisely what a mutation is also was not understood.

2. The one-gene–one-polypeptide hypothesis (described in Chapter 6) postulated that genes control the structure of proteins and other polypeptides.

3. Genes were known to be carried on chromosomes.

4. The chromosomes were found to consist of DNA and protein.

5. The results of a series of experiments beginning in the 1920s revealed that DNA is the genetic material. These experiments, described next, showed that bacterial cells that express one phenotype can be transformed into cells that express a different phenotype and that the transforming agent is DNA.

Discovery of transformation

Frederick Griffith made a puzzling observation in the course of experiments performed in 1928 on the bacterium *Streptococcus pneumoniae*. This bacterium, which causes pneumonia in humans, is normally lethal in mice. However, some strains of this bacterial species have evolved to be less virulent (less able to cause disease or death). Griffith's experiments are summarized in

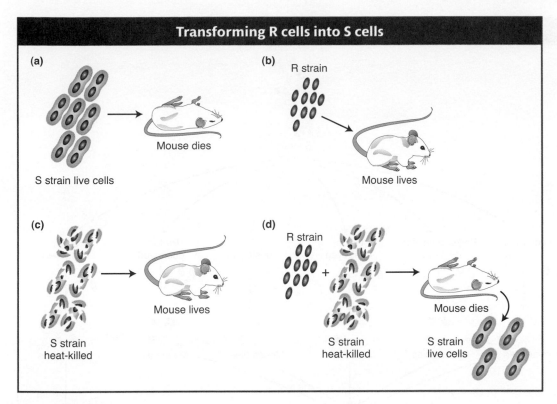

Transforming R cells into S cells

(a) S strain live cells → Mouse dies

(b) R strain → Mouse lives

(c) S strain heat-killed → Mouse lives

(d) R strain + S strain heat-killed → Mouse dies → S strain live cells

Figure 7-2 The presence of heat-killed S cells transforms live R cells into live S cells. (a) Mouse dies after injection with the virulent S strain. (b) Mouse survives after injection with the R strain. (c) Mouse survives after injection with heat-killed S strain. (d) Mouse dies after injection with a mixture of heat-killed S strain and live R strain. Live S cells were isolated from the dead mouse, indicating that the heat-killed S strain somehow transforms the R strain into the virulent S strain. [*After G. S. Stent and R. Calendar, Molecular Genetics, 2nd ed. Copyright 1978 by W. H. Freeman and Company. After R. Sager and F. J. Ryan, Cell Heredity. Wiley, 1961.*]

Figure 7-2. In these experiments, Griffith used two strains that are distinguishable by the appearance of their colonies when grown in laboratory cultures. One strain was a normal virulent type deadly to most laboratory animals. The cells of this strain are enclosed in a polysaccharide capsule, giving colonies a smooth appearance; hence, this strain is identified as S. Griffith's other strain was a mutant nonvirulent type that grows in mice but is not lethal. In this strain, the polysaccharide coat is absent, giving colonies a rough appearance; this strain is called *R*.

Griffith killed some virulent cells by boiling them. He then injected the heat-killed cells into mice. The mice survived, showing that the carcasses of the cells do not cause death. However, mice injected with a mixture of heat-killed virulent cells and live nonvirulent cells did die. Furthermore, live cells could be recovered from the dead mice; these cells gave smooth colonies and were virulent on subsequent injection. Somehow, the cell debris of the boiled S cells had converted the live R cells into live S cells. The process, already discussed in Chapter 5, is called *transformation.*

The next step was to determine which chemical component of the dead donor cells had caused this transformation. This substance had changed the genotype of the recipient strain and therefore might be a candidate for the hereditary material. This problem was solved by experiments conducted in 1944 by Oswald Avery and two colleagues, Colin MacLeod and Maclyn McCarty

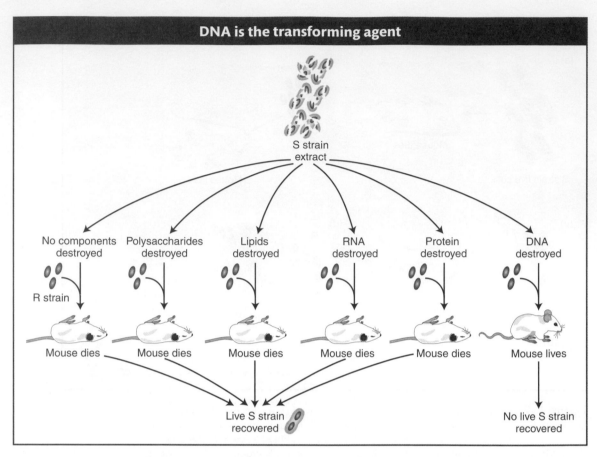

Figure 7-3 DNA is the agent transforming the R strain into virulence. If the DNA in an extract of heat-killed S-strain cells is destroyed, then mice injected with a mixture of the heat-killed cells and the live nonvirulent R-strain cells are no longer killed.

(Figure 7-3). Their approach to the problem was to chemically destroy all the major categories of chemicals in an extract of dead cells one at a time and find out if the extract had lost the ability to transform. The virulent cells had a smooth polysaccharide coat, whereas the nonvirulent cells did not; hence, polysaccharides were an obvious candidate for the transforming agent. However, when polysaccharides were destroyed, the mixture could still transform. Proteins, fats, and ribonucleic acids (RNAs) were all similarly shown not to be the transforming agent. The mixture lost its transforming ability only when the donor mixture was treated with the enzyme deoxyribonuclease (DNase), which breaks up DNA. These results strongly implicate DNA as the genetic material. It is now known that fragments of the transforming DNA that confer virulence enter the bacterial chromosome and replace their counterparts that confer nonvirulence.

> **Message** The demonstration that DNA is the transforming principle was the first demonstration that genes (the hereditary material) are composed of DNA.

Hershey–Chase experiment

The experiments conducted by Avery and his colleagues were definitive, but many scientists were very reluctant to accept DNA (rather than proteins) as the genetic material. After all, how could such a low-complexity molecule as DNA encode the diversity of life on this planet? Alfred Hershey and Martha Chase provided additional evidence in 1952 in an experiment that made use of phage T2, a virus that infects bacteria. They reasoned that infecting phage must inject into the bacterium the specific information that dictates the reproduction of new viral particles. If they could find out what material the phage was injecting into the bacterial host, they would have determined the genetic material of phages.

The phage is relatively simple in molecular constitution. Most of its structure is protein, with DNA contained inside the protein sheath of its "head." Hershey and Chase decided to give the DNA and protein distinct labels by using radioisotopes so that they could track the two materials during infection. Phosphorus is not found in the amino acids building blocks of proteins but is an integral part of DNA; conversely, sulfur is present in proteins but never in DNA. Hershey and Chase incorporated the radioisotope of phosphorus (^{32}P) into phage DNA and that of sulfur (^{35}S) into the proteins of a separate phage culture. As shown in Figure 7-4, they then infected two *E. coli* cultures with many virus particles per cell: one *E. coli* culture received phage labeled with ^{32}P and the other received phage labeled with ^{35}S. After allowing sufficient time for infection to take place, they sheared the empty phage carcasses (called *ghosts*) off the bacterial cells by agitation in a kitchen blender. They separated the bacterial cells from the phage ghosts in a centrifuge and then measured the radioactivity in the two fractions. When the ^{32}P-labeled phages were used to infect *E. coli,* most of the radioactivity ended up inside the bacterial cells, indicating that the phage DNA entered the cells. When the ^{35}S-labeled phages were used, most of the radioactive material ended up in the phage ghosts, indicating that the phage protein never entered the bacterial cell. The conclusion is inescapable: DNA is the hereditary material. The phage proteins are mere structural packaging that is discarded after delivering the viral DNA to the bacterial cell.

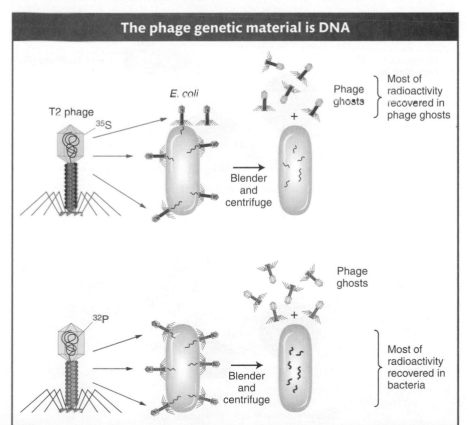

The phage genetic material is DNA

Figure 7-4 The Hershey–Chase experiment demonstrated that the genetic material of phages is DNA, not protein. The experiment uses two sets of T2 bacteriophage. In one set, the protein coat is labeled with radioactive sulfur (^{35}S), not found in DNA. In the other set, the DNA is labeled with radioactive phosphorus (^{32}P), not found in amino acids. Only the ^{32}P is injected into the *E. coli,* indicating that DNA is the agent necessary for the production of new phages.

7.2 The DNA Structure

Even before the structure of DNA was elucidated, genetic studies indicated that the hereditary material must have three key properties:

1. Because essentially every cell in the body of an organism has the same genetic makeup, faithful replication of the genetic material at every cell division is crucial. Thus, the structural features of DNA *must allow faithful replication*. These structural features will be considered later in this chapter.

2. Because it must encode the constellation of proteins expressed by an organism, the genetic material *must have informational content*. How the information coded in DNA is deciphered to produce proteins will be the subject of Chapters 8 and 9.

3. Because hereditary changes, called mutations, provide the raw material for evolutionary selection, the genetic material *must be able to change* on rare occasion. Nevertheless, the structure of DNA must be stable so that organisms can rely on its encoded information. We will consider the mechanisms of mutation in Chapter 16.

Figure 7-5 These nucleotides, two with purine bases and two with pyrimidine bases, are the fundamental building blocks of DNA. The sugar is called *deoxyribose* because it is a variation of a common sugar, ribose, that has one more oxygen atom (position indicated by the red arrow).

DNA structure before Watson and Crick

Consider the discovery of the double-helical structure of DNA by Watson and Crick as the solution to a complicated three-dimensional puzzle. To solve this puzzle, Watson and Crick used a process called "model building" in which they assembled the results of earlier and ongoing experiments (the puzzle pieces) to

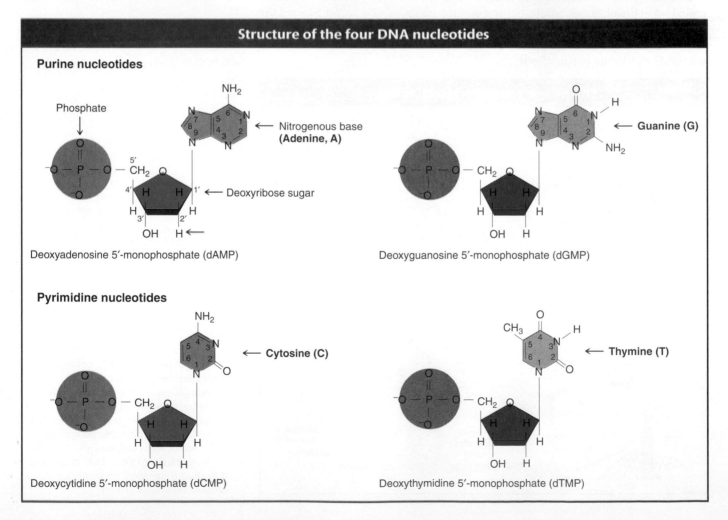

Structure of the four DNA nucleotides

Purine nucleotides

Phosphate

Nitrogenous base
(Adenine, A)

Deoxyribose sugar

Deoxyadenosine 5′-monophosphate (dAMP)

Guanine (G)

Deoxyguanosine 5′-monophosphate (dGMP)

Pyrimidine nucleotides

Cytosine (C)

Deoxycytidine 5′-monophosphate (dCMP)

Thymine (T)

Deoxythymidine 5′-monophosphate (dTMP)

Table 7-1   Molar Properties of Bases* in DNAs from Various Sources

Organism	Tissue	Adenine	Thymine	Guanine	Cytosine	$\dfrac{A + T}{G + C}$
Escherichia coli (K12)	–	26.0	23.9	24.9	25.2	1.00
Diplococcus pneumoniae	–	29.8	31.6	20.5	18.0	1.59
Mycobacterium tuberculosis	–	15.1	14.6	34.9	35.4	0.42
Yeast	–	31.3	32.9	18.7	17.1	1.79
Paracentrotus lividus (sea urchin)	Sperm	32.8	32.1	17.7	18.4	1.85
Herring	Sperm	27.8	27.5	22.2	22.6	1.23
Rat	Bone marrow	28.6	28.4	21.4	21.5	1.33
Human	Thymus	30.9	29.4	19.9	19.8	1.52
Human	Liver	30.3	30.3	19.5	19.9	1.53
Human	Sperm	30.7	31.2	19.3	18.8	1.62

*Defined as moles of nitrogenous constituents per 100 g-atoms phosphate in hydrolysate.
Source: E. Chargaff and J. Davidson, eds., *The Nucleic Acids*. Academic Press, 1955.

form the three-dimensional puzzle (the double-helix model). To understand how they did so, we first need to know what pieces of the puzzle were available to Watson and Crick in 1953.

The building blocks of DNA   The first piece of the puzzle was knowledge of the basic building blocks of DNA. As a chemical, DNA is quite simple. It contains three types of chemical components: (1) **phosphate,** (2) a sugar called **deoxyribose,** and (3) four nitrogenous **bases**—adenine, guanine, cytosine, and thymine. The sugar in DNA is called "deoxyribose" because it has only a hydrogen atom (H) at the 2′-carbon atom, unlike ribose (a component of RNA), which has a hydroxyl (OH) group at that position. Two of the bases, adenine and guanine, have a double-ring structure characteristic of a type of chemical called a **purine.** The other two bases, cytosine and thymine, have a single-ring structure of a type called a **pyrimidine.**

The carbon atoms in the bases are assigned numbers for ease of reference. The carbon atoms in the sugar group also are assigned numbers—in this case, the number is followed by a prime (1′, 2′, and so forth).

The chemical components of DNA are arranged into groups called **nucleotides,** each composed of a phosphate group, a deoxyribose sugar molecule, and any one of the four bases (Figure 7-5). It is convenient to refer to each nucleotide by the first letter of the name of its base: A, G, C, or T. The nucleotide with the adenine base is called deoxyadenosine 5′-monophosphate, where the 5′ refers to the position of the carbon atom in the sugar ring to which the single (mono) phosphate group is attached.

Chargaff's rules of base composition   The second piece of the puzzle used by Watson and Crick came from work done several years earlier by Erwin Chargaff. Studying a large selection of DNAs from different organisms (Table 7-1), Chargaff established certain empirical rules about the amounts of each type of nucleotide found in DNA:

1.   The total amount of pyrimidine nucleotides (T + C) always equals the total amount of purine nucleotides (A + G).

2.   The amount of T always equals the amount of A, and the amount of C always equals the amount of G. But the amount of A + T is not necessarily equal to the amount of G + C, as can be seen in the right-hand column of Table 7-1. This ratio varies among different organisms but is virtually the same in different tissues of the same organism.

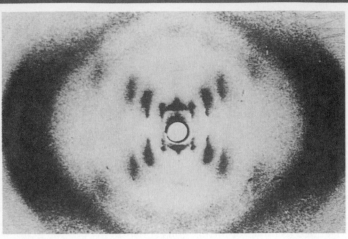

Rosalind Franklin's critical experimental result

Figure 7-6 Rosalind Franklin (*left*) and her X-ray diffraction pattern of DNA (*right*). [*(Left) Science Source; (right) Rosalind Franklin/Science Source/ Photo Researchers.*]

X-ray diffraction analysis of DNA   The third and most controversial piece of the puzzle came from X-ray diffraction data on DNA structure that were collected by Rosalind Franklin when she was in the laboratory of Maurice Wilkins (Figure 7-6). In such experiments, X rays are fired at DNA fibers, and the scatter of the rays from the fibers is observed by catching the rays on photographic film, on which the X rays produce spots. The angle of scatter represented by each spot on the film gives information about the position of an atom or certain groups of atoms in the DNA molecule. This procedure is not simple to carry out (or to explain), and the interpretation of the spot patterns requires complex mathematical treatment that is beyond the scope of this text. The available data suggested that DNA is long and skinny and that it has two similar parts that are parallel to each other and run along the length of the molecule. The X-ray data showed the molecule to be helical (spiral-like). Unknown to Rosalind Franklin, her best X-ray picture was shown to Watson and Crick by Maurice Wilkins, and it was this crucial piece of the puzzle that allowed them to deduce the three-dimensional structure that could account for the X-ray spot patterns.

The first model of DNA

Figure 7-7 James Watson and Francis Crick with their DNA model. [*A. Barrington Brown / Science Source.*]

The double helix

A 1953 paper by Watson and Crick in the journal *Nature* began with two sentences that ushered in a new age of biology: "We wish to suggest a structure for the salt of deoxyribose nucleic acid (D.N.A.). This structure has novel features which are of considerable biological interest." The structure of DNA had been a subject of great debate since the experiments of Avery and co-workers in 1944. As we have seen, the general composition of DNA was known, but how the parts fit together was not known. The structure had to fulfill the main requirements for a hereditary molecule: the ability to store information, the ability to be replicated, and the ability to mutate.

The three-dimensional structure derived by Watson and Crick is composed of two side-by-side chains ("strands") of nucleotides twisted into the shape of a **double helix** (Figure 7-7). The two nucleotide strands are held together by hydrogen bonds between the bases of each strand, forming a structure like a spiral staircase (Figure 7-8a). The backbone of each strand is formed of alternating phosphate and deoxyribose sugar units that are connected by phosphodiester linkages (Figure 7-8b). We can use these linkages to describe how a nucleotide chain is organized. As already mentioned,

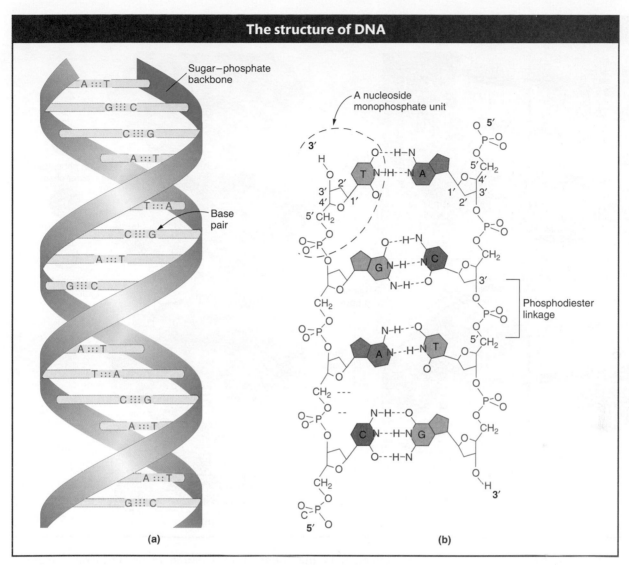

The structure of DNA

Figure 7-8 (a) A simplified model showing the helical structure of DNA. The sticks represent base pairs, and the ribbons represent the sugar–phosphate backbones of the two antiparallel chains. (b) An accurate chemical diagram of the DNA double helix, unrolled to show the sugar–phosphate backbones (blue) and base-pair rungs (red). The backbones run in opposite directions; the 5′ and 3′ ends are named for the orientation of the 5′ and 3′ carbon atoms of the sugar rings. Each base pair has one purine base, adenine (A) or guanine (G), and one pyrimidine base, thymine (T) or cytosine (C), connected by hydrogen bonds (dashed lines). [*From R. E. Dickerson, "The DNA Helix and How It Is Read." Copyright 1983 by Scientific American, Inc. All rights reserved.*]

the carbon atoms of the sugar groups are numbered 1′ through 5′. A phosphodiester linkage connects the 5′-carbon atom of one deoxyribose to the 3′-carbon atom of the adjacent deoxyribose. Thus, each sugar–phosphate backbone is said to have a 5′-to-3′ polarity, or direction, and understanding this polarity is essential in understanding how DNA fulfills its roles. In the double-stranded DNA molecule, the two backbones are in opposite, or **antiparallel,** orientation (see Figure 7-8b).

Each base is attached to the 1′-carbon atom of a deoxyribose sugar in the backbone of each strand and faces inward toward a base on the other strand. Hydrogen bonds between pairs of bases hold the two strands of the DNA molecule together. The hydrogen bonds are indicated by dashed lines in Figure 7-8b.

Two representations of the DNA double helix

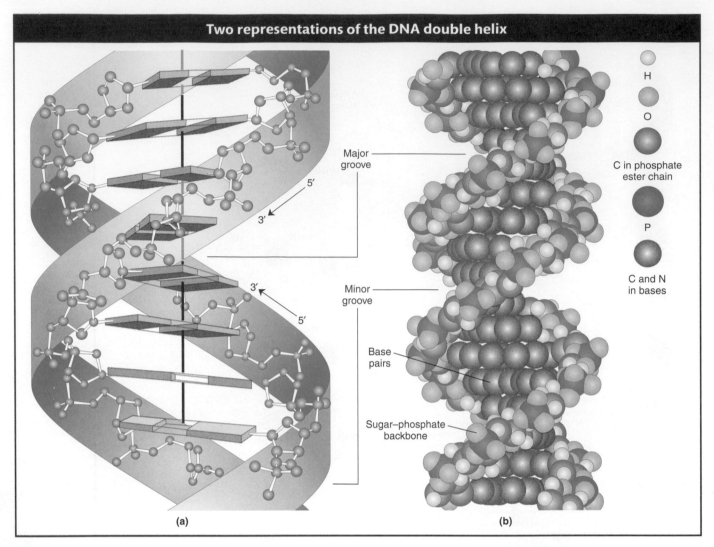

Major groove

Minor groove

Base pairs

Sugar–phosphate backbone

H

O

C in phosphate ester chain

P

C and N in bases

(a)

(b)

Figure 7-9 The ribbon diagram (a) highlights the stacking of the base pairs, whereas the space-filling model (b) shows the major and minor grooves. [*(b) From C. Yanofsky, "Gene Structure and Protein Structure." Copyright 1967 by Scientific American, Inc. All rights reserved.*]

Two complementary nucleotide strands paired in an antiparallel manner automatically assume a double-helical conformation (Figure 7-9), mainly through the interaction of the base pairs. The base pairs, which are flat planar structures, stack on top of one another at the center of the double helix (see Figure 7-9a). Stacking adds to the stability of the DNA molecule by excluding water molecules from the spaces between the base pairs. The most stable form that results from base stacking is a double helix with two distinct sizes of grooves running in a spiral: the **major groove** and the **minor groove,** which can be seen in both the ribbon and the space-filling models of Figure 7-9a and 7-9b. Most DNA–protein associations are in major grooves. A single strand of nucleotides has no helical structure; the helical shape of DNA depends entirely on the pairing and stacking of the bases in the antiparallel strands. DNA is a right-handed helix; in other words, it has the same structure as that of a screw that would be screwed into place by using a clockwise turning motion.

Base pairing in DNA

Pyrimidine + pyrimidine: DNA too thin

Purine + purine: DNA too thick

Purine + pyrimidine: thickness compatible with X-ray data

Figure 7-10 The pairing of purines with pyrimidines accounts exactly for the diameter of the DNA double helix determined from X-ray data. That diameter is indicated by the vertical dashed lines. [*From R. E. Dickerson, "The DNA Helix and How It Is Read." Copyright 1983 by Scientific American, Inc. All rights reserved.*]

The double helix accounted nicely for the X-ray data and successfully accounted for Chargaff's data. By studying models that they made of the structure, Watson and Crick realized that the observed radius of the double helix (known from the X-ray data) would be explained if a purine base always pairs (by hydrogen bonding) with a pyrimidine base (Figure 7-10). Such pairing would account for the $(A + G) = (T + C)$ regularity observed by Chargaff, but it would predict four possible pairings: T···A, T···G, C···A, and C···G. Chargaff's data, however, indicate that T pairs only with A, and C pairs only with G. Watson and Crick concluded that each base pair consists of one purine base and one pyrimidine base, paired according to the following rule: G pairs with C, and A pairs with T.

Note that the G–C pair has three hydrogen bonds, whereas the A–T pair has only two (see Figure 7-8b). We would predict that DNA containing many G–C pairs would be more stable than DNA containing many A–T pairs. In fact, this prediction is confirmed. Heat causes the two strands of DNA double helix to separate (a process called DNA melting or DNA denaturation); DNAs with higher $G + C$ content can be shown to require higher temperatures to melt because of the greater attraction of the G–C pairing.

> **Message** DNA is a double helix composed of two nucleotide chains held together by complementary pairing of A with T and G with C.

Watson and Crick's discovery of the structure of DNA is considered by some to be the most important biological discovery of the twentieth century and led to their being awarded the Nobel Prize with Maurice Wilkins in 1962 (Rosalind Franklin died of cancer in 1958 and the prize is not awarded posthumously). The reason that this discovery is considered so important is that the double helix, in addition to being consistent with earlier data about DNA structure, fulfilled the three requirements for a hereditary substance.

1. The double-helical structure suggested how the genetic material might determine the structure of proteins. Perhaps the *sequence* of nucleotide pairs in DNA dictates the sequence of amino acids in the protein specified by that gene. In other words, some sort of **genetic code** may write information in DNA as a sequence of nucleotides and then translate it into a different language of amino acid sequences in protein. Just how it is done is the subject of Chapter 9.

2. If the base sequence of DNA specifies the amino acid sequence, then mutation is possible by the substitution of one type of base for another at one or more positions. Mutations will be discussed in Chapter 16.

3. As Watson and Crick stated in the concluding words of their 1953 *Nature* paper that reported the double-helical structure of DNA: "It has not escaped our notice that the specific pairing we have postulated immediately suggests a possible copying mechanism for the genetic material." To geneticists at the time, the meaning of this statement was clear, as we see in the next section.

7.3 Semiconservative Replication

The copying mechanism to which Watson and Crick referred is called semiconservative replication and is diagrammed in Figure 7-11. The sugar–phosphate backbones are represented by thick

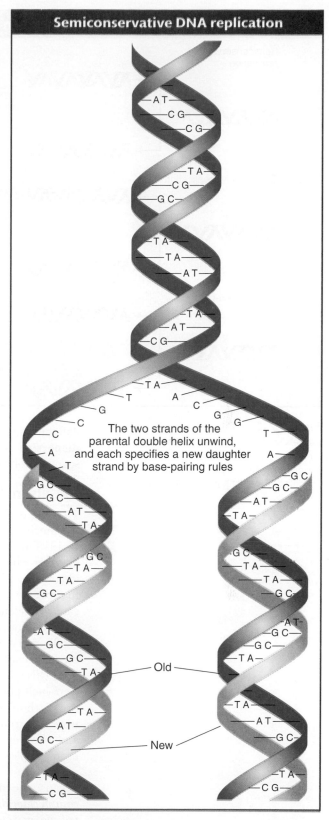

Semiconservative DNA replication

The two strands of the parental double helix unwind, and each specifies a new daughter strand by base-pairing rules

Old

New

Figure 7-11 The semiconservative model of DNA replication proposed by Watson and Crick is based on the hydrogen-bonded specificity of the base pairs. Parental strands, shown in blue, serve as templates for polymerization. The newly polymerized strands, shown in gold, have base sequences that are complementary to their respective templates.

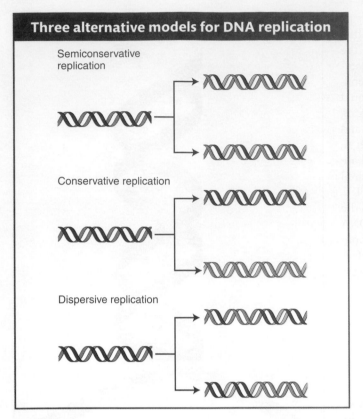

Three alternative models for DNA replication

Semiconservative replication

Conservative replication

Dispersive replication

Figure 7-12 Of three alternative models for DNA replication, the Watson–Crick model of DNA structure would produce the first (semiconservative) model. Gold lines represent the newly synthesized strands.

ribbons, and the sequence of base pairs is random. Let's imagine that the double helix is analogous to a zipper that unzips, starting at one end (see Figure 7-11). We can see that, if this zipper analogy is valid, the unwinding of the two strands will expose single bases on each strand. Each exposed base has the potential to pair with free nucleotides in solution. Because the DNA structure imposes strict pairing requirements, each exposed base will pair only with its **complementary base,** A with T and G with C. Thus, each of the two single strands will act as a **template,** or mold, to direct the assembly of complementary bases to re-form a double helix identical with the original. The newly added nucleotides are assumed to come from a pool of free nucleotides that must be present in the cell.

If this model is correct, then each daughter molecule should contain one parental nucleotide chain and one newly synthesized nucleotide chain. However, a little thought shows that there are at least three different ways in which a parental DNA molecule might be related to the daughter molecules. These hypothetical modes of replication are called semiconservative (the Watson–Crick model), conservative, and dispersive (Figure 7-12). In **semiconservative replication,** the double helix of each daughter DNA molecule contains one strand from the original DNA molecule and one newly synthesized strand. However, in **conservative replication,** the parent DNA molecule is conserved, and a single daughter double helix is produced consisting of two newly synthesized strands. In **dispersive replication,** daughter molecules consist of strands *each* containing segments of *both* parental DNA and newly synthesized DNA.

Meselson–Stahl experiment

The first problem in understanding DNA replication was to figure out whether the mechanism of replication was semiconservative, conservative, or dispersive. In 1958, two young scientists, Matthew Meselson and Franklin Stahl, set out to discover which of these possibilities correctly described DNA replication. Their idea was to allow parental DNA molecules containing nucleotides of one density to replicate in medium containing nucleotides of different density. If DNA replicated semiconservatively, the daughter molecules should be half old and half new and therefore of intermediate density.

To carry out their experiment, Meselson and Stahl grew *E. coli* cells in a medium containing the heavy isotope of nitrogen (^{15}N) rather than the normal light (^{14}N) form. This isotope was inserted into the nitrogen bases, which then were incorporated into newly synthesized DNA strands. After many cell divisions in ^{15}N, the DNA of the cells were well labeled with the heavy isotope. The cells were then removed from the ^{15}N medium and put into a ^{14}N medium; after one and two cell divisions, samples were taken and the DNA was isolated from each sample.

Meselson and Stahl were able to distinguish DNA of different densities because the molecules can be separated from one another by a procedure called *cesium chloride gradient centrifugation.* If cesium chloride (CsCl) is spun in a centrifuge at tremendously high speeds (50,000 rpm) for many hours, the cesium and chloride ions tend to be pushed by centrifugal force toward the bottom of the tube. Ultimately, a gradient of ions is established in the tube, with the highest ion concentration, or density, at the bottom. DNA centrifuged with the cesium chlo-

Figure 7-13 The Meselson–Stahl experiment demonstrates that DNA is copied by semiconservative replication. DNA centrifuged in a cesium chloride (CsCl) gradient will form bands according to its density. (a) When the cells grown in ^{15}N are transferred to a ^{14}N medium, the first generation produces a single intermediate DNA band and the second generation produces two bands: one intermediate and one light. This result matches the predictions of the semiconservative model of DNA replication. (b and c) The results predicted for conservative and dispersive replication, shown here, were *not* found.

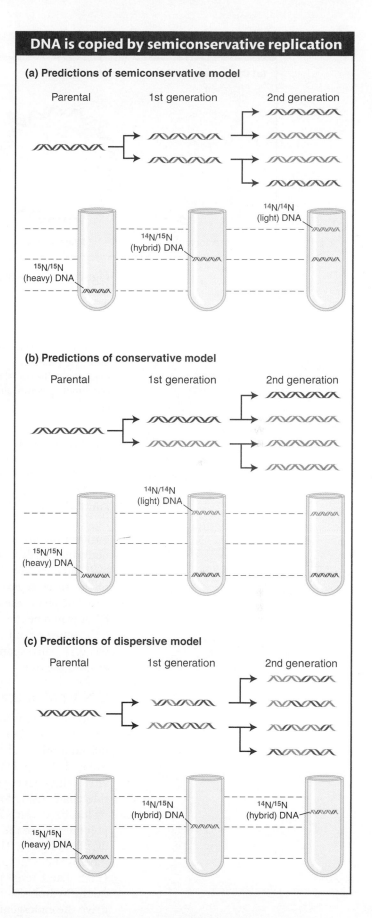

DNA is copied by semiconservative replication

(a) Predictions of semiconservative model

Parental   1st generation   2nd generation

^{14}N/^{14}N (light) DNA
^{14}N/^{15}N (hybrid) DNA
^{15}N/^{15}N (heavy) DNA

(b) Predictions of conservative model

Parental   1st generation   2nd generation

^{14}N/^{14}N (light) DNA
^{15}N/^{15}N (heavy) DNA

(c) Predictions of dispersive model

Parental   1st generation   2nd generation

^{14}N/^{15}N (hybrid) DNA
^{14}N/^{15}N (hybrid) DNA
^{15}N/^{15}N (heavy) DNA

ride forms a band at a position identical with its density in the gradient (Figure 7-13). DNA of different densities will form bands at different places. Cells initially grown in the heavy isotope ^{15}N showed DNA of high density. This DNA is shown in blue at the left-hand side of Figure 7-13a. After growing these cells in the light isotope ^{14}N for one generation, the researchers found that the DNA was of intermediate density, shown half blue (^{15}N) and half gold (^{14}N) in the middle of Figure 7-13a. Note that Meselson and Stahl continued the experiment through two *E. coli* generations so that they could distinguish semiconservative replication from dispersive. After two generations, both intermediate- and low-density DNA was observed (right-hand side of Figure 7-13a), precisely as predicted by Watson–Crick's semiconservative replication model.

Message DNA is replicated by the unwinding of the two strands of the double helix and the building up of a new complementary strand on each of the separated strands of the original double helix.

The replication fork

Another prediction of the Watson–Crick model of DNA replication is that a replication zipper, or *fork,* will be found in the DNA molecule during replication. This fork is the location at which the double helix is unwound to produce the two single strands that serve as templates for copying. In 1963, John Cairns tested this prediction by allowing replicating DNA in bacterial cells to incorporate tritiated thymidine ([^{3}H]thymidine)—the thymine nucleotide labeled with a radioactive hydrogen isotope called tritium. Theoretically, each newly synthesized daughter molecule should then contain one radioactive ("hot") strand (with ^{3}H) and another nonradioactive ("cold") strand. After varying intervals and varying numbers of replication cycles in a "hot" medium, Cairns carefully lysed the bacteria and allowed the cell contents to settle onto grids designed for electron microscopy. Finally, Cairns covered the grid with photographic emulsion and exposed it in the dark for 2 months. This procedure, called autoradiography, allowed Cairns to develop a picture of the location of ^{3}H in the cell material. As ^{3}H decays, it emits a beta particle (an energetic electron). The photo-

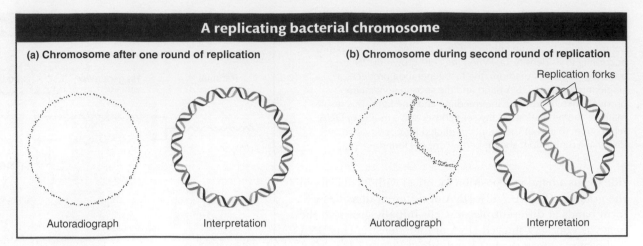

Figure 7-14 A replicating bacterial chromosome has two replication forks. (a) *Left:* Autoradiograph of a bacterial chromosome after one replication in tritiated thymidine. According to the semiconservative model of replication, one of the two strands should be radioactive. *Right:* Interpretation of the autoradiograph. The gold helix represents the tritiated strand. (b) *Left:* Autoradiograph of a bacterial chromosome in the second round of replication in tritiated (^{3}H) thymidine. In this molecule, the newly replicated double helix that crosses the circle could consist of two radioactive strands (if the parental strand were the radioactive one). *Right:* The double thickness of the radioactive tracing on the autoradiogram appears to confirm the interpretation shown here.

graphic emulsion detects a chemical reaction that takes place wherever a beta particle strikes the emulsion. The emulsion can then be developed like a photographic print so that the emission track of the beta particle appears as a black spot or grain.

After one replication cycle in [^{3}H]thymidine, a ring of dots appeared in the autoradiograph. Cairns interpreted this ring as a newly formed radioactive strand in a circular daughter DNA molecule, as shown in Figure 7-14a. It is thus apparent that the bacterial chromosome is circular—a fact that also emerged from genetic analysis described earlier (see Chapter 5). In the second replication cycle, the forks predicted by the model were indeed seen. Furthermore, the density of grains in the three segments was such that the interpretation shown in Figure 7-14b could be made: the thick curve of dots cutting through the interior of the circle of DNA would be the newly synthesized daughter strand, this time consisting of *two* radioactive strands. Cairns saw all sizes of these moon-shaped, autoradiographic patterns, corresponding to the progressive movement of the replication forks, around the ring.

DNA polymerases

A problem confronted by scientists was to understand just how the bases are brought to the double-helix template. Although scientists suspected that enzymes played a role, that possibility was not proved until 1959, when Arthur Kornberg isolated DNA polymerase from *E. coli* and demonstrated its enzymatic activity in vitro. This enzyme adds deoxyribonucleotides to the 3′ end of a growing nucleotide chain, using for its template a single strand of DNA that has been exposed by localized unwinding of the double helix (Figure 7-15). The substrates for DNA polymerase are the triphosphate forms of the deoxyribonucleotides, dATP, dGTP, dCTP, and dTTP. The addition of each base to the growing polymer is accompanied by the removal of two of the three phosphates in the form of pyrophosphate (PP$_i$). The energy produced by cleaving this high-energy bond and the subsequent hydrolysis of pyrophosphate to two inorganic phosphate molecules helps drive the endergonic process of building a DNA polymer.

Reaction catalyzed by DNA polymerase

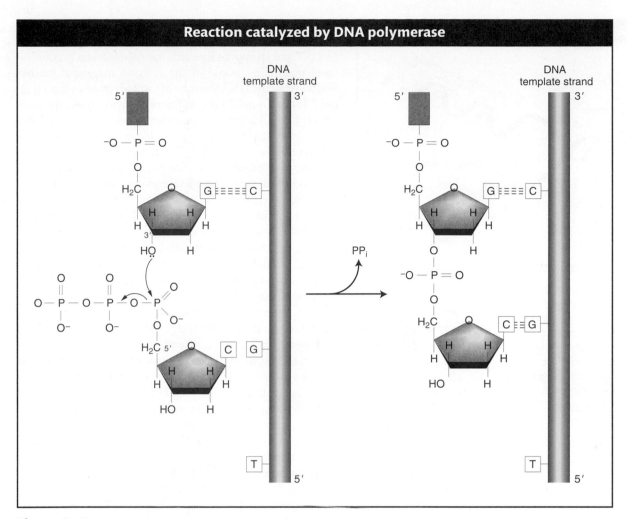

Figure 7-15 DNA polymerase catalyzes the chain-elongation reaction. Energy for the reaction comes from breaking the high-energy phosphate bond of the triphosphate substrate.

GENETICS*PORTAL* **ANIMATED ART: The nucleotide polymerization process**

There are now known to be five DNA polymerases in *E. coli*. The first enzyme that Kornberg purified is now called DNA polymerase I, or pol I. This enzyme has three activities, which appear to be located in different parts of the molecule:

1. a polymerase activity, which catalyzes chain growth in the 5′-to-3′ direction;

2. a 3′-to-5′ exonuclease activity, which removes mismatched bases; and

3. a 5′-to-3′ exonuclease activity, which degrades single strands of DNA or RNA.

We will return to the significance of the two exonuclease activities later in this chapter.

Although pol I has a role in DNA replication (see next section), some scientists suspected that it was not responsible for the majority of DNA synthesis because it was too slow (~20 nucleotides/second) and too abundant (~400 molecules/cell) and because it dissociated from the DNA after incorporating from only 20 to 50 nucleotides. In 1969, John Cairns and Paula DeLucia settled this matter when they demonstrated that an *E. coli* strain harboring a mutation in the gene that encodes DNA pol I was still able to grow normally and replicate its DNA. They concluded that another DNA polymerase, now called pol III, catalyzes DNA synthesis at the replication fork.

DNA replication at the growing fork

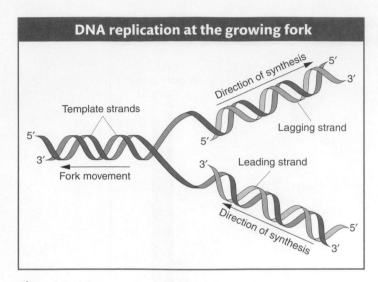

Figure 7-16 The replication fork moves in DNA synthesis as the double helix continuously unwinds. Synthesis of the leading strand can proceed smoothly without interruption in the direction of movement of the replication fork, but synthesis of the lagging strand must proceed in the opposite direction, away from the replication fork.

7.4 Overview of DNA Replication

As DNA pol III moves forward, the double helix is continuously unwinding ahead of the enzyme to expose further lengths of single DNA strands that will act as templates (Figure 7-16). DNA pol III acts at the **replication fork,** the zone where the double helix is unwinding. However, because DNA polymerase always adds nucleotides at the 3′ *growing tip,* only one of the two antiparallel strands can serve as a template for replication in the direction of the replication fork. For this strand, synthesis can take place in a smooth continuous manner in the direction of the fork; the new strand synthesized on this template is called the **leading strand.**

Synthesis on the other template also takes place at 3′ growing tips, but this synthesis is in the "wrong" direction, because, for this strand, the 5′-to-3′ direction of synthesis is away from the replication fork (see Figure 7-16). As we will see, the nature of the replication machinery requires that synthesis of both strands take place in the region of the replication fork. Therefore, synthesis moving away from the growing fork cannot go on for long. It must be in short segments: polymerase synthesizes a segment, then moves back to the segment's 5′ end, where the growing fork has exposed new template, and begins the process again. These short (1000–2000 nucleotides) stretches of newly synthesized DNA are called **Okazaki fragments.**

Another problem in DNA replication arises because DNA polymerase can extend a chain but cannot start a chain. Therefore, synthesis of both the leading strand and each Okazaki fragment must be initiated by a **primer,** or short chain of nucleotides, that binds with the template strand to form a segment of duplex nucleic acid. The primer in DNA replication can be seen in Figure 7-17. The primers are synthesized by a set of proteins called a **primosome,** of which a central component is an enzyme called **primase,** a type of RNA polymerase. Primase synthesizes a short (~8-12 nucleotides) stretch of RNA complementary to a specific region of the chromosome. On the leading strand, only one initial primer is needed because, after the initial priming, the growing DNA strand serves as the primer for continuous addition. However, on the lagging strand, every Okazaki fragment needs its own primer. The RNA chain composing the primer is then extended as a DNA chain by DNA pol III.

Synthesizing the lagging strand

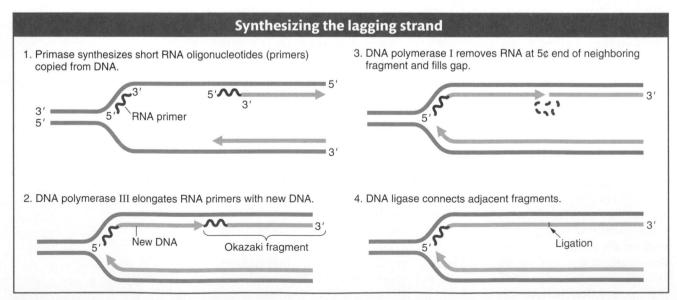

1. Primase synthesizes short RNA oligonucleotides (primers) copied from DNA.

3. DNA polymerase I removes RNA at 5¢ end of neighboring fragment and fills gap.

2. DNA polymerase III elongates RNA primers with new DNA.

4. DNA ligase connects adjacent fragments.

Figure 7-17 Steps in the synthesis of the lagging strand. DNA synthesis proceeds by continuous synthesis on the leading strand and discontinuous synthesis on the lagging strand.

A different DNA polymerase, pol I, removes the RNA primers with its 5′ to 3′ exonuclease activity and fills in the gaps with its 5′-to-3′ polymerase activity. As mentioned earlier, pol I is the enzyme originally purified by Kornberg. Another enzyme, **DNA ligase,** joins the 3′ end of the gap-filling DNA to the 5′ end of the downstream Okazaki fragment. The new strand thus formed is called the **lagging strand.** DNA ligase joins broken pieces of DNA by catalyzing the formation of a phosphodiester bond between the 5′-phosphate end of one fragment and the adjacent 3′-OH group of another fragment.

A hallmark of DNA replication is its accuracy, also called fidelity: overall, less than one error per 10^{10} nucleotides is inserted. Part of the reason for the accuracy of DNA replication is that both DNA pol I and DNA pol III possess 3′-to-5′ exonuclease activity, which serves a "proofreading" function by excising erroneously inserted mismatched bases.

Given the importance of proofreading, let's take a closer look at how it works. A mismatched base pair occurs when the 5′-to-3′ polymerase activity inserts, for example, an A instead of a G next to a C. The addition of an incorrect base is often due to a process called **tautomerization.** Each of the bases in DNA can appear in one of several forms, called tautomers, which are isomers that differ in the positions of their atoms and in the bonds between the atoms. The forms are in equilibrium. The **keto** form of each base is normally present in DNA, but in rare instances a base may shift to the **imino** or **enol** form. The imino and keto forms may pair with the wrong base, forming a mispair (Figure 7-18). When a

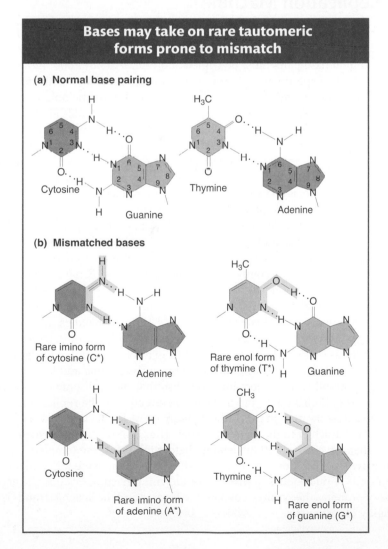

Figure 7-18 Normal base pairing compared with mismatched bases. (a) Pairing between the normal (keto) forms of the bases. (b) Rare tautomeric forms of bases result in mismatches.

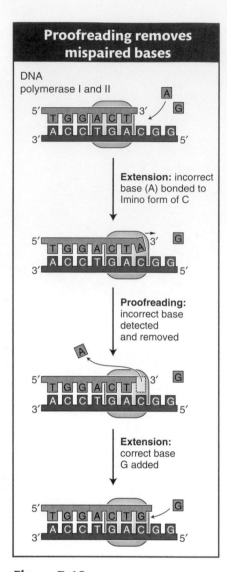

Proofreading removes mispaired bases

DNA polymerase I and II

Extension: incorrect base (A) bonded to Imino form of C

Proofreading: incorrect base detected and removed

Extension: correct base G added

Figure 7-19 DNA polymerase backs up to remove the A-C mismatch using its 3′-to-5′ exonuclease activity.

C shifts to its rare imino form, the polymerase adds an A rather than a G (Figure 7-19). Fortunately, such a mismatch is usually detected and removed by the 3′-to-5′ exonuclease activity. Once the mismatched base is removed, the polymerase has another chance to add the correct complementary G base.

As you would expect, mutant strains lacking a functional 3′-to-5′ exonuclease have a higher rate of mutation. In addition, because primase lacks a proofreading function, the RNA primer is more likely than DNA to contain errors. The need to maintain the high fidelity of replication is one reason that the RNA primers at the ends of Okazaki fragments must be removed and replaced with DNA. Only after the RNA primer is gone does DNA pol I catalyze DNA synthesis to replace the primer. The subject of DNA repair will be covered in detail in Chapter 16.

> **Message** DNA replication takes place at the replication fork, where the double helix is unwinding and the two strands are separating. DNA replication proceeds continuously in the direction of the unwinding replication fork on the leading strand. DNA is synthesized in short segments, in the direction away from the replication fork, on the lagging strand. DNA polymerase requires a primer, or short chain of nucleotides, to be already in place to begin synthesis.

7.5 The Replisome: A Remarkable Replication Machine

Another hallmark of DNA replication is speed. The time needed for *E. coli* to replicate its chromosome can be as short as 40 minutes. Therefore, its genome of about 5 million base pairs must be copied at a rate of about *2000 nucleotides per second*. From the experiment of Cairns, we know that *E. coli* uses only two replication forks to copy its entire genome. Thus, each fork must be able to move at a rate of as many as *1000 nucleotides per second*. What is remarkable about the entire process of DNA replication is that it does not sacrifice speed for accuracy. How can it maintain both speed and accuracy, given the complexity of the reactions at the replication fork? The answer is that DNA polymerase is part of a large "nucleoprotein" complex that coordinates the activities at the replication fork. This complex, called the **replisome,** is an example of a "molecular machine." You will encounter other examples in later chapters. The discovery that most of the major functions of cells—replication, transcription, and translation, for example—are carried out by large multisubunit complexes has changed the way that we think about the cell. To begin to understand why, let's look at the replisome more closely.

Some of the interacting components of the replisome in *E. coli* are shown in Figure 7-20. At the replication fork, the catalytic core of DNA pol III is part of a much larger complex, called the **pol III holoenzyme,** which consists of two catalytic cores and many **accessory proteins.** One of the catalytic cores handles the synthesis of the leading strand while the other handles lagging-strand synthesis. Some of the accessory proteins (not visible in Figure 7-20) form a connection that bridges the two catalytic cores, thus coordinating the synthesis of the leading and lagging strands. The lagging strand is shown looping around so that the replisome can coordinate the synthesis of both strands and move in the direction of the replication fork. An important accessory protein called the β **clamp** encircles the DNA like a donut and keeps pol III attached to the DNA molecule. Thus, pol III is transformed from an enzyme that can add only 10 nucleotides before falling off the template (termed a **distributive enzyme**) into an enzyme that stays at the moving fork and adds tens of thousands of nucleotides (a **processive enzyme**). In sum, through the action of accessory proteins, the synthesis of both the leading and the lagging strands is rapid and highly coordinated.

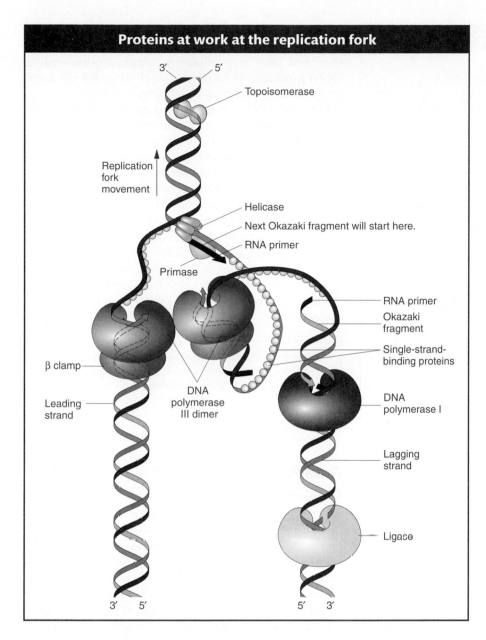

Proteins at work at the replication fork

3′ 5′

Topoisomerase

Replication fork movement

Helicase

Next Okazaki fragment will start here.

RNA primer

Primase

RNA primer

Okazaki fragment

Single-strand-binding proteins

β clamp

DNA polymerase III dimer

DNA polymerase I

Leading strand

Lagging strand

Ligase

3′ 5′   5′ 3′

Figure 7-20 The replisome and accessory proteins carry out a number of steps at the replication fork. Topoisomerase and helicase unwind and open the double helix in preparation for DNA replication. When the double helix has been unwound, single-strand-binding proteins prevent the double helix from re-forming. The illustration is a representation of the so-called trombone model (named for its resemblance to a trombone owing to the looping of the lagging strand) showing how the two catalytic cores of the replisome are envisioned to interact to coordinate the numerous events of leading- and lagging-strand replication. [*After Geoffrey Cooper, The Cell. Sinauer Associates, 2000.*]

GENETICSPORTAL ANIMATED ART: **Leading- and lagging-strand synthesis**

Note that primase, the enzyme that synthesizes the RNA primer, is not touching the clamp protein. Therefore, primase acts as a distributive enzyme—it adds only a few ribonucleotides before dissociating from the template. This mode of action makes sense because the primer need be only long enough to form a suitable duplex starting point for DNA pol III.

Unwinding the double helix

When the double helix was proposed in 1953, a major objection was that the replication of such a structure would require the unwinding of the double helix at the replication fork and the breaking of the hydrogen bonds that hold the strands together. How could DNA be unwound so rapidly and, even if it could, wouldn't that overwind the DNA behind the fork and make it hopelessly tangled? We now know that the replisome contains two classes of proteins that open the helix and prevent overwinding: they are **helicases** and **topoisomerases,** respectively.

Helicases are enzymes that disrupt the hydrogen bonds that hold the two strands of the double helix together. Like the clamp protein, the helicase fits like

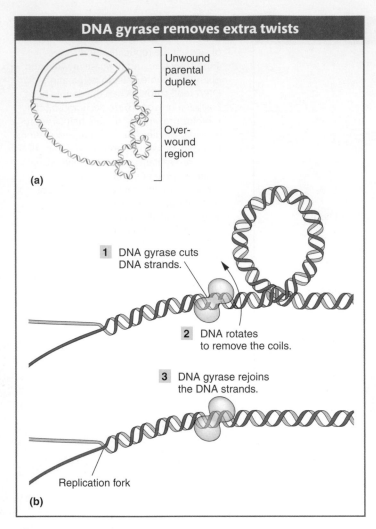

DNA gyrase removes extra twists

Unwound parental duplex

Over-wound region

(a)

1 DNA gyrase cuts DNA strands.

2 DNA rotates to remove the coils.

3 DNA gyrase rejoins the DNA strands.

Replication fork

(b)

Figure 7-21 DNA gyrase, a topoisomerase, removes extra twists during replication. (a) Extra-twisted (positively supercoiled) regions accumulate ahead of the fork as the parental strands separate for replication. (b) A topoisomerase such as DNA gyrase removes these regions, by cutting the DNA strands, allowing them to rotate, and then rejoining the strands. [*(a) From A. Kornberg and T. A. Baker, DNA Replication, 2nd ed. Copyright 1992 by W. H. Freeman and Company.*]

a donut around the DNA; from this position, it rapidly unzips the double helix ahead of DNA synthesis. The unwound DNA is stabilized by **single-strand-binding (SSB) proteins,** which bind to single-stranded DNA and prevent the duplex from re-forming.

Circular DNA can be twisted and coiled, much like the extra coils that can be introduced into a rubber band. The unwinding of the replication fork by helicases causes extra twisting at other regions, and coils called supercoils form to release the strain of the extra twisting. Both the twists and the supercoils must be removed to allow replication to continue. This supercoiling can be created or relaxed by enzymes termed topoisomerases, of which an example is DNA gyrase (Figure 7-21). Topoisomerases relax supercoiled DNA by breaking either a single DNA strand or both strands, which allows DNA to rotate into a relaxed molecule. Topoisomerases finish by rejoining the strands of the now relaxed DNA molecule.

> **Message** A molecular machine called the replisome carries out DNA synthesis. It includes two DNA polymerase units to handle synthesis on each strand and coordinates the activity of accessory proteins required for priming, unwinding the double helix, and stabilizing the single strands.

Assembling the replisome: replication initiation

Assembly of the replisome is an orderly process that begins at precise sites on the chromosome (called **origins**) and takes place only at certain times in the life of the cell. *E. coli* replication begins from a fixed origin (called *oriC*) and then proceeds in both directions (with moving forks at both ends, as shown in Figure 7-14) until the forks merge. Figure 7-22 shows the process of replisome assembly. The first step is the binding of a protein called DnaA to a specific 13-base-pair (bp) sequence (called a "DnaA box") that is repeated five times in *oriC*. In response to the binding of DnaA, the origin is unwound at a cluster of A and T nucleotides. Recall that AT base pairs are held together with only two hydrogen bonds, whereas GC base pairs are held together with three. Thus, it is easier to separate (melt) the double helix at stretches of DNA that are enriched in A and T bases.

After unwinding begins, additional DnaA proteins bind to the newly unwound single-stranded regions. With DnaA coating the origin, two helicases (the DnaB protein) now bind and slide in a 5′-to-3′ direction to begin unzipping the helix at the replication fork. Primase and DNA pol III holoenzyme are now recruited to the replication fork by protein–protein interactions, and DNA synthesis begins. You may be wondering why DnaA is not present in Figure 7-20, showing the replisome machine. The answer is that, although it is necessary for the assembly of the replisome, it is not part of the replication machinery. Rather, its job is to bring the replisome to the correct place in the circular chromosome for the initiation of replication.

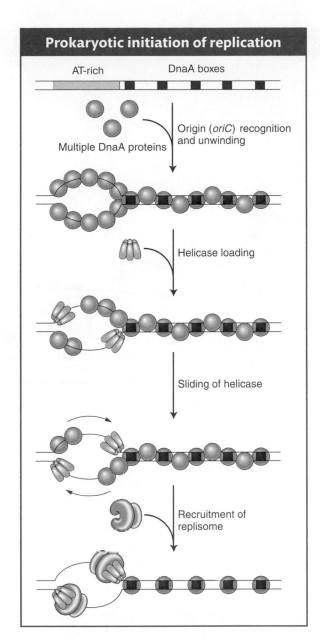

Prokaryotic initiation of replication

AT-rich   DnaA boxes

Multiple DnaA proteins

Origin (*oriC*) recognition and unwinding

Helicase loading

Sliding of helicase

Recruitment of replisome

Figure 7-22 DNA synthesis is initiated at origins of replication in prokaryotes. Proteins bind to the origin (*oriC*), where they separate the two strands of the double helix and recruit replisome components to the two replication forks.

7.6 Replication in Eukaryotic Organisms

DNA replication in both prokaryotes and eukaryotes uses a semiconservative mechanism and employs leading- and lagging-strand synthesis. For this reason, it should not come as a surprise that the components of the prokaryotic replisome and those of the eukaryotic replisome are very similar. However, as organisms increase in complexity, the number of replisome components also increases.

The eukaryotic replisome

Replication is far more complex in eukaryotes with their larger genomes that are packed into the DNA with histones (see Chapter 1). As such, it is not surprising that there are now known to be 13 components of the *E. coli* replisome and at least 27 in the replisomes of yeast and mammals. One reason for the added complexity of the eukaryotic replisome is the higher complexity of the eukaryotic template. Recall that, unlike the bacterial chromosome, eukaryotic chromosomes exist in

Assembling nucleosomes during DNA replication

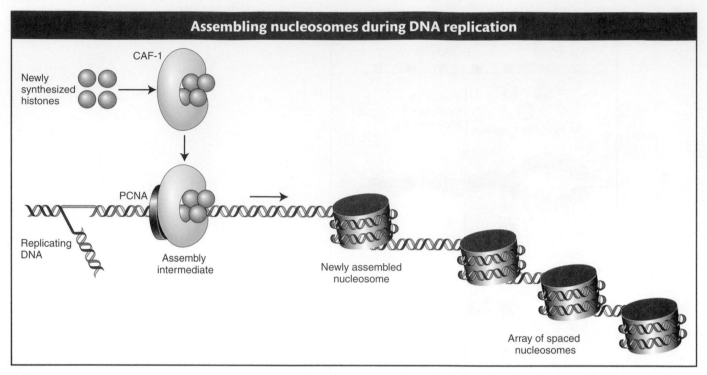

Figure 7-23 In eukaryotes, the protein CAF-1 (chromatin assembly factor 1) brings histones to the replication fork, where they are assembled to form nucleosomes. PCNA, proliferating cell nuclear antigen. For simplicity, nucleosome assembly is shown on only one of the replicated DNA strands.

the nucleus as chromatin. As described in Chapter 1, the basic unit of chromatin is the **nucleosome,** which consists of DNA wrapped around histone proteins. Thus, the replisome has not only to copy the parental strands but also to disassemble the nucleosomes in the parental strands and reassemble them in the daughter molecules. This maneuver is done by randomly distributing the old histones (from the existing nucleosomes) to daughter molecules and delivering new histones in association with a protein called chromatin assembly factor 1 (CAF-1) to the replisome. CAF-1 binds to histones and targets them to the replication fork, where they can be assembled together with newly synthesized DNA. CAF-1 and its cargo of histones arrive at the replication fork by binding to the eukaryotic version of the β clamp, called **proliferating cell nuclear antigen (PCNA)** (Figure 7-23).

> **Message** The eukaryotic replisome performs all the functions of the prokaryotic replisome; in addition, it must disassemble and reassemble the protein–DNA complexes called nucleosomes.

Eukaryotic origins of replication

Bacteria such as *E. coli* usually complete a replication–division cycle in 20 to 40 minutes but, in eukaryotes, the cycle can vary from 1.4 hours in yeast to 24 hours in cultured animal cells and may last from 100 to 200 hours in some cells. Eukaryotes have to solve the problem of coordinating the replication of more than one chromosome.

To understand eukaryotic replication origins, we will first turn our attention to the simple eukaryote yeast. Many eukaryotic proteins having roles at replication origins were first identified in yeast because of the ease of genetic analysis in yeast research (see the yeast Model Organism box in Chapter 12). The origins of replication in yeast are very much like *oriC* in *E. coli*. The 100- to 200-bp origins have a conserved DNA sequence that includes an AT-rich region that melts when an initiator protein binds to adjacent binding sites. Unlike prokaryotic chromosomes, each eukaryotic chromosome has many replication origins to replicate the much larger eukaryotic genomes quickly. Approximately 400 replication origins are dispersed

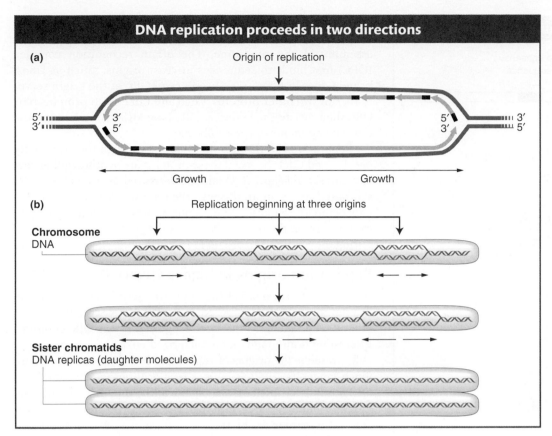

Figure 7-24 DNA replication proceeds in both directions from an origin of replication. Black arrows indicate the direction of growth of daughter DNA molecules. (a) Starting at the origin, DNA polymerases move outward in both directions. Long yellow arrows represent leading strands and short joined yellow arrows represent lagging strands. (b) How replication proceeds at the chromosome level. Three origins of replication are shown in this example.

GENETICS*PORTAL* ANIMATED ART: DNA replication: replication of a chromosome

throughout the 16 chromosomes of yeast, and there are estimated to be thousands of growing forks in the 23 chromosomes of humans. Thus, in eukaryotes, replication proceeds in both directions from multiple points of origin (Figure 7-24). The double helices that are being produced at each origin of replication elongate and eventually join one another. When replication of the two strands is complete, two identical **daughter molecules** of DNA result.

Message Where and when replication takes place are carefully controlled by the ordered assembly of the replisome at a precise site called the origin. Replication proceeds in both directions from a single origin on the circular prokaryotic chromosome. Replication proceeds in both directions from hundreds or thousands of origins on each of the linear eukaryotic chromosomes.

DNA replication and the yeast cell cycle

DNA synthesis takes place in the S (synthesis) phase of the eukaryotic cell cycle (Figure 7-25). How is the onset of DNA synthesis limited to this single stage? In yeast, the method of

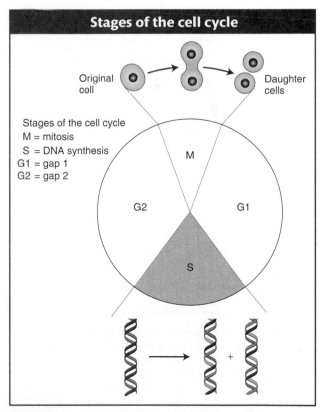

Figure 7-25 DNA is replicated during the S phase of the cell cycle.

Eukaryotic initiation of replication

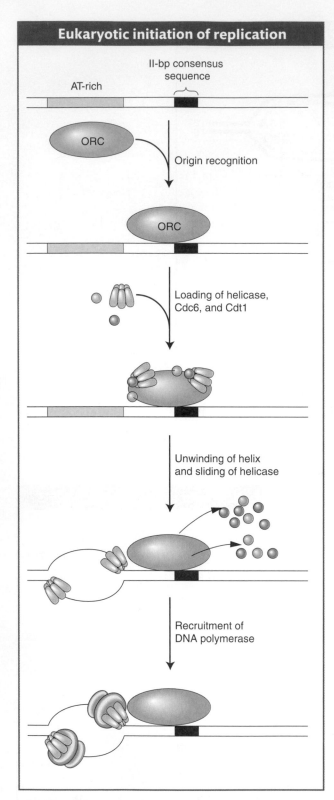

Figure 7-26 This example from yeast shows the initiation of DNA synthesis at an origin of replication in a eukaryote. As with prokaryotic initiation (see Figure 7-20), proteins of the origin recognition complex (ORC) bind to the origin, where they separate the two strands of the double helix and recruit replisome components at the two replication forks. Replication is linked to the cell cycle through the availability of two proteins: Cdc6 and Cdt1.

control is to link replisome assembly to the cell cycle. Figure 7-26 shows the process. In yeast, three proteins are required to begin assembly of the replisome. The origin recognition complex (ORC) first binds to sequences in yeast origins, much as DnaA protein does in *E. coli.* The presence of ORC at the origin serves to recruit two other proteins, Cdc6 and Cdt1. Both proteins plus ORC then recruit the helicase, called the MCM complex, and the other components of the replisome.

Replication is linked to the cell cycle through the availability of Cdc6 and Cdt1. In yeast, these proteins are synthesized during late mitosis and gap 1 (G1) and are destroyed by proteolysis after synthesis has begun. In this way, the replisome can be assembled only before the S phase. When replication begins, new replisomes cannot form at the origins, because Cdc6 and Cdt1 are degraded during the S phase and are no longer available.

Replication origins in higher eukaryotes

As already stated, most of the approximately 400 origins of replication in yeast are composed of similar DNA sequence motifs (100–200 bp in length) that are recognized by the ORC subunits. Interestingly, although all characterized eukaryotes have similar ORC proteins, the origins of replication in higher eukaryotes are much longer, possibly as long as tens of thousands or hundreds of thousands of nucleotides. Significantly, they have limited sequence similarity. Thus, although the yeast ORC recognizes specific DNA sequences in yeast chromosomes, what the related ORCs of higher eukaryotes recognize is not clear at this time, but the feature recognized is probably not a specific DNA sequence. What this uncertainty means in practical terms is that it is much harder to isolate origins from humans and other higher eukaryotes because scientists cannot use an isolated DNA sequence of one human origin, for example, to perform a computer search of the entire human genome sequence to find other origins.

If the ORCs of higher eukaryotes do not interact with a specific sequence scattered throughout the chromosomes, then how do they find the origins of replication? These ORCs are thought to interact indirectly with origins by associating with other protein complexes that are bound to chromosomes. Such a recognition mechanism may have evolved so that higher eukaryotes can regulate the timing of DNA replication during S phase (see Chapter 12 for more about euchromatin and heterochromatin). Gene-rich regions of the chromosome (the euchromatin) have been known for some time to replicate early in S phase, whereas gene-poor regions, including the densely packed heterochromatin, replicate late in S phase. DNA replication could not be timed by region if ORCs were to bind to related sequences scattered throughout the chromosomes. Instead, ORCs may, for example, have a higher affinity for origins in open chromatin and bind to these origins first and then bind to condensed chromatin only after the gene-rich regions have been replicated.

> **Message** The yeast origin of replication, like the origin in prokaryotes, contains a conserved DNA sequence that is recognized by the ORC and other proteins needed to assemble the replisome. In contrast, the origins of higher eukaryotes have been difficult to isolate and study because they are long and complex and do not contain a conserved DNA sequence.

7.7 Telomeres and Telomerase: Replication Termination

Replication of the linear DNA molecule in a eukaryotic chromosome proceeds in both directions from numerous replication origins, as shown in Figure 7-24. This process replicates most of the chromosomal DNA, but there is an inherent problem in replicating the two ends of linear DNA molecules, the regions called **telomeres.** Continuous synthesis on the leading strand can proceed right up to the very tip of the template. However, lagging-strand synthesis requires primers ahead of the process; so, when the last primer is removed, sequences are missing at the end of that strand. As a consequence, a single-stranded tip remains in one of the daughter DNA molecules (Figure 7-27). If the daughter chromosome with this DNA molecule were replicated again, the strand missing sequences at the end would become a shortened double-stranded molecule after replication. At each subsequent replication cycle, the telomere would continue to shorten, until eventually essential coding information would be lost.

Cells have evolved a specialized system to prevent this loss. The solution involves the addition of multiple copies of a simple noncoding sequence to the DNA at the chromosome tips. Thus, every time a chromosome is duplicated, it is shortened and only these repeating sequences, which contain no information, are lost. The lost repeats are then added back to the chromosome ends.

The discovery that the ends of chromosomes are made up of sequences repeated in tandem was made in 1978 by Elizabeth Blackburn and Joe Gall, who were studying the DNA in the unusual macronucleus of the single-celled ciliate *Tetrahymena*. Like other ciliates, *Tetrahymena* has a conventional micronucleus and an unusual macronucleus in which the chromosomes are fragmented into thousands of gene-size pieces with new ends added to each piece. With so many chromosome ends, *Tetrahymena* has about 40,000 telomeres and, as such, was the perfect choice to determine telomere composition. Blackburn and Gall were able to isolate the fragments containing the genes for ribosomal RNA (fragments called rDNA; see Chapter 9 for more on ribosomes) by using CsCl gradient centrifugation, the technique developed by Meselson and Stahl to isolate newly replicated *E. coli* DNA (see page 000). The ends of rDNA fragments contained tandem arrays of the sequence TTGGGG. We now know that virtually all eukaryotes have short tandem repeats at their chromosome ends; however, the sequence is not exactly the same. Human chromosomes, for example, end in about 10 to 15 kb of tandem repeats of the sequence TTAGGG.

The question of how these repeats are actually added to chromosome ends after each round of replication was addressed by Elizabeth Blackburn and Carol Grieder. They hypothesized that an enzyme catalyzed the process. Working again with extracts from the *Tetrahymena* macronucleus, they identified an enzyme, which they called **telomerase,** that adds the short repeats to the 3′ ends of DNA molecules. Interestingly, the telomerase protein carries a small RNA molecule, part of which acts as a template for the synthesis of the telomeric repeat unit. In all vertebrates, including humans, the RNA sequence 3′-AAUCCC-5′ acts as the

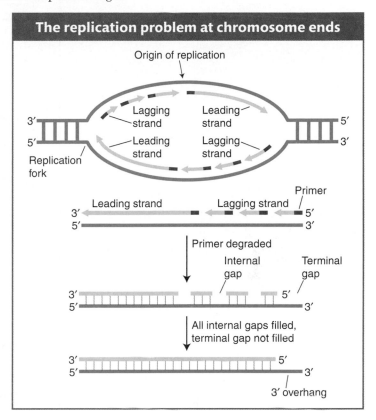

The replication problem at chromosome ends

Figure 7-27 *Top:* The replication of each Okazaki fragment on the lagging strand begins with the insertion of a primer. *Bottom:* The fate of the bottom strand in the transcription bubble. When the primer for the last Okazaki fragment of the lagging strand is removed, there is no way to fill the gap by conventional replication. A shortened chromosome would result when the chromosome containing the gap was replicated.

Figure 7-28 Telomerase carries a short RNA molecule (red letters) that acts as a template for the addition of a complementary DNA sequence, which is added to the 3′ overhang (blue letters). To add another repeat, the telomerase translocates to the end of the repeat that it just added. The extended 3′ overhang can then serve as template for conventional DNA replication. [*After Lin Kah Wai, "Telomeres, Telomerase, and Tumorigenesis: A Review," Medscape Gen. Med. 2004, 6(3):19. Copyright 2004 Medscape.*]

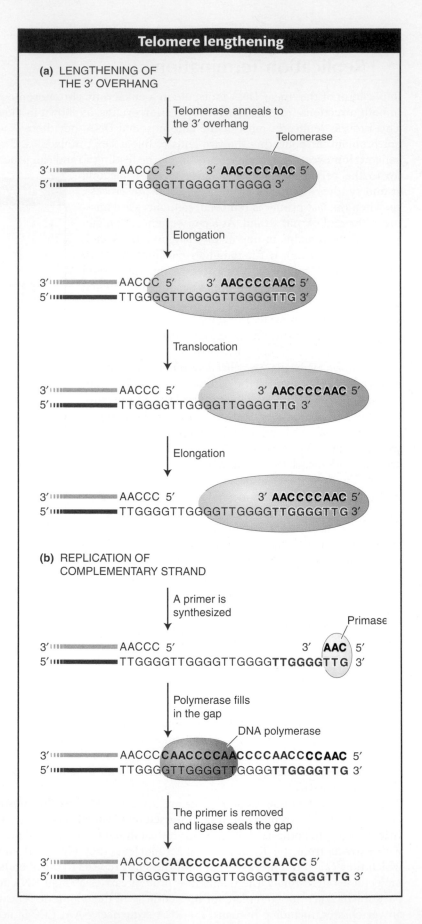

template for the 5′-TTAGGG-3′ repeat unit by a mechanism shown in Figure 7-28. Briefly, the telomerase RNA first anneals to the 3′ DNA overhang, which is then extended with the use of the telomerase's two components: the small RNA (as template) and the protein (as polymerase activity). After the addition of a few nucleotides to the 3′ overhang, the telomerase RNA moves along the DNA so that the 3′ end can be further extended by its polymerase activity. The 3′ end continues to be extended by repeated movement of the telomerase RNA. Primase and DNA polymerase then use the very long 3′ overhang as a template to fill in the end of the other DNA strand. Working with Elizabeth Blackburn, a third researcher, Jack Szostak, went on to show that telomeres also exist in the less unusual eukaryote yeast. For contributing to the discovery of how telomeres protect chromosomes from shortening, Blackburn, Grieder, and Szostak were awarded the 2009 Nobel Prize in Medicine or Physiology.

One notable feature of this reaction is that RNA is serving as the template for the synthesis of DNA. As you saw in Chapter 1 (and will revisit in Chapter 8), DNA normally serves as the template of RNA synthesis in the process called transcription. It is for this reason that the polymerase of telomerase is said to have reverse transcriptase activity. We will revisit reverse transcriptase in Chapters 10 and 15.

In addition to preventing the erosion of genetic material after each round of replication, telomeres preserve chromosomal integrity by associating with proteins to form protective caps. These caps sequester the 3′ single-stranded overhang, which can be as much as 100 nucleotides long (Figure 7-29). Without this protective cap, the double-stranded ends of chromosomes would be mistaken for double-stranded breaks by the cell and dealt with accordingly. As you will see later, in Chapter 16, double-stranded breaks are potentially very dangerous because they can result in chromosomal instability that can lead to cancer and a variety of phenotypes associated with aging. For this reason, when a double-stranded break is detected, the ccll responds in a variety of ways, depending, in part, on the cell type and the extent of the damage. For example, the double-stranded break can

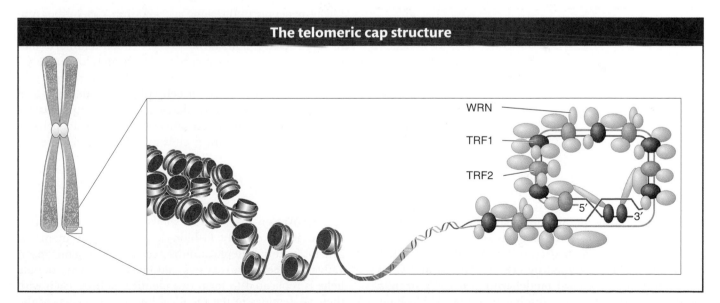

The telomeric cap structure

WRN

TRF1

TRF2

5′

3′

Figure 7-29 A "cap" protects the telomere at the end of a chromosome. The 3′ overhang is "hidden" when it displaces a DNA strand in a region where the telomeric repeats are double stranded. The proteins TRF1 and TRF2 bind to the telomeric repeats, and other proteins, including WRN, bind to TRF1 and TRF2, thus forming the protective telomeric cap.

be fused to another break or the cell can limit the damage to the organism by stopping further cell division (called senescence) or by initiating a cell-death pathway (called apoptosis).

> **Message** Telomeres are specialized structures at the ends of chromosomes that contain tandem repeats of a short DNA sequence that is added to the 3′ end by the enzyme telomerase. Telomeres stabilize chromosomes by preventing the loss of genomic information after each round of DNA replication and by associating with proteins to form a cap that "hides" the chromosome ends from the cell's DNA-repair machinery.

Surprisingly, although most germ cells have ample telomerase, somatic cells produce very little or no telomerase. For this reason, the chromosomes of proliferating somatic cells get progressively shorter with each cell division until the cell stops all divisions and enters a senescence phase. This observation led many investigators to suspect that there was a link between telomere shortening and aging. Geneticists studying human diseases that lead to a premature-aging phenotype have recently uncovered evidence that supports such a connection. People with Werner syndrome experience the early onset of many age-related events, including wrinkling of the skin, cataracts, osteoporosis, graying of the hair, and cardiovascular disease (Figure 7-30). Genetic and biochemical studies have found that afflicted people have shorter telomeres than those of normal people owing to a mutation in a gene called WRN, which encodes a protein (a helicase) that is part of the telomere cap structure (see Figure 7-29). This mutation is hypothesized to disrupt the normal telomere, resulting in chromosomal instability and the premature-aging phenotype. Patients with another premature-aging syndrome called dyskeratosis congenital also have shorter telomeres than those of healthy people of the same age, and they, too, harbor mutations in genes required for telomerase activity.

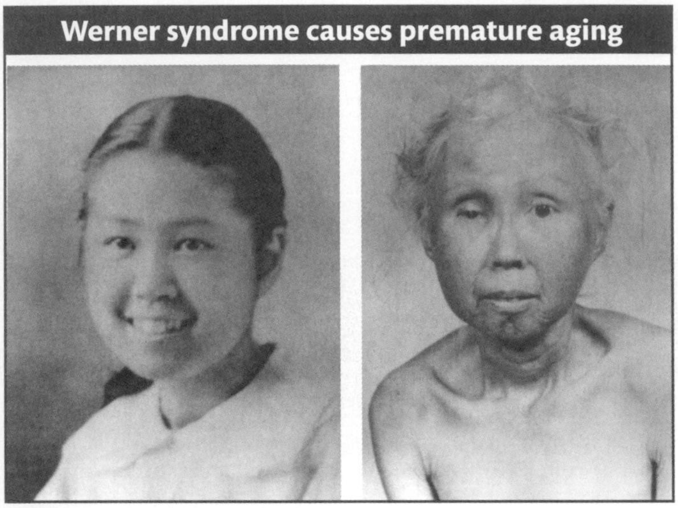

Figure 7-30 A woman with Werner syndrome at ages 15 and 48. [*International Registry of Werner Syndrome, www.wernersyndrome.org.*]

Geneticists are also very interested in connections between telomeres and cancer. Unlike normal somatic cells, most cancerous cells have telomerase activity. The ability to maintain functional telomeres may be one reason that cancer cells, but not normal cells, can grow in cell culture for decades and are considered to be immortal. As such, many pharmaceutical companies are seeking to capitalize on this difference between cancerous and normal cells by developing drugs that selectively target cancer cells by inhibiting telomerase activity.

Summary

Experimental work on the molecular nature of hereditary material has demonstrated conclusively that DNA (not protein, lipids, or carbohydrates) is indeed the genetic material. Using data obtained by others, Watson and Crick deduced a double-helical model with two DNA strands, wound around each other, running in antiparallel fashion. The binding of the two strands together is based on the fit of adenine (A) to thymine (T) and guanine (G) to cytosine (C). The former pair is held by two hydrogen bonds; the latter, by three.

The Watson–Crick model shows how DNA can be replicated in an orderly fashion—a prime requirement for genetic material. Replication is accomplished semiconservatively in both prokaryotes and eukaryotes. One double helix is replicated to form two identical helices, each with their nucleotides in the identical linear order; each of the two new double helices is composed of one old and one newly polymerized strand of DNA.

The DNA double helix is unwound at a replication fork, and the two single strands thus produced serve as templates for the polymerization of free nucleotides. Nucleotides are polymerized by the enzyme DNA polymerase, which adds new nucleotides only to the 3′ end of a growing DNA chain. Because addition is only at 3′ ends, polymerization on one template is continuous, producing the leading strand, and, on the other, it is discontinuous in short stretches (Okazaki fragments), producing the lagging strand. Synthesis of the leading strand and of every Okazaki fragment is primed by a short RNA primer (synthesized by primase) that provides a 3′ end for deoxyribonucleotide addition.

The multiple events that have to occur accurately and rapidly at the replication fork are carried out by a biological machine called the replisome. This protein complex includes two DNA polymerase units, one to act on the leading strand and one to act on the lagging strand. In this way, the more time-consuming synthesis and joining of the Okazaki fragments into a continuous strand can be temporally coordinated with the less complicated synthesis of the leading strand. Where and when replication takes place is carefully controlled by the ordered assembly of the replisome at certain sites on the chromosome called origins. Eukaryotic genomes may have tens of thousands of origins. The assembly of replisomes at these origins can take place only at a specific time in the cell cycle.

The ends of linear chromosomes (telomeres) present a problem for the replication system because there is always a short stretch on one strand that cannot be primed. The enzyme telomerase adds a number of short, repetitive sequences to maintain length. Telomerase carries a short RNA that acts as the template for the synthesis of the telomeric repeats. These noncoding telomeric repeats associate with proteins to form a telomeric cap. Telomeres shorten with age in somatic cells because telomerase is not made in those cells. Individuals who have defective telomeres experience premature aging.

KEY TERMS

accessory protein (p. 268)
antiparallel orientation (p. 259)
base (p. 257)
β clamp (p. 268)
complementary base (p. 262)
conservative replication (p. 262)
daughter molecule (p. 273)
deoxyribose (p. 257)
dispersive replication (p. 262)
distributive enzyme (p. 268)
DNA ligase (p. 267)
double helix (p. 258)
enol (p. 267)
genetic code (p. 261)
helicase (p. 269)
imino (p. 267)

keto (p. 267)
lagging strand (p. 267)
leading strand (p. 266)
major groove (p. 260)
minor groove (p. 260)
nucleosome (p. 272)
nucleotide (p. 257)
Okazaki fragment (p. 266)
origin (p. 270)
phosphate (p. 257)
polymerase III (pol III)
    holoenzyme (p. 268)
primase (p. 266)
primer (p. 266)
primosome (p. 266)
processive enzyme (p. 268)

proliferating cell nuclear antigen
    (PCNA) (p. 272)
purine (p. 257)
pyrimidine (p. 257)
replication fork (p. 266)
replisome (p. 268)
semiconservative replication (p. 262)
single-strand-binding (SSB) protein
    (p. 270)
tautomerization (p. 267)
telomerase (p. 275)
telomere (p. 275)
template (p. 262)
topoisomerase (p. 269)

SOLVED PROBLEMS

SOLVED PROBLEM 1. Mitosis and meiosis were presented in Chapter 2. Considering what has been covered in this chapter concerning DNA replication, draw a graph showing DNA content against time in a cell that undergoes mitosis and then meiosis. Assume a diploid cell.

Solution

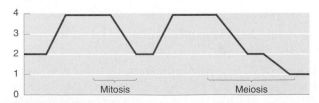

SOLVED PROBLEM 2. If the GC content of a DNA molecule is 56 percent, what are the percentages of the four bases (A, T, G, and C) in this molecule?

Solution

If the GC content is 56 percent, then, because G = C, the content of G is 28 percent and the content of C is 28 percent. The content of AT is 100 − 56 = 44 percent. Because A = T, the content of A is 22 percent and the content of T is 22 percent.

SOLVED PROBLEM 3. Describe the expected pattern of bands in a CsCl gradient for *conservative* replication in the Meselson–Stahl experiment. Draw a diagram.

Solution

Refer to Figure 7-13 for an additional explanation. In conservative replication, if bacteria are grown in the presence of ^{15}N and then shifted to ^{14}N, one DNA molecule will be all ^{15}N after the first generation and the other molecule will be all ^{14}N, resulting in one heavy band and one light band in the gradient. After the second generation, the ^{15}N DNA will yield one molecule with all ^{15}N and one molecule with all ^{14}N, whereas the ^{14}N DNA will yield only ^{14}N DNA. Thus, only all ^{14}N or all ^{15}N DNA is generated, again yielding a light band and a heavy band:

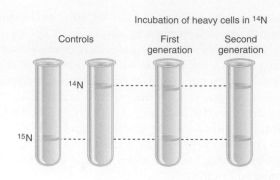

Incubation of heavy cells in ^{14}N

Controls   First generation   Second generation

^{14}N

^{15}N

PROBLEMS

Most of the problems are also available for review/grading through the GENETICSPORTAL www.yourgeneticsportal.com.

WORKING WITH THE FIGURES

1. In Table 7-1, why are there no entries for the first four tissue sources? For the last three entries, what is the most likely explanation for the slight differences in the composition of human DNA from the three tissue sources?

2. In Figure 7-7, do you recognize any of the components used to make Watson and Crick's DNA model? Where have you seen them before?

3. Referring to Figure 7-20, answer the following questions:

   a. What is the DNA polymerase I enzyme doing?

   b. What other proteins are required for the DNA polymerase III on the left to continue synthesizing DNA?

   c. What other proteins are required for the DNA polymerase III on the right to continue synthesizing DNA?

4. What is different about the reaction catalyzed by the green helicase in Figure 7-20 and the yellow gyrase in Figure 7-21?

5. In Figure 7-24(a), label all the leading and lagging strands.

BASIC PROBLEMS

6. Describe the types of chemical bonds in the DNA double helix.

7. Explain what is meant by the terms *conservative* and *semiconservative replication*.

8. What is meant by a *primer*, and why are primers necessary for DNA replication?

9. What are helicases and topoisomerases?

10. Why is DNA synthesis continuous on one strand and discontinuous on the opposite strand?

11. If the four deoxynucleotides showed nonspecific base pairing (A to C, A to G, T to G, and so on), would the unique information contained in a gene be maintained through round after round of replication? Explain.

12. If the helicases were missing during replication, what would happen to the replication process?

13. Both strands of a DNA molecule are replicated simultaneously in a continuous fashion on one strand and a discontinuous one on the other. Why can't one strand be replicated in its entirety (from end to end) before replication of the other is initiated?

14. What would happen if, in the course of replication, the topoisomerases were unable to reattach the DNA fragments of each strand after unwinding (relaxing) the DNA molecule?

15. Which of the following would happen if DNA synthesis were discontinuous on both strands?

    a. The DNA fragments from the two new strands could become mixed, producing possible mutations.

    b. DNA synthesis would not take place, because the appropriate enzymes to carry out discontinuous replication on both strands would not be present.

    c. DNA synthesis might take longer, but otherwise there would be no noticeable difference.

    d. DNA synthesis would not take place, because the entire length of the chromosome would have to be unwound before both strands could be replicated in a discontinuous fashion.

16. Which of the following is *not* a key property of hereditary material?

    a. It must be capable of being copied accurately.

    b. It must encode the information necessary to form proteins and complex structures.

    c. It must occasionally mutate.

    d. It must be able to adapt itself to each of the body's tissues.

17. It is essential that RNA primers at the ends of Okazaki fragments be removed and replaced by DNA because otherwise which of the following events would result?

a. The RNA might not be accurately read during transcription, thus interfering with protein synthesis.

b. The RNA would be more likely to contain errors because primase lacks a proofreading function.

c. The stretches of RNA would destabilize and begin to break up into ribonucleotides, thus creating gaps in the sequence.

d. The RNA primers would be likely to hydrogen bond to each other, forming complex structures that might interfere with the proper formation of the DNA helix.

18. Polymerases usually add only about 10 nucleotides to a DNA strand before dissociating. However, during replication, DNA pol III can add tens of thousands of nucleotides at a moving fork. How is this addition accomplished?

19. At each origin of replication, DNA synthesis proceeds bidirectionally from two replication forks. Which of the following would happen if a mutant arose having only one functional fork per replication bubble? (See diagram.)

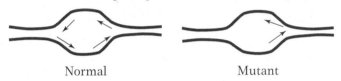

Normal	Mutant

a. No change at all in replication.

b. Replication would take place only on one half of the chromosome.

c. Replication would be complete only on the leading strand.

d. Replication would take twice as long.

20. In a diploid cell in which $2n = 14$, how many telomeres are there in each of the following phases of the cell cycle?
(a) G1   **(c)** mitotic prophase
(b) G2   **(d)** mitotic telophase

21. If thymine makes up 15 percent of the bases in a specific DNA molecule, what percentage of the bases is cytosine?

GENETICS P♦RTAL **Unpacking the Problem   22.** If the GC content of a DNA molecule is 48 percent, what are the percentages of the four bases (A, T, G, and C) in this molecule?

23. Bacteria called extremophiles are able to grow in hot springs such as Old Faithful at Yellowstone National Park in Wyoming. Do you think that the DNA of extremophiles would have a higher content of GC or AT base pairs? Justify your answer.

24. Assume that a certain bacterial chromosome has one origin of replication. Under some conditions of rapid cell division, replication could start from the origin before the preceding replication cycle is complete. How many replication forks would be present under these conditions?

25. A molecule of composition

$$5'\text{-AAAAAAAAAAA-}3'$$

$$3'\text{-TTTTTTTTTTTTT-}5'$$

is replicated in a solution of adenine nucleoside triphosphate with all its phosphorus atoms in the form of the radioactive isotope ^{32}P. Will both daughter molecules be radioactive? Explain. Then repeat the question for the molecule

$$5'\text{-ATATATATATATAT-}3'$$

$$3'\text{-TATATATATATATA-}5'$$

26. Would the Meselson and Stahl experiment have worked if diploid eukaryotic cells had been used instead?

27. Consider the following segment of DNA, which is part of a much longer molecule constituting a chromosome:

$$5'....\text{ATTCGTACGATCGACTGACTGACAGTC}....3'$$

$$3'....\text{TAAGCATGCTAGCTGACTGACTGTCAG}....5'$$

If the DNA polymerase starts replicating this segment from the right,

a. which will be the template for the leading strand?

b. Draw the molecule when the DNA polymerase is halfway along this segment.

c. Draw the two complete daughter molecules.

d. Is your diagram in part *b* compatible with bidirectional replication from a single origin, the usual mode of replication?

28. The DNA polymerases are positioned over the following DNA segment (which is part of a much larger molecule) and moving from right to left. If we assume that an Okazaki fragment is made from this segment, what will be the fragment's sequence? Label its $5'$ and $3'$ ends.

$$5'....\text{CCTTAAGACTAACTACTTACTGGGATC}....3'$$

$$3'....\text{GGAATTCTGATTGATGAATGACCCTAG}....5'$$

29. *E. coli* chromosomes in which every nitrogen atom is labeled (that is, every nitrogen atom is the heavy isotope ^{15}N instead of the normal isotope ^{14}N) are allowed to replicate in an environment in which all the nitrogen is ^{14}N. Using a solid line to represent a heavy polynucleotide chain and a dashed line for a light chain, sketch each of the following descriptions:

a. The heavy parental chromosome and the products of the first replication after transfer to a ^{14}N medium, assuming that the chromosome is one DNA double helix and that replication is semiconservative.

b. Repeat part *a*, but now assume that replication is conservative.

c. Repeat part *a*, but assume that the chromosome is in fact two side-by-side double helices, each of which replicates semiconservatively.

d. Repeat part *c*, but assume that each side-by-side double helix replicates conservatively and that the overall *chromosome* replication is semiconservative.

e. If the daughter chromosomes from the first division in ^{14}N are spun in a cesium chloride density gradient and a single band is obtained, which of the possibilities in parts *a* through *d* can be ruled out? Reconsider the Meselson and Stahl experiment: What does it *prove*?

CHALLENGING PROBLEMS

30. If a mutation that inactivated telomerase occurred in a cell (telomerase activity in the cell = zero), what do you expect the outcome to be?

31. On the planet Rama, the DNA is of six nucleotide types: A, B, C, D, E, and F. Types A and B are called *marzines,*

C and D are *orsines,* and E and F are *pirines.* The following rules are valid in all Raman DNAs:

Total marzines = total orsines = total pirines

$$A = C = E$$
$$B = D = F$$

a. Prepare a model for the structure of Raman DNA.

b. On Rama, mitosis produces three daughter cells. Bearing this fact in mind, propose a replication pattern for your DNA model.

c. Consider the process of meiosis on Rama. What comments or conclusions can you suggest?

32. If you extract the DNA of the coliphage ϕX174, you will find that its composition is 25 percent A, 33 percent T, 24 percent G, and 18 percent C. Does this composition make sense in regard to Chargaff's rules? How would you interpret this result? How might such a phage replicate its DNA?

RNA: Transcription and Processing 8

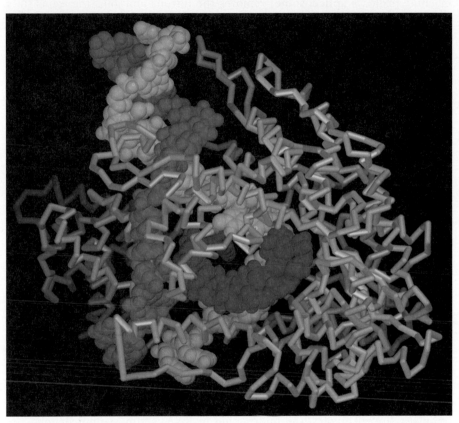

RNA polymerase in action. A very small RNA polymerase (blue), made by the bacteriophage T7, transcribes DNA into a strand of RNA (red). The enzyme separates the DNA double helix (yellow, orange), exposing the template strand to be copied into RNA. [*David S. Goodsell, Scripps Research Institute.*]

Using their newly acquired knowledge of the DNA sequences of entire genomes, scientists have been able to determine the approximate number of genes in several organisms, both simple and complex. At first there were no surprises: the bacterium *Escherichia coli* has about 3200 genes, the unicellular eukaryote yeast *Saccharomyces cerevisiae* has about 6300 genes, and the multicellular fruit fly *Drosophila melanogaster* has about 13,600 genes. Scientists assumed that more complex organisms would require more genes, and so early estimates were that our genome would have 100,000 genes. At a conference focused on genome research in 2000, scientists started an informal betting pool called GeneSweep that would be won by the person who most closely predicted the actual number of genes in the human genome. The entries ranged from ~26,000 to ~150,000 genes.

With the release of the first draft sequence, a winner was announced. Surprisingly, the winner was the entrant with the very lowest estimate, 25,947 genes. How could *Homo sapiens* with their complex brains and sophisticated immune systems have only twice as many genes as the roundworm and the same number of genes as the first sequenced plant genome, the mustard weed *Arabidopsis thaliana*? Part of the answer to this question has to do with a remarkable discovery made in the late 1970s. At that time, the proteins of higher organisms were found to be encoded in DNA not as continuous stretches (as they are in bacteria) but in pieces. Thus, the genes of higher eukaryotes are usually composed of pieces called **exons** (for *ex*pressed regi*on*) that encode parts of proteins and pieces called **introns** (for *in*tervening regi*on*) that separate exons. As you will learn in this chapter, an RNA copy containing both exons and introns is synthesized from a gene. A biological machine (called a **spliceosome**) removes the introns and joins the exons (in a process called **RNA splicing**) to produce a mature RNA that contains the continuous information needed to synthesize a protein.

What do exons and introns have to do with the low human gene count? For now, suffice it to say that the RNA transcribed from a gene can be spliced in alternative ways. Although we have only about 20,000 genes, these genes encode more than 100,000 proteins, thanks to the process of **alternative splicing** of RNA.

Even more surprising is the finding that only a small fraction of the genome actually codes for proteins (a little more than 2 percent for most complex multicellular organisms). The content of genomes will be the subject of future chapters. For now it is important to note that despite having such a small proportion of coding DNA, most of the genome still encodes RNA. The story of this aptly named **non-protein-coding RNA (ncRNA)** is a work in progress. That story will be introduced in this chapter and developed in succeeding chapters.

In this chapter, we see the first steps in the transfer of information from genes to gene products. Within the DNA sequence of any organism's genome is encoded information specifying each of the gene products that the organism can make. These DNA sequences also contain information specifying when, where, and how much of the product is made. However, this information is static, embedded in the sequence of the DNA. To utilize the information, an intermediate molecule that is a copy of a discrete gene must be synthesized with the use of the DNA sequence as a guide. This molecule is RNA, and the process of its synthesis from DNA is called *transcription.*

The transfer of information from gene to gene product takes place in several steps. The first step, which is the focus of this chapter, is to copy (*transcribe*) the information into a strand of RNA with the use of DNA as an alignment guide, or template. In prokaryotes, the information in RNA is almost immediately converted into an amino acid chain (polypeptide) by a process called *translation.* This second step is the focus of Chapter 9. In eukaryotes, transcription and translation are spatially separated: transcription takes place in the nucleus and translation in the cytoplasm. However, before RNAs are ready to be transported into the cytoplasm for translation or other uses, they undergo extensive processing, including the removal of introns and the addition of a special 5′ cap and a 3′ tail of adenine nucleotides. One fully processed type of RNA, called *messenger RNA (mRNA)*, is the intermediary in the synthesis of proteins. In addition, in both prokaryotes and eukaryotes, there are other types of RNAs that are never translated. These ncRNAs perform many essential roles.

DNA and RNA function is based on two principles:

1. Complementarity of bases is responsible for determining the sequence of a new DNA strand in replication and of the RNA transcript in transcription. Through the matching of complementary bases, DNA is replicated, the information encoded in the DNA passes into RNA, and protein complexes associated with ncRNAs are guided to specific regions in the DNA or RNA to regulate their expression.

2. Certain proteins recognize particular base sequences in DNA. These nucleic-acid-binding proteins bind to these sequences and act on them.

We will see these principles at work throughout the detailed discussions of transcription and translation that follow in this chapter and in chapters to come.

> **Message** The transactions of DNA and RNA take place through the matching of complementary bases and the binding of various proteins to specific sites on the DNA or RNA.

8.1 RNA

Early investigators had good reason for thinking that information is not transferred directly from DNA to protein. In a eukaryotic cell, DNA is found in the nucleus, whereas protein is synthesized in the cytoplasm. An intermediate is needed.

Early experiments suggest an RNA intermediate

In 1957, Elliot Volkin and Lawrence Astrachan made a significant observation. They found that one of the most striking molecular changes that takes place when *E. coli* is infected with the phage T2 is a rapid burst of RNA synthesis. Furthermore, this phage-induced RNA "turns over" rapidly; that is, its lifetime is brief, on the order of minutes. Its rapid appearance and disappearance suggested that RNA might play some role in the expression of the T2 genome necessary to make more virus particles.

Volkin and Astrachan demonstrated the rapid turnover of RNA by using a protocol called a **pulse–chase experiment.** To conduct a pulse–chase experiment, the infected bacteria are first fed (pulsed with) radioactive uracil (a molecule needed for the synthesis of RNA but not DNA). Any RNA synthesized in the bacteria from then on is "labeled" with the readily detectable radioactive uracil. After a short period of incubation, the radioactive uracil is washed away and replaced (chased) by uracil that is not radioactive. This procedure "chases" the label out of the RNA because, as the RNA breaks down, only the unlabeled precursors are available to synthesize new RNA molecules (the labeled nucleotides are "diluted" by the huge excess of unlabeled uracil added in the chase). The RNA recovered shortly after the pulse is labeled, but that recovered somewhat longer after the chase is unlabeled, indicating that the RNA has a very short lifetime.

A similar experiment can be done with eukaryotic cells. Cells are first pulsed with radioactive uracil and, after a short time, they are transferred to medium with unlabeled uracil. In samples taken after the pulse, most of the label is in the nucleus. In samples taken after the chase, the labeled RNA is found in the cytoplasm (Figure 8-1). Apparently, in eukaryotes, the RNA is synthesized in the nucleus and then moves into the cytoplasm, where proteins are synthesized. Thus, RNA is a good candidate for an information-transfer intermediary between DNA and protein.

Properties of RNA

Let's consider the general features of RNA. Although both RNA and DNA are nucleic acids, RNA differs from DNA in several important ways:

1. RNA is usually a single-stranded nucleotide chain, not a double helix like DNA. A consequence is that RNA is more flexible and can form a much greater variety of complex three-dimensional molecular shapes than can double-stranded DNA. An RNA strand can bend in such a way that some of its own bases pair with each other. Such *intramolecular* base pairing is an important determinate of RNA shape.

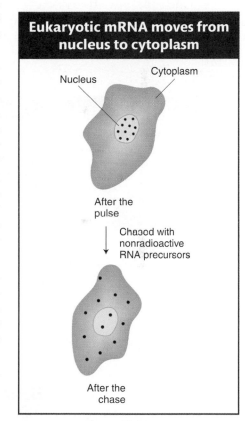

Eukaryotic mRNA moves from nucleus to cytoplasm

Figure 8-1 The pulse–chase experiment showed that mRNA moves into the cytoplasm. Cells are grown briefly in radioactive uracil to label newly synthesized RNA (pulse). Cells are washed to remove the radioactive uracil and are then grown in excess nonradioactive uracil (chase). The red dots indicate the location of the RNA containing radioactive uracil over time.

2. RNA has **ribose** sugar in its nucleotides, rather than the deoxyribose found in DNA. As the names suggest, the two sugars differ in the presence or absence of just one oxygen atom. The RNA sugar contains a hydroxyl group (OH) bound to the 2′-carbon atom, whereas the DNA sugar has only a hydrogen atom bound to the 2′-carbon atom.

As you will see later in this chapter, the presence of the hydroxyl group at the 2′-carbon atom facilitates the action of RNA in many important cellular processes.

Like an individual DNA strand, a strand of RNA is formed of a sugar–phosphate backbone, with a base covalently linked at the 1′ position on each ribose. The sugar–phosphate linkages are made at the 5′ and 3′ positions of the sugar, just as in DNA; so an RNA chain will have a 5′ end and a 3′ end.

3. RNA nucleotides (called ribonucleotides) contain the bases adenine, guanine, and cytosine, but the pyrimidine base **uracil** (abbreviated **U**) is present instead of thymine.

Uracil forms two hydrogen bonds with adenine just as thymine does. Figure 8-2 shows the four ribonucleotides found in RNA.

In addition, uracil is capable of base pairing with G. The bases U and G form base pairs only during RNA folding and not during transcription. The two hydrogen bonds that can form between U and G are weaker than the two that form between U and A. The ability of U to pair with both A and G is a major reason why RNA can form extensive and complicated structures, many of which are important in biological processes.

4. RNA—like protein, but unlike DNA—can catalyze biological reactions. The name **ribozyme** was coined for the RNA molecules that function like protein enzymes.

Ribose

Deoxyribose

Uracil

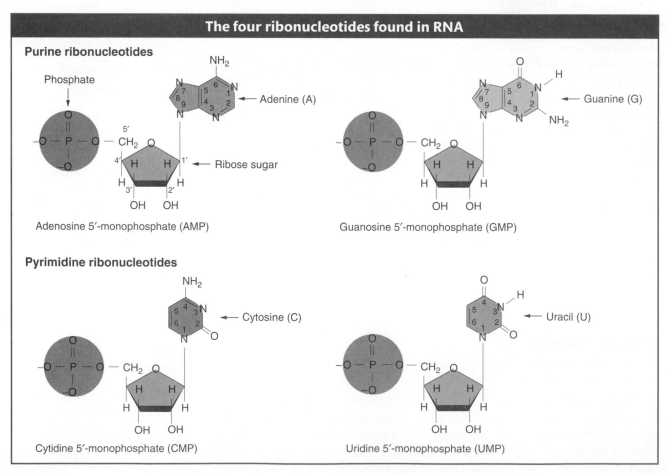

The four ribonucleotides found in RNA

Purine ribonucleotides

Phosphate

Adenine (A)

Ribose sugar

Adenosine 5′-monophosphate (AMP)

Guanine (G)

Guanosine 5′-monophosphate (GMP)

Pyrimidine ribonucleotides

Cytosine (C)

Cytidine 5′-monophosphate (CMP)

Uracil (U)

Uridine 5′-monophosphate (UMP)

Figure 8-2

Classes of RNA

RNAs can be grouped into two general classes. One class of RNA encodes the information necessary to make polypeptide chains (proteins). We refer to this class as **messenger RNA (mRNA)** because, like a messenger, these RNAs serve as the intermediary that passes information from DNA to protein. We refer to the other class as **functional RNA** because the RNA does not encode information to make protein. Instead, the RNA itself is the final functional product.

Messenger RNA   The steps through which a gene influences phenotype are called *gene expression*. For the vast majority of genes, the RNA transcript is only an intermediate necessary for the synthesis of a protein, which is the ultimate functional product that influences phenotype.

Functional RNA   As more is learned about the intimate details of gene expression and regulation, it becomes apparent that functional RNAs fall into a variety of classes and play diverse roles. Again, it is important to emphasize that functional RNAs are active as RNA; they are never translated into polypeptides.

The main classes of functional RNAs contribute to various steps in the transfer of information from DNA to protein, in the processing of other RNAs, and in the regulation of RNA and protein levels in the cell. Two such classes of functional RNAs are found in both prokaryotes and eukaryotes: transfer RNAs and ribosomal RNAs.

- **Transfer RNA (tRNA)** molecules are responsible for bringing the correct amino acid to the mRNA in the process of translation.

- **Ribosomal RNA (rRNA)** molecules are the major components of ribosomes, which are large macromolecular machines that guide the assembly of the amino acid chain by the mRNAs and tRNAs.

The entire collection of tRNAs and rRNAs are encoded by a small number of genes (a few tens to a few hundred at most). However, though the genes that encode them are few in number, rRNAs account for a very large percentage of the RNA in the cell because they are both stable and transcribed into many copies.

Another class of functional RNAs participate in the processing of RNA and are specific to eukaryotes:

- **Small nuclear RNAs (snRNAs)** are part of a system that further processes RNA transcripts in eukaryotic cells. Some snRNAs unite with several protein subunits to form the ribonucleoprotein processing complex (the *spliceosome*) that removes introns from eukaryotic mRNAs.

Finally, a large group of functional RNAs suppress the expression of genes at many levels and also maintain genome stability. Three classes of these functional RNAs may be encoded by large parts of eukaryotic genomes: microRNAs, small interfering RNAs, and piwi-interacting RNAs.

- **MicroRNAs (miRNAs)** have recently been recognized by scientists to have a widespread role in regulating the amount of protein produced by many eukaryotic genes.

- **Small interfering RNAs (siRNAs)** and **piwi-interacting RNAs (piRNAs)** help protect the integrity of plant and animal genomes. siRNAs inhibit the production of viruses, while both siRNAs and piRNAs prevent the spread of transposable elements to other chromosomal loci. siRNAs restrain transposable elements in plants, and piRNAs perform the same function in animals.

Long noncoding RNAs (lncRNAs, or sometimes just abbreviated **ncRNAs)** were recently found to be transcribed from most regions of the genomes of humans and other animals and plants. While a few lncRNAs play a role in classic genetic phenomena such as dosage compensation (see Chapter 12), the function, if any, of most lncRNAs is currently unknown.

Because protein synthesis and mRNA processing occur throughout the lifetime of most cells, tRNA, rRNA, and snRNAs are always needed. As such, these RNAs are continuously synthesized (their transcription is said to be **constitutive**). In contrast, miRNAs, siRNAs, piRNAs, and lncRNAs are transcribed and/or processed from larger transcripts intermittently, only when they are needed to fulfill their roles in protecting the genome and regulating gene expression.

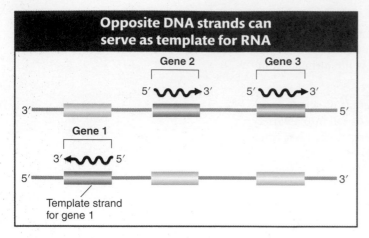

Figure 8-3 Only one strand of DNA is the template for gene transcription, but which strand varies with the gene. The direction of transcription is always the same for any gene and starts from the 3′ end of the DNA template and the 5′ end of the RNA transcript. Hence, genes transcribed in different directions use opposite strands of the DNA as templates.

> **Message** There are two general classes of RNAs, those that encode proteins (mRNA) and those that are functional as RNA (ncRNAs). Functional RNAs participate in a variety of cellular processes including protein synthesis (tRNAs, rRNAs), RNA processing (snRNAs), the regulation of gene expression (miRNAs), and genome defense (siRNAs, piRNAs).

8.2 Transcription

The first step in the transfer of information from gene to protein is to produce an RNA strand whose base sequence matches the base sequence of a DNA segment, sometimes followed by modification of that RNA to prepare it for its specific cellular roles. Hence, RNA is produced by a process that copies the nucleotide sequence of DNA. Because this process is reminiscent of transcribing (copying) written words, the synthesis of RNA is called **transcription.** The DNA is said to be transcribed into RNA, and the RNA is called a **transcript.**

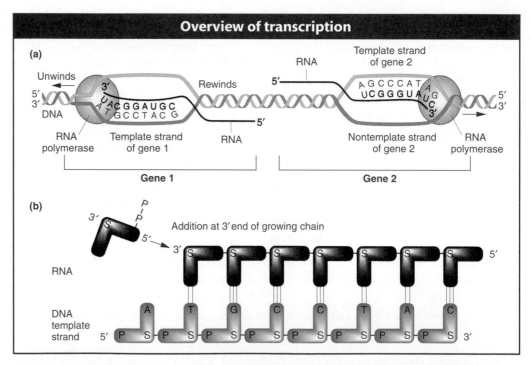

Figure 8-4 (a) Transcription of two genes in opposite directions. Genes 1 and 2 from Figure 8-3 are shown. Gene 1 is transcribed from the bottom strand. The RNA polymerase migrates to the left, reading the template strand in a 3′-to-5′ direction and synthesizing RNA in a 5′-to-3′ direction. Gene 2 is transcribed in the opposite direction, to the right, because the top strand is the template. As transcription proceeds, the 5′ end of the RNA is displaced from the template as the transcription bubble closes behind the polymerase. (b) As gene 1 is transcribed, the phosphate group on the 5′ end of the entering ribonucleotide (U) attaches to the 3′ end of the growing RNA chain. S = sugar.

GENETICS*PORTAL*   ANIMATED ART: Animated Art Transcription

Overview: DNA as transcription template

How is the information encoded in the DNA molecule transferred to the RNA transcript? Transcription relies on the complementary pairing of bases. Consider the transcription of a chromosomal segment that constitutes a gene. First, the two strands of the DNA double helix separate locally, and one of the separated strands acts as a **template** for RNA synthesis. In the chromosome overall, both DNA strands are used as templates, *but, in any one gene, only one strand is used,* and, in that gene, it is always the same strand (Figure 8-3). Next, ribonucleotides that have been chemically synthesized elsewhere in the cell form stable pairs with their complementary bases in the template. The ribonucleotide A pairs with T in the DNA, G with C, C with G, and U with A. Each ribonucleotide is positioned opposite its complementary base by the enzyme **RNA polymerase.** This enzyme attaches to the DNA and moves along it, linking the aligned ribonucleotides to make an ever-growing RNA molecule, as shown in Figure 8-4a. Hence, we already see the two principles of base complementarity and nucleic-acid–protein binding in action (in this case, the binding of RNA polymerase).

We have seen that RNA has a 5′ end and a 3′ end. During synthesis, RNA growth is always in the 5′-to-3′ direction; in other words, nucleotides are always added at a 3′ growing tip, as shown in Figure 8-4b. Because complementary nucleic acid strands are oppositely oriented, the fact that RNA is synthesized from 5′ to 3′ means that the template strand must be oriented from 3′ to 5′.

As an RNA polymerase molecule moves along the gene, it unwinds the DNA double helix ahead of it and rewinds the DNA that has already been transcribed. As the RNA molecule progressively lengthens, the 5′ end of the RNA is displaced from the template and the transcription bubble closes behind the polymerase. "Trains" of RNA polymerases, each synthesizing an RNA molecule, move along the gene (Figure 8-5).

We have also seen that the bases in transcript and template are complementary. Consequently, the nucleotide sequence in the RNA must be the same as that in the nontemplate strand of the DNA, except that the T's are replaced by U's, as shown in Figure 8-6. When DNA base sequences are cited in scientific literature, the sequence of the nontemplate strand is conventionally given, because this sequence is the same as that found in the RNA. For this reason, the nontemplate strand of the DNA is referred to as the **coding strand.** This distinction is extremely important to keep in mind when transcription is discussed.

> Message Transcription is asymmetrical: only one strand of the DNA of a gene is used as a template for transcription. This strand is in the 3′-to-5′ orientation, and RNA is synthesized in the 5′-to-3′ direction.

Stages of transcription

The protein-encoding sequence in a gene is a relatively small segment of DNA embedded in a much longer DNA molecule (the chromosome). How is the appropriate segment transcribed into a single-stranded RNA molecule of correct length

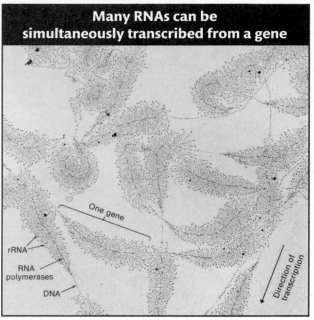

Many RNAs can be simultaneously transcribed from a gene

Figure 8-5 This electromicrograph shows the transcription of ribosomal RNA genes repeated in tandem in the nucleus of the amphibian *Triturus viridiscens.* Along each gene, many RNA polymerases are transcribing in one direction. The growing RNA transcripts appear as threads extending outward from the DNA backbone. The shorter transcripts are close to the start of transcription; the longer ones are near the end of the gene. The "Christmas tree" appearance is the result. [*Photograph from O. L. Miller, Jr., and Barbara A. Hamkalo.*]

Figure 8-6 The mRNA sequence is complementary to the DNA template strand from which it is transcribed and therefore matches the sequence of the nontemplate strand (except that the RNA has U where the DNA has T). This sequence is from the gene for the enzyme β-galactosidase.

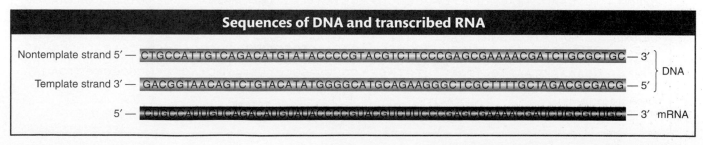

Sequences of DNA and transcribed RNA

Nontemplate strand 5′ — CTGCCATTGTCAGACATGTATACCCCGTACGTCTTCCCGAGCGAAAACGATCTGCGCTGC — 3′ ⎤
Template strand 3′ — GACGGTAACAGTCTGTACATATGGGGCATGCAGAAGGGCTCGCTTTTGCTAGACGCGACG — 5′ ⎦ DNA

5′ — CUGCCAUUGUCAGACAUGUAUACCCCGUACGUCUUCCCGAGCGAAAACGAUCUGCGCUGC — 3′ mRNA

and nucleotide sequence? Because the DNA of a chromosome is a continuous unit, the transcriptional machinery must be directed to the start of a gene to begin transcribing at the right place, continue transcribing the length of the gene, and finally stop transcribing at the other end. These three distinct stages of transcription are called **initiation, elongation,** and **termination.** Although the overall process of transcription is remarkably similar in prokaryotes and eukaryotes, there are important differences. For this reason, we will follow the three stages first in prokaryotes (by using the gut bacterium *E. coli* as an example) and then in eukaryotes.

Initiation in prokaryotes How does RNA polymerase find the correct starting point for transcription? In prokaryotes, RNA polymerase usually binds to a specific DNA sequence called a **promoter,** located close to the start of the transcribed region. A promoter is an important part of the regulatory region of a gene. Remember that, because the synthesis of an RNA transcript begins at its 5′ end and continues in the 5′-to-3′ direction, the convention is to draw and refer to the orientation of the gene in the 5′-to-3′ direction, too. Generally, the 5′ end is drawn at the left and the 3′ at the right. With this view, because the promoter must be near the end of the gene where transcription begins, it is said to be at the 5′ end of the gene; thus, the promoter region is also called the 5′ regulatory region (Figure 8-7a).

The first transcribed base is always at the same location, designated the *initiation site.* The promoter is referred to as **upstream** of the initiation site because it is located ahead of the initiation site, in the direction opposite the direction of transcription. A **downstream** site would be located later in the direction of transcription. By convention, the first DNA base to be transcribed is numbered +1. Nucleotide positions upstream of the initiation site are indicated by a negative (−) sign and those downstream by a positive (+) sign.

Figure 8-7b shows the promoter sequences of seven different genes in the *E. coli* genome. Because the same RNA polymerase binds to the promoter sequences of these different genes, the similarities among the promoters are not surprising. In particular, two regions of great similarity appear in virtually every case. These regions have been termed the −35 (minus 35) and −10 *regions* because they are located 35 base pairs and 10 base pairs, respectively, upstream of the first transcribed base. They are shown in yellow in Figure 8-7b. As you can see, the −35 and −10 regions from different genes do not have to be identical to perform a similar function. Nonetheless, it is possible to arrive at a sequence of nucleotides, called a

Figure 8-7 (a) The promoter lies "upstream" (toward the 5′ end) of the initiation point and coding sequences. (b) Promoters have regions of similar sequences, as indicated by the yellow shading in seven different promoter sequences in *E. coli.* Spaces (dots) are inserted in the sequences to optimize the alignment of the common sequences. Numbers refer to the number of bases before (−) or after (+) the RNA synthesis initiation point. The consensus sequence for most *E. coli* promoters is at the bottom. [*After H. Lodish, D. Baltimore, A. Berk, S. L. Zipursky, P. Matsudaira, and J. Darnell, Molecular Cell Biology, 3rd ed. Copyright 1995 by Scientific American Books, Inc. All rights reserved. See W. R. McClure, Annu. Rev. Biochem. 54, 1985, 171, Consensus Sequences.*]

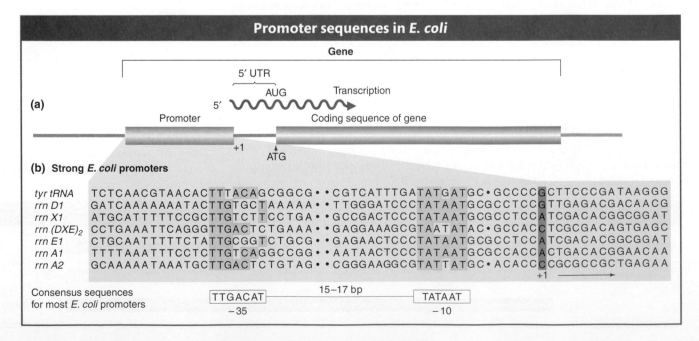

consensus sequence, that is in agreement with most sequences. The *E. coli* promoter consensus sequence is shown at the bottom of Figure 8-7b. An RNA polymerase holoenzyme (see next paragraph) binds to the DNA at this point, then unwinds the DNA double helix and begins the synthesis of an RNA molecule. Note in Figure 8-7a that the protein-encoding part of the gene usually begins at an ATG sequence, but the initiation site, where transcription begins, is usually well upstream of this sequence. The intervening part is referred to as the **5′ untranslated region (5′ UTR).**

The bacterial RNA polymerase that scans the DNA for a promoter sequence is called the **RNA polymerase holoenzyme** (Figure 8-8). This multisubunit complex is composed of the five subunits of the core enzyme (two subunits of α, one of β, one of β′, and one of ω) plus a subunit called **sigma factor (σ).** The two α subunits help assemble the enzyme and promote interactions with regulatory proteins, the β subunit is active in catalysis, the β′ subunit binds DNA, and the ω subunit has roles in enzyme assembly and the regulation of gene expression. The σ subunit binds to the −10 and −35 regions, thus positioning the holoenzyme to initiate transcription correctly at the start site (see Figure 8-8a). The σ subunit also has a role in separating (melting) the DNA strands around the −10 region so that the core enzyme can bind tightly to the DNA in preparation for RNA synthesis. After the core enzyme is bound, transcription begins and the σ subunit dissociates from the rest of the complex (see Figure 8-8b).

E. coli, like most other bacteria, has several different σ factors. One, called σ⁷⁰ because its mass in kilodaltons is 70, is the primary σ subunit used to initiate the transcription of the vast majority of *E. coli* genes. Other σ factors recognize different promoter sequences. Thus, by associating with different σ factors, the same core enzyme can recognize different promoter sequences and transcribe different sets of genes.

Elongation As the RNA polymerase moves along the DNA, it unwinds the DNA ahead of it and rewinds the DNA that has already been transcribed. In this way, it maintains a region of single-stranded DNA, called a **transcription bubble,** within which the template strand is exposed. In the bubble, polymerase monitors the binding of a free ribonucleoside triphosphate to the next exposed base on the DNA template and, if there is a complementary match, adds it to the chain. The energy for the addition of a nucleotide is derived from splitting the high-energy triphosphate and releasing inorganic diphosphate, according to the following general formula:

$$\text{NTP} + (\text{NMP})_n \xrightarrow[\substack{\text{Mg}^{2+} \\ \text{RNA polymerase}}]{\text{DNA}} (\text{NMP})_{n+1} + PP_i$$

Figure 8-9a gives a physical picture of elongation. Inside the bubble, the last eight or nine nucleotides added to the RNA chain form an RNA–DNA hybrid by complementary base pairing with the template strand. As the RNA chain

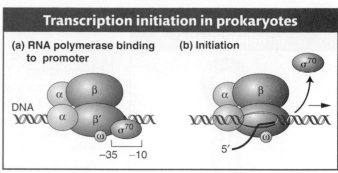

Transcription initiation in prokaryotes

(a) RNA polymerase binding to promoter

(b) Initiation

Figure 8-8 The σ subunit positions prokaryotic RNA polymerase for transcription initiation. (a) Binding of the σ subunit to the –10 and –35 regions positions the other subunits for correct initiation. (b) Shortly after RNA synthesis begins, the σ subunit dissociates from the other subunits, which continue transcription. [*After B. M. Turner, Chromatin and Gene Regulation. Copyright 2001 by Blackwell Science Ltd.*]

Figure 8-9 The five subunits of RNA polymerase are shown as a single ellipselike shape surrounding the transcription bubble. (a) *Elongation:* Synthesis of an RNA strand complementary to the single-strand region of the DNA template strand is in the 5′-to-3′ direction. DNA that is unwound ahead of RNA polymerase is rewound after it has been transcribed. (b) *Termination:* The intrinsic mechanism shown here is one of two ways used to end RNA synthesis and release the completed RNA transcript and RNA polymerase from the DNA. In this case, the formation of a hairpin loop sets off their release. For both the intrinsic and the rho-mediated mechanism, termination first requires the synthesis of certain RNA sequences.

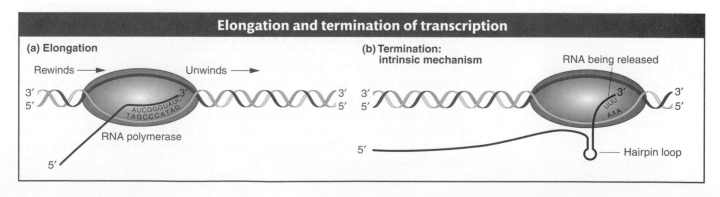

Elongation and termination of transcription

(a) Elongation

Rewinds →   Unwinds →

RNA polymerase

(b) Termination: intrinsic mechanism

RNA being released

Hairpin loop

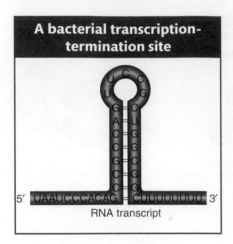

A bacterial transcription-termination site

RNA transcript

Figure 8-10 The structure of a termination site for RNA polymerase in bacteria. The hairpin structure forms by complementary base pairing within a GC-rich RNA strand. Most of the RNA base pairing is between G and C, but there is one A–U pair.

lengthens at its 3′ end, the 5′ end is further extruded from the polymerase. The complementary base pairs are broken at the point of exit, leaving the extruding strand single stranded.

Termination   The transcription of an individual gene continues beyond the protein-encoding segment of the gene, creating a **3′ untranslated region (3′ UTR)** at the end of the transcript. Elongation proceeds until RNA polymerase recognizes special nucleotide sequences that act as a signal for chain termination. The encounter with the signal nucleotides initiates the release of the nascent RNA and the enzyme from the template (Figure 8-9b). The two major mechanisms for termination in *E. coli* (and other bacteria) are called intrinsic and rho dependent.

In the intrinsic mechanism, the termination is direct. The terminator sequences contain about 40 base pairs, ending in a GC-rich stretch that is followed by a string of six or more A's. Because G and C in the template will give C and G, respectively, in the transcript, the RNA in this region also is GC rich. These C and G bases are able to form complementary hydrogen bonds with each other, resulting in a hairpin loop (Figure 8-10). Recall that the G–C base pair is more stable than the A–T pair because it is hydrogen bonded at three sites, whereas the A–T (or A–U) pair is held together by only two hydrogen bonds. Hairpin loops that are largely G–C pairs are more stable than loops that are largely A–U pairs. The loop is followed by a string of about eight U's that correspond to the A residues on the DNA template.

Normally, in the course of transcription elongation, RNA polymerase will pause if the short DNA–RNA hybrid in the transcription bubble is weak and will backtrack to stabilize the hybrid. Like that of hairpins, the strength of the hybrid is determined by the relative number of G–C base pairs compared with A–U base pairs (or A–T base pairs in RNA–DNA hybrids). In the intrinsic mechanism, the polymerase is believed to pause after synthesizing the U's (which form a weak DNA–RNA hybrid). However, the backtracking polymerase encounters the hairpin loop. This roadblock sets off the release of RNA from the polymerase and the release of the polymerase from the DNA template.

The second type of termination mechanism requires the help of a protein called the rho factor. This protein recognizes the termination signals for RNA polymerase. RNAs with rho-dependent termination signals do not have the string of U residues at their 3′ end and usually do not have hairpin loops. Instead, they have a sequence of about 40 to 60 nucleotides that is rich in C residues and poor in G residues and includes an upstream segment called the *rut* (*r*ho *ut*ilization) site. Rho is a hexamer consisting of six identical subunits that bind a nascent RNA chain at the *rut* site. These sites are located just upstream from (recall that upstream means 5′ of) sequences at which the RNA polymerase tends to pause. After binding, rho facilitates the release of the RNA from RNA polymerase. Thus, rho-dependent termination entails the binding of rho to *rut*, the pausing of polymerase, and rho-mediated dissociation of the RNA from the RNA polymerase.

8.3 Transcription in Eukaryotes

As described in Chapter 7, the replication of DNA in eukaryotes, although more complicated, is very similar to the replication of DNA in prokaryotes. In some ways, the same can be said for transcription, because eukaryotes retain many of the events associated with initiation, elongation, and termination in prokaryotes. DNA replication is more complex in eukaryotes in large part because there is a lot more DNA to copy. Transcription is more complicated in eukaryotes for three primary reasons.

1. The larger eukaryotic genomes have many more genes to be recognized and transcribed. Whereas bacteria usually have a few thousand genes, eukaryotes have tens of thousands of genes. Furthermore, there is much more noncoding DNA in eukaryotes. Noncoding DNA originates by a variety of mechanisms that will be discussed in Chapter 15. So, even though eukaryotes have more genes than prokaryotes do, their genes are, on average, farther apart. For example, whereas the gene density (average number of genes per length of DNA) in *E. coli* is 1 gene per 1400 bp, that number drops to 1 gene per 9000 bp for the fruit fly *Drosophila,* and it is only 1 gene per 100,000 bp for humans. This low gene density makes the initiation step of transcription a much more complicated process. In the genomes of multicellular eukaryotes, finding the start of a gene can be like finding a needle in a haystack.

As you will see, eukaryotes deal with this situation in several ways. First, they have divided the job of transcription among three different polymerases.

   a. RNA polymerase I transcribes rRNA genes (excluding 5S rRNA).

   b. RNA polymerase II transcribes all protein-encoding genes, for which the ultimate transcript is mRNA, and transcribes some snRNAs.

   c. RNA polymerase III transcribes the small functional RNA genes (such as the genes for tRNA, some snRNAs, and 5S rRNA). In this section, we will focus our attention on RNA polymerase II.

Second, eukaryotes require the assembly of many proteins at a promoter before RNA polymerase II can begin to synthesize RNA. Some of these proteins, called **general transcription factors (GTFs),** bind before RNA polymerase II binds, whereas others bind afterward. The role of the GTFs and their interaction with RNA polymerase II will be described in the next section, on transcription initiation in eukaryotes.

2. A significant difference between eukaryotes and prokaryotes is the presence of a nucleus in eukaryotes. In prokaryotes, the information in RNA is almost immediately translated into an amino acid chain (polypeptide), as we will see in Chapter 9. In eukaryotes, transcription and translation are spatially separated—transcription takes place in the nucleus and translation in the cytoplasm (Figure 8-11). In eukaryotes, RNA is synthesized in the nucleus where the DNA is located and exported out of the nucleus into the cytoplasm for translation.

Before the RNA leaves the nucleus, it must be modified in several ways. These modifications are collectively referred to as **RNA processing.** To distinguish the RNA before and after processing, newly synthesized RNA is called the **primary**

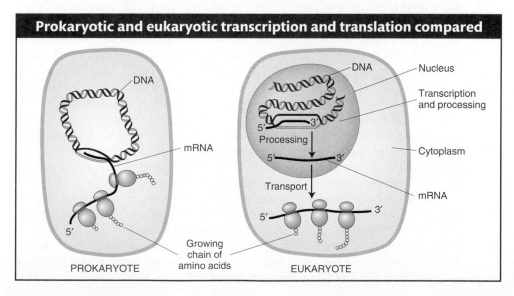

Prokaryotic and eukaryotic transcription and translation compared

PROKARYOTE

EUKARYOTE

Figure 8-11 Transcription and translation take place in the same cellular compartment in prokaryotes but in different compartments in eukaryotes. Moreover, unlike prokaryotic RNA transcripts, eukaryotic transcripts undergo extensive processing before they can be translated into proteins. [*After J. Darnell, H. Lodish, and D. Baltimore, Molecular Cell Biology, 2nd ed. Copyright 1990 by Scientific American Books, Inc. All rights reserved.*]

transcript or pre-mRNA, and the term mRNA is reserved for the fully processed transcript that can be exported out of the nucleus. As you will see, the 5′ half of the RNA undergoes processing while the 3′ half is still being synthesized. Thus, RNA polymerase II must synthesize RNA while simultaneously coordinating a diverse array of processing events. For this reason, among others, RNA polymerase II is a more complicated multisubunit enzyme than prokaryotic RNA polymerase. In fact, it is considered to be another molecular machine. The coordination of RNA processing and synthesis by RNA polymerase II will be discussed in the section on transcription elongation in eukaryotes.

3. Finally, the template for transcription, genomic DNA, is organized into chromatin in eukaryotes (see Chapter 1), whereas it is virtually "naked" in prokaryotes. As you will learn in Chapter 12, certain chromatin structures can block the access of RNA polymerase to the DNA template. This feature of chromatin has evolved into a very sophisticated mechanism for regulating eukaryotic gene expression. However, a discussion of the influence of chromatin on the ability of RNA polymerase II to initiate transcription will be put aside until Chapter 12 as we focus on the events that take place *after* RNA polymerase II gains access to the DNA template.

Transcription initiation in eukaryotes

As stated earlier, transcription starts in prokaryotes when the σ subunit of the RNA polymerase holoenzyme recognizes the −10 and −35 regions in the promoter of a gene. After transcription begins, the σ subunit dissociates and the core polymerase continues to synthesize RNA within a transcription bubble that moves along the DNA. Similarly, in eukaryotes, the core of RNA polymerase II also cannot recognize promoter sequences on its own. However, unlike bacteria, where σ factor is an integral part of the polymerase holoenzyme, eukaryotes require GTFs to bind to regions in the promoter *before* the binding of the core enzyme.

The initiation of transcription in eukaryotes has some features that are reminiscent of the initiation of replication at origins of replication. Recall from Chapter 7 that proteins that are not part of the replisome initiate the assembly of the replication machine. DnaA in *E. coli* and the origin recognition complex (ORC) in yeast, for example, first recognize and bind to origin DNA sequences. These proteins serve to attract replication proteins, including DNA polymerase III, through protein–protein interactions. Similarly, GTFs, which do not take part in RNA synthesis, recognize and bind to sequences in the promoter or to other GTFs and serve to attract the RNA polymerase II core and position it at the correct site to start transcription. The GTFs are designated TFIIA, TFIIB, and so forth (for *t*ranscription *f*actor of RNA polymerase *II*).

The GTFs and the RNA polymerase II core constitute the **preinitiation complex (PIC).** This complex is quite large: it contains six GTFs, *each* of which is a multiprotein complex, plus the RNA polymerase II core, which is made up of a dozen or more protein subunits. The sequence of amino acids of some of the RNA polymerase II core subunits is conserved from yeast to humans. This conservation can be dramatically demonstrated by replacing some yeast RNA polymerase II subunits with their human counterparts to form a *chimeric* RNA polymerase II complex (named after a fire-breathing creature from Greek mythology that had a lion's head, a goat's body, and a serpent's tail). This chimeric RNA polymerase II complex is fully functional in yeast.

Like prokaryotic promoters, eukaryotic promoters are located on the 5′ side (upstream) of the transcription start site. When eukaryotic promoter regions from different species are aligned, the sequence TATA can often be seen to be located about 30 base pairs (−30 bp) from the transcription start site (Figure 8-12). This sequence, called the **TATA box,** is the site of the first event in transcription: the binding of the **TATA-binding protein (TBP).** The TBP is part of the TFIID complex, one of the six GTFs. When bound to the TATA box, TBP attracts other GTFs and the RNA polymerase II core to the promoter, thus forming the preinitiation complex. After transcription has been initiated, RNA polymerase II dissociates from most of the GTFs to elongate the primary RNA transcript. Some of the GTFs remain at the promoter to attract the next RNA polymerase core. In this way, multiple RNA polymerase II enzymes can simultaneously synthesize transcripts from a single gene.

How is the RNA polymerase II core able to separate from the GTFs and start transcription? Although the details of this process are still being worked out, what is known is that the β subunit of RNA polymerase II contains a protein tail, called the **carboxyl tail domain (CTD),** that plays a key role. The CTD is strategically located near the site at which nascent RNA will emerge from the polymerase. The initiation phase ends and the elongation phase begins after the CTD has been phosphorylated by one of the GTFs. This phosphorylation is thought to somehow weaken the connection of RNA polymerase II to the other proteins of the preinitiation complex and permit elongation. The CTD also participates in several other critical phases of RNA synthesis and processing.

> **Message** Eukaryotic promoters are first recognized by general transcription factors. The function of GTFs is to attract the core RNA polymerase II so that it is positioned to begin RNA synthesis at the transcription start site.

Elongation, termination, and pre-mRNA processing in eukaryotes

Elongation takes place inside the transcription bubble essentially as described for the synthesis of prokaryotic RNA. However, nascent RNA has very different fates in prokaryotes and eukaryotes. In prokaryotes, translation begins at the 5′ end of the nascent RNA while the 3′ half is still being synthesized. In contrast, the RNA of eukaryotes must undergo further processing before it can be translated. This processing includes (1) the addition of a cap at the 5′ end, (2) splicing to eliminate introns, and (3) the addition of a 3′ tail of adenine nucleotides (polyadenylation).

Like DNA replication, the synthesis and processing of pre-mRNA to mRNA requires that many steps be performed rapidly and accurately. At first, most of the processing of eukaryotic pre-mRNA was thought to take place after RNA synthesis was complete. Processing after RNA synthesis is complete is said to be **posttranscriptional.** However, experimental evidence now indicates that processing actually takes place during RNA synthesis; it is **cotranscriptional.** Therefore, the partly synthesized (nascent) RNA is undergoing processing reactions as it emerges from the RNA polymerase II complex.

The CTD of eukaryotic RNA polymerase II plays a central role in coordinating all processing events. The CTD is composed of many repeats of a sequence of seven amino acids. These repeats serve as binding sites for some of the enzymes and other proteins that are required for RNA capping, splicing, and cleavage followed by polyadenylation. The CTD is located near the site where nascent RNA emerges from the polymerase, and so it is in an ideal place to orchestrate the binding and release of proteins needed to process the nascent RNA transcript while

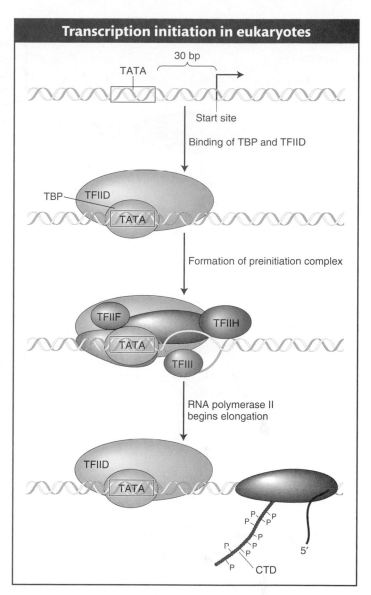

Transcription initiation in eukaryotes

Figure 8-12 Formation of the preinitiation complex usually begins with the binding of the TATA-binding protein (TBP), which then recruits the other general transcription factors (TFs) and RNA polymerase II to the transcription start site. Transcription begins after phosphorylation of the carboxyl tail domain (CTD) of RNA polymerase II. [*After "RNA Polymerase II Holoenzyme and Transcription Factors." Copyright 2001 Macmillan Publishing Group Ltd./Nature Publishing Group.*]

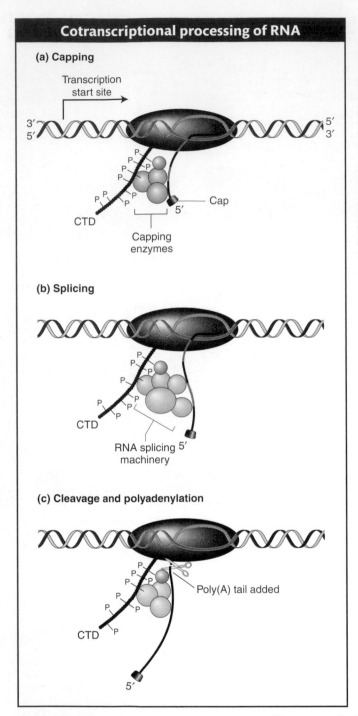

Figure 8-13 Cotranscriptional processing of RNA is coordinated by the carboxyl tail domain (CTD) of the β subunit of RNA polymerase II. Reversible phosphorylation of the amino acids of the CTD (indicated by the P's) creates binding sites for the different processing enzymes and factors required for (a) capping, (b) splicing, and (c) cleavage and polyadenylation. [*Parts (a) and (b) after R. I. Drapkin and D. F. Reinberg, "RNA Synthesis," Encyclopedia of Life Sciences. Copyright 2002, Macmillan Publishing Group Ltd./Nature Publishing Group.*]

RNA synthesis continues. In the various phases of processing, the amino acids of the CTD are reversibly modified—usually through the addition and removal of phosphate groups (called phosphorylation and dephosphorylation, respectively). The phosphorylation state of the CTD determines which processing proteins can bind. In this way, the CTD determines the task to be performed on the RNA as it emerges from the polymerase. The processing events and the role of CTD in executing them are shown in Figure 8-13 and considered next.

Processing 5′ and 3′ ends   Figure 8-13a depicts the processing of the 5′ end of the transcript of a protein-encoding gene. When the nascent RNA first emerges from RNA polymerase II, a special structure, called a **cap,** is added to the 5′ end by several proteins that interact with the CTD. The cap consists of a 7-methylguanosine residue linked to the transcript by three phosphate groups. The cap has two functions. First, it protects the RNA from degradation in its long journey to the site of translation. Second, as you will see in Chapter 9, the cap is required for translation of the mRNA.

RNA elongation continues until the conserved sequence AAUAAA or AUUAAA near the 3′ end is reached. An enzyme recognizes that sequence and cuts off the end of the RNA approximately 20 bases farther down. To this cut end, a stretch of 150 to 200 adenine nucleotides called a **poly(A) tail** is added (see Figure 8-13c). Hence, the AAUAAA sequence of the mRNA from protein-encoding genes is called a *polyadenylation signal.*

RNA splicing, the removal of introns   In 1977, a scientific study appeared titled "An amazing sequence arrangement at the 5′ end of adenovirus 2 messenger RNA." Scientists are usually understated, at least in their publications, and the use of the word "amazing" indicated that something truly unexpected had been found. The laboratories of Richard Roberts and Phillip Sharp had independently discovered that the information encoded by eukaryotic genes (in their case, the gene of a virus that infects eukaryotic cells) can be fragmented into pieces of two types, exons and introns. As stated earlier, pieces that encode parts of proteins are exons, and pieces that separate exons are introns. Introns are present not only in protein-encoding genes but also in some rRNA and even tRNA genes.

Introns are removed from the primary transcript while RNA is still being transcribed and after the cap has been added but before the transcript is transported into the cytoplasm. The removal of introns and the joining of exons is called **splicing,** because it is reminiscent of the way in which videotape or movie film can be cut and rejoined to delete a specific segment. Splicing brings together the coding regions, or exons, so that the mRNA now contains a coding sequence that is completely colinear with the protein that it encodes.

The number and size of introns vary from gene to gene and from species to species. For example, only about 200 of the 6300 genes in yeast have introns, whereas typical genes in mammals, including humans, have several. The average size of a mammalian intron is about 2000 nucleotides and the average exon is about 200 nucleotides; thus, a larger percentage of the DNA in mammals encodes introns than exons. An extreme example is the human Duchenne muscular dystrophy gene. This gene has 79 exons and 78 introns spread across

2.5 million base pairs. When spliced together, its 79 exons produce an mRNA of 14,000 nucleotides, which means that introns account for the vast majority of the 2.5 million base pairs.

Alternative splicing   At this point, you might be wondering about the utility of having genes organized into exons and introns. Recall that this chapter began with a discussion of the number of genes in the human genome. This number (now estimated at ~21,000 genes) is less than twice the number of genes in the round-worm, yet the spectrum of human proteins (called the **proteome;** see Chapter 9) is in excess of 70,000. That proteins so outnumber genes indicates that a gene can encode the information for more than one protein. One way that a gene can encode multiple proteins is through a process called **alternative splicing.** In this process, different mRNAs and, subsequently, different proteins are produced from the same primary transcript by splicing together different combinations of exons. For reasons that are currently unknown, the proportion of alternatively spliced genes varies from species to species. Although alternative splicing is rare in plants, more than 70 percent of human genes are alternatively spliced. Many mutations with serious consequences for the organism are due to splicing defects.

The consequences of alternative splicing on protein structure and function will be presented later in the book. For now, suffice it to say that proteins produced by alternative splicing are usually related (because they usually contain subsets of the same exons from the primary transcript) and that they are often used in different cell types or at different stages of development. Figure 8-14 shows the myriad combinations produced by alternative splicing of the primary RNA transcript of the α-tropomyosin gene. The mechanism of splicing is considered in the next section.

> **Message** Eukaryotic pre-mRNA is extensively processed before being transported as mRNA to the cytoplasm for translation into protein. A 5′ cap and 3′ poly(A) tail are added; introns are removed and exons spliced together. One gene can encode more than one polypeptide when its pre-mRNA is alternatively spliced.

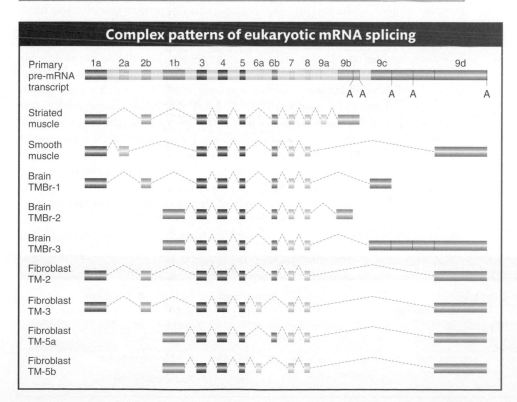

Figure 8-14 The pre-mRNA transcript of the rat α-tropomyosin gene is alternatively spliced in different cell types. The light green boxes represent introns; the other colors represent exons. Polyadenylation signals are indicated by an A. Dashed lines in the mature mRNAs indicate regions that have been removed by splicing. TM, tropomyosin. [*After J. P. Lees et al., Mol. Cell. Biol. 10, 1990, 1729–1742.*]

8.4 Intron Removal and Exon Splicing

Because RNA is such a versatile molecule, it participates in a variety of cellular processes. In Chapter 9, you will learn more about the role of **functional RNAs** as important components of the ribosome, the biological machine that is the site of protein synthesis. In this section and the next, you will see that functional RNAs also have prominent roles in both the processing of mRNA and the regulation of its level in the cell.

Small nuclear RNAs (snRNAs): The mechanism of exon splicing

After the discovery of exons and introns, scientists turned their attention to the mechanism of RNA splicing. Because introns must be precisely removed and exons precisely joined, the first approach was to compare the sequences of pre-mRNAs for clues to how introns and exons are recognized. Figure 8-15 shows the exon–intron junctions of pre-mRNAs. These junctions are the sites at which the splicing reactions take place. At these junctions, certain specific nucleotides were found to be nearly identical across genes and across species; they have been highly conserved because they participate in the splicing reactions. Each intron is cut at each end, and these intron ends almost always have GU at the 5′ end and AG at the 3′ end (the **GU–AG rule**). Another invariant site is an A residue between 15 and 45 nucleotides upstream of the 3′ splice site. The nucleotides flanking the highly conserved ones also are conserved, but to a lesser degree. The existence of conserved nucleotide sequences at splice junctions suggested that there must be cellular machinery that recognizes these sequences and carries out splicing. As is often the case in scientific research, the splicing machinery was found by accident and the mechanism of splicing was entirely unexpected.

A serendipitous finding in the laboratory of Joan Steitz led to the discovery of components of the splicing machinery. Patients with a variety of autoimmune diseases, including systemic lupus erythematosis, produce antibodies against their own proteins. In the course of analyzing blood samples from patients with lupus, Steitz and colleagues identified antibodies that could bind to a large molecular complex of small RNAs and proteins. Because this riboprotein complex was localized in the nucleus, the RNA components were named *small nuclear RNAs*. The snRNAs were found to be complementary to the consensus sequences at splice junctions, leading scientists to hypothesize a role for the snRNAs in the splicing reaction. The conserved nucleotides in the transcript are now known to be recognized by five small nuclear ribonucleoproteins (snRNPs), which are complexes of protein and one of five snRNAs (U1, U2, U4, U5, and U6). These snRNPs and more than 100 additional proteins are part of the spliceosome, the large biological machine that removes introns and joins exons. Components of the spliceosome interact with the CTD, as suggested in Figure 8-13b.

These components of the spliceosome attach to intron and exon sequences, as shown in Figure 8-16. The snRNPs help to align the splice sites at either end of an intron by forming hydrogen bonds to the conserved intron and exon sequences. Then the snRNPs recruit other complexes and form the spliceosome, which catalyzes the removal of the intron through two consecutive splicing

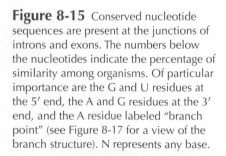

Figure 8-15 Conserved nucleotide sequences are present at the junctions of introns and exons. The numbers below the nucleotides indicate the percentage of similarity among organisms. Of particular importance are the G and U residues at the 5′ end, the A and G residues at the 3′ end, and the A residue labeled "branch point" (see Figure 8-17 for a view of the branch structure). N represents any base.

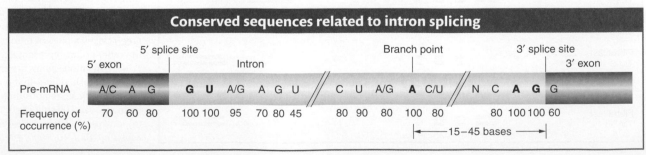

steps (see Figure 8-16). The first step attaches one end of the intron to the conserved internal adenine, forming a structure having the shape of a cowboy's lariat. The second step releases the lariat and joins the two adjacent exons. Figure 8-17 portrays the chemistry behind intron removal and exon

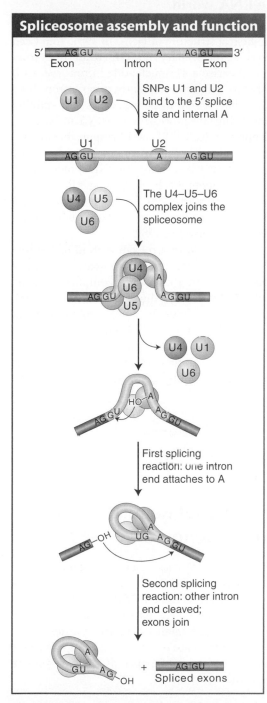

Figure 8-16 The spliceosome is composed of several snRNPs that attach sequentially to the RNA, taking up positions roughly as shown. Alignment of the snRNPs results from hydrogen bonding of their snRNA molecules to the complementary sequences of the intron. In this way, the reactants are properly aligned and the two splicing reactions can take place. The chemistry of these reactions can be seen in more detail in Figure 8-17.

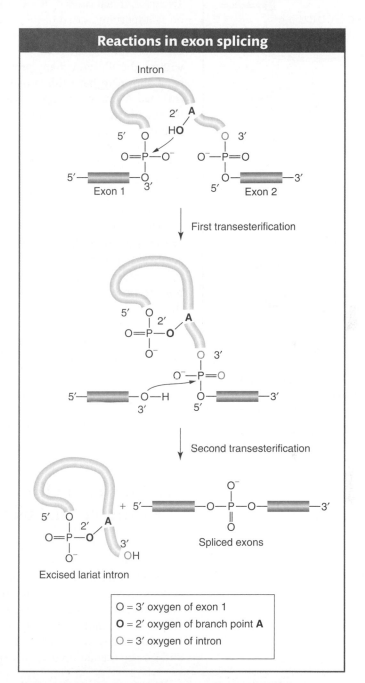

Figure 8-17 Two transesterification reactions take place in the splicing of RNA: first, to join the 5' donor end to the internal branch point (first reaction in Figure 8-16) and, second, to join the two exons together (second reaction in Figure 8-16). [*After H. Lodish, A. Berk, S. O. Zipurski, P. Matsudaira, D. Baltimore, and J. Darnell, Molecular Cell Biology, 4th ed. Copyright 2000 by W. H. Freeman and Company.*]

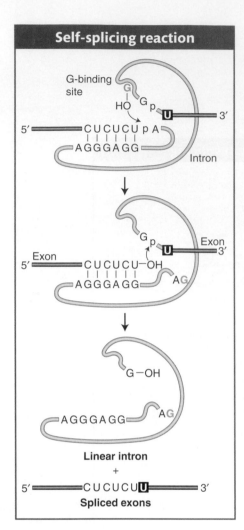

Self-splicing reaction

G-binding site

Intron

Exon

Exon

Linear intron
+
Spliced exons

Figure 8-18 The self-splicing intron from *Tetrahymena* executes two transesterification reactions to excise itself from RNA.

splicing. Chemically, the two steps are transesterification reactions between the conserved nucleotides. Hydroxyl groups at the 2′ and 3′ positions of ribonucleotides are key reaction participants.

Self-splicing introns and the RNA world

Two exceptional cases of RNA processing led to a discovery considered by some to be as important as that of the double-helical structure of DNA. In 1981, Tom Cech and co-workers reported that, in a test tube, the primary transcript of an rRNA from the ciliate protozoan *Tetrahymena* could excise a 413-nucleotide intron *from itself* without the addition of any protein (Figure 8-18). Subsequently, other introns have been shown to have this property and have come to be known as **self-splicing introns.** A few years earlier while studying the processing of tRNA in bacteria, Sidney Altman identified a ribonucleoprotein (called RNase P) responsible for cutting the pre-tRNA molecule at a specific site. The big surprise came when they determined that the catalytic activity of RNase P resided in the RNA component of the enzyme rather than in the protein component. Cech and Altman's findings are considered landmark discoveries because they marked the first time that biological molecules other than protein were shown to catalyze reactions. As such, it was fitting that they received the Nobel Prize in Chemistry in 1989.

The discovery of self-splicing introns has led to a reexamination of the role of the snRNAs in the spliceosome. The most recent studies indicate that intron removal is catalyzed by the snRNAs and not by the protein component of the spliceosome. As you will see in Chapter 9, the RNAs in the ribosome (the rRNAs), not the ribosomal proteins, are now thought to have the central role in most of the important events of protein synthesis. The numerous examples of ribozymes have provided solid evidence for a theory called the **RNA world,** which holds that RNA must have been the genetic material in the first cells because only RNA is known to both encode genetic information and catalyze biological reactions.

Message Intron removal and exon joining are catalyzed by RNA molecules. In eukaryotes, the snRNAs of the spliceosome catalyze the removal of introns from pre-mRNA. Some introns are self-splicing; in these cases, the intron itself catalyzes its own removal. RNAs capable of catalysis are called ribozymes.

8.5 Small Functional RNAs That Regulate and Protect the Eukaryotic Genome

In 2002, one of the leading science journals, *Science* magazine, named "Small RNA" as their Breakthrough of the Year (Figure 8-19). The RNAs to which they were referring were not the previously described small RNAs such as snRNAs or tRNAs, which are considered to have a housekeeping role and, as such, are synthesized constitutively. Instead, these other small RNAs are synthesized in response to changes in a cell's developmental state or its surroundings. We now know that they are critically important for the maintenance of a stable genome and for the regulation of gene expression.

miRNAs are important regulators of gene expression

The first of this type of small RNA was discovered in 1993 by Victor Ambros and colleagues while they were studying the *lin-4* gene of the roundworm *C. elegans*. Because mutations in *lin-4* resulted in abnormal larvae, it was hypothesized that this gene encoded a protein that was required for normal larval development. Thus, it came as a surprise when the group isolated the *lin-4* gene and reported that rather than encode a protein, it produced two small RNAs of 22 nucleotides and 61 nucleotides. They then found that the 22-nucleotide RNA was produced by processing the larger 61-nucleotide RNA. Finally, they found that the 22-nucleotide RNA repressed the expression of certain other genes by base pairing with their mRNAs.

The 22-nucleotide RNA product of the *lin-4* gene was the first member to be discovered of a very large class of RNAs called **microRNAs (miRNAs)** now known to be present in the genomes of plants and animals. Most miRNAs act to repress the expression of genes. In fact, it is estimated that plant and animal genomes each have up to a thousand miRNAs that in turn regulate the expression of thousands of genes. Like the product of *lin-4*, many miRNAs are initially transcribed by RNA polymerase II as a longer RNA from a gene that produces only an RNA product. The longer RNA assumes a double-stranded stem loop structure with a mismatched base in the stem (Figure 8-20). The RNA is processed in the nucleus to a smaller but not yet final form, then exported to the cytoplasm. There, two biological machines, both with the ability to cleave RNA, take part in a two-step process. One machine, called **Dicer,** recognizes **double-stranded RNA (dsRNA)** molecules and cleaves them into ~22-nucleotide products. A second machine, called **RISC (RNA-induced silencing complex),** binds to a short dsRNA and unwinds it into the

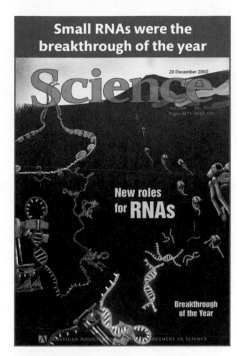

Figure 8-19 Cover of *Science* magazine, December 2002. [*Science, vol. 298, no. 5602, December 20, 2002.*]

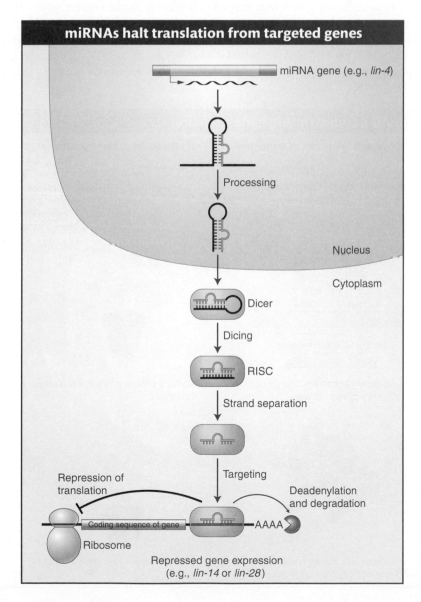

Figure 8-20 miRNAs are synthesized by polII as longer RNAs that are processed in several steps to their mature form. Once fully processed, miRNAs bind to RISC and direct its activities to reduce the expression of complementary mRNAs by either repressing their translation or promoting their degradation.

biologically active single-stranded miRNA. The miRNA, still bound to RISC, binds to complementary mRNAs. RISC then represses the translation of these mRNAs into protein or removes the polyA tail, which hastens mRNA degradation. In the example shown in Figure 8-20, the *lin-4* miRNA binds to *lin-14* and *lin-28* mRNAs and represses their translation.

You will learn more about the function of miRNAs in Chapters 12 and 13. The key point to remember is that part of an miRNA is complementary to the RNA of the gene it regulates. When the regulated gene needs to be shut down or its expression reduced, the miRNA gene is transcribed into RNA and that RNA binds to the RNA of the regulated gene, interfering with translation into protein or promoting its degradation.

> **Message** miRNAs are processed from longer RNA pol II transcripts by Dicer, which binds to double-stranded RNAs. The biologically active single-stranded miRNA binds to RISC and guides it to complementary sequences in protein-coding mRNAs, where RISC either represses translation or promotes mRNA degradation.

siRNAs ensure genome stability

Scientists soon found a different case of dsRNA that could repress gene expression prior to translation. This finding led to the discovery of a second type of short RNA, siRNAs. This second type of short RNA has a very different origin and function from miRNAs. In contrast to miRNAs, an siRNA silences the gene that produces it. Thus, it is not used to regulate other genes, but rather to shut off

Figure 8-21 Three experiments reveal key features of gene silencing. (a) Fire and Mello demonstrated that dsRNA copies can selectively silence genes in *C. elegans*. (b) Jorgensen discovered that a transgene can silence an endogenous petunia gene necessary for floral color. (c) Baulcombe showed that plants with a copy of a viral transgene were resistant to viral infection and produced siRNAs complementary to the viral genome.

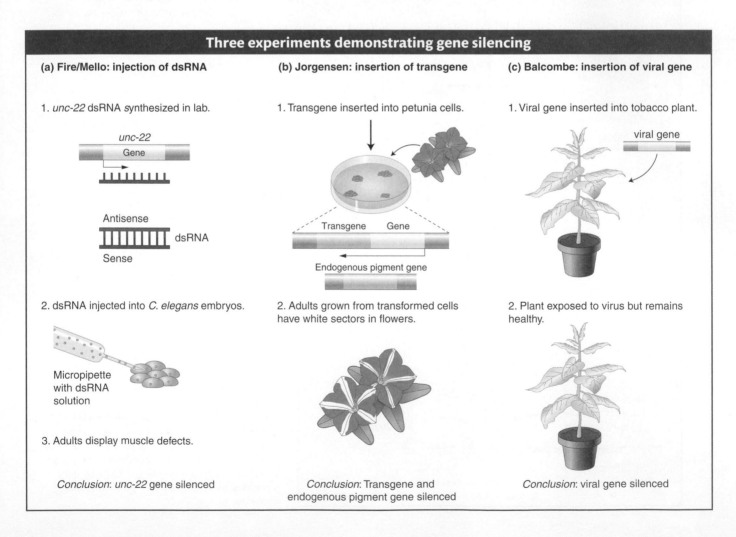

Three experiments demonstrating gene silencing

(a) Fire/Mello: injection of dsRNA

1. *unc-22* dsRNA synthesized in lab.

unc-22
Gene

Antisense
dsRNA
Sense

2. dsRNA injected into *C. elegans* embryos.

Micropipette with dsRNA solution

3. Adults display muscle defects.

Conclusion: *unc-22* gene silenced

(b) Jorgensen: insertion of transgene

1. Transgene inserted into petunia cells.

Transgene   Gene

Endogenous pigment gene

2. Adults grown from transformed cells have white sectors in flowers.

Conclusion: Transgene and endogenous pigment gene silenced

(c) Balcombe: insertion of viral gene

1. Viral gene inserted into tobacco plant.

viral gene

2. Plant exposed to virus but remains healthy.

Conclusion: viral gene silenced

undesirable genetic elements that insert into the genome. Such undesirable elements could be the genes in an infecting virus or they could be internal genetic elements called transposons that you will learn about in Chapter 15.

In 1998, five years after the discovery of miRNAs, Andrew Fire and Craig Mello reported that they had found a potent way to selectively turn off genes, also in the roundworm *C. elegans*. Fire and Mello discovered that, by injecting dsRNA copies of a *C. elegans* gene into *C. elegans* embryos, they were able to block the synthesis of the protein product of that gene (Figure 8-21a). The selective shutting off of the gene by this procedure is called **gene silencing.** The dsRNA had been synthesized in the laboratory and was composed of a sense (coding) RNA strand and a complementary **antisense** RNA strand. In their initial experiment, Fire and Mello injected dsRNA copies of the *unc-22* gene into *C. elegans* embryos and watched as the embryos grew into adults that twitched and had muscle defects. This result was exciting because *unc-22* was known to encode a muscle protein and null mutants of *unc-22* displayed the same twitching and muscle defects. Taken together, these observations indicated that the injected dsRNA prevented the production of the Unc-22 protein. For their discovery of a new way to silence genes, Fire and Mello were awarded the Nobel Prize in Medicine or Physiology in 2004.

If instead of a dsRNA copy of a gene, what would happen if a DNA copy of a gene normally found in an organism were inserted into its genome? In such an experiment the introduced gene would be an example of a **transgene,** which is short for "transformed gene." A transgene is a gene that has been introduced into the chromosomes of an organism in the laboratory. An organism containing a transgene in its genome is called either a **transgenic** organism or **genetically modified organism** (or the popular abbreviated term **GMO**). This experiment was actually done in 1990 by Rich Jorgensen, a plant scientist studying the color of flowers in petunias.

One of the greatest joys of doing scientific research is observing a completely unexpected result. This is precisely what happened to Jorgensen after he inserted a petunia gene that encodes an enzyme necessary for the synthesis of purple-blue floral pigment into a normal petunia plant having purple-blue flowers (Figure 8-21b). He expected that the floral color of this transgenic plant would be unchanged. After all, the transgenic plant had two good genes necessary for pigment production—one at its usual locus in the petunia genome (called the **endogenous gene**; in Figure 8-21b it is called the pigment gene) plus the introduced transgene that was inserted elsewhere in the genome. However, instead of purple flowers, the transgenic plants displayed the unusual floral pat-

Figure 8-22 (a) The wild-type (no transgene) phenotype. (b and c) So-called cosuppression phenotypes resulting from the transformation of the wild-type petunia shown in part *a* with a petunia gene required for pigmentation. In the colorless regions, both the transgene and the chromosomal copy of the same gene have been inactivated. [*Photographs courtesy of Richard A. Jorgensen, Department of Plant Sciences, University of Arizona.*]

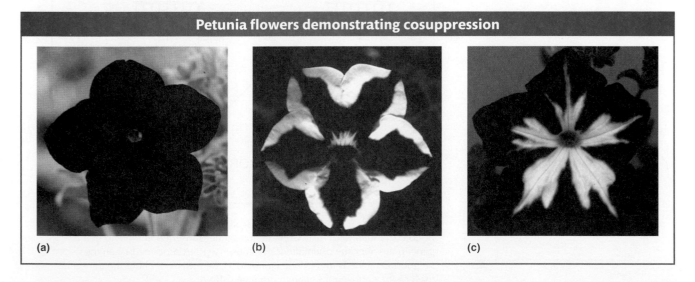

Petunia flowers demonstrating cosuppression

(a)　　　　(b)　　　　(c)

terns shown in Figure 8-22. In a totally unexpected outcome, the transgene triggered suppression of both the transgene and the endogenous pigment gene, resulting in white flowers or, more commonly, white floral sectors. This phenomenon is called **cosuppression** because the expression of both the introduced transgene and the endogenous copy is suppressed.

To review, introduction of either a dsRNA copy of a gene or the gene itself into an organism can silence that gene. To understand why these different experiments led to the same result, scientists hypothesized that the insertion of the transgene led to the synthesis of antisense RNA, which could complement with sense RNA to produce dsRNA. Because scientists cannot control where transgenes insert, some transgenes will end up next to genes in an opposite orientation (Figure 8-23). Transcription initiated at the gene promoter can "read through" into the transgene and produce a very long "chimeric" RNA containing both the sense strand of the gene and the antisense strand of the transgene. Double-stranded RNA will then form when the antisense part of the long RNA hybridizes with sense RNA produced either by the transgene or the endogenous gene.

Thus, dsRNA is a common feature of this form of gene silencing. However, the function of this process is clearly not to shut off genes introduced by scientists. What is the normal role of this form of gene silencing in the cell? An important clue came from the experiments conducted by another plant scientist, David Baulcombe, who was investigating the reason why tobacco plants that were engineered to express a viral gene were resistant to subsequent infection by the virus. In this experiment the viral gene is another example of a transgene, which was, in this case, introduced into the tobacco genome (see Figure 8-21c). A key difference between this and the petunia experiment was that tobacco plants do not normally have a viral gene in their genome. So this experiment suggested that this form of gene silencing functions to silence invading viruses.

Baulcombe and his co-workers found that the resistant plants, and only the resistant plants, produced large amounts of short RNAs, 25 nucleotides in length, that were complementary to the viral genome. Significantly, short RNAs related

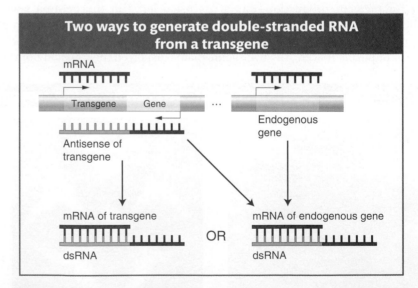

Figure 8-23 The insertion of a transgene can lead to the production of double-stranded RNA (dsRNA) if the transgene is inserted at the end of a gene in the opposite orientation. The antisense RNA produced when the neighboring gene is transcribed can bind to the mRNA of either the transgene itself or the endogenous gene to produce dsRNA.

to the endogenous genes have also been found to be present during gene silencing in the worm and in the petunia. The short RNAs generated during viral resistance and gene silencing associated with either injected dsRNAs or transgenes are now collectively called **small interfering RNAs (siRNAs).** The phenomena that results in gene silencing and viral resistance through the production of siRNAs is called **RNA interference (RNAi).**

The short RNAs (21–31 nucleotides in length) are now classified as one of three types depending on their biogenesis: miRNAs or siRNAs (both 21–25 nucleotides) or the recently discovered **piwi-interacting RNAs (piRNAs,** 24–31 nucleotides). Because the mechanism of piRNA synthesis is still under investigation, we will focus on the better characterized miRNAs and siRNAs.

Similar mechanisms generate siRNA and miRNA

As we have seen, siRNAs can arise from an antisense copy of any source of mRNA in the genome: from endogenous genes to transgenes to invading viruses. However, the most likely source of antisense RNA is not an organism's own genes, but rather foreign DNA that inserts into the genome. In this regard, it would be correct to think of siRNAs as the product of a genome immune system that detects the insertion of foreign DNA by, in some cases, promoting the synthesis of antisense mRNA. Complementarity between sense and antisense RNAs produces dsRNAs, which, as in the miRNA pathway, are recognized by Dicer and cleaved into short double-stranded products that are bound by RISC (Figure 8-24). As with miRNAs, RISC unwinds the product into the biologically active single-stranded siRNA that targets RISC to complementary mRNAs so they can be degraded. Unlike miRNAs, complementarity between siRNAs and mRNAs is perfect; there are no mismatches. This difference is probably responsible for the different outcomes: miRNAs direct RISC to repress the translation of a mRNA, while siRNAs direct RISC to degrade the mRNA.

As discussed above, the production of siRNAs probably plays an important role in viral defense. However, its most important role may be to protect the hereditary material of an organism from genetic elements in its own genome. In Chapter 15, you will learn about the transposable elements that constitute a huge fraction of the genomes of multicellular eukaryotes, including humans. These elements can amplify themselves and move to new locations, creating an obvious threat to the integrity of the genome. Just like the introduction of transgenes by scientists, the movement of transposable elements into new chromosomal locations can trigger the production of siRNAs by generating dsRNA. The siRNAs eventually inactivate the transposable elements in part by preventing the production of the protein products needed for their movement and amplification.

> **Message** Antisense RNA is frequently formed in response to the insertion of foreign DNA into the genome. Dicer detects double-stranded RNA that forms between antisense and sense RNA and processes it into short RNAs. RISC binds a short RNA and unwinds it to form biologically active siRNA. The siRNA targets RISC to a perfectly complementary mRNA, which is degraded, thus silencing the expression of the foreign DNA.

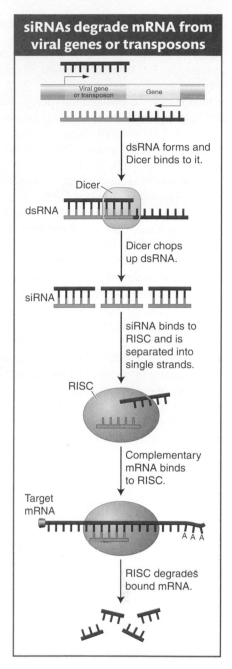

Figure 8-24 In the RNA interference pathway, double-stranded RNA (dsRNA) specifically interacts with the Dicer complex, which chops the dsRNA up. The RNA-induced silencing complex (RISC) uses the small dsRNAs to find and destroy homologous mRNA transcribed from the target DNA, thereby repressing gene expression.

Summary

We know that information is not transferred directly from DNA to protein, because, in a eukaryotic cell, DNA is in the nucleus, whereas protein is synthesized in the cytoplasm. Information transfer from DNA to protein requires an intermediate. That intermediate is RNA.

Although DNA and RNA are nucleic acids, RNA differs from DNA in that (1) it is usually single stranded rather than a double helix, (2) its nucleotides contain the sugar ribose rather than deoxyribose, (3) it has the pyrimidine base uracil rather than thymine, and (4) it can serve as a biological catalyst.

The similarity of RNA to DNA suggests that the flow of information from DNA to RNA relies on the complementarity of bases, which is also the key to DNA replication. A template DNA strand is copied, or transcribed, into either a functional RNA (such as transfer RNA or ribosomal RNA), which is never translated into polypeptides, or a messenger RNA, from which proteins are synthesized.

In prokaryotes, all classes of RNA are transcribed by a single RNA polymerase. This multisubunit enzyme initiates transcription by binding to the DNA at promoters that contain specific sequences at -35 and -10 bases before the transcription start site at $+1$. After being bound, RNA polymerase locally unwinds the DNA and begins incorporating ribonucleotides that are complementary to the template DNA strand. The chain grows in the 5′-to-3′ direction until one of two mechanisms, intrinsic or rho dependent, leads to the dissociation of the polymerase and the RNA from the DNA template. As we will see in Chapter 9, in the absence of a nucleus, prokaryotic RNAs that encode proteins are translated while they are being transcribed.

In eukaryotes, there are three different RNA polymerases; only RNA polymerase II transcribes mRNAs. Overall, the phases of initiation, elongation, and termination of RNA synthesis in eukaryotes resemble those in prokaryotes. However, there are important differences. RNA polymerase II does not bind directly to promoter DNA, but rather to general transcription factors, one of which recognizes the TATA sequence in most eukaryotic promoters. RNA polymerase II is a much larger molecule than its prokaryotic counterpart. It contains numerous subunits that function not only to elongate the primary RNA transcript, but also to coordinate the extensive processing events that are necessary to produce the mature mRNA. These processing events are 5′ capping, intron removal and exon joining by spliceosomes, and 3′ cleavage followed by polyadenylation. Part of the RNA polymerase II core, the carboxyl tail domain, is positioned ideally to interact with the nascent RNA as it emerges from polymerase. Through the CTD, RNA polymerase II coordinates the numerous events of RNA synthesis and processing.

Discoveries of the past 20 years have revealed the importance of new classes of functional RNAs. Once thought to be a lowly messenger, RNA is now recognized as a versatile and dynamic participant in many cellular processes. The discovery of self-splicing introns demonstrated that RNA can function as a catalyst, much like proteins. Since the discovery of these ribozymes, the scientific community has begun to pay more attention to RNA. Small nuclear RNAs, the noncoding RNAs in the spliceosome, are now recognized to provide the catalytic activity to remove introns and join exons. The twentieth century ended with the discovery that two other classes of functional RNA, miRNA and siRNA, associate with RNA-induced silencing complexes (RISC) and target complementary cellular mRNA for repression (in the case of miRNA) or for destruction (in the case of siRNA).

KEY TERMS

alternative splicing (p. 284, 297)
antisense RNA strand (p. 303)
cap (p. 296)
carboxyl tail domain (CTD) (p. 295)
coding strand (p. 289)
consensus sequence (p. 291)
constitutive (p. 288)
cosuppression (p. 304)
cotranscriptional processing (p. 295)
Dicer (p. 301)
double-stranded RNA (dsRNA) (p. 301)
downstream (p. 290)
elongation (p. 290)
endogenous gene (p. 303)
exon (p. 284)
functional RNA (p. 287, 298)
general transcription factor (GTF) (p. 293)
gene silencing (p. 303)
genetically modified organism (GMO) (p. 303)
GU–AG rule (p. 298)
initiation (p. 290)
intron (p. 284)

long noncoding RNAs (lncRNAs) (p. 287)
messenger RNA (mRNA) (p. 287)
microRNA (miRNA) (p. 287, 301)
non-protein-coding RNA (ncRNA) (p. 284)
piwi-interacting RNAs (piRNAs) (p. 287, 305)
poly(A) tail (p. 296)
posttranscriptional processing (p. 295)
preinitiation complex (PIC) (p. 294)
primary transcript (pre-mRNA) (pp. 293–294)
promoter (p. 290)
proteome (p. 297)
pulse–chase experiment (p. 285)
ribose (p. 286)
ribosomal RNA (rRNA) (p. 287)
ribozyme (p. 286)
RISC (RNA-induced silencing complex) (p. 301)
RNA interference (RNAi) (p. 305)
RNA polymerase (p. 289)
RNA polymerase holoenzyme (p. 291)

RNA processing (p. 293)
RNA splicing (p. 284)
RNA world (p. 300)
self-splicing intron (p. 300)
sigma factor (σ) (p. 291)
small interfering RNA (siRNA) (p. 287, 305)
small nuclear RNA (snRNA) (p. 287)
spliceosome (p. 284)
splicing (p. 296)
TATA-binding protein (TBP) (p. 294)
TATA box (p. 294)
template (p. 289)
termination (p. 290)
transcript (p. 288)
transcription (p. 288)
transcription bubble (p. 291)
transfer RNA (tRNA) (p. 287)
transgene (p. 303)
transgene silencing (p. 000)
3′ untranslated region (3′ UTR) (p. 292)
5′ untranslated region (5′ UTR) (p. 291)
upstream (p. 290)
uracil (U) (p. 286)

PROBLEMS

Most of the problems are also available for review/grading through the ᴳᴱᴺᴱᵀᴵᶜˢ**PORTAL** www.yourgeneticsportal.com.

WORKING WITH THE FIGURES

1. In Figure 8-3, why are the arrows for genes 1 and 2 pointing in opposite directions?

2. In Figure 8-5, draw the "one gene" at much higher resolution with the following components: DNA, RNA polymerase(s), RNA(s).

3. In Figure 8-6, describe where the gene promoter is located.

4. In Figure 8-9b, write a sequence that could form the hairpin loop structure.

5. How do you know that the events in Figure 8-13 are occurring in the nucleus?

6. In Figure 8-15, what do you think would be the effect of a G to A mutation in the first G residue of the intron?

7. In Figure 8-23, show how the double-stranded RNA is able to silence the transgene. What would have to happen for the transgene to also silence the flanking cellular gene (in yellow)?

BASIC PROBLEMS

8. The two strands of λ phage DNA differ from each other in their GC content. Owing to this property, they can be separated in an alkaline cesium chloride gradient (the alkalinity denatures the double helix). When RNA synthesized by λ phage is isolated from infected cells, it is found to form DNA–RNA hybrids with both strands of λ DNA. What does this finding tell you? Formulate some testable predictions.

9. In both prokaryotes and eukaryotes, describe what else is happening to the RNA while RNA polymerase is synthesizing a transcript from the DNA template.

10. List three examples of proteins that act on nucleic acids.

11. What is the primary function of the sigma factor? Is there a protein in eukaryotes analogous to the sigma factor?

12. You have identified a mutation in yeast, a unicellular eukaryote, that prevents the capping of the 5′ end of the RNA transcript. However, much to your surprise, all the enzymes required for capping are normal. You determine that the mutation is, instead, in one of the subunits of RNA polymerase II. Which subunit is mutant and how does this mutation result in failure to add a cap to yeast RNA?

13. Why is RNA produced only from the template DNA strand and not from both strands?

14. A linear plasmid contains only two genes, which are transcribed in opposite directions, each one from the end, toward the center of the plasmid. Draw diagrams of

    a. the plasmid DNA, showing the 5′ and 3′ ends of the nucleotide strands.

    b. the template strand for each gene.

    c. the positions of the transcription-initiation site.

    d. the transcripts, showing the 5′ and 3′ ends.

15. Are there similarities between the DNA replication bubbles and the transcription bubbles found in eukaryotes? Explain.

16. Which of the following statements are true about eukaryotic mRNA?

    a. The sigma factor is essential for the correct initiation of transcription.

    b. Processing of the nascent mRNA may begin before its transcription is complete.

    c. Processing takes place in the cytoplasm.

    d. Termination is accomplished by the use of a hairpin loop or the use of the rho factor.

    e. Many RNAs can be transcribed simultaneously from one DNA template.

17. A researcher was mutating prokaryotic cells by inserting segments of DNA. In this way, she made the following mutation:

    Original   TTGACAT <u>15 to 17 bp</u> TATAAT

    Mutant   TATAAT <u>15 to 17 bp</u> TTGACAT

    a. What does this sequence represent?

    b. What do you predict will be the effect of such a mutation? Explain.

18. You will learn more about genetic engineering in Chapter 10, but for now, put on your genetic engineer's cap and try to solve this problem. *E. coli* is widely used in laboratories to produce proteins from other organisms.

    a. You have isolated a yeast gene that encodes a metabolic enzyme and want to produce this enzyme in *E. coli*. You suspect that the yeast promoter will not work in *E. coli*. Why?

    b. After replacing the yeast promoter with an *E. coli* promoter, you are pleased to detect RNA from the yeast gene but are confused because the RNA is almost twice the length of the mRNA from this gene isolated from yeast. Explain why this result might have occurred.

19. Draw a prokaryotic gene and its RNA product. Be sure to include the promoter, transcription start site, transcription termination site, untranslated regions, and labeled 5′ and 3′ ends.

20. Draw a two-intron eukaryotic gene and its pre-mRNA and mRNA products. Be sure to include all the features of the prokaryotic gene included in your answer to Problem 19, plus the processing events required to produce the mRNA.

21. A certain *Drosophila* protein-encoding gene has one intron. If a large sample of null alleles of this gene is examined, will any of the mutant sites be expected

   a. in the exons?

   b. in the intron?

   c. in the promoter?

   d. in the intron–exon boundary?

22. What are self-splicing introns and why does their existence support the theory that RNA evolved before protein?

23. Antibiotics are drugs that selectively kill bacteria without harming animals. Many antibiotics act by selectively binding to certain proteins that are critical for bacterial function. Explain why some of the most successful antibiotics target bacterial RNA polymerase.

CHALLENGING PROBLEMS

24. The following data represent the base compositions of double-stranded DNA from two different bacterial species and their RNA products obtained in experiments conducted in vitro:

Species	$(A + T)/$ $(G + C)$	$(A + U)/$ $(G + C)$	$(A + G)/$ $(U + C)$
Bacillus subtilis	1.36	1.30	1.02
E. coli	1.00	0.98	0.80

   a. From these data, can you determine whether the RNA of these species is copied from a single strand or from both strands of the DNA? How? Drawing a diagram will make it easier to solve this problem.

   b. Explain how you can tell whether the RNA itself is single stranded or double stranded.

   (Problem 24 is reprinted with the permission of Macmillan Publishing Co., Inc., from M. Strickberger, *Genetics.* Copyright 1968, Monroe W. Strickberger.)

25. A human gene was initially identified as having three exons and two introns. The exons are 456, 224, and 524 bp, whereas the introns are 2.3 kb and 4.6 kb.

   a. Draw this gene, showing the promoter, introns, exons, and transcription start and stop sites.

   b. Surprisingly, this gene is found to encode not one but two mRNAs that have only 224 nucleotides in common. The original mRNA is 1204 nucleotides, and the new mRNA is 2524 nucleotides. Use your drawing to show how this one region of DNA can encode these two transcripts.

26. While working in your laboratory, you isolate an mRNA from *C. elegans* that you suspect is essential for embryos to develop successfully. With the assumption that you are able to turn mRNA into double-stranded RNA, design an experiment to test your hypothesis.

27. Glyphosate is an herbicide used to kill weeds. It is the main component of a product made by the Monsanto Company called Roundup. Glyphosate kills plants by inhibiting an enzyme in the shikimate pathway called EPSPS. This herbicide is considered safe because animals do not have the shikimate pathway. To sell even more of their herbicide, Monsanto commissioned its plant geneticists to engineer several crop plants, including corn, to be resistant to glyphosate. To do so, the scientists had to introduce an EPSPS enzyme that was resistant to inhibition by glyphosate into crop plants and then test the transformed plants for resistance to the herbicide.

Imagine that you are one of these scientists and that you have managed to successfully introduce the resistant EPSPS gene into the corn chromosomes. You find that some of the transgenic plants are resistant to the herbicide, whereas others are not. Your supervisor is very upset and demands an explanation of why some of the plants are not resistant even though they have the transgene in their chromosomes. Draw a picture to help him understand.

28. Many human cancers result when a normal gene mutates and leads to uncontrolled growth (a tumor). Genes that cause cancer when they mutate are called oncogenes. Chemotherapy is effective against many tumors because it targets rapidly dividing cells and kills them. Unfortunately, chemotherapy has many side effects, such as hair loss or nausea, because it also kills many of our normal cells that are rapidly dividing, such as those in the hair follicles or stomach lining.

Many scientists and large pharmaceutical companies are excited about the prospects of exploiting the RNAi pathway to selectively inhibit oncogenes in life-threatening tumors. Explain in very general terms how gene-silencing therapy might work to treat cancer and why this type of therapy would have fewer side effects than chemotherapy.

Proteins and Their Synthesis 9

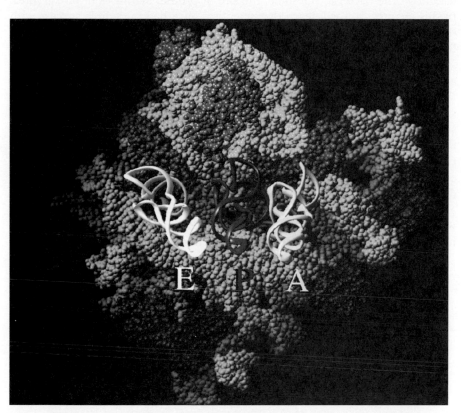

This image shows at atomic resolution a surface of the ribosome from the bacterium *Haloarcula marismortui,* deduced from X-ray crystallography. The part of the ribosome consisting of RNA is shown in blue; that consisting of protein is shown in purple. The white, red, and yellow structures in the center are tRNAs at the E, P, and A binding sites, their acceptor stems disappearing into a cleft in the ribosome. [*From P. Nissen, J. Hansen, N. Ban, P. B. Moore, and T. A. Steitz, "The Structural Basis of Ribosome Activity in Peptide Bond Synthesis," Science 289, 2000, 920–930, Fig. 10A on p. 926.*]

In an address to Congress in 1969, William Stewart, Surgeon General of the United States, said, "It is time to close the book on infectious diseases. The war against pestilence is over." At the time, his claim of victory was not an unreasonable boast. In the preceding two decades, three infectious diseases that had plagued humankind for centuries—polio, smallpox, and tuberculosis—had been virtually eliminated throughout the world. A major contributing factor to the eradication of some infectious diseases was the discovery and widespread use of antibiotics, a diverse group of chemical compounds that kill specific bacterial pathogens without harming the animal host. Antibiotics such as penicillin, tetracycline, ampicillin, and chloramphenicol, to name but a few, have saved hundreds of millions of lives.

KEY QUESTIONS

- How are the sequences of a gene and its protein related?

- Why is it said that the genetic code is nonoverlapping and degenerate?

- What is the evidence that the ribosomal RNA, not the ribosomal proteins, carries out the key steps in translation?

- What is posttranslational processing, and why is it important for protein function?

OUTLINE

9.1 Protein structure

9.2 The genetic code

9.3 tRNA: the adapter

9.4 Ribosomes

9.5 The proteome

Unfortunately, William Stewart's claim of victory in the battle against infectious disease was premature. The overuse of antibiotics worldwide has spurred the evolution of resistant bacterial strains. For example, each year, more than 2 million hospital patients in the United States acquire an infection that is resistant to antibiotics and 90,000 die as a result. How did resistance develop so quickly? Will infectious disease be, once again, a significant cause of human mortality? Or will scientists be able to use their understanding of resistance mechanisms to develop more durable antibiotics?

To answer these questions, scientists have focused on the cellular machinery that is targeted by antibiotics. More than half of all antibiotics currently in use target the bacterial ribosome, the site of protein synthesis in prokaryotes. In this chapter, you will learn that scientists have had incredible success with the use of a technique called X-ray crystallography to visualize the ribosomal RNAs (rRNAs) and the ~100 proteins that make up the large and small ribosomal subunits of bacterial ribosomes. Although the ribosomes of prokaryotes and eukaryotes are very similar, there are still subtle differences. Because of these differences, antibiotics are able to target bacterial ribosomes but leave eukaryotic ribosomes untouched. Using X-ray crystallography, scientists have also succeeded in visualizing antibiotics bound to the ribosome (Figure 9-1). From these studies, they have determined that mutations in bacterial rRNA and/or ribosomal proteins are responsible for antibiotic resistance. With this knowledge of the points of contact between certain antibiotics and the ribosome, drug designers are attempting to design a new generation of antibiotics that, for example, will be able to bind to multiple nearby sites. Resistance to such a drug would be less likely to evolve because it would require the occurrence of two mutations, which is a very unlikely event even for bacteria.

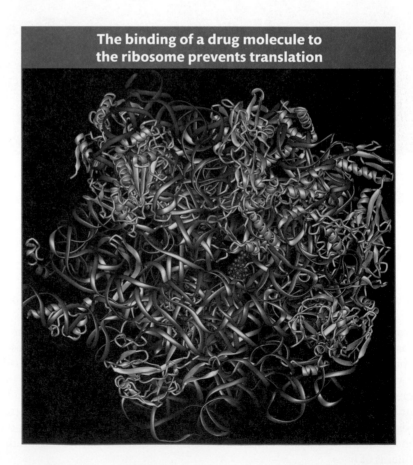

The binding of a drug molecule to the ribosome prevents translation

Figure 9-1 The drug erythromycin (red) blocks the tunnel from which a newly synthesized protein emerges from the ribosome. The image is a top view of the 50S ribosomal subunit in the bacterium *Deinococcus radiodurans.* Ribosomal RNAs are shown in blue, and ribosomal proteins in gold. [*Dr. Joerg Harms, MPI for Molecular Genetics, Berlin, Germany.*]

Chapters 7 and 8 described how DNA is copied from generation to generation and how RNA is synthesized from specific regions of DNA. We can think of these processes as two stages of information transfer: *replication* (the synthesis of DNA) and *transcription* (the synthesis of an RNA copy of a part of the DNA). In this chapter, you will learn about the final stage of information transfer: *translation* (the synthesis of a polypeptide directed by the RNA sequence).

As you learned in Chapter 8, RNA transcribed from genes is classified as either messenger RNA (mRNA) or functional RNA. In this chapter, we will see the fate of both RNA classes. The vast majority of genes encode mRNAs whose function is to serve as an intermediate in the synthesis of the ultimate gene product, protein. In contrast, recall that *functional* RNAs are active as RNAs; they are never translated into proteins. The main classes of functional RNAs are important actors in protein synthesis. They include transfer RNAs and ribosomal RNAs.

- **Transfer RNA (tRNA)** molecules are the adapters that translate the three-nucleotide codon in the mRNA into the corresponding amino acid, which is brought by the tRNA to the ribosome in the process of translation. The tRNAs are general components of the translation machinery; a tRNA molecule can bring an amino acid to the ribosome for the purpose of translating *any* mRNA.

- **Ribosomal RNAs (rRNAs)** are the major components of **ribosomes,** which are large macromolecular complexes that assemble amino acids to form the protein whose sequence is encoded in a specific mRNA. Ribosomes are composed of several types of rRNA and scores of different proteins. Like tRNA, ribosomes are general in function in the sense that they can be used to translate the mRNAs of *any* protein-coding gene.

Although most genes encode mRNAs, functional RNAs make up, by far, the largest fraction of total cellular RNA. In a typical actively dividing eukaryotic cell, rRNA and tRNA account for almost 95 percent of the total RNA, whereas mRNA accounts for only about 5 percent. Two factors explain the abundance of rRNAs and tRNAs. First, they are much more stable than mRNAs, and so these molecules remain intact much longer. Second, the transcription of rRNA and tRNA genes constitutes more than half of the total nuclear transcription in active eukaryotic cells and almost 80 percent of transcription in yeast cells.

The components of the translational machinery and the process of translation are very similar in prokaryotes and eukaryotes. The major feature that distinguishes translation in prokaryotes from that in eukaryotes is the location where transcription and translation take place in the cell: the two processes take place in the same compartment in prokaryotes, whereas they are physically separated in eukaryotes. After extensive processing, eukaryotic mRNAs are exported from the nucleus for translation on ribosomes that reside in the cytoplasm. In contrast, transcription and translation are coupled in prokaryotes: translation of an RNA begins at its 5′ end while the rest of the mRNA is still being transcribed.

9.1 Protein Structure

When a primary transcript has been fully processed into a mature mRNA molecule, translation into protein can take place. Before considering how proteins are made, we need to understand protein structure.

Proteins are the main determinants of biological form and function. These molecules heavily influence the shape, color, size, behavior, and physiology of organisms. Because genes function by encoding proteins, understanding the nature of proteins is essential to understanding gene action.

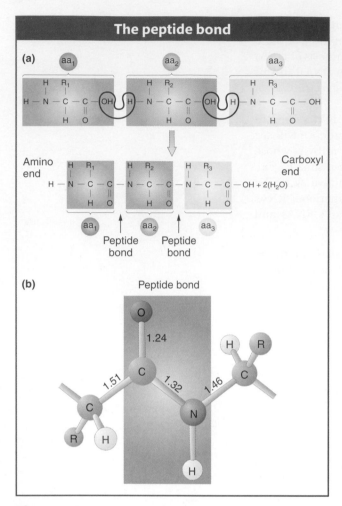

The peptide bond

(a)

Amino end

Carboxyl end

aa₁   aa₂   aa₃

Peptide bond   Peptide bond

(b)

Peptide bond

O

1.24

1.51   1.32   1.46

C   C   N

R   H   H   H

R

H

Figure 9-2 (a) A polypeptide is formed by the removal of water between amino acids to form peptide bonds. Each aa indicates an amino acid. R_1, R_2, and R_3 represent R groups (side chains) that differentiate the amino acids. (b) The peptide bond is a rigid planar unit with the R groups projecting out from the C–N backbone. Standard bond distances (in angstroms) are shown. [*(b) From L. Stryer, Biochemistry, 4th ed. Copyright 1995 by Lubert Stryer.*]

GENETICS**PORTAL** **ANIMATED ART: Translation: peptide-bond formation**

A protein is a polymer composed of monomers called **amino acids.** In other words, a protein is a chain of amino acids. Because amino acids were once called *peptides*, the chain is sometimes referred to as a **polypeptide.** Amino acids all have the general formula

$$H_2N - \underset{\underset{R}{|}}{\overset{\overset{H}{|}}{C}} - COOH$$

All amino acids have a side chain, or R (reactive) group. There are 20 amino acids known to exist in proteins, each having a different R group that gives the amino acid its unique properties. The side chain can be anything from a hydrogen atom (as in the amino acid glycine) to a complex ring (as in the amino acid tryptophan). In proteins, the amino acids are linked together by covalent bonds called peptide bonds. A peptide bond is formed by the linkage of the **amino end** (NH_2) of one amino acid with the **carboxyl end** (COOH) of another amino acid. One water molecule is removed during the reaction (Figure 9-2). Because of the way in which the peptide bond forms, a polypeptide chain always has an amino end (NH_2) and a carboxyl end (COOH), as shown in Figure 9-2a.

Proteins have a complex structure that has four levels of organization, illustrated in Figure 9-3. The linear sequence of the amino acids in a polypeptide chain constitutes the **primary structure** of the protein. Local regions of the polypeptide chain fold into specific shapes, called the protein's **secondary structure.** Each shape arises from the bonding forces between amino acids that are close together in the linear sequence. These forces include several types of weak bonds, notably hydrogen bonds, electrostatic forces, and van der Waals forces. The most common secondary structures are the α helix and the pleated sheet. Different proteins show either one or the other or sometimes both within their structures. **Tertiary structure** is produced by the folding of the secondary structure. Some proteins have **quaternary structure:** such a protein is composed of two or more separate folded polypeptides, also called **subunits,** joined by weak bonds. The quaternary association can be between different types of polypeptides (resulting in a heterodimer if there are two subunits) or between identical polypeptides (making a homodimer). Hemoglobin is an example of a heterotetramer, a four-subunit protein; it is composed of two copies each of two different polypeptides, shown in green and purple in Figure 9-3d.

Many proteins are compact structures; they are called **globular proteins.** Enzymes and antibodies are among the best-known globular proteins. Proteins with linear shape, called **fibrous proteins,** are important components of such structures as skin, hair, and tendons.

Shape is all-important to a protein because a protein's specific shape enables it to do its specific job in the cell. A protein's shape is determined by its primary amino acid sequence and by conditions in the cell that promote the folding and bonding necessary to form higher-level structures. The folding of proteins into their correct conformation will be discussed at the end of this chapter. The amino acid sequence also determines which R groups are present at specific positions and thus available to bind with other cellular components. The active sites of enzymes are good illustrations of the precise interactions of R groups. Each enzyme has a pocket called the **active site** into which its substrate or substrates

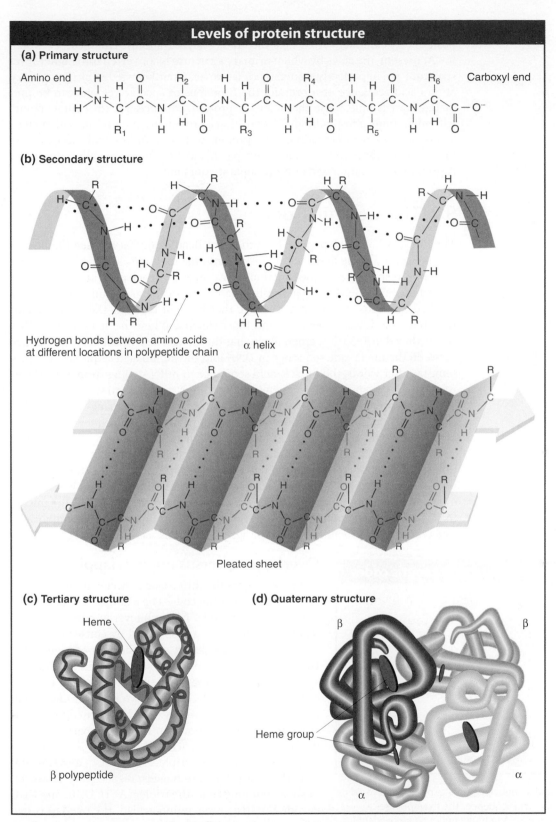

Figure 9-3 A protein has four levels of structure. (a) Primary structure. (b) Secondary structure. The polypeptide can form a helical structure (an α helix) or a zigzag structure (a pleated sheet). The pleated sheet has two polypeptide segments arranged in opposite polarity, as indicated by the arrows. (c) Tertiary structure. The heme group is a nonprotein ring structure with an iron atom at its center. (d) Quaternary structure illustrated by hemoglobin, which is composed of four polypeptide subunits: two α subunits and two β subunits.

can fit. Within the active site, the R groups of certain amino acids are strategically positioned to interact with a substrate and catalyze a specific chemical reaction.

At present, the rules by which primary structure is converted into higher-level structure are imperfectly understood. However, from knowledge of the primary amino acid sequence of a protein, the functions of specific regions can be predicted. For example, some characteristic protein sequences are the contact points with membrane phospholipids that position a protein in a membrane. Other characteristic sequences act to bind the protein to DNA. Amino acid sequences or protein folds that are associated with particular functions are called **domains.** A protein may contain one or more separate domains.

9.2 The Genetic Code

The one-gene–one-polypeptide hypothesis of Beadle and Tatum (see Chapter 6) was the source of the first exciting insight into the functions of genes: genes were somehow responsible for the function of enzymes, and each gene apparently controlled one enzyme. This hypothesis became one of the great unifying concepts in biology because it provided a bridge that brought together the concepts and research techniques of genetics and biochemistry. When the structure of DNA was deduced in 1953, it seemed likely that there must be a linear correspondence between the nucleotide sequence in DNA and the amino acid sequence in a protein. In other words, the nucleic acid sequence in mRNA going from 5′ to 3′ corresponds to the amino acid sequence going from N-terminus to C-terminus.

If genes are segments of DNA and if a strand of DNA is just a string of nucleotides, then the sequence of nucleotides must somehow dictate the sequence of amino acids in proteins. How does the DNA sequence dictate the protein sequence? The analogy to a code springs to mind at once. Simple logic tells us that, if the nucleotides are the "letters" in a code, then a combination of letters can form "words" representing different amino acids. First, we must ask how the code is read. Is it overlapping or nonoverlapping? Then we must ask how many letters in the mRNA make up a word, or **codon,** and which codon or codons represent each amino acid. The cracking of the genetic code is the story told in this section.

Overlapping versus nonoverlapping codes

Figure 9-4 shows the difference between an overlapping and a nonoverlapping code. The example shows a three-letter, or **triplet,** code. For a nonoverlapping code, consecutive amino acids are specified by consecutive code words (codons), as shown at the bottom of Figure 9-4. For an overlapping code, consecutive amino acids are specified by codons that have some consecutive bases in common; for example, the last two bases of one codon may also be the first two bases of the next codon. Overlapping codons are shown in the upper part of Figure 9-4. Thus, for the sequence AUUGCUCAG in a nonoverlapping code, the three triplets AUU, GCU, and CAG encode the first three amino acids, respectively. However, in an overlapping code, the triplets AUU, UUG, and UGC encode the first three amino acids if the overlap is two bases, as shown in Figure 9-4.

By 1961, it was already clear that the genetic code was nonoverlapping. Analyses of mutationally altered proteins showed that only a single amino acid changes at one time in one region of the protein. This result is

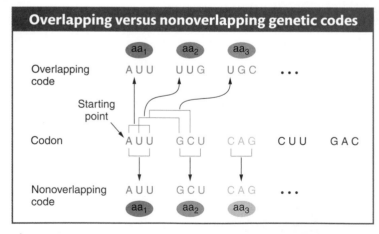

Overlapping versus nonoverlapping genetic codes

Figure 9-4 An overlapping and a nonoverlapping genetic code would translate differently into an amino acid sequence. The example uses a codon with three nucleotides in the RNA (a triplet code). In an overlapping code, single nucleotides occupy positions in multiple codons. In this illustration, the third nucleotide in the RNA, U, is found in three codons. In a nonoverlapping code, a protein is translated by reading nucleotides sequentially in sets of three. A nucleotide is found in only one codon. In this example, the third U in the RNA is only in the first codon.

predicted by a nonoverlapping code. As you can see in Figure 9-4, an overlapping code predicts that a single base change will alter as many as three amino acids at adjacent positions in the protein.

Number of letters in the codon

If an mRNA molecule is read from one end to the other, only one of four different bases, A, U, G, or C, can be found at each position. Thus, if the words encoding amino acids were one letter long, only four words would be possible. This vocabulary cannot be the genetic code, because we must have a word for each of the 20 amino acids commonly found in cellular proteins. If the words were two letters long, then $4 \times 4 = 16$ words would be possible; for example, AU, CU, or CC. This vocabulary is still not large enough.

If the words are three letters long, then $4 \times 4 \times 4 = 64$ words are possible; for example, AUU, GCG, or UGC. This vocabulary provides more than enough words to describe the amino acids. We can conclude that the code word must consist of at least three nucleotides. However, if all words are "triplets," then the possible words are in considerable excess of the 20 needed to name the common amino acids. We will come back to these excess codons later in the chapter.

Use of suppressors to demonstrate a triplet code

Convincing proof that a codon is, in fact, three letters long (and no more than three) came from beautiful genetic experiments first reported in 1961 by Francis Crick, Sidney Brenner, and their co-workers. These experiments used mutants in the *rII* locus of T4 phage. The use of *rII* mutations in recombination analysis was discussed in Chapter 5. Phage T4 is usually able to grow on two different *E. coli* strains, called B and K. However, mutations in the *rII* gene change the host range of the phage: mutant phages can still grow on an *E. coli* B host, but they cannot grow on an *E. coli* K host. Mutations causing this rII phenotype were induced by using a chemical called proflavin, which was thought to act by the addition or deletion of single nucleotide pairs in DNA. (This assumption is based on experimental evidence not presented here.) The following examples illustrate the action of proflavin on double-stranded DNA.

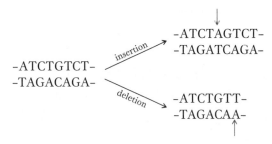

Starting with one particular proflavin-induced mutation called FCO, Crick and his colleagues found "reversions" (reversals of the mutation) that were able to grow on *E. coli* strain K. Genetic analysis of these plaques revealed that the "revertants" were not identical with true wild types. In fact, the reversion was found to be due to the presence of a *second mutation* at a different site from that of FCO, although in the same gene.

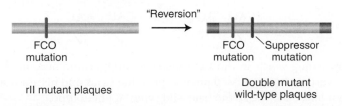

This second mutation "suppressed" mutant expression of the original FCO. Recall from Chapter 6 that a suppressor mutation counteracts or suppresses the effects of another mutation so that the bacterium is more like wild type.

How can we explain these results? If we assume that the gene is read from one end only, then the original addition or deletion induced by proflavin could result in a mutation because it interrupts a normal reading mechanism that establishes the group of bases to be read as words. For example, if each three bases on the resulting mRNA make a word, then the "reading frame" might be established by taking the first three bases from the end as the first word, the next three as the second word, and so forth. In that case, a proflavin-induced addition or deletion of a single pair on the DNA would shift the reading frame on the mRNA from that corresponding point on, causing all following words to be misread. Such a frame-shift mutation could reduce most of the genetic message to gibberish. However, the proper reading frame could be restored by a compensatory insertion or deletion somewhere else, leaving only a short stretch of gibberish between the two. Consider the following example in which three-letter English words are used to represent the codons:

	THE	FAT	CAT	ATE	THE	BIG	RAT
Delete C:	THE	FAT	ATA	TET	HEB	IGR	AT

	THE	FAT	ATA	ATE	THE	BIG	RAT
Insert A:							

The insertion suppresses the effect of the deletion by restoring most of the sense of the sentence. By itself, however, the insertion also disrupts the sentence:

THE FAT CAT AAT ETH EBI GRA T

If we assume that the FCO mutant is caused by an addition, then the second (suppressor) mutation would have to be a deletion because, as we have seen, only a deletion would restore the reading frame of the resulting message (a second insertion would not correct the frame). In the following diagrams, we use a hypothetical nucleotide chain to represent RNA for simplicity. We also assume that the code words are three letters long and are read in one direction (from left to right in our diagrams).

1. Wild-type message

CAU CAU CAU CAU CAU

2. rII_a message: Words after the addition are changed ($\times$) by frameshift mutation (words marked ✓ are unaffected).

Addition

CAU ACA UCA UCA UCA U
✓ × × × ×

3. $rII_a rII_b$ message: Few words are wrong, but reading frame is restored for later words.

Deletion

CAU ACA UCU CAU CAU
✓ × × ✓ ✓

The few wrong words in the suppressed genotype could account for the fact that the "revertants" (suppressed phenotypes) that Crick and his associates recovered did not look exactly like the true wild types in phenotype.

We have assumed here that the original frameshift mutation was an addition, but the explanation works just as well if we assume that the original FCO mutation is a deletion and the suppressor is an addition. You might want to verify it on your own. Very interestingly, combinations of *three* additions or *three* deletions have been shown to act together to restore a wild-type phenotype. This observation provided the first experimental confirmation that a word in the genetic code consists of three successive nucleotides, or a triplet. The reason is that three additions or three deletions within a gene automatically restore the reading frame in the mRNA if the words are triplets.

Degeneracy of the genetic code

As already stated, with four letters from which to choose at each position, a three-letter codon could make $4 \times 4 \times 4 = 64$ words. With only 20 words needed for the 20 common amino acids, what are the other words used for, if anything? Crick's work suggested that the genetic code is **degenerate,** meaning that each of the 64 triplets must have some meaning within the code. For the code to be degenerate, some of the amino acids must be specified by at least two or more different triplets.

The reasoning goes like this. If only 20 triplets were used, then the other 44 would be nonsense in that they would not encode any amino acid. In that case, most frameshift mutations could be expected to produce nonsense words, which presumably stop the protein-building process, and the suppression of frameshift mutations would rarely, if ever, work. However, if all triplets specified some amino acid, then the changed words would simply result in the insertion of incorrect amino acids into the protein. Thus, Crick reasoned that many or all amino acids must have several different names in the base-pair code; this hypothesis was later confirmed biochemically.

Message The discussion so far demonstrates that

1. The linear sequence of nucleotides in a gene determines the linear sequence of amino acids in a protein.

2. The genetic code is nonoverlapping.

3. Three bases encode an amino acid. These triplets are termed codons.

4. The code is read from a fixed starting point and continues to the end of the coding sequence. We know that the code is read sequentially because a single frameshift mutation anywhere in the coding sequence alters the codon alignment for the rest of the sequence.

4. The code is degenerate in that some amino acids are specified by more than one codon.

Cracking the code

The deciphering of the genetic code—determining the amino acid specified by each triplet—was one of the most exciting genetic breakthroughs of the past 50 years. After the necessary experimental techniques became available, the genetic code was broken in a rush.

One breakthrough was the discovery of how to make synthetic mRNA. If the nucleotides of RNA are mixed with a special enzyme (polynucleotide phosphorylase), a single-stranded RNA is formed in the reaction. Unlike transcription, no DNA template is needed for this synthesis, and so the nucleotides are incorporated at random. The ability to synthesize RNA offered the exciting prospect of creating specific mRNA sequences and then seeing which amino acids they would specify. The first synthetic messenger obtained was made by mixing only uracil nucleotides with the RNA-synthesizing enzyme, producing . . . UUUU . . . [poly(U)]. In 1961, Marshall Nirenberg and Heinrich Matthaei mixed poly(U) with the protein-synthesizing machinery of *E. coli* in vitro and *observed the formation of a protein*. The main excitement centered on the question of the amino acid sequence of this protein. It proved to be polyphenylalanine—a string of phenylalanine molecules attached to form a polypeptide. Thus, the triplet UUU must code for phenylalanine:

–UUUUUUUUUUUUUUUUUU–

–Phe–Phe–Phe–Phe–Phe–Phe–

For this discovery, Nirenberg was awarded the Nobel Prize.

Next, mRNAs containing two types of nucleotides in repeating groups were synthesized. For instance, synthetic mRNA having the sequence (AGA)*n*, which is a long sequence of AGAAGAAGAAGAAGA, was used to stimulate polypeptide synthesis in vitro (in a test tube that also contained a cell extract with all the components necessary for translation). The sequence of the resulting polypeptides was observed from a variety of such tests, with the use of different triplets residing in other synthetic RNAs. From such tests, many code words could be verified. (This kind of experiment is detailed in Problem 44 at the end of this chapter. In solving it, you can put yourself in the place of H. Gobind Khorana, who received a Nobel Prize for directing the experiments.)

Additional experimental approaches led to the assignment of each amino acid to one or more codons. Recall that the code was proposed to be degenerate, meaning that some amino acids had more than one codon assignment. This degeneracy can be seen clearly in Figure 9-5, which gives the codons and the amino acids that they specify. Virtually all organisms on Earth use this same genetic code. (There are just a few exceptions in which a small number of the codons have different meanings—for example, in mitochondrial genomes.)

Stop codons

You may have noticed in Figure 9-5 that some codons do not specify an amino acid at all. These codons are stop, or termination, codons. They can be regarded as being similar to periods or commas punctuating the message encoded in the DNA.

The genetic code

		Second letter				
		U	C	A	G	
First letter	U	UUU UUC } Phe UUA UUG } Leu	UCU UCC UCA UCG } Ser	UAU UAC } Tyr UAA Stop UAG Stop	UGU UGC } Cys UGA Stop UGG Trp	U C A G
	C	CUU CUC CUA CUG } Leu	CCU CCC CCA CCG } Pro	CAU CAC } His CAA CAG } Gln	CGU CGC CGA CGG } Arg	U C A G
	A	AUU AUC AUA } Ile AUG Met	ACU ACC ACA ACG } Thr	AAU AAC } Asn AAA AAG } Lys	AGU AGC } Ser AGA AGG } Arg	U C A G
	G	GUU GUC GUA GUG } Val	GCU GCC GCA GCG } Ala	GAU GAC } Asp GAA GAG } Glu	GGU GGC GGA GGG } Gly	U C A G

Third letter

Figure 9-5 The genetic code designates the amino acids specified by each codon.

One of the first indications of the existence of stop codons came in 1965 from Brenner's work with the T4 phage. Brenner analyzed certain mutations (m_1–m_6) in a single gene that controls the head protein of the phage. He found that the head protein of each mutant was a shorter polypeptide chain than that of the wild type. Brenner examined the ends of the shortened proteins and compared them with the wild-type protein. For each mutant, he recorded the next amino acid that *would* have been inserted to continue the wild-type chain. The amino acids for the six mutations were glutamine, lysine, glutamic acid, tyrosine, tryptophan, and serine. These results present no immediately obvious pattern, but Brenner deduced that certain codons for each of these amino acids are similar. Specifically, each of these codons can mutate to the codon UAG by a single change in a DNA nucleotide pair. He therefore postulated that UAG is a stop (termination) codon—a signal to the translation mechanism that the protein is now complete.

UAG was the first stop codon deciphered; it is called the amber codon (amber is the English translation of the last name of the codon's discoverer, Bernstein). Mutants that are defective owing to the presence of an abnormal amber codon are called *amber mutants*. Two other stop codons are UGA and UAA. Analogously to the amber codon, and continuing the theme of naming for colors and gems, UGA is called the opal codon and UAA is called the ochre codon. Mutants that are defective because they contain abnormal opal or ochre codons are called opal and ochre mutants, respectively. Stop codons are often called nonsense codons because they designate no amino acid.

In addition to a shorter head protein, Brenner's phage mutants had another interesting feature in common: the presence of a suppressor mutation ($su-$) in the host chromosome would cause the phage to develop a head protein of normal (wild-type) chain length despite the presence of the m mutation. We will consider stop codons and their suppressors further after we have dealt with the process of protein synthesis.

9.3 tRNA: The Adapter

Once the genetic code was deciphered, scientists began to wonder how the sequence of amino acids of a protein was determined by the triplet codons of the mRNA. An early model, quickly dismissed as naive and unlikely, proposed that the mRNA codons could fold up and form 20 distinct cavities that directly bind specific amino acids in the correct order. Instead, in 1958, Crick recognized:

> It is therefore a natural hypothesis that the amino acid is carried to the template by an adapter molecule, and that the adapter is the part which actually fits on to the RNA. In its simplest form [this hypothesis] would require twenty adapters, one for each amino acid.

He speculated that the adapter "might contain nucleotides. This would enable them to join on the RNA template by the same 'pairing' of bases as is found in DNA." Furthermore, "a separate enzyme would be required to join each adapter to its own amino acid."

We now know that Crick's "adapter hypothesis" is largely correct. Amino acids are in fact attached to an adapter (recall that adapters constitute a special class of stable RNAs called *transfer RNAs*). Each amino acid becomes attached to a specific tRNA, which then brings that amino acid to the ribosome, the molecular complex that will attach the amino acid to a growing polypeptide.

Codon translation by tRNA

The structure of tRNA holds the secret of the specificity between an mRNA codon and the amino acid that it designates. The single-stranded tRNA molecule has a

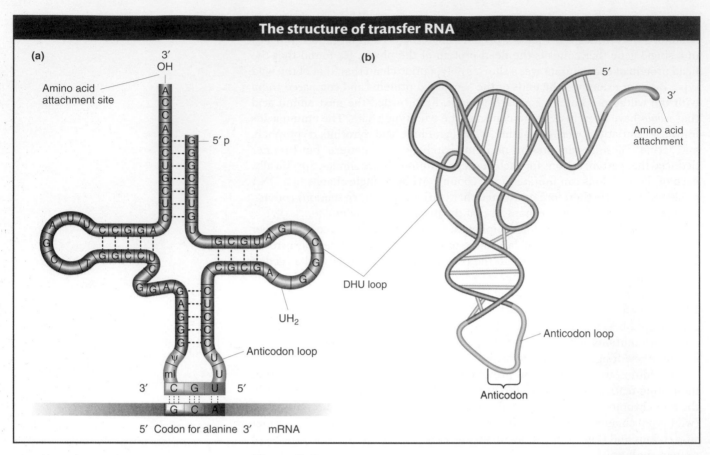

The structure of transfer RNA

(a)

3'
OH

Amino acid
attachment site

5' p

DHU loop

UH₂

Anticodon loop

ml

3'   C G   U   5'
         G C A

5' Codon for alanine 3'   mRNA

(b)

5'

3'

Amino acid
attachment

DHU loop

Anticodon loop

Anticodon

Figure 9-6 (a) The structure of yeast alanine tRNA, showing the anticodon of the tRNA binding to its complementary codon in mRNA. (b) Diagram of the actual three-dimensional structure of yeast phenylalanine tRNA. [(a) *After S. Arnott, "The Structure of Transfer RNA," Prog. Biophys. Mol. Biol. 22, 1971, 186; (b) after L. Stryer, Biochemistry, 4th ed. Copyright 1995 by Lubert Stryer. Part b is based on a drawing by Sung-Hou Kim.*]

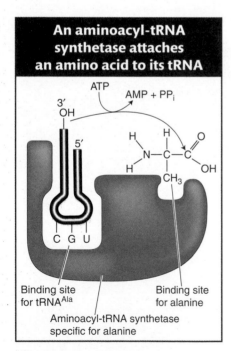

An aminoacyl-tRNA synthetase attaches an amino acid to its tRNA

ATP

AMP + PP$_i$

3'
OH

5'

$$\begin{array}{c} H \\ | \\ H-N-C-C \\ | \quad | \quad \backslash \\ H \quad CH_3 \quad OH \end{array}$$

C G U

Binding site
for tRNA^{Ala}

Binding site
for alanine

Aminoacyl-tRNA synthetase
specific for alanine

Figure 9-7 Each aminoacyl-tRNA synthetase has binding pockets for a specific amino acid and its cognate tRNA. By this means, an amino acid is covalently attached to the tRNA with the corresponding anticodon.

cloverleaf shape consisting of four double-helical stems and three single-stranded loops (Figure 9-6a). The middle loop of each tRNA is called the anticodon loop because it carries a nucleotide triplet called an **anticodon.** This sequence is complementary to the codon for the amino acid carried by the tRNA. The anticodon in tRNA and the codon in the mRNA bind by specific RNA-to-RNA base pairing. (Again, we see the principle of nucleic acid complementarity at work, this time in the binding of two different RNAs.) Because codons in mRNA are read in the $5' \rightarrow 3'$ direction, anticodons are oriented and written in the $3' \rightarrow 5'$ direction, as Figure 9-6a shows.

Amino acids are attached to tRNAs by enzymes called **aminoacyl-tRNA synthetases.** There are 20 of these enzymes in the cell, one for each of the 20 amino acids. Each amino acid has a specific synthetase that links it only to those tRNAs that recognize the codons for that particular amino acid. To catalyze this reaction, synthetases have two binding sites, one for the amino acid and the other for its cognate tRNA (Figure 9-7). An amino acid is attached at the free 3' end of its tRNA, the amino acid alanine in the case shown in Figures 9-6a and 9-7. The tRNA with an attached amino acid is said to be **charged.**

A tRNA normally exists as an L-shaped folded cloverleaf, as shown in Figure 9-6b, rather than the "flattened" cloverleaf shown in Figure 9-6a. The three-dimensional structure of tRNA was determined with the use of X-ray crystallography. In the years since this technique was used to deduce the double-helical structure of DNA, it has been refined so that it can now be used to determine

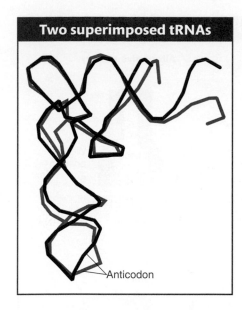

Two superimposed tRNAs

Anticodon

Figure 9-8 When folded into their correct three-dimensional structures, the yeast tRNA for glutamine (blue) almost completely overlaps the yeast tRNA for phenylalanine (red) except for the anticodon loop and aminoacyl end. [*After M. A. Rould, J. J. Perona, D. Soll, and T. A. Steitz, "Structure of E. coli Glutaminyl–tRNA Synthetase Complexed with tRNA(Gln) and ATP at 2.8 Å Resolution," Science 246, 1989, 1135–1142.*]

the structure of very complex macromolecules such as the ribosome. Although tRNAs differ in their primary nucleotide sequence, all tRNAs fold into virtually the same L-shaped conformation except for differences in the anticodon loop and aminoacyl end. This similarity of structure can be easily seen in Figure 9-8, which shows two different tRNAs superimposed. Conservation of structure tells us that shape is important for tRNA function.

What would happen if the wrong amino acid were covalently attached to a tRNA? A convincing experiment answered this question. The experiment used cysteinyl-tRNA (tRNA^{Cys}), the tRNA specific for cysteine. This tRNA was "charged" with cysteine, meaning that cysteine was attached to the tRNA. The charged tRNA was treated with nickel hydride, which converted the cysteine (while still bound to tRNA^{Cys}) into another amino acid, alanine, without affecting the tRNA:

$$\text{cysteine–tRNA} \xrightarrow{\text{nickel hydride}} \text{alanine–tRNA}^{Cys}$$

Protein synthesized with this hybrid species had alanine wherever we would expect cysteine. The experiment demonstrated that the amino acids are "illiterate"; they are inserted at the proper position because the tRNA "adapters" recognize the mRNA codons and insert their attached amino acids appropriately. Thus, the attachment of the correct amino acid to its cognate tRNA is a critical step in ensuring that a protein is synthesized correctly. If the wrong amino acid is attached, there is no way to prevent it from being incorporated into a growing protein chain.

Degeneracy revisited

As can be seen in Figure 9-5, the number of codons for a single amino acid varies, ranging from one codon (UGG for tryptophan) to as many as six (UCC, UCU, UCA, UCG, AGC, or AGU for serine). Why the genetic code contains this variation is not exactly clear, but two facts account for it:

1. Most amino acids can be brought to the ribosome by several alternative tRNA types. Each type has a different anticodon that base-pairs with a different codon in the mRNA.

2. Certain charged tRNA species can bring their specific amino acids to any one of several codons. These tRNAs recognize and bind to several alternative codons, not just the one with a complementary sequence, through a loose kind of base pairing at the 3' end of the codon and the 5' end of the anticodon. This loose pairing is called **wobble.**

Figure 9-9 In the third site (5′ end) of the anticodon, G can take either of two wobble positions, thus being able to pair with either U or C. This ability means that a single tRNA species carrying an amino acid (in this case, serine) can recognize two codons—UCU and UCC—in the mRNA.

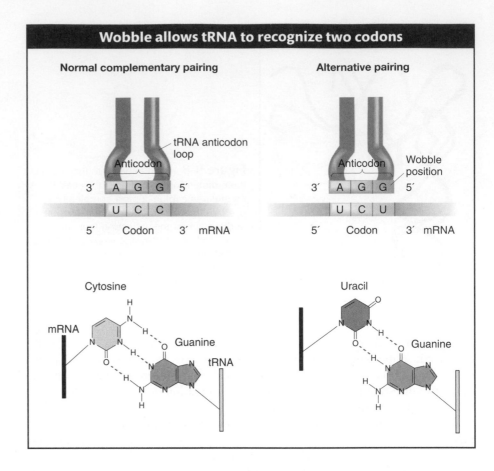

Wobble is a situation in which the third nucleotide of an anticodon (at the 5′ end) can form either of two alignments (Figure 9-9). This third nucleotide can form hydrogen bonds either with its normal complementary nucleotide in the third position of the codon or with a different nucleotide in that position. "Wobble rules" dictate which nucleotides can and cannot form hydrogen bonds with alternative nucleotides through wobble (Table 9-1). In Table 9-1, the letter *I* stands for inosine, one of the rare bases found in tRNA, often in the anticodon.

Table 9-1   Codon–Anticodon Pairings Allowed by the Wobble Rules

5′ end of anticodon	3′ end of codon
G	C or U
C	G only
A	U only
U	A or G
I	U, C, or A

> **Message** The genetic code is said to be degenerate because, in many cases, more than one codon is assigned to a single amino acid; in addition, several codons can pair with more than one anticodon (wobble).

9.4 Ribosomes

Protein synthesis takes place when tRNA and mRNA molecules associate with *ribosomes*. The task of the tRNAs and the ribosome is to translate the sequence of nucleotide codons in mRNA into the sequence of amino acids in protein. The term *biological machine* was used in preceding chapters to characterize multisubunit complexes that perform cellular functions. The replisome, for example, is a biological machine that can replicate DNA with precision and speed. The site of protein synthesis, the ribosome, is much larger and more complex than the machines described thus far. Its complexity is due to the fact that it has to perform several jobs with precision and speed. For this reason, it is better to think of the ribosome as a factory containing many machines that act in concert. Let's see how this factory is organized to perform its numerous functions.

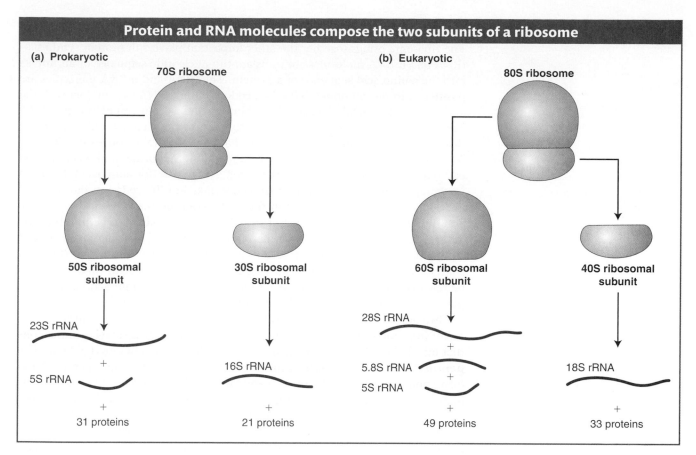

Figure 9-10 A ribosome contains a large and a small subunit. Each subunit contains both rRNA of varying lengths and a set of proteins. There are two principal rRNA molecules in all ribosomes. Prokaryotic ribosomes also contain one 120-base-long rRNA that sediments at 5S, whereas eukaryotic ribosomes have two small rRNAs: a 5S RNA molecule similar to the prokaryotic 5S and a 5.8S molecule 160 bases long.

In all organisms, the ribosome consists of one small and one large subunit, each made up of RNA (called ribosomal RNA or rRNA) and protein. Each subunit is composed of one to three rRNA types and as many as 50 proteins. Ribosomal subunits were originally characterized by their rate of sedimentation when spun in an ultracentrifuge, and so their names are derived from their sedimentation coefficients in Svedberg (S) units, which is an indication of molecular size. In prokaryotes, the small and large subunits are called 30S and 50S, respectively, and they associate to form a 70S particle (Figure 9-10a). The eukaryotic counterparts are called 40S and 60S, and the complete ribosome is called 80S (Figure 9-10b). Although eukaryotic ribosomes are bigger owing to their larger and more numerous components, the components and the steps in protein synthesis are similar overall. The similarities clearly indicate that translation is an ancient process that originated in the common ancestor of eukaryotes and prokaryotes.

When ribosomes were first studied, the fact that almost two-thirds of their mass is RNA and only one-third is protein was surprising. For decades, rRNAs had been assumed to function as the scaffold or framework necessary for the correct assembly of the ribosomal proteins. That role seemed logical because rRNAs fold up by intramolecular base pairing into stable secondary structures (Figure 9-11). According to this model, the ribosomal proteins were solely responsible for carrying out the important steps in protein synthesis. This view changed with the discovery in the 1980s of catalytic RNAs (see Chapter 8). As you will see, scientists now believe that the rRNAs, assisted by the ribosomal proteins, carry out most of the important steps in protein synthesis.

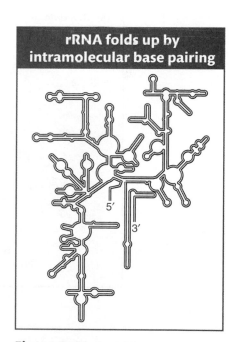

Figure 9-11 The folded structure of the prokaryotic 16S ribosomal RNA of the small ribosomal subunit.

Ribosome features

The ribosome brings together the other important players in protein synthesis—tRNA and mRNA molecules—to translate the nucleotide sequence of an mRNA into the amino acid sequence of a protein. The tRNA and mRNA molecules are positioned in the ribosome so that the codon of the mRNA can interact with the anticodon of the tRNA. The key sites of interaction are illustrated in Figure 9-12. The binding site for mRNA is completely within the small subunit. There are three binding sites for tRNA molecules. Each bound tRNA bridges the 30S and 50S subunits, positioned with its anticodon end in the former and its aminoacyl end (carrying the amino acid) in the latter. The **A site** (for aminoacyl) binds an incoming aminoacyl-tRNA whose anticodon matches the codon in the A site of the 30S subunit. As we proceed in the 5′ direction on the mRNA, the next codon interacts with the anticodon of the tRNA in the **P site** (for peptidyl) of the 30S subunit. The tRNA in the P site binds the growing peptide chain, part of which fits into a tunnel-like structure in the 50S subunit. The **E site** (for exit) contains a deacylated tRNA (it no longer carries an amino acid) that is ready to be released from the ribosome. Whether codon–anticodon interactions also take place between the mRNA and the tRNA in the E site is not clear.

Two additional regions in the ribosome are critical for protein synthesis. The **decoding center** in the 30S subunit ensures that only tRNAs carrying anticodons that match the codon (called *cognate* tRNAs) will be accepted into the A site. The **peptidyltransferase center** in the 50S subunit is the site where peptide-bond formation is catalyzed. Recently, many laboratories, especially that of Thomas Steitz, Venkatraman Ramakrishnan, and Ada Yonath, have used X-ray crystallography to "solve" the structure of the ribosome at the atomic level. For this accomplishment these three scientists received the Nobel Prize in Chemistry in 2009. The results of their elegant studies clearly show that both the decoding and peptidyltransferase centers are composed entirely of regions of rRNA; that is, the

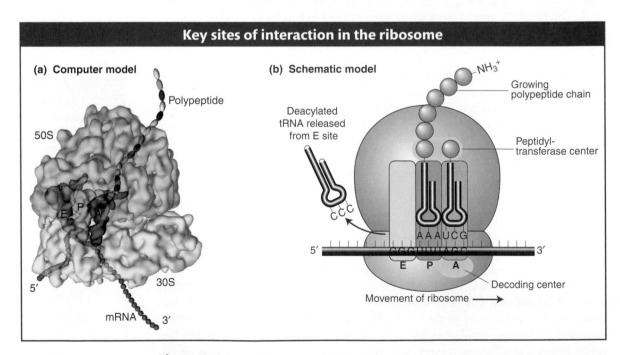

Key sites of interaction in the ribosome

(a) **Computer model**

Polypeptide

50S

E P A

5′

30S

mRNA   3′

(b) **Schematic model**

NH₃⁺

Growing polypeptide chain

Deacylated tRNA released from E site

Peptidyl-transferase center

CCC

AAAUCG

5′   CCGUUUAGC   3′

E   P   A

Decoding center

Movement of ribosome ⟶

Figure 9-12 Key sites of interaction in a ribosome in the elongation phase of translation. (a) A computer model of the three-dimensional structure of the ribosome including mRNA, tRNAs, and the nascent polypeptide chain as it emerges from the large ribosomal subunit. (b) A schematic model of the ribosome during translation elongation. See text for details. *[(a) From J. Frank, Bioessays 23, 2001, 725–732, Fig. 2.]*

important contacts in these centers are tRNA–rRNA contacts. Peptide-bond formation is even thought to be catalyzed by an active site in the ribosomal RNA and only assisted by ribosomal proteins. In other words, the large ribosomal subunit functions as a ribozyme to catalyze peptide-bond formation.

Similar structural studies have examined the large ribosomal subunit complexed with several different antibiotics. These studies identified the contact points between the antibiotic and the ribosome and, in doing so, have provided an explanation for why certain antibiotics inactivate only bacterial ribosomes. For example, the macrolides are a family of structurally similar compounds that includes the popular antibiotics erythromycin and Zithromax. These antibiotics inhibit protein synthesis by stalling the ribosome on the mRNA. They do so by binding to a specific region on the 23S RNA in the large ribosomal subunit and blocking the so-called exit tunnel where the nascent polypeptide emerges from the large subunit (see Figure 9-1). Because of minor sequence differences between in the rRNAs of prokaryotes and eukaryotes, macrolides only inhibit bacterial translation. Interestingly, pathogenic bacteria that have evolved resistance to some of these antibiotics appear to have ribosomal mutations that make the exit tunnel larger. Thus, knowledge of how antibiotics bind to ribosomes helps scientists understand how the ribosome works and how to design new antibiotics that can be active against resistant mutants. The use of basic information about cellular machinery to develop new antibiotics and other drugs has been dubbed **structure-based drug design.**

Translation initiation, elongation, and termination

The process of translation can be divided into three phases: initiation, elongation, and termination. Aside from the ribosome, mRNA, and tRNAs, additional proteins are required for the successful completion of each phase. Because certain steps in initiation differ significantly in prokaryotes and eukaryotes, initiation is described separately for the two groups. The elongation and termination phases are described largely as they take place in bacteria, which have been the focus of many recent studies of translation.

Translation initiation   The main task of initiation is to place the first aminoacyl-tRNA in the P site of the ribosome and, in this way, establish the correct reading frame of the mRNA. In most prokaryotes and all eukaryotes, the first amino acid in any newly synthesized polypeptide is methionine, specified by the codon AUG. It is inserted not by tRNA^{Met} but by a special tRNA called an **initiator,** symbolized tRNA$^{Met}_i$. In bacteria, a formyl group is added to the methionine while the amino acid is attached to the initiator, forming *N*-formylmethionine. (The formyl group on *N*-formylmethionine is removed later.)

How does the translation machinery know where to begin? In other words, how is the initiation AUG codon selected from among the many AUG codons in an mRNA molecule? Recall that, in both prokaryotes and eukaryotes, mRNA has a 5′ untranslated region consisting of the sequence between the transcriptional start site and the translational start site. As you will see below, the nucleotide sequence of the 5′ UTR adjacent to the AUG initiator is critical for ribosome binding in prokaryotes but not in eukaryotes.

Initiation in prokaryotes Initiation codons are preceded by special sequences called **Shine–Dalgarno sequences** that pair with the 3′ end of an rRNA, called the 16S rRNA, in the 30S ribosomal subunit. This pairing correctly positions the initiator codon in the P site where the initiator tRNA will bind (Figure 9-13). The mRNA can pair only with a 30S subunit that is dissociated from the rest of the ribosome. Note again that rRNA performs the key function in ensuring that the ribosome is at the right place to start translation.

**What Geneticists
Are Doing Today**

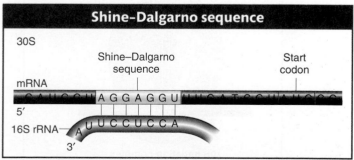

Figure 9-13 In bacteria, base complementarity between the 3′ end of the 16S rRNA of the small ribosomal subunit (30S) and the Shine–Dalgarno sequence of the mRNA positions the ribosome to correctly initiate translation at the downstream AUG codon.

Three proteins—IF1, IF2, and IF3 (for **initiation factor**)—are required for correct initiation (Figure 9-14). IF3 is necessary to keep the 30S subunit dissociated from the 50S subunit, and IF1 and IF2 act to ensure that only the initiator tRNA enters the P site. The 30S subunit, mRNA, and initiator tRNA constitute the initiation complex. The complete 70S ribosome is formed by the association of the 50S large subunit with the initiation complex and the release of the initiation factors.

Because a prokaryote lacks a nuclear compartment that separates transcription and translation, the prokaryotic initiation complex is able to form at a Shine–Dalgarno

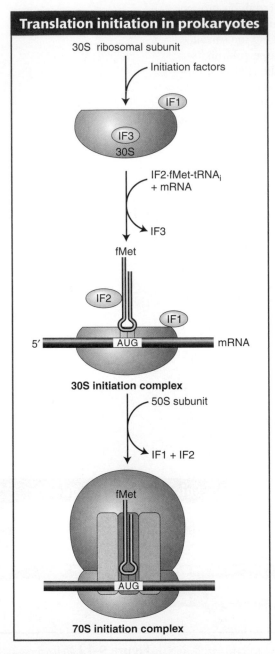

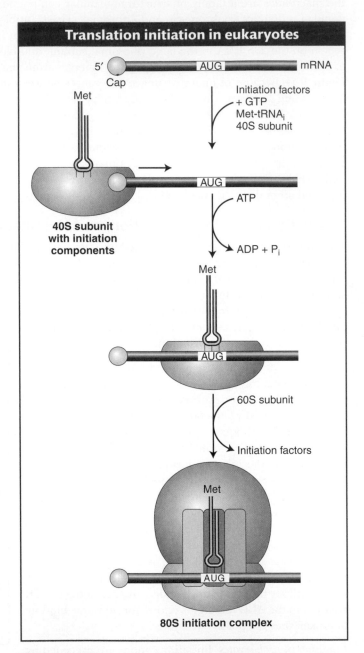

Figure 9-14 Initiation factors assist the assembly of the ribosome at the translation start site and then dissociate before translation. [*After J. Berg, J. Tymoczko, and L. Stryer, Biochemistry, 5th ed. Copyright 2002 by W. H. Freeman and Company.*]

Figure 9-15 The initiation complex forms at the 5′ end of the mRNA and then scans in the 3′ direction in search of a start codon. Recognition of the start codon triggers the assembly of the complete ribosome and the dissociation of initiation factors (not shown). The hydrolysis of ATP provides energy to drive the scanning process. [*After J. Berg, J. Tymoczko, and L. Stryer, Biochemistry, 5th ed. Copyright 2002 by W. H. Freeman and Company.*]

sequence near the 5′ end of an RNA that is still being transcribed. Thus, translation can begin on prokaryotic RNAs even before they are completely transcribed.

Initiation in eukaryotes Transcription and translation take place in separate compartments of the eukaryotic cell. As discussed in Chapter 8, eukaryotic mRNAs are transcribed and processed in the nucleus before export to the cytoplasm for translation. On arrival in the cytoplasm, the mRNA is usually covered with proteins, and regions may be double helical due to intramolecular base pairing. These regions of secondary structure must be removed to expose the AUG initiator codon. This removal is accomplished by eukaryotic initiation factors called eIF4A, B, and G. These initiation factors associate with the cap structure (found at the 5′ end of virtually all eukaryotic mRNAs) and with the 40S subunit and initiator tRNA to form an initiation complex. Once in place, the complex moves in the 5′-to-3′ direction and unwinds the base-paired regions (Figure 9-15). At the same time, the exposed sequence is "scanned" for an AUG codon where translation can begin. After the AUG codon is properly aligned with the initiator tRNA, the initiation complex is joined by the 60S subunit to form the 80S ribosome. As in prokaryotes, the eukaryotic initiation factors dissociate from the ribosome before the elongation phase of translation begins.

Elongation It is during the process of elongation that the ribosome most resembles a factory. The mRNA acts as a blueprint specifying the delivery of cognate tRNAs, each carrying as cargo an amino acid. Each amino acid is added to the growing polypeptide chain while the deacylated tRNA is recycled by the addition of another amino acid. Figure 9-16 details the steps in elongation. Two protein factors called elongation factor Tu (EF-Tu) and elongation factor G (EF-G) assist the elongation process.

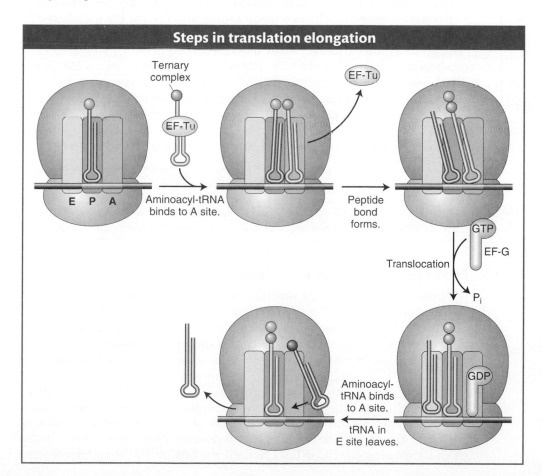

Steps in translation elongation

Ternary complex

EF-Tu

E P A   Aminoacyl-tRNA binds to A site.

EF-Tu

Peptide bond forms.

GTP
EF-G

Translocation

P_i

GDP

Aminoacyl-tRNA binds to A site.

tRNA in E site leaves.

Figure 9-16 A ternary complex consisting of an aminoacyl-tRNA attached to an EF-Tu factor binds to the A site. When its amino acid has joined the growing polypeptide chain, an EF-G factor binds to the A site while nudging the tRNAs and their mRNA codons into the E and P sites. See text for details. GENETICS*PORTAL* **ANIMATED ART: The three steps of translation**

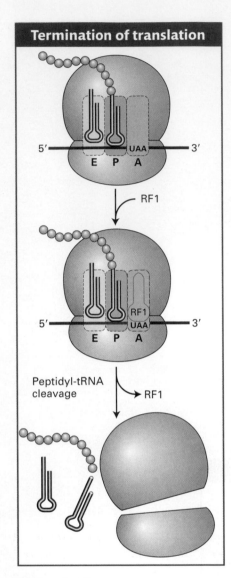

Termination of translation

5′ ——— UAA ——— 3′
E P A

↓ RF1

5′ ——— RF1 UAA ——— 3′
E P A

Peptidyl-tRNA cleavage   → RF1

Figure 9-17 Translation is terminated when release factors recognize stop codons in the A site of the ribosome. [*From H. Lodish et al., Molecular Cell Biology, 5th ed. Copyright 2004 by W. H. Freeman and Company.*]

As described earlier in this chapter, an aminoacyl-tRNA is formed by the covalent attachment of an amino acid to the 3′ end of a tRNA that contains the correct anticodon. Before aminoacyl-tRNAs can be used in protein synthesis, they associate with the protein factor EF-Tu to form a ternary complex composed of tRNA, amino acid, and EF-Tu. The elongation cycle commences with an initiator tRNA (and its attached methionine) in the P site and with the A site ready to accept a ternary complex (see Figure 9-16). Which of the 20 different ternary complexes to accept is determined by codon–anticodon recognition in the decoding center of the small subunit (see Figure 9-12b). When the correct match has been made, the ribosome changes shape, the EF-Tu leaves the ternary complex, and the two aminoacyl ends are juxtaposed in the peptidyltransferase center of the large subunit (see Figure 9-12b). There, a peptide bond is formed with the transfer of the methionine in the P site to the amino acid in the A site. At this point, the second protein factor, EF-G, plays its part. The EF-G factor appears to fit into the A site. Its entry into that site shifts the tRNAs in the A and P sites to the P and E sites, respectively, and the mRNA moves through the ribosome so that the next codon is positioned in the A site (see Figure 9-16). When EF-G leaves the ribosome, the A site is open to accept the next ternary complex.

In subsequent cycles, the A site is filled with a new ternary complex as the deacylated tRNA leaves the E site. As elongation progresses, the number of amino acids on the peptidyl-tRNA (at the P site) increases. Eventually, the amino-terminal end of the growing polypeptide emerges from the tunnel in the 50S subunit and protrudes from the ribosome.

Termination The cycle continues until the codon in the A site is one of the three stop codons: UGA, UAA, or UAG. Recall that no tRNAs recognize these codons. Instead, proteins called **release factors** (RF1, RF2, and RF3 in bacteria) recognize stop codons (Figure 9-17). In bacteria, RF1 recognizes UAA or UAG, whereas RF2 recognizes UAA or UGA; both are assisted by RF3. The interaction between release factors 1 and 2 and the A site differs from that of the ternary complex in two important ways. First, the stop codons are recognized by tripeptides in the RF proteins, not by an anticodon. Second, release factors fit into the A site of the 30S subunit but do not participate in peptide-bond formation. Instead, a water molecule gets into the peptidyltransferase center, and its presence leads to the release of the polypeptide from the tRNA in the P site. The ribosomal subunits separate, and the 30S subunit is now ready to form a new initiation complex.

> **Message** Translation is carried out by ribosomes moving along mRNA in the 5′ → 3′ direction. A set of tRNA molecules bring amino acids to the ribosome, and their anticodons bind to mRNA codons exposed on the ribosome. An incoming amino acid becomes bonded to the amino end of the growing polypeptide chain in the ribosome.

Nonsense suppressor mutations

It is interesting to consider the suppressors of the nonsense mutations defined by Brenner and co-workers. Recall that mutations in phages called amber mutants replaced wild-type codons with stop codons but that suppressor mutations in the host chromosome counteracted the effects of the amber mutations. We can now say more specifically where the suppressor mutations were located and how they worked.

Many of these suppressors are mutations in genes encoding tRNAs and are known as tRNA suppressors. These mutations alter the anticodon loops of specific tRNAs in such a way that a tRNA becomes able to recognize a stop codon in mRNA. In Figure 9-18, the amber mutation replaces a wild-type codon with the chain-terminating stop codon UAG. By itself, the UAG would cause the protein to

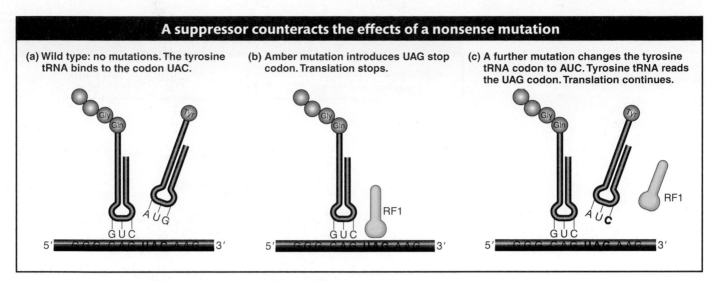

A suppressor counteracts the effects of a nonsense mutation

(a) Wild type: no mutations. The tyrosine tRNA binds to the codon UAC.

(b) Amber mutation introduces UAG stop codon. Translation stops.

(c) A further mutation changes the tyrosine tRNA codon to AUC. Tyrosine tRNA reads the UAG codon. Translation continues.

be prematurely cut off at the corresponding position. The suppressor mutation in this case produces a tRNA^{Tyr} with an anticodon that recognizes the mutant UAG stop codon. Thus, in the suppressed mutant, tRNA^{Tyr} competes with the release factor for access to the UAG stop codon. As a result, if tyrosine is inserted, translation continues past that triplet.

Could tRNA suppressors (like tRNA^{Tyr}) also bind to normal termination signals and result in the synthesis of abnormally long proteins? Now that many genomes have been sequenced, it is known that one of the three stop codons, UAA (ochre), is used much more often to terminate protein synthesis. As such, it is not surprising that cells with ochre suppressors are usually sicker than cells with amber or opal suppressor mutations.

9.5 The Proteome

Chapter 8 began with a discussion of the number of genes in the human genome and how that number (about 20,000) was much lower than the actual number of proteins in a human cell (more than 100,000). Now that you are familiar with how information encoded in DNA is transcribed into RNA and how RNA is translated into protein, it is a good time to revisit this matter and look more closely at the sources of protein diversification. First, let's review a few old terms and add a new one that will be useful in this discussion. You already know that the *genome* is the entire set of genetic material in an organism. You will learn in Chapter 14 that the *transcriptome* is the complete set of coding and noncoding transcripts in an organism, organ, tissue, or cell. Another term is the **proteome,** which was briefly introduced in Chapter 8 but is defined here as the complete set of proteins in an organism, organ, tissue, or cell. In the remainder of this chapter, you will see how the proteome is enriched by two cellular processes: the alternative splicing of pre-mRNA and the posttranslational modification of proteins.

Alternative splicing generates protein isoforms

As you recall, alternative splicing of pre-mRNA allows one gene to encode more than one protein. Proteins are made up of functional domains that are often encoded by different exons. Thus, the alternative splicing of a pre-mRNA can lead to the synthesis of multiple proteins (called **isoforms**) with different combinations of functional domains. This concept is illustrated by *FGFR2*, a human gene that encodes the receptor that binds fibroblast growth factors and then transduces

Figure 9-18 A suppressor allows translation to continue when otherwise a mutation would have stopped it. (a) In the wild type, a tRNA reads the codon UAC and translation continues. (b) Termination of translation. Here, the translation apparatus cannot go past a stop codon (UAG in this case), because no tRNA can recognize the UAG triplet. Instead, a release factor binds with the codon and protein synthesis ends, with the subsequent release of the polypeptide fragment. (c) A mutation alters the anticodon of a tyrosine tRNA so that this tRNA can now read the UAG codon. The suppression of the UAG codon by the altered tRNA now permits chain elongation. [*After D. Watson, J. Tooze, and D. T. Kurtz, Recombinant DNA: A Short Course. Copyright 1983 by W. H. Freeman and Company.*]

GENETICS P**O**RTAL **ANIMATED ART: Nonsense suppression of the rod^{ns} allele, The tRNA nonsense suppressor, The rod^{ns} nonsense mutation**

Figure 9-19 Messenger RNAs produced by alternative splicing of the pre-mRNA of the human *FGFR2* gene encode two protein isoforms that bind to different ligands (the growth factors).

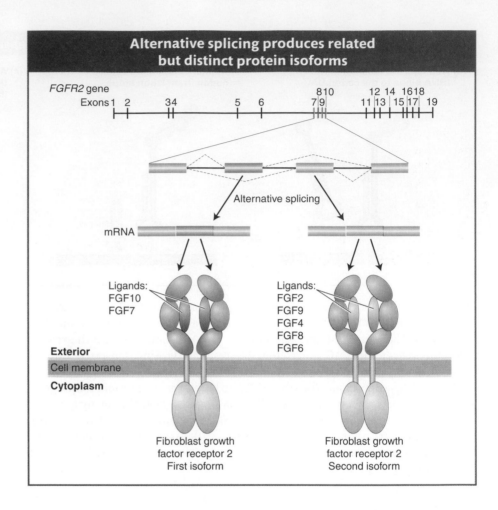

a signal inside the cell (Figure 9-19). The FGFR2 protein is made up of several domains, including an extracellular ligand-binding domain. Alternative splicing results in two isoforms that differ in their extracellular domains. Because of this difference, each isoform binds different growth factors. For many genes that are alternatively spliced, different isoforms are made in different tissues.

Posttranslational events

When released from the ribosome, most newly synthesized proteins are unable to function. This fact may come as a surprise to those who believe that the protein sequences encoded in DNA and transcribed into mRNAs are all that is needed to explain how organisms work. As you will see in this section and in subsequent chapters of this book, DNA sequence is only part of the story. In this case, all newly synthesized proteins need to fold up correctly and the amino acids of some proteins need to be chemically modified. Because some protein folding and modification take place after protein synthesis, they are called posttranslational events.

Protein folding inside the cell   The most important posttranslational event is the folding of the nascent (newly synthesized) protein into its correct three-dimensional shape. A protein that is folded correctly is said to be in its native conformation (in contrast with an unfolded or misfolded protein that is nonnative). As we saw at the beginning of this chapter, proteins exist in a remarkable diversity of structures. The distinct structures of proteins are essential for their enzymatic activity, for their ability to bind to DNA, or for their structural roles in the cell. Although it has been known since the 1950s that the amino acid sequence of a protein determines its three-dimensional structure, it is also known that the aqueous environment inside the cell does not favor the correct folding of most proteins. Given that proteins do in fact fold correctly in the cell, a long-standing question has been, How is this correct folding accomplished?

The answer seems to be that nascent proteins are folded correctly with the help of chaperones—a class of proteins found in all organisms from bacteria to plants to humans. One family of chaperones, called the GroE chaperonins, form large multisubunit complexes called chaperonin folding machines. Although the precise mechanism is not yet understood, newly synthesized, unfolded proteins are believed to enter a chamber in the folding machine that provides an electrically neutral microenvironment within which the nascent protein can successfully fold into its native conformation.

Posttranslational modification of amino acid side chains As already stated, proteins are polymers of amino acids made from any of the 20 different types. However, biochemical analysis of many proteins reveals that a variety of molecules can be covalently attached to amino acid side chains. More than 300 modifications of amino acid side chains are possible after translation. Two of the more commonly encountered posttranslational modifications—phosphorylation and ubiquitinylation—are considered next.

Phosphorylation Enzymes called *kinases* attach phosphate groups to the hydroxyl groups of the amino acids serine, threonine, and tyrosine, whereas enzymes called *phosphatases* remove these phosphate groups. Because phosphate groups are negatively charged, their addition to a protein usually changes protein conformation. The addition and removal of phosphate groups serves as a reversible switch to control a variety of cellular events, including enzyme activity, protein–protein interactions, and protein–DNA interactions (Figure 9-20).

One measure of the importance of protein phosphorylation is the number of genes encoding kinase activity in the genome. Even a simple organism such as yeast has hundreds of kinase genes, whereas the mustard plant *Arabidopsis thaliana* has more than 1000. Another measure of the significance of protein phosphorylation is that most of the numerous protein–protein interactions that take place in a typical cell are regulated by phosphorylation.

Recent analyses of the protein–protein interactions of the proteome indicate that most proteins function by interacting with other proteins. The **interactome**

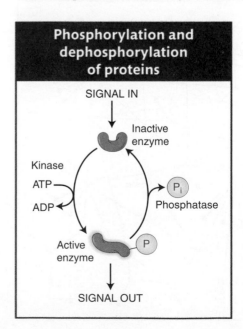

Figure 9-20 Proteins can be activated through the enzymatic attachment of phosphate groups to their amino acid side groups and inactivated by the removal of those phosphate groups.

Figure 9-21 Proteins (represented by circles) interact with other proteins (connected by lines) to form simple or large protein complexes. This interactome shows 3186 interactions among 1705 human proteins. [*Ulrich Steizl et al., Max Delbrück Center for Molecular Medicine (MDC) Berlin-Buch. Copyright MDC.*]

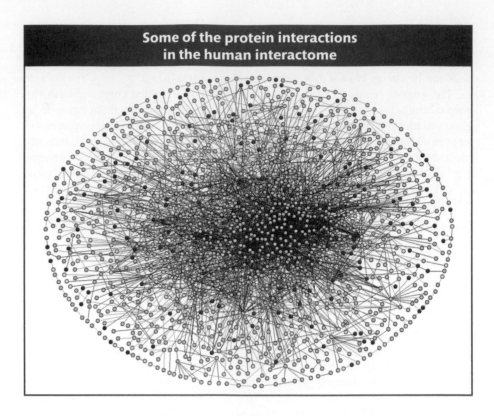

is the name given to the complete set of protein–protein interactions in an organism, organ, tissue, or cell. One way to display the network of protein–protein interactions that constitute an interactome is shown in Figure 9-21. To generate this figure, researchers determined the 3186 protein interactions among 1705 human proteins. However, these interactions constitute only a tiny fraction of the protein–protein interactions that are taking place in all human cells under all growth conditions.

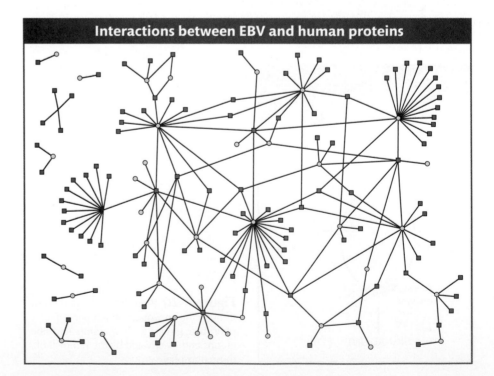

Figure 9-22 The web of 173 interactions among 40 proteins from Epstein–Barr virus (EBV) and 112 human proteins. Virus proteins are shown as yellow circles and human proteins as blue squares. Interactions are shown as red lines. [*Calderwood et al., Proceedings of the National Academy of Sciences 104, 2007, 7606–7611. Copyright 2007 by National Academy of Sciences.*]

What is the biological significance of these interactions? In this chapter and preceding ones, you have seen that protein–protein interactions are central to the function of large biological machines such as the replisome, the spliceosome, and the ribosome. Another set of significant interactions is the associations between human proteins and the proteins of human pathogens. For example, the interactome of 40 Epstein–Barr virus (EBV) proteins and 112 human proteins consists of 173 interactions (Figure 9-22). Understanding of this web of interactions may lead to new therapies for mononucleosis, a disease caused by EBV infection.

Ubiquitination Surprisingly, one of the most common posttranslational modifications is not a subtle one like the addition of a phosphate group. Instead, this modification targets the protein for degradation by a biological machine and protease called the 26S proteasome (Figure 9-23). The modification targeting a protein for degradation is the addition of chains of multiple copies of a protein called **ubiquitin** to the ε-amine of lysine residues (called **ubiquitination**). Ubiquitin contains 76 amino acids and is found only in eukaryotes, where it is highly conserved in plants and animals. Two broad classes of proteins are targeted for destruction by ubiquitinization: short-lived proteins such as cell-cycle regulators or proteins that have become damaged or mutated.

Protein targeting   In eukaryotes, all proteins are synthesized on ribosomes in the cytoplasm. However, some of these proteins end up in the nucleus, others in the mitochondria, and still others anchored in the membrane or secreted from the cell. How do these proteins "know" where they are supposed to go? The answer to this seemingly complex problem is actually quite simple: a newly synthesized protein contains a short sequence that targets the protein to the correct place or cellular compartment. For example, a newly synthesized membrane protein or a protein destined for an organelle has a short leader peptide, called a **signal sequence,** at its amino-terminal end. For membrane proteins, this stretch of 15 to 25 amino acids directs the protein to channels in the endoplasmic reticulum membrane where the signal sequence is cleaved by a peptidase (Figure 9-24). From the endoplasmic reticulum, the protein is directed to its ultimate destination. A similar phenomenon exists for certain bacterial proteins that are secreted.

Proteins destined for the nucleus include the RNA and DNA polymerases and transcription factors discussed in Chapters 7 and 8. Amino acid sequences embedded in the interiors of such nucleus-bound proteins are necessary for transport from the cytoplasm into the nucleus. These **nuclear localization sequences (NLSs)** are recognized by cytoplasmic receptor proteins that transport newly synthesized proteins through nuclear pores—sites in the membrane through which large molecules are able to pass into and out of the nucleus. A protein not normally found in the nucleus will be directed to the nucleus if an NLS is attached to it.

Why are signal sequences cleaved during targeting, whereas an NLS, located in a protein's interior, remains after the protein moves into the nucleus? One explanation might be that, in the nuclear disintegration that accompanies

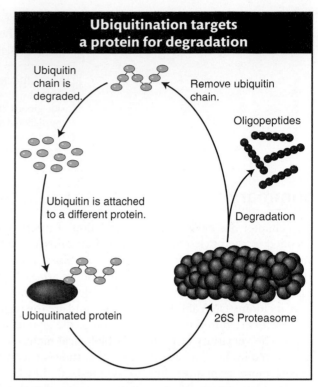

Figure 9-23 The major steps in ubiquitin-mediated protein degradation are shown. Ubiquitin is first conjugated to another protein and then degraded by the proteasome. Ubiquitin and oligopeptides are then recycled.

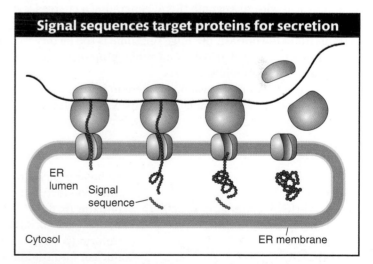

Figure 9-24 Proteins destined to be secreted from the cell have an amino-terminal sequence that is rich in hydrophobic residues. This signal sequence binds to proteins in the endoplasmic reticulum (ER) membrane that draw the remainder of the protein through the lipid bilayer. The signal sequence is cleaved from the protein in this process by an enzyme called *signal peptidase* (not shown). Once inside the endoplasmic reticulum, the protein is directed to the cell membrane, from which it will be secreted.

mitosis (see Chapter 2), proteins localized to the nucleus may find themselves in the cytoplasm. Because such a protein contains an NLS, it can relocate to the nucleus of a daughter cell that results from mitosis.

> **Message** Most eukaryotic proteins are inactive unless modified after translation. Some posttranslational events, such as phosphorylation or ubiquitination, modify amino acid side groups, thus promoting protein activation or degradation, respectively. Other post-translational mechanisms recognize amino acid signatures in a protein sequence and target those proteins to places where their activity is required inside or outside the cell.

Summary

This chapter has dealt with the translation of information encoded in the nucleotide sequence of an mRNA into the amino acid sequence of a protein. Our proteins, more than any other macromolecule, determine who we are and what we are. They are the enzymes responsible for cell metabolism, including DNA and RNA synthesis, and are the regulatory factors required for the expression of the genetic program. The versatility of proteins as biological molecules is manifested in the diversity of shapes that they can assume. Furthermore, even after they are synthesized, they can be modified in a variety of ways by the addition of molecules that can alter their function.

Given the central role of proteins in life, it is not surprising that both the genetic code and the machinery for translating this code into protein have been highly conserved from bacteria to humans. The major components of translation are three classes of RNA: tRNA, mRNA, and rRNA. The accuracy of translation depends on the enzymatic linkage of an amino acid with its cognate tRNA, generating a charged tRNA molecule. As adapters, tRNAs are the key molecules in translation. In contrast, the huge ribosome is the factory where mRNA, charged tRNAs, and other protein factors come together for protein synthesis.

The key decision in translation is where to initiate translation. In prokaryotes, the initiation complex assembles on mRNA at the Shine–Dalgarno sequence, just upstream of the AUG start codon. The initiation complex in eukaryotes is assembled at the 5′ cap structure of the mRNA and moves in a 3′ direction until the start codon is recognized. The longest phase of translation is the elongation cycle; in this phase, the ribosome moves along the mRNA, revealing the next codon that will interact with its cognate-charged tRNA so that the charged tRNA's amino acid can be added to the growing polypeptide chain. This cycle continues until a stop codon is encountered. Release factors facilitate translation termination.

In the past few years, new imaging techniques have revealed ribosomal interactions at the atomic level. With these new "eyes," we can now see that the ribosome is an incredibly dynamic machine that changes shape in response to the contacts made with tRNAs and with proteins. Furthermore, imaging at atomic resolution has revealed that the ribosomal RNAs, not the ribosomal proteins, are intimately associated with the functional centers of the ribosome.

The proteome is the complete set of proteins that can be expressed by the genetic material of an organism. Whereas a typical multicellular eukaryote has about 20,000 genes, the typical proteome is probably 10- to 50-fold larger. This difference is in part the result of posttranslational modifications such as phosphorylation and ubiquitination, which influence protein activity and stability.

KEY TERMS

active site (p. 312)
amino acid (p. 312)
aminoacyl-tRNA synthetase (p. 320)
amino end (p. 312)
anticodon (p. 320)
A site (p. 324)
carboxyl end (p. 312)
charged tRNA (p. 320)
codon (p. 314)
decoding center (p. 324)
degenerate code (p. 317)
domain (p. 314)
E site (p. 324)
fibrous protein (p. 312)

globular protein (p. 312)
initiation factor (p. 326)
initiator (p. 325)
interactome (p. 331)
isoform (p. 329)
nuclear localization sequence (NLS) (p. 333)
peptidyltransferase center (p. 324)
polypeptide (p. 312)
primary structure (p. 312)
proteome (p. 329)
P site (p. 324)
quaternary structure (p. 312)
release factor (RF) (p. 328)

ribosomal RNA (rRNA) (p. 311)
ribosome (p. 311)
secondary structure (p. 312)
Shine–Dalgarno sequence (p. 325)
signal sequence (p. 333)
structure-based drug design (p. 325)
subunit (p. 312)
tertiary structure (p. 312)
transfer RNA (tRNA) (p. 311)
triplet (p. 314)
ubiquitin (p. 333)
ubiquitination (p. 333)
wobble (p. 321)

SOLVED PROBLEMS

SOLVED PROBLEM 1. Using Figure 9-5, show the consequences on subsequent translation of the addition of an adenine base to the beginning of the following coding sequence:

Ⓐ
↓

-CGA-UCG-GAA-CCA-CGU-GAU-AAG-CAU-
- Arg - Ser - Glu - Pro - Arg - Asp - Lys - His -

Solution

With the addition of A at the beginning of the coding sequence, the reading frame shifts, and a different set of amino acids is specified by the sequence, as shown here (note that a set of nonsense codons is encountered, which results in chain termination):

-ACG-AUC-GGA-ACC-ACG-UGA-UAA-GCA-
- Thr - Ile - Gly - Thr - Thr - stop - stop

SOLVED PROBLEM 2. A single nucleotide addition followed by a single nucleotide deletion approximately 20 bp apart in DNA causes a change in the protein sequence from

-His-Thr-Glu-Asp-Trp-Leu-His-Gln-Asp-

to

-His-Asp-Arg-Gly-Leu-Ala-Thr-Ser-Asp-

Which nucleotide has been added and which nucleotide has been deleted? What are the original and the new mRNA sequences? (**Hint:** Consult Figure 9-5.)

Solution

We can draw the mRNA sequence for the original protein sequence (with the inherent ambiguities at this stage):

- His - Thr - Glu - Asp - Trp - Leu - His - Gln - Asp

$$
-CA^U_C-ACC-GA^A_{\substack{A\\G}}-GA^U_C-UGG-CUC-CA^U_{\substack{A\\G\\UUA\\G}}-CA^A_G-GA^U_C
$$

Because the protein-sequence change given to us at the beginning of the problem begins after the first amino acid (His) owing to a single nucleotide addition, we can deduce that a Thr codon must change to an Asp codon. This change must result from the addition of a G directly before the Thr codon (indicated by a box), which shifts the reading frame, as shown here:

$$
-CA^U_C-\boxed{G}AC-UGA-{}^A_{\substack{©\\Ⓐ\\G}}GA-ⓊUG-G©U-UCA-Ⓤ CA\uparrow{}^{\substack{A\\G}}-GA^U_C-
$$

- His - Asp - Arg - Gly - Leu - Ala - Thr - Ser - Asp -

Additionally, because a deletion of a nucleotide must restore the final Asp codon to the correct reading frame, an A or G must have been deleted from the end of the original next-to-last codon, as shown by the arrow. The original protein sequence permits us to draw the mRNA with a number of ambiguities. However, the protein sequence resulting from the frameshift allows us to determine which nucleotide was in the original mRNA at most of these points of ambiguity. The nucleotide that must have appeared in the original sequence is circled. In only a few cases does the ambiguity remain.

PROBLEMS

Most of the problems are also available for review/grading through the ɢᴇɴᴇᴛɪᴄsᴘ⦿ʀᴛᴀʟ www.yourgeneticsportal.com.

WORKING WITH THE FIGURES

1. The primary protein structure is shown in Figure 9-3(a). Where in the mRNA (near the 5′ or 3′ end) would a mutation in R_2 be encoded?

2. In this chapter you were introduced to nonsense suppressor mutations in tRNA genes. However, suppressor mutations also occur in protein-coding genes. Using the tertiary structure of the β subunit of hemoglobin shown in Figure 9-3(c), explain in structural terms how a mutation could cause the loss of globin protein function. Now explain how a mutation at a second site in the same protein could suppress this mutation and lead to a normal or near-normal protein.

3. Using the quarternary structure of hemoglobin shown in Figure 9-3(d), explain in structural terms how a mutation in the β subunit protein could be suppressed by a mutation in the α subunit gene.

4. Transfer RNAs (tRNAs) are examples of RNA molecules that do not encode protein. Based on Figures 9-6 and 9-8, what is the significance of the sequence of tRNA molecules? What do you predict would be the impact on translation of a mutation in one of the bases of one of the stems in the tRNA structure? On the mutant organism?

5. Ribosomal RNAs (rRNAs) are another example of a functional RNA molecule. Based on Figure 9-11, what

do you think is the significance of the secondary structure of rRNA?

6. The components of prokaryotic and eukaryotic ribosomes are shown in Figure 9-10. Based on this figure, do you think that the large prokaryotic ribosomal RNA (23S rRNA) would be able to substitute for the eukaryotic 28S rRNA? Justify your answer.

7. In Figure 9-12, is the terminal amino acid emerging from the ribosome encoded by the 5′ or 3′ end of the mRNA?

8. In Figure 9-12(b), what do you think happens to the tRNA that is released from the E site?

9. In Figure 9-17, what do you think happens next to the ribosomal subunits after they are finished translating that mRNA?

10. Based on Figure 9-19, can you predict the position of a mutation that would affect the synthesis of one isoform but not the other?

11. Based on Figure 9-24, can you predict the position of a mutation that would produce an active protein that was not directed to the correct location?

BASIC PROBLEMS

Unpacking Problem 12

a. Use the codon dictionary in Figure 9-5 to complete the following table. Assume that reading is from left to right and that the columns represent transcriptional and translational alignments.

C									
			T	G	A				DNA double helix
	C	A		U					mRNA transcribed
						G	C	A	Appropriate tRNA anticodon
		Trp							Amino acids incorporated into protein

b. Label the 5′ and 3′ ends of DNA and RNA, as well as the amino and carboxyl ends of the protein.

13. Consider the following segment of DNA:

5′ GCTTCCCAA 3′

3′ CGAAGGGTT 5′

Assume that the top strand is the template strand used by RNA polymerase.

a. Draw the RNA transcribed.

b. Label its 5′ and 3′ ends.

c. Draw the corresponding amino acid chain.

d. Label its amino and carboxyl ends.

Repeat parts *a* through *d*, assuming the bottom strand to be the template strand.

14. A mutational event inserts an extra nucleotide pair into DNA. Which of the following outcomes do you expect? (1) No protein at all; (2) a protein in which one amino acid is changed; (3) a protein in which three amino acids are changed; (4) a protein in which two amino acids are changed; (5) a protein in which most amino acids after the site of the insertion are changed.

15. Before the true nature of the genetic coding process was fully understood, it was proposed that the message might be read in overlapping triplets. For example, the sequence GCAUC might be read as GCA CAU AUC:

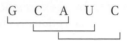

Devise an experimental test of this idea.

16. In protein-synthesizing systems in vitro, the addition of a specific human mRNA to the *E. coli* translational apparatus (ribosomes, tRNA, and so forth) stimulates the synthesis of a protein very much like that specified by the mRNA. What does this result show?

17. Which anticodon would you predict for a tRNA species carrying isoleucine? Is there more than one possible answer? If so, state any alternative answers.

18. a. In how many cases in the genetic code would you fail to know the amino acid specified by a codon if you knew only the first two nucleotides of the codon?

b. In how many cases would you fail to know the first two nucleotides of the codon if you knew which amino acid is specified by it?

19. Deduce what the six wild-type codons may have been in the mutants that led Brenner to infer the nature of the amber codon UAG.

20. If a polyribonucleotide contains equal amounts of randomly positioned adenine and uracil bases, what proportion of its triplets will encode (a) phenylalanine, (b) isoleucine, (c) leucine, (d) tyrosine?

21. You have synthesized three different messenger RNAs with bases incorporated in random sequence in the following ratios: **(a)** 1 U:5 C's, **(b)** 1 A:1 C:4 U's, **(c)** 1 A:1 C:1 G:1 U. In a protein-synthesizing system in vitro, indicate the identities and proportions of amino acids that will be incorporated into proteins when each of these mRNAs is tested. (Refer to Figure 9-5.)

22. In the fungus *Neurospora*, some mutants were obtained that lacked activity for a certain enzyme. The mutations were found, by mapping, to be in either of two unlinked genes. Provide a possible explanation in reference to quaternary protein structure.

23. A mutant is found that lacks all detectable function for one specific enzyme. If you had a labeled antibody that detects this protein in a Western blot (see Chapter 1), would you expect there to be any protein detectable by the antibody in the mutant? Explain.

24. In a Western blot (see Chapter 1), the enzyme tryptophan synthetase usually shows two bands of different mobility on the gel. Some mutants with no enzyme activity showed exactly the same bands as the wild type. Other mutants with no activity showed just the slow band; still others, just the fast band.

   a. Explain the different types of mutants at the level of protein structure.

   b. Why do you think there were no mutants that showed no bands?

25. In the Crick-Brenner experiments described in this chapter, three "insertions" or three "deletions" restored the normal reading frame and the deduction was that the code was read in groups of three. Is this deduction really proved by the experiments? Could a codon have been composed of six bases, for example?

26. A mutant has no activity for the enzyme isocitrate lyase. Does this result prove that the mutation is in the gene encoding isocitrate lyase?

27. A certain nonsense suppressor corrects a nongrowing mutant to a state that is near, but not exactly, wild type (it has abnormal growth). Suggest a possible reason why the reversion is not a full correction.

28. In bacterial genes, as soon as any partial mRNA transcript is produced by the RNA polymerase system, the ribosome jumps on it and starts translating. Draw a diagram of this process, identifying 5′ and 3′ ends of mRNA, the COOH and NH₂ ends of the protein, the RNA polymerase, and at least one ribosome. (Why couldn't this system work in eukaryotes?)

29. In a haploid, a nonsense suppressor *su1* acts on mutation 1 but not on mutation 2 or 3 of gene *P*. An unlinked nonsense suppressor *su2* works on *P* mutation 2 but not on 1 or 3. Explain this pattern of suppression in regard to the nature of the mutations and the suppressors.

30. In vitro translation systems have been developed in which specific RNA molecules can be added to a test tube containing a bacterial cell extract that includes all the components needed for translation (ribosomes, tRNAs, amino acids). If a radioactively labeled amino acid is included, any protein translated from that RNA can be detected and displayed on a gel. If a eukaryotic mRNA is added to the test tube, would radioactive protein be produced? Explain.

31. An in vitro translation system contains a eukaryotic cell extract that includes all the components needed for translation (ribosomes, tRNAs, amino acids). If bacterial RNA is added to the test tube, would a protein be produced? If not, why not?

32. Would a chimeric translation system containing the large ribosomal subunit from *E. coli* and the small ribosomal subunit from yeast (a unicellular eukaryote) be able to function in protein synthesis? Explain why or why not.

33. Mutations that change a single amino acid in the active site of an enzyme can result in the synthesis of wild-type amounts of an inactive enzyme. Can you think of other regions in a protein where a single amino acid change might have the same result?

34. What evidence supports the view that ribosomal RNAs are a more important component of the ribosome than the ribosomal proteins?

35. Explain why antibiotics, such as erythromycin and Zithromax, that bind the large ribosomal subunit do not harm us.

36. Why do multicellular eukaryotes need to have hundreds of kinase-encoding genes?

37. Our immune system makes many different proteins that protect us from viral and bacterial infection. Biotechnology companies must produce large quantities of these immune proteins for human testing and eventual sale to the public. To this end, their scientists engineer bacterial or human cell cultures to express these immune proteins. Explain why proteins isolated from bacterial cultures are often inactive, whereas the same proteins isolated from human cell cultures are active (functional).

CHALLENGING PROBLEMS

38. A single nucleotide addition and a single nucleotide deletion approximately 15 sites apart in the DNA cause a protein change in sequence from

   Lys-Ser-Pro-Ser-Leu-Asn-Ala-Ala-Lys

   to

   Lys-Val-His-His Leu-Met-Ala-Ala-Lys

   a. What are the old and new mRNA nucleotide sequences? (Use the codon dictionary in Figure 9-5.)

   b. Which nucleotide has been added and which has been deleted?

   (Problem 38 is from W. D. Stansfield, *Theory and Problems of Genetics.* McGraw-Hill, 1969.)

39. You are studying an *E. coli* gene that specifies a protein. A part of its sequence is

   –Ala-Pro-Trp-Ser-Glu-Lys-Cys-His–

   You recover a series of mutants for this gene that show no enzymatic activity. By isolating the mutant enzyme products, you find the following sequences:

Mutant 1:
   –Ala-Pro-Trp-Arg-Glu-Lys-Cys-His–
Mutant 2:
   –Ala-Pro–
Mutant 3:
   –Ala-Pro-Gly-Val-Lys-Asn-Cys-His–
Mutant 4:
   –Ala-Pro-Trp-Phe-Phe-Thr-Cys-His–

What is the molecular basis for each mutation? What is the DNA sequence that specifies this part of the protein?

40. Suppressors of frameshift mutations are now known. Propose a mechanism for their action.

41. Consider the gene that specifies the structure of hemoglobin. Arrange the following events in the most likely sequence in which they would take place.

   a. Anemia is observed.

   b. The shape of the oxygen-binding site is altered.

   c. An incorrect codon is transcribed into hemoglobin mRNA.

   d. The ovum (female gamete) receives a high radiation dose.

   e. An incorrect codon is generated in the DNA of the hemoglobin gene.

   f. A mother (an X-ray technician) accidentally steps in front of an operating X-ray generator.

   g. A child dies.

   h. The oxygen-transport capacity of the body is severely impaired.

   i. The tRNA anticodon that lines up is one of a type that brings an unsuitable amino acid.

   j. Nucleotide-pair substitution occurs in the DNA of the gene for hemoglobin.

42. An induced cell mutant is isolated from a hamster tissue culture because of its resistance to α-amanitin (a poison derived from a fungus). Electrophoresis shows that the mutant has an altered RNA polymerase; *just one* electrophoretic band is in a position different from that of the wild-type polymerase. The cells are presumed to be diploid. What do the results of this experiment tell you about ways in which to detect recessive mutants in such cells?

43. A double-stranded DNA molecule with the sequence shown here produces, in vivo, a polypeptide that is five amino acids long.

   TACATGATCATTTCACGGAATTTCTAGCATGTA

   ATGTACTAGTAAAGTGCCTTAAAGATCGTACAT

   a. Which strand of DNA is transcribed and in which direction?

   b. Label the 5′ and the 3′ ends of each strand.

   c. If an inversion occurs between the second and the third triplets from the left and right ends, respectively, and the same strand of DNA is transcribed, how long will the resultant polypeptide be?

   d. Assume that the original molecule is intact and that the bottom strand is transcribed from left to right. Give

the base sequence, and label the 5′ and 3′ ends of the anticodon that inserts the *fourth* amino acid into the nascent polypeptide. What is this amino acid?

44. One of the techniques used to decipher the genetic code was to synthesize polypeptides in vitro, with the use of synthetic mRNA with various repeating base sequences—for example, (AGA)$_n$ which can be written out as AGAAGAAGAAGAAGA. . . . Sometimes the synthesized polypeptide contained just one amino acid (a homopolymer), and sometimes it contained more than one (a heteropolymer), depending on the repeating sequence used. Furthermore, sometimes different polypeptides were made from the same synthetic mRNA, suggesting that the initiation of protein synthesis in the system in vitro does not always start on the end nucleotide of the messenger. For example, from (AGA)n, three polypeptides may have been made: aa$_1$ homopolymer (abbreviated aa$_1$-aa$_1$), aa$_2$ homopolymer (aa$_2$-aa$_2$), and aa$_3$ homopolymer (aa$_3$-aa$_3$). These polypeptides probably correspond to the following readings derived by starting at different places in the sequence:

AGA AGA AGA AGA . . .

GAA GAA GAA GAA . . .

AAG AAG AAG AAG . . .

The following table shows the actual results obtained from the experiment done by Khorana.

Synthetic mRNA	Polypeptide(s) synthesized
(UC)$_n$	(Ser–Leu)
(UG)$_n$	(Val–Cys)
(AC)$_n$	(Thr–His)
(AG)$_n$	(Arg–Glu)
(UUC)$_n$	(Ser–Ser) and (Leu–Leu) and (Phe–Phe)
(UUG)$_n$	(Leu–Leu) and (Val–Val) and (Cys–Cys)
(AAG)$_n$	(Arg–Arg) and (Lys–Lys) and (Glu–Glu)
(CAA)$_n$	(Thr–Thr) and (Asn–Asn) and (Gln–Gln)
(UAC)$_n$	(Thr–Thr) and (Leu–Leu) and (Tyr–Tyr)
(AUC)$_n$	(Ile–Ile) and (Ser–Ser) and (His–His)
(GUA)$_n$	(Ser–Ser) and (Val–Val)
(GAU)$_n$	(Asp–Asp) and (Met–Met)
(UAUC)$_n$	(Tyr–Leu–Ser–Ile)
(UUAC)$_n$	(Leu–Leu–Thr–Tyr)
(GAUA)$_n$	None
(GUAA)$_n$	None

Note: The order in which the polypeptides or amino acids are listed in the table is not significant except for (UAUC)$_n$ and (UUAC)$_n$.

a. Why do $(GUA)_n$ and $(GAU)_n$ each encode only two homopolypeptides?

b. Why do $(GAUA)_n$ and $(GUAA)_n$ fail to stimulate synthesis?

c. Assign an amino acid to each triplet in the following list. Bear in mind that there are often several codons for a single amino acid and that the first two letters in a codon are usually the important ones (but that the third letter is occasionally significant). Also remember that some very different-looking codons sometimes encode the same amino acid. Try to carry out this task without consulting Figure 9-5.

AUG	GAU	UUG	AAC
GUG	UUC	UUA	CAA
GUU	CUC	AUC	AGA
GUA	CUU	UAU	GAG
UGU	CUA	UAC	GAA
CAC	UCU	ACU	UAG
ACA	AGU	AAG	UGA

To solve this problem requires both logic and trial and error. Don't be disheartened: Khorana received a Nobel Prize for doing it. Good luck!

(Problem 44 is from J. Kuspira and G. W. Walker, *Genetics: Questions and Problems.* McGraw-Hill, 1973.)

Gene Isolation and Manipulation

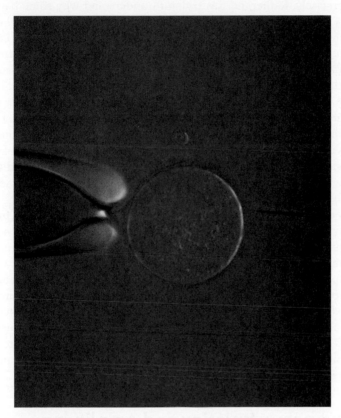

Injection of foreign DNA into an animal cell. The microneedle used for injection is shown at the right, and a cell-holding pipette is shown at the left. [*Copyright M. Baret/Rapho/Photo Researchers.*]

KEY QUESTIONS

- How is a gene isolated and amplified by cloning?

- How are libraries used to identify specific DNA molecules?

- How is DNA amplified without cloning?

- How is DNA analyzed after cloning?

- How are eukaryotic genomes modified in the laboratory?

OUTLINE

Genes are the central focus of genetics, and so, clearly, it is desirable to be able to isolate a gene of interest (or any DNA region) from the genome and amplify it to obtain a working amount to study. Isolating individual genes and producing enough of them to analyze can be a daunting task because a single gene is a tiny fraction of an entire genome. For example, the human genome contains about 3 billion base pairs, whereas the coding region of an average gene contains only a few thousand base pairs. How do scientists find the proverbial needle in the haystack, the gene, and then produce quantities of it for analysis?

Just like a construction worker, a genetic engineer needs tools. Most toolboxes that we are familiar with are filled with tools like hammers, screwdrivers, and wrenches that are designed by people and manufactured in factories. In contrast, the tools of the genetic engineer are molecules isolated from cells. Most of these tools were the product of scientific discovery—where the objective was to answer a biological question. Only later did some scientists appreciate the potential practical value of some of these molecules and invent ways to integrate them

341

into protocols with the goal of isolating and amplifying DNA fragments. We have already been introduced to some of these molecules in previous chapters, and in this chapter you will see how they have become the foundation of the biotechnology revolution.

The first step in isolating a gene from the genome is usually to cut up the chromosome into gene-size fragments so that any scientist in any laboratory can begin with the DNA from the same organism and by applying "molecular scissors" end up with the virtually the same array of fragments. Werner Arber discovered these molecular scissors, and for this discovery he was awarded the Nobel Prize in Physiology or Medicine in 1978. However, Arber was not looking for a tool to cut DNA precisely. Rather, he was trying to understand why some bacteria are resistant to infection by bacterial viruses. By answering this biological question, he discovered that resistant bacteria possess a previously unknown enzyme, a restriction endonuclease that cuts DNA at specific sequences. The enzyme he discovered, *Eco*RI, became the first commercially available molecular scissors.

As another example, it is unlikely that anyone would have predicted that DNA polymerase, the enzyme discovered by Arthur Kornberg, a discovery for which he received the Nobel Prize in Physiology or Medicine in 1959, could be fashioned into two powerful tools for DNA isolation and analysis. To this day, most of the techniques used to determine the nucleotide sequence of DNA rely on synthesizing DNA with DNA polymerase. Similarly, most of the protocols used to isolate and amplify specific regions of DNA from sources as disparate as a crime scene to a fossil embedded in amber rely on the activity of DNA polymerase.

DNA technology is a term that describes the collective techniques for obtaining, amplifying, and manipulating specific DNA fragments. Since the mid-1970s, the development of DNA technology has revolutionized the study of biology, opening many areas of research to molecular investigation. **Genetic engineering,** the application of DNA technology to specific biological, medical, or agricultural problems, is now a well-established branch of technology. **Genomics** is the ultimate extension of the technology to the global analysis of the nucleic acids present in a nucleus, a cell, an organism, or a group of related species (see Chapter 14). Later in this chapter, we will see how the techniques of DNA technology and genomics, along with methods presented in Chapters 2 and 4, can be used together to isolate and identify a gene.

10.1 Overview: Isolating and Amplifying Specific DNA Fragments

How can a specific segment of DNA be isolated from an entire genome? Furthermore, how can it be isolated in quantities sufficient to analyze features of the DNA such as its DNA sequence and its protein product? A crucial insight was that researchers could create the large samples of DNA that they needed to isolate a gene by tricking the DNA replication machinery (see Chapter 7) to replicate the DNA segment in question. Such replication is called **amplification.** It can be done either within live bacterial cells (in vivo) or in a test tube (in vitro).

In the in vivo approach (Figure 10-1a), an investigator begins with a sample of DNA molecules containing the gene of interest. This sample is called the **donor DNA,** and most often it is an entire genome. Fragments of the donor DNA are inserted into a specially designed plasmid or bacterial virus that will "carry" and amplify the gene of interest and are hence called **vectors.** First, the donor DNA molecules are cut up by using enzymes called restriction endonucleases as molecular "scissors." They cut long chromosome-size DNA molecules into hundreds or thousands of fragments of more manageable size. Next, each fragment is inserted into a cut vector chromosome to form **recombinant DNA** molecules.

The recombinant DNA molecules are transferred into bacterial cells, and, generally, only one recombinant molecule is taken up by each cell. The recombinant molecule is amplified along with the vector during the division of the bacterial cell. This process results in a *clone* of identical cells, each containing the recombinant DNA molecule, and so this technique of amplification is called **DNA cloning.** The next stage is to find the rare clone containing the DNA of interest.

In the in vitro approach, called the **polymerase chain reaction (PCR)** (Figure 10-1b), a specific gene or DNA region of interest is isolated and amplified by DNA polymerase extracted from a heat-tolerant bacteria. PCR "finds" the DNA region of interest (called the target DNA) by the complementary binding of specific short primers to the ends of that sequence. These primers then guide the replication process, which cycles exponentially, resulting in the production of large quantities of the target DNA as an isolated DNA fragment. Even larger quantities of target DNA can be obtained by inserting the PCR product into a plasmid, thus generating a recombinant DNA molecule like that described above.

We will see repeatedly that DNA technology depends on two basic foundations of molecular biology research:

- The ability of specific proteins to recognize and bind to specific base sequences within the DNA double helix (examples are shown in yellow in Figure 10-1).

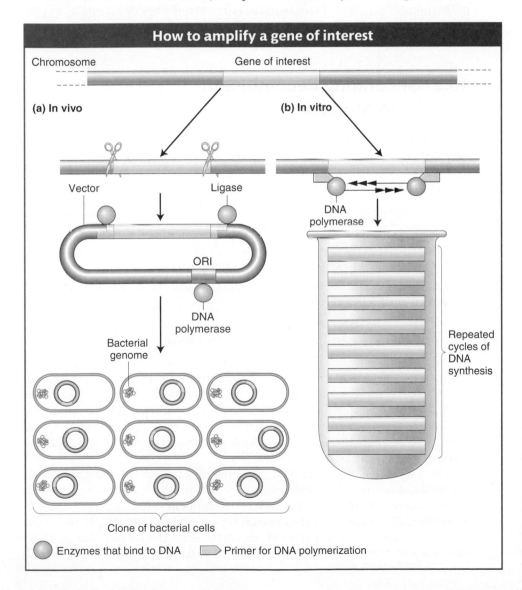

How to amplify a gene of interest

Chromosome   Gene of interest

(a) In vivo   (b) In vitro

Vector   Ligase

DNA
polymerase

ORI

DNA
polymerase

Bacterial
genome

Repeated
cycles of
DNA
synthesis

Clone of bacterial cells

● Enzymes that bind to DNA   ▷ Primer for DNA polymerization

Figure 10-1 Two methods of isolating and amplifying a gene are (a) in vivo, by tricking the replication machinery of a bacterium into amplifying recombinant DNA containing the gene, and (b) in vitro, in the test tube using the polymerase-chain-reaction technique. Both methods employ the basic principles of molecular biology: the ability of specific proteins (yellow) to bind to DNA and the ability of complementary single-stranded nucleic acid segments to hybridize together (the primer used in the test-tube method).

- The ability of complementary single-stranded DNA or RNA sequences to anneal to form double-stranded molecules (an example is the binding of the primers shown in yellow in Figure 10-1).

The remainder of the chapter will explore some of the uses to which we put amplified DNA. These uses range from routine gene isolation for basic biological research to gene therapy to treat human disease to the production of herbicides and pesticides by crop plants. To illustrate how recombinant DNA is made, let's consider the cloning of the gene for human insulin, a protein hormone used in the treatment of diabetes. Diabetes is a disease in which blood sugar levels are abnormally high either because the body does not produce enough insulin (type I diabetes) or because cells are unable to respond to insulin (type II diabetes). Mild forms of type I diabetes can be treated by dietary restrictions, but, for many patients, daily insulin treatments are necessary. Until about 30 years ago, cows were the major source of insulin protein. The protein was harvested from the pancreases of animals slaughtered in meatpacking plants and purified on a large scale to eliminate the majority of proteins and other contaminants in the pancreas extracts. Then, in 1982, the first recombinant human insulin came on the market. Human insulin could be made in purer form, at lower cost, and on an industrial scale because it was produced in bacteria by recombinant DNA techniques. The recombinant insulin is a higher proportion of the proteins in the bacterial cell than is the bovine insulin; hence, the protein purification is much easier. We will follow the general steps necessary for making any recombinant DNA and apply them to insulin.

10.2 Generating Recombinant DNA Molecules

Recombinant DNA molecules usually contain a DNA fragment inserted into a bacterial vector. In this section, you will see that there are many types of recombinant DNA molecules that can be constructed from a variety of donor DNAs and vectors. We begin by discussing sources of donor DNA:

- If the experimenter wants a collection of inserts that represents the entire genome of an organism, the genomic DNA can be cut up before cloning.

- Alternatively, if the goal is to isolate a single gene, the polymerase chain reaction can be used to amplify selected regions of DNA in vitro.

- Finally, DNA copies of the mRNA products, called cDNA, can be synthesized and inserted into a vector.

Genomic DNA can be cut up before cloning

Genomic DNA is obtained directly from the chromosomes of the organism under study, usually by grinding up fresh tissue and purifying the DNA. Chromosomal DNA can be used as the starting point for both in vivo and PCR methods to isolate genes. For the in vivo method, genomic DNA needs to be cut up before cloning is possible. As you will see later in this section, to perform PCR, genomic DNA does not have to be cut up because the specific short primers that anneal to it guide the replication of the intervening DNA.

The long chromosome-size DNA molecules of genomic DNA must be cut into fragments of much smaller size before they can be inserted into a vector. Most cutting is done with the use of bacterial **restriction enzymes.** These enzymes cut at specific DNA sequences, called *restriction sites,* and this property is one of the key features that make restriction enzymes suitable for DNA manipulation. These enzymes are examples of endonucleases that cleave a phosphodiester bond.

Purely by chance, any DNA molecule, whether it is derived from virus, fly, or human, contains restriction-enzyme recognition sites. Thus, a restriction enzyme will cut the DNA into a set of **restriction fragments** determined by the locations of the restriction sites.

Another key property of some restriction enzymes is that they make "sticky ends." Let's look at an example. The restriction enzyme *Eco*RI (from *E. coli*) recognizes the following sequence of six nucleotide pairs in the DNA of any organism:

5'-GAATTC-3'

3'-CTTAAG-5'

This type of segment is called a **DNA palindrome,** which means that both strands have the same nucleotide sequence but in antiparallel orientation. Different restriction enzymes cut at different palindromic sequences. Sometimes the cuts are in the same position on each of the two antiparallel strands. However, the most useful restriction enzymes make cuts that are offset, or staggered. The enzyme *Eco*RI makes cuts only between the G and the A nucleotides on each strand of the palindrome:

5'-G↓AATTC-3'

3'-CTTAA↑G-5'

These staggered cuts leave a pair of identical sticky ends, each a single strand five bases long. The ends are called *sticky* because, being single stranded, they can base-pair (that is, stick) to a complementary sequence. Single-strand pairing of this type is sometimes called **hybridization.** Figure 10-2 illustrates the restriction enzyme *Eco*RI making staggered double-strand cuts in a circular DNA molecule such as a plasmid; the cut opens up the circle, and the resulting linear molecule has two sticky ends. It can now hybridize with a fragment of a different DNA molecule having the same complementary sticky ends.

Digesting human genomic DNA with *Eco*RI generates approximately 500,000 fragments. You will see later in this section how scientists sift through all of these fragments to find the needle in the haystack—the one or two fragments that contain the gene of interest.

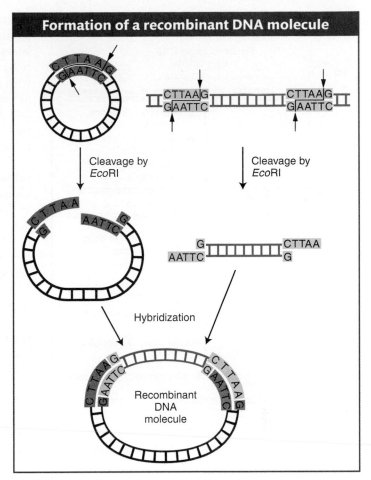

Figure 10-2 To form a recombinant DNA molecule, the restriction enzyme *Eco*RI cuts a circular DNA molecule bearing one target sequence, resulting in a linear molecule with single-stranded sticky ends. Because of complementarity, other linear molecules with *Eco*RI-cut sticky ends can hybridize with the linearized circular DNA, forming a recombinant DNA molecule.

> **Message** Genomic DNA can be used directly for cloning genes. As a first step, restriction enzymes cut DNA into fragments of manageable size, and many of them generate single-stranded sticky ends suitable for making recombinant DNA.

The polymerase chain reaction amplifies selected regions of DNA in vitro

If we know the sequence of at least some parts of the gene or sequence of interest, it is not necessary to search for the gene among the hundreds of thousands of genomic fragments. Instead, we can simply amplify it in a test tube using a procedure called the *polymerase chain reaction (PCR)*. The basic strategy of PCR

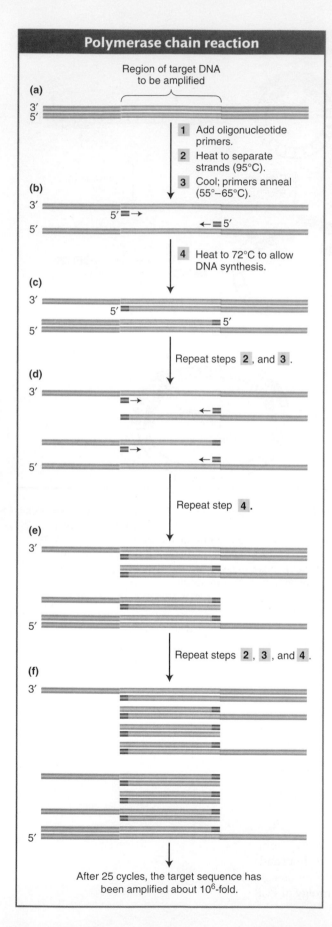

Polymerase chain reaction

Region of target DNA
to be amplified

(a)

3′
5′

1 Add oligonucleotide
primers.

2 Heat to separate
strands (95°C).

3 Cool; primers anneal
(55°–65°C).

(b)

3′

5′ ■→

←■ 5′

5′

4 Heat to 72°C to allow
DNA synthesis.

(c)

3′

5′ ■

■ 5′

5′

Repeat steps 2 , and 3 .

(d)

3′

■→

←■

■→

←■

5′

Repeat step 4 .

(e)

3′

5′

Repeat steps 2 , 3 , and 4 .

(f)

3′

5′

After 25 cycles, the target sequence has
been amplified about 10⁶-fold.

is outlined in Figure 10-3. The process uses multiple copies of a pair of short chemically synthesized primers, from 15 to 20 bases long, each binding to a different end of the gene or region to be amplified. The two primers bind to opposite DNA strands, with their 3′ ends pointing at each other. Polymerases add bases to these primers, and the polymerization process shuttles back and forth between them, forming an exponentially growing number of double-stranded DNA molecules. The details are as follows.

We start with a solution containing the DNA source, the primers, the four deoxyribonucleotide triphosphates, and an unusual heat-tolerant DNA polymerase. The DNA is denatured by heat (95°C), resulting in single-stranded DNA molecules. When the solution is cooled (to between 50 and 65°C) the primers hybridize (they are said to *anneal*) to their complementary sequences in the single-stranded DNA molecules. After the temperature is raised to 72°C, the heat-tolerant DNA polymerase replicates the single-stranded DNA segments extending from a primer. The DNA polymerase *Taq* polymerase, from the bacterium *Thermus aquaticus,* is one such enzyme commonly used. (This bacterium normally grows in thermal vents and so has evolved proteins that are extremely heat resistant. Thus, it is able to survive the high temperatures required to denature the DNA duplex, which would denature and inactivate DNA polymerase from most species.) Complementary new strands are synthesized as in normal DNA replication in cells, forming two double-stranded DNA molecules identical with the parental double-stranded molecule. These steps leading to a single replication of the segment between the two primers represent one cycle.

Figure 10-3 The polymerase chain reaction quickly copies a target DNA sequence. (a) Double-stranded DNA containing the target sequence. (b) Two chosen or created primers have sequences complementing primer-binding sites at the 3′ ends of the target gene on the two strands. The strands are separated by heating; then cooled to allow the two primers to anneal to the primer-binding sites. Together, the primers thus flank the targeted sequence. (c) After the temperature is raised, *Taq* polymerase then synthesizes the first set of complementary strands in the reaction. These first two strands are of varying length, because they do not have a common stop signal. They extend beyond the ends of the target sequence as delineated by the primer-binding sites. (d) The two duplexes are heated again, exposing four binding sites. After cooling, the two primers again bind to their respective strands at the 3′ ends of the target region. (e) After the temperature is raised, *Taq* polymerase synthesizes four complementary strands. Although the template strands at this stage are variable in length, two of the four strands just synthesized from them are precisely the length of the target sequence desired. This precise length is achieved because each of these strands begins at the primer-binding site, at one end of the target sequence, and proceeds until it runs out of template, at the other end of the sequence. (f) The process is repeated for many cycles, each time creating more double-stranded DNA molecules identical with the target sequence. [*After D. L. Nelson and M. M. Cox, Lehninger Principles of Biochemistry, 4th ed. Copyright 2005 by W. H. Freeman and Company.*]

GENETICS*PORTAL* **ANIMATED ART: Polymerase chain reaction**

After the replication of the segment between the two primers is completed, the two new duplexes are again heat-denatured to generate single-stranded templates, and a second cycle of replication is carried out by lowering the temperature in the presence of all the components necessary for the polymerization. Repeated cycles of denaturation, annealing, and synthesis result in an exponential increase in the number of segments replicated. Amplifications by as much as a million-fold can be readily achieved within 1 to 2 hours. As you will see later in this section, the PCR products can be further amplified by cloning them in bacterial cells.

PCR is a powerful technique that is routinely used to isolate specific genes or DNA fragments when there is prior knowledge of the sequence to be amplified. In fact, if the sequences corresponding to the primers are each present only once in the genome and are sufficiently close together, the *only* DNA segment that can be amplified is the one between the two primers. PCR is a very sensitive technique with numerous applications in biology. It can amplify target sequences that are present in extremely low copy numbers in a sample as long as primers specific to this rare sequence are used. For example, crime investigators can amplify segments of human DNA from the few follicle cells surrounding a single pulled-out hair.

It would not be an overstatement to say that PCR has revolutionized the study of many fields of biology where DNA analysis is required. In recognition of its importance to science, the person who made PCR a viable protocol, Kary Mullis, was awarded the Nobel Prize in Chemistry in 1993.

> **Message** The polymerase chain reaction uses specially designed primers for direct isolation and amplification of specific regions of DNA in a test tube.

DNA copies of mRNA can be synthesized

Genomic DNA may not be a good starting point if the goal is to isolate and analyze genes that encode proteins. As we have seen in Chapter 8, eukaryotic genes often contain several introns that disrupt the coding regions. Further, as we will see in Chapters 14 and 15, protein-coding genes are often less than 5 percent of the genomic DNA of multicellular eukaryotes. Thus, in higher eukaryotes, collections of mRNA are a more useful starting point for the isolation of protein-coding genes. In addition, mRNA is a better predictor of a polypeptide sequence than is a genomic sequence, because the introns have been spliced out (see Chapter 8). Thus, the sequence of the mRNA can be virtually "translated" into the amino acid sequence of the protein by simply reading the triplet codons.

Complementary DNA (cDNA)   is a DNA version of an mRNA molecule. Researchers use cDNA rather than mRNA itself because RNAs are inherently less stable than DNA. Moreover, techniques for routinely amplifying and purifying individual RNA molecules do not exist. The cDNA is made from mRNA with the use of a special enzyme called *reverse transcriptase*, originally isolated from retroviruses (see Chapter 15). Retroviruses have RNA genomes that are copied into DNA that inserts into the host chromosome. Can you think of why it is called *reverse* transcriptase? To make cDNA, a researcher purifies mRNA (usually from a particular tissue such as pancreas or plant roots) and adds it to a test tube containing reverse transcriptase, the four dNTPs, and a short primer of polymerized dTTP residues (called an oligo-dT primer). The oligo-dT primer anneals to the poly(A) tail of the mRNA molecule being copied. Using this mRNA molecule as a template, reverse transcriptase catalyzes the synthesis of a single-stranded DNA molecule starting from the oligo-dT primer. DNA polymerase then copies the cDNA into a double-stranded DNA

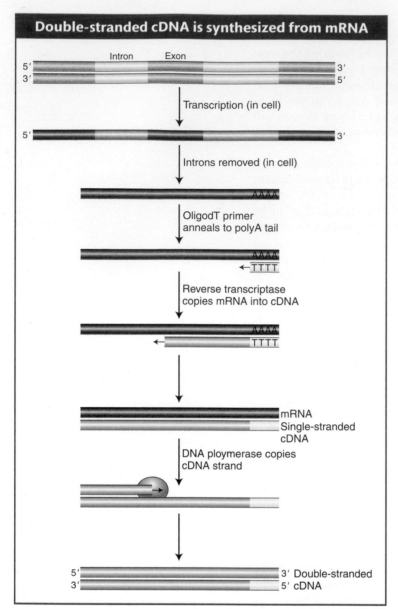

Double-stranded cDNA is synthesized from mRNA

Intron   Exon

Transcription (in cell)

Introns removed (in cell)

OligodT primer
anneals to polyA tail

Reverse transcriptase
copies mRNA into cDNA

DNA ploymerase copies
cDNA strand

mRNA
Single-stranded
cDNA

3′ Double-stranded
5′ cDNA

Figure 10-4 The formation of cDNA for the insulin gene. The insulin gene (with its two introns) is transcribed in the pancreas into pre-mRNA. The introns are removed by splicing, and A residues are added to the 3′ end to form polyadenylated mRNA. In the laboratory, mRNAs are isolated from pancreatic cells and a short oligo(dT) primer is hybridized to the poly(A) tail of all mRNAs to prime synthesis of complementary DNA from the RNA template by reverse transcriptase. When the mRNA strand has been degraded (by treatment with NaOH or with RNAseH), addition of a second primer (step not shown) permits initiation by DNA polymerase and completes the synthesis of double-stranded cDNA.

molecule (Figure 10-4). Like fragments of genomic DNA or PCR products, double-stranded cDNA can be inserted into recombinant DNA molecules for further amplification (see below).

The human insulin gene contains two introns. To create bacteria that synthesize human insulin, cDNA was the initial choice because bacteria do not have the ability to splice out introns present in natural genomic DNA.

> **Message** mRNA is often a preferable starting point in the isolation of a gene. Enzymatic conversion of mRNA into cDNA allows for the isolation of a gene copy without introns.

Attaching donor and vector DNA

As we have seen, there are diverse sources of donor DNA, including genomic DNA fragments, PCR products, and double-stranded cDNA. The first step in the production of large quantities of these DNAs is to construct recombinant DNA molecules by inserting donor DNA into vector DNA.

Cloning DNA fragments with sticky ends To make recombinant DNA molecules containing donor genomic DNA fragments, both donor and vector DNAs are digested by a restriction enzyme that produces complementary sticky ends. The resulting fragments are then mixed in a test tube to allow the sticky ends of vector and donor DNA to hybridize with each other and form recombinant molecules. Figure 10-5a shows a bacterial plasmid DNA that carries a single *Eco*RI restriction site; so digestion with the restriction enzyme *Eco*RI converts the circular DNA into a single linear molecule with sticky ends. Donor DNA from any other source, such as human DNA, also is treated with the *Eco*RI enzyme to produce a population of fragments carrying the same sticky ends. When the two populations are mixed under the proper physiological conditions, DNA fragments from the two sources can hybridize, because double helices form between their sticky ends (Figure 10-5b). There are many opened-up plasmid molecules in the solution, as well as many different *Eco*RI fragments of donor DNA. Therefore, a diverse array of plasmids recombined with different donor fragments will be produced. At this stage, the hybridized molecules do not have covalently joined sugar–phosphate backbones. However, the backbones can be sealed by the addition of the enzyme **DNA ligase,** which creates phosphodiester linkages at the junctions (Figure 10-5c).

Cloning DNA fragments with blunt ends Some restriction enzymes produce blunt ends rather than staggered cuts. In addition, cDNA and the DNA fragments that arise from PCR have blunt or near-blunt ends. While blunt end fragments from all these sources can be joined to the vector with the use of ligase alone, this is a very inefficient reaction. One alternative method is to create PCR products with sticky ends by using specially designed PCR primers that contain restriction endonuclease recognition sequences at their 5′ ends (Figure 10-6). Digestion of the final PCR product with the restriction enzyme (*Eco*RI in this case) produces a fragment that is ready to be inserted into a vector (see Figure 10-5b).

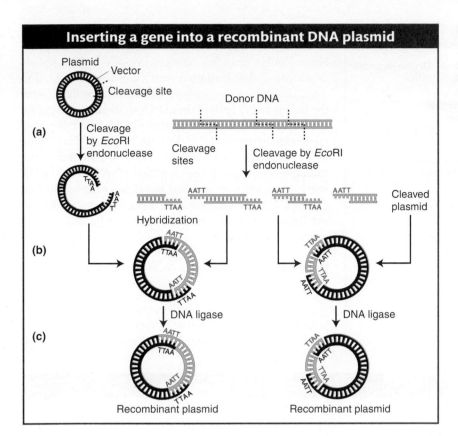

Inserting a gene into a recombinant DNA plasmid

Figure 10-5 Method for generating a collection of recombinant DNA plasmids containing genes derived from restriction enzyme digestion of donor DNA. [*After S. N. Cohen, "The Manipulation of Genes." Copyright 1975 by Scientific American, Inc. All rights reserved.*]

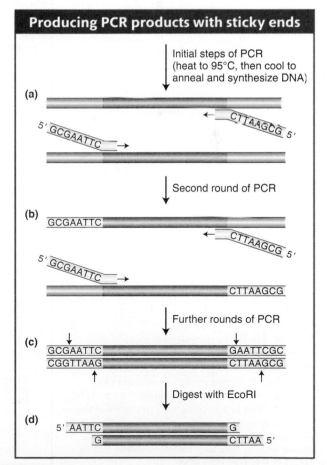

Producing PCR products with sticky ends

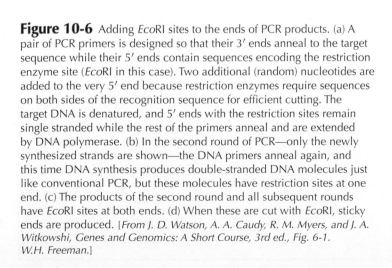

Figure 10-6 Adding *Eco*RI sites to the ends of PCR products. (a) A pair of PCR primers is designed so that their 3′ ends anneal to the target sequence while their 5′ ends contain sequences encoding the restriction enzyme site (*Eco*RI in this case). Two additional (random) nucleotides are added to the very 5′ end because restriction enzymes require sequences on both sides of the recognition sequence for efficient cutting. The target DNA is denatured, and 5′ ends with the restriction sites remain single stranded while the rest of the primers anneal and are extended by DNA polymerase. (b) In the second round of PCR—only the newly synthesized strands are shown—the DNA primers anneal again, and this time DNA synthesis produces double-stranded DNA molecules just like conventional PCR, but these molecules have restriction sites at one end. (c) The products of the second round and all subsequent rounds have *Eco*RI sites at both ends. (d) When these are cut with *Eco*RI, sticky ends are produced. [*From J. D. Watson, A. A. Caudy, R. M. Myers, and J. A. Witkowshi, Genes and Genomics: A Short Course, 3rd ed., Fig. 6-1. W.H. Freeman.*]

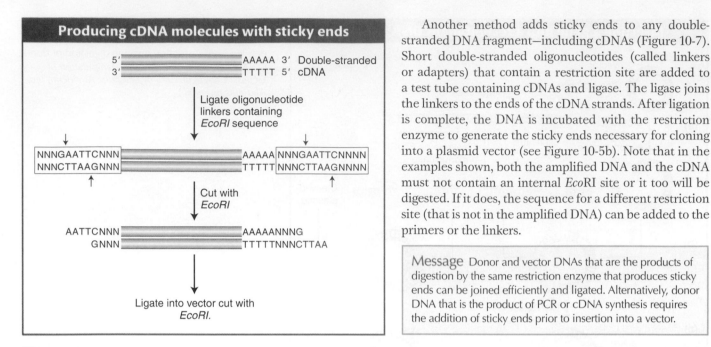

Producing cDNA molecules with sticky ends

5′ ———————————— AAAAA 3′   Double-stranded
3′ ———————————— TTTTT 5′   cDNA

Ligate oligonucleotide
linkers containing
*Eco*RI sequence

| NNNGAATTCNNN |————————————| AAAAA | NNNGAATTCNNNN |
| NNNCTTAAGNNN |————————————| TTTTT | NNNCTTAAGNNNN |

Cut with
*Eco*RI

AATTCNNN ———————————— AAAAANNNG
    GNNN ———————————— TTTTTNNNCTTAA

Ligate into vector cut with
*Eco*RI.

Figure 10-7 Adding *Eco*RI sites to the ends of cDNA molecules. The cDNA molecules come from the last step in Figure 10-4. Adapters (boxed region) are added at both ends of the cDNA molecules. These adapters are double-stranded oligonucleotides that contain a restriction site (*Eco*RI is shown in red) and random DNA sequence at both ends (represented by N).

Another method adds sticky ends to any double-stranded DNA fragment—including cDNAs (Figure 10-7). Short double-stranded oligonucleotides (called linkers or adapters) that contain a restriction site are added to a test tube containing cDNAs and ligase. The ligase joins the linkers to the ends of the cDNA strands. After ligation is complete, the DNA is incubated with the restriction enzyme to generate the sticky ends necessary for cloning into a plasmid vector (see Figure 10-5b). Note that in the examples shown, both the amplified DNA and the cDNA must not contain an internal *Eco*RI site or it too will be digested. If it does, the sequence for a different restriction site (that is not in the amplified DNA) can be added to the primers or the linkers.

Message Donor and vector DNAs that are the products of digestion by the same restriction enzyme that produces sticky ends can be joined efficiently and ligated. Alternatively, donor DNA that is the product of PCR or cDNA synthesis requires the addition of sticky ends prior to insertion into a vector.

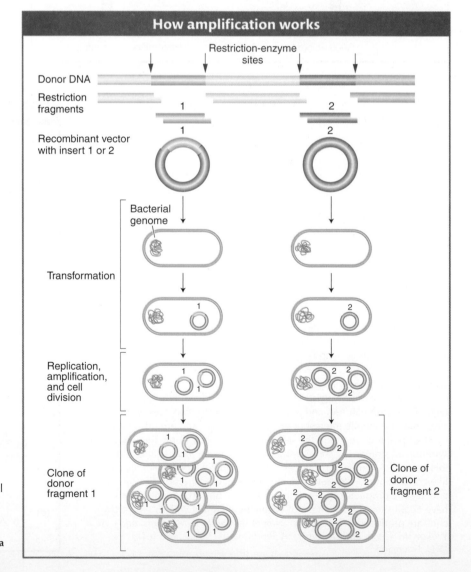

How amplification works

Restriction-enzyme
sites

Donor DNA

Restriction
fragments

Recombinant vector
with insert 1 or 2

Bacterial
genome

Transformation

Replication,
amplification,
and cell
division

Clone of
donor
fragment 1

Clone of
donor
fragment 2

Figure 10-8 The general strategy used to clone a gene. Restriction-enzyme treatment of donor DNA and vector allows the insertion of single fragments into vectors. A single vector enters a bacterial host, where replication and cell division result in a large number of copies of the donor fragment.

GENETICS*PORTAL* **ANIMATED ART: Finding specific cloned genes by functional complementation: making a library of wild-type yeast DNA**

Amplification of donor DNA inside a bacterial cell

Amplification of the recombinant DNA molecules takes advantage of prokaryotic genetic processes, including those of bacterial transformation, plasmid replication, and bacteriophage growth, all discussed in Chapter 5. Figure 10-8 illustrates the cloning of a donor DNA segment. A single recombinant vector enters a bacterial cell and is amplified by the same machinery that replicates the bacterial chromosome. One basic requirement is the presence of an origin of DNA replication (as described in Chapter 7). There are generally many copies of each vector in each bacterial cell. Hence, after amplification, a colony of bacteria will typically contain billions of copies of the single-donor DNA insert fused to its vector. This set of amplified copies of the single-donor DNA fragment within the cloning vector is the recombinant DNA *clone*.

The amplification of donor DNA inside a bacterial cell entails the following steps:

- Choosing a cloning vector and introducing the insert (see the preceding section for a discussion of the latter)

- Introducing the recombinant DNA molecule inside a bacterial cell

- Recovering the amplified recombinant molecules

Choice of cloning vectors Vectors must be small molecules for convenient manipulation. Some need to be capable of prolific replication in a living cell in order to amplify the inserted donor fragment. In contrast, others are designed to be present in only a single copy to maintain the integrity of the inserted DNA (see below). All vectors must have convenient restriction sites at which the DNA to be cloned may be inserted. Ideally, the restriction site should be present only once in the vector because then restriction fragments of donor DNA will insert only at that one location in the vector. Having a way to identify and recover the recombinant molecule quickly also is important. Numerous cloning vectors are in current use, suitable for different sizes of DNA insert or for different uses of the clone. Some general classes of cloning vectors follow.

Plasmid vectors As described earlier, bacterial plasmids are small circular DNA molecules that replicate their DNA independent of the bacterial chromosome. The plasmids routinely used as vectors carry a gene for drug resistance and a gene to distinguish plasmids with and without DNA inserts. These drug-resistance genes provide a convenient way to select for bacterial cells transformed by plasmids: those cells still alive after exposure to the drug must carry the plasmid vectors. However, not all the plasmids in these transformed cells will contain DNA inserts. For this reason, it is desirable to be able to identify bacterial colonies with plasmids containing DNA inserts. Such a feature is part of the pUC18 plasmid vector shown in Figure 10-9; DNA inserts disrupt a gene (*lacZ*) in the plasmid that encodes an enzyme (β-galactosidase) necessary to cleave a compound added to the agar (X-gal) so that it produces a

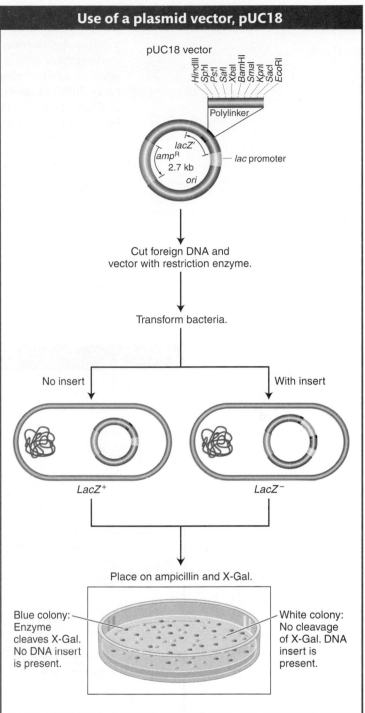

Figure 10-9 The plasmid vector pUC18 has been designed for use as a vector for DNA cloning. Insertion of DNA into pUC18 is detected by inactivation of the β-galactosidase function of *lacZ*, resulting in an inability to convert the artificial substrate X-Gal into a blue dye. The polylinker has several alternative restriction sites into which donor DNA can be inserted.

blue pigment. Thus, the colonies that contain the plasmids with the DNA insert will be white rather than blue (they cannot cleave X-gal because they do not produce β-galactosidase).

Bacteriophage vectors A bacteriophage vector harbors DNA as an insert "packaged" inside the phage particle. Different classes of bacteriophage vectors can carry different sizes of donor DNA insert. Bacteriophage λ (lambda) (discussed in Chapters 5 and 11) is an effective cloning vector for double-stranded DNA inserts as long as about 15 kb. Lambda phage heads can package DNA molecules no larger than about 50 kb in length (the size of a normal λ chromosome). The central part of the phage genome is not required for replication or packaging of λ DNA molecules in *E. coli* and so can be cut out by using restriction enzymes and discarded. The deleted central part is then replaced by inserts of donor DNA. An insert will be from 10 to 15 kb in length because an insert of this size brings the total chromosome size back to its normal 50 kb.

As Figure 10-10 shows, the recombinant molecules can be directly packaged into phage heads in vitro and then introduced into the bacterium. The ends of the linear phage genome have 12-bp sticky ends (called the COS site). When digested with *Bam*HI, the left and right arms will each have one sticky end and one *Bam*HI end. Ligation of the isolated arms to genomic DNA fragments will produce long complex molecules (called *concatomers*) because the sticky ends of the arms can

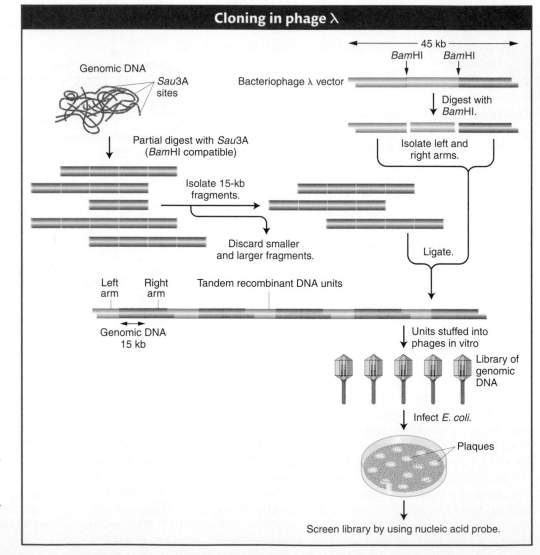

Figure 10-10 To clone in phage λ, a nonessential central region of the phage chromosome is discarded and the ends are ligated to random 15-kb fragments of donor DNA. A linear multimer (concatenate) forms, which is then stuffed into phage heads one monomer at a time by using an in vitro packaging system. [*After J. D. Watson, M. Gilman, J. Witkowski, and M. Zoller, Recombinant DNA, 2nd ed. Copyright 1992 by Scientific American Books.*]

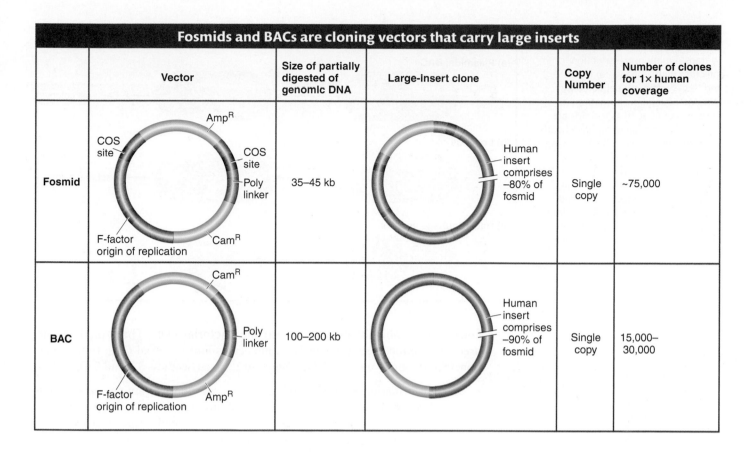

Fosmids and BACs are cloning vectors that carry large inserts

	Vector	Size of partially digested of genomic DNA	Large-Insert clone	Copy Number	Number of clones for 1× human coverage
Fosmid	COS site, AmpR, COS site, Poly linker, CamR, F-factor origin of replication	35–45 kb	Human insert comprises −80% of fosmid	Single copy	~75,000
BAC	CamR, Poly linker, AmpR, F-factor origin of replication	100–200 kb	Human insert comprises −90% of fosmid	Single copy	15,000–30,000

anneal to each other at one end and to the genomic DNA fragments at their other end. When mixed in a test tube with phage heads and tails, the concatomers are cut to unit lengths (including a left and right arm flanking the genomic *Bam*HI fragment) while they are being packaged into viral particles. The phages are now ready to be introduced into bacterial cells (see below).

Vectors for larger DNA inserts The standard plasmid and phage λ vectors just described can accept donor DNA of sizes as large as 25 to 30 kb. However, many experiments require inserts well in excess of this upper limit. To meet these needs, special vectors have been engineered. In each case, after the DNAs have been delivered into the bacterium, they replicate as large plasmids.

Fosmids are vectors that can carry 35- to 45-kb inserts (Figure 10-11). They are engineered hybrids of λ phage DNA (COS sites contain the 12-bp sticky ends; see the previous section) and bacterial plasmid DNA. Fosmids are packaged into λ phage particles, which act as the syringes that introduce these big pieces of recombinant DNA into recipient *E. coli* cells. After they are in the cell, these hybrids, just like the λ phage, form circular molecules that replicate extrachromosomally in a manner similar to plasmids. However, because of the presence of sequences from the F plasmid (see Chapter 5), very few copies of fosmids accumulate in a cell.

The most popular vector for cloning very large DNA inserts is the **bacterial artificial chromosome (BAC).** Derived from the F plasmid, it can carry inserts ranging from 100 to 200 kb, although the vector itself is only ~7 kb (see Figure 10-11). The DNA to be cloned is inserted into the plasmid, and this large circular recombinant DNA is introduced into the bacterium. BACs were the "workhorse" vectors for the extensive cloning required by large-scale genome-sequencing projects, including the public project to sequence the human genome (discussed in Chapter 14).

Figure 10-11 Features of some large-insert cloning vectors. The number of clones needed to cover the human genome once (1×) is based on a genome size of 3000 Mb (3 billion base pairs).

Figure 10-12 Recombinant DNA can be delivered into bacterial cells by transformation, transduction, or infection with a phage. (a) Plasmid and BAC vectors are delivered by DNA-mediated transformation. (b) Certain vectors such as fosmids are delivered within bacteriophage heads (transduction); however, after having been injected into the bacterium, they form circles and replicate as large plasmids. (c) Bacteriophage vectors such as phage l infect and lyse the bacterium, releasing a clone of progeny phages, all carrying the identical recombinant DNA molecule within the phage genome.

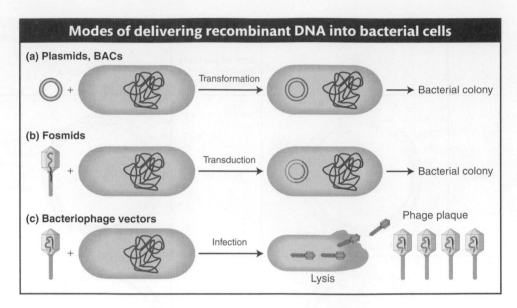

Entry of recombinant molecules into the bacterial cell   Three methods are used to introduce recombinant DNA molecules into bacterial cells: transformation, transduction, and infection (Figure 10-12) (see Sections 5.3 and 5.4).

- In transformation, bacteria are bathed in a solution containing the recombinant DNA molecule. Because bacterial cells are not able to take up molecules as large as recombinant plasmids, they must be made *competent* by either incubation in a calcium solution or exposure to a high-voltage electrical pulse (*electroporation*). After entering a competent cell through membrane pores, the recombinant molecule becomes a plasmid chromosome (Figure 10-12a). Electroporation is the method of choice for introducing BACs into bacterial cells.

- In transduction, the recombinant molecule is combined with phage head and tail proteins. These engineered phages are then mixed with bacteria and they inject their DNA cargo into the bacterial cells. Fosmids are introduced into cells by transduction (Figure 10-12b).

- In contrast to transduction, which produces plasmids and bacterial colonies, infection produces recombinant phage particles (Figure 10-12c). Through repeated rounds of re-infection, a plaque full of phage particles forms from each initial bacterium that was infected. Each phage particle in a plaque contains a copy of the original recombinant λ chromosome.

Recovery of amplified recombinant molecules   The recombinant DNA packaged into phage particles is easily obtained by collecting phage lysate and isolating the DNA that they contain. To obtain the recombinant DNA packaged in plasmids, fosmids, or BACs, the bacteria are chemically or mechanically broken apart. The recombinant DNA plasmid is separated from the much larger main bacterial chromosome by centrifugation, electrophoresis, or other selective techniques.

> **Message**   Gene cloning is carried out through the introduction of single recombinant vectors into recipient bacterial cells, followed by the amplification of these molecules as either plasmid chromosomes or phages.

Making genomic and cDNA libraries

We have seen how to make and amplify individual recombinant DNA molecules. Any one clone represents a small part of the genome of an organism or only one of thousands of mRNA molecules that the organism can synthesize. To ensure that we have cloned the DNA segment of interest, we have to make large collections of DNA segments that are all-inclusive. For example, we take all the DNA from a genome, break it up into segments of the right size for our cloning vector, and insert each segment into a different copy of the vector, thereby creating a collection of recombinant DNA molecules that, taken together, represent the entire genome. We then transform or infect these molecules into separate bacterial recipient cells, where they are amplified. The resulting collection of recombinant-DNA-bearing bacteria or bacteriophages is called a **genomic library.** If we are using a cloning vector that accepts an average insert size of 10 kb and if the entire genome is 100,000 kb in size (the approximate size of the genome of the nematode *Caenorhabditis elegans*), then 10,000 independent recombinant clones will represent one genome's worth of DNA. To ensure that all sequences of the genome that can be cloned are contained within a collection, genomic libraries typically represent an average segment of the genome at least five times (and so, in our example, there will be 50,000 independent clones in the genomic library). This multifold representation makes it highly unlikely that, by chance, a sequence is not represented at least once in the library.

Similarly, representative collections of cDNA inserts require tens or hundreds of thousands of independent cDNA clones; these collections are **cDNA libraries** and represent only the protein-coding regions of the genome. A comprehensive cDNA library is based on mRNA samples from different tissues, different developmental stages, or from organisms grown in different environmental conditions.

Whether we choose to construct a genomic DNA library or a cDNA library depends on the situation. If we are seeking a specific gene that is active in a specific type of tissue in a plant or animal, then it makes sense to construct a cDNA library from a sample of that tissue. For example, suppose we want to identify cDNAs corresponding to insulin mRNAs. The β-islet cells of the pancreas are the most abundant source of insulin, and so mRNAs from pancreas cells are the appropriate source for a cDNA library because these mRNAs should be enriched for the gene in question. A cDNA library represents a subset of the transcribed regions of the genome; so it will inevitably be smaller than a complete genomic library. Although genomic libraries are bigger, they do have the benefit of containing genes in their native form, including introns and untranscribed regulatory sequences. A genomic library is necessary at some stage as a prelude to cloning an entire gene or an entire genome.

> **Message** The task of isolating a clone of a specific gene begins with making a library of genomic DNA or cDNA—if possible, enriched for sequences containing the gene in question.

10.3 Finding a Specific Clone of Interest

The production of a library as just described is sometimes referred to as "shotgun" cloning because the experimenter clones a large sample of fragments and hopes that one of the clones will contain a "hit"—the desired gene. The task then is to find that particular clone, considered next.

Finding specific clones by using probes

A library might contain as many as hundreds of thousands of cloned fragments. This huge collection of fragments must be screened to find the recombinant DNA molecule containing the gene of interest to a researcher. Such screening is accomplished by using a specific **probe** that will find and mark only the desired clone. There are two types of probes: (1) those that recognize a specific nucleic acid sequence and (2) those that recognize a specific protein.

Probes for finding DNA   Probing for DNA makes use of the power of base complementarity. Two single-stranded nucleic acids with full or partial complementary base sequence will "find" each other in solution by random collision. After being united, the double-stranded hybrid so formed is stable. This approach provides a powerful means of finding specific sequences of interest. Probing for DNA requires that all molecules be made single stranded by heating. A single-stranded probe labeled radioactively or chemically is sent out to find its complementary target sequence in a population of DNAs such as a library. Probes as small as 15 to 20 base pairs will hybridize to specific complementary sequences within much larger cloned DNAs. Thus, probes can be thought of as "bait" for identifying much larger "prey."

The identification of a specific clone in a library is a multistep procedure. In Figure 10-13, these steps are shown for a library cloned into a fosmid vector. The steps are similar for libraries of plasmids or BACs. For libraries of phages, plaques are screened rather than colonies. First, colonies of the library on a petri dish are transferred to an absorbent membrane by simply laying the membrane on the surface of the medium. The membrane is peeled off, colonies clinging to the surface are lysed in situ, and the DNA is denatured. Second, the membrane is bathed with a solution of a single-stranded probe that is specific for the DNA being sought. Generally, the probe is itself a cloned piece of DNA that has a sequence that is complementary to that of the desired gene. The probe must be labeled with either a radioactive isotope or a fluorescent dye. Thus, the position of a positive clone will become clear from the position of the concentrated radioactive or fluorescent label. For radioactive labels, the membrane is placed on a piece of X-ray film, and the decay of the radioisotope produces subatomic particles that "expose" the film, producing a dark spot on the film adjacent to the location of the radioisotope concentration. Such an exposed film is called an **autoradiogram.** If a fluorescent dye is used as a label, the membrane is exposed to the correct wavelength of light to activate the dye's fluorescence, and a photograph is taken of the membrane to record the location of the fluorescing dye.

Where does the DNA to make a probe come from? The DNA can come from one of several sources.

- One can use a homologous gene or a cDNA from a related organism. This method depends on the fact that organisms descended from a recent common ancestor will have similar DNA sequences. Even though the probe DNA and the DNA of the desired clone might not be identical, they are often similar enough to promote hybridization.

- One can use the protein product of the gene of interest. If part or all of the protein sequence is known, one can back-translate, by using the table of the genetic code in reverse (from amino acid to codon), to obtain the DNA sequence that encoded it. A synthetic DNA probe that matches that sequence is then designed. Recall, however, that the genetic code is degenerate—that is, most amino acids are encoded by multiple codons. Thus, several possible DNA sequences could in theory encode the protein in question, but only one of these DNA sequences is present in the gene that actually encodes the protein. To get around

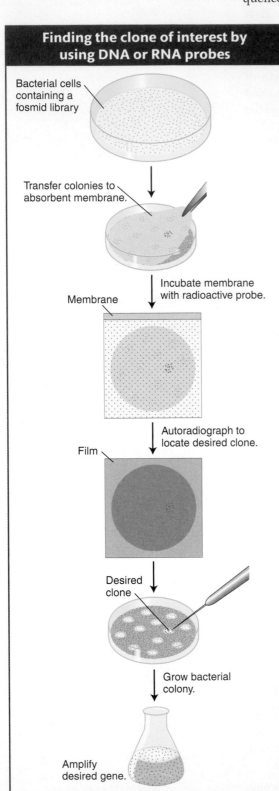

Finding the clone of interest by using DNA or RNA probes

Bacterial cells containing a fosmid library

Transfer colonies to absorbent membrane.

Membrane

Incubate membrane with radioactive probe.

Film

Autoradiograph to locate desired clone.

Desired clone

Grow bacterial colony.

Amplify desired gene.

Figure 10-13 The clone carrying a gene of interest is identified by probing a genomic library, in this case made by cloning genes in a fosmid vector, with DNA or RNA known to be related to the desired gene. A radioactive probe hybridizes with any recombinant DNA incorporating a matching DNA sequence, and the position of the clone having the DNA is revealed by autoradiography. Now the desired clone can be selected from the corresponding spot on the petri dish and transferred to a fresh bacterial host so that a pure gene can be manufactured. [*Modified from R. A. Weinberg, "A Molecular Basis of Cancer," and P. Leder, "The Genetics of Antibody Diversity." Copyright 1983, 1982 by Scientific American, Inc. All rights reserved.*]

this problem, a short stretch of amino acids with minimal degeneracy is selected. A mixed set of probes is then designed containing all possible DNA sequences that can encode this amino acid sequence. This "cocktail" of oligonucleotides is used as a probe. The correct strand within this cocktail finds the gene of interest. About 20 nucleotides embody enough specificity to hybridize to one unique complementary DNA sequence in the library.

Probes for finding proteins If the protein product of a gene is known and isolated in pure form, then this protein can be used to detect the clone of the corresponding gene in a library. The process, described in Figure 10-14, requires two components. First, it requires an expression library, made by using expression

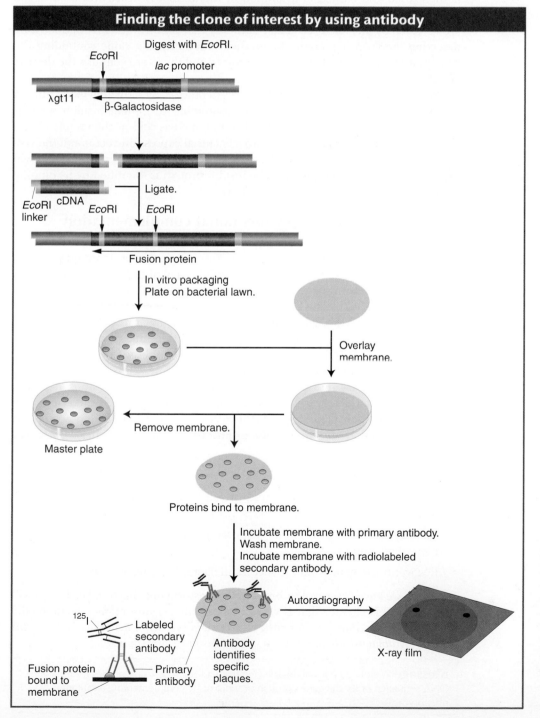

Finding the clone of interest by using antibody

Digest with *Eco*RI.

*Eco*RI

lac promoter

λgt11

β-Galactosidase

Ligate.

*Eco*RI   cDNA   *Eco*RI   *Eco*RI
linker

Fusion protein

In vitro packaging
Plate on bacterial lawn.

Overlay membrane.

Remove membrane.

Master plate

Proteins bind to membrane.

Incubate membrane with primary antibody.
Wash membrane.
Incubate membrane with radiolabeled secondary antibody.

Autoradiography

125I
Labeled secondary antibody

Antibody identifies specific plaques.

X-ray film

Fusion protein bound to membrane

Primary antibody

Figure 10-14 To find the clone of interest, an expression library made with special phage λ vector called λgt11 is screened with a protein-specific antibody. After the unbound antibodies have been washed off the membrane, the bound antibodies are visualized through the binding of a radioactive secondary antibody. [*After J. D. Watson, M. Gilman, J. Witkowski, and M. Zoller, Recombinant DNA, 2nd ed. Copyright 1992 by Scientific American Books.*]

vectors that will produce the protein. To make the library, cDNA is inserted into the vector in the correct triplet reading frame with a bacterial protein (in this case, β-galactosidase), and cells containing the vector and its insert produce a "fusion" protein that is partly a translation of the cDNA insert and partly a part of the normal β-galactosidase. Second, the process requires an antibody to the specific protein product of the gene of interest. (An **antibody** is a protein made by an animal's immune system that binds with high affinity to a given molecule.) The antibody is used to screen the expression library for that protein. A membrane is lain over the surface of the medium and removed so that some of the cells of each colony are now attached to the membrane at locations that correspond to their positions on the original petri dish (see Figure 10-14). The imprinted membrane is then dried and bathed in a solution of the antibody, which will bind to the imprint of any colony that contains the fusion protein of interest. Positive clones are revealed by a labeled secondary antibody that binds to the first antibody. By detecting the correct protein, the antibody identifies the clone containing the gene that must have synthesized that protein and therefore contains the desired cDNA.

We can see how this type of probe works in practice by returning to the human insulin example. To clone a cDNA corresponding to human insulin, we first synthesize cDNA using mRNA isolated from pancreas cells as the template. The cDNA molecules are then inserted into a bacterial expression vector and the vector is transformed into bacteria. Bacterial colonies containing insulin cDNA will express an insulin fusion protein. The insulin protein is identified by its binding with an insulin antibody as described above.

Finding specific clones by functional complementation

In many cases, we don't have a probe for the gene to start with, but we do have a recessive mutation in the gene of interest. This gene could be a mutant gene in a bacterium or yeast or even a plant or mouse. The goal of this approach is to identify the clone containing the gene of interest by the fact that it will restore the function eliminated by the recessive mutation. In practice, one first generates a genomic or cDNA library from an organism that has the wild-type allele of the gene of interest. The gene of interest is one of thousands represented in the library. However, only the gene of interest has the ability to complement the mutant organism and restore the wild-type phenotype. Thus, if we are able to introduce the library into the species bearing the recessive mutation (see Section 10.6), we can detect specific clones in the library through their ability to restore the function eliminated by the recessive mutation. This procedure is called **functional complementation** or **mutant rescue.** The general outline of the procedure is as follows:

- Make a library containing wild-type a^+ recombinant-donor DNA inserts.

- Transform cells of recessive-mutant-cell-line a^- with this library of DNA inserts.

- Identify clones from the library that produce transformed cells with the dominant a^+ phenotype.

- Recover the a^+ gene from the successful bacterial or phage clone.

So far, we have described techniques to transform only bacterial cells. You will see later in this chapter that DNA can be introduced into many genetic model organisms, including *Saccharomyces cerevisiae* (yeast), *Caenorhabditis elegans* (worm*)*, *Arabidopsis thaliana* (plant), and *Mus musculus* (mouse).

> **Message** A cloned gene can be selected from a library by using probes for the gene's DNA sequence or for the gene's protein product or by complementing a mutant phenotype.

Southern- and Northern-blot analysis of DNA

After you have amplified your PCR product or selected a clone of interest from a genomic or cDNA library, the next step is to find out more about the DNA. Let's say that you have recovered the insulin cDNA from an expression vector and want to determine the restriction sites in the genomic copy of the insulin gene. Perhaps you want to see whether these sites differ among diverse human populations. You might also want to know whether the size of the insulin mRNA varies among human populations. Alternatively, you might want to determine whether a similar gene is present in the genome of a related organism. In the section below you will see that these important questions can be answered by using relatively simple techniques. In these techniques, complex mixtures of DNA or RNA are sorted by size and then probed by hybridization to detect DNA molecules related to some other DNA molecule.

The most extensively used method for detecting a molecule within a mixture is *blotting*, which starts with **gel electrophoresis** to separate the molecules in the mixture. A mixture of linear DNA molecules is placed into a well cut into an agarose gel. The gel is oriented so that the wells are at the cathode end (negatively charged) and the DNA migrates to the anode end (positively charged) because of its negative charge. The speed of migration of DNA molecules in the gel is inversely dependent on their size (Figure 10-15). Therefore, the fragments in distinct size classes will form distinct bands on the gel. The bands can be visualized by staining the DNA with ethidium bromide, which causes the DNA to fluoresce in ultraviolet light. The absolute size of each fragment in the mixture can be determined by comparing its migration distance with a set of standard fragments of known sizes. If the bands are well separated, an individual band can be cut from the gel, and the DNA sample can be purified from the gel matrix. Therefore, DNA electrophoresis can be either diagnostic (showing sizes and relative amounts of the DNA fragments present) or preparative (useful in isolating specific DNA fragments).

Genomic DNA digested by restriction enzymes generally yields so many fragments that electrophoresis produces a continuous smear of DNA and no discrete bands. A probe can identify one fragment in this mixture, with the use of a technique

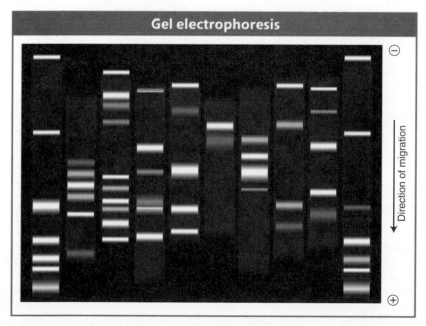

Figure 10-15 Mixtures of different-size DNA fragments have been separated electrophoretically on an agarose gel. The mixtures are applied to wells near the top of the gel, and fragments move from the negative to the positive end under the influence of an electrical field to different positions dependent on size. [*Ingram Publishing/Thinkstock.*]

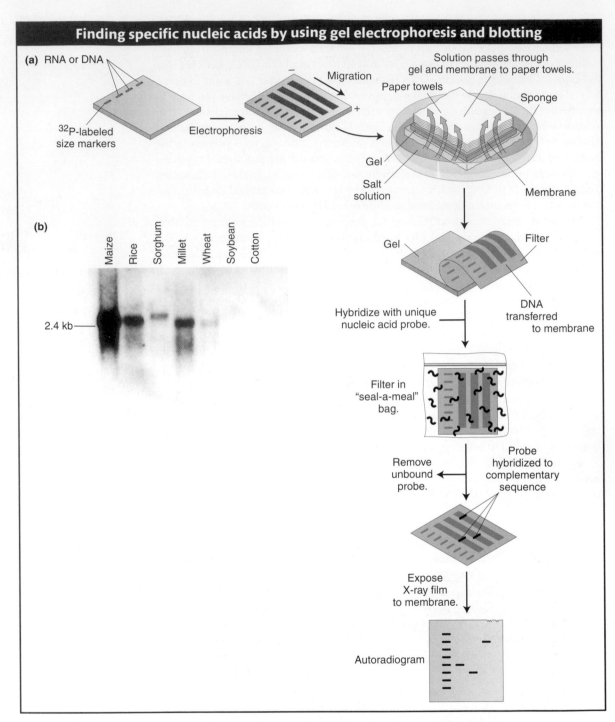

Figure 10-16 In this example, a radioactive probe is used to identify specific nucleic acids separated by gel electrophoresis. (a) RNA or DNA restriction fragments are applied to an agarose gel and undergo electrophoresis. The various fragments migrate at differing rates according to their respective sizes. The gel is placed in buffer and covered by a membrane and a stack of paper towels. The fragments are denatured to single strands so that they can stick to the membrane. They are carried to the membrane by the buffer, which is wicked up by the towels. The membrane is then removed and incubated with a radioactively labeled single-stranded probe that is complementary to the targeted sequence. Unbound probe is washed away, and X-ray film is exposed to the membrane. Because the radioactive probe has hybridized only with its complementary restriction fragments, the film will be exposed only in bands corresponding to those fragments. Comparison of these bands with labeled markers reveals the number and size of the fragments in which the targeted sequences are found. This procedure is termed *Southern blotting* when DNA is transferred to the membrane and *Northern blotting* when RNA is transferred. (b) An actual Northern blot, run with RNA isolated from the seeds of various plants. A single RNA probe is used to identify the presence of a single locus. The results show that maize is more closely related to rice, sorghum, and millet than it is to soybean or cotton. [(a) *After J. D. Watson, M. Gilman, J. Witkowski, and M. Zoller, Recombinant DNA, 2nd ed. Copyright 1992 by Scientific American Books; (b) Susan Wessler.*]

developed by E. M. Southern called **Southern blotting** (Figure 10-16a). Like clone identification (see Figure 10-13), this technique entails getting an imprint of DNA molecules on a membrane by using the membrane to blot the gel after electrophoresis is complete. The DNA must be denatured first, which allows it to stick to the membrane. Then the membrane is hybridized with labeled probe. An autoradiogram or a photograph of fluorescent bands will reveal the presence of any bands on the gel that are complementary to the probe. To detect the insulin gene, we can apply this protocol to human genomic DNA digested with restriction enzymes on the membrane, using insulin cDNA as the labeled probe.

The Southern-blotting technique can be modified to detect a specific *RNA* molecule from a mixture of RNAs fractionated on a gel. This technique is called **RNA blotting** or more commonly **Northern blotting** (thanks to some scientist's sense of humor) to contrast it with the Southern-blotting technique used for *DNA* analysis. The RNA separated by electrophoresis can be a sample of the total RNA isolated from a tissue or from an entire organism. In the example shown in Figure 10-16b, the gel was run with RNA isolated from the seeds of various plants. Unlike DNA that is loaded onto a gel, there is no need to digest the RNA sample as it is already in discrete transcript-size molecules. RNA gels are blotted onto a membrane and probed in the same way as DNA is blotted and probed for Southern blotting. One application of Northern analysis is to determine whether a specific gene is transcribed in a certain tissue or under certain environmental conditions. Another is to determine the size of the mRNA and whether an RNA of similar size can be detected in closely related plants (as in Figure 10-16b).

We started this section on blot analysis by posing questions about the insulin gene and its mRNA in human populations and in related species. Based on the techniques above, can you design Southern- and Northern-blot experiments to answer these questions? You can assume that you have access to samples of the required genomic DNAs and RNAs.

Hence, we see that cloned DNA finds widespread application as a probe used for detecting a specific clone, a DNA fragment, or an RNA molecule. In all these cases, note that the technique again exploits the ability of nucleic acids with *complementary* nucleotide sequences to find and bind to each other.

> **Message** Recombinant-DNA techniques that depend on complementarity to a cloned DNA probe include blotting and hybridization systems for the identification of specific clones, restriction fragments, or mRNAs for measurement of the size of specific DNAs or RNAs.

10.4 Determining the Base Sequence of a DNA Segment

After we have cloned our desired gene or have amplified it by using PCR, the task of trying to understand its function begins. The ultimate language of the genome is composed of strings of the nucleotides A, T, C, and G. Obtaining the complete nucleotide sequence of a segment of DNA is often an important part of understanding the organization of a gene and its regulation, its relation to other genes, or the function of its encoded RNA or protein. Indeed, the DNA sequence can be used to determine the protein primary structure since for the most part, translating the nucleic acid sequence of a cDNA molecule to discover the amino sequence of its encoded polypeptide chain is simpler than directly sequencing the polypeptide itself. In this section, we consider the techniques used to read the nucleotide sequence of DNA.

As with recombinant-DNA technologies and PCR, DNA sequencing exploits base-pair complementarity together with an understanding of the basic biochemistry of DNA replication. Several techniques have been developed, but one of them has been the predominant method used to date to sequence large genomes, such as the human genome. For this reason it will be described in detail. However, as you will see in Chapter 14, new sequencing technology has largely supplanted

The structure of 2′,3′-dideoxynucleotides

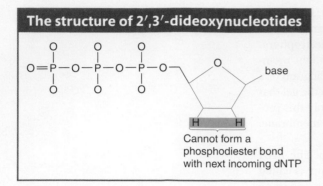

Cannot form a
phosphodiester bond
with next incoming dNTP

Figure 10-17 2′,3′-Dideoxynucleotides, which are employed in the Sanger DNA-sequencing method, are missing the ribose hydroxyl group present in DNA.

this technique when the goal is to determine the sequence of entire genomes. However, the technique we will describe in this section is still the most widely used technique and the one researchers turn to when sequencing more-limited regions of the genome.

This sequencing technique is called **dideoxy sequencing** or, sometimes, **Sanger sequencing** after its inventor. The term *dideoxy* comes from a special modified nucleotide, called a dideoxynucleotide triphosphate (generically, a ddNTP). This modified nucleotide is key to the Sanger technique because of its ability to block continued DNA synthesis. What is a dideoxynucleotide triphosphate? And how does it block DNA synthesis? A dideoxynucleotide lacks the 3′-hydroxyl group as well as the 2′-hydroxyl group, which is also absent in a deoxynucleotide (Figure 10-17). For DNA synthesis to take place, the DNA polymerase must catalyze a reaction between the 3′-hydroxyl group of the last nucleotide and the 5′-phosphate group of the next nucleotide to be added. Because a dideoxynucleotide lacks the 3′-hydroxyl group, this reaction cannot take place, and therefore DNA synthesis is blocked at the point of addition.

The logic of dideoxy sequencing is straightforward. Suppose we want to read the sequence of a cloned DNA segment of up to 800 base pairs. This DNA segment could be a plasmid insert or even a PCR product. First, we denature the two strands of this segment. Next, we create a primer for DNA synthesis that will hybridize to exactly one location on the cloned DNA segment and then add a special "cocktail" of DNA polymerase, normal deoxynucleotide triphosphates (dATP, dCTP, dGTP, and dTTP), and a small amount of a special dideoxynucleotide for one of the four bases (for example, dideoxyadenosine triphosphate, abbreviated ddATP). The polymerase will begin to synthesize the complementary DNA strand, starting from the primer, but will stop at any point at which the dideoxynucleotide triphosphate is incorporated into the growing DNA chain in place of the normal deoxynucleotide triphosphate. Suppose the DNA sequence of the DNA segment that we're trying to sequence is

5′ ACGGGATAGCTAATTGTTTACCGCCGGAGCCA 3′

We would then start DNA synthesis from a complementary primer:

5′ ACGGGATAGCTAATTGTTTACCGCCGGAGCCA 3′
3′   CGGCC TCGGT 5′

◄——— Direction of DNA synthesis

Using the special DNA-synthesis cocktail "spiked" with ddATP, for example, we will create a nested set of DNA fragments that have the same starting point but different end points because the fragments stop at whatever point the insertion of ddATP instead of dATP halted DNA replication. The array of different ddATP-arrested DNA chains looks like the list of sequences below. (*A indicates the dideoxynucleotide.)

5′	ATGGGATAGCTAATTGTTTACCGCCGGAGCCA 3′	Template DNA clone
3′	CGGCC TCGGT 5′	Primer for synthesis
	◄———————	Direction of DNA synthesis
3′	*ATGGCGGCC TCGGT 5′	Dideoxy fragment 1
3′	*AATGGCGGCC TCGGT 5′	Dideoxy fragment 2
3′	*AAATGGCGGCC TCGGT 5′	Dideoxy fragment 3
3′	*ACAAATGGCGGCC TCGGT 5′	Dideoxy fragment 4
3′	*AACAAATGGCGGCC TCGGT 5′	Dideoxy fragment 5
3′	*ATTAACAAATGGCGGCC TCGGT 5′	Dideoxy fragment 6
3′	*ATCGATTAACAAATGGCGGCC TCGGT 5′	Dideoxy fragment 7
3′	*ACCCTATCGATTAACAAATGGCGGCC TCGGT 5′	Dideoxy fragment 8

We can generate an array of such fragments for each of the four possible dideoxynucleotide triphosphates in four separate cocktails (one spiked with ddATP, one with ddCTP, one with ddGTP, and one with ddTTP). Each will produce a different array of fragments, with no two spiked cocktails producing fragments of the same size. Next, the DNA fragments generated in the four cocktails are separated and displayed in order by using gel electrophoresis. By running the fragments in four adjacent lanes of a very long gel, we see that the fragments can be ordered by length with the lengths increasing by one base at a time.

The newly synthesized strands must be labeled in some way to make the bands visible on the gel. Strands are labeled as they are made either by using a primer that is radioactively labeled or having one of the regular dNTPs carry a radioactive label. Fluorescent labels can also be used, and in this case they are carried by each ddNTP (see below).

The products of such dideoxy sequencing reactions are shown in Figure 10-18. That result is a ladder of labeled DNA chains increasing in length by one, and so

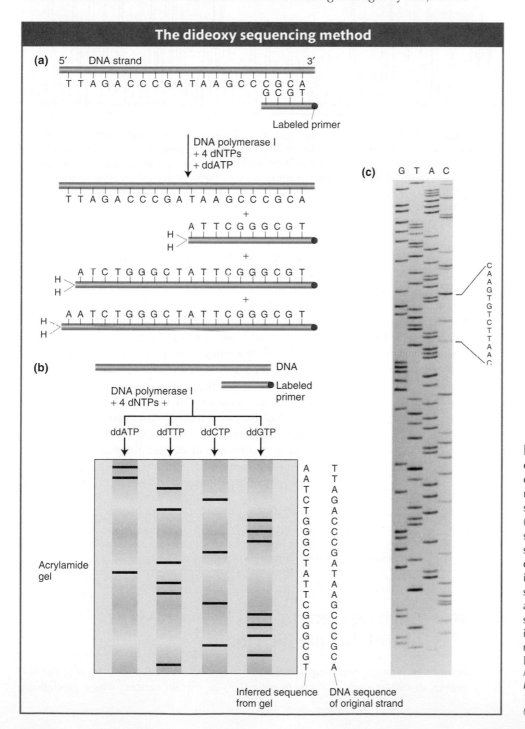

Figure 10-18 DNA is efficiently sequenced by including dideoxynucleotides among the nucleotides used to copy a DNA segment. (a) A labeled primer (designed from the flanking vector sequence) is used to initiate DNA synthesis. The addition of four different dideoxynucleotides (ddATP is shown here) randomly arrests synthesis. (b) The resulting fragments are separated electrophoretically and subjected to autoradiography. The inferred sequence is shown at the right. (c) Sanger sequencing gel. [(a and b) From J. D. Watson, M. Gilman, J. Witkowski, and M. Zoller, Recombinant DNA, 2nd ed. Copyright 1992 by Scientific American Books. (c) From Loida Escote-Carlson.]

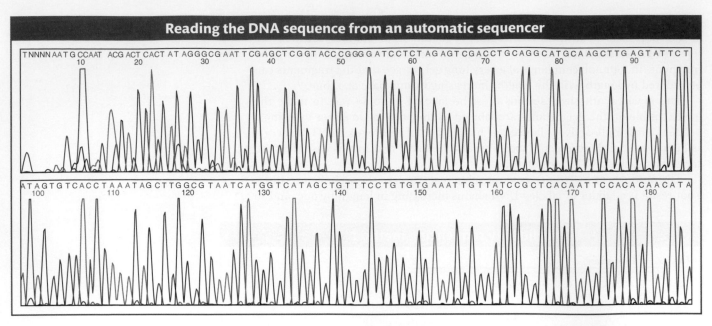

Figure 10-19 Printout from an automatic sequencer that uses fluorescent dyes. Each of the four colors represents a different base. The letter N represents a base that cannot be assigned, because peaks are too low. Note that, if this were a gel as in Figure 10-18c, each of these peaks would correspond to one of the dark bands on the gel; in other words, these colored peaks represent a different readout of the same sort of data as are produced on a sequencing gel.

all we need do is read up the gel to read the DNA sequence of the synthesized strand in the 5′-to-3′ direction.

If the tag is a fluorescent dye and a different fluorescent color emitter is used for each of the four ddNTP reactions, then the four reactions can take place in the same test tube and the four sets of nested DNA chains can undergo electrophoresis together. Thus, four times as many sequences can be produced in the same amount of time as can be produced by running the reactions separately. This logic is used in fluorescence detection by automated DNA-sequencing machines. Thanks to these machines, DNA sequencing can proceed at a massive level, and sequences of whole genomes have been obtained by scaling up the procedures described in this section. Figure 10-19 illustrates a readout of automated sequencing. Each colored peak represents a different-size fragment of DNA, ending with a fluorescent base that was detected by the fluorescent scanner of the automated DNA sequencer; the four different colors represent the four bases of DNA. Applications of automated sequencing technology on a genome-wide scale is a major focus of Chapter 14.

> **Message** A cloned DNA segment can be sequenced by characterizing the end bases of a serial set of truncated synthetic DNA fragments, each terminated at different positions corresponding to the incorporation of a dideoxynucleotide.

10.5 Aligning Genetic and Physical Maps to Isolate Specific Genes

Before complete genome sequences were available, molecular cloning of genes for genetic disorders such as cystic fibrosis (CF) or certain cancers was a prodigious undertaking. Cloning these genes often required a major research project involving the collaborative efforts of several laboratories. The process is known as **positional cloning,** and its strategy is to use the genetic position to isolate the gene underlying the trait. Even with the availability of entire genome sequences, it is still often necessary to first map traits that have not been associated with a gene product.

To initiate positional cloning, researchers need to first map the gene responsible for a particular trait. To map the gene, researchers can test for linkage with landmarks of known location, as described in Chapter 4. Landmarks might be RFLPs (restriction fragment length polymorphisms), SNPs (single nucleotide polymorphisms), or other molecular polymorphisms (see Chapters 4 and 14), or they might be well-mapped chromosomal break points (see Chapter 17). Landmarks on either side of the gene of interest are best, because they delimit the possible location of that gene.

It is important to keep in mind that mapping a gene only serves to locate the chromosomal "neighborhood" of the gene. Thus, regions delimited by the molecular landmarks usually contain many genes spread over hundreds of thousands or even millions of base pairs. To identify the correct gene responsible for a particular trait, researchers need to be able to analyze the whole neighborhood for the gene of interest. In model organisms for which the entire genome sequence is available, the local neighborhood with its numerous genes can simply be obtained from a computer database (see Chapter 14). From these genes, candidates are chosen that might represent the gene being sought.

Below, we will briefly discuss how positional cloning was used before the availability of the human genome sequence to isolate the gene responsible for the devastating human disease cystic fibrosis (the CF gene). While this technique is no longer necessary for identifying human genes (because of the availability of databases containing the entire genome sequence), it is still used to identify specific genes in organisms where a genome sequence is yet to be determined.

Using positional cloning to identify a human-disease gene

Let's follow the methods used to identify the genomic sequence of the cystic fibrosis gene. No primary biochemical defect was known at the time that the gene was isolated, and so it was very much a gene in search of a function.

Genetic screens can be used to dissect any biological process. However, genetic screens cannot be used with human beings, because we do not want to intentionally create human mutants. So, pedigree analyses of large families with the disease trait are performed (pedigree analysis is described in Chapter 2) when such information is available to determine the position of the genetic defect causing a disease such as cystic fibrosis. Members of a family carrying the disease are found to have one or more molecular markers in common that are not found in other families (see a discussion of molecular markers in Chapter 4). Linkage to molecular markers had located the CF gene to the long arm of chromosome 7, between bands 7q22 and 7q31.1. The CF gene was thought to be inside this region, but between these markers lay 1.5 centimorgans (map units) of chromosome, a vast uncharted terrain of more than 1 million bases. To get closer, it was necessary to generate more molecular markers within this region. The general method for isolating molecular markers (described in Chapter 4) is to identify a region of DNA that is polymorphic in individuals or populations that differ for the trait of interest. By finding additional molecular markers linked to the CF gene, geneticists narrowed down the region containing the CF gene to about 500 kbp, still a considerable distance.

A physical map was created of this entire region; that is, a random set of clones from this region was placed into the correct order. This was done in part

Figure 10-20 This chromosome walk begins with a recombinant phage or BAC clone obtained from a library that contains large inserts representing an entire eukaryotic genome. In the example shown, the molecular marker 7q22 is used to probe a human genomic library. Only the insert DNAs are shown. The insert DNA selected by the probe is then used to isolate another recombinant phage or BAC containing a neighboring segment of eukaryotic DNA. This walk illustrates how to start at molecular landmark 7q22 and get to marker 7q31.1, which is on the other side of the CF gene.

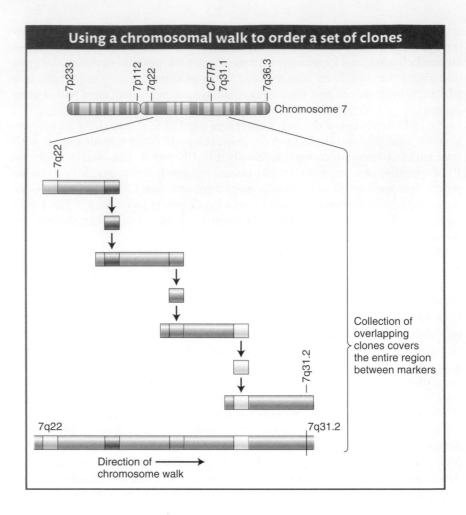

by using a technique called a **chromosome walk** (Figure 10-20). The basic idea is to use the sequence of the nearby landmark as a probe to identify a second set of clones that overlaps the marker clone containing the landmark but extends out from it in one of two directions (toward the target or away from the target). End fragments from the new sets of clones can be used as probes for identifying a third set of overlapping clones from the genomic library. In this tedious way, geneticists identified the clone containing the molecular markers that are most tightly linked to the CF trait and sequenced that clone. With the sequence in hand, the hunt for any genes along this stretch of DNA, containing both genes and noncoding sequences, could begin.

In the CF example, candidate genes were identified by noting features, such as start and stop signals, common to genes. The sequences of candidate genes and cDNAs were then compared between normal individuals and CF patients. A mutation in a candidate gene was found that appeared in all CF patients analyzed but in none of the normal individuals. This mutation was a deletion of three base pairs, eliminating a phenylalanine from the protein. In turn, from the inferred sequence, the three-dimensional structure of the protein was predicted. This protein is structurally similar to ion-transport proteins in other systems, suggesting that a transport defect is the primary cause of CF. When used to transform mutant cell lines from CF patients, the wild-type gene restored normal function; this phenotypic "rescue" was the final confirmation that the isolated sequence was in fact the CF gene.

Other human genes isolated by positional cloning include those involved in several heritable diseases, including Huntington's disease, breast cancer, Werner syndrome (see Chapter 7), and susceptibility to asthma. Because of the relative ease of crossing plants, positional cloning has been a very powerful technique to isolate genes involved in many processes, including the identification of genes that contributed to crop domestication.

Using fine mapping to identify genes

Today, the extremely tedious process of "walking" to the gene is no longer necessary for any organism for which a genome sequence is available. Researchers still begin the gene hunt by identifying two molecular markers that closely flank the gene of interest (see Figure 10-20). In Figure 10-21a, the two initial flanking markers are labeled starting marker 1 and starting marker 2. In the interval between the two starting markers, there are seven genes that are known to exist. Which one of these genes is responsible for the trait of interest?

Researchers try to narrow the interval enclosing the gene of interest. To accomplish this, they select additional markers, located in between the starting

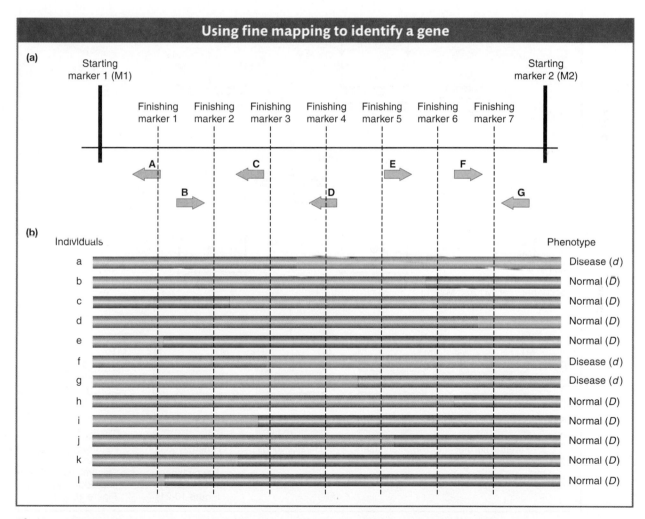

Figure 10-21 A modern gene hunt uses markers and sequences obtained from online databases to determine the genotypes within a region of DNA for large numbers of individuals with and without the disease trait. The individuals shown here were derived from the F$_2$ progeny in Figure 10-22. The target gene is the gene allele shared by all with the disease, gene *D*. Red is homozygous for the dominant *D* allele of Parent 1 (normal); blue is homozygous for the recessive *d* allele of Parent 2 (disease, mutant); gray is heterozygous.

markers, from the online genome sequence and marker databases. They must select markers that have one allele in individuals with the trait phenotype and a different allele in individuals without the phenotype. These additional markers are called finishing markers; in Figure 10-21a they are finishing markers 1 to 7. The goal of the researchers is to find the markers most tightly linked to the gene of interest.

The next step is to look for individuals in which a rare crossover event occurred within the region bounded by the two starting markers. Usually, the starting markers are 1 or more centimorgans apart. As such, there will be, on average, a single crossover in an interval of this size in 100 progeny. Because of the large number of progeny required, this type of mapping has been applied successfully in many model genetic organisms (such as the fruit fly and *C. elegans*) but is especially successful in plants because plant crosses can produce hundreds or thousands of progeny that can be obtained and analyzed. Figure 10-22 shows the series of crosses used to produce progeny containing recombinants between starting markers M1 and M2 (see Figure 10-21).

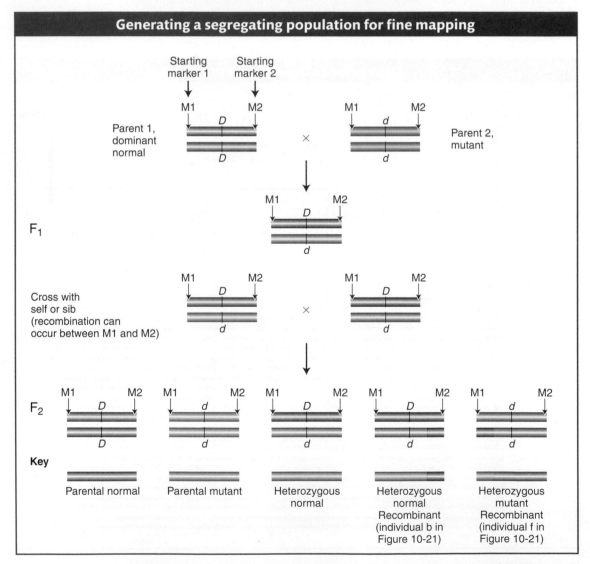

Generating a segregating population for fine mapping

Figure 10-22 A modern gene hunt often begins with a cross between parents with contrasting traits. In the example shown, Parent 1 carries the wild type allele (*D*) and is normal, while Parent 2 carries the mutant allelel (*d*) and as a result has a disease. All F$_1$ progeny are heterozygous at all loci and are normal. F$_2$ progeny are segregaing for the disease. Most progeny with the disease have the Parent 2 genotype (Parental mutant). Rare mutant individuals have experienced a recombination event in one of the parents in the chromosomal interval between markers M1 and M2 (heterozygous mutant, bottom right).

The marker alleles reveal whether the markers, and the genes within the same region, were inherited from the mother or the father. In the recombinants, part of the stretch in between the starting markers on one chromosome will have been inherited from the mother and part from the father (Parent 2 in Figure 10-22). In Figure 10-21, the trait is a disease inherited from the father. In individuals with the disease, the crossover has created a region, colored blue in Figure 10-21b, that is homozygous for the disease gene. Thus, the gene must lie within the blue region in all individuals with the disease. Comparing all individuals with the disease, you can see that the only blue region shared by all of them is the region where gene D resides.

In other words, D is the one gene present in regions of the genome that are homozygous for the paternal allele in individuals with the disease. If the number of individuals in the pedigrees or study populations is sufficiently large, then it may be possible to identify not only the gene in question but also the disease lesion, which is the polymorphic site within the gene that controls the trait difference. Notice that this process does not involve either cloning bits of DNA into BAC libraries or screening such libraries. As such, it is more appropriately referred to as **fine mapping** rather than positional cloning.

Investigators still must work very hard and overcome many hurdles to isolate the genes that control disease conditions or other traits. First, they need to have large samples of individuals to ensure that they can identify rare crossover events between all the genes. Typically, this means having thousands of individuals. Without large populations, investigators might recover crossovers only every several genes or so and consequently would not be certain which of these was the causative gene. For example, with only individuals a and b in Figure 10-21b, an investigator could only narrow the search to four genes (D, E, F, G). Second, although online databases contain lists of DNA-based markers such as SNPs, not all of these have alleles that are only present in individuals with the trait of interest in a particular pedigree or cross, so researchers must first screen a large number of markers to find those that do. Finally, investigators need to determine the complete DNA sequence of the disease allele of the gene to identify the causative lesion. In most cases, the online genome sequence will contain the wild-type allele. The disease allele is most easily sequenced using PCR to amplify that allele from the DNA of affected individuals and without actually cloning the DNA. The precise mutation can then be deduced from the DNA sequence of the PCR product.

> **Message** Even with access to the sequences of entire genomes, the isolation of defective disease-causing genes begins with the genetic mapping of the disease trait. With tightly linked markers flanking the trait in hand, investigators can use fine mapping to narrow the search for the gene of interest.

The preceding sections have introduced the fundamental techniques that have revolutionized genetics. The final section of this chapter will focus on the application of these techniques to genetic engineering.

10.6 Genetic Engineering

Thanks to recombinant-DNA technology, genes can be isolated in a test tube and characterized as specific nucleotide sequences. But even this achievement is not the end of the story. We will see next that knowledge of a sequence is often the beginning of a fresh round of genetic manipulation. When characterized, a sequence can be manipulated to alter an organism's genotype. The introduction of an altered gene into an organism has become central to basic genetic research, but it also finds wide commercial application. Two examples of the latter are (1) goats that secrete in their milk antibiotics derived from a fungus and (2) plants kept from freezing by the incorporation of arctic-fish "antifreeze" genes into their genomes. The use of recombinant-DNA techniques to alter an organism's genotype and phenotype is termed *genetic engineering*.

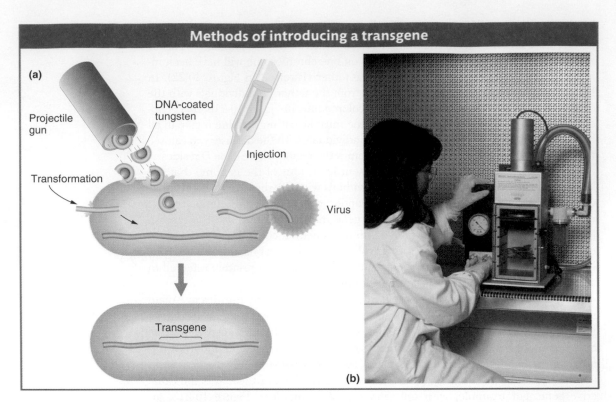

Figure 10-23 (a) Some of the different ways to introduce foreign DNA into a cell. (b) A gene gun. [*Matt Meadows/Peter Arnold/Photolibrary.*]

The techniques of genetic engineering described in the first part of this chapter were originally developed in bacteria. Thus, these techniques needed to be extended to model eukaryotes, which constitute a large proportion of model research organisms. Eukaryotic genes are still typically cloned and sequenced in bacterial hosts, but eventually they are introduced into a eukaryote, either the original donor species or a completely different one. The gene transferred is called a **transgene**, and the engineered product is called a **transgenic organism.**

The transgene can be introduced into a eukaryotic cell by a variety of techniques, including transformation, injection, bacterial or viral infection, and bombardment with DNA-coated tungsten or gold particles using a gene gun like the one shown (Figure 10-23). When the transgene enters a cell, it is able to travel to the nucleus. Once in the nucleus, it must become a stable part of the genome. Thus, the transgene must insert into a chromosome or (in a few species only) replicate as part of a plasmid. If insertion occurs, the transgene can either replace the resident gene or insert **ectopically**—that is, at other locations in the genome. Transgenes from other species typically insert ectopically.

> **Message** Transgenesis can introduce new or modified genetic material into eukaryotic cells.

We now turn to some examples in fungi, plants, and animals.

Genetic engineering in *Saccharomyces cerevisiae*

It is fair to say that *S. cerevisiae* is the most sophisticated eukaryotic genetic model. Most of the techniques typically used for eukaryotic genetic engineering were developed in yeast; so let's consider the general routes for transgenesis in yeast.

The simplest yeast vectors are yeast integrative plasmids (YIps), derivatives of bacterial plasmids into which the yeast DNA of interest has been inserted. When

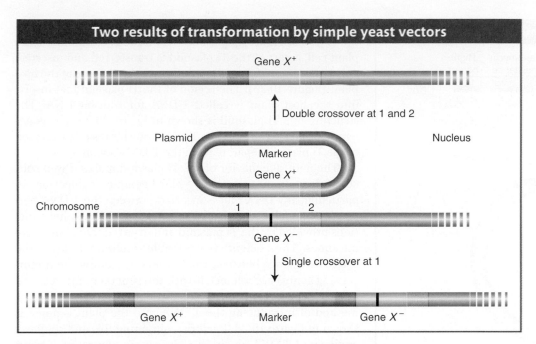

Figure 10-24 A plasmid bearing an active allele (gene X^+) inserts into a recipient yeast strain bearing a defective gene (X^-) by homologous recombination. The result can be replacement of the defective gene X^- by X^+ (*top*) or its retention along with the new allele (bottom). The mutant site of gene X^- is represented as a vertical black bar. Single crossovers at position 2 also are possible but are not shown.

transformed into yeast cells, these plasmids insert into yeast chromosomes, generally by homologous recombination with the resident gene, by either a single or a double crossover (Figure 10-24). As a result, either the entire plasmid is inserted or the targeted allele is replaced by the allele on the plasmid. The latter is an example of *gene replacement*—in this case, the substitution of an engineered gene for the gene originally in the yeast cell. Gene replacement can be used to delete a gene or substitute a mutant allele for its wild-type counterpart or, conversely, to substitute a wild-type allele for a mutant. Such substitutions can be detected by plating cells on a medium that selects for a marker allele on the plasmid.

The bacterial origin of replication is different from eukaryotic origins, and so bacterial plasmids do not replicate in yeast. Therefore, the only way in which such vectors can generate a stable modified genotype is if they are integrated into the yeast chromosome.

Genetic engineering in plants

Recombinant DNA technology has introduced a new dimension to the effort to develop improved crop varieties. No longer is genetic diversity achieved solely by selecting variants within a given species. DNA can now be introduced from other species of plants, animals, or even bacteria, producing **genetically modified organisms (GMOs).** The genome modifications made possible by this technology are almost limitless. In response to new possibilities, a sector of the public has expressed concern that the introduction of GMOs into the food supply may produce unexpected health problems. The concern about GMOs is one facet of an ongoing public debate about complex public health, safety, ethical, and educational issues raised by the new genetic technologies.

A vector routinely used to produce transgenic plants is derived from the **Ti plasmid,** a natural plasmid from a soil bacterium called *Agrobacterium tumefaciens.* This bacterium causes what is known as *crown gall disease,* in which the infected plant produces uncontrolled growths called tumors or galls. The key to

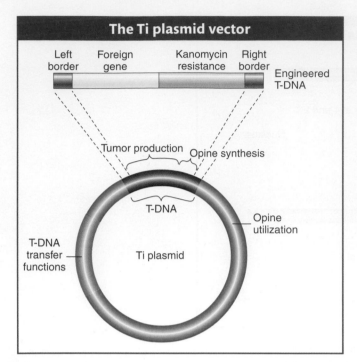

The Ti plasmid vector

Left border — Foreign gene — Kanomycin resistance — Right border — Engineered T-DNA

Tumor production — Opine synthesis

T-DNA

Opine utilization

T-DNA transfer functions

Ti plasmid

Figure 10-25 Simplified representation of the major regions of the Ti plasmid of *A. tumefaciens* containing an engineered T-DNA.

tumor production is a large (200-kb) circular DNA plasmid— the *Ti (tumor-inducing) plasmid.* When the bacterium infects a plant cell, a part of the Ti plasmid is transferred and inserted, apparently more or less at random, into the genome of the host plant (Figure 10-25). The region of the Ti plasmid that inserts into the host plant is called *T-DNA* for transfer DNA. The structure of a Ti plasmid is shown in Figure 10-25. The genes whose products catalyze this T-DNA transfer reside in a region of the Ti plasmid separate from the T-DNA region itself.

The natural behavior of the Ti plasmid makes it well suited to the role of a vector for plant genetic engineering. In particular, any DNA that is inserted between the T-DNA border (24-bp ends) sequences can be mobilized by other functions provided by the Ti plasmid and inserted into plant chromosomes. Thus, scientists were able to eliminate all of the T-DNA sequence between the borders (including the tumor-causing genes) and replace it with the gene(s) of interest and a selectable marker (for example, kanamycin resistance). One method of introducing the T-DNA into the plant genome is shown in Figure 10-26. Bacteria containing this and similarly engineered T-DNA are used to infect cut segments of plant tissue, such as punched-out leaf disks. If the leaf disks are placed on a medium containing kanamycin, the only plant cells that will undergo cell division are those that have acquired the *kan*^R gene engineered into the T-DNA. The transformed cells grow into a clump, or callus, that can be induced to form shoots and roots. These calli are transferred to soil, where they develop into transgenic plants (see Figure 10-26). Typically, only a single copy of the T-DNA region inserts into a given plant genome, where it segregates at meiosis like a regular Mendelian allele (Figure 10-27). The presence of the insert can be verified by screening the transgenic tissue for transgenic genetic markers or by screening purified DNA with a T-DNA probe in a Southern hybridization.

Transgenic plants carrying any one of a variety of foreign genes are in current use, including crop plants carrying genes that confer resistance to certain bacterial or fungal pests, and many more are in development. Not only are the qualities of plants themselves being manipulated, but, like microorganisms, plants are also being used as convenient "factories" to produce proteins encoded by foreign genes.

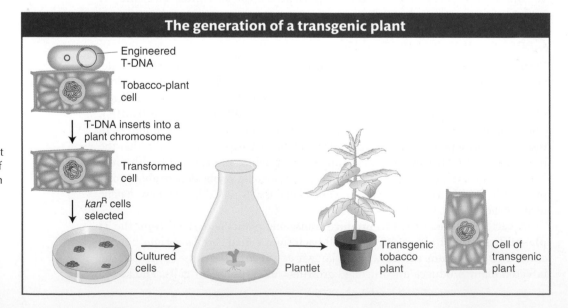

The generation of a transgenic plant

Engineered T-DNA

Tobacco-plant cell

T-DNA inserts into a plant chromosome

Transformed cell

kan^R cells selected

Cultured cells

Plantlet

Transgenic tobacco plant

Cell of transgenic plant

Figure 10-26 The insertion of T-DNA into plant chromosomes. Incubation of leaf disks with the bacterium *A. tumefaciens* containing an engineered T-DNA leads to leaf cells with the T-DNA in their genome, which are able to grow on agar plates and can be coaxed to differentiate into transgenic tobacco plants.

Genetic engineering in animals

Transgenic technologies are now being employed with many animal-model systems. We will focus on the two animal models heavily used for basic genetic research: the nematode *Caenorhabditis elegans* and the mouse *Mus musculus*. A commonly used method to transform a third model organism, the fruit fly *Drosophila melanogaster*, is described in Chapter 15. Versions of many of the techniques considered so far can also be applied in these animal systems.

Transgenesis in *C. elegans*   The method used to introduce trangenes into *C. elegans* is simple: transgenic DNAs are injected directly into the organism, typically as plasmids, fosmids, or other DNAs cloned in bacteria. The injection strategy is determined by the worm's reproductive biology. The gonads of the worm are syncitial, meaning that there are many nuclei within the same gonadal cell. One syncitial cell is a large proportion of one arm of the gonad, and the other syncitial cell is the bulk of the other arm (Figure 10-28a). These nuclei do not form individual cells until meiosis, when they begin their transformation into individual eggs or sperm. A solution of DNA is injected into the syncitial region of one of the arms, thereby exposing more than 100 nuclei to the transforming DNA. By chance, a few of these nuclei will incorporate the DNA (remember, the nuclear membrane breaks down in the course of division, and so the cytoplasm into which the DNA is injected becomes continuous with the nucleoplasm). Typically, the transgenic DNA forms multicopy *extrachromosomal arrays* (Figure 10-28b) that exist as independent units outside the chromosomes. More rarely, the transgenes will become integrated into an ectopic position in a chromosome, still as a multicopy array. Unfortunately, sequences may become scrambled within the arrays, complicating the work of the researcher.

Transgenesis in *M. musculus*   Mice are the most important models for mammalian genetics. Most exciting, much of the technology developed in mice is potentially applicable to humans. There are two strategies for transgenesis in mice, each having its advantages and disadvantages:

- *Ectopic insertions.* Transgenes are inserted randomly in the genome, usually as multicopy arrays.

- *Gene targeting.* The transgene sequence is inserted into a location occupied by a homologous sequence in the genome. That is, the transgene replaces its normal homologous counterpart.

Ectopic insertions   To insert transgenes in random locations, the procedure is simply to inject a solution of bacterially cloned DNA into the nucleus of a fertilized egg (Figure 10-29a). Several injected eggs are inserted into the female oviduct, where some will develop into baby mice. At some later stage, the transgene becomes integrated into

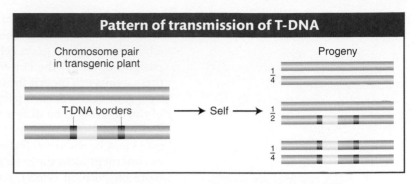

Pattern of transmission of T-DNA

Figure 10-27 The T-DNA region and any DNA inserted into a plant chromosome in a transgenic plant are transmitted in a Mendelian pattern of inheritance.

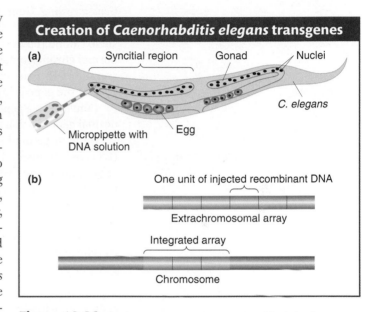

Figure 10-28 *C. elegans* transgenes are created by injecting transgenic DNA directly into a gonad. (a) Method of injection. (b) The two main types of transgenic results: extrachromosomal arrays and arrays integrated in ectopic chromosomal locations.

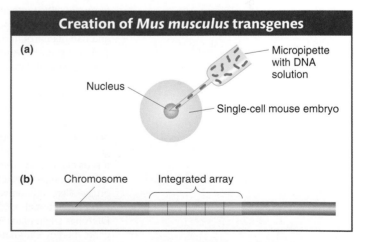

Figure 10-29 *M. musculus* transgenes are created by injection of cloned DNA into fertilized eggs and subsequent insertion in ectopic chromosomal locations. (a) Method of injection. (b) A typical ectopic integrant, with multiple copies of the recombinant transgene inserted in an array.

the chromosomes of random nuclei. On occasion, the transgenic cells form part of the germ line, and, in these cases, an injected embryo will develop into an adult mouse whose germ cells contain the transgene inserted at some random position in one of the chromosomes (Figure 10-29b). Some of the progeny of these adults will inherit the transgene in all cells. There will be an array of multiple gene copies at each point of insertion, but the location, size, and structure of the arrays will be different for each integration event. The technique does give rise to some problems: (1) the expression pattern of the randomly inserted genes may be abnormal (called a **position effect**) because the local chromosome environment lacks the gene's normal regulatory sequences (see Chapter 12 for more on position effect), and (2) DNA rearrangements can occur inside the multicopy arrays (in essence, mutating the sequences). Nonetheless, this technique is much more efficient and less laborious than gene targeting.

Gene targeting Gene targeting enables researchers to eliminate a gene or modify the function it encodes. In one application, called **gene replacement,** a mutant allele can be repaired by substituting a wild-type allele for the mutant one in its normal chromosomal location. Gene replacement avoids both the position effect and the DNA rearrangements associated with ectopic insertion, because a single copy of the gene is inserted in its normal chromosomal environment.

Conversely, a gene may be inactivated by substituting an inactive gene for the normal gene. Such a targeted inactivation is called a **gene knockout.**

Gene targeting in the mouse is carried out in cultured embryonic stem cells (ES cells). In general, a stem cell is an undifferentiated cell in a given tissue or organ that divides asymmetrically to produce a progeny stem cell and a cell that will differentiate into a terminal cell type. ES cells are special stem cells that can differentiate to form any cell type in the body—including, most importantly, the germ line.

To illustrate the process of gene targeting, we look at how it achieves one of its typical outcomes—namely, the substitution of an inactive gene for the normal gene, or gene knockout. The process requires two stages:

1. An inactive gene is targeted to replace the functioning gene in a culture of ES cells, producing ES cells containing a gene knockout (Figure 10-30).

Figure 10-30 Producing cells that contain a mutation in one specific gene, known as a targeted mutation or a gene knockout. (a) Copies of a cloned gene are altered in vitro to produce the targeting vector. The gene shown here was inactivated by the insertion of the neomycin-resistance gene (*neo*^R) into a protein-coding region (exon 2) of the gene and had been inserted into a vector. The *neo*^R gene will serve later as a marker to indicate that the vector DNA took up residence in a chromosome. The vector was also engineered to carry a second marker at one end: the herpes *tk* gene. These markers are standard, but others could be used instead. When a vector, with its dual markers, is complete, it is introduced into cells isolated from a mouse embryo. (b) When homologous recombination occurs (*left*), the homologous regions on the vector, together with any DNA in between but excluding the marker at the tip, take the place of the original gene. This event is important because the vector sequences serve as a useful tag for detecting the presence of this mutant gene. In many cells, though, the full vector (complete with the extra marker at the tip) inserts ectopically (*middle*) or does not become integrated at all (*bottom*). (c) To isolate cells carrying a targeted mutation, all the cells are put into a medium containing selected drugs—here, a neomycin analog (G418) and ganciclovir. G418 is lethal to cells unless they carry a functional *neo*^R gene, and so it eliminates cells in which no integration of vector DNA has taken place (yellow). Meanwhile, ganciclovir kills any cells that harbor the *tk* gene, thereby eliminating cells bearing a randomly integrated vector (red). Consequently, virtually the only cells that survive and proliferate are those harboring the targeted insertion (green). [*After M. R. Capecchi, "Targeted Gene Replacement." Copyright 1994 by Scientific American, Inc. All rights reserved.*]

Producing cells containing a targeted gene knockout

(a) Production of ES cells with a gene knockout

(b) Targeted insertion of vector DNA by homologous recombination

(c) Selection of cells with gene knockout

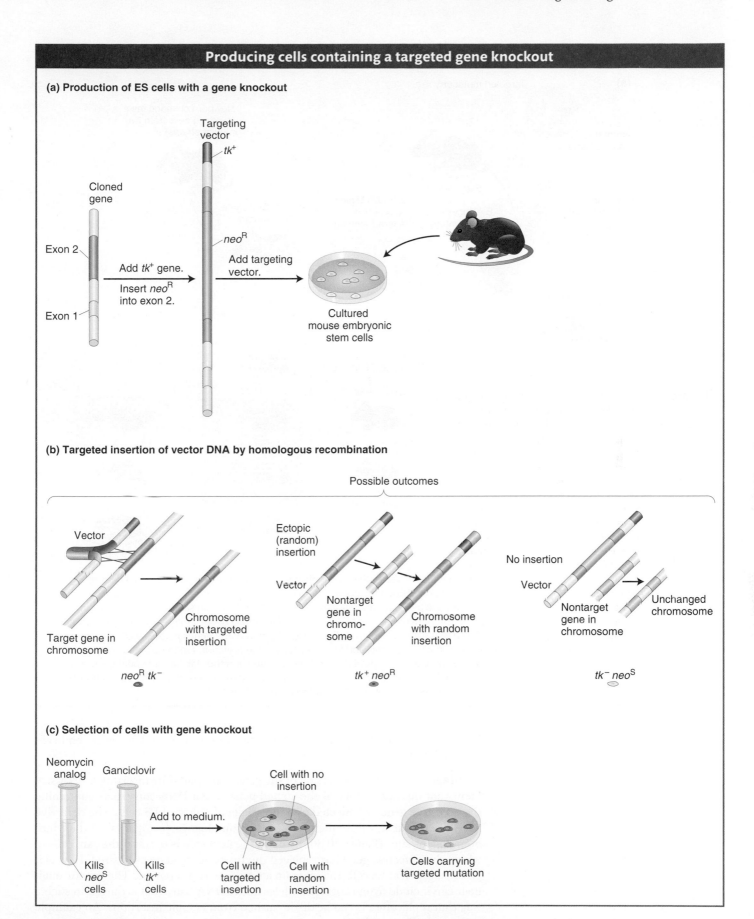

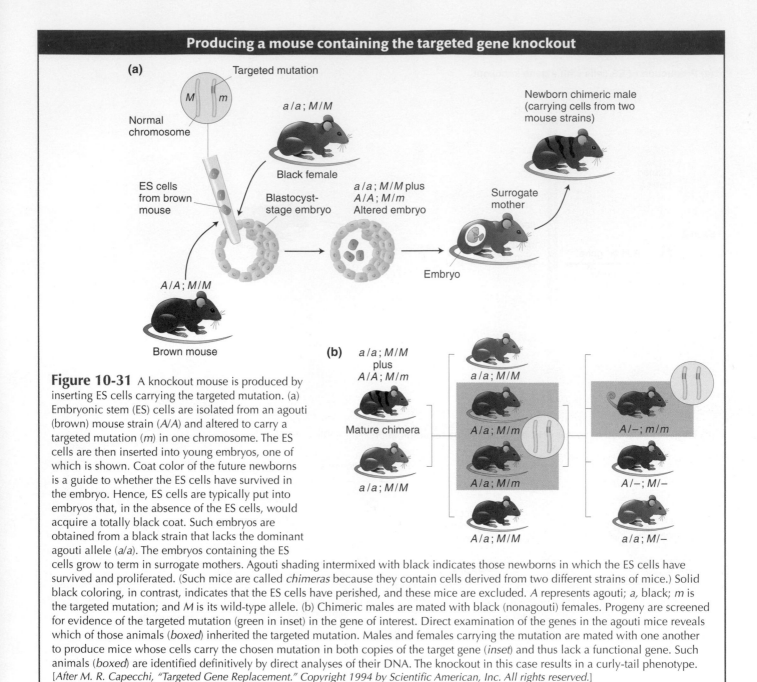

Producing a mouse containing the targeted gene knockout

(a)

Targeted mutation

Normal chromosome

M m

a/a; M/M

Black female

a/a; M/M plus A/A; M/m
Altered embryo

ES cells from brown mouse

Blastocyst-stage embryo

A/A; M/M

Brown mouse

Newborn chimeric male (carrying cells from two mouse strains)

Surrogate mother

Embryo

(b)

a/a; M/M plus A/A; M/m

Mature chimera

a/a; M/M

a/a; M/M

A/a; M/m

A/a; M/m

A/a; M/M

$A/-$; m/m

$A/-$; $M/-$

a/a; $M/-$

Figure 10-31 A knockout mouse is produced by inserting ES cells carrying the targeted mutation. (a) Embryonic stem (ES) cells are isolated from an agouti (brown) mouse strain (A/A) and altered to carry a targeted mutation (m) in one chromosome. The ES cells are then inserted into young embryos, one of which is shown. Coat color of the future newborns is a guide to whether the ES cells have survived in the embryo. Hence, ES cells are typically put into embryos that, in the absence of the ES cells, would acquire a totally black coat. Such embryos are obtained from a black strain that lacks the dominant agouti allele (a/a). The embryos containing the ES cells grow to term in surrogate mothers. Agouti shading intermixed with black indicates those newborns in which the ES cells have survived and proliferated. (Such mice are called *chimeras* because they contain cells derived from two different strains of mice.) Solid black coloring, in contrast, indicates that the ES cells have perished, and these mice are excluded. A represents agouti; a, black; m is the targeted mutation; and M is its wild-type allele. (b) Chimeric males are mated with black (nonagouti) females. Progeny are screened for evidence of the targeted mutation (green in inset) in the gene of interest. Direct examination of the genes in the agouti mice reveals which of those animals (*boxed*) inherited the targeted mutation. Males and females carrying the mutation are mated with one another to produce mice whose cells carry the chosen mutation in both copies of the target gene (*inset*) and thus lack a functional gene. Such animals (*boxed*) are identified definitively by direct analyses of their DNA. The knockout in this case results in a curly-tail phenotype. [*After M. R. Capecchi, "Targeted Gene Replacement." Copyright 1994 by Scientific American, Inc. All rights reserved.*]

2. ES cells containing the inactive gene are transferred to mice embryos (Figure 10-31).

Stage 1: The inactive version of the gene is prepared by inserting a DNA segment that disrupts copies of the cloned gene. Then DNA constructs containing the defective gene are injected into the nuclei of cultured ES cells. The defective gene inserts far more frequently into nonhomologous (ectopic) sites than into homologous sites (Figure 10-30b), and so the next step is to select the rare cells in which the defective gene has replaced the functioning gene as desired. How is it possible to select ES cells that contain a rare gene replacement? The genetic engineer can include drug-resistance alleles in the DNA construct arranged in such a way that replacements can be distinguished from ectopic insertions. An example is shown in Figure 10-30c.

Stage 2: The ES cells that contain one copy of the disrupted gene of interest (that is, gene knockout) are injected into a blastocyst-stage embryo, which is then implanted in a surrogate mother (Figure 10-31a). Some of the ES cells may become incorporated into the host embryo, and if that happens, the mouse that develops will be chimeric—that is, it will contain cells from two different mouse strains. When the chimeric mouse reaches adulthood, it is mated with a normal mouse. If the chimeric mouse had taken up ES cells (with the knockout gene) into germ-line cells, then some of the resulting offspring will inherit the gene knockout in all their cells. Sibling mice that are identified as being heterozygous for the knockout version of the gene of interest are then mated in order to produce mice that are homozygous for the knockout allele (if it does not cause lethality in homozygotes) (Figure 10-31b).

> **Message** Germ-line transgenic techniques have been developed for all well-studied eukaryotic species. These techniques depend on an understanding of the reproductive biology of the recipient species.

Summary

Recombinant DNA is constructed in the laboratory to allow researchers to amplify and analyze DNA segments (donor DNA) from any genome or from DNA copies of mRNAs. Three sources of donor DNA are (1) the entire genome digested with a restriction enzyme, (2) PCR products of specific DNA regions defined by the flanking primer sequences, and (3) cDNA copies of mRNAs.

The polymerase chain reaction is a powerful method for the direct amplification of a relatively small sequence of DNA from within a complex mixture of DNA, without the need for a host cell or very much starting material. The key is to have primers that are complementary to flanking regions on each of the two DNA strands. These regions act as sites for polymerization. Multiple rounds of denaturation, priming, and polymerization amplify the sequence of interest exponentially.

To insert donor DNA into vectors, donor DNA and vector DNA are cut by the same restriction endonuclease at specific sequences. Vector and donor DNA are joined in a test tube by annealing the sticky ends that result from digestion, followed by ligation to covalently join the molecules. PCR and cDNA molecules are inserted into vectors by first adding restriction-endonuclease-recognition sequences to the 5' end of PCR primers or by ligating short adapters containing restriction sites to their ends before insertion into the vector.

There are a wide variety of bacterial vectors. The choice of vector depends largely on the size of the DNA fragment to be cloned. Plasmids are used to clone small restriction fragments, PCR molecules, or cDNA molecules (from ~100 bp to up to 10 kb). Intermediate-size fragments, such as those resulting from the digestion of genomic DNA, can be cloned into modified versions of λ bacteriophage (for inserts of 10–15 kb) or into phage–plasmid hybrids called fosmids (for inserts of 35–45 kb). Finally, bacterial artificial chromosomes (BACs) are used routinely to clone very large genomic fragments (~100–200 kb).

The vector–donor DNA construct is amplified inside host cells as extrachromosomal molecules that are replicated when the host is replicating its genome. For this to happen, the vector must contain all the necessary signals for proper replication and segregation in that host cell. For plasmid-based systems (including BACs), the vector must include an origin of replication and selectable markers such as drug-resistance genes that can be used to ensure that the plasmid is not lost from the host cell. In bacteriophage, the vector must include all sequences necessary for carrying the bacteriophage (and the hitchhiking foreign DNA) through the lytic growth cycle. Finally, as vectors with a hybrid lifestyle, fosmids must contain sequences needed for their packaging into phage heads plus all of the plasmid-based systems mentioned above. The result of amplification of plasmids, phages, and BACs is clones containing multiple copies of each recombinant DNA construct. In contrast, only a single fosmid is present in each bacterial cell.

Often, finding a specific clone with a gene of interest requires screening a full genomic library. A genomic library is a set of clones, ligated in the same vector, that together represent all regions of the genome of the organism in question. The number of clones that constitute a genomic library depends on (1) the size of the genome in question and (2) the insert size tolerated by the particular cloning-vector system. Similarly, a cDNA library is a representation of the total mRNA set produced by a given tissue or developmental stage in a given organism. A comparison of a genomic region and its cDNA can be a source of insight into the locations of transcription start and stop sites and boundaries between introns and exons.

Labeled single-stranded DNA or RNA probes are important "bait" for fishing out similar or identical sequences from complex mixtures of molecules, either in genomic or cDNA libraries or in Southern (DNA) and Northern (RNA) blotting. The general principle of the technique for identifying clones or gel fragments is to create an "image" of the colonies or plaques on an agar petri-dish culture or of the nucleic acids that have been separated in an electric field passed through a gel matrix. The DNA or RNA is then denatured and mixed with a denatured probe that has been labeled with a fluorescent dye or a radioactive label. After unbound probe has been washed off, the location of the probe is detected either by observing its fluorescence or, if radioactive, by exposing the sample to X-ray film. The locations of the probe correspond to the locations of the relevant DNA or RNA in the original petri dish or electrophoresis gel.

The vast genomic resources are making it increasingly possible to isolate genes solely from knowledge of their position on a genetic map. Two overall procedures are forward genetic strategies called positional cloning and fine-structure mapping. With the sequencing of the human genome and the availability of families with inherited disorders, fine-structure-mapping strategies have led to isolation of genes that when mutated produce human disease.

Transgenes are engineered DNA molecules that are introduced and expressed in eukaryotic cells. They can be used to engineer a novel mutation or to study the regulatory sequences that constitute part of a gene. Transgenes can be introduced as extrachromosomal molecules or they can be integrated into a chromosome, either in random (ectopic) locations or in place of the homologous gene, depending on the system. Typically, the mechanisms used to introduce a transgene depend on an understanding and exploitation of the reproductive biology of the organism.

KEY TERMS

amplification (p. 342)
antibody (p. 358)
autoradiogram (p. 356)
bacterial artificial chromosome (BAC) (p. 353)
cDNA library (p. 355)
chromosome walk (p. 366)
complementary DNA (cDNA) (p. 347)
dideoxy (Sanger) sequencing (p. 362)
DNA cloning (p. 343)
DNA ligase (p. 348)
DNA palindrome (p. 345)
DNA technology (p. 342)
donor DNA (p. 342)
ectopically (p. 370)

fine mapping (p. 369)
fosmid (p. 353)
functional complementation (mutant rescue) (p. 358)
gel electrophoresis (p. 359)
gene knockout (p. 374)
gene replacement (p. 374)
genetically modified organism (GMO) (p. 371)
genetic engineering (p. 342)
genomic library (p. 355)
genomics (p. 342)
hybridization (p. 345)
mutant rescue (functional complementation) (p. 358)
Northern blotting (p. 361)

polymerase chain reaction (PCR) (p. 343)
positional cloning (p. 365)
position effect (p. 374)
probe (p. 355)
recombinant DNA (p. 342)
restriction enzyme (p. 344)
restriction fragment (p. 345)
RNA blotting (p. 361)
Sanger (dideoxy) sequencing (p. 362)
Southern blotting (p. 361)
Ti plasmid (p. 371)
transgene (p. 370)
transgenic organism (p. 370)
vector (p. 342)

SOLVED PROBLEMS

SOLVED PROBLEM 1. In Chapter 9, we studied the structure of tRNA molecules. Suppose that you want to clone a fungal gene that encodes a certain tRNA. You have a sample of the purified tRNA and an *E. coli* plasmid that contains a single *Eco*RI cutting site in a *tet*^{R} (tetracycline-resistance) gene, as well as a gene for resistance to ampicillin (*amp*^{R}). How can you clone the gene of interest?

Solution

You can use the tRNA itself or a cloned cDNA copy of it to probe for the DNA containing the gene. One method is to digest the genomic DNA with *Eco*RI and then mix it with the plasmid, which you also have cut with *Eco*RI. After transformation of an *amp*^{S} *tet*^{S} recipient, select Amp^R colonies, indicating successful transformation. Of these Amp^R colonies, select the colonies that are Tet^S. These Tet^S colonies will contain vectors with inserts in the *tet*^{R} gene, and a great number of them are needed to make the library. Test the library by using the tRNA as the probe. Those clones that hybridize to the probe will contain the gene of interest.

Alternatively, you can subject *Eco*RI-digested genomic DNA to gel electrophoresis and then identify the correct band by probing with the tRNA. This region of the gel can be cut out and used as a source of enriched DNA to clone into the plasmid cut with *Eco*RI. You then probe these clones with the tRNA to confirm that these clones contain the gene of interest.

PROBLEMS

Most of the problems are also available for review/grading through the GENETICSPORTAL www.yourgeneticsportal.com.

WORKING WITH THE FIGURES

1. Figure 10-1 shows that specific DNA fragments can be synthesized in vitro prior to cloning. What are two ways to synthesize DNA inserts for recombinant DNA in vitro?

2. In Figure 10-4, why is cDNA made only from mRNA and not also from tRNAs and ribosomal RNAs?

3. Redraw Figure 10-6 with the goal of adding one *Eco*RI end and one *Xho*I end. On the next page is the *Xho*I recognition sequence.

Recognition sequence:	After cut:
...CTCGAG... ...GAGCTC...	...C TCGAG... ...GAGCT C...

4. Redraw Figure 10-7 so that the cDNA can insert into an *XhoI* site of a vector rather than into an *Eco*RI site as shown.

5. In Figure 10-11, determine approximately how many BAC clones are needed to provide 1× coverage of

  a. the yeast genome (12 Mbp).

  b. the *E. coli* genome (4.1 Mbp).

  c. the fruit-fly genome (130 Mbp).

6. In Figure 10-15, why does DNA migrate to the anode (+ pole)?

7. In Figure 10-18a, why are DNA fragments of different length and all ending in an A residue synthesized?

8. As you will see in Chapter 15, most of the genomes of higher eukaryotes (plants and animals) are filled with DNA sequences that are present in hundreds, even thousands of copies throughout the chromosomes. In the chromosome-walking procedure shown in Figure 10-20, how would the experimenter know whether the fragment he or she is using to "walk" to the next BAC or phage is repetitive? Can repetitive DNA be used in a chromosome walk?

9. Redraw Figure 10-24 to include the positions of the single and double crossovers.

10. In Figure 10-26, why do only plant cells that have T-DNA inserts in their chromosomes grow on the agar plates? Do all of the cells of a transgenic plant grown from one clump of cells contain T-DNA? Justify your answer.

11. In Figure 10-28, what is the difference between extra-chromosomal DNA and integrated arrays of DNA? Are the latter ectopic? What is distinctive about the syncitial region that makes it a good place to inject DNA?

BASIC PROBLEMS

12. From this chapter, make a list of all the examples of **(a)** the hybridization of single-stranded DNAs and **(b)** proteins that bind to DNA and then act on it.

13. Compare and contrast the use of the word *recombinant* as used in the phrases **(a)** "recombinant DNA" and **(b)** "recombinant frequency."

14. Why is ligase needed to make recombinant DNA? What would be the immediate consequence in the cloning process if someone forgot to add it?

15. In the PCR process, if we assume that each cycle takes 5 minutes, how manyfold amplification would be accomplished in 1 hour?

16. The position of the gene for the protein actin in the haploid fungus *Neurospora* is known from the com-plete genome sequence. If you had a slow-growing mutant that you suspected of being an actin mutant and you wanted to verify that it was one, would you **(a)** clone the mutant by using convenient restriction sites flanking the actin gene and then sequence it or **(b)** amplify the mutant sequence by using PCR and then sequence it?

17. You obtain the DNA sequence of a mutant of a 2-kb gene in which you are interested and it shows base differences at three positions, all in different codons. One is a silent change, but the other two are missense changes (they encode new amino acids). How would you demonstrate that these changes are real mutations and not sequencing errors? (Assume that sequencing is about 99.9 percent accurate.)

18. In a T-DNA transformation of a plant with a transgene from a fungus (not found in plants), the presumptive transgenic plant does not express the expected phenotype of the transgene. How would you demonstrate that the transgene is in fact present? How would you demonstrate that the transgene was expressed?

19. How would you produce a mouse that is homozygous for a rat growth-hormone transgene?

20. Why was cDNA and not genomic DNA used in the commercial cloning of the human insulin gene?

21. After *Drosophila* DNA has been treated with a restriction enzyme, the fragments are inserted into plasmids and selected as clones in *E. coli*. With the use of this "shotgun" technique, every DNA sequence of *Drosophila* in a library can be recovered.

  a. How would you identify a clone that contains DNA encoding the protein actin, whose amino acid sequence is known?

  b. How would you identify a clone encoding a specific tRNA?

22. In any particular transformed eukaryotic cell (say, of *Saccharomyces cerevisiae*), how could you tell if the transforming DNA (carried on a circular bacterial vector)

  a. replaced the resident gene of the recipient by double crossing over or single crossing over?

  b. was inserted ectopically?

23. In an electrophoretic gel across which is applied a powerful electrical alternating pulsed field, the DNA of the haploid fungus *Neurospora crassa* ($n = 7$) moves slowly but eventually forms seven bands, which represent DNA fractions that are of different sizes and hence have moved at different speeds. These bands are presumed to be the seven chromosomes. How would you show which band corresponds to which chromosome?

24. The protein encoded by the cystic-fibrosis gene is 1480 amino acids long, yet the gene spans 250 kb. How is this difference possible?

25. In yeast, you have sequenced a piece of wild-type DNA and it clearly contains a gene, but you do not know what gene it is. Therefore, to investigate further, you would like to find out its mutant phenotype. How would you use the cloned wild-type gene to do so? Show your experimental steps clearly.

CHALLENGING PROBLEMS

26. Prototrophy is often the phenotype selected to detect transformants. Prototrophic cells are used for donor DNA extraction; then this DNA is cloned and the clones are added to an auxotrophic recipient culture. Successful transformants are identified by plating the recipient culture on minimal medium and looking for colonies. What experimental design would you use to make sure that a colony that you hope is a transformant is not, in fact,

a. a prototrophic cell that has entered the recipient culture as a contaminant?

b. a revertant (mutation back to prototrophy by a second mutation in the originally mutated gene) of the auxotrophic mutation?

27. A cloned fragment of DNA was sequenced by using the dideoxy chain-termination method. A part of the autoradiogram of the sequencing gel is represented here.

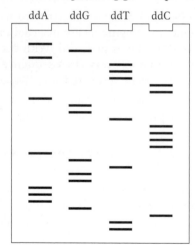

a. Deduce the nucleotide sequence of the DNA nucleotide chain synthesized from the primer. Label the 5′ and 3′ ends.

b. Deduce the nucleotide sequence of the DNA nucleotide chain used as the template strand. Label the 5′ and 3′ ends.

c. Write out the nucleotide sequence of the DNA double helix (label the 5′ and 3′ ends).

d. How many of the six reading frames are "open" as far as you can tell?

28. The cDNA clone for the human gene encoding tyrosinase was radioactively labeled and used in a Southern analysis of *Eco*RI-digested genomic DNA of wild-type mice. Three mouse fragments were found to be radioactive (were bound by the probe). When albino mice were used in this Southern analysis, no genomic fragments bound to the probe. Explain these results in relation to the nature of the wild-type and mutant mouse alleles.

29. Transgenic tobacco plants were obtained in which the vector Ti plasmid was designed to insert the gene of interest plus an adjacent kanamycin-resistance gene. The inheritance of chromosomal insertion was followed by testing progeny for kanamycin resistance. Two plants typified the results obtained generally. When plant 1 was backcrossed with wild-type tobacco, 50 percent of the progeny were kanamycin resistant and 50 percent were sensitive. When plant 2 was backcrossed with the wild type, 75 percent of the progeny were kanamycin resistant and 25 percent were sensitive. What must have been the difference between the two transgenic plants? What would you predict about the situation regarding the gene of interest?

30. A cystic-fibrosis mutation in a certain pedigree is due to a single nucleotide-pair change. This change destroys an *Eco*RI restriction site normally found in this position. How would you use this information in counseling members of this family about their likelihood of being carriers? State the precise experiments needed. Assume that you find that a woman in this family is a carrier, and it transpires that she is married to an unrelated man who also is a heterozygote for cystic fibrosis, but, in his case, it is a different mutation in the same gene. How would you counsel this couple about the risks of a child's having cystic fibrosis?

31. Bacterial glucuronidase converts a colorless substance called X-Gluc into a bright blue indigo pigment. The gene for glucuronidase also works in plants if given a plant promoter region. How would you use this gene as a reporter gene to find the tissues in which a plant gene that you have just cloned is normally active? (Assume that X-Gluc is easily taken up by the plant tissues.)

32. The plant *Arabidopsis thaliana* was transformed by using the Ti plasmid into which a kanamycin-resistance gene had been inserted in the T-DNA region. Two kanamycin-resistant colonies (A and B) were selected, and plants were regenerated from them. The plants were allowed to self-pollinate, and the results were as follows:

Plant A selfed → $\frac{3}{4}$ progeny resistant to kanamycin

$\frac{1}{4}$ progeny sensitive to kanamycin

Plant B selfed → $\frac{15}{16}$ progeny resistant to kanamycin

$\frac{1}{16}$ progeny sensitive to kanamycin

a. Draw the relevant plant chromosomes in both plants.

b. Explain the two different ratios.

Regulation of Gene Expression in Bacteria and Their Viruses

11

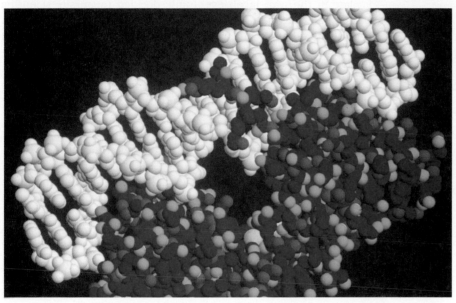

The control of gene expression is governed primarily by DNA-binding proteins that recognize specific control sequences of genes. Here, the binding of the λ repressor protein to DNA is modeled. [*Kenneth Eward/Science Source.*]

I n December 1965, the king of Sweden presented the Nobel Prize in Physiology or Medicine to François Jacob, Jacques Monod, and André Lwoff of the Pasteur Institute for their discoveries of how gene expression is regulated (Figure 11-1). The prizes were the fruit of an exceptional collaboration among three superb scientists. They were also triumphs over great odds. The chances that each of these three men would live to see that day, let alone earn such honors, were poor.

Twenty-five years earlier, Monod was a doctoral student at the Sorbonne in Paris, working on a phenomenon in bacteria called "enzymatic adaptation" that seemed so obscure to some that the director of the zoological laboratory where he worked stated, "What Jacques Monod is doing is of no interest whatever to the Sorbonne." Jacob was a 19-year-old medical student intent on becoming a surgeon. Lwoff was by that time a well-established member of the Pasteur, chief of its department of microbial physiology.

Then came World War II.

As France was invaded and quickly defeated, Jacob raced for the coast to join the Free French forces assembling in England. He served as a medic in North Africa and in Normandy until badly wounded. Monod joined the French Resistance while continuing his work. After a Gestapo raid on his Sorbonne laboratory, Monod decided that working there was dangerous (his predecessor in the Resistance was arrested and executed), and André Lwoff offered him space at the

Pioneers of gene regulation

Figure 11-1 François Jacob, Jacques Monod, and André Lwoff were awarded the 1965 Nobel Prize in Physiology or Medicine for their pioneering work on how gene expression is regulated. [*The Pasteur Institute.*]

Pasteur. Monod, in turn, helped Lwoff to join the Resistance, which put the senior scientist in double jeopardy: as a Jew, Lwoff also risked deportation (several Pasteur scientists were deported and died during the occupation of France).

After the liberation of Paris, Monod served in the French army and happened on, in a mobile U.S. Army library, an article by Oswald Avery and colleagues demonstrating that DNA is the hereditary material in bacteria (see Chapter 7). His interest in genetics was rekindled, and he rejoined Lwoff after the war. Jacob's injuries were too severe for him to pursue a career in surgery. Inspired by the enormous impact of antibiotics introduced late in the war, Jacob eventually decided to pursue scientific research. Jacob approached Lwoff several times for a position in his laboratory but was declined. He made one last try and caught Lwoff in a jovial mood. The senior scientist told Jacob, "You know, we have just found the induction of the prophage. Would you be interested in working on the phage?" Jacob had no idea what Lwoff was talking about. He stammered, "That's just what I would like to do."

The cast was set. What unfolded in the subsequent decade was one of the most creative and productive collaborations in the history of genetics, whose discoveries still reverberate throughout biology today.

One of the most important insights arrived not in the laboratory but in a movie theater. Struggling with a lecture that he had to prepare, Jacob opted instead to take his wife, Lise, to a Sunday matinee. Bored and daydreaming, Jacob drew a connection between the work he had been doing on the induction of prophage and that of Monod on the induction of enzyme synthesis. Jacob became "involved by a sudden excitement mixed with a vague pleasure. . . . The astonishment of the obvious. How could I not have thought of it sooner? Both experiments . . . on the phage . . . and that done with Pardee and Monod on the lactose system . . . are the same! Same situation. Same result . . . In both cases, a gene governs the formation of a cytoplasmic product, of a repressor blocking the expression of other genes and so preventing either the synthesis of the galactosidase or the multiplication of the virus. . . . Where can the repressor act to stop everything at once? The only simple answer . . . is on the DNA itself!" (F. Jacob, *The Statue Within: An Autobiography*, 1988).

And so was born the concept of a repressor acting on DNA to repress the induction of genes. It would take many years before the hypothesized repressors were isolated and characterized biochemically. The concepts worked out by Jacob and Monod and explained in this chapter—messenger RNA, promoters, operators, regulatory genes, operons, and allosteric proteins—were deduced entirely from genetic evidence, and these concepts shaped the future field of molecular genetics.

Walter Gilbert, who isolated the first repressor and was later awarded a Nobel Prize in Chemistry for co-inventing a method of sequencing DNA, explained the effect of Jacob and Monod's work at that time: "Most of the crucial discoveries in science are of such a simplifying nature that they are very hard even to conceive without actually have gone through the experience involved in the discovery. . . . Jacob's and Monod's suggestion made things that were utterly dark, very simple" (H. F. Judson, *The Eighth Day of Creation: Makers of the Revolution in Biology*, 1979).

The concepts that Jacob and Monod illuminated went far beyond bacterial enzymes and viruses. They understood, and were able to articulate with exceptional eloquence, how their discoveries about gene regulation pertained to the general mysteries of cell differentiation and embryonic development in animals. Jacques Monod once quipped, "What is true for *E. coli* is also true for the elephant."

In the next three chapters, we will see to what degree that is true. We'll start in this chapter with bacterial examples that illustrate key themes and mechanisms in the regulation of gene expression. We will largely focus on single regulatory proteins and the genetic "switches" on which they act. Then, in Chapter 12, we'll tackle gene regulation in eukaryotic cells, which entails more complex biochemical and genetic machinery. And, finally, in Chapter 13, we'll examine the role of gene regulation in the development of multicellular animals. There we will see how sets of regulatory proteins act on arrays of genetic switches to control gene expression in time and space and choreograph the building of bodies and body parts.

11.1 Gene Regulation

Despite their simplicity of form, bacteria have in common with larger and more complex organisms the need to regulate expression of their genes. One of the main reasons is that they are nutritional opportunists. Consider how bacteria obtain the many important compounds, such as sugars, amino acids, and nucleotides, needed for metabolism. Bacteria swim in a sea of potential nutrients. They can either acquire the compounds that they need from the environment or synthesize them by enzymatic pathways. But synthesizing these compounds also requires expending energy and cellular resources to produce the necessary enzymes for these pathways. Thus, given the choice, bacteria will take compounds from the environment instead. Natural selection favors efficiency and selects against the waste of resources and energy. To be economical, bacteria will synthesize the enzymes necessary to produce compounds only when there is no other option—in other words, when compounds are unavailable in their local environment.

Bacteria have evolved regulatory systems that couple the expression of gene products to sensor systems that detect the relevant compound in a bacterium's local environment. The regulation of enzymes taking part in sugar metabolism provides an example. Sugar molecules can be broken down to provide energy or they can be used as building blocks for a great range of organic compounds. However, there are many different types of sugar that bacteria could use, including lactose, glucose, galactose, and xylose. A different import protein is required to allow each of these sugars to enter the cell. Further, a different set of enzymes is required to process each of the sugars. If a cell were to simultaneously synthesize all the enzymes that it might possibly need, the cell would expend much more energy and materials to produce the enzymes than it could ever derive from breaking down prospective carbon sources. The cell has devised mechanisms to shut down (repress) the transcription of all genes encoding enzymes that are not needed at a given time and to turn on (activate) those genes encoding enzymes that are needed. For example, if only lactose is in the environment, the cell will shut down the transcription of the genes encoding enzymes needed for the import and metabolism of glucose, galactose, xylose, and other sugars. Conversely, *E. coli* will initiate the transcription of the genes encoding enzymes needed for the import and metabolism of lactose. In sum, cells need mechanisms that fulfill two criteria:

1. They must be able to recognize environmental conditions in which they should activate or repress the transcription of the relevant genes.

2. They must be able to toggle on or off, like a switch, the transcription of each specific gene or group of genes.

Let's preview the current model for prokaryotic transcriptional regulation and then use a well-understood example—the regulation of the genes in the metabolism of the sugar lactose—to examine it in detail. In particular, we will focus on how this regulatory system was dissected with the use of the tools of classical genetics and molecular biology.

The basics of prokaryotic transcriptional regulation: genetic switches

The regulation of transcription depends mainly on two types of protein–DNA interactions. Both take place near the site at which gene transcription begins.

One of these DNA–protein interactions determines where transcription begins. The DNA that participates in this interaction is a DNA segment called the **promoter,** and the protein that binds to this site is RNA polymerase. When RNA polymerase binds to the promoter DNA, transcription can start a few bases away from the promoter site. Every gene must have a promoter or it cannot be transcribed.

The other type of DNA–protein interaction determines whether promoter-driven transcription takes place. DNA segments near the promoter serve as binding sites for sequence-specific regulatory proteins called **activators** and **repressors.** In bacteria, most binding sites for repressors are termed **operators.** For some genes, an activator protein must bind to its target DNA site as a necessary prerequisite for transcription to begin. Such instances are sometimes referred to as *positive regulation* because the *presence* of the bound protein is required for transcription (Figure 11-2). For other genes, a repressor protein must be prevented from binding to its target site as a necessary prerequisite for transcription to begin. Such cases are sometimes termed *negative regulation* because the *absence* of the bound repressor allows transcription to begin. How do activators and repressors regulate transcription? Often, a DNA-bound activator protein physically helps tether RNA polymerase to its nearby promoter so that polymerase may begin transcribing. A DNA-bound repressor protein typically acts either by physically interfering with the binding of RNA polymerase to its promoter (blocking transcription initiation) or by impeding the movement of RNA polymerase along the DNA chain (blocking transcription). Together, these regulatory proteins and their binding sites constitute **genetic switches** that control the efficient changes in gene expression that occur in response to environmental conditions.

> **Message** Genetic switches control gene transcription. The on/off function of the switches depends on the interactions of several proteins with their binding sites on DNA. RNA polymerase interacts with the promoter to begin transcription. Activator or repressor proteins bind to sites in the vicinity of the promoter to control its accessibility to RNA polymerase.

Figure 11-2 The binding of regulatory proteins can either activate or block transcription.

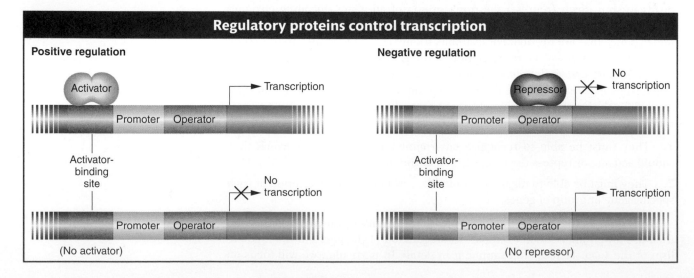

Regulatory proteins control transcription

Positive regulation

Negative regulation

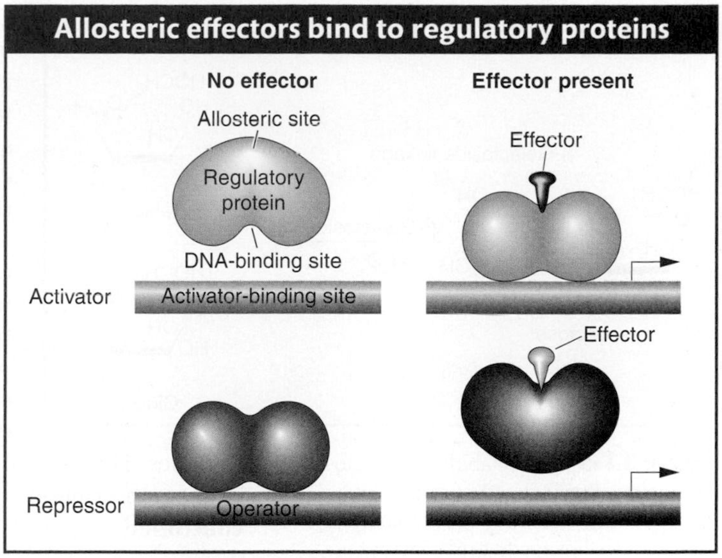

Both activator and repressor proteins must be able to recognize when environmental conditions are appropriate for their actions and act accordingly. Thus, for activator or repressor proteins to do their job, each must be able to exist in two states: one that can bind its DNA targets and another that cannot. The binding state must be appropriate to the set of physiological conditions present in the cell and its environment. For many regulatory proteins, DNA binding is effected through the interaction of two different sites in the three-dimensional structure of the protein. One site is the **DNA-binding domain.** The other site, the **allosteric site,** acts as a sensor that sets the DNA-binding domain in one of two modes: functional or nonfunctional. The allosteric site interacts with small molecules called *allosteric effectors.* In lactose metabolism, an isomer of the sugar lactose (called allolactose) is an allosteric effector: the sugar binds to a regulatory protein that inhibits the expression of genes needed for lactose metabolism. In general, an **allosteric effector** binds to the allosteric site of the regulatory protein in such a way as to change its activity. In this case, allolactose changes the shape and structure of the DNA-binding domain of a regulatory protein. Some activator or repressor proteins must bind to their allosteric effectors before they can bind DNA. Others can bind DNA only in the absence of their allosteric effectors. Two of these situations are shown in Figure 11-3.

Message Allosteric effectors control the ability of activator or repressor proteins to bind to their DNA target sites.

Allosteric effectors bind to regulatory proteins

Figure 11-3 Allosteric effectors influence the DNA-binding activities of activators and repressors.

A first look at the *lac* regulatory circuit

The pioneering work of François Jacob and Jacques Monod in the 1950s showed how lactose metabolism is genetically regulated. Let's examine the system under two conditions: the presence and the absence of lactose. Figure 11-4 is a simplified view of the components of this system. The cast of characters for *lac* operon regulation includes protein-coding genes and sites on the DNA that are targets for DNA-binding proteins.

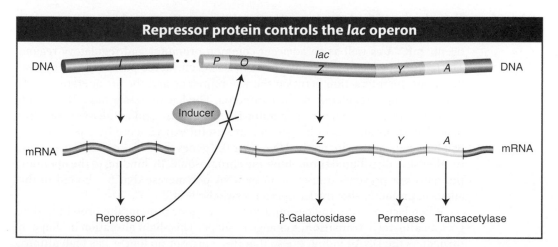

Repressor protein controls the *lac* operon

Figure 11-4 A simplified *lac* operon model. Coordinate expression of the *Z, Y,* and *A* genes is under the negative control of the product of the *I* gene, the repressor. When the inducer binds the repressor, the operon is fully expressed.

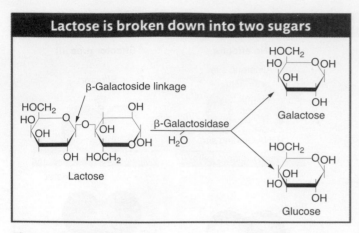

Figure 11-5 The metabolism of lactose. The enzyme β-galactosidase catalyzes a reaction in which water is added to the β-galactoside linkage to break lactose into separate molecules of glucose and galactose.

The *lac* structural genes   The metabolism of lactose requires two enzymes: (1) a permease to transport lactose into the cell and (2) β-galactosidase to cleave the lactose molecule to yield glucose and galactose (Figure 11-5). The structures of the β-galactosidase and permease proteins are encoded by two adjacent sequences, *Z* and *Y*, respectively. A third contiguous sequence encodes an additional enzyme, termed *transacetylase,* which is not required for lactose metabolism. We will call *Z, Y,* and *A structural genes*—in other words, segments encoding proteins—while reserving judgment on this categorization until later. We will focus mainly on the *Z* and *Y* genes. All three genes are transcribed into a single messenger RNA molecule. Regulation of the production of this mRNA coordinates the synthesis of all three enzymes. That is, either all or none of the three enzymes are synthesized. Genes whose transcription is controlled by a common means are said to be **coordinately controlled.**

> **Message** If the genes encoding proteins constitute a single transcription unit, the expression of all these genes will be coordinately regulated.

Regulatory components of the *lac* system   Key regulatory components of the lactose metabolic system include a gene encoding a transcription regulatory protein and two binding sites on DNA: one site for the regulatory protein and another site for RNA polymerase.

1.   *The gene for the Lac repressor.* A fourth gene (besides the structural genes, *Z, Y,* and *A*), the *I* gene, encodes the Lac repressor protein. It is so named because it can block the expression of the *Z, Y,* and *A* genes. The *I* gene happens to map close to the *Z, Y,* and *A* genes, but this proximity is not important to its function because it encodes a diffusible protein.

2.   *The lac promoter site.* The promoter (*P*) is the site on the DNA to which RNA polymerase binds to initiate transcription of the *lac* structural genes (*Z, Y,* and *A*).

3.   *The lac operator site.* The operator (*O*) is the site on the DNA to which the Lac repressor binds. It is located between the promoter and the *Z* gene near the point at which transcription of the multigenic mRNA begins.

The induction of the *lac* system   The *P, O, Z, Y,* and *A* segments (shown in Figure 11-6) together constitute an **operon,** defined as a segment of DNA that encodes a multigenic mRNA as well as an adjacent common promoter and regulatory region. The *lacI* gene, encoding the Lac repressor, is *not* considered part of the *lac* operon itself, but the interaction between the Lac repressor and the *lac* operator site is crucial to proper regulation of the *lac* operon. The Lac repressor has a *DNA-binding* site that can recognize the operator DNA sequence and an *allosteric site* that binds lactose or analogs of lactose that are useful experimentally. The repressor will bind only to the site on the DNA near the genes that it is controlling and not to other sites distributed throughout the chromosome. By binding to the operator, the repressor prevents transcription by RNA polymerase that has bound to the adjacent promoter site; the *lac* operon is switched "off."

When lactose or its analogs bind to the repressor protein, the protein undergoes an **allosteric transition,** a change in shape. This slight alteration in shape in turn alters the DNA-binding site so that the repressor no longer has high affinity for the operator. Thus, in response to binding lactose, the repressor falls off the DNA: the *lac* operon is switched "on." The repressor's response to lactose satisfies

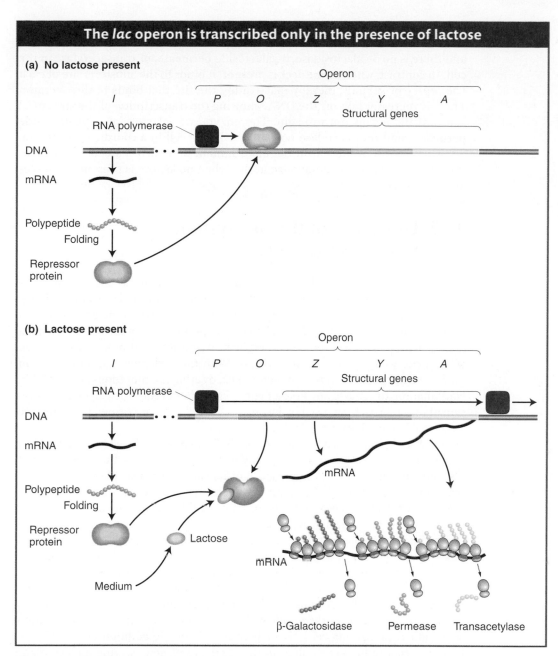

The *lac* operon is transcribed only in the presence of lactose

(a) No lactose present

Operon

I   P   O   Z   Y   A
Structural genes

RNA polymerase

DNA

mRNA

Polypeptide
Folding

Repressor
protein

(b) Lactose present

Operon

I   P   O   Z   Y   A
Structural genes

RNA polymerase

DNA

mRNA

Polypeptide
Folding

Repressor
protein

Lactose

Medium

mRNA

mRNA

β-Galactosidase   Permease   Transacetylase

Figure 11-6 Regulation of the *lac* operon. The *I* gene continually makes repressor. (a) In the absence of lactose, the repressor binds to the *O* (operator) region and blocks transcription. (b) The binding of lactose changes the shape of the repressor so that the repressor no longer binds to *O* and falls off the DNA. The RNA polymerase is then able to transcribe the *Z, Y,* and *A* structural genes, and so the three enzymes are produced.

GENETICS**PORTAL** **ANIMATED ART:** Assaying lactose presence or absence through the Lac repressor

one requirement for such a control system—that the presence of lactose stimulates the synthesis of genes needed for its processing. The relief of repression for systems such as *lac* is termed **induction.** Lactose and its analogs that allosterically inactivate the repressor, leading to the expression of the *lac* genes, are termed **inducers.**

Let's summarize how the *lac* switch works. In the absence of an inducer (lactose or an analog), the Lac repressor binds to the *lac* operator site and prevents transcription of the *lac* operon by blocking the movement of RNA polymerase.

In this sense, the Lac repressor acts as a roadblock on the DNA. Consequently, all the structural genes of the *lac* operon (the *Z, Y,* and *A* genes) are repressed, and there is no β-galactosidase, β-galactoside permease, or transacetylase in the cell. In contrast, when an inducer is present, it binds to the allosteric site of each Lac repressor subunit, thereby inactivating the site that binds to the operator. The Lac repressor falls off the DNA, allowing the transcription of the structural genes of the *lac* operon to begin. The enzymes β-galactosidase, β-galactoside permease, and transacetylase now appear in the cell in a coordinated fashion. So, when lactose is present in the environment of a bacterial cell, it produces the enzymes needed to metabolize it. But when no lactose is present, resources are not wasted.

11.2 Discovery of the *lac* System: Negative Control

To study gene regulation, ideally we need three ingredients: a biochemical assay that lets us measure the amount of mRNA or expressed protein or both, reliable conditions in which the levels of expression differ in a wild-type genotype, and genetic mutations that perturb the levels of expression. In other words, we need a way of describing wild-type gene regulation and we need mutations that can disrupt the wild-type regulatory process. With these elements in hand, we can analyze the expression in mutant genotypes, treating the mutations singly and in combination, to unravel any kind of gene-regulation event. The classical application of this approach was used by Jacob and Monod, who performed the definitive studies of bacterial gene regulation.

Jacob and Monod used the lactose metabolism system of *E. coli* (see Figure 11-4) to genetically dissect the process of enzyme induction—that is, the appearance of a specific enzyme only in the presence of its substrates. This phenomenon had been observed in bacteria for many years, but how could a cell possibly "know" precisely which enzymes to synthesize? How could a particular substrate induce the appearance of a specific enzyme?

In the *lac* system, the presence of the inducer lactose causes cells to produce more than 1000 times as much of the enzyme β-galactosidase as they produced when grown in the absence of lactose. What role did the inducer play in the induction phenomenon? One idea was that the inducer was simply activating a precursor form of β-galactosidase that had accumulated in the cell. However, when Monod and co-workers followed the fate of radioactively labeled amino acids added to growing cells either before or after the addition of an inducer, he found that induction resulted in the synthesis of new enzyme molecules, as indicated by the presence of the radioactive amino acids in the enzymes. These new molecules could be detected as early as 3 minutes after the addition of an inducer. Additionally, withdrawal of the inducer brought about an abrupt halt in the synthesis of the new enzyme. Therefore, it became clear that the cell has a rapid and effective mechanism for turning gene expression on and off in response to environmental signals.

Genes controlled together

When Jacob and Monod induced β-galactosidase, they found that they also induced the enzyme permease, which is required to transport lactose into the cell. The analysis of mutants indicated that each enzyme was encoded by a different gene. The enzyme transacetylase (with a dispensable and as yet unknown function) also was induced together with β-galactosidase and permease and was later shown to be encoded by a separate gene. Therefore, Jacob and Monod could identify three coordinately controlled genes. Recombination mapping showed that the *Z*, *Y*, and *A* genes were very closely linked on the chromosome.

Genetic evidence for the operator and repressor

Now we come to the heart of Jacob's and Monod's work: How did they deduce the mechanisms of gene regulation in the *lac* system? Their approach was a classic genetic approach: to examine the physiological consequences of mutations. Thus, they induced mutations in the structural genes and regulatory elements of the *lac* operon. As we will see, the properties of mutations in these different components of the *lac* operon are quite different, providing important clues for Jacob and Monod.

Natural inducers, such as lactose, are not optimal for these experiments, because they are broken down by β-galactosidase. The inducer concentration decreases during the experiment, and so the measurements of enzyme induction become quite complicated. Instead, for such experiments, Jacob and Monod used synthetic inducers, such as isopropyl-β-D-thiogalactoside (IPTG; Figure 11-7). IPTG is not hydrolyzed by β-galactosidase.

Jacob and Monod found that several different classes of mutations can alter the expression of the structural genes of the *lac* operon. They were interested in assessing the interactions between the new alleles, such as which alleles exhibited dominance. But to perform such tests, one needs diploids, and bacteria are haploid. However, Jacob and Monod were able to produce bacteria that are partially diploid by inserting F′ factors (see Chapter 5) carrying the *lac* region of the genome. They could then create strains that were heterozygous for selected *lac* mutations. These **partial diploids** allowed Jacob and Monod to distinguish mutations in the regulatory DNA site (the *lac* operator) from mutations in the regulatory protein (the Lac repressor encoded by the *I* gene).

We begin by examining mutations that inactivate the structural genes for β-galactosidase and permease (designated *Z⁻* and *Y⁻*, respectively). The first thing that we learn is that *Z⁻* and *Y⁻* are recessive to their respective wild-type alleles (*Z⁺* and *Y⁺*). For example, strain 2 in Table 11-1 can be induced to synthesize β-galactosidase (like the wild-type haploid strain 1 in this table) even though it is heterozygous for mutant and wild-type *Z* alleles. This demonstrates that the *Z⁺* allele is dominant over its *Z⁻* counterpart.

Jacob and Monod first identified two classes of regulatory mutations, called *O^C* and *I⁻*. These were called **constitutive mutations** because they caused the *lac* operon structural genes to be expressed regardless of whether inducer was present. Jacob and Monod identified the existence of the operator on the basis of their analysis of the *O^C* mutations. These mutations make the operator incapable of binding to repressor; they damage the switch such that the operon is always "on" (Table 11-1, strain 3). Importantly, the constitutive effects of *O^C* mutations were

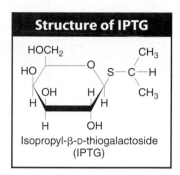

Figure 11-7 IPTG is an inducer of the lac operon.

Table 11-1 Synthesis of β-Galactosidase and Permease in Haploid and Heterozygous Diploid Operator Mutants

Strain	Genotype	β-*Galactosidase* (*Z*)		*Permease* (*Y*)		Conclusion
		Noninduced	Induced	Noninduced	Induced	
1	*O⁺ Z⁺ Y⁺*	−	+	−	+	Wild type is inducible
2	*O⁺ Z⁺ Y⁺/F′ O⁺ Z⁻ Y⁺*	−	+	−	+	*Z⁺* is dominant to *Z⁻*
3	*O^C Z⁺ Y⁺*	+	+	+	+	*O^C* is constitutive
4	*O⁺ Z⁻ Y⁺/F′ O^C Z⁺ Y⁻*	+	+	−	+	Operator is cis-acting

Note: Bacteria were grown in glycerol (no glucose present) with and without the inducer IPTG. The presence or absence of enzyme is indicated by + or −, respectively. All strains are *I⁺*.

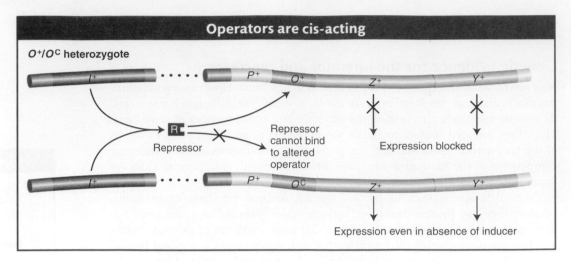

Figure 11-8 O^+/O^C heterozygotes demonstrate that operators are cis-acting. Because a repressor cannot bind to O^C operators, the lac structural genes linked to an O^C operator are expressed even in the absence of an inducer. However, the lac genes adjacent to an O^+ operator are still subject to repression.

GENETICS*PORTAL* **ANIMATED ART:** O^C *lac* operator mutations

restricted solely to those *lac* structural genes *on the same chromosome* as the O^C mutation. For this reason, the operator mutant was said to be **cis-acting,** as demonstrated by the phenotype of strain 4 in Table 11-1. Here, because the wild-type permease (Y^+) gene is cis to the wild-type operator, permease is expressed only when lactose or an analog is present. In contrast, the wild-type β-galactosidase (Z^+) gene is cis to the O^C mutant operator; hence, β-galactosidase is expressed constitutively. This unusual property of cis action suggested that the operator is a segment of DNA that influences only the expression of the structural genes linked to it (Figure 11-8). The operator thus acts simply as a protein-binding site and makes *no* gene product.

Jacob and Monod did comparable genetic tests with the I^- mutations (Table 11-2). A comparison of the inducible wild-type I^+ (strain 1) with I^- strains shows that I^- mutations are constitutive (strain 2). That is, they cause the structural genes to be expressed at all times. Strain 3 demonstrates that the inducible phenotype of I^+ is dominant over the constitutive phenotype of I^-. This finding showed Jacob and Monod that the amount of wild-type protein encoded by one copy of the gene is sufficient to regulate both copies of the operator in a diploid cell. Most significantly, strain 4 showed them that the I^+ gene product is **trans-acting,** meaning that the gene product can regulate *all* structural *lac* operon genes, whether residing on the same DNA molecule or on different ones (in cis or in trans, respectively). Unlike the operator, the I gene behaves like a standard

Table 11-2   Synthesis of β-Galactosidase and Permease in Haploid and Heterozygous Diploid Strains Carrying I^+ and I^-

| Strain | Genotype | β-Galactosidase (Z) | | Permease (Y) | | Conclusion |
		Noninduced	Induced	Noninduced	Induced	
1	$I^+ Z^+ Y^+$	−	+	−	+	I^+ is inducible
2	$I^- Z^+ Y^+$	+	+	+	+	I^- is constitutive
3	$I^+ Z^- Y^+/F'\ I^-\ Z^+ Y^+$	−	+	−	+	I^+ is dominant to I^-
4	$I^- Z^- Y^+/F'\ I^+ Z^+ Y^-$	−	+	−	+	I^+ is trans-acting

Note: Bacteria were grown in glycerol (no glucose present) and induced with IPTG. The presence of the maximal level of the enzyme is indicated by a plus sign; the absence or very low level of an enzyme is indicated by a minus sign. (All strains are O^+.)

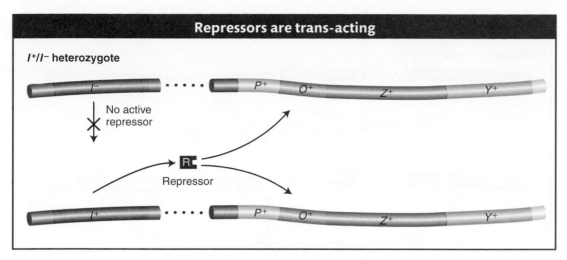

Figure 11-9 The recessive nature of I^- mutations demonstrates that the repressor is trans-acting. Although no active repressor is synthesized from the I^- gene, the wild-type (I^+) gene provides a functional repressor that binds to both operators in a diploid cell and blocks *lac* operon expression (in the absence of an inducer).
GENETICS P🔵RTAL **ANIMATED ART: I^- Lac repressor mutations**

protein-coding gene. The protein product of the *I* gene is able to diffuse throughout a cell and act on both operators in the partial diploid (Figure 11-9).

> **Message** Operator mutations reveal that such a site is cis-acting; that is, it regulates the expression of an adjacent transcription unit on the same DNA molecule. In contrast, mutations in the gene encoding a repressor protein reveal that this protein is trans-acting; that is, it can act on any copy of the target DNA site in the cell.

Genetic evidence for allostery

Finally, Jacob and Monod were able to demonstrate allostery through the analysis of another class of repressor mutations. Recall that the Lac repressor inhibits transcription of the *lac* operon in the absence of an inducer but permits transcription when the inducer is present. This regulation is accomplished through a second site on the repressor protein, the allosteric site, which binds to the inducer. When bound to the inducer, the repressor undergoes a change in overall structure such that its DNA-binding site can no longer function.

Jacob and Monod isolated another class of repressor mutation, called superrepressor (I^S) mutations. I^S mutations cause repression to persist even in the presence of an inducer (compare strain 2 in Table 11-3 with the inducible wild-type strain 1). Unlike I^- mutations, I^S mutations are dominant over I^+ (see Table 11-3, strain 3). This key observation led Jacob and Monod to speculate that I^S mutations

Table 11-3   Synthesis of β-Galactosidase and Permease by the Wild Type and by Strains Carrying Different Alleles of the *I* Gene

| Strain | Genotype | β-*Galactosidase* (Z) | | Permease (Y) | | Conclusion |
		Noninduced	Induced	Noninduced	Induced	
1	$I^+ Z^+ Y^+$	−	+	−	+	I^+ is inducible
2	$I^S Z^+ Y^+$	−	−	−	−	I^S is always repressed
3	$I^S Z^+ Y^+/F'\ I^+ Z^+ Y^+$	−	−	−	−	I^S is dominant to I^+

Note: Bacteria were grown in glycerol (no glucose present) with and without the inducer IPTG. Presence of the indicated enzyme is represented by +; absence or low levels, by −.

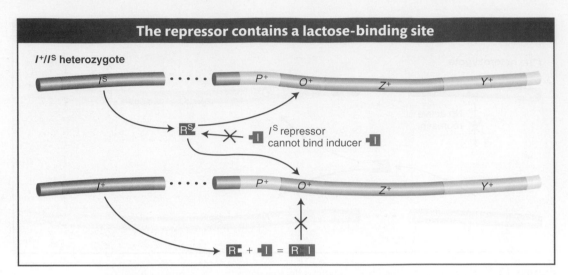

Figure 11-10 The dominance of I^S mutation is due to the inactivation of the allosteric site on the Lac repressor. In an I^S/I^- diploid cell, none of the lac structural genes are transcribed. The I^S repressor lacks a functional lactose-binding site (the allosteric site) and thus is not inactivated by an inducer. Therefore, even in the presence of an inducer, the I^S repressor binds irreversibly to all operators in a cell, thereby blocking transcription of the lac operon.
GENETICS**PORTAL**   ANIMATED ART: I^S Lac superrepressor mutations

alter the allosteric site so that it can no longer bind to an inducer. As a consequence, I^S-encoded repressor protein continually binds to the operator—preventing transcription of the *lac* operon even when the inducer is present in the cell. On this basis, we can see why I^S is dominant over I^+. Mutant I^S protein will bind to both operators in the cell, even in the presence of an inducer and regardless of the fact that I^+-encoded protein may be present in the same cell (Figure 11-10).

Genetic analysis of the *lac* promoter

Mutational analysis also demonstrated that an element essential for *lac* transcription is located between the gene for the repressor *I* and the operator site *O*. This element, termed the *promoter* (*P*), serves as the initiation site for transcription by RNA polymerase, as described in Chapter 8. There are two binding regions for RNA polymerase in a typical prokaryotic promoter, shown in Figure 11-11 as the two highly conserved regions at −35 and −10. Promoter mutations are cis-acting in that they affect the transcription of all adjacent structural genes in the operon. Like operators and other cis-acting elements, promoters are sites on the DNA molecule that are bound by proteins and themselves produce no protein product.

Molecular characterization of the Lac repressor and the *lac* operator

Walter Gilbert and Benno Müller-Hill provided a decisive demonstration of the *lac* system in 1966 by monitoring the binding of the radioactively labeled inducer IPTG to purified repressor protein. They first showed that the repressor consists of four identical subunits, and hence contains four IPTG- (and hence lactose-) binding sites. Second, they showed that, in the test tube, repressor protein binds to DNA containing the operator and comes off the DNA in the presence of IPTG. (A more detailed description of how the repressor and other DNA-binding proteins work is given later, at the end of Section 11.6.)

Gilbert and his co-workers showed that the repressor can protect specific bases in the operator from chemical reagents. This information allowed them to isolate

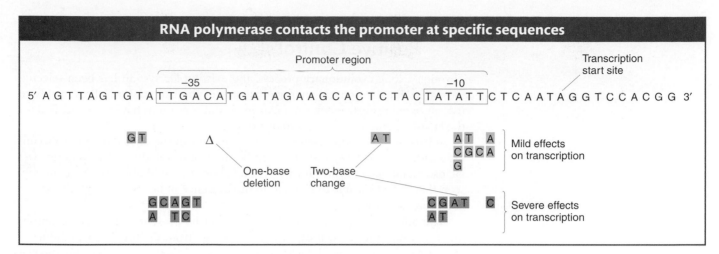

Figure 11-11 Specific DNA sequences are important for the efficient transcription of *E. coli* genes by RNA polymerase. The boxed sequences are highly conserved in all *E. coli* promoters, an indication of their role as contact sites on the DNA for RNA polymerase binding. Mutations in these regions have mild (gold) and severe (brown) effects on transcription. The mutations may be changes of single nucleotides or pairs of nucleotides or a deletion (Δ) may occur. [*From J. D. Watson, M. Gilman, J. Witkowski, and M. Zoller, Recombinant DNA, 2nd ed. Copyright 1992 by James D. Watson, Michael Gilman, Jan Witkowski, and Mark Zoller.*]

the DNA segment constituting the operator and to determine its sequence. They took operon DNA to which repressor was bound and treated it with the enzyme DNase, which breaks up DNA. They were able to recover short DNA strands that had been shielded from the enzyme activity by the repressor molecule. These short strands presumably constituted the operator sequence. The base sequence of each strand was determined, and each operator mutation was shown to be a change in the sequence (Figure 11-12). These results showed that the operator locus is a specific sequence of 17 to 25 nucleotides situated just before (5′ to) the structural Z gene. They also showed the incredible specificity of repressor–operator recognition, which can be disrupted by a single base substitution. When the sequence of bases in the *lac* mRNA (transcribed from the *lac* operon) was determined, the first 21 bases on the 5′ initiation end proved to be complementary to the operator sequence that Gilbert had determined, showing that the operator sequence is transcribed.

The results of these experiments provided crucial confirmation of the mechanism of repressor action formulated by Jacob and Monod.

Polar mutations

Some of the mutations that mapped to the Z and Y genes were found to be polar—that is, affecting genes "downstream" in the operon. For example, polar Z mutations resulted in null function not only for Z but also for Y and A. Polar mutations in Y also affected A but not Z. These polar mutations suggested to Jacob and Monod that the three genes were transcribed from one end as a unit. The polar mutations created stop codons that cause the ribosomes to fall off the transcript. This left a naked stretch of mRNA that was degraded, thereby inactivating downstream genes. (The normal stop and start codons that cause ribosomes to exit and enter the mRNA between the structural genes do not trigger this degradation.)

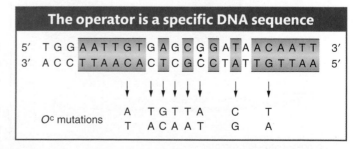

Figure 11-12 The DNA base sequence of the lactose operator and the base changes associated with eight O^C mutations. Regions of twofold rotational symmetry are indicated by color and by a dot at their axis of symmetry. [*From W. Gilbert, A. Maxam, and A. Mirzabekov, in N. O. Kjeldgaard and O. Malløe, eds., Control of Ribosome Synthesis. Academic Press, 1976. Used by permission of Munksgaard International Publishers, Ltd., Copenhagen.*]

11.3 Catabolite Repression of the *lac* Operon: Positive Control

Through a long evolutionary process, the existing *lac* system has been selected to operate for the optimal energy efficiency of the bacterial cell. Presumably to maximize energy efficiency, two environmental conditions have to be satisfied for the lactose metabolic enzymes to be expressed.

One condition is that lactose must be present in the environment. This condition makes sense, because it would be inefficient for the cell to produce the lactose metabolic enzymes if there is no lactose to metabolize. We have already seen that the cell is able to respond to the presence of lactose through the action of a repressor protein.

The other condition is that glucose cannot be present in the cell's environment. Because the cell can capture more energy from the breakdown of glucose than it can from the breakdown of other sugars, it is more efficient for the cell to metabolize glucose rather than lactose. Thus, mechanisms have evolved that prevent the cell from synthesizing the enzymes for lactose metabolism when both lactose and glucose are present together. The repression of the transcription of lactose-metabolizing genes in the presence of glucose is an example of **catabolite repression** (glucose is a breakdown product, or a *catabolite*, of lactose). The transcription of genes encoding proteins necessary for the metabolism of many different sugars is similarly repressed in the presence of glucose. We will see that catabolite repression works through an *activator protein*.

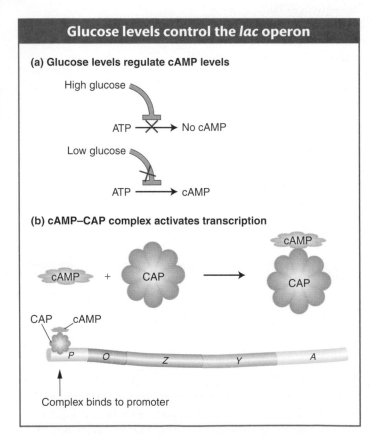

Glucose levels control the *lac* operon

(a) Glucose levels regulate cAMP levels

High glucose

ATP ⟶✕⟶ No cAMP

Low glucose

ATP ⟶ cAMP

(b) cAMP–CAP complex activates transcription

cAMP + CAP ⟶ cAMP / CAP

CAP   cAMP

P O Z Y A

Complex binds to promoter

Figure 11-13 Catabolite control of the *lac* operon. (a) Only under conditions of low glucose is cAMP (cyclic adenosine monophosphate) formed. (b) When cAMP is present, it forms a complex with CAP (catabolite activator protein) that activates transcription by binding to a region within the *lac* promoter.

The basics of *lac* catabolite repression: choosing the best sugar to metabolize

If both lactose and glucose are present, the synthesis of β-galactosidase is not induced until all the glucose has been metabolized. Thus, the cell conserves its energy by metabolizing any existing glucose before going through the energy-expensive process of creating new machinery to metabolize lactose. There are multiple mechanisms that bacteria have evolved to ensure the preferential use of a carbon source and optimal growth. One mechanism is to exclude the inducer lactose from the cell. A second mechanism is to regulate operon expression via catabolites.

The results of studies indicate that the breakdown product of glucose prevents activation of the *lac* operon by lactose—the catabolite repression just mentioned. The identity of this breakdown product is as yet unknown. However, the glucose breakdown product is known to modulate the level of an important cellular constituent—**cyclic adenosine monophosphate (cAMP).** When glucose is present in high concentrations, the cell's cAMP concentration is low. As the glucose concentration decreases, the cell's concentration of cAMP increases correspondingly. A high concentration of cAMP is necessary for activation of the *lac* operon. Mutants that cannot convert ATP into cAMP cannot be induced to produce β-galactosidase, because the concentration of cAMP is not great enough to activate the *lac* operon.

What is the role of cAMP in *lac* activation? A study of a different set of mutants provided an answer. These mutants make cAMP but cannot activate the Lac enzymes, because they lack yet another protein, called **catabolite activator**

protein (CAP), encoded by the *crp* gene. CAP binds to a specific DNA sequence of the *lac* operon (the CAP-binding site; see Figure 11-14b). The DNA-bound CAP is then able to interact physically with RNA polymerase and increases that enzyme's affinity for the *lac* promoter. By itself, CAP cannot bind to the CAP-binding site of the *lac* operon. However, by binding to cAMP, its allosteric effector, CAP is able to bind to the CAP-binding site and activate transcription by RNA polymerase. By inhibiting CAP when glucose is available, the catabolite-repression system ensures that the *lac* operon will be activated only when glucose is scarce (Figure 11-13).

> **Message** The *lac* operon has an added level of control so that the operon is inactive in the presence of glucose even if lactose also is present. An allosteric effector, cAMP, binds to the activator CAP to permit the induction of the *lac* operon. However, high concentrations of glucose catabolites produce low concentrations of cAMP, thus failing to produce cAMP–CAP and thereby failing to activate the *lac* operon.

The structures of target DNA sites

The DNA sequences to which the CAP–cAMP complex binds are now known. These sequences (Figure 11-14) are very different from the sequences to which

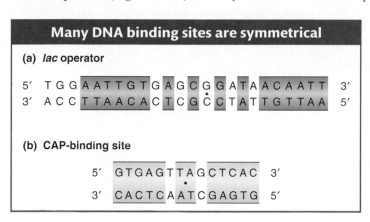

Figure 11-14 The DNA base sequences of (a) the *lac* operator, to which the Lac repressor binds, and (b) the CAP-binding site, to which the CAP–cAMP complex binds. Sequences exhibiting twofold rotational symmetry are indicated by the colored boxes and by a dot at the center point of symmetry. [(a) From W. Gilbert, A. Maxam, and A. Mirzabekov, in N. O. Kjeldgaard and O. Malløe, eds., Control of Ribosome Synthesis. Academic Press, 1976. Used by permission of Munksgaard International Publishers, Ltd., Copenhagen.]

the Lac repressor binds. These differences underlie the specificity of DNA binding by these very different regulatory proteins. One property that these sequences do have in common and that is common to many other DNA-binding sites is rotational twofold symmetry. In other words, if we rotate the DNA sequence shown in Figure 11-14 by 180 degrees within the plane of the page, the sequence of the highlighted bases of the binding sites will be identical. The highlighted bases are thought to constitute the important contact sites for protein–DNA interactions. This rotational symmetry corresponds to symmetries within the DNA-binding proteins, many of which are composed of two or four identical subunits. We will consider the structures of some DNA-binding proteins later in the chapter.

How does the binding of the cAMP-CAP complex to the operon further the binding of RNA polymerase to the *lac* promoter? In Figure 11-15, the DNA is shown as being bent when CAP is bound. This bending of DNA may aid the binding of RNA polymerase to the promoter. There is also evidence that CAP

Figure 11-15 (a) When CAP binds the promoter, it creates a bend greater than 90 degrees in the DNA. (b) Image derived from the structural analysis of the CAP–DNA complex. [(a) Redrawn from B. Gartenberg and D. M. Crothers, Nature 333, 1988, 824. (See H. N. Lie-Johnson et al., Cell 47, 1986, 995.) After H. Lodish, D. Baltimore, A. Berk, S. L. Zipursky, P. Matsudaira, and J. Darnell, Molecular Cell Biology, 3rd ed. Copyright 1995 by Scientific American Books. (b) From S. Schultz and T. A. Steitz.]

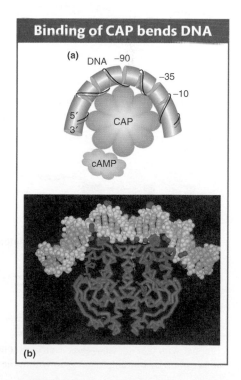

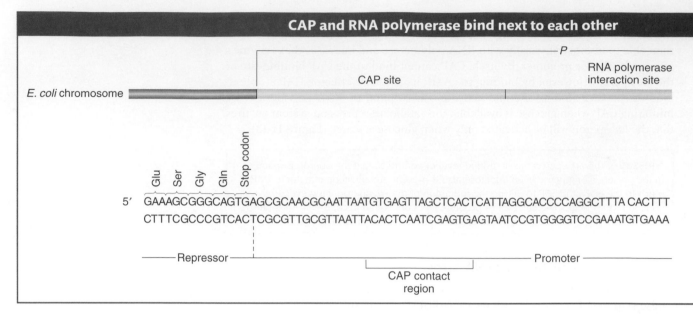

Figure 11-16 The control region of the *lac* operon. The base sequence and the genetic boundaries of the control region of the *lac* operon, with partial sequences for the structural genes. [*After R. C. Dickson, J. Abelson, W. M. Barnes, and W. S. Reznikoff, "Genetic Regulation: The Lac Control Region," Science 187, 1975, 27. Copyright 1975 by the American Association for the Advancement of Science.*]

makes direct contact with RNA polymerase that is important for the CAP activation effect. The base sequence shows that CAP and RNA polymerase bind directly adjacent to each other on the *lac* promoter (Figure 11-16).

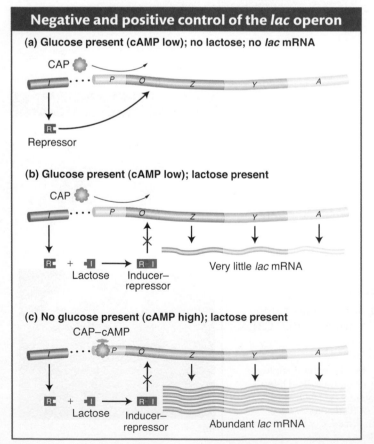

> **Message** Generalizing from the *lac* operon model, we can envision DNA as occupied by regulatory proteins binding to the operator sites that they control. The exact pattern of binding will depend on which genes are turned on or off and whether activators or repressors regulate particular operons.

A summary of the *lac* operon

We can now fit the CAP-cAMP- and RNA-polymerase-binding sites into the detailed model of the *lac* operon, as shown in Figure 11-17. The presence of glucose prevents lactose metabolism because a glucose breakdown product inhibits maintenance of the high cAMP levels necessary for formation of the CAP–cAMP complex, which in turn is required for the RNA polymerase to attach at the *lac* promoter site. Even when there is a shortage of glucose catabolites and CAP–cAMP forms, the mechanism for lactose

Figure 11-17 The *lac* operon is controlled jointly by the Lac repressor (negative control) and the catabolite activator protein (CAP) (positive control). Large amounts of mRNA are produced only when lactose is present to inactivate the repressor, and low glucose levels promote the formation of the CAP–cAMP complex, which positively regulates transcription. [*Redrawn from B. Gartenberg and D. M. Crothers, Nature 333, 1988, 824. (See H. N. Lie-Johnson et al., Cell 47, 1986, 995.) After H. Lodish, D. Baltimore, A. Berk, S. L. Zipursky, P. Matsudaira, and J. Darnell, Molecular Cell Biology, 3rd ed. Copyright 1995 by Scientific American Books.*]

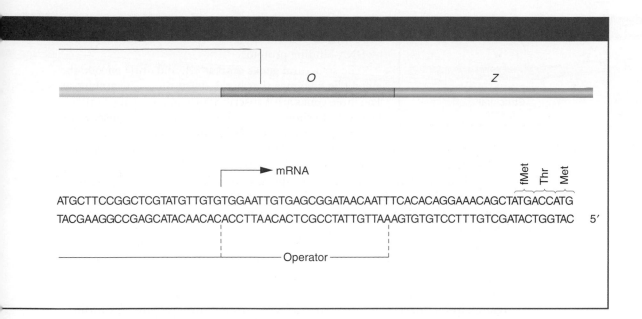

metabolism will be implemented only if lactose is present. This level of control is accomplished because lactose must bind to the repressor protein to remove it from the operator site and permit transcription of the *lac* operon. Thus, the cell conserves its energy and resources by producing the lactose-metabolizing enzymes only when they are both needed and useful.

Inducer–repressor control of the *lac* operon is an example of repression, or **negative control,** in which expression is normally blocked. In contrast, the CAP–cAMP system is an example of activation, or **positive control,** because it acts as a signal that activates expression—in this case, the activating signal is the interaction of the CAP–cAMP complex with the CAP-binding site on DNA. Figure 11-18 outlines these two basic types of control systems.

> **Message** The *lac* operon is a cluster of structural genes that specify enzymes taking part in lactose metabolism. These genes are controlled by the coordinated actions of cis-acting promoter and operator regions. The activity of these regions is, in turn, determined by repressor and activator molecules specified by separate regulator genes.

11.4 Dual Positive and Negative Control: The Arabinose Operon

As with the *lac* system, the control of transcription in bacteria is neither purely positive nor purely negative; rather, both positive and negative regulation may govern individual operons. The regulation of the arabinose operon provides an example in which a single DNA-binding

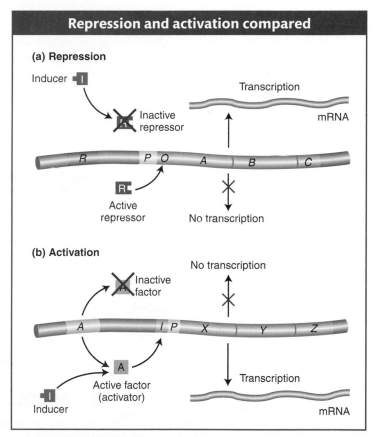

Figure 11-18 (a) In repression, an active repressor (encoded by the *R* gene in this example) blocks expression of the *A, B, C* operon by binding to an operator site (*O*). (b) In activation, a functional activator is required for gene expression. A nonfunctional activator results in no expression of genes *X, Y, Z*. Small molecules can convert a nonfunctional activator into a functional one that then binds to the control region of the operon, termed *I* in this case. The positions of both *O* and *I* with respect to promoter *P* in the two examples are arbitrarily drawn, inasmuch as their positions differ in different operons.

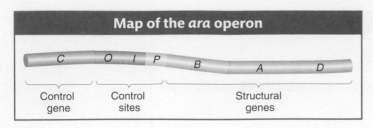

Figure 11-19 The *B, A,* and *D* genes together with the *I* and *O* sites constitute the *ara* operon. *O* is *araO* and *I* is *araI*.

protein may act as *either* a repressor *or* an activator—a twist on the general theme of transcriptional regulation by DNA-binding proteins.

The structural genes *araB, araA,* and *araD* encode the metabolic enzymes that break down the sugar arabinose. The three genes are transcribed in a unit as a single mRNA. Figure 11-19 shows a map of the *ara* operon. Transcription is activated at *araI*, the **initiator** region, which contains a binding site for an activator protein. The *araC* gene, which maps nearby, encodes an activator protein. When bound to arabinose, this protein binds to the *araI* site and activates transcription of the *ara* operon, perhaps by helping RNA polymerase bind to the promoter. In addition, the same CAP–cAMP catabolite repression system that prevents *lac* operon expression in the presence of glucose also prevents expression of the *ara* operon.

In the presence of arabinose, both the CAP–cAMP complex and the AraC–arabinose complex must bind to *araI* in order for RNA polymerase to bind to the promoter and transcribe the *ara* operon (Figure 11-20a). In the absence of arabinose, the AraC protein assumes a different conformation and represses the *ara* operon by binding both to *araI* and to a second distant site, *araO*, thereby forming a loop (Figure 11-20b) that prevents transcription. Thus, the AraC protein has two conformations, one that acts as an activator and another that acts as a repressor. The on/off switch of the operon is "thrown" by arabinose. The two conformations, dependent on whether the allosteric effector arabinose has bound to the protein, differ in their abilities to bind a specific target site in the *araO* region of the operon.

> **Message** Operon transcription can be regulated by both activation and repression. Operons regulating the metabolism of similar compounds, such as sugars, can be regulated in quite different ways.

Figure 11-20 Dual control of the *ara* operon. (a) In the presence of arabinose, the AraC protein binds to the *araI* region. The CAP–cAMP complex binds to a site adjacent to *araI*. This binding stimulates the transcription of the *araB, araA,* and *araD* genes. (b) In the absence of arabinose, the AraC protein binds to both the *araI* and the *araO* regions, forming a DNA loop. This binding prevents transcription of the *ara* operon.

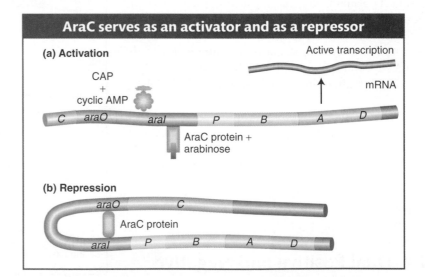

11.5 Metabolic Pathways and Additional Levels of Regulation: Attenuation

Coordinate control of genes in bacteria is widespread. In the preceding section, we looked at examples illustrating the regulation of pathways for the breakdown of specific sugars. In fact, most coordinated genes in bacteria are coordinated

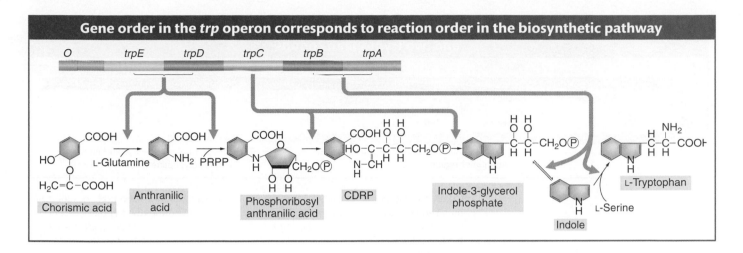

through operon mechanisms. In many pathways that synthesize essential molecules from simple inorganic building blocks, the genes that encode the enzymes are organized into operons, complete with multigenic mRNAs. Furthermore, in cases for which the sequence of catalytic activity is known, there is a remarkable congruence between the order of operon genes on the chromosome and the order in which their products act in the metabolic pathway. This congruence is strikingly illustrated by the organization of the tryptophan operon in *E. coli* (Figure 11-21). The tryptophan operon contains five genes (*trpE*, *trpD*, *trpC*, *trpB*, *trpA*) that encode enzymes that contribute to the synthesis of the amino acid tryptophan.

> **Message** In bacteria, genes that encode enzymes that are in the same metabolic pathways are generally organized into operons.

There are two mechanisms for regulating transcription of the tryptophan operon and some other operons functioning in amino acid biosynthesis. One provides global control of operon mRNA expression and the other provides fine-tuned control.

The level of *trp* operon gene expression is governed by the level of tryptophan. When tryptophan is absent from the growth medium, *trp* gene expression is high; when levels of tryptophan are high, the *trp* operon is repressed. One mechanism for controlling the transcription of the *trp* operon is similar to the mechanism that we have already seen controls the *lac* operon: a repressor protein binds an operator, preventing the initiation of transcription. This repressor is the Trp repressor, the product of the *trpR* gene. The Trp repressor binds tryptophan when adequate levels of the amino acid are present, and only after binding tryptophan will the Trp repressor bind to the operator and switch off transcription of the operon. This simple mechanism ensures that the cell does not waste energy producing tryptophan when the amino acid is sufficiently abundant. *E. coli* strains with mutations in *trpR* continue to express the *trp* mRNA and thus continue to produce tryptophan when the amino acid is abundant.

In studying these *trpR* mutant strains, Charles Yanofsky discovered that, when tryptophan was removed from the medium, the production of *trp* mRNA further increased several-fold. This finding was evidence that, in addition to the Trp repressor, a second control mechanism existed to negatively regulate transcription. This mechanism is called **attenuation** because mRNA production is normally *attenuated*, meaning "decreased," when tryptophan is plentiful. Unlike the other bacterial control mechanisms described thus far, attenuation acts at a step *after* transcription initiation.

Figure 11-21 The chromosomal order of genes in the *trp* operon of *E. coli* and the sequence of reactions catalyzed by the enzyme products of the *trp* structural genes. The products of genes *trpD* and *trpE* form a complex that catalyzes specific steps, as do the products of genes *trpB* and *trpA*. Tryptophan synthetase is a tetrameric enzyme formed by the products of *trpB* and *trpA*. It catalyzes a two-step process leading to the formation of tryptophan. Abbreviations: PRPP, phosphoribosylpyrophosphate; CDRP, 1-(*o*-carboxyphenylamino)-1-deoxyribulose 5-phosphate. [*After S. Tanemura and R. H. Bauerle, Genetics 95, 1980, 545.*]

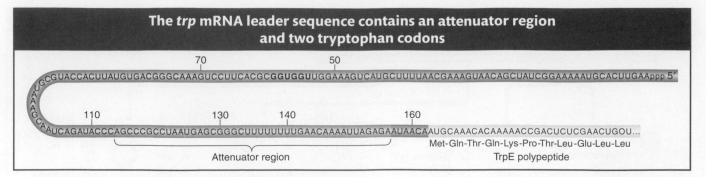

The *trp* mRNA leader sequence contains an attenuator region and two tryptophan codons

Figure 11-22 In the *trp* mRNA leader sequence, the attenuator region precedes the *trpE* coding sequence. Farther upstream, at bases 54 through 59, are the two tryptophan codons (shown in red) of the leader peptide. [*After G. S. Stent and R. Calendar, Molecular Genetics, 2nd ed. Copyright 1978 by W. H. Freeman and Company. Based on unpublished data provided by Charles Yanofsky.*]

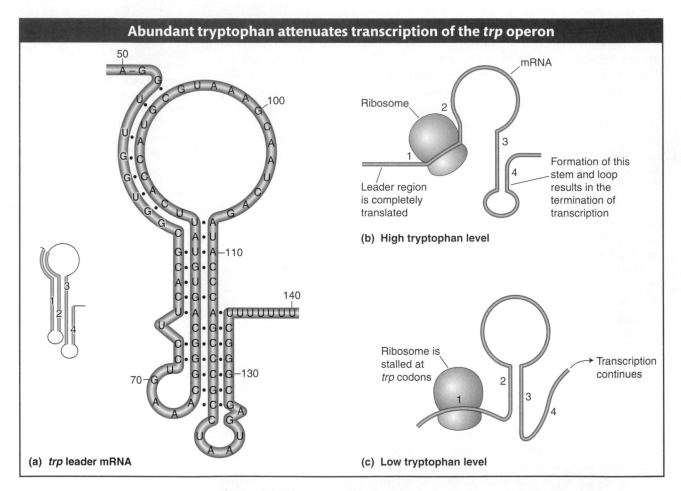

Abundant tryptophan attenuates transcription of the *trp* operon

(a) *trp* leader mRNA

(b) High tryptophan level

(c) Low tryptophan level

Figure 11-23 Model for attenuation in the *trp* operon. (a) Proposed secondary structures in the conformation of *trp* leader mRNA that favors termination of transcription. Four regions can base-pair to form three stem-and-loop structures. (b) When tryptophan is abundant, segment 1 of the *trp* mRNA is translated. Segment 2 enters the ribosome (although it is not translated), which enables segments 3 and 4 to base-pair. This base-paired region causes RNA polymerase to terminate transcription. (c) In contrast, when tryptophan is scarce, the ribosome is stalled at the codons of segment 1. Segment 2 interacts with segment 3 instead of being drawn into the ribosome, and so segments 3 and 4 cannot pair. Consequently, transcription continues. [*After D. L. Oxender, G. Zurawski, and C. Yanofsky, Proc. Natl. Acad. Sci. USA 76, 1979, 5524.*]

The mechanisms governing attenuation were discovered by identifying mutations that reduced or abolished attenuation. Strains with these mutations produce *trp* mRNA at maximal levels even in the presence of tryptophan. Yanofsky mapped the mutations to a region between the *trp* operator and the *trpE* gene; this region, termed the **leader sequence,** is at the 5′ end of the *trp* operon mRNA before the first codon of the *trpE* gene (Figure 11-22). The *trp* leader sequence is unusually long for a prokaryotic mRNA, 160 bases, and detailed analyses have revealed how a part of this sequence works as an **attenuator** that governs the further transcription of *trp* mRNA.

The key observations are that, in the absence of the TrpR repressor protein, the presence of tryptophan halts transcription after the first 140 bases or so, whereas, in the absence of tryptophan, transcription of the operon continues. The mechanism for terminating or continuing transcription consists of two key elements. First, the *trp* mRNA leader sequence encodes a short, 14-amino-acid peptide that includes two adjacent tryptophan codons. Tryptophan is one of the least abundant amino acids in proteins, and it is encoded by a single codon. This pair of tryptophan codons is therefore an unusual feature. Second, parts of the *trp* mRNA leader form stem-and-loop structures that are able to alternate between two conformations. One of these conformations favors the termination of transcription (Figure 11-23a).

The regulatory logic of the operon pivots on the abundance of tryptophan. When tryptophan is abundant, there is a sufficient supply of tRNA^{Trp} to allow translation of the 14-amino-acid peptide. Recall that transcription and translation in bacteria are coupled; so ribosomes can engage mRNA transcripts and initiate translations before transcription is complete. The engagement of the ribosome alters *trp* mRNA conformation to the form that favors termination of transcription (Figure 11-23b). However, when tryptophan is scarce, the ribosome is stalled at the tryptophan codons and transcription is able to continue (Figure 11-23c).

Other operons for enzymes in biosynthetic pathways have similar attenuation controls. One signature of amino acid biosynthesis operons is the presence of multiple codons for the amino acid being synthesized in a separate peptide encoded by the 5′ leader sequence. For instance, the *phe* operon has seven phenylalanine codons in a leader peptide and the *his* operon has seven tandem histidine codons in its leader peptide (Figure 11-24).

> **Message** A second level of regulation in amino acid biosynthesis operons is attenuation of transcription mediated by the abundance of the amino acid and translation of a leader peptide.

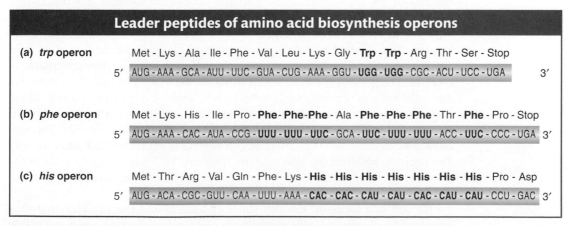

Leader peptides of amino acid biosynthesis operons

(a) *trp* operon

Met - Lys - Ala - Ile - Phe - Val - Leu - Lys - Gly - **Trp** - **Trp** - Arg - Thr - Ser - Stop

5′ AUG - AAA - GCA - AUU - UUC - GUA - CUG - AAA - GGU - **UGG** - **UGG** - CGC - ACU - UCC - UGA   3′

(b) *phe* operon

Met - Lys - His - Ile - Pro - **Phe-Phe-Phe** - Ala - **Phe-Phe-Phe** - Thr - **Phe** - Pro - Stop

5′ AUG - AAA - CAC - AUA - CCG - **UUU - UUU - UUC** - GCA - **UUC - UUU - UUU** - ACC - **UUC** - CCC - UGA 3′

(c) *his* operon

Met - Thr - Arg - Val - Gln - Phe - Lys - **His - His - His - His - His - His - His** - Pro - Asp

5′ AUG - ACA - CGC - GUU - CAA - UUU - AAA - **CAC - CAC - CAU - CAU - CAC - CAU - CAU** - CCU - GAC 3′

Figure 11-24 (a) The translated part of the *trp* leader region contains two consecutive tryptophan codons, (b) the *phe* leader sequence contains seven phenylalanine codons, and (c) the *his* leader sequence contains seven consecutive histidine codons.

11.6 Bacteriophage Life Cycles: More Regulators, Complex Operons

In that Paris movie theater, François Jacob had a flash of insight that the phenomenon of prophage induction might be closely analogous to the induction of β-galactosidase synthesis. He was right. Here, we are going to see how the life cycle of the bacteriophage λ is regulated. Although its regulation is more complex than that of individual operons, it is controlled by now-familiar modes of gene regulation.

Bacteriophage λ is a so-called temperate phage that has two alternative life cycles (Figure 11-25). When a normal bacterium is infected by a wild-type λ phage, two possible outcomes may follow: (1) the phage may replicate and eventually lyse the cell (the **lytic cycle**) or (2) the phage genome may be integrated into the bacterial chromosome as an inert prophage (the **lysogenic cycle**). In the lytic state, most of the phage's 71 genes are expressed at some point, whereas in the lysogenic state, most genes are inactive.

What decides which of these two pathways is taken? The physiological control of the decision between the lytic or lysogenic pathway depends on the resources available in the host bacterium. If resources are abundant, the lytic cycle is preferred because then there are sufficient nutrients to make many copies of the virus. If resources are limited, the lysogenic pathway is taken. The virus then remains present as a *prophage* until conditions improve. The inert prophage can be induced by ultraviolet light to enter the lytic cycle—the phenomenon studied by Jacob. The lytic and lysogenic states are characterized by very distinct programs of gene expression that must be regulated. Which alternative state is selected is determined by a complex genetic switch comprising several DNA-binding regulatory proteins and a set of operator sites.

Just as they were for the *lac* and other regulatory systems, genetic analyses of mutants were sources of crucial insights into the components and logic of the λ genetic switch. Jacob used simple phenotypic screens to isolate mutants that were defective in either the lytic or the lysogenic pathway. Mutants of each type could be recognized by the appearance of infected plaques on a lawn of bacteria. When wild-type phage particles are placed on a lawn of sensitive bacteria, clearings (plaques) appear where bacteria are infected and lysed, but these plaques are turbid because bacteria that are lysogenized grow within them. Mutant phages that form clear plaques are unable to lysogenize cells.

Such *clear* mutants (designated by *c*) turn out to be analogous to the *I* and *O* mutants of the *lac* system. These mutants were often isolated as temperature-sensitive mutants that had *clear* phenotypes at higher temperatures but wild-type phenotypes at lower temperatures. Three classes of mutants led to the identification of the key regulatory features of phage λ. In the first class, mutants for the *cI, cII,* and *cIII* genes form clear plaques; that is, they are unable to establish lysogeny. A second class of mutants were isolated that do not lysogenize cells but can replicate and enter the lytic cycle in a lysogenized cell. These mutants turn out to be analogous to the operator-constitutive mutants of the *lac* system. A third key mutant can lysogenize but is unable to lyse cells. The mutated gene in this case is the *cro* gene (for control of repressor and other things). The decision between the lytic and the lysogenic pathways hinges on the activity of the proteins encoded by the four genes *cI, cII, cIII,* and *cro,* three of which are DNA-binding proteins.

We will first focus on the two genes *cI* and *cro* and the proteins that they encode. The *cI* gene encodes a repressor, often referred to as λ repressor, that represses lytic growth and promotes lysogeny. The *cro* gene encodes a repressor

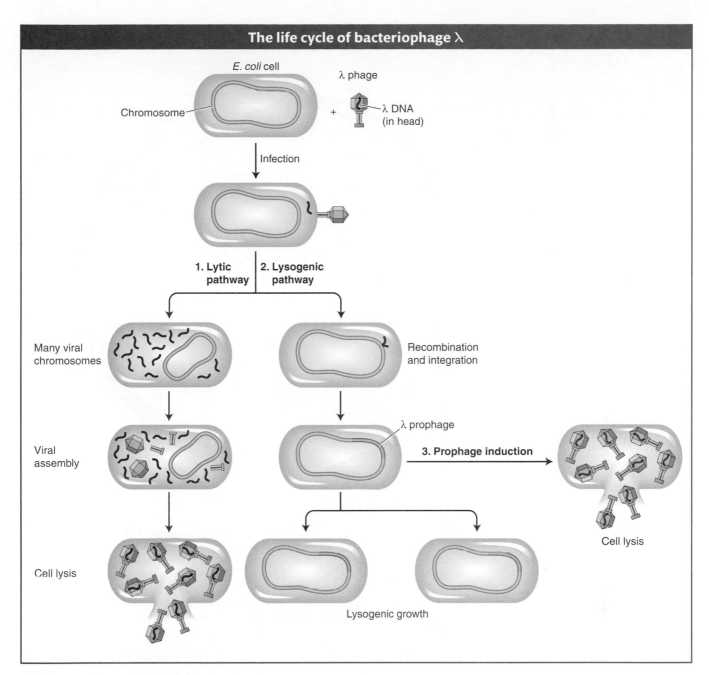

Figure 11-25 Whether bacteriophage l enters the lytic cycle immediately or enters the lysogenic pathway depends on the availability of resources. The lysogenic virus inserts its genome into the bacterial chromosome, where it remains quiescent until conditions are favorable.

that represses lysogeny, thereby permitting lytic growth. The genetic switch controlling the two λ phage life cycles has two states: in the lysogenic state, *cI* is on, but *cro* is off, and in the lytic cycle, *cro* is on, but *cI* is off. Therefore, λ repressor and Cro are in competition, and whichever repressor prevails determines the state of the switch and of the expression of the λ genome.

The race between λ repressor and Cro is initiated when phage λ infects a normal bacterium. The sequence of events in the race is critically determined by the organization of genes in the λ genome and of promoters and operators between the *cI* and the *cro* genes. The roughly 50-kb λ genome encodes proteins having

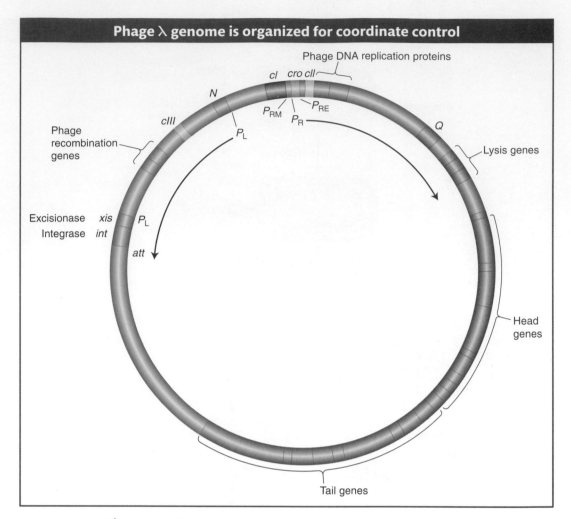

Figure 11-26 Map of phage λ in the circular form. The genes for recombination, integration and excision, replication, head and tail assembly, and cell lysis are clustered together and coordinately regulated. Transcription of the right side of the genome begins at P_R, and that of the leftward genes begins at P_L. Key regulatory interactions governing the lysogenic-versus-lytic decision take place at operators between the *cro* and the *cI* genes. [*After Fig. 16.25, p. 513, in J. Watson et al., Molecular Biology of the Gene, 5th edition. Copyright 2004 by Pearson Education, Inc. Reprinted by permission.*]

roles in DNA replication, recombination, assembly of the phage particle, and cell lysis (Figure 11-26). These proteins are expressed in a logical sequence such that copies of the genome are made first, these copies are then packaged into viral particles, and, finally, the host cell is lysed to release the virus and begin the infection of other host cells (see Figure 11-25). The order of viral gene expression flows from the initiation of transcription at two promoters, P_L and P_R (for *l*eftward and *r*ightward promoter with respect to the genetic map). On infection, RNA polymerase initiates transcription at both promoters. Looking at the genetic map (see Figure 11-26), we see that from P_R, *cro* is the first gene transcribed, and from P_L, *N* is the first gene transcribed.

The *N* gene encodes a positive regulator, but the mechanism of this protein differs from those of other regulators that we have considered thus far. Protein N works by enabling RNA polymerase to continue to transcribe through regions of DNA that would otherwise cause transcription to terminate. A regulatory protein that acts by preventing transcription termination is called an **antiterminator.** Thus, *N* allows the transcription of *cIII* and other genes to the left of *N*, as well as *cII* and other genes to the right of *cro*. The *cII* gene encodes an activator protein that binds to a site that promotes transcription leftward from a different promoter, P_{RE} (for promoter of *r*epressor *e*stablishment), which activates transcription of the *cI* gene. Recall that the *cI* gene encodes λ repressor, which will prevent lytic growth.

Before the expression of the rest of the viral genes takes place, a "decision" must be made—whether to continue with viral-gene expression and lyse the cell or to repress that pathway and lysogenize the cell. The decision whether to lyse or lysogenize a cell pivots on the activity of the cII protein. The cII protein is unstable because it is sensitive to bacterial proteases—enzymes that degrade proteins. These proteases respond to environmental conditions: they are more active when resources are abundant but less active when cells are starved.

Let's look at what happens to cII when resources are abundant and not abundant. When resources are abundant, cII is degraded and little λ repressor is produced. The genes transcribed from P_L and P_R continue to be expressed and the lytic cycle prevails. However, if resources are limited, cII is more active and more repressor is produced. In this case, the genes transcribed from P_L and P_R do not continue to be expressed and the lysogenic pathway is entered. The cII protein is also responsible for activating the transcription of *int,* a gene that encodes an additional protein required for lysogeny—an integrase required for the λ genome to integrate into the host chromosome. The cIII protein shields cII from degradation; so it, too, contributes to the lysogenic decision.

Let's briefly recap the sequence of events and the decision points in the bacteriophage λ life cycle:

On infection:

Host RNA polymerase transcribes the *cro* and *N* genes.

Then:

Antiterminator protein N enables transcription of the *cIII* gene and recombination genes (see Figure 11-26, left) and the *cII* gene and other genes (see Figure 11-26, right).

Then:

The cII protein, protected by the cIII protein, turns on *cI* and *int.*

Then:

If resources and proteases are abundant, cII is degraded, Cro represses *cI,* and the lytic cycle continues.

If resources and proteases are not abundant, cII is active, *cI* transcription proceeds at a high level, Int protein integrates the phage chromosome, and cI (λ repressor) shuts off all genes except itself.

Molecular anatomy of the genetic switch

To see how the decision is executed at the molecular level, let's turn to the activities of λ repressor and Cro. The O_R operator lies between the two genes encoding these proteins and contains three sites, O_{R1}, O_{R2}, and O_{R3}, that overlap two opposing promoters: P_R, which promotes transcription of lytic genes, and P_{RM} (for *r*epressor *m*aintenance), which directs transcription of the *cI* gene. Recall that the *cI* gene encodes the λ repressor. The three operator sites are similar but not identical in sequence, and although Cro and λ repressor can each bind to any one of the operators, they do so with different affinities: λ repressor binds to O_{R1} with the highest affinity, whereas Cro binds to O_{R3} with the highest affinity. The λ repressor's occupation of O_{R1} blocks transcription from P_R and thus blocks the transcription of genes for the lytic cycle. Cro's occupation of O_{R3} blocks transcription from P_{RM} and thus blocks maintenance of *cI* transcription. Hence,

Figure 11-27 The binding of λ repressor and Cro to operator sites. Lysogeny is promoted by λ repressor binding to O_{R1} and O_{R2}, which prevents transcription from P_R. On induction or in the lytic cycle, the binding of Cro to O_{R3} prevents transcription of the *cl* gene. [*After M. Ptashne and A. Gann, Genes and Signals, p. 30, Fig. 1-13. © Cold Spring Laboratory Press, 2002.*]

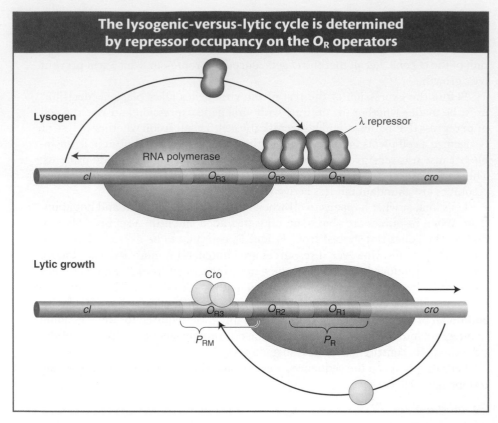

no λ repressor is produced, and transcription of genes for the lytic cycle can continue. The occupation of the operator sites therefore determines the lytic-versus-lysogenic patterns of λ gene expression (Figure 11-27).

After a lysogen has been established, it is generally stable. But the lysogen can be induced to enter the lytic cycle by various environmental changes. Ultraviolet light induces the expression of host genes. One of the host genes encodes a protein, RecA, that stimulates cleavage of the λ repressor, thus crippling maintenance of lysogeny and resulting in lytic growth. Prophage induction, just as Jacob and Monod surmised, requires the release of a repressor from DNA. The physiological role of ultraviolet light in lysogen induction makes sense in that this type of radiation damages host DNA and stresses the bacteria; the phage replicates and leaves the damaged, stressed cell for another host.

> **Message** The phage λ genetic switch illustrates how a few DNA-binding regulatory proteins, acting through a few sites, control the expression of a much larger number of genes in the virus in a "cascade" mechanism. Just as in the *lac, ara, trp,* and other systems, the alternative states of gene expression are determined by physiological signals.

Sequence-specific binding of regulatory proteins to DNA

How do λ repressor and Cro recognize different operators with different affinities? This question directs our attention to a fundamental principle in the control of gene transcription—the regulatory proteins bind to specific DNA sequences. For individual proteins to bind to certain sequences and not others requires specificity in the interactions between the side chains of the protein's amino acids and the chemical groups of DNA bases. Detailed structural studies of λ repressor, Cro, and other bacterial regulators have revealed how the three-dimensional structures of regulators and DNA interact and how the arrangement of particular amino acids enables them to recognize specific base sequences.

Crystallographic analysis has identified a common structural feature of the DNA-binding domains of λ and Cro. Both proteins make contact with DNA through a *helix-turn-helix* domain that consists of two α helices joined by a short flexible linker region (Figure 11-28). One helix, the recognition helix, fits into the major groove of DNA. In that position, amino acids on the helix's outer face are able to interact with chemical groups on the DNA bases. The specific amino acids in the recognition helix determine the affinity of a protein for a specific DNA sequence.

The recognition helices of the λ repressor and Cro have similar structures and some identical amino acid residues. Differences between the helices in key amino acid residues determine their DNA-binding properties. For example, in the λ repressor and Cro proteins, glutamine and serine side chains contact the same bases, but an alanine residue in the λ repressor and lysine and asparagine residues in the Cro protein impart different binding affinities for sequences in O_{R1} and O_{R3} (Figure 11-29).

The Lac and TrpR repressors, as well as the AraC activator and many other proteins, also bind to DNA through helix-turn-helix motifs of differing specificities, depending on the primary amino acid sequences of their recognition helices. In general, other domains of these proteins, such as those that bind their respective allosteric effectors, are dissimilar.

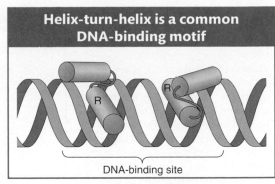

Figure 11-28 The binding of a helix-turn-helix motif to DNA. The purple cylinders are alpha helices. Many regulatory proteins bind as dimers to DNA. In each monomer, the recognition helix (R) makes contact with bases in the major groove of DNA. [*After Fig. 6-11, p. 494, in J. Watson et al., Molecular Biology of the Gene, 5th ed. Copyright 2004 by Pearson Education, Inc. Reprinted by permission.*]

> **Message** The biological specificity of gene regulation is due to the chemical specificity of amino acid–base interactions between individual regulatory proteins and discrete DNA sequences.

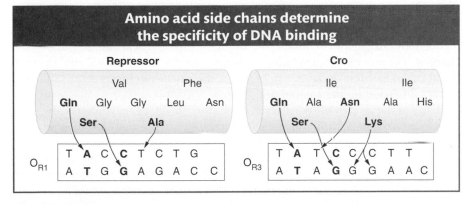

Figure 11-29 Interactions between amino acids and bases determine the specificity and affinity of DNA-binding proteins. The amino acid sequences of the recognition helices of the λ repressor and Cro proteins are shown. Interactions between the glutamine (Gln), serine (Ser), and alanine (Ala) residues of the λ repressor and bases in the O_R operator determine the strength of binding. Similarly, interactions between the glutamine, serine, asparagine (Asn), and lysine (Lys) residues of the Cro protein mediate binding to the O_{R3} operator. Each DNA sequence shown is that bound by an individual monomer of the respective repressor; it is half of the operator site occupied by the repressor dimer. [*From M. Ptashne, A Genetic Switch: Phage l and Higher Organisms, 2nd ed. Copyright 1992 by Cell Press and Blackwell Scientific.*]

11.7 Alternative Sigma Factors Regulate Large Sets of Genes

Thus far, we have seen how single switches can control the expression of single operons or two operons containing as many as a couple of dozen genes. Some physiological responses, such as the formation of spores in certain Gram-positive

Figure 11-30 Sporulation in *Bacillus subtilis* is regulated by cascades of σ factors. (a) In vegetative cells, σ^A and σ^H are active. On initiation of sporulation, σ^F is active in the forespore and σ^E is active in the mother cell. These σ factors are then superseded by σ^G and σ^K, respectively. The mother cell eventually lyses and releases the mature spore. (b) Factors σ^E and σ^F control the regulons of many genes (*ybaN*, and so forth, in this illustration). Three examples of the large number of promoters regulated by each σ factor are shown. Each σ factor has a distinct sequence-specific binding preference at the −35 and −10 sequences of target promoters. [*Based on data from P. Eichenberger et al., J. Mol. Biol. 327, 2003, 945–972; and S. Wang et al., J. Mol. Biol. 358, 2006, 16–37.*]

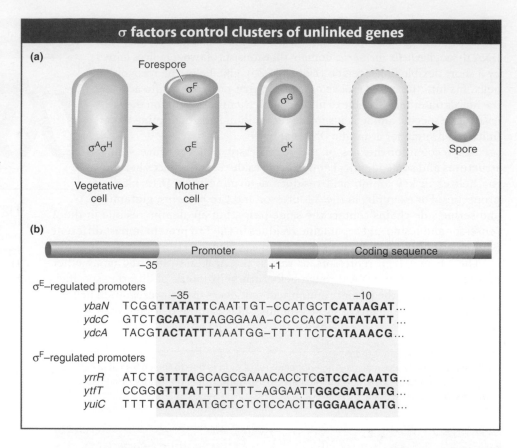

bacteria, require the coordinated expression of large sets of unlinked genes located throughout the genome to bring about dramatic physiological and morphological changes. The process of sporulation in *Bacillus subtilis* has been analyzed in great detail in the past few decades. Under stress, the bacterium forms spores that are remarkably resistant to heat and desiccation.

Early in the process of sporulation, the bacterium divides asymmetrically, generating two components of unequal size that have very different fates. The smaller compartment, the forespore, develops into the spore. The larger compartment, the mother cell, nurtures the developing spore and lyses when spore morphogenesis is complete to liberate the spore (Figure 11-30a). Genetic dissection of this process has entailed the isolation of many mutants that cannot sporulate. Detailed investigations have led to the characterization of several key regulatory proteins that directly regulate programs of gene expression that are specific to either the forespore or the mother cell. Four of these proteins are alternative σ factors.

Recall that transcription initiation in bacteria includes the binding of the σ subunit of RNA polymerase to the −35 and −11 regions of gene promoters. The σ factor disassociates from the complex when transcription begins and is recycled. In *B. subtilis*, two σ factors, σ^A and σ^H, are active in vegetative cells. During sporulation, a different σ factor, σ^F, becomes active in the forespore and activates a group of more than 40 genes. One gene activated by σ^F is a secreted protein that in turn triggers the proteolytic processing of the inactive precursor pro-σ^E, a distinct σ factor in the mother cell. The σ^E factor is required to activate sets of genes in the mother cell. Two additional σ factors, σ^K and σ^G, are subsequently activated in the mother cell and forespore, respectively (Figure 11-30a). The expression of distinct σ factors allows for the coordinated transcription of different sets of genes, or **regulons,** by a single RNA polymerase.

How do these alternative σ factors control different aspects of the sporulation process? The answers have become crystal clear with the advent of new approaches for characterizing the expression of all genes in a genome (see Section 14.6).

It is now possible to monitor the transcription of each *B. subtilis* gene during vegetative growth and spore formation and in different compartments of the spore. Several hundred genes have been identified in this fashion that are transcriptionally activated or repressed during spore formation.

How are the different sets of genes controlled by each σ factor? Each σ factor has different sequence-specific DNA-binding properties. The operons or individual genes regulated by particular σ factors have characteristic sequences in the −35 and −11 regions of their promoters that are bound by one σ factor and not others (Figure 11-30b). For example, σ^E binds to at least 121 promoters, within 34 operons and 87 individual genes, to regulate more than 250 genes, and σ^F binds to at least 36 promoters to regulate 48 genes.

> **Message** Sequential expression of alternative σ factors that recognize alternative promoter sequences provides for the coordinated expression of large numbers of independent operons and unlinked genes during the developmental program of sporulation.

Alternative σ factors also play important roles in the virulence of human pathogens. For example, bacteria of the genus *Clostridium* produce potent toxins that are responsible for severe diseases such as botulism, tetanus, and gangrene. Key toxin genes of *C. botulinum*, *C. tetani*, and *C. perfringens* have recently been discovered to be controlled by related, alternative σ factors that recognize similar sequences in the −35 and −10 regions of the toxin genes. Understanding the mechanisms of toxin-gene regulation may lead to new means of disease prevention and therapy.

Summary

Gene regulation is often mediated by proteins that react to environmental signals by raising or lowering the transcription rates of specific genes. The logic of this regulation is straightforward. In order for regulation to operate appropriately, the regulatory proteins have built-in sensors that continually monitor cellular conditions. The activities of these proteins would then depend on the right set of environmental conditions.

In bacteria and their viruses, the control of several structural genes may be coordinated by clustering the genes together into operons on the chromosome so that they are transcribed into multigenic mRNAs. Coordinated control simplifies the task for bacteria because one cluster of regulatory sites per operon is sufficient to regulate the expression of all the operon's genes. Alternatively, coordinate control can also be achieved through discrete σ factors that regulate dozens of independent promoters simultaneously.

In negative regulatory control, a repressor protein blocks transcription by binding to DNA at the operator site. Negative regulatory control is exemplified by the *lac* system. Negative regulation is one very straightforward way for the *lac* system to shut down genes in the absence of appropriate sugars in the environment. In positive regulatory control, protein factors are required to activate transcription. Some prokaryotic gene control, such as that for catabolite repression, operates through positive gene control.

Many regulatory proteins are members of families of proteins that have very similar DNA-binding motifs, such as the helix-turn-helix domain. Other parts of the proteins, such as their protein–protein interaction domains, tend to be less similar. The specificity of gene regulation depends on chemical interactions between the side chains of amino acids and chemical groups on DNA bases.

KEY TERMS

activator (p. 384)
allosteric effector (p. 385)
allosteric site (p. 385)
allosteric transition (p. 386)
antiterminator (p. 404)
attenuation (p. 399)
attenuator (p. 401)
catabolite activator protein (CAP) (pp. 394–395)
catabolite repression (p. 394)
cis-acting (p. 390)

constitutive mutation (p. 389)
coordinately controlled genes (p. 386)
cyclic adenosine monophosphate (cAMP) (p. 394)
DNA-binding domain (p. 385)
genetic switch (p. 384)
inducer (p. 387)
induction (p. 387)
initiator (p. 398)
leader sequence (p. 401)

lysogenic cycle (p. 402)
lytic cycle (p. 402)
negative control (p. 397)
operator (p. 384)
operon (p. 386)
partial diploid (p. 389)
positive control (p. 397)
promoter (p. 384)
regulon (p. 408)
repressor (p. 384)
trans-acting (p. 390)

SOLVED PROBLEMS

This set of four solved problems, which are similar to Problem 10 in the Basic Problems at the end of this chapter, is designed to test understanding of the operon model. Here, we are given several diploids and are asked to determine whether Z and Y gene products are made in the presence or absence of an inducer. Use a table similar to the one in Problem 10 as a basis for your answers, except that the column headings will be as follows:

	Z gene		Y gene	
Genotype	No inducer	Inducer	No inducer	Inducer

SOLVED PROBLEM 1.

$$\frac{I^- \, P^- \, O^C \, Z^+ \, Y^+}{I^+ \, P^+ \, O^+ \, Z^- \, Y^-}$$

Solution

One way to approach these problems is first to consider each chromosome separately and then to construct a diagram. The following illustration diagrams this diploid:

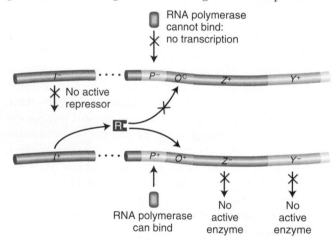

The first chromosome is P^-, and so transcription is blocked and no Lac enzyme can be synthesized from it. The second chromosome (P^+) can be transcribed, and thus transcription is repressible (O^+). However, the structural genes linked to the good promoter are defective; thus, no active Z product or Y product can be generated. The symbols to add to your table are "$-, -, -, -$."

SOLVED PROBLEM 2.

$$\frac{I^+ \, P^- \, O^+ \, Z^+ \, Y^+}{I^- \, P^+ \, O^+ \, Z^+ \, Y^-}$$

Solution

The first chromosome is P^-, and so no enzyme can be synthesized from it. The second chromosome is O^+, and so transcription is repressed by the repressor sup-

plied from the first chromosome, which can act in trans through the cytoplasm. However, only the Z gene from this chromosome is intact. Therefore, in the absence of an inducer, no enzyme is made; in the presence of an inducer, only the Z gene product, β-galactosidase, is generated. The symbols to add to the table are "$-, +, -, -$."

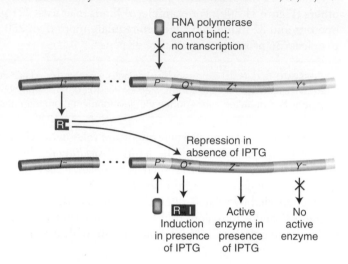

SOLVED PROBLEM 3.

$$\frac{I^+ \, P^+ \, O^C \, Z^- \, Y^+}{I^+ \, P^- \, O^+ \, Z^+ \, Y^-}$$

Solution

Because the second chromosome is P^-, we need consider only the first chromosome. This chromosome is O^C, and so enzyme is made in the absence of an inducer, although, because of the Z^- mutation, only active permease is generated. The entries in the table should be "$-, -, +, +$."

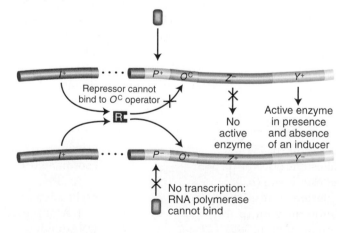

SOLVED PROBLEM 4.

$$\frac{I^S \, P^+ \, O^+ \, Z^+ \, Y^-}{I^- \, P^+ \, O^C \, Z^- \, Y^+}$$

Solution

In the presence of an I^S repressor, all wild-type operators are shut off, both with and without an inducer. Therefore, the first chromosome is unable to produce any enzyme. However, the second chromosome has an altered (O^C) operator and can produce enzyme in both the absence and the presence of an inducer. Only the Y gene is wild type on the O^C chromosome, and so only permease is produced constitutively. The entries in the table should be "−, −, +, +."

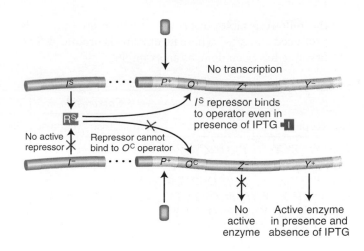

PROBLEMS

Most of the problems are also available for review/grading through the GENETICSP⊙RTAL www.yourgeneticsportal.com.

WORKING WITH THE FIGURES

1. Compare the structure of IPTG shown in Figure 11-7 with the structure of galactose shown in Figure 11-5. Why is IPTG bound by the lac repressor but not broken down by β-galactosidase?

2. Looking at Figure 11-9, why were partial diploids essential for establishing the trans-acting nature of the Lac repressor? Could one distinguish cis-acting from trans-acting genes in haploids?

3. Why do promoter mutations cluster at positions −10 and −35 as shown in Figure 11-11?

4. Looking at Figure 11-16, how large is the overlap between the operator and the *lac* transcription unit?

5. Examining Figure 11-21, what effect do you predict *trpA* mutations will have on tryptophan levels?

6. Examining Figure 11-21, what effect do you predict *trpA* mutations have on *trp* mRNA expression?

Basic Problems

7. Explain why I^- alleles in the *lac* system are normally recessive to I^+ alleles and why I^+ alleles are recessive to I^S alleles.

8. What do we mean when we say that O^C mutations in the *lac* system are cis-acting?

GENETICSP⊙RTAL **Unpacking the Problem** **9.** The genes shown in the following table are from the *lac* operon system of *E. coli*. The symbols *a*, *b*, and *c* represent the repressor (*I*) gene, the operator (*O*) region, and the structural gene (*Z*) for β-galactosidase, although not necessarily in that order. Furthermore, the order in which the symbols are written in the genotypes is not necessarily the actual sequence in the *lac* operon.

Activity (+) or inactivity (−) of Z gene

Genotype	Inducer absent	Inducer present
$a^- \, b^+ \, c^+$	+	+
$a^+ \, b^+ \, c^-$	+	+
$a^+ \, b^- \, c^-$	−	−
$a^+ \, b^- \, c^+ / a^- \, b^+ \, c^-$	+	+
$a^+ \, b^+ \, c^+ / a^- \, b^- \, c^-$	−	+
$a^+ \, b^+ \, c^- / a^- \, b^- \, c^+$	−	+
$a^- \, b^+ \, c^+ / a^+ \, b^- \, c^-$	+	+

a. State which symbol (*a*, *b*, or *c*) represents each of the *lac* genes *I*, *O*, and *Z*.

b. In the table, a superscript minus sign on a gene symbol merely indicates a mutant, but you know that some mutant behaviors in this system are given special mutant designations. Use the conventional gene symbols for the *lac* operon to designate each genotype in the table.

(Problem 9 is from J. Kuspira and G. W. Walker, *Genetics: Questions and Problems*. Copyright 1973 by McGraw-Hill.)

10. The map of the *lac* operon is

POZY

The promoter (*P*) region is the start site of transcription through the binding of the RNA polymerase molecule before actual mRNA production. Mutationally altered promoters (P^-) apparently cannot bind the RNA polymerase molecule. Certain predictions can be made about the effect of P^- mutations. Use your predictions and your knowledge of the lactose system to complete the following table. Insert a "+" where an enzyme is produced and a "−" where no enzyme is produced. The first one has been done as an example.

Genotype	β-*Galactosidase*		*Permease*	
	No lactose	Lactose	No lactose	Lactose
$I^+ P^+ O^+ Z^+ Y^+/I^+ P^+ O^+ Z^+ Y^+$	−	+	−	+
a. $I^- P^+ O^C Z^+ Y^-/ I^+ P^+ O^+ Z^- Y^+$				
b. $I^+ P^- O^C Z^- Y^+/I^- P^+ O^C Z^+ Y^-$				
c. $I^S P^+ O^+ Z^+ Y^-/I^+ P^+ O^+ Z^- Y^+$				
d. $I^S P^+ O^+ Z^+ Y^+/I^- P^+ O^+ Z^+ Y^+$				
e. $I^- P^+ O^C Z^+ Y^-/I^- P^+ O^+ Z^- Y^+$				
f. $I^- P^- O^+ Z^+ Y^+/I^- P^+ O^C Z^+ Y^-$				
g. $I^+ P^+ O^+ Z^- Y^+/I^- P^+ O^+ Z^+ Y^-$				

11. Explain the fundamental differences between negative control and positive control.

12. Mutants that are *lacY*⁻ retain the capacity to synthesize β-galactosidase. However, even though the *lacI* gene is still intact, β-galactosidase can no longer be induced by adding lactose to the medium. Explain.

13. What are the analogies between the mechanisms controlling the *lac* operon and those controlling phage λ genetic switches?

14. Compare the arrangement of cis-acting sites in the control regions of the *lac* operon and phage λ.

CHALLENGING PROBLEMS

15. An interesting mutation in *lacI* results in repressors with 110-fold increased binding to both operator and nonoperator DNA. These repressors display a "reverse" induction curve, allowing β-galactosidase synthesis in the absence of an inducer (IPTG) but partly repressing β-galactosidase expression in the presence of IPTG. How can you explain this? (Note that, when IPTG binds a repressor, it does not completely destroy operator affinity, but rather it reduces affinity 1100-fold. Additionally, as cells divide and new operators are generated by the synthesis of daughter strands, the repressor must find the new operators by searching along the DNA, rapidly binding to nonoperator sequences and dissociating from them.)

16. In *Neurospora*, all mutants affecting the enzymes carbamyl phosphate synthetase and aspartate transcarbamylase map at the *pyr-3* locus. If you induce *pyr-3* mutations by ICR-170 (a chemical mutagen), you find that either both enzyme functions are lacking or only the transcarbamylase function is lacking; in no case is the synthetase activity lacking when the transcarbamylase activity is present. (ICR-170 is assumed to induce frameshifts.) Interpret these results in regard to a possible operon.

17. Certain *lacI* mutations eliminate operator binding by the Lac repressor but do not affect the aggregation of subunits to make a tetramer, the active form of the repressor. These mutations are partly dominant over wild type. Can you explain the partly dominant *I*⁻ phenotype of the *I*⁻/*I*⁺ heterodiploids?

18. You are examining the regulation of the lactose operon in the bacterium *Escherichia coli*. You isolate seven new independent mutant strains that lack the products of all three structural genes. You suspect that some of these mutations are *lacI*ˢ mutations and that other mutations are alterations that prevent the binding of RNA polymerase to the promoter region. Using whatever haploid and partial diploid genotypes that you think are necessary, describe a set of genotypes that will permit you to distinguish between the *lacI* and *lacP* classes of uninducible mutations.

19. You are studying the properties of a new kind of regulatory mutation of the lactose operon. This mutation, called *S*, leads to the complete repression of the *lacZ*, *lacY*, and *lacA* genes, regardless of whether inducer (lactose) is present. The results of studies of this mutation in partial diploids demonstrate that this mutation is completely dominant over wild type. When you treat bacteria of the *S* mutant strain with a mutagen and select for mutant bacteria that can express the enzymes encoded by *lacZ*, *lacY*, and *lacA* genes in the presence of lactose, some of the mutations map to the *lac* operator region and others to the *lac* repressor gene. On the basis of your knowledge of the lactose operon, provide a molecular genetic explanation for all these properties of the *S* mutation. Include an explanation of the constitutive nature of the "reverse mutations."

20. The *trp* operon in *E. coli* encodes enzymes essential for the biosynthesis of tryptophan. The general mechanism for controlling the *trp* operon is similar to that observed with the *lac* operon: when the repressor binds

to the operator, transcription is prevented; when the repressor does not bind the operator, transcription proceeds. The regulation of the *trp* operon differs from the regulation of the *lac* operon in the following way: the enzymes encoded by the *trp* operon are not synthesized when tryptophan is present but rather when it is absent. In the *trp* operon, the repressor has two binding sites: one for DNA and the other for the effector molecule, tryptophan. The *trp* repressor must first bind to a molecule of tryptophan before it can bind effectively to the *trp* operator.

a. Draw a map of the tryptophan operon, indicating the promoter (*P*), the operator (*O*), and the first structural gene of the tryptophan operon (*trpA*). In your drawing, indicate where on the DNA the repressor protein binds when it is bound to tryptophan.

b. The *trpR* gene encodes the repressor; *trpO* is the operator; *trpA* encodes the enzyme tryptophan synthetase. A *trpR²* repressor cannot bind tryptophan, a *trpO²* operator cannot be bound by the repressor, and the enzyme encoded by a *trpA²* mutant gene is completely inactive. Do you expect to find active tryptophan synthetase in each of the following mutant strains when the cells are grown in the presence of tryptophan? In its absence?

i. $R^+ O^+ A^+$ (wild type)

ii. $R^- O^+ A^+/R^+ O^+ A^-$

iii. $R^+ O^- A^+/R^+ O^+ A^-$

21. The activity of the enzyme β-galactosidase produced by wild-type cells grown in media supplemented with different carbon sources is measured. In relative units, the following levels of activity are found:

Glucose	Lactose	Lactose + glucose
0	100	1

Predict the relative levels of β-galactosidase activity in cells grown under similar conditions when the cells are $lacI^-$, $lacI^S$, $lacO$, and crp^-.

22. A bacteriophage λ is found that is able to lysogenize its *E. coli* host at 30°C but not at 42°C. What genes may be mutant in this phage?

23. What would happen to the ability of bacteriophage λ to lyse a host cell if it acquired a mutation in the O_R binding site for the Cro protein? Why?

24. Contrast the effects of mutations in genes encoding sporulation-specific σ factors with mutations in the −35 and −10 regions of the promoters of genes in their regulons.

a. Would functional mutations in the σ-factor genes or in the individual promoters have the greater effect on sporulation?

b. On the basis of the sequences shown in Figure 11-30b, would you expect all point mutations in −35 or −10 regions to affect gene expression?

Regulation of Gene Expression in Eukaryotes

12

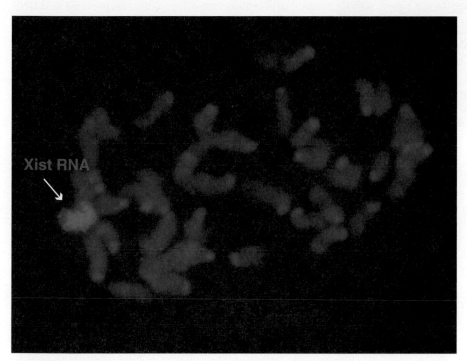

Xist RNA (labeled by a red rhodamine dye) covers one of the two copies of the X chromosome. The expression of Xist will lead to the chromosome's inactivation. The image is from an RNA fluorescent in situ hybridization (FISH) experiment performed on a metaphase chromosome spread taken from a female fibroblast cell line. [*Permission granted from Genes & Development 23, 2009, 1831–1842. J. T. Lee, Lessons from X Chromosome Inactivation. Cold Spring Harbor Laboratory Press. Photography by Jeannie Lee.*]

KEY QUESTIONS

- What are the molecular mechanisms of gene regulation in eukaryotes?

- How do eukaryotes generate many different patterns of gene expression with a limited number of regulatory proteins?

- What role does chromatin play in eukaryotic gene regulation?

- What are epigenetic marks and how do they influence gene expression?

- What roles do RNA molecules play in repressing eukaryotic gene expression?

OUTLINE

12.1 Transcriptional regulation in eukaryotes: an overview

12.2 Lessons from yeast: the GAL system

12.3 Dynamic chromatin

12.4 Short-term activation of genes in a chromatin environment

12.5 Long-term inactivation of genes in a chromatin environment

12.6 Gender-specific silencing of genes and whole chromosomes

12.7 Post-transcriptional gene repression by miRNAs

The cloning of Dolly, a sheep, was reported worldwide in 1996. Dolly developed from adult somatic nuclei that had been implanted into enucleated eggs (eggs with the nuclei removed). More recently, cows, pigs, mice, and other mammals have been cloned as well with the use of similar technology (Figure 12-1). The successful cloning of Dolly was a great surprise to the scientific community because the cloning of mammals from somatic cells was thought to be impossible. A reason for the initial skepticism was that the formation of male and female gametes (sperm and egg cells) was known to include sex-specific modifications to the respective genomes that resulted in sex-specific patterns of gene expression. As such, Dolly is symbolic of how far we have progressed in understanding aspects of eukaryotic gene regulation such as the global control of gene expression exemplified by gamete development. However, for every successful clone, including Dolly, there are many more, perhaps hundreds of embryos that fail to develop into viable progeny. The extremely high failure rate underscores how much remains to be deciphered about eukaryotic gene regulation.

415

The first cloned mammal

Figure 12-1 The first cloned mammal was a sheep named Dolly. [*PHOTOTAKE/Alamy.*]

In this chapter, we will examine gene regulation in eukaryotes. In many ways, our look at gene regulation will be a study of contrasts. In bacteria, you learned how the activities of genetic switches were often governed by single activator or repressor proteins and how the control of sets of genes was achieved by their organization into operons or by the activity of specific factors (see Chapter 11). Initial expectations were that eukaryotic gene expression would be regulated by similar means. In eukaryotes, however, most genes are not found in operons. Furthermore, we will see that the proteins and DNA sequences participating in eukaryotic gene regulation are more numerous. Often, many DNA-binding proteins act on a single switch, with many separate switches per gene, and the regulatory sequences of these switches are often located far from promoters. A key additional difference between bacteria and eukaryotes is that the access to eukaryotic gene promoters is restricted by chromatin. Gene regulation in eukaryotes requires the activity of large protein complexes that promote or restrict access to gene promoters by RNA polymerase. This chapter will provide an essential foundation for understanding the spatiotemporal regulation of gene expression that choreographs the process of development described in Chapter 13.

12.1 Transcriptional Regulation in Eukaryotes: An Overview

The biological properties of each eukaryotic cell type are largely determined by the proteins expressed within it. This constellation of expressed proteins determines much of the cell's architecture, its enzymatic activities, its interactions with its environment, and many other physiological properties. However, at any given time in a cell's life history, only a fraction of the RNAs and proteins encoded in its genome are expressed. At different times, the profile of expressed gene products can differ dramatically, both in regard to which proteins are expressed and at what levels. How are these specific profiles generated?

As one might expect, if the final product is a protein, regulation could be achieved by controlling the transcription of DNA into RNA or the translation of RNA into protein. In fact, gene regulation takes place at many levels, including at the mRNA level (through alterations in splicing or the stability of the mRNA) and after translation (by modifications of proteins). The varied ways that eukaryotic genes can be regulated have been divided into two general categories that reflect when they act: transcriptional gene regulation and post-transcriptional gene regulation. While the former will be the primary focus of this chapter, the latter is receiving increasing attention. In particular, the role of RNA to act post-transcriptionally to repress gene expression (called gene silencing; see Chapter 8) is one of the hottest areas of current research. Three of the RNA players that participate in silencing genes, miRNA, ncRNA, and siRNA, were introduced in Chapter 8. Later in this chapter we will explore the mechanisms of gene regulation mediated by miRNA and ncRNA.

Most regulation characterized to date takes place at the level of gene transcription; so, in this chapter, the primary focus is on transcriptional gene regulation. The basic mechanism at work is that molecular signals from outside or inside the cell lead to the binding of regulatory proteins to specific DNA sites outside of protein-encoding regions, and the binding of these proteins modulates the rate of transcription. These proteins may directly or indirectly assist RNA polymerase in binding to its transcription initiation site—the *promoter*—or they may repress transcription by preventing the binding of RNA polymerase.

Although bacteria and eukaryotes have much of the logic of gene regulation in common, there are some fundamental differences in the underlying mechanisms and machinery. Both use sequence-specific DNA-binding proteins to modulate the level of transcription. However, eukaryotic genomes are bigger, and their range of properties is larger than that of bacteria. Inevitably, the regulation of eukaryotic genomes is more complex, requiring more types of regulatory proteins and more types of interactions with the adjacent regulatory regions in DNA. The most important difference is that eukaryotic DNA is packaged into *nucleosomes,* forming *chromatin,* whereas bacterial DNA lacks nucleosomes. In eukaryotes, chromatin structure is dynamic and is an essential ingredient in gene regulation.

In general, the ground state of a bacterial gene is "on." Thus, RNA polymerase can usually bind to a promoter when no other regulatory proteins are around to bind to the DNA. In bacteria, transcription initiation is prevented or reduced if the binding of RNA polymerase is blocked, usually through the binding of a repressor regulatory protein. Activator regulatory proteins increase the binding of RNA polymerase to promoters where a little help is needed. In contrast, the ground state in eukaryotes is "off." Therefore, the transcriptional machinery (including RNA polymerase II and associated general transcription factors) cannot bind to the promoter in the absence of other regulatory proteins (Figure 12-2). In many cases, the binding of the transcriptional apparatus is not possible, because nucleosomes are positioned to block the promoter. Thus, chromatin structure usually has to be changed to activate eukaryotic transcription. Those changes generally depend on the binding of sequence-specific DNA-binding regulatory proteins.

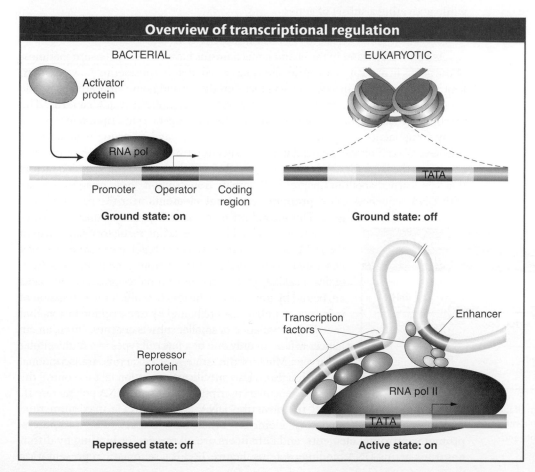

Figure 12-2 In bacteria, RNA polymerase can usually begin transcription unless a repressor protein blocks it. In eukaryotes, however, the packaging of DNA with nucleosomes prevents transcription unless other regulatory proteins are present. These regulatory proteins expose promoter sequences by altering nucleosome density or position. They may also recruit RNA polymerase II more directly through binding.

The structure of chromatin around activated or repressed genes within cells can be quite stable and inherited by daughter cells. The inheritance of chromatin states is a form of inheritance that does not directly entail DNA sequence.

The unique features of eukaryotic transcriptional regulation are the focus of this chapter. Some differences from transcriptional regulation in bacteria were already noted in Chapter 8:

1. In bacteria, all genes are transcribed into RNA by the same RNA polymerase, whereas three RNA polymerases function in eukaryotes. RNA polymerase II, which transcribes DNA into mRNA, was the focus of Chapter 8 and will be the only polymerase discussed in this chapter.

2. RNA transcripts are extensively processed during transcription in eukaryotes; the 5′ and 3′ ends are modified and introns are spliced out.

3. RNA polymerase II is much larger and more complex than its bacterial counterpart. One reason it is more complex is that RNA polymerase II must synthesize RNA *and* coordinate the special processing events unique to eukaryotes.

Multicellular eukaryotes may have as many as 25,000 genes, several-fold more than the average bacterium. Moreover, patterns of eukaryotic gene expression can be extraordinarily complex. That is, there is great variation among genes in when a gene is on (transcribed) or off (not transcribed) and in how much transcript needs to be made. For example, one gene may be transcribed only during one stage of development and another only in the presence of a viral infection. Finally, the majority of the genes in a eukaryotic cell are off at any one time. On the basis of these considerations alone, eukaryotic gene regulation must be able to

1. Ensure that the expression of most genes in the genome is off at any one time while activating a subset of genes

2. Generate thousands of patterns of gene expression

As you will see later in the chapter, mechanisms have evolved to ensure that most of the genes in a eukaryotic cell are not transcribed. Before considering how genes are kept transcriptionally inactive, we will focus on the second point: How are eukaryotic genes able to exhibit an enormous number and diversity of expression patterns? The machinery required for generating so many patterns of gene transcription in vivo has many components, including both regulatory proteins and cis-acting regulatory sequences. We can divide the regulatory proteins into two sets. The first set of proteins comprises the large RNA polymerase II complex and the general transcription factors that you learned about in Chapter 8. To initiate transcription, these proteins interact with DNA sequences called **promoter-proximal elements** near the promoter of a gene. The second set of regulatory proteins consists of transcription factors that bind to cis-acting regulatory sequences in the DNA called **enhancers.** These regulatory sequences may be located a considerable distance from gene promoters. Generally speaking, promoters and promoter-proximal elements are bound by transcription factors that affect the expression of many genes. Enhancers are bound by transcription factors that control the regulation of smaller subsets of genes. Often, an enhancer will act in only one or a few cell types in a multicellular eukaryote. Much of the strategy of eukaryotic transcriptional control hinges on how specific transcription factors control the access of general transcription factors and RNA polymerase II.

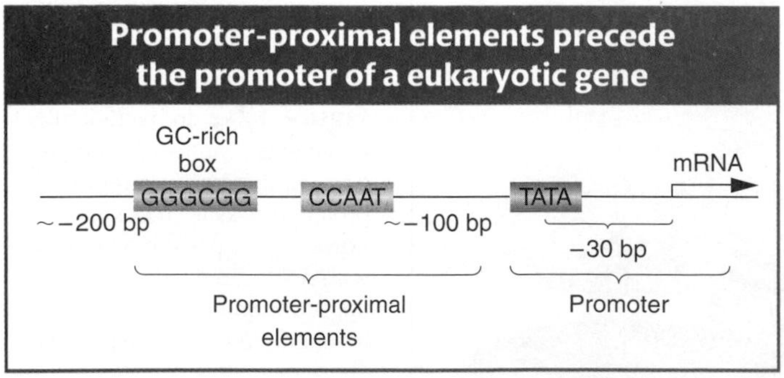

Promoter-proximal elements precede the promoter of a eukaryotic gene

Figure 12-3 The region upstream of the transcription start site in higher eukaryotes contains promoter-proximal elements and the promoter.

For RNA polymerase II to transcribe DNA into RNA at a maximum rate, multiple cis-acting regulatory elements must play a part. The promoters, promoter-proximal elements, and enhancers are all targets for binding by different trans-acting DNA binding proteins. Figure 12-3 is a schematic representation

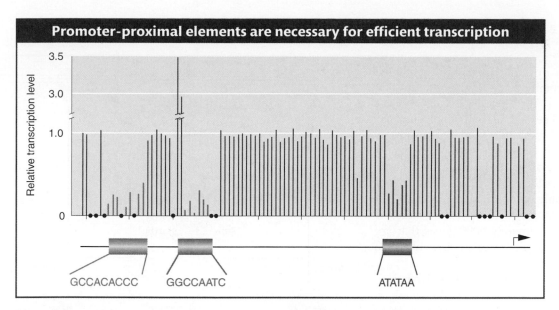

Figure 12-4 Point mutations in the promoter and promoter-proximal elements hinder transcription of the β-globin gene. Point mutations throughout the promoter region were analyzed for their effects on transcription rates. The height of each line represents the transcription level relative to a wild-type promoter or promoter-proximal element (1.0). Only the base substitutions that lie within the three elements shown change the level of transcription. Positions with black dots were not tested. [*From T. Maniatis, S. Goodbourn, and J. A. Fischer, "Regulation of Inducible and Tissue-Specific Gene Expression," Science 236, 1987, 1237.*]

of the promoter and promoter-proximal sequence elements. The binding of RNA polymerase II to the promoter does not produce efficient transcription by itself. Transcription requires the binding of general transcription factors to additional promoter-proximal elements that are commonly found within 100 bp of the transcription initiation site of many (but not all) genes. One of these elements is the CCAAT box, and often another is a GC-rich segment farther upstream. The general transcription factors that bind to the promoter-proximal elements are expressed in most cells, and so they are available to initiate transcription at any time. Mutations in these sites can have a dramatic effect on transcription, demonstrating how important they are. If these sequence elements are mutated, the level of transcription is generally reduced, as shown in Figure 12-4.

To modulate transcription, regulatory proteins possess one or more of the following *functional domains:*

1. A domain that recognizes a DNA regulatory sequence (the protein's DNA-binding site)

2. A domain that interacts with one or more proteins of the transcriptional apparatus (RNA polymerase or a protein associated with RNA polymerase)

3. A domain that interacts with proteins bound to nearby regulatory sequences on DNA such that they can act cooperatively to regulate transcription

4. A domain that influences chromatin condensation either directly or indirectly

5. A domain that acts as a sensor of physiological conditions within the cell

Eukaryotic gene regulatory mechanisms have been discovered through both biochemical and genetic approaches. The latter has been advanced in particular by studies of the single-celled yeast *Saccharomyces cerevisiae* (see the Model Organism box on page 420). Several decades of research have been a source of many insights into general principles of how eukaryotic transcriptional regulatory proteins work

Model Organism *Yeast*

Saccharomyces cerevisiae, or budding yeast, has emerged in recent years as the premier eukaryotic genetic system. Humans have grown yeast for centuries because it is an essential component of beer, bread, and wine. Yeast has many features that make it an ideal model organism. As a unicellular eukaryote, it can be grown on agar plates and, with yeast's life cycle of just 90 minutes, large quantities of it can be cultured in liquid media. It has a very compact genome with only about 12 mega–base pairs of DNA (compared with almost 3000 mega–base pairs for humans) containing approximately 6000 genes that are distributed on 16 chromosomes. It was the first eukaryote to have its genome sequenced.

The yeast life cycle makes it very versatile for laboratory studies. Cells can be grown as either diploid or haploid. In both cases, the mother cell produces a bud containing an identical daughter cell. Diploid cells either continue to grow by budding or are induced to undergo meiosis, which produces four haploid spores held together in an *ascus* (also called a *tetrad*). Haploid spores of opposite mating type (**a** or **α**) will fuse and form a diploid. Spores of the same mating type will continue growth by budding.

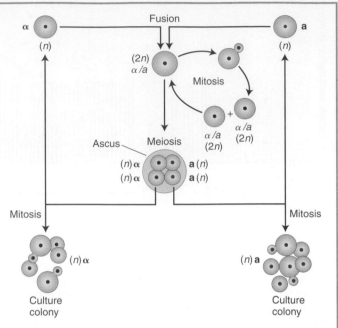

The life cycle of baker's yeast. The nuclear alleles *MATa* and *MATα* determine mating type.

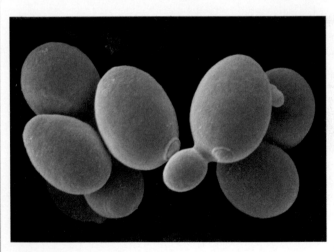

Electron micrograph of budding yeast cells. [*SciMAT/Photo Researchers.*]

Yeast has been called the *E. coli* of eukaryotes because of the ease of forward and reverse mutant analysis. To isolate mutants by using a forward genetic approach, haploid cells are mutagenized (with X rays, for example) and screened on plates for mutant phenotypes. This procedure is usually done by first plating cells on a rich medium on which all cells grow and by copying, or *replica plating*, the colonies from this master plate onto replica plates containing selective media or special growth conditions. (See also Chapter 16.) For example, temperature-sensitive mutants will grow on the master plate at the permissive temperature but not on a replica plate at a restrictive temperature. Comparison of the colonies on the master and replica plates will reveal the temperature-sensitive mutants. Using reverse genetics, scientists can also replace any yeast gene (of known or unknown function) with a mutant version (synthesized in a test tube) to understand the nature of the gene product.

and how different cell types are generated. We'll examine two yeast gene regulatory systems in detail: the first concerns the galactose-utilization pathway; the second is the control of mating type.

12.2 Lessons from Yeast: The GAL System

To make use of extracellular galactose, yeast imports the sugar and converts it into a form of glucose that can be metabolized. Several genes—*GAL1, GAL2, GAL7,* and *GAL10*—in the yeast genome encode enzymes that catalyze steps in the biochemical pathway that converts galactose into glucose (Figure 12-5). Three additional genes—*GAL3, GAL4,* and *GAL80*—encode proteins that regulate the expression of the enzyme genes. Just as in the *lac* system, the abundance of the sugar determines the level of gene expression in the biochemical pathway. In yeast cells growing in media lacking galactose, the *GAL* genes are largely silent. But, in the presence of galactose (and the absence of glucose), the *GAL* genes are induced. Just as for the *lac* operon, genetic and molecular analyses of mutants have been key to understanding how the expression of the genes in the galactose pathway is controlled.

The key regulator of *GAL* gene expression is the Gal4 protein, a sequence-specific DNA-binding protein. Gal4 is perhaps the best-studied transcriptional regulatory protein in eukaryotes. The detailed dissection of its regulation and activity has been a source of several key insights into the control of transcription in eukaryotes.

Gal4 regulates multiple genes through upstream activation sequences

In the presence of galactose, the expression levels of the *GAL1, GAL2, GAL7,* and *GAL10* genes are 1000-fold or more higher than in its absence. In *GAL4* mutants, however, they remain silent. Each of these four genes has two or more Gal4-binding sites located 5′ (upstream) of its promoter. Consider the *GAL10* and *GAL1* genes, which are adjacent to each other and transcribed in opposite directions. Between the *GAL1* transcription start site and the *GAL10* transcription start site is a single 118-bp region that contains four Gal4-binding sites (Figure 12-6). Each Gal4-binding site is 17 base pairs long and is bound by one Gal4 protein dimer. There are two Gal4-binding sites upstream of the *GAL2* gene as well, and another two upstream of the *GAL7* gene. These binding sites are required for gene activation in vivo. If they are deleted, the genes are silent, even in the presence of galactose. These regulatory sequences are enhancers. The presence of enhancers located at a considerable linear distance from a eukaryotic gene's promoter is typical. Because the *GAL4* enhancers are located upstream of the genes they regulate, they are also called **upstream activation sequences (UAS)**.

> **Message** The binding of sequence-specific DNA-binding proteins to regions outside the promoters of target genes is a common feature of eukaryotic transcriptional regulation.

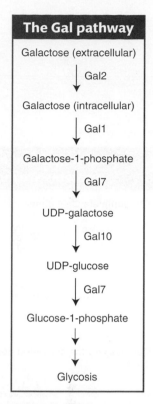

The Gal pathway

Galactose (extracellular)
⬇ Gal2
Galactose (intracellular)
⬇ Gal1
Galactose-1-phosphate
⬇ Gal7
UDP-galactose
⬇ Gal10
UDP-glucose
⬇ Gal7
Glucose-1-phosphate
⬇
⬇
Glycosis

Figure 12-5 Galactose is converted into glucose-1-phosphate in a series of steps. These steps are catalyzed by enzymes (Gal1, and so forth) encoded by the structural genes *GAL1, GAL2, GAL7,* and *GAL10.*

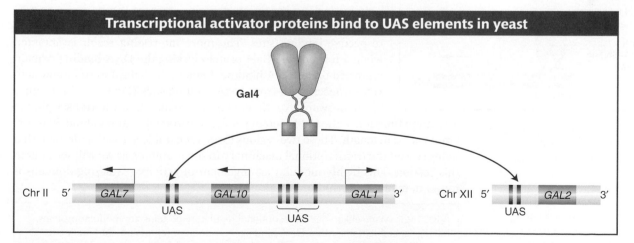

Transcriptional activator proteins bind to UAS elements in yeast

Gal4

Chr II 5′ *GAL7*   UAS   *GAL10*   UAS   *GAL1* 3′   Chr XII 5′ UAS *GAL2* 3′

Figure 12-6 The Gal4 protein activates target genes through upstream-activating-sequence (UAS) elements. The Gal4 protein has two functional domains: a DNA-binding domain (pink square) and an activation domain (orange oval). The protein binds to specific sequences upstream of the promoters of Gal-pathway genes. Some of the *GAL* genes are adjacent (*GAL1, GAL10*), whereas others are on different chromosomes. The *GAL1* UAS element contains four Gal4-binding sites.

The Gal4 protein has separable DNA-binding and activation domains

After Gal4 is bound to the UAS element, how is gene expression induced? A distinct domain of the Gal4 protein, the **activation domain,** is required for regulatory activity. Thus, the Gal4 protein has at least two domains: one for DNA binding and another for activating transcription. A similar modular organization has been found to be a common feature of other DNA-binding transcription factors as well.

The modular organization of the Gal4 protein was demonstrated in a series of simple, elegant experiments. The strategy was to test the DNA binding and gene activation of mutant forms of the protein in which parts had been either deleted or fused to other proteins. By this means, investigators could determine whether a part of the protein was necessary for a particular function. To carry out these studies, experimenters needed a simple means of assaying the expression of the enzymes encoded by the *GAL* genes.

The expression of *GAL* genes and other targets of transcription factors is typically monitored by using a **reporter gene** whose level of expression is easily measured. In reporter-gene constructs, the reporter gene is linked to the regulatory sequences that govern the expression of the gene being investigated. The expression of the reporter gene reflects the activity of the regulatory element being investigated. Often, the reporter gene is the *lacZ* gene of *E. coli*. LacZ is an effective reporter gene because the products of its activity are easily measured. Another common reporter gene is the gene that encodes the green fluorescent protein (GFP) of jellyfish. As its name suggests, the concentration of reporter protein is easily measured by the amount of light that it emits. To investigate the control of *GAL* gene expression, the coding region of one of these reporter genes and a promoter are placed downstream of a UAS element from a *GAL* gene. Reporter expression is then a readout of Gal4 activity in cells.

Let's see what happens when a form of the Gal4 protein lacking the activation domain is expressed in yeast. In this case, the binding sites of the UAS element are occupied, but no transcription is stimulated (Figure 12-7b). The same is true when other regulatory proteins lacking activation domains, such as the bacterial repressor LexA, are expressed in cells bearing reporter genes with their respective binding sites. The more interesting result is obtained when a form of the Gal4 protein lacking the DNA-binding domain is grafted to the DNA-binding domain; the hybrid protein now activates transcription from LexA binding sites (Figure 12-7). Further "domain-swap" experiments have revealed that the transcriptional activation function of the Gal4 protein resides in two small regions about 50 to 100 amino acids in length. These two regions form a separable activation domain that helps recruit the transcriptional machinery to the promoter, as we will see later in this section. This highly modular arrangement of activity-regulating domains is found in many transcription factors.

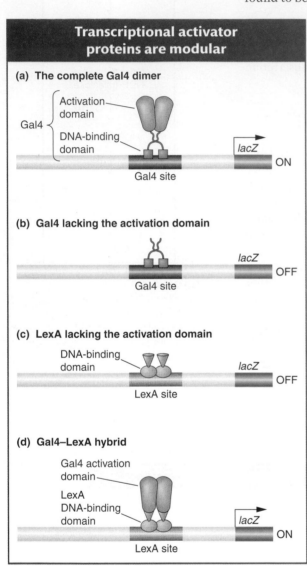

Figure 12-7 Transcriptional activator proteins have multiple, separable domains. (a) The Gal4 protein has two domains and forms a dimer. (b) The experimental removal of the activation domain shows that DNA binding is not sufficient for gene activation. (c) Similarly, the bacterial LexA protein cannot activate transcription on its own, but, when fused to the Gal4 activation domain (d), it can activate transcription through LexA-binding sites. [*After J. Watson et al., Molecular Biology of the Gene, 5th ed., ©2004, Benjamin Cummings.*]

> **Message** Many eukaryotic transcriptional regulatory proteins are modular proteins, having separable domains for DNA binding, activation or repression, and interaction with other proteins.

Gal4 activity is physiologically regulated

How does Gal4 become active in the presence of galactose? Key clues came from analyzing mutations in the *GAL80* and *GAL3* genes. In *GAL80* mutants, the *GAL* structural genes are active even in the absence of galactose. This result suggests that the normal function of the Gal80 protein is to somehow inhibit *GAL* gene expression. Conversely, in *GAL3* mutants, the *GAL* structural genes are not active in the presence of galactose, suggesting that Gal3 normally promotes expression of the *GAL* genes.

Extensive biochemical analyses have revealed that the Gal80 protein binds to the Gal4 protein with high affinity and directly inhibits Gal4 activity. Specifically, Gal80 binds to a region within one of the Gal4 activation domains, blocking its ability to promote the transcription of target genes. The Gal80 protein is expressed continuously, so it is always acting to repress transcription of the *GAL* structural genes unless stopped. The role of the Gal3 protein is to release the GAL structural genes from their repression by Gal80 when galactose is present.

Gal3 is thus both a sensor and inducer. When Gal3 binds galactose and ATP, it undergoes an allosteric change that promotes binding to Gal80, which in turn causes Gal80 to release Gal4, which is then able to activate transcription of its target genes. Thus, Gal3, Gal80, and Gal4 are all part of a switch whose state is determined by the presence or absence of galactose (Figure 12-8). In this switch, DNA binding by the transcriptional regulator is not the physiologically regulated step (as is the case in the *lac* operon and bacteriophage l); rather, the activity of the activation domain is regulated.

> **Message** The activity of eukaryotic transcriptional regulatory proteins is often controlled by interactions with other proteins.

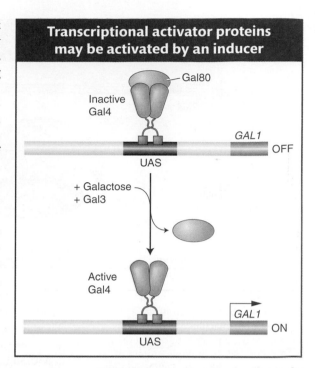

Figure 12-8 Gal4 activity is regulated by the Gal80 protein. (*Top*) In the absence of galactose, the Gal4 protein is inactive, even though it can bind to sites upstream of the *GAL1* target gene. Gal4 activity is suppressed by the binding of the Gal80 protein. (*Bottom*) In the presence of galactose and the Gal3 protein, Gal80 undergoes a conformational change and releases the Gal4 activation domain, permitting target gene transcription.

Gal4 functions in most eukaryotes

In addition to its action in yeast cells, Gal4 has been shown to be able to activate transcription in insect cells, human cells, and many other eukaryotic species. This versatility suggests that biochemical machinery and mechanisms of gene activation are common to a broad array of eukaryotes and that features revealed in yeast are generally present in other eukaryotes and vice versa. Furthermore, because of their versatility, Gal4 and its UAS elements have become favored tools in genetic analysis for manipulating gene expression and function in a wide variety of model systems.

> **Message** The ability of Gal4, as well as other eukaryotic regulators, to function in a variety of eukaryotes indicates that eukaryotes generally have the transcriptional regulatory machinery and mechanisms in common.

Now we look at how activators and other regulatory proteins interact with the transcriptional machinery to control gene expression.

Activators recruit the transcriptional machinery

In bacteria, activators commonly stimulate transcription by interacting directly with DNA and with RNA polymerase. In eukaryotes, activators generally work indirectly. Eukaryotic activators recruit RNA polymerase II to gene promoters through two major mechanisms. First, activators can interact with subunits of the protein complexes having roles in transcription initiation. Second, activators can recruit proteins that modify chromatin structure, allowing RNA polymerase II and other proteins access to the DNA. Many activators, including Gal4, have both activities. We'll examine the recruitment of parts of the transcriptional initiation complex first.

Recall from Chapter 8 that the eukaryotic transcriptional machinery contains many proteins that are parts of various subcomplexes within the transcriptional apparatus that is assembled on gene promoters. One subcomplex, transcription

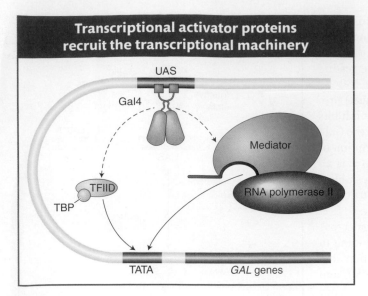

Transcriptional activator proteins recruit the transcriptional machinery

UAS

Gal4

TFIID

TBP

Mediator

RNA polymerase II

TATA

GAL genes

Figure 12-9 Gal4 recruits the transcriptional machinery. The Gal4 protein, and many other transcriptional activators, binds to multiple protein complexes, including the TFIID and mediator complexes, that recruit RNA polymerase II to gene promoters. The interactions facilitate gene activation through binding sites that are distant from gene promoters. [*After J. Watson et al., Molecular Biology of the Gene, 5th ed., ©2004, Benjamin Cummings.*]

factor IID (TFIID), binds to the TATA box of eukaryotic promoters through the TATA-binding protein (TBP; see Figure 8-12). One way that Gal4 works to activate gene expression is by binding to TBP at a site in its activation domain. Through this binding interaction, it recruits the TFIID complex and, in turn, RNA polymerase II to the promoter (Figure 12-9). The strength of this interaction between Gal4 and TBP correlates well with Gal4's potency as an activator.

A second way that Gal4 works to activate gene expression is by interacting with the **mediator complex,** a large multiprotein complex that, in turn, directly interacts with RNA polymerase II to recruit it to gene promoters. The mediator complex is an example of a **co-activator,** a term applied to a protein or protein complex that facilitates gene activation by a transcription factor but that itself is neither part of the transcriptional machinery nor a DNA-binding protein.

The ability of transcription factors to bind to upstream DNA sequences and to interact with proteins that bind directly or indirectly to promoters helps to explain how transcription can be stimulated from more distant regulatory sequences (see Figure 12-9).

> **Message** Eukaryotic transcriptional activators often work by recruiting parts of the transcriptional machinery to gene promoters.

The control of yeast mating type: combinatorial interactions

Thus far, we have focused in this chapter on the regulation of single genes or a few genes in one pathway. In multicellular organisms, distinct cell types differ in the expression of hundreds of genes. The expression or repression of sets of genes must therefore be coordinated in the making of particular cell types. One of the best-understood examples of cell-type regulation in eukaryotes is the regulation of mating type in yeast. This regulatory system has been dissected by an elegant combination of genetics, molecular biology, and biochemistry. Mating type serves as an excellent model for understanding the logic of gene regulation in multicellular animals.

The yeast *Saccharomyces cerevisiae* can exist in any of three different cell types known as **a,** α**,** and **a/**α (see Chapter 2). The two cell types **a** and α are haploid and contain only one copy of each chromosome. The **a/**α cell is diploid and contains two copies of each chromosome. Although the two haploid cell types cannot be distinguished by their appearance in the microscope, they can be differentiated by a number of specific cellular characteristics, principally their mating type (see the Model Organism box on page 420). An α cell mates only with an **a** cell, an **a** cell mates only with an α cell. An α cell secretes an oligopeptide *pheromone,* or sex hormone, called α factor that arrests **a** cells in the cell cycle. Similarly, an **a** cell secretes a pheromone, called **a** factor, that arrests α cells. Cell arrest of both participants is necessary for successful mating. The diploid **a/**α cell does not mate, is larger than the α and **a** cells, and does not respond to the mating hormones.

Genetic analysis of mutants defective in mating has shown that cell type is controlled by a single genetic locus, the mating-type locus, *MAT.* There are two alleles of the *MAT* locus: haploid **a** cells have the *MATa* allele, and the haploid α cells have the *MAT*α allele. The **a/**α diploid has both alleles. Although mating type is under genetic control, certain strains switch their mating type, sometimes as frequently as every cell division. We will examine the basis of switching later in this chapter, but first, let's see how each cell type expresses the right set of genes. We will see that different combinations of DNA-binding proteins regulate the expression of sets of genes specific to different cell types.

How does the *MAT* locus control cell type? Genetic analyses of mutants that cannot mate have identified a number of structural genes that are separate from the *MAT* locus but whose protein products are required for mating. One group of structural genes is expressed only in the α cell type (α-specific genes), and another set is expressed only in the **a** cell type (**a**-specific genes). The *MAT* locus controls which of these sets of structural genes is expressed in each cell type. The *MATa* allele causes the structural genes of the **a**-type cell to be expressed, whereas the *MAT*α allele causes the structural genes of the α-type cell to be expressed. These two alleles activate different sets of genes because they encode different regulatory proteins. In addition, a regulatory protein not encoded by the *MAT* locus, called MCM1, plays a key role in regulating cell type.

The simplest case is the **a** cell type (Figure 12-10a). The *MATa* locus encodes a single regulatory protein, a1. However, this regulatory protein has no effect in haploid cells, only in diploid cells. In a haploid **a** cell, the regulatory protein MCM1 turns on the expression of the structural genes needed by an **a** cell, by binding to regulatory sequences within promoters for **a**-specific genes.

In an α cell, the α-specific structural genes must be transcribed, but, in addition, the MCM1 protein must be prevented from activating the **a**-specific genes. The DNA sequence of the *MAT*α allele encodes two proteins, α1 and α2, that are produced by separate transcription units. These two proteins have different regulatory roles in the cell, as can be demonstrated by analyzing their DNA-binding properties in vitro (Figure 12-10b). The α1 protein is an activator of α-specific gene expression. It binds in concert with the MCM1 protein to a discrete DNA

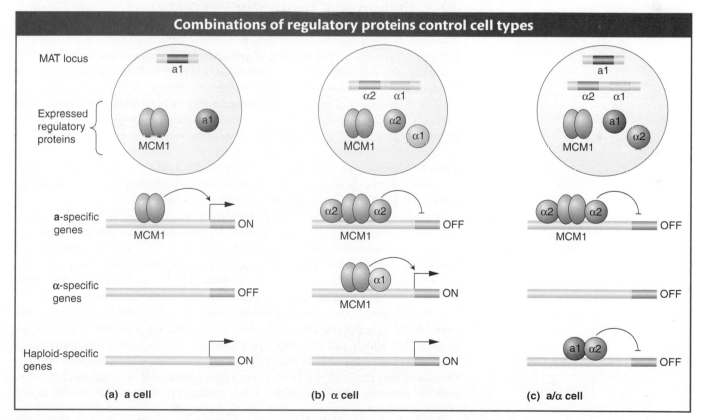

Figure 12-10 Control of cell-type-specific gene expression in yeast. The three cell types of *S. cerevisiae* are determined by the regulatory proteins a1, α1, and α2, which regulate different subsets of target genes. The MCM1 protein acts in all three cell types and interacts with α1 and α2.

sequence controlling several α-specific genes. The α2 protein represses transcription of the **a**-specific genes. It binds as a dimer, with MCM1, to sites in DNA sequences located 5′ of a group of **a**-specific genes and acts as a repressor.

In a diploid yeast cell, regulatory proteins encoded by each *MAT* locus are expressed (Figure 12-10c). What is the result? All the structural genes involved in cell mating are shut down, as are a separate set of genes, called haploid specific, that are expressed in haploid cells but not diploid cells. How does this happen? The a1 protein encoded by *MATa* has a part to play at last. The a1 protein can bind to some of the α2 protein present and alter its binding specificity such that the a1–α2 complex does not bind to **a**-specific genes. Rather, the a1–α2 complex binds to a different sequence found upstream of the haploid-specific genes. In diploid cells, then, α2 exists in two forms: (1) as an α2–MCM1 complex that represses **a**-specific genes and (2) in a complex with a1 that represses haploid-specific genes. Moreover, the a1–α2 complex also represses expression of α1, which is thus no longer present to turn on α-specific genes. The different binding partners determine which specific DNA sequences are bound and which genes are regulated by each α2-containing complex. The regulation of different sets of target genes by the association of the same transcription factor with different binding partners plays a major role in the generation of different patterns of gene expression in different cell types within multicellular eukaryotes.

> **Message** In yeast and in multicellular eukaryotes, cell-type-specific patterns of gene expression are governed by combinations of interacting transcription factors.

12.3 Dynamic Chromatin

A second mechanism for influencing gene transcription in eukaryotes modifies the local chromatin structure around gene regulatory sequences. To fully understand how this mechanism works, we need to first review chromatin structure and then consider how it can change and how these changes affect gene expression.

The recruitment of transcriptional machinery by activators may appear to be somewhat similar in eukaryotes and bacteria, with the major difference being the number of interacting proteins in the transcriptional machinery. Indeed, a little more than a decade ago, many biologists pictured eukaryotic regulation simply as a biochemically more complicated version of what had been discovered in bacteria. However, this view has changed dramatically as biologists have considered the effect of the organization of genomic DNA in eukaryotes.

Compared with eukaryotic DNA, bacterial DNA is relatively "naked," making it readily accessible to RNA polymerase. In contrast, eukaryotic chromosomes are packaged into chromatin, which is composed of DNA and proteins (mostly histones). As mentioned briefly in Chapter 1, the basic unit of chromatin is the nucleosome. The nucleosome contains about 150 bp of DNA wrapped about 1.8 times around a core of eight histones called the histone octomer (Figure 12-11). The histone core is composed of two subunits of each of the four histones: histone 2A, 2B, 3, and 4 (Figure 12-12). Surrounding the nucleosome core is a linker histone, H1, which can compact the nucleosomes into higher-order structures that further condense the DNA. The packaging of eukaryotic DNA into chromatin means that much of the DNA is not readily accessible to regulatory proteins and the transcriptional apparatus. Thus, whereas prokaryotic genes are generally accessible and "on" unless repressed, eukaryotic genes are inaccessible and "off" unless activated. Therefore, the modification of chromatin structure is a distinctive feature of eukaryotic gene regulation.

For now, there is one feature about chromatin structure that is important to keep in mind—that it can be inherited. This form of inheritance is given a name—

The structure of chromatin

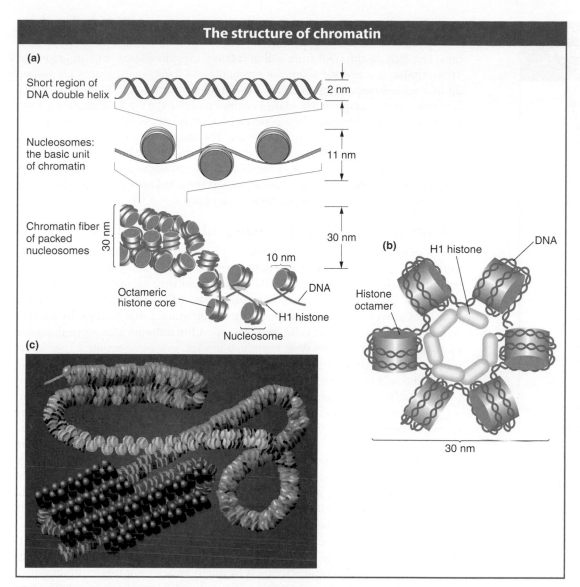

(a)

Short region of DNA double helix — 2 nm

Nucleosomes: the basic unit of chromatin — 11 nm

Chromatin fiber of packed nucleosomes — 30 nm

30 nm

10 nm

Octameric histone core

DNA

H1 histone

Nucleosome

(b)

H1 histone

DNA

Histone octamer

30 nm

(c)

Figure 12-11 (a) The nucleosome in decondensed and condensed chromatin. (b) End view of the coiled chain of nucleosomes. (c) Chromatin structure varies along the length of a chromosome. The least-condensed chromatin (euchromatin) is shown in yellow, regions of intermediate condensation are in orange and blue, and heterochromatin coated with special proteins (purple) is in red. [(b) From P. J. Horn and C. L. Peterson, "Chromatin Higher Order Folding: Wrapping Up Transcription," Science 297, 2002, 1827, Fig. 3. Copyright 2002, AAAS.]

epigenetic inheritance—and defined operationally as the inheritance of chromatin states from one cell generation to the next. What this inheritance means is that, in DNA replication, both the DNA sequence and the chromatin structure are faithfully passed on to the next cell generation. However, unlike the sequence of DNA, chromatin structure can change in the course of the cell cycle.

One way to alter chromatin structure might be to simply move the histone octamer along the DNA. In the 1980s, biochemical techniques were developed that allowed researchers to determine the position of nucleosomes in and around specific genes. In these studies, chromatin was isolated from tissues or cells in which a gene was on and compared with chromatin from tissue where the same gene was off. The result for most genes analyzed was

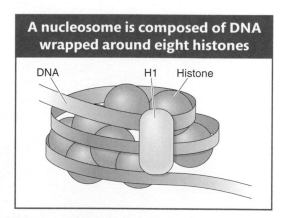

A nucleosome is composed of DNA wrapped around eight histones

DNA   H1   Histone

Figure 12-12 The components of a nucleosome unit showing the core histones (H2A, H2B, H3, H4), the surrounding DNA, and the H1 linker histone.

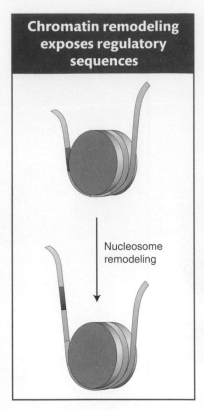

Chromatin remodeling exposes regulatory sequences

Nucleosome remodeling

Figure 12-13 The histone octamer slides in response to chromatin-remodeling activity (such as that of the SWI–SNF complex), in this case exposing the DNA marked in red. (See Figure 12-19 for details on how SWI–SNF is recruited to a particular DNA region). [*After J. Watson et al., Molecular Biology of the Gene, 5th ed., ©2004, Benjamin Cummings.*]

that nucleosome positions changed, especially in a gene's regulatory regions. Thus, which DNA regions are wrapped up in nucleosomes can change: nucleosome positions can shift on the DNA from cell to cell and over the life cycle of an organism. Transcription is repressed when the promoter and flanking sequences are wound up in a nucleosome, which prevents the initiation of transcription by RNA pol II. Activation of transcription would thus require nudging the nucleosomes away from the promoter. Conversely, when gene repression is necessary, nucleosomes shift into a position that prevents transcription. The changing of nucleosome position is referred to as **chromatin remodeling.** Chromatin remodeling is known to be an integral part of eukaryotic gene expression, and great advances are being made in determining the underlying mechanism(s) and the regulatory proteins taking part. Here, again, genetic studies in yeast have been pivotal.

Chromatin-remodeling proteins and gene activation

Two genetic screens in yeast for mutants in seemingly unrelated processes led to the discovery of the same gene whose product plays a key role in chromatin remodeling. In both cases, yeast cells were treated with agents that would cause mutations. In one screen, these mutagenized yeast cells were screened for cells that could not grow well on sucrose (sugar *non*fermenting mutants, *snf*). In another screen, mutagenized yeast cells were screened for mutants that were defective in switching their mating type (*switch* mutants, *swi;* see Section 12.5). Many mutants for different loci were recovered in each screen, but one mutant gene was found to cause both phenotypes. Mutants at the so-called *swi2/snf2* locus ("switch–sniff") could neither utilize sucrose effectively nor switch mating type.

What was the connection between the ability to utilize sugar and the ability to switch mating types? The Snf2–Swi2 protein was purified and discovered to be part of a large, multisubunit complex called the SWI–SNF complex that can reposition nucleosomes in a test-tube assay if ATP is provided as an energy source (Figure 12-13). In some situations, the multisubunit SWI–SNF complex activates transcription by moving nucleosomes that are covering the TATA sequences. In this way, the complex facilitates the binding of RNA polymerase II. The SWI–SNF complex is thus a co-activator.

Gal4 also binds to the SWI–SNF chromatin-remodeling complex and recruits it to activated promoters. Yeast strains containing a defective SWI–SNF complex show a reduced level of Gal4 activity. Why might an activator use multiple activation mechanisms? There are at least two reasons understood at present. The first is that target promoters may become less accessible at certain stages of the cell cycle or in certain cell types (in multicellular eukaryotes). For example, genes are less accessible during mitosis, when chromatin is more condensed. At that stage, Gal4 must recruit the chromatin-remodeling complex, whereas at other times, it might not be necessary to use the complex.

A second reason is that many transcription factors act in combinations to control gene expression synergistically. We will see shortly that this combinatorial synergy is a result of the fact that chromatin-remodeling complexes and the transcriptional machinery are recruited more efficiently when multiple transcription factors act together.

> **Message** Chromatin can be dynamic; nucleosomes are not necessarily in fixed positions on the chromosome. Chromatin remodeling changes nucleosome density or position and is an integral part of eukaryotic gene regulation.

Histones and chromatin remodeling

Let's look at the nucleosome more closely to see if any part of this structure could carry the information necessary to influence nucleosome position, nucleosome density, or both.

Histone modifications   As already stated, most nucleosomes are composed of an octamer made up of two copies each of the four core histones. Histones are known to be the most conserved proteins in nature; that is, histones are almost identical in all eukaryotic organisms from yeast to plants to animals. This conservation contributed to the view that histones could not take part in anything more complicated than the packaging of DNA to fit in the nucleus. However, recall that DNA with its four bases also was considered too simple a molecule to carry the blueprint for all organisms on Earth.

Figure 12-14 shows a model of nucleosome structure that represents contributions from many studies. Of note is that the histone proteins are organized into the core octamer with their amino-terminal ends making electrostatic contacts with the phosphate backbone of the surrounding DNA. These protruding ends are called **histone tails.** Since the early 1960s, it has been known that specific basic amino acid residues (lysine and arginine) in the histone tails can be covalently modified by the attachment of acetyl and methyl groups. These reactions take place after the histone protein has been translated and even after the histone has been incorporated into a nucleosome.

There are now known to be at least 150 different histone modifications that require a wide variety of molecules in addition to the acetyl and methyl groups already mentioned. The role of histone acetylation in gene expression is described below. Later in this chapter, the involvement of histone methylation in gene activity and repression will be discussed.

Histone acetylation, deacetylation, and gene expression   The acetylation reaction is one of the best-characterized histone modifications:

Note that the reaction is reversible, which means that acetyl groups can be added and removed from the same histone residue. There are 44 histone lysine residues available to accept acetyl groups, so the presence or absence of these groups can carry a tremendous amount of information. For this reason, the covalent modification of histone tails is said to be a **histone code.** Scientists coined the expression histone code because the covalent modification of histone tails is reminiscent of the genetic code. For the histone code, information is stored in the patterns of histone modification rather than in the sequence of nucleotides. With more than 150 known histone modifications, there are a huge number of possible patterns, and scientists are just

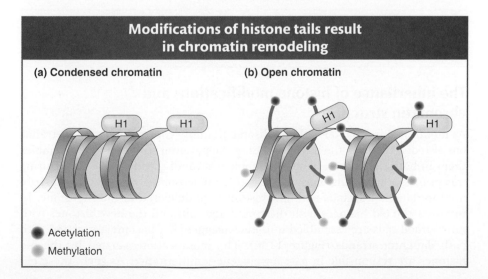

Figure 12-14 Removing and adding acetyl and methyl groups to histone tails causes the nucleosomes to slide apart, exposing DNA to the activity of proteins that regulate transcription.

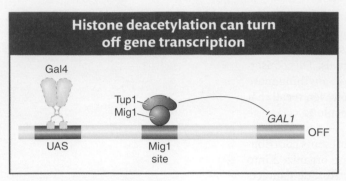

Histone deacetylation can turn off gene transcription

Gal4

Tup1
Mig1

GAL1

OFF

UAS

Mig1
site

Figure 12-15 Recruitment of a repressing complex leads to repression of transcription. In the presence of glucose, *GAL1* transcription is repressed by the Mig1 protein, which binds to a site between the UAS and the promoter of the *GAL1* gene. Mig1 recruits the Tup1 repressing complex, which recruits a histone deacetylase, turning gene transcription off. *[After J. Watson et al., Molecular Biology of the Gene, 5th ed., copyright © 2004, Benjamin Cummings.]*

beginning to decipher their effects on chromatin structure and transcriptional regulation. To add to this complexity, the code is likely not interpreted in precisely the same way in all organisms. For now, let's see how the acetylation of histone amino acids influences chromatin structure and gene expression.

Evidence had been accumulating for years that the histones associated with the nucleosomes of active genes are rich in acetyl groups (said to be **hyperacetylated**), whereas inactive genes are underacetylated (**hypoacetylated**). The enzyme responsible for adding acetyl groups, histone acetyltransferase (HAT), proved very difficult to isolate. When it was finally isolated and its protein sequence deduced, it was found to be an *ortholog* of a yeast transcriptional activator called GCN5 (meaning that it was encoded by the same gene in a different organism). Thus, the conclusion was that GCN5 is a histone acetyltransferase. It binds to the DNA in the regulatory regions of some genes and activates transcription by acetylating nearby histones. Various protein complexes that are recruited by transcriptional activators are now understood to possess a HAT activity.

How does histone acetylation facilitate changes in gene expression? There appear to be at least three mechanisms for doing so. First, the addition of acetyl groups to specific histone residues can alter the interaction in a nucleosome between the DNA and a histone octamer so that the octamer is more likely to slide along the DNA to a new position. Second, the addition of acetyl groups can alter the interaction between adjacent nucleosomes, resulting in a more open chromatin (Figure 12-14). Third, histone acetylation, in conjunction with other histone modifications, influences the binding of regulatory proteins to the DNA. The bound regulatory protein may take part in one of several functions that either directly or indirectly increase the frequency of transcription initiation.

Like other histone modifications, acetylation is reversible, and **histone deacetylases (HDACs)** also have been identified. Such proteins play key roles in gene repression. For example, in the presence of galactose and glucose, the activation of *GAL* genes is prevented by the Mig1 protein. Mig1 is a sequence-specific DNA-binding repressor that binds to a site between the UAS element and the promoter of the *GAL1* gene (Figure 12-15). Mig1 recruits a protein complex called Tup1 that contains a histone deacetylase and that represses gene transcription. The Tup1 complex is an example of a **corepressor,** which facilitates gene repression but is not itself a DNA-binding repressor. The Tup1 complex is also recruited by other yeast repressors, such as MATα2 (see page 425), and counterparts of this complex are found in all eukaryotes.

> **Message** In most cases examined, histone acetylation and deacetylation promote and repress gene transcription, respectively. These activities are recruited to genes by sequence-specific activators and repressors.

The inheritance of histone modifications and chromatin structure

As mentioned in Chapter 7, the replisome not only copies the parental strands but also disassembles the nucleosomes in the parental strands and reassembles them in both the parental and the daughter strands (see Figure 7-23). During this process, the old histones from existing nucleosomes are randomly distributed to daughter strands and new histones are delivered to the replisome. In this way, the old histones with their modified tails and the new histones with unmodified tails are assembled into nucleosomes that become associated with both daughter strands (Figure 12-16). The modifications carried by the old histones are responsible in part for epigenetic inheritance. As such, these old

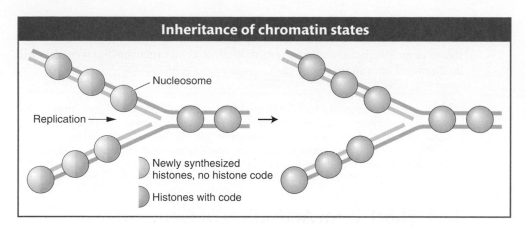

Figure 12-16 In replication, old histones (purple) with their histone codes are distributed randomly to the daughter strands, where they direct the coding of adjacent newly assembled histones (orange) to form complete nucleosomes.

modifications are called epigenetic marks because they guide the modification of the new histones.

DNA methylation: another heritable mark that influences chromatin structure

There is another important epigenetic mark in most (but not all) eukaryotes. This mark is not a histone modification; rather, it is the addition of methyl groups to DNA residues after replication. An enzyme usually attaches these methyl groups to the carbon-5 position of a specific cytosine residue.

In mammals, the methyl group is usually added to the cytosine in a CG dinucleotide. The pattern of methylation is called symmetric methylation because the methyl groups are present on both strands in the same context:

$$C*G$$
$$G\ C*$$

A remarkable number of C residues are methylated in mammals: 70 to 80 percent of all CG dinucleotides are methylated genome-wide. Interestingly, most of the unmethylated CG dinucleotides are found in clusters near gene promoters. These regions are called **CpG islands,** where the "p" represents the phosphodiester bond. Given this distribution, do you think C methylation would be associated with active or inactive regions of the genome? If you said inactive, you would be correct.

Like histone modifications, DNA methylation marks can be stably inherited from one cell generation to the next. The inheritance of DNA methylation is better understood than the inheritance of histone modifications. Semiconservative replication generates daughter helices that are methylated on one of their two strands (the parental strand). DNA molecules methylated on only one strand are termed **hemimethylated.** Methyl groups are added to unmethylated strands by DNA methyltransferases that have a high affinity for these hemimethylated substrates. These enzymes are guided by the methylation pattern on the parental strand (Figure 12-17). As you

Figure 12-17 After replication, the hemimethylated dinucleotide CG (shown as CpG) residues are fully methylated. The parental strands are black, and the daughter strand is red. The letter "M" represents the methyl group on the C nucleotide. [*After Y. H. Jiang, J. Bressler, and A. L. Beaudet, "Epigenetics and Human Disease," Annu. Rev. Genomics Hum. Genet. 5, 2004, 479–510.*]

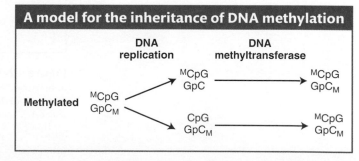

will see later in the chapter, because DNA methylation is more stable than histone modifications, it is often associated with regions of the genome that are maintained in an inactive state for the entire lifetime of an organism. Such regions will be discussed later in this chapter.

> **Message** Chromatin structure is inherited from cell generation to cell generation because mechanisms exist to replicate the DNA along with the associated epigenetic marks. In this way, the information inherent in the histone modifications and the existing DNA methylation patterns serve to reconstitute the local chromatin structure that existed before DNA synthesis and mitosis.

12.4 Short-Term Activation of Genes in a Chromatin Environment

As you have seen in this chapter, the transcription of eukaryotic genes has to be turned on and off during the lifetime of an organism. To understand how eukaryotes regulate genes during their lifetime, it is necessary to see how chromatin changes during transcriptional activation. In addition, the development of a complex organism requires that transcription levels be regulated over a wide range. Think of a regulation mechanism as more like a rheostat than an on-or-off switch: rather than a gene producing either many or no proteins, it may produce a number anywhere in between depending on the transcription level. In eukaryotes, transcription levels are made finely adjustable in a chromatin environment by clustering binding sites into enhancers. Several different transcription factors or several molecules of the same transcription factor may bind to adjacent sites. The binding of these factors to sites that are the correct distance apart leads to an amplified, or superadditive, effect on activating transcription. When an effect is greater than additive, it is said to be **synergistic.**

The binding of multiple regulatory proteins to the multiple binding sites in an enhancer can catalyze the formation of an **enhanceosome,** a large protein

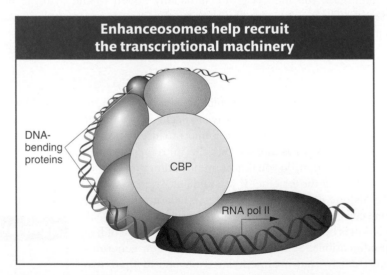

Figure 12-18 The β-interferon enhanceosome. In this case, the transcription factors recruit a co-activator (CBP), which binds both to the transcription factors and to RNA polymerase II, initiating transcription. [*After A. J. Courey, "Cooperativity in Transcriptional Control," Curr. Biol. 7, 2001, R250–R253, Fig. 1.*]

complex that acts synergistically to activate transcription. In Figure 12-18 you can see how architectural proteins bend the DNA to promote cooperative interactions between the other DNA-binding proteins. In this mode of enhanceosome action, transcription is activated to very high levels only when all the proteins are present and touching one another in just the right way.

To better understand what an enhanceosome is and how it acts synergistically, let's look at a specific example.

The β-interferon enhanceosome

The human β-interferon gene, which encodes the antiviral protein interferon, is one of the best-characterized genes in eukaryotes. It is normally switched off but is activated to very high levels of transcription on viral infection. The key to the activation of this gene is the assembly of transcription factors into an enhanceosome about 100 bp upstream of the TATA box and transcription start site. The regulatory proteins of the β-interferon enhanceosome all bind to the same face of the DNA double helix. Binding to the other side of the helix are several architectural proteins that bend the DNA and allow the different regulatory proteins to touch one another and form an activated complex. When all of the regulatory proteins are bound and interacting correctly, they form a "landing pad," a high-affinity binding site for the protein CBP, a co-activator protein that also recruits the transcriptional machinery. The large CBP protein also contains an intrinsic histone acetylase activity that modifies nucleosomes and facilitates high levels of transcription.

Although the β-interferon promoter is shown without nucleosomes in Figure 12-18, the enhanceosome is actually surrounded by two nucleosomes, called nuc 1 and nuc 2 in Figure 12-19. One of them, nuc 2, is strategically positioned over the TATA box and transcription start site. However, the binding of GCN5, another co-activator, is now known to actually precede CBP binding. GCN5 acetylates the two nucleosomes. After acetylation, the activating transcription factors recruit the co-activator CBP, the RNA pol II holoenzyme, and the SWI–SNF chromatin-remodeling complex. SWI–SNF is then positioned to nudge the nucleosome 37 bp off the TATA box, making the TATA box accessible to the TATA-binding protein and allowing transcription to be initiated.

Cooperative interactions help to explain several perplexing observations about enhancers. For example, they explain why mutating any one transcription factor or binding site dramatically reduces enhancer activity. They also explain why the distance between binding sites within the enhancer is such a critical feature. Furthermore, enhancers do not have to be close to the start site of transcription, as is the example shown in Figure 12-19. One characteristic of enhancers is that they can activate transcription when they are located at great distances from the promoter (>50 kb), either upstream or downstream from a gene or even in an intron.

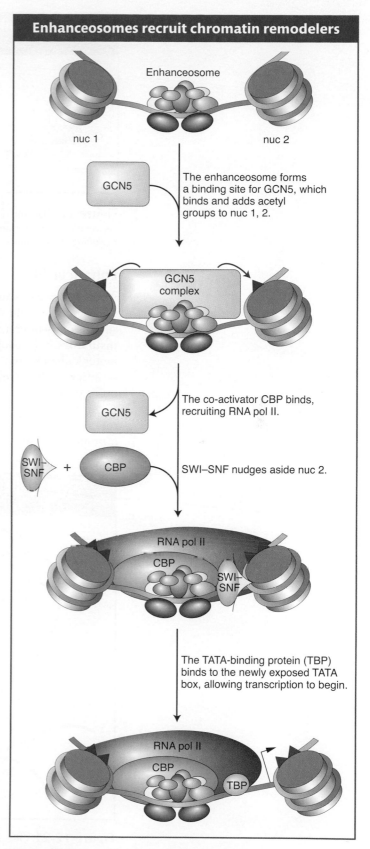

Enhanceosomes recruit chromatin remodelers

The enhanceosome forms a binding site for GCN5, which binds and adds acetyl groups to nuc 1, 2.

The co-activator CBP binds, recruiting RNA pol II.

SWI–SNF nudges aside nuc 2.

The TATA-binding protein (TBP) binds to the newly exposed TATA box, allowing transcription to begin.

Figure 12-19 The β-interferon enhanceosome acts to move nucleosomes by recruiting the SWI–SNF complex.

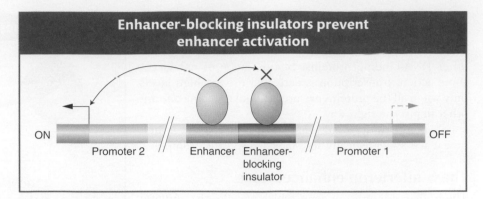

Figure 12-20 Enhancer-blocking insulators prevent gene activation when placed between an enhancer and a promoter. [*After M. Gaszner and G. Felsenfeld, "Insulators: Exploiting Transcriptional and Epigenetic Mechanisms," Nat. Rev. Genet. 7, 2006, 703–713.*]

Enhancer-blocking insulators

A regulatory element, such as an enhancer, that can act over tens of thousands of base pairs could interfere with the regulation of nearby genes. To prevent such promiscuous activation, regulatory elements called **enhancer-blocking insulators** have evolved. When positioned between an enhancer and a promoter, enhancer-blocking insulators prevent the enhancer from activating transcription at that promoter. Such insulators have no effect on the activation of other promoters that are not separated from their enhancers by the insulator (Figure 12-20). Several models have been proposed to explain how an insulator could block enhancer activity only when placed between an enhancer and a promoter. Many of the models, like the one shown in Figure 12-21, propose

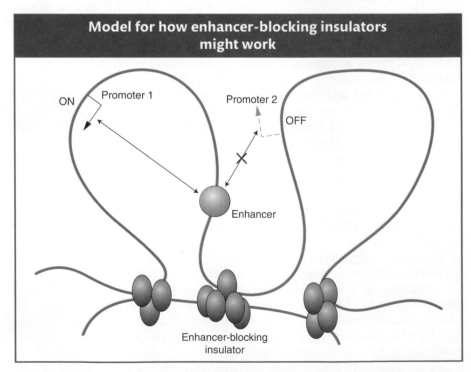

Figure 12-21 One proposal is that enhancer-blocking insulators (EB) create new loops that physically separate a promoter from its enhancer (E). [*After M. Gaszner and G. Felsenfeld, "Insulators: Exploiting Transcriptional and Epigenetic Mechanisms," Nat. Rev. Genet. 7, 2006, 703–713.*]

that the DNA is organized into loops containing active genes. According to this model, insulators act by moving a promoter into a new loop, where it is shielded from the enhancer.

As you will see later in the chapter, enhancer-blocking insulators are a fundamental component of a phenomenon called genomic imprinting.

> **Message** Eukaryotic enhancers can act at great distances to modulate the activity of the transcriptional apparatus. Enhancers contain binding sites for many transcription factors, which bind and interact cooperatively. These interactions result in a variety of responses, including the recruitment of additional co-activators and the remodeling of chromatin.

12.5 Long-Term Inactivation of Genes in a Chromatin Environment

Thus far, we have looked at how genes are activated in a chromatin environment. However, as stated at the beginning of this chapter, most of the genes in eukaryotic genomes are off at any one time. One of the most surprising findings of the genomics era is that many eukaryotic genes are inactive for the life of the organism. This leads to two questions that will be addressed in this section. First, why do organisms have genes that are always inactive? Second, how do organisms keep genes in an inactive state for their entire lifetime?

One of the most useful models for understanding mechanisms that maintain the long-term inactivity of genes concerns the control of mating-type switching in yeast. The components of the yeast mating-type locus were introduced at the end of Section 12.2. Here we continue the story by focusing on the mechanism of mating-type switching, which requires each yeast cell to maintain inactive copies of a and α genes elsewhere in their genome.

Mating-type switching and gene silencing

Haploid yeast cells are able to switch their mating type, sometimes as often as every cell cycle. In this way, a yeast haploid cell of one mating type (say **a**) will form a colony of both **a** and α cells that can mate to form diploid (a/α) cells. During times of crisis such as periods of nutrient scarcity, each diploid cell can undergo meiosis and produce four haploid spores. This process is advantageous to the survival of the species because spores can survive adverse environment conditions better than haploid cells.

Genetic analyses of certain mutants that either could not switch or could not mate (they were sterile) were sources of key insights into mating-type switching. Among the switch mutants were several mutant loci, including the *HO* gene and the *HMRa* and *HMLα* genes. Further study revealed that the *HO* gene encodes an endonuclease, an enzyme that cleaves DNA (see Chapter 10), required for the initiation of switching. It was also found that the *HMRa* and *HMLα* loci, which are on the same chromosome as the *MAT* locus, contain "cassettes" of the *MATa* and *MATα* alleles, respectively, that are not expressed. The *HMR* and *HML* loci are thus "silent" cassettes. Recall from Chapter 8 that one form of gene silencing occurs when dsRNA targets the RISC complex to destroy complementary RNA. This is an example of **post-transcriptional gene silencing.** In contrast, *HMRa* and *HMRα* cannot be transcribed, and, as such, they are examples of **transcriptional gene silencing.**

Two features of the mating-type switch were of interest to geneticists: how do cells switch their mating type, and why are *HMRa* and *HMRα* transcriptionally silent? The key to switching is the HO endonuclease, which initiates the

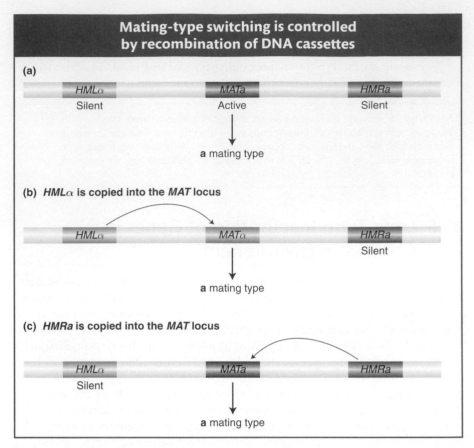

Mating-type switching is controlled by recombination of DNA cassettes

(a)

HMLα — Silent   MATa — Active   HMRa — Silent

↓

a mating type

(b) *HML*α is copied into the *MAT* locus

HMLα   MATα   HMRa — Silent

↓

a mating type

(c) *HMRa* is copied into the *MAT* locus

HMLα — Silent   MATa   HMRa

↓

a mating type

Figure 12-22 *S. cerevisiae* chromosome III encodes three mating-type loci, but only the genes at the MAT locus are expressed. *HML* encodes a silent cassette of the α genes, and *HMR* encodes a silent cassette of the **a** genes. Copying of a silent cassette and insertion through recombination at the MAT locus switches mating type.

mating-type switch by inserting a double-strand break at the *MAT* locus. The interconversion of mating type then takes place by a type of recombination between the segment of DNA (a cassette) from one of the two unexpressed loci and the *MAT* locus. The result is the replacement of the old cassette at the *MAT* locus with a new cassette from either *HMRa* or *HMR*α. The resulting mating type is either the MATa or the MATα type, depending on which gene is at the *MAT* locus (Figure 12-22). The inserted cassette is actually copied from the *HML* or *HMR* locus. In this manner, the switch is reversible because the information for the *a* and α cassettes is always present at the *HMR* and *HML* loci and never lost. Thus, the mating switch provides one example of genes that need to be silenced for the entire lifetime of an organism. As you will see later in this chapter, the silencing of genes for an entire lifetime also occurs in humans and all other mammals.

The second feature of mating-type switching of interest to geneticists is the mechanism underlying gene silencing. Why are genes in the *HMR* and *HML* cassettes not expressed? Normally, these cassettes are "silent." However, in *SIR* mutants (silent information regulators), silencing is compromised such that both *a* and α information is expressed. The resulting mutants are sterile. This means that in normal, nonmutant yeast, genes at the *HMR* and *HML* cassettes are capable of being expressed but are not because of the action of the Sir proteins. The Sir2, Sir3, and Sir4 proteins form a complex that plays a key role in gene silencing. Sir2 is a histone deacetylase that facilitates the condensation of chromatin and helps lock up *HMR* and *HML* in chromatin domains where transcription cannot be initiated.

Gene silencing is a very different process from gene repression: silencing is a position effect that depends on the neighborhood in which genetic information is located. For example, a normally active gene inserted into the *HMR* or *HML* loci would be silenced. You will learn more about position effects later, in the section on position-effect variegation in the fruit fly *Drosophila melanogaster*.

Heterochromatin and euchromatin compared

Let's examine why long-term gene silencing, of the type that silences *HML* and *HMR*, is a different process from gene repression. To do so, it is important to note that chromatin is not uniform over all chromosomes: certain regions of chromosomes are bundled in highly condensed chromatin called **heterochromatin.** Other domains are packaged in less-condensed chromatin called **euchromatin** (see Figure 12-11b). Chromatin condensation changes in the course of the cell cycle. The chromatin of cells entering mitosis becomes highly condensed as the chromosomes align in preparation for cell division. After cell division, regions forming heterochromatin remain condensed, especially around the centromeres and telomeres (called **constitutive heterochromatin**), whereas the regions forming euchromatin become less condensed. As seen for the β-interferon example (see Figure 12-19), the chromatin of active genes can change in response to, for example, developmental stage or environmental conditions.

The major distinction between heterochromatin and euchromatin is that the former contains few genes while the latter is rich in genes. But what is heterochromatin if not genes? Most of the eukaryotic genome is composed of repetitive sequences that do not make protein or structural RNA—sometimes called junk DNA (see Chapter 4). Thus, the densely packed nucleosomes of heterochromatin (organized into 30-nm chromatin fibers; see Figure 12-11a) are said to form a "closed" structure that is largely inaccessible to regulatory proteins and inhospitable to gene activity. In contrast, euchromatin, with its more widely spaced nucleosomes (organized into 10-nm fibers; see Figure 12-11a), assumes an "open" structure that permits transcription.

> **Message** The chromatin of eukaryotes is not uniform. Highly condensed heterochromatic regions have fewer genes and lower recombination frequencies than do the less-condensed euchromatic regions.

Position-effect variegation in *Drosophila* reveals genomic neighborhoods

Long before the silent-mating loci of yeast were described, geneticist Hermann Muller discovered an interesting genetic phenomenon while studying *Drosophila:* chromosomal neighborhoods exist that can silence genes that are experimentally "relocated" to adjacent regions of the chromosome. In these experiments, flies were irradiated with X rays to induce mutations in their germ cells. The progeny of the irradiated flies were screened for unusual phenotypes. A mutation in the *white* gene, near the tip of the X chromosome, will result in progeny with white eyes instead of the wild-type red color. Some of the progeny had very unusual eyes with patches of white and red color. Cytological examination revealed a chromosomal rearrangement in the mutant flies: present in the X chromosome was an inversion of a piece of the chromosome carrying the

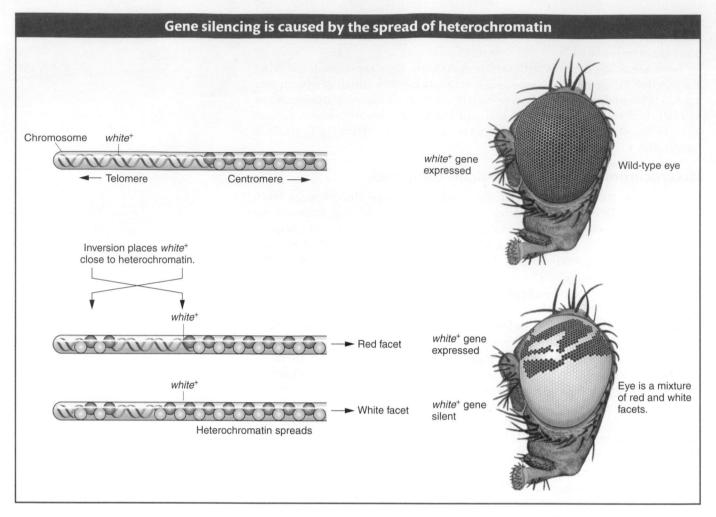

Gene silencing is caused by the spread of heterochromatin

Chromosome   *white*⁺

← Telomere   Centromere →

Inversion places *white*⁺
close to heterochromatin.

white⁺

→ Red facet

white⁺

→ White facet

Heterochromatin spreads

white⁺ gene
expressed

Wild-type eye

white⁺ gene
expressed

white⁺ gene
silent

Eye is a mixture
of red and white
facets.

Figure 12-23 Chromosomal rearrangement produces position-effect variegation. Chromosomal inversion places the wild-type white allele close to heterochromatin. The spread of heterochromatin silences the allele. Eye facets are white instead of the wild-type red wherever the allele has been silenced. [*After J. C. Eissenberg and S. Elgin, Encyclopedia of Life Sciences. Nature Publishing Group, 2001, p. 3, Fig. 1.*]

white gene (Figure 12-23). Inversions and other chromosomal rearrangements will be discussed in Chapter 17. In this rearrangement, the *white* gene, which is normally located in a euchromatic region of the X chromosome, now finds itself near the heterochromatic centromere. In some cells, the heterochromatin can "spread" to the neighboring euchromatin and silences the *white* gene. Patches of white tissue in the eye are derived from the descendants of a single cell in which the *white* gene has been silenced and remains silenced through future cell divisions. In contrast, the red patches arise from cells in which heterochromatin has not spread to the *white* gene, and so this gene remains active in all its descendants. The existence of red and white patches of cells in the eye of a single organism dramatically illustrates two features of epigenetic silencing. First, that the expression of a gene can be repressed by virtue of its position in the chromosome rather than by a mutation in its DNA sequence. Second, that epigenetic silencing can be inherited from one cell generation to the next.

Findings from subsequent studies in *Drosophila* and yeast demonstrated that many active genes are silenced in this mosaic fashion when they are relocated to neighborhoods (near centromeres or telomeres) that are heterochromatic. Thus, the ability of heterochromatin to spread into euchromatin and silence genes is a feature common to many organisms. This phenomenon has been called **position-effect variegation (PEV).** It provides powerful evidence that chromatin structure is able to regulate the expression of genes—in this case, determining whether genes with identical DNA sequence will be active or silenced.

> **Message** Active genes that are relocated to genomic neighborhoods that are heterochromatic may be silenced if the heterochromatin spreads to the genes.

Genetic analysis of PEV reveals proteins necessary for heterochromatin formation

Geneticists reasoned that PEV could be exploited to identify the proteins necessary for forming heterochromatin. To this end, they isolated mutations at a second chromosomal locus that either suppressed or enhanced the variegated pattern (Figure 12-24). Suppressors of variegation [called *Su(var)*] are genes that, when mutated, reduce the spread of heterochromatin, meaning that the wild-type products of these genes are required for spreading. In fact, the *Su(var)* alleles have proved to be a treasure trove for scientists interested in the proteins that are required to establish and maintain the inactive, heterochromatic state. Among more than 50 *Drosophila* gene products identified by these screens was **heterochromatin protein-1 (HP-1),** which had previously been found associated with the heterochromatic telomeres and centromeres. Thus, it makes sense that a mutation in the gene encoding HP-1 will show up as a *Su(var)* allele because the protein is required in some way to generate the higher-order chromatin structure associated with heterochromatin.

But why does HP-1 bind to some DNA regions and not others? The answer to that came with the discovery that another *Su(var)* gene encoded a methyltransferase that adds methyl groups to a specific amino acid residue (lysine 9) in the

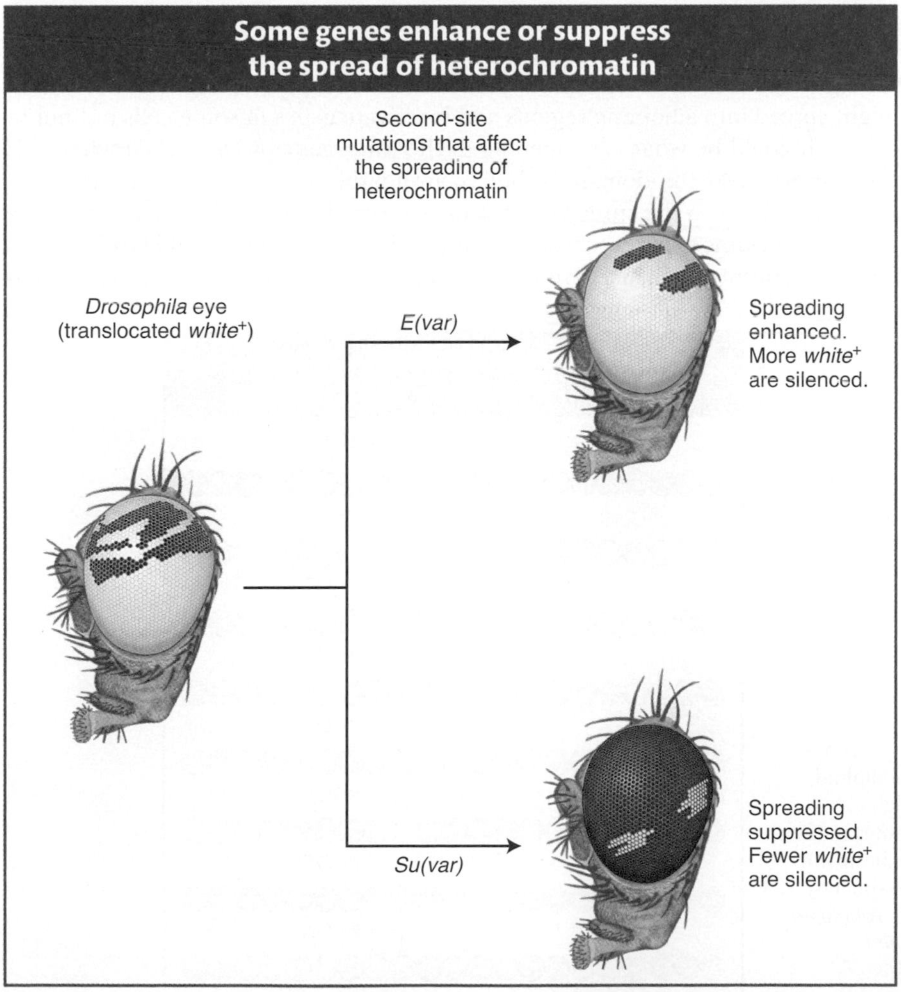

What Geneticists Are Doing Today

Figure 12-24 Mutations were used to identify genes that suppress, *Su(var)*, or enhance, *E(var)*, position-effect variegation. [*After J. C. Eissenberg and S. Elgin, Encyclopedia of Life Sciences. Nature Publishing Group, 2001, p. 3, Fig. 1.*]

tail of histone H3 (called histone H3 methyltransferase or HMTase). One of the reactions catalyzed by HMTase is shown here:

| Lysine | Monomethyl lysine | Dimethyl lysine | Trimethyl lysine |

The amino acid lysine is abbreviated with a "K." As such, these epigenetic marks are referred to as H3K9me1, H3K9me2, and H3K9me3, respectively. Chromatin modified in this way binds HP-1 proteins, which then associate to form heterochromatin. Proteins similar to HP-1 and HMTase have been isolated in diverse taxa, suggesting the conservation of an important eukaryotic function.

We have seen that actively transcribed regions are associated with nucleosomes whose histone tails are hyperacetylated and that transcriptional activators such as *GCN5* encode a histone acetyltransferase activity. As already discussed, acetyl marks can also be removed from histones by histone deacetylases. Similarly, chromatin made up of nucleosomes that are methylated at H3meK9 and bound up with HP-1 protein contain epigenetic marks that are associated with heterochromatin. Scientists are now able to separate heterochromatin and euchromatin and analyze differences in histone modifications and bound proteins. The procedure used, *ch*romatin *i*mmuno*p*recipitation (ChIP), is described in Chapter 14.

Figure 12-25 illustrates that, in the absence of any barriers, heterochromatin might spread into adjoining regions and inactivate genes in some cells but not in others. It could be what is happening to the *white* gene of *Drosophila* when it is translocated near the domain of heterochromatin associated with the chromosome ends. But can the spread of heterochromatin be stopped? One can imagine that the spreading of heterochromatin into active gene regions could be disastrous for an organism because active genes would be silenced as they are converted

Figure 12-25 The spread of heterochromatin into adjacent euchromatin is variable. In four genetically identical diploid cells, heterochromatin spreads enough to knock out a gene in some chromosomes but not others. Heterochromatin and euchromatin are represented by orange and green spheres, respectively. [*After M. Gaszner and G. Felsenfeld, "Insulators: Exploiting Transcriptional and Epigenetic Mechanisms," Nat. Rev. Genet. 7, 2006, 703–713.*]

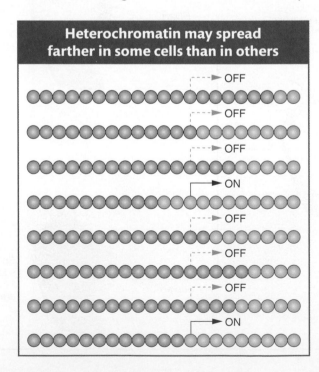

Heterochromatin may spread farther in some cells than in others

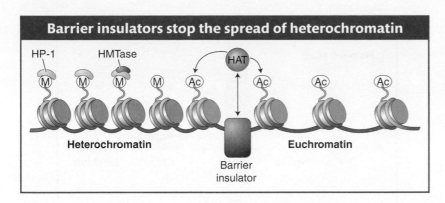

Figure 12-26 In this model, barrier insulators recruit enzymatic activities such as histone acetyltransferase (HAT) that promote euchromatin formation. The letter "M" stands for methylation and the letters "Ac" for acetylation. [*After M. Gaszner and G. Felsenfeld, "Insulators: Exploiting Transcriptional and Epigenetic Mechanisms," Nat. Rev. Genet. 7, 2006, 703–713.*]

into heterochromatin. To avert this potential disaster, the genome contains DNA elements called **barrier insulators** that prevent the spreading of heterochromatin by creating a local environment that is not favorable to heterochromatin formation. For example, a barrier insulator could bind HATs and, in doing so, make sure that the adjacent histones are hyperacetylated. A model for how a barrier insulator might act to "protect" a region of euchromatin from being converted into heterochromatin is shown in Figure 12-26.

> **Message** The isolation of critical proteins necessary for the formation of heterochromatin, including HP-1 and HMTase, was made possible by the isolation of mutant strains of *Drosophila* that suppressed or enhanced PEV.

12.6 Gender-Specific Silencing of Genes and Whole Chromosomes

Thus far, we have discussed chromosomal domains that are open or condensed in all members of a species. In this section we consider two widespread genetic phenomena in mammals that depend on the sex of the individual. In these cases, specific genes or even a whole chromosome are silenced for the entire lifetime of an organism. However, unlike the prior examples, these genes or chromosomes are silenced in males or females but not both.

Genomic imprinting explains some unusual patterns of inheritance

The phenomenon of **genomic imprinting** was discovered about 20 years ago in mammals. In genomic imprinting, certain autosomal genes have unusual inheritance patterns. For example, an *igf2* allele is expressed in a mouse only if it is inherited from the mouse's father—an example of **maternal imprinting** because the copy of the gene derived from the mother is inactive. Conversely, a mouse *H19* allele is expressed only if it is inherited from the mother; *H19* is an example of **paternal imprinting** because the paternal copy is inactive. The consequence of parental imprinting is that imprinted genes are expressed as if there were only one copy of the gene present in the cell even though there are two. Importantly, no changes are observed in the DNA sequences of imprinted genes; that is, the identical gene can be active or inactive in the progeny, depending on whether it was inherited from mom or dad.

If the DNA sequence of the gene does not correlate with activity, what does? The answer is that that during the development of gametes, methyl groups are added to the DNA in the regulatory regions of imprinted genes in one sex only. We saw earlier that DNA of genes that are shut down for an entire lifetime are usually highly methylated. However, it is important to note that DNA methylation

Figure 12-27 Genomic imprinting in the mouse. The imprinting control region (ICR) is unmethylated in female gametes and can bind a CTCF dimer, forming an insulator that blocks enhancer activation of *Igf2*. Methylation (M) of the ICR in male germ cells prevents CTCF binding, but it also prevents the binding of other proteins to the *H19* promoter.

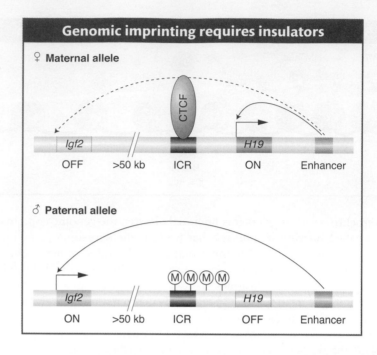

Genomic imprinting requires insulators

♀ **Maternal allele**

Igf2 OFF   >50 kb   ICR   *H19* ON   Enhancer

♂ **Paternal allele**

Igf2 ON   >50 kb   ICR   *H19* OFF   Enhancer

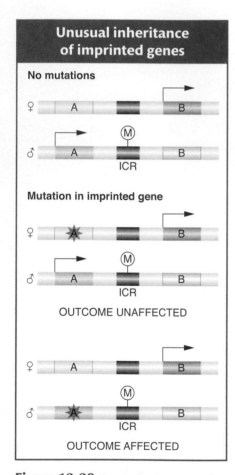

Unusual inheritance of imprinted genes

No mutations

♀ A   B

♂ A   ⓂICR   B

Mutation in imprinted gene

♀ A   B

♂ A   ⓂICR   B

OUTCOME UNAFFECTED

♀ A   B

♂ A   ⓂICR   B

OUTCOME AFFECTED

Figure 12-28 A mutation (represented by an orange star) in gene A will have no effect if inherited from the male. Abbreviations: M, methylation; ICR, imprinting control region. [*After S. T. da Rocha and A. C. Ferguson-Smith, "Genomic Imprinting," Curr. Biol. 14, 2004, R646–R649.*]

is one of several epigenetic marks associated with the long-term inactivation of genes. Other marks include methylation of specific histone amino acids, including H3K27me1.

Let's turn again to the mouse *ifg2* and *H19* genes to see how imprinting works at the molecular level. These two genes are located in a cluster of imprinted genes on mouse chromosome 7. There are an estimated 100 imprinted genes in the mouse, and most are found in clusters comprising from 3 to 11 imprinted genes. (Humans have most of the same clustered imprinted genes as those in the mouse.) In all cases examined, there is a specific pattern of DNA methylation for each gene copy of an imprinted gene. For the *ifg2–H19* cluster, a specific region of DNA lying between the two genes (Figure 12-27) is methylated in male germ cells and unmethylated in female germ cells. This region is called the imprinting control region (ICR). Thus, methylation of the ICR leads to *ifg2* being active and *H19* being inactive, whereas lack of methylation leads to the reverse.

How does methylation control which of the two genes is active? Methylation acts as a block to the binding of proteins needed for transcription. Only the unmethylated (female) ICR can bind a regulatory protein called CTCF. When bound, CTCF acts as an enhancer-blocking insulator that prevents enhancer activation of *Igf2* transcription. However, the enhancer in females can still activate *H19* transcription. In males, CTCF cannot bind to the ICR and the enhancer can activate *Igf2* transcription (recall that enhancers can act at great distances). The enhancer cannot activate *H19*, however, because the methylated region extends into the *H19* promoter. The methylated promoter cannot bind proteins needed for the transcription of *H19*.

Thus, we see how an enhancer-blocking insulator (in this case, CTCF bound to part of the ICR) prevents the enhancer from activating a distant gene (in this case, *Igf2*). Furthermore, we see that the CTCF-binding site is methylated only in chromosomes derived from the male parent. The methylation of the CTCF-binding site prevents CTCF binding in males and permits the enhancer to activate *Igf2*.

Note that parental imprinting can greatly affect pedigree analysis. Because the inherited allele from one parent is inactive, a mutation in the allele inherited from the other parent will appear to be dominant, whereas, in fact, the allele is expressed because only one of the two homologs is active for this gene. Figure 12-28 shows how a mutation in an imprinted gene can have different outcomes on the phenotype of the organism if inherited from the male or from the female parent.

Many steps are required for imprinting (Figure 12-29). Soon after fertilization, mammals set aside cells that will become their germ cells. Imprints are erased before the germ cells form. Without their distinguishing mark of DNA methylation, these genes are now said to be *epigenetically equivalent*. As these primordial germ cells become fully formed gametes, imprinted genes receive the sex-specific mark that will determine whether the gene will be active or silent after fertilization.

But what about Dolly and other cloned mammals?

Many thought that genomic imprinting would lead to a requirement that both male and female germ cells participate in mammalian embryo development. That is, male and female gametes contain different subsets of imprinted genes; so germ cells of both sexes must participate for the embryo to have a full complement of active imprinted genes. Why, then, are mammals such as Dolly and, more recently, cloned pigs, cats, dogs, and cows that were derived from somatic nuclei able to survive and even flourish? After all, as already noted, the mutation of even a single imprinted gene can be lethal or can lead to serious disease.

At this point, scientists do not understand why the cloning of many mammalian species has been successful. However, despite these successes, cloning is extremely inefficient in all species tested. For most experiments, a successful clone is an exceedingly rare event, requiring hundreds, even thousands of attempts. One could argue that the failure of most cloned embryos to develop into viable organisms is a testament to the importance of the epigenetic mechanisms of gene regulation in eukaryotes. As such, it illustrates how knowledge of the complete DNA sequence of all genes in an organism is only a first step in understanding how eukaryotic genes are regulated.

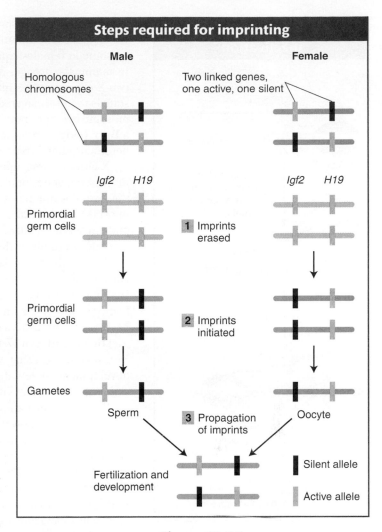

Figure 12-29 How *Igf2* and *H19* are differentially imprinted in males and females.

Silencing an entire chromosome: X-chromosome inactivation

The epigenetic phenomenon called X-chromosome inactivation has intrigued scientists for decades. In Chapter 17, you will learn about the effects of gene copy number on the phenotype of an organism. For now, you need only know that the number of transcripts produced by a gene is usually proportional to the number of copies of that gene in a cell. Mammals, for example, are diploid and have two copies of each gene located on their autosomes. For the vast majority of genes, both alleles are expressed. Therefore, all individuals are producing about the same number of transcripts for these genes, proportional to two gene copies.

There is an exception to this generalization, however. All individuals would not produce the same number of transcripts of genes located on the sex chromosomes if both X chromosomes were expressed in females. As discussed in Chapter 2, the number of the X and Y sex chromosomes differs between the sexes, with female mammals having two X chromosomes and males having only one. The mammalian X chromosome is thought to contain about 1000 genes. Females have twice as many copies of these X-linked genes as males and would express twice as many transcripts from these genes as males do if there were not a mechanism to correct this imbalance. (The absence of a Y chromosome is not a problem for females, because the very few genes on this chromosome are required only for the development of males.) We say that the females produce two doses of transcripts for every one dose produced by males.

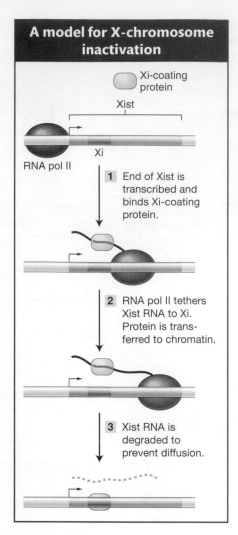

A model for X-chromosome inactivation

Xi-coating protein

Xist

RNA pol II

Xi

1 End of Xist is transcribed and binds Xi-coating protein.

2 RNA pol II tethers Xist RNA to Xi. Protein is transferred to chromatin.

3 Xist RNA is degraded to prevent diffusion.

Figure 12-30 A model showing how Xist RNA might act in cis to bind proteins that inactivate one X chromosome by forming heterochromatin.

This dosage imbalance is corrected by a process called **dosage compensation,** which makes the amount of most gene products from the two copies of the X chromosome in females equivalent to the single dose of the X chromosome in males. In mammals, dosages are made equivalent by randomly inactivating one of the two X chromosomes in each cell at an early stage in development. This inactive state is then propagated to all progeny cells. (In the germ line, the second X chromosome becomes reactivated in oogenesis). The inactivated chromosome, called a **Barr body,** can be seen in the nucleus as a darkly staining, highly condensed, heterochromatic structure.

X-chromosome inactivation is an example of epigenetic inheritance. First, most of the genes on the inactivated X chromosome (called Xi) are silenced, and the chromosome has epigenetic marks associated with heterochromatin, including H3K9me, hypoacetylation of histones, and hypermethylation of its DNA. Second, most but not all of the genes on the inactivated chromosome remain inactive in all descendants of these cells, yet the DNA sequence itself is unchanged.

The mechanism that converts a fully functional X chromosome into heterochromatin is the subject of current investigations. The process is well characterized in the mouse, and X-chromosome inactivation in that organism shares many features with X-chromosome inactivation in human female somatic cells. Both have a locus on the X chromosome called the X-inactivation center (abbreviated *Xic*) that produces a 17-kb non-protein-coding RNA (ncRNA; see Chapter 8) called Xist. It is thought that Xist is transcribed from only one chromosome early in the development of female mouse embryos. The chromosome producing Xist becomes inactivated as Xist specifically coats the central region of that chromosome, leading to the formation of heterochromatin. Neither how Xist is localized to one chromosome nor how it triggers the conversion to heterochromatin is understood.

One interesting model for how transcription of an ncRNA could influence chromatin structure in Xi is shown in Figure 12-30. According to this model, as an ncRNA is transcribed by RNA Pol II, proteins bind specifically to its sequences and catalyze the histone modifications that initiate heterochromatin formation. In this way, ncRNAs act as tethers to recruit chromatin-modifying proteins to the X chromosome from which it is transcribed.

Message For most diploid organisms, both alleles of a gene are expressed independently. Genomic imprinting and X inactivation are examples of only a single allele being available for expression. In these cases, epigenetic mechanisms silence a single chromosomal locus or one copy of an entire chromosome, respectively.

12.7 Post-Transcriptional Gene Repression by miRNAs

Xist is one example of the rapidly growing class of functional RNAs (see Chapter 8). Functional RNAs do not encode proteins: rather, they perform a variety of tasks that exploit the complementarity of RNA and RNA and of RNA and DNA. The functional RNAs discussed here contain specific sequences that direct proteins or protein complexes to places in the cell where their services are needed. For example, Xist acts to direct proteins involved in heterochromatin formation to one of the two X chromosomes.

Two types of small functional RNAs were introduced in Chapter 8: siRNAs and miRNAs. In this section we explore how miRNAs assist in the regulation of eukaryotic gene expression. The function of siRNAs will be considered further in Chapter 15.

Recall from Chapter 8 that miRNAs are synthesized by RNA pol II as longer RNAs that are processed into the smaller (~22 nt) biologically active miRNAs (see Figure 8-20). Organisms contain hundreds of miRNAs that regulate thousands

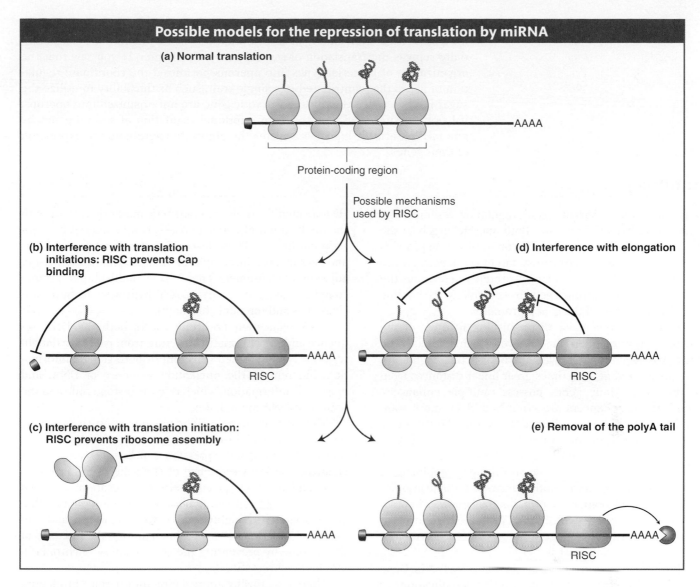

Figure 12-31 See text for details.

of genes. Of these, about 1/3 are organized into clusters that are transcribed into a single transcript, which is later processed to form several miRNAs. In contrast, about 1/4 of all miRNAs are processed from transcripts derived from spliced introns. The final steps in the processing of miRNAs occurs in the cytoplasm.

In Chapter 8 we saw how the active single-stranded miRNAs bind to the RNA-inducing silencing complex (RISC) and hybridize to mRNAs that are complementary to the miRNAs. Specifically, the binding region of the miRNA consists of nucleotides 2 through 8 of the ~22-nt miRNA, called the *seed region*. The nucleotides of the seed region bind to the 3′UTR of an mRNA that is being translated by ribosomes (Figure 12-31). While the miRNA-RISC complex is known to inhibit translation, the precise mechanism is still under investigation. Models for how translation may be repressed include interference with translation initiation or elongation or the removal of the poly(A) tail, which would hasten mRNA degradation (Figure 12-31).

Although miRNAs were discovered almost 20 years ago, scientists are only beginning to decipher the extent and complexity of miRNA control of eukaryotic gene expression. Consider that in mammals, sequences complementary to the seed regions of miRNAs are found in the 3′UTR of several hundred genes. Furthermore, the 3′UTRs of some genes contain sequences complementary to several miRNAs, while many miRNAs contain sequences complementary to the

3'UTRs of several genes. Thus, one gene can potentially be repressed by several miRNAs (either individually or in combination) while one miRNA can potentially repress the translation of several genes. In Chapter 11 you saw that the organization of bacterial genes into operons permitted the coordinate regulation of genes that contributed to a single trait, such as the ability to utilize the sugar lactose. Given that most eukaryotic gene are not organized into operons, it has been suggested that post-transcriptional regulation of several genes by one miRNA affords higher organisms the ability to coordinate the expression of their genes.

Summary

Many aspects of eukaryotic gene regulation resemble the regulation of bacterial operons. Both operate largely at the level of transcription, and both rely on trans-acting proteins that bind to cis-acting regulatory target sequences on the DNA molecule. These regulatory proteins determine the level of transcription from a gene by controlling the binding of RNA polymerase to the gene's promoter.

There are three major distinguishing features of the control of transcription in eukaryotes. First, eukaryotic genes possess enhancers, which are cis-acting regulatory elements located at sometimes great linear distances from the promoter. Many genes possess multiple enhancers. Second, these enhancers are often bound by more transcription factors than are bacterial operons. Multicellular eukaryotes must generate thousands of patterns of gene expression with a limited number of regulatory proteins (transcription factors). They do so through combinatorial interactions among transcription factors. Enhanceosomes are complexes of regulatory proteins that interact in a cooperative and synergistic fashion to promote high levels of transcription through the recruitment of RNA polymerase II to the transcription start site.

Third, eukaryotic genes are packaged in chromatin. Gene activation and repression require specific modifications to chromatin. The vast majority of the tens of thousands of genes in a typical eukaryotic genome are turned off at any one time. Genes are maintained in a transcriptionally inactive state through the participation of nucleosomes, which serve to compact the chromatin and prevent the binding of RNA polymerase II. The position of nucleosomes and the extent of chromatin condensation are instructed by the pattern of post-translational modifications of the histone tails.

Histone modifications are epigenetic marks that, along with the methylation of cytosine bases, can be altered by transcription factors. These factors bind to regulatory regions and recruit protein complexes that enzymatically modify adjacent nucleosomes. These large multisubunit protein complexes use the energy of ATP hydrolysis to move nucleosomes and remodel chromatin.

DNA replication faithfully copies both the DNA sequence and the chromatin structure from parent to daughter cells. Newly formed cells inherit both genetic information, inherent in the nucleotide sequence of DNA, and epigenetic information, which is in the histone code and the pattern of DNA methylation.

The existence of epigenetic phenomena such as genetic imprinting and X-chromosome inactivation demonstrates that eukaryotic gene expression can be silenced without changing the DNA sequence of the gene. Another epigenetic phenomenon, position-effect variegation, revealed the existence of repressive heterochromatic domains that are associated with highly condensed nucleosomes and contain few genes. Barrier insulators maintain the integrity of the genome by preventing the conversion of euchromatin into heterochromatin.

There is a growing appreciation for the role of functional RNAs, such as ncRNAs and miRNAs, in the regulation of eukaryotic gene expression. These RNAs serve to target protein complexes to complementary DNA or RNA in the cell. For some RNAs (like Xist), the act of transcription may tether the RNA to a chromosomal region where proteins will bind and alter chromatin. In contrast, the translation of hundreds of mRNAs is repressed when RISC bound to complementary miRNAs is targeted to their 3′ UTRs.

KEY TERMS

activation domain (p. 422)
Barr body (p. 444)
barrier insulator (p. 441)
chromatin remodeling (p. 428)
co-activator (p. 424)
constitutive heterochromatin (p. 437)
corepressor (p. 430)
CpG island (p. 431)

dosage compensation (p. 444)
enhanceosome (p. 432)
enhancer (p. 418)
enhancer-blocking insulator (p. 434)
epigenetic inheritance (p. 427)
euchromatin (p. 437)
genomic imprinting (p. 441)
hemimethylation (p. 431)

heterochromatin (p. 437)
heterochromatin protein-1 (HP-1) (p. 439)
histone code (p. 429)
histone deacetylase (HDACs) (p. 430)
histone tail (p. 429)
hyperacetylation (p. 430)
hypoacetylation (p. 430)
maternal imprinting (p. 441)

mediator complex (p. 424)
paternal imprinting (p. 441)
position-effect variegation (PEV)
  (p. 438)

post-transcriptional gene silencing
  (p. 435)
promoter-proximal element (p. 418)
reporter gene (p. 422)

synergistic effect (p. 432)
transcriptional gene silencing (p. 435)
upstream activation sequence (UAS)
  (p. 421)

PROBLEMS

Most of the problems are also available for review/grading through the GENETICSPORTAL www.yourgeneticsportal.com.

WORKING WITH THE FIGURES

1. In Figure 12-4, certain mutations decrease the relative transcription rate of the β-globin gene. Where are these mutations located, and how do they exert their effects on transcription?

2. Based on the information in Figure 12-6, how does Gal4 regulate four different *GAL* genes at the same time? Contrast this mechanism with how the Lac repressor controls the expression of three genes.

3. In any experiment, controls are essential in order to determine the specific effect of changing some parameter. In Figure 12-7, which constructs are the "controls" that serve to establish the principle that activation domains are modular and interchangeable?

4. Contrast the role of the MCM1 protein in different yeast cell types shown in Figure 12-10. How are the **a**-specific genes controlled differently in different cell types?

5. In Figure 12-11b, in what chromosomal region are you likely to find the most H1 histone protein?

6. What is the conceptual connection between Figures 12-13 and 12-19?

7. In Figure 12-19, where is the TATA box located before the enhanceosome forms at the top of the figure?

8. Let's say that you have incredible skill and can isolate the white and red patches of tissue from the *Drosophila* eyes shown in Figure 12-24 in order to isolate mRNA from each tissue preparation. Using your knowledge of DNA techniques from Chapter 10, design an experiment that would allow you to determine whether RNA is transcribed from the white gene in the red tissue or the white tissue or both. If you need it, you have access to radioactive white-gene DNA.

9. In Figure 12-26, provide a biochemical mechanism for why HP-1 can bind to the DNA only on the left side of the barrier insulator. Similarly, why can HMTase bind only to the DNA on the left of the barrier insulator?

10. In reference to Figure 12-28, draw the outcome if there is a mutation in gene B.

BASIC PROBLEMS

11. What analogies can you draw between transcriptional trans-acting factors that activate gene expression in eukaryotes and the corresponding factors in bacteria? Give an example.

12. Contrast the states of genes in bacteria and eukaryotes with respect to gene activation.

13. Predict and explain the effect on *GAL1* transcription, in the presence of galactose alone, of the following mutations:

    a. Deletion of one Gal4-binding site in the *GAL1* UAS element.

    b. Deletion of all four Gal4-binding sites in the *GAL1* UAS element.

    c. Deletion of the Mig1-binding site upstream of *GAL1*.

    d. Deletion of the Gal4 activation domain.

    e. Deletion of the *GAL80* gene.

    f. Deletion of the *GAL1* promoter.

    g. Deletion of the *GAL3* gene.

14. How is the activation of the *GAL1* gene prevented in the presence of galactose and glucose?

15. What are the roles of histone deacetylation and histone acetylation in gene regulation, respectively?

16. An α strain of yeast that cannot switch mating type is isolated. What mutations might it carry that would explain this phenotype?

17. What genes are regulated by the α1 and α2 proteins in an α cell?

18. What are Sir proteins? How do mutations in *SIR* genes affect the expression of mating-type cassettes?

19. What is meant by the term *epigenetic inheritance*? What are two examples of such inheritance?

20. What is an enhanceosome? Why could a mutation in any one of the enhanceosome proteins severely reduce the transcription rate?

21. Why are mutations in imprinted genes usually dominant?

22. What has to happen for the expression of two different genes on two different chromosomes to be regulated by the same miRNA?

23. What mechanisms are thought to be responsible for the inheritance of epigenetic information?

24. What is the fundamental difference in how bacterial and eukaryotic genes are regulated?

25. Why is it said that transcriptional regulation in eukaryotes is characterized by combinatorial interactions?

26. The following diagram represents the structure of a gene in *Drosophila melanogaster;* blue segments are exons, and yellow segments are introns.

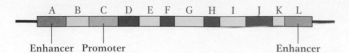

Enhancer   Promoter   Enhancer

a. Which segments of the gene will be represented in the initial RNA transcript?

b. Which segments of the gene will be removed by RNA splicing?

c. Which segments would most likely bind proteins that interact with RNA polymerase?

CHALLENGING PROBLEMS

27. The transcription of a gene called *YFG* (your *f*avorite *g*ene) is activated when three transcription factors (TFA, TFB, TFC) interact to recruit the co-activator CRX. TFA, TFB, TFC, and CRX and their respective binding sites constitute an enhanceosome located 10 kb from the transcription start site. Draw a diagram showing how you think the enhanceosome functions to recruit RNA polymerase to the promoter of *YFG*.

28. A single mutation in one of the transcription factors in Problem 27 results in a drastic reduction in *YFG* transcription. Diagram what this mutant interaction might look like.

29. Diagram the effect of a mutation in the binding site for one of the transcription factors in Problem 27.

30. How does an epigenetically silenced gene differ from a mutant gene (a null allele of the same gene)?

31. What are epigenetic marks? Which are associated with heterochromatin? How are epigenetic marks thought to be interpreted into chromatin structure?

32. You receive four strains of yeast in the mail and the accompanying instructions state that each strain contains a single copy of transgene *A*. You grow the four strains and determine that only three strains express the protein product of transgene *A*. Further analysis reveals that transgene *A* is located at a different position in the yeast genome in each of the four strains. Provide a hypothesis to explain this result.

33. In *Neurospora*, all mutants affecting the enzymes carbamyl phosphate synthetase and aspartate transcarbamylase map at the *pyr*-3 locus. If you induce *pyr*-3 mutations by ICR-170 (a chemical mutagen), you find that either both enzyme functions are lacking or only the transcarbamylase function is lacking; in no case is the synthetase activity lacking when the transcarbamylase activity is present. (ICR-170 is assumed to induce frameshifts.) Interpret these results in regard to a possible operon.

34. You wish to find the cis-acting regulatory DNA elements responsible for the transcriptional responses of two genes,

c-fos and *globin*. Transcription of the *c-fos* gene is activated in response to fibroblast growth factor (FGF), but it is inhibited by cortisol (Cort). On the other hand, transcription of the *globin* gene is not affected by either FGF or cortisol, but it is stimulated by the hormone erythropoietin (EP). To find the cis-acting regulatory DNA elements responsible for these transcriptional responses, you use the following clones of the *c-fos* and *globin* genes, as well as two "hybrid" combinations (fusion genes), as shown in the diagram below. The letter A represents the intact *c-fos* gene, D represents the intact *globin* gene, and B and C represent the *c-fos–globin* gene fusions. The *c-fos* and *globin* exons (E) and introns (I) are numbered. For example, E3(f) is the third exon of the *c-fos* gene and I2(g) is the second intron of the *globin* gene. (These labels are provided to help you make your answer clear.) The transcription start sites (black arrows) and polyadenylation sites (red arrows) are indicated.

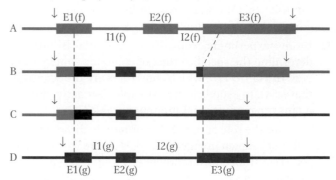

You introduce all four of these clones simultaneously into tissue-culture cells and then stimulate individual aliquots of these cells with one of the three factors. Gel analysis of the RNA isolated from the cells gives the following results.

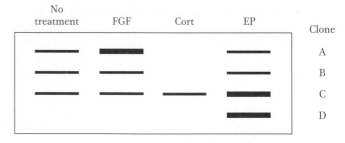

The levels of transcripts produced from the introduced genes in response to various treatments are shown; the intensity of these bands is proportional to the amount of transcript made from a particular clone. (The failure of a band to appear indicates that the level of transcript is undetectable.)

a. Where is the DNA element that permits activation by FGF?

b. Where is the DNA element that permits repression by Cort?

c. Where is the DNA element that permits induction by EP? Explain your answer.

The Genetic Control 13
of Development

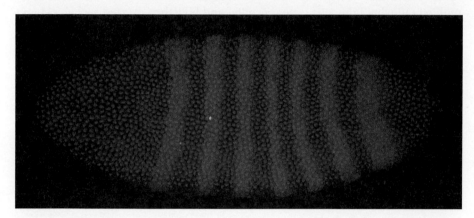

Gene expression in a developing fruit-fly embryo. The seven magenta stripes mark the cells expressing the mRNA of a gene encoding a regulatory protein that controls segment number in the *Drosophila* embryo. The spatial regulation of gene expression is central to the control of animal development. [*Photograph by Dave Kosman, Ethan Bier, and Bill McGinnis.*]

O f all the phenomena in biology, few if any inspire more awe than the formation of a complex animal from a single-celled egg. In this spectacular transformation, unseen forces organize the dividing mass of cells into a form with a distinct head and tail, various appendages, and many organs. The great geneticist Thomas Hunt Morgan was not immune to its aesthetic appeal:

> A transparent egg as it develops is one of the most fascinating objects in the world of living beings. The continuous change in form that takes place from hour to hour puzzles us by its very simplicity. The geometric patterns that present themselves at every turn invite mathematical analysis. . . . This pageant makes an irresistible appeal to the emotional and artistic sides of our nature.
> [T. H. Morgan, *Experimental Embryology.* Columbia University Press, 1927.]

Yet, for all its beauty and fascination, biologists were stumped for many decades concerning how biological form is generated during development. Morgan also said that "if the mystery that surrounds embryology is ever to come within our comprehension, we must . . . have recourse to other means than description of the passing show."

The long drought in embryology lasted well beyond Morgan's heyday in the 1910s and 1920s, but it was eventually broken by geneticists working very much in the tradition of Morgan-style genetics and with his favorite, most productive genetic model, the fruit fly *Drosophila melanogaster.*

The key catalysts to understanding the making of animal forms were the discoveries of genetic monsters—mutant fruit flies with dramatic alterations of body structures (Figure 13-1). In the early days of *Drosophila* genetics, rare mutants arose

KEY QUESTIONS

- Which genes control development, and how are they identified?
- Where and when are these genes active in the course of development?
- How are pattern-regulating genes controlled?
- How do pattern-regulating genes affect animal form?
- Do different taxa have pattern-regulating genes and processes in common?

OUTLINE

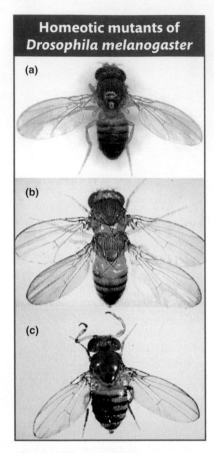

Homeotic mutants of Drosophila melanogaster

(a)

(b)

(c)

Figure 13-1 In homeotic mutants, the identity of one body structure has been changed into another. (a) Normal fly with one pair of forewings on the second thoracic segment and one pair of small hind wings on the third thoracic segment. (b) Triple mutant for three mutations in the *Ultrabithorax* gene. *Ubx* function is lost in the posterior thorax, which causes the development of forewings in place of the hind wings. (c) *Antennapedia* mutant in which the antennae are transformed into legs. [*Courtesy of Sean Carroll.*]

spontaneously or as by-products of other experiments with spectacular transformations of body parts. In 1915, Calvin Bridges, then Morgan's student, isolated a fly having a mutation that caused the tiny hind wings (halteres) of the fruit fly to resemble the large forewings. He dubbed the mutant *bithorax*. The transformation in *bithorax* mutants is called *homeotic* (Greek *homeos*, meaning same or similar) because one part of the body (the hind wing) is transformed to resemble another (the forewing), as shown in Figure 13-1b. Subsequently, several more homeotic mutants were identified in *Drosophila*, such as the dramatic *Antennapedia* mutant in which legs develop in place of the antennae (Figure 13-1c).

The spectacular effects of homeotic mutants inspired what would become a revolution in embryology, once the tools of molecular biology became available to understand what homeotic genes encoded and how they exerted such enormous influence on the development of entire body parts. Surprisingly, these strange fruit-fly genes turned out to be a passport to the study of the entire animal kingdom, as counterparts to these genes were discovered that played similar roles in almost all animals.

The study of development is a large and still-growing discipline. In this chapter, we will focus on a few concepts that illustrate the logic of the genetic control of the development of animal body plans and we will see how the information for building complex organisms is encoded in the genome. The genetic control of body formation and body patterning is fundamentally a matter of gene regulation in three-dimensional space and over time. We will see that the principles governing the genetic control of development are connected to those already presented in Chapters 11 and 12 governing the physiological control of gene expression in bacteria and single-celled eukaryotes.

13.1 The Genetic Approach to Development

For many decades, the study of embryonic development largely entailed the physical manipulation of embryos, cells, and tissues. Several key concepts were established about the properties of developing embryos through experiments in which one part of an embryo was transplanted into another part of the embryo. For example, the transplantation of a part of a developing amphibian embryo to another site in a recipient embryo was shown to induce the surrounding tissue to form a second complete body axis (Figure 13-2a). Similarly, transplantation of the posterior part of a developing chick limb bud to the anterior could induce extra digits, but with reversed polarity with respect to the normal digits (Figure 13-2b). These transplanted regions of the amphibian embryo and chick limb bud were termed *organizers* because of their remarkable ability to organize the development of surrounding tissues. The cells in the organizers were postulated to produce *morphogens*, molecules that induced various responses in surrounding tissue in a concentration-dependent manner.

Although these experimental results were spectacular and fascinating, further progress in understanding the nature of organizers and morphogens stalled after their discovery in the first half of the 1900s. It was essentially impossible to isolate the molecules responsible for these activities by using biochemical separation techniques. Embryonic cells make thousands of substances—proteins, glycoconjugates, hormones, and so forth. A morphogen could be any one of these molecules but would be present in minuscule quantities—one needle in a haystack of cellular products.

The long impasse in defining embryology in molecular terms was broken by genetic approaches—mainly the systematic isolation of mutants with discrete

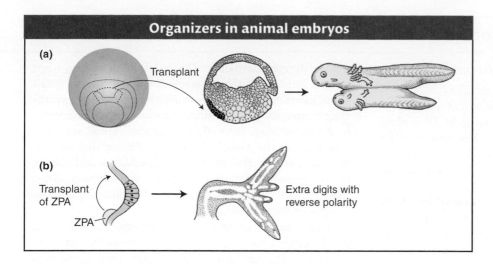

Organizers in animal embryos

(a) Transplant

(b) Transplant of ZPA

ZPA

Extra digits with reverse polarity

Figure 13-2 Transplantation experiments played a central role in early embryology and demonstrated the long-range organizing activity of embryonic tissues. (a) The Spemann organizer. The dorsal blastopore "lip" of an early amphibian embryo can induce a second embryonic axis and embryo when transplanted to the ventral region of a recipient embryo. (b) In the developing chick vertebral limb bud, the zone of polarizing activity (ZPA) organizes pattern along the anteroposterior axis. Transplantation of the ZPA from a posterior to anterior position induces extra digits with reverse polarity.

defects in development and the subsequent characterization and study of the gene products that they encoded. The genetic approach to studying development presented many advantages over alternative, biochemical strategies. First, the geneticist need not make any assumptions about the number or nature of molecules required for a process. Second, the (limited) quantity of a gene product is no impediment: all genes can be mutated regardless of the amount of product made by a gene. And, third, the genetic approach can uncover phenomena for which there is no biochemical or other bioassay.

From the genetic viewpoint, there are four key questions concerning the number, identity, and function of genes taking part in development:

1. Which genes are important in development?

2. Where in the developing animal and at what times are these genes active?

3. How is the expression of developmental genes regulated?

4. Through what molecular mechanisms do gene products affect development?

To address these questions, strategies had to be devised to identify, catalog, and analyze genes that control development. One of the first considerations in the genetic analysis of animal development was which animal to study. Of the millions of living species, which offered the most promise? The fruit fly *Drosophila melanogaster* emerged as the leading genetic model of animal development because its ease of rearing, rapid life cycle, cytogenetics, and decades of classical genetic analysis (including the isolation of many very dramatic mutants) provided important experimental advantages (see the Model Organism box on *Drosophila* on pages 452–453). The nematode worm *Caenorhabditis elegans* also presented many attractive features, most particularly its simple construction and well-studied cell lineages. Among vertebrates, the development of targeted gene disruption techniques opened up the laboratory mouse to more systematic genetic study, and the zebrafish *Danio rerio* has recently become a favorite model owing to the transparency of the embryo and to advances in its genetic study.

Through systematic and targeted genetic analysis, as well as comparative genomic studies, much of the *genetic toolkit* for the development of the bodies, body parts, and cell types of several different species has been defined. We will first focus on the genetic toolkit of *Drosophila melanogaster* because its identification was a source of major insights into the genetic control of development; its discovery catalyzed the identification of the genetic toolkit of other animals, including humans.

Model Organism *Drosophila*

Mutational Analysis of Early *Drosophila* Development

The initial insights into the genetic control of pattern formation emerged from studies of the fruit fly *Drosophila melanogaster*. *Drosophila* development has proved to be a gold mine to researchers because developmental problems can be approached by the use of genetic and molecular techniques simultaneously.

The *Drosophila* embryo has been especially important in understanding the formation of the basic animal body plan. One important reason is that an abnormality in the body plan of a mutant is easily identified in the larval exoskeleton in the *Drosophila* embryo. The larval exoskeleton is a noncellular structure, made of a polysaccharide polymer called chitin that is produced as a secretion of the epidermal cells of the embryo. Each structure of the exoskeleton is formed from epidermal cells or cells immediately underlying that structure. With its intricate pattern of hairs, indentations, and other structures, the exoskeleton provides numerous landmarks to serve as indicators of the fates assigned to the many epidermal cells. In particular, there are many distinct anatomical structures along the antero-posterior (A–P) and dorsoventral (D–V) axes. Furthermore, because all the nutrients necessary to develop to the larval stage are prepackaged in the egg, mutant embryos in which the A–P or D–V cell fates are drastically altered can nonetheless develop to the end of embryogenesis and produce a mutant larva in about 1 day (see diagram). The exoskeleton of such a mutant larva mirrors the mutant fates assigned to subsets of the epidermal cells and can thus identify genes worthy of detailed analysis.

The development of the *Drosophila* adult body pattern takes a little more than a week (see illustration). Small populations of cells set aside during embryogenesis proliferate during three larval stages (instars) and differentiate in the pupal stage into adult structures. These set-aside cells include the *imaginal disks,* which are disk-shaped regions that give rise to specific appendages and tissues in each segment as the leg, wing, eye, and antennal disks. Imaginal disks are easy to remove for analysis of gene expression (see Figure 13-7).

Genes that contribute to the *Drosophila* body plan can be cloned and characterized at the molecular level with ease. The analysis of the cloned genes often provides valuable information on the function of the protein product—usually by identifying close relatives in amino acid sequence of the encoded polypeptide through comparisons with all the protein sequences stored in public databases. In addition, one can investigate the spatial and temporal patterns of expression of (1) an mRNA, by using histochemically tagged single-stranded DNA sequences complementary to the mRNA to perform RNA in situ hybridization, or (2) a protein, by using histochemically tagged antibodies that bind specifically to that protein.

Using Knowledge from One Model Organism to Fast-Track Developmental Gene Discovery in Others

With the discovery that there are numerous homeobox genes within the *Drosophila* genome, similarities among the DNA sequences of these genes could be exploited in treasure hunts for other members of the homeotic-gene

13.2 The Genetic Toolkit for *Drosophila* Development

Animal genomes typically contain about 13,000 to 22,000 genes. Many of these genes encode proteins that function in essential processes in all cells of the body (for example, in cellular metabolism or the biosynthesis of macromolecules). Such genes are often referred to as **housekeeping genes.** Other genes encode proteins that carry out the specialized tasks of various organ systems, tissues, and cells of the body such as the globin proteins in oxygen transport or antibody proteins that mediate immunity. Here, we are interested in a different set of genes, those concerned with the building of organs and tissues and the specification of cell types—the toolkit that determines the overall body plan and the number, identity, and pattern of body parts.

Toolkit genes of the fruit fly have generally been identified through the monstrosities or catastrophes that arise when they are mutated. Toolkit-gene mutations from two sources have yielded most of our knowledge. The first source consists of spontaneous mutations that arise in laboratory populations.

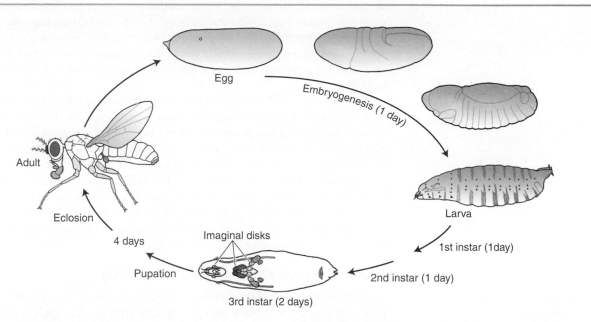

Overview of *Drosophila* development. The larva forms in 1 day and then undergoes several stages of growth during which the imaginal disks and other precursors of adult structures proliferate. These structures differentiate during pupation, and the adult fly hatches (eclosion) and begins the cycle again.

family. These hunts depend on DNA base-pair complementarity. For this purpose, DNA hybridizations were carried out under *moderate stringency conditions,* in which there could be some mismatch of bases between the hybridizing strands without disrupting the proper hydrogen bonding of nearby base pairs. Some of these treasure hunts were carried out in the *Drosophila* genome itself, in looking for more family members. Others searched for homeobox genes in other animals, by means of *zoo blots* (Southern blots of restriction-enzyme-digested DNA from different animals), by using radioactive *Drosophila* homeobox DNA as the probe. This approach led to the discovery of homologous homeobox sequences in many different animals, including humans and mice. (Indeed, it is a very powerful approach for "fishing" for relatives of almost any gene in your favorite organism.)

The second source comprises mutations induced at random by treatment with mutagens (such as chemicals or radiation) that greatly increase the frequency of damaged genes throughout the genome. Elegant refinements of the latter approach have made possible systematic searches for mutants that have identified many members of the fly's genetic toolkit. The members of this toolkit constitute only a small fraction, perhaps several hundred genes, of the roughly 14,000 genes in the fly genome.

> Message The genetic toolkit for animal development is composed of a small fraction of all genes. Only a small subset of the entire complement of genes in the genome affect development in discrete ways.

Classification of genes by developmental function

One of the first tasks following the execution of a genetic screen for mutations is to sort out those of interest. Many mutations are lethal when hemi- or homozygous because cells cannot survive without products affected by these mutations. The more interesting mutations are those that cause some discrete defect in either the embryonic or the adult body pattern or both. It has proved useful to group the genes affected by mutations into several categories based on the nature of their mutant phenotypes. Many toolkit genes can be classified according to their function in controlling the identity of body parts (for example, of different segments or appendages), the formation of body parts (for example, of organs and appendages), the number of body parts, the formation of cell types, and the organization of the primary body axes (the anteroposterior, or A–P, and dorsoventral, or D–V, axes).

We will begin our inventory of the *Drosophila* toolkit by examining the genes that control the identity of segments and appendages. We do so for both historical and conceptual purposes. The genes controlling segmental and appendage identity were among the very first toolkit genes identified. Subsequent discoveries about their nature were sources of profound insights into not just how their products work, but also the content and workings of the toolkits of most animals. Furthermore, their spectacular mutant phenotypes indicate that they are among the most globally acting genes that affect animal form. Learning about these genes should whet our appetites for learning more about the whole toolkit that controls the development of animal form.

Homeotic genes and segmental identity

Among the most fascinating abnormalities to be described in animals are those in which one normal body part is replaced by another. Such homeotic transformations have been observed in many species in nature, including sawflies in which a leg forms in place of an antenna and frogs in which a thoracic vertebra forms in place of a cervical vertebra (Figure 13-3). Whereas only one member of a bilateral pair of structures is commonly altered in many naturally occurring variants, both members of a bilateral pair of structures are altered in homeotic mutants of fruit flies. Such homeotic mutants breed true from generation to generation.

The scientific fascination with homeotic mutants stems from three properties. First, it is amazing that a single gene mutation can alter a developmental pathway so dramatically. Second, it is striking that the structure formed in the mutant is a well-developed likeness of another body part. And, third, it is important to note

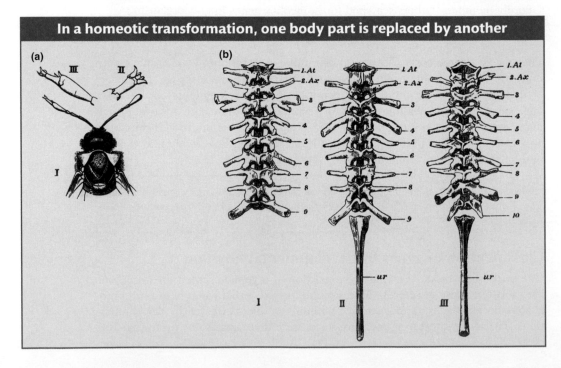

In a homeotic transformation, one body part is replaced by another

Figure 13-3 A late-nineteenth-century drawing from one of the first studies of homeotic transformations in nature. (a) Homeosis in a sawfly, with the left antenna transformed into a leg. (b) Homeosis in a frog. The middle specimen is normal. The specimen on the left has extra structures growing out of the top of the vertebral column. The specimen on the right has an extra set of vertebrae. [*From W. Bateson, Material for the Study of Variation. Macmillan, 1894.*]

that homeotic mutations transform the identity of **serially reiterated structures.** Insect and many animal bodies are made of repeating parts of similar structure, like building blocks, arranged in a series. The forewings and hind wings, the segments, and the antennae, legs, and mouthparts of insects are sets of serially reiterated body parts. Homeotic mutations transform identities within these sets.

A mutation may cause a loss of homeotic gene function where the gene normally acts or it may cause a gain of homeotic function where the homeotic gene does not normally act. For example, the *Ultrabithorax* (*Ubx*) gene acts in the developing hind wing to promote hind-wing development and to repress forewing development. Loss-of-function mutations in *Ubx* transform the hind wing into a forewing. Dominant gain-of-function mutations in *Ubx* transform the forewing into a hind wing. Similarly, the antenna-to-leg transformations of *Antennapedia* (*Antp*) mutants are caused by the dominant gain of *Antp* function in the antenna. In addition to these transformations in appendage identity, homeotic mutations can transform segment identity, causing one body segment of the adult or larva to resemble another.

Although homeotic genes were first identified through spontaneous mutations affecting adult flies, they are required throughout most of a fly's development. Systematic searches for homeotic genes have led to the identification of eight loci, now referred to as ***Hox* genes,** that affect the identity of segments and their associated appendages in *Drosophila*. Generally, the complete loss of any *Hox*-gene function is lethal in early development. The dominant mutations that transform adults are viable in heterozygotes because the wild-type allele provides normal gene function to the developing animal.

Organization and expression of *Hox* genes

A most intriguing feature of *Hox* genes is that they are clustered together in two **gene complexes** that are located on the third chromosome of *Drosophila*. The *Bithorax* complex contains three *Hox* genes, and the *Antennapedia* complex contains five *Hox* genes. Moreover, the order of the genes in the complexes and on the chromosome corresponds to the order of body regions, from head to tail, that are influenced by each *Hox* gene (Figure 13-4).

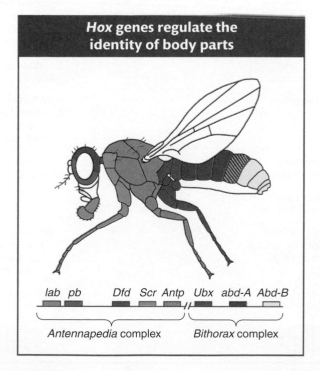

Figure 13-4 The *Hox* genes of *Drosophila*. Eight *Hox* genes regulate the identity of regions within the adult. The color coding identifies the segments and structures that are affected by mutations in the various *Hox* genes. [*After S. B. Carroll, J. K. Grenier, and S. D. Weatherbee, From DNA to Diversity: Molecular Genetics and the Evolution of Animal Design, 2nd ed. Blackwell, 2005.*]

The relation between the structure of the *Hox*-gene complexes and the phenotypes of *Hox*-gene mutants was illuminated by the molecular characterization of the genes. Molecular cloning of the sequences encompassing each *Hox* locus provided the means to analyze where in the developing animal each gene is expressed. These spatial aspects of gene expression and gene regulation are crucial to understanding the logic of the genetic control of development. In regard to the *Hox* genes and other toolkit genes, the development of technology that made possible the visualization of gene and protein expression was crucial to understanding the relation among gene organization, gene function, and mutant phenotypes.

Two principal technologies for the visualization of gene expression in embryos or other tissues are (1) the expression of RNA transcripts visualized by in situ hybridization and (2) the expression of Hox proteins visualized by immunological methods. Each technology depends on the isolation of cDNA clones representing the mature mRNA transcript and protein (Figure 13-5).

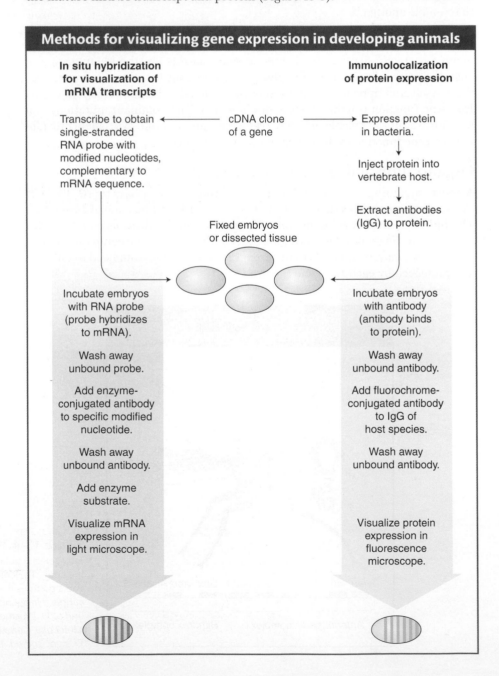

Figure 13-5 The two principal technologies for visualizing where a gene is transcribed or where the protein that it encodes is expressed are (*left*) in situ hybridization of complementary RNA probe to mRNA and (*right*) immunolocalization of protein expression. The procedures for each method are outlined. Expression patterns may be visualized as the product of an enzymatic reaction or of a chromogenic substrate or with fluorescently labeled compounds.

In the developing embryo, the *Hox* genes are expressed in spatially restricted, sometimes overlapping domains within the embryo (Figure 13-6). The genes are also expressed in the larval and pupal tissues that will give rise to the adult body parts.

The patterns of *Hox*-gene expression (and other toolkit genes) generally correlate with the regions of the animal affected by gene mutations. For example, the purple shading in Figure 13-6 indicates where the *Ubx* gene is expressed. This *Hox* gene is expressed in the posterior thoracic and most of the abdominal segments of the embryo. The development of these segments is altered in *Ubx* mutants. *Ubx* is also expressed in the developing hind wing but not in the developing forewing (Figure 13-7), as one would expect knowing that *Ubx* promotes hind-wing development and represses forewing development in this appendage.

> **Message** The spatial expression of toolkit genes is usually closely correlated with the regions of the animal affected by gene mutations.

It is crucial to distinguish the role of *Hox* genes in determining the *identity* of a structure from that governing its *formation*. In the absence of function of all *Hox* genes, segments form, but they all have the same identity; limbs also can form, but they have antennal identity; and, similarly, wings can form, but they have forewing identity. Other genes control the formation of segments, limbs, and wings and will be described later. First, we must understand how *Hox* genes exert their dramatic effects on fly development.

The homeobox

Because *Hox* genes have large effects on the identities of entire segments and other body structures, the nature and function of the proteins that they encode are of special interest. Edward Lewis, a pioneer in the study of homeotic genes, noted early on that the clustering of *Bithorax* complex genes suggested that the

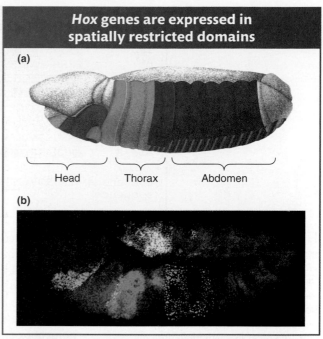

Hox genes are expressed in spatially restricted domains

(a)

Head   Thorax   Abdomen

(b)

Figure 13-6 Expression of *Hox* genes in the *Drosophila* embryo. (a) Schematic representation of *Drosophila* embryo showing regions where eight individual *Hox* genes are expressed. (b) Actual image of the expression of seven *Hox* genes visualized by in situ hybridization. Colors indicate expression of *labial* (turquoise), *Deformed* (lavender), *Sex combs reduced* (green), *Antennapedia* (orange), *Ultrabithorax* (purple), *Abdominal-A* (red), and *Abdominal-B* (yellow). The embryo is folded so that the posterior end (yellow) appears near the top center. [*(b) Photomicrograph by Dave Kosman, Ethan Bier, and Bill McGinnis.*]

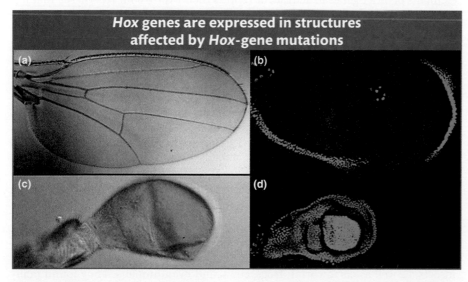

Hox genes are expressed in structures affected by Hox-gene mutations

(a)   (b)

(c)   (d)

Figure 13-7 An example of *Hox*-gene expression. (a) The adult wing of *D. melanogaster*. (b) Ubx protein is not expressed in cells of the developing imaginal disk that will form the forewing. Cells enriched in Hox proteins are stained green; in this image, the green-stained cells are cells that do *not* form the wing. (c) The adult hind wing (haltere). (d) The Ubx protein is expressed at high levels in all cells of the developing hind-wing imaginal disk. [*Photographs by Scott Weatherbee.*]

Hox proteins have a sequence in common

lab	NNSGRTNFTNKQLTELEKEFHFNRYLTRARRIEIANTLQLNETQVKIWFQNRRMKQKKRV
pb	PRRLRTAYTNTQLLELEKEFHFNKYLCRPRRIEIAASLDLTERQVKVWFQNRRMKHKRQT
Dfd	PKRQRTAYTRHQILELEKEFHYNRYLTRRRRIEIAHTLVLSERQIKIWFQNRRMKWKKDN
Scr	TKRQRTSYTRYQTLELEKEFHFNRYLTRRRRIEIAHALCLTERQIKIWFQNRRMKWKKEH
Antp	RKRGRQTYTRYQTLELEKEFHFNRYLTRRRRIEIAHALCLTERQIKIWFQNRRMKWKKEN
Ubx	RRRGRQTYTRYQTLELEKEFHTNHYLTRRRRIEMAHALCLTERQIKIWFQNRRMKLKKEI
abd-A	RRRGRQTYTRFQTLELEKEFHFNHYLTRRRRIEIAHALCLTERQIKIWFQNRRMKLKKEL
abd-B	VRKKRKPYSKFQTLELEKEFLFNAYVSKQKRWELARNLQLTERQVKIWFQNRRMKNKKNS

Consensus -RRGRT-YTR-QTLELEKEFHFNRYLTRRRRIEIAHALCLTERQIKIWFQNRRMK-KKE-
sequence   Helix 1   Helix 2   Helix 3

Figure 13-8 Sequences of fly homeodomains. All eight *Drosophila Hox* genes encode proteins containing a highly conserved 60 amino acid domain, the homeodomain, composed of three α helices. Helices 2 and 3 form a helix-turn-helix motif similarly to the Lac repressor, Cro, and other DNA-binding proteins. Residues common to the *Hox* genes are shaded in yellow; divergent residues are shaded in red; those common to subsets of proteins are shaded in blue or green. [*From S. B. Carroll, J. K. Grenier, and S. D. Weatherbee, From DNA to Diversity: Molecular Genetics and the Evolution of Animal Design, 2nd ed. Blackwell, 2005.*]

multiple loci had arisen by tandem duplication of an ancestral gene. This idea led researchers to search for similarities in the DNA sequences of *Hox* genes. They found that all eight *Hox* genes of the two complexes were similar enough to hybridize to each other. This hybridization was found to be due to a short region of sequence in each gene, 180 bp in length. This stretch of DNA sequence similarity, because of its presence in homeotic genes, was dubbed the **homeobox.** The homeobox encodes a protein domain, the **homeodomain,** containing 60 amino acids. The amino acid sequence of the homeodomain is very similar among the Hox proteins (Figure 13-8).

Although the discovery of a common protein motif in each of the Hox proteins was very exciting, further analysis of the structure of the homeodomain revealed that it forms a helix-turn-helix motif—the structure common to the Lac repressor, the λ repressor, Cro, and the α2 and a1 regulatory proteins of the yeast mating-type loci. This similarity suggested immediately (and it was subsequently borne out) that Hox proteins are sequence-specific DNA-binding proteins and that they exert their effects by controlling the expression of genes within developing segments and appendages. Thus, the products of these remarkable genes function through principles that are already familiar from Chapters 11 and 12—by binding to regulatory elements of other genes to activate or repress their expression. We will see that it is also true of many other toolkit genes: a significant fraction of these genes encode transcription factors that control the expression of other genes.

> **Message** Many toolkit genes encode transcription factors that regulate the expression of other genes.

We will examine how Hox proteins and other toolkit proteins orchestrate gene expression in development a little later. First, there is one more huge discovery to describe, which revealed that what we learn from fly *Hox* genes has very general implications for the animal kingdom.

Clusters of *Hox* genes control development in most animals

When the homeobox was discovered in fly *Hox* genes, it raised the question whether this feature was some peculiarity of these bizarre fly genes or was more widely distributed, in other insects or segmented animals, for example. To address this possibility, researchers searched for homeoboxes in the genomes of other insects, as well as earthworms, frogs, cows, and even humans. They found many homeoboxes in each of these animal genomes.

The similarities in the homeobox sequences from different species were astounding. Over the 60 amino acids of the homeodomain, some mouse and frog Hox proteins were identical with the fly sequences at as many as 59 of the 60 positions (Figure 13-9). In light of the vast evolutionary distances between

Drosophila and vertebrate Hox protein show striking similarities

Fly *Dfd*	PKRQRTAYTRHQILELEKEFHYNRYLTRRRRIEIAHTLVLSERQIKIWFQNRRMKWKKDN	KLPNTKNVR
Amphibian *Hox4*	TKRSRTAYTRQQVLELEKEFHFNRYLTRRRRIEIAHSLGLTERQIKIWFQNRRMKWKKDN	RLPNTKTRS
Mouse *HoxB4*	PKRSRTAYTRQQVLELEKEFHYNRYLTRRRRVEIAHALCLSERQIKIWFQNRRMKWKKDH	KLPNTKIRS
Human *HoxB4*	PKRSRTAYTRQQVLELEKEFHYNRYLTRRRRVEIAHALCLSERQIKIWFQNRRMKWKKDH	KLPNTKIRS
Chick *HoxB4*	PKRSRTAYTRQQVLELEKEFHYNRYLTRRRRVEIAHSLCLSERQIKIWFQNRRMKWKKDH	KLPNTKIRS
Frog *HoxB4*	AKRSRTAYTRQQVLELEKEFHYNRYLTRRRRVEIAHTLRLSERQIKIWFQNRRMKWKKDH	KLPNTKIKS
Fugu *HoxB4*	PKRSRTAYTRQQVLELEKEFHYNRYLTRRRRVEIAHTLCLSERQIKIWFQNRRMKWKKDH	KLPNTKVRS
Zebrafish *HoxB4*	AKRSRTAYTRQQVLELEKEFHYNRYLTRRRRVEIAHTLRLSERQIKIWFQNRRMKWKKDH	KLPNTKIKS

these animals, more than 500 million years since they last had a common ancestor, the extent of sequence similarity indicates very strong pressure to maintain the sequence of the homeodomain.

The existence of *Hox* genes with homeoboxes throughout the animal kingdom was entirely unexpected. Why different types of animals would possess the same regulatory genes was not obvious, which is why biologists were further surprised by the results when the organization and expression of *Hox* genes was examined in other animals. In vertebrates, such as the laboratory mouse, the *Hox* genes also are clustered together in four large gene complexes on four different chromosomes. Each cluster contains from 9 to 11 *Hox* genes, a total of 39 *Hox* genes altogether. Furthermore, the order of the genes in the mouse *Hox* complexes parallels the order of their most related counterparts in the fly *Hox* complexes, as well as in each of the other mouse *Hox* clusters (Figure 13-10a). This correspondence indicates that the *Hox* complexes of insects and vertebrates are related and that some form of *Hox* complex existed in their distant common ancestor. The four *Hox* complexes in the mouse arose by duplications of entire *Hox* complexes (perhaps of entire chromosomes) in vertebrate ancestors.

Figure 13-9 The sequences of the *Drosophila* Deformed protein homeodomain and of several members of the vertebrate *Hox* group 4 genes are strikingly similar. Residues in common are shaded in yellow; divergent residues are shaded in red; residues common to subsets of proteins are shaded in blue. The very similar C-terminal flanking regions outside of the homeodomain are shaded in green. [*From S. B. Carroll, J. K. Grenier, and S. D. Weatherbee, From DNA to Diversity: Molecular Genetics and the Evolution of Animal Design, 2nd ed., Blackwell, 2005.*]

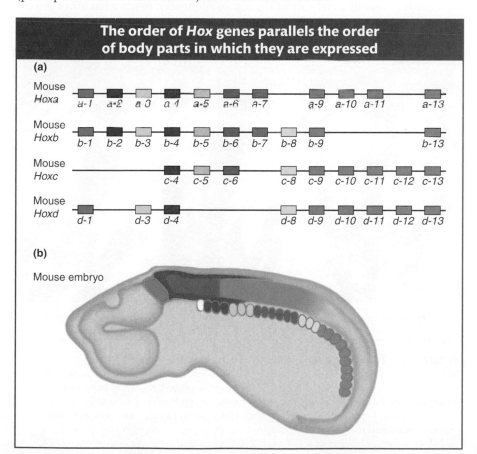

The order of *Hox* genes parallels the order of body parts in which they are expressed

(a)

Mouse *Hoxa*   a-1 a-2 a-3 a-4 a-5 a-6 a-7   a-9 a-10 a-11   a-13

Mouse *Hoxb*   b-1 b-2 b-3 b-4 b-5 b-6 b-7 b-8 b-9   b-13

Mouse *Hoxc*   c-4 c-5 c-6   c-8 c-9 c-10 c-11 c-12 c-13

Mouse *Hoxd*   d-1   d-3 d-4   d-8 d-9 d-10 d-11 d-12 d-13

(b)

Mouse embryo

Figure 13-10 Like those of the fruit fly, vertebrate *Hox* genes are organized in clusters and expressed along the anteroposterior axis. (a) In the mouse, four complexes of *Hox* genes, comprising 39 genes in all, are present on four different chromosomes. Not every gene is represented in each complex; some have been lost in the course of evolution. (b) The *Hox* genes are expressed in distinct domains along the anteroposterior axis of the mouse embryo. The color shading represents the different groups of genes shown in part *a*. [*From S. B. Carroll, "Homeotic Genes and the Evolution of Arthropods and Chordates," Nature 376, 1995, 479–485.*]

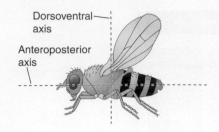

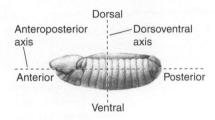

The relationship between adult and embryonic body axes.

Why would such different animals have these sets of genes in common? Their deep, common ancestry indicates that *Hox* genes play some fundamental role in the development of most animals. That role is apparent from analyses of how the *Hox* genes are expressed in different animals. In vertebrate embryos, adjacent *Hox* genes also are expressed in adjacent or partly overlapping domains along the anteroposterior body axis. Furthermore, the order of the *Hox* genes in the complexes corresponds to the head-to-tail order of body regions in which the genes are expressed (Figure 13-10b).

The *Hox*-gene expression patterns of vertebrates suggested that they also specify the identity of body regions, and subsequent analyses of *Hox*-gene mutants have borne this suggestion out. For example, mutations in the *Hoxa11* and *Hoxd11* genes cause the homeotic transformation of sacral vertebrae to lumbar vertebrae (Figure 13-11). Thus, as in the fly, the loss or gain of function of *Hox* genes in vertebrates causes transformation of the identity of serially repeated structures. Such results have been obtained in several classes, including mammals, birds, amphibians, and fish. Furthermore, clusters of *Hox* genes have been shown to govern the patterning of other insects and to be deployed in regions along the anteroposterior axis in annelids, molluscs, nematodes, various arthropods, primitive chordates, flatworms, and other animals. Therefore, despite enormous differences in anatomy, the possession of one or more clusters of *Hox* genes that are deployed in regions along the main body axis is a common, fundamental feature of at least all bilateral animals. Indeed, the surprising lessons from the *Hox* genes portended what turned out to be a general trend among toolkit genes; that is, most toolkit genes are common to different animals.

> **Message** Despite great differences in anatomy, many toolkit genes are common to a broad array of different animal phyla.

Now let's take an inventory of the rest of the toolkit to see what other general principles emerge.

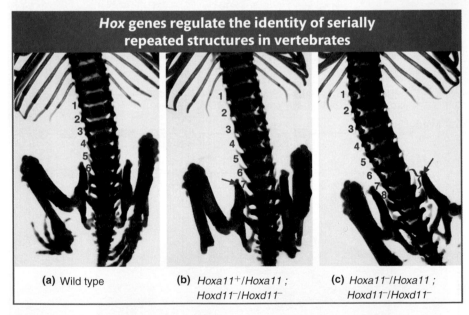

Figure 13-11 The morphologies of different regions of the vertebral column are regulated by *Hox* genes. (a) In the mouse, six lumbar vertebrae form just anterior to the sacral vertebrae (numbers in red). (b) In mice lacking the function of the posteriorly acting *Hoxd11* gene and possessing one functional copy of the *Hoxa11* gene, seven lumbar vertebrae form and one sacral vertebra is lost. (c) In mice lacking both *Hoxa11* and *Hoxd11* function, eight lumbar vertebrae form and two sacral vertebrae are lost. [*Photographs courtesy of Dr. Anne Boulet, HHMI, University of Utah; from S. B. Carroll, J. K. Grenier, and S. D. Weatherbee, From DNA to Diversity: Molecular Genetics and the Evolution of Animal Design, 2nd ed. Blackwell, 2005.*]

13.3 Defining the Entire Toolkit

The *Hox* genes are perhaps the best-known members of the toolkit, but they are just a small family in a much larger group of genes required for the development of the proper numbers, shapes, sizes, and kinds of body parts. Little was known about the rest of the toolkit until the late 1970s and early 1980s, when Christine Nüsslein-Volhard and Eric Wieschaus, working at the Max Planck Institute in Tübingen, Germany, set out to find the genes required for the formation of the segmental organization of the *Drosophila* embryo and larva.

Until their efforts, most work on fly development focused on viable adult phenotypes and not the embryo. Nüsslein-Volhard and Wieschaus realized that the sorts of genes that they were looking for were probably lethal to embryos or larvae in homozygous mutants. So, they came up with a scheme to search for genes that were required in the **zygote** (the product of fertilization; Figure 13-12, bottom). They also developed screens to identify those genes with products that function in the egg, before the zygotic genome is active, and that are required for the proper patterning of the embryo. Genes with products provided by the female to the egg are called **maternal-effect genes.** Mutant phenotypes of strict maternal-effect genes depend only on the genotype of the female (Figure 13-12, top).

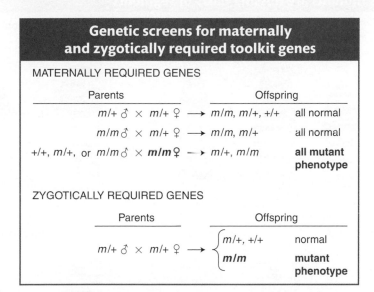

Figure 13-12 Genetic screens identify whether a gene product functions in the egg or in the zygote. The phenotypes of offspring depend on either (*top*) the maternal genotype for maternal-effect genes or (*bottom*) the offspring (zygotic) genotype for zygotically required genes (*m*, mutant; +, wild type).

In these screens, genes were identified that were necessary to make the proper number and pattern of larval segments, to make its three tissue layers (ectoderm, mesoderm, and endoderm), and to pattern the fine details of an animal's anatomy. The power of the genetic screens was their systematic nature. By saturating each of a fly's chromosomes (except the small fourth chromosome) with chemically induced mutations, the researchers were able to identify most genes that were required for the building of the fly. For their pioneering efforts, Nüsslein-Volhard, Wieschaus, and Lewis shared the 1995 Nobel Prize in Physiology or Medicine.

The most striking and telling features of the newly identified mutants were that they showed dramatic but discrete defects in embryo organization or patterning. That is, the dead larva was not an amorphous carcass but exhibited specific, often striking patterning defects. The *Drosophila* larval body has various features whose number, position, or pattern can serve as landmarks to diagnose or classify the abnormalities in mutant animals. Each locus could thus be classified according to the body axis that it affected and the pattern of defects caused by mutations. Genetic crosses reveal whether the locus is active in the maternal egg or

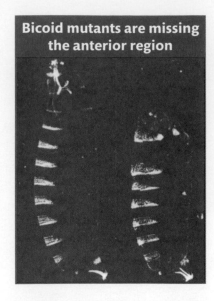

Figure 13-13 The *Bicoid* (*bcd*) maternal-effect gene affects the anterior part of the developing larva. These photomicrographs are of *Drosophila* larvae that have been prepared to show their hard exoskeletons. Dense structures, such as the segmental denticle bands, appear white. (*Left*) A normal larva. (*Right*) A larva from a homozygous *bcd* mutant female. Head and anterior thoracic structures are missing. [*From C. H. Nüsslein-Volhard, G. Frohnhöfer, and R. Lehmann, "Determination of Anteroposterior Polarity in Drosophila," Science 238, 1987, 1678.*]

the zygote. Each class of genes appeared to represent different steps in the progressive refinement of the embryonic body plan—from those that affect large regions of the embryo to those with more limited realms of influence.

For any toolkit gene, three pieces of information are key toward understanding gene function: (1) the mutant phenotype, (2) the pattern of gene expression, and (3) the nature of the gene product. Extensive study of a few dozen genes has led to a fairly detailed picture of how each body axis is established and subdivided into segments or germ layers.

The anteroposterior and dorsoventral axes

A few dozen genes are required for proper organization of the anteroposterior body axis of the fly embryo. The

Segmentation-gene mutants are missing parts of segments

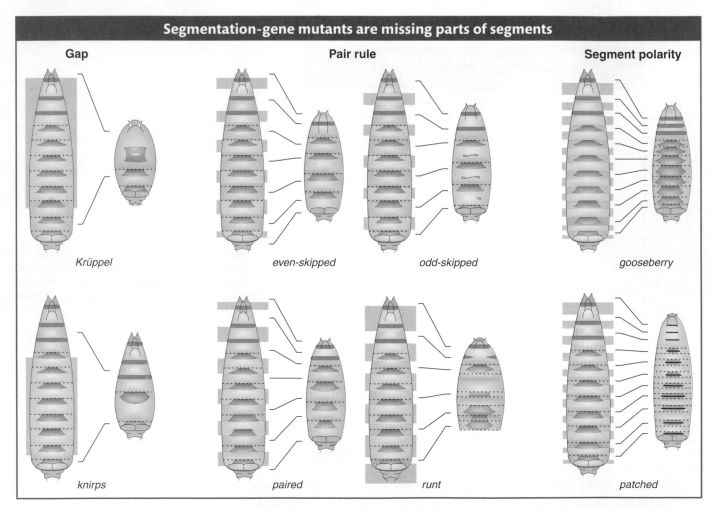

Gap   **Pair rule**   **Segment polarity**

Krüppel   *even-skipped*   *odd-skipped*   *gooseberry*

knirps   *paired*   *runt*   *patched*

Figure 13-14 Classes of *Drosophila* segmentation-gene mutants. These diagrams depict representative gap, pair-rule, and segment-polarity mutants. The red trapezoids are the dense bands of exoskeleton seen in Figure 13-13. The boundary of each segment is indicated by a dotted line. The left-hand diagram of each pair depicts a wild-type larva, and the right-hand diagram depicts the pattern formed in a given mutant. The shaded light orange regions on the wild-type diagrams indicate the domains of the larva that are missing or affected in the mutant.

genes are grouped into five classes on the basis of their realm of influence on embryonic pattern.

- The first class sets up the anteroposterior axis and consists of the maternal-effect genes. A key member of this class is the *Bicoid* gene. Embryos from *Bicoid* mutant mothers are missing the anterior region of the embryo (Figure 13-13), telling us that the gene is required for the development of that region.

     The next three classes are zygotically active genes required for the development of the segments of the embryo.

- The second class contains the **gap genes.** Each of these genes affects the formation of a contiguous block of segments; mutations in gap genes lead to large gaps in segmentation (Figure 13-14, left).

- The third class comprises the **pair-rule genes,** which act at a double-segment periodicity. Pair-rule mutants are missing part of each pair of segments, but different pair-rule genes affect different parts of each double segment. For example, the *even-skipped* gene affects one set of segmental boundaries, and the *odd-skipped* gene affects the complementary set of boundaries (Figure 13-14, middle).

- The fourth class consists of the **segment-polarity genes,** which affect patterning within each segment. Mutants of this class display defects in segment polarity and number (Figure 13-14, right).

     The fifth class of genes determines the fate of each segment.

- The fifth class includes the *Hox* genes already discussed; *Hox* mutants do not affect segment number, but they alter the appearance of one or more segments.

Expression of toolkit genes

To understand the relation between genes and mutant phenotype, we must know the timing and location of gene-expression patterns and the molecular nature of the gene products. The patterns of expression of the toolkit genes turn out to vividly correspond to their phenotypes, inasmuch as they are often precisely correlated with the parts of the developing body that are altered in mutants. Each gene is expressed in a region that can be mapped to specific coordinates along either axis of the embryo. For example, the maternal-effect Bicoid protein is expressed in a graded pattern emanating from the anterior pole of the early embryo, the section of the embryo missing in mutants (Figure 13-15a). Similarly, the gap proteins are expressed in blocks of cells that correspond to the future positions of the segments that are missing in respective gap-gene mutants (Figure 13-15b). The pair-rule proteins are expressed in striking striped patterns: one transverse stripe is expressed per every 2 segments, in a total of 7 stripes covering the 14 future body segments (the position and periodicity of the stripes correspond to the periodicity of defects in mutant larvae), as shown in Figure 13-15c. Many segment-polarity genes are expressed in stripes of cells within each segment, 14 stripes in all corresponding to 14 body segments (Figure 13-15d). Note that the domains of gene expression become progressively more refined as development proceeds: genes are expressed first in large regions (gap proteins), then in stripes from three to four cells wide (pair-rule proteins), and then in stripes from one to two cells wide (segment-polarity proteins).

In addition to what we have learned from the spatial patterns of toolkit-gene expression, the order of toolkit-gene expression over time is logical. The maternal-effect protein Bicoid appears before the zygotic gap proteins, which are expressed before the 7-striped patterns of pair-rule proteins appear, which in turn precede the 14-striped patterns of segment-polarity proteins. The order of gene expression and the progressive refinement of domains within the embryo reveal that the making of the body plan is a step-by-step process, with major subdivisions of the body outlined first and then refined until a fine-grain pattern is established.

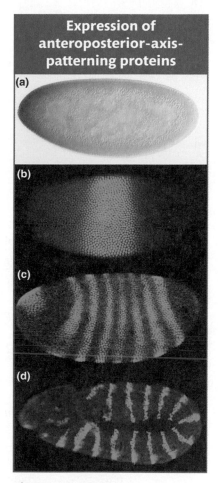

Expression of anteroposterior-axis-patterning proteins

(a)
(b)
(c)
(d)

Figure 13-15 Patterns of toolkit-gene expression correspond to mutant phenotypes. *Drosophila* embryos have been stained with antibodies to the (a) maternally derived Bicoid protein, (b) Krüppel gap protein, (c) Hairy pair-rule protein, and (d) Engrailed segment-polarity protein and visualized by immunoenzymatic (staining is brown) (a) or immunofluorescence (staining is green) (b–d) methods. Each protein is localized to nuclei in regions of the embryo that are affected by mutations in the respective genes. [*Photomicrographs courtesy of (a) Ruth Lehmann and (b–d) James Langeland.*]

The order of gene action further suggests that the expression of one set of genes might govern the expression of the succeeding set of genes.

One clue that this progression is indeed the case comes from analyzing the effects of mutations in toolkit genes on the expression of other toolkit genes. For example, in embryos from *Bicoid* mutant mothers, the expression of several gap genes is altered, as well as that of pair-rule and segment-polarity genes. This finding suggests that the Bicoid protein somehow (directly or indirectly) influences the regulation of gap genes.

Another clue that the expression of one set of genes might govern the expression of the succeeding set of genes comes from examining the protein products. Inspection of the Bicoid protein sequence reveals that it contains a homeodomain, related to but distinct from those of Hox proteins. Thus, Bicoid has the properties of a DNA-binding transcription factor. Each gap gene also encodes a transcription factor, as does each pair-rule gene, several segment-polarity genes, and, as described earlier, all *Hox* genes. These transcription factors include representatives of most known families of sequence-specific DNA-binding proteins; so, although there is no restriction concerning to which family they may belong, many early-acting toolkit proteins are transcription factors. Those that are not transcription factors tend to be components of signaling pathways (Table 13-1). These pathways, shown in generic form in Figure 13-16, mediate ligand-induced signaling processes between cells, and their output generally leads to gene activation or repression. Thus, most toolkit proteins either directly (as transcription factors) or indirectly (as components of signaling pathways) affect gene regulation.

> **Message** Most toolkit proteins are transcription factors or components of ligand-mediated signal-transduction pathways.

Table 13-1 Examples of *Drosophila* A–P Axis Genes That Contribute to Pattern Formation

Gene symbol	Gene name	Protein function	Role(s) in early development
hb-z	*hunchback-zygotic*	Transcription factor—zinc-finger protein	Gap gene
Kr	*Krüppel*	Transcription factor—zinc-finger protein	Gap gene
kni	*knirps*	Transcription factor—steroid receptor-type protein	Gap gene
eve	*even-skipped*	Transcription factor—homeodomain protein	Pair-rule gene
ftz	*fushi tarazu*	Transcription factor—homeodomain protein	Pair-rule gene
opa	*odd-paired*	Transcription factor—zinc-finger protein	Pair-rule gene
prd	*paired*	Transcription factor—PHOX protein	Pair-rule gene
en	*engrailed*	Transcription factor—homeodomain protein	Segment-polarity gene
wg	*wingless*	Signaling WG protein	Segment-polarity gene
hh	*hedgehog*	Signaling HH protein	Segment-polarity gene
ptc	*patched*	Transmembrane protein	Segment-polarity gene
lab	*labial*	Transcription factor—homeodomain protein	Segment-identity gene
Dfd	*Deformed*	Transcription factor—homeodomain protein	Segment-identity gene
Antp	*Antennapedia*	Transcription factor—homeodomain protein	Segment-identity gene
Ubx	*Ultrabithorax*	Transcription factor—homeodomain protein	Segment-identity gene

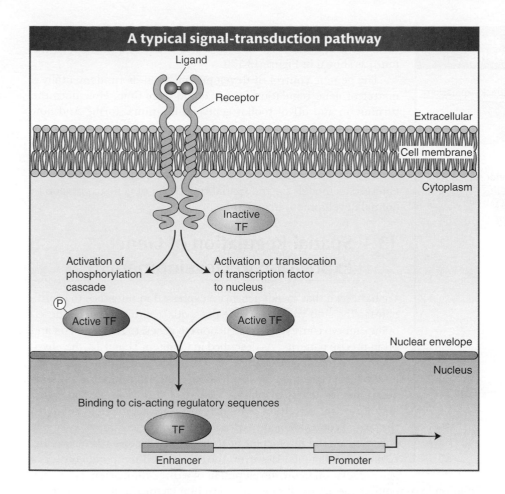

A typical signal-transduction pathway

Ligand

Receptor

Extracellular

Cell membrane

Cytoplasm

Inactive TF

Activation of phosphorylation cascade

Activation or translocation of transcription factor to nucleus

Ⓟ Active TF

Active TF

Nuclear envelope

Nucleus

Binding to cis-acting regulatory sequences

TF

Enhancer   Promoter

Figure 13-16 Most signaling pathways operate through similar logic but have different protein components and signal-transduction mechanisms. Signaling begins when a ligand binds to a membrane-bound receptor, leading to the release or activation of intracellular proteins. Receptor activation often leads to the modification of inactive transcription factors (TF). The modified transcription factors are translocated to the cell nucleus, where they bind to cis-acting regulatory DNA sequences or to DNA-binding proteins and regulate the level of target-gene transcription. [*From S. B. Carroll, J. K. Grenier, and S. D. Weatherbee, From DNA to Diversity: Molecular Genetics and the Evolution of Animal Design, 2nd ed. Blackwell, 2005.*]

The same principles apply to the making of the dorsoventral body axis as apply to the anteroposterior axis. The dorsoventral axis also is subdivided into regions. Several maternal-effect genes, such as *dorsal*, are required to establish these regions at distinct positions from the dorsal (top) to ventral (bottom) side of the embryo. *Dorsal* mutants are "dorsalized" and lack ventral structures (such as the mesoderm and nervous system). A handful of genes that are activated in the zygote are also required for the subdivision of the dorsoventral axis.

The product of the *dorsal* maternal-effect gene is a transcription factor—the Dorsal protein. This protein is expressed in a gradient along the dorsoventral axis, with its highest level of accumulation in ventral cells (Figure 13-17a). The gradient establishes subregions of differing Dorsal concentration. In each subregion, a different set of zygotic genes are expressed that contribute to dorsoventral patterning. The sets of zygotic genes expressed define regions that give rise to

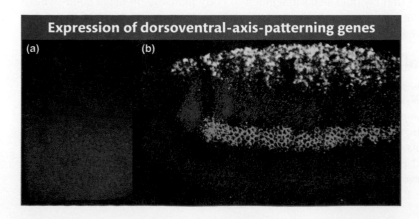

Expression of dorsoventral-axis-patterning genes

(a)   (b)

Figure 13-17 Expression domains of specific dorsoventral-axis-patterning genes correspond to particular future tissue layers. (a) The maternally derived Dorsal protein is expressed in a gradient, with the highest concentration of Dorsal in the nuclei of ventral cells (*bottom of photograph*). (b) The expression of four zygotic dorsoventral-axis-patterning genes revealed by in situ hybridization to RNA. In this lateral view, the domains of the *decapentaplegic* (yellow), *muscle segment homeobox* (red), *intermediate* neuroblasts *defective* (green), and *ventral neuroblasts defective* (blue) genes are revealed. [*Photographs courtesy of (a) Michael Levine and (b) David Kosman, Bill McGinnis, and Ethan Bier.*]

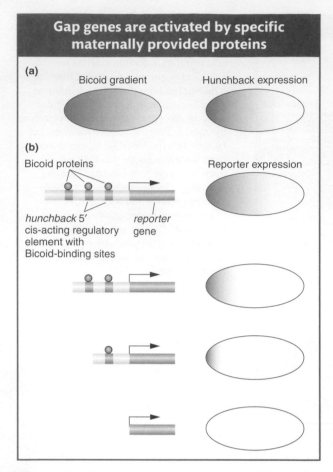

Gap genes are activated by specific maternally provided proteins

(a)

Bicoid gradient   Hunchback expression

(b)

Bicoid proteins   Reporter expression

hunchback 5′   *reporter*
cis-acting regulatory   *gene*
element with
Bicoid-binding sites

Figure 13-18 The Bicoid protein activates zygotic expression of the *hunchback* gene. (a) Bicoid protein expression is graded along the anteroposterior axis. The *hunchback* gap gene is expressed in the anterior half of the zygote. (b) The Bicoid protein (blue) binds to three sites 5′ of the *hunchback* gene. When this 5′ DNA is placed upstream of a reporter gene, reporter-gene *expression* recapitulates the pattern of *hunchback* expression (*top right*). However, progressive deletion of one, two, or all three Bicoid-binding sites either leads to more restricted expression of the reporter gene or abolishes it altogether. These observations show that the level and pattern of *hunchback* expression are controlled by Bicoid through its binding to *hunchback* DNA regulatory sequences.

particular tissue layers, such as the mesoderm and neuroectoderm (the part of the ectoderm that gives rise to the ventral nervous system), as shown in Figure 13-17b.

The genetic control of development, then, is fundamentally a matter of gene regulation in space and over time. How does the turning on and off of toolkit genes build animal form? And how is it choreographed during development? To answer these questions, we will examine the interactions among fly toolkit proteins and genes in more detail. The mechanisms that we will see for controlling toolkit-gene expression in the *Drosophila* embryo have emerged as models for the spatial regulation of gene expression in animal development in general.

13.4 Spatial Regulation of Gene Expression in Development

We have seen that toolkit genes are expressed in reference to coordinates in the embryo. But how are the spatial coordinates of the developing embryo conveyed as instructions to genes, to turn them on and off in precise patterns? As described in Chapters 11 and 12, the physiological control of gene expression in bacteria and simple eukaryotes is ultimately governed by sequence-specific DNA-binding proteins acting on cis-acting regulatory elements (for example, operators and upstream-activation-sequence, or UAS, elements). Similarly, the spatial control of gene expression during development is largely governed by the interaction of transcription factors with cis-acting regulatory elements. However, the spatial and temporal control of gene regulation in the development of a three-dimensional multicellular embryo requires the action of more transcription factors on more numerous and more complex cis-acting regulatory elements.

To define a position in an embryo, regulatory information must exist that distinguishes that position from adjacent regions. If we picture a three-dimensional embryo as a globe, then **positional information** must be specified that indicates longitude (location along the anteroposterior axis), latitude (location along the dorsoventral axis), and altitude or depth (position in the germ layers). We will illustrate the general principles of how the positions of gene expression are specified with three examples. These examples should be thought of as just a few snapshots of the vast number of regulatory interactions that govern fly and animal development. Development is a continuum in which every pattern of gene activity has a preceding causal basis. The entire process includes tens of thousands of regulatory interactions and outputs.

We will focus on a few connections between genes in different levels of the hierarchies that lay out the basic segmental body plan and on *nodal points* where key genes integrate multiple regulatory inputs and respond by producing simpler gene-expression outputs.

Maternal gradients and gene activation

The Bicoid protein is a homeodomain-type transcription factor that is translated from mRNA deposited in the egg and localized at the anterior pole. Because the early *Drosophila* embryo is a syncytium, lacking any cell membranes that would impede the diffusion of protein molecules, Bicoid can diffuse through the cytoplasm. This diffusion establishes a protein concentration gradient (Figure 13-18a): the Bicoid protein is highly concentrated at the anterior end, and this concentration gradually decreases as distance from that end increases, until there is very little Bicoid protein beyond the middle of the embryo. This concentration gradient provides positional information about the location along the anteroposterior axis. A high concentration means anterior end, a lower concentration means middle, and so on. Thus, a way to ensure that a gene is activated in only one location along the axis is to link gene expression to the concentration level. A case in point is the gap genes, which must be activated in specific regions along the axis.

Several zygotic genes, including gap genes, are regulated by different levels of the Bicoid protein. For example, the *hunchback* gene is a gap gene activated in the zygote in the anterior half of the embryo. This activation is through direct binding of the Bicoid protein to three sites 5′ of the promoter of the *hunchback* gene. Bicoid binds to these sites *cooperatively;* that is, the binding of one Bicoid protein molecule to one site facilitates the binding of other Bicoid molecules to nearby sites.

How the activation of *hunchback* depends on the concentration gradient can be seen by performing some tests in vivo. These tests require linking gene regulatory sequences to a reporter gene (an enzyme-encoding gene such as the *LacZ* gene or the green fluorescent protein of jellyfish), introducing the DNA construct into the fly germ line, and monitoring reporter expression in the embryo offspring of transgenic flies (Figure 13-19). Although the wild-type sequences

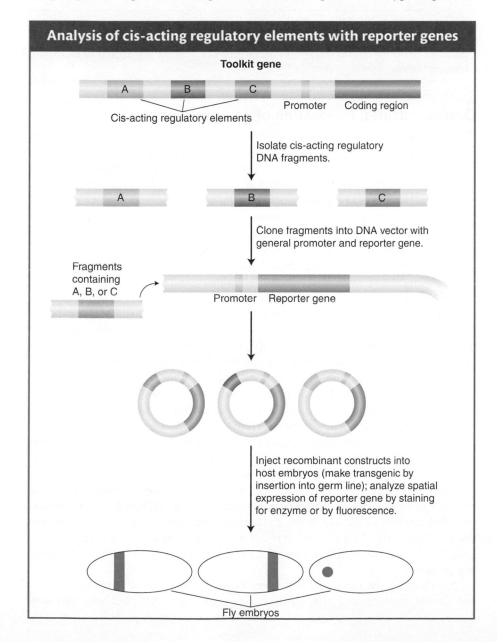

Analysis of cis-acting regulatory elements with reporter genes

Toolkit gene

Cis-acting regulatory elements

Promoter   Coding region

Isolate cis-acting regulatory DNA fragments.

Clone fragments into DNA vector with general promoter and reporter gene.

Fragments containing A, B, or C

Promoter   Reporter gene

Inject recombinant constructs into host embryos (make transgenic by insertion into germ line); analyze spatial expression of reporter gene by staining for enzyme or by fluorescence.

Fly embryos

Figure 13-19 Toolkit loci often contain multiple independent cis-acting regulatory elements that control gene expression in different places or at different times during development or both (for example, A, B, C, here). These elements are identified by their ability, when placed in cis to a reporter gene and inserted back into a host genome, to control the pattern, timing, or level, or all three, of reporter-gene expression. In this example, each element drives a different pattern of gene expression in a fly embryo. Most reporter genes encode enzymes or fluorescent proteins that can be easily visualized.

5′ of the *hunchback* gene are sufficient to drive reporter expression in the anterior half of the embryo, deletions of Bicoid-binding sites in this cis-acting regulatory element reduce or abolish reporter expression (see Figure 13-18b). More than one Bicoid site must be occupied to generate a sharp boundary of reporter expression, which indicates that a threshold concentration of Bicoid protein is required to occupy multiple sites before gene expression is activated. A gap gene with fewer binding sites will not be activated at locations with lower concentration of Bicoid protein.

Each gap gene contains cis-acting regulatory elements with different arrangements of binding sites, and these binding sites may have different affinities for the Bicoid protein. Consequently, each gap gene is expressed in a unique distinct domain in the embryo, in response to different levels of Bicoid and other transcription-factor gradients. A similar theme is found in the patterning of the dorsoventral axis: cis-acting regulatory elements contain different numbers and arrangements of binding sites for Dorsal and other dorsoventral transcription factors. Consequently, genes are activated in discrete domains along the dorsoventral axis.

> **Message** The concentration-dependent response of genes to graded inputs is a crucial feature of gene regulation in the early *Drosophila* embryo. The cis-acting regulatory elements governing distinct responses contain different numbers and arrangements of transcription-factor-binding sites.

Drawing stripes: Integration of gap-protein inputs

The expression of each pair-rule gene in seven stripes is the first sign of the periodic organization of the embryo and future animal. How are such periodic patterns generated from prior aperiodic information? Before the molecular analysis of pair-rule-gene regulation, several models were put forth to explain stripe formation. Every one of these ideas viewed all seven stripes as identical outputs in response to identical inputs. However, the actual way in which the patterns of a few key pair-rule genes are encoded and generated is one stripe at a time. The solution to the mystery of stripe generation highlights one of the most important concepts concerning the spatial control of gene regulation in developing animals; namely, the distinct cis-acting regulatory elements of individual genes are controlled independently.

The key discovery was that each of the seven stripes that make up the expression patterns of the *even-skipped* and *hairy* pair-rule genes is controlled independently. Consider the second stripe expressed by the *even-skipped* gene (Figure 13-20a). This stripe lies within the broad region of *hunchback* expression and on the edges of the regions of expression of two other gap proteins, Giant and Krüppel (Figure 13-20b). Thus, within the area of the future stripe, there will be large amounts of Hunchback protein and small amounts of Giant protein and Krüppel protein. There will also be a certain concentration of the maternal-effect Bicoid protein. No other stripe of the embryo will contain these proteins in these proportions. The formation of stripe 2 is controlled by a specific cis-acting regulatory element, an enhancer, that contains a number of binding sites for these four proteins (Figure 13-20c). Thus, the entire seven-striped periodic pattern is the sum of different sets of inputs into separate cis-acting regulatory elements. Detailed analysis of the *eve* stripe 2 cis-acting regulatory element revealed that the position of this "simple" stripe is controlled by no fewer than four aperiodically distributed transcription factors, including one maternal protein and three gap proteins.

Specifically, the *eve* stripe 2 element contains multiple sites for the maternal Bicoid protein and the Hunchback, Giant, and Krüppel gap proteins

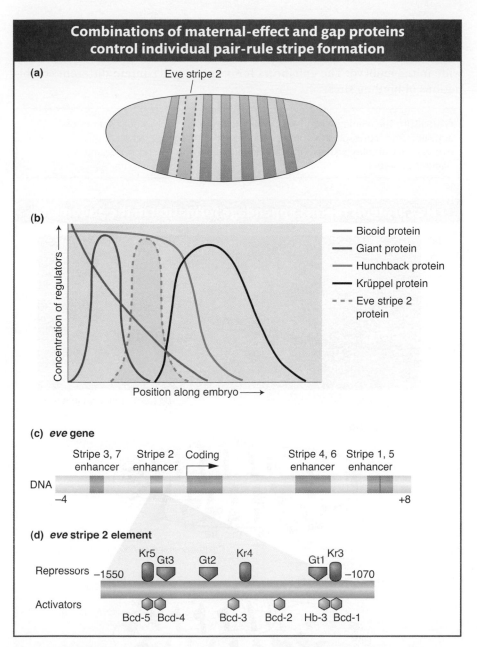

Figure 13-20 Regulation of a pair-rule stripe: combinatorial control of an independent cis-acting regulatory element. (a) The regulation of the *eve* stripe 2 cis-acting regulatory element controls the formation of the second stripe of *eve* expression in the early embryo, just one of seven stripes of *eve* expression. (b) The stripe forms within the domains of the Bicoid and Hunchback proteins and at the edge of the Giant and Krüppel gap proteins. Bcd and Hb are activators, Gt and Kr are repressors of the stripe. (c) The *eve* stripe 2 element is just one of several cis-acting regulatory elements of the *eve* gene, each of which controls different parts of *eve* expression. The *eve* stripe 2 element spans from about 1 to 1.7 kb upstream of the *eve* transcription unit. (d) Within the *eve* stripe 2 element, several binding sites exist for each transcription factor (repressors are shown above the element, activators below). The net output of this combination of activators and repressors is expression of the narrow *eve* stripe. [*After J. Gerhart and M. Kirshner, Cells, Embryos, and Evolution. Blackwell Science, 1997.*]

(Figure 13-20d). Mutational analyses of different combinations of binding sites revealed that Bicoid and Hunchback activate the expression of the *eve* stripe 2 element over a broad region. The Giant and Krüppel proteins are repressors that sharpen the boundaries of the stripe to just a few cells wide. The *eve* stripe 2 element acts, then, as a genetic switch, integrating multiple regulatory protein activities to produce one stripe from three to four cells wide in the embryo. The enhancers for other stripes contain different combinations of binding sites.

> **Message** The regulation of cis-acting regulatory elements by combinations of activators and repressors is a common theme in the spatial regulation of gene expression. Complex patterns of inputs are often integrated to produce simpler patterns of outputs.

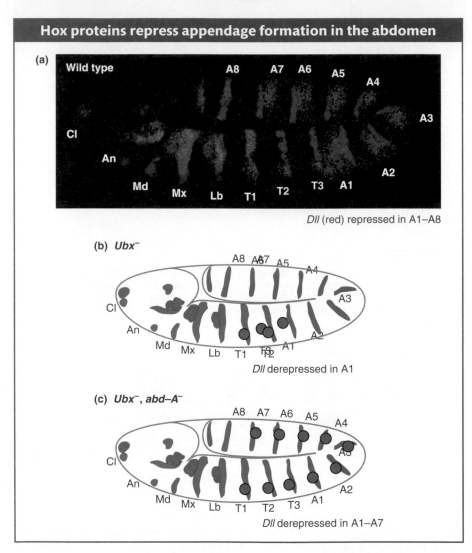

Figure 13-21 The absence of limbs in the abdomen is controlled by *Hox* genes. (a) The expression of the *Distal-less* (*Dll*) gene (red) marks the position of future appendages, expression of the *Hox* gene *Ultrabithorax* (purple) marks the position of the abdominal segments A1 through A7, and expression of the *engrailed* gene (blue) marks the posterior of each segment. (b) Schematic representation of *Ubx⁻* embryo showing that *Dll* expression (red circles) is derepressed in segment A1. (c) Schematic representation of *Ubx⁻ abd-A⁻* embryo showing that *Dll* expression (red circles) is derepressed in the first seven abdominal segments. [(a) Photomicrograph by Dave Kosman, Ethan Bier, and Bill McGinnis; (b and c) based on B. Gebelein, D. J. McKay, and R. S. Mann, "Direct Integration of Hox and Segmentation Gene Inputs During Drosophila Development," Nature 431, 2004, 653–659.]

Making segments different: integration of Hox inputs

The combined and sequential activity of the maternal-effect, gap, pair-rule, and segment-polarity proteins establishes the basic segmented body plan of the embryo and larva. How are the different segmental identities established by Hox proteins? This process has two aspects. First, the *Hox* genes are expressed in different domains along the anteroposterior axis. *Hox*-gene expression is largely controlled by segmentation proteins, especially gap proteins, through mechanisms that are similar to those already described herein (as well as some cross-regulation by Hox proteins of other *Hox* genes). The regulation of *Hox* genes will not be considered in depth here. The second aspect of Hox control of segmental identity is the regulation of target genes by Hox proteins. We will examine one example that nicely illustrates how a major feature of the fruit fly's body plan is controlled through the integration of many inputs by a single cis-acting regulatory element.

The paired limbs, mouthparts, and antennae of *Drosophila* each develop from initially small populations of about 20 cells. Different structures develop from the different segments of the head and thorax, whereas the abdomen is limbless. The first sign of the development of these structures is the activation of regulatory genes within small clusters of cells, which are called the appendage *primordia*. The expression of the *Distal-less* (*Dll*) gene marks the start of the development of the appendages. This gene is one of the key targets of the *Hox* genes, and its function is required for the subsequent development of the distal parts of each of these appendages. The small clusters of cells expressing *Distal-less* arise in several head segments and in each of the three thoracic segments but not in the abdomen (Figure 13-21a).

How is *Distal-less* expression restricted to the more anterior segments? By repressing its expression in the abdomen. Several lines of evidence have revealed that the *Distal-less* gene is repressed by two Hox proteins—the Ultrabithorax and Abdominal-A proteins—working in collaboration with two segmentation proteins. Notice in Figure 13-6 that Ultrabithorax is expressed in abdominal segments one through seven, and Abdominal-A is expressed in abdominal segments two through seven, overlapping with all but the first segment covered by Ultrabithorax. In *Ultrabithorax* mutant embryos, *Distal-less* expression expands to the first abdominal segment (Figure 13-21b), and in *Ultrabithorax/Abdominal-A* double-mutant embryos, *Distal-less* expression extends through the first seven abdominal segments (Figure 13-21c), indicating that both proteins are required for the repression of *Distal-less* expression in the abdomen.

The cis-acting regulatory element responsible for *Distal-less* expression in the embryo has been identified and characterized in detail (Figure 13-22a). It contains two binding sites for the Hox proteins. If these two binding sites are mutated such that the Hox proteins cannot bind, *Distal-less* expression is derepressed in the abdomen (Figure 13-22b). Several additional proteins collaborate with the Hox proteins in repressing *Distal-less*. Two are proteins encoded by segment-polarity genes, *Sloppy-paired* (*Slp*) and *Engrailed* (*En*). The Sloppy-paired and Engrailed proteins are expressed in stripes that mark the anterior and posterior compartments of each segment, respectively. Each protein also binds to the *Distal-less* cis-acting regulatory element. When the Sloppy-paired-binding site is mutated in the cis-acting regulatory element, reporter-gene expression is derepressed in the anterior compartments of abdominal segments (Figure 13-22c). When the Engrailed-binding site is mutated, reporter expression is derepressed in the posterior compartments of each abdominal segment (Figure 13-22d). And, when the binding sites for both proteins are mutated, reporter-gene expression is derepressed in both compartments of each abdominal segment, just as when the Hox-binding sites are mutated (Figure 13-22e). Two other proteins, called

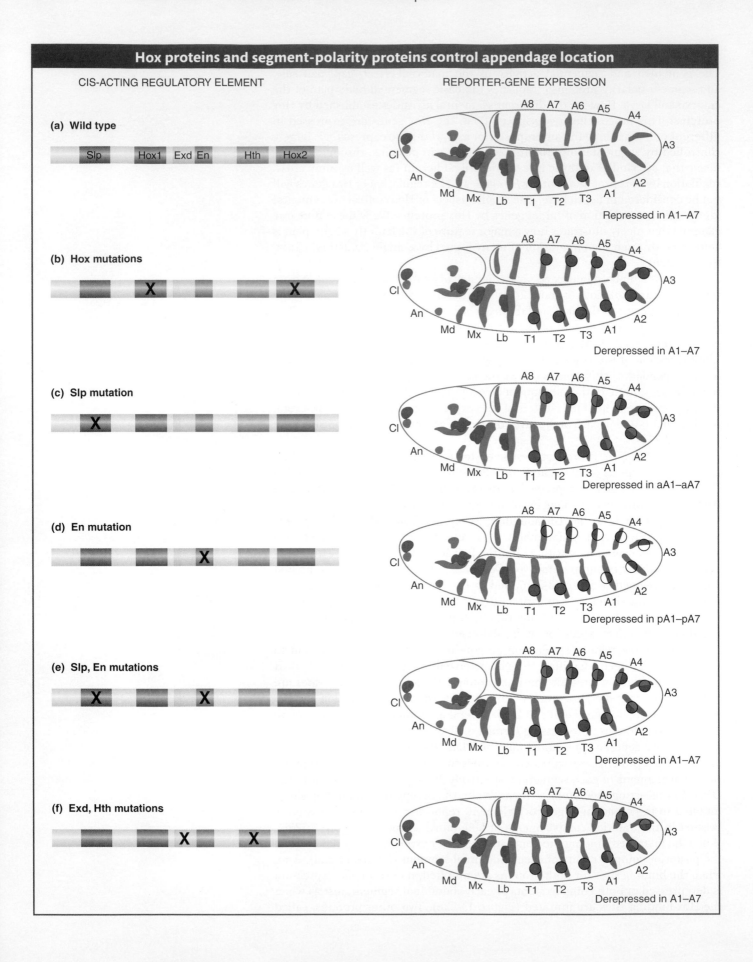

Hox proteins and segment-polarity proteins control appendage location

CIS-ACTING REGULATORY ELEMENT

REPORTER-GENE EXPRESSION

(a) Wild type

Repressed in A1–A7

(b) Hox mutations

Derepressed in A1–A7

(c) Slp mutation

Derepressed in aA1–aA7

(d) En mutation

Derepressed in pA1–pA7

(e) Slp, En mutations

Derepressed in A1–A7

(f) Exd, Hth mutations

Derepressed in A1–A7

Figure 13-22 Integration of Hox and segmentation-protein inputs by a cis-acting regulatory element. (a, *left*) A cis-acting regulatory element of the *Dll* gene governs the repression of *Dll* expression in the abdomen by a set of transcription factors. (a, *right*) *Dll* expression (red) extends to the thorax but not into the abdomen in a wild-type embryo. (b–f) Mutations in the respective binding sites shown derepress *Dll* expression in various patterns in the abdomen. Binding sites are: Slp, Sloppy-paired; Hox1 and Hox2, Ultrabithorax and Abdominal-A; Exd, Extradenticle; En, Engrailed; Hth, Homothorax. [*Based on data of B. Gebelein, D. J. McKay, and R. S. Mann, "Direct Integration of Hox and Segmentation Gene Inputs During Drosophila Development," Nature 431, 2004, 653–659.*]

Extradenticle and Homothorax, which are broadly expressed in every segment, also bind to the *Distal-less* cis-acting regulatory element and are required for transcriptional repression in the abdomen (Figure 13-22f).

Thus, altogether, two Hox proteins and four other transcription factors bind within a span of 57 base pairs and act together to repress *Distal-less* expression and, hence, appendage formation in the abdomen. The repression of *Distal-less* expression is a clear demonstration of how Hox proteins regulate segment identity and the number of reiterated body structures. It is also a good illustration of how diverse regulatory inputs converge and act combinatorially on cis-acting regulatory elements. In this instance, the presence of Hox-binding sites is not sufficient for transcriptional repression: collaborative and cooperative interactions are required among several proteins to fully repress gene expression in the abdomen.

> **Message** Combinatorial and cooperative regulation of gene transcription imposes greater specificity on spatial patterns of gene expression and allows for their greater diversity.

Although evolutionary diversity has not been explicitly addressed in this chapter, the presence of multiple independent cis-acting regulatory elements for each toolkit gene has profound implications for the evolution of form. Specifically, the modularity of these elements allows for changes in one aspect of gene expression independent of other gene functions. The evolution of gene regulation plays a major role in the evolution of development and morphology. We will return to this topic in Chapter 20.

13.5 Posttranscriptional Regulation of Gene Expression in Development

Although transcriptional regulation is a major means of restricting the expression of gene products to defined areas during development, it is not at all the exclusive means of doing so. Alternative RNA splicing also contributes to gene regulation, and so does the regulation of mRNA translation by proteins and microRNAs (miRNAs). In each case, regulatory sequences in RNA are recognized—by splicing factors, mRNA-binding proteins, or miRNAs—and govern the structure of the protein product, its amount, or the location where the protein is produced. We will look at one example of each type of regulatory interaction at the RNA level.

RNA splicing and sex determination in *Drosophila*

A fundamental developmental decision in sexually reproducing organisms is the specification of sex. In animals, the development of many tissues follows different paths, depending on the sex of the individual animal. In *Drosophila*, many genes have been identified that govern *sex determination* through the analysis of mutant phenotypes in which sexual identity is altered or ambiguous.

Figure 13-23 Three pre-mRNAs of major *Drosophila* sex-determining genes are alternatively spliced. The female-specific pathway is shown on the left and the male-specific pathway shown on the right. The pre-mRNAs are identical in both sexes and shown in the middle. In the male *Sex-lethal* and *transformer* mRNAs, there are stop codons that terminate translation. These sequences are removed by splicing to produce functional proteins in the female. The Transformer and Tra-2 proteins then splice the female *doublesex* pre-mRNA to produce the female-specific isoform of the Dsx protein, which differs from the male-specific isoform by the alternative splicing of several exons. [*After S. S. Gilbert, Developmental Biology, 7th ed. Sinauer, 2003.*]

The *doublesex* (*dsx*) gene plays a central role in governing the sexual identity of somatic (non-germ-line) tissue. Null mutations in *dsx* cause females and males to develop as intermediate *intersexes*, which have lost the distinct differences between male and female tissues. Although *dsx* function is required in both sexes, different gene products are produced from the locus in different sexes. In males, the product is a specific, longer isoform, Dsx^M, that contains a unique C-terminal region of 150 amino acids not found in the female-specific isoform Dsx^F, which instead contains a unique 30 amino acid sequence at its carboxyl terminus. Each form of the Dsx protein is a DNA-binding transcription factor that apparently binds the same DNA sequences. However, the activities of the two isoforms differ: Dsx^F activates certain target genes in females that Dsx^M represses in males.

The alternative forms of the Dsx protein are generated by alternative splicing of the primary *dsx* RNA transcript. Thus, in this case, the choice of splice sites must be regulated to produce mature mRNAs that encode different proteins. The various genetic factors that influence Dsx expression and sex determination have been identified by mutations that affect the sexual phenotype.

One key regulator is the product of the *transformer* (*tra*) gene. Whereas null mutations in *tra* have no effect on males, XX female flies bearing *tra* mutations are transformed into the male phenotype. The Tra protein is an alternative splicing factor that affects the splice choices in the *dsx* RNA transcript. In the presence of Tra (and a related protein Tra2), a splice occurs that incorporates exon 4 of the *dsx* gene into the mature *dsx*^{F} transcript (Figure 13-23), but not exons 5 and 6. Males lack Tra; so this splice does not occur, and exons 5 and 6 are incorporated into the *dsx*^{M} transcript, but not exon 4 (Figure 13-23).

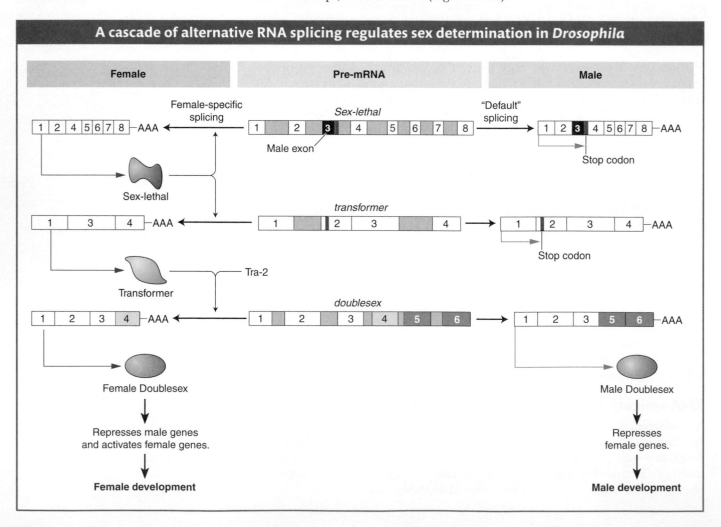

A cascade of alternative RNA splicing regulates sex determination in *Drosophila*

The Tra protein explains how alternative forms of Dsx are expressed, but how is Tra expression itself regulated to differ in females and males? The *tra* RNA itself is alternatively spliced. In females, a splicing factor encoded by the *Sex-lethal* (*Sxl*) gene is present. This splicing factor binds to the *tra* RNA and prevents a splicing event that would otherwise incorporate an exon that contains a stop codon. In males, no Tra protein is made, because this stop codon is present.

The production of the Sex-lethal protein is, in turn, regulated both by RNA splicing and by factors that alter the level of transcription. The level of *Sxl* transcription is governed by activators on the X chromosome and repressors on the autosomes. In females, *Sxl* activation prevails and the Sxl protein is produced, which regulates *tra* RNA splicing and feeds back to regulate the splicing of *Sxl* RNA itself. In females, a stop codon is spliced out so that Sxl protein production can continue. However, in males, where no Sxl protein is present, the stop codon is still present in the unspliced *Sxl* RNA transcript and no Sxl protein can be produced.

This cascade of sex-specific RNA splicing in *D. melanogaster* illustrates one way that the sex-chromosome genotype leads to different forms of regulatory proteins being expressed in one sex and not the other. Interestingly, the genetic regulation of sex determination differs greatly between animal species, in that sexual genotype can lead to differential expression of regulatory genes through distinctly different paths. However, proteins related to Dsx do play roles in sexual differentiation in a wide variety of animals, including humans. Thus, although there are many ways to generate differential expression of transcription factors, a family of similar proteins appear to underlie much sexual differentiation.

Regulation of mRNA translation and cell lineage in *C. elegans*

In many animal species, the early development of the embryo entails the partitioning of cells or groups of cells into discrete lineages that will give rise to distinct tissues in the adult. This process is best understood in the nematode worm *C. elegans,* in which the adult animal is composed of just about 1000 somatic cells (a third of which are nerve cells) and a similar number of germ cells in the gonad. The simple construction, rapid life cycle, and transparency of *C. elegans* has made it a powerful model for developmental analysis (see the Model Organism box on *C. elegans* on page 476). All of this animal's cell lineages were mapped out in a series of elegant studies led by John Sulston at the Medical Research Council (MRC) Laboratory in Cambridge, England. Systematic genetic screens for mutations that disrupt or extend cell lineages have provided a bounty of information about the genetic control of lineage decisions. *C. elegans* genetics has been especially important in understanding the role of posttranscriptional regulation at the RNA level, and we will examine two mechanisms here: (1) control of translation by mRNA-binding proteins and (2) miRNA control of gene expression.

Translational control in the early embryo

We first look at how a cell lineage begins. After two cell divisions, the *C. elegans* embryo contains four cells, called blastomeres. Each cell will begin a distinct lineage, and the descendants of the separate lineages will have different fates. Already at this stage, differences are observed in the proteins present in the four blastomeres. Not surprisingly from what we have learned, many of these proteins are toolkit proteins that determine which genes will be expressed in descendant cells. What is surprising, though, is that the mRNAs encoding some worm toolkit proteins are present in *all* cells of the early embryo. However, in a specific cell, only some of these mRNAs will be translated into proteins. Thus, in the *C. elegans* embryo, posttranscriptional regulation is critical for the proper specification of early cell fates. During the very first cell division, polarity within the zygote leads to the partitioning of regulatory molecules to specific embryonic cells. For

Model Organism *Caenorhabditis elegans*

The Nematode *Caenorhabditis elegans* as a Model for Cell-Lineage-Fate Decisions

In the past 20 years, studies of the nematode worm *Caenorhabditis elegans* (see Diagram 1) have greatly advanced our understanding of the genetic control of cell-lineage decisions. The transparency and simple construc-tion of this animal led Sydney Brenner to advance its use as a model organism. The adult worm contains about 1000 somatic cells, and researchers, led by John Sulston, have carefully mapped out the entire series of somatic-cell decisions that produce the adult animal.

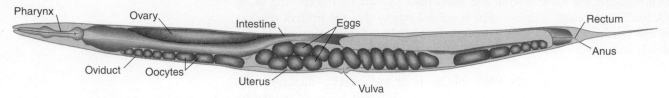

Diagram 1 An adult hermaphrodite *Caenorhabtis elegans*, showing various organs.

Some of the lineage decisions, such as the formation of the vulva, have been key models of so-called *inductive interactions* in development, where signaling between cells induces cell-fate changes and organ formation (see Diagram 2). Exhaustive genetic screens have identified many components participating in signaling and signal transduction in the formation of the vulva.

(a) Tissue derived from 1°, 2°, and 3° cells

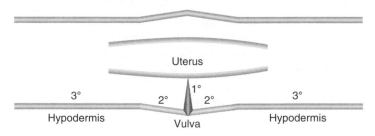

(b) Pedigrees of cells

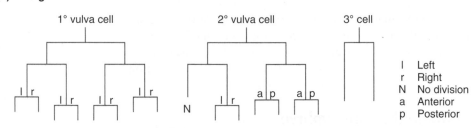

l	Left
r	Right
N	No division
a	Anterior
p	Posterior

Diagram 2 Production of the vulval-cell lineages. (a) The parts of the vulval anatomy that are occupied by so-called primary (1°), secondary (2°), and tertiary (3°) cells. (b) The lineages or pedigrees of the primary, secondary, and tertiary cells are distinguished by their cell-division patterns.

For some of the embryonic and larval cell divisions, particularly those that will contribute to a worm's nervous system, a progenitor cell gives rise to two progeny cells, one of which then undergoes programmed cell death. Analysis of mutants in which programmed cell death is aberrant, led by Robert Horvitz, has revealed many components of programmed-cell-death pathways common to most animals. Sydney Brenner, John Sulston, and Robert Horvitz shared the 2002 Nobel Prize in Physiology or Medicine for their pioneering work based on *C. elegans*.

example, the *glp-1* gene encodes a transmembrane receptor protein (related to the Notch receptor of flies and other animals). Although the *glp-1* mRNA is present in all cells at the four-cell stage, the GLP-1 protein is translated only in the two anterior cells ABa and ABp (Figure 13-24a). This localized expression of GLP-1 is critical for establishing distinct fates. Mutations that abolish *glp-1* function at the four-cell stage alter the fates of ABp and ABa descendants.

GLP-1 is localized to the anterior cells by repressing its translation in the posterior cells. The repression of GLP-1 translation requires sequences in the 3′ UTR of the *glp-1* mRNA—specifically, a 61-nucleotide region called the spatial control region (SCR). The importance of the SCR has been demonstrated by linking mRNA transcribed from reporter genes to different variants of the SCR. Deletion of this region or mutation of key sites within it causes the reporter gene to be expressed in all four blastomeres of the early embryo (Figure 13-24b).

On the basis of how we have seen transcription controlled, we might guess that a protein binds to the SCR to repress translation of the *glp-1* mRNA. To identify these repressor proteins, researchers isolated proteins that bound to the SCR. One protein, GLD-1, binds specifically to a region of the SCR. Furthermore, the GLD-1 protein is enriched in posterior blastomeres, just where the expression of

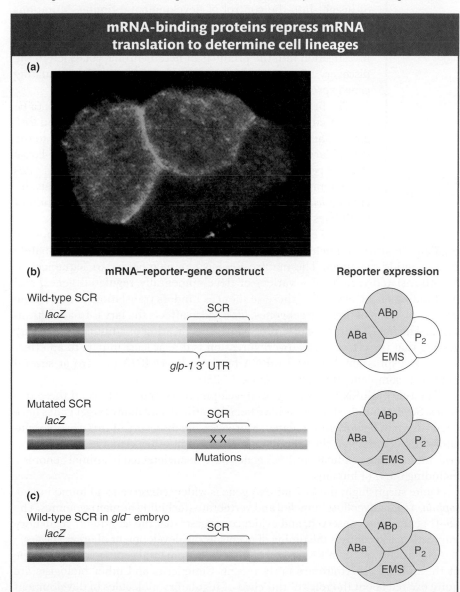

mRNA-binding proteins repress mRNA translation to determine cell lineages

Figure 13-24 Translational regulation and cell-lineage decisions in the early *C. elegans* embryo. (a) At the four-cell stage of the *C. elegans* embryo, the GLP-1 protein is expressed in two anterior cells (bright green) but not in other cells. Translation of the *glp-1* mRNA is regulated by the GLD-1 protein in posterior cells. (b) Fusion of the *glp-1* 3′ UTR to the *lacZ* reporter gene leads to reporter expression in the ABa and ABp cells of the four-cell stage of the *C. elegans* embryo (shaded, *right*). Mutations in GLD-1-binding sites in the spatial control region (SCR) cause derepression of translation in the EMS and P_2 lineages, as does (c) loss of *gld* function. [*(a) Courtesy of Thomas Evans, University of Colorado Health Sciences Center.*]

glp-1 is repressed. Finally, when GLD-1 expression is inhibited by using RNA interference, the GLP-1 protein is expressed in posterior blastomeres (Figure 13-24c). This evidence suggests that GLD-1 is a translational repressor protein controlling the expression of *glp-1*.

The spatial regulation of GLP-1 translation is but one example of translational control in development or by GLD-1. Many other mRNAs are translationally regulated, and GLD-1 binds to other target mRNAs in embryonic and germ-line cells.

> **Message** Sequence-specific RNA-binding proteins act through cis-acting RNA sequences to regulate the spatial pattern of protein translation.

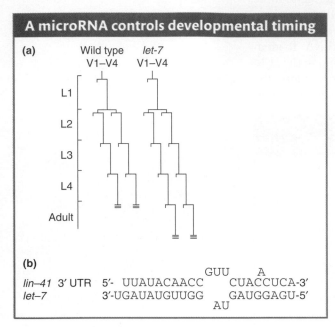

Figure 13-25 Normally, *C. elegans* develops into an adult after four larval stages, and hypodermal cell lineages conclude their development at L4 (double-hatched lines at ends of V1–V4 lineages). (a) In *let-7* mutants, the transition from the L4 larval stage to adult is delayed and the cell lineages of lateral hypodermal cells (V) are reiterated. (b) *let-7* encodes an miRNA that is complementary to sequences in the 3′ UTR of *lin-41* mRNA.

miRNA control of developmental timing in *C. elegans* and other species

Development is a temporally as well as spatially ordered process. When events take place is just as important as where. Mutations in the *heterochronic* genes of *C. elegans* have been sources of insight into the control of developmental timing. Mutations in these genes alter the timing of events in cell-fate specification, causing such events to be either reiterated or omitted. Detailed investigation into the products of heterochronic genes led to the discovery of an entirely unexpected mechanism for regulating gene expression, through microRNAs.

The first members of this class of regulatory molecules to be discovered in *C. elegans* are RNAs produced by the *lin-4* and *let-7* genes. The *lin-4* gene governs the transition from the first to the second larval stage; *let-7* regulates the transition from late-larval to adult cell fates. In *let-7* mutants, for example, larval cell fates are reiterated in the adult stage (Figure 13-25a). Conversely, increased *let-7* gene dosage causes the precocious specification of adult fates in larval stages.

Neither *let-7* nor *lin-4* encodes proteins. The *let-7* gene encodes a temporally regulated mature 22-nucleotide RNA that is processed from an approximately 70-nucleotide precursor. The mature RNA is complementary to sequences in 3′ untranslated regions of a variety of developmentally regulated genes, and the binding of the miRNA to these sequences hinders translation of these gene transcripts. One of these target genes, *lin-41*, also affects the larval-to-adult transition. The *lin-41* mutants cause precocious specification of adult cell fates, suggesting that the effect of *let-7* overexpression is due at least in part to an effect on *lin-41* expression. The *let-7* mRNA binds to *lin-41* RNA in vitro at several imperfect complementary sites (Figure 13-25b).

The role of miRNAs in *C. elegans* development extends far beyond these two genes. Several hundred miRNAs have been identified, and many target genes have been shown to be miRNA regulated. Moreover, the discovery of this class of regulatory RNAs prompted the search for such genes in other genomes, and, in general, hundreds of candidate miRNA genes have been detected in animal genomes, including those of humans.

Quite surprisingly, the *let-7* miRNA gene is widely conserved and found in *Drosophila*, ascidian, mollusc, annelid, and vertebrate (including human) genomes. The *lin-41* gene also is conserved, and evidence suggests that the *let-7–lin-41* regulatory interaction also controls the timing of events in the development of other species.

The discoveries of miRNA regulation of developmental genes and of the scope of the miRNA repertoire are fairly recent. Geneticists and other biologists are quite excited about the roles of this class of regulatory molecules in development and physiology, leading to a very vigorous, fast-paced area of new research.

13.6 From Flies to Fingers, Feathers, and Floor Plates: The Many Roles of Individual Toolkit Genes

We have seen that toolkit proteins and regulatory RNAs have multiple roles in development. For example, recall that the Ultrabithorax protein represses limb formation in the fly abdomen and promotes hind-wing development in the fly thorax. Similarly, Sloppy-paired and Engrailed participate in the generation of the basic segmental organization of the embryo and collaborate with Hox proteins to suppress limb formation. These roles are just a few of the many roles played by these toolkit genes in the entire course of fly development. Most toolkit genes function at more than one time and place, and most may influence the formation or patterning of many different structures that are formed in different parts of the larval or adult body. Those that regulate gene expression may directly regulate scores to hundreds of different genes. The function of an individual toolkit protein (or RNA) is almost always context dependent, which is why the toolkit analogy is perhaps so fitting. As with a carpenter's toolkit, a common set of tools can be used to fashion many structures.

To illustrate this principle more vividly, we will look at the role of one toolkit protein in the development of many vertebrate features, including features present in humans. This toolkit protein is the vertebrate homolog of the *Drosophila hedgehog* gene. The *hedgehog* gene was first identified by Nüsslein-Volhard and Wieschaus as a segment-polarity gene. It has been characterized as encoding a signaling protein secreted from cells in *Drosophila*.

As the evidence grew that toolkit genes are common to different animal phyla, the discovery and characterization of fly toolkit genes such as *hedgehog* became a common springboard to the characterization of genes in other taxa, particularly vertebrates. The cloning of homologous genes based on sequence similarity (see Chapter 14) was a fast track to the identification of vertebrate toolkit genes. The application of this strategy to the *hedgehog* gene illustrates the power and payoffs of using homology to discover important genes. Several distinct homologs of *hedgehog* were isolated from zebrafish, mice, chickens, and humans. In the whimsical spirit of the *Drosophila* gene nomenclature, the three vertebrate homologs were named *Sonic hedgehog* (after the video-game character), *Indian hedgehog*, and *Desert hedgehog*.

One of the first means of characterizing the potential roles of these genes in development was to examine where they are expressed. *Sonic hedgehog* (*Shh*) was found to be expressed in several parts of the developing chicken and other vertebrates. Most intriguing was its expression in the posterior part of the developing limb bud (Figure 13-26a). This part of the limb bud was known for decades to be the *zone of polarizing activity* (ZPA), because it is an organizer responsible for establishing the anteroposterior polarity of the limb and its digits (Figure 13-26b). To test whether *Shh* might play a role in ZPA function, Cliff Tabin and his colleagues at Harvard Medical School caused the Shh protein to be expressed in the *anterior* region of developing chick limb buds. They observed the same effect as transplantation of the ZPA—the induction of extra digits with reversed polarity. Their results were stunning evidence that Shh was the long-sought morphogen produced by the ZPA.

Shh is also expressed in other intriguing patterns in the chicken and other vertebrates. For example, Shh is expressed in developing feather buds, where it plays a role in establishing the pattern and polarity of feather formation (see Figure 13-26b). Shh is also expressed in the developing neural tube of vertebrate embryos, in a region called the *floor plate* (see Figure 13-26a). Subsequent experiments have shown that Shh signaling from these floor-plate cells is critical for the subdivision of the brain hemispheres and the subdivision of the developing eye into the left

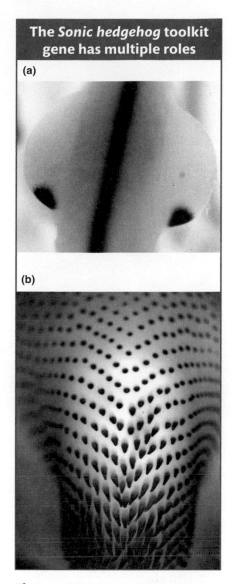

The *Sonic hedgehog* toolkit gene has multiple roles

(a)

(b)

Figure 13-26 The *Shh* gene is expressed in many different parts of the developing chick embryo (indicated by dark staining), including (a) the zone of polarizing activity in each of the two developing limb buds and the long neural tube and (b) the developing feather buds. *Shh* mRNA is visualized by in situ hybridization. [*Photomicrographs courtesy of (a) Cliff Tabin and (b) Matthew Harris and John Fallon.*]

and right sides. When the function of the *Shh* gene is eliminated by mutation in the mouse, these hemispheres and eye regions do not separate, and the resulting embryo is *cyclopic,* with one central eye and a single forebrain (it also lacks limb structures).

The dramatic and diverse roles of *Shh* are a striking example of the different roles played by toolkit genes at different places and times in development. The outcomes of Shh signaling are different in each case: the Shh signaling pathway will induce the expression of one set of genes in the developing limb, a different set in the feather bud, and yet another set in the floor plate. How are different cell types and tissues able to respond differently to the same signaling molecule? The outcome of Shh signaling depends on the context provided by other toolkit genes that are acting at the same time.

> **Message** Most toolkit genes have multiple roles in different tissues and cell types. The specificity of their action is determined by the context provided by the other toolkit genes that act in combination with them.

13.7 Development and Disease

The discovery of the fly, vertebrate, and human toolkits for development has also had a profound effect on the study of the genetic basis of human diseases, particularly of birth defects and cancer. A large number of toolkit-gene mutations have been identified that affect human development and health. We will focus here on just a few examples that illustrate how understanding gene function and regulation in model animals has translated into better understanding of human biology.

Polydactyly

A fairly common syndrome in humans is the development of extra partial or complete digits on the hands and feet. This condition, called *polydactyly,* arises in about 5 to 17 of every 10,000 live births. In the most dramatic cases, the condition is present on both hands and feet (Figure 13-27). Polydactyly occurs widely throughout vertebrates—in cats, chickens, mice, and other species.

The discovery of the role of Shh in digit patterning led geneticists to investigate whether the *Shh* gene was altered in polydactylous humans and other species. In fact, certain polydactyly mutations are mutations of the *Shh* gene. Im-

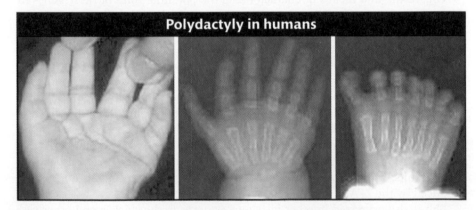

Figure 13-27 This person has six fingers on each hand and seven toes on each foot owing to a regulatory mutation in the *Sonic hedgehog* gene. [*Photographs courtesy of Dr. Robert Hill, MRC Human Genetics Unit, Edinburgh, Scotland; from L. A. Lettice et al., "Disruption of a Long-Range Cis-Acting Regulator for Shh Causes Preaxial Polydactyly," Proc. Natl. Acad. Sci. USA 99, 2002, 7548.*]

portantly, the mutations are not in the coding region of the *Shh* gene; rather, they lie in a cis-acting regulatory element, far from the coding region, that controls *Shh* expression in the developing limb bud. The extra digits are induced by the expression of *Shh* in a part of the limb where the gene is not normally expressed. Mutations in cis-acting regulatory elements have two important properties that are distinct from mutations in coding regions. First, because they affect regulation in cis, the phenotypes are often dominant. Second, because only one of several cis-acting regulatory elements may be affected, other gene functions may be completely normal. Polydactyly can occur without any collateral developmental problems. Coding mutations in *Shh*, however, tell a different story, as we will see in the next section.

Holoprosencephaly

Mutations in the human *Shh* coding region also have been identified. The consequent alterations in the Shh protein are associated with a syndrome termed *holoprosencephaly*, in which abnormalities occur in brain size, in the formation of the nose, and in other midline structures. These abnormalities appear to be less severe counterparts of the developmental defects observed in homozygous *Shh* mutant mice. Indeed, the affected children seen in clinics are heterozygous. One copy of a normal *Shh* gene appears to be insufficient for normal midline development (the gene is *haploinsufficient*). Human fetuses homozygous for loss-of-function *Shh* mutations very likely die in gestation with more severe defects.

Holoprosencephaly is not caused exclusively by *Shh* mutations. Shh is a ligand in a signal-transduction pathway. As might be expected, mutations in genes encoding other components of the pathway affect the efficiency of Shh signaling and are also associated with holoprosencephaly. Several components of the human Shh pathway were first identified as homologs of members of the fly pathway, demonstrating once again both the conservation of the genetic toolkit and the power of model systems for biomedical discovery.

Cancer as a developmental disease

In long-lived animals, such as ourselves and other mammals, development does not cease at birth or at the end of adolescence. Tissues and various cell types are constantly being replenished. The maintenance of many organ functions depends on the controlled growth and differentiation of cells that replace those that are sloughed off or otherwise die. Tissue and organ maintenance is generally controlled by signaling pathways. Inherited or spontaneous mutations in genes encoding components of these pathways can disrupt tissue organization and contribute to the loss of control of cell proliferation. Because unchecked cell proliferation is a characteristic of cancer, the formation of cancers may be a consequence. Cancer, then, is a developmental disease, a product of normal developmental processes gone awry.

Some of the genes associated with types of human cancers are shared members of the animal toolkit. For example, the *patched* gene encodes a receptor for the Hedgehog signaling proteins. In addition to causing inherited developmental disorders such as polydactyly and holoprosencephaly, mutations in the human *patched* gene are associated with the formation of a variety of cancers. About 30 to 40 percent of patients with a dominant genetic disorder called *basal cell nevus syndrome* (BCNS) carry *patched* mutations. These persons are strongly disposed to develop a type of skin cancer called basal-cell carcinoma. They also have a greatly increased incidence of medulloblastoma, a very deadly form of brain tumor. A growing list of cancers are now associated with disruptions of signal-transduction

Table 13-2   Some Toolkit Genes Having Roles in Cancer

	Fly gene	Mammalian gene	Cancer type
Signaling-Pathway Components			
Wingless	*armadillo*	*β-catenin*	Colon and skin
	D.TCF	*TLF*	Colon
Hedgehog	*cubitus interruptus*	*Gli1*	Basal-cell carcinoma
	patched	*patched*	Basal-cell-carcinoma, medulloblastoma
	smoothened	*smoothened*	Basal-cell carcinoma
Notch	*Notch*	*hNotch1*	Leukemia, lymphoma
EGF receptor	*torpedo*	*C-erbB-2*	Breast and colon
Decapentaplegic/TFG-*β*	*Medea*	*DPC4*	Pancreatic and colon
Toll	*dorsal*	*NF-κB*	Lymphoma
Other	*extradenticle*	*Pbx1*	Acute pre-B-cell leukemia

pathways—pathways that were first elucidated by these early systematic genetic screens for patterning mutants in fruit flies (Table 13-2).

The discoveries of links between mutations of signal-transduction-pathway genes and human cancer have greatly facilitated the study of the biology of cancer and the development of new therapies. For example, about 30 percent of mice heterozygous for a targeted mutation in the *patched* gene develop medulloblastoma. These mice therefore serve as an excellent model for the biology of human disease and a testing platform for therapy. Many of the newest anticancer agents employed today are in fact targeted toward components of signal-transduction pathways that are disrupted in certain types of tumors.

It is fair to say that even the most optimistic and farsighted researchers did not expect that the discovery of the genetic toolkit for building a fly would have such far-ranging effects on understanding human development and disease. But such huge unforeseen dividends are familiar in the recent history of basic genetic research. The advent of genetically engineered medicines, monoclonal antibodies for diagnosis and therapy, and forensic DNA testing all had similar origins in seemingly unrelated investigations.

Summary

In Chapter 11, we mentioned the quip from Jacques Monod that "what is true for *E. coli* is also true for the elephant." Now that we have seen the regulatory processes that build worms, flies, mice, and elephants, would we say he was right? If Monod was referring to the principle that gene transcription is controlled by sequence-specific regulatory proteins, we have seen that the bacterial Lac repressor and the fly Hox proteins do indeed act similarly. Moreover, their DNA-binding proteins have the same type of motif. The fundamental insights that François Jacob and Jacques Monod had concerning the central role of the control of gene transcription in bacterial physiology and that they expected would apply to cell differentiation and development in complex multicellular organisms have been borne out in many respects in the genetic control of animal development.

Many features in single-celled and multicellular eukaryotes, however, are not found in bacteria and their viruses. Geneticists and molecular biologists have discovered the functions of introns, RNA splicing, distant and multiple cis-acting regulatory elements, chromatin, alternative splicing, and, more recently, miRNAs. Still, central to the genetic control of development is the control of differential gene expression.

This chapter has presented an overview of the logic and mechanisms for the control of gene expression and development in a few model species. We have concentrated on the toolkit of animal genes for developmental processes and the mechanisms that control the organization of major features of the body plan—the establishment of body axes, segmentation, and segment identity. Although we explored only a modest number of regulatory mechanisms in depth, and just a few species, similarities in regulatory logic and mechanisms allow us to identify some general themes concerning the genetic control of development.

1.   *Despite vast differences in appearance and anatomy, animals have in common a toolkit of genes that govern development.* This toolkit is a small fraction of all genes in the genome, and most of these toolkit genes control transcription factors and components of signal-transduction pathways. Individual toolkit genes typically have multiple functions and affect the development of different structures at different stages.

2.   *The development of the growing embryo and its body parts takes place in a spatially and temporally ordered progression.* Domains within the embryo are established by the expression of toolkit genes that mark out progressively finer subdivisions along both embryonic axes.

3.   *Spatially restricted patterns of gene expression are products of combinatorial regulation.* Each pattern of gene expression has a preceding causal basis. New patterns are generated by the combined inputs of preceding patterns. In the examples presented in this chapter, the positioning of pair-rule stripes and the restriction of appendage-regulatory-gene expression to individual segments requires the integration of numerous positive and negative regulatory inputs by cis-acting regulatory elements.

Posttranscriptional regulation at the RNA level adds another layer of specificity to the control of gene expression. Alternative RNA splicing and translational control by proteins and miRNAs also contribute to the spatial and temporal control of toolkit-gene expression.

Combinatorial control is key to both the *specificity* and the *diversity* of gene expression and toolkit-gene function. In regard to specificity, combinatorial mechanisms provide the means to localize gene expression to discrete cell populations by using inputs that are not specific to cell type or tissue type. The actions of toolkit proteins can thus be quite specific in different contexts. In regard to diversity, combinatorial mechanisms provide the means to generate a virtually limitless variety of gene-expression patterns.

4.   *The modularity of cis-acting regulatory elements allows for independent spatial and temporal control of toolkit-gene expression and function.* Just as the operators and UAS elements of prokaryotes and simple eukaryotes act as switches in the physiological control of gene expression, the cis-acting regulatory elements of toolkit genes act as switches in the developmental control of gene expression. The distinguishing feature of toolkit genes is the typical presence of numerous independent cis-acting regulatory elements that govern gene expression in different spatial domains and at different stages of development. The independent spatial and temporal regulation of gene expression enables individual toolkit genes to have different but specific functions in different contexts. In this light, it is not adequate or accurate to describe a given toolkit-gene function solely in relation to the protein (or miRNA) that it encodes, because the function of the gene product almost always depends on the context in which it is expressed.

KEY TERMS

gap gene (p. 463)
gene complex (p. 455)
homeobox (p. 458)
homeodomain (p. 458)

housekeeping gene (p. 452)
Hox gene (p. 455)
maternal-effect gene (p. 461)
pair-rule gene (p. 463)

positional information (p. 466)
segment-polarity gene (p. 463)
serially reiterated structure (p. 455)
zygote (p. 461)

SOLVED PROBLEMS

SOLVED PROBLEM 1. The *Bicoid* gene (*bcd*) is a maternal-effect gene required for the development of the *Drosophila* anterior region. A mother heterozygous for a *bcd* deletion has only one copy of the *bcd* gene. With the use of *P* elements to insert copies of the cloned *bcd*$^+$ gene into the genome by transformation, it is possible to produce mothers with extra copies of the gene. The early *Drosophila* embryo develops an indentation called the cephalic furrow that is more or less perpendicular to the longitudinal, anteroposterior (A–P) body axis. In the progeny of mothers with only a single copy of *bcd*$^+$, this furrow is very close to the anterior tip, lying at a position one-sixth of the distance from the anterior to the posterior tip. In the progeny of standard wild-type diploids (having two copies of *bcd*$^+$), the cephalic furrow arises more posteriorly, at a position one-fifth of the distance from the anterior to the posterior tip of the em-

bryo. In the progeny of mothers with three copies of *bcd*$^+$, it is even more posterior. As additional gene doses are added, the cephalic furrow moves more and more posteriorly, until, in the progeny of mothers with six copies of *bcd*$^+$, it is midway along the A–P axis of the embryo. Explain the gene-dosage effect of *bcd*$^+$ on the formation of the cephalic furrow in light of the contribution that *bcd* makes to A–P pattern formation.

Solution

The determination of anterior–posterior parts of the embryo is governed by a concentration gradient of Bicoid protein. The furrow develops at a critical concentration of *bcd*. As *bcd*$^+$ gene dosage (and, therefore, Bicoid protein concentration) decreases, the furrow shifts anteriorly; as the gene dosage increases, the furrow shifts posteriorly.

PROBLEMS

Most of the problems are also available for review/grading through the GENETICSPORTAL www.yourgeneticsportal.com.

WORKING WITH THE FIGURES

1. In Figure 13-2, the transplantation of certain regions of embryonic tissue induces the development of structures in new places. What are these special regions called, and what are the substances they are proposed to produce?

2. In Figure 13-5, two different methods are illustrated for visualizing gene expression in developing animals. Which method would allow one to detect where within a cell a protein is localized?

3. Figure 13-7 illustrates the expression of the Ultrabithorax (Ubx) Hox protein in developing flight appendages. What is the relationship between where the protein is expressed and the phenotype resulting from the loss of its expression (shown in Figure 13-1)?

4. In Figure 13-11, what is the evidence that vertebrate *Hox* genes govern the identity of serially repeated structures?

5. As shown in Figure 13-14, what is the fundamental distinction between a pair-rule gene and a segment-polarity gene?

6. In Table 13-1, what is the most common function of proteins that contribute to pattern formation? Why is this the case?

7. In Figure 13-20, which gap protein regulates the posterior boundary of *eve* stripe 2? Describe how it does so in molecular terms.

8. As shown in Figure 13-22, how many different transcription factors govern where the *Distal-less (Dll)* gene will be expressed?

9. As shown in Figure 13-26, the *Sonic hedgehog* gene is expressed in many places in a developing chicken. Is the identical Sonic hedgehog protein expressed in each tissue? If so, how do the tissues develop into different structures? If not, how are different Sonic hedgehog proteins produced?

BASIC PROBLEMS

10. *Gooseberry, runt, knirps,* and *Antennapedia.* To a *Drosophila* geneticist, what are they? How do they differ?

11. Describe the expression pattern of the *Drosophila* gene *eve* in the early embryo.

12. Contrast the function of homeotic genes with that of pair-rule genes.

13. When an embryo is homozygous mutant for the gap gene *Kr,* the fourth and fifth stripes of the pair-rule gene *ftz* (counting from the anterior end) do not form normally. When the gap gene *kni* is mutant, the fifth

and sixth *ftz* stripes do not form normally. Explain these results in regard to how segment number is established in the embryo.

14. Some of the mammalian *Hox* genes have been shown to be more similar to one of the insect *Hox* genes than to the others. Describe an experimental approach that would enable you to demonstrate this finding in a functional context.

15. The three homeodomain proteins ABD-B, ABD-A, and UBX are encoded by genes within the *Bithorax* complex of *Drosophila.* In wild-type embryos, the *Abd-B* gene is expressed in the posterior abdominal segments, *Abd-A* in the middle abdominal segments, and *Ubx* in the anterior abdominal and posterior thoracic segments. When the *Abd-B* gene is deleted, *Abd-A* is expressed in both the middle and the posterior abdominal segments. When *Abd-A* is deleted, *Ubx* is expressed in the posterior thorax and in the anterior and middle abdominal segments. When *Ubx* is deleted, the patterns of *Abd-A* and *Abd-B* expression are unchanged from wild type. When both *Abd-A* and *Abd-B* are deleted, *Ubx* is expressed in all segments from the posterior thorax to the posterior end of the embryo. Explain these observations, taking into consideration the fact that the gap genes control the initial expression patterns of the homeotic genes.

16. How can you tell if a gene is required zygotically and if it has a maternal effect?

17. In considering the formation of the A–P and D–V axes in *Drosophila,* we noted that, for mutations such as *bcd,* homozygous mutant mothers uniformly produce mutant offspring with segmentation defects. This outcome is always true regardless of whether the offspring themselves are *bcd*[+]/*bcd* or *bcd*/*bcd.* Some other maternal-effect lethal mutations are different, in that the mutant phenotype can be "rescued" by introducing a wild-type allele of the gene from the father. In other words, for such rescuable maternal-effect lethals, *mut*[+]/*mut* animals are normal, whereas *mut*/*mut* animals have the mutant defect. Explain the difference between rescuable and nonrescuable maternal-effect lethal mutations.

18. Suppose you isolate a mutation affecting A–P patterning of the *Drosophila* embryo in which every other segment of the developing mutant larva is missing.

    a. Would you consider this mutation to be a mutation in a gap gene, a pair-rule gene, a segment-polarity gene, or a segment-identity gene?

    b. You have cloned a piece of DNA that contains four genes. How could you use the spatial-expression pattern of their mRNA in a wild-type embryo to identify which represents a candidate gene for the mutation described?

c. Assume that you have identified the candidate gene. If you now examine the spatial-expression pattern of its mRNA in an embryo that is homozygous mutant for the gap gene *Krüppel,* would you expect to see a normal expression pattern? Explain.

19. How does the Bicoid protein gradient form?

20. In an embryo from a homozygous *Bicoid* mutant female, which class(es) of gene expression is (are) abnormal?

   a. Gap genes

   b. Pair-rule genes

   c. Segment-polarity genes

   d. *Hox* genes

CHALLENGING PROBLEMS

21. a. The *eyeless* gene is required for eye formation in *Drosophila.* It encodes a homeodomain. What would you predict about the biochemical function of the Eyeless protein?

   b. Where would you predict that the *eyeless* gene is expressed in development? How would you test your prediction?

   c. The *Small eye* and *Aniridia* genes of mice and humans, respectively, encode proteins with very strong sequence similarity to the fly Eyeless protein, and they are named for their effects on eye development. Devise one test to examine whether the mouse and human genes are functionally equivalent to the fly *eyeless* gene.

22. Gene *X* is expressed in the developing brain, heart, and lungs of mice. Mutations that selectively affect gene *X* function in these three tissues map to three different regions (A, B, and C, respectively) 5′ of the X coding region.

   a. Explain the nature of these mutations.

   b. Draw a map of the *X* locus consistent with the preceding information.

   c. How would you test the function of the A, B, and C regions?

23. Why are regulatory mutations at the *Sonic hedgehog* gene dominant and viable? Why do coding mutations cause more widespread defects?

24. A mutation occurs in the *Drosophila doublesex* gene that prevents Tra from binding to the *dsx* RNA transcript. What would be the consequences of this mutation for Dsx protein expression in males? In females?

25. You isolate a *glp-1* mutation of *C. elegans* and discover that the DNA region encoding the spatial control region (SCR) has been deleted. What will the GLP-1 protein expression pattern be in a four-cell embryo in mutant heterozygotes? In mutant homozygotes?

Genomes and Genomics

14

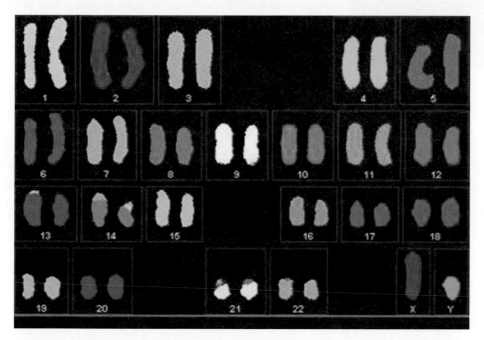

The human nuclear genome viewed as a set of labeled DNA. The DNA of each chromosome has been labeled with a dye that emits fluorescence at one specific wavelength (producing a specific color). [*Nallasivam Palanisamy, Michigan Center for Translational Pathology, University of Michigan.*]

In 1997, a research team at the University of Munich led by Svante Pääbo reported the sequence of a 379-bp region of mitochondrial DNA from the thigh bone of the original Neanderthal fossil discovered in 1856 (Figure 14-1). This sequencing was an astounding technical achievement. DNA molecules break down and accumulate chemical modifications with the passage of time, and so only a series of very short sequences could be deciphered and stitched together. The amount of mitochondrial DNA present in the sample was very small, and the amount of nuclear DNA was even less. Furthermore, the scientists had to take great care to be certain that the sequence that they obtained was not a contaminant from either modern humans or some other source. Most exciting, the sequence of the mitochondrial DNA fragment indicated that Neanderthals became extinct without contributing mitochondrial DNA to modern humans.

Thirteen years later, the Pääbo team, now at the Max Planck Institute for Evolutionary Anthropology in Leipzig, reported that they had obtained more than 4 *billion* bases of *nuclear* DNA sequence from three Neanderthal individuals.

487

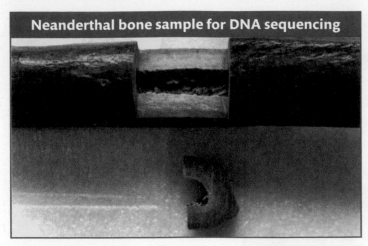

Neanderthal bone sample for DNA sequencing

Figure 14-1 A sample (*bottom*) was removed from the right thigh bone (*top*) of the original Neanderthal specimen for use in DNA sequencing. [*From M. Krings et al., "Neanderthal DNA Sequences and the Origin of Modern Humans," Cell 90, 1997, 19–30, Fig. 1.*]

These advances in Neanderthal genetics illustrate the dramatic advances in the technology and ambitions of **genomics**—the study of genomes in their entirety. What began as a trickle is today a flood of data. In 1995, the 1.8-Mb (1.8-megabase) genome of the bacterium *Hemophilus influenzae* was the first genome of a free-living organism to be sequenced. In 1996 came the 12-Mb genome of *Saccharomyces cerevisiae;* in 1998, the 100-Mb genome of *C. elegans;* in 2000, the 180-Mb genome of *Drosophila melanogaster;* in 2001, the first draft of the 3000-Mb human genome; and, in 2005, the first draft of our closest living relative, the chimpanzee. These species are just a small sample. By the end of 2009, we had the sequences of almost 2000 bacterial genomes, 88 fungal genomes, 14 plants (*Arabidopsis* and rice, for example), 31 mammals (rat, dog, opossum), and many other animal species. The genomes of more than 1000 species are currently in the process of being deciphered.

Genomics has revolutionized how genetic analysis is performed and has opened avenues of inquiry that were not conceivable just a few years ago. Most of the genetic analyses that we have so far considered employ a forward approach to analyzing genetic and biological processes. That is, the analysis begins by first screening for mutants that affect some observable phenotype, and the characterization of these mutants eventually leads to the identification of the gene and the function of DNA, RNA, and protein sequences. In contrast, having the entire DNA sequences of an organism's genome allows geneticists to work in both directions—forward from phenotype to gene, and in reverse from gene to phenotype. Without exception, genome sequences reveal many genes that were not detected from classical mutational analysis. Using so-called reverse genetics, geneticists can now systematically study the roles of such formerly unidentified genes. Moreover, a lack of prior classical genetic study is no longer an impediment to the genetic investigation of organisms. The frontiers of experimental analysis are growing far beyond the bounds of the modest number of long-explored model organisms.

Analyses of whole genomes now contribute to every corner of biological research. In human genetics, genomics is providing new ways to locate genes that contribute to the many genetic diseases determined by complex combinations of genetic factors. For long-studied model organisms, the availability of genome sequences for these species and their relatives has dramatically accelerated gene identification, the analysis of gene function, and the characterization of noncoding elements of the genome. New technologies for the global, genome-wide analysis of the physiological role of all gene products are driving the development of the new field called *systems biology.* From an evolutionary perspective, genomics provides a detailed view of how genomes and organisms have diverged and adapted over geological time. In ecological research, biologists are developing new tools for assaying the distribution of organisms based on detecting the presence and concentration of different genomes in natural samples. And in human medicine, the day is soon approaching when a person's genome sequence is a standard part of his or her medical record.

The DNA sequence of the genome is the starting point for a whole new set of analyses aimed at understanding the structure, function, and evolution of the genome and its components. In this chapter, we will focus on three major aspects of genomic analysis:

- *Bioinformatics,* the analysis of the information content of entire genomes. This information includes the numbers and types of genes and gene products as well as the location, number, and types of binding sites on DNA and RNA that allow functional products to be produced at the correct time and place.

- *Comparative genomics,* which considers the genomes of closely and distantly related species for evolutionary insight and enables conserved sequences to be used as a guide to analyzing gene function.

- *Functional genomics,* the use of an expanding variety of methods, including reverse genetics, to understand gene function and to delineate networks of interacting genes and proteins in biological processes.

14.1 The Genomics Revolution

After the development of recombinant DNA technology in the 1970s, research laboratories typically undertook the cloning and sequencing of one gene at a time, and then only after having had first found out something interesting about that gene from a classic mutational analysis. The steps in proceeding from the classical genetic map of a locus to isolating the DNA encoding a gene (*cloning*) to determining its sequence were often numerous and time consuming. In the 1980s, some scientists realized that a large team of researchers making a concerted effort could clone and sequence the *entire* genome of a selected organism. Such **genome projects** would then make the clones and the sequence publicly available resources. One appeal of having these resources available is that, when researchers become interested in a gene of a species whose genome has been sequenced, they need only find out where that gene is located on the map of the genome to be able to zero in on its sequence and potentially its function. By this means, a gene could be characterized much more rapidly than by cloning and sequencing it from scratch, a project that at the time could take several years to carry out. This quicker approach is now a reality for all model organisms. In a similar way, in human genetics, the genome sequence can aid in identifying disease-causing genes.

From a broader perspective, the genome projects had the appeal that they could provide some glimmer of the principles on which genomes are built. The human genome contains 3 billion base pairs of DNA. Having the entire sequence raised questions such as: How many genes does it contain? How are they distributed and why? What fraction of the genome is coding sequence? What fraction is regulatory sequence? Although we might convince ourselves that we understand a single gene of interest, the major challenge of genomics today is genomic literacy: How do we read the storehouse of information enciphered in the sequence of complete genomes?

The basic techniques needed for sequencing entire genomes were already available in the 1980s (see Chapter 10). But the scale that was needed to sequence a complex genome was, as an engineering project, far beyond the capacity of the research community then. Genomics in the late 1980s and the 1990s evolved out of large research centers that could integrate these elemental technologies into an industrial-level production line. These centers developed robotics and automation to carry out the many thousands of cloning steps and millions of sequencing reactions necessary to assemble the sequence of a complex organism. With these centers in place, the late 1990s and 2000s have been the golden age of genome sequencing.

The rate of genome sequencing has continued to accelerate, and a variety of powerful sequencing methodologies have been invented that are providing greater throughput (the amount of sequence obtained per unit time per instrument) at much lower cost. New technologies can now obtain more than 500 million bases of sequence in a working day on a single instrument. This figure represents an approximately 1000-fold increase in throughput over earlier instruments used to obtain the first human genome sequence.

Genomics, aided by the explosive growth in information technology, has encouraged researchers to develop ways of experimenting on the genome as a whole rather than simply one gene at a time. Genomics has also demonstrated the value of collecting large-scale data sets in advance so that they can be used later to address specific research problems. Genomics has also changed the sociology of biological research, demonstrating the value of large collaborative research networks as a complement to the small, independent research laboratories (which still flourish). These effects will only increase as more information, technology and insight emerge. In the last section of this chapter, we will explore some ways that genomics now drives basic and applied genetics research. In subsequent chapters, we will see how genomics is catalyzing advances in understanding the dynamics of mutation, recombination, and evolution.

> **Message** Characterizing whole genomes is fundamental to understanding the entire body of genetic information underlying the physiology and development of living organisms and to the discovery of new genes such as those having roles in human genetic disease.

14.2 Obtaining the Sequence of a Genome

When people encounter new territory, one of their first activities is to create a map. This practice has been true for explorers, geographers, oceanographers, and astronomers, and it is equally true for geneticists. Geneticists use many kinds of maps to explore the terrain of a genome. Examples are linkage maps based on inheritance patterns of gene alleles and cytogenetic maps based on the location of microscopically visible features such as rearrangement break points.

The highest-resolution map is the complete DNA sequence of the genome—that is, the complete sequence of nucleotides A, T, C, and G of each double helix in the genome. Because obtaining the complete sequence of a genome is such a massive undertaking, of a sort not seen before in biology, new strategies must be used, all based on automation.

Turning sequence reads into an assembled sequence

You've probably seen a magic act in which the magician cuts up a newspaper page into a great many pieces, mixes it in his hat, says a few magic words, and *voila!* an intact newspaper page reappears. Basically, that's how genomic sequences are obtained. The approach is to (1) break the DNA molecules of a genome up into thousands to millions of more or less random, overlapping small segments, (2) read the sequence of each small segment, (3) computationally find the overlap among the small segments where their sequences are identical, and (4) continue overlapping ever larger pieces until all the small segments are linked (Figure 14-2). At that point, the sequence of a genome is assembled.

Why does this process require automation? To understand why, let's consider the human genome, which contains about 3×10^9 bp of DNA, or 3 billion base pairs (3 gigabase pairs = 3 Gbp). Suppose we could purify the DNA intact from each of the 24 human chromosomes (X, Y, and the 22 autosomes), separately put each of these 24 DNA samples into a sequencing machine, and read their sequences directly from one telomere to the other. Obtaining a complete sequence would be utterly straightforward, like reading a book with 24 chapters—albeit a very, very long book with 3 billion characters (about the length of 3000 novels). Unfortunately, such a sequencing machine does not exist.

Rather, automated sequencing is the current state of the art in DNA sequencing technology. Initially based on the pioneering Sanger dideoxy chain-termination method (discussed in Chapter 10), automated sequencing now employs new chemistries and optical-detection methods. A variety of methods are now available that vary in the length of DNA sequence obtained, the bases determined per

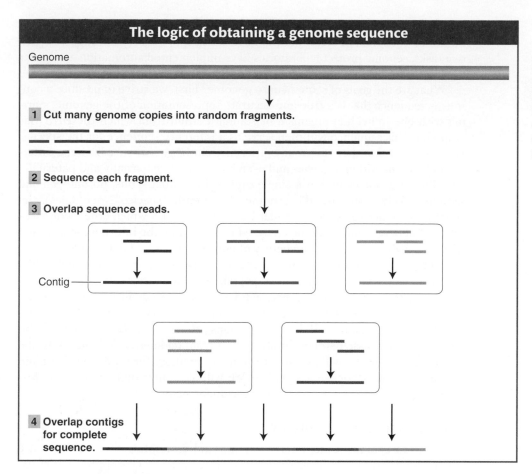

The logic of obtaining a genome sequence

Genome

1 **Cut many genome copies into random fragments.**

2 **Sequence each fragment.**

3 **Overlap sequence reads.**

Contig

4 **Overlap contigs for complete sequence.**

Figure 14-2 To obtain a genome sequence, multiple copies of the genome are cut into small pieces that are sequenced. The resulting sequence reads are overlapped by matching identical sequences in different fragments until a consensus sequence of each DNA double helix in the genome is produced.

second, and raw accuracy. For large-scale sequencing projects that seek to analyze large individual genomes or the genomes of many different individuals or species, choosing a method requires balancing speed, cost, and accuracy.

Individual sequencing reactions (called *sequencing reads*) provide letter strings that, depending on the sequencing technique employed, are generally about 300 to 600 bases long. Such lengths are tiny compared with the DNA of a single chromosome. For example, an individual read of 300 bases is only 0.0001 percent of the longest human chromosome (about 3×10^8 bp of DNA) and only about 0.00001 percent of the entire human genome. Thus, one major challenge facing a genome project is **sequence assembly**—that is, building up all of the individual reads into a **consensus sequence,** a sequence for which there is consensus (or agreement) that it is an authentic representation of the sequence for each of the DNA molecules in that genome.

Let's look at these numbers in a somewhat different way to understand the scale of the problem. As with any experimental observation, automated sequencing machines do not always give perfectly accurate sequence reads. The error rate is not constant; it depends on such factors as the dyes that are attached to the sequenced molecules, the purity and homogeneity of the starting DNA sample, and the specific sequence of base pairs in the DNA sample. Thus, to ensure accuracy, genome projects conventionally obtain multiple (as many as 10) independent sequence reads of each base pair in a genome. Tenfold coverage (denoted $10\times$) ensures that chance errors in the reads do not give a false reconstruction of the consensus sequence. Given an average sequence read of about 300 bases of DNA and a human genome of 3 billion base pairs, 100 million successful independent reads are required to give 10-fold average coverage of each base pair. However, not all reads are successful, and so the number of attempted reads is larger. Thus, the amount of information and material to be tracked is enormous. To try to minimize both human error and the need for people to carry out highly repetitive tasks, genome-project laboratories have implemented automation, computer tracking with the use of bar coding, and computer-analysis systems.

What are the goals of sequencing a genome? First, we strive to produce a consensus sequence that is a true and accurate representation of the genome, starting with one individual organism or standard strain from which the DNA was obtained. This sequence will then serve as a reference sequence for the species. We now know that there are many differences in DNA sequence between different individuals within a species and even between the maternally and paternally contributed genomes within a single diploid individual. Thus, no one genome sequence truly represents the genome of the entire species. Nonetheless, the genome sequence serves as a standard or reference with which other sequences can be compared, and it can be analyzed to determine the information encoded within the DNA, such as the inventory of encoded RNAs and polypeptides.

Like written manuscripts, genome sequences can range from *draft* quality (the general outline is there, but there are typographical errors, grammatical errors, gaps, sections that need rearranging, and so forth), to *finished* quality (a very low rate of typographical errors, some missing sections but everything that is currently possible has been done to fill in these sections), to truly *complete* (no typographical errors, every base pair absolutely correct from telomere to telomere). In the following sections, we will examine the current methods for producing draft and finished genome-sequence assemblies. We will also encounter some of the features of genomes that challenge genome-sequencing projects.

Whole-genome sequencing

The current general strategy for obtaining and assembling the sequence of a genome is called *whole-genome shotgun (WGS) sequencing.* This approach is based on determining the sequence of many segments of genomic DNA that have been generated by breaking the long chromosomes of DNA into many short segments. Two approaches to whole-genome shotgun sequencing are responsible for most genome sequences obtained to date. The fundamental differences between them are in how the short segments of DNA are obtained and prepared for sequencing and the sequencing chemistry employed. The first method, used to sequence the first human genome, relied on the cloning of DNA in microbial cells and employed the Sanger dideoxy sequencing technique. We will refer to this approach as "traditional WGS." The second group of methods are generally cell-free methods that employ new techniques for sequencing and are designed for very high throughput. We will refer to this group of methods as "next-generation WGS."

Traditional WGS

The traditional WGS approach begins with the construction of **genomic libraries,** which are collections of these short segments of DNA, representing the entire genome. The short DNA segments in such a library have been inserted into one of a number of types of *accessory* chromosomes (nonessential elements such as plasmids, modified bacterial viruses, or artificial chromosomes) and propagated in microbes, usually bacteria or yeast. These accessory chromosomes carrying DNA inserts are called **vectors.**

To generate a genomic library, a researcher first uses *restriction enzymes,* which cleave DNA at specific sequences, to cut up purified genomic DNA. Some enzymes cut the DNA at many places, whereas others cut it at fewer places; so the researcher can control whether the DNA is cut, on average, into longer or shorter pieces. The resulting fragments have short single strands of DNA at both ends. Each fragment is then joined to the DNA molecule of the accessory chromosome, which also has been cut with a restriction enzyme and which has ends that are complementary to those of the genomic fragments. In order for the entire genome to be represented, multiple copies of the genomic DNA are cut into fragments. By this means, thousands to millions of different fragment-vector recombinant molecules are generated.

The resulting pool of recombinant DNA molecules is then propagated, typically by introducing the molecules into bacterial cells. Each cell takes up one recombinant molecule. Then each recombinant molecule is replicated in the normal growth and division of its host so that many identical copies of the inserted fragment are produced for use in analyzing the fragment's DNA sequence. Because each recombinant molecule is amplified from an individual cell, each cell is a distinct *clone*. (More details about DNA cloning are provided in Chapter 10.) The resulting library of clones is called a *shotgun library* because sequence reads are obtained from clones randomly selected from the whole-genome library without any information on where these clones map in the genome.

Next, the genome fragments in clones from the shotgun library are partially sequenced. The sequencing reaction must start from a primer of known sequence. Because the sequence of a cloned insert is not known (and is the goal of the exercise), primers are based on the sequence of adjacent vector DNA. These primers are used to guide the sequencing reaction into the insert. Hence, short regions at one or both ends of the genomic inserts can be sequenced (Figure 14-3). After sequencing, the output is a large collection of random short sequences, some of them overlapping. These sequence reads are assembled into a consensus sequence covering the whole genome by matching homologous sequences shared by reads from overlapping clones. The sequences of overlapping reads are assembled into units called **sequence contigs** (sequences that are contiguous, or touching).

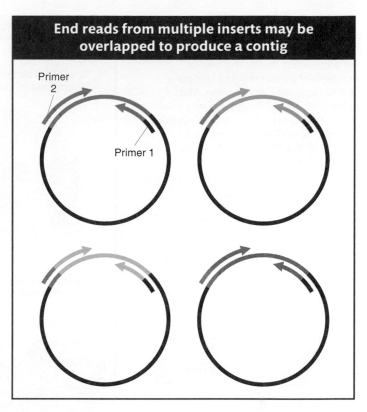

End reads from multiple inserts may be overlapped to produce a contig

Primer 2

Primer 1

Figure 14-3 Sequencing reads are taken only of the ends of cloned inserts. The use of two different sequence priming sites, one at each end of the vector, makes possible the sequencing of as many as 600 base pairs at each end of the genomic insert. If both ends of the same clone are sequenced, the two resulting sequence reads are called *paired-end reads*.

Next-generation whole-genome shotgun sequencing

The goal of next-generation WGS is the same as that of traditional WGS—to obtain a large number of overlapping sequence reads that can be assembled into contigs. However, the methodologies used differ in several substantial ways from traditional WGS. Several different systems have been developed that, while they differ in certain details, each employ three strategies that have dramatically increased throughput:

1. DNA molecules are prepared for sequencing in cell-free reactions, without cloning in microbial hosts.

2. Millions of individual DNA fragments are isolated and sequenced in parallel during each machine run.

3. Advanced fluid-handling technologies, cameras, and software make it possible to detect the products of sequencing reactions in extremely small reaction volumes.

Since the field of genomic technology is evolving rapidly, we will not describe all next-generation systems. However, we will examine one widely used approach that employs all of these features. One of the first next-generation systems was developed by the 454 Life Sciences Corporation. This approach illustrates the gains that have been made in throughput and what such gains enable geneticists to do. The approach can be considered to have three stages:

Figure 14-4 (a) In the 454 sequencing system, single strands of DNA are replicated on tiny beads in preparation for sequencing. (b) The sequencing reactions of pyrosequencing take place in tiny wells arranged on plates. The many wells in a plate, and the very small reaction volumes, allow massively parallel sequencing of DNA at modest cost. [*Part (b) by Roy Kaltschmidt, Lawrence Berkeley National Laboratory.*]

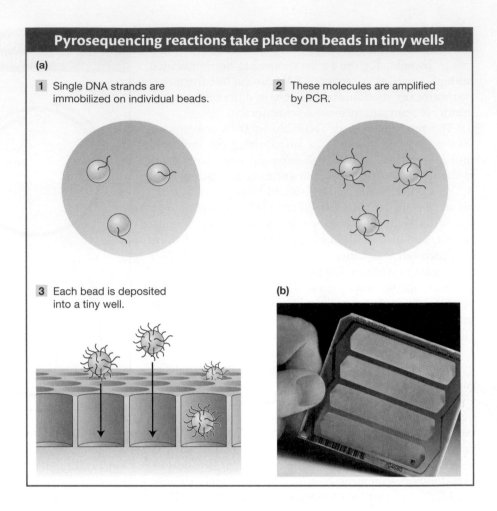

Pyrosequencing reactions take place on beads in tiny wells

(a)

1 Single DNA strands are immobilized on individual beads.

2 These molecules are amplified by PCR.

3 Each bead is deposited into a tiny well.

(b)

Stage 1. A **DNA template library** of single-stranded DNA molecules is constructed.

Stage 2. The DNA molecules in the template library are amplified into many copies, not by growing colonies as for traditional genomic libraries, but by using the polymerase chain reaction (PCR) (see Chapter 10). First, single molecules are immobilized on individual beads. The molecules are then amplified by PCR such that single-stranded DNA molecules remain attached to the beads. Thus, each bead contains many identical DNA fragments. Each bead is then deposited individually into wells of a very small volume in a device that hosts the sequencing reactions (Figure 14-4).

Stage 3. The sequencing of each bead is performed using a novel "sequencing-by-synthesis" chemistry termed **pyrosequencing** (Figure 14-5). DNA polymerase and a primer are added to the wells to prime the synthesis of a complementary DNA strand. Each of the four deoxyribonucleotides dATP, dGTP, dTTP, and dCTP are made to flow through all of the wells, one at a time, in a specific order. When a nucleotide is added that is complementary to the next base in the template strand in a given well, it is incorporated and the reaction releases a pyrophosphate molecule. Two enzymes, *sulfurylase* and *luciferase,* which are also present, then act to convert the pyrophosphate signal to a visible-light signal (Figure 14-5). The light is detected by a special camera. Hence, growing DNA strands that have A as the first base after the primer will yield a signal only when dATP is made to flow through the well and not when the other deoxynucleotides are made to flow through. The reaction is repeated for at least 100 cycles, and the signals from each well over all of the cycles are integrated to generate the sequence reads from each well.

The high throughput of this approach is the product of the massively parallel sequencing: several hundred thousand to more than 1 million reactions can be run simultaneously. Sequencing machines of the previous generation were able to achieve just 384 sequencing reactions per run.

Whole-genome-sequence assembly

Whichever method of obtaining raw sequence is used, the challenge remains to assemble the contigs into the entire genome sequence. The difficulty of that process depends a great deal on the size and complexity of the genome.

For instance, the genomes of bacterial species are relatively easy to assemble. Bacterial DNA is essentially *single-copy* DNA, with no repeating sequences. Therefore, any given DNA sequence read from a bacterial genome will come from one unique place in that genome. Owing to these properties, contigs within bacterial genomes can often be assembled into larger contigs representing most or all of the genome sequence in a relatively straightforward manner. In addition, a typical bacterial genome is only a few megabase pairs of DNA in size. As of May 2010, 674 bacterial genomes had been fully sequenced and the sequencing of more than 1000 additional species was in progress.

For eukaryotes, genome assembly often presents some difficulties. A big stumbling block is the existence of numerous classes of repeated sequences, some arranged in tandem and others dispersed. Why are they a problem for genome sequencing? In short, because a sequencing read of repetitive DNA fits into many places in the draft of the genome. Not infrequently, a tandem repetitive sequence is in total longer than the length of a maximum sequence read. In that case, there is no way to bridge the gap between adjacent unique sequences. Dispersed repetitive elements can cause reads from different chromosomes or different parts of the same chromosome to be mistakenly aligned together.

> **Message** The landscape of eukaryotic chromosomes includes a variety of repetitive DNA segments. These segments are difficult to align as sequence reads.

Whole-genome shotgun sequencing is particularly good at producing draft-quality sequences of complex genomes with many repetitive sequences. As an example, we will consider the genome of the fruit fly *D. melanogaster*, which was initially sequenced by the traditional WGS method. The project began with the sequencing of libraries of genomic clones of different sizes (2 kb, 10 kb, 150 kb). Sequence reads were obtained from *both* ends of genomic-clone inserts and aligned by a logic identical to that used for bacterial WGS sequencing. Through this logic, sequence overlaps were identified and clones were placed in order, producing sequence contigs—consensus sequences for these single-copy stretches of the genome. However, unlike the situation in bacteria, the contigs eventually ran into a repetitive DNA segment that prevented unambiguous assembly of the contigs into a whole genome. The sequence contigs had an average size of about 150 kb. The challenge then was how to glue the thousands of such sequence contigs together in their correct order and orientation.

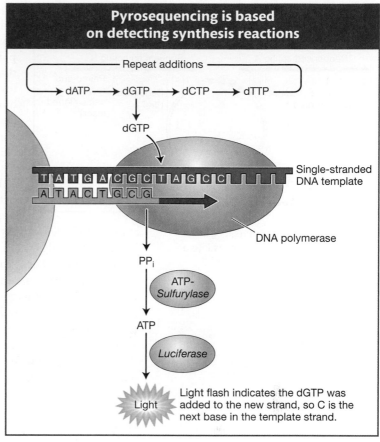

Pyrosequencing is based on detecting synthesis reactions

Repeat additions

dATP → dGTP → dCTP → dTTP

dGTP

Single-stranded DNA template

DNA polymerase

PP$_i$

ATP-Sulfurylase

ATP

Luciferase

Light

Light flash indicates the dGTP was added to the new strand, so C is the next base in the template strand.

Figure 14-5 In the pyrosequencing process, nucleotides are sequentially added to form the complementary strand of the single-stranded template, to which a sequencing primer has been annealed. The reactions are carried out in the presence of the enzymes DNA polymerase, sulfurylase, and luciferase. One molecule of pyrophosphate (PPi) is released for every nucleotide incorporated into the growing strand by the DNA polymerase and is converted to ATP by sulfurylase. Visible light is produced from luciferin in a luciferase-catalyzed reaction that utilizes the ATP produced by sulfurylase.

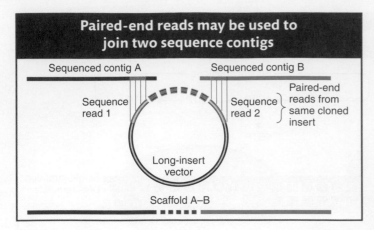

Figure 14-6 Paired-end reads can be used to join two sequence contigs into a single ordered and oriented scaffold.

The solution to this problem was to make use of the pairs of sequence reads from opposite ends of the genomic inserts in the same clone—these reads are called **paired-end reads.** The idea was to find paired-end reads that spanned the gaps between two sequence contigs (Figure 14-6). In other words, if one end of an insert was part of one contig and the other end was part of a second contig, then this insert must span the gap between two contigs, and the two contigs were clearly near each other. Indeed, because the size of each clone was known (that is, it came from a library containing genomic inserts of uniform size, either the 2-kb, 100-kb, or 150-kb library), the distance between the end reads was known. Further, aligning the sequences of the two contigs by using paired-end reads automatically determines the relative orientation of the two contigs. In this manner, single-copy contigs could be joined together, albeit with gaps where the repetitive elements reside. These gapped collections of joined-together sequence contigs are called **scaffolds** (sometimes also referred to as **supercontigs**). Because most *Drosophila* repeats are large (3–8 kb) and widely spaced (one repeat approximately every 150 kb), this technique was extremely effective at producing a correctly assembled draft sequence of the single-copy DNA. A summary of the logic of this approach is shown in Figure 14-7.

Next-generation WGS does not circumvent the problem of repetitive sequences and gaps. Since this approach is intended to circumvent the construction of libraries, which would otherwise facilitate the bridging of gaps between contigs via paired-end reads, next-generation WGS researchers had to devise a way to bridge these gaps without building genomic libraries in vectors. One solution was to build a library of circularized genomic DNA fragments of desired sizes. The circularization allows for short segments of previously distant sequences located at the ends of each fragment to be juxtaposed on either side of a linker sequence. Shearing of these circular molecules and amplification and sequencing of linker-containing fragments produces paired-end reads equivalent to those obtained from sequencing of traditional genomic-library inserts (Figure 14-8).

In both traditional and next-generation whole-genome shotgun sequencing, some gaps usually remain. Occasional gaps are encountered whenever a region of the genome is by chance not found in the traditional shotgun library—for

Figure 14-7 In whole-genome shotgun sequencing, first, the unique sequence overlaps between sequence reads are used to build contigs. Paired-end reads are then used to span gaps and to order and orient the contigs into larger units, called *scaffolds.*

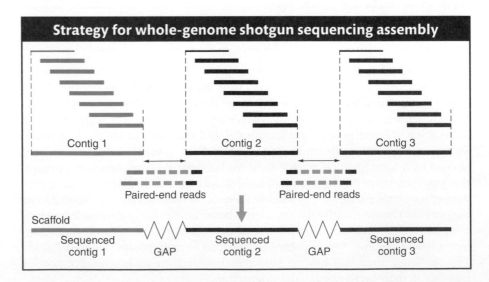

example, some DNA fragments do not replicate well in particular cloning vectors. Specific procedures targeted to individual gaps must be used to fill the missing data in the sequence assemblies. If the gaps are short, missing fragments can be generated by using the known sequences at the ends of the assemblies as primers to amplify and analyze the genomic sequence in between. If the gaps are longer, attempts can be made to clone these sequences in a different host, such as yeast. If cloning in a different host fails, then the gaps in the sequence may remain.

Whether a genome is sequenced to "draft" or "finished" standards is a cost–benefit judgment. It is relatively easy to create a draft but very hard to complete a finished sequence.

Toward the personalized genome

Once the sequence of a given species genome is completed, that accomplishment opens the door to much more rapid and less costly analysis of other individuals of the species. The reason is that with a known assembly as a reference, it is much easier to align the raw sequence reads of additional individuals.

The proof of this principle was demonstrated in 2008 by the massively parallel sequencing of the genome of James D. Watson, the co-discoverer of the double-helical structure of DNA. Over 100 million sequence reads comprising approximately 24.5 billion DNA bases were analyzed, which represented 7.4-fold coverage of the genome. The sequence was obtained at a reported cost of less than $1 million, less than 1/100th the estimated cost of the first human genome sequenced by traditional WGS methods. It is hoped that technological improvements will soon bring this cost to perhaps $10,000 per human genome and eventually to as little as $1000.

As the genome sequences of more human individuals are obtained, we will be able to capture detailed pictures of genetic variation among individuals (see Section 14.5) and make progress toward a "personal genomics" that may allow for a comprehensive and detailed assessment of disease risks in individuals and their relatives.

14.3 Bioinformatics: Meaning from Genomic Sequence

The genomic sequence is a highly encrypted code containing the information for building and maintaining a functional organism. The study of the information content of genomes is called **bioinformatics.** We are far from being able to read this information from beginning to end in the way that we would read a book. Even though we know which triplets encode which amino acids in the protein-coding segments, much of the information contained in a genome is not decipherable from mere inspection.

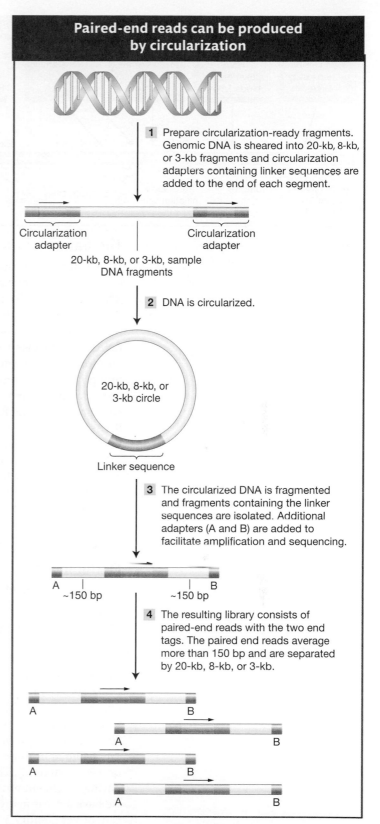

Paired-end reads can be produced by circularization

1 Prepare circularization-ready fragments. Genomic DNA is sheared into 20-kb, 8-kb, or 3-kb fragments and circularization adapters containing linker sequences are added to the end of each segment.

Circularization adapter   Circularization adapter

20-kb, 8-kb, or 3-kb, sample DNA fragments

2 DNA is circularized.

20-kb, 8-kb, or 3-kb circle

Linker sequence

3 The circularized DNA is fragmented and fragments containing the linker sequences are isolated. Additional adapters (A and B) are added to facilitate amplification and sequencing.

A ~150 bp   B ~150 bp

4 The resulting library consists of paired-end reads with the two end tags. The paired end reads average more than 150 bp and are separated by 20-kb, 8-kb, or 3-kb.

A   B
A   B
A   B
A   B

Figure 14-8 Paired-end reads for high-throughput sequencing can be produced without genomic-library construction. The figure is based on the paired-end protocol of the Roche GS FLX Titanium Series, Roche Applied Science, Mannheim, Germany.

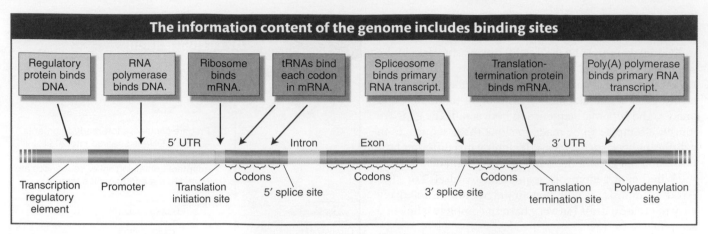

The information content of the genome includes binding sites

| Regulatory protein binds DNA. | RNA polymerase binds DNA. | Ribosome binds mRNA. | tRNAs bind each codon in mRNA. | Spliceosome binds primary RNA transcript. | Translation-termination protein binds mRNA. | Poly(A) polymerase binds primary RNA transcript. |

5′ UTR   Intron   Exon   3′ UTR

Transcription regulatory element   Promoter   Translation initiation site   Codons   5′ splice site   Codons   3′ splice site   Codons   Translation termination site   Polyadenylation site

Figure 14-9 A gene within DNA may be viewed as a series of binding sites for proteins and RNAs.

The nature of the information content of DNA

DNA contains information, but in what way is it encoded? Conventionally, the information is thought of as the sum of all the gene products, both proteins and RNAs. However, the information content of the genome is more complex than that. The genome also contains binding sites for different proteins and RNAs. Many proteins bind to sites located in the DNA itself, whereas other proteins and RNAs bind to sites located in mRNA (Figure 14-9). The sequence and relative positions of those sites permit genes to be transcribed, spliced, and translated properly, at the appropriate time in the appropriate tissue. For example, regulatory protein-binding sites determine when, where, and at what level a gene will be expressed. At the RNA level in eukaryotes, the locations of binding sites for the RNAs and proteins of spliceosomes will determine the 5′ and 3′ splice sites where introns are removed. Regardless of whether a binding site actually functions as such in DNA or RNA, the site must be encoded in the DNA. The information in the genome can be thought of as the sum of all the sequences that encode proteins and RNAs, plus the binding sites that govern the time and place of their actions. As a genome draft continues to be improved, the principal objective is the identification of all of the functional elements of the genome. This process is referred to as **annotation.**

Deducing the protein-encoding genes from genomic sequence

Because the proteins present in a cell largely determine its morphology and physiological properties, one of the first orders of business in genome analysis and annotation is to try to determine an inventory of all of the polypeptides encoded by an organism's genome. This inventory is termed the organism's **proteome.** It can be considered a "parts list" for the cell. To determine the list of polypeptides, the sequence of each mRNA encoded by the genome must be deduced. Because of intron splicing, this task is particularly challenging in multicellular eukaryotes, where introns are the norm. In humans, for example, an average gene has about 10 exons. Furthermore, many genes encode alternative exons; that is, some exons are included in some versions of a processed mRNA but are not included in others (see Chapter 8). The alternatively processed mRNAs can encode polypeptides having much, but not all, of their amino acid sequences in common. Even though we have a great many examples of completely sequenced genes and mRNAs, we cannot yet identify 5′ and 3′ splice sites merely from DNA sequence with a high degree of accuracy. Therefore, we cannot be certain which sequences are introns. Predictions of alternatively used exons are even more error prone. For such reasons, deducing the total polypeptide parts list in higher eukaryotes is a large problem. Some approaches follow.

ORF detection   The main approach to producing a polypeptide list is to use the computational analysis of the genome sequence to predict mRNA and polypeptide sequences, an important part of bioinformatics. The procedure is to look for sequences that have the characteristics of genes. These sequences would be gene size and composed of sense codons after possible introns had been removed. The appropriate 5′- and 3′-end sequences would be present, such as start and stop codons. Sequences with these characteristics typical of genes are called **open reading frames (ORFs).** To find candidate ORFs, the computer scans the DNA sequence on both strands in each reading frame. Because there are three possible reading frames on each strand, there are six possible reading frames in all.

Direct evidence from cDNA sequences   Another means of identifying ORFs and exons is through the analysis of mRNA expression. This analysis is done by creating libraries of DNA molecules that are complementary to mRNA sequences, called cDNA. Complementary DNA sequences are extremely valuable in two ways. First, they are direct evidence that a given segment of the genome is expressed and may thus encode a gene. Second, because the cDNA is complementary to the mature mRNA, the introns of the primary transcript have been removed, which greatly facilitates the identification of the exons and introns of a gene (Figure 14-10). The alignment of cDNAs with their corresponding genomic sequence clearly delineates the exons, and hence introns are revealed as the regions falling between the exons. In the cDNA, the ORF should be continuous from initiation codon through stop codon. Thus, cDNA sequences can greatly assist in identifying the correct reading frame, including the initiation and stop codons. Full-length cDNA evidence is taken as the gold-standard proof that one has identified the sequence of a transcription unit, including its exons and its location in the genome.

In addition to full-length cDNA sequences, there are large data sets of cDNAs for which only the 5′ or the 3′ ends or both have been sequenced. These short cDNA sequence reads are called **expressed sequence tags (ESTs).** Expressed sequence tags can be aligned with genomic DNA and thereby used to determine the 5′ and 3′ ends of transcripts—in other words, to determine the boundaries of the transcript as shown in Figure 14-10.

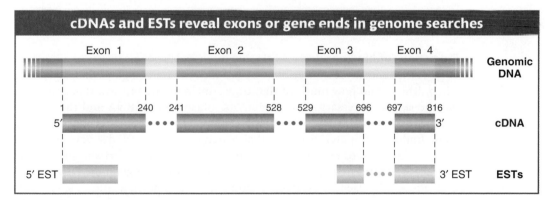

Figure 14-10 Alignment of fully sequenced complementary DNAs (cDNAs) and expressed sequence tags (ESTs) with genomic DNA. The dashed lines indicate regions of alignment; for the cDNA, these regions are the exons of the gene. The dots between segments of cDNA or ESTs indicate regions in the genomic DNA that do not align with cDNA or EST sequences; these regions are the locations of the introns. The numbers above the cDNA line indicate the base coordinates of the cDNA sequence, where base 1 is the 5′-most base and base 816 is the 3′-most base of the cDNA. For the ESTs, only a short sequence read is obtained from each end (5′ and 3′) of the corresponding cDNA. These sequence reads establish the boundaries of the transcription unit, but they are not informative about the internal structure of the transcript unless the EST sequences cross an intron (as is true for the 3′ EST depicted here).

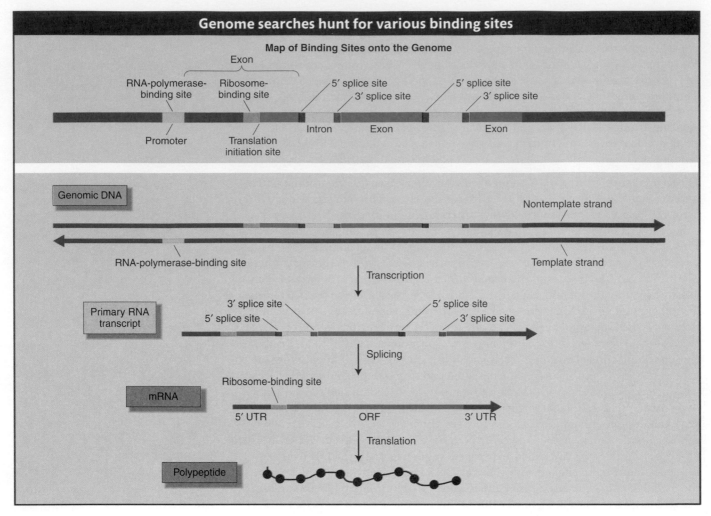

Figure 14-11 Eukaryotic information transfer from gene to polypeptide chain. Note the DNA and RNA "binding sites" that are bound by protein complexes to initiate the events of transcription, splicing, and translation.

Predictions of binding sites   As already discussed, a gene consists of a segment of DNA that encodes a transcript as well as the regulatory signals that determine when, where, and how much of that transcript is made. In turn, that transcript has the signals necessary to determine its splicing into mRNA and the translation of that mRNA into a polypeptide (Figure 14-11). There are now statistical "gene-finding" computer programs that search for the predicted sequences of the various binding sites used for promoters, for transcription start sites, for 3' and 5' splice sites, and for translation initiation codons within genomic DNA. These predictions are based on consensus motifs for such known sequences, but they are by no means perfect.

Using polypeptide and DNA similarity   Because organisms have common ancestors, they also have many genes with similar sequences in common. Hence, a gene will likely have relatives among the genes isolated and sequenced in other organisms, especially in the closely related ones. Candidate genes predicted by the preceding techniques can often be verified by comparing them with all the other gene sequences that have ever been found. A candidate sequence is submitted as a "query sequence" to public databases containing a record of all known gene sequences. This procedure is called a BLAST search (BLAST stands for Basic Local Alignment Search Tool). The sequence can be submitted as a nucleotide sequence (a BLASTn search) or as a translated amino acid sequence (BLASTp). The computer scans the database and returns a list of full or partial "hits," starting with the closest matches. If the candidate sequence closely resembles that of a gene previously identified from another organism, then this resemblance provides a strong indication that the candidate gene is a real gene. Less-close matches are still useful. For example, an amino acid identity of only 35 percent, but at identical positions, is a strong indicator that two proteins have a common three-dimensional structure.

BLAST searches are used in many other ways, but always the goal is to find out more about some identified sequence of interest.

Predictions based on codon bias   Recall from Chapter 9 that the triplet code for amino acids is degenerate; that is, most amino acids are encoded by two or more codons (see Figure 9-5). The multiple codons for a single amino acid are termed *synonymous codons.* In a given species, not all synonymous codons for an amino acid are used with equal frequency. Rather, certain codons are present much more frequently in mRNAs (and hence in the DNA that encodes them). For example, in *D. melanogaster*, of the two codons for cysteine, UGC is used 73 percent of the time, whereas UGU is used 27 percent. This usage is a diagnostic for *Drosophila* because, in other organisms, this "codon bias" pattern is quite different. Codon biases are thought to be due to the relative abundance of the tRNAs complementary to these various codons in a given species. If the codon usage of a predicted ORF matches that species' known pattern of codon usage, then this match is supporting evidence that the proposed ORF is genuine.

Putting it all together   A summary of how different sources of information are combined to create the best-possible mRNA and gene predictions is depicted in Figure 14-12. These different kinds of evidence are complementary

Figure 14-12 The different forms of gene-product evidence—cDNAs, ESTs, BLAST-similarity hits, codon bias, and motif hits—are integrated to make gene predictions. Where multiple classes of evidence are found to be associated with a particular genomic DNA sequence, there is greater confidence in the likelihood that a gene prediction is accurate.

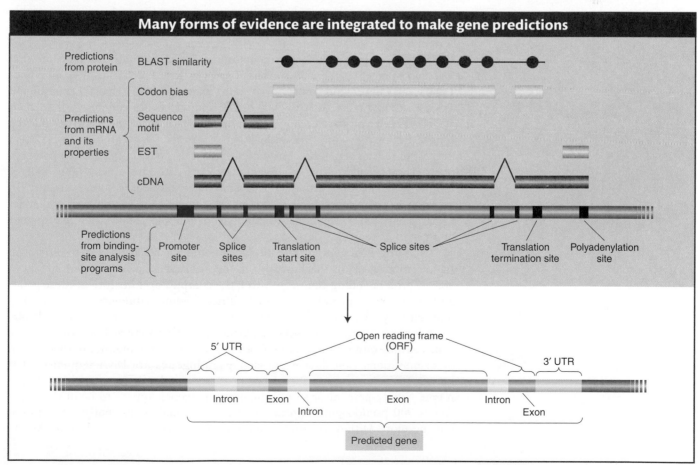

Many forms of evidence are integrated to make gene predictions

Predictions from protein — BLAST similarity

Predictions from mRNA and its properties — Codon bias / Sequence motif / EST / cDNA

Predictions from binding-site analysis programs — Promoter site / Splice sites / Translation start site / Splice sites / Translation termination site / Polyadenylation site

5′ UTR   Open reading frame (ORF)   3′ UTR

Intron   Exon   Intron   Exon   Intron   Exon

Predicted gene

and can cross-validate one another. For example, the structure of a gene may be inferred from evidence of protein similarity within a region of genomic DNA bounded by 5' and 3' ESTs. Useful predictions are possible even without a cDNA sequence or evidence of protein similarities. A binding-site-prediction program can propose a hypothetical ORF, and proper codon bias would be supporting evidence.

> **Message** Predictions of mRNA and polypeptide structure from genomic DNA sequence depend on the integration of information from cDNA sequence, binding-site predictions, polypeptide similarities, and codon bias.

Let's consider some of the insights from our first view of the overall genome structures and global parts lists of a few species whose genomes have been sequenced. We will start with ourselves. What can we learn by looking at the human genome by itself? Then we will see what we can learn by comparing our genome with others.

14.4 The Structure of the Human Genome

In describing the overall structure of the human genome, we must first confront its repeat structure. A considerable fraction of the human genome, about 45 percent, is repetitive. Much of this repetitive DNA is composed of copies of transposable elements. Indeed, even within the remaining single-copy DNA, a fraction has sequences suggesting that they might be descended from ancient transposable elements that are now immobile and have accumulated random mutations, causing them to diverge in sequence from the ancestral transposable elements. Thus, much of the human genome appears to be composed of genetic "hitchhikers."

Only a small part of the human genome encodes polypeptides; that is, somewhat less than 3 percent of it encodes exons of mRNAs. Exons are typically small (about 150 bases), whereas introns are large, many extending more than 1000 bases and some extending more than 100,000 bases. Transcripts are composed of an average of 10 exons, although many have substantially more. Finally, introns may be spliced out of the same gene in locations that vary. This variation in the location of splice sites generates considerable added diversity in mRNA and polypeptide sequence. On the basis of current cDNA and EST data, at least 60 percent of human protein-coding genes are likely to have two or more splice variants. On average, there are several splice variants per gene. Hence, the number of distinct proteins encoded by the human genome is several-fold greater than the number of recognized genes.

The number of genes in the human genome has not been easy to pin down. In the initial draft of the human genome, there were an estimated 30,000 to 40,000 protein-coding genes. However, the complex architecture of these genes and the genome can make annotation difficult. Some sequences scored as genes may actually be exons of larger genes. In addition, there are more than 19,000 **pseudogenes,** which are ORFs or partial ORFs that may at first appear to be genes but are either nonfunctional or inactive due to the manner of their origin or to mutations. So-called **processed pseudogenes** are DNA sequences that have been reverse-transcribed from RNA and randomly inserted into the genome. Ninety percent or so of human pseudogenes appear to be of this type. About 900 pseudogenes appear to be conventional genes that have acquired one or more ORF-disrupting mutations in the course of evolution. As the

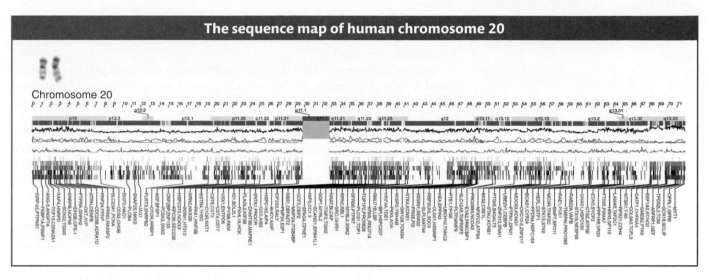

Figure 14-13 Numerous genes have been identified on human chromosome 20. The recombinational and cytogenetic map coordinates are shown in the top lines of the figure. Various graphics depicting gene density and different DNA properties are shown in the middle sections. The identifiers of the predicted genes are shown at the bottom of the panel. [*Courtesy of Jim Kent, Ewan Birney, Darryl Leja, and Francis Collins. After the International Human Genome Sequencing Consortium, "Initial Sequencing and Analysis of the Human Genome," Nature 409, 2001, 860–921.*]

challenges in annotation have been overcome, the estimated number of genes in the human genome has dropped steadily. A recent estimate is that there are about 20,500 protein-coding genes.

The annotation of the human genome has progressed as the sequences of each chromosome were finished one by one. These sequences then became the searching ground in the hunt for candidate genes. An example of gene predictions for a chromosome from the human genome is shown in Figure 14-13. Such predictions are being revised continually as new data and new computer programs become available. The current state of the predictions can be viewed at many Web sites, most notably at the public DNA databases in the United States and Europe (see Appendix B). These predictions are the current best inferences of the protein-coding genes present in the sequenced species and, as such, are works in progress.

Proteins can be grouped into families of related proteins, similar in structure and function, on the basis of similarity in amino acid sequence. For a given protein family that is known in many organisms, the number of proteins in the family is generally larger in humans than in nonvertebrates whose genomes have been sequenced. Proteins are composed of modular domains that are mixed and matched to carry out different roles. Many domains are associated with specific biological functions. The number of modular domains per protein also seems to be higher in humans than in nonvertebrate organisms.

As more refined information on the human genome emerges, additional features can be gleaned. A recent example is the finished sequence of one of the best-studied human chromosomes—chromosome 7. Initially, this chromosome was intensively studied because it contains the gene that, when mutated, causes cystic fibrosis. The location of the cystic fibrosis gene was identified in the early days of the Human Genome Project by overlaying the linkage map on the physical map and sequence, as described in Chapters 4 and 10. Groups have continued to study this chromosome in detail, and about 1700 genes are known or predicted to reside on chromosome 7. About 800 physical-map clones have been mapped to human chromosome 7.

One use of physical-map clones is in mapping rearrangement break points associated with human disease. Chromosomal rearrangements are a class of mutations that result from the breaking of a chromosome at one location—the rearrangement break point—and its rejoining with another similarly severed site on the same chromosome or a different one. These breaks cause mutations when a gene happens to reside at the break point. With the use of physical clones, about 1600 rearrangement break points associated with human disease have been mapped on chromosome 7, creating a high-density cytogenetic map (Figure 14-14). Of these

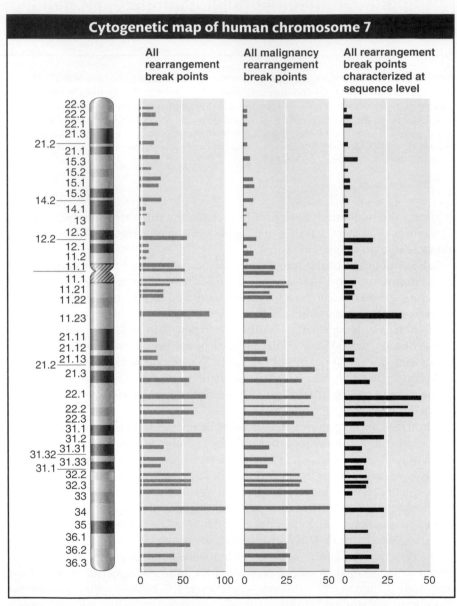

Figure 14-14 Rearrangement break points from patients with genetic disorders have been mapped in chromosome 7, creating a cytogenetic map. [*After W. S. Scherer et al., "Human Chromosome 7: DNA Sequence and Biology," Science 300, 2003, 769 and 771, Figs. 2 and 5.*]

break points, 440 have been sequenced, allowing the association of mutant pheno-types and genes found in the DNA sequences.

14.5 Comparative Genomics

One of the most powerful means to advance the analysis of our or any other genome is to compare genome structure and sequence among related species. Because natural selection generally culls mutations that decrease fitness, genes and other functional DNA sequences are conserved over long periods of evolution. A tract of DNA sequence that is common to divergent species is likely to perform a nec-essary function, and such common tracts can be used to guide studies aimed at discovering these functions. Furthermore, genes identified in one model species are likely to be identifiable, on the bases of sequence and genome location, in related species.

In addition to the identification of conserved regions, **comparative ge-nomics** has the potential to reveal how species diverge. Species evolve and traits change through changes in DNA sequence. Comparisons among species genomes can reveal events unique to particular lineages that may contribute to differences in physiology, behavior, or anatomy. Such events could include, for example, the gain and loss of individual genes or groups of genes. Here we will explore the key principles underlying comparative genomics and look at a few examples of how comparisons among relevant genomes reveal what is similar and different among species and among different individuals of the same species.

Phylogenetic inference

The first step in comparing genomes is to decide which species' genomes to com-pare. In order for comparisons to be informative, it is crucial to understand the evolutionary relationships among the species to be compared. The evolutionary history of a group is called a **phylogeny.** Phylogenies are useful because they allow us to infer how species genomes have changed over time.

The second step in comparing genomes is the identification of the most closely related genes, called **homologs.** Genes that are homologs can be recog-nized by similarities in their DNA sequences and in the amino acid sequences of the proteins they encode. It is important to distinguish here two classes of homologous genes. Some homologs are genes at the same genetic locus in dif-ferent species. These genes would have been inherited from a common ancestor and are referred to as **orthologs.** However, many homologous genes belong to families that have expanded (and contracted) in number in the course of evolu-tion. These homologous genes are at different genetic loci in the same organ-ism. They arose when genes within a genome were duplicated. Genes that are related by gene-duplication events in a genome are called **paralogs.** In genome comparisons, it is not always possible to identify the relationships between par-alogs in gene families, but the presence of homologs can nonetheless be quite informative.

For example, suppose we would like to know how the mammalian genome has evolved over the history of the group. We would like to know whether mammals as a group might have acquired some unique genes, whether mammals with dif-ferent lifestyles might possess different sets of genes, and what the fate was of genes that existed in mammalian ancestors.

Fortunately, we now have a large and expanding set of mammal genome sequences to compare that includes representatives of the three main branches of mammals—monotremes (e.g., platypus), marsupials (e.g., wallaby, opossum), and eutherian mammals (e.g., dog, cat, mouse, human). The relationships between these groups, some members within these groups, and other amniote vertebrates are shown in Figure 14-15.

To illustrate the importance of understanding phylogenies and how to utilize them, we consider the platypus genome. Monotremes differ from other mammals in that they lay eggs. Inspection of the platypus genome revealed that it contains one egg-yolk gene called vitellogenin. Analyses of marsupial and eutherian genomes revealed no such genes. The presence of vitellogenin in the platypus and its absence from other mammals could be explained in two ways: (1) vitellogenin is a novel invention of the platypus, or (2) vitellogenin existed in a common ancestor of monotremes, marsupials, and eutherians but was subsequently lost from marsupials and eutherians. The direction of evolutionary change is opposite in these two alternatives.

A simple pair-wise comparison between the platypus and another mammal does not distinguish between these alternatives. To do that, first we have to infer whether vitellogenin was likely to be present in the last common ancestor of the platypus, marsupials, and eutherians. We make this **phylogenetic inference** by examining whether vitellogenin is found in taxa outside of this entire group of mammals, what is referred to as an evolutionary **outgroup.** Indeed, three homologous vitellogenin genes exist in the chicken. Next, we consider the relationship of the chicken to mammals. Chickens belong to another major branch of the amniotes (amniotes are mostly land-dwelling vertebrates that have a terrestrially adapted egg). Looking at the evolutionary tree in Figure 14-15, we can explain the presence of vitellogenins in chickens and the platypus

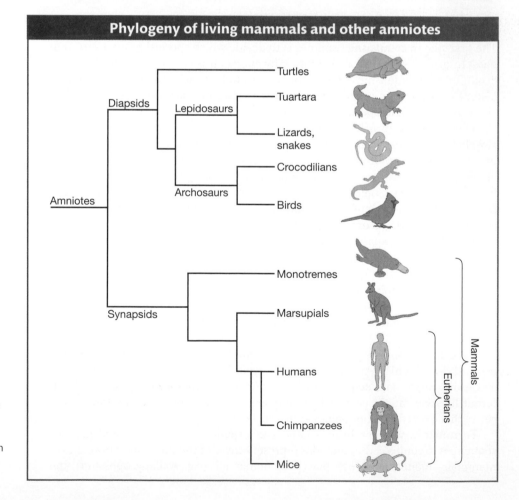

Figure 14-15 Phylogeny of living mammals and other amniotes. The phylogenetic tree depicts the evolutionary relationships among the three major groups of mammals (monotremes, marsupials, and eutherians) and other amniotes, including birds and various reptiles. By mapping the presence or absence of genes in particular groups onto known phylogenies, one can infer the direction of evolutionary change (gain or loss) in particular lineages.

as the result of two independent acquisitions (in the platypus lineage and the chicken lineage, respectively) or as the result of just one acquisition in a common ancestor of the platypus and chicken (which, based on the tree, would be a common ancestor of all amniotes) followed by the loss of vitellogenin genes in marsupials and eutherians.

How do we finally decide between these alternatives? When studying infrequent events such as the invention of a gene, evolutionary biologists prefer to rely on the principle of **parsimony,** that is, to favor the simplest explanation involving the smallest number of evolutionary changes. Therefore, the preferred explanation for the pattern of vitellogenin evolution in mammals is that this egg-yolk protein and corresponding gene were present in some egg-laying amniote ancestor and were retained in the egg-laying platypus and lost from non-egg-laying mammals.

Further inspection of the platypus genome also reveals the platypus's kinship with other mammals. For example, milk production is a shared trait among all mammals. Genes encoding the casein milk proteins are tightly clustered together in the platypus genome as they are in other mammals. Just this brief glance at a few mammalian genomes informs us that, yes indeed, some mammals have genes that others do not, some genes are shared by all mammals, and the presence or absence of certain genes correlates with mammals' lifestyle. The latter is a pervasive finding in comparative genomics.

> **Message** Determining which genomic elements have been gained or lost during evolution requires knowledge of the phylogeny of the species being compared. The presence or absence of genes often correlates with organism lifestyles.

Let's look at a few more examples that illuminate the evolutionary history of our genome and how we are different from, and similar to, other mammals.

Of mice and humans

The sequence of the mouse genome has been particularly informative for understanding the human genome because of the mouse's long-standing role as a model genetic species, the vast knowledge of its classical genetics, and the mouse's evolutionary relationship to humans. The mouse and human lineages diverged approximately 75 million years ago, which is sufficient time for mutations to cause their genomes to differ, on average, at about one of every two nucleotides. Thus, sequences common to the mouse and human genomes are likely to indicate common functions.

Homologs are identified because they have similar DNA sequences. Analysis of the mouse genome indicates that the number of protein-coding genes that it contains is similar to that of the human genome. Further inspection of the mouse genes reveals that at least 99 percent of all mouse genes have some homolog in the human genome and that at least 99 percent of all human genes have some homolog in the mouse genome. Thus, the kinds of proteins encoded in each genome are essentially the same. Furthermore, about 80 percent of all mouse and human genes are clearly identifiable orthologs.

The similarities between the genomes extend well beyond the inventory of protein-coding genes to overall genome organization. More than 90 percent of the mouse and human genomes can be partitioned into corresponding regions of conserved **synteny,** where the order of genes within variously sized blocks is the same as their order in the most recent common ancestor of the two species. This synteny is very helpful in relating the maps of the two genomes. For example, human chromosome 17 is orthologous to a single mouse chromosome (chromosome 11).

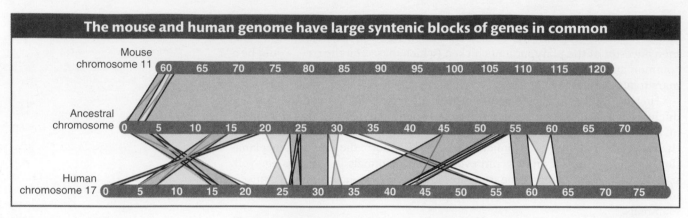

Figure 14-16 Synteny between human chromosome 17 and mouse chromosome 11. Large conserved syntenic blocks 100 kb or greater in size are shown in human chromosome 17, mouse chromosome 11, and the inferred chromosome of their last common ancestor (reconstructed by analysis of other mammalian genomes). Direct blocks of synteny are shown in light purple, inverted blocks are shown in green. Chromosome sizes are indicated in megabases (Mb). [*After M. C. Zody et al., "DNA Sequence of Human Chromosome 17 and Analysis of Rearrangement in the Human Lineage," Nature 440, 2006, 1045–1049, Fig. 2.*]

Although there have been extensive intrachromosomal rearrangements in the human chromosome, there are 23 segments of colinear sequences more than 100 kb in size (Figure 14-16).

> **Message** The mouse and human genomes contain similar sets of genes, often arranged in similar order.

There are some detectable differences between the inventories of mouse and human genes. In one family of genes involved in color vision, the opsins, humans possess one additional paralog. The presence of this opsin has equipped humans with so-called trichromatic vision, so that we can perceive colors across the entire spectrum of visible light—violet, blue, green, red—whereas mice cannot. But again, the presence of this additional paralog in humans and its absence in mice does not alone tell us whether it was gained in the human lineage or lost in the mouse lineage. Analysis of other primate and mammalian genomes has revealed that Old World primates such as chimpanzees, gorillas, and the colobus monkey possess this gene but that all nonprimate mammals lack it. We can safely infer from this phylogenetic distribution of the additional opsin gene that it evolved in an ancestor of Old World primates (that includes humans).

On the other hand, the mouse genome contains more functional copies of some genes that reflect its lifestyle. Mice have about 1400 genes involved in olfaction—this is the largest single functional category of genes in its genome. Dogs, too, have a large number of olfactory genes. This certainly makes sense for the species' lifestyles. Mice and dogs rely heavily on their sense of smell, and they encounter different odors from those encountered by humans. And the set of human olfactory genes, compared to that of mice and dogs, is strikingly inferior. We have a lot of olfactory genes, but a very large fraction of them are pseudogenes that bear inactivating mutations. For example, in just one class of olfactory genes called *V1r* genes, mice have about 160 functional genes, but just 5 out of the 200 or so *V1r* genes in the human genome are functional.

Still, these differences in gene content are relatively modest in light of the vast differences in anatomy and behavior. The overall similarity in the mouse and human genomes corresponds to the picture we get from examining the genetic tool kit controlling development in different taxa (see Chapter 13)—that great differences can evolve from genomes containing similar sets of genes. This same theme is illustrated by comparing our genome with that of our closest living relative, the chimpanzee.

Comparative genomics of chimpanzees and humans

Chimpanzees and humans last had a common ancestor about 5 to 6 million years ago. Since that time, genetic differences have accumulated by mutations that have occurred in each lineage. Genome sequencing has revealed that there are about 35 million single-nucleotide differences between chimpanzees and humans, corresponding to about a 1.06 percent degree of divergence. In addition, about 5 million insertions and deletions, ranging in length from just a single nucleotide to more than 15 kb, contribute a total of about 90 Mb of divergent DNA sequence (about 3 percent of the overall genome). Most of these insertions or deletions lie outside of coding regions.

Overall, the proteins encoded by the human and chimpanzee genomes are extremely similar. Twenty-nine percent of all orthologous proteins are *identical* in sequence. Most proteins that differ do so by only about two amino acid replacements. There are some detectable differences between chimpanzees and humans in the sets of functional genes. About 80 or so genes that were functional in their common ancestor are no longer functional in humans, owing to their deletion or to the accumulation of mutations. Some of these changes may contribute to differences in physiology.

In addition to changes in particular genes, duplications of chromosome segments in a single lineage have contributed to genome divergence. More than 170 genes in the human genome and more than 90 genes in the chimpanzee genome are present in large duplicated segments. These duplications are responsible for a greater amount of the total genome divergence than all single-nucleotide mutations combined. However, whether they contribute to major phenotypic differences is not yet clear.

Of course, all genetic differences between species originate as variations within species. The sequencing of the human genome and the advent of faster and less expensive high-throughput sequencing methods have opened the door to the detailed analysis of human genetic variation.

Comparative genomics of humans

The human species, *Homo sapiens,* originated in Africa approximately 200,000 years ago. Around 60,000 years ago, populations left Africa and migrated across the world, eventually populating five additional continents. These migrating populations encountered different climates, adopted different diets, and combated different pathogens in different parts of the world. Much of the recent evolutionary history of our species is recorded in our genomes, as are the genetic differences that make individuals or populations more or less susceptible to disease.

Overall, any two unrelated humans' genomes are 99.9 percent identical. That difference of just 0.1 percent still corresponds to roughly 3 million bases. The challenge today is to decipher which of those base differences are meaningful with respect to physiology, development, or disease.

One of the first and greatest surprises that has emerged from comparing individual human genomes is the degree to which humans differ not merely at one base in a thousand, but in the number of copies of parts of individual genes, entire genes, or sets of genes. These **copy number variations (CNVs)** include repeats and duplications that increase copy number and deletions that reduce copy number. Between any two unrelated individuals, there may be 1000 or more segments of DNA greater than 500 bp in length that differ in copy number. Some CNVs can be quite large and span over 1 million base pairs.

How such copy numbers may play a role in human evolution and disease is of intense interest. One case where increased copy number appears to have been adaptive concerns diet. People with high-starch diets have, on average, more copies of a salivary amylase (an enzyme that breaks down starch) gene than people with traditionally low-starch diets.

Conserved and ultraconserved noncoding elements

The discussion thus far has focused exclusively on the protein-coding regions of the genome. This emphasis is due more to analytical ease than to biological importance. Because of the simplicity and universality of the genetic code, the detection of ORFs and exons is much easier than the detection of functional noncoding sequences. As stated earlier, only about 3 percent of the human genome encodes exons of mRNAs, and fewer than half of these exon sequences, about 1 to 2 percent of the total genome DNA, encode protein sequences. So, more than 98 percent of our genome does not encode proteins. How do we identify other functional parts of the genome?

Other than in gene promoter regions, which contain some typical sequence motifs (see Chapter 12), it is difficult to assign function to much noncoding sequence. However, one way to locate potentially functional noncoding elements is to look for conserved sequences, which have not changed much over millions of years of evolution. Comparisons of the mouse and human genomes reveal that about 5 percent of all sequence is conserved. About one-third of this amount consists of protein-coding sequences and the remaining two-thirds consists of sequences that do not encode proteins. Thus, the proportion of the genome that governs how our genes are regulated may be greater than that encoding proteins.

The methods of comparative genomics can expedite identification of functional noncoding elements. For example, one can search for very highly conserved sequences of modest length among a few species or for less perfectly conserved sequences of greater length among a larger number of species. Comparisons of the human, rat, and mouse genomes have led to the identification of so-called *ultraconserved elements,* which are sequences that are perfectly conserved among the three species. Searches of these genomes have found more than 5000 sequences of more than 100 bp and 481 sequences of more than 200 bp that are absolutely conserved.

Extending this analysis to include the dog genome, researchers have found more than 140,000 highly conserved elements 50 bp or greater in length outside of protein-coding sequences. Although 50 percent of these elements are found in gene-poor regions, they are most richly concentrated near regulatory genes important for development. The majority of highly conserved noncoding elements may largely take part in regulating the expression of the genetic tool kit for the development of mammals and other vertebrates.

How can we verify that such conserved elements play a role in gene regulation? These elements can be tested in the same manner as the transcriptional cis-acting regulatory elements examined in earlier chapters, with the use of reporter genes (see Section 12.2). A researcher places candidate regulatory regions adjacent to a promoter and reporter gene and introduces the reporter gene into a host species. One such example is shown in Figure 14-17. An element that is highly conserved among mammalian, chicken, and a frog species lies 488 kb from the 3′ end of the human *ISL1* gene, which encodes a protein required for motor-neuron differentiation. This element was placed upstream of a promoter and the β-galactosidase (*lacZ*) reporter gene, and the construct was injected into the pronuclei of fertilized mouse oocytes. The reporter gene is then seen to be expressed along the spinal cord and in the head, as one would expect for the location of future motor neurons (see Figure 14-17). Most significantly, the expression pattern corresponds to part of the expression pattern of the native mouse *ISL1* gene (presumably other noncoding elements control the other features of *ISL1* expression). The expression pattern strongly suggests that the conserved element is a regulatory region

Testing the role of a conserved element in gene regulation

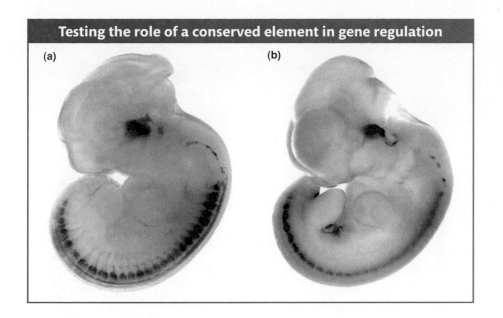

(a) (b)

Figure 14-17 A transcriptional cis-acting regulatory element is identified in an ultraconserved element of the human genome. An ultraconserved element lying near the human *ISL1* gene was coupled to a reporter gene and injected into fertilized mouse oocytes. The regions where the gene is expressed are stained dark blue or black. (a) The reporter gene is expressed in the head and spinal cord of a transgenic mouse, as seen here on day 11.5 of gestation. This expression pattern corresponds to (b) the native pattern of expression of the mouse *ISL1* gene on day 11.5 of gestation. This experiment demonstrates how functional noncoding elements can be identified by comparative genomics and tested in a model organism. [*From G. Bejerono et al., "A Distal Enhancer and an Ultraconserved Exon Are Derived from a Novel Retroposon," Nature 441, 2006, 87–90, Fig. 3.*]

for the *ISL1* gene. Many thousands of human noncoding regulatory elements will likely be identified on the basis of sequence conservation and the activity of those elements in reporter assays.

Comparisons of the mouse, human, and chimpanzee proteomes, as well as the identification and analysis of their common noncoding elements, underscore the conservative nature of genome evolution. Nevertheless, some very dramatic differences in genome content have been revealed by comparative genomics, with great implications for human medicine. We will look at one such case next.

Comparative genomics of nonpathogenic and pathogenic *E. coli*

Escherichia coli are found in our mouths and intestinal tracts in vast numbers, and this species is generally a benign symbiont. Because of its central role in genetics research, it was one of the first bacterial genomes sequenced. The *E. coli* genome is about 4.6 Mb in size and contains 4405 genes. However, calling it "the *E. coli* genome" is really not accurate. The first genome sequenced was derived from the common laboratory *E. coli* strain K-12. Many other *E. coli* strains exist, including several important to human health.

In 1982, a multistate outbreak of human disease was traced to the consumption of undercooked ground beef. The *E. coli* strain O157:H7 was identified as the culprit, and it has since been associated with a number of large-scale outbreaks of infection. In fact, there are an estimated 75,000 cases of *E. coli* infection annually in the United States. Although most people recover from the infection, a fraction develop hemolytic uremia syndrome, a potentially life-threatening kidney disease.

To understand the genetic bases of pathogenicity, the genome of an *E. coli* O157:H7 strain has been sequenced. The O157 and K-12 strains have a backbone of 3574 protein-coding genes in common, and the average nucleotide identity among orthologous genes is 98.4 percent, comparable to that of human and chimpanzee orthologs. About 25 percent of the *E. coli* orthologs encode identical proteins, similar to the 29 percent for human and chimpanzee orthologs.

Despite the similarities in many proteins, the genomes and proteomes differ enormously in content. The *E. coli* O157 genome encodes 5416 genes, whereas the *E. coli* K-12 genome encodes 4405 genes. The *E. coli* O157 genome contains 1387 genes that are not found in the K-12 genome, and the K-12 genome contains 528 genes not found in the O157 genome. Comparison of the genome maps reveals

Figure 14-18 The circular genome maps of *E. coli* strains K-12 and O157:H7. The circle depicts the distribution of sequences specific to each strain. The colinear backbone common to both strains is shown in blue. The positions of O157:H7-specific sequences are shown in red. The positions of K-12-specific sequences are shown in green. The positions of O157:H7- and K-12-specific sequences at the same location are shown in tan. Hypervariable sequences are shown in purple. [*After N. T. Perna et al., "Genome Sequence of Enterohaemorrhagic Escherichia coli O157:H7," Nature 409, 2001, 7529–7533. Courtesy of Guy Plunkett III and Frederick Blattner.*]

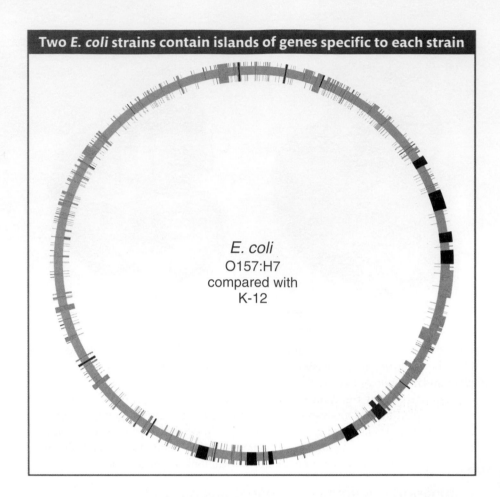

Two *E. coli* strains contain islands of genes specific to each strain

E. coli
O157:H7
compared with
K-12

that the backbones common to the two strains are interspersed with islands of genes specific to either K-12 or O157 (Figure 14-18).

Among the 1387 genes specific to *E. coli* O157 are many genes that are suspected to encode virulence factors, including toxins, cell-invasion proteins, adherence proteins, and secretion systems for toxins, as well as possible metabolic genes that may be required for nutrient transport, antibiotic resistance, and other activities that may confer the ability to survive in different hosts. Most of these genes were not known before sequencing and would not be known today had researchers relied solely on *E. coli* K-12 as a guide to all *E. coli*.

The surprising level of diversity between two members of the same species shows how dynamic genome evolution can be. Most new genes in *E. coli* strains are thought to have been introduced by horizontal transfer from the genomes of viruses and other bacteria. Differences can also evolve owing to gene deletion. Other pathogenic *E. coli* and bacterial species also exhibit many differences in gene content from their non-pathogenic cousins. The identification of genes that may contribute directly to pathogenicity opens new avenues to the prevention and treatment of disease.

14.6 Functional Genomics and Reverse Genetics

Geneticists have been studying the expression and interactions of gene products for the past several decades. However, these studies were small scale: they considered just one gene or a few genes at a time. With the advent of genomics, we have an opportunity to expand these studies to a global level by using genome-wide approaches to study most or all gene products systematically and simultaneously. This global approach to the study of the function, expression, and interaction of gene products is termed **functional genomics.**

Ome, sweet ome

In addition to the genome, other global data sets are of interest. Following the example of the term *genome,* for which "gene" plus "ome" becomes a word for "all genes," genomics researchers have coined a number of terms to describe other global data sets on which they are working. This *-ome* wish list includes

The transcriptome. The sequence and expression patterns of all transcripts (where, when, how much).

The proteome. The sequence and expression patterns of all proteins (where, when, how much).

The interactome. The complete set of physical interactions between proteins and DNA segments, between proteins and RNA segments, and between proteins.

We will not consider all of these *-omes* in this section but will focus on some of the global techniques that are beginning to be exploited to obtain these data sets.

Using DNA microarrays to study the transcriptome   Suppose we want to answer the question, What genes are active in a particular cell under certain conditions? Those conditions could be one or more stages in development or they could be the presence or absence of a pathogen or a hormone. Active genes are transcribed into RNA, and so the set of RNA transcripts present in the cell can tell us what genes are active. Here is where the new technology of DNA chips used to assay RNA transcripts is so powerful.

DNA chips are samples of DNA laid out as a series of microscopic spots bound to a glass "chip" the size of a microscope cover slip. The set of DNAs so displayed is called a **microarray.** A typical type of microarray contains short synthetic oligonucleotides representing most or all of the genes in a genome (Figure 14-20). DNA microarrays have powered molecular genetics by permitting the assay of RNA transcripts for all genes simultaneously in a single experiment. Let's see how this process works in more detail.

Microarrays are exposed to cDNA probes—for example, one set of probes used as a control and one set of probes representing a specific condition. The set used as a control might be made from the total set of RNA molecules extracted from a particular cell type grown under typical conditions. The second set of probes might be made from RNA extracted from cells grown under some experimental condition. Fluorescent labels are attached to the probes and the probes are hybridized to the microarray. The relative binding of the probe molecules to the microarray is monitored automatically with the use of a laser-beam-illuminated microscope (Figure 14-19). In this manner, genes whose levels of expression are increased or decreased under the given experimental condition are identified. Simi-

Figure 14-19 The key steps in a microarray analysis are (1) extraction of mRNA from cells or tissues, (2) synthesis of fluorescent-dye-labeled cDNA probes, (3) hybridization to the microarray, (4) detection of the fluorescent signal from hybridized probes, and (5) image analysis to identify relative levels of hybridized probe. The relative levels reveal those genes whose expression is increased or decreased under the conditions analyzed. [*After G. Gibson and S. Muse, A Primer of Genome Science 3rd ed., Fig. 4.3, p. 199. Sinauer Associates, Sunderland, Mass., 2009.*]

GENETICS PORTAL **ANIMATED ART: DNA microarrays: using an oligonucleotide array to analyze patterns of gene expression**

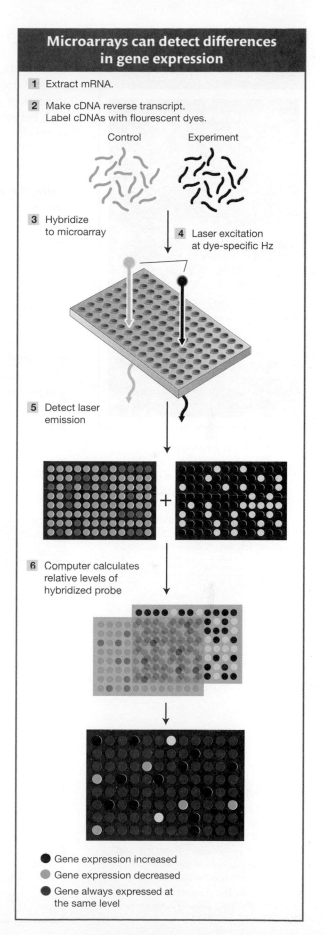

Microarrays can detect differences in gene expression

1 Extract mRNA.

2 Make cDNA reverse transcript. Label cDNAs with flourescent dyes.

Control   Experiment

3 Hybridize to microarray

4 Laser excitation at dye-specific Hz

5 Detect laser emission

6 Computer calculates relative levels of hybridized probe

● Gene expression increased
● Gene expression decreased
● Gene always expressed at the same level

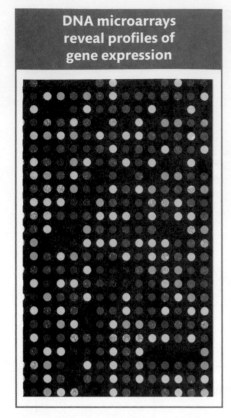

DNA microarrays reveal profiles of gene expression

Figure 14-20 Display of gene-expression patterns detected by DNA microarrays. Each position in the array is a separate gene. Samples of unbound cDNA, to which are attached differently colored fluorescent labels, bind to sites on the array that have a complementary sequence. [*Pasieka/Science Photo Library.*]

larly, genes that are active in a given cell type or at a given stage of development can be identified.

With an understanding of which genes are active or inactive at a given developmental stage, in a particular cell type, or in various environmental conditions, the sets of genes that may respond to similar regulatory inputs can be identified. Furthermore, gene-expression profiles can paint a picture of the differences between normal and diseased cells. By identifying genes whose expression is altered by mutations, in cancer cells, or by a pathogen, researchers may be able to devise new therapeutic strategies.

Using the two-hybrid test to study the protein–protein interactome One of the most important activities of proteins is their interaction with other proteins. Because of the large number of proteins in any cell, biologists have sought ways of systematically studying all of the interactions of individual proteins in a cell. One of the most common ways of studying the interactome uses an engineered system in yeast cells called the **two-hybrid test,** which detects physical interactions between two proteins. The basis for the test is the transcriptional activator encoded by the yeast *GAL4* gene (see Chapter 12).

Recall that this protein has two domains: (1) a DNA-binding domain that binds to the transcriptional start site and (2) an activation domain that will activate transcription but cannot itself bind to DNA. Thus, the two domains must be in close proximity in order for transcriptional activation to take place. Suppose that you are investigating whether two proteins interact. The strategy of the two-hybrid system is to separate the two domains of the activator encoded by *GAL4*, making activation of a reporter gene impossible. Each domain is connected to a different protein. If the two proteins interact, they will join the two domains together. The activator will become active and start transcription of the reporter gene.

How is this scheme implemented in practice? The *GAL4* gene is divided between two plasmids so that one plasmid contains the part encoding the DNA-binding

Figure 14-21 The system uses the binding of two proteins, a "bait" protein and a "target" protein to restore the function of the Gal4 protein, which activates a reporter gene. *Cam,* Trp, and Leu are components of the selection systems for moving the plasmids around between cells. The reporter gene is *lacZ,* which resides on a yeast chromosome (shown in blue).

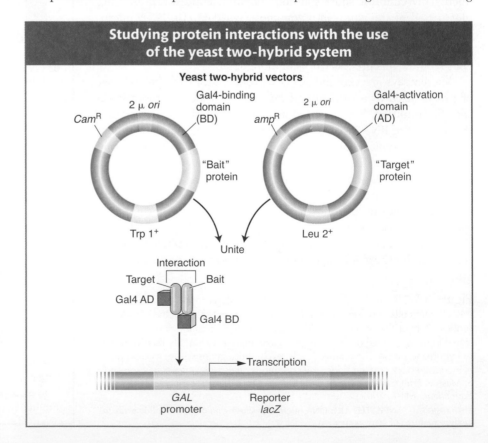

Studying protein interactions with the use of the yeast two-hybrid system

Yeast two-hybrid vectors

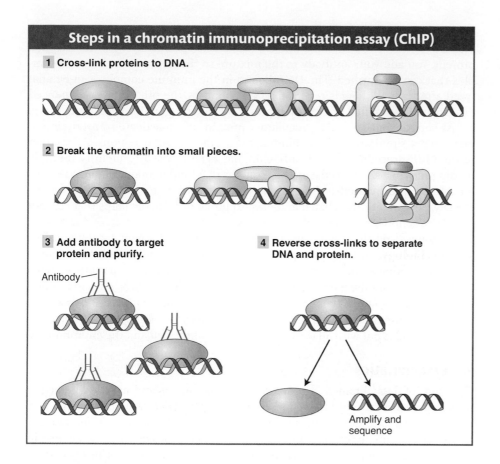

Figure 14-22 ChIP is a technique for isolating the DNA and its associated proteins in a specific region of chromatin so that both can be analyzed together.

domain and the other plasmid contains the part encoding the activation domain. On one plasmid, a gene for one protein under investigation is spliced next to the DNA-binding domain, and this fusion protein acts as "bait." On the other plasmid, a gene for another protein under investigation is spliced next to the activation domain and this fusion protein is said to be the "target" (Figure 14-21). The two hybrid plasmids are then introduced into the same yeast cell—perhaps by mating haploid cells containing bait and target plasmids. The final step is to look for activation of transcription by a Gal4-regulated reporter gene construct, which would be proof that bait and target bind to each other. The two-hybrid system can be automated to make it possible to hunt for protein interactions throughout the proteome.

Studying the protein–DNA interactome using chromatin immunoprecipitation assay (ChIP)   The sequence-specific binding of proteins to DNA is critical for correct gene expression. For example, regulatory proteins bind to promoters and activate or repress transcription in both bacteria and eukaryotes (see Chapters 11, 12, and 13). In the case of eukaryotes, chromosomes are organized into chromatin, in which the fundamental unit, the nucleosome, contains DNA wrapped around histones. Posttranslational modification of histones often dictates what proteins bind and where (see Chapter 12). A variety of technologies have been developed that allow researchers to isolate specific regions of chromatin so that DNA and its associated proteins can be analyzed together. The most widely used method is called **ChIP** (for **chromatin immunoprecipitation**), and its application is described below (Figure 14-22).

Let's say that you have isolated a gene from yeast and suspect that it encodes a protein that binds to DNA when yeast is grown at high temperature. You want to know whether this protein binds to DNA and, if so, to what yeast sequence. One way to address this question is first to treat yeast cells that have been grown at high temperature with a chemical that will cross-link proteins to the DNA. In this way proteins bound to the DNA at the time of chromatin isolation will remain bound through subsequent treatments. The next step is to break the chromatin into small pieces. To separate the fragment containing your protein–DNA complex from others, you use an antibody that reacts specifically with the encoded protein. You add your antibody to the mixture so that it forms an immune complex that can be purified. The DNA bound in the immune complex can be analyzed after cross-linking is reversed. DNA bound by the protein may sequenced directly or be amplified into many copies by PCR to prepare for DNA sequencing.

As you saw in Chapter 12, regulatory proteins often activate transcription of many genes simultaneously by binding to several promoter regions. A variation of the ChIP procedure, called ChIP-chip, has been devised to identify multiple binding sites in a sequenced genome. Proteins that bind to many genomic regions are immunoprecipitated as described above. Then, after cross-linking is reversed, the DNA fragments are labeled and used to probe microarray chips that contain the entire genomic sequence of the species under study.

Genomics and the other "omics" areas have spawned a new discipline called **systems biology.** Whereas the approach of genetics has traditionally been reductionist, dissecting an organism with mutations to see what the parts are, systems biology tries to put the parts together to understand the whole as a system. A biological system comprises gene-regulation networks, signal-transduction cascades, cell-to-cell communication, and many forms of interactions not only between "genetic" molecules but with all the other molecules of the cell and the environment.

Reverse genetics

The kinds of data obtained from microarray experiments and protein-interaction screens are suggestive of interactions within the genome and proteome, but they do not allow one to draw firm conclusions about gene functions and interactions in vivo. For example, finding out that the expression of certain genes is lost in some cancers is not proof of cause and effect. It is necessary to specifically disrupt gene function and to understand phenotypes in native conditions. Starting from the available gene sequences, researchers can now use a variety of methods to disrupt the function of a specific gene. These methods are referred to as **reverse genetics.** Reverse-genetic analysis starts with a known molecule—a DNA sequence, an mRNA, or a protein—and then attempts to disrupt this molecule to assess the role of the normal gene product in the biology of the organism.

There are several approaches to reverse genetics. One approach is to introduce random mutations into the genome but then home in on the gene of interest by molecular identification of mutations in the gene. A second approach is to conduct a targeted mutagenesis that produces mutations directly in the gene of interest. A third approach is to create *phenocopies*—effects comparable to mutant phenotypes—usually by treatment with agents that interfere with the mRNA transcript of the gene.

Each approach has its advantages. Random mutagenesis is well established, but it often requires substantial time to sift through all the mutations to find the small proportion that includes the gene of interest. Targeted mutagenesis also is labor intensive but, after the targeted mutation has been obtained, its characterization is more straightforward. Creating phenocopies can be very efficient, especially as libraries of tools have been developed for particular model species. We will consider examples of each of these approaches.

Reverse genetics through random mutagenesis   Random mutagenesis for reverse genetics employs the same kinds of general mutagens that are used for forward genetics: chemical agents, radiation, or transposable genetic elements (see page 199). However, instead of screening the genome at large for mutations that exert a particular phenotypic effect, reverse genetics focuses on the gene in question, which can be done in one of two general ways.

One approach is to focus on the map location of the gene. Only mutations falling in the region of the genome where the gene is located are retained for further detailed molecular analysis. Thus, in this approach, the recovered mutations must be mapped. One straightforward way is to cross a new mutant with a mutant containing a known deletion or mutation of the gene of interest. Symbolically, the pairing is *new mutant/known mutant*. Only the pairings that result in progeny with a mutant phenotype (showing lack of complementation) are saved for study.

In another approach, the gene of interest is identified in the mutagenized genome and checked for the presence of mutations. For example, if the mutagen causes small deletions, then, after PCR amplification, genes from the parental and mutagenized genomes can be compared, looking for a mutagenized genome in which the gene of interest is reduced in size. Similarly, transposable-element insertions into the gene of interest can be readily detected because they increase its size. Techniques for recognizing single-base-pair substitutions also are available. In these ways, a set of genomes containing random mutations can be effectively screened to identify the small fraction of mutations that are of interest to a researcher.

Reverse genetics by targeted mutagenesis   For most of the twentieth century, researchers viewed the ability to direct mutations to a specific gene as the unattainable "holy grail" of genetics. However, now several such techniques are available. After a gene has been inactivated in an individual, geneticists can evaluate the phenotype exhibited for clues to the gene's function. Generally speaking, the tools for targeted gene mutations rely on genetic techniques developed for model organisms. So, although the disruption of yeast, fly, or mouse genes in an efficient, directed fashion is feasible, such disruption cannot be done in many nonmodel species.

Gene-specific mutagenesis usually requires the replacement of a resident wild-type copy of an entire gene by a mutated version of that gene. The mutated gene inserts into the chromosome by a mechanism resembling homologous recombination, replacing the normal sequence with the mutant (Figure 14-23). This approach can be used for targeted gene knockout, in which a null allele replaces the wild-type copy. Some techniques are so efficient that, in *E. coli* and *S. cerevisiae*, for example, it has been possible to mutate every gene in the genome to try to ascertain its biological function.

Disrupting gene function with the use of targeted mutagenesis

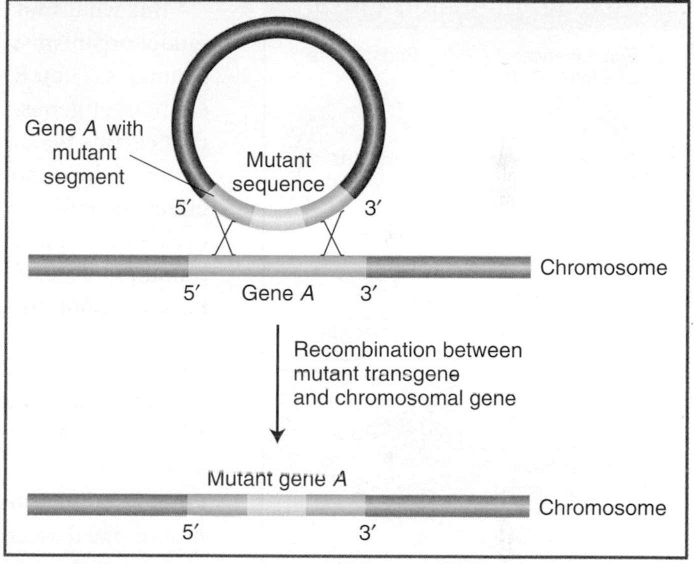

Figure 14-23 The basic molecular event in targeted gene replacement. A transgene containing sequences from two ends of a gene but with a selectable segment of DNA in between is introduced into a cell. Double recombination between the transgene and a normal chromosomal gene produces a recombinant chromosomal gene that has incorporated the abnormal segment.

> **Message**   Targeted mutagenesis is the most precise means of obtaining mutations in a specific gene and can now be practiced in a variety of model systems, including mice and flies.

Reverse genetics by phenocopying   The advantage of inactivating a gene itself is that mutations will be passed on from one generation to the next, and so, once obtained, a line of mutants is always available for future study. However, only organisms well developed as molecular models can be used for such manipulations. On the other hand, phenocopying can be applied to a great many organisms regardless of how well developed the genetic technology is for a given species.

One of the most exciting discoveries of the past decade or so has been the discovery of a widespread mechanism whose natural function seems to be to protect a cell from foreign DNA. This mechanism is called **RNA interference (RNAi),** described on page 305. Researchers have capitalized on this cellular mechanism to make a powerful method for inactivating specific genes. The inactivation is achieved as follows. A double-stranded RNA is made with sequences homologous to part of the gene under study and is introduced into a

Figure 14-24 Three ways to create and introduce double-stranded RNA (dsRNA) into a cell. The dsRNA will then stimulate RNAi, degrading sequences that match those in the dsRNA. [*Reprinted with permission from S. Hammond, A. Caudy, and G. Hannon, Nat. Rev. Genet. 2, 2001, 116.*]

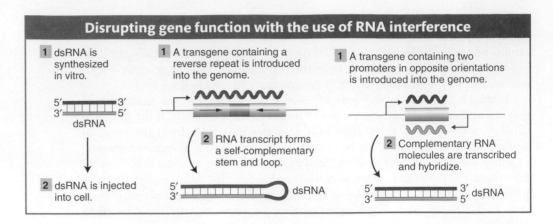

Disrupting gene function with the use of RNA interference

1 dsRNA is synthesized in vitro.

5′ ⌶⌶⌶⌶⌶ 3′
3′ ⌶⌶⌶⌶⌶ 5′
dsRNA

2 dsRNA is injected into cell.

1 A transgene containing a reverse repeat is introduced into the genome.

2 RNA transcript forms a self-complementary stem and loop.

5′ ⌶⌶⌶⌶⌶⌶
3′ ⌶⌶⌶⌶⌶⌶ dsRNA

1 A transgene containing two promoters in opposite orientations is introduced into the genome.

2 Complementary RNA molecules are transcribed and hybridize.

5′ ⌶⌶⌶⌶⌶ 3′
3′ ⌶⌶⌶⌶⌶ 5′ dsRNA

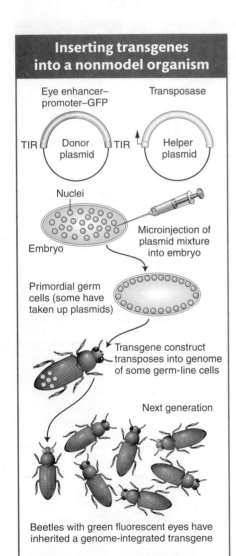

Inserting transgenes into a nonmodel organism

Eye enhancer–promoter–GFP

Transposase

TIR — Donor plasmid — TIR

Helper plasmid

Nuclei

Embryo

Microinjection of plasmid mixture into embryo

Primordial germ cells (some have taken up plasmids)

Transgene construct transposes into genome of some germ-line cells

Next generation

Beetles with green fluorescent eyes have inherited a genome-integrated transgene

Figure 14-25 Creation of transgenic beetles expressing a green fluorescent protein. TIR, terminal inverted repeat. [*From E. A. Wimmer, "Applications of Insect Transgenesis," Nat. Rev. Genet. 4, 2003, 225–232.*]

cell (Figure 14-24). The RNA-induced silencing complex, or RISC, then degrades native mRNA that is complementary to the double-stranded RNA. The net result is a complete or considerable reduction of mRNA levels that lasts for hours or days, thereby nullifying expression of that gene. The technique has been widely applied successfully in model systems, including *C. elegans, Drosophila,* zebrafish, and several plant species.

But what makes RNAi especially attractive is that it can be applied to nonmodel organisms. First, target genes of interest can be identified by comparative genomics. Then RNAi sequences are produced to target the inhibition of the specific target genes. This technique has been applied, for example, to a mosquito that carries malaria (*Anopheles gambiae*). Using these techniques, scientists can better understand the biological mechanisms relating to the medical or economic effect of such species. For example, the genes that control the complicated life cycle of the malaria parasite, partly inside a mosquito host and partly inside the human body, can be better understood, revealing new ways to control the single most common infectious disease in the world.

> **Message** RNAi-based methods provide general ways of experimentally interfering with the function of a specific gene without changing its DNA sequence (generally called *phenocopying*).

Functional genomics with nonmodel organisms   Much of our consideration of mutational dissection and phenocopying has focused on genetic model organisms. One current focus of many geneticists is the broader application of these techniques, including to species that have negative effects on human society, such as parasites, disease carriers, or agricultural pests. Classical genetic techniques are not readily applicable to most of these species, but the roles of specific genes can be assessed by either transgenesis or phenocopying.

The first approach—inserting transgenes—is shown in Figure 14-25. This example concerns beetles, many of which are agricultural pests. In this case, transgenes were inserted randomly into the beetle genome. Transgenic beetles can be produced by using methodology similar to that used to produce transgenic *Drosophila* (see Chapter 10). However, some way is needed to identify successful transgenesis. Therefore, the technique depends on using a reporter gene that can be expressed in a wild-type recipient. The green fluorescent protein (GFP), originally isolated from a jellyfish, is a useful reporter for this application. As in *Drosophila,* transgenes are inserted as parts of transposons, and a helper plasmid encoding a transposase facilitates insertion of the transposon bearing the transgene. Figure 14-25 shows the use of GFP transgenes driven by an enhancer element that drives expression in the insect eye. This method has also been effectively used to create GFP-expressing transgenes in mosquitos and the silkworm moth (*Bombyx mori*) (Figure 14-26).

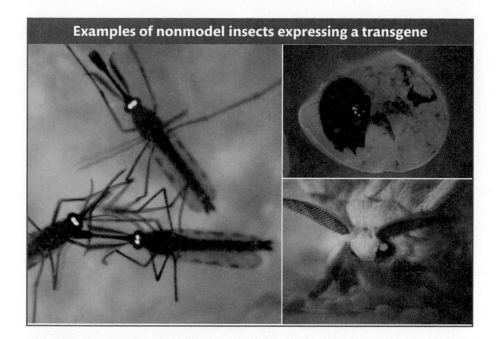

Figure 14-26 Examples of a transgenic green fluorescent protein reporter expressed in the eyes of some nonmodel insects. Expression is driven from one single promoter active in the eye. The insects are mosquito (left) and silkworm moth (*Bombyx mori*) (right). [*Mosquito: Jacquelyn Martin/AP/Corbis; silkworm moth: Marek Jindra.*]

Summary

Genomic analysis takes the approaches of genetic analysis and applies them to the collection of global data sets to fulfill goals such as the mapping and sequencing of whole genomes and the characterization of all transcripts and proteins. Genomic techniques require the rapid processing of large sets of experimental material, all dependent on extensive automation.

The key problem in compiling an accurate sequence of a genome is to take short sequence reads and relate them to one another by sequence identity to build up a consensus sequence of an entire genome. This can be done straightforwardly for bacterial or archaeal genomes by aligning overlapping sequences from different sequence reads to compile the entire genome, because there are few or no DNA segments that are present in more than one copy in prokaryotes. The problem is that complex genomes are replete with such repetitive sequences. These repetitive sequences interfere with accurate sequence-contig production. The problem is resolved in whole-genome shotgun (WGS) sequencing with the use of paired-end reads.

Having a genomic sequence map provides the raw, encrypted text of the genome. The job of bioinformatics is to interpret this encrypted information. For the analysis of gene products, computational techniques are used to identify ORFs and noncoding RNAs, then to integrate these results with available experimental evidence for transcript structures (cDNA sequences), protein similarities, and knowledge of characteristic sequence motifs.

One of the most powerful means to advance the analysis and annotation of genomes is by comparing with the genomes of related species. Conservation of sequences among species is a reliable guide to identifying functional sequences in the complex genomes of many animals and plants. Comparative genomics can also reveal how genomes have changed in the course of evolution and how these changes may relate to differences in physiology, anatomy, or behavior among species. In bacterial genomics, comparisons of pathogenic and nonpathogenic strains have revealed many differences in gene content that may contribute to pathogenicity.

Functional genomics attempts to understand the working of the genome as a whole system. Two key elements are the transcriptome, the set of all transcripts produced, and the interactome, the set of interacting gene products and other molecules that together enable a cell to be produced and to function. The function of individual genes and gene products for which classical mutations are not available can be tested through reverse genetics—by targeted mutation or phenocopying.

KEY TERMS

annotation (p. 498)

bioinformatics (p. 497)

ChIP (chromatin immunoprecipitation assay) (p. 515)

comparative genomics (p. 505)

consensus sequence (p. 491)

copy number variation (CNVs) (p. 509)

DNA template library (p. 494)

expressed sequence tag (EST) (p. 499)

functional genomics (p. 512)

genome project (p. 489)

genomic library (p. 492)

genomics (p. 488)

homolog (p. 505)

microarray (p. 513)

<table>
<tr><td>

open reading frame (ORF)
(p. 499)
ortholog (p. 505)
outgroup (p. 506)
paired-end read (p. 496)
paralog (p. 505)
parsimony (p. 507)
phylogeny (p. 505)

</td><td>

phylogenetic inference (p. 506)
processed pseudogene (p. 502)
proteome (p. 498)
pseudogene (p. 502)
pyrosequencing (p. 494)
reverse genetics (p. 516)
RNA interference (RNAi) (p. 517)
scaffold (p. 496)

</td><td>

sequence assembly (p. 491)
sequence contig (p. 493)
supercontig (p. 496)
synteny (p. 507)
systems biology (p. 516)
two-hybrid test (p. 514)
vector (p. 492)

</td></tr>
</table>

SOLVED PROBLEM

SOLVED PROBLEM 1. You want to study the development of the olfactory (smell-reception) system in the mouse. You know that the cells that sense specific chemical odors (odorants) are located in the lining of the nasal passages of the mouse. Describe some approaches for using reverse genetics to study olfaction.

Solution

There are many approaches that can be imagined. For reverse genetics, you would want to identify candidate genes that are expressed in the lining of the nasal passages. Given the techniques of functional genomics, this identification could be accomplished by purifying RNA from isolated nasal-passage-lining cells and using this RNA as a probe of DNA chips containing sequences that correspond to all known mRNAs in the mouse. For example, you may choose to first examine mRNAs that are expressed in the nasal-passage lining but nowhere else in the mouse as important candidates for a specific role in olfaction. (Many of the important molecules may also have other jobs elsewhere in the body, but you have to start somewhere.) Alternatively, you may choose to start with those genes whose protein products are candidate proteins for binding the odorants themselves. Regardless of your choice, the next step would be to engineer a targeted knockout of the gene that encodes each mRNA or protein of interest or to use RNA interference to attempt to phenocopy the loss-of-function phenotype of each of the candidate genes.

PROBLEMS

Most of the problems are also available for review/grading through the GENETICSPORTAL www.yourgeneticsportal.com.

WORKING WITH THE FIGURES

1. Based on Figure 14-2, why must the DNA fragments sequenced overlap in order to obtain a genome sequence?

2. Filling gaps in draft genome sequences is a major challenge. Based on Figure 14-6, can paired-end reads from a library of 2-kb fragments fill a 10-kb gap?

3. In Figure 14-9, how are the positions of codons determined?

4. In Figure 14-10, expressed sequence tags (ESTs) are aligned with genomic sequence. How are ESTs helpful in genome annotation?

5. In Figure 14-10, cDNA sequences are aligned with genomic sequence. How are cDNA sequences helpful in genome annotation? Are cDNAs more important for bacterial or eukaryotic genome annotations?

6. Figure 14-16 shows syntenic regions of mouse chromosome 11 and human chromosome 17. What do these syntenic regions reveal about the genome of the last common ancestor of mice and humans?

7. Based on Figure 14-17 and the features of ultraconserved elements, what would you predict you'd observe if you injected a reporter-gene construct of the rat ortholog of the *ISL1* ultraconserved element into fertilized mouse oocytes and examined reporter gene expression in the developing embryo?

8. The genomes of two *E. coli* strains are compared in Figure 14-18. Would you expect any third strain to contain more of the blue, tan, or red regions shown in Figure 14-18? Explain.

9. Figure 14-21 depicts the Gal4-based two-hybrid system. Why don't the "bait" proteins fused to the Gal4 DNA-binding protein activate reporter-gene expression?

BASIC PROBLEMS

10. The word *contig* is derived from the word *contiguous*. Explain the derivation.

11. Explain the approach that you would apply to sequencing the genome of a newly discovered bacterial species.

12. Terminal-sequencing reads of clone inserts are a routine part of genome sequencing. How is the central part of the clone insert ever obtained?

13. What is the difference between a contig and a scaffold?

14. Two particular contigs are suspected to be adjacent, possibly separated by repetitive DNA. In an attempt to link them, end sequences are used as primers to try to

bridge the gap. Is this approach reasonable? In what situation will it not work?

15. A segment of cloned DNA containing a protein-encoding gene is radioactively labeled and used in an in situ hybridization to chromosomes. Radioactivity was observed over five regions on different chromosomes. How is this result possible?

16. In an in situ hybridization experiment, a certain clone bound to only the X chromosome in a boy with no disease symptoms. However, in a boy with Duchenne muscular dystrophy (X-linked recessive disease), it bound to the X chromosome and to an autosome. Explain. Could this clone be useful in isolating the gene for Duchenne muscular dystrophy?

17. In a genomic analysis looking for a specific disease gene, one candidate gene was found to have a single-base-pair substitution resulting in a nonsynonymous amino acid change. What would you have to check before concluding that you had identified the disease-causing gene?

18. Is a bacterial operator a binding site?

19. A certain cDNA of size 2 kb hybridized to eight genomic fragments of total size 30 kb and contained two short ESTs. The ESTs were also found in two of the genomic fragments each of size 2 kb. Sketch a possible explanation for these results.

20. A sequenced fragment of DNA in *Drosophila* was used in a BLAST search. The best (closest) match was to a kinase gene from *Neurospora*. Does this match mean that the *Drosophila* sequence contains a kinase gene?

21. In a two-hybrid test, a certain gene *A* gave positive results with two clones, M and N. When M was used, it gave positives with three clones, A, S, and Q. Clone N gave only one positive (with A). Develop a tentative interpretation of these results.

22. You have the following sequence reads from a genomic clone of the *Drosophila melanogaster* genome:

Read 1: TGGCCGTGATGGGCAGTTCCGGTG

Read 2: TTCCGGTGCCGGAAAGA

Read 3: CTATCCGGGCGAACTTTTGGCCG

Read 4: CGTGATGGGCAGTTCCGGTG

Read 5: TTGGCCGTGATGGGCAGTT

Read 6: CGAACTTTTGGCCGTGATGGGCAGTTCC

Use these six sequence reads to create a sequence contig of this part of the *D. melanogaster* genome.

23. Sometimes, cDNAs turn out to be "monsters"; that is, fusions of DNA copies of two different mRNAs accidentally inserted adjacently to each other in the same clone. You suspect that a cDNA clone from the nematode *Caenorhabditis elegans* is such a monster because the sequence of the cDNA insert predicts a protein with two structural domains not normally observed in the same protein. How would you use the availability of the entire genomic sequence to assess if this cDNA clone is a monster or not?

24. In browsing through the human genome sequence, you identify a gene that has an apparently long coding region, but there is a two-base-pair deletion that disrupts the reading frame.

a. How would you determine whether the deletion was correct or an error in the sequencing?

b. You find that the exact same deletion exists in the chimpanzee homolog of the gene but that the gorilla gene reading frame is intact. Given the phylogeny of great apes below, what can you conclude about when in ape evolution the mutation occurred?

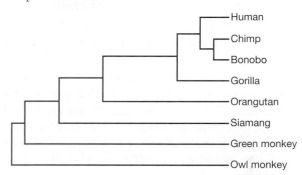

25. In browsing through the chimpanzee genome, you find that it has three homologs of a particular gene, whereas humans have only two.

a. What are two alternative explanations for this observation?

b. How could you distinguish between these two possibilities?

26. The platypus is one of the few venomous mammals. The male platypus has a spur on the hind foot through which it can deliver a mixture of venom proteins. Looking at the phylogeny in Figure 14-15, how would you go about determining whether these venom proteins are unique to the platypus?

27. You have sequenced the genome of the bacterium *Salmonella typhimurium,* and you are using BLAST analysis to identify similarities within the *S. typhimurium* genome to known proteins. You find a protein that is 100 percent identical in the bacterium *Escherichia coli.* When you compare nucleotide sequences of the *S. typhimurium* and *E. coli* genes, you find that their nucleotide sequences are only 87 percent identical.

a. Explain this observation.

b. What do these observations tell you about the merits of nucleotide- versus protein-similarity searches in identifying related genes?

28. To inactivate a gene by RNAi, what information do you need? Do you need the map position of the target gene?

29. What is the purpose of generating a phenocopy?

30. What is the difference between forward and reverse genetics?

CHALLENGING PROBLEMS

31. You have the following sequence reads from a genomic clone of the *Homo sapiens* genome:

Read 1: ATGCGATCTGTGAGCCGAGTCTTTA

Read 2: AACAAAAATGTTGTTATTTTTATTTCAGATG

Read 3: TTCAGATGCGATCTGTGAGCCGAG

Read 4: TGTCTGCCATTCTTAAAAACAAAAATGT

Read 5: TGTTATTTTTATTTCAGATGCGA

Read 6: AACAAAAATGTTGTTATT

a. Use these six sequence reads to create a sequence contig of this part of the *H. sapiens* genome.

b. Translate the sequence contig in all possible reading frames.

c. Go to the BLAST page of the National Center for Biotechnology Information, or NCBI (http://www.ncbi.nlm.nih.gov/BLAST/, Appendix B) and see if you can identify the gene of which this sequence is a part by using each of the reading frames as a query for protein–protein comparison (BLASTp).

32. Some sizable regions of different chromosomes of the human genome are more than 99 percent nucleotide identical with one another. These regions were overlooked in the production of the draft genome sequence of the human genome because of their high level of similarity. Of the techniques discussed in this chapter, which would allow genome researchers to identify the existence of such duplicate regions?

33. Some exons in the human genome are quite small (less than 75 bp long). Identification of such "microexons" is difficult because these distances are too short to reliably use ORF identification or codon bias to determine if small genomic sequences are truly part of an mRNA and a polypeptide. What techniques of "gene finding" can be used to try to assess if a given region of 75 bp constitutes an exon?

34. You are studying proteins having roles in translation in the mouse. By BLAST analysis of the predicted proteins of the mouse genome, you identify a set of mouse genes that encode proteins with sequences similar to those of known eukaryotic translation-initiation factors. You are interested in determining the phenotypes associated with loss-of-function mutations of these genes.

a. Would you use forward- or reverse-genetics approaches to identify these mutations?

b. Briefly outline two different approaches that you might use to look for loss-of-function phenotypes in one of these genes.

35. The entire genome of the yeast *Saccharomyces cerevisiae* has been sequenced. This sequencing has led to the identification of all the open reading frames (ORFs, gene-size sequences with appropriate translational initiation and termination signals) in the genome. Some of these ORFs are previously known genes with established functions; however, the remainder are unassigned reading frames (URFs). To deduce the possible functions of the URFs, they are being systematically, one at a time, converted into null alleles by in vitro knockout techniques. The results are as follows:

15 percent are lethal when knocked out.

25 percent show some mutant phenotype (altered morphology, altered nutrition, and so forth).

60 percent show no detectable mutant phenotype at all and resemble wild type.

Explain the possible molecular-genetic basis of these three mutant categories, inventing examples where possible.

36. Different strains of *E. coli* are responsible for enterohemorrhagic and urinary tract infections. Based on the differences between the benign K-12 strain and the enterohemorrhagic O157:H7 strain, would you predict that there are obvious genomic differences:

a. Between K-12 and uropathogenic strains?

b. Between O157:H7 and uropathogenic strains?

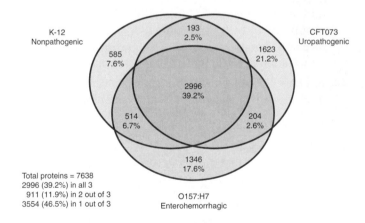

c. What might explain the observed pair-by-pair differences in genome content?

d. How might the function of strain-specific genes be tested?

The Dynamic Genome: Transposable Elements

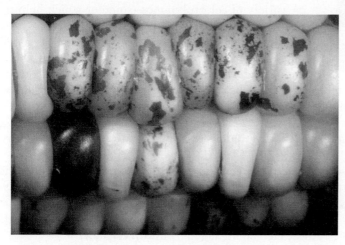

Kernels on an ear of corn. The spotted kernels on this ear of corn result from the interaction of a mobile genetic element (a transposable element) with a corn gene whose product is required for pigmentation. [*Cliff Weil and Susan Wessler.*]

KEY QUESTIONS

- Why were transposable elements first discovered genetically in maize but first isolated molecularly from *E. coli*?

- How do transposable elements participate in the spread of antibiotic-resistant bacteria?

- What are the two major mechanisms used by elements to transpose?

- How can humans survive, given that as much as 50 percent of the human genome is derived from transposable elements?

- Why can the host repress the spread of some transposable elements but not others?

A boy is born with a disease that makes his immune system ineffective. Diagnostic testing determines that he has a recessive genetic disorder called SCID (severe combined immunodeficiency disease), more commonly known as *bubble-boy disease*. This disease is caused by a mutation in the gene encoding the blood enzyme adenosine deaminase (ADA). As a result of the loss of this enzyme, the precursor cells that give rise to one of the cell types of the immune system are missing. Because this boy has no ability to fight infection, he has to live in a completely isolated and sterile environment—that is, a bubble in which the air is filtered for sterility (Figure 15-1). No pharmaceutical or other conventional therapy is available to treat this disease. Giving the boy a tissue transplant containing the precursor cells from another person would not work in the vast majority of cases, because a precise tissue match between donor and patient is extremely rare. Consequently, the donor cells would end up creating an immune response against the boy's own tissues (graft-versus-host disease).

In the past two decades, techniques have been developed that offer the possibility of a different kind of transplantation therapy—**gene therapy**—that could help people with SCID and other incurable diseases. In regard to SCID, a normal ADA gene is "transplanted" into cells of a patient's immune system, thereby permitting these cells to survive and function normally. In the earliest human gene-therapy trials, scientists modified a type of virus called a retrovirus in the laboratory ("engineered") so that it could insert itself and a normal ADA gene into chromosomes of the immune cells taken from patients with SCID. In this chapter, you will see that retroviruses have many biological properties in common with a type of mobile element called a *retrotransposon*, which is present in our genome and the genomes of most eukaryotes. Lessons learned about the behavior

OUTLINE

15.1 Discovery of transposable elements in maize

15.2 Transposable elements in prokaryotes

15.3 Transposable elements in eukaryotes

15.4 The dynamic genome: more transposable elements than ever imagined

15.5 Epigenetic regulation of transposable elements by the host

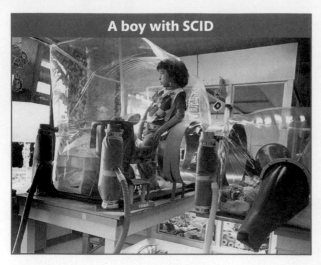

Figure 15-1 A patient with SCID must live in a protective bubble. [*UPI/Bettmann/CORBIS.*]

of retrotransposons and other mobile elements from model organisms such as yeast are sources of valuable insights into the design of a new generation of biological vectors for human gene therapy.

Starting in the 1930s, genetic studies of maize yielded results that greatly upset the classical genetic picture of genes residing only at fixed loci on the main chromosomes. The research literature began to carry reports suggesting that certain genetic elements present in the main chromosomes can somehow move from one location to another. These findings were viewed with skepticism for many years, but it is now clear that such mobile elements are widespread in nature.

A variety of colorful names (some of which help to describe their respective properties) have been applied to these genetic elements: controlling elements, jumping genes, mobile genes, mobile elements, and transposons. Here we use the terms *transposable elements* and *mobile elements,* which embrace the entire family of types. Transposable elements can move to new positions within the same chromosome or even to a different chromosome. They have been detected genetically in model organisms such as *E. coli,* maize, yeast, *C. elegans,* and *Drosophila* through the mutations that they produce when they insert into and inactivate genes.

DNA sequencing of genomes from a variety of microbes, plants, and animals indicates that transposable elements exist in virtually all organisms. Surprisingly, they are by far the largest component of the human genome, accounting for almost 50 percent of our chromosomes. Despite their abundance, the normal genetic role of these elements is not known with certainty.

In their studies, scientists are able to exploit the ability of transposable elements to insert into new sites in the genome. Transposable elements engineered in the test tube are valuable tools, both in prokaryotes and in eukaryotes, for genetic mapping, creating mutants, cloning genes, and even producing transgenic organisms. Let us reconstruct some of the steps in the evolution of our present understanding of transposable elements. In doing so, we will uncover the principles guiding these fascinating genetic units.

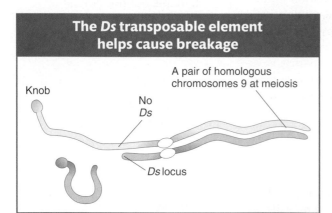

Figure 15-2 Chromosome 9 of corn breaks at the *Ds* locus, where the *Ds* transposable element has inserted.

15.1 Discovery of Transposable Elements in Maize

McClintock's experiments: the *Ds* element

In the 1940s, Barbara McClintock made an astonishing discovery while studying the colored kernels of so-called Indian corn, known as maize (see the Model Organism box on the next page). Maize has 10 chromosomes, numbered from largest (1) to smallest (10). While analyzing the breakage of maize chromosomes, McClintock noticed some unusual phenomena. She found that, in one strain of maize, chromosome 9 broke very frequently and at one particular site, or locus (Figure 15-2). Breakage of the chromosome at this locus, she determined, was due to the presence of two genetic factors. One factor that she called **Ds** (for **Dissociation**) was located at the site of the break. Another, unlinked genetic factor was required to "activate" the breakage of chromosome 9 at the *Ds* locus. Thus, McClintock called this second factor **Ac** (for **Activator**).

McClintock began to suspect that *Ac* and *Ds* were actually mobile genetic elements when she found it impossible to map *Ac*. In some plants, it mapped to one position; in other plants of the same line, it mapped to different positions.

Model Organism *Maize*

Maize, also known as corn, is actually *Zea mays*, a member of the grass family. Grasses—also including rice, wheat, and barley—are the most important source of calories for humanity. Maize was domesticated from the wild grass teosinte by Native Americans in Mexico and Central America and was first introduced to Europe by Columbus on his return from the New World.

In the 1920s, Rollins A. Emerson set up a laboratory at Cornell University to study the genetics of corn traits, including kernel color, which were ideal for genetic analysis. In addition, the physical separation of male and female flowers into the tassel and ear, respectively, made controlled genetic crosses easy to accomplish. Among the outstanding geneticists attracted to the Emerson laboratory were Marcus Rhoades, Barbara McClintock, and George Beadle (see Chapter 6). Before the advent of molecular biology and the rise of microorganisms as model organisms, geneticists performed microscopic analyses of chromosomes and related their behavior to the segregation of traits. The large pachytene chromosomes of maize and the salivary-gland chromosomes of *Drosophila* made them the organisms of choice for cytogenetic analyses. The results of these early studies led to an understanding of chromosome behavior during meiosis and mitosis, including such events as recombination and the consequences of chromosome breakage such as inversions, translocations, and duplication.

The maize laboratory of Rollins A. Emerson at Cornell University, 1929. Standing from left to right: Charles Burnham, Marcus Rhoades, Rollins Emerson, and Barbara McClintock. Kneeling is George Beadle. Both McClintock and Beadle were awarded a Nobel Prize. [*Courtesy of the Department of Plant Breeding, Cornell University.*]

Maize still serves as a model genetic organism. Molecular biologists continue to exploit its beautiful pachytene chromosomes with new antibody probes (see photograph *b* below) and have used its wealth of genetically well-characterized transposable elements as tools to identify and isolate important genes.

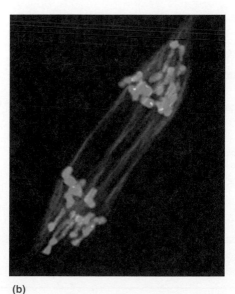

(a) (b)

Analysis of maize chromosomes, then and now. Maize chromosomes are large and easily visualized by light microscopy. (a) An image from Marcus Rhoades (1952). (b) This image is comparable to that in part *a* except that the spindle is shown in blue (stained with antibodies to tubulin), the centromeres are shown in red (stained with antibodies to a centromere-associated protein), and the chromosomes are shown in green. [*(a) From M. M. Rhoades, "Preferential Segregation in Maize," in J. W. Gowen, ed., Heterosis, pp. 66–80. Iowa State College Press, 1952. (b) From R. K. Dawe, L. Reed, H.-G. Yu, M. G. Muszynski, and E. N. Hiatt, "A Maize Homolog of Mammalian CENPC Is a Constitutive Component of the Inner Kinetochore," Plant Cell 11, 1999, 1227–1238.*]

As if this variable mapping were not enough of a curiosity, rare kernels with dramatically different phenotypes could be derived from the original strain that had frequent breaks in chromosome 9. One such phenotype was a rare colorless kernel containing pigmented spots.

Figure 15-3 compares the phenotype of the chromosome-breaking strain with the phenotype of one of these derivative strains. For the chromosome-breaking strain, a chromosome that breaks at or near *Ds* loses its end containing wild-type alleles of the *C, Sh,* and *Wx* genes. In the example shown in Figure 15-3a, a break occurred in a single cell, which divided mitotically to produce the large sector of mutant tissue (*c sh wx*). Breakage can happen many times in a single kernel, but each sector of tissue will display the loss of expression of all three genes. In contrast, each new derivative affected the expression of only a single gene. One derivative that affected the expression of only the pigment gene *C* is shown in Figure 15-3b. In this example, pigmented spots appeared on a colorless kernel background. Although the expression of *C* was altered in this strange way, the

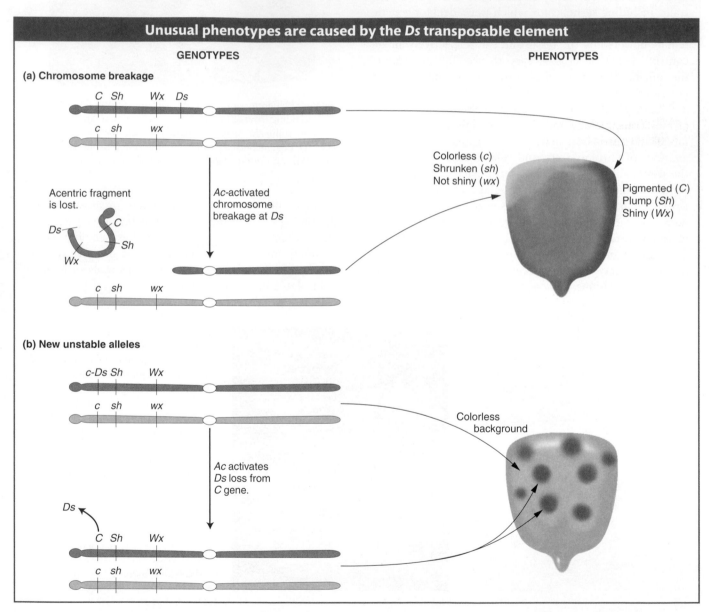

Figure 15-3 New phenotypes in corn are produced through the movement of the *Ds* transposable element on chromosome 9. (a) A chromosome fragment is lost through breakage at the *Ds* locus. Recessive alleles on the homologous chromosome are expressed, producing the colorless sector in the kernel. (b) Insertion of *Ds* in the *C* gene (top) creates colorless corn-kernel cells. Excision of *Ds* from the *C* gene through the action of *Ac* in cells and their mitotic descendants allows color to be expressed again, producing the spotted phenotype.

expression of *Sh* and *Wx* was normal and chromosome 9 no longer sustained frequent breaks.

To explain the new derivatives, McClintock hypothesized that *Ds* had moved from a site near the centromere into the *C* gene located close to the telomeric end. In its new location, *Ds* prevents the expression of *C*. The inactivation of the *C* gene explains the colorless parts of the kernel, but what explains the appearance of the pigmented spots? The spotted kernel is an example of an **unstable phenotype.** McClintock concluded that such unstable phenotypes resulted from the movement or transposition of *Ds* away from the *C* gene. That is, the kernel begins development with a *C* gene that has been mutated by the insertion of *Ds*. However, in some cells of the kernel, *Ds* leaves the *C* gene, allowing the mutant phenotype to revert to wild type and produce pigment in the original cell and in all its mitotic descendants. There are big spots of color when *Ds* leaves the *C* gene early in kernel development (because there are more mitotic descendants), whereas there are small spots when *Ds* leaves the *C* gene later in kernel development. Unstable mutant phenotypes that revert to wild type are a clue to the participation of mobile elements.

Autonomous and nonautonomous elements

What is the relation between *Ac* and *Ds*? How do they interact with genes and chromosomes to produce these interesting and unusual phenotypes? These questions were answered by further genetic analysis. Interactions between *Ds, Ac,* and the pigment gene *C* are used as an example in Figure 15-4. There, *Ds* is shown as a piece of DNA that has inactivated the *C* gene by inserting into its coding region. The allele carrying the insert is called *c-mutable(Ds)* or *c-m(Ds)* for short. A strain with *c-m(Ds)* and no *Ac* has colorless kernels because *Ds* cannot move; it is stuck in the *C* gene. A strain with *c-m(Ds)* and *Ac* has spotted kernels because *Ac* activates *Ds* in some cells to leave the *C* gene, thereby restoring gene function. The leaving element is said to **excise** from the chromosome or **transpose.**

Other strains were isolated in which the *Ac* element itself had inserted into the *C* gene [called *c-m(Ac)*]. Unlike the *c-m(Ds)* allele, which is unstable only when *Ac* is in the genome, *c-m(Ac)* is always unstable. Furthermore, McClintock found that, on rare occasions, an allele of the *Ac* type could be transformed into an allele of the *Ds* type. This transformation was due to the spontaneous generation of a *Ds* element from the inserted *Ac* element. In other words, *Ds* is, in all likelihood, an incomplete, mutated version of *Ac* itself.

Several systems like *Ac/Ds* were found by McClintock and other geneticists working with maize. Two other systems are *Dotted* [(*Dt*), discovered by Marcus Rhoades] and *Suppressor/mutator* [(*Spm*), independently discovered by McClintock and Peter Peterson, who called it *Enhancer/Inhibitor* (*En/In*)]. In addition, as you will see in the sections that follow, elements with similar genetic behavior have been isolated from bacteria, plants, and animals.

The common genetic behavior of these elements led geneticists to propose new categories for all the elements. *Ac* and elements with similar genetic properties are now called **autonomous elements** because they require no other elements for their mobility. Similarly, *Ds* and elements with similar genetic properties are called **nonautonomous elements.** An element *family* is composed of one or more

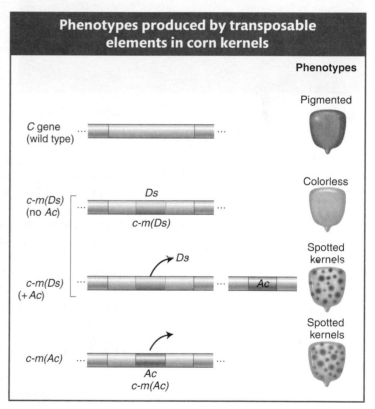

Phenotypes produced by transposable elements in corn kernels

Phenotypes

C gene (wild type) — Pigmented

c-m(Ds) (no *Ac*) — *Ds* / *c-m(Ds)* — Colorless

c-m(Ds) (+*Ac*) — *Ds* / *Ac* — Spotted kernels

c-m(Ac) — *Ac* / *c-m(Ac)* — Spotted kernels

Figure 15-4 Kernel spotting is controlled by the insertion and excision of *Ds* or *Ac* elements in the *C* gene controlling pigment.

Transposable elements at work in a rose

Figure 15-5 Mosaicism is caused by the excision of transposable elements in roses. The insertion of a transposable element disrupts pigment production, resulting in white flowers. The excision of the transposable element restores pigment production, resulting in red floral-tissue sectors. [*Photo courtesy of Susan Wessler.*]

autonomous elements and the nonautonomous members that can be mobilized. Autonomous elements encode the information necessary for their own movement and for the movement of unlinked nonautonomous elements in the genome. Because nonautonomous elements do not encode the functions necessary for their own movement, they cannot move unless an autonomous element from their family is present somewhere else in the genome.

Figure 15-5 shows an example of the effects of transposons in a rose.

> **Message** Transposable elements in maize can inactivate a gene in which they reside, cause chromosome breaks, and transpose to new locations within the genome. Autonomous elements can perform these functions unaided; nonautonomous elements can transpose only with the help of an autonomous element elsewhere in the genome.

Transposable elements: only in maize?

Although geneticists accepted McClintock's discovery of transposable elements in maize, many were reluctant to consider the possibility that similar elements resided in the genomes of other organisms. Their existence in all organisms would imply that genomes are inherently unstable and dynamic. This view was inconsistent with the fact that the genetic maps of members of the same species were the same. After all, if genes can be genetically mapped to a precise chromosomal location, doesn't this mapping indicate that they are not moving around?

Because McClintock was a highly respected geneticist, her results were explained by saying that maize is not a natural organism: it is a crop plant that is the product of human selection and domestication. This view was held by some until the 1960s, when the first transposable elements were isolated from the *E. coli* genome and studied at the DNA-sequence level. Transposable elements were subsequently isolated from the genomes of many organisms, including *Drosophila* and yeast. When it became apparent that transposable elements are a significant component of the genomes of most and perhaps all organisms, Barbara McClintock was recognized for her seminal discovery by being awarded the 1983 Nobel Prize in Medicine or Physiology.

15.2 Transposable Elements in Prokaryotes

The genetic discovery of transposable elements led to many questions about what such elements might look like at the DNA-sequence level and how they are able to move from one site to another in the genome. Did all organisms have them? Did all elements look alike or were there different classes of transposable elements? If there were many classes of elements, could they coexist in one genome? Did the number of transposable elements in the genome vary from species to species? The molecular nature of transposable genetic elements was first understood in bacteria. Therefore, we will continue this story by examining the original studies performed with prokaryotes.

There are two broad types of transposable elements in bacteria:

- Short sequences called *IS elements* that can move themselves to new positions but do not carry genes other than those needed for their movement.

- Longer sequences called *transposons* that not only carry the genes they need for their movement but also carry other genes.

Bacterial insertion sequences

Insertion sequences, or **insertion-sequence (IS) elements,** are segments of bacterial DNA that can move from one position on a chromosome to a different position on the same chromosome or on a different chromosome. When IS elements appear in the middle of genes, they interrupt the coding sequence and inactivate the expression of that gene. Owing to their size and in some cases the presence of transcription- and translation-termination signals in the IS element, IS elements can also block the expression of other genes in the same operon if those genes are downstream of the insertion site. IS elements were first found in *E. coli* in the *gal* operon—a cluster of three genes taking part in the metabolism of the sugar galactose.

Identification of discrete IS elements   Several *E. coli gal⁻* mutants were found to contain large insertions of DNA into the *gal* operon. This finding led naturally to the next question: Are the segments of DNA that insert into genes merely random DNA fragments or are they distinct genetic entities? The answer to this question came from the results of hybridization experiments showing that many different insertion mutations are caused by a small set of insertion sequences. These experiments are performed with the use of λd*gal* phages that contain the *gal⁻* operon from several independently isolated *gal* mutant strains. Individual phage particles from the strains are isolated, and their DNA is used to synthesize radioactive RNA in vitro. Certain fragments of this RNA are found to hybridize with the DNA from other *gal⁻* mutations containing large DNA insertions but not with wild-type DNA. These results were interpreted to mean that independently isolated *gal* mutants contain the same extra piece of DNA. These particular RNA fragments also hybridize to DNA from other mutants containing IS insertions in other genes, showing that the same bit of DNA can insert in different places in the bacterial chromosome.

Structure of IS elements   On the basis of their patterns of cross-hybridization, a number of distinct IS elements have been identified. One sequence, termed IS1, is the 800-bp segment identified in *gal.* Another sequence, termed IS2, is 1350 bp long. Although IS elements differ in DNA sequence, they have several features in common. For example, all IS elements encode a protein, called a **transposase,** which is an enzyme required for the movement of IS elements from one site in the chromosome to another. In addition, all IS elements begin and end with short inverted repeat sequences that are required for their mobility. The transposition of IS elements and other mobile genetic elements will be considered later in the chapter.

The genome of the standard wild-type *E. coli* is rich in IS elements: it contains eight copies of IS1, five copies of IS2, and copies of other less well-studied IS types. Because IS elements are regions of identical sequence, they are sites where crossovers may take place. For example, recombination between the F-factor plasmid and the *E. coli* chromosome to form *Hfr* strains is the result of a single crossover between an IS element located on the plasmid and an IS element located on the chromosome (see Chapter 5, Figure 5-18). Because there are multiple IS elements, the F factor can insert at multiple sites.

> **Message** The bacterial genome contains segments of DNA, termed IS elements, that can move from one position on the chromosome to a different position on the same chromosome or on a different chromosome.

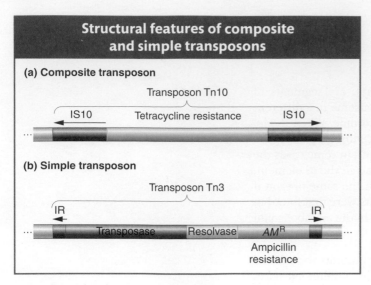

Figure 15-6 (a) Tn10, an example of a composite transposon. The IS elements are inserted in opposite orientation and form inverted repeats (IRs). Each IS element carries a transposase, but only one is usually functional. (b) Tn3, an example of a simple transposon. Short inverted repeats contain no transposase. Instead, simple transposons encode their own transposase. The resolvase is a protein that promotes recombination and resolves the cointegrates (see Figure 15-9).

Prokaryotic transposons

In Chapter 5, you learned about **R factors,** which are plasmids carrying genes that encode resistance to several antibiotics. These R factors (for resistance) are transferred rapidly on cell conjugation, much like the F factor in *E. coli.*

The R factors proved to be just the first of many similar F-like factors to be discovered. R factors have been found to carry many different kinds of genes in bacteria. In particular, R factors pick up genes conferring resistance to different antibiotics. How do they acquire their new genetic abilities? It turns out that the drug-resistance genes reside on a mobile genetic element called a **transposon (Tn).** There are two types of bacterial transposons. **Composite transposons** contain a variety of genes that reside between two nearly identical IS elements that are oriented in opposite direction (Figure 15-6a) and, as such, form what is called an **inverted repeat** sequence. Transposase encoded by one of the two IS elements is necessary to catalyze the movement of the entire transposon. An example of a composite transposon is Tn10, shown in Figure 15-6a. Tn10 carries a gene that confers resistance to the antibiotic tetracycline and is flanked by two IS10 elements in opposite orientation. The IS elements that make up composite transposons are not capable of transposing on their own without the rest of the transposon, because of mutations in their inverted repeats.

Simple transposons also consist of bacterial genes flanked by inverted repeat sequences, but these sequences are short (<50 bp) and do not encode the transposase enzyme that is necessary for transposition. Thus, their mobility is not due to an association with IS elements. Instead, simple transposons encode their own transposase in the region between the inverted repeat sequences in addition to carrying bacterial genes. An example of a simple transposon is Tn3, shown in Figure 15-6b.

To review, IS elements are short mobile sequences that encode only those proteins necessary for their mobility. Composite transposons and simple transposons contain additional genes that confer new functions to bacterial cells. Whether composite or simple, transposons are usually just called transposons, and different transposons are designated Tn1, Tn2, Tn505, and so forth.

A transposon can jump from a plasmid to a bacterial chromosome or from one plasmid to another plasmid. In this manner, multiple-drug-resistant plasmids are generated. Figure 15-7 is a composite diagram of an R factor, indicating the various places at which transposons can be located. We next consider the question of how such **transposition** or mobilization events occur.

> **Message** Transposons were originally detected as mobile genetic elements that confer drug resistance. Many of these elements consist of IS elements flanking a gene that encodes drug resistance. This organization promotes the spread of drug-resistant bacteria by facilitating movement of the resistance gene from the chromosome of a resistant bacterium to a plasmid that can be conjugated into another (susceptible) bacterial strain.

Mechanism of transposition

As already stated, the movement of a transposable element depends on the action of a transposase. This enzyme plays key roles in the two stages of transposition: excision (leaving) from the original location and inserting into the new location.

Excision from the original location   Most transposable elements in prokaryotes (and in eukaryotes) employ one of two mechanisms of transposition, called

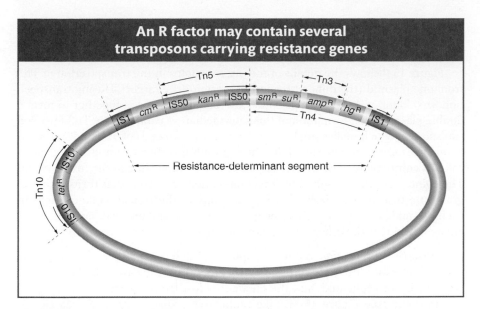

Figure 15-7 A schematic map of a plasmid with several insertions of simple and composite transposons carrying resistance genes. Plasmid sequences are in blue. Genes encoding resistance to the antibiotics tetracycline (*tet*R), kanamycin (*kan*R), streptomycin (*sm*R), sulfonamide (*su*R), and ampicillin (*amp*R) and to mercury (*hg*R) are shown. The resistant-determinant segment can move as a cluster of resistance genes. Tn3 is within Tn4. Each transposon can be transferred independently. [*Simplified from S. N. Cohen and J. A. Shapiro, "Transposable Genetic Elements." Copyright 1980 by Scientific American, Inc. All rights reserved.*]

replicative and **conservative** (nonreplicative), as illustrated in Figure 15-8. In the replicative pathway (as shown for Tn3), a new copy of the transposable element is generated in the transposition event. The results of the transposition are that one copy appears at the new site and one copy remains at the old site. In the conservative pathway (as shown for Tn10), there is no replication. Instead, the element is excised from the chromosome or plasmid and is integrated into the new site. The conservative pathway is also called **"cut and paste."**

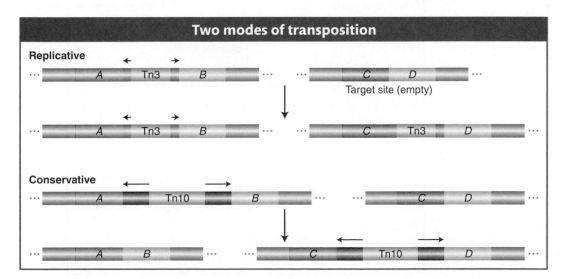

Figure 15-8 Mobile-element transposition may be either replicative or conservative. See text for details. [*Adapted with permission from Nature Reviews: Genetics 1, no. 2, p. 138, Fig. 3, November 2000, "Mobile Elements and the Human Genome," E. T. Luning Prak and H. H. Kazazian, Jr. Copyright 2000 by Macmillan Magazines Ltd.*]

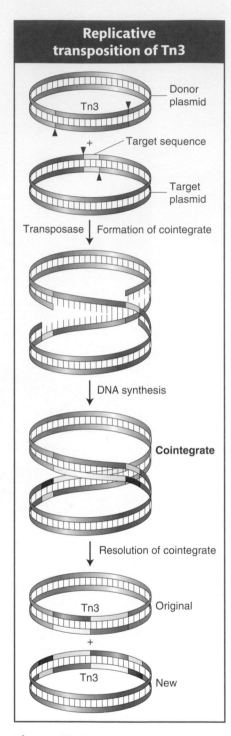

Figure 15-9 Replicative transposition of Tn3 takes place through a cointegrate intermediate. [*After Robert J. Brooker, Genetics: Analysis and Principles, Fig. 18-14. Benjamin-Cummings, 1999.*]

GENETICS**PORTAL** ANIMATED ART: **Replicative transposition**

Replicative transposition Because this mechanism is a bit complicated, it will be described here in more detail. As Figure 15-8 illustrates, one copy of Tn3 is produced from an initial single copy, yielding two copies of Tn3 altogether.

Figure 15-9 shows the details of the intermediates in the transposition of Tn3 from one plasmid (the donor) to another plasmid (the target). During transposition, the donor and recipient plasmids are temporarily fused together to form a double plasmid. The formation of this intermediate is catalyzed by Tn3-encoded transposase, which makes single-strand cuts at the two ends of Tn3 and staggered cuts at the target sequence and joins the free ends together, forming a fused circle called a **cointegrate.** The transposable element is duplicated in the fusion event. The cointegrate then resolves by a recombination-like event that turns a cointegrate into two smaller circles, leaving one copy of the transposable element in each plasmid. The result is that one copy remains at the original location of the element, whereas the other is integrated at a new genomic position.

Conservative transposition Some transposons, such as Tn10, excise from the chromosome and integrate into the target DNA. In these cases, the DNA of the element is not replicated, and the element is lost from the site of the original chromosome (see Figure 15-8). Like replicative transposition, this reaction is initiated by the element-encoded transposase, which cuts at the ends of the transposon. However, in contrast with replicative transposition, the transposase

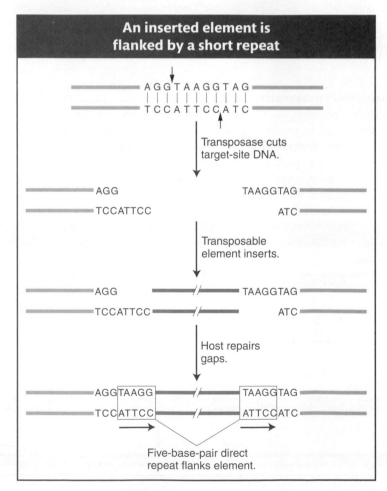

Figure 15-10 A short sequence of DNA is duplicated at the transposon insertion site. The recipient DNA is cleaved at staggered sites (a 5-bp staggered cut is shown), leading to the production of two copies of the five-base-pair sequence flanking the inserted element.

cuts the element out of the donor site. It then makes a staggered cut at a target site and inserts the element into the target site. We will revisit this mechanism in more detail in a discussion of the transposition of eukaryotic transposable elements, including the *Ac/Ds* family of maize.

Insertion into a new location   We have seen that transposase continues to play an important role in insertion. In one of the first steps of insertion, the transposase makes a staggered cut in the target-site DNA (not unlike the staggered breaks catalyzed by restriction endonucleases in the sugar–phosphate backbone of DNA). Figure 15-10 shows the steps in the insertion of a generic transposable element. In this case, the transposase makes a five-base-pair staggered cut. The transposable element inserts between the staggered ends, and the host DNA repair machinery (see Chapter 16) fills in the gap opposite each single-strand overhang by using the bases in the overhang as a template. There are now two duplicate sequences, each five base pairs in length, at the sites of the former overhangs. These sequences are called a **target-site duplication.** Virtually all transposable elements (in both prokaryotes and eukaryotes) are flanked by a target-site duplication, indicating that all use a mechanism of insertion similar to that shown in Figure 15-10. What differs is the length of the duplication; a particular type of transposable element has a characteristic length for its target-site duplication—as small as two base pairs for some elements. It is important to keep in mind that the transposable elements have *inverted repeats* at their ends and that the inverted repeats are flanked by the target-site duplication—which is a *direct repeat.*

> **Message** In prokaryotes, transposition occurs by at least two different pathways. Some transposable elements can replicate a copy of the element into a target site, leaving one copy behind at the original site. In other cases, transposition consists of the direct excision of the element and its reinsertion into a new site.

15.3 Transposable Elements in Eukaryotes

Although transposable elements were first discovered in maize, the first eukaryotic elements to be molecularly characterized were isolated from mutant yeast and *Drosophila* genes. Eukaryotic transposable elements fall into two classes: class 1 retrotransposons and class 2 DNA transposons. The first class to be isolated, the retrotransposons, are not at all like the prokaryotic IS elements and transposable elements.

Class I: retrotransposons

The laboratory of Gerry Fink was among the first to use yeast as a model organism to study eukaryotic gene regulation. Through the years, he and his colleagues isolated thousands of mutations in the *HIS4* gene, which encodes one of the enzymes in the pathway leading to the synthesis of the amino acid histidine.

They isolated more than 1500 spontaneous *HIS4* mutants and found that 2 of them had an unstable mutant phenotype. The unstable mutants (called pseudorevertants) were more than 1000 times as likely to revert to a phenotype that was similar to wild type as the other *HIS4* mutants. Symbolically, we say that these unstable mutants reverted from His^- to His^+ (wild types have a superscript plus sign, whereas mutants have a superscript minus sign). Like the *E. coli* gal^- mutants, these yeast mutants were found to harbor a large DNA insertion in the *HIS4* gene. The insertion turned out to be very similar to one of a group of transposable elements already characterized in yeast, called the **Ty elements.** There are, in fact, about 35 copies of the inserted element, called *Ty1*, in the yeast genome.

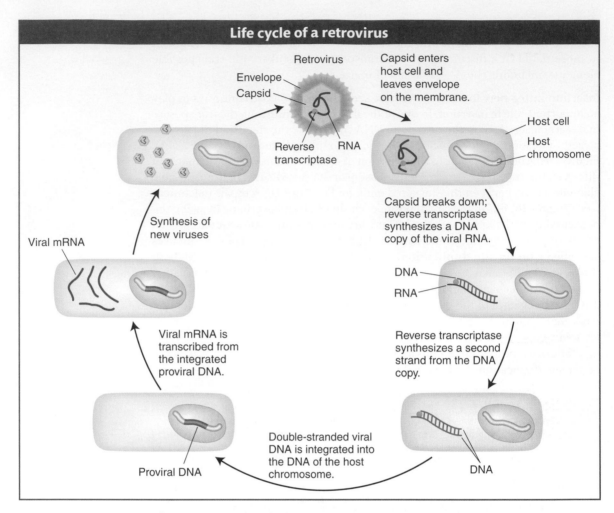

Figure 15-11 The retrovirus RNA genome undergoes reverse transcription into double-stranded DNA inside the host cell.

Cloning of the elements from these mutant alleles led to the surprising discovery that the insertions did not look at all like bacterial IS elements or transposons. Instead, they resembled a well-characterized class of animal viruses called retroviruses. A **retrovirus** is a single-stranded RNA virus that employs a double-stranded DNA intermediate for replication. The RNA is copied into DNA by the enzyme **reverse transcriptase.** The double-stranded DNA is integrated into host chromosomes, from which it is transcribed to produce the RNA viral genome and proteins that form new viral particles. When integrated into host chromosomes as double-stranded DNA, the double-stranded DNA copy of the retroviral genome is called a **provirus.** The life cycle of a typical retrovirus is shown in Figure 15-11. Some retroviruses, such as mouse mammary tumor virus (MMTV) and Rous sarcoma virus (RSV), are responsible for the induction of cancerous tumors. This happens because they insert randomly into the genome and, in the process, may insert next to a gene whose altered expression leads to cancer.

Figure 15-12 shows the similarity in structure and gene content of a retrovirus and the *Ty1* element isolated from the *HIS4* mutants. Both are flanked by **long terminal repeat (LTR)** sequences that are several hundred base pairs long. Both contain the genes *gag* and *pol*.

Retroviruses encode at least three proteins that take part in viral replication: the products of the *gag, pol,* and *env* genes. The *gag*-encoded protein has a role in the maturation of the RNA genome, *pol* encodes the all-important reverse tran-

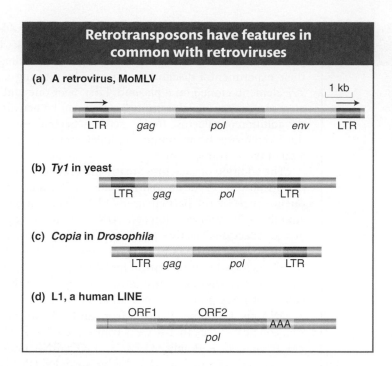

Figure 15-12 Structural comparison of a retrovirus with retrotransposons found in eukaryotic genomes. (a) A retrovirus, Moloney murine leukemia virus (MoMLV), of mice. (b) A retrotransposon, *Ty1*, in yeast. (c) A retrotransposon, *copia*, in *Drosophila*. (d) A long interspersed element (LINE) in humans. Abbreviations: LTR, long terminal repeat; ORF, open reading frame.

scriptase, and *env* encodes the structural protein that surrounds the virus. This protein is necessary for the virus to leave the cell to infect other cells. Interestingly, *Ty1* elements have genes related to *gag* and *pol* but not *env*. These features led to the hypothesis that, like retroviruses, *Ty1* elements are transcribed into RNA transcripts that are copied into double-stranded DNA by the reverse transcriptase. However, unlike retroviruses, *Ty1* elements cannot leave the cell, because they do not encode *env*. Instead, the double-stranded DNA copies are inserted back into the genome of the same cell. These steps are diagrammed in Figure 15-13.

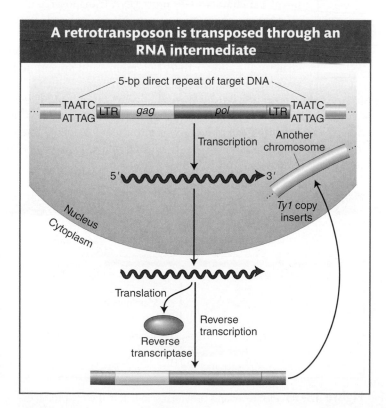

Figure 15-13 An RNA transcript from the retrotransposon undergoes reverse transcription into DNA, by a reverse transcriptase encoded by the retrotransposon. The DNA copy is inserted at a new location in the genome.

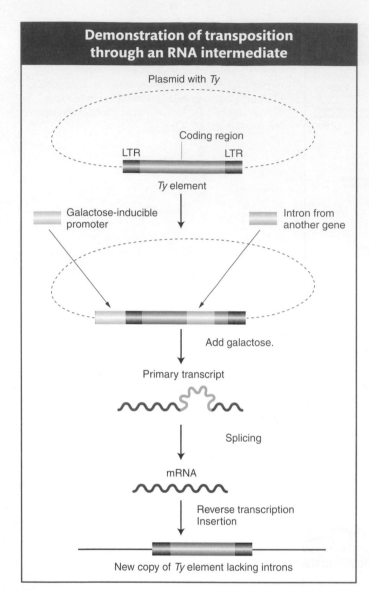

Figure 15-14 A *Ty* element is altered by adding an intron and a promoter that can be activated by the addition of galactose. The intron sequences are spliced before reverse transcription. [*After H. Lodish, D. Baltimore, A. Berk, S. L. Zipursky, P. Matsudaira, and J. Darnell, Molecular Cell Biology, 3rd ed., p. 332. Copyright 1995 by Scientific American Books.*]

In 1985, David Garfinkel, Jef Boeke, and Gerald Fink showed that, like retroviruses, *Ty* elements do in fact transpose through an RNA intermediate. Figure 15-14 diagrams their experimental design. They began by altering a yeast *Ty1* element, cloned on a plasmid. First, near one end of an element, they inserted a promoter that can be activated by the addition of galactose to the medium. Second, they introduced an intron from another yeast gene into the coding region of the *Ty* transposon.

The addition of galactose greatly increases the frequency of transposition of the altered *Ty* element. This increased frequency suggests the participation of RNA, because galactose stimulates the transcription of *Ty* DNA into RNA, beginning at the galactose-sensitive promoter. The key experimental result, however, is the fate of the transposed *Ty* DNA. The researchers found that the intron had been removed from the transposed *Ty* DNA. Because introns are spliced only in the course of RNA processing (see Chapter 8), the transposed *Ty* DNA must have been copied from an RNA intermediate. The conclusion was that RNA is transcribed from the original *Ty* element and spliced. The spliced mRNA undergoes reverse transcription back into double-stranded DNA, which is then integrated into the yeast chromosome. Transposable elements that employ reverse transcriptase to transpose through an RNA intermediate are termed **retrotransposons.** They are also known as **class 1 transposable elements.** Retrotransposons such as *Ty1* that have *long terminal repeats* at their ends are called **LTR-retrotransposons,** and the mechanism they use to transpose is called **"copy and paste"** to distinguish them from the cut-and-paste mechanism that characterizes most DNA transposable elements.

Several spontaneous mutations isolated through the years in *Drosophila* also were shown to contain retrotransposon insertions. The ***copia*-like elements** of *Drosophila* are structurally similar to *Ty1* elements and appear at 10 to 100 positions in the *Drosophila* genome (see Figure 15-12c). Certain classic *Drosophila* mutations result from the insertion of *copia*-like and other elements. For example, the *white-apricot* (w^a) mutation for eye color is caused by the insertion of an element of the *copia* family into the *white* locus. The insertion of LTR-retrotransposons into plant genes (including maize) also has been shown to contribute to spontaneous mutations in this kingdom.

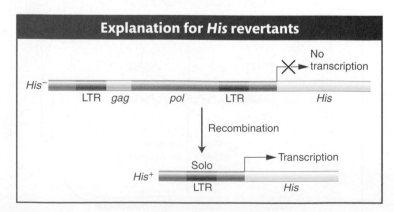

Figure 15-15 *His⁺* revertants contain a solo LTR in the *His* promoter.

Before we leave retrotransposons (we will return to them later in this chapter), there is one question that needs to be answered. Recall that the first LTR-retrotransposon was discovered in an unstable *His⁻* strain of yeast that reverted frequently to *His⁺*. However, we have just seen that LTR retrotransposons, unlike most DNA transposable elements, do not excise when they transpose. What, then, is responsible for this allele's ~1000-fold increase in reversion frequency when compared to other *His⁻* alleles? The answer is shown in Figure 15-15, which shows that the *Ty1* element in the *His⁻* allele is located in the promoter region of the *His* gene, where it prevents gene transcription. In contrast, the revertants contain a single copy of the LTR, called a **solo LTR.** This much smaller insertion does not interfere with the transcription of the *His* gene. The solo LTR is the product of recombination between the identical LTRs, which results in the deletion of the rest of the element (see Chapters 4 and 16 for more on recombination). Solo LTRs are a very common feature in the genomes of virtually all eukaryotes, indicating the importance of this process. The sequenced yeast genome contains more than fivefold as many solo LTRs as complete *Ty1* elements.

> **Message** Transposable elements that transpose through RNA intermediates predominate in eukaryotes. Retrotransposons, also known as class 1 elements, encode a reverse transcriptase that produces a double-stranded DNA copy (from an RNA intermediate) that is capable of integrating at a new position in the genome.

Class 2: DNA transposons

Some mobile elements found in eukaryotes appear to transpose by mechanisms similar to those in bacteria. As illustrated in Figure 15-8 for IS elements and transposons, the entity that inserts into a new position in the genome is either the element itself or a copy of the element. Elements that transpose in this manner are called **class 2 elements,** or **DNA transposons.** The first transposable elements discovered by McClintock in maize are now known to be DNA transposons. However, the first DNA transposons to be molecularly characterized were the *P* elements in *Drosophila.*

***P* elements** Of all the transposable elements in *Drosophila,* the most intriguing and useful to the geneticist are the *P* **elements.** The full-size *P* element resembles the simple transposons of bacteria in that its ends are short (31-bp) inverted repeats and it encodes a single protein—the transposase that is responsible for its mobilization (Figure 15-16). The *P* elements vary in size, ranging from 0.5 to 2.9 kb in length. This size difference is due to the presence of many defective *P* elements from which parts of the middle of the element—encoding the transposase gene—have been deleted.

P elements were discovered by Margaret Kidwell, who was studying **hybrid dysgenesis**—a phenomenon that occurs when females from laboratory strains of

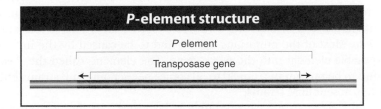

Figure 15-16 DNA sequence analysis of the 2.9-kb *P* element reveals a gene that encodes transposase. A perfect 31-bp inverted repeat resides at each of the element's termini. [*From G. Robin, in J. A. Shapiro, ed., Mobile Genetic Elements, pp. 329–361. Academic Press, 1983.*]

Figure 15-17 In hybrid dysgenesis, a cross between a female from laboratory stock and a wild male yields defective progeny. See text for details.

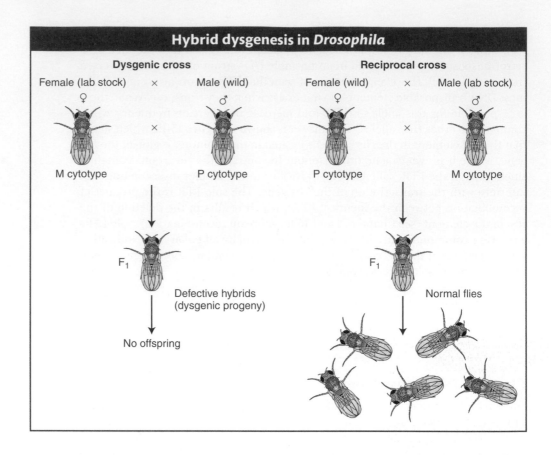

D. melanogaster are mated with males derived from natural populations. In such crosses, the laboratory stocks are said to possess an **M cytotype** (cell type), and the natural stocks are said to possess a **P cytotype.** In a cross of M (female) × P (male), the progeny show a range of surprising phenotypes that are manifested in the germ line, including sterility, a high mutation rate, and a high frequency of chromosomal aberration and nondisjunction (Figure 15-17). These hybrid progeny are *dysgenic,* or biologically deficient (hence, the expression *hybrid dysgenesis*). Interestingly, the reciprocal cross, P (female) × M (male), produces no dysgenic offspring. An important observation is that a large percentage of the dysgenically induced mutations are unstable; that is, they revert to wild type or to other mutant alleles at very high frequencies. This instability is generally restricted to the germ line of an individual fly possessing an M cytotype by a mechanism explained below.

The unstable *Drosophila* mutants had similarities to the unstable maize mutants characterized by McClintock. Investigators hypothesized that the dysgenic mutations are caused by the insertion of transposable elements into specific genes, thereby rendering them inactive. According to this view, reversion would usually result from the excision of these inserted sequences. This hypothesis has been critically tested by isolating unstable dysgenic mutations at the eye-color locus *white.* Most of the mutations were found to be caused by the insertion of a transposable element into the *white+* gene. The element, called the *P element,* was found to be present in from 30 to 50 copies per genome in P strains but to be completely absent in M strains.

Why do *P* elements not cause trouble in P strains? The simple answer is that *P*-element transposition is repressed in P strains. At first it was thought that repression was due to a polypeptide repressor that was in P but not M strains. This model is no longer favored. Instead, geneticists now think that all the transposase genes in *P* elements are silenced in P strains. The genes are activated in the F$_1$ generation as shown in Figure 15-18. Gene silencing has been discussed previously (see Chapters 8 and 12) and will be revisited at the end of this chapter. For some reason, most laboratory strains have no *P* elements, and consequently the silencing mechanism is not activated. In hybrids from the cross M (female, no *P* elements) × P (male, *P* elements), the *P* elements in the newly formed zygote are in a silencing-free environment. The *P* elements derived from the male genome can now transpose throughout the diploid genome, causing a variety of damage as they insert into genes and cause mutations. These molecular events are expressed as the various manifestations of hybrid dysgencsis. On the other hand, P (female) × M (male) crosses do not result in dysgenesis, because, presumably, the egg cytoplasm contains the components required for silencing the *P*-element transposase.

An intriguing question remains unanswered: Why do laboratory strains lack *P* elements, whereas strains in the wild have *P* elements? One hypothesis is that most of the current laboratory strains descended from the original isolates taken from the wild by Morgan and his students almost a century ago. At some point between the capture of those original strains and the present, *P* elements spread

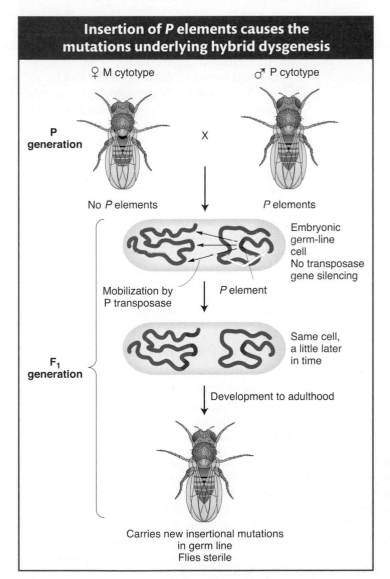

Figure 15-18 Molecular events underlying hybrid dysgenesis. Crosses of male *Drosophila* bearing P transposase with female *Drosophila* that do not have functional *P* elements produce mutations in the germ line of F$_1$ progeny caused by *P*-element insertions. *P* elements are able to move, causing mutations, because the egg does not silence the transposase gene.

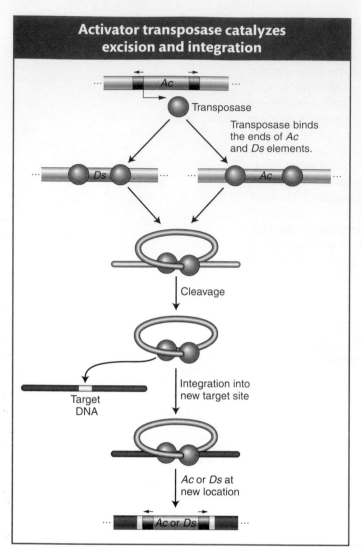

Activator transposase catalyzes excision and integration

Transposase

Transposase binds the ends of *Ac* and *Ds* elements.

Ds   *Ac*

Cleavage

Target DNA

Integration into new target site

Ac or *Ds* at new location

Ac or *Ds*

Figure 15-19 The *Ac* element in maize encodes a transposase that binds its own ends or those of a *Ds* element, excising the element, cleaving the target site, and allowing the element to insert elsewhere in the genome.

through natural populations but not through laboratory strains. This difference was not noticed until wild strains were again captured and mated with laboratory strains.

Although the exact scenario of how *P* elements have spread throughout wild populations is not clear, what is clear is that transposable elements can spread rapidly from a few individual members of a population. In this regard, the spread of *P* elements resembles the spread of transposons carrying resistance genes to formerly susceptible bacterial populations.

Maize transposable elements revisited Although the causative agent responsible for unstable mutants was first shown genetically to be transposable elements in maize, it was almost 50 years before the maize *Ac* and *Ds* elements were isolated and shown to be related to DNA transposons in bacteria and in other eukaryotes. Like the *P* element of *Drosophila*, *Ac* has terminal inverted repeats and encodes a single protein, the transposase. The nonautonomous *Ds* element does not encode transposase and thus cannot transpose on its own. When *Ac* is in the genome, its transposase can bind to both ends of *Ac* or *Ds* elements and promote their transposition (Figure 15-19).

As noted earlier in the chapter, *Ac* and *Ds* are members of a single transposon family, and there are other families of transposable elements in maize. Each family contains an autonomous element encoding a transposase that can move elements in the same family but cannot move elements in other families, because the transposase can bind only to the ends of family members.

Although some organisms such as yeast have no DNA transposons, elements structurally similar to the *P* and *Ac* elements have been isolated from many plant and animal species.

> **Message** The first known transposable elements in maize are DNA transposons that structurally resemble DNA transposons in bacteria and other eukaryotes. DNA transposons encode a transposase that cuts the transposon from the chromosome and catalyzes its reinsertion at other chromosomal locations.

Utility of DNA transposons for gene discovery

Quite apart from their interest as a genetic phenomenon, DNA transposons have become major tools used by geneticists working with a variety of organisms. Their mobility has been exploited to tag genes for cloning and to insert transgenes. The *P* element in *Drosophila* provides one of the best examples of how geneticists exploit the properties of transposable elements in eukaryotes.

Using *P* elements to tag genes for cloning *P* elements can be used to create mutations by insertion, to mark the position of genes, and to facilitate the cloning of genes. *P* elements inserted into genes in vivo disrupt genes at random, creating mutants with different phenotypes. Fruit flies with interesting mutant phenotypes can be selected for cloning of the mutant gene, which is marked by the presence of the *P* element, a method termed **transposon tagging.** After the interrupted gene has been cloned, fragments from the mutant gene can be used as a probe to isolate the wild-type gene.

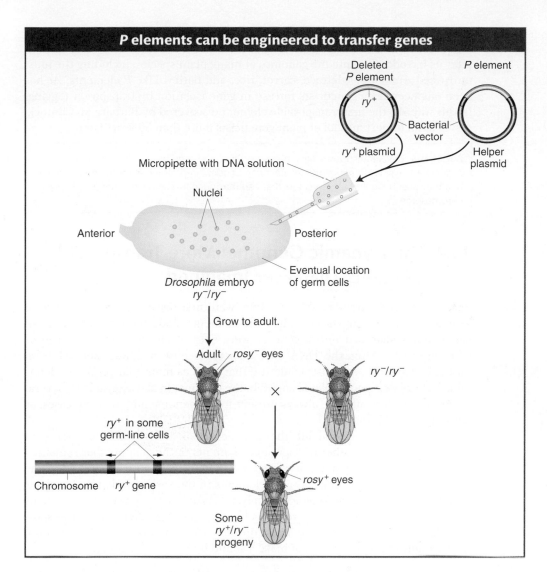

P elements can be engineered to transfer genes

Figure 15-20 *P*-element-mediated gene transfer in *Drosophila*. The *rosy⁺* (*ry⁺*) eye-color gene is engineered into a deleted *P* element carried on a bacterial vector. At the same time, a helper plasmid bearing an intact *P* element is used. Both are injected into an *ry⁻* embryo, where *ry⁺* transposes with the *P* element into the chromosomes of the germ-line cells.

Using *P* elements to insert genes Gerald Rubin and Allan Spradling showed that *P*-element DNA can be an effective vehicle for transferring donor genes into the germ line of a recipient fly. They devised the following experimental procedure (Figure 15-20). Suppose the goal is to transfer the allele *ry⁺*, which confers a characteristic eye color, into the fly genome. The recipient genotype is homozygous for the *rosy* (*ry⁻*) mutation. From this strain, embryos are collected at the completion of about nine nuclear divisions. At this stage, the embryo is one multinucleate cell, and the nuclei destined to form the germ cells are clustered at one end. (*P* elements mobilize only in germ-line cells.) Two types of DNA are injected into embryos of this type. The first is a bacterial plasmid carrying a defective *P* element into which the *ry⁺* gene has been inserted. The defective *P* element resembles the maize *Ds* element in that it does not encode transposase but still has the ends that bind transposase and allow transposition. This deleted element is not able to transpose, and so, as mentioned earlier, a helper plasmid encoding transposase also is injected. Flies developing from these embryos are phenotypically still *rosy* mutants, but their offspring include a large proportion of *ry⁺* flies. In situ hybridization confirmed that the *ry⁺* gene, together with the deleted *P* element, was inserted into one of several distinct chromosome locations. None appeared exactly at the normal locus of the *rosy* gene. These new *ry⁺* genes are found to be inherited in a stable, Mendelian fashion.

Because the *P* element can transpose only in *Drosophila*, these applications are restricted to these flies. In contrast, the maize *Ac* element is able to transpose after its introduction into the genomes of many plant species, including the mustard weed *Arabidopsis*, lettuce, carrot, rice, and barley. Like *P* elements, *Ac* has been engineered by geneticists for use in gene isolation by transposon tagging. In this way, *Ac*, the first transposable element discovered by Barbara McClintock, serves as an important tool of plant geneticists more than 50 years later.

> **Message** DNA transposons have been modified and used by scientists in two important ways: (1) to make mutants that can be identified molecularly by the presence of a transposon tag and (2) as vectors that can introduce foreign genes into a chromosome.

15.4 The Dynamic Genome: More Transposable Elements Than Ever Imagined

As you have seen, transposable elements were first discovered with the use of genetic approaches. In these studies, the elements made their presence known when they transposed into a gene or were sites of chromosome breakage or rearrangement. After the DNA of transposable elements was isolated from unstable mutations, scientists could use that DNA as molecular probes to determine if there were more related copies in the genome. In all cases, at least several copies of the element were always present in the genome and, in some cases, as many as several hundred.

Scientists wondered about the prevalence of transposable elements in genomes. Were there other transposable elements in the genome that remained unknown because they had not caused a mutation that could be studied in the laboratory? Were there transposable elements in the vast majority of organisms that were not amenable to genetic analysis? Asked another way, do organisms without mutations induced by transposable elements nonetheless have transposable elements in their genomes? These questions are reminiscent of the question, If a tree falls in the forest, does it make a sound if no one is listening?

Large genomes are largely transposable elements

Long before the advent of DNA-sequencing projects, scientists using a variety of biochemical techniques discovered that DNA content (called **C-value**) varied dramatically in eukaryotes and did not correlate with biological complexity. For example, the genomes of salamanders are 20 times as large as the human genome, whereas the genome of barley is more than 10 times as large as the genome of rice, a related grass. The lack of correlation between genome size and the biological complexity of an organism is known as the **C-value paradox.**

Barley and rice are both cereal grasses, and, as such, their gene content should be similar. However, if genes are a relatively constant component of the genomes of multicellular organisms, what is responsible for the additional DNA in the larger genomes? On the basis of the results of additional experiments, scientists were able to determine that DNA sequences that are repeated thousands, even hundreds of thousands, of times make up a large fraction of eukaryotic genomes and that some genomes contain much more repetitive DNA than others.

Thanks to many recent projects to sequence the genomes of a wide variety of taxa (including *Drosophila*, humans, the mouse, *Arabidopsis*, and rice), we now know that there are many classes of repetitive sequences in the genomes of higher organisms and that some are similar to the DNA transposons and retrotransposons shown to be responsible for mutations in plants, yeast, and insects. Most

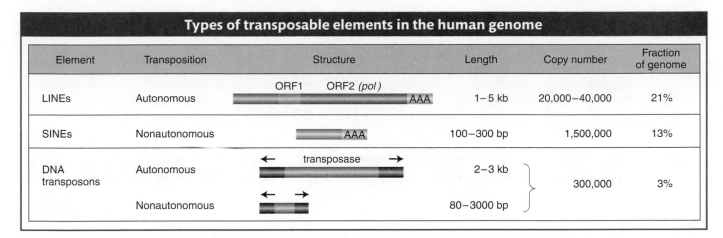

Figure 15-21 Several general classes of transposable elements are found in the human genome. [*Reprinted by permission from Nature 409, 880 (15 February 2001), "Initial Sequencing and Analysis of the Human Genome," The International Human Genome Sequencing Consortium. Copyright 2001 by Macmillan Magazines Ltd.*]

remarkably, these sequences make up most of the DNA in the genomes of multicellular eukaryotes.

Rather than correlating with gene content, genome size frequently correlates with the amount of DNA in the genome that is derived from transposable elements. Organisms with big genomes have lots of sequences that resemble transposable elements, whereas organisms with small genomes have many fewer. Two examples, one from the human genome and the other from a comparison of the grass genomes, illustrate this point. The structural features of the transposable elements that are found in human genomes are summarized in Figure 15-21 and will be referred to in the next section.

> **Message** The *C*-value paradox is the lack of correlation between genome size and biological complexity. Genes make up only a small proportion of the genomes of multicellular organisms. Genome size usually corresponds to the amount of transposable-element sequences rather than to gene content.

Transposable elements in the human genome

Almost half of the human genome is derived from transposable elements. The vast majority of these transposable elements are two types of retrotransposons called **long interspersed elements,** or **LINEs,** and **short interspersed elements,** or **SINEs** (see Figure 15-21). LINEs move like a retrotransposon with the help of an element-encoded reverse transcriptase but lack some structural features of retrovirus-like elements, including LTRs (see Figure 15-12d). SINEs can be best described as nonautonomous LINEs, because they have the structural features of LINEs but do not encode their own reverse transcriptase. Presumably, they are mobilized by reverse transcriptase enzymes that are encoded by LINEs residing in the genome.

The most abundant SINE in humans is called *Alu,* so named because it contains a target site for the Alu restriction enzyme. The human genome contains more than 1 million whole and partial *Alu* sequences, scattered between genes and within introns. These *Alu* sequences make up more than 10 percent of the human genome. The full *Alu* sequence is about 200 nucleotides long and bears remarkable resemblance to 7SL RNA, an RNA that is part of a complex by which newly synthesized polypeptides are secreted through the endoplasmic reticulum. Presumably, the *Alu* sequences originated as reverse transcripts of these RNA molecules.

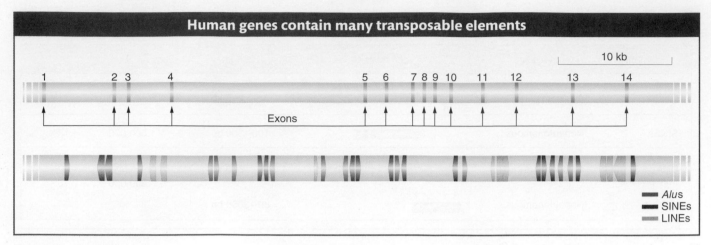

Human genes contain many transposable elements

Figure 15-22 Numerous repetitive elements are found in the human gene (*HGO*) encoding homogentisate 1,2-dioxygenase, the enzyme whose deficiency causes alkaptonuria. The upper row diagrams the positions of the *HGO* exons. The locations of *Alu* (blue), other SINEs (purple), and LINEs (yellow) in the *HGO* sequence are indicated in the lower row. [*After B. Granadino, D. Beltrán-Valero de Bernabé, J. M. Fernández-Cañón, M. A. Peñalva, and S. Rodríguez de Córdoba, "The Human Homogentisate 1,2-dioxygenase (HGO) Gene," Genomics 43, 1997, 115.*]

There is about 20 times as much DNA in the human genome derived from transposable elements as there is DNA encoding all human proteins. Figure 15-22 illustrates the number and diversity of transposable elements present in the human genome, using as an example the positions of individual *Alu*s, other SINEs, and LINEs in the vicinity of a typical human gene.

The human genome seems to be typical for a multicellular organism in the abundance and distribution of transposable elements. Thus, an obvious question is, How do plants and animals survive and thrive with so many insertions in genes and so much mobile DNA in the genome? First, with regard to gene function, all of the elements shown in Figure 15-22 are inserted into introns. Thus, the mRNA produced by this gene will not include any sequences from transposable elements, because they will have been spliced out of the pre-mRNA with the surrounding intron. Presumably, transposable elements insert into both exons and introns, but only the insertions into introns will remain in the population because they are less likely to cause a deleterious mutation. Insertions into exons are said to be subjected to **negative selection.** Second, humans, as well as all other multicellular organisms, can survive with so much mobile DNA in the genome because the vast majority is inactive and cannot move or increase in copy number. Most transposable-element sequences in a genome are relics that have accumulated inactivating mutations through evolutionary time. Others are still capable of movement but are rendered inactive by host regulatory mechanisms (see Section 15.5). There are, however, a few active LINEs and *Alu*s that have managed to escape host control and have inserted into important genes, causing several human diseases. Three separate insertions of LINEs have disrupted the factor VIII gene, causing hemophilia A. At least 11 *Alu* insertions into human genes have been shown to cause several diseases, including hemophilia B (in the factor IX gene), neurofibromatosis (in the *NF1* gene), and breast cancer (in the *BRCA2* gene).

The overall frequency of spontaneous mutation due to the insertion of class 2 elements in humans is quite low, accounting for less than 0.2 percent (1 in 500) of all characterized spontaneous mutations. Surprisingly, retrotransposon insertions account for about 10 percent of spontaneous mutations in another mammal, the mouse. The approximately 50-fold increase in this type of mutation in the mouse most likely corresponds to the much higher activity of these elements in the mouse genome than in the human genome.

Message Transposable elements compose the largest fraction of the human genome, with LINEs and SINEs being the most abundant. The vast majority of transposable elements are ancient relics that can no longer move or increase their copy number. A few elements remain active, and their movement into genes can cause disease.

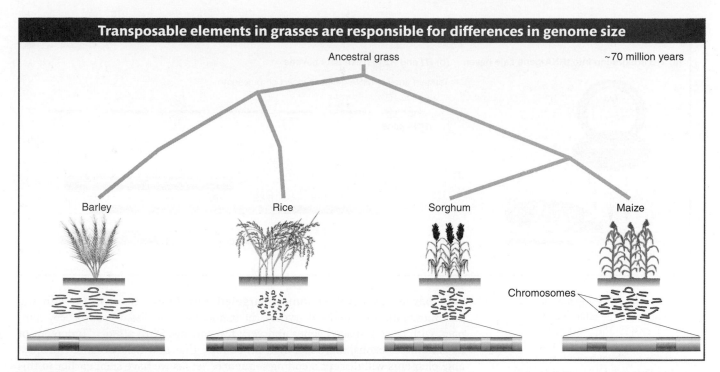

Figure 15-23 The grasses, including barley, rice, sorghum, and maize, diverged from a common ancestor about 70 million years ago. Since that time, the transposable elements have accumulated to different levels in each species. Chromosomes are larger in maize and barley, whose genomes contain large amounts of LTR retrotransposons. Green in the partial genome at the bottom represents a cluster of transposons, whereas orange represents genes.

The grasses: LTR retrotransposons thrive in large genomes

As already mentioned, the *C*-value paradox is the lack of correlation between genome size and biological complexity. How can organisms have very similar gene content but differ dramatically in the size of their genomes? This situation has been investigated in the cereal grasses. Differences in the genome sizes of these grasses have been shown to correlate primarily with the number of one class of elements, the LTR retrotransposons. The cereal grasses are evolutionary relatives that have arisen from a common ancestor in the past 70 million years. As such, their genomes are still very similar with respect to gene content and organization (called **synteny;** see Chapter 14), and regions can be compared directly. These comparisons reveal that linked genes in the small rice genome are physically closer together than are the same genes in the larger maize and barley genomes. In the maize and barley genomes, genes are separated by large clusters of retrotransposons (Figure 15-23).

Safe havens

The abundance of transposable elements in the genomes of multicellular organisms led some investigators to postulate that successful transposable elements (those that are able to attain very high copy numbers) have evolved mechanisms to prevent harm to their hosts by not inserting into host genes. Instead, successful transposable elements insert into so-called **safe havens** in the genome. For the grasses, a safe haven for new insertions appears to be into other retrotransposons. Another safe haven is the heterochromatin of centromeres, where there are very few genes but lots of repetitive DNA (see Chapter 12 for more on heterochromatin). Many classes of transposable elements in both plant and animal species tend to insert into the centric heterochromatin.

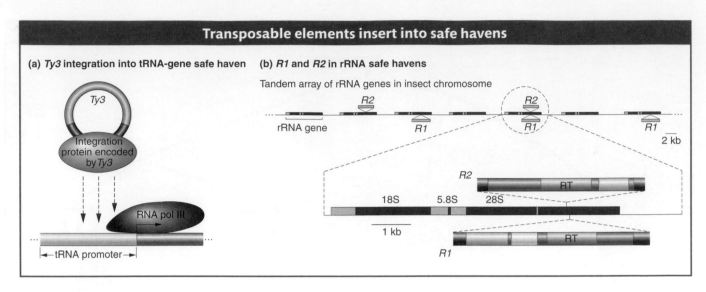

Figure 15-24 Some transposable elements are targeted to specific safe havens. (a) The yeast *Ty3* retrotransposon inserts into the promoter region of transfer RNA genes. (b) The *Drosophila R1* and *R2* non-LTR retrotransposons (LINEs) insert into the genes encoding ribosomal RNA that are found in long tandem arrays on the chromosome. Only the reverse transcriptase (RT) genes of *R1* and *R2* are noted. [(a) Inspired by D. F. Voytas and J. D. Boeke, "Ty1 and Ty5 of Saccharomyces cerevisiae," in N. L. Craig et al., eds., Mobile DNA II, Chap. 26, Fig. 15, p. 652. ASM Press, 2002. (b) After T. H. Eickbush, "R2 and Related Site-Specific Non-Long Terminal Inverted Repeat Retrotransposons," in N. L. Craig et al., eds., Mobile DNA II, Chap. 34, Fig. 1, p. 814. ASM Press, 2002.]

Safe havens in small genomes: targeted insertions In contrast with the genomes of multicellular eukaryotes, the genome of unicellular yeast is very compact, with closely spaced genes and very few introns. With almost 70 percent of its genome as exons, there is a high probability that new insertions of transposable elements will disrupt a coding sequence. Yet, as we have seen earlier in this chapter, the yeast genome supports a collection of LTR-retrotransposons called *Ty* elements.

How are transposable elements able to spread to new sites in genomes with few safe havens? Investigators have identified hundreds of *Ty* elements in the sequenced yeast genome and have determined that they are not randomly distributed. Instead, each family of *Ty* elements inserts into a particular genomic region. For example, the *Ty3* family inserts almost exclusively near but not in tRNA genes, at sites where they do not interfere with the production of tRNAs and, presumably, do not harm their hosts. *Ty* elements have evolved a mechanism that allows them to insert into particular regions of the genome: *Ty* proteins necessary for integration interact with specific yeast proteins bound to genomic DNA. *Ty3* proteins, for example, recognize and bind to subunits of the RNA polymerase complex that have assembled at tRNA promoters (Figure 15-24a).

The ability of some transposons to insert preferentially into certain sequences or genomic regions is called **targeting.** A remarkable example of targeting is illustrated by the *R1* and *R2* elements of arthropods, including *Drosophila. R1* and *R2* are LINEs (see Figure 15-21) that insert only into the genes that produce ribosomal RNA. In arthropods, several hundred rRNA genes are organized in tandem arrays (Figure 15-24b). With so many genes encoding the same product, the host tolerates insertion into a subset. However, too many insertions of *R1* and *R2* have been shown to decrease insect viability, presumably by interfering with ribosome assembly.

Gene therapy revisited This chapter began with a description of a recessive genetic disorder called SCID (severe combined immunodeficiency disease). The immune systems of persons afflicted with SCID are severely compromised owing to a mutation in a gene encoding the enzyme adenosine deaminase. To correct this genetic defect, bone-marrow cells from SCID patients were collected and treated with a retrovirus vector containing a good ADA gene. The transformed cells were then infused back into the patients. The immune systems of most of the patients showed significant improvement. However, the therapy had a very serious side effect: two of the patients developed leukemia. In both patients, the retroviral vector had inserted (integrated) near a cellular gene whose aberrant expression is associated with leukemia. A likely scenario is that insertion of the retroviral vector near the cellular gene altered its expression and, indirectly, caused the leukemia.

Clearly, the serious risk associated with this form of gene therapy might be greatly improved if doctors were able to control where the retroviral vector integrates into the human genome. We have already seen that there are many similarities between LTR retrotransposons and retroviruses. It is hoped that, by understanding *Ty* targeting in yeast, we can learn how to construct retroviral vectors that insert themselves and their transgene cargo into safe havens in the human genome.

> **Message** A successful transposable element increases copy number without harming its host. One way in which an element safely increases copy number is to target new insertions into safe havens, regions of the genome where there are few genes.

15.5 Epigenetic Regulation of Transposable Elements by the Host

Genetic analysis is a very powerful tool used to dissect complex biological processes through the isolation of mutants and, ultimately, mutant genes (for example, see Chapter 13 on development). Many laboratories around the world are using genetic analyses to identify the host genes responsible for repressing the movement of transposable elements and, in this way, maintaining the stability of the genome.

The repression of transposable elements was first investigated in the laboratory of Ron Plasterk in the late 1990s, using the model organism *C. elegans* (a nematode; see the Model Organism box in Chapter 13). This story starts with the observation of a striking difference between the mobility of a transposable element called *Tc*1 in two different cell types of the model organism. *Tc*1 is a DNA transposon that, like the *Ac* element of maize, can lead to an unstable mutant phenotype when it excises from a gene with a visible phenotype (see Figure 15-4). There are 32 *Tc*1 elements in the sequenced genome of the common laboratory strain called Bristol. Significantly, *Tc*1 transposes in somatic but not germ-line cells. That observation suggested to Plasterk that transposition is repressed in the germ line by the host. Evidently, germ-line repression results from the silencing of the transposase genes of all 32 *Tc*1 copies in germ-line cells. Can you propose an explanation for why it makes biological sense for a host to repress transposition in the germ line and not in the soma?

Plasterk and his co-workers set out to identify *C. elegans* genes responsible for silencing the transposase gene. They began with a *C. elegans* strain that had *Tc*1 inserted in the *unc-22* gene (designated *unc-22/Tc*1; Figure 15-25). This was the same gene that was silenced in the experiment of Fire and Mello that led to their sharing the Nobel Prize (see Chapter 8). Whereas wild-type *C. elegans* glides smoothly on the surface of the agar in a petri dish, worms with the mutant *unc-22/Tc*1 gene have a twitching movement that can be easily observed with a microscope. Because *Tc*1 cannot normally transpose in the germ line, it remains inserted in the *unc-22*

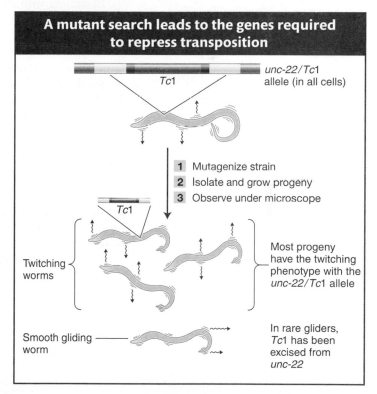

A mutant search leads to the genes required to repress transposition

*unc-22/Tc*1 allele (in all cells)

*Tc*1

1 Mutagenize strain
2 Isolate and grow progeny
3 Observe under microscope

*Tc*1

Twitching worms

Most progeny have the twitching phenotype with the *unc-22/Tc*1 allele

Smooth gliding worm

In rare gliders, *Tc*1 has been excised from *unc-22*

Figure 15-25 Experimental design used to identify genes required to repress transposition. Investigators look for mutants that have regained normal movement, because mutations in these individuals would have disabled the repression mechanism that prevents the transposition of the *Tc*1 element from the *unc-22* gene.

A single *Tc*1 element can repress transposition

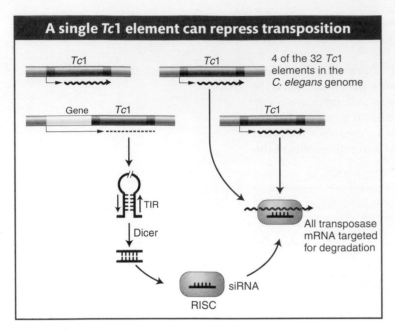

4 of the 32 *Tc*1 elements in the *C. elegans* genome

All transposase mRNA targeted for degradation

Figure 15-26 The production of dsRNA from only a single *Tc*1 element is sufficient to silence all of the *Tc*1 transposase genes and thereby repress transposition in the germ line. The siRNA derived from *Tc*1 dsRNA is bound to RISC and targets all complementary RNA for degradation.

gene and continues to disrupt its function. Thus, the strain with the mutant *unc-22/Tc*1 gene should express a twitching phenotype from generation to generation. However, Plasterk and co-workers reasoned that mutations that inactivated *C. elegans* genes required for repression would allow *Tc*1 to excise from the *unc-22/Tc*1 allele in the germ line and revert the twitching phenotype to wild type (*unc-22*). To this end, they exposed the mutant *unc-22/Tc*1 strain to a chemical that greatly increased the frequency of mutation (called a mutagen; see Chapter 16) and examined their progeny under a microscope, searching for rare worms that no longer twitched.

This and subsequent genetic screens identified over 25 *C. elegans* genes that, when mutated, allowed the host to excise *Tc*1 in the germ line. Significantly, many of the products of these genes are integral components of the RNAi silencing pathway, including proteins found in Dicer and RISC (see Chapters 8 and 12). Recall from Chapter 8 that Dicer binds to long dsRNAs and cleaves them into small dsRNA fragments. These fragments are then unwound so that one strand, the siRNA, can target RISC to chop up complementary mRNAs (see Figure 8-24).

Beginning with this elegant genetic screen, many years of experimentation have led to the following model for the repression of transposable elements in the germ line of *C. elegans*. With 32 *Tc*1 elements scattered throughout the *C. elegans* genome, a few elements near genes are transcribed along with the nearby gene by "read-through" transcription (see Chapter 8). Because the ends of *Tc*1 are 54-bp terminal inverted repeats, the *Tc*1 RNA spontaneously forms dsRNA (Figure 15-26). Like all dsRNAs produced in most eukaryotes, this RNA is recognized by Dicer and ultimately siRNA is produced, which directs RISC to chop up complementary *Tc*1 transcripts. Because all *Tc*1 RNA is efficiently chopped up in the germ line, the element-encoded transposase gene is silenced. Without transposase, the element cannot excise. It has been hypothesized that *Tc*1 can transpose in somatic cells because RNAi is not as efficient and some transposase can be produced.

Over the past decade, numerous laboratories working with both plants and animals have discovered that mutations that disrupt the RNAi pathway often lead to the activation of transposable elements in their respective genomes. Because of the abundance of transposable elements in eukaryotic genomes, it has been suggested that the natural function of the RNAi pathway is to maintain genome stability by repressing the movement of transposable elements.

> **Message** Eukaryotic hosts use RNAi to repress the expression of active transposable elements in their genomes. In this way, a single element that inserts near a gene can be transcribed to produce dsRNA that will trigger the silencing of all copies of the element in the genome.

What Geneticists Are Doing Today

The RNAi silencing pathway is akin to radar in that the host is able to detect new insertions into the genome. The host can then respond by silencing the transposase gene and thus prevent the movement of all family members. However, much like planes that evade radar by flying close to the ground, some transposons have evolved mechanisms that allow them to evade the RNAi silencing pathway. These transposons can attain very high copy numbers. Evidence for these mechanisms can be found in the genomes of all characterized plants and animals containing transposon families (such as *Alu* in humans) that have thousands of members.

How do some transposons avoid detection by the RNAi silencing pathway? The short answer is that in most cases, we do not know. To understand how a transposon avoids detection, it is necessary to study actively transposing elements. To date, scientists have detected very few transposon families with high copy numbers that are still actively transposing. One of the best-characterized elements among this small group is a special type of nonautonomous DNA transposon called a **miniature inverted repeat transposable element** (abbreviated **MITE**). Like other nonautonomous elements, MITEs can form by deletion of the transposase gene from an autonomous element. However, unlike most nonautonomous elements, MITEs can attain very high copy numbers, particularly in the genomes of some grasses (see Figure 15-23). Some MITEs in grasses have been amplified to thousands of copies.

The only actively transposing MITE isolated to date is the *mPing* element of rice, which is formed from the autonomous *Ping* element by deletion of the entire transposase gene (Figure 15-27). This element was discovered in the laboratory of Susan Wessler by Ning Jiang. Another member of the Wessler laboratory, Ken Naito, documented that in individuals of some rice strains there are only 3 to 7 copies of *Ping* and over 1000 copies of *mPing*. Remarkably, the copy number of *mPing* in these strains is increasing by almost 40 new insertions per plant per generation.

Two questions about the rapid increase in *mPing* copy number immediately come to mind. First, how does a rice strain survive a transposable-element burst of this magnitude? To address this question, the Wessler laboratory used next-generation sequencing technology (see Chapter 14) to determine the insertion sites of over 1700 *mPing* elements in the rice genome. Surprisingly, they found that the element avoided inserting into exons, thus minimizing the impact of insertion on rice gene expression. The mechanism underlying this preference is currently being investigated.

The second question is, why does the rice host apparently fail to repress *mPing* transposition? While this question is also an active area of current research, a reasonable hypothesis is that *mPing* can fly under the hosts' RNAi radar because it does not contain any part of the transposase gene that resides on the *Ping* element (Figure 15-27). Thus, read-through transcription into *mPing* elements inserted throughout the rice genome will produce lots of dsRNA and siRNA. However, because siRNAs derived from *mPing* share no sequence with the source of transposase, siRNAs produced from *mPing* will not induce silencing mechanisms aimed at transposase. Instead, the transposase gene will remain active and will continue to catalyze the movement of *mPing*. According to this hypothesis, *mPing* transposition will be repressed only when a much rarer *Ping* insertion generates dsRNA that triggers the silencing of its transposase gene.

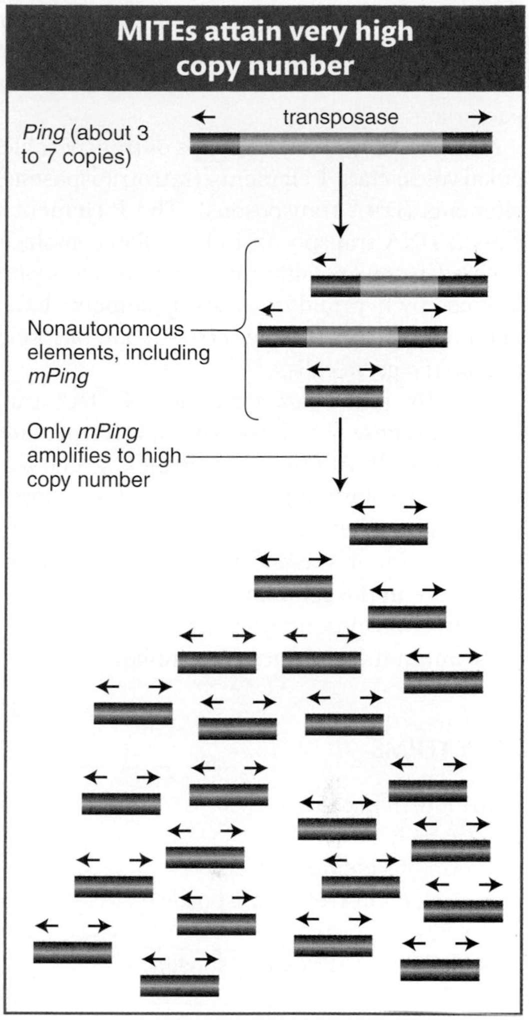

> **Message** MITEs are nonautonomous DNA transposons that can attain high copy numbers. While MITEs can utilize the transposase of autonomous elements, they probably evade host repression because their amplification does not lead to the silencing of the transposase gene.

MITEs attain very high copy number

Figure 15-27 MITEs are nonautonomous DNA transposons that can attain very high copy number because they do not encode the transposase necessary for their transposition. The active MITE *mPing* is the only deletion derivative of the autonomous *Ping* element that has attained high copy number in certain strains of rice.

Summary

Transposable elements were discovered in maize by Barbara McClintock as the cause of several unstable mutations. An example of a nonautonomous element is *Ds*, the transposition of which requires the presence of the autonomous *Ac* element in the genome.

Bacterial insertion-sequence elements were the first transposable elements isolated molecularly. There are many different types of IS elements in *E. coli* strains, and they are usually present in at least several copies. Composite transposons contain IS elements flanking one or more genes, such as genes conferring resistance to antibiotics. Transposons with resistance genes can insert into plasmids and are then transferred by conjugation to nonresistant bacteria.

There are two major groups of transposable elements in eukaryotes: class 1 elements (retrotransposons) and class 2 elements (DNA transposons). The *P* element was the first class 2 DNA transposon to be isolated molecularly. It was isolated from unstable mutations in *Drosophila* that were induced by hybrid dysgenesis. *P* elements have been developed into vectors for the introduction of foreign DNA into *Drosophila* germ cells.

Ac, Ds, and *P* are examples of DNA transposons, so named because the transposition intermediate is the DNA element itself. Autonomous elements such as *Ac* encode a transposase that binds to the ends of autonomous and nonautonomous elements and catalyzes excision of the element from the donor site and reinsertion into a new target site elsewhere in the genome.

Retrotransposons were first molecularly isolated from yeast mutants, and their resemblance to retroviruses was immediately apparent. Retrotransposons are class 1 elements, as are all transposable elements that use RNA as their transposition intermediate.

The active transposable elements isolated from such model organisms as yeast, *Drosophila, E. coli,* and maize constitute a very small fraction of all the transposable elements in the genome. DNA sequencing of whole genomes, including the human genome, has led to the remarkable finding that almost half of the human genome is derived from transposable elements. Despite having so many transposable elements, eukaryotic genomes are extremely stable as transposition is relatively rare because of two factors. First, most of the transposable elements in eukaryotic genomes cannot move because inactivating mutations prevent the production of normal transposase and reverse transcriptase. Second, expression of the vast majority of the remaining elements is silenced by the host RNAi pathway. Some high-copy-number elements, such as MITEs, may evade silencing because they do not trigger the silencing of the transposase that catalyzes their transposition.

KEY TERMS

Activator (*Ac*) (p. 524)

Alu (p. 543)

autonomous element (p. 527)

class 1 element (retrotransposon) (p. 536)

class 2 element (DNA transposon) (p. 537)

cointegrate (p. 532)

composite transposon (p. 530)

conservative transposition (p. 531)

copia-like element (p. 536)

"copy and paste" (p. 536)

"cut and paste" (p. 531)

C-value (p. 542)

C-value paradox (p. 542)

Dissociation (*Ds*) (p. 524)

DNA transposon (p. 537)

excise (p. 527)

gene therapy (p. 523)

hybrid dysgenesis (p. 537)

insertion-sequence (IS) element (p. 529)

inverted repeat (p. 530)

long interspersed element (LINE) (p. 543)

long terminal repeat (LTR) (p. 534)

LTR-retrotransposon (p. 536)

M cytotype (p. 538)

miniature inverted repeat transposable element (MITE) (p. 549)

negative selection (p. 544)

nonautonomous element (p. 527)

P cytotype (p. 538)

P element (p. 537)

provirus (p. 534)

replicative transposition (p. 531)

retrotransposon (p. 536)

retrovirus (p. 530, 534)

reverse transcriptase (p. 534)

R factor (p. 530)

safe haven (p. 545)

short interspersed element (SINE) (p. 543)

simple transposon (p. 530)

solo LTR (p. 537)

synteny (p. 545)

targeting (p. 546)

target-site duplication (p. 533)

transposase (p. 529)

transpose (p. 527)

transposition (p. 530)

transposon (Tn) (p. 530)

transposon tagging (p. 540)

Ty element (p. 533)

unstable phenotype (p. 527)

SOLVED PROBLEMS

SOLVED PROBLEM 1. Transposable elements have been referred to as "jumping genes" because they appear to jump from one position to another, leaving the old locus and appearing at a new locus. In light of what we now know concerning the mechanism of transposition, how appropriate is the term "jumping genes" for bacterial transposable elements?

Solution

In bacteria, transposition takes place by two different modes. The conservative mode results in true jumping genes, because, in this case, the transposable element excises from its original position and inserts at a new position. The other mode is the replicative mode. In this pathway, a transposable element moves to a new location by replicating into the target DNA, leaving behind a copy of the transposable element at the original site. When operating by the replicative mode, transposable elements are not really jumping genes, because a copy does remain at the original site.

SOLVED PROBLEM 2. Following from the question above, in light of what we now know concerning the mechanism of transposition, how appropriate is the term "jumping genes" for the vast majority of transposable elements in the human genome and in the genomes of most other mammals?

Solution

The vast majority of transposable elements in the char-acterized mammalian genomes are retrotransposons. In humans, two retrotransposons (the LINE called *L1* and the SINE called *Alu*) account for a whopping one-third of our entire genome. Like bacterial elements, retrotrans-posons do not excise from the original site, so they are not really jumping genes. Instead, the element serves as a template for the transcription of RNAs that can be reverse-transcribed by the enzyme reverse transcriptase into double-stranded cDNA. Each cDNA can potentially insert into target sites throughout the genome. Note that while both bactcrial elements and retrotransposons do not leave the original site, their respective mechanisms of transposition are dramatically different. Finally, while LTR retrotransposons do not excise, they can become much shorter insertions due to the production of solo LTRs by recombination.

PROBLEMS

Most of the problems are also available for review/grading through the GENETICS PORTAL www.yourgeneticsportal.com.

WORKING WITH THE FIGURES

1. In the chapter-opening photograph of kernels on an ear of corn, what is the genetic basis of the following (**Hint:** Refer to Figure 15-4 for some clues):

   a. the fully pigmented kernel?

   b. the unpigmented kernels? Note that they can arise in two different ways.

2. In Figure 15-3a, what would the kernel phenotype be if the strain was homozygous for all dominant markers on chromosome 9?

3. For Figure 15-7, draw out a series of steps that could explain the origin of this large plasmid containing many transposable elements.

4. For Figure 15-8, draw a figure for the third mode of transposition, retrotransposition.

5. In Figure 15-10, show where the transposase would have to cut to generate a 6-bp target-site duplication. Also show the location of the cut to generate a 4-bp target-site duplication.

6. If the transposable element in Figure 15-14 were a DNA transposon that had an intron in its transposase gene, would the intron be removed following transposition? Justify your answer.

7. For Figure 15-22, draw the pre-mRNA that is tran-scribed from this gene and then draw its mRNA.

BASIC PROBLEMS

8. Describe the generation of multiple-drug-resistant plasmids.

9. Briefly describe the experiment that demonstrates that the transposition of the *Ty1* element in yeast takes place through an RNA intermediate.

10. Explain how the properties of *P* elements in *Drosophila* make gene-transfer experiments possible in this organism.

11. Although class 2 elements are abundant in the ge-nomes of multicellular eukaryotes, class 1 elements usually make up the largest fraction of very large ge-nomes such as those from humans (~2500 Mb), maize (~2500 Mb), and barley (~5000 Mb). Given what you know about class 1 and class 2 elements, what is it about their distinct mechanisms of transposition that would account for this consistent difference in abundance?

12. As you saw in Figure 15-22, the genes of multicellular eukaryotes often contain many transposable elements. Why do most of these elements not affect the expres-sion of the gene?

13. What are safe havens? Are there any places in the much more compact bacterial genomes that might be a safe haven for insertion elements?

14. Nobel prizes are usually awarded many years after the actual discovery. For example, James Watson, Francis Crick, and Maurice Wilkens were awarded the Nobel Prize in Medicine or Physiology in 1962, almost a de-cade after their discovery of the double-helical struc-ture of DNA. However, Barbara McClintock was award-ed the Nobel Prize in 1983, almost four decades after her discovery of transposable elements in maize. Why do you think it took this long?

CHALLENGING PROBLEMS

15. The insertion of transposable elements into genes can alter the normal pattern of expression. In the following situations, describe the possible consequences on gene expression.

a. A LINE inserts into an enhancer of a human gene.

b. A transposable element contains a binding site for a transcriptional repressor and inserts adjacent to a promoter.

c. An *Alu* element inserts into the 3′ splice (AG) site of an intron.

d. A *Ds* element that was inserted into the exon of a gene excises imperfectly and leaves 3 base pairs behind in the exon.

e. Another excision by that same *Ds* element leaves 2 base pairs behind in the exon.

f. A *Ds* element that was inserted into the middle of an intron excises imperfectly and leaves 5 base pairs behind in the intron.

16. Before the integration of a transposon, its transposase makes a staggered cut in the host target DNA. If the staggered cut is at the sites of the arrows below, draw what the sequence of the host DNA will be after the transposon has been inserted. Represent the transposon as a rectangle.

↓

AATTTGGCCTAGTACTAATTGGTTGG
TTAAACCGGATCATGATTAACCAACC

↑

17. In *Drosophila*, M. Green found a *singed* allele (*sn*) with some unusual characteristics. Females homozygous for this X-linked allele have singed bristles, but they have numerous patches of *sn*⁺(wild-type) bristles on their heads, thoraxes, and abdomens. When these flies are mated with *sn* males, some females give only singed progeny, but others give both singed and wild-type progeny in variable proportions. Explain these results.

18. Consider two maize plants:

a. Genotype *C*/*c*ᵐ ; *Ac*/*Ac*⁺, where *c*ᵐ is an unstable allele caused by *Ds* insertion

b. Genotype *C*/*c*ᵐ, where *c*ᵐ is an unstable allele caused by *Ac* insertion

What phenotypes would be produced and in what proportions when (1) each plant is crossed with a base-pair-substitution mutant *c*/*c* and (2) the plant in part *a* is crossed with the plant in part *b*? Assume that *Ac* and *c* are unlinked, that the chromosome-breakage frequency is negligible, and that mutant *c*/*C* is *Ac*⁺.

19. You meet your friend, a scientist, at the gym and she begins telling you about a mouse gene that she is studying in the lab. The product of this gene is an enzyme required to make the fur brown. The gene is called *FB* and the enzyme is called FB enzyme. When *FB* is mutant and cannot produce the FB enzyme, the fur is white.

The scientist tells you that she has isolated the gene from two mice with brown fur and that, surprisingly, she found that the two genes differ by the presence of a 250-bp SINE (like the human *Alu* element) in the *FB* gene of one mouse but not in the gene of the other. She does not understand how this difference is possible, especially given that she determined that both mice make the FB enzyme. Can you help her formulate a hypothesis that explains why the mouse can still produce FB enzyme with a transposable element in its *FB* gene?

20. The yeast genome has class 1 elements (*Ty1*, *Ty2*, and so forth) but no class 2 elements. Can you think of a possible reason why DNA elements have not been successful in the yeast genome?

21. In addition to *Tc1*, the *C. elegans* genome contains other families of DNA transposons such as *Tc2*, *Tc3*, *Tc4*, and *Tc5*. Like *Tc1*, their transposition is repressed in the germ line but not in somatic cells. Predict the behavior of these elements in the mutant strains where *Tc1* is no longer repressed due to mutations in the RNAi pathway. Justify your answer.

22. Based on the mechanism of gene silencing, what features of transposable elements does the RNAi pathway exploit to ensure that the host's own genes are not also silenced?

23. What are the similarities and differences between retroviruses and retrotransposons? It has been hypothesized that retroviruses evolved from retrotransposons. Do you agree with this model? Justify your answer.

24. You have isolated a transposable element from the human genome and have determined its DNA sequence. How would you use this sequence to determine the copy number of the element in the human genome if you just had a computer with an Internet connection? (**Hint:** see Chapter 14.)

25. Following up on the previous question, how would you determine whether other primates had a similar element in their genomes?

26. Of all the genes in the human genome, the ones with the most characterized *Alu* insertions are those that cause hemophilia, including several insertions in the factor VIII and factor IX genes. Based on this fact, your colleague hypothesizes that the *Alu* element prefers to insert into these genes. Do you agree? What other reason can you provide that also explains these data?

27. If all members of a transposable element family can be silenced by dsRNA synthesized from a single family member, how is it possible for one element family (like *Tc1*) to have 32 copies in the *C. elegans* genome while another family (*Tc2*) has fewer than 5 copies?

Mutation, Repair, and Recombination

16

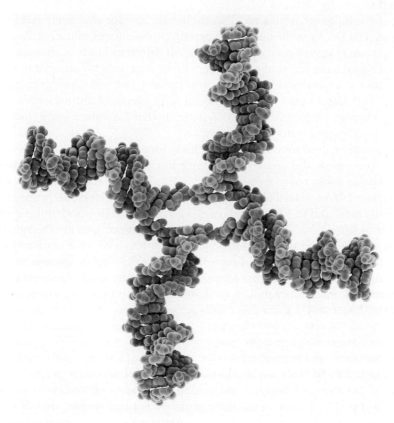

A molecular model of two chromosomes undergoing a crossover. [*Laguna Design/Science Photo Library.*]

A young patient develops a great many small, frecklelike, precancerous skin growths and is extremely sensitive to sunlight (Figure 16-1). A family history is taken, and the patient is diagnosed with an autosomal recessive disease called xeroderma pigmentosum. Throughout her life, she will be prone to developing pigmented skin cancers. Several different genes can be mutated to generate this disease phenotype. In a person without the disease, each of these genes contributes to the biochemical processes in the cell that respond to chemical damage to DNA and repair this damage before it leads to the formation of new mutations. Later in this chapter, we will see how mutations in the repair systems lead to genetic diseases such as xeroderma pigmentosum.

Persons with this disease are examples of genetic *variants*—individuals that show phenotypic differences in one or more particular characters. Because genetics is the study of inherited differences, genetic analysis would not be possible without variants. In preceding chapters you saw many analyses of the inheritance of such variants; now, we consider their origin. How do genetic variants arise?

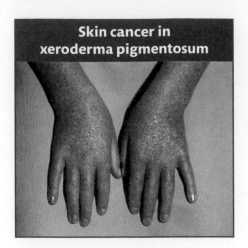

Figure 16-1 The recessive hereditary disease xeroderma pigmentosum is caused by a deficiency in an enzyme that helps correct damaged DNA. This enzyme deficiency leads to the formation of skin cancers on exposure of the skin to ultraviolet rays in sunlight. [*KOKEL/BSIP/SuperStock.*]

Two major processes are responsible for genetic variation: *mutation* and *recombination*. We have seen that mutation is a change in the DNA sequence of a gene. Mutation is especially significant because it is the ultimate source of evolutionary change; new alleles arise in all organisms, some spontaneously and others resulting from exposure to radiation or chemicals in the environment. The new alleles produced by mutation become the raw material for a second level of variation, effected by recombination. As its name suggests, recombination is the outcome of cellular processes that cause alleles of different genes to become grouped in new combinations (see Chapter 4). To use an analogy, mutation occasionally produces a new playing card, but it is recombination that shuffles the cards and deals them out as different hands.

In the cellular environment, DNA molecules are not absolutely stable: each base pair in a DNA double helix has a certain probability of mutating. As we will see, the term *mutation* covers a broad array of different kinds of changes. These changes range from the simple swapping of one base pair for another to the disappearance of an entire chromosome. In Chapter 17, we will consider mutational changes that affect entire chromosomes or large pieces of chromosomes. In the present chapter, we focus on mutational events that take place *within* individual genes. We call such events *gene mutations*.

Cells have evolved sophisticated systems to identify and repair damaged DNA, thereby preventing the occurrence of most but not all mutations. We can view DNA as being subjected to a dynamic tug-of-war between the chemical processes that damage DNA and lead to new mutations and the cellular repair processes that constantly monitor DNA for such damage and correct it. However, this tug-of-war is not straightforward. As already mentioned, mutations provide the raw material for evolution and thus the introduction of a low level of mutation must be tolerated. We will see that DNA-replication and DNA-repair systems can actually introduce mutations. Others turn potentially devastating mutations (such as double-strand breaks) into mutations that may affect only a single gene product.

We will see that the most potentially serious class of DNA damage, a double-strand break, is also an intermediate step in the normal cellular process of recombination through meiotic crossing over. Thus, we can draw parallels between mutation and recombination at two levels. First, as mentioned earlier, mutation and recombination are the major sources of variation. Second, mechanisms of DNA repair and recombination have some features in common, including the use of some of the same proteins. For this reason, we will explore mechanisms of DNA repair first and then compare them with the mechanism of DNA recombination.

16.1 The Phenotypic Consequences of DNA Mutations

The term **point mutation** typically refers to the alteration of a single base pair of DNA or of a small number of adjacent base pairs. In this section, we will consider the effects of such changes at the phenotypic level. Point mutations are classified in molecular terms in Figure 16-2, which shows the main types of DNA changes and their effects on protein function when they occur within the protein-coding region of a gene.

Types of point mutation

The two main types of point mutation in DNA are *base substitutions* and *base insertions* or *deletions*. Base substitutions are mutations in which one base pair is replaced by another. Base substitutions can be divided into two subtypes: transitions and transversions. To describe these subtypes, we consider how a mutation

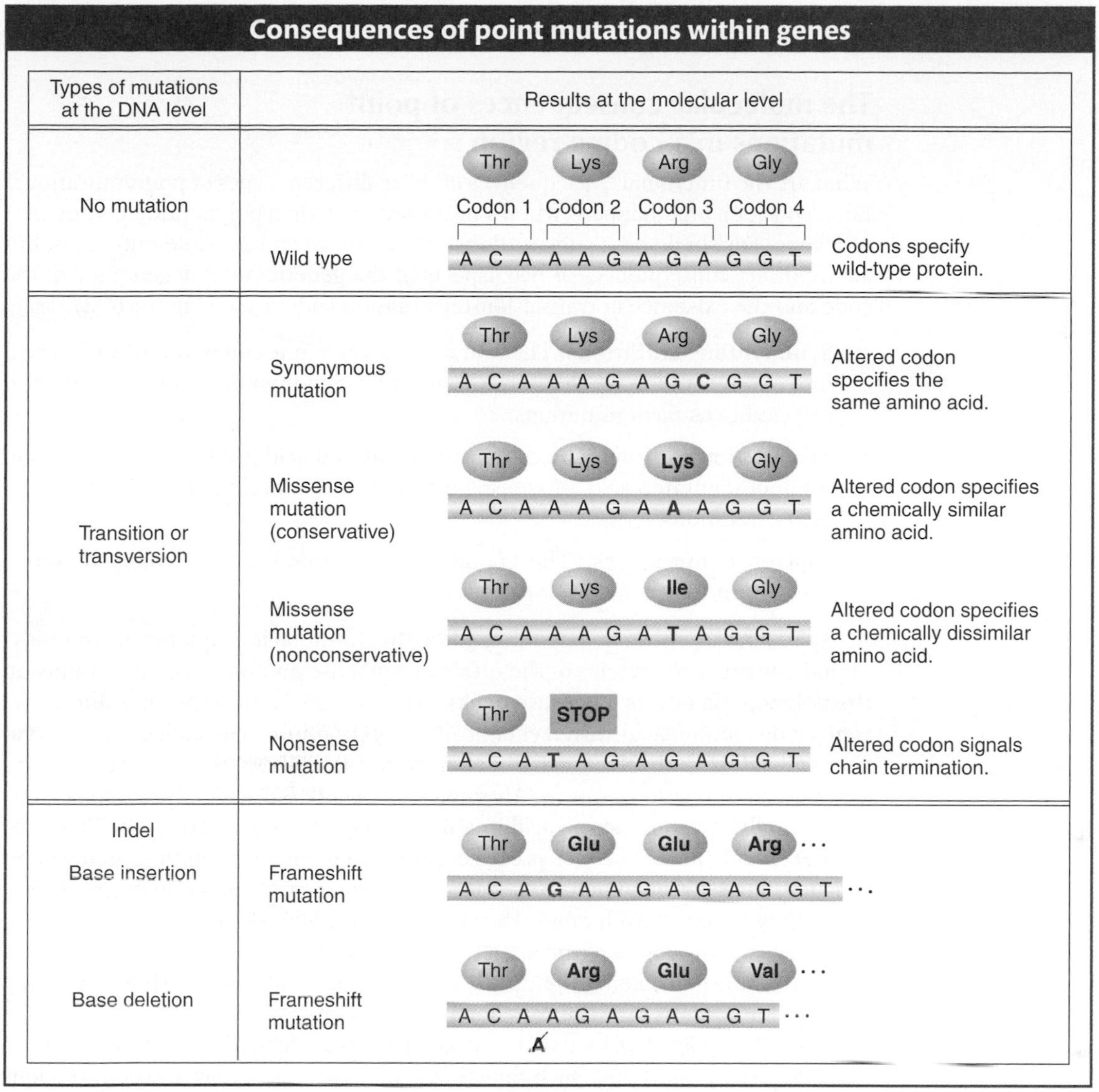

Consequences of point mutations within genes

Figure 16-2 Point mutations within the coding region of a gene vary in their effects on protein function. Proteins with synonymous and missense mutations are usually still functional.

alters the sequence on one DNA strand (the complementary change will take place on the other strand). A **transition** is the replacement of a base by the other base of the same chemical category. Either a purine is replaced by a purine (from A to G or from G to A) or a pyrimidine is replaced by a pyrimidine (from C to T or from T to C). A **transversion** is the opposite—the replacement of a base of one chemical category by a base of the other. Either a pyrimidine is replaced by a purine (from C to A, C to G, T to A, or T to G) or a purine is replaced by a pyrimidine (from A to C, A to T, G to C, or G to T). In describing the same changes at the double-stranded level of DNA, we must represent both members of a base pair in the same relative location. Thus, an example of a transition is $G \cdot C \rightarrow A \cdot T$; that of a transversion is $G \cdot C \rightarrow T \cdot A$. Insertion or deletion mutations are actually insertions or deletions of *nucleotide* pairs; nevertheless, the convention is to call them *base*-pair insertions or deletions. Collectively, they are termed *indel mutations* (for *in*sertion-*del*etion). The simplest of these mutations is the addition or deletion of a single base pair. Mutations sometimes arise through the simultaneous addition or deletion of multiple base pairs at once. As we will see later in this chapter, mechanisms that selectively produce certain kinds of multiple-base-pair additions or deletions are the cause of certain human genetic diseases.

The molecular consequences of point mutations in a coding region

What are the functional consequences of these different types of point mutations? First, consider what happens when a mutation arises in a polypeptide-coding part of a gene. For single-base substitutions, there are several possible outcomes, but all are direct consequences of two aspects of the genetic code: degeneracy of the code and the existence of translation-termination codons (see Figure 16-2).

- **Synonymous mutations.** The mutation changes one codon for an amino acid into another codon for that same amino acid. Synonymous mutations are also referred to as *silent* mutations.

- **Missense mutations.** The codon for one amino acid is changed into a codon for another amino acid. Missense mutations are sometimes called *nonsynonymous* mutations.

- **Nonsense mutations.** The codon for one amino acid is changed into a translation-termination (stop) codon.

Synonymous substitutions never alter the amino acid sequence of the polypeptide chain. The severity of the effect of missense and nonsense mutations on the polypeptide differs from case to case. For example, a missense mutation may replace one amino acid with a chemically similar amino acid, called a **conservative substitution.** In this case, the alteration is less likely to affect the protein's structure and function severely. Alternatively, one amino acid may be replaced by a chemically different amino acid in a **nonconservative substitution.** This type of alteration is more likely to produce a severe change in protein structure and function. Nonsense mutations will lead to the premature termination of translation. Thus, they have a considerable effect on protein function. The closer a nonsense mutation is to the 3′ end of the open reading frame (ORF), the more plausible it is that the resulting protein might possess some biological activity. However, many nonsense mutations produce completely inactive protein products.

Single-base-pair changes that inactivate proteins are often due to splice site mutations. As seen in Figure 16-3, such changes can dramatically change the coding regions by leading to large insertions or deletions that may or may not be in frame.

Like nonsense mutations, indel mutations may have consequences on polypeptide sequence that extend far beyond the site of the mutation itself (see Figure 16-2). Recall that the sequence of mRNA is "read" by the translational apparatus in register ("in frame"), three bases (one codon) at a time. The addition or deletion of a single base pair of DNA changes the reading frame for the remain-

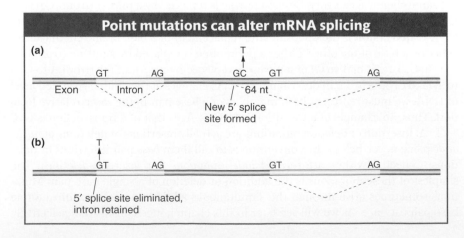

Figure 16-3 Two examples show the consequences of point mutations at splice sites. (a) A C to T transition mutation leads to a GT dinucleotide in the exon, forming a new 5′ splice site. As a result, 64 nucleotides at the end of an exon are spliced out. (b) A G to T transversion mutation would eliminate the 5′ splice site so the intron would be retained in the mRNA.

der of the translation process, from the site of the base-pair mutation to the next stop codon in the new reading frame. Hence, these lesions are called **frameshift mutations.** These mutations cause the entire amino acid sequence translationally downstream of the mutant site to bear no relation to the original amino acid sequence. Thus, frameshift mutations typically result in complete loss of normal protein structure and function.

The molecular consequences of point mutations in a noncoding region

Now let's turn to mutations that occur in regulatory and other noncoding sequences. Those parts of a gene that do not directly encode a protein contain many crucial DNA binding sites for proteins interspersed among sequences that are nonessential to gene expression or gene activity. At the DNA level, the binding sites include the sites to which RNA polymerase and its associated factors bind, as well as sites to which specific transcription-regulating proteins bind. At the RNA level, additional important binding sites include the ribosome-binding sites of bacterial mRNAs, the 5′ and 3′ splice sites for exon joining in eukaryotic mRNAs, and sites that regulate translation and localize the mRNA to particular areas and compartments within the cell.

The ramifications of mutations in parts of a gene other than the polypeptide-coding segments are harder to predict than are those of mutations in coding segments. In general, the functional consequences of any point mutation in such a region depend on whether the mutation disrupts (or creates) a binding site. Mutations that disrupt these sites have the potential to change the expression pattern of a gene by altering the amount of product expressed at a certain time or in a certain tissue or by altering the response to certain environmental cues. Such regulatory mutations will alter the amount of the protein product produced but *not* the structure of the protein. Alternatively, some binding-site mutations might completely obliterate a required step in normal gene expression (such as the binding of RNA polymerase or splicing factors) and hence totally inactivate the gene product or block its formation. Figure 16-4 shows some examples of how different types of mutations affect mRNA and protein.

Figure 16-4 Point mutations in coding regions can alter protein structure with or without altering mRNA size. Point mutations in regulatory regions can prevent the synthesis of mRNA (and protein).

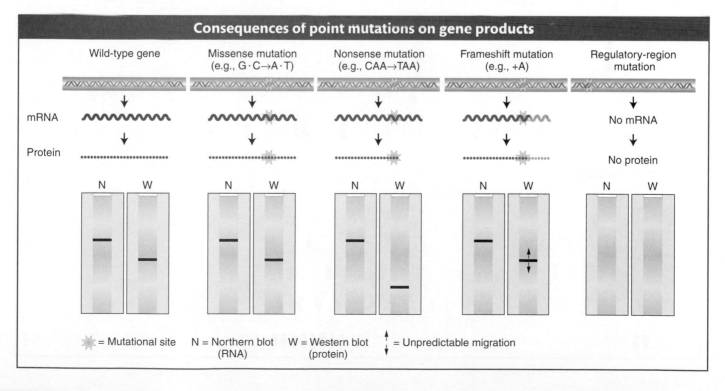

Consequences of point mutations on gene products

It is important to keep in mind the distinction between the occurrence of a gene mutation—that is, a change in the DNA sequence of a given gene—and the detection of such an event at the phenotypic level. Many point mutations within noncoding sequences elicit little or no phenotypic change; these mutations are located between DNA binding sites for regulatory proteins. Such sites may be functionally irrelevant, or other sites within the gene may duplicate their function.

16.2 The Molecular Basis of Spontaneous Mutations

Gene mutations can arise spontaneously or they can be induced. **Spontaneous mutations** are naturally occurring mutations and arise in all cells. **Induced mutations** arise through the action of certain agents, called **mutagens,** that increase the rate at which mutations occur. In this section, we consider the nature of spontaneous mutations.

Luria and Delbrück fluctuation test

The origin of spontaneous hereditary change has always been a topic of considerable interest. Among the first questions asked by geneticists was, Do spontaneous mutations occur in response to the selecting agent or are variants present at a low frequency in most populations? An ideal experimental system to address this important question was the analysis of mutations in bacteria that confer resistance to specific environmental agents not normally tolerated by wild-type cells.

One experiment by Salvador Luria and Max Delbrück in 1943 was particularly influential in shaping our understanding of the nature of mutation,

Figure 16-5 These cell pedigrees illustrate the expectations from two contrasting hypotheses about the origin of resistant cells. [*From G. S. Stent and R. Calendar, Molecular Genetics, 2nd ed. W. H. Freeman and Company, 1978.*]

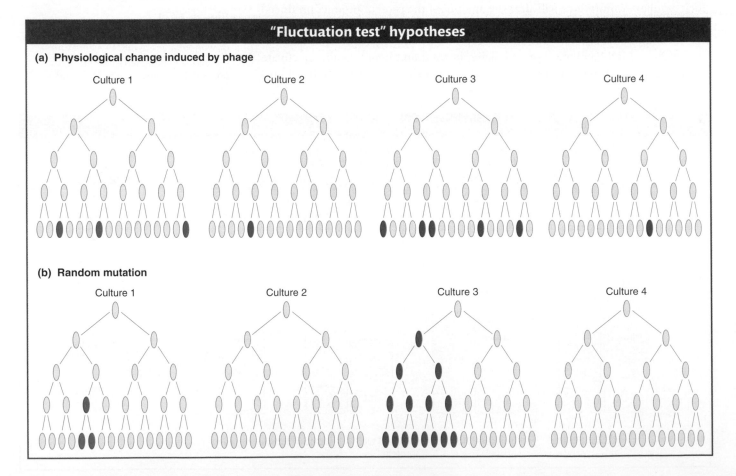

"Fluctuation test" hypotheses

(a) Physiological change induced by phage

Culture 1   Culture 2   Culture 3   Culture 4

(b) Random mutation

Culture 1   Culture 2   Culture 3   Culture 4

not only in bacteria, but in organisms generally. It was known at the time that, if *E. coli* bacteria are spread on a plate of nutrient medium in the presence of phage T1, the phages soon infect and kill the bacteria. However, rarely but regularly, colonies were seen that were resistant to phage attack; these colonies were stable and so appeared to be genuine mutants. However, whether these mutants were produced spontaneously but randomly in time or the presence of the phage induced a physiological change that caused resistance was not known.

Luria reasoned that, if mutations occurred spontaneously, then the mutations might be expected to occur at different times in different cultures. In this case, the numbers of resistant colonies per culture should show high variation (or "fluctuation" in his word). He later claimed that the idea came to him as he watched the fluctuating returns obtained by colleagues gambling on a slot machine at a faculty dance in a local country club; hence the origin of the term "*jackpot*" mutation.

Luria and Delbrück designed their **"fluctuation test"** as follows. They inoculated 20 small cultures, each with a few cells, and incubated them until there were 10^8 cells per milliliter. At the same time, a much larger culture also was inoculated and incubated until there were 10^8 cells per milliliter. The 20 individual cultures and 20 samples of the same size from the large culture were plated in the presence of phage. The 20 individual cultures showed high variation in the number of resistant colonies: 11 plates had 0 resistant colonies, and the remainder had 1, 1, 3, 5, 5, 6, 35, 64, and 107 per plate (Figure 16-5a). The 20 samples from the large culture showed much less variation from plate to plate, all in the range of 14 to 26. If the phage were inducing mutations, there was no reason why fluctuation should be higher on the individual cultures, because all were exposed to phage similarly. The best explanation was that mutation was occurring randomly in time: the early mutations gave the higher numbers of resistant cells because the mutant cells had time to produce many resistant descendants. The later mutations produced fewer resistant cells (Figure 16-5b). This result led to the reigning "paradigm" of mutation; that is, whether in viruses, bacteria, or eukaryotes, mutations can occur in any cell at any time and their occurrence is random. For this and other work, Luria and Delbrück were awarded the Nobel Prize in Physiology or Medicine in 1969. Interestingly, this was after Luria's first graduate student, James Watson, won his Nobel Prize (with Frances Crick in 1964) for the discovery of the DNA double-helix structure.

This elegant analysis suggests that the resistant cells are selected by the environmental agent (here, phage) rather than produced by it. Can the existence of mutants in a population before selection be demonstrated directly? This demonstration was made possible by the use of a technique called **replica plating,** developed by Joshua and Esther Lederberg in 1952. A population of bacteria was plated on nonselective medium—that is, medium containing no phage—and from each cell a colony grew. This plate was called the *master plate*. A sterile piece of velvet was pressed down lightly on the surface of the master plate, and the velvet picked up cells wherever there was a colony (Figure 16-6). In this way, the velvet picked up a colony "imprint" from the whole plate. The velvet was then touched to replica plates containing selective medium (that is, containing T1 phage). On touching velvet to plates, cells clinging to the velvet are inoculated onto the replica plates in the same relative positions as those of the colonies on the original master plate. As expected, rare resistant mutant colonies were found on the replica plates, but the multiple replica plates showed identical

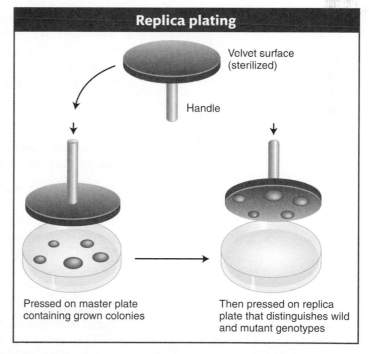

Replica plating

Velvet surface (sterilized)

Handle

Pressed on master plate containing grown colonies

Then pressed on replica plate that distinguishes wild and mutant genotypes

Figure 16-6 Replica plating reveals mutant colonies on a master plate through their behavior on selective replica plates. [*From G. S. Stent and R. Calendar, Molecular Genetics, 2nd ed. W. H. Freeman and Company, 1978.*]

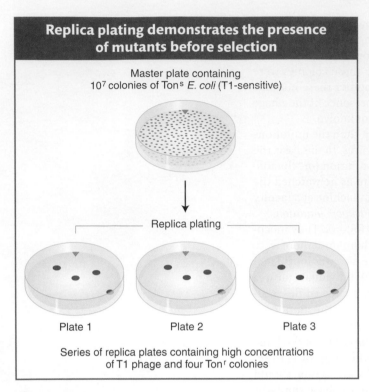

Replica plating demonstrates the presence of mutants before selection

Master plate containing 10^7 colonies of Ton^s *E. coli* (T1-sensitive)

Replica plating

Plate 1   Plate 2   Plate 3

Series of replica plates containing high concentrations of T1 phage and four Ton^r colonies

Figure 16-7 The identical patterns on the replicas show that the resistant colonies are from the master. [*From G. S. Stent and R. Calendar, Molecular Genetics, 2nd ed. W. H. Freeman and Company, 1978.*]

patterns of resistant colonies (Figure 16-7). If the mutations had occurred *after* exposure to the selective agents, the patterns for each plate would have been as random as the mutations themselves. The mutation events must have occurred before exposure to the selective agent. Again, these results confirm that mutation is occurring randomly all the time, rather than in response to a selective agent.

> **Message** Mutation is a random process. Any allele in any cell may mutate at any time.

Mechanisms of spontaneous mutations

Spontaneous mutations arise from a variety of sources. One source is the DNA-replication process. Although DNA replication is a remarkably accurate process, mistakes are made in the copying of the millions, even billions, of base pairs in a genome. Spontaneous mutations also arise in part because DNA is a very labile molecule and the cellular environment itself can damage it. As described in Chapter 15, mutations can even be caused by the insertion of a transposable element from elsewhere in the genome. In this chapter, we focus on mutations that are not caused by transposable elements.

Errors in DNA replication An error in DNA replication can result when an illegitimate nucleotide pair (say, A–C) forms in DNA synthesis, leading to a base substitution that may be either a transition or a transversion. Other errors may add or subtract base pairs such that a frameshift mutation is created.

Transitions You saw in Chapter 7 that each of the bases in DNA can appear in one of several tautomeric forms that can pair to the wrong base. Mismatches can also result when one of the bases becomes *ionized*. This type of mismatch may occur more frequently than mismatches due to tautomerization. These errors are frequently corrected by the proofreading (editing) function of bacterial DNA pol III (see Figure 7-18). If proofreading does not occur, all the mismatches described so far lead to transition mutations, in which a purine substitutes for a purine or a pyrimidine for a pyrimidine (see Figure 16-2). Other repair systems (described later in this chapter) correct many of the mismatched bases that escape correction by the polymerase editing function.

Transversions In transversion mutations, a pyrimidine substitutes for a purine, or vice versa (see Figure 16-2). The creation of a transversion by a replication error would require, at some point in the course of replication, the mispairing of a purine with a purine or a pyrimidine with a pyrimidine. Although the dimensions of the DNA double helix render such mismatches energetically unfavorable, we now know from X-ray diffraction studies that G–A pairs, as well as other purine–purine pairs, can form.

Frameshift mutations Replication errors can also lead to frameshift mutations. Recall from Chapter 9 that such mutations result in greatly altered proteins.

Certain kinds of replication errors can lead to **indel mutations**—that is, insertions or deletions of one or more base pairs. These insertions or deletions produce frameshift mutations when they add or subtract a number of bases not divisible by three (the size of a codon) in the protein-coding regions. The prevailing model (Figure 16-8) proposes that indels arise when loops in single-stranded regions are stabilized by the "slipped mispairing" of repeated sequences in the course of replication. This mechanism is sometimes called *replication slippage*.

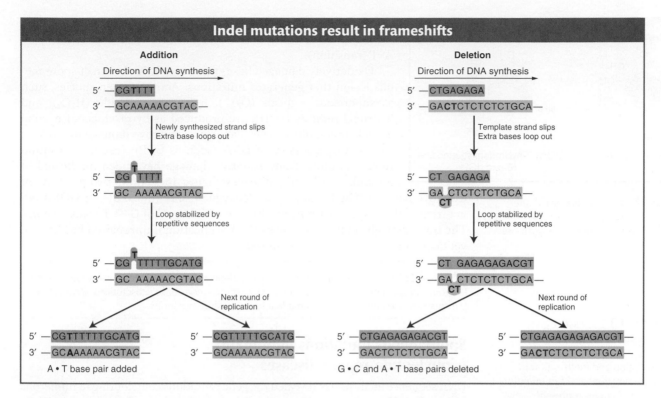

Indel mutations result in frameshifts

Addition

Direction of DNA synthesis →

5′ — CG**TTTT**
3′ — GC**AAAAA**CGTAC —

Newly synthesized strand slips
Extra base loops out ↓

5′ — CG T TTTT
3′ — GC AAAAACGTAC —

Loop stabilized by
repetitive sequences ↓

5′ — CG T TTTTGCATG
3′ — GC AAAAACGTAC —

Next round of replication

5′ — CG**T**TTTTTGCATG —
3′ — GC**A**AAAAACGTAC —

A • T base pair added

5′ — CGTTTTTGCATG —
3′ — GCAAAAACGTAC —

Deletion

Direction of DNA synthesis →

5′ — CTGAGAGA
3′ — GA**CT**CTCTCTCTGCA —

Template strand slips
Extra bases loop out ↓

5′ — CT GAGAGA
3′ — GA CTCTCTCTGCA —
    CT

Loop stabilized by
repetitive sequences ↓

5′ — CT GAGAGAGACGT
3′ — GA CTCTCTCTGCA —
    CT

Next round of replication

5′ — CTGAGAGAGACGT —
3′ — GACTCTCTCTGCA —

G • C and A • T base pairs deleted

5′ — CTGAGAGAGAGACGT —
3′ — GA**CT**CTCTCTCTGCA —

Spontaneous lesions In addition to replication errors, **spontaneous lesions**, naturally occurring damage to the DNA, can generate mutations. Two of the most frequent spontaneous lesions result from depurination and deamination.

Depurination, the more common of the two, is the loss of a purine base. Depurination consists of the interruption of the glycosidic bond between the base and deoxyribose and the subsequent loss of a guanine or an adenine residue from the DNA.

Figure 16-8 Base additions and deletions (indel mutations) cause frameshift mutations through the slipped mispairing of repeated sequences in the course of replication.
GENETICSP**O**RTAL **ANIMATED ART: Molecular mechanism of mutation**

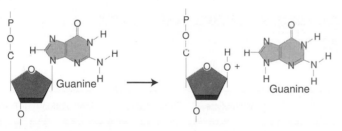

A mammalian cell spontaneously loses about 10,000 purines from its DNA in a 20-hour cell-cycle period at 37°C. If these lesions were to persist, they would result in significant genetic damage because, in replication, the resulting **apurinic sites** cannot specify a base complementary to the original purine. However, as we will see later in the chapter, efficient repair systems remove apurinic sites. Under certain conditions (to be described later), a base can be inserted across from an apurinic site; this insertion will frequently result in a mutation.

The deamination of cytosine yields uracil.

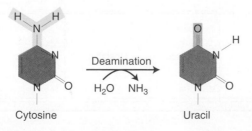

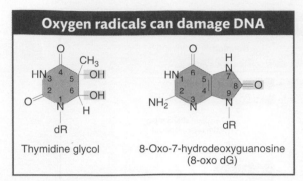

Figure 16-9 Products formed after DNA has been attacked by oxygen radicals. Abbreviation: dR, deoxyribose.

Unrepaired uracil residues will pair with adenine in replication, resulting in the conversion of a G·C pair into an A·T pair (a G·C → A·T transition).

Oxidatively damaged bases represent a third type of spontaneous lesion that generates mutations. Active oxygen species, such as superoxide radicals ($O_2 \cdot {}^-$), hydrogen peroxide (H_2O_2), and hydroxyl radicals ($\cdot OH$), are produced as by-products of normal aerobic metabolism. They can cause oxidative damage to DNA, as well as to precursors of DNA (such as GTP), resulting in mutation. Mutations from oxidative damage have been implicated in a number of human diseases. Figure 16-9 shows two products of oxidative damage. The 8-oxo-7-hydrodeoxyguanosine (8-oxo dG, or GO) product frequently mispairs with A, resulting in a high level of G → T transversions. The thymidine glycol product blocks DNA replication if unrepaired but has not yet been implicated as a cause of mutation.

> **Message** Spontaneous mutations can be generated by different processes. Replication errors and spontaneous lesions generate most spontaneous base substitutions. Replication errors can also cause deletions that lead to frameshift mutations.

Figure 16-10 The *FMR-1* gene in fragile X syndrome. (a) Exon structure and upstream CGG repeat. (b) Transcription and methylation in normal, premutation, and full mutation alleles. The red circles represent methyl groups. [*From W. T. O'Donnell and S. T. Warren, Annu. Rev. Neurosci. 25, 2002, 315–338, Fig. 1.*]

Spontaneous mutations in humans: trinucleotide-repeat diseases

DNA sequence analysis has revealed the gene mutations contributing to numerous human hereditary diseases. Many are of the expected base-substitution or single-base-pair indel type. However, some mutations are more complex. A number of these human disorders are due to duplications of short repeated sequences.

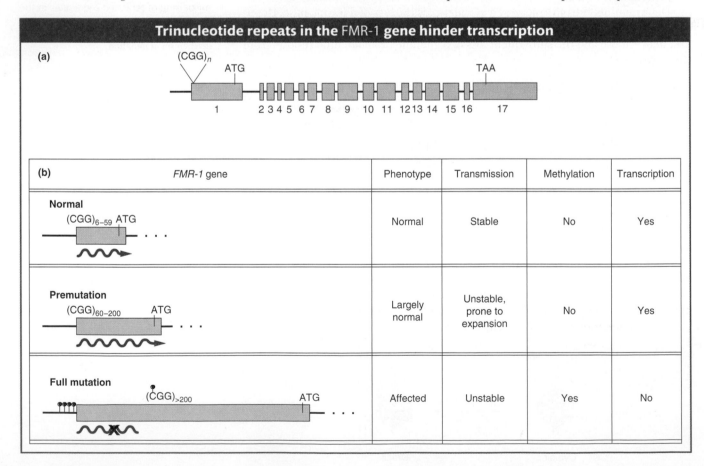

A common mechanism responsible for a number of genetic diseases is the expansion of a three-base-pair repeat. For this reason, they are termed **trinucleotide-repeat** diseases. An example is the human disease called *fragile X syndrome.* This disease is the most common form of inherited mental impairment, occurring in close to 1 of 1500 males and 1 of 2500 females. It is manifested cytologically by a fragile site in the X chromosome that results in breaks in vitro (but this does not lead to the disease phenotype). Fragile X syndrome results from changes in the number of a $(CGG)_n$ repeat in a region of the *FMR-1* gene that is transcribed but not translated (Figure 16-10a).

How does repeat number correlate with the disease phenotype? Humans normally show considerable variation in the number of CGG repeats in the *FMR-1* gene, ranging from 6 to 54, with the most frequent allele containing 29 repeats. Sometimes, unaffected parents and grandparents give rise to several offspring with fragile X syndrome. The offspring with the symptoms of the disease have enormous repeat numbers, ranging from 200 to 1300 (Figure 16-10b). The unaffected parents and grandparents also have been found to contain increased copy numbers of the repeat, but ranging from only 50 to 200. For this reason, these ancestors have been said to carry *premutations*. The repeats in these premutation alleles are not sufficient to cause the disease phenotype, but they are much more unstable (that is, readily expanded) than normal alleles, and so they lead to even greater expansion in their offspring. (In general, the more expanded the repeat number, the greater the instability appears to be.)

The proposed mechanism for the generation of these repeats is replication slippage that occurs in the course of DNA synthesis (Figure 16-11). However, the extraordinarily high frequency of mutation at the trinucleotide repeats in fragile X syndrome suggests that in human cells, after a threshold level of about 50 repeats, the replication machinery cannot faithfully replicate the correct sequence and large variations in repeat numbers result.

Other diseases, such as Huntington disease (see Chapter 2), also have been associated with the expansion of trinucleotide repeats in a gene. Several general themes apply to these diseases. In Huntington disease, for example, the wild-type *HD* gene includes a repeated sequence, often within the protein-coding region, and mutation correlates with a considerable expansion of this repeat region. The severity of the disease correlates with the number of repeat copies.

Huntington disease and Kennedy disease (also called *X-linked spinal and bulbar muscular atrophy*) result from the amplification of a three-base-pair repeat, CAG. Unaffected persons have an average of 19 to 21 CAG repeats, whereas affected patients have an average of about 46. In Kennedy disease, which is characterized by progressive muscle weakness and atrophy, the expansion of the trinucleotide repeat is in the gene that encodes the androgen receptor.

Properties common to some trinucleotide-repeat diseases suggest a common mechanism by which the abnormal phenotypes are produced. First, many of these diseases seem to include neurodegeneration—that is, cell death within the nervous system. Second, in such diseases, the trinucleotide repeats fall within the open reading frames of the transcripts of the mutated gene, leading to expansions or contractions of the number of repeats of a single amino acid in the polypeptide (for example, CAG repeats encode a polyglutamine repeat). Thus, it is easy to understand why these diseases result from expansions of codon-size units three base pairs in length.

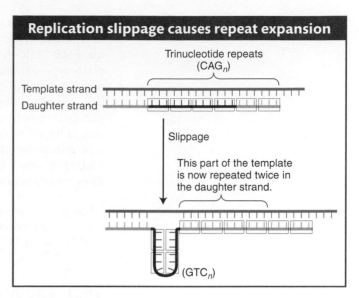

Replication slippage causes repeat expansion

Figure 16-11 Regions of trinucleotide repeats are prone to slipping during replication (red loop). As a consequence, the same region of trinucleotide repeats may be duplicated twice in the course of replication.

However, this explanation cannot hold for all trinucleotide-repeat diseases. After all, in fragile X syndrome, the trinucleotide expansion is near the 5′ end of the *FMR-1* mRNA, *before* the translation start site. Thus, we cannot ascribe the phenotypic abnormalities of the *FMR-1* mutations to an effect on protein structure. One clue to the problem with the mutant *FMR-1* genes is that they, unlike the normal gene, are hypermethylated, a feature associated with transcriptionally silenced genes (see Figure 16-10b). On the basis of these findings, repeat expansion is hypothesized to lead to changes in chromatin structure that silence the transcription of the mutant gene (see Chapter 12). In support of this model is the finding that the *FMR-1* gene is deleted in some patients with fragile X syndrome. These observations support a loss-of-function mutation.

> **Message** Trinucleotide-repeat diseases arise through the expansion of the number of copies of a three-base-pair sequence normally present in several copies, often within the coding region of a gene.

16.3 The Molecular Basis of Induced Mutations

Whereas some mutations are spontaneously produced inside the cell, other sources of mutation are present in the environment, whether intentionally applied in the laboratory or accidentally encountered in the course of everyday life. The production of mutations in the laboratory through exposure to mutagens is called **mutagenesis,** and the organism is said to be *mutagenized.*

Mechanisms of mutagenesis

Mutagens induce mutations by at least three different mechanisms. They can *replace* a base in the DNA, *alter* a base so that it specifically mispairs with another base, or *damage* a base so that it can no longer pair with any base under normal conditions. Mutagenizing genes and observing the phenotypic consequences is one of the primary experimental strategies used by geneticists.

Incorporation of base analogs   Some chemical compounds are sufficiently similar to the normal nitrogen bases of DNA that they occasionally are incorporated into DNA in place of normal bases; such compounds are called **base analogs.** After they are in place, these analogs have pairing properties unlike those of the normal bases; thus, they can produce mutations by causing incorrect nucleotides to be inserted opposite them in replication. The original base analog exists in only a single strand, but it can cause a nucleotide-pair substitution that is replicated in all DNA copies descended from the original strand.

One base analog widely used in research is 2-amino-purine (2-AP). This analog of adenine can pair with thymine but can also mispair with cytosine when protonated, as shown in Figure 16-12. Therefore, when 2-AP is incorporated into DNA

Figure 16-12 (a) An analog of adenine, 2-aminopurine (2-AP) can pair with thymine. (b) In its protonated state, 2-AP can pair with cytosine.

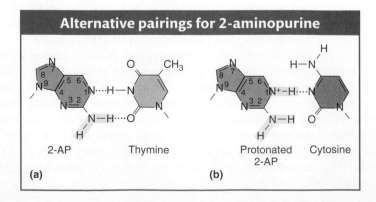

Alternative pairings for 2-aminopurine

2-AP   Thymine   Protonated 2-AP   Cytosine

(a)   (b)

by pairing with thymine, it can generate $A \cdot T \rightarrow G \cdot C$ transitions by mispairing with cytosine in subsequent replications. Or, if 2-AP is incorporated by mispairing with cytosine, then $G \cdot C \rightarrow A \cdot T$ transitions will result when it pairs with thymine. Genetic studies have shown that 2-AP causes transitions almost exclusively.

Specific mispairing   Some mutagens are not incorporated into the DNA but instead alter a base in such a way that it will form a specific mispair. Certain alkylating agents, such as ethylmethanesulfonate (EMS) and the widely used nitrosoguanidine (NG), operate by this pathway.

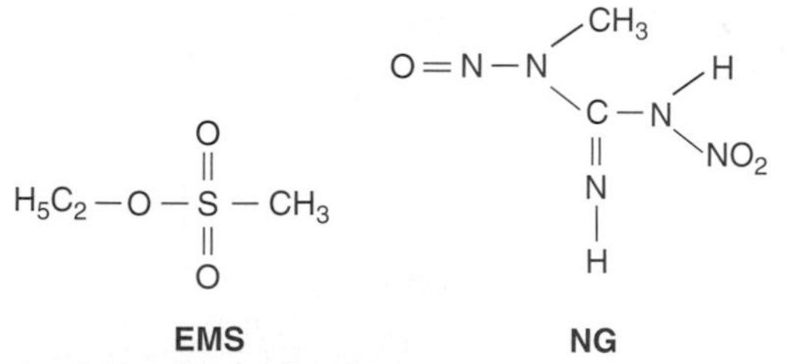

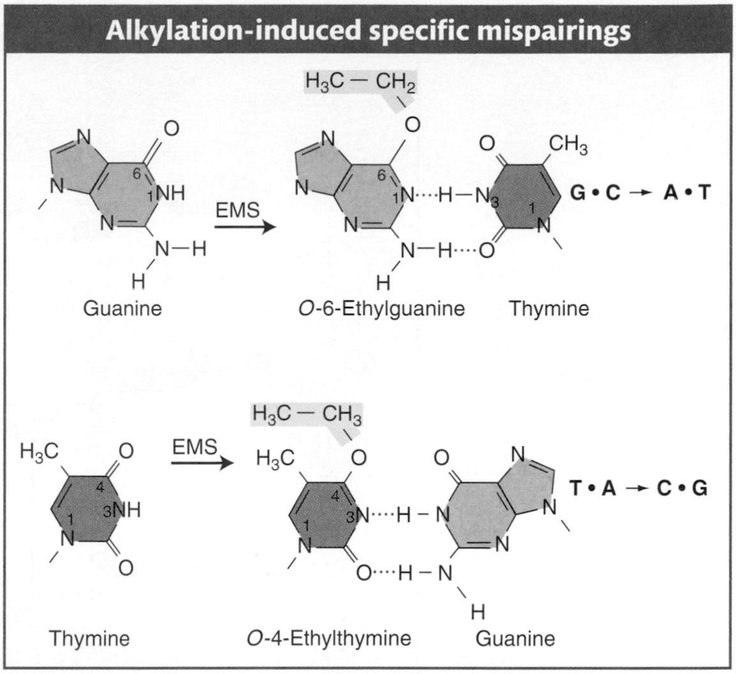

Such agents add alkyl groups (an ethyl group in EMS and a methyl group in NG) to many positions on all four bases. However, the formation of a mutation is best correlated with an addition to the oxygen at position 6 of guanine to create an *O*-6-alkylguanine. This addition leads to direct mispairing with thymine, as shown in Figure 16-13, and would result in $G \cdot C \rightarrow A \cdot T$ transitions at the next round of replication.

Alkylating agents can also modify the bases in dNTPs (where N is any base), which are precursors in DNA synthesis.

Intercalating agents   The **intercalating agents** form another important class of DNA modifiers. This group of compounds includes proflavin, acridine orange, and a class of chemicals termed ICR compounds (Figure 16-14a). These agents are planar molecules that mimic base pairs and are able to slip themselves in (intercalate) between the stacked nitrogen bases at the core of the DNA double helix (Figure 16-14b). In this intercalated position, such an agent can cause an insertion or deletion of a single nucleotide pair.

Figure 16-13 Treatment with EMS alters the structure of guanine and thymine and leads to mispairings.

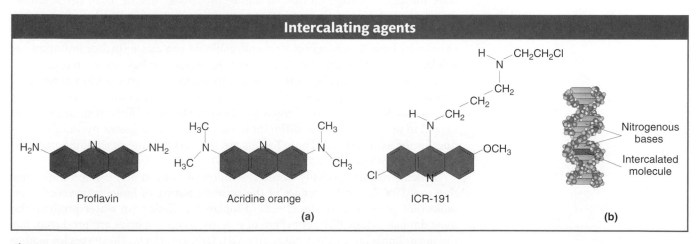

Figure 16-14 Structures of common intercalating agents (a) and their interaction with DNA (b). [*From L. S. Lerman, "The Structure of the DNA–Acridine Complex," Proc. Natl. Acad. Sci. USA 49, 1963, 94.*]

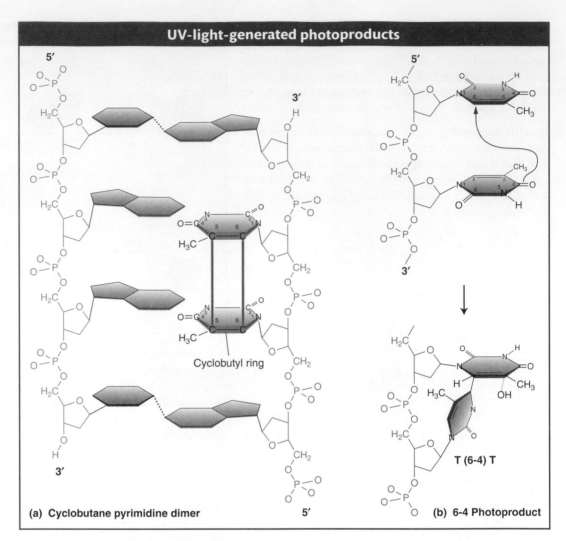

Figure 16-15 Photoproducts that unite adjacent pyrimidines in DNA are strongly correlated with mutagenesis. [*(Left) After E. C. Friedberg, DNA Repair. Copyright 1985 by W. H. Freeman and Company. (Right) From J. S. Taylor et al.*]

Base damage   A large number of mutagens damage one or more bases, and so no specific base pairing is possible. The result is a replication block, because DNA synthesis will not proceed past a base that cannot specify its complementary partner by hydrogen bonding. Replication blocks can cause further mutation—as will be explained later in the chapter (see the section on base excision repair).

Ultraviolet light usually causes damage to nucleotide bases in most organisms. Ultraviolet light generates a number of distinct types of alterations in DNA, called *photoproducts* (from the word *photo,* for "light"). The most likely of these products to lead to mutations are two different lesions that unite adjacent pyrimidine residues in the same strand. These lesions are the cyclobutane pyrimidine photodimer and the 6-4 photoproduct (Figure 16-15).

Ionizing radiation results in the formation of ionized and excited molecules that can damage DNA. Because of the aqueous nature of biological systems, the molecules generated by the effects of ionizing radiation on water produce the most damage. Many different types of reactive oxygen species are produced, but the most damaging to DNA bases are $\cdot$OH, O_2^-, and H_2O_2. These species lead to the formation of different adducts and degradation products. Among the most

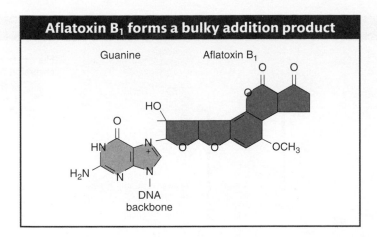

Figure 16-16 Metabolically activated aflatoxin B_1 binds to DNA.

prevalent, pictured in Figure 16-9, are thymine glycol and 8-oxo dG, both of which can result in mutations.

Ionizing radiation can also damage DNA directly rather than through reactive oxygen species. Such radiation may cause breakage of the N-glycosydic bond, leading to the formation of apurinic or apyrimidinic sites, and it can cause strand breaks. In fact, strand breaks are responsible for most of the lethal effects of ionizing radiation.

Aflatoxin B_1 is a powerful carcinogen that attaches to guanine at the N-7 position (Figure 16-16). The formation of this addition product leads to the breakage of the bond between the base and the sugar, thereby liberating the base and generating an apurinic site. Aflatoxin B_1 is a member of a class of chemical carcinogens known as bulky addition products when they bind covalently to DNA. Other examples include the diol epoxides of benzo(a)pyrene, a compound produced by internal combustion engines. All compounds of this class induce mutations, although by what mechanisms is not always clear.

> **Message** Mutagens induce mutations by a variety of mechanisms. Some mutagens mimic normal bases and are incorporated into DNA, where they can mispair. Others damage bases and either cause specific mispairing or destroy pairing by causing nonrecognition of bases.

The Ames test: evaluating mutagens in our environment

A huge number of chemical compounds have been synthesized, and many have possible commercial applications. We have learned the hard way that the potential benefits of these applications have to be weighed against health and environmental risks. Thus, having efficient screening techniques to assess some of the risks of a large number of compounds is essential.

Many compounds are potential cancer-causing agents (carcinogens), and so having valid model systems in which the carcinogenicity of compounds can be efficiently and effectively evaluated is very important. However, using a model mammalian system such as the mouse is very slow, time consuming, and expensive.

In the 1970s, Bruce Ames recognized that there is a strong correlation between the ability of compounds to cause cancer and their ability to cause mutations. He surmised that measurement of mutation rates in bacterial systems would be an effective model for evaluating the mutagenicity of compounds as a first level of detection of potential carcinogens. However, it became clear that not all carcinogens were themselves mutagenic; rather, some carcinogens' metabolites produced

Figure 16-17 Summary of the procedure used for the Ames test. Solubilized liver enzymes (S9) are added to a suspension of auxotrophic bacteria in a solution of the potential carcinogen (X). The mixture is plated on a medium containing no histidine. The presence of revertants indicates that the chemical is a mutagen and possibly a carcinogen as well.

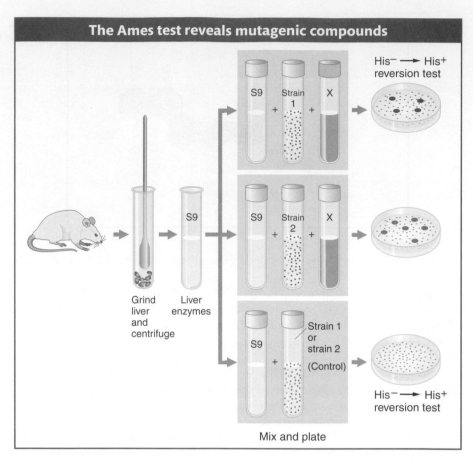

The Ames test reveals mutagenic compounds

His⁻ → His⁺ reversion test

His⁻ → His⁺ reversion test

Mix and plate

Figure 16-18 TA100, TA1538, and TA1535 are strains of *Salmonella* bearing different histidine auxotrophic mutations. The TA100 strain is highly sensitive to reversion through base-pair substitution. The TA1535 and TA1538 strains are sensitive to reversion through frameshift mutation. The test results show that aflatoxin B₁ is a potent mutation that causes base-pair substitutions but not frameshifts. [*From J. McCann and B. N. Ames, in W. G. Flamm and M. A. Mehlman, eds. Advances in Modern Technology, vol. 5. Copyright by Hemisphere Publishing Corporation, Washington, D.C.*]

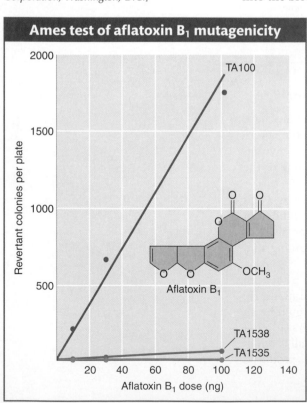

Ames test of aflatoxin B₁ mutagenicity

in the body are actually the mutagenic agents. Typically, these metabolites are produced in the liver, and the enzymatic reactions that convert the carcinogens into the bioactive metabolites did not take place in bacteria.

Ames realized that he could overcome this problem by treating special strains of the bacterium *Salmonella typhimurium* with extracts of rat livers containing metabolic enzymes (Figure 16-17). The special strains of *S. typhimurium* had one of several mutant alleles of a gene responsible for histidine synthesis that were known to "revert" (that is, return to wild-type phenotype) only by certain kinds of additional mutational events. For example, an allele called TA100 could be reverted to wild type only by a base-substitution mutation, whereas TA1538 and 1535 could be reverted only by indel mutations resulting in a protein frameshift (Figure 16-18).

The treated bacteria of each of these strains were exposed to the test compound, then grown on petri plates containing medium lacking histidine. The absence of this nutrient ensured that only revertant individuals containing the appropriate base substitution or frameshift mutation would grow. The number of colonies on each plate and the total number of bacteria tested were determined, allowing Ames to measure the frequency of reversion. Compounds that yielded metabolites inducing elevated levels of reversion relative to untreated control liver extracts would then clearly be mutagenic and would be possible carcinogens. The **Ames test** thus provided an important way of screening thousands of compounds and evaluating one aspect of their risk to health and the environment. It is still in use today as an important tool for the evaluation of the safety of chemical compounds.

16.4 Biological Repair Mechanisms

After surveying the numerous ways that DNA can be damaged—from sources both inside the cell (replication, reactive oxygen, and so forth) and outside (environmental: UV light, ionizing radiation, mutagens)—you might be wondering how life has managed to survive and thrive for billions of years. The fact is that organisms ranging from bacteria to humans to plants can efficiently repair their DNA. These organisms make use of a variety of repair mechanisms that together employ as many as 100 known proteins. In fact, our current understanding is that DNA is the only molecule that organisms repair rather than replace. As you will see, failure of these repair systems is a significant cause of many inherited human diseases.

The most important repair mechanism was briefly mentioned in Chapter 7—the proofreading function of the DNA polymerases that replicate DNA as part of the replisome. As noted there, both DNA polymerase I and DNA polymerase III are able to excise mismatched bases that have been inserted erroneously. Let's now examine some of the other repair pathways, beginning with error-free repair.

Direct reversal of damaged DNA

The most straightforward way to repair a lesion is to reverse it directly, thereby regenerating the normal base (Figure 16-19). Although most types of damage are essentially irreversible, lesions can be repaired by direct reversal in a few cases. One case is a mutagenic photodimer caused by UV light. The cyclobutane pyrimidine dimer (CPD) can be repaired by an enzyme called *CPD photolyase*. The enzyme binds to the photodimer and splits it to regenerate the original bases. This repair mechanism is called *photoreactivation* because the enzyme requires light to function. Other repair pathways are required to remove UV damage in the absence of light of the appropriate wavelength (>300 nm).

Alkyltransferases are enzymes that also directly reverse lesions. They remove certain alkyl groups that have been added to position O-6 of guanine (see Figure 16-13) by such mutagens as nitrosoguanidine and ethylmethanesulfonate. The methyltransferase from *E. coli* has been well studied. This enzyme transfers the methyl group from O-6-methylguanine to a cysteine residue in the enzyme's active site. However, the transfer inactivates the enzyme, and so this repair system can be saturated if the level of alkylation is high enough.

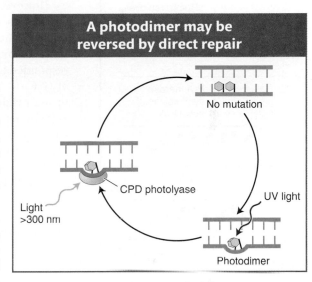

Figure 16-19 The enzyme CPD photolyase splits a cyclobutane pyrimidine photodimer to repair this mutation. GENETICS*PORTAL* ANIMATED ART: **UV-induced photodimers and excision repair**

Base-excision repair

An overarching principle guiding cellular genetic systems is the power of nucleotide sequence complementarity. (Recall that genetic analysis also depends heavily on this principle.) Important repair systems exploit the properties of antiparallel complementarity to restore damaged DNA segments to their initial, undamaged state. In these systems, a base or longer segment of a DNA chain is removed and replaced with a newly synthesized nucleotide segment complementary to the opposite template strand.

Because these systems depend on the complementarity, or homology, of the template strand to the strand being repaired, they are called **homology-dependent repair systems.** Unlike the examples of reversal of damage described in the preceding section, these pathways include the removal and replacement of one or more bases.

The first homology-dependent repair system we will examine is **base-excision repair.** After DNA proofreading by DNA polymerase, base-excision repair is the most important mechanism used to remove incorrect or damaged bases. The main target of base-excision repair is nonbulky damage to bases. This type of

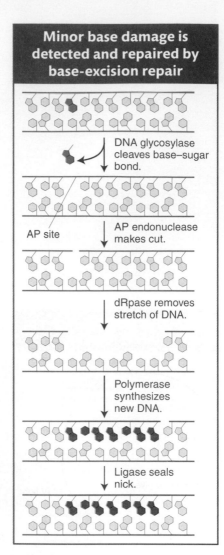

DNA glycosylase cleaves base–sugar bond.

AP site

AP endonuclease makes cut.

dRpase removes stretch of DNA.

Polymerase synthesizes new DNA.

Ligase seals nick.

Figure 16-20 In base-excision repair, damaged bases are removed and repaired through sequential action of a DNA glycosylase, AP endonuclease, deoxyribophosphordiesterase (dRpase), DNA polymerase, and ligase.

damage can result from the variety of causes mentioned in preceding sections, including methylation, deamination, oxidation, or the spontaneous loss of a DNA base. Base-excision repair (Figure 16-20) is carried out by DNA glycosylases that cleave base–sugar bonds, thereby liberating the altered bases and generating apurinic or apyrimidinic (AP) sites. An enzyme called AP endonuclease then nicks the damaged strand upstream of the AP site. A third enzyme, deoxyribophosphodiesterase, cleans up the backbone by removing a stretch of neighboring sugar–phosphate residues so that a DNA polymerase can fill the gap with nucleotides complementary to the other strand. DNA ligase then seals the new nucleotide into the backbone (see Figure 16-20).

Numerous DNA glycosylases exist. One, uracil-DNA glycosylase, removes uracil from DNA. Uracil residues, which result from the spontaneous deamination of cytosine (see page 561), can lead to a C-to-T transition if unrepaired. One advantage of having thymine (5-methyluracil) rather than uracil as the natural pairing partner of adenine in DNA is that spontaneous cytosine deamination events can be recognized as abnormal and then excised and repaired. If uracil were a normal constituent of DNA, such repair would not be possible.

However, deamination does pose other problems for both bacteria and eukaryotes. By analyzing a large number of mutations in the *lacI* gene, Jeffrey Miller identified places in the gene where one or more bases were prone to frequent mutation. Miller found that these so-called mutational hotspots corresponded to deaminations at certain cytosine residues. DNA sequence analysis of $G \cdot C \rightarrow A \cdot T$ transition hot spots in the *lacI* gene showed that 5-methylcytosine residues are present at each hotspot. Recall from Chapter 12 that eukaryotic DNA may be methylated to inactive genes. Similarly, *E. coli* and other bacteria also methylate their DNA, although for different purposes. Some of the data from this *lacI* study are shown in Figure 16-21. The height of each bar on the graph represents the frequency of mutations at each site. The positions of 5-methylcytosine residues can be seen to correlate nicely with the most mutable sites.

Why are 5-methylcytosines hot spots for mutations? The deamination of 5-methylcytosine generates thymine (5-methyluracil):

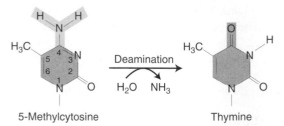

5-Methylcytosine   Thymine

Thymine is not recognized by the enzyme uracil-DNA glycosylase and thus is not repaired. Therefore, C → T transitions generated by deamination are seen more frequently at 5-methylcytosine sites because they escape this repair system. A consequence of the frequent mutation of 5-methylcytosine to thymine is that methylated regions of the genome (which are usually transcriptionally inactive; see Chapter 12) are converted, over evolutionary time, to AT-rich regions. In contrast, coding and regulatory regions, which are less methylated, remain GC rich.

> **Message** In base-excision repair, nonbulky damage to the DNA is recognized by one of several enzymes called DNA glycosylases that cleave the base–sugar bonds, releasing the incorrect base. Repair consists of the removal of the site that now lacks a base and the insertion of the correct base as guided by the complementary base in the undamaged strand.

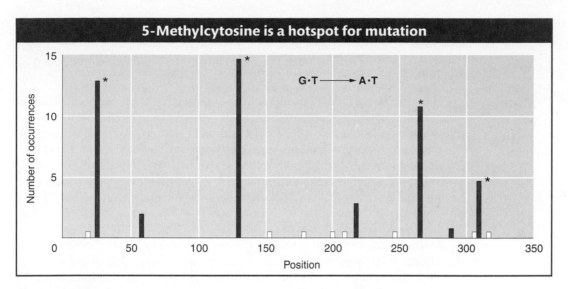

Figure 16-21 Methylcytosine hotspots in *E. coli*. Nonsense mutations at 15 different sites in *lacI* were scored. All resulted in G·C → A·T transitions. The asterisk (*) marks the positions of 5-methylcytosines, and the white bars mark sites where transitions known to occur were not isolated in this group. [*From C. Coulondre, J. H. Miller, P. J. Farabaugh, and W. Gilbert, "Molecular Basis of Base Substitution Hotspots in Escherichia coli," Nature 274, 1978, 775.*]

Nucleotide-excision repair

Although the vast majority of the damage sustained by an organism is minor base damage that can be handled by base-excision repair, this mechanism can neither correct bulky adducts that distort the DNA helix, adducts such as the cyclobutane pyrimidine dimers caused by UV light (see Figure 16-15), nor correct damage to more than one base. A DNA polymerase cannot continue DNA synthesis past such lesions, and so the result is a replication block. A blocked replication fork can cause cell death. Similarly, an abnormal or damaged base can stall the transcription complex. To cope with both of these situations, prokaryotes and eukaryotes utilize an extremely versatile pathway called **nucleotide-excision repair (NER)** that is able to relieve replication and transcription blocks and repair the damage.

Interestingly, two autosomal recessive diseases in humans, **xeroderma pigmentosum (XP)** and **Cockayne syndrome,** are caused by defects in nucleotide-excision repair. Although patients with either XP or Cockayne syndrome are exceptionally sensitive to UV light, other symptoms are dramatically different. Xeroderma pigmentosum was introduced at the beginning of this chapter and is characterized by the early development of cancers, especially skin cancer and, in some cases, neurological defects. In contrast, patients afflicted with Cockayne syndrome have a variety of developmental disorders including dwarfism, deafness, and retardation. In broad terms, XP patients get early cancer, whereas Cockayne syndrome patients age prematurely. How can defects in the same repair pathway lead to such different disease symptoms? Although there is no simple answer to this question, work on the genetic basis of these diseases has led to the identification of important proteins in the NER pathway.

Nucleotide-excision repair is a complex process that requires dozens of proteins. Despite this complexity, the repair process can be divided into four phases:

1. Recognition of damaged base(s)
2. Assembly of a multiprotein complex at the site

3. Cutting of the damaged strand several nucleotides upstream and downstream of the damage site and removal of the nucleotides (~30) between the cuts

4. Use of the undamaged strand as a template for DNA polymerase followed by strand ligation

The fact that, as already mentioned, both stalled replication forks and stalled transcription complexes activate this repair pathway implies that there are two types of nucleotide-excision repair that differ in damage recognition (step 1). We now know that one type repairs transcribed regions of DNA and is called, not surprisingly, **transcription-coupled nucleotide-excision repair (TC-NER).** The other type, **global genomic repair (GGR),** corrects lesions anywhere in the genome and is activated by stalled replication forks. As can be seen in Figure 16-22, although the recognition step differs, both TC-NER and GGR share the last four steps.

At this point, some of you might be asking yourselves, What if differences in the disease symptoms of XP and Cockayne syndrome were due to mutations in different classes of recognition proteins? You would be on the right track in asking that question. Patients with Cockayne syndrome have a mutation in one of two proteins called CSA and CSB, which are thought to recognize stalled transcription complexes. A model for how CSA and CSB may bind to the stalled polymerase and then attract the multiprotein complex needed for DNA repair is shown in Figure 16-22.

For GGR, the first step is the recognition of a base that (1) has been damaged by a heterodimeric complex and so distorts the double helix and (2) is not part of an actively transcribed template strand. CSA and CSB bind at this site to form a recognition complex that attracts the multiprotein complex, which contains the 10 subunits of the general transcription factor TFIIH. Two of the 10 subunits, XPB and XPD, are helicases that unwind the DNA, and the entire structure is stabilized by RPA, a single-stranded-DNA-binding protein. Subsequent steps that mediate the cleavage and excision of the damaged base and as many as 30 adjacent nucleotides followed by DNA synthesis to fill the gap are common to both the TC-NER and GGR pathways. The steps that resynthesize DNA to fill in the gap (so-called gap-repair synthesis) will be discussed in more detail later, inasmuch as other repair pathways share this step.

Many of the human counterparts of nuclear excision-repair proteins (XPA, XPB, and so forth) were first identified in patients with XP. Both XP and Cockayne syndrome are examples of **heterogeneous genetic disorders** in which patients have mutations in one of several genes having roles in a particular process—in this case, nucleotide-excision repair.

Can the molecular differences between TC-NER and GGR provide an explanation for the different symptoms displayed by patients with XP and Cockayne syndrome? Recall that XP patients develop early cancers, whereas Cockayne syndrome patients have a variety of symptoms associated with premature aging. We have seen that the repair system of Cockayne syndrome patients cannot recognize stalled transcription complexes. A consequence of this defect is that the cell is more likely to activate the apoptosis suicide pathway. In a healthy person, cell death is often preferable to the propagation of a cell that has sustained DNA damage. However, according to this theory, the cell-death pathway would be activated more frequently in a Cockayne syndrome patient, thus leading to a variety of premature-aging symptoms. In contrast, XP patients can recognize stalled transcription complexes (they have normal CSA and CSB proteins) and prevent cell death when transcription is restarted. However, they cannot repair the original damage, because of mutations in one of their XP proteins. Thus, mutations will accumulate in the cells of patients with XP and, as stated earlier in this chapter, the presence of mutations, whether

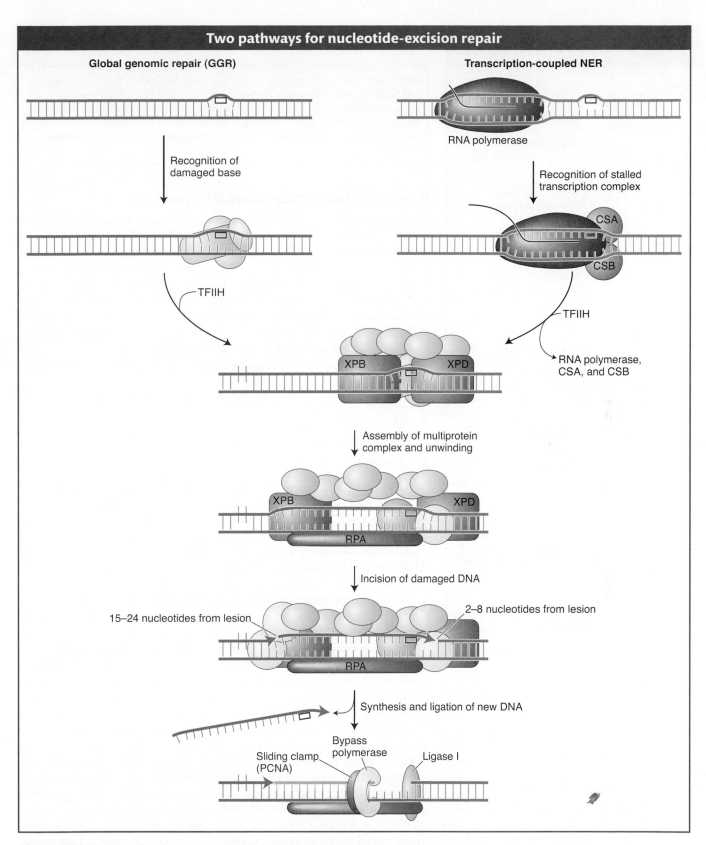

Figure 16-22 The nucleotide-excision repair pathway is activated when bulky adducts or multiple damaged bases are recognized in nontranscribed (GGR) or transcribed (TC-NER) regions of the genome. A multiprotein complex excises several bases and resynthesizes them using the opposite strand as a template. See text for details.

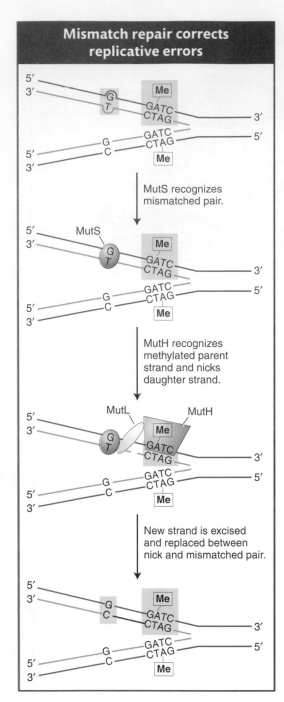

Figure 16-23 Model for mismatch repair in *E. coli*. DNA is methylated (Me) at the A residue in the sequence GATC. DNA replication yields a hemimethylated duplex that exists until methylase can modify the newly synthesized strand. The mismatch-repair system makes any necessary corrections based on the sequence found on the methylated strand (original template). MutS, MutH, and MutL are proteins.

caused by mutagens or the failure of repair pathways, increases the risk of developing many types of cancer.

> **Message** Nucleotide-excision repair is a versatile pathway that recognizes and corrects DNA lesions due largely to UV damage and, in doing so, relieves stalled replication forks and transcription complexes. Patients with xeroderma pigmentosum and Cockayne syndrome are UV sensitive owing to mutations in key nucleotide-excision-repair proteins that recognize or repair the damaged bases.

Postreplication repair: mismatch repair

You learned in the first half of this chapter that many errors occur in DNA replication. In fact, the error rate is about 10^{-5}. Correction by the 3′-to-5′ proofreading function of the replicative polymerase reduces the error rate to less than 10^{-7}. The major pathway that corrects the remaining replicative errors is called **mismatch repair.** This repair pathway reduces the error rate to less than 10^{-9} by recognizing and repairing mismatched bases and small loops caused by the insertion and deletion of nucleotides (indels) in the course of replication. From these values, you can see that mutations leading to the loss of the mismatch-repair pathway could increase the mutation frequency 100-fold. In fact, loss of mismatch repair is associated with hereditary forms of colon cancer.

Mismatch-repair systems have to do at least three things:

1. Recognize mismatched base pairs
2. Determine which base in the mismatch is the incorrect one
3. Excise the incorrect base and carry out repair synthesis

Most of what is known about mismatch repair comes from decades of genetic and biochemical analysis with the use of the model bacterium *E. coli* (see the *E. coli* Model Organism box on page 173). Especially noteworthy was the reconstitution of the mismatch-repair system in the test tube in the laboratory of Paul Modrich. Conservation of many of the mismatch-repair proteins from bacteria to yeast to human indicates that this pathway is both ancient and important in all living organisms. Recently, the human mismatch-repair system also was reconstituted in the test tube in the Modrich laboratory. The ability to study the details of the reaction will spur future studies of the human pathway. However, for now we will focus on the very well characterized *E. coli* system (Figure 16-23).

The first step in mismatch repair is the recognition of the damage in newly replicated DNA by the MutS protein. The binding of this protein to distortions in the DNA double helix caused by mismatched bases initiates the mismatch-repair pathway by attracting three other proteins to the site of the lesion (MutL, MutH, and UvrD [not shown]). The key protein is MutH, which performs the crucial function of cutting the strand containing the incorrect base. Without this ability to discriminate between the correct and the incorrect bases, the mismatch-repair system could not determine which base to excise to prevent a mutation from arising. If, for example, a G–T mismatch occurs as a replication error, how can the system determine whether G or T is incorrect? Both are normal bases in DNA. But replication errors produce mismatches on the newly synthesized strand, and so the mismatch-repair system replaces the base on that strand.

How does mismatch repair distinguish the newly synthesized strand from the old one? Recall from Chapter 12 that cytosine bases are often methylated in eukaryotes and that this so-called epigenetic mark is propagated from parent to daughter strand soon after replication. *E. coli* DNA also is methylated, but the methyl groups relevant to mismatch repair are added to adenine bases. To distinguish the old, template strand from the newly synthesized strand, the bacterial repair system takes advantage of a delay in the methylation of the following sequence:

$$5'\text{-G-A-T-C-}3'$$
$$3'\text{-C-T-A-G-}5'$$

The methylating enzyme is adenine methylase, which creates 6-methyladenine on each strand. However, adenine methylase requires several minutes to recognize and modify the newly synthesized GATC stretches. In that interval, the MutH protein nicks the methylation site on the strand containing the A that has not yet been methylated. This site can be several hundred base pairs away from the mismatched base. After the site has been nicked, the UrvD protein binds at the nick and uses its helicase activity to unwind the DNA. A protective single-strand-binding protein coats the unwound parental strand while the part of the new strand between the mismatch and the nick is excised.

Many of the proteins in *E. coli* mismatch repair are conserved in human mismatch repair. Nonetheless, how eukaryotes recognize and repair only the newly replicated strand is still unknown. The problem is particularly perplexing in organisms that lack most or all DNA methylation such as yeast, *Drosophila,* and *C. elegans.* A popular model proposes that discrimination is based on the recognition of free 3' ends that characterize the newly synthesized leading and lagging strands.

An important target of the human mismatch system is short repeat sequences that can be expanded or deleted in replication by the slipped-mispairing mechanism described previously (see Figure 16-8). Mutations in some of the components of this pathway have been shown to be responsible for several human diseases, especially cancers. There are thousands of short repeats (microsatellites) located throughout the human genome (see Chapter 4). Although most are located in noncoding regions (given that most of the genome is noncoding), a few are located in genes that are critical for normal growth and development.

Therefore, defects in the human mismatch-repair pathway would be predicted to have very serious disease consequences. This prediction has turned out to be true, a case in point being a syndrome called hereditary nonpolyposis colorectal cancer (HNPCC), which, despite its name, is not a cancer itself but increases cancer risk. One of the most common inherited predispositions to cancer, the disease affects as many as 1 in 200 people in the Western world. Studies have shown that HNPCC results from a loss of the mismatch-repair system due in large part to inherited mutations in genes that encode the human counterparts (and homologs) of the bacterial MutS and MutL proteins (see Figure 16-23). The inheritance of HNPCC is autosomal dominant. Cells with one functional copy of the mismatch-repair genes have normal mismatch-repair activity, but tumor cell lines arise from cells that have lost the one functional copy and are thus mismatch deficient. These cells display high mutation rates owing in part to an inability to correct the formation of indels in replication.

> **Message** The mismatch-repair system corrects errors in replication that are not corrected by the proofreading function of the replicative DNA polymerase. Repair is restricted to the newly synthesized strand, which is recognized by the repair machinery in prokaryotes because it lacks a methylation marker.

Error-prone repair: translesion DNA synthesis

Thus far, all of the repair mechanisms that we have encountered are error free, inasmuch as they either reverse the damage directly or use base complementarity to insert the correct base. Yet, there are repair pathways that are themselves a significant source of mutation. These mechanisms appear to have evolved to prevent the occurrence of potentially more serious outcomes such as cell death

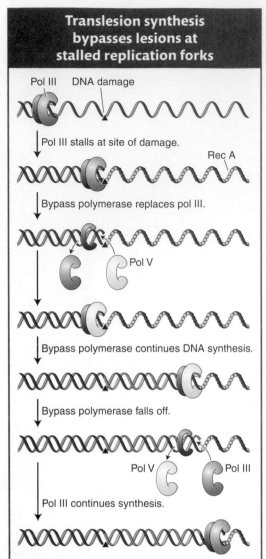

Translesion synthesis bypasses lesions at stalled replication forks

Pol III   DNA damage

Pol III stalls at site of damage.

Rec A

Bypass polymerase replaces pol III.

Pol V

Bypass polymerase continues DNA synthesis.

Bypass polymerase falls off.

Pol V   Pol III

Pol III continues synthesis.

Figure 16-24 A model for translesion synthesis in *E. coli*. In the course of replication, DNA polymerase III is temporarily replaced by a bypass polymerase (pol V) that can continue replicating past a lesion. Bypass polymerases are error prone. [*After E. C. Friedberg, A. R. Lehmann, and R. P. Fuchs, "Trading Places: How Do DNA Polymerases Switch During Translesion DNA Synthesis?" Molec. Cell 18, 2005, 499–505.*]

or cancer. As already mentioned, a stalled replication fork can initiate a cell-death pathway. In both prokaryotes and eukaryotes, such replication blocks can be *bypassed* by the insertion of nonspecific bases. In *E. coli*, this process requires the activation of the **SOS system.** The name *SOS* comes from the idea that this system is induced as an emergency response to prevent cell death in the presence of significant DNA damage. As such, SOS induction is a mechanism of last resort, a form of damage tolerance that allows the cell to trade death for a certain level of mutagenesis.

It has taken more than 30 years to figure out how the SOS system generates mutations while allowing DNA polymerase to bypass lesions at stalled replication forks. We are already familiar with DNA damage induced by UV light (see Figure 16-15). An unusual class of *E. coli* mutants that survived UV exposure without sustaining additional mutations was isolated in the 1970s. The fact that such mutants even existed suggested that some *E. coli* genes function to generate mutations when exposed to UV light. UV-induced mutation will not occur if the *DinB, UmuC,* or *UmuD′* genes are mutated.

Figure 16-24 shows the steps in the SOS mechanism. In the first step, UV light induces the synthesis of a protein called RecA. We will see more of the RecA protein later in the chapter because it is a key player in key mechanisms of DNA repair and recombination. When the replicative polymerase (DNA polymerase III) stalls at a site of DNA damage, the DNA ahead of the polymerase continues to be unwound, exposing regions of single-stranded DNA that become bound by single-strand-binding proteins. Next, RecA proteins join the single-strand-binding proteins and form a protein–DNA filament. The RecA filament is the biologically active form of this protein. In this situation, RecA acts as a signal that leads to the induction of several genes that are now known to encode members of a newly discovered family of DNA polymerases that can bypass the replication block and are distinct from replicative polymerases. DNA polymerases that can bypass replication stalls have also been found in diverse taxa of eukaryotes ranging from yeast to human. These eukaryotic polymerases contribute to a damage-tolerance mechanism called **translesion DNA synthesis** that resembles the SOS bypass system in *E. coli*.

These **translesion,** or **bypass, polymerases,** as they have come to be known, differ from the main replicative polymerases in several ways. First, they can tolerate unusually large adducts on the bases. Whereas the replicative polymerase stalls if a base does not fit into an active site, the bypass polymerases have much larger pockets that can accommodate damaged bases. Second, in some situations, the bypass polymerases have a much higher error rate, in part because they lack the 3′-to-5′ proofreading activity of the main replicative polymerases. Third, they can only add a few nucleotides before falling off. This feature is attractive because the main function of an error-prone polymerase is to unblock the replication fork, not to synthesize long stretches of DNA that could contain many mismatches.

Several bypass polymerases that appear to be always present in eukaryotic cells are now known. Because they are always present, their access to DNA must be regulated so that they are used only when needed. The cell has evolved a neat solution to this problem. Recall that an integral part of the replisome is the PCNA (proliferating cell nuclear antigen) protein that functions as a sliding clamp to orchestrate the myriad events at the replication fork (see Figure 7-20). One critical protein present at a stalled replication fork is Rad6, which, curiously, is an enzyme that adds ubiquitin to proteins (Figure 16-25). As described in Chapter 9, the addition of chains of many ubiquitin monomers serves to target a protein for degradation (see Figure 9-23). In contrast, the binding of a single ubiquitin monomer to PCNA changes its conformation so that it can now bind the bypass polymerase and orchestrate transle-sion synthesis. Enzymatic removal of the ubiquitin tag on PCNA leads to the dissociation of the bypass poly-merase and the eventual restoration of normal replica-tion. Any base mismatch due to translesion synthesis still has a chance of detection and correction by the mismatch-repair pathway.

The regulation of PCNA function by the addition and removal of ubiquitin monomers illustrates the impor-tance of post-translational modifications in eukaryotes. If base damage in the template strand is not corrected quickly, the stalled replication fork will signal the activa-tion of the cell-death pathway. A eukaryotic cell cannot wait for the de novo synthesis of bypass polymerases following transcription and translation as occurs in the *E. coli* SOS system. Instead, eukaryotic bypass polymerases are constitutively transcribed and are always present; their access to the replication fork is controlled by rapid and reversible post-translational modifications.

> **Message** In translesion synthesis, bypass polymerases are recruited to replication forks that have stalled because of damage in the template strand. Bypass polymerases may introduce errors in the course of synthesis that may persist and lead to mutation or that can be corrected by other mechanisms such as mismatch repair.

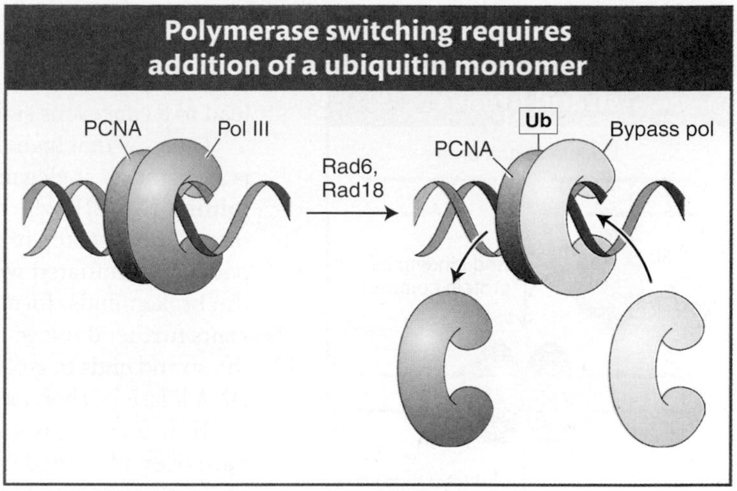

Figure 16-25 The addition of a single ubiquitin (Ub) monomer to the sliding clamp (PCNA) allows the bypass polymerase to bind to PCNA and begin replicating.

Repair of double-strand breaks

As we have seen, many correction systems exploit DNA complementarity to make error-free repairs. Such error-free repair is characterized by two stages: (1) removal of the damaged bases, perhaps along with nearby DNA, from one strand of the double helix and (2) use of the other strand as a template for the DNA synthesis needed to fill the single-strand gap. However, what would happen if *both* strands of the double helix were damaged in such a way that complementarity could not be exploited? For example, exposure to X-rays often causes both strands of the double helix to break at sites that are close together. This type of mutation is called a **double-strand break.** If left unrepaired, double-strand breaks can cause a variety of chromosomal aberrations resulting in cell death or a precancerous state.

Interestingly, the generation of double-strand breaks is an integral feature of some normal cellular processes that require DNA rearrangements. One example is meiotic recombination. As will be seen in the remainder of this chapter, the cell uses many of the same proteins and pathways to repair double-strand breaks and to carry out meiotic recombination. For this reason, we begin by focusing on the molecular mechanisms that repair double-strand breaks before turning our atten-tion to the mechanism of meiotic recombination.

Double-strand breaks can arise spontaneously (for example, in response to reactive oxygen species produced as a by-product of cellular metabolism) or they can be induced by ionizing radiation. Several mechanisms are known to repair double-strand breaks, and new mechanisms are still being discovered. Two dis-tinct mechanisms are described in the following section: nonhomologous end joining and homologous recombination.

Nonhomologous end joining   Many of the previously described repair mecha-nisms are called on in the S phase of the cell cycle, when the DNA is replicating in preparation for mitosis or meiosis. However, unlike the cells of most prokaryotes and lower eukaryotes, the cells of higher eukaryotes are usually not replicating their DNA, because they are either in a resting phase of the cell cycle or have ceased dividing entirely. What happens when double-strand breaks occur in cells

Error-prone nonhomologous end joining repairs double-strand breaks

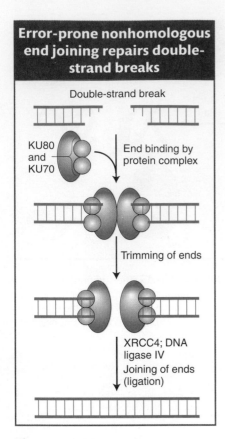

Double-strand break

KU80 and KU70 — End binding by protein complex

Trimming of ends

XRCC4; DNA ligase IV
Joining of ends (ligation)

Figure 16-26 Mechanism of nonhomologous end joining (NHEJ). This mechanism is error prone. See text for details.

What Geneticists Are Doing Today

where undamaged strands or sister chromatids are not present? The answer is that these ends must be repaired, either perfectly or imperfectly, because broken ends can initiate potentially harmful chromosomal rearrangements that could lead to a cancerous state (see Chapter 17).

One way that higher eukaryotes put double-stranded broken ends back together is by a rather inelegant but important mechanism called **nonhomologous end joining (NHEJ),** which is shown in Figure 16-26. Like that of other repair mechanisms, the first step in the NHEJ pathway is recognition of the damage. The NHEJ pathway is initiated when two very abundant proteins, KU70 and KU80, bind to the broken ends, forming a heterodimer that serves two functions. First, it prevents further damage to the ends, and, second, it recruits other proteins that trim the strand ends to generate the 5′-P and 3′-OH ends that are required for ligation. DNA ligase IV then joins the two ends.

How do scientists know when all of the components of a biological pathway have been identified? As it turns out, this problem is difficult. The identification of a new component of the NHEJ pathway provides a recent example.

For several reasons, all of the components of the NHEJ pathway were thought to have been identified. However, geneticists analyzing a cell line (called 2BN) derived from a child with a rare inherited disorder were in for a surprise. Although they were able to demonstrate that cell line 2BN was defective for double-strand-break repair, they were not able to restore the repair system and produce the wild-type phenotype by genetic complementation with any of the genes encoding NHEJ proteins. That is, when they introduced wild-type genes encoding known NHEJ proteins (for example, KU70, KU80, ligase IV) into the 2BN line, the cell line was still defective in the repair of double-strand breaks. This negative result indicated that cell line 2BN carried a mutation in an unknown NHEJ protein.

In the era of genomics, the identification of proteins linked to diseases is becoming more common because of the wide availability of cell lines from persons with disease phenotypes. When we humans have a health problem, we go to a doctor and tell him or her about our symptoms, including information about relatives with similar problems. Such information is of increasing importance in the genomics era with its ever-expanding genetic toolbox that can often be used to identify mutant genes associated with inherited disorders (see Chapters 10 and 14).

What can geneticists do in the laboratory to find a protein, such as the unknown NHEJ protein, that has not yet been identified? Two laboratories using very different approaches succeeded in identifying the new NHEJ component; one approach will be described here because it has been successfully employed to discover several other proteins. As noted in preceding chapters, many cellular proteins perform their jobs by interacting with other proteins. Chapter 14, for example, described the yeast two-hybrid test used to identify proteins that interact with a protein of interest. In the case under consideration now, the protein of interest was the NHEJ component XRCC4 (see Figure 16-26), and the two-hybrid test identified a 33-kD interacting protein that was encoded by an uncharacterized human open reading frame. That two proteins interact in the yeast two-hybrid test does not necessarily mean that these proteins interact in human cells. To establish a connection between the 33-kD protein and the NHEJ pathway, the geneticists used another valuable technique from their toolbox, RNAi (see Chapter 8). In this case, they demonstrated that normal cells expressing antisense RNA from the ORF that encodes the 33-kD protein, which would prevent translation of this gene into protein, were now defective in the execution of the NHEJ pathway.

This story came full circle when the 2BN cells defective in double-strand repair were shown to lack the 33-kD protein. Expression of this protein corrected the cellular defects.

Message NHEJ is an error-prone pathway that repairs double-strand breaks in higher eukaryotes by ligating the free ends back together. The identification of genes responsible for inherited disorders is an important way used by geneticists to isolate formerly unknown components of repair and other biological pathways.

Homologous recombination   If a double-strand break occurs after replication of a chromosomal region in a dividing cell, the damage can be corrected by an error-free mechanism called **synthesis-dependent strand annealing (SDSA).** This mechanism is depicted in Figure 16-27. It uses the sister chromatids available in mitosis as the templates to ensure correct repair.

The first steps in SDSA are the binding of the broken ends by specialized proteins and enzymes, the trimming of the 5′ ends by an endonuclease to expose single-stranded regions, and the coating of these regions with proteins that include the RecA homolog, Rad51. Recall that in the SOS response, RecA monomers associate with regions of single-stranded DNA to form nucleoprotein filaments. Similarly, Rad51 forms long filaments as it associates with the exposed single-stranded region. The Rad51–DNA filament then takes part in a remarkable search of the undamaged sister chromatid for the complementary sequence that will be used as a template for DNA synthesis. This process is called strand invasion. The 3′ end of the invading strand displaces one of the undamaged sister chromatids, which forms a D-loop (for displacement), and primes DNA synthesis from its free 3′ end. New DNA synthesis continues from both 3′ ends until both strands unwind from their templates and anneal. Ligation seals the nicks, leaving a repaired patch of DNA that has one very distinctive feature: it has been replicated by a conservative process. That is, both strands are newly synthesized, which stands in marked contrast to the semiconservative replication of most DNA (see Chapter 7).

Message Synthesis-dependent strand annealing is an error-free mechanism that repairs double-strand breaks in dividing cells in which a sister chromatid is available to serve as template for repair synthesis.

The involvement of DSB repair in meiotic recombination

Our consideration of the repair of double-strand breaks in dividing cells leads naturally to the topic of crossing over at meiosis because a double-strand break initiates the crossover event. Although the breaks are a normal and essential part of meiosis, they are, if not processed correctly and efficiently, as dangerous as the accidental breaks discussed so far.

Crossing over is a remarkably precise process that takes place between two homologous chromosomes (Figure 16-28). That process was described in Section 4.8. Recall that recombination takes place after the replication fork has passed

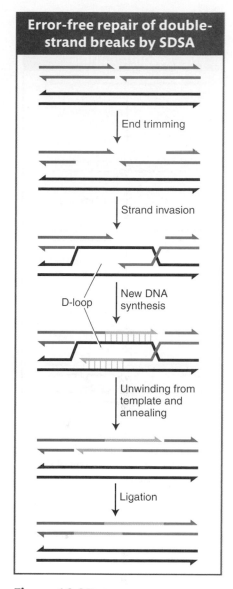

Figure 16-27 The error-free mechanism of synthesis-dependent strand annealing (SDSA) repairs double-strand breaks in dividing cells.

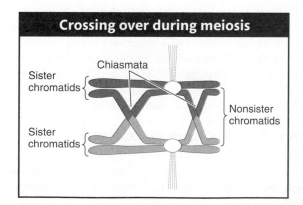

Figure 16-28 Exchange of chromosome arms between nonsister chromatids during meiosis yields a chiasma, the location of crossovers. Circles represent centromeres that are attached to the spindle fibers. [*Adapted from Mathew J. Neale and Scott Keeney, Nature 442, 2006, 153–158.*]

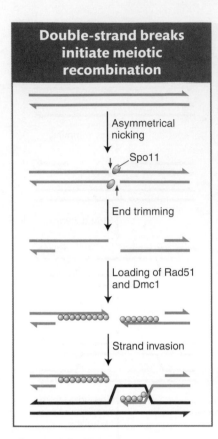

Double-strand breaks initiate meiotic recombination

Asymmetrical nicking

Spo11

End trimming

Loading of Rad51 and Dmc1

Strand invasion

Figure 16-29 Meiotic recombination is initiated when the enzyme Spo11 makes staggered nicks in a pair of DNA strands in a chromatid.

through a chromosomal region, forming two chromatids from each homologous chromosome. One chromatid from one homologous chromosome will recombine with a nonsister chromatid from the other homologous chromosome. For meiotic segregation to work correctly, every pair of homologs must have at least one crossover. Recombination is initiated when an enzyme called Spo11 makes DNA double-strand cuts in one of the chromatids that will recombine (Figure 16-29). Although first discovered in yeast, the Spo11 protein is widely conserved in eukaryotes, indicating that this mechanism to initiate recombination is widely employed.

After making its cuts, the Spo11 enzyme remains attached to the now free 5′ ends, where it appears to serve two purposes. First, it protects the ends from further damage, including spurious recombination with other free ends. Second, it may attract other proteins that are needed for the next step in recombination. That step is actually very similar to what happens in the repair of double-strand breaks in dividing cells. The 5′ ends are trimmed back (resected), and a protein complex binds to the single-stranded 3′ ends (see Figure 16-29). That complex includes the Rad51 protein, which, as already mentioned, is a homolog to the RecA protein that takes part in that remarkable search for complementarity in the sister chromatid.

At this time, meiotic recombination takes a dramatically different path from double-strand-break repair. In meiosis, Rad51 associates with another protein, Dmc1, which is present only during meiosis (see Figure 16-29). (It should be noted that the model organisms *Drosophila* and *C. elegans* do not have Dmc1 homologs.) Somehow, by an incompletely understood mechanism, the filament containing Rad51–Dmc1 conducts a search for a complementary sequence. However, in contrast with double-strand-break repair, the filament searches a nonsister chromatid from the homologous chromosome, not the sister chromatid. The search culminates in strand invasion and D-loop formation, just as in double-strand-break repair. These events are necessary for chiasma formation in meiosis I. That is, the homologs become connected as a result of recombination.

> **Message** Meiotic recombination is initiated by the Spo11 enzyme, which introduces double-strand cuts into chromosomes after they have replicated but before homologs separate.

16.5 Cancer: An Important Phenotypic Consequence of Mutation

Why do so many mutagenic agents cause cancer? What is the connection between cancer and mutation? In this section, we explore the mutation–cancer connection.

It has become clear that virtually all cancers of somatic cells arise owing to a series of special mutations that accumulate in a cell. Some of these mutations alter the activity of a gene; others simply eliminate the gene's activity. Cancer-promoting mutations fall into one of a few major categories: those that increase the ability of a cell to proliferate; those that decrease the susceptibility of a cell to a suicide pathway, called **apoptosis;** or those that increase the general mutation rate of the cell or its longevity so that all mutations, including those that encourage proliferation or apoptosis, are more likely to occur.

How cancer cells differ from normal cells

A malignant tumor, or **cancer,** is an aggregate of cells, all descended from an initial aberrant founder cell. In other words, the malignant cells are all members of a single clone, even in advanced cancers having multiple tumors at many sites in the body. Cancer cells typically differ from their normal neighbors by a host

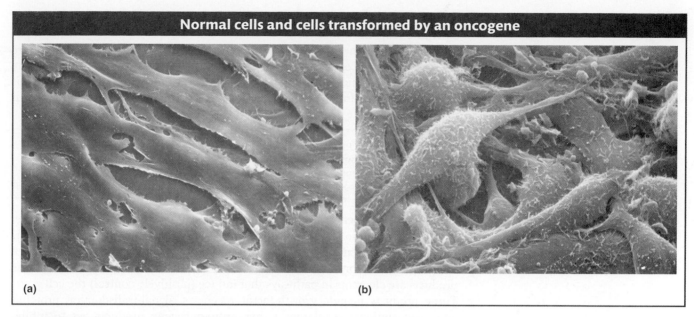

Normal cells and cells transformed by an oncogene

(a)

(b)

Figure 16-30 Scanning electron micrographs of (a) normal cells and (b) cells transformed by Rous sarcoma virus, which infects cells with the *src* oncogene. (a) A normal cell line called 3T3. Note the organized monolayer structure of the cells. (b) A transformed derivative of 3T3. Note how the cells are rounder and piled up on one another. [*Courtesy of L.-B. Chen.*]

of phenotypic characters, such as rapid division rate, ability to invade new cellular territories, high metabolic rate, and abnormal shape. For example, when cells from normal epithelial cell sheets are placed in cell culture, they can grow only when anchored to the culture dish itself. In addition, normal epithelial cells in culture divide only until they form a single continuous layer (Figure 16-30a). At that point, they somehow recognize that they have formed a single epithelial sheet and stop dividing. In contrast, malignant cells derived from epithelial tissue continue to proliferate, piling up on one another (Figure 16-30b).

Clearly, the factors regulating normal cellular physiology have been altered. What, then, is the underlying cause of cancer? Many different cell types can be converted into a malignant state. Is there a common theme? Or does each arise in a quite different way? We can think about cancer in a general way as being due to the accumulation of multiple mutations in a single cell that cause it to proliferate out of control. Some of those mutations may be transmitted from the parents through the germ line. But most arise de novo in the somatic-cell lineage of a particular cell.

Mutations in cancer cells

Several lines of evidence point to a genetic origin for the transformation of cells from the benign into the cancerous state. First, as already discussed in this chapter, many mutagenic agents such as chemicals and radiation cause cancer, suggesting that they produce cancer by introducing mutations into genes. Second, and most importantly, mutations that are frequently associated with particular kinds of cancers have been identified.

Two general kinds are associated with tumors: oncogene mutations and mutations in tumor-suppressor genes. **Oncogene** mutations act in the cancer cell as gain-of-function dominant mutations (see Chapter 6 for a discussion of dominant mutations). That statement suggests two key characteristics of oncogene mutations. First, the proteins encoded by oncogenes are usually *activated* in tumor cells, and, second, the mutation need be present in only one allele to contribute to tumor formation. The gene in its normal, unmutated form is called a **proto-oncogene.**

Mutations in **tumor-suppressor genes** that promote tumor formation are loss-of-function recessive mutations. That is, this type of mutation causes the encoded gene products to lose much or all of their activity (that is, the mutation is a null mutation). Moreover, for cancer to develop, the mutation must be present in both alleles of the gene.

> **Message** Oncogenes encode mutated forms of normal cellular proteins that result in dominant mutations, usually owing to their inappropriate activation. In contrast, tumor-suppressor genes encode proteins whose loss of activity can contribute to a cancerous state. As such, they are recessive mutations.

Classes of oncogenes Roughly a hundred different oncogenes have been identified. How do their normal counterparts, proto-oncogenes, function? Proto-oncogenes generally encode a class of proteins that are active only when the proper regulatory signals allow them to be activated. Many proto-oncogene products are elements in pathways that induce (positively control) the cell cycle. These products include growth-factor receptors, signal-transduction proteins, and transcriptional regulators. Other proto-oncogene products act to inhibit (negatively control) the apoptotic pathway that destroys damaged cells. In both types of oncogene mutation, the activity of the mutant protein has been uncoupled from its normal regulatory pathway, leading to its continuous unregulated expression. The continuously expressed protein product of an oncogene is called an **oncoprotein.** Several categories of oncogenes have been identified according to the different ways in which the regulatory functions have been uncoupled.

The *ras* oncogene can be used to illustrate what happens when a normal gene sustains a tumor-promoting mutation. As is often the case, the change from normal protein to oncoprotein entails structural modifications of the protein itself—in this case, caused by a simple point mutation. A single base-pair substitution that converts glycine into valine at amino acid number 12 of the Ras protein, for example, creates the oncoprotein found in human bladder cancer (Figure 16-31a).

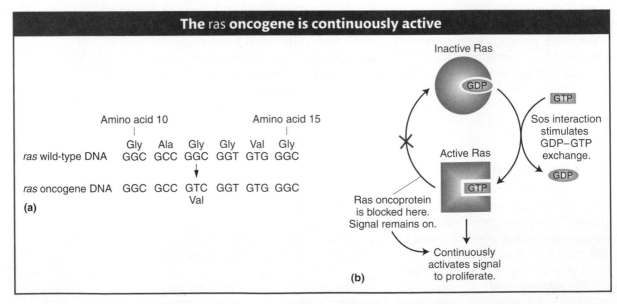

Figure 16-31 Formation and effect of the Ras oncoprotein. (a) The *ras* oncogene differs from the wild type by a single base pair, producing a Ras oncoprotein that differs from wild type in one amino acid, at position 12 in the *ras* open reading frame. (b) The Ras oncoprotein cannot hydrolyze GTP to GDP. Because of this defect, the Ras oncoprotein remains in the active Ras-GTP complex and continuously activates the signal to proliferate.

Table 16-1   Functions of Wild-Type Proteins and Properties of Tumor-Promoting Mutations in the Corresponding Genes

Wild-type protein function	Properties of tumor-promoting mutations
Promotes cell-cycle progression	Oncogene (gain of function)
Inhibits cell-cycle progression	Tumor-suppressor mutation (loss of function)
Promotes apoptosis	Tumor-suppressor mutation (loss of function)
Inhibits apoptosis	Oncogene (gain of function)
Promotes DNA repair	Tumor-suppressor mutation (loss of function)

The normal Ras protein is a G-protein subunit that takes part in signal transduction. It normally functions by cycling between the active GTP-bound state and the inactive GDP-bound state. The missense mutation in the *ras* oncogene produces an oncoprotein that always binds GTP (Figure 16-31b), even in the absence of normal signals. As a consequence, the Ras oncoprotein continuously propagates a signal that promotes cell proliferation.

Tumor-suppressor genes   The normal functions of tumor-suppressor genes fall into categories complementary to those of proto-oncogenes (Table 16-1). Some tumor-suppressor genes encode negative regulators whose normal function is to inhibit the cell cycle. Others encode positive regulators that normally activate apoptosis, or cell death, of a damaged cell. Still others are indirect players in cancer, with a normal role in the repair of damaged DNA or in controlling cellular longevity. We will consider one example here.

Mutations in the *p53* gene are associated with many types of tumors. In fact, estimates are that 50 percent of human tumors lack a functional *p53* gene. The active p53 protein is a transcriptional regulator that is activated in response to DNA damage. Activated wild-type p53 serves double duty: it prevents the progression of the cell cycle until the DNA damage is repaired, and, under some circumstances, it induces apoptosis. If no functional *p53* gene is present, the cell cycle progresses even if damaged DNA has not been repaired. The progression of the cell cycle into mitosis elevates the overall frequency of mutations, chromosomal rearrangements, and aneuploidy and thus increases the chances that other mutations that promote cell proliferation or block apoptosis will arise.

It is now clear that mutations able to elevate the mutation rate are important contributors to the progression of tumors in humans. These mutations are recessive mutations in tumor-suppressor genes that normally function in DNA-repair pathways. Mutations in these genes thus interfere with DNA repair. They promote tumor growth indirectly by elevating the mutation rate, which makes it much more likely that a series of oncogene and tumor-suppressor mutations will arise, corrupting the normal regulation of the cell cycle and programmed cell death. Large numbers of such tumor-suppressor-gene mutations have been identified, including some associated with heritable forms of cancer in specific tissues. Examples are the *BRCA1* and *BRCA2* mutations and breast cancer.

> **Message** Mutagenic agents can cause some cancers because cancer is, in part, caused by mutant versions of normal genes that lead to uncontrolled growth.

Summary

DNA change within a gene (point mutation) generally entails one or a few base pairs. Single-base-pair substitutions can create missense codons or nonsense (transcription termination) codons. A purine replaced by the other purine (or a pyrimidine replaced by the other pyrimidine) is a transition. A purine replaced by a pyrimidine (or vice versa) is a transversion. Single-base-pair additions or deletions (indels) produce frameshift mutations. Certain human genes that contain trinucleotide repeats—especially those that are expressed in neural tissue—become mutated through the expansion of these repeats and can thus cause disease. The formation of monoamino acid repeats within the polypeptides encoded by these genes is responsible for the mutant phenotypes.

Mutations can occur spontaneously as a by-product of normal cellular processes such as DNA replication or metabolism or they can be induced by mutagenic radiation or chemicals. Mutagens often result in a specific type of change because of their chemical specificity. For example, some produce exclusively $G \cdot C \rightarrow A \cdot T$ transitions; others, exclusively frameshifts.

Although mutations are necessary to generate diversity, many mutations are associated with inherited genetic diseases such as xeroderma pigmentosum. In addition, mutations that occur in somatic cells are the source of many human cancers. Many biological pathways have evolved to correct the broad spectrum of spontaneous and induced mutations. Some pathways, such as base- and nucleotide-excision repair and mismatch repair, use the information inherent in base complementarity to execute error-free repair. Other pathways that use bypass polymerases to correct damaged bases can introduce errors in the DNA sequence.

The correction of double-strand breaks is particularly important because these lesions can lead to destabilizing chromosomal rearrangements. Nonhomologous end joining is a pathway that ligates broken ends back together so that a stalled replication fork does not result in cell death. In replicating cells, double-strand breaks can be repaired in an error-free manner by the synthesis-dependent strand-annealing pathway, which utilizes the sister chromatid to repair the break.

Hundreds of programmed double-strand breaks initiate meiotic crossing over between nonsister chromatids. Just like other double-strand breaks, the meiotic breaks must be processed quickly and efficiently to prevent serious consequences such as cell death and cancer. Just how this repair is done is still being explored.

KEY TERMS

Ames test (p. 568)
apoptosis (p. 580)
apurinic site (p. 561)
base analog (p. 564)
base-excision repair (p. 569)
bypass (translesion) polymerase
    (p. 576)
cancer (p. 580)
Cockayne syndrome (p. 571)
conservative substitution (p. 556)
double-strand break (p. 577)
fluctuation test (p. 559)
frameshift mutation (p. 557)
global genomic repair (GGR) (p. 572)
heterogeneous genetic disorder
    (p. 572)
homology-dependent repair system
    (p. 569)

indel mutation (p. 560)
induced mutation (p. 558)
intercalating agent (p. 565)
mismatch repair (p. 574)
missense mutation (p. 556)
mutagen (p. 558)
mutagenesis (p. 564)
nonconservative substitution
    (p. 556)
nonhomologous end joining (NHEJ)
    (p. 578)
nonsense mutation (p. 556)
nucleotide-excision repair (NER)
    (p. 571)
oncogene (p. 581)
oncoprotein (p. 582)
point mutation (p. 554)
proto-oncogene (p. 582)

replica plating (p. 559)
SOS system (p. 576)
spontaneous lesion (p. 561)
spontaneous mutation (p. 558)
synonymous mutation (p. 556)
synthesis-dependent strand
    annealing (SDSA) (p. 579)
transcription-coupled nucleotide-
    excision repair (TC-NER)
    (p. 572)
transition (p. 555)
translesion (bypass) polymerase
    (p. 576)
translesion DNA synthesis (p. 576)
transversion (p. 555)
trinucleotide repeat (p. 563)
tumor-suppressor gene (p. 582)
xeroderma pigmentosum (XP) (p. 571)

SOLVED PROBLEMS

SOLVED PROBLEM 1. In Chapter 9, we learned that UAG and UAA codons are two of the chain-terminating nonsense triplets. On the basis of the specificity of aflatoxin B_1 and ethylmethanesulfonate (EMS), describe whether each mutagen would be able to revert these codons to wild type.

Solution

EMS induces primarily $G \cdot C \rightarrow A \cdot T$ transitions. UAG codons could not be reverted to wild type, because only the

UAG → UAA change would be stimulated by EMS and that generates a nonsense (ochre) codon. UAA codons would not be acted on by EMS. Aflatoxin B_1 induces primarily $G \cdot C \rightarrow T \cdot A$ transversions. Only the third position of UAG codons would be acted on, resulting in a UAG → UAU change (on the mRNA level), which produces tyrosine. Therefore, if tyrosine were an acceptable amino acid at the corresponding site in the protein, aflatoxin B_1 could revert UAG codons. Aflatoxin B_1 would not revert UAA codons, because no G·C base pairs appear at the corresponding position in the DNA.

SOLVED PROBLEM 2. Explain why mutations induced by acridines in phage T4 or by ICR-191 in bacteria cannot be reverted by 5-bromouracil.

Solution

Acridines and ICR-191 induce mutations by deleting or adding one or more base pairs, which results in a frameshift. However, 5-bromouracil induces mutations by causing the substitution of one base for another. This substitution cannot compensate for the frameshift resulting from ICR-191 and acridines.

PROBLEMS

Most of the problems are also available for review/grading through the GENETICSPORTAL www.yourgeneticsportal.com.

WORKING WITH THE FIGURES

1. In Figure 16-3a, what is the consequence of the new 5′ splice site on the open reading frame? In 16-3b, how big could the intron be to maintain the reading frame (let's say between 75 and 100 bp)?

2. Using Figure 16-4 as an example, compare the migration of RNA and protein for the wild-type gene and the mutation shown in Figure 16-3b. Assume that the retained intron maintains the reading frame.

3. In the Ames test shown in Figure 16-17, what is the reason for adding the liver extract to each sample?

4. Based on the mode of action of aflatoxin (Figure 16-16), propose a scenario that explains its response in the Ames test (Figure 16-18).

5. In Figure 16-22, point out the mutant protein(s) in patients with Cockayne syndrome. What protein(s) is/are mutant in patients with XP? How are these different mutations thought to account for the different disease symptoms?

6. The MutH protein nicks the newly synthesized strand (Figure 16-23). How does it "know" which strand this is?

7. What features of the bypass polymerase make it ideal for its role in translesion synthesis, shown in Figure 16-24?

BASIC PROBLEMS

8. Consider the following wild-type and mutant sequences:

   Wild-type ....CTTGCAAGCGAATC....

   Mutant   ....CTTGCTAGCGAATC....

   The substitution shown *seems* to have created a stop codon. What further information do you need to be confident that it has done so?

9. What type of mutation is depicted by the following sequences (shown as mRNA)?

   Wild type ....5′ AAUCCUUACGGA 3′....

   Mutant   ....5′ AAUCCUACGGA 3′....

10. Can a missense mutation of proline to histidine be made with a G·C → A·T transition-causing mutagen? What about a proline-to-serine missense mutation?

11. By base-pair substitution, what are all the synonymous changes that can be made starting with the codon CGG?

12. **a.** What are all the transversions that can be made starting with the codon CGG?

    b. Which of these transversions will be missense? Can you be sure?

13. **a.** Acridine orange is an effective mutagen for producing null alleles by mutation. Why does it produce null alleles?

    b. A certain acridine-like compound generates only single insertions. A mutation induced with this compound is treated with the same compound, and some revertants are produced. How is this outcome possible?

14. Defend the statement "Cancer is a genetic disease."

15. Give an example of a DNA-repair defect that leads to cancer.

16. In mismatch repair in *E. coli*, only a mismatch in the newly synthesized strand is corrected. How is *E. coli* able to recognize the newly synthesized strand? Why does this ability make biological sense?

17. A mutational lesion results in a sequence containing a mismatched base pair:

   5′ AGCT**G**CCTT 3′

   3′ ACG**AT**GGAA 5′

   Codon

   If mismatch repair occurs in either direction, which amino acids could be found at this site?

18. Under what circumstances could nonhomologous end joining be said to be error prone?

19. Why are many chemicals that test positive by the Ames test also classified as carcinogens?

20. The Spo11 protein is conserved in eukaryotes. Do you think it is also conserved in bacterial species? Justify your answer.

21. Differentiate between the elements of the following pairs:

    a. Transitions and transversions

    b. Synonymous and neutral mutations

c. Missense and nonsense mutations

d. Frameshift and nonsense mutations

22. Describe two spontaneous lesions that can lead to mutations.

23. What are bypass polymerases? How do they differ from the replicative polymerases? How do their special features facilitate their role in DNA repair?

24. In adult cells that have stopped dividing, what types of repair systems are possible?

25. A certain compound that is an analog of the base cytosine can become incorporated into DNA. It normally hydrogen bonds just as cytosine does, but it quite often isomerizes to a form that hydrogen bonds as thymine does. Do you expect this compound to be mutagenic, and, if so, what types of changes might it induce at the DNA level?

26. Two pathways, homologous recombination and nonhomologous end joining (NHEJ), can repair double-strand breaks in DNA. If homologous recombination is an error-free pathway whereas NHEJ is not always error free, why is NHEJ used most of the time in eukaryotes?

27. Which repair pathway recognizes DNA damage during transcription? What happens if the damage is not repaired?

CHALLENGING PROBLEMS

28. a. Why is it impossible to induce nonsense mutations (represented at the mRNA level by the triplets UAG, UAA, and UGA) by treating wild-type strains with mutagens that cause only A·T → G·C transitions in DNA?

b. Hydroxylamine (HA) causes only G·C → A·T transitions in DNA. Will HA produce nonsense mutations in wild-type strains?

c. Will HA treatment revert nonsense mutations?

29. Several auxotrophic point mutants in *Neurospora* are treated with various agents to see if reversion will take place. The following results were obtained (a plus sign indicates reversion; HA causes only G·C → A·T transitions).

Mutant	5-BU	HA	Proflavin	Spontaneous reversion
1	−	−	−	−
2	−	−	+	+
3	+	−	−	+
4	−	−	−	+
5	+	+	−	+

a. For each of the five mutants, describe the nature of the original mutation event (not the reversion) at the molecular level. Be as specific as possible.

b. For each of the five mutants, name a possible mutagen that could have caused the original mutation event. (Spontaneous mutation is not an acceptable answer.)

c. In the reversion experiment for mutant 5, a particularly interesting prototrophic derivative is obtained. When this type is crossed with a standard wild-type strain, the progeny consist of 90 percent prototrophs and 10 percent auxotrophs. Give a full explanation for these results, including a precise reason for the frequencies observed.

30. You are using nitrosoguanidine to "revert" mutant *nic-2* (nicotinamide-requiring) alleles in *Neurospora*.

You treat cells, plate them on a medium without nicotinamide, and look for prototrophic colonies. You obtain the following results for two mutant alleles. Explain these results at the molecular level, and indicate how you would test your hypotheses.

a. With *nic-2* allele 1, you obtain no prototrophs at all.

b. With *nic-2* allele 2, you obtain three prototrophic colonies A, B, and C, and you cross each separately with a wild-type strain. From the cross prototroph A × wild type, you obtain 100 progeny, all of which are prototrophic. From the cross prototroph B × wild type, you obtain 100 progeny, of which 78 are prototrophic and 22 are nicotinamide requiring. From the cross prototroph C × wild type, you obtain 1000 progeny, of which 996 are prototrophic and 4 are nicotinamide requiring.

31. You are working with a newly discovered mutagen, and you wish to determine the base change that it introduces into DNA. Thus far, you have determined that the mutagen chemically alters a single base in such a way that its base-pairing properties are altered permanently. To determine the specificity of the alteration, you examine the amino acid changes that take place after mutagenesis. A sample of what you find is shown here:

Original:	Gln–His–Ile–Glu–Lys
Mutant:	Gln–His–Met–Glu–Lys
Original:	Ala–Val–Asn–Arg
Mutant:	Ala–Val–Ser–Arg
Original:	Arg–Ser–Leu
Mutant:	Arg–Ser–Leu–Trp–Lys–Thr–Phe

What is the base-change specificity of the mutagen?

32. You now find an additional mutant from the experiment in Problem 31:

Original:	Ile–Leu–His–Gln
Mutant:	Ile–Pro–His–Gln

Could the base-change specificity in your answer to Problem 31 account for this mutation? Why or why not?

33. You are an expert in DNA-repair mechanisms. You receive a sample of a human cell line derived from a woman who has symptoms of xeroderma pigmentosum. You determine that she has a mutation in a gene that has not been previously associated with XP. How is this possible?

34. Ozone (O_3) is an important naturally occurring component in our atmosphere, where it forms a layer that absorbs UV radiation. A hole in the ozone layer was discovered in the 1970s over Antarctica and Australia. The hole appears seasonally and was found to be due to human activity. Specifically, ozone is destroyed by a class of chemicals (called CFCs for chlorofluorocarbons) that are found in refrigerants, air-conditioning systems, and aerosols.

As a scientist working on DNA-repair mechanisms, you discover that there has been a significant increase in skin cancer in the beach communities in Australia. A newspaper reporter friend offers to let you publish a short note (a paragraph) in which you are to describe the possible connection between the ozone hole and the increased skin cancers. On the basis of what you have learned about DNA repair in this chapter, write a paragraph that explains the mechanistic connection.

Large-Scale Chromosomal Changes

17

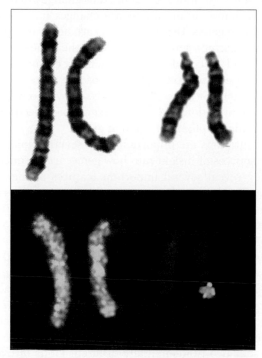

A reciprocal translocation demonstrated by chromosome painting. A suspension of chromosomes from many cells is passed through an electronic device that sorts the chromosomes by size. DNA is extracted from individual chromosomes, denatured, bound to one of several fluorescent dyes, and then added to partly denatured chromosomes on a slide. The fluorescent DNA "finds" its own chromosome and binds along its length by base complementarity, thus "painting" it. In this example, a red and a green dye have been used to paint different chromosomes. The figure shows unpainted (above) and painted (below) preparations. The painted preparation shows one normal green chromosome, one normal red, and two that have exchanged segments. [*Addenbrookes Hospital/Photo Researchers.*]

KEY QUESTIONS

- How do polyploids (3*n*, 4*n*, and so forth) arise, and what are their properties?

- How do aneuploids (2*n* − 1, 2*n* + 1, and so forth) arise, and what are their properties?

- How do duplications and deletions arise, and what are their properties?

- How do inversions and translocations arise, and what are their properties?

- What is the relevance of such changes to human beings?

A young couple is planning to have children. The husband knows that his grandmother had a child with Down syndrome by a second marriage. Down syndrome is a set of physical and mental disorders caused by the presence of an extra chromosome 21 (Figure 17-1). No records of the birth, which occurred early in the twentieth century, are available, but the couple knows of no other cases of Down syndrome in their families.

The couple has heard that Down syndrome results from a rare chance mistake in egg production and therefore decide that they stand only a low chance of having such a child. They decide to have children. Their first child is unaffected, but the next conception aborts spontaneously (a miscarriage), and their second child is born with Down syndrome. Was their having a Down syndrome child

OUTLINE

17.1 Changes in chromosome number

17.2 Changes in chromosome structure

17.3 Overall incidence of human chromosome mutations

589

Figure 17-1 Down syndrome results from having an extra copy of chromosome 21. [*Bob Daemmrich/The Image Works.*]

a coincidence or did a connection between the genetic makeup of the child's father and that of his grandmother lead to their both having Down syndrome children? Was the spontaneous abortion significant? What tests might be necessary to investigate this situation? The analysis of such questions is the topic of this chapter.

We have seen throughout the book that gene mutations are an important source of change in the genomic sequence. However, the genome can also be remodeled on a larger scale by alterations to chromosome structure or by changes in the number of copies of chromosomes in a cell. These large-scale variations are termed **chromosome mutations** to distinguish them from gene mutations. Broadly speaking, gene mutations are defined as changes that take place within a gene, whereas chromosome mutations are changes in a chromosome region encompassing multiple genes. Gene mutations are never detectable microscopically; a chromosome bearing a gene mutation looks the same under the microscope as one carrying the wild-type allele. In contrast, many chromosome mutations are detectable with the use of microscopy. Chromosome mutations can be detected by microscopy, by genetic or molecular analysis, or by a combination of all techniques. Chromosome mutations have been best characterized in eukaryotes, and all the examples in this chapter are from that group.

Chromosome mutations are important from several biological perspectives. First, they can be sources of insight into how genes act in concert on a genomic scale. Second, they reveal several important features of meiosis and chromosome architecture. Third, they constitute useful tools for experimental genomic manipulation. Fourth, they are sources of insight into evolutionary processes. Fifth, chromosomal mutations are regularly found in humans, and some of these mutations cause genetic disease.

Many chromosome mutations cause abnormalities in cell and organismal function. Most of these abnormalities stem from changes in *gene number* or *gene position*. In some cases, a chromosome mutation results from chromosome breakage. If the break occurs within a gene, the result is functional *disruption* of that gene.

For our purposes, we will divide chromosome mutations into two groups: changes in chromosome *number* and changes in chromosome *structure*. These two groups represent two fundamentally different kinds of events. Changes in chromosome number are not associated with structural alterations of any of the DNA molecules of the cell. Rather, it is the *number* of these DNA molecules that is changed, and this change in number is the basis of their genetic effects. Changes

Figure 17-2 The illustration is divided into three colored regions to depict the main types of chromosome mutations that can occur: the loss, gain, or relocation of entire chromosomes or chromosome segments. The wild-type chromosome is shown in the center.

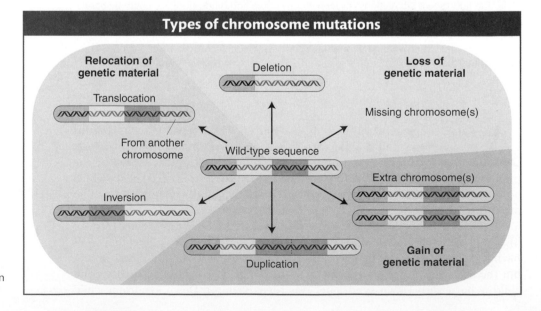

Types of chromosome mutations

Relocation of genetic material
Translocation
From another chromosome
Inversion

Deletion

Wild-type sequence

Duplication

Loss of genetic material
Missing chromosome(s)
Extra chromosome(s)
Gain of genetic material

in chromosome structure, on the other hand, result in novel sequence arrangements within one or more DNA double helices. These two types of chromosome mutations are illustrated in Figure 17-2, which is a summary of the topics of this chapter. We begin by exploring the nature and consequences of changes in chromosome number.

17.1 Changes in Chromosome Number

In genetics as a whole, few topics impinge on human affairs quite so directly as that of changes in the number of chromosomes present in our cells. Foremost is the fact that a group of common genetic disorders results from the presence of an abnormal number of chromosomes. Although this group of disorders is small, it accounts for a large proportion of the genetically determined health problems that afflict humans. Also of relevance to humans is the role of chromosome mutations in plant breeding: plant breeders have routinely manipulated chromosome number to improve commercially important agricultural crops.

Changes in chromosome number are of two basic types: changes in *whole* chromosome sets, resulting in a condition called *aberrant euploidy,* and changes in *parts* of chromosome sets, resulting in a condition called *aneuploidy.*

Aberrant euploidy

Organisms with multiples of the basic chromosome set (genome) are referred to as **euploid.** You learned in earlier chapters that familiar eukaryotes such as plants, animals, and fungi carry in their cells either one chromosome set (haploidy) or two chromosome sets (diploidy). In these species, both the haploid and the diploid states are cases of normal euploidy. Organisms that have more or fewer than the normal number of sets are aberrant euploids. **Polyploids** are individual organisms that have more than two chromosome sets. They can be represented by $3n$ **(triploid),** $4n$ **(tetraploid),** $5n$ **(pentaploid),** $6n$ **(hexaploid),** and so forth. (The number of chromosome sets is called the ploidy or ploidy level.) An individual member of a normally diploid species that has only one chromosome set (n) is called a **monoploid** to distinguish it from an individual member of a normally haploid species (also n). Examples of these conditions are shown in the first four rows of Table 17-1.

Monoploids   Male bees, wasps, and ants are monoploid. In the normal life cycles of these insects, males develop by **parthenogenesis** (the development

Table 17-1   Chromosome Constitutions in a Normally Diploid Organism with Three Chromosomes (Identified as A, B, and C) in the Basic Set

Name	Designation	Constitution	Number of chromosomes
Euploids			
Monoploid	n	A B C	3
Diploid	$2n$	AA BB CC	6
Triploid	$3n$	AAA BBB CCC	9
Tetraploid	$4n$	AAAA BBBB CCCC	12
Aneuploids			
Monosomic	$2n - 1$	A BB CC	5
		AA B CC	5
		AA BB C	5
Trisomic	$2n + 1$	AAA BB CC	7
		AA BBB CC	7
		AA BB CCC	7

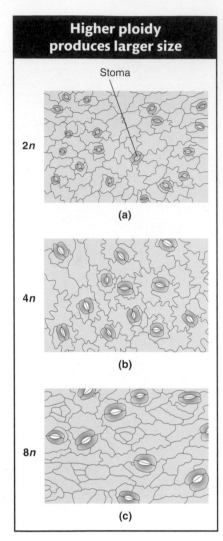

Higher ploidy produces larger size

Stoma

2n

(a)

4n

(b)

8n

(c)

Figure 17-3 Epidermal leaf cells of tobacco plants with increasing ploidy. Cell size increases, particularly evident in stoma size, with an increase in ploidy. (a) Diploid; (b) tetraploid; (c) octoploid. [*From W. Williams, Genetic Principles and Plant Breeding. Blackwell Scientific Publications, Ltd., 1964.*]

of a specialized type of unfertilized egg into an embryo without the need for fertilization). In most other species, however, monoploid zygotes fail to develop. The reason is that virtually all members of a diploid species carry a number of deleterious recessive mutations, together called a **"genetic load."** The deleterious recessive alleles are masked by wild-type alleles in the diploid condition but are automatically expressed in a monoploid derived from a diploid. Monoploids that do develop to advanced stages are abnormal. If they survive to adulthood, their germ cells cannot proceed through meiosis normally, because the chromosomes have no pairing partners. Thus, monoploids are characteristically sterile. (Male bees, wasps, and ants bypass meiosis; in these groups, gametes are produced by *mitosis*.)

Polyploids   Polyploidy is very common in plants but rarer in animals (for reasons that we will consider later). Indeed, an increase in the number of chromosome sets has been an important factor in the origin of new plant species. The evidence for this benefit is that above a haploid number of about 12, even numbers of chromosomes are much more common than odd numbers. This pattern is a consequence of the polyploid origin of many plant species, because doubling and redoubling of a number can give rise only to even numbers. Animal species do not show such a distribution, owing to the relative rareness of polyploid animals.

In aberrant euploids, there is often a correlation between the number of copies of the chromosome set and the size of the organism. A tetraploid organism, for example, typically looks very similar to its diploid counterpart in its proportions, except that the tetraploid is bigger, both as a whole and in its component parts. The higher the ploidy level, the larger the size of the organism (Figure 17-3).

> **Message** Polyploids are often larger and have larger component parts than their diploid relatives.

In the realm of polyploids, we must distinguish between **autopolyploids,** which have multiple chromosome sets originating from within one species, and **allopolyploids,** which have sets from two or more different species. Allopolyploids form only between closely related species; however, the different chromosome sets are only **homeologous** (partly homologous), not fully homologous as they are in autopolyploids.

Autopolyploids   Triploids ($3n$) are usually autopolyploids. They arise spontaneously in nature, but they can be constructed by geneticists from the cross of a $4n$ (tetraploid) and a $2n$ (diploid). The $2n$ and the n gametes produced by the tetraploid and the diploid, respectively, unite to form a $3n$ triploid. Triploids are characteristically sterile. The problem (which is also true of monoploids) lies in the presence of unpaired chromosomes at meiosis. The molecular mechanisms for synapsis, or true pairing, dictate that, in a triploid, pairing can take place between only two of the three chromosomes of each type (Figure 17-4). Paired homologs (**bivalents**) segregate to opposite poles, but the unpaired homologs (**univalents**) pass to either pole randomly. In a **trivalent,** a paired group of three, the paired centromeres segregate as a bivalent and the unpaired one as a univalent. These segregations take place for every chromosome threesome; so, for any chromosomal type, the gamete could receive either one or two chromosomes. It is unlikely that a gamete will receive *two* for *every* chromosomal type or that it will receive *one* for *every* chromosomal type. Hence, the likelihood is that gametes will have chromosome numbers intermediate between the haploid and the diploid number; such genomes are of a type called **aneuploid** ("not euploid").

Aneuploid gametes do not generally give rise to viable offspring. In plants, aneuploid pollen grains are generally inviable and hence unable to fertilize the

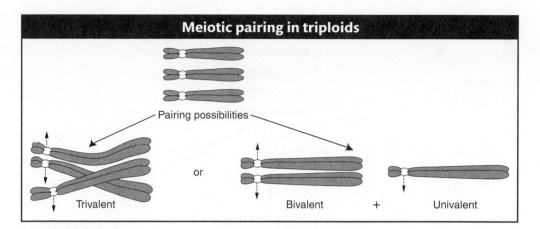

Meiotic pairing in triploids

Pairing possibilities

or

Trivalent   Bivalent   +   Univalent

Figure 17-4 The three homologous chromosomes of a triploid may pair in two ways at meiosis, as a trivalent or as a bivalent plus a univalent.

female gamete. In any organism, zygotes that might arise from the fusion of a haploid and an aneuploid gamete will themselves be aneuploid, and typically these zygotes also are inviable. We will examine the underlying reason for the inviability of aneuploids when we consider gene balance later in the chapter.

Message Polyploids with odd numbers of chromosome sets, such as triploids, are sterile or highly infertile because their gametes and offspring are aneuploid.

Autotetraploids arise by the doubling of a $2n$ complement to $4n$. This doubling can occur spontaneously, but it can also be induced artificially by applying chemical agents that disrupt microtubule polymerization. As stated in Chapter 2, chromosome segregation is powered by spindle fibers, which are polymers of the protein tubulin. Hence, disruption of microtubule polymerization blocks chromosome segregation. The chemical treatment is normally applied to somatic tissue during the formation of spindle fibers in cells undergoing division. The resulting polyploid tissue (such as a polyploid branch of a plant) can be detected by examining stained chromosomes from the tissue under a microscope. Such a branch can be removed and used as a cutting to generate a polyploid plant or allowed to produce flowers, which, when selfed, would produce polyploid offspring. A commonly used antitubulin agent is colchicine, an alkaloid extracted from the autumn crocus. In colchicine-treated cells, the S phase of the cell cycle takes place, but chromosome segregation or cell division does not. As the treated cell enters telophase, a nuclear membrane forms around the entire doubled set of chromosomes. Thus, treating diploid ($2n$) cells with colchicine for one cell cycle leads to tetraploids ($4n$) with exactly four copies of each type of chromosome (Figure 17-5). Treatment for an additional cell cycle produces octoploids ($8n$), and so forth. This method works in both plant and animal cells, but, generally, plants seem to be much

Figure 17-5 Colchicine may be applied to generate a tetraploid from a diploid. Colchicine added to mitotic cells during metaphase and anaphase disrupts spindle-fiber formation, preventing the migration of chromatids after the centromere has split. A single cell is created that contains pairs of identical chromosomes that are homozygous at all loci.

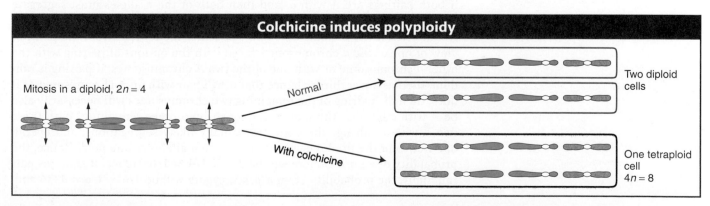

Colchicine induces polyploidy

Mitosis in a diploid, $2n = 4$

Normal → Two diploid cells

With colchicine → One tetraploid cell $4n = 8$

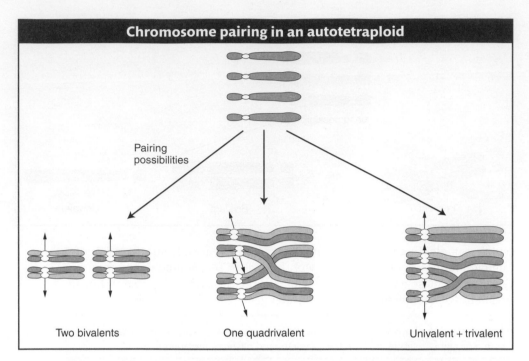

Figure 17-6 There are three different pairing possibilities at meiosis in tetraploids. The four homologous chromosomes may pair as two bivalents or as a quadrivalent, and each can yield functional gametes. A third possibility, a trivalent plus a univalent, yields nonfunctional gametes.

GENETICS*PORTAL*   ANIMATED ART: **Autotetraploid meiosis**

more tolerant of polyploidy. Note that all alleles in the genotype are doubled. Therefore, if a diploid cell of genotype *A/a* ; *B/b* is doubled, the resulting autotetraploid will be of genotype *A/A/a/a* ; *B/B/b/b*.

Because four is an even number, autotetraploids can have a regular meiosis, although this result is by no means always the case. The crucial factor is how the four chromosomes of each set pair and segregate. There are several possibilities, as shown in Figure 17-6. If the chromosomes pair as bivalents or quadrivalents, the chromosomes segregate normally, producing diploid gametes. The fusion of gametes at fertilization regenerates the tetraploid state. If trivalents form, segregation leads to nonfunctional aneuploid gametes and, hence, sterility.

What genetic ratios are produced by an autotetraploid? Assume for simplicity that the tetraploid forms only bivalents. If we start with an *A/A/a/a* tetraploid plant and self it, what proportion of progeny will be *a/a/a/a*? We first need to deduce the frequency of *a/a* gametes because this type is the only one that can produce a recessive homozygote. The *a/a* gametes can arise only if both pairings are *A* with *a*, and then both of the *a* alleles must segregate to the same pole. Let's use the following thought experiment to calculate the frequencies of the possible outcomes. Consider the options from the point of view of one of the *a* chromosomes faced with the options of pairing with the other *a* chromosome or with one of the two *A* chromosomes; if pairing is random, there is a two-thirds chance that it will pair with an *A* chromosome. If it does, then the pairing of the remaining two chromosomes will necessarily also be *A* with *a* because those are the only chromosomes remaining. With these two *A*-with-*a* pairings there are two equally likely segregations, and overall one-fourth of the products will contain both *a* alleles at one pole. Hence, the probability of an *a/a* gamete will be $2/3 \times 1/4 = 1/6$. Hence, if gametes pair randomly, the probability of an *a/a/a/a* zygote will be $1/6 \times 1/6 = 1/36$ and, by subtraction, the probability of $A/-/-/-$ will be 35/36. Therefore, a 35:1 phenotypic ratio is expected.

Allopolyploids An allopolyploid is a plant that is a hybrid of two or more species, containing two or more copies of each of the input genomes. The prototypic allopolyploid was an allotetraploid synthesized by Georgi Karpechenko in 1928. He wanted to make a fertile hybrid that would have the leaves of the cabbage (*Brassica*) and the roots of the radish (*Raphanus*), because they were the agriculturally important parts of each plant. Each of these two species has 18 chromosomes, and so $2n_1 = 2n_2 = 18$, and $n_1 = n_2 = 9$. The species are related closely enough to allow intercrossing. Fusion of an n_1 and an n_2 gamete produced a viable hybrid progeny individual of constitution $n_1 + n_2 = 18$. However, this hybrid was functionally sterile because the 9 chromosomes from the cabbage parent were different enough from the radish chromosomes that pairs did not synapse and segregate normally at meiosis, and thus the hybrid could not produce functional gametes.

Eventually, one part of the hybrid plant produced some seeds. On planting, these seeds produced fertile individuals with 36 chromosomes. All these individuals were allopolyploids. They had apparently been derived from spontaneous, accidental chromosome doubling to $2n_1 + 2n_2$ in one region of the sterile hybrid, presumably in tissue that eventually became a flower and underwent meiosis to produce gametes. In $2n_1 + 2n_2$ tissue, there is a pairing partner for each chromosome, and functional gametes of the type $n_1 + n_2$ are produced. These gametes fuse to give $2n_1 + 2n_2$ allopolyploid progeny, which also are fertile. This kind of allopolyploid is sometimes called an **amphidiploid,** or doubled diploid (Figure 17-7). Treating a sterile hybrid with colchicine greatly increases the chances that the chromosome sets will double. Amphidiploids are now synthesized routinely in this manner. (Unfortunately for Karpechenko, his amphidiploid had the roots of a cabbage and the leaves of a radish.)

When Karpechenko's allopolyploid was crossed with either parental species—the cabbage or the radish—sterile offspring resulted. The offspring of the cross with cabbage were $2n_1 + n_2$, constituted from an $n_1 + n_2$ gamete from the allopolyploid and an n_1 gamete from the cabbage. The n_2 chromosomes had no pairing partners; hence, a normal meiosis could not take place, and the offspring were sterile. Thus, Karpechenko had effectively created a new species, with no possibility of gene exchange with either cabbage or radish. He called his new plant *Raphanobrassica.*

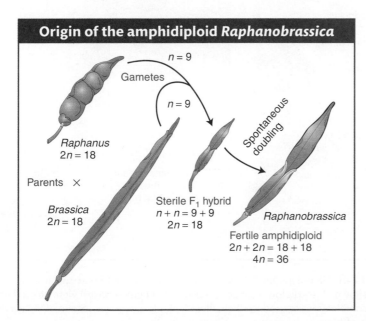

Figure 17-7 In the progeny of a cross of cabbage (*Brassica*) and radish (*Raphanus*), the fertile amphidiploid arose from spontaneous doubling in the $2n = 18$ sterile hybrid. [*From A. M. Srb, R. D. Owen, and R. S. Edgar, General Genetics, 2nd ed. Copyright 1965 by W. H. Freeman and Company. After G. Karpechenko, Z. Indukt. Abst. Vererb. 48, 1928, 27.*]

In nature, allopolyploidy seems to have been a major force in the evolution of new plant species. One convincing example is shown by the genus *Brassica,* as illustrated in Figure 17-8. Here, three different parent species have hybridized in all possible pair combinations to form new amphidiploid species. Natural polyploidy was once viewed as a somewhat rare occurrence, but recent work has shown that it is a recurrent event in many plant species. The use of DNA markers has made it possible to show that polyploids in any population or area that appear to be the same are the result of many independent past fusions between genetically distinct individuals of the same two parental species. An estimated 50 percent of all angiosperm plants are polyploids, resulting from auto- or allopolyploidy. As a result of multiple polyploidizations, the amount of allelic variation within a polyploid species is much higher than formerly thought, perhaps contributing to its potential for adaptation.

A particularly interesting natural allopolyploid is bread wheat, *Triticum aestivum* ($6n = 42$). By studying its wild relatives, geneticists have reconstructed a probable evolutionary history of this plant. Figure 17-9 shows that bread wheat is composed of two sets each of three ancestral genomes. At meiosis, pairing is

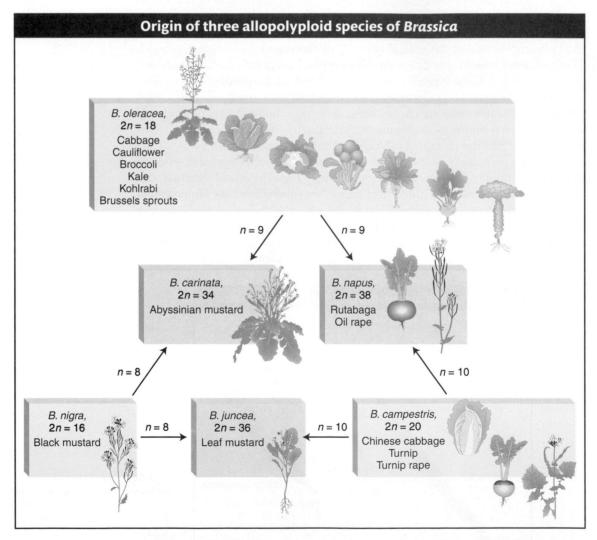

Figure 17-8 Allopolyploidy is important in the production of new species. In the example shown, three diploid species of *Brassica* (light green boxes) were crossed in different combinations to produce their allopolyploids (tan boxes). Some of the agricultural derivatives of some of the species are shown within the boxes.

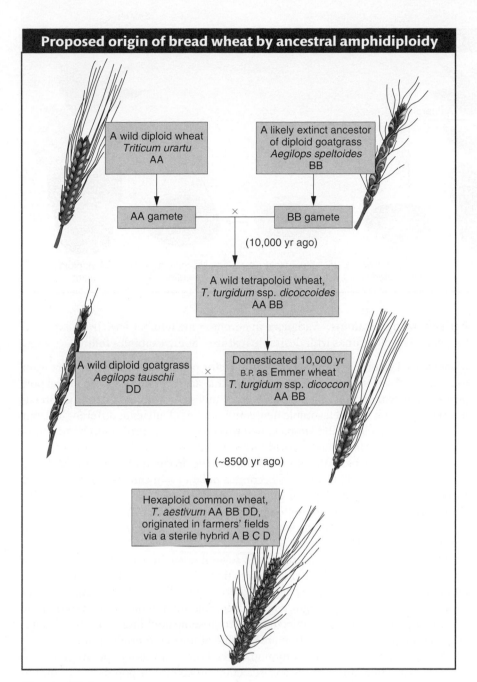

Proposed origin of bread wheat by ancestral amphidiploidy

A wild diploid wheat
Triticum urartu
AA

A likely extinct ancestor
of diploid goatgrass
Aegilops speltoides
BB

AA gamete × BB gamete

(10,000 yr ago)

A wild tetraploid wheat,
T. turgidum ssp. *dicoccoides*
AA BB

A wild diploid goatgrass
Aegilops tauschii
DD
×
Domesticated 10,000 yr
B.P. as Emmer wheat
T. turgidum ssp. *dicoccon*
AA BB

(~8500 yr ago)

Hexaploid common wheat,
T. aestivum AA BB DD,
originated in farmers' fields
via a sterile hybrid A B C D

Figure 17-9 Modern wheat arose from two ancestral cases of amphidiploidy, first by unreduced gametes, second via a sterile intermediate.

always between homologs from the same ancestral genome. Hence, in bread-wheat meiosis, there are always 21 bivalents.

Allopolyploid plant cells can also be produced artificially by fusing diploid cells from different species. First, the walls of two diploid cells are removed by treatment with an enzyme, and the membranes of the two cells fuse and become one. The nuclei often fuse, too, resulting in the polyploid. If the cell is nurtured with the appropriate hormones and nutrients, it divides to become a small allopolyploid plantlet, which can then be transferred to soil.

Message Allopolyploid plants can be synthesized by crossing related species and doubling the chromosomes of the hybrid or by fusing diploid cells.

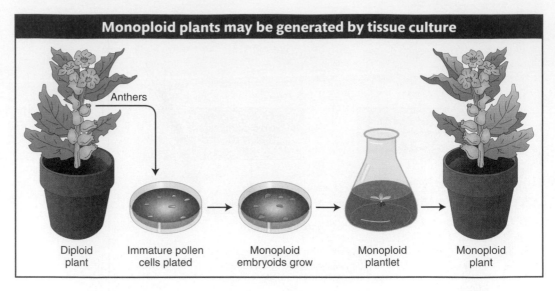

Monoploid plants may be generated by tissue culture

Anthers

Diploid plant | Immature pollen cells plated | Monoploid embryoids grow | Monoploid plantlet | Monoploid plant

Agricultural applications   Variations in chromosome number have been exploited to create new plant lines with desirable features. Some examples follow.

Monoploids   Diploidy is an inherent nuisance for plant breeders. When they want to induce and select new recessive mutations that are favorable for agricultural purposes, the new mutations cannot be detected unless they are homozygous. Breeders may also want to find favorable new combinations of alleles at different loci, but such favorable allele combinations in heterozygotes will be broken up by recombination at meiosis. Monoploids provide a way around some of these problems.

Monoploids can be artificially derived from the products of meiosis in a plant's anthers. A haploid cell destined to become a pollen grain can instead be induced by cold treatment (subjected to low temperatures) to grow into an **embryoid,** a small dividing mass of monoploid cells. The embryoid can be grown on agar to form a monoploid plantlet, which can then be potted in soil and allowed to mature (Figure 17-10).

Plant monoploids can be exploited in several ways. In one approach, they are first examined for favorable allelic combinations that have arisen from the recombination of alleles already present in a heterozygous diploid parent. Hence, from a parent that is A/a ; B/b might come a favorable monoploid combination a ; b. The monoploid can then be subjected to chromosome doubling to produce homozygous diploid cells, a/a ; b/b, that are capable of normal reproduction.

Another approach is to treat monoploid cells basically as a population of haploid organisms in a mutagenesis-and-selection procedure. A population of monoploid cells is isolated, their walls are removed by enzymatic treatment, and they are exposed to a mutagen. They are then plated on a medium that selects for some desirable phenotype. This approach has been used to select for resistance to toxic compounds produced by a plant parasite as well as to select for resistance to herbicides being used by farmers to kill weeds. Resistant plantlets eventually grow into monoploid plants, whose chromosome number can then be doubled with the use of colchicine, leading to a resistant homozygous diploid. These powerful techniques can circumvent the normally slow process of meiosis-based plant breeding. They have been successfully applied to important crop plants such as soybeans and tobacco.

> **Message**   Geneticists can create new plant lines by producing monoploids with favorable genotypes and then doubling their chromosomes to form fertile, homozygous diploids.

Autotriploids The bananas that are widely available commercially are sterile triploids with 11 chromosomes in each set ($3n = 33$). The most obvious expression of the sterility of bananas is the absence of seeds in the fruit that we eat. (The black specks in bananas are not seeds; banana seeds are rock hard—real tooth breakers.) Seedless watermelons are another example of the commercial exploitation of triploidy in plants.

Autotetraploids Many autotetraploid plants have been developed as commercial crops to take advantage of their increased size (Figure 17-11). Large fruits and flowers are particularly favored.

Allopolyploids Allopolyploidy (formation of polyploids between different species) has been important in the production of modern crop plants. New World cotton is a natural allopolyploid that arose spontaneously, as is wheat. Allopolyploids are also synthesized artificially to combine the useful features of parental species into one type. Only one synthetic amphidiploid has ever been widely used commercially, a crop known as *Triticale.* It is an amphidiploid between wheat (*Triticum*, $6n = 42$) and rye (*Secale*, $2n = 14$). Hence, for *Triticale*, $2n = 2 \times (21 + 7) = 56$. This novel plant combines the high yields of wheat with the ruggedness of rye.

Polyploid animals   As noted earlier, polyploidy is more common in plants than in animals, but there are cases of naturally occurring polyploid animals. Polyploid species of flatworms, leeches, and brine shrimps reproduce by parthenogenesis. Triploid and tetraploid *Drosophila* have been synthesized experimentally. However, examples are not limited to these so-called lower forms. Naturally occurring polyploid amphibians and reptiles are surprisingly common. They have several modes of reproduction: polyploid species of frogs and toads participate in sexual reproduction, whereas polyploid salamanders and lizards are parthenogenetic. The Salmonidae (the family of fishes that includes salmon and trout) provide a familiar example of the numerous animal species that appear to have originated through ancestral polyploidy.

The sterility of triploids has been commercially exploited in animals as well as in plants. Triploid oysters have been developed because they have a commercial advantage over their diploid relatives. The diploids go through a spawning season, when they are unpalatable, but the sterile triploids do not spawn and are palatable year-round.

Figure 17-11 Diploid (*left*) and tetraploid (*right*) watermelon leaves and flowers. [*Michael E. Compton, University of Wisconsin-Platteville.*]

Aneuploidy

Aneuploidy is the second major category of chromosomal aberrations in which the chromosome number is abnormal. An aneuploid is an individual organism whose chromosome number differs from the wild type by part of a chromosome set. Generally, the aneuploid chromosome set differs from the wild type by only one chromosome or by a small number of chromosomes. An aneuploid can have a chromosome number either greater or smaller than that of the wild type. Aneuploid nomenclature (see Table 17-1) is based on the number of copies of the specific chromosome in the aneuploid state. For autosomes in diploid organisms, the aneuploid $2n + 1$ is **trisomic,** $2n - 1$ is **monosomic,** and $2n - 2$ (the "$- 2$" represents the loss of both homologs of a chromosome) is **nullisomic.** In haploids, $n + 1$ is **disomic.** Special notation is used to describe sex-chromosome aneuploids because it must deal with the two different chromosomes. The notation merely lists the copies of each sex chromosome, such as XXY, XYY, XXX, or XO (the "O" stands for absence of a chromosome and is included to show that the single X symbol is not a typographical error).

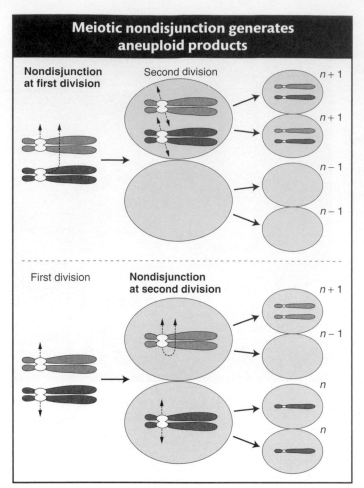

Figure 17-12 Aneuploid products of meiosis (that is, gametes) are produced by nondisjunction at the first or second meiotic division. Note that all other chromosomes are present in normal number, including in the cells in which no chromosomes are shown.

GENETICS PORTAL   **ANIMATED ART:** Meiotic nondisjunction

Nondisjunction   The cause of most aneuploidy is **nondisjunction** in the course of meiosis or mitosis. *Disjunction* is another word for the normal segregation of homologous chromosomes or chromatids to opposite poles at meiotic or mitotic divisions. *Nondisjunction* is a failure of this process, in which two chromosomes or chromatids incorrectly go to one pole and none to the other.

Mitotic nondisjunction can occur as cells divide during development. Sections of the body will be aneuploid (aneuploid *sectors*) as a result. *Meiotic* nondisjunction is more commonly encountered. In this case, the products of meiosis are aneuploid, leading to descendants in which the entire organism is aneuploid. In meiotic nondisjunction, the chromosomes may fail to disjoin at either the first or the second meiotic division (Figure 17-12). Either way, $n - 1$ and $n + 1$ gametes are produced. If an $n - 1$ gamete is fertilized by an n gamete, a monosomic ($2n - 1$) zygote is produced. The fusion of an $n + 1$ and an n gamete yields a trisomic $2n + 1$.

> **Message** Aneuploid organisms result mainly from nondisjunction in a parental meiosis.

Nondisjunction occurs spontaneously. Like most gene mutations, it is an example of a chance failure of a basic cellular process. The precise molecular processes that fail are not known but, in experimental systems, the frequency of nondisjunction can be increased by interference with microtubule polymerization, thereby inhibiting normal chromosome movement. Disjunction appears to be more likely to go awry in meiosis I. This failure is not surprising, because normal anaphase I disjunction requires that the homologous chromatids of the tetrad remain paired during prophase I and metaphase I, and it requires crossovers. In contrast, proper disjunction at anaphase II or at mitosis requires that the centromere split properly but does not require chromosome pairing or crossing over.

Crossovers are a necessary component of the normal disjunction process. Somehow the formation of a chiasma helps to hold a bivalent together and ensures that the two dyads will go to opposite poles. In most organisms, the amount of crossing over is sufficient to ensure that all bivalents will have at least one chiasma per meiosis. In *Drosophila*, many of the nondisjunctional chromosomes seen in disomic ($n + 1$) gametes are nonrecombinant, showing that they arise from meioses in which there is no crossing over on that chromosome. Similar observations have been made in human trisomies. In addition, in several different experimental organisms, mutations that interfere with recombination have the effect of massively increasing the frequency of meiosis I nondisjunction. All these observations provide evidence for the role of crossing over in maintaining chromosome pairing; in the absence of these associations, chromosomes are vulnerable to anaphase I nondisjunction.

> **Message** Crossovers are needed to keep bivalents paired until anaphase I. If crossing over fails for some reason, first-division nondisjunction occurs.

Monosomics ($2n - 1$)   Monosomics are missing one copy of a chromosome. In most diploid organisms, the absence of one chromosome copy from a pair is

deleterious. In humans, monosomics for any of the autosomes die in utero. Many X-chromosome monosomics also
die in utero, but some are viable. A human chromosome
complement of 44 autosomes plus a single X produces a
condition known as **Turner syndrome,** represented as XO.
Affected persons have a characteristic phenotype: they are
sterile females, short in stature, and often have a web of
skin extending between the neck and shoulders (Figure
17-13). Although their intelligence is near normal, some of
their specific cognitive functions are defective. About 1 in
5000 female births show Turner syndrome.

Geneticists have used viable plant monosomics to map
newly discovered recessive mutant alleles to a specific chromosome. For example, one can make a set of monosomic lines,
each known to lack a different chromosome. Homozygotes for
the new mutant allele are crossed with each monosomic line,
and the progeny of each cross are inspected for the recessive
phenotype. The appearance of the recessive phenotype identifies the chromosome that has one copy missing as the one on
which the gene is normally located. The test works because half
the gametes of a fertile $2n - 1$ monosomic will be $n - 1$, and,
when an $n - 1$ gamete is fertilized by a gamete bearing a new
mutation on the homologous chromosome, the mutant allele
will be the only allele of that gene present and hence will be
expressed.

As an illustration, let's assume that a gene A/a is on chromosome 2. Crosses
of a/a and monosomics for chromosome 1 and chromosome 2 are predicted to
produce different results (chromosome 1 is abbreviated chr1):

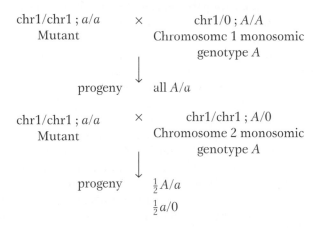

$$\text{chr1/chr1}\,;\,a/a \qquad \times \qquad \text{chr1/0}\,;\,A/A$$
$$\text{Mutant} \qquad\qquad \text{Chromosome 1 monosomic}$$
$$\text{genotype } A$$

$$\downarrow$$
$$\text{progeny} \qquad \text{all } A/a$$

$$\text{chr1/chr1}\,;\,a/a \qquad \times \qquad \text{chr1/chr1}\,;\,A/0$$
$$\text{Mutant} \qquad\qquad \text{Chromosome 2 monosomic}$$
$$\text{genotype } A$$

$$\downarrow$$
$$\text{progeny} \qquad \tfrac{1}{2}\,A/a$$
$$\tfrac{1}{2}\,a/0$$

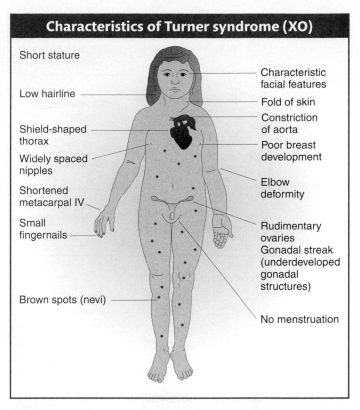

Characteristics of Turner syndrome (XO)

Short stature

Low hairline

Shield-shaped
thorax

Widely spaced
nipples

Shortened
metacarpal IV

Small
fingernails

Brown spots (nevi)

Characteristic
facial features

Fold of skin

Constriction
of aorta

Poor breast
development

Elbow
deformity

Rudimentary
ovaries
Gonadal streak
(underdeveloped
gonadal
structures)

No menstruation

Figure 17-13 Turner syndrome
results from the presence of a single X
chromosome (XO). [*After F. Vogel and A. G.
Motulsky, Human Genetics. Springer-Verlag,
1982.*]

Trisomics ($2n + 1$)   Trisomics contain an extra copy of one chromosome. In diploid organisms generally, the chromosomal imbalance from the trisomic condition
can result in abnormality or death. However, there are many examples of viable
trisomics. Furthermore, trisomics can be fertile. When cells from some trisomic
organisms are observed under the microscope at the time of meiotic chromosome
pairing, the trisomic chromosomes are seen to form an associated group of three
(a trivalent), whereas the other chromosomes form regular bivalents.

What genetic ratios might we expect for genes on the trisomic chromosome?
Let's consider a gene A that is close to the centromere on that chromosome, and
let's assume that the genotype is $A/a/a$. Furthermore, let's postulate that, at anaphase I, the two paired centromeres in the trivalent pass to opposite poles and that
the other centromere passes randomly to either pole. Then we can predict the three

Meiotic products of a trisomic

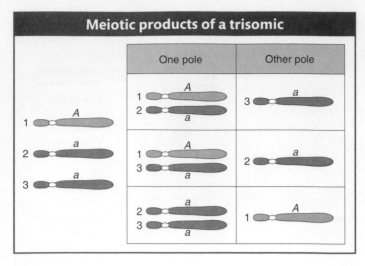

Figure 17-14 Three equally likely segregations may take place in the meiosis of an *A/a/a* trisomic, yielding the genotypes shown.

Characteristics of Klinefelter syndrome (XXY)

Tall stature

Slightly feminized physique

Mildly impaired IQ (15 points less than average)

Tendency to lose chest hairs

Female-type pubic hair pattern

Frontal baldness absent

Poor beard growth

Breast development (in 30% of cases)

Osteoporosis

Small testes

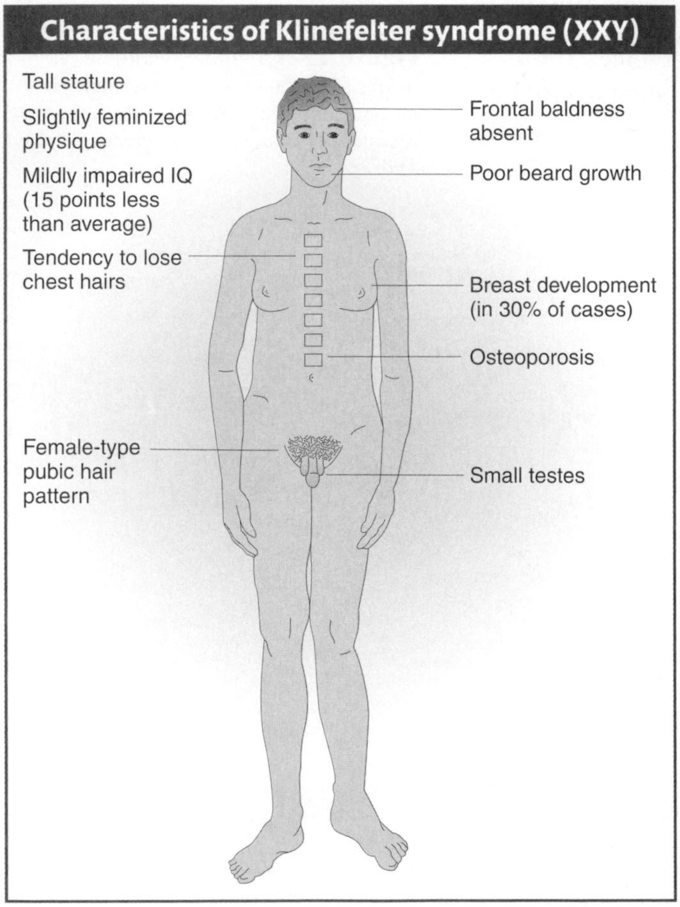

Figure 17-15 Klinefelter syndrome results from the presence of two X chromosomes and a Y chromosome. [*After F. Vogel and A. G. Motulsky, Human Genetics. Springer-Verlag, 1982.*]

equally frequent segregations shown in Figure 17-14. These segregations result in an overall gametic ratio as shown in the six compartments of Figure 17-14; that is,

$$\frac{1}{6}A$$

$$\frac{2}{6}a$$

$$\frac{2}{6}A/a$$

$$\frac{1}{6}a/a$$

If a set of lines is available, each carrying a different trisomic chromosome, then a gene mutation can be located to a chromosome by determining which of the lines gives a trisomic ratio of the preceding type.

There are several examples of viable human trisomies. Several types of sex-chromosome trisomics can live to adulthood. Each of these types is found at a frequency of about 1 in 1000 births of the relevant sex. (In considering human sex-chromosome trisomies, recall that mammalian sex is determined by the presence or absence of the Y chromosome.) The combination XXY results in **Klinefelter syndrome.** Persons with this syndrome are males who have lanky builds and a mildly impaired IQ and are sterile (Figure 17-15). Another abnormal combination, XYY, has a controversial history. Attempts have been made to link the XYY condition with a predisposition toward violence. However, it is now clear that an XYY condition in no way guarantees such behavior. Most males with XYY are fertile. Meioses show normal pairing of the X with one of the Y's; the other Y does not pair and is not transmitted to gametes. Therefore, the gametes contain either X or Y, never YY or XY. Triplo-X trisomics (XXX) are phenotypically normal and fertile females. Meiosis shows pairing of only two X chromosomes; the third does not pair. Hence, eggs bear only one X and, like that of XYY males, the condition is not passed on to progeny.

Of human trisomies, the most familiar type is **Down syndrome** (Figure 17-16), discussed briefly at the beginning of the chapter. The frequency of Down syndrome is about 0.15 percent of all live births. Most affected persons have an extra copy of chromosome 21 caused by nondisjunction of chromosome 21 in a parent who is chromosomally normal. In this *sporadic* type of Down syndrome, there is no family history of aneuploidy. Some rarer types of Down syndrome arise from translocations (a type of chromosomal rearrangement discussed later in the chapter); in these cases, as we will see, Down syndrome recurs in the pedigree because the translocation may be transmitted from parent to child.

The combined phenotypes that make up Down syndrome include mental retardation (with an IQ in the 20 to 50 range); a broad, flat face; eyes with an epicanthic fold; short stature; short hands with a crease across the middle; and a large, wrinkled tongue. Females may be fertile and may produce normal or trisomic progeny, but males are sterile with very few exceptions. Mean life expectancy is about 17 years, and only 8 percent of persons with Down syndrome survive past age 40.

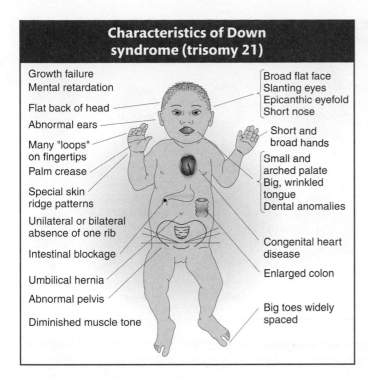

Figure 17-16 Down syndrome results from the presence of an extra copy of chromosome 21. [*After F. Vogel and A. G. Motulsky, Human Genetics. Springer-Verlag, 1982.*]

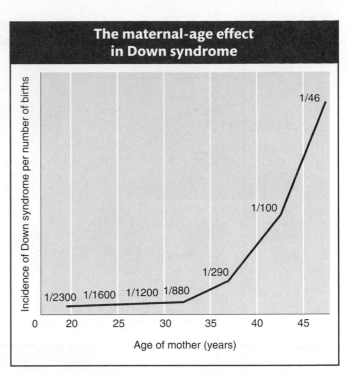

Figure 17-17 Older mothers have a higher proportion of babies with Down syndrome than younger mothers do. [*From L. S. Penrose and G. F. Smith, Down's Anomaly. Little, Brown and Company, 1966.*]

The incidence of Down syndrome is related to maternal age: older mothers run a greatly elevated risk of having a child with Down syndrome (Figure 17-17). For this reason, fetal chromosome analysis (by amniocentesis or by chorionic villus sampling) is now recommended for older expectant mothers. A less-pronounced paternal-age effect also has been demonstrated.

Even though the maternal-age effect has been known for many years, its cause is still not known. Nonetheless, there are some interesting biological correlations. With age, possibly the chromosome bivalent is less likely to stay together during prophase I of meiosis. Meiotic arrest of oocytes (female meiocytes) in late prophase I is a common phenomenon in many animals. In female humans, all oocytes are arrested at diplotene before birth. Meiosis resumes at each menstrual period, which means that the chromosomes in the bivalent must remain properly associated for as long as five or more decades. If we speculate that these associations have an increasing probability of breaking down by accident as time passes, we can envision a mechanism contributing to increased maternal nondisjunction with age. Consistent with this speculation, most nondisjunction related to the effect of maternal age is due to nondisjunction at anaphase I, not anaphase II.

The only other human autosomal trisomics to survive to birth are those with trisomy 13 (Patau syndrome) and trisomy 18 (Edwards syndrome). Both have severe physical and mental abnormalities. The phenotypic syndrome of trisomy 13 includes a harelip; a small, malformed head; "rocker bottom" feet; and a mean life expectancy of 130 days. That of trisomy 18 includes "faunlike" ears, a small jaw, a narrow pelvis, and rocker-bottom feet; almost all babies with trisomy 18 die within the first few weeks after birth. All other trisomics die in utero.

The concept of gene balance

In considering aberrant euploidy, we noted that an increase in the number of full chromosome sets correlates with increased organism size but that the general

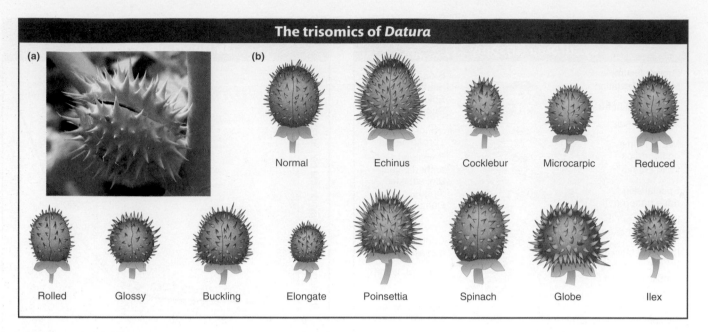

The trisomics of *Datura*

(a)

(b)

Normal   Echinus   Cocklebur   Microcarpic   Reduced

Rolled   Glossy   Buckling   Elongate   Poinsettia   Spinach   Globe   Ilex

Figure 17-18 Each of the 12 possible trisomics of *Datura* is disproportionate in a different way. (a) *Datura* fruit. (b) Each drawing is of the fruit of a different trisomic, each of which has been named. [*(a) iStockphoto/Thinkstock; (b) After E. W. Sinnott, L. C. Dunn, and T. Dobzhansky, Principles of Genetics, 5th ed. McGraw-Hill Book Company, 1958.*]

shape and proportions of the organism remain very much the same. In contrast, autosomal aneuploidy typically alters the organism's shape and proportions in characteristic ways.

Plants tend to be somewhat more tolerant of aneuploidy than are animals. Studies in jimsonweed (*Datura stramonium*) provide a classic example of the effects of aneuploidy and polyploidy. In jimsonweed, the haploid chromosome number is 12. As expected, the polyploid jimsonweed is proportioned like the normal diploid, only larger. In contrast, each of the 12 possible trisomics is disproportionate but in ways different from one another, as exemplified by changes in the shape of the seed capsule (Figure 17-18). The 12 different trisomies lead to 12 different and characteristic shape changes in the capsule. Indeed, these characteristics and others of the individual trisomics are so reliable that the phenotypic syndrome can be used to identify plants carrying a particular trisomy. Similarly, the 12 monosomics are themselves different from one another and from each of the trisomics. In general, a monosomic for a particular chromosome is more severely abnormal than is the corresponding trisomic.

We see similar trends in aneuploid animals. In the fruit fly *Drosophila*, the only autosomal aneuploids that survive to adulthood are trisomics and monosomics for chromosome 4, which is the smallest *Drosophila* chromosome, representing only about 1 to 2 percent of the genome. Trisomics for chromosome 4 are only very mildly affected and are much less abnormal than are monosomics for chromosome 4. In humans, no autosomal monosomic survives to birth, but, as already stated, three types of autosomal trisomics can do so. As is true of aneuploid jimsonweed, each of these three trisomics shows unique phenotypic syndromes because of the special effects of altered dosages of each of these chromosomes.

Why are aneuploids so much more abnormal than polyploids? Why does aneuploidy for each chromosome have its own characteristic phenotypic effects? And why are monosomics typically more severely affected than are the corresponding trisomics? The answers seem certain to be a matter of **gene balance.** In a euploid, the ratio of genes on any one chromosome to the genes on other chromosomes is always 1:1, regardless of whether we are considering a monoploid, diploid, triploid, or tetraploid. For example, in a tetraploid, for gene *A* on chromosome 1 and gene *B* on chromosome 2, the ratio is 4 *A*:4 *B*, or 1:1. In contrast, in an aneuploid, the ratio of genes on the aneuploid chromosome to genes on the other chromosomes differs from the wild type by 50 percent: 50 percent for monosomics; 150 percent for trisomics. Using the same example as before, in a trisomic for chromosome 2, we find that the ratio of the *A* and *B* genes is 2 *A* : 3 *B*. Thus, we can see that the aneuploid genes are out of balance. How does their being out of balance help us answer the questions raised?

In general, the amount of transcript produced by a gene is directly proportional to the number of copies of that gene in a cell. That is, for a given gene, the rate of transcription is directly related to the number of DNA templates available. Thus, the more copies of the gene, the more transcripts are produced and the more of the corresponding protein product is made. This relation between the number of copies of a gene and the amount of the gene's product made is called a **gene-dosage effect.**

We can infer that normal physiology in a cell depends on the proper ratio of gene products in the euploid cell. This ratio is the normal gene balance. If the relative dosage of certain genes changes—for example, because of the removal of one of the two copies of a chromosome (or even a segment thereof)—physiological imbalances in cellular pathways can arise.

In some cases, the imbalances of aneuploidy result from the effects of a few "major" genes whose dosage has changed, rather than from changes in the dosage of all the genes on a chromosome. Such genes can be viewed as *haplo-abnormal* (resulting in an abnormal phenotype if present only once) or *triplo-abnormal* (resulting in an abnormal phenotype if present in three copies) or both. They contribute significantly to the aneuploid phenotypic syndromes. For example, the study of persons trisomic for only part of chromosome 21 has made it possible to localize genes contributing to Down syndrome to various regions of chromosome 21; the results hint that some aspects of the phenotype might be due to triplo-abnormality for single major genes in these chromosome regions. In addition to these major-gene effects, other aspects of aneuploid syndromes are likely to result from the cumulative effects of aneuploidy for numerous genes whose products are all out of balance. Undoubtedly, the entire aneuploid phenotype results from a combination of the imbalance effects of a few major genes, together with a cumulative imbalance of many minor genes.

However, the concept of gene balance does not tell us why having too few gene products (monosomy) is much worse for an organism than having too many gene products (trisomy). In a parallel manner, we can ask why there are many more haplo-abnormal genes than triplo-abnormal ones. A key to explaining the extreme abnormality of monosomics is that any deleterious recessive alleles present on a monosomic autosome will be automatically expressed.

How do we apply the idea of gene balance to cases of sex-chromosome aneuploidy? Gene balance holds for sex chromosomes as well, but we also have to take into account the special properties of the sex chromosomes. In organisms with XY sex determination, the Y chromosome seems to be a degenerate X chromosome in which there are very few functional genes other than some concerned with sex determination itself, in sperm production, or in both. The X chromosome, on the other hand, contains many genes concerned with basic cellular processes ("housekeeping genes") that just happen to reside on the chromosome that eventually evolved into the X chromosome. XY sex-determination mechanisms have probably evolved independently from 10 to 20 times in different taxonomic groups. For example, there appears to be one sex-determination mechanism for all mammals, but it is completely different from the mechanism governing XY sex determination in fruit flies.

In a sense, X chromosomes are naturally aneuploid. In species with an XY sex-determination system, females have two X chromosomes, whereas males have only one. Nonetheless, the X chromosome's housekeeping genes are expressed to approximately equal extents per cell in females and in males. In other words, there is **dosage compensation.** How is this compensation accomplished? The answer depends on the organism. In fruit flies, the male's X chromosome appears to be hyperactivated, allowing it to be transcribed at twice the rate of either X chromosome in the female. As a result, the XY male *Drosophila* has an X gene dosage equivalent to that of an XX female. In mammals, in contrast, the rule is that no matter how many X chromosomes are present, there is only one transcriptionally active X chromosome in each somatic cell. This rule gives the XX female mammal an X gene dosage equivalent to that of an XY male. Dosage compensation in mammals is achieved by X-chromosome inactivation. A female with two X chromosomes, for example, is a mosaic of two cell types in which one or the other X chromosome is active. We examined this phenomenon in Chapter 12. Thus, XY and XX individuals produce the same amounts of X-chromosome housekeeping-gene products. X-chromosome inactivation also explains why triplo-X humans are phenotypically normal: only one of the three X chromosomes is transcriptionally active in a given cell. Similarly, an XXY male is only moderately affected because only one of his two X chromosomes is active in each cell.

Why are XXY individuals abnormal at all, given that triplo-X individuals are phenotypically normal? It turns out that a few genes scattered throughout an "inactive X" are still transcriptionally active. In XXY males, these genes are transcribed at twice the level that they are in XY males. In XXX females, on the other hand, the few transcribed genes are active at only 1.5 times the level that they are in XX females. This lower level of "functional aneuploidy" in XXX than in XXY, plus the fact that the active X genes appear to lead to feminization, may explain the feminized phenotype of XXY males. The severity of Turner syndrome (XO) may be due to the deleterious effects of monosomy and to the lower activity of the transcribed genes of the X (compared with XX) females. As is usually observed for aneuploids, monosomy for the X chromosome produces a more abnormal phenotype than does having an extra copy of the same chromosome (triplo-X females or XXY males).

Gene dosage is also important in the phenotypes of polyploids. Human polyploid zygotes do arise through various kinds of mistakes in cell division. Most die in utero. Occasionally, triploid babies are born, but none survive. This fact seems to violate the principle that polyploids are more normal than aneuploids. The explanation for this contradiction seems to lie with X-chromosome dosage compensation. Part of the rule for gene balance in organisms that have a single active X seems to be that there must be one active X for every two copies of the autosomal chromosome complement. Thus, some cells in triploid mammals are found to have one active X, whereas others, surprisingly, have two. Neither situation is in balance with autosomal genes.

> **Message** Aneuploidy is nearly always deleterious because of gene imbalance: the ratio of genes is different from that in euploids, and this difference interferes with the normal function of the genome.

17.2 Changes in Chromosome Structure

Changes in chromosome structure, called **rearrangements,** encompass several major classes of events. A chromosome segment can be lost, constituting a **deletion,** or doubled, to form a **duplication.** The orientation of a segment within the chromosome can be reversed, constituting an **inversion.** Or a segment can be moved to a different chromosome, constituting a **translocation.** DNA breakage is a major cause of each of these events. Both DNA strands must break at two different locations, followed by a rejoining of the broken ends to produce a new chromosomal arrangement (Figure 17-19, left side). Chromosomal rearrangements by breakage can be induced artificially by using ionizing radiation. This kind of radiation, particularly X rays and gamma rays, is highly energetic and causes numerous double-stranded breaks in DNA.

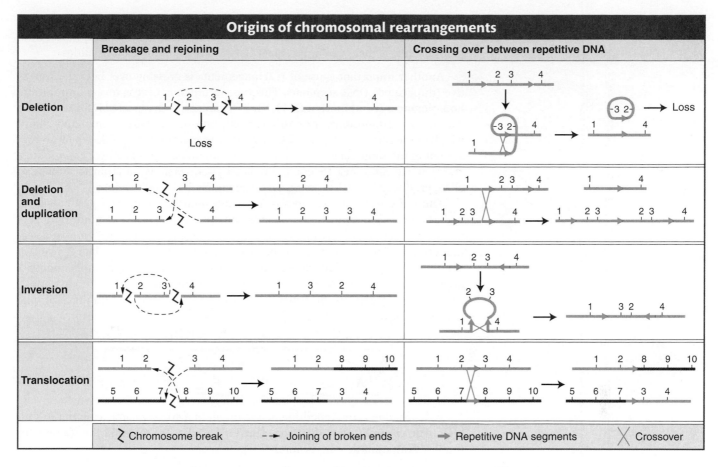

Figure 17-19 Each of the four types of chromosomal rearrangements can be produced by either of two basic mechanisms: chromosome breakage and rejoining or crossing over between repetitive DNA. Chromosome regions are numbered 1 through 10. Homologous chromosomes are the same color.

To understand how chromosomal rearrangements are produced by breakage, several points should be kept in mind:

1. Each chromosome is a single double-stranded DNA molecule.

2. The first event in the production of a chromosomal rearrangement is the generation of two or more double-stranded breaks in the chromosomes of a cell (see Figure 17-19, top row at left).

3. Double-stranded breaks are potentially lethal, unless they are repaired.

4. Repair systems in the cell correct the double-stranded breaks by joining broken ends back together (see Chapter 16 for a detailed discussion of DNA repair).

5. If the two ends of the same break are rejoined, the original DNA order is restored. If the ends of two different breaks are joined, however, one result is one or another type of chromosomal rearrangement.

6. The only chromosomal rearrangements that survive meiosis are those that produce DNA molecules that have one centromere and two telomeres. If a rearrangement produces a chromosome that lacks a centromere, such an **acentric** chromosome will not be dragged to either pole at anaphase of mitosis or meiosis and will not be incorporated into either progeny nucleus. Therefore, acentric chromosomes are not inherited. If a rearrangement produces a chromosome with two centromeres (a **dicentric**), it will often be pulled simultaneously to opposite poles at anaphase, forming an **anaphase bridge**. Anaphase-bridge chromosomes typically will not be incorporated into either progeny cell. If a chromosome break produces a chromosome lacking a telomere, that chromosome cannot replicate properly. Recall from Chapter 7 that telomeres are needed to prime proper DNA replication at the ends (see Figure 7-27).

7. If a rearrangement duplicates or deletes a segment of a chromosome, gene balance may be affected. The larger the segment that is lost or duplicated, the more likely it is that gene imbalance will cause phenotypic abnormalities.

Another important cause of rearrangements is crossing over between repetitive (duplicated) DNA segments. This type of crossing over is termed **nonallelic homologous recombination (NAHR).** In organisms with repeated DNA sequences within one chromosome or on different chromosomes, there is ambiguity about which of the repeats will pair with each other at meiosis. If sequences pair up that are not in the same relative positions on the homologs, crossing over can produce aberrant chromosomes. Deletions, duplications, inversions, and translocations can all be produced by such crossing over (see Figure 17-19, right side).

There are two general types of rearrangements: unbalanced and balanced. **Unbalanced rearrangements** change the gene dosage of a chromosome segment. As with aneuploidy for whole chromosomes, the loss of one copy of a segment or the addition of an extra copy can disrupt normal gene balance. The two simple classes of unbalanced rearrangements are deletions and duplications. A *deletion* is the loss of a segment within one chromosome arm and the juxtaposition of the two segments on either side of the deleted segment, as in this example, which shows loss of segment C–D:

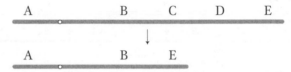

A *duplication* is the repetition of a segment of a chromosome arm. In the simplest type of duplication, the two segments are adjacent to each other (a tandem duplication), as in this duplication of segment C:

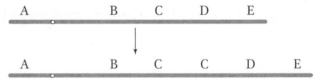

However, the duplicate segment can end up at a different position on the same chromosome or even on a different chromosome.

Balanced rearrangements change the chromosomal gene order but do not remove or duplicate any DNA. The two simple classes of balanced rearrangements are inversions and reciprocal translocations. An *inversion* is a rearrangement in which an internal segment of a chromosome has been broken twice, flipped 180 degrees, and rejoined.

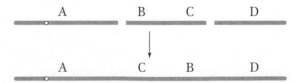

A *reciprocal translocation* is a rearrangement in which two nonhomologous chromosomes are each broken once, creating acentric fragments, which then trade places:

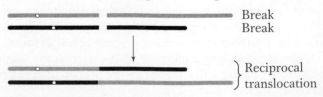

Sometimes the DNA breaks that precede the formation of a rearrangement occur *within* genes. When they do, they disrupt gene function because part of the gene moves to a new location and no complete transcript can be made. In addition, the DNA sequences on either side of the rejoined ends of a rearranged chromosome are sequences that are not normally juxtaposed. Sometimes the junction occurs in such a way that fusion produces a nonfunctional hybrid gene composed of parts of two other genes.

The following sections consider the properties of these balanced and unbalanced rearrangements.

Deletions

A deletion is simply the loss of a part of one chromosome arm. The process of deletion requires two chromosome breaks to cut out the intervening segment. The deleted fragment has no centromere; consequently, it cannot be pulled to a spindle pole in cell division and is lost. The effects of deletions depend on their size. A small deletion *within* a gene, called an **intragenic deletion,** inactivates the gene and has the same effect as that of other null mutations of that gene. If the homozygous null phenotype is viable (as, for example, in human albinism), the homozygous deletion also will be viable. Intragenic deletions can be distinguished from mutations caused by single nucleotide changes because genes with such deletions never revert to wild type.

For most of this section, we will be dealing with **multigenic deletions,** in which several to many genes are missing. The consequences of these deletions are more severe than those of intragenic deletions. If such a deletion is made homozygous by inbreeding (that is, if both homologs have the same deletion), the combination is always lethal. This fact suggests that all regions of the chromosomes are essential for normal viability and that complete elimination of any segment from the genome is deleterious. Even an individual organism heterozygous for a multigenic deletion—that is, having one normal homolog and one that carries the deletion—may not survive. Principally, this lethal outcome is due to disruption of normal gene balance. Alternatively, the deletion may "uncover" deleterious recessive alleles, allowing the single copies to be expressed.

> **Message** The lethality of large heterozygous deletions can be explained by gene imbalance and the expression of deleterious recessives.

Small deletions are sometimes viable in combination with a normal homolog. Such deletions may be identified by examining meiotic chromosomes under the microscope. The failure of the corresponding segment on the normal homolog to pair creates a visible **deletion loop** (Figure 17-20a). In *Drosophila*, deletion loops are also visible in the **polytene chromosomes.** These chromosomes are found in the cells of salivary glands and other specific tissues of certain insects. In these cells, the homologs pair and replicate many times, and so each chromosome is represented by a thick bundle of replicates. These polytene chromosomes are easily visible, and each has a set of dark-staining bands of fixed position and number. These bands act as useful chromosomal landmarks. An example of a polytene chromosome in which one original homolog carried a deletion is shown in Figure 17-20b. A deletion can be assigned to a specific chromosome location by examining polytene chromosomes microscopically and determining the position of the deletion loop.

Another clue to the presence of a deletion is that the deletion of a segment on one homolog sometimes unmasks recessive alleles present on the other homolog,

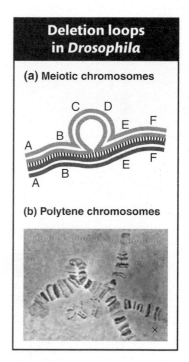

Deletion loops in *Drosophila*

(a) Meiotic chromosomes

(b) Polytene chromosomes

Figure 17-20 In meiosis, the chromosomes of a deletion heterozygote form a looped configuration. (a) In meiotic pairing, the normal homolog forms a loop. The genes in this loop have no alleles with which to synapse. (b) Because *Drosophila* polytene chromosomes (found in salivary glands and other specific locations) have specific banding patterns, we can infer which bands are missing from the homolog with the deletion by observing which bands appear in the loop of the normal homolog. [*(b) From William M. Gelbart.*]

leading to their unexpected expression. Consider, for example, the deletion shown in the following diagram:

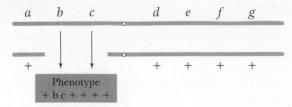

If there is no deletion, none of the seven recessive alleles is expected to be expressed; however, if *b* and *c* are expressed, then a deletion spanning the b^+ and c^+ genes has probably occurred on the other homolog. Because recessive alleles seem to be showing dominance in such cases, the effect is called **pseudodominance.**

In the reverse case—if we already know the location of the deletion—we can apply the pseudodominance effect in the opposite direction to map the positions of mutant alleles. This procedure, called **deletion mapping,** pairs mutations against a set of defined overlapping deletions. An example from *Drosophila* is shown in **Figure 17-21.** In this diagram, the recombination map is shown at the top, marked with distances in map units from the left end. The horizontal red bars below the chromosome show the extent of the deletions listed at the left. Each deletion is paired with each mutation under test, and the phenotype is observed to see if the mutation is pseudodominant. The mutation *pn* (prune), for example, shows pseudodominance only with deletion 264-38, and this result determines its location in the 2D-4 to 3A-2 region. However, *fa* (facet) shows pseudodominance with all but two deletions (258-11 and 258-14); so its position can be pinpointed to band 3C-7, which is the region that all but two deletions have in common.

> **Message** Deletions can be recognized by deletion loops and pseudodominance.

Clinicians regularly find deletions in human chromosomes. The deletions are usually small, but they do have adverse effects, even though heterozygous. Deletions of specific human chromosome regions cause unique syndromes of

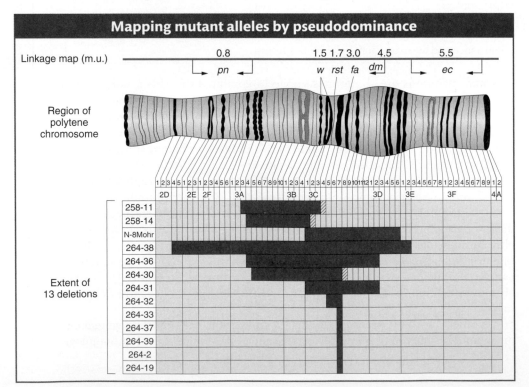

Figure 17-21 A *Drosophila* strain heterozygous for deletion and normal chromosomes may be used to map mutant alleles. The red bars show the extent of the deleted segments in 13 deletions. All recessive alleles in the same region that is deleted in a homologous chromosome will be expressed.

phenotypic abnormalities. One example is cri du chat syndrome, caused by a heterozygous deletion of the tip of the short arm of chromosome 5 (Figure 17-22). The specific bands deleted in cri du chat syndrome are 5p15.2 and 5p15.3, the two most distal bands identifiable on 5p. (The short and long arms of human chromosomes are traditionally called p and q, respectively.) The most characteristic phenotype in the syndrome is the one that gives it its name, the distinctive catlike mewing cries made by affected infants. Other manifestations of the syndrome are microencephaly (abnormally small head) and a moonlike face. Like syndromes caused by other deletions, cri du chat syndrome includes mental retardation. Fatality rates are low, and many persons with this deletion reach adulthood.

Another instructive example is Williams syndrome. This syndrome is autosomal dominant and is characterized by unusual development of the nervous system and certain external features. Williams syndrome is found at a frequency of about 1 in 10,000 people. Patients often have pronounced musical or singing ability. The syndrome is almost always caused by a 1.5-Mb deletion on one homolog of chromosome 7. Sequence analysis showed that this segment contains 17 genes of known and unknown function. The abnormal phenotype is thus caused by haploinsufficiency of one or more of these 17 genes. Sequence analysis also reveals the origin of this deletion because the normal sequence is bounded by repeated copies of a gene called *PMS*, which happens to encode a DNA-repair protein. As we have seen, repeated sequences can act as substrates for unequal crossing over. A crossover between flanking copies of *PMS* on opposite ends of the 17-gene segment leads to a duplication (not found) and a Williams syndrome deletion, as shown in Figure 17-23.

Most human deletions, such as those that we have just considered, arise spontaneously in the gonads of a normal parent of an affected person; thus, no signs of the deletions are usually found in the chromosomes of the parents. Less commonly, deletion-bearing individuals appear among the offspring of an individual having an undetected balanced rearrangement of chromosomes. For example, cri du chat syndrome can result from a parent heterozygous for a reciprocal

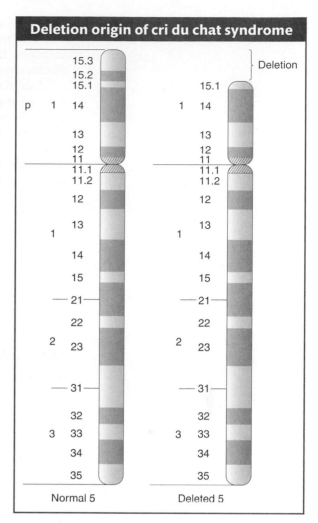

Deletion origin of cri du chat syndrome

Normal 5   Deleted 5

Figure 17-22 Cri du chat syndrome is caused by the loss of the tip of the short arm of one of the homologs of chromosome 5.

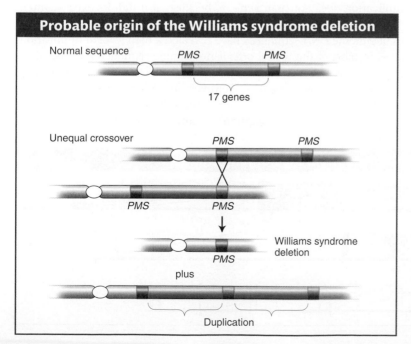

Probable origin of the Williams syndrome deletion

Normal sequence   *PMS*   *PMS*

17 genes

Unequal crossover   *PMS*   *PMS*

PMS   *PMS*

PMS   Williams syndrome deletion

plus

Duplication

Figure 17-23 A crossover between left and right repetitive flanking genes results in two reciprocal rearrangements, one of which corresponds to the Williams syndrome deletion.

translocation, because (as we will see) segregation produces deletions. Deletions may also result from recombination within a heterozygote having a pericentric inversion (an inversion spanning the centromere) on one chromosome. Both mechanisms will be detailed later in the chapter.

Animals and plants show differences in the survival of gametes or offspring that bear deletions. A male animal with a deletion in one chromosome produces sperm carrying one or the other of the two chromosomes in approximately equal numbers. These sperm seem to function to some extent regardless of their genetic content. In diploid plants, on the other hand, the pollen produced by a deletion heterozygote is of two types: functional pollen carrying the normal chromosome and nonfunctional (aborted) pollen carrying the deficient homolog. Thus, pollen cells seem to be sensitive to changes in the amount of chromosomal material, and this sensitivity might act to weed out deletions. This effect is analogous to the sensitivity of pollen to whole-chromosome aneuploidy, described earlier in this chapter. Unlike animal sperm cells, whose metabolic activity relies on enzymes that have already been deposited in them during their formation, pollen cells must germinate and then produce a long pollen tube that grows to fertilize the ovule. This growth requires that the pollen cell manufacture large amounts of protein, thus making it sensitive to genetic abnormalities in its own nucleus. Plant ovules, in contrast, are quite tolerant of deletions, presumably because they receive their nourishment from the surrounding maternal tissues.

Duplications

The processes of chromosome mutation sometimes produce an extra copy of some chromosome region. The duplicate regions can be located adjacent to each other—called a **tandem duplication**—or the extra copy can be located elsewhere in the genome—called an **insertional duplication.** A diploid cell containing a duplication will have three copies of the chromosome region in question: two in one chromosome set and one in the other—an example of a duplication heterozygote. In meiotic prophase, tandem-duplication heterozygotes show a loop consisting of the unpaired extra region.

Synthetic duplications of known coverage can be used for gene mapping. In haploids, for example, a chromosomally normal strain carrying a new recessive mutation m may be crossed with strains bearing a number of duplication-generating rearrangements (for example, translocations and pericentric inversions). In any one cross, if some duplication progeny have the recessive phenotype, the duplication does not span gene m, because, if it did, its extra segment would mask the recessive m allele.

Analyses of genome DNA sequences have revealed a high level of duplications in humans and in most of the model organisms. Simple sequence repeats, which are extensive throughout the genome and useful as molecular markers in mapping, were discussed in earlier chapters. However, another class of duplications is based on duplicated units that are much bigger than the simple sequence repeats. Duplications in this class are termed **segmental duplications.** The duplicated units

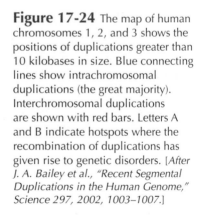

Figure 17-24 The map of human chromosomes 1, 2, and 3 shows the positions of duplications greater than 10 kilobases in size. Blue connecting lines show intrachromosomal duplications (the great majority). Interchromosomal duplications are shown with red bars. Letters A and B indicate hotspots where the recombination of duplications has given rise to genetic disorders. [*After J. A. Bailey et al., "Recent Segmental Duplications in the Human Genome," Science 297, 2002, 1003–1007.*]

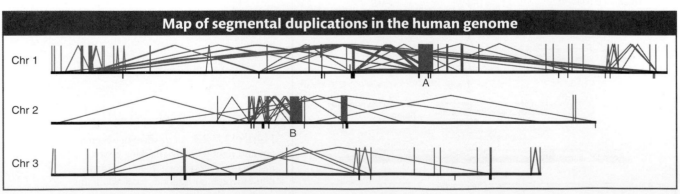

Map of segmental duplications in the human genome

Chr 1

Chr 2

Chr 3

in segmental duplications range from 10 to 50 kilobases in length and encompass whole genes and the regions in between. The extent of segmental duplications is shown in Figure 17-24, in which most of the duplications are dispersed, but there are some tandem cases. Another property shown in Figure 17-24 is that the dispersion of the duplicated units is mostly within the same chromosome, not between chromosomes. The origin of segmental duplications is still not known.

Segmental duplications are thought to have an important role as substrates for nonallelic homologous recombination, as shown in Figure 17-19. Crossing over between segmental duplications can lead to various chromosomal rearrangements. These rearrangements seem to have been important in evolution, inasmuch as some major inversions that are key differences between human and ape sequences have almost certainly come from NAHR. It also seems likely that NAHR has been responsible for rearrangements that cause some human diseases. The loci of such diseases are at segmental-duplication hotspots; examples of such loci are shown in Figure 17-24.

We have seen that, in some organisms such as polyploids, the present-day genome evolved as a result of an ancestral whole-genome duplicating. When whole-genome duplication has taken place, every gene is doubled. These doubled genes are a source of some of the segmental duplications found in genomes. A well-studied case is baker's yeast, *Saccharomyces cerevisiae*. The evolution of this genome has been analyzed by comparing the whole-genome sequence of *S. cerevisiae* with that of another yeast, *Kluyveromyces*, whose genome is similar to that of the ancestral genome of yeast. Apparently, in the course of the evolution of *Saccharomyces*, the *Kluyveromyces*-like ancestral genome doubled, and so there were two sets, each containing the whole genome. After doubling occurred, many gene copies were lost from one set or the other, and the remaining sets were rearranged, resulting in the present *Saccharomyces* genome. This process is reconstructed in Figure 17-25.

Figure 17-25 A common ancestor similar to the modern *Kluyveromyces* yeast duplicated its genome (1). Some genes were lost (2). Duplicate genes such as 3 and 13 are in the same relative order. The bottom panel compares the two modern genomes. [*After Figure 1, Manolis Kellis, Bruce W. Birren, and Eric S. Lander, "Proof and Evolutionary Analysis of Ancient Genome Duplication in the Yeast Saccharomyces cerevisiae," Nature, vol. 428, April 8, 2004, copyright Nature Publishing Group.*]

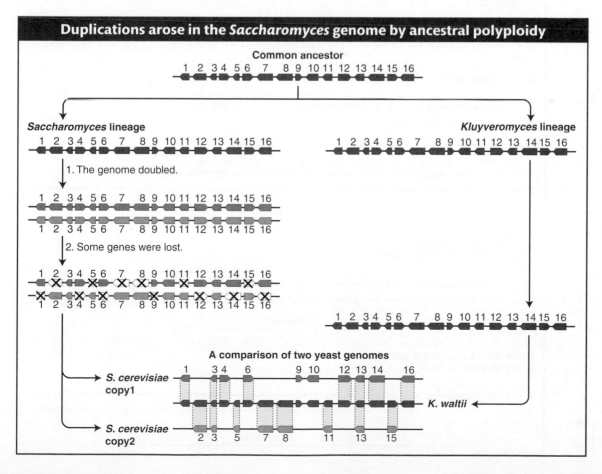

Duplications arose in the *Saccharomyces* genome by ancestral polyploidy

Inversions

We have seen that, to create an inversion, a segment of a chromosome is cut out, flipped, and reinserted. Inversions are of two basic types. If the centromere is outside the inversion, the inversion is said to be **paracentric.** Inversions spanning the centromere are **pericentric.**

	A	B		C	D	E	F
Normal sequence	A	B		C	D	E	F
Paracentric	A	B		C	E	D	F
Pericentric	A	D	C		B	E	F

Because inversions are balanced rearrangements, they do not change the overall amount of genetic material, and so they do not result in gene imbalance. Individuals with inversions are generally normal, if there are no breaks within genes. A break that disrupts a gene produces a mutation that may be detectable as an abnormal phenotype. If the gene has an essential function, then the break point acts as a lethal mutation linked to the inversion. In such a case, the inversion cannot be bred to homozygosity. However, many inversions can be made

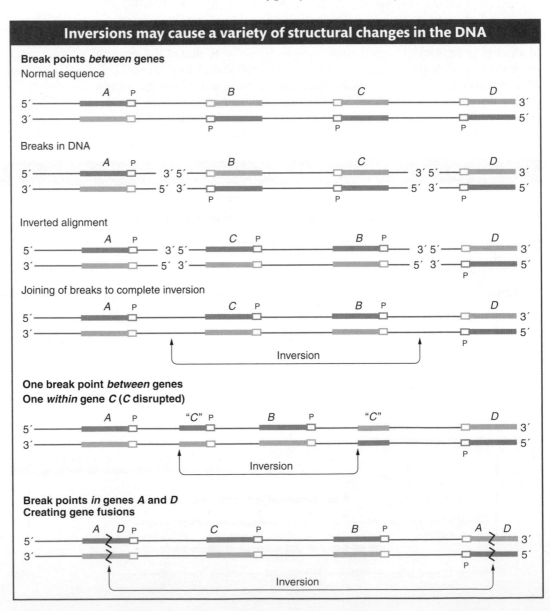

Figure 17-26 An inversion may have no effect on genes, may disrupt a gene, or may fuse parts of two genes, depending on the location of the break point. Genes are represented by *A, B, C,* and *D.* Template strand is dark green; nontemplate strand is light green; jagged red lines indicate where breaks in the DNA produced gene fusions (*A* with *D*) after inversion and rejoining. The letter P stands for promoter; arrows indicate the positions of the break points.

homozygous, and, furthermore, inversions can be detected in haploid organisms. In these cases, the break points of the inversion are clearly not in essential regions. Some of the possible consequences of inversion at the DNA level are shown in Figure 17-26.

Most analyses of inversions are carried out on diploid cells that contain one normal chromosome set plus one set carrying the inversion. This type of cell is called an **inversion heterozygote,** but note that this designation does not imply that any gene locus is heterozygous; rather, it means that one normal and one abnormal chromosome set are present. The location of the inverted segment can often be detected microscopically. In meiosis, one chromosome twists once at the ends of the inversion to pair with the its untwisted homolog; in this way, the paired homologs form a visible **inversion loop** (Figure 17-27).

In a *paracentric* inversion, crossing over within the inversion loop at meiosis connects homologous centromeres in a **dicentric bridge** while also producing an **acentric fragment** (Figure 17-28). Then,

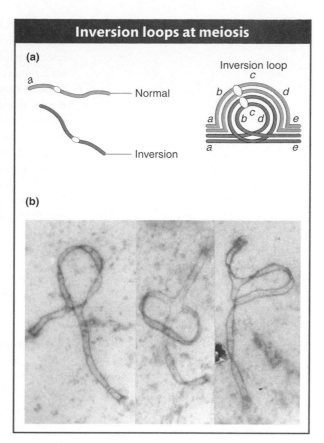

Inversion loops at meiosis

(a)

Normal

Inversion

Inversion loop

(b)

Figure 17-27 The chromosomes of inversion heterozygotes pair in a loop at meiosis. (a) Diagrammatic representation. (b) Electron micrographs of synaptonemal complexes at prophase I of meiosis in a mouse heterozygous for a paracentric inversion. Three different meiocytes are shown. [*(b) From M. J. Moses, Department of Anatomy, Duke Medical Center.*]
GENETICS PORTAL **ANIMATED ART: Chromosome rearrangements: formation of paracentric inversions**

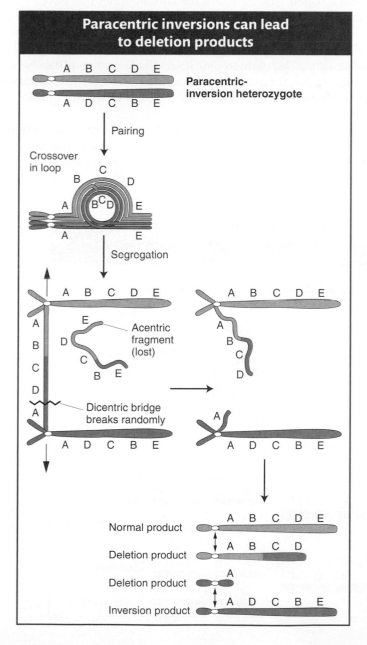

Paracentric inversions can lead to deletion products

A B C D E

Paracentric-inversion heterozygote

A D C B E

Pairing

Crossover in loop

Segregation

Acentric fragment (lost)

Dicentric bridge breaks randomly

Normal product   A B C D E

Deletion product   A B C D

Deletion product   A

Inversion product   A D C B E

Figure 17-28 A crossover in the loop of a paracentric-inversion heterozygote gives rise to chromosomes containing deletions.
GENETICS PORTAL **ANIMATED ART: Chromosome rearrangements: meiotic behavior of paracentric inversions**

as the chromosomes separate in anaphase I, the centromeres remain linked by the bridge. The acentric fragment cannot align itself or move; consequently, it is lost. Tension eventually breaks the dicentric bridge, forming two chromosomes with terminal deletions. Either the gametes containing such chromosomes or the zygotes that they eventually form will probably be inviable. Hence, a crossover event, which normally generates the recombinant class of meiotic products, is instead lethal to those products. The overall result is a drastically lower frequency of viable recombinants. In fact, for genes within the inversion, the recombinant frequency is close to zero. (It is not exactly zero, because rare double crossovers between only two chromatids are viable.) For genes flanking the inversion, the RF is reduced in proportion to the size of the inversion because, for a longer inversion, there is a greater probability of a crossover occurring within it and producing an inviable meiotic product.

In a heterozygous *pericentric* inversion, the net genetic effect is the same as that of a paracentric inversion—crossover products are not recovered—but the reasons are different. In a pericentric inversion, the centromeres are contained within the inverted region. Consequently, the chromosomes that have engaged in crossing over separate in the normal fashion, without the creation of a bridge (Figure 17-29). However, the crossover produces chromatids that contain a duplication and a deletion for different parts of the chromosome. In this case, if a gamete carrying a crossover chromosome is fertilized, the zygote dies because of gene imbalance. Again, the result is that only noncrossover chromatids are present in viable progeny. Hence, the RF value of genes within a pericentric inversion also is zero.

Inversions affect recombination in another way, too. Inversion heterozygotes often have mechanical pairing problems in the region of the inversion. The inversion loop causes a large distortion that can extend beyond the loop itself. This distortion reduces the opportunity for crossing over in the neighboring regions.

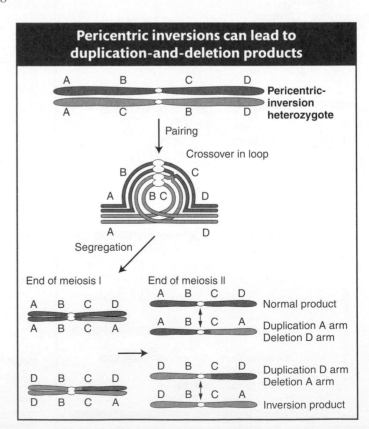

Figure 17-29 A crossover in the loop of a pericentric-inversion heterozygote gives rise to chromosomes containing duplications and deletions.

Let us consider an example of the effects of an inversion on recombinant frequency. A wild-type *Drosophila* specimen from a natural population is crossed with a homozygous recessive laboratory stock *dp cn/dp cn*. (The *dp* allele encodes dumpy wings and *cn* encodes cinnabar eyes. The two genes are known to be 45 map units apart on chromosome 2.) The F_1 generation is wild type. When an F_1 female is crossed with the recessive parent, the progeny are

250	wild type	$+ +/dp\ cn$
246	dumpy cinnabar	$dp\ cn/dp\ cn$
5	dumpy	$dp +/dp\ cn$
7	cinnabar	$+ cn/dp\ cn$

In this cross, which is effectively a dihybrid testcross, 45 percent of the progeny are expected to be dumpy or cinnabar (they constitute the crossover classes), but only 12 of 508, about 2 percent, are obtained. Something is reducing crossing over in this region, and a likely explanation is an inversion spanning most of the *dp–cn* region. Because the expected RF was based on measurements made on laboratory strains, the wild-type fly from nature was the most likely source of the inverted chromosome. Hence, chromosome 2 in the F_1 can be represented as follows:

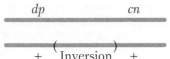

Pericentric inversions also can be detected microscopically through new arm ratios. Consider the following pericentric inversion:

Normal ●━━○━━━━━━━● Arm ratio, long : short ~ 4:1

Inversion ●━(━━○━━)━━● Arm ratio, long : short ~ 1:1

Note that the length ratio of the long arm to the short arm has been changed from about 4:1 to about 1:1 by the inversion. Paracentric inversions do not alter the arm ratio, but they may be detected microscopically by observing changes in banding or other chromosomal landmarks, if available.

> **Message** The main diagnostic features of heterozygous inversions are inversion loops, reduced recombinant frequency, and reduced fertility because of unbalanced or deleted meiotic products.

In some model experimental systems, notably *Drosophila* and the nematode *Caenorhabditis elegans*, inversions are used as balancers. A **balancer** chromosome contains *multiple* inversions; so, when it is combined with the corresponding wild-type chromosome, there can be no viable crossover products. In some analyses, it is important to keep stock with all the alleles on one chromosome together. The geneticist creates individuals having genomes that combine such a chromosome with a balancer. This combination eliminates crossovers, and so only parental combinations appear in the progeny. For convenience, balancer chromosomes are marked with a dominant morphological mutation. The marker allows the geneticist to track the segregation of the entire balancer or its normal homolog by noting the presence or absence of the marker.

Reciprocal translocations

There are several types of translocations, but here we consider only reciprocal translocations, the simplest type. Recall that, to form a reciprocal translocation,

Figure 17-30 The segregating chromosomes of a reciprocal-translocation heterozygote form a cross-shaped pairing configuration. The two most commonly encountered segregation patterns that result are the often inviable "adjacent-1" and the viable "alternate." N_1 and N_2, normal nonhomologous chromosomes; T_1 and T_2, translocated chromosomes. Up and Down designate the opposite poles to which homologs migrate in anaphase I.
GENETICS*PORTAL* ANIMATED ART: **Chromosome rearrangements: reciprocal translocation**

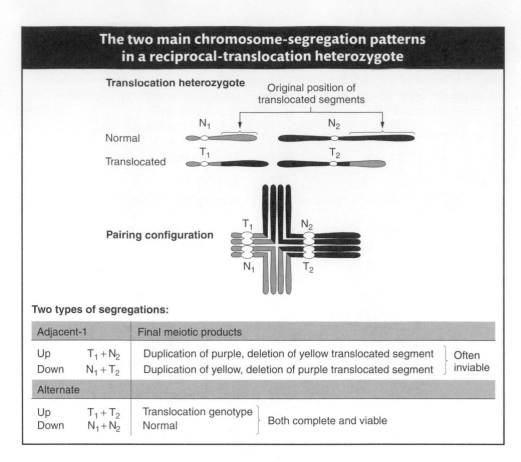

The two main chromosome-segregation patterns in a reciprocal-translocation heterozygote

Translocation heterozygote

Original position of translocated segments

Normal N_1 N_2

Translocated T_1 T_2

Pairing configuration T_1 N_2 N_1 T_2

Two types of segregations:

Adjacent-1		Final meiotic products	
Up	$T_1 + N_2$	Duplication of purple, deletion of yellow translocated segment	Often inviable
Down	$N_1 + T_2$	Duplication of yellow, deletion of purple translocated segment	
Alternate			
Up	$T_1 + T_2$	Translocation genotype	Both complete and viable
Down	$N_1 + N_2$	Normal	

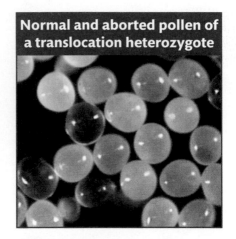

Normal and aborted pollen of a translocation heterozygote

Figure 17-31 Pollen of a semisterile corn plant. The clear pollen grains contain chromosomally unbalanced meiotic products of a reciprocal-translocation heterozygote. The opaque pollen grains, which contain either the complete translocation genotype or normal chromosomes, are functional in fertilization and development. [*William Sheridan.*]

two chromosomes trade acentric fragments created by two simultaneous chromosome breaks. As with other rearrangements, meiosis in heterozygotes having two translocated chromosomes and their normal counterparts produces characteristic configurations. Figure 17-30 illustrates meiosis in an individual that is heterozygous for a reciprocal translocation. Note the cross-shaped pairing configuration. Because the law of independent assortment is still in force, there are two common patterns of segregation. Let us use N_1 and N_2 to represent the normal chromosomes and T_1 and T_2 the translocated chromosomes. The segregation of each of the structurally normal chromosomes with one of the translocated ones ($T_1 + N_2$ and $T_2 + N_1$) is called **adjacent-1 segregation.** Each of the two meiotic products is deficient for a different arm of the cross and has a duplicate of the other. These products are inviable. On the other hand, the two normal chromosomes may segregate together, as will the reciprocal parts of the translocated ones, to produce $N_1 + N_2$ and $T_1 + T_2$ products. This segregation pattern is called **alternate segregation.** These products are both balanced and viable.

Adjacent-1 and alternate segregations are equal in number, and so half the overall population of gametes will be nonfunctional, a condition known as **semisterility** or "half sterility." Semisterility is an important diagnostic tool for identifying translocation heterozygotes. However, semisterility is defined differently for plants and animals. In plants, the 50 percent of meiotic products that are from the adjacent-1 segregation generally abort at the gametic stage (Figure 17-31). In animals, these products are viable as gametes but lethal to the zygotes that they produce on fertilization.

Remember that heterozygotes for inversions also may show some reduction in fertility but by an amount dependent on the size of the affected region. The precise 50 percent reduction in viable gametes or zygotes is usually a reliable diagnostic clue for a translocation.

Genetically, genes on translocated chromosomes act as though they are linked if their loci are close to the translocation break point. Figure 17-32 shows a translocation heterozygote that has been established by crossing an a/a ; b/b individual with a translocation homozygote bearing the wild-type alleles. When the heterozygote is testcrossed, recombinants are created but do not survive, because they carry unbalanced genomes (duplication-and-deletions). The only viable progeny are those bearing the parental genotypes; so linkage is seen between loci that were originally on different chromosomes. The apparent linkage of genes normally known to be on separate nonhomologous chromosomes—sometimes called **pseudolinkage**—is a genetic diagnostic clue to the presence of a translocation.

> **Message** Heterozygous reciprocal translocations are diagnosed genetically by semisterility and by the apparent linkage of genes whose normal loci are on separate chromosomes.

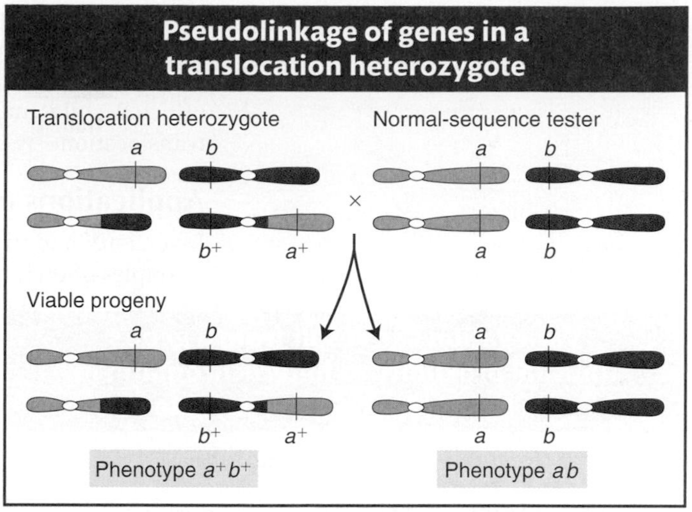

Pseudolinkage of genes in a translocation heterozygote

Figure 17-32 When a translocated fragment carries a marker gene, this marker can show linkage to genes on the other chromosome.
GENETICS*PORTAL* **ANIMATED ART: Chromosome rearrangements: pseudolinkage of genes**

Robertsonian translocations

Let's return to the family with the Down syndrome child, introduced at the beginning of the chapter. The birth can indeed be a coincidence—after all, coincidences do happen. However, the miscarriage gives a clue that something else might be going on. A large proportion of spontaneous abortions carry chromosomal abnormalities, so perhaps that is the case in this example. If so, the couple may have had two conceptions with chromosome mutations, which would be very unlikely unless there was a common cause. However, a small proportion of Down syndrome cases are known to result from a translocation in one of the parents. We have seen that translocations can produce progeny that have extra material from part of the genome, and so a translocation concerning chromosome 21 can produce progeny that have extra material from that chromosome. In Down syndrome, the translocation responsible is of a type called a *Robertsonian translocation*. It produces progeny carrying an almost complete extra copy of chromosome 21. The translocation and its segregation are illustrated in Figure 17-33. Note that, in addition to complements causing Down syndrome, other aberrant

Figure 17-33 In a small minority of cases, the origin of Down syndrome is a parent heterozygous for a Robertsonian translocation concerning chromosome 21. Meiotic segregation results in some gametes carrying a chromosome with a large additional segment of chromosome 21. In combination with a normal chromosome 21 provided by the gamete from the opposite sex, the symptoms of Down syndrome are produced even though there is not full trisomy 21.

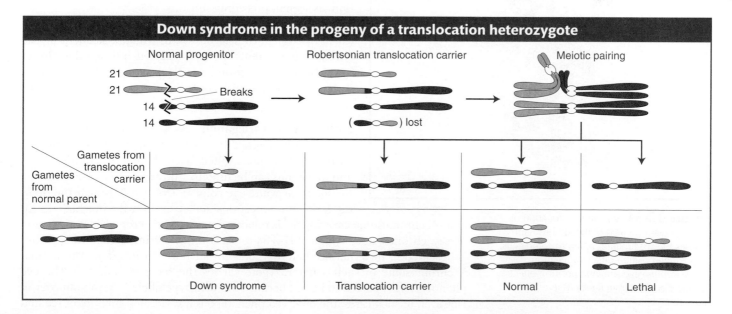

Down syndrome in the progeny of a translocation heterozygote

chromosome complements are produced, most of which abort. In our example, the man may have this translocation, which he may have inherited from his grandmother. To confirm this possibility, his chromosomes are checked. His unaffected child might have normal chromosomes or might have inherited his translocation.

Applications of inversions and translocations

Inversions and translocations have proved to be useful genetic tools; some examples of their uses follow.

Gene mapping Inversions and translocations are useful for the mapping and subsequent isolation of specific genes. The gene for human neurofibromatosis was isolated in this way. The critical information came from people who not only had the disease, but also carried chromosomal translocations. All the translocations had one break point in common, in a band close to the centromere of chromosome 17. Hence, this band appeared to be the locus of the neurofibromatosis gene, which had been disrupted by the translocation break point. Subsequent analysis showed that the chromosome 17 break points were not at identical positions; however, because they must have been within the gene, the range of their positions revealed the segment of the chromosome that constituted the neurofibromatosis gene. The isolation of DNA fragments from this region eventually led to the recovery of the gene itself.

Synthesizing specific duplications or deletions Translocations and inversions are routinely used to delete or duplicate specific chromosome segments. Recall, for example, that pericentric inversions as well as translocations generate products of meiosis that contain a duplication *and* a deletion (see Figures 17-29 and 17-30). If the duplicated or the deleted segment is very small, then the duplication-and-deletion meiotic products are tantamount to duplications or deletions, respectively. Duplications and deletions are useful for a variety of experimental applications, including the mapping of genes and the varying of gene dosage for the study of regulation, as seen in preceding sections.

Another approach to creating duplications uses unidirectional *insertional* translocations, in which a segment of one chromosome is removed and inserted into another. In an insertional-translocation heterozygote, a duplication results if the chromosome with the insertion segregates along with the normal copy.

Position-effect variegation As we saw in Chapter 12, gene action can be blocked by proximity to the densely staining chromosome regions called *heterochromatin*. Translocations and inversions can be used to study this effect. For example, the locus for white eye color in *Drosophila* is near the tip of the X chromosome. Consider a translocation in which the tip

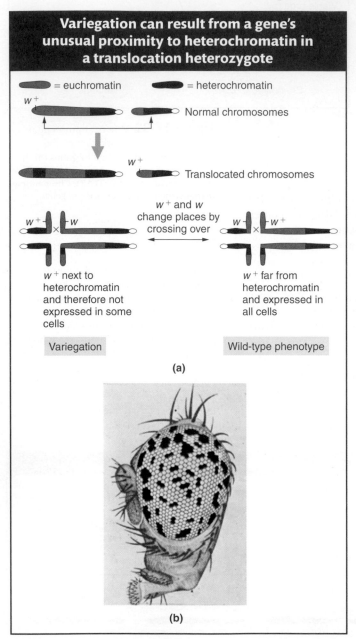

Variegation can result from a gene's unusual proximity to heterochromatin in a translocation heterozygote

= euchromatin   = heterochromatin

w^+ 〔 Normal chromosomes

w^+ 〔 Translocated chromosomes

w^+ and w change places by crossing over

w^+ next to heterochromatin and therefore not expressed in some cells

Variegation

w^+ far from heterochromatin and expressed in all cells

Wild-type phenotype

(a)

(b)

Figure 17-34 (a) The translocation of w^+ to a position next to heterochromatin causes the w^+ function to fail in some cells, producing position-effect variegation. (b) A *Drosophila* eye showing position-effect variegation. [*(b) From Randy Mottus.*]

of an X chromosome carrying w^+ is relocated next to the heterochromatic region of, say, chromosome 4 (Figure 17-34a, top section). **Position-effect variegation** is observed in flies that are heterozygotes for such a translocation. The normal X chromosome in such a heterozygote carries the recessive allele w. The eye phenotype is expected to be red because the wild-type allele is dominant over w. However, in such cases, the observed phenotype is a variegated mixture of red and white eye facets (Figure 17-34b). How can we explain the white areas? The w^+ allele is not always expressed, because the heterochromatin boundary is somewhat variable: in some cells, it engulfs and inactivates the w^+ gene, thereby allowing the expression of w. If the positions of the w^+ and w alleles are exchanged by a crossover, then position-effect variegation is not detected (see Figure 17-34a, bottom section).

Rearrangements and cancer

Cancer is a disease of abnormal cell proliferation. As a result of some insult inflicted on it, a cell of the body divides out of control to form a population of cells called a cancer. A localized knot of proliferated cells is called a tumor, whereas cancers of mobile cells such as blood cells disperse throughout the body. Cancer is most often caused by a mutation in the coding or regulatory sequence of a gene whose normal function is to regulate cell division. Such genes are called *proto-oncogenes*. However, chromosomal rearrangements, especially translocations, also can interfere with the normal function of such proto-oncogenes.

There are two basic ways in which translocations can alter the function of proto-oncogenes. In the first mechanism, the translocation relocates a proto-oncogene next to a new regulatory element. A good example is provided by Burkitt lymphoma. The proto-oncogene in this cancer encodes the protein MYC, a transcription factor that activates genes required for cell proliferation. Normally, the *myc* gene is transcribed only when a cell needs to undergo proliferation, but, in cancerous cells, the proto-oncogene *MYC* is relocated next to the regulatory region of immunoglobulin (Ig) genes (Figure 17-35a). These immunoglobulin genes are constitutively transcribed; that is, they are on all the time. Consequently, the *myc* gene is transcribed at all times, and the cell-proliferation genes are continuously activated.

The other mechanism by which translocations can cause cancer is the formation of a hybrid gene. An example is provided by the disease chronic myelogenous

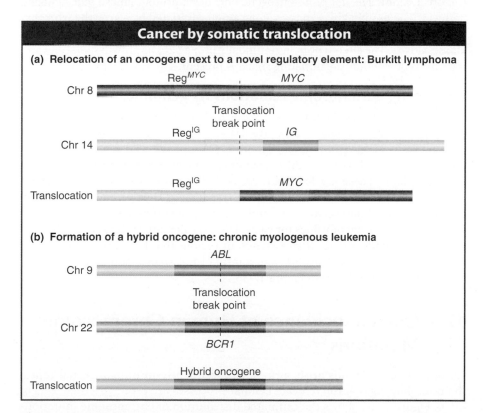

Figure 17-35 The two main ways that translocations can cause cancer in a body (somatic) cell are illustrated by the cancers Burkitt lymphoma (a) and chronic myelogenous leukemia (b). The genes *MYC, BCR1,* and *ABL* are proto-oncogenes.

Figure 17-36 To detect chromosomal rearrangements, mutant and wild-type genomic DNA is tagged with dyes that fluoresce at different wavelengths. These tagged DNAs are added to cDNA clones arranged in chromosomally ordered microarrays, and the ratio of bound fluorescence at each wavelength is calculated for each clone. The expected results for a normal genome and three types of mutants are illustrated.

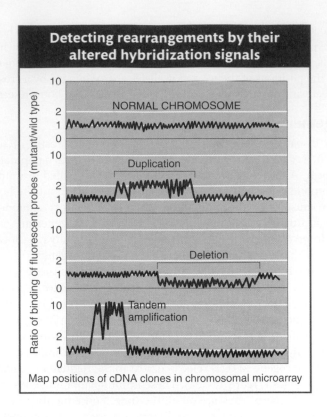

leukemia (CML), a cancer of white blood cells. This cancer can result from the formation of a hybrid gene between the two proto-oncogenes *BCR1* and *ABL* (Figure 17-35b). The *abl* proto-oncogene encodes a protein kinase in a signaling pathway. The protein kinase passes along a signal initiated by a growth factor that leads to cell proliferation. The Bcr1-Abl fusion protein has a permanent protein kinase activity. The altered protein continually propagates its growth signal onward, regardless of whether the initiating signal is present.

Identifying chromosome mutations by genomics

DNA microarrays (see Figure 14-19) have made it possible to detect and quantify duplications or deletions of a given DNA segment. The technique is called *comparative genomic hybridization*. The total DNA of the wild type and that of a mutant are labeled with two different fluorescent dyes that emit distinct wavelengths of light. These labeled DNAs are added to a cDNA microarray together, and both of them hybridize to the array. The array is then scanned by a detector tuned to one fluorescent wavelength and is then scanned again for the other wavelength. The ratio of values for each cDNA is calculated. Mutant-to-wild-type ratios substantially greater than 1 represent regions that have been amplified. A ratio of 2 points to a duplication, and a ratio of less than 1 points to a deletion. Some examples are shown in Figure 17-36.

17.3 Overall Incidence of Human Chromosome Mutations

Chromosome mutations arise surprisingly frequently in human sexual reproduction, showing that the relevant cellular processes are prone to a high level of error. Figure 17-37 shows the estimated distribution of chromosome mutations among human conceptions that develop sufficiently to implant in the uterus. Of the estimated 15 percent of conceptions that abort spontaneously (pregnancies that terminate naturally), fully half show chromosomal abnormalities.

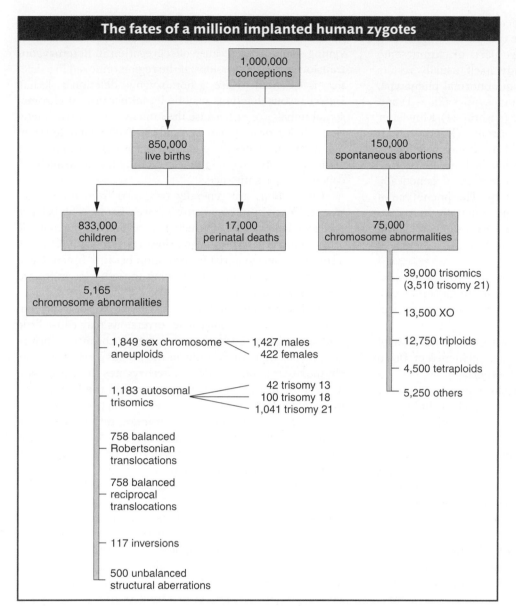

Figure 17-37 The proportion of chromosomal mutations is much higher in spontaneous abortions. [*From K. Sankaranarayanan, Mutat. Res. 61, 1979, 249–257.*]

Some medical geneticists believe that even this high level is an underestimate, because many cases are never detected. Among live births, 0.6 percent have chromosomal abnormalities, resulting from both aneuploidy and chromosomal rearrangements.

Summary

Polyploidy is an abnormal condition in which there is a larger-than-normal number of chromosome sets. Polyploids such as triploids ($3n$) and tetraploids ($4n$) are common among plants and are represented even among animals. Organisms with an odd number of chromosome sets are sterile because not every chromosome has a partner at meiosis. Unpaired chromosomes attach randomly to the poles of the cell in meiosis, leading to unbalanced sets of chromosomes in the resulting gametes. Such unbalanced gametes do not yield viable progeny. In polyploids with an even number of sets, each chromosome has a potential pairing partner and hence can produce balanced gametes and progeny. Polyploidy can result in an organism of larger dimensions; this discovery has permitted important advances in horticulture and in crop breeding.

In plants, allopolyploids (polyploids formed by combining chromosome sets from different species) can be made by crossing two related species and then doubling the progeny chromosomes through the use of colchicine or through somatic cell fusion. These techniques have potential applications in crop breeding because allopolyploids combine the features of the two parental species.

When cellular accidents change parts of chromosome sets, aneuploids result. Aneuploidy itself usually results in an unbalanced genotype with an abnormal phenotype. Examples of aneuploids include monosomics $(2n - 1)$ and trisomics $(2n + 1)$. Down syndrome (trisomy 21), Klinefelter syndrome (XXY), and Turner syndrome (XO) are well-documented examples of aneuploid conditions in humans. The spontaneous level of aneuploidy in humans is quite high and accounts for a large proportion of genetically based ill health in human populations. The phenotype of an aneuploid organism depends very much on the particular chromosome affected. In some cases, such as human trisomy 21, there is a highly characteristic constellation of associated phenotypes.

Most instances of aneuploidy result from accidental chromosome missegregation at meiosis (nondisjunction). The error is spontaneous and can occur in any particular meiocyte at the first or second division. In humans, a maternal-age effect is associated with nondisjunction of chromosome 21, resulting in a higher incidence of Down syndrome in the children of older mothers.

The other general category of chromosome mutations comprises structural rearrangements, which include deletions, duplications, inversions, and translocations. These changes result either from breakage and incorrect reunion or from crossing over between repetitive elements (nonallelic homologous recombination). Chromosomal rearrangements are an important cause of ill health in human populations and are useful in engineering special strains of organisms for experimental and applied genetics. In organisms with one normal chromosome set plus a rearranged set (heterozygous rearrangements), there are unusual pairing structures at meiosis resulting from the strong pairing affinity of homologous chromosome regions. For example, heterozygous inversions show loops, and reciprocal translocations show cross-shaped structures. Segregation of these structures results in abnormal meiotic products unique to the rearrangement.

A deletion is the loss of a section of chromosome, either because of chromosome breaks followed by loss of the intervening segment or because of segregation in heterozygous translocations or inversions. If the region removed in a deletion is essential to life, a homozygous deletion is lethal. Heterozygous deletions may be lethal because of chromosomal imbalance or because they uncover recessive deleterious alleles, or they may be nonlethal. When a deletion in one homolog allows the phenotypic expression of recessive alleles in the other, the unmasking of the recessive alleles is called pseudodominance.

Duplications are generally produced from other rearrangements or by aberrant crossing over. They also unbalance the genetic material, producing a deleterious phenotypic effect or death of the organism. However, duplications can be a source of new material for evolution because function can be maintained in one copy, leaving the other copy free to evolve new functions.

An inversion is a 180-degree turn of a part of a chromosome. In the homozygous state, inversions may cause little problem for an organism unless heterochromatin brings about a position effect or one of the breaks disrupts a gene. On the other hand, inversion heterozygotes show inversion loops at meiosis, and crossing over within the loop results in inviable products. The crossover products of pericentric inversions, which span the centromere, differ from those of paracentric inversions, which do not, but both show reduced recombinant frequency in the affected region and often result in reduced fertility.

A translocation moves a chromosome segment to another position in the genome. A simple example is a reciprocal translocation, in which parts of nonhomologous chromosomes exchange positions. In the heterozygous state, translocations produce duplication-and-deletion meiotic products, which can lead to unbalanced zygotes. New gene linkages can be produced by translocations. The random segregation of centromeres in a translocation heterozygote results in 50 percent unbalanced meiotic products and, hence, 50 percent sterility (semisterility).

KEY TERMS

acentric chromosome (p. 607)
acentric fragment (p. 615)
adjacent-1 segregation (p. 618)
allopolyploid (p. 592)
alternate segregation (p. 618)
amphidiploid (p. 595)
anaphase bridge (p. 607)
aneuploid (p. 592)
autopolyploid (p. 592)
balanced rearrangement (p. 608)
balancer (p. 617)
bivalent (p. 592)
chromosome mutation (p. 590)
deletion (p. 606)

deletion loop (p. 609)
deletion mapping (p. 610)
dicentric bridge (p. 615)
dicentric chromosome (p. 607)
disomic (p. 599)
dosage compensation (p. 605)
Down syndrome (p. 602)
duplication (p. 606)
embryoid (p. 598)
euploid (p. 591)
gene balance (p. 604)
gene-dosage effect (p. 605)
genetic load (p. 592)
hexaploid (p. 591)

homeologous chromosomes (p. 592)
insertional duplication (p. 612)
intragenic deletion (p. 609)
inversion (p. 606)
inversion heterozygote (pp. 606, 615)
inversion loop (p. 615)
Klinefelter syndrome (p. 602)
monoploid (p. 591)
monosomic (p. 599)
multigenic deletion (p. 609)
nonallelic homologous
     recombination (NAHR) (p. 608)
nondisjunction (p. 600)
nullisomic (p. 599)

SOLVED PROBLEMS

SOLVED PROBLEM 1. A corn plant is heterozygous for a reciprocal translocation and is therefore semisterile. This plant is crossed with a chromosomally normal strain that is homozygous for the recessive allele brachytic (*b*), located on chromosome 2. A semisterile F₁ plant is then backcrossed to the homozygous brachytic strain. The progeny obtained show the following phenotypes:

Nonbrachytic		Brachytic	
Semisterile	Fertile	Semisterile	Fertile
334	27	42	279

a. What ratio would you expect to result if the chromosome carrying the brachytic allele does not take part in the translocation?

b. Do you think that chromosome 2 takes part in the translocation? Explain your answer, showing the conformation of the relevant chromosomes of the semisterile F₁ and the reason for the specific numbers obtained.

SOLUTION

a. We should start with the methodical approach and simply restate the data in the form of a diagram, where

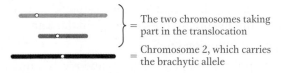

To simplify the diagram, we do not show the chromosomes divided into chromatids (although they would be at this stage of meiosis). We then diagram the first cross:

Translocation strain

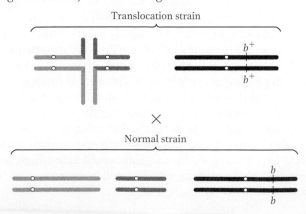

×

Normal strain

All the progeny from this cross will be heterozygous for the chromosome carrying the brachytic allele, but what about the chromosomes taking part in the translocation? In this chapter, we have seen that only alternate-segregation products survive and that half of these survivors will be chromosomally normal and half will carry the two rearranged chromosomes. The rearranged combination will regenerate a translocation heterozygote when it combines with the chromosomally normal complement from the normal parent. These latter types—the semisterile F₁'s—are diagrammed as part of the backcross to the parental brachytic strain:

Semisterile F₁

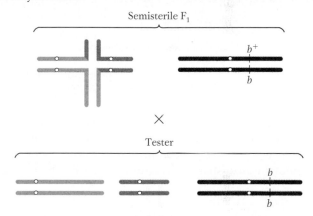

×

Tester

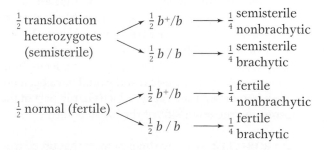

In calculating the expected ratio of phenotypes from this cross, we can treat the behavior of the translocated chromosomes independently of the behavior of chromosome 2. Hence, we can predict that the progeny will be

$\frac{1}{2}$ translocation heterozygotes (semisterile) → $\frac{1}{2}\,b^+/b$ → $\frac{1}{4}$ semisterile nonbrachytic
↘ $\frac{1}{2}\,b/b$ → $\frac{1}{4}$ semisterile brachytic

$\frac{1}{2}$ normal (fertile) → $\frac{1}{2}\,b^+/b$ → $\frac{1}{4}$ fertile nonbrachytic
↘ $\frac{1}{2}\,b/b$ → $\frac{1}{4}$ fertile brachytic

This predicted 1:1:1:1 ratio is quite different from that obtained in the actual cross.

b. Because we observe a departure from the expected ratio based on the independence of the brachytic phenotype and semisterility, chromosome 2 likely *does* take part in the translocation. Let's assume that the brachytic locus (*b*) is on the orange chromosome. But where? For the purpose of the diagram, it doesn't matter where we put it, but it does matter genetically because the position of the *h* locus affects the ratios in the progeny. If we assume that the *b* locus is near the tip of the piece that is translocated, we can redraw the pedigree:

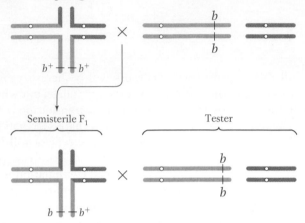

If the chromosomes of the semisterile F₁ segregate as diagrammed here, we could then predict

$\frac{1}{2}$ fertile, brachytic

$\frac{1}{2}$ semisterile, nonbrachytic

Most progeny are certainly of this type, and so we must be on the right track. How are the two less-frequent types produced? Somehow, we have to get the *b⁺* allele onto the normal yellow chromosome and the *b* allele onto the translocated chromosome. This positioning must be achieved by crossing over between the translocation break point (the center of the cross-shaped structure) and the brachytic locus.

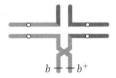

The recombinant chromosomes produce some progeny that are fertile and nonbrachytic and some that are semisterile and brachytic (these two classes together constitute 69 progeny of a total of 682, or a frequency of about 10 percent). We can see that this frequency is really a measure of the map distance (10 m.u.) of the brachytic locus from the breakpoint. (The same basic result would have been obtained if we had drawn the brachytic locus in the part of the chromosome on the other side of the break point.)

SOLVED PROBLEM 2. We have lines of mice that breed true for two alternative behavioral phenotypes that we know are determined by two alleles at a single locus: *v* causes a mouse to move with a "waltzing" gait, whereas *V* determines a normal gait. After crossing the true-breeding waltzers and normals, we observe that most of the F₁ is normal,

but, unexpectedly, there is one waltzer female. We mate the F₁ waltzer with two different waltzer males and note that she produces only waltzer progeny. When we mate her with normal males, she produces normal progeny and no waltzers. We mate three of her normal female progeny with two of their brothers, and these mice produce 60 progeny, all normal. When, however, we mate one of these same three females with a third brother, we get six normals and two waltzers in a litter of eight. By thinking about the parents of the F₁ waltzer, we can consider some possible explanations of these results:

a. A dominant allele may have mutated to a recessive allele in her normal parent.

b. In one parent, there may have been a dominant mutation in a second gene to create an epistatic allele that acts to prevent the expression of *V*, leading to waltzing.

c. Meiotic nondisjunction of the chromosome carrying *V* in her normal parent may have given a viable aneuploid.

d. There may have been a viable deletion spanning *V* in the meiocyte from her normal parent.

Which of these explanations are possible, and which are eliminated by the genetic analysis? Explain in detail.

SOLUTION

The best way to answer the question is to take the explanations one at a time and see if each fits the results given.

a. Mutation *V* to *v*

This hypothesis requires that the exceptional waltzer female be homozygous *v/v*. This assumption is compatible with the results of mating her both with waltzer males, which would, if she is *v/v*, produce all waltzer offspring (*v/v*), and with normal males, which would produce all normal offspring (*V/v*). However, brother–sister matings within this normal progeny should then produce a 3 : 1 normal-to-waltzer ratio. Because some of the brother–sister matings actually produced no waltzers, this hypothesis does not explain the data.

b. Epistatic mutation *s* to *S*

Here the parents would be *V/V · s/s* and *v/v · s/s*, and a germinal mutation in one of them would give the F₁ waltzer the genotype *V/v · S/s*. When we crossed her with a waltzer male, who would be of the genotype *v/v · s/s*, we would expect some *V/v · S/s* progeny, which would be phenotypically normal. However, we saw no normal progeny from this cross, and so the hypothesis is already overthrown. Linkage could save the hypothesis temporarily if we assumed that the mutation was in the normal parent, giving a gamete *VS*. Then the F₁ waltzer would be *VS/vs*, and, if linkage were tight enough, few or no *V s* gametes would be produced, the type that are necessary to combine with the *v s* gamete from the male to give *Vs/vs* normals. However, if the linkage hypothesis were true, the cross with the normal males would be *VS/v s × Vs/Vs*, and this would give a high percentage of *VS/Vs* progeny, which would be waltzers, none of which were seen.

c. Nondisjunction in the normal parent

This explanation would give a nullisomic gamete that would combine with v to give the F_1 waltzer the hemizygous genotype v. The subsequent matings would be

- $v \times v/v$, which gives v/v and v progeny, all waltzers. This fits.
- $v \times V/V$, which gives V/v and V progeny, all normals. This also fits.
- First intercrosses of normal progeny: $V \times V$. These intercrosses give V and V/V, which are normal. This fits.
- Second intercrosses of normal progeny: $V \times V/v$. These intercrosses give 25 percent each of V/V, V/v, V (all normals), and v (waltzers). This also fits.

This hypothesis is therefore consistent with the data.

d. Deletion of V in normal parent

Let's call the deletion D. The F_1 waltzer would be D/v, and the subsequent matings would be

- D/v × v/v, which gives v/v and D/v, which are waltzers. This fits.
- D/v × V/V, which gives V/v and D/V, which are normal. This fits.
- First intercrosses of normal progeny: D/V × D/V, which give D/V and V/V, all normal. This fits.
- Second intercrosses of normal progeny: D/V × V/v, which give 25 percent each of V/V, V/v, D/V (all normals), and D/v (waltzers). This also fits.

Once again, the hypothesis fits the data provided; so we are left with two hypotheses that are compatible with the results, and further experiments are necessary to distinguish them. One way of doing so would be to examine the chromosomes of the exceptional female under the microscope: aneuploidy should be easy to distinguish from deletion.

PROBLEMS

Most of the problems are also available for review/grading through the GENETICS PORTAL www.yourgeneticsportal.com.

WORKING WITH THE FIGURES

1. Based on Table 17-1, how would you categorize the following genomes? (Letters H through J stand for four different chromosomes.)

    HH II J KK

    HH II JJ KKK

    HHHH IIII JJJJ KKKK

2. Based on Figure 17-4, how many chromatids are in a trivalent?

3. Based on Figure 17-5, if colchicine is used on a plant in which $2n = 18$, how many chromosomes would be in the abnormal product?

4. Basing your work on Figure 17-7, use colored pens to represent the chromosomes of the fertile amphidiploid.

5. If Emmer wheat (Figure 17-9) is crossed to another wild wheat CC (not shown), what would be the constitution of a sterile product of this cross? What amphidiploid could arise from the sterile product? Would the amphidiploid be fertile?

6. In Figure 17-12, what would be the constitution of an individual formed from the union of a monosomic from a first-division nondisjunction in a female and a disomic from a second-division nondisjunction in a male, assuming the gametes were functional?

7. In Figure 17-14, what would be the expected percentage of each type of segregation?

8. In Figure 17-19, is there any difference between the inversion products formed from breakage and those formed from crossing over?

9. Referring to Figure 17-19, draw a diagram showing the process whereby an inversion formed from crossing over could generate a normal sequence.

10. In Figure 17-21, would the recessive fa allele be expressed when paired with deletion 264-32? 265-11?

11. Look at Figure 17-22 and state which bands are missing in the cri du chat deletion.

12. In Figure 17-25, which species is most closely related to the ancestral yeast strain? Why are genes 3 and 13 referred to as duplicate?

13. Referring to Figure 17-26, draw the product if breaks occurred within genes A and B.

14. In Figure 17-26, the bottom panel shows that genes B and C are oriented in a different direction (note the promoters). Do you think this difference in orientation would affect their functionality?

15. In Figure 17-28, what would be the consequence of a crossover between the centromere and locus A?

16. Based on Figure 17-30, are normal genomes ever formed from the two types of segregation? Are normal genomes ever formed from an adjacent-1 segregation?

17. Referring to Figure 17-32, draw an *inviable* product from the same meiosis.

18. Based on Figure 17-35, write a sentence stating how translocation can lead to cancer. Can you think of another genetic cause of cancer?

19. Looking at Figure 17-36, why do you think the signal ratio is so much higher in the bottom panel?

20. Using Figure 17-37, calculate what percentage of conceptions are triploid. The same figure shows XO in the spontaneous-abortion category; however, we know that many XO individuals are viable. In which of the viable categories would XO be grouped?

BASIC PROBLEMS

21. In keeping with the style of Table 17-1, what would you call organisms that are MM N OO; MM NN OO; MMM NN PP?

22. A large plant arose in a natural population. Qualitatively, it looked just the same as the others, except much larger. Is it more likely to be an allopolyploid or an autopolyploid? How would you test that it was a polyploid and not just growing in rich soil?

23. Is a trisomic an aneuploid or a polyploid?

24. In a tetraploid $B/B/b/b$, how many quadrivalent possible pairings are there? Draw them (see Figure 17-5).

25. Someone tells you that cauliflower is an amphidiploid. Do you agree? Explain.

26. Why is *Raphanobrassica* fertile, whereas its progenitor wasn't?

27. In the designation of wheat genomes, how many chromosomes are represented by the letter B?

28. How would you "re-create" hexaploid bread wheat from *Triticum tauschii* and Emmer?

29. How would you make a monoploid plantlet by starting with a diploid plant?

30. A disomic product of meiosis is obtained. What is its likely origin? What other genotypes would you expect among the products of that meiosis under your hypothesis?

31. Can a trisomic $A/A/a$ ever produce a gamete of genotype a?

32. Which, if any, of the following sex-chromosome aneuploids in humans are fertile: XXX, XXY, XYY, XO?

33. Why are older expectant mothers routinely given amniocentesis or CVS?

34. In an inversion, is a 5′ DNA end ever joined to another 5′ end? Explain.

35. If you observed a dicentric bridge at meiosis, what rearrangement would you predict had taken place?

36. Why do acentric fragments get lost?

37. Diagram a translocation arising from repetitive DNA. Repeat for a deletion.

38. From a large stock of *Neurospora* rearrangements available from the fungal genetics stock center, what type would you choose to synthesize a strain that had a duplication of the right arm of chromosome 3 and a deletion for the tip of chromosome 4?

39. You observe a very large pairing loop at meiosis. Is it more likely to be from a heterozygous inversion or heterozygous deletion? Explain.

40. A new recessive mutant allele doesn't show pseudodominance with any of the deletions that span *Drosophila* chromosome 2. What might be the explanation?

41. Compare and contrast the origins of Turner syndrome, Williams syndrome, cri du chat syndrome, and Down syndrome. (Why are they called *syndromes*?)

42. List the diagnostic features (genetic or cytological) that are used to identify these chromosomal alterations:

 a. Deletions

 b. Duplications

 c. Inversions

 d. Reciprocal translocations

43. The normal sequence of nine genes on a certain *Drosophila* chromosome is 123 · 456789, where the dot represents the centromere. Some fruit flies were found to have aberrant chromosomes with the following structures:

 a. 123 · 476589

 b. 123 · 46789

 c. 1654 · 32789

 d. 123 · 4566789

Name each type of chromosomal rearrangement, and draw diagrams to show how each would synapse with the normal chromosome.

44. The two loci *P* and *Bz* are normally 36 m.u. apart on the same arm of a certain plant chromosome. A paracentric inversion spans about one-fourth of this region but does not include either of the loci. What approximate recombinant frequency between *P* and *Bz* would you predict in plants that are

 a. heterozygous for the paracentric inversion?

 b. homozygous for the paracentric inversion?

45. As stated in Solved Problem 2, certain mice called *waltzers* have a recessive mutation that causes them to execute bizarre steps. W. H. Gates crossed waltzers with homozygous normals and found, among several hundred normal progeny, a single waltzing female mouse. When mated with a waltzing male, she produced all waltzing offspring. When mated with a homozygous normal male, she produced all normal progeny. Some males and females of this normal progeny were intercrossed, and there were no waltzing offspring among their progeny. T. S. Painter examined the chromosomes of waltzing mice that were derived from some of Gates's crosses and that showed a breeding behavior similar to that of the original, unusual waltzing female. He found that these mice had 40 chromosomes, just as in normal mice or the usual waltzing mice. In the unusual waltzers, however, one member of a chromosome pair was abnormally short. Interpret these observations as completely as possible, both genetically and cytologically. (Problem 45 is from A. M. Srb, R. D. Owen, and R. S. Edgar, *General Genetics,* 2nd ed. W. H. Freeman and Company, 1965.)

46. Six bands in a salivary-gland chromosome of *Drosophila* are shown in the following illustration, along with the extent of five deletions (Del 1 to Del 5):

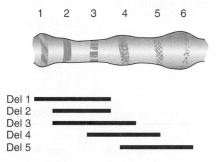

Recessive alleles *a, b, c, d, e,* and *f* are known to be in the region, but their order is unknown. When the deletions are combined with each allele, the following results are obtained:

	a	b	c	d	e	f
Del 1	−	−	−	+	+	+
Del 2	−	+	−	+	+	+
Del 3	−	+	−	+	−	+
Del 4	+	+	−	−	−	+
Del 5	+	+	+	−	−	−

In this table, a minus sign means that the deletion is missing the corresponding wild-type allele (the deletion uncovers the recessive), and a plus sign means that the corresponding wild-type allele is still present. Use these data to infer which salivary band contains each gene. (Problem 46 is from D. L. Hartl, D. Friefelder, and L. A. Snyder, *Basic Genetics.* Jones and Bartlett, 1988.)

47. A fruit fly was found to be heterozygous for a paracentric inversion. However, obtaining flies that were homozygous for the inversion was impossible even after many attempts. What is the most likely explanation for this inability to produce a homozygous inversion?

48. Orangutans are an endangered species in their natural environment (the islands of Borneo and Sumatra), and so a captive-breeding program has been established using orangutans currently held in zoos throughout the world. One component of this program is research into orangutan cytogenetics. This research has shown that all orangutans from Borneo carry one form of chromosome 2, as shown in the accompanying diagram, and all orangutans from Sumatra carry the other form. Before this cytogenetic difference became known, some matings were carried out between animals from different islands, and 14 hybrid progeny are now being raised in captivity.

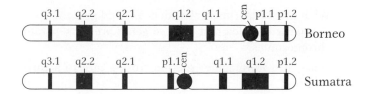

a. What term or terms describe the differences between these chromosomes?

b. Draw the chromosomes 2, paired in the first meiotic prophase, of such a hybrid orangutan. Be sure to show all the landmarks indicated in the accompanying diagram, and label all parts of your drawing.

c. In 30 percent of meioses, there will be a crossover somewhere in the region between bands p1.1 and q1.2. Draw the gamete chromosomes 2 that would result from a meiosis in which a single crossover occurred within band q1.1.

d. What fraction of the gametes produced by a hybrid orangutan will give rise to viable progeny, if these chromosomes are the only ones that differ between the parents? (Problem 48 is from Rosemary Redfield.)

49. In corn, the genes for tassel length (alleles *T* and *t*) and rust resistance (alleles *R* and *r*) are known to be on separate chromosomes. In the course of making routine crosses, a breeder noticed that one *T/t ; R/r* plant gave unusual results in a testcross with the double-recessive pollen parent *t/t ; r/r.* The results were

Progeny:	*T/t ; R/r*	98
	t/t ; r/r	104
	T/t ; r/r	3
	t/t ; R/r	5

Corncobs: Only about half as many seeds as usual

a. What key features of the data are different from the expected results?

b. State a concise hypothesis that explains the results.

c. Show genotypes of parents and progeny.

d. Draw a diagram showing the arrangement of alleles on the chromosomes.

e. Explain the origin of the two classes of progeny having three and five members.

Unpacking Problem 49

1. What do a "gene for tassel length" and a "gene for rust resistance" mean?

2. Does it matter that the precise meaning of the allelic symbols *T, t, R,* and *r* is not given? Why or why not?

3. How do the terms *gene* and *allele,* as used here, relate to the concepts of locus and gene pair?

4. What prior experimental evidence would give the corn geneticist the idea that the two genes are on separate chromosomes?

5. What do you imagine "routine crosses" are to a corn breeder?

6. What term is used to describe genotypes of the type T/t ; R/r?

7. What is a "pollen parent"?

8. What are testcrosses, and why do geneticists find them so useful?

9. What progeny types and frequencies might the breeder have been expecting from the testcross?

10. Describe how the observed progeny differ from expectations.

11. What does the approximate equality of the first two progeny classes tell you?

12. What does the approximate equality of the second two progeny classes tell you?

13. What were the gametes from the unusual plant, and what were their proportions?

14. Which gametes were in the majority?

15. Which gametes were in the minority?

16. Which of the progeny types seem to be recombinant?

17. Which allelic combinations appear to be linked in some way?

18. How can there be linkage of genes supposedly on separate chromosomes?

19. What do these majority and minority classes tell us about the genotypes of the parents of the unusual plant?

20. What is a corncob?

21. What does a normal corncob look like? (Sketch one and label it.)

22. What do the corncobs from this cross look like? (Sketch one.)

23. What exactly is a kernel?

24. What effect could lead to the absence of half the kernels?

25. Did half the kernels die? If so, was the female or the male parent the reason for the deaths?

Now try to solve the problem.

50. A yellow body in *Drosophila* is caused by a mutant allele *y* of a gene located at the tip of the X chromosome (the wild-type allele causes a gray body). In a radiation experiment, a wild-type male was irradiated with X rays and then crossed with a yellow-bodied female. Most of the male progeny were yellow, as expected, but the scanning of thousands of flies revealed two gray-bodied (phenotypically wild-type) males. These gray-bodied males were crossed with yellow-bodied females, with the following results:

	Progeny
gray male 1 × yellow female	females all yellow
	males all gray
gray male 2 × yellow female	$\frac{1}{2}$ females yellow
	$\frac{1}{2}$ females gray
	$\frac{1}{2}$ males yellow
	$\frac{1}{2}$ males gray

a. Explain the origin and crossing behavior of gray male 1.

b. Explain the origin and crossing behavior of gray male 2.

51. In corn, the allele *Pr* stands for green stems, *pr* for purple stems. A corn plant of genotype *pr/pr* that has standard chromosomes is crossed with a *Pr/Pr* plant that is homozygous for a reciprocal translocation between chromosomes 2 and 5. The F₁ is semisterile and phenotypically Pr. A backcross with the parent with standard chromosomes gives 764 semi-sterile Pr, 145 semisterile pr, 186 normal Pr, and 727 normal pr. What is the map distance between the *Pr* locus and the translocation point?

52. Distinguish among Klinefelter, Down, and Turner syndromes. Which syndromes are found in both sexes?

53. Show how you could make an allotetraploid between two related diploid plant species, both of which are $2n = 28$.

54. In *Drosophila*, trisomics and monosomics for the tiny chromosome 4 are viable, but nullisomics and tetrasomics are not. The *b* locus is on this chromosome. Deduce the phenotypic proportions in the progeny of the following crosses of trisomics.

a. $b^+/b/b \times b/b$

b. $b^+/b^+/b \times b/b$

c. $b^+/b^+/b \times b^+/b$

55. A woman with Turner syndrome is found to be colorblind (an X-linked recessive phenotype). Both her mother and her father have normal vision.

a. Explain the simultaneous origin of Turner syndrome and color blindness by the abnormal behavior of chromosomes at meiosis.

b. Can your explanation distinguish whether the abnormal chromosome behavior occurred in the father or the mother?

c. Can your explanation distinguish whether the abnormal chromosome behavior occurred at the first or second division of meiosis?

d. Now assume that a color-blind Klinefelter man has parents with normal vision, and answer parts *a*, *b*, and *c*.

56. a. How would you synthesize a pentaploid?

b. How would you synthesize a triploid of genotype $A/a/a$?

c. You have just obtained a rare recessive mutation a^* in a diploid plant, which Mendelian analysis tells you is A/a^*. From this plant, how would you synthesize a tetraploid ($4n$) of genotype $A/A/a^*/a^*$?

d. How would you synthesize a tetraploid of genotype $A/a/a/a$?

57. Suppose you have a line of mice that has cytologically distinct forms of chromosome 4. The tip of the chromosome can have a knob (called 4^K) or a satellite (4^S) or neither (4). Here are sketches of the three types:

You cross a $4^K/4^S$ female with a 4/4 male and find that most of the progeny are $4^K/4$ or $4^S/4$, as expected. However, you occasionally find some rare types as follows (all other chromosomes are normal):

a. $4^K/4^K/4$

b. $4^K/4^S/4$

c. 4^K

Explain the rare types that you have found. Give, as precisely as possible, the stages at which they originate, and state whether they originate in the male parent, the female parent, or the zygote. (Give reasons briefly.)

58. A cross is made in tomatoes between a female plant that is trisomic for chromosome 6 and a normal diploid male plant that is homozygous for the recessive allele for potato leaf (p/p). A trisomic F_1 plant is back-crossed to the potato-leaved male.

a. What is the ratio of normal-leaved plants to potato-leaved plants when you assume that p is located on chromosome 6?

b. What is the ratio of normal-leaved to potato-leaved plants when you assume that p is not located on chromosome 6?

59. A tomato geneticist attempts to assign five recessive mutations to specific chromosomes by using trisomics. She crosses each homozygous mutant ($2n$) with each of three trisomics, in which chromosomes 1, 7, and 10 take part. From these crosses, the geneticist selects trisomic progeny (which are less vigorous) and back-crosses them to the appropriate homozygous recessive. The *diploid* progeny from these crosses are examined. Her results, in which the ratios are wild type:mutant, are as follows:

Trisomic chromosome	Mutation				
	d	*y*	*c*	*h*	*cot*
1	48:55	72:29	56:50	53:54	32:28
7	52:56	52:48	52:51	58:56	81:40
10	45:42	36:33	28:32	96:50	20:17

Which of the mutations can the geneticist assign to which chromosomes? (Explain your answer fully.)

60. A petunia is heterozygous for the following autosomal homologs:

A	B	C	D	E	F	G	H	I
a	b	c	d	h	g	f	e	i

a. Draw the pairing configuration that you would see at metaphase I, and identify all parts of your diagram. Number the chromatids sequentially from top to bottom of the page.

b. A three-strand double crossover occurs, with one crossover between the C and D loci on chromatids 1 and 3, and the second crossover between the G and H loci on chromatids 2 and 3. Diagram the results of these recombination events as you would see them at anaphase I, and identify all parts of your diagram.

c. Draw the chromosome pattern that you would see at anaphase II after the crossovers described in part *b*.

d. Give the genotypes of the gametes from this meiosis that will lead to the formation of viable progeny. Assume that all gametes are fertilized by pollen that has the gene order $A\ B\ C\ D\ E\ F\ G\ H\ I$.

61. Two groups of geneticists, in California and in Chile, begin work to develop a linkage map of the medfly. They both independently find that the loci for body color (B = black, b = gray) and eye shape (R = round, r = star) are linked 28 m.u. apart. They send strains to each other and perform crosses; a summary of all their findings is shown here:

Cross	F_1	Progeny of $F_1 \times$ any $b\,r/b\,r$	
$B\,R/B\,R$ (Calif.)	$B\,R/b\,r$	$B\,R/b\,r$	36%
$\times\ b\,r/b\,r$ (Calif.)		$b\,r/b\,r$	36
		$B\,r/b\,r$	14
		$b\,R/b\,r$	14
$B\,R/B\,R$ (Chile)	$B\,R/b\,r$	$B\,R/b\,r$	36
$\times\ b\,r/b\,r$ (Chile)		$b\,r/b\,r$	36
		$B\,r/b\,r$	14
		$b\,R/b\,r$	14
$B\,R/B\,R$ (Calif.)	$B\,R/b\,r$	$B\,R/b\,r$	48
$\times\ b\,r/b\,r$ (Chile) or		$b\,r/b\,r$	48
$b\,r/b\,r$ (Calif.)		$B\,r/b\,r$	2
$\times\ B\,R/B\,R$ (Chile)		$b\,R/b\,r$	2

a. Provide a genetic hypothesis that explains the three sets of testcross results.

b. Draw the key chromosomal features of meiosis in the F_1 from a cross of the Californian and Chilean lines.

62. An aberrant corn plant gives the following RF values when testcrossed:

	Interval				
	$d–f$	$f–b$	$b–x$	$x–y$	$y–p$
Control	5	18	23	12	6
Aberrant plant	5	2	2	0	6

(The locus order is centromere–d–f–b–x–y–p.) The aberrant plant is a healthy plant, but it produces far fewer normal ovules and pollen than does the control plant.

a. Propose a hypothesis to account for the abnormal recombination values and the reduced fertility in the aberrant plant.

b. Use diagrams to explain the origin of the recombinants according to your hypothesis.

63. The following corn loci are on one arm of chromosome 9 in the order indicated (the distances between them are shown in map units):

$$c–bz–wx–sh–d–\text{centromere}$$
$$12\quad 8\quad 10\quad 20\ 10$$

C gives colored aleurone; c, white aleurone.

Bz gives green leaves; bz, bronze leaves.

Wx gives starchy seeds; wx, waxy seeds.

Sh gives smooth seeds; sh, shrunken seeds.

D gives tall plants; d, dwarf.

A plant from a standard stock that is homozygous for all five recessive alleles is crossed with a wild-type plant from Mexico that is homozygous for all five dominant alleles. The F_1 plants express all the dominant alleles and, when backcrossed to the recessive parent, give the following progeny phenotypes:

colored, green, starchy, smooth, tall	360
white, bronze, waxy, shrunk, dwarf	355
colored, bronze, waxy, shrunk, dwarf	40
white, green, starchy, smooth, tall	46
colored, green, starchy, smooth, dwarf	85
white, bronze, waxy, shrunk, tall	84
colored, bronze, waxy, shrunk, tall	8
white, green, starchy, smooth, dwarf	9
colored, green, waxy, smooth, tall	7
white, bronze, starchy, shrunk, dwarf	6

Propose a hypothesis to explain these results. Include

a. a general statement of your hypothesis, with diagrams if necessary;

b. why there are 10 classes;

c. an account of the origin of each class, including its frequency; and

d. at least one test of your hypothesis.

64. Chromosomally normal corn plants have a p locus on chromosome 1 and an s locus on chromosome 5.

P gives dark green leaves; p, pale green leaves.

S gives large ears; s, shrunken ears.

An original plant of genotype P/p ; S/s has the expected phenotype (dark green, large ears) but gives unexpected results in crosses as follows:

- On selfing, fertility is normal, but the frequency of p/p ; s/s types is 1/4 (not 1/16 as expected).

- When crossed with a normal tester of genotype p/p; s/s, the F_1 progeny are $\frac{1}{2}$; P/p ; S/s and $\frac{1}{2}$; p/p ; s/s; fertility is normal.

- When an F_1 P/p ; S/s plant is crossed with a normal p/p ; s/s tester, it proves to be semisterile, but, again, the progeny are $\frac{1}{2}$; P/p ; S/s and $\frac{1}{2}$; p/p ; s/s.

Explain these results, showing the full genotypes of the original plant, the tester, and the F_1 plants. How would you test your hypothesis?

65. A male rat that is phenotypically normal shows reproductive anomalies when compared with normal male rats, as shown in the table on the top of the next page. Propose a genetic explanation of these unusual results, and indicate how your idea could be tested.

Mating	Embryos (mean number)			Degeneration (%)
	Implanted in the uterine wall	Degeneration after implantation	Normal	
exceptional ♂ × normal ♀	8.7	5.0	3.7	37.5
normal ♂ × normal ♀	9.5	0.6	8.9	6.5

66. A tomato geneticist working on *Fr,* a dominant mutant allele that causes rapid fruit ripening, decides to find out which chromosome contains this gene by using a set of lines of which each is trisomic for one chromosome. To do so, she crosses a homozygous diploid mutant with each of the wild-type trisomic lines.

a. A trisomic F_1 plant is crossed with a diploid wild-type plant. What is the ratio of fast- to slow-ripening plants in the diploid progeny of this second cross if *Fr* is on the trisomic chromosome? Use diagrams to explain.

b. What is the ratio of fast- to slow-ripening plants in the diploid progeny of this second cross if *Fr* is not located on the trisomic chromosome? Use diagrams to explain.

c. Here are the results of the crosses. On which chromosome is *Fr,* and why?

Trisomic chromosome	Fast ripening:slow ripening in diploid progeny
1	45:47
2	33:34
3	55:52
4	26:30
5	31:32
6	37:41
7	44:79
8	49:53
9	34:34
10	37:39

(Problem 66 is from Tamara Western.)

CHALLENGING PROBLEMS

67. The *Neurospora un-3* locus is near the centromere on chromosome 1, and crossovers between *un-3* and the centromere are very rare. The *ad-3* locus is on the other side of the centromere of the same chromosome, and crossovers occur between *ad-3* and the centromere in about 20 percent of meioses (no multiple crossovers occur).

a. What types of linear asci (see Chapter 4) do you predict, and in what frequencies, in a normal cross of *un-3 ad-3* × wild type? (Specify genotypes of spores in the asci.)

b. Most of the time such crosses behave predictably, but, in one case, a standard *un-3 ad-3* strain was crossed with a wild type isolated from a field of sugarcane in Hawaii. The results follow:

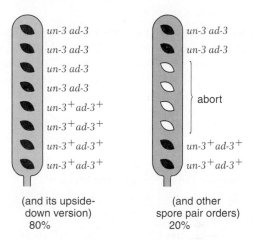

(and its upside-down version)
80%

(and other spore pair orders)
20%

Explain these results, and state how you could test your idea. (**Note:** In *Neurospora,* ascospores with extra chromosomal material survive and are the normal black color, whereas ascospores lacking any chromosome region are white and inviable.)

68. Two mutations in *Neurospora, ad-3* and *pan-2,* are located on chromosomes 1 and 6, respectively. An unusual *ad-3* line arises in the laboratory, giving the results shown in the table below. Explain all three results with the aid of clearly labeled diagrams. (**Note:** In *Neurospora,* ascospores with extra chromosomal material survive and are the normal black color, whereas ascospores lacking any chromosome region are white and inviable.)

	Ascospore appearance	RF between *ad-3* and *pan-2*
1. Normal *ad-3* × normal p*an-2*	All black	50%
2. Abnormal *ad-3* × normal p*an-2*	About $\frac{1}{2}$ black and $\frac{1}{2}$ white (inviable)	1%

3. Of the black spores from cross 2, about half were completely normal and half repeated the same behavior as the original abnormal ad-3 strain

69. Deduce the phenotypic proportions in the progeny of the following crosses of autotetraploids in which the a^+/a locus is very close to the centromere. (Assume that the four homologous chromosomes of any one type pair randomly two by two and that only one copy of the a^+ allele is necessary for the wild-type phenotype.)

a. $a^+/a^+/a/a \times a/a/a/a$

b. $a^+/a/a/a \times a/a/a/a$

c. $a^+/a/a/a \times a^+/a/a/a$

d. $a^+/a^+/a/a \times a^+/a/a/a$

70. The New World cotton species *Gossypium hirsutum* has a $2n$ chromosome number of 52. The Old World species *G. thurberi* and *G. herbaceum* each have a $2n$ number of 26. Hybrids between these species show the following chromosome pairing arrangements at meiosis:

Hybrid	Pairing arrangement
G. hirsutum	13 small bivalents
$\times$ *G. thurberi*	+ 13 large univalents
G. hirsutum	13 large bivalents
$\times$ *G. herbaceum*	+ 13 small univalents
G. thurberi	13 large univalents
$\times$ *G. herbaceum*	+ 13 small univalents

Draw diagrams to interpret these observations phylogenetically, clearly indicating the relationships between the species. How would you go about proving that your interpretation is correct? (Problem 70 is adapted from A. M. Srb, R. D. Owen, and R. S. Edgar, *General Genetics*, 2nd ed. W. H. Freeman and Company, 1965.)

71. There are six main species in the *Brassica* genus: *B. carinata*, *B. campestris*, *B. nigra*, *B. oleracea*, *B. juncea*, and *B. napus*. You can deduce the interrelationships among these six species from the following table:

Species or F$_1$ hybrid	Chromosome number	Number of bivalents	Number of univalents
B. juncea	36	18	0
B. carinata	34	17	0
B. napus	38	19	0
B. juncea × *B. nigra*	26	8	10
B. napus × *B. campestris*	29	10	9
B. carinata × *B. oleracea*	26	9	8
B. juncea × *B. oleracea*	27	0	27
B. carinata × *B. campestris*	27	0	27
B. napus × *B. nigra*	27	0	27

a. Deduce the chromosome number of *B. campestris*, *B. nigra*, and *B. oleracea*.

b. Show clearly any evolutionary relationships between the six species that you can deduce at the chromosomal level.

72. Several kinds of sexual mosaicism are well documented in humans. Suggest how each of the following examples may have arisen by nondisjunction at *mitosis*:

a. XX/XO (that is, there are two cell types in the body, XX and XO)

b. XX/XXYY

c. XO/XXX

d. XX/XY

e. XO/XX/XXX

73. In *Drosophila*, a cross (cross 1) was made between two mutant flies, one homozygous for the recessive mutation bent wing (*b*) and the other homozygous for the recessive mutation eyeless (*e*). The mutations *e* and *b* are alleles of two different genes that are known to be very closely linked on the tiny autosomal chromosome 4. All the progeny had a wild-type phenotype. One of the female progeny was crossed with a male of genotype *b e/b e*; we will call this cross 2. Most of the progeny of cross 2 were of the expected types, but there was also one rare female of wild-type phenotype.

a. Explain what the common progeny are expected to be from cross 2.

b. Could the rare wild-type female have arisen by (1) crossing over or (2) nondisjunction? Explain.

c. The rare wild-type female was testcrossed to a male of genotype *b e/b e* (cross 3). The progeny were

$\frac{1}{6}$ wild type

$\frac{1}{6}$ bent, eyeless

$\frac{1}{3}$ bent

$\frac{1}{3}$ eyeless

Which of the explanations in part *b* is compatible with this result? Explain the genotypes and phenotypes of the progeny of cross 3 and their proportions.

Unpacking Problem 73

1. Define *homozygous, mutation, allele, closely linked, recessive, wild type, crossing over, nondisjunction, testcross, phenotype,* and *genotype.*

2. Does this problem concern sex linkage? Explain.

3. How many chromosomes does *Drosophila* have?

4. Draw a clear pedigree summarizing the results of crosses 1, 2, and 3.

5. Draw the gametes produced by both parents in cross 1.

6. Draw the chromosome 4 constitution of the progeny of cross 1.

7. Is it surprising that the progeny of cross 1 are wild-type phenotype? What does this outcome tell you?

8. Draw the chromosome 4 constitution of the male tester used in cross 2 and the gametes that he can produce.

9. With respect to chromosome 4, what gametes can the female parent in cross 2 produce in the absence of nondisjunction? Which would be common and which rare?

10. Draw first- and second-division meiotic nondisjunction in the female parent of cross 2, as well as in the resulting gametes.

11. Are any of the gametes from part 10 aneuploid?

12. Would you expect aneuploid gametes to give rise to viable progeny? Would these progeny be nullisomic, monosomic, disomic, or trisomic?

13. What progeny phenotypes would be produced by the various gametes considered in parts 9 and 10?

14. Consider the phenotypic ratio in the progeny of cross 3. Many genetic ratios are based on halves and quarters, but this ratio is based on thirds and sixths. To what might this ratio point?

15. Could there be any significance to the fact that the crosses concern genes on a very small chromosome? When is chromosome size relevant in genetics?

16. Draw the progeny expected from cross 3 under the two hypotheses, and give some idea of relative proportions.

74. In the fungus *Ascobolus* (similar to *Neurospora*), ascospores are normally black. The mutation *f*, producing fawn-colored ascospores, is in a gene just to the right of the centromere on chromosome 6, whereas mutation *b*, producing beige ascospores, is in a gene just to the left of the same centromere. In a cross of fawn and beige parents ($+f \times b+$), most octads showed four fawn and four beige ascospores, but three rare exceptional octads were found, as shown in the accompanying illustration. In the sketch, black is the wild-type phenotype, a vertical line is fawn, a horizontal line is beige, and an empty circle represents an aborted (dead) ascospore.

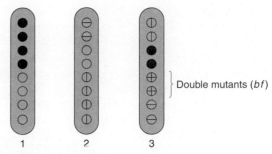

a. Provide reasonable explanations for these three exceptional octads.

b. Diagram the meiosis that gave rise to octad 2.

75. The life cycle of the haploid fungus *Ascobolus* is similar to that of *Neurospora*. A mutational treatment produced two mutant strains, 1 and 2, both of which when crossed with wild type gave unordered tetrads, all of the following type (fawn is a light brown color; normally, crosses produce all black ascospores):

| spore pair 1 black | spore pair 3 fawn |
| spore pair 2 black | spore pair 4 fawn |

a. What does this result show? Explain.

The two mutant strains were crossed. Most of the unordered tetrads were of the following type:

| spore pair 1 fawn | spore pair 3 fawn |
| spore pair 2 fawn | spore pair 4 fawn |

b. What does this result suggest? Explain.

When large numbers of unordered tetrads were screened under the microscope, some rare ones that contained black spores were found. Four cases are shown here:

	Case A	Case B	Case C	Case D
spore pair 1	black	black	black	black
spore pair 2	black	fawn	black	abort
spore pair 3	fawn	fawn	abort	fawn
spore pair 4	fawn	fawn	abort	fawn

(**Note:** Ascospores with extra genetic material survive, but those with less than a haploid genome abort.)

c. Propose reasonable genetic explanations for each of these four rare cases.

d. Do you think the mutations in the two original mutant strains were in one single gene? Explain.

Population Genetics

18

Artist Lynn Fellman's conception of the "Eurasian Adam," an African man with a Y chromosome belonging to a haplotype group that was ancestral to all Y chromosomes of men outside of Africa and arose within Africa approximately 70,000 years ago. [*Lynn Fellman.*]

KEY QUESTIONS

- What types of and how much genetic variation exist within populations?

- What are the forces that influence the amount of variation within populations?

- How can population genetic principles be used to address questions in human health, environmental issues, and evolution?

I n 2009, Sean Hodgson was released from a British prison after serving 27 years behind bars for the murder of Teresa De Simone, a clerk and part-time barmaid. Hodgson, who suffers from mental illness, initially confessed to the crime but withdrew his confession during the trial. Throughout his years in prison, he maintained his innocence. More than two decades after the crime, the courts analyzed DNA of the assailant found at the crime scene and determined that it did not come from Mr. Hodgson. His conviction was overturned, and the police have now reopened the investigation of Ms. De Simone's murder. As you will learn in this chapter, the DNA-based analysis used to exonerate Mr. Hodgson was dependent on population genetic analysis.

The principles of population genetics are at the heart of many questions facing society today. What are the risks that a couple will have a child with a genetic disease? Have the practices of plant and animal breeding caused a loss of genetic diversity on the farm and does this loss of diversity place our food supply at risk? As the human population continues to expand and wildlife retreats

OUTLINE

18.1 Detecting genetic variation

18.2 The gene-pool concept and the Hardy–Weinberg law

18.3 Mating systems

18.4 Genetic variation and its measurement

18.5 The modulation of genetic variation

18.6 Biological and social applications

into smaller and smaller parts of the earth, will wildlife species be able to avoid inbreeding and survive? The principles of population genetics are also fundamental to understanding many historical and evolutionary questions. How are human populations from different regions of the world related to one another? How has the human genome responded as humans have spread out across the globe and become adapted to different environments and lifestyles? How do populations and species evolve over time?

A **population** is a group of individuals of the same species. **Population genetics** analyzes the amount and distribution of genetic variation in populations and the forces that control this variation. It has its roots in the early 1900s, when geneticists began to study how Mendel's laws could be extended to understand genetic variation within whole populations of organisms. While Mendel's laws explain how genes are passed from parent to offspring in the cases of controlled crosses and known pedigrees, these laws are insufficient to understand the transmission of genes from one generation to the next in natural populations, in which not all individuals produce offspring and not all offspring survive. Geneticists began developing the principles of population genetics in the early 1900s, but at the time, they had rather limited tools to actually measure genetic variation. With the development of DNA-based technologies over the past two decades, geneticists now have the ability to observe directly differences between the DNA sequences of individuals throughout their genomes, and they can measure these differences in large samples of individuals in many species. The result has been a revolution in our understanding of genetic variation in populations.

In this chapter, we will consider the concept of the gene pool and how geneticists estimate allele and genotype frequencies in populations. Next, we will examine the impact that mating systems have on the frequencies of genotypes in a population. We will also discuss how geneticists measure variation using DNA-based technologies. We will then discuss the forces that modulate the levels of genetic variation within populations. Finally, we will look at some case studies involving the application of population genetics to questions of interest to society.

18.1 Detecting Genetic Variation

The methods of population genetics can be used to analyze any variable or polymorphic locus in the DNA sequences of a population of organisms. Historically, geneticists lacked the molecular tools needed to observe differences in the DNA sequences among individuals directly, and so most population genetic analyses looked at differences in proteins or phenotypes. For example, differences in the protein encoded by the *ABO* glycocyltransferase gene controlling the ABO blood group in humans can be detected using antibody probes. From these protein differences, investigators can infer differences in the DNA sequence of this gene among individuals. Over the past two decades, new technologies, such as DNA sequencing, DNA microarrays, and PCR (see Chapter 10), have been developed that allow geneticists to observe differences in the DNA sequences directly. As a result, population genetic analyses are no longer confined to a small set of genes such as *ABO* but have expanded to include every nucleotide in the genome.

In population genetics, a **locus** is simply a location in the genome; it can be a single nucleotide site or a stretch of many nucleotides. The simplest form of variation one might observe at a locus is a difference in the nucleotide present at a single nucleotide site whether adenine, cytosine, guanine, or thymine. These types of variants are called **single nucleotide polymorphisms (SNPs)** and they are the most widely studied variants in human population genetics (Figure 18-1; see also Chapter 4). Population genetics also makes extensive use of **microsatellite** loci (see Chapter 4). These loci have a short sequence motif, 2 to 6 base

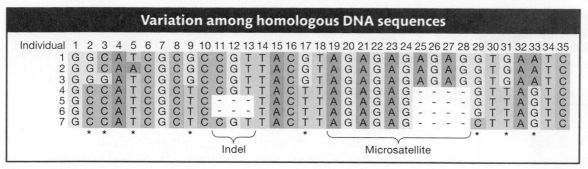

Figure 18-1 Variation in the aligned DNA sequences of seven individual chromosomes. The *'s show the location of SNPs. The location of an indel and a microsatellite are also indicated.

pairs long, that is repeated multiple times with different alleles having different numbers of repeats. For example, the 2-bp-sequence motif AG at a locus might be tandemly repeated five times in one allele (AGAGAGAGAG) but three times in another (AGAGAG) (see Figure 18-1).

Single nucleotide polymorphisms (SNPs)

SNPs are the most prevalent types of polymorphism in most genomes. Most SNPs have just two alleles—for example, A and C. SNPs are usually considered **common SNPs** in a population if the less common allele occurs at a frequency of about 5 percent or greater. SNPs for which the less common allele occurs at a frequency below 5 percent are considered **rare SNPs.** For humans, there is a common SNP about every 300 to 1000 bp in the genome. Of course, there are a far greater number of rare SNPs.

SNPs occur within genes, including within exons, introns, and regulatory regions. SNPs within protein-coding regions can be classified into one of three groups: *synonymous* if the different alleles encode the same amino acid, *nonsynonymous* if the two alleles encode different amino acids, and *nonsense* if one allele encodes a stop codon and the other an amino acid. Thus, it is sometimes possible to associate an SNP with functional variation in proteins and any associated change in phenotype. SNPs located outside of coding sequences are called *silent* SNPs. These SNPs can be very useful in population genetics even if they are found in nonfunctional DNA since they can be used as markers to address questions about population-genetic processes such as gene flow between populations.

To study SNP variation in a population, we first need to determine which nucleotide sites in the genome are variable—that is, constitute an SNP. This first step is called SNP discovery. SNPs are often discovered by sequencing the genomes of a small sample of individuals of a species, then comparing these sequences. For example, SNP discovery in humans began by partially sequencing the genomes of a **discovery panel** of 48 individuals from around the world. Variable nucleotide sites were discovered by comparing the partial genome sequences of these 48 individuals with one another. This initial effort led to the discovery of more than 1 million SNPs.

Once SNPs have been discovered, the genotype (allelic composition) of different individuals in the population at each SNP can be determined. DNA microarrays are a widely used technology for this purpose (Figure 18-2). The microarrays used for SNP assays can contain thousands of probes corresponding to known SNPs. Biotechnologists have developed several different methods to detect SNP variants using microarrays. In one method, DNA from an individual is labeled with fluorescent tags and hybridized to the microarray. Each spot (SNP) on the microarray will fluoresce red for one homozygous class, green for the other homozygote, and yellow for a heterozygote (see Figure 18-2). The entire procedure

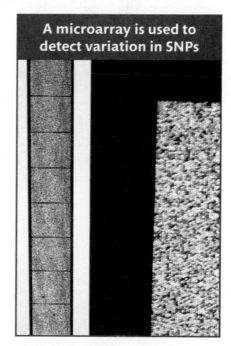

Figure 18-2 Detecting variation in DNA: SNPs. View of a scan of a single individual's genome using a microarray (left) with a closeup of a small section of the microarray (right). Each dot represents one SNP, with red and green for the homozygous classes and yellow for heterozygous. [*Stephen Ausmus/U.S Department of Agriculture/Photo Researchers, Inc.*]

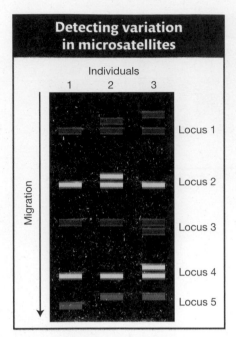

Figure 18-3 Detecting variation in DNA: microsatellites. Pseudo-gel image of the loci for five microsatellites scored simultaneously. The three vertical lanes correspond to three individuals. Notice that there are three alleles present for Locus 1 and that individuals 2 and 3 are both heterozygous for this locus.

has been enhanced with robotics to allow rapid *genotyping,* or assignment of genotypes (for example, A/A versus A/C) on a large-scale basis.

Microsatellites

Microsatellites are powerful loci for population genetic analysis for several reasons. First, unlike SNPs, which typically have only two alleles per locus and can never have more than four alleles, the number of alleles at a microsatellite is often very large (20 or more). Second, they have a high mutation rate, typically in the range of 10^{-3} to 10^{-4} mutations per locus per generation as compared to 10^{-8} to 10^{-9} mutations per site per generation for SNPs. The high mutation rate means that levels of variation are higher: more alleles per locus and a greater chance that any two individuals will have different genotypes. Third, microsatellites are very abundant in most genomes. Humans have over a million microsatellites.

Microsatellites are found throughout the genomes of most organisms and may be present in exons, introns, regulatory regions, and nonfunctional DNA sequences. Microsatellites with tri-nucleotide repeats are found in the coding sequences of some genes; these encode strings of a single amino acid. The Huntington disease gene (*HD*) (see Chapter 16) contains a repeat of CAG, which encodes a string of glutamines. Individuals carrying alleles with more than 30 glutamines are predisposed to develop the disease. In general, however, most microsatellites are located outside of coding sequences, and variation in the number of repeats is not associated with differences in phenotype.

Two main methods are used to discover microsatellite loci in the genome. If a complete genomic sequence is available for an organism, one can simply conduct a search to find them using a computer. For species without genome sequences (most nonmodel organisms), considerable laboratory work is required to discover microsatellites. Typically, one creates a genomic library, screens the library with a probe for the motif of interest (for example, AG repeats), and determines the DNA sequence of the selected clones to identify the microsatellites and the sequences that flank them. The molecular methods for doing this type of work were discussed in Chapter 10.

Once a microsatellite and its flanking sequences have been identified, DNA samples from a set of individuals in the population can be analyzed to determine the number of repeats that are present in each individual. To carry out the analysis, oligonucleotide primers are designed that match the flanking sequences for use in PCR. If the primers are labeled with a fluorescent tag, then the sizes of the PCR products can be determined on the same apparatus used to determine the sequence of DNA molecules (Figure 18-3). These sizes reveal the number of repeats in a microsatellite allele. For example, the PCR product of a microsatellite allele containing seven AG repeats will be 8 bp longer than an allele containing three AG repeats. Heterozygous individuals will possess products of two different sizes. Since PCR, the sizing of PCR products, and scoring of the alleles can all be automated, it is possible to determine the genotypes of large samples of individuals for large numbers of microsatellites relatively rapidly.

Haplotypes

For some questions in population genetics, it is important to consider the genotypes of linked loci as a group rather than individually. Geneticists use the term **haplotype** to refer to the combination of alleles at multiple loci on the same chromosomal homolog. Two homologous chromosomes that share the same allele at each of the loci under consideration have the same haplotype. If two chromosomes have different genotypes at even one of the loci in question, then they have different haplotypes. If the *A* locus with alleles *A* and *a* is linked to the *B* locus with alleles *B* and *b*, then there are four possible haplotypes for the chromosomal segment on which these two loci are located:

A	*B*
A	*b*
a	*B*
a	*b*

In Figure 18-4a, there are seven chromosome segments but only six haplotypes because segments 5 and 6 have the same haplotype (E).

Haplotypes are most often used in population genetics for loci that are physically close. For example, the variable-nucleotide sites in a single gene can be used to define haplotypes for that gene. However, the haplotype concept works for larger regions when there is little or no recombination over the region. It can even be applied to an entire chromosome such as the human Y chromosome. Finally, it is sometimes useful to group haplotypes into classes. As shown in Figure 18-4a, there are two major classes of haplotypes (I and II) that differ at five nucleotide sites plus a microsatellite. However, each classes contains several subtypes (I-a, I-b, . . .). The **haplotype network** shows the relationships among the haplotypes, placing each mutation on one of the branches (Figure 18-4b).

What insights can we gain from haplotype analysis? Population geneticists studying the human Y chromosome among Asian men discovered one highly

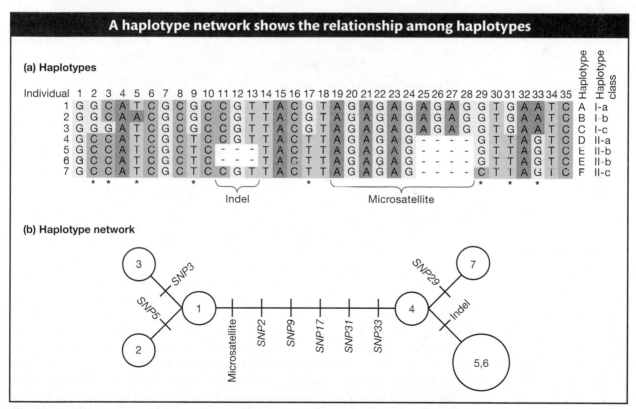

Figure 18-4 (a) There are a total of six haplotypes (A–F) in the aligned DNA sequences from seven individual chromosomes. (b) These six haplotypes are joined in a haplotype network showing the relationships among the haplotypes. Each circle represents one of the six haplotypes. The individual(s) carrying a specific haplotype are noted within each circle. Any two haplotypes differ at the loci noted on all of the branches connecting them.

Figure 18-5 (a) Haplotype network for the Y chromosomes of Asian men showing the predominance of the star-cluster haplotype thought to trace back to Genghis Khan. The area of the circle is proportional to the number of individuals with the specific haplotype that the circle represents. (b) Geographical distribution of the star-cluster haplotype. Populations are shown as circles with an area proportional to sample size; the proportion of individuals in the sample carrying star-cluster chromosomes is indicated by green sectors. No star-cluster chromosomes were found in populations having no green sector in the circle. The shaded area represents the extent of Genghis Khan's empire. [*From T. Zerjal et al., Am. J. Hum. Genet. 72, 2003, 717–721.*]

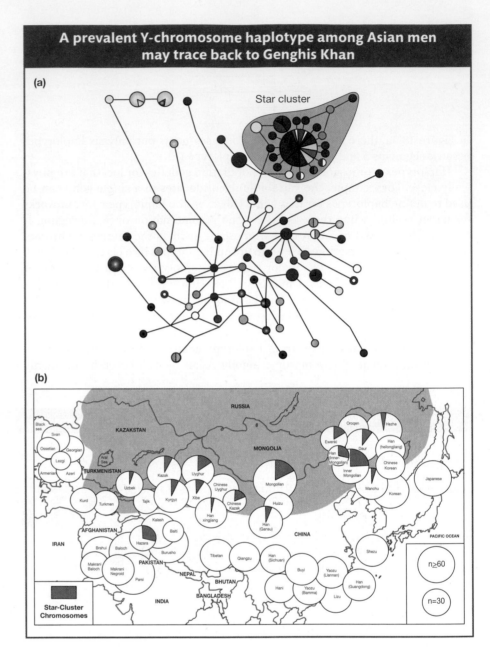

prevalent haplotype, termed the "star-cluster" haplotype (Figure 18-5a). Typically, most men have a rare Y chromosome haplotype, but the "star-cluster" haplotype is present in 8 percent of Asian men. Using the known mutation rate, the researchers estimated that this common haplotype arose between 700 and 1300 years ago. (Later in this chapter, we will discuss mutation rates and their use in population genetics.) This haplotype is most common in Mongolia, suggesting that it arose there. The researchers inferred that the "star-cluster" haplotype traces back to one man in Mongolia about 1000 years ago. Remarkably, the present-day distribution of this haplotype follows the geographic boundaries of the Mongolian Empire established by Genghis Khan about 1200 years ago (Figure 18-5b). It appears that contemporary men with this haplotype are all descendants of Genghis Khan (or his male-lineage relatives).

Other sources and forms of variation

Beyond SNP and microsatellites, any variation in the DNA sequence of the chromosomes in a population is amenable to population genetic analysis. Variations

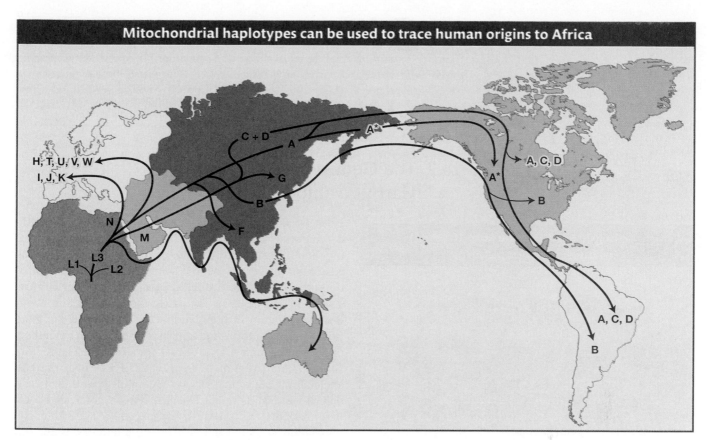

Figure 18-6 The haplotype network for human mtDNA haplotype groups drawn onto a world map. The ancestral L haplotype group appears in Africa, and the derived groups (A, B, and so on) are dispersed throughout the world. [*From www.mitomap.org.*]

that can be analyzed include inversions, translocations, deletions or duplications, and the presence or absence of a transposable element at a particular locus in the genome. Another common form of variation is insertion-deletion polymorphism, or *indel* for short (see Chapter 16). This type of polymorphism involves the presence or absence of one or more nucleotides at a locus in one allele relative to another. In Figure 18-1, individuals 5 and 6 differ from the other five individuals by a 3-bp indel. Unlike microsatellites, indels do not contain repeat motifs such as AGAGAGAG.

Thus far, our discussion of SNP and microsatellites has focused on the nuclear genome. However, interesting genetic variation can also be found in the mitochondrial (mtDNA) and chloroplast (cpDNA) genomes of eukaryotes. Both SNP and microsatellites are found in these organelle genomes. Since mtDNA and cpDNA are usually maternally inherited, their analysis can be used to follow the history of female lineages. In 1987, a prominent study of the human mitochondrial lineage traced the history of the human mtDNA haplotypes and determined that the mitochondrial genomes of all modern humans trace back to a single woman who lived in Africa about 150,000 years ago (Figure 18-6). She was dubbed the "mitochondrial Eve" in the popular press. This study of mtDNA was the first thorough genetic analysis to show that all modern humans came from Africa.

The HapMap Project

A major advance in human population genetics over the past decade was the creation of a genome-wide haplotype map, or **HapMap.** A consortium of scientists around the world genotyped thousands of people representing the diversity of our species for hundreds of thousands of SNPs and microsatellites. The result is a highly detailed picture of variation in our species. The data are available to the public at several Web sites, including that of the International HapMap Project (www.hapmap.org) and the Human Genome Diversity Project (hgdp.uchicago.edu). In this chapter, we will use these data to present the principles of population genetics.

Although first developed for humans, HapMaps have since been developed for several other species, including *Drosophila*, mouse, *Arabidopsis*, rice, and maize.

> **Message** Genomes are replete with diverse types of variation suitable for population genetic analysis. SNPs and microsatellites are the two most commonly studied types of polymorphism in population genetics. High-throughput technologies allow hundreds of thousands of polymorphisms to be scored in tens of thousands of individuals.

18.2 The Gene-Pool Concept and Hardy–Weinberg Law

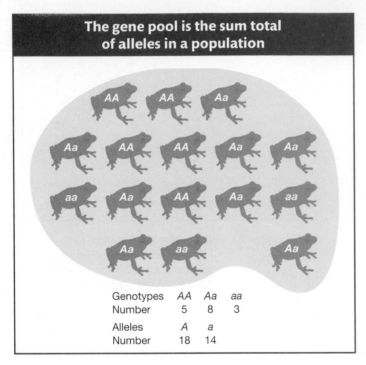

The gene pool is the sum total of alleles in a population

Genotypes	*AA*	*Aa*	*aa*
Number	5	8	3

Alleles	*A*	*a*
Number	18	14

Figure 18-7   A frog gene pool.

Perhaps you have watched someone performing a death-defying stunt and thought that they were at risk of eliminating themselves from the "gene pool." If so, you were using a concept, the gene pool, that comes straight out of population genetics and has worked its way into popular culture. The gene-pool concept is a basic tool for thinking about genetic variation in populations. We can define the **gene pool** as the sum total of all alleles in the breeding members of a population at a given time. For example, Figure 18-7 shows a population of 16 frogs, each of which carries two alleles at the autosomal locus *A*. By simple counting, we can determine that there are five *A/A* homozygotes, eight *A/a* heterozygotes, and three *a/a* homozygotes. The size of the population, usually symbolized by the letter *N*, is 16, and there is a total of 32 or *2N* alleles in this diploid population. With this simple set of numbers, we have described the gene pool.

Typically, population geneticists do not care about the absolute counts of the different genotypes in a population but about the **genotype frequencies.** We can calculate the frequency of the *A/A* genotype simply by dividing the number of *A/A* individuals by the total number of individuals in the population (*N*) to get 0.31. The frequency of *A/a* heterozygotes is 0.50, and the frequency of *a/a* homozygotes is 0.19. Since these are frequencies, they sum to 1.0. Frequencies are a more practical measurement than absolute counts because rarely are population geneticists able to study every individual in a population. Rather, population geneticists will draw a random or unbiased sample of individuals from a population and use the sample to infer the genotype frequencies in the entire population.

We can make a simpler description of this frog gene pool if we calculate the **allele frequencies** rather than the genotype frequencies (Box 18-1). In Figure 18-7, 18 of the 32 alleles are *A*, so the frequency of *A* is 18/32 = 0.56. The frequency of the *A* allele is typically symbolized by the letter *p*, and in this case *p* = 0.56. The frequency of the *a* allele is symbolized by the letter *q*, and in this case *q* = 14/32 = 0.44. Again, since these are frequencies, they sum to 1.0: $p + q = 0.56 + 0.44 = 1.0$. We now have a description of our frog gene pool using only two numbers, *p* and *q*.

> **Message** The gene pool is a fundamental concept for the study of genetic variation in populations: it is the sum total of all alleles in the breeding members of a population at a given time. We can describe the variation in a population in terms of genotype and allele frequencies.

Box 18-1	Calculation of Allele Frequencies

At a locus with two alleles A and a, let's define the frequencies of the three genotypes A/A, A/a, and a/a as $f_{A/A}, f_{A/a}$, and $f_{a/a}$, respectively. We can use these genotype frequencies to calculate the allele frequencies: p is the frequency of the A allele, and q is the frequency of the a allele. Because each homozygote A/A consists only of A alleles and because half the alleles of each heterozygote A/a are A alleles, the total frequency p of A alleles in the population is calculated as

$$p = f_{A/A} + \tfrac{1}{2} f_{A/a} = \text{ frequency of } A$$

Similarly, the frequency q of the a allele is given by

$$q = f_{a/a} + \tfrac{1}{2} f_{A/a} = \text{ frequency of } a$$

Therefore,

$$p + q = f_{A/A} + f_{A/a} + f_{a/a} = 1.0$$

and

$$q = 1 - p$$

If there are more than two different allelic forms, the frequency for each allele is simply the frequency of its homozygote plus half the sum of the frequencies for all the heterozygotes in which it appears.

As mentioned above, an important goal of population genetics is to understand the transmission of alleles from one generation to the next in natural populations. In this section, we will begin to look at how this works. We will see how we can use the allele frequencies in the gene pool to make predictions about the genotype frequencies in the next generation.

The frequency of an allele in the gene pool is equal to the probability that the allele will be chosen when randomly picking an allele from the gene pool to form an egg or a sperm. Knowing this, we can calculate the probability that a frog in the next generation will be an A/A homozygote. If we reach into the frog gene pool (Figure 18-7) and pick the first allele, the probability that it will be an A is $p = 0.56$, and similarly the probability that the second allele we pick is also an A is $p = 0.56$. The product of these two probabilities, or $p^2 = 0.3136$, is the probability that a frog in the next generation will be A/A. The probability that a frog in the next generation will be a/a is $q^2 = 0.44 \times 0.44 = 0.1936$. There are two ways to make a heterozygote. We might first pick an A with probability p and then pick an a with probability q, or we might pick the a first and the A second. Thus, the probability that a frog in the next generation will be heterozygous A/a is $pq + qp = 2pq = 0.4928$. Overall, the frequencies (f) of the genotypes are

$$f_{A/A} = p^2$$
$$f_{a/a} = q^2$$
$$f_{A/a} = 2pq$$

Finally, as expected, the sum of the probability of being A/A plus the probability of being A/a plus the probability of being a/a is 1.0:

$$p^2 + 2pq + q^2 = 1.0$$

This simple equation is the **Hardy–Weinberg law,** and it is part of the foundation for the theory of population genetics.

The process of reaching into the gene pool to pick an allele is called *sampling* the gene pool. Since any individual that contributes to the gene can produce many eggs or sperm that carry exactly the same copy of an allele, it is possible to pick a particular copy and then reach back into the gene pool and pick exactly the same copy again. There is also an element of chance involved when sampling the gene pool. Just by chance, some copies may be picked more than once and other copies may not be picked at all. Later in the chapter, we will look at how these properties of sampling the gene pool can lead to changes in the gene pool over time.

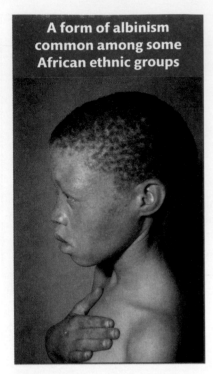

Figure 18-8 Individual of African ancestry with brown oculocutaneous albinism (BOCA), a condition defined by light tan skin and beige to light brown hair. [*Dr. Michele Ramsay, Department of Human Genetics School of Pathology the National Health Laboratory Service University of Witwatersrand.*]

We used the Hardy-Weinberg law to calculate genotype frequencies in the next generation from the allele frequencies in the current generation. We can also use the Hardy-Weinberg law to calculate allele frequencies from the genotype frequencies within a single generation. For example, some forms of albinism in humans are due to recessive alleles at the *OCA2* locus. In Africa, a form of albinism called brown oculocutaneous albinism results from a recessive allele of *OCA2* (Figure 18-8). Individuals with this condition are present at frequencies as high as 1 in 1100 among some ethnic groups in Africa. We can use the Hardy-Weinberg law to calculate the allele frequencies:

$$f_{a/a} = q^2 = 1/1100 = 0.0009$$

so

$$q = \sqrt{0.0009} = 0.03$$

and

$$p = 1 - q = 0.97$$

Using the allele frequencies, we can also calculate the frequency of heterozygotes in the population as

$$2pq = 2 \times 0.97 \times 0.03 = 0.06$$

The latter number predicts that about 6 percent of this population are heterozygotes, or carriers of the recessive allele at *OCA2*.

When we use the Hardy-Weinberg law to calculate allele or genotype frequencies, we make some critical assumptions.

- *First,* we assume that mating is random in the population with respect to the gene in question. Deviation from random mating violates this assumption, making it inappropriate to apply Hardy-Weinberg. For example, a tendency for individuals who are phenotypically similar to mate with each other violates the Hardy-Weinberg law. If albinos mated more frequently with other albinos than with non-albinos, then the Hardy-Weinberg law would overestimate the frequency of the recessive allele.

- *Second,* if one of the genotypes has reduced viability such that some individuals with that genotype die before the genotype frequencies are counted, then the estimate of the gene frequencies will be inaccurate.

- *Third,* for the Hardy-Weinberg law to apply, the population must not be divided into subpopulations that are partially or fully genetically isolated. If there are separate subpopulations, alleles may be present at different frequencies in the different subpopulations. If so, using genotypic counts from the overall population may not give an accurate estimate of the overall allele frequencies.

- *Finally,* the Hardy-Weinberg law strictly applies only to infinitely large populations. For finite populations, there will be deviations from the frequencies predicted by the Hardy-Weinberg law due to chance when sampling the gene pool to produce the next generation.

We have seen how we can use the Hardy-Weinberg law and the gene frequencies in the current generation (t_0) to calculate genotype frequencies in the next generation (t_1) by randomly sampling the gene pool for the production of eggs and sperm. Similarly, the predicted genotype frequencies for generation t_1 can be used in turn to calculate gene frequencies for the next generation (t_2). The gene frequencies in generation t_2 will remain the same as generation t_1. Under the Hardy-Weinberg law, neither gene nor genotype frequencies change from one generation to the next when an infinitely large population is randomly sampled for the formation of eggs and sperm. Thus, an important lesson from the Hardy-Weinberg law is that, in large populations, genetic variation is neither created nor destroyed by the process of transmitting genes from one generation to the next. Populations that adhere to this principle are said to be at **Hardy–Weinberg equilibrium.**

Generation	Genotype frequencies			Gene frequencies	
	A/A	A/a	a/a	A	a
t_0	0.64	0.32	0.04	0.8	0.2
t_1	0.64	0.32	0.04	0.8	0.2
$\vdots$	$\vdots$	$\vdots$	$\vdots$	$\vdots$	$\vdots$
t_n	0.64	0.32	0.04	0.8	0.2

Here are a few more points about the Hardy–Weinberg law.

1. For any allele that exists at a very low frequency, homozygous individuals will only very rarely be found. If allele a has a frequency of 1 in a thousand ($q = 0.001$), then only 1 in a million (q^2) individuals will be homozygous for that allele. As a consequence, recessive alleles for genetic disorders can occur in the heterozygous state in many more individuals than there are individuals that actually express the genetic disorder in question.

2. The Hardy–Weinberg law still applies where there are more than two alleles per locus. If there are n alleles, $A_1, A_2, \ldots A_n$ with frequencies $p_1, p_2, \ldots p_n$, then the sum of all the individual frequencies equals 1.0. The frequencies of each of the homozygous genotypes are simply the square of the frequencies of the alleles, and the frequencies of the different heterozygous classes are two times the product of the frequencies of the first and second allele. Table 18-1 gives an example with $p_1 = 0.5$, $p_2 = 0.3$, and $p_3 = 0.2$.

Table 18-1 **Hardy–Weinberg Genotype Frequencies for a Locus with Three Alleles A_1, A_2, and A_3 with Frequencies 0.5, 0.3, and 0.2, Respectively**

Genotype	Expectation	Frequency
A_1A_1	p_1^2	0.25
A_2A_2	p_2^2	0.09
A_3A_3	p_3^2	0.04
A_1A_2	p_1p_2	0.30
A_1A_3	p_1p_3	0.20
A_2A_3	p_2p_3	0.12
Sum		1.00

3. Hardy–Weinberg logic applies to X-linked loci as well. Males are hemizygous for X-linked genes, meaning that a male has a single copy of these genes. Thus, for X-linked genes in males, the genotype frequencies are equal to the allele frequencies. The frequency of the A genotype will be p and the frequency of the a genotype will be q. For females, genotype frequencies for X-linked genes follow normal Hardy–Weinberg expectations: $f_{A/A} = p^2$, $f_{A/a} = 2pq$, and $f_{a/a} = q^2$.

Male pattern baldness is an X-linked trait (Figure 18-9). *AR* (for androgen receptor) is an X-linked gene involved in male development. There is an *AR* haplotype called *Eur-H1* that is strongly associated with pattern baldness. Male pattern baldness is common in Europe, where the *Eur-H1* haplotype occurs at a frequency of 0.71, meaning that 71 percent of European men carry it. Using the Hardy–Weinberg law, we can calculate that 50 percent of European women are *Eur-H1* homozygotes and 41 percent are heterozygous. The inheritance of baldness is complex and is affected by multiple genes, and so not all men who have *Eur-H1* go bald.

Male pattern baldness

Figure 18-9 Individual showing male pattern baldness, an X-chromosome-linked condition. [*B2M Productions/Getty Images.*]

4. One can test whether the observed genotype frequencies at a locus fit Hardy–Weinberg predictions using the χ^2 test (see Chapter 3). An example is provided by the human leukocyte antigen gene, *HLA-DQA1*, of the major histocompatibility complex (MHC). MHC is a cluster of genes on chromosome 6 that play roles in the immune system. Table 18-2 has genotype frequencies for an SNP (rs9272426) in the *HLA-DQA1* for 84 natives of Tuscany, Italy. This SNP has alleles A and G. From the genotype frequencies in Table 18-2, we can calculate the allele frequencies: $f(A) = p = 0.53$ and $f(G) = q = 0.47$. Next, we can calculate expected genotype frequencies under the Hardy–Weinberg law: $p^2 = 0.281$, $2pq = 0.498$, and $q^2 = 0.221$. Multiplying the expected genotype frequencies times the sample size ($N = 84$) gives us the expected number of individuals for each genotype. Now we can calculate the χ^2 statistic to be 8.29. Using Table 3-1, we see that the probability under the null hypothesis that the observed data fit Hardy–Weinberg predictions is $P < 0.005$ with df = 1. [We have only one degree of freedom because we have three genotypic categories and we used two numbers from the data (N and p) to calculate the expected values ($3 - 2$ leaves 1 degree of freedom). We did not need to use q since $q = p - 1$.] This analysis makes us strongly suspect that Tuscans do not conform to Hardy–Weinberg expectations with regard to *HLA-DQA1*. We will look further at the population genetics of MHC in Section 18.3 on mating systems and Section 18.5 on natural selection.

Table 18-2   Frequencies of SNP rs9272426 Genotypes in *HLA-DQA1* of the MHC Locus for People from Tuscany, Italy

| | *Genotypes* | | | |
	A/A	A/G	G/G	Sum
Observed number	17	55	12	84
Observed frequency	0.202	0.655	0.143	1
Expected frequency	0.281	0.498	0.221	1
Expected number	23.574	41.851	18.574	84
(Observed – expected)²/expected	1.833	4.131	2.327	8.29

Source: International HapMap Project (www.hapmap.org).

The Hardy–Weinberg law is part of the foundation of population genetics. It applies to an idealized population that is infinite in size and in which mating is random. It also assumes that all genotypes are equally fit—that is, that they are all equally viable and have the same success at reproduction. Real populations deviate from this idealized one. In the rest of chapter, we will examine how factors such as nonrandom mating, finite population size, and the unequal fitness of different genotypes cause deviations from Hardy–Weinberg expectations. We will also see how the Hardy–Weinberg law can be modified to compensate for these factors.

> **Message** The Hardy–Weinberg law describes the relationship between allele and genotype frequencies. This law informs us that genetic variation is neither created nor destroyed by the process of transmitting genes from one generation to the next. The Hardy–Weinberg law only strictly applies in infinitely large and randomly mating populations.

18.3 Mating Systems

Random mating is a critical assumption of the Hardy–Weinberg law. The assumption of random mating is met if all individuals in the population are equally likely as a choice when a mate is chosen. However, if a relative, a neighbor, or a phenotypically similar individual is a more likely mate than a random individual, then the assumption of random mating has been violated. Populations that are not random mating will not exhibit exact Hardy–Weinberg proportions for the genotypes at some or all genes. Three types of bias in mate choice that violate the assumption of random mating are assortative mating, isolation by distance, and inbreeding.

Assortative mating

Assortative mating occurs if individuals choose mates based on resemblance to themselves. **Positive assortative mating** occurs when similar types mate; for example, if tall individuals preferentially mate with other tall individuals and short individuals mate with other short individuals. In these cases, genes controlling the difference in height will not follow the Hardy–Weinberg law. Rather, we'd expect to see an excess of homozygotes for the "tall" alleles among the progeny of tall mating pairs and an excess of homozygotes for "short" alleles among the progeny of short mating pairs. In humans, there is positive assortative mating for height.

Negative assortative or **disassortative mating** occurs when unlike individuals mate—that is, when opposites attract. One example of negative assortative mating is provided by the self-incompatibility, or S, locus in plants such as *Brassica* (broccoli and its relatives). There are numerous alleles at the S locus, S_1, S_2, S_3, and so forth. The stigma of a plant will not be receptive to pollen that carries either of its own two alleles (Figure 18-10). For example, the stigma of an S_1/S_2 heterozygote will not allow pollen grains carrying either an S_1 or S_2 allele to germinate and fertilize its ovules, although pollen grains carrying the S_3 or S_4 alleles can do so. This mechanism blocks self-fertilization, thereby enforcing cross-pollination. The S locus violates the Hardy–Weinberg law since homozygous genotypes at S are not formed.

A second example of negative assortative mating is provided by the major histocompatibility complex (MHC), which is known to influence mate choice in vertebrates. MHC affects body odor in mice and rats, providing a basis for mate choice. In what are known as the "sweaty T-shirt experiments," researchers asked a group of men to wear T-shirts for two days. Then they asked a group of women to smell the T-shirts and rate them for "pleasantness." Women preferred the scent of men whose MHC haplotypes were different from their own. Data from the human HapMap project have since confirmed that American couples are significantly more heterozygous at the MHC than expected by chance. The MHC plays a central role in our immune response to pathogens, and heterozygotes may be more resistant to pathogens. Therefore, our offspring benefit if we mate disassortatively with respect to our MHC genotype. This mechanism may explain why the SNP in the MHC gene *HLA-DQA1* that we discussed above does not follow the Hardy–Weinberg law among residents of Tuscany. Look back at Table 18-2 and you will notice that there are more heterozygotes than expected, 55 versus 42. Tuscans appear to be practicing disassortative mating with respect to this SNP.

Isolation by distance

Another form of bias in mate choice arises from the amount of geographic distance between individuals. Individuals are more apt to mate with a neighbor than another member of their species on the opposite side of the continent—that is, individuals can show **isolation by distance.** As a consequence, allele and genotype frequencies often differ between fish in separate lakes or between pine trees in different regions of a continent. Species or populations exhibiting such patterning of genetic variation are said to show **population structure.** A species can be divided into a series of subpopulations such as frogs in different ponds or people in different cities.

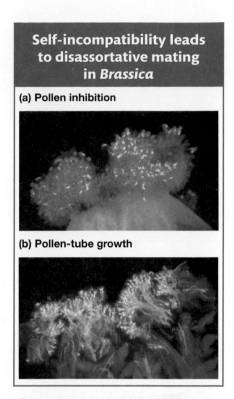

Self-incompatibility leads to disassortative mating in *Brassica*

(a) Pollen inhibition

(b) Pollen-tube growth

Figure 18-10 Disassortative mating caused by the self-incompatibility locus (S) of the flowering plant genus *Brassica*. (a) A self-pollinated S_1/S_2 stigma shows no pollen-tube growth. (b) There is pollen-tube growth for an S_1/S_2 stigma cross-pollinated with pollen from an S_3/S_4 heterozygote. [*June Bowman Nasrallah.*]

If a species has population structure, the proportion of homozygotes will be greater species-wide than expected under the Hardy–Weinberg law. Consider a species of wild sunflowers distributed across Kansas with a gradient in the frequency of the A allele from 0.9 near Kansas City to 0.1 near Elkhart (Figure 18-11a). We sample 100 sunflower plants from each of these two cities plus 100 from Hutchinson, in the middle of the state, and we calculate allele frequencies. Each city represents a subpopulation. For any of the three cities, the Hardy–Weinberg law works fine. For example, in Elkhart, we expect $Nq^2 = 100 \times (0.9)^2 = 81$ a/a homozygotes, and that is what we observe. However, state-wide, we'd predict $Nq^2 = 300 \times (0.5)^2 = 75$ a/a homozygotes, yet we observed 107. Because of population structure, there are more homozygous sunflower plants than expected.

		Number of individuals				
	N	A/A	A/a	a/a	p	q
Kansas City	100	81	18	1	0.90	0.10
Hutchinson	100	25	50	25	0.50	0.50
Elkhart	100	1	18	81	0.10	0.90
State-wide (observed)	300	107	86	107	0.50	0.50
State-wide (expected)	300	75	150	75	—	—

Here is an example of population structure from our own species. In Africa, the FY^{null} allele of the Duffy blood group shows a gradient with a low frequency in eastern and northern Africa, moderate frequency in southern Africa and high frequency across central Africa (Figure 18-11b). This allele is rare outside of Africa. Because of this gradient, we cannot use overall allele frequencies in Africa to calculate genotype frequencies using the Hardy–Weinberg law. Later in the chapter and in Chapter 20, we will discuss the relationship between FY^{null} and malaria.

> **Message** Assortative mating and isolation by distance violate the Hardy–Weinberg law and can cause genotype frequencies to deviate from Hardy–Weinberg expectations.

Inbreeding

The third type of bias in mating is **inbreeding,** or mating between relatives. Long before anyone knew about deleterious recessive alleles, many societies recognized that disorders such as "dumbness, deafness, and blindness" were more frequent among the children of marriages between relatives. Accordingly, brother–sister and first-cousin marriages are either outlawed or discouraged in many societies. Nevertheless, many famous individuals have married a cousin, including Charles Darwin, Albert Einstein, J. S. Bach, Edgar Allan Poe, Jesse James, and Queen Victoria. As we will see, the offspring of marriages between relatives are at higher risk of having an inherited disorder.

Progeny of inbreeding are more likely to be homozygous at any locus than progeny of non-inbred matings. Thus, they are more likely to be homozygous for deleterious recessive alleles. For this reason, inbreeding can lead to a reduction in vigor and reproductive success called **inbreeding depression.** However, inbreeding can have advantages too. Many plant species are highly self-pollinating and highly inbred. These include the model plant *Arabidopsis,* a successful weed, and the productive cereal

Figure 18-11 (a) Allele frequency variation across Kansas for a hypothetical species of wild sunflower. (b) Frequency variation for the FY^{null} allele of the Duffy blood group locus in Africa. [*From P. C. Sabeti et al., Science 312, 2006, 1614–1620.*]

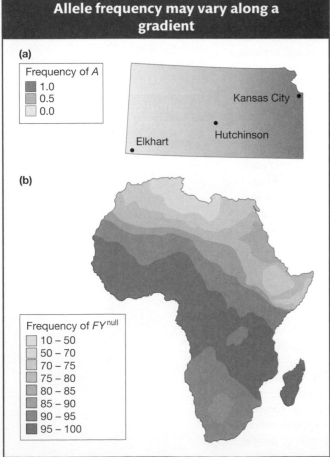

Allele frequency may vary along a gradient

(a)

Frequency of A
- 1.0
- 0.5
- 0.0

Kansas City

Hutchinson

Elkhart

(b)

Frequency of FY^{null}
- 10 – 50
- 50 – 70
- 70 – 75
- 75 – 80
- 80 – 85
- 85 – 90
- 90 – 95
- 95 – 100

crops rice and wheat. Since most plant species bear male and female organs on the same individual, self-pollination can be accomplished more easily than outcrossing. Another advantage of self-pollination is that when a single seed is dispersed to a new location, the plant that grows from the seed has a ready mate—itself, enabling a new population to be established from a single seed. Finally, if an individual plant has a beneficial combination of alleles at different loci, then inbreeding preserves that combination. In selfing plant species, benefits such as these offer advantages that outweigh the cost associated with inbreeding depression.

The inbreeding coefficient

Inbreeding increases the risk that an individual will be homozygous for a recessive deleterious allele and exhibit a genetic disease. The amount that risk increases depends on two factors: (1) the frequency of the deleterious allele in the population and (2) the degree of inbreeding. To measure the degree of inbreeding, geneticists use the **inbreeding coefficient (F),** which is the probability that two alleles in an individual trace back to the same copy in a common ancestor. Let's first consider how to calculate F using pedigrees and then examine how F can be used to determine the increase in risk of inheriting a recessive disease condition.

Consider a simple pedigree for a mating between half-sibs, individuals who have one parent in common (Figure 18-12a). In the figure, B and C are half-sibs who have the same mother, A, but different fathers; B and C have a daughter, I. Notice that there is a closed loop from I through B and A and back to I through C. The presence of a closed loop in the pedigree informs us that I is inbred. The two copies of the gene in A are colored blue and pink—the blue from A's father and pink from her mother. As drawn, I has inherited the pink copy both through her father (B) and her mother (C). Since I's two copies of the gene trace back to the same copy in her grandmother, her two copies are **identical by descent (IBD).** More generally, if the two copies of a gene in an individual trace back to the same copy in an ancestor, then the copies are IBD. We'd like a way to calculate the probability that I's two alleles will be IBD. This probability is the inbreeding coefficient for I, which is in symbol form as F_I.

First, since we are only interested in tracing the path of IBD alleles, we can simplify the pedigree to contain only the individuals in the closed loop and still follow the transmission of any IBD alleles (Figure 18-12b). Also, since the sex of the individual doesn't matter, we use circles for both sexes. The alleles transmitted with each mating are labeled w, x, y, and z. We use "~" to symbolize IBD. We'd like to calculate the probability that w and x are IBD, but let's take this calculation step by step. First, what is the probability that x and y are IBD or, symbolically, what is $P(x \sim y)$? This is the probability that C transmits the copy inherited from A to I, which is 1/2, or $P(x \sim y) = 1/2$. Similarly, the probability that B transmits the copy inherited from A to I is 1/2, or $P(w \sim z) = 1/2$.

Now we need to calculate the probability that z and y are IBD. There are two ways that z and y can be IBD. The first way is when z and y are both the same copy (both pink or both blue). This happens 1/2 of the time since 1/4 of the time they are both blue and 1/4 both pink. The second way is when z and y are different copies (one pink and the other blue) but individual A was inbred. If individual A is inbred, then there is a probability that her two copies of the gene are IBD. The probability that A's two copies are IBD is the inbreeding coefficient of A, F_A. The probability that z and y are different copies (one pink, the other blue) is 1/2. So, the probability that z and y are different copies that are IBD is 1/2 multiplied by the inbreeding coefficient (F_A) to give $\frac{1}{2}F_A$. Altogether, the probability that z and y are IBD is the probability that they are the same copy (1/2) plus the probability that they are different copies that are IBD ($\frac{1}{2}F_A$). Symbolically, we write

$$P(z \sim y) = \tfrac{1}{2} + \tfrac{1}{2}F_A$$

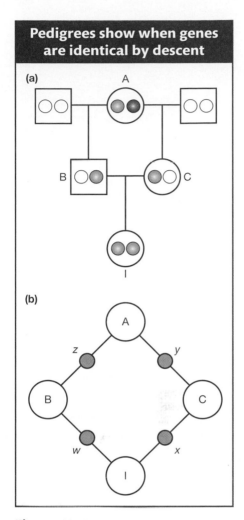

Pedigrees show when genes are identical by descent

(a)

(b)

Figure 18-12 (a) Pedigree for a half-sib mating drawn in the standard format. Small colored balls represent a single copy of a gene. Within individual A, the pink and blue copies represent the copies of the gene that she inherited from her mother and father, respectively. (b) Pedigree for a half-sib mating drawn in the simplified format used for the analysis of inbreeding. Only lines connecting parent to offspring are drawn, and only individuals in the "closed inbreeding loop" are included.

$P(x \sim y)$, $P(w \sim z)$ and $P(z \sim y)$ are independent probabilities, so can use the product rule and put it all together to obtain

$$F_I = P(x \sim y) \times P(w \sim z) \times P(z \sim y)$$

$$= \tfrac{1}{2} \times \tfrac{1}{2} \times \left(\tfrac{1}{2} + \tfrac{1}{2} F_A\right)$$

$$= \left(\tfrac{1}{2}\right)^3 (1 + F_A)$$

In the analysis of inbred pedigrees, we can substitute the value of F_A into the equation above if it is known. Otherwise, we can assume F_A is zero if there is no information to suggest that individual A is inbred. In the current example, if we assume $F_A = 0$, then

$$F_I = \left(\tfrac{1}{2}\right)^3 = \tfrac{1}{8}$$

This calculation tells us that the offspring of half-sib matings will be homozygous for alleles that are IBD for at least 1/8 of their genes. It could be more than 1/8 if F_A is greater than zero. Additional inbred pedigrees and a general formula for calculating F can be found in Box 18-2.

When there is inbreeding in a population, the random-mating assumption of Hardy–Weinberg will be violated. However, Hardy–Weinberg can be modified to correct the predicted genotypic proportions for different degrees of inbreeding

---

Box 18-2	Calculating Inbreeding Coefficients from Pedigrees

In the main text, we saw that the inbreeding coefficient (F_I) for the offspring of a mating between half-sibs is

$$F_I = \left(\tfrac{1}{2}\right)^3 (1 + F_A)$$

where F_A is the inbreeding coefficient of the ancestor. This expression includes the term 1/2 to the third power $\left(\tfrac{1}{2}\right)^3$. In Figure 18-12, you'll see there are three individuals in the inbreeding loop, not counting I. The general formula for computing inbreeding coefficients from pedigrees is

$$F_I = \left(\tfrac{1}{2}\right)^n (1 + F_A)$$

where n is the number of individuals in the inbreeding loop not counting I. Let's look at another pedigree, one in which the grandparents of I are half-sibs:

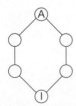

There are five individuals in the inbreeding loop other than I, so if we assume that the ancestor was not inbred ($F_A = 0$), then

$$F_I = \left(\tfrac{1}{2}\right)^5 (1 + F_A) = 0.03125$$

In some pedigrees, there is more than one inbreeding loop. Here's a pedigree in which I is the offspring of a mating between full sibs:

For pedigrees with multiple inbreeding loops, you sum the contribution over all of the loops where F_A is the inbreeding coefficient of the ancestor (A) of the given loop:

$$F_I = \sum_{\text{loops}} \left(\tfrac{1}{2}\right)^n (1 + F_A)$$

Thus, for the pedigree where I is the offspring of a mating between full sibs, we get

$$F_I = \left(\tfrac{1}{2}\right)^3 (1 + F_{A_1}) + \left(\tfrac{1}{2}\right)^3 (1 + F_{A_2}) = \tfrac{1}{4}$$

assuming that the inbreeding coefficients for both ancestors are 0.

Table 18-3 Number of Homozygous Recessives per 10,000 Individuals for Different Allele Frequencies (q)

Mating	F	$q = 0.01$	$q = 0.005$	$q = 0.001$
Unrelated parents	0.0	1.00	0.25	0.01
Parent–offspring or brother–sister	1/4	25.75	12.69	2.51
Half-sib	1/8	13.38	6.47	1.26
First cousin	1/16	7.19	3.36	0.63
Second cousin	1/64	2.55	1.03	0.17

using F, the mean inbreeding coefficient for the population. The modified Hardy–Weinberg frequencies are

$$f_{A/A} = p^2 + pqF$$
$$f_{A/a} = 2pq - 2pqF$$
$$f_{a/a} = q^2 + pqF$$

These modified Hardy–Weinberg proportions make intuitive sense, showing how inbreeding reduces the frequency of heterozygotes by $2pqF$ and adds half this amount to each of the homozygous classes.

How much does inbreeding increase the risk that offspring will exhibit a recessive disease condition? Table 18-3 shows the inbreeding coefficients for offspring of some different inbred matings and the predicted number of homozygous recessives for different frequencies (q) of the recessive allele. When $q = 0.01$, there is a 7-fold (7.19/1.0) increase in homozygous recessive offspring for first-cousin matings as compared to matings between unrelated individuals. The increase in risk jumps 13-fold (3.36/0.25) when $q = 0.005$ and 63-fold (0.63/0.01) when $q = 0.001$. In other words, the degree of risk jumps dramatically for rare alleles. Brother–sister and parent–offspring matings are the riskiest: when $q = 0.001$, they show a 250-fold (2.51/0.01) greater risk compared to matings between unrelated individuals.

The impact of inbreeding on the frequency of genetic disorders in human populations can be seen in Figure 18-13. In Japan, approximately 5 percent of marriages are between first cousins; however, the proportion of first-cousin marriages among families with children having microcephaly is over 75 percent, a 15-fold increase. Genetic disorders are more common among children of first-cousin marriages in Europe as well. Historical records suggest that the risks of inbreeding were understood long before the field of genetics existed.

Population size and inbreeding

Population size is a major factor contributing to the level of inbreeding in populations. In small populations, individuals are more likely to mate with a relative than in large ones. The phenomenon is seen in small human populations such the one on the Tristan de Cunha Islands in the South Atlantic, which has fewer than 300 people. Let's look at the effect of population size on the overall level of inbreeding in a population as measured by F.

Consider a population with F_t being the level of inbreeding at generation t. To form an individual in the next generation $t + 1$, we select the first allele from the gene pool. Suppose the population size is N. After the first allele is selected, the probability that the second allele we pick will be exactly the same copy is $1/2N$ and the inbreeding coefficient for this individual is 1.0. The

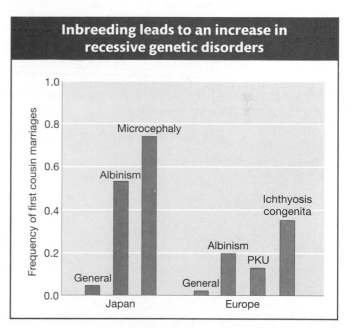

Figure 18-13 Frequency of first-cousin marriages for parents of children showing several recessive disorders compared to the frequency of first-cousin marriages for families in general in Japan and Europe. [*From P. W. Hedrick, Genetics of Populations, Jones & Bartlett, 2005.*]

probability that the second allele we pick will be a different copy from the first allele is $1 - 1/2N$ and the level of inbreeding for the resulting individual would be F_t, the average inbreeding coefficient for the initial population at generation t. The level of inbreeding in the next generation is the sum of these two possible outcomes or

$$F_{t+1} = \left(\frac{1}{2N}\right)1 + \left(1 - \frac{1}{2N}\right)F_t$$

This equation informs us that F will increase over time as a function of population size. When N is large, F increases slowly over time. When N is small, F increases rapidly over time. For example, suppose F_t in the initial population is 0.1 and $N = 10,000$. Then F_{t+1} would be 0.10005, just a slightly higher value. However if $N = 10$, then F_{t+1} would be 0.145, a much higher value. We can also use this equation recursively to calculate F_{t+2} by using F_{t+1} in place of F_t on the right side. The result with $N = 10$ and $F_t = 0.1$ would be $F_{t+2} = 0.188$. The effects of population size on inbreeding in populations are further explored in Box 18-3.

A consequence of the increased inbreeding is that individuals in small populations are more likely to be homozygous for deleterious alleles just as the offspring of first-cousin marriages are more likely to be homozygous for such alleles. This effect is seen in ethnic groups that live in small, reproductively

---

Box 18-3	**Inbreeding in Finite Populations**

In the main text, we derived the formula for the increase in inbreeding between generations in finite populations as

$$F_{t+1} = \left(\frac{1}{2N}\right)1 + \left(1 - \frac{1}{2N}\right)F_t$$

which can be rewritten as

$$(1 - F_{t+1}) = \left(1 - \frac{1}{2N}\right)(1 - F_t)$$

We also presented the formula for the frequency of heterozygotes (H) with inbreeding as

$$H = f_{A/a} = 2pq - 2pqF$$

which can be rewritten as

$$(1 - F) = H/2pq$$

Combining these two equations, we obtain

$$H_{t+1}/2pq = \left(1 - \frac{1}{2N}\right)H_t/2pq$$

and then

$$H_{t+1} = \left(1 - \frac{1}{2N}\right)H_t$$

Thus, each generation, the level of heterozygosity is reduced by the fraction $(1 - 1/2N)$. The reduction in H over t generations is

$$H_t = \left(1 - \frac{1}{2N}\right)^t H_0$$

and the change in F over t generations is given by

$$F_t = 1 - \left(1 - \frac{1}{2N}\right)^t (1 - F_0)$$

As shown in the figure below, inbreeding will increase with time in a finite population even when there is no inbreeding in the initial population.

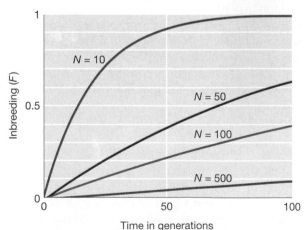

Increase in inbreeding (F) over time for several different population sizes.

isolated communities. For example, a form of dwarfism in which affected individuals have six fingers occurs at a frequency of more than 1 in 200 among a population of about 13,000 Amish in Lancaster County, Pennsylvania, although its frequency in the general U.S. population is only 1 in 60,000.

> **Message** Inbreeding increases the frequency of homozygotes in a population, and can result in a higher frequency of recessive genetic disorders. The inbreeding coefficient (F) is the probability that two alleles in an individual trace back to the same copy in a common ancestor.

18.4 Genetic Variation and Its Measurement

To study the amount and distribution of genetic variation in populations, we need some ways to quantify variation. To describe how we can quantify variation, we will use data for the glucose-6-phosphate dehydrogenase (*G6PD*) gene from humans. *G6PD* is an X-linked gene that encodes an enzyme that catalyzes a step in glycolysis. *G6PD* is of special interest because one of its alleles (A^-) leads to strongly reduced enzyme activity and individuals who carry this allele develop hemolytic anemia. However, this allele also confers a 50 percent reduction in the risk of severe malaria in carriers. In regions of Africa where malaria is endemic, the A^- allele reaches frequencies near 20 percent, although this allele is absent or rare elsewhere.

Figure 18-14 shows SNPs at 18 polymorphic sites that were identified by sequencing a 5102-bp segment of *G6PD* from a worldwide sample of 47 men. The remaining 5084 sites were **fixed,** or invariant: only a single allele (nucleotide) exists in the entire sample for each of these sites. By sampling only males, we observe just one allele and one haplotype for each individual because the gene is X linked. The wild-type allele (*B*) of *G6PD* has full enzyme activity. Another allele (A^+) leads to only modestly reduced enzyme activity. The A^+ allele differs from *B* by a single amino acid change (aspartic acid in place of asparagine) at *SNP3* in Figure 18-14. Unlike individuals carrying the A^- allele, individuals carrying only the A^+ or *B* alleles do not develop hemolytic anemia. The A^- allele differs from the *B* allele at two amino acids: it contains both the "aspartic acid in place of asparagine" change found in the A^+ allele and a second amino acid difference (methionine in place of valine) at *SNP2*.

How can we quantify variation at the *G6PD* locus? One simple measure is the number of polymorphic or **segregating sites (S).** For the *G6PD* data, *S* is 18 for the total sample, 14 for the African sample, and 7 for the non-African sample. Africans contain twice the number of segregating sites despite the fact that our sample has fewer Africans. Another simple measure is the **number of haplotypes (NH).** The value of *NH* is 12 for the total sample, 9 for the African sample, and 6 for the non-African sample. Again, the African sample has greater variation. One shortcoming of measures such as *S* and *NH* is that the values we observe depend heavily on sample size. If one samples more individuals, then the values of *S* and *NH* are apt to increase. For example, our sample has 16 Africans compared to 31 non-Africans. Although *S* is twice as large in Africans as non-Africans, the difference would likely be even greater if we had an equal number (31) of Africans and non-Africans.

In place of *S* and *NH*, we can calculate allele frequencies, which are not biased by differences in sample size. For the *G6PD* data, *B*, A^-, and A^+ have worldwide frequencies of 0.83, 0.13, and 0.04, respectively. However, you'll note that A^- has a frequency of 0.0 outside of Africa and 0.38 in our African sample, which is a substantial difference. We can use allele frequency data to calculate a statistic called **gene diversity (GD),** which is the probability that two alleles drawn at random

Nucleotide variation at the *G6PDH* gene in humans

Individual	Origin	Allele	1	2	3	4	5	6	7	8	9	10	11	12	13	14	15	16	17	18	Haplotype
			A	G	A	C	C	G	C	C	C	C	C	G	G	C	T	C	A	C	
1	Southern African	A–	G	**A**	G	•	G	•	•	T	•	T	•	•	•	•	C	•	G	•	1
2	Central African	A–	G	**A**	G	•	G	•	•	T	•	T	•	•	•	•	C	•	G	•	1
3	Central African	A–	G	**A**	G	•	G	•	•	T	•	T	•	•	•	•	C	•	G	•	1
4	African American	A–	G	**A**	G	•	G	•	•	T	•	T	•	•	•	•	C	•	G	•	1
5	African American	A–	G	**A**	G	•	G	•	•	T	•	T	•	•	•	•	C	•	G	•	1
6	Central African	A–	G	**A**	G	•	G	•	•	T	•	T	•	•	•	•	C	•	G	•	1
7	Central African	A+	G	•	G	•	•	•	•	•	•	T	•	•	•	•	C	•	G	•	2
8	Central African	A+	G	•	G	•	•	•	•	•	•	T	•	•	•	•	C	•	G	•	2
9	Central African	B	•	•	•	•	•	•	•	•	•	•	•	•	•	•	C	•	G	•	3
10	Southern African	B	•	•	•	•	•	•	•	•	•	•	•	•	A	•	C	T	G	•	4
11	Southern African	B	•	•	•	•	•	•	•	•	•	•	•	•	A	•	C	T	G	•	4
12	Southern African	B	•	•	•	T	•	•	•	•	•	•	•	•	•	•	C	T	G	•	5
13	Southern African	B	•	•	•	•	•	•	•	•	•	•	•	•	•	•	C	T	G	•	6
14	Southern African	B	•	•	•	•	•	•	•	•	•	•	T	•	A	•	C	•	G	•	7
15	Central African	B	•	•	•	•	•	•	•	•	•	•	•	•	•	T	C	•	G	•	8
16	European	B	•	•	•	•	•	•	•	•	•	•	•	•	•	T	C	•	G	•	8
17	European	B	•	•	•	•	•	•	•	•	•	•	•	•	•	T	C	•	G	•	8
18	European	B	•	•	•	•	•	•	•	•	•	•	•	•	•	T	C	•	G	•	8
19	Southwest Asian	B	•	•	•	•	•	•	•	•	•	•	•	•	•	T	C	•	G	•	8
20	East Asian	B	•	•	•	•	•	•	•	•	•	•	•	•	•	•	C	•	G	•	3
21	Native American	B	•	•	•	•	•	A	T	•	•	•	•	•	•	•	C	•	G	•	9
22	Southern African	B	•	•	•	•	•	•	•	•	•	•	•	•	•	•	•	•	•	•	10
23	Native American	B	•	•	•	•	•	•	•	•	•	•	•	•	•	•	•	•	•	•	10
24	Native American	B	•	•	•	•	•	•	•	•	•	•	•	•	•	•	•	•	•	•	10
25	Native American	B	•	•	•	•	•	•	•	•	•	•	•	•	•	•	•	•	•	•	10
26	Native American	B	•	•	•	•	•	•	•	•	•	•	•	•	•	•	•	•	•	•	10
27	Native American	B	•	•	•	•	•	•	•	•	•	•	•	•	•	•	•	•	•	•	10
28	Native American	B	•	•	•	•	•	•	•	•	•	•	•	•	•	•	•	•	•	•	10
29	Native American	B	•	•	•	•	•	•	•	•	•	•	•	•	•	•	•	•	•	•	10
30	Native American	B	•	•	•	•	•	•	•	•	•	•	•	•	•	•	•	•	•	•	10
31	Native American	B	•	•	•	•	•	•	•	•	•	•	•	•	•	•	•	•	•	•	10
32	European	B	•	•	•	•	•	•	•	•	•	•	•	•	•	•	•	•	•	•	10
33	European	B	•	•	•	•	•	•	•	•	•	•	•	•	•	•	•	•	•	•	10
34	European	B	•	•	•	•	•	•	•	•	•	•	•	•	•	•	•	•	•	•	10
35	European	B	•	•	•	•	•	•	•	•	•	•	•	•	•	•	•	•	•	•	10
36	European	B	•	•	•	•	•	•	•	•	•	•	•	•	•	•	•	•	•	•	10
37	European	B	•	•	•	•	•	•	•	•	•	•	•	•	•	•	•	•	•	•	10
38	Southwest Asian	B	•	•	•	•	•	•	•	•	•	•	•	•	•	•	•	•	•	•	10
39	East Asian	B	•	•	•	•	•	•	•	•	•	•	•	•	•	•	•	•	•	•	10
40	East Asian	B	•	•	•	•	•	•	•	•	•	•	•	•	•	•	•	•	•	•	10
41	East Asian	B	•	•	•	•	•	•	•	•	•	•	•	•	•	•	•	•	•	•	10
42	East Asian	B	•	•	•	•	•	•	•	•	•	•	•	•	•	•	•	•	•	•	10
43	East Asian	B	•	•	•	•	•	•	•	•	•	•	•	•	•	•	•	•	•	•	10
44	East Asian	B	•	•	•	•	•	•	•	•	•	•	•	•	•	•	•	•	•	•	10
45	East Asian	B	•	•	•	•	•	•	•	•	•	•	•	•	•	•	•	•	•	•	10
46	Pacific Islander	B	•	•	•	•	•	•	•	•	T	•	•	•	•	•	•	•	•	•	11
47	East Asian	B	•	•	•	•	•	•	•	•	•	•	•	•	•	•	•	•	•	T	12

Figure 18-14 Nucleotide variation for 5102 bp of the *G6PD* gene for a worldwide sample of 47 men. Only the 18 variable sites are shown. The functional allele class (*A⁻*, *A⁺*, or *B*) is shown for each sequence. *SNP2* is a nonsynonymous SNP that causes a valine-to-methionine change that underlies differences in enzyme activity associated with the *A⁻* allele. *SNP3* is a nonsynonymous SNP that causes an aspartic-acid-to-asparagine amino acid change. [*From M. A. Saunders et al., Genetics 162, 2002, 1849–1861.*]

from the gene pool will be different. The probability of drawing two different alleles is equal to 1 minus the probability of drawing two copies of the same allele summed over all alleles at the locus. Thus,

$$GD = 1 - \sum p_i^2$$
$$= 1 - (p_1^2 + p_2^2 + p_3^2 + \cdots p_n^2)$$

where p_i is the frequency of the ith allele and Σ is the summation sign, indicating that we add the squares of all n observed values of p for $i = 1, 2$, through the nth allele. The value of *GD* can vary from 0 to 1. It will approach 1 when there is a large number of alleles of roughly equal frequencies. It is 0 when there is a single allele, and it is near 0 whenever there is a single very common allele with a frequency of 0.99 or higher. Table 18-4 shows that gene diversity is quite high in Africans (0.59). Since non-Africans have only the *B* allele, gene diversity is 0.0.

The value of *GD* is equal to the expected proportion of heterozygotes under Hardy–Weinberg equilibrium, **heterozygosity (*H*)**. However, *H* as a concept applies only to diploids, and it would not apply to X-linked loci in males. Thus, conceptually gene diversity (*GD*) is more appropriate even if it is mathematically the same quantity as *H* for populations of diploids under Hardy–Weinberg equilibrium.

Gene diversity can be calculated for a single nucleotide site. It can be averaged over all the nucleotide sites in a gene, in which case it is referred to as **nucleotide diversity.** Since the vast majority of nucleotides in any two copies of a gene from a species are typically the same, values for nucleotide diversity for genes are typically very small. For *G6PD*, there are only 18 polymorphic nucleotide sites but 5084 invariant sites. The average nucleotide diversity for the entire *G6PD* gene sequence is 0.0008 in Africans, 0.0002 in non-Africans, and 0.0006 for the entire sample. These values tell us that Africans have four times as much nucleotide diversity at *G6PD* as non-Africans.

Figure 18-15 shows the level of nucleotide diversity in several organisms. Unicellular eukaryotes are the most diverse, followed by plants and then invertebrates. Vertebrates are the least diverse group; however, most vertebrates still possess a lot of nucleotide diversity. For humans, nucleotide diversity is about 0.001, meaning that two randomly chosen human

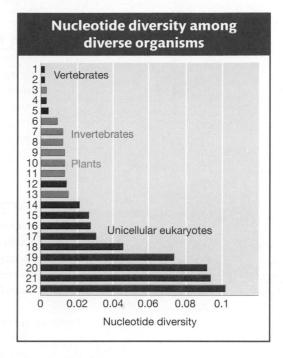

Figure 18-15 Levels of nucleotide diversity at synonymous and silent sites in some different organisms. (1) *Mus musculus*, (2) *Homo sapiens*, (3) *Oryza sativa*, (4) *Plasmodium falciparum*, (5) *Fugu rubripes*, (6) *Strongylocentrotus purpuratus*, (7) *Anopheles gambiae*, (8) *Ciona intestinalis*, (9) *Arabidopsis thaliana*, (10) *Caenorhabditis elegans*, (11) *Zea mays*, (12) *Encephalitozoon cuniculi*, (13) *Drosophila melanogaster*, (14) *Leishmania major*, (15) *Trypanosoma* species, (16) *Toxoplasma gondii*, (17) *Giardia lamblia*, (18) *Neurospora crassa*, (19) *Dictyostelium discoideum*, (20) *Saccharomyces cerevisiae*, (21) *Cryptosporidium parvum*, (22) *Cryptococcus neoformans*. [*From M. Lynch and J. S. Conery, Science 302, 2003, 1401–1404.*]

Table 18-4 **Diversity Data for Glucose-6-Phosphate Dehydrogenase (*G6PD*) in Humans**

	Total sample	Africans	Non-Africans
Sample size	47	16	31
Number of segregating sites	18	14	7
Number of haplotypes	12	9	6
Gene diversity (*GD*) at *SNP2*	0.22	0.47	0.00
Nucleotide diversity	0.0006	0.0008	0.0002

chromosomes will differ at about 1 bp per thousand. With 3 billion bp in our genome, that adds up to a total of about 3 million differences between the set of chromosomes inherited from a person's mother and the set inherited from a person's father for non-inbred individuals.

> **Message** Biological populations are often rich in genetic variation. This diversity can be quantified by different statistics to compare levels of variation among populations and species.

18.5 The Modulation of Genetic Variation

What are the forces that modulate the amount of genetic variation in a population? How do new alleles enter the gene pool? What forces remove alleles from the gene pool? How can genetic variants be recombined to create novel combinations of alleles? Answers to these questions are at the heart of understanding the process of evolution. In this section, we will examine the roles of mutation, migration, recombination, genetic drift (chance), and selection in sculpting the genetic composition of populations.

New alleles enter the population: mutation and migration

Mutation is the ultimate source of all genetic variation. In Chapter 16, we discussed the molecular mechanisms that underlie small-scale mutations such as point mutations, indels, and changes in the number of repeat units in microsatellites. Population geneticists are particularly interested in the **mutation rate,** which is the probability that a copy of an allele changes to some other allelic form in one generation. The mutation rate is typically symbolized by the Greek letter μ. As we will see below, if we know the mutation rate and the number of nucleotide differences between two sequences, then we can estimate how long ago the two sequences diverged.

How can geneticists estimate the mutation rate? Geneticists can estimate mutation rates by starting with a single homozygous individual and following the pedigree of its descendants for several generations. Then they can compare the DNA sequence of the founding individual to the DNA sequences of the descendants several generations later and record any new mutations that have occurred. The number of observed mutations per genome per generation provides an estimate of the rate. Because one is looking for rather rare events, it is necessary to sequence billions of nucleotides to find just a few SNP mutations. In 2009, the SNP mutation rate for a part of the human Y chromosome was estimated by this approach to be 3.0×10^{-8} mutations/nucleotide/generation, or about one mutation every 30 million bp. If we extrapolate to the entire human genome (3 billion bp), then each of us has inherited 100 new mutations from each of our parents. Luckily, the vast majority of mutations are not detrimental since they occur in regions of the genome that are not critical.

Table 18-5 lists the mutation rates for SNPs and microsatellites in several model organisms. The SNP mutation rate is several orders of magnitude lower than the microsatellite rate. Their higher mutation rate and greater variation make microsatellites particularly useful in population genetics and DNA forensics. The SNP mutation rate per generation appears to be lower for unicellular organisms than for large multicellular organisms. This difference can be explained at least partially by the number of cell divisions per generation. There are about 200 cell divisions from zygote to gamete in humans but only 1 in *E. coli*. If the human rate is

Table 18-5 Approximate Mutation Rates per Generation per Haploid Genome

Organism	SNP mutations (per bp)	Microsatellite
Arabidopsis	7×10^{-9}	9×10^{-4}
Maize	3×10^{-8}	8×10^{-4}
E. coli	5×10^{-10}	—
Yeast	5×10^{-10}	4×10^{-5}
C. elegans	3×10^{-9}	4×10^{-3}
Drosophila	4×10^{-9}	9×10^{-6}
Mouse	4×10^{-9}	3×10^{-4}
Human	3×10^{-8}	6×10^{-4}

Note: Microsatellite rate is for di- or trinucleotide repeat microsatellites.
Source: Data from multiple published studies.

divided by 200, then the rate per cell division in humans is remarkably close to the rate in *E. coli.*

Other than mutation, the only other means for new variation to enter a population is through **migration** or **gene flow,** the movement of individuals (or gametes) between populations. Most species are divided into a set of small local populations or subpopulations. Physical barriers such as oceans, rivers, or mountains may reduce gene flow between subpopulations, but often some degree of gene flow occurs despite such barriers. Within subpopulations, an individual may have a chance to mate with any other member of the opposite sex; however, individuals from different subpopulations cannot mate unless there is migration.

Isolated subpopulations tend to diverge as each accumulates its own unique mutations. Gene flow limits genetic divergence between subpopulations. One of the genetic consequences of migration is **genetic admixture,** the mix of genes that results when individuals have ancestry from more than one subpopulation. This phenomenon is common in human populations. It is readily observed in South Africa, where migrants from around the world were brought together. As shown in Figure 18-16, the genomes of South Africans of mixed ancestry are complex and include parts from the indigenous people of southern Africa plus contributions of migrants from western Africa, Europe, India, East Asia, and other regions.

> **Message** Mutation is the ultimate source of all genetic variation. Migration can add genetic variation to a population via gene flow from another population of the same species.

Recombination and linkage disequilibrium

Recombination is a critical force sculpting patterns of genetic variation in populations. In this case, alleles are not gained or lost; rather, recombination creates new haplotypes. Let's look at how this works. Consider linked loci *A* and *B*. There could be a population in which only two haplotypes are found at generation t_0: *AB* and *ab*. Suppose an individual in this population is heterozygous for these two haplotypes:

$$\frac{A \qquad\qquad B}{a \qquad\qquad b}$$

If a crossover occurs in this individual, then gametes with two new haplotypes, *Ab* and *aB*, could be formed and enter the population in generation t_1.

$$\overline{A \qquad\qquad b} \qquad \overline{a \qquad\qquad B}$$

Thus, recombination can create variation that takes the form of new haplotypes. The new haplotypes can have unique properties that alter protein function. For example, suppose an amino acid variant in a protein on one haplotype increases the enzyme activity of the protein twofold and a second amino acid variant on another haplotype also increases activity twofold. A recombination event that combines these two variants would yield a protein with fourfold higher activity.

Let's now consider the observed and expected frequencies of the four possible haplotypes for two loci, each with two alleles. Linked loci, *A* and *B*, have alleles *A* and *a* and *B* and *b* with frequencies p_A, p_a, p_B, and p_b, respectively. The four possible haplotypes are *AB*, *Ab*, *aB*, and *ab* with observed frequencies P_{AB}, P_{Ab}, P_{aB}, and P_{ab}. At what frequency do we expect to find each of these four haplotypes? If there is a random relationship between the alleles at the two loci, then

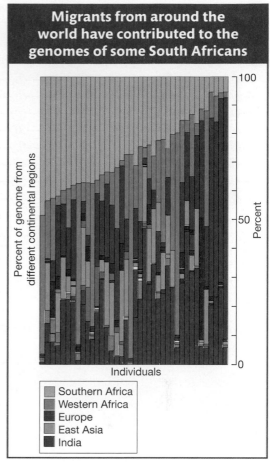

Migrants from around the world have contributed to the genomes of some South Africans

Legend:
- Southern Africa
- Western Africa
- Europe
- East Asia
- India

Figure 18-16 Graphical representation of genetic admixture for 39 people of mixed ancestry from South Africa. Each column represents one person's genome, and the colors represent the parts of their genome contributed by their ancestors, who came from many regions of the world. The figure is based on the population genetic analysis of over 800 microsatellites and 500 indels that were scored for nearly 4000 people from around the world, including the 39 of mixed ancestry from South Africa. [*From S. A. Tishkoff et al., Science 324, 2009, 1035–1044.*]

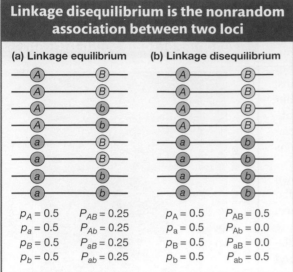

Linkage disequilibrium is the nonrandom association between two loci

(a) Linkage equilibrium

$p_A = 0.5$ $P_{AB} = 0.25$
$p_a = 0.5$ $P_{Ab} = 0.25$
$p_B = 0.5$ $P_{aB} = 0.25$
$p_b = 0.5$ $P_{ab} = 0.25$

(b) Linkage disequilibrium

$p_A = 0.5$ $P_{AB} = 0.5$
$p_a = 0.5$ $P_{Ab} = 0.0$
$p_B = 0.5$ $P_{aB} = 0.0$
$p_b = 0.5$ $P_{ab} = 0.5$

Figure 18-17 (a) Linkage equilibrium and (b) linkage disequilibrium for two loci (*A* and *B*).

the frequency of any haplotype will be the product of the frequencies of the two alleles that compose that haplotype:

$$P_{AB} = p_A \times p_B$$
$$P_{Ab} = p_A \times p_b$$
$$P_{aB} = p_a \times p_B$$
$$P_{ab} = p_a \times p_b$$

For example, suppose that the frequency of each of the alleles is 0.5; that is, $p_A = p_a = p_B = p_b = 0.5$. When we sample the gene pool, the probability of drawing a chromosome with an A allele is 0.5. If the relationship between the alleles at locus A and the alleles at locus B is random, then the probability that the selected chromosome has the B allele is also 0.5. Thus, the probability that we draw a chromosome with the AB haplotype is

$$P_{AB} = p_A \times p_B = 0.5 \times 0.5 = 0.25$$

If the association between the alleles at two loci is random as just described, then the two loci are said to be at **linkage equilibrium.** In this case, the observed and expected frequencies will be the same. Figure 18-17a diagrams a case of two loci at linkage equilibrium.

If the association between the alleles at two loci is nonrandom, then the loci are said to be in **linkage disequilibrium (LD).** In this case, a specific allele at the first locus is associated with a specific allele at the second locus more often than expected by chance. Figure 18-17b diagrams a case of complete LD between two loci. The A allele is always associated with the B allele, while the a allele is always associated with the b allele. There are no chromosomes with haplotypes Ab or aB. In this case, the observed and expected frequencies will not be the same.

We can quantify the level of LD between two loci as the difference (D) between the observed frequency of a haplotype and the expected frequency given a random association among alleles at the two loci. If both loci involved have just two alleles, then

$$D = P_{AB} - p_A p_B$$

In Figure 18-17a, $D = 0$ since there is no LD, and in Figure 18-17b, $D = 0.25$, which is greater than 0, indicating the presence of LD.

How does LD arise? Whenever a new mutation occurs at a locus, the mutation appears on a single specific chromosome and so it is instantly linked to (or associated with) the specific alleles at any neighboring loci on that chromosome. Consider a population in which there are just two haplotypes: AB and Ab. If a new mutation (a) arises at the A locus on a chromosome that already possesses the b allele at the B locus, then a new ab haplotype would be formed. Over time, this new ab haplotype rises in frequency in the population. Other chromosomes in the population would possess the AB or Ab haplotypes at these two loci, but no chromosomes would possess aB. Thus, the loci would be in LD. Migration can also cause LD when one subpopulation possesses only the AB haplotype and another only the ab haplotype. Any migrants between the subpopulations would give rise to LD within the subpopulation that receives the migrants.

LD between two loci will decline over time as crossovers between them randomize the relationship between their alleles. The rate at which this happens depends on the frequency of recombinants (RF) between the two loci among the gametes that form the next generation (see Chapter 4). In population genetics, this frequency is referred to as the recombination frequency and it is symbolized by the lowercase letter r. If D_0 is the value for linkage disequilibrium between two loci in the current generation, then the value in the next generation (D_1) is given by this equation:

$$D_1 = D_0(1 - r)$$

In other words, linkage disequilibrium as measured by D declines at a rate of $(1 - r)$ per generation. When r is small, D declines slowly over time. When r is at its maximum (0.5), then D declines by 1/2 each generation.

Since LD decays as a function of time and the recombination fraction, population geneticists can use the level of LD between a mutation and the loci surrounding it to estimate the time in generations since the mutation first arose in the population. Older mutations have little LD with neighboring loci, while recent mutations show a high level of LD with neighboring loci. If you look again at Figure 18-14, you'll notice that there is considerable LD between *SNP2* in *G6PD* and the neighboring SNPs. *SNP2* encodes the amino acid change of valine to methionine in the A^- allele that confers resistance to malaria. Population geneticists have used LD at *G6PD* to estimate that the A^- allele arose about 10,000 years ago. Malaria is not thought to have been prevalent in Africa until then. Thus, the A^- arose by random mutation but was maintained in the population because it provided protection against malaria.

> **Message** Linkage disequilibrium is the outcome of the fact that new mutations arise on a single haplotype. Linkage disequilibrium will decay over time because of recombination.

Genetic drift and population size

The Hardy–Weinberg law tells us that allele frequencies remain the same from one generation to the next in an *infinitely large population*. However, actual populations of organisms in nature are *finite* rather than infinite in size. In finite populations, allele frequencies may change from one generation to the next as the result of chance (sampling error) when gametes are drawn from the gene pool to form the next generation. Change in allele frequencies between generations due to sampling error is called **random genetic drift** or just drift for short.

Let's consider a simple but extreme case—a population composed of a single heterozygous (A/a) individual ($N = 1$) at generation t_0. We will allow self-fertilization. In this case, the gene pool can be described as having two alleles, A and a, each present at a frequency of $p = q = 0.5$. The size of the population remains the same, $N = 1$, in the subsequent generation, t_1. What is the probability that the allele frequencies will change ("drift") to $p = 1$ and $q = 0$ at generation t_1? In other words, what is the probability that the population will become fixed for the A allele, so that it consists of a single homozygous A/A individual? Since $N = 1$, we need to draw just two gametes from the gene pool to form a single individual. The probability of drawing two A's is $p^2 = 0.5^2 = 0.25$. Thus, 25 percent of the time this population will "drift" away from the initial allele frequencies and become fixed for the A allele after just one generation.

What happens if we increase the population size to $N = 2$ and the initial gene pool still has $p = q = 0.5$? The allele frequencies will change to $p = 1$ and $q = 0$ in the next generation only if the population consists of two A/A individuals. For this to happen, we need to draw four A alleles, each with a probability of $p = 0.5$, so the probability that the next generation will have $p = 1$ and $q = 0.0$ is $p^4 = (0.5)^4 = 0.0625$, or just over 6 percent. Thus, an $N = 2$ population is less likely to drift to fixation of the A allele than an $N = 1$ population. More generally, the probability of a population drifting to the fixation of the A allele in a single generation is p^{2N}, and thus this probability gets progressively smaller as the population size (N) gets larger. Drift is a weaker force in large populations.

Drift means any change in allele frequencies due to sampling error, not just loss or fixation of an allele. In a population of $N = 500$ with two alleles at a frequency of $p = q = 0.5$, there are 500 copies of A and 500 copies of a. If the next generation has 501 copies of A ($p = 0.501$) and 499 copies of a ($q = 0.499$), then there has been genetic drift, albeit a very modest level of drift. A general formula for calculating the probability of observing a specific number of copies of an allele in the next generation, given the frequencies in the current generation, is presented in Box 18-4.

When drift is operating in a finite population, one can calculate the probabilities of different outcomes, but one cannot accurately predict the specific outcome that will occur. The process is like rolling dice. At any locus, drift can continue from one generation to the next until one allele has become fixed. Also, in a particular population, the frequency of the A allele may increase from generation t_0 to t_1 but then decrease from generation t_1 to t_2. Drift does not proceed in a specific direction toward loss or fixation of an allele.

Figures 18-18a and 18-18b show computer-simulated random trials (rolls of the dice) for six populations of size $N = 10$ and $N = 500$. Each population starts having two alleles at a frequency of $p = q = 0.5$, then the random trials proceed for 30 generations. First, notice the randomness of the process from one generation to the next. For example, the frequency of A in the population depicted by the yellow line in Figure 18-18a bounces up and down from one generation to the next, hitting a low of $p = 0.15$ at t_{16} but then rebounding to $p = 0.75$ at t_{30}. Second, whether $N = 10$ or $N = 500$, notice that no two populations have exactly the same trajectory. Drift is a random process, and we are not likely to observe exactly the same outcome with different populations over many generations except when N is very small. Third, notice that when $N = 10$, the populations became fixed (either $p = 1$ or $p = 0$) before generation 20 in five of the six trials. However, when $N = 500$, the populations retained both alleles in all six trials even after 30 generations.

Box 18-4	Allele Frequency Changes Under Drift

Consider a population of N diploid individuals segregating for two alleles A and a at the A locus with frequencies p and q, respectively. The population is random mating, and the size of the population remains the same (N) in each generation. When the gene pool is sampled to create the next generation, the exact number of copies of the A allele that are drawn cannot be strictly predicted because of sampling error. However, the probability that a specific number of copies of A will be drawn can be calculated using the binomial formula. Let k be a specific number of copies of the A allele. The probability of drawing k copies is

$$\text{Prob}(k) = \left(\frac{2N!}{k!\,(2N-k)!} \right) p^k\, q^{(2N-k)}$$

If we set $N = 10$ and $p = q = 0.5$, then the probability of drawing 10 copies of the A allele is

$$\text{Prob}(10) = \left(\frac{20!}{10!\,(20-10)!} \right) 0.5^{10}\, 0.5^{(20-10)} = 0.176$$

Thus, only 17.6 percent of the time will the next generation have the same frequency of A and a as the original generation. We can use this formula to calculate the out-

comes for all possible values of k and obtain a probability distribution, shown in the figure below.

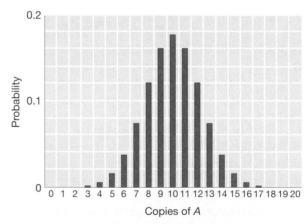

Probability distribution showing the likelihood that different numbers of A will be present after one generation.

The most probable single outcome is no drift, with $k = 10$ and a probability of 0.176. However, the other outcomes all involve some drift, and so the probability that the population will experience some drift is 0.824.

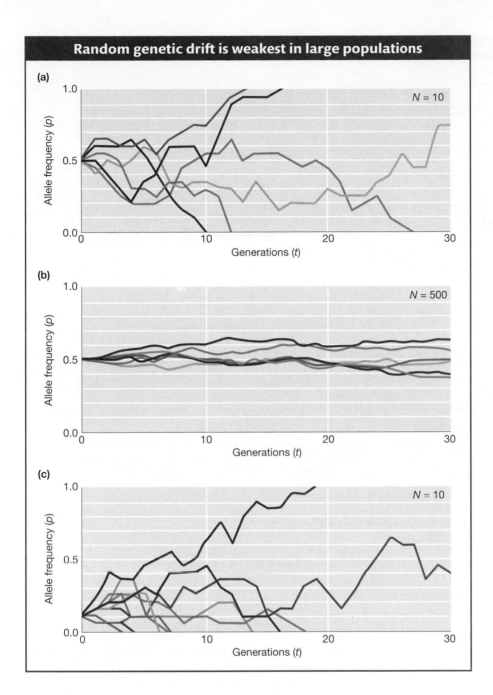

Random genetic drift is weakest in large populations

(a) *N* = 10, *p* = *q* = 0.5

(b) *N* = 500

(c) *N* = 10

Figure 18-18 Computer simulations of random genetic drift. Each colored line represents a simulated population over 30 generations. (a) $N = 10$, $p = q = 0.5$. (b) $N = 500$, $p = q = 0.5$. (c) $N = 10$, $p = 0.1$, $q = 0.9$.

In addition to population size, the fate of an allele is determined by its frequency in the population. Specifically, the probability that an allele will drift to fixation in a future generation is equal to its frequency in the present generation. An allele that is at a frequency of 0.5 has a 50 : 50 chance of fixation or loss from the population in a future generation. You can see the effect of allele frequency on the fate of an allele in Figure 18-18c. For ten populations with an initial frequency of $p = 0.1$, eight populations experienced the loss of the *A* allele, one its fixation, and one population retained both alleles after 30 generations. That's very close to the expectation that *A* will go to fixation 10 percent of the time when $p = 0.1$.

The fact that the frequency of an allele is equal to its probability of fixation means that most newly arising mutations will ultimately be lost from a population because of drift. The initial frequency of a new mutation in the gene pool is

$$\frac{1}{2N}$$

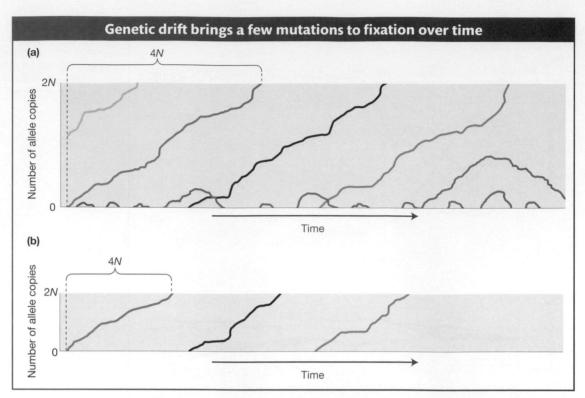

Figure 18-19 (a) Graphical representation of the appearance, loss, and eventual incorporation of new mutations in a population over time under the action of genetic drift. Gray lines show the fate of most new mutations, which appear and then are lost from the population within a few generations. Colored lines show the fate of the few "lucky" mutations that continue to rise in frequency until they reach fixation. (b) A population that is 1/2 the size of the population in part a. In this population, 4N generations is 1/2 as long and the lucky new mutations are fixed more rapidly. [*Adapted from J. Crow and M. Kimura, An Introduction to Population Genetics Theory, Harper and Row, 1970.*]

If N is even modestly large, such as 10,000, then the probability that a new mutation will ultimately reach fixation is extremely small: $1/2N = 1/20{,}000 = 5 \times 10^{-5}$. The probability that a new mutation will ultimately be lost from the population is

$$\frac{2N - 1}{2N} = 1 - \frac{1}{2N}$$

which is close to 1.0 in large populations. It is 0.99995 in a population of 10,000.

Figure 18-19a shows a graphical representation of the fate of new mutations in a population. The *x*-axis represents time and the *y*-axis the number of copies of an allele. The black lines show the fate of most new mutations. They appear and then are soon lost from the population. The colored lines show the few "lucky" new mutations that become fixed. From population genetic theory, it can be shown that the average time required for a lucky mutation to become fixed is 4N generations. Figure 18-19b shows a population that is 1/2 the size of the population in Figure 18-19a. Thus, 4N generations is 1/2 as long and the lucky new mutations are fixed more rapidly.

An important consequence of drift is that slightly deleterious alleles can be brought to fixation or advantageous alleles lost by this random process. Consider a new allele that arises in a population and endows the individual carrying it with a stronger immune system. This individual can pass the advantageous allele to his or her offspring, but those offspring might die before reproducing because of a random event such as being struck by lightning. Or if the individual carrying the favorable allele is heterozygous, he or she may pass only the less favorable allele to his or her offspring by chance.

In calculating the probabilities of different outcomes under genetic drift, we are assuming that the *A* and *a* alleles do not confer differences in viability or reproductive success to the individuals that carry them. We assume that *A/A*, *A/a*, and *a/a* individuals are all equally likely to survive and reproduce. In this case, *A* and *a* would be termed **neutral alleles** (or variants) relative to each other. Change in the frequencies of neutral alleles over time due to drift is called **neutral evolution.** The process of neutral evolution is the foundation for the **molecular clock,** the constant rate of substitution of newly arising allelic variants for preexisting ones over long periods (Box 18-5). Neutral evolution is distinct from Darwinian evolution, in which favorable alleles rise in frequency because the individuals that carry them leave more offspring. We will discuss Darwinian evolution in the next section of this chapter and in Chapter 20.

Up until now, we have been considering drift in the context of populations that remain the same size from one generation to the next. In reality, populations often contract or expand in size over time. For example, a new population of much smaller size can suddenly form when a relatively small number of the

---

Box 18-5	The Molecular Clock

As species diverge over time, their DNA sequences become increasingly different as mutations arise and become fixed in the population. At what rate do sequences diverge? To answer this question, consider a population at generation t_0. The number of mutations that will appear in generation t_1 is the product of the number of copies of the sequence in the gene pool ($2N$) times the rate at which they mutate (μ); that is, $2N\mu$. If a mutation is neutral, then the probability that it drifts to fixation is $1/2N$. So each generation, $2N\mu$ new mutations enter the gene pool, and $1/2N$ of these will become fixed. The product of these two numbers is the rate (k) at which sequences evolve:

$$k = 2N\mu \times \frac{1}{2N} = \mu$$

The value k is called the substitution rate, and it is equal to the mutation rate for neutral mutations. If the mutation rate remains constant over time, then the substitution rate will "tick" regularly like a clock, the molecular clock.

Consider two species A and B and their common ancestor. Let's define d as the number of neutral substitutions at nucleotide sites in the DNA sequence of a gene that have occurred since the divergence of A and B from their ancestor.

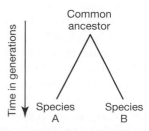

The expected value for d will be the product of the rate (k) at which substitutions occur and two times the time in generations ($2t$) during which substitution accumulated. The 2 is required because there are two lineages leading away from the common ancestor. Thus, we have

$$d = 2tk$$

This equation can be rewritten as

$$t = \frac{d}{2k}$$

showing how we can calculate the time in generations since the divergence of two species if we know d and k. The SNP mutation rate per generation (μ) is known for many groups of organisms (Table 18-5), and it is the same as the substitution rate (k) for neutral mutations. One can sequence one or more genes from two species and determine the proportion of silent (neutral) nucleotide sites at which they differ and use this proportion as an estimate for d. Thus, one can calculate the time since two sequences (two species) diverged using the molecular clock. Between humans and chimps, there are about 0.018 base differences at synonymous sites in coding sequences. The SNP mutation rate for humans is 3×10^{-8}, and the generation time is about 20 years. Using these values and the equation above, the estimated divergence time for humans and chimps is 6.0 million years ago. These calculations assume that the substitutions are neutral and that the rate of substitution has been constant over time.

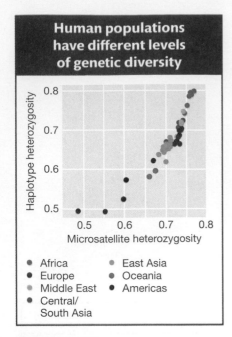

Human populations have different levels of genetic diversity

- Africa
- Europe
- Middle East
- Central/South Asia
- East Asia
- Oceania
- Americas

Figure 18-20 Plot of haplotype heterozygosity versus microsatellite heterozygosity shows genetic diversity for different geographical groups of humans. Genetic diversity is lowest for Native Americans because of the founder effect. [*From D. F. Conrad et al., Nat. Genet. 38, 2006, 1251–1260.*]

members of a population migrate to a new location and establish a new population. The migrants, or "founders," of the new population may not carry all the alleles present in the original population or they may carry the same alleles but at different frequencies. Genetic drift caused by random sampling of the original population to create the new population is known as the **founder effect.** One of many founder events in human history occurred when people crossed the Bering land bridge from Asia to the Americas during the ice age about 15,000 to 30,000 years ago. As a result, genetic diversity among Native Americans is lower than among people in other regions of the world (Figure 18-20).

Population size can also change within a single location. A period of one or several consecutive generations of contraction in population size is known as a population **bottleneck.** Bottlenecks occur in natural populations because of environmental fluctuations such as a reduction in the food supply or increase in predation. The gray wolf, American bison, bald eagle, California condor, whooping crane, and many whale species are some familiar examples of species that have experienced recent bottlenecks because of hunting by humans or encroachment by humans on their habitat. The reduction in population size during a bottleneck increases the level of drift in a population. As explained earlier in the chapter, the level of inbreeding in populations is also dependent on population size. Thus, bottlenecks also cause an increase in the level of inbreeding.

The California condor presents a remarkable example of a bottleneck. This species was once wide ranging but in the 1980s declined to a breeding population of only 14 captive birds. The population is now above 300 individuals, but the average heterozygosity in the genome decreased by 8 percent during the initial bottleneck. Furthermore, a deleterious recessive allele for a lethal form of dwarfism occurs at a frequency of about 9 percent among the surviving animals, presumably as a result of drift from a lower frequency in the pre-bottleneck population. To manage these problems, conservation biologists set up matings of captive animals to minimize further inbreeding and to purge deleterious alleles from the population.

Box 18-6 discusses the well-characterized bottleneck that occurred during the domestication of crop species. This bottleneck explains why our crop plants possess much less genetic diversity than their wild ancestors.

> **Message** Population size is a key factor affecting genetic variation in populations. Genetic drift is a stronger force in small populations than in large ones. The probability that an allele will become fixed in (or lost from) a population by drift is a function of its frequency in the population and population size. Most new neutral mutations are lost from populations by drift.

Selection

So far, we have considered how new alleles enter a population through mutation and migration and how these alleles can become fixed in (or lost from) a population by random drift. But mutation, migration, and drift cannot explain why organisms seem so well adapted to their environments. They cannot explain **adaptations,** features of an organism's form or physiology that allow it to better cope with the environmental conditions under which it lives. To explain the origin of adaptations, Charles Darwin, in 1859 in his historic book *The Origin of Species,* proposed that adaptations arise through the action of another process, which he called "natural selection." In this section, we will explore the role of natural selection in modulating genetic variation within populations. Later, in Chapter 20, we will consider the effects of natural selection on the evolution of genes and traits over extended periods.

Let's define **natural selection** as the process by which individuals with certain heritable features are more likely to survive and reproduce than are other individuals that lack these features. As outlined by Darwin, the process works like this. In each generation more offspring are produced than can survive and

Box 18-6	The Domestication Bottleneck

Before 10,000 years ago, our ancestors around the world provided for themselves by hunting wild animals and collecting wild plant foods. At about that time, human societies began to develop farming. People took local wild plants and animals and bred them into crop plants and domesticated animals. Some of the major crops that were domesticated at this time include wheat in the Middle East, rice in Asia, sorghum in Africa, and maize in Mexico.

When the first farmers collected seeds from the wild to begin domestication, they drew a sample of the wild gene pool. This sample possessed only a subset of the genetic variation found in the wild. The domesticated populations were put through a bottleneck. As a consequence, crop plants and domesticated animals typically have less genetic variation than their wild progenitors.

Modern scientific plant breeding aimed at crop improvement has created a second bottleneck. By sampling the gene pool of the traditional crop varieties, modern plant breeders have created elite varieties with traits of commercial value such as high yield and suitability for mechanical harvesting and processing. As a consequence, elite or modern varieties have even less genetic variation than traditional varieties.

The loss of genetic variation resulting from the domestication and improvement bottlenecks can pose a threat. Since there are fewer alleles per locus, crops have a smaller repertoire of alleles at disease-resistance genes and potentially greater susceptibility to emerging pathogens. To reduce this vulnerability, breeders make crosses between modern varieties and the wild relatives (or traditional varieties) to reintroduce critically important alleles into modern crops.

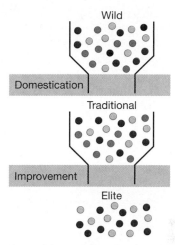

Crop domestication and improvement bottlenecks. Colored dots represent different alleles. [*M. Yamasaki et al., Plant Cell 17, 2005, 2859-2872.*]

reproduce in the environment. Nature has a mechanism (mutation) to generate new heritable forms or variants. Individuals with particular variants of some features are more likely to survive and reproduce. Individuals with features that enhance their ability to survive and reproduce will transmit these features to their offspring. Over time, these features will rise in frequency in the population. Thus, populations will change over time (*evolve*) as the environment (*nature*) favors (*selects*) features that enhance the ability to survive and reproduce. This is Darwin's theory of evolution by means of natural selection.

Darwinian evolution is often described using the phrase *"survival of the fittest."* This phrase can be misleading. An individual who is physically strong, resistant to disease, and lives a long life but has no offspring is not fit in the Darwinian sense. **Darwinian fitness** refers to the ability to survive *and reproduce*. It considers both viability and fecundity. One measure of Darwinian fitness is simply the number of offspring that an individual has. This measure is called **absolute fitness,** and we will symbolize it with an uppercase W. For an individual with no offspring, W equals 0, for an individual with one offspring, W equals 1, for an individual with two offspring, W equals 2, and so forth. W is also the number of alleles at a locus that an individual contributes to the gene pool.

Absolute fitness confounds population size and differences in reproductive success among individuals. Population geneticists are primarily interested in the latter, and so they use a measure called **relative fitness** (symbolized by a lowercase w), which is the fitness of an individual relative to some other individual, usually the most fit individual in the population. If individual X has two offspring and the most fit individual, Y, has 10 offspring, then the relative fitness of X is $w = 2/10 = 0.2$. The relative fitness of Y is $w = 10/10 = 1$. For every 10 alleles Y contributes to the next generation, X will contribute 2.

The concept of fitness applies to genotypes as well as to individuals. The absolute fitness for the A/A genotype ($W_{A/A}$) is the average number of offspring left by individuals with that genotype. If we know the absolute fitnesses for all genotypes at a locus, we can calculate the relative fitnesses for each of the genotypes.

Let's now look at how allele frequencies can change over time when different genotypes have different fitnesses; that is, when natural selection is at work. Below are the fitnesses and genotype frequencies for the three genotypes at the A locus in a population. In this case, A is a favored dominant allele since the fitnesses of the A/A and A/a individuals are the same and superior to the fitness of the a/a individuals. We are assuming that this population follows the Hardy–Weinberg law, with $p = 0.1$ and $q = 0.9$.

	A/A	A/a	a/a
Average number of offspring (W)	10	10	5
Relative fitness (w)	1.0	1.0	0.5
Genotype frequency	0.01	0.18	0.81

The relative contribution of each genotype to the gene pool is determined by the product of its fitness and its frequency. The more fit and the higher the frequency of a genotype, then the more it contributes.

Genotype	A/A	A/a	a/a	Sum
Relative contribution	1×0.01 $= 0.01$	1×0.18 $= 0.18$	0.5×0.81 $= 0.405$	0.595

The relative contributions do not sum to 1, so we need to rescale them by dividing each by the sum of all three (0.595) to get the expected frequencies of the genotypes that contribute to the gene pool.

Genotype	A/A	A/a	a/a	Sum
Genotype frequencies	0.02	0.30	0.68	1.0

Using these expected genotype frequencies and the Hardy–Weinberg law, we can calculate the frequencies of the alleles in the next generation:

$$p' = 0.02 + \left(\tfrac{1}{2} \times 0.3\right) = 0.17$$

and

$$q' = 0.68 + \left(\tfrac{1}{2} \times 0.3\right) = 0.83$$

The difference between p' and p ($\Delta p = p' - p$) is $0.17 - 0.1 = 0.07$, so we conclude that the A allele has climbed 7 percent in one generation due to natural selection. Box 18-7 presents the standard equations for calculating changes in allele frequencies over time due to natural selection.

We could go through this process recursively, using the allele frequencies from the first generation to calculate those in the second generation, then using those from the second to calculate the third, and so forth. If we then plotted p by time measured in number of generations (t), we'd have a picture of the tempo with which allele frequencies change under the force of natural selection. Figure 18-21 shows such a plot for both a favored dominant and a favored recessive allele. The dominant allele rises rapidly to start but then hits a plateau and only slowly approaches fixation. Once the favored dominant allele is at a high frequency, the unfavored recessive allele occurs mostly in heterozygotes and rarely as homozygotes with reduced fitness, so selection is ineffective at purging it from the population. The favored recessive behaves in the opposite manner—it rises slowly in frequency at first since

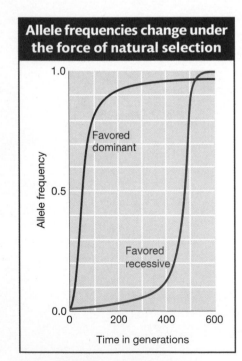

Allele frequencies change under the force of natural selection

Favored dominant

Favored recessive

Allele frequency

Time in generations

Figure 18-21 Change in allele frequency of a favored dominant allele (red) and a favored recessive allele (blue) driven by natural selection over the course of 600 generations.

Box 18-7	The Effect of Selection on Allele Frequencies

Selection causes change in allele frequencies between generations because some genotypes contribute more alleles to the gene pool than others. Let's describe a set of equations to predict gene frequencies in the next generation when selection is operating. The genotype frequencies and absolute fitnesses are symbolized as follows:

genotype	A/A	A/a	a/a
frequency	p^2	$2pq$	q^2
absolute fitness	$W_{A/A}$	$W_{A/a}$	$W_{a/a}$

The average number of alleles contributed by individuals of a given genotype is the frequency of the genotype times the absolute fitness. If N is the population size, the total number of alleles contributed by all individuals of a given genotype is N multiplied by the average number of alleles contributed by individuals of a given genotype:

average number	$p^2 W_{A/A}$	$2pq W_{A/a}$	$q^2 W_{a/a}$
total number	$N(p^2) W_{A/A}$	$N(2pq) W_{A/a}$	$N(q^2) W_{a/a}$

Thus, the gene pool will have

number of A alleles $= N(p^2)W_{A/A} + \frac{1}{2}[N(2pq)W_{A/a}]$

number of a alleles $= N(q^2)W_{a/a} + \frac{1}{2}[N(2pq)W_{A/a}]$

The mean fitness of the population is

$$\overline{W} = p^2 W_{A/A} + 2pq W_{A/a} + q^2 W_{a/a}$$

which is the average number of alleles contributed to the gene pool by an individual. $N\overline{W}$ is the total number of alleles in the gene pool.

We can now calculate the proportion of A alleles in the gene pool for the next generation as

$$p' = \frac{Np^2 W_{A/A} + Npq W_{A/a}}{N\overline{W}}$$

This equation reduces to

$$p' = p\frac{pW_{A/A} + qW_{A/a}}{\overline{W}}$$

Notice the expression $pW_{A/A} + qW_{A/a}$. This is called the allelic fitness or mean fitness of A alleles (W_A):

$$W_A = pW_{A/A} + qW_{A/a}$$

From the Hardy–Weinberg law, we know that a proportion p of all A alleles are present in homozygotes with another A, in which case they have a fitness of $W_{A/A}$, whereas a proportion q of all the A alleles are present in heterozygotes with a and have a fitness of $W_{A/a}$. Substituting W_A into the equation above, we obtain

$$p' = p\frac{W_A}{\overline{W}}$$

This equation can be used calculate the frequency of A in the next generation and used recursively to follow the change in p over time.

Although we derived these formulas using absolute fitness, generally we are not interested in population size, so we use forms of these equations with relative fitness:

$$\overline{w} = p^2 w_{A/A} + 2pq w_{A/a} + q^2 w_{a/a}$$

$$w_A = pw_{A/A} + qw_{A/a}$$

$$p' = p\frac{w_A}{\overline{w}}$$

Finally, we can express change in allele frequency between generations as

$$\Delta p = p' - p = p\frac{w_A}{\overline{w}} - p$$

$$= \frac{p(w_A - \overline{w})}{\overline{w}}$$

But $\overline{w}$, the mean relative fitness of the population, is the average of w_A and w_a, which are the allelic fitnesses of the A and a alleles, respectively:

$$\overline{w} = pw_A + qw_a$$

Substituting this expression for $\overline{w}$ in the formula for Δp and remembering that $q = 1 - p$, we obtain

$$\Delta p = \frac{pq(w_A - w_a)}{\overline{w}}$$

a/a homozygotes with enhanced fitness are rare but proceeds more rapidly to fixation later. Since the heterozygous class has reduced fitness, the unfavored dominant allele can eventually be purged from the population.

Forms of selection

Natural selection can operate in several different ways. **Directional selection,** which we have been discussing, moves the frequency of an allele in one direction until it

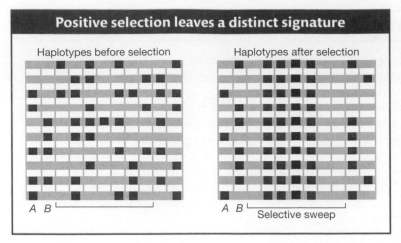

Figure 18-22 Schematic of haplotypes found in a population before and after a favored allele (red) is swept to fixation. There are 11 loci altogether. There are two alleles (red and gray) at the locus that was the target of selection. There are two alleles (black and gray) at each locus that is linked to the target locus. After selection, the target and some neighboring sites have all been swept to fixation.

reaches fixation or loss. Directional selection can be either *positive* or *purifying*. **Positive selection** works to bring a new, favorable mutation or allele to a higher frequency. This type of selection is at work when new adaptations evolve. A *selective sweep* occurs when a favorable allele reaches fixation. Directional selection can also work to remove deleterious mutations from the population. This form of selection is called **purifying selection,** and it prevents existing adaptive features from being degraded or lost. Selection does not always proceed directionally until loss or fixation of an allele. If the heterozygous class has a higher fitness than either of the homozygous classes, then natural selection will favor the maintenance of both alleles in the population. In this case, the locus is under **balancing selection** and natural selection will move the population to an equilibrium point at which both alleles are maintained in the population (see Chapter 20).

The different forms of selection each leave a distinct signature on the DNA sequence near the target locus in a population. For example, positive selection can be detected in DNA sequences by its effects on genetic diversity and linkage disequilibrium. Figure 18-22 shows schematic haplotypes before and after an episode of positive selection. In the panel showing the haplotypes before selection, the bracketed region has many polymorphisms and multiple haplotypes. However, after selection, there is only a single haplotype in this region and thus no polymorphism. When selection is applied to the target site (shown in red), the target and neighboring sites can all be swept to fixation before recombination breaks up the haplotype in which the favorable mutation first occurred. The result is lower diversity and higher LD near the target. As distance from the target increases, there is more opportunity for recombination, and so diversity goes gradually back up.

Figure 18-23 shows the pattern of diversity in the region surrounding the *SLC24A5* gene in humans. This gene influences the deposition of the melanin in the skin. When people migrated from Africa to Europe, a selective sweep at *SLC24A5* caused a loss of all diversity at this locus. As a consequence, there is a single allele and a single haplotype at this locus in Europe. The single allele that was selected for in Europe produces lighter skin color. Moving away from the gene in either direction, the number of haplotypes rises in European populations since recombination disrupted the linkage disequilibrium between *SLC25A5* and more distance sites. Light skin may be adaptive in northern latitudes. People are able to synthesize vitamin D, but to do so they need to absorb UV radiation through the skin. In the equatorial latitudes, people are exposed to high levels of UV light and can synthesize vitamin D even with heavily pigmented skin. At more distance from the equator, people are exposed to less UV light, and lighter skin color may facilitate vitamin D synthesis at these latitudes.

Table 18-6 lists a few of the genes that show evidence for natural selection in modern humans. These genes fall into a few basic categories. One group strengthens resistance to pathogens. The genes *G6PD*, *FY*null, and *Hb* (*hemoglobin B*, the sickle-cell-anemia gene) all help adapt humans to the threat of malaria. Figure 18-11b shows that the frequency of *FY*null is highest in central Africa. Central Africa also has the highest prevalence of malaria, suggesting that selection has driven *FY*null to its highest frequency in the region where selection pressure is greatest. Recently, medical geneticists have uncovered the gene *CCR5* (*chemokine receptor 5*), having an allele (*CCR5-Δ32*) that provides resistance to AIDS. This allele is

Figure 18-23 Gene diversity in human continental groups along a 2-million-bp segment of human chromosome 15 surrounding the *SLC24A5* gene. [*From Human Diversity Genome Project, www.hgdp.uchicago.edu.*]

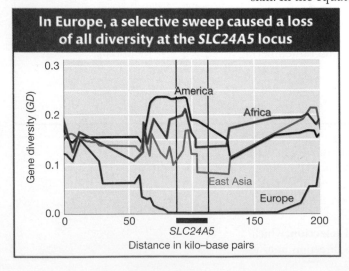

Table 18-6 Some Genes Showing Evidence for Natural Selection in Specific Human Populations

Gene	Presumed Trait	Population
EDA2R (ectodysplasin A2 receptor)	Male pattern baldness	Europeans
EDAR (ectodysplasin A receptor)	Hair morphology	East Asians
FY^{null} (Duffy antigen)	Resistance to malaria	Africans
G6PD (glucose-6-phosphate dehydrogenase)	Resistance to malaria	Africans
Hb (hemoglobin B)	Resistance to malaria	Africans
KITLG (KIT ligand)	Skin pigmentation	East Asians and Europeans
LARGE (glycosyltransferase)	Resistance to Lassa fever	Africans
LCT (lactase)	Lactase persistence; ability to digest milk sugar as an adult	Africans, Europeans
LPR (leptin receptor)	Processing of dietary fats	East Asians
MC1R (melanocortin receptor)	Hair and skin pigmentation	East Asians
MHC (major histocompatibility complex)	Infectious disease resistance	Multiple populations
OCA2 (oculocutaneous albinism)	Skin pigmentation and eye color	Europeans
PPARD (peroxisome proliferator-activated receptor delta)	Processing of dietary fats	Europeans
SI (sucrase-isomaltase)	Sucrose metabolism	East Asians
SLC24A5 (solute carrier family 24)	Skin pigmentation	Europeans and West Asians
TYRP1 (tyrosinase-related protein 1)	Skin pigmentation	Europeans

Source: P. C. Sabeti et al., *Science* 312, 2006, 1614–1620; P. C. Sabeti et al., *Nature* 449, 2007, 913–919; B. F. Voight et al., *PLoS Biology* 4, 2006, 446–458; J. K. Pickrell et al., *Genome Research* 19, 2009, 826–837.

now a target of natural selection. As long as there are pathogens, natural selection will continue to operate in human populations.

Another group of selected genes in Table 18-6 adapts people to regional diets. Before 10,000 years ago, all humans were hunter–gatherers. More recently, most humans switched to agricultural foods, but there are regional differences in diet. In northern Europe and parts of Africa, milk products are a substantial part of the diet. In most populations, the lactase enzyme for digesting milk sugar (lactose) is expressed during childhood but is switched off in adults. In parts of Europe and Africa where adults drink milk, however, special alleles of the *lactase* gene that continue to express the lactase enzyme during adulthood have risen in frequency due to natural selection. Finally, Table 18-6 includes some genes for physiological adaption to climate. Among these are the genes for skin pigmentation such as *SLC24A5*, discussed above.

Whereas directional selection causes a loss of genetic variation in the region surrounding the target locus, balancing selection can prevent the loss of diversity by random genetic drift, leading to regions of unusually high genetic diversity in the genome. One region of high genetic diversity surrounds the major histocompatibility complex (MHC) gene complex on chromosome 6. Figure 18-24 shows a distinct spike in the number of SNPs at the MHC. This complex includes the human leukocyte antigen (HLA) genes, which are involved in immune system recognition of (and response to) pathogens. Balancing selection is one hypothesis proposed to explain the high diversity observed at the MHC. Since heterozygotes have two alleles, they may be resistant to a greater repertoire of pathogen types, giving heterozygotes a fitness advantage.

Finally, selection can be imposed by an agent other than nature. Humans have imposed selection in the process of domesticating and improving cultivated plants and animals. This form of selection is called **artificial selection.** In this case, individuals with traits that humans prefer contribute more alleles to the gene pool than individuals with unfavored

Figure 18-24 Number of segregating sites (S) or SNPs in 20-kilo-basepair windows along the short arm of human chromosome 6. There is a spike of high diversity at the MHC locus. [*From International HapMap Project, www.hapmap.org.*]

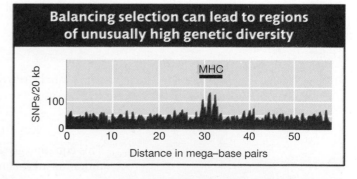

traits. Over time, the alleles that confer the favored traits rise in frequency in the population. The many breeds of dogs and dairy cows and varieties of garden vegetables and cereal crops are all the products of artificial selection.

> **Message** Natural selection is a force that can both drive favorable alleles at a locus to fixation or maintain multiple alleles at a locus in a population. Selection leaves a signature in the genome in the form of the pattern of genetic diversity surrounding the target of selection. Population geneticists have identified a number of genes that have been targets of selection in humans.

Balance between mutation and drift

We have considered the forces that regulate variation in populations individually. Let's now consider the opposing effects of mutation and drift, the former adding variation and the latter removing it from populations. When these two forces are in balance, a population can reach an equilibrium at which the loss and gain of variation are equal. We will use heterozygosity (H) as a measure of variation. Remember that H will be near 0 when a population is near fixation for a single allele (low variation), and H approaches 1 when there are many alleles of equal frequency (high variation).

Let's use H with a "hat," $\hat{H}$, as the symbol for the equilibrium value of H. To find $\hat{H}$, we start with two mathematical equations: one equation that relates change in H to population size (drift) and another equation that relates change in H to the mutation rate. We can then set these equations equal to each other and solve for $\hat{H}$.

First, we need an equation for the decline in variation (H) between generations as a function of population size (drift). We developed such an equation in Box 18-3 when discussing inbreeding:

$$H' = \left(1 - \frac{1}{2N}\right)H$$

This equation applies to the effects of drift as well as those of inbreeding. From this equation, it follows that the change in H between generations due to drift is

$$\Delta H = H - H' = \frac{1}{2N}H$$

Second, we need an equation for the increase in variation, as measured by H, between generations due to mutation. Any new mutation will increase heterozygosity at a rate proportional to the frequency of homozygotes in the population $(1 - H)$ times the rate at which mutation converts them to heteroygotes (2μ). (The 2 is necessary because there are two alleles that could mutate in a diploid.) Thus, the change in H between generations due to mutation is

$$\Delta H = 2\mu(1 - H)$$

When the population reaches an equilibrium, the loss of heterozygosity by drift will be equal to the gain from mutation. Thus, we have

$$\frac{1}{2N}\hat{H} = 2\mu(1 - \hat{H})$$

which can be rewritten as

$$\hat{H} = \frac{4N\mu}{4N\mu + 1}$$

This equation gives the equilibrium value of $\hat{H}$ when the loss by drift and gain by mutation are balanced. This equation applies only to neutral variation; that is, we are assuming selection is not at work. We are also assuming that each new mutation yields a unique allele.

Expressions such as this are useful when we have estimates for two of the variables and would like to know the third. For example, nucleotide diversity (H at the nucleotide level) for noncoding sequences, which are largely neutral, is about 0.0013 in humans, and μ for humans is 3×10^{-8} (Table 18-5). Using these values and solving the equation above for N yields an estimate of the human population size of 10,498 humans. This estimate is far below the 6.8 billion of us alive today. What's up? This is an estimate for the cquilibrium value. Modern humans are a young group, only about 150,000 years old. Over the last 150,000 years, our population has grown dramatically as we filled the globe, but mutation is a slow process, so genetic diversity has not kept up and the human population is not at equilibrium. The population size of 10,498 represents an estimate of our historical size, or how many breeding members there were about 150,000 years ago.

Balance between mutation and selection

Allelic frequencies may also reach a stable equilibrium when the introduction of new alleles by repeated mutation is balanced by their removal by natural selection. This balance probably explains the persistence of genetic diseases as low-level polymorphisms in human populations. New deleterious mutations are constantly arising spontaneously. These mutations may be completely recessive or partly dominant. Selection removes them from the population, but there is an equilibrium between their appearance and removal.

Let's begin with the simplest case—the frequency for a deleterious recessive when an equilibrium is reached between mutation and selection. For this purpose, it is convenient to express the relative fitnesses in terms of the **selection coefficient (s)**, which is the selective disadvantage of (or loss of fitness in) a genotype:

$$
\begin{array}{ccc}
W_{A/A} & W_{A/a} & w_{a/a} \\
1 & 1 & 1-s
\end{array}
$$

Then, as shown in Box 18-8, the equation for equilibrium frequency of a deleterious recessive allele is

$$\hat{q} = \sqrt{\frac{\mu}{s}}$$

Box 18-8	The Balance Between Selection and Mutation

If we let q be the frequency of the deleterious allele a and $p = 1 - q$ be the frequency of the normal allele A, then the change in allele frequency due to the mutation rate μ is

$$\Delta q_{mut} = \mu p$$

A simple way to express the fitnesses of the genotypes in the case of a recessive deleterious allele a is $w_{A/A} = w_{A/a} = 1.0$ and $w_{a/a} = 1 - s$, where s, the selection coefficient, is the loss of fitness in the recessive homozygotes. We now can substitute these fitnesses in our general expression for allele frequency change (see Box 18-7) and obtain

$$\Delta q_{sel} = \frac{-pq(sq)}{1 - sq^2} = \frac{-spq^2}{1 - sq^2}$$

Equilibrium means that the increase in the allele frequency due to mutation exactly balances the decrease in the allele frequency due to selection, so

$$\mu\hat{p} = \frac{-s\hat{p}\hat{q}^2}{1 - s\hat{q}^2}$$

The frequency of a recessive deleterious allele ($\hat{q}$) at equilibrium will be quite small, so $1 - s\hat{q}^2 \approx 1$, and we have

$$\mu\hat{p} = -s\hat{p}\hat{q}^2$$

$$\hat{q} = \sqrt{\frac{\mu}{s}}$$

at equilibrium.

This equation shows that the frequency at equilibrium depends on the ratio μ/s. When the mutation rate for $A \rightarrow a$ gets larger and the selective disadvantage smaller, then the equilibrium frequency ($\hat{q}$) of a recessive deleterious allele will rise. As an example, a recessive lethal allele ($s = 1$) that arises by mutation from the wild-type allele at the rate of $\mu = 10^{-6}$ will have an equilibrium frequency of 10^{-3}.

Let's consider the equilibrium between selection and mutation for the slightly more complicated case of a partially dominant deleterious allele—that is, an allele with some deleterious effect in heterozygotes as well as its effect in homozygotes. We'll define h as the degree of dominance of the deleterious allele. When h is 1, the deleterious allele is fully dominant, and when h is 0, the deleterious allele is fully recessive. Then, the fitnesses are

$$W_{A/A} \qquad W_{A/a} \qquad w_{a/a}$$
$$1 \qquad 1 - hs \qquad 1 - s$$

where a is a partially dominant deleterious allele. A derivation similar to the one in Box 18-8 gives us

$$\hat{q} = \frac{\mu}{hs}$$

Here is an example. If $\mu = 10^{-6}$ and the lethal allele is not totally recessive but causes a 5 percent reduction in fitness in heterozygotes ($s = 1.0$, $h = 0.05$), then

$$\hat{q} = \frac{\mu}{hs} = 2 \times 10^{-5}$$

This result is smaller by two orders of magnitude than the equilibrium frequency for the purely recessive case described above. In general, then, we can expect deleterious, completely recessive alleles to have frequencies much higher than those of partly dominant alleles, because the recessive alleles are protected in heterozygotes.

> **Message** The amount of genetic variation in populations represents a balance between opposing forces: mutation and migration, which add new variation, versus drift and selection, which remove variation. Balancing selection also serves to maintain variation in populations. As a result of these processes, allele frequencies can reach equilibrium values, explaining why populations often maintain high levels of genetic variation.

18.6 Biological and Social Applications

Just as the principles of physics guide engineers who design bridges and jet airliners, so the principles of population genetics touch all of our lives in many, if unseen, ways. In Chapter 19, you'll see how population genetics figures prominently in the search for genes that contribute to disease risk in people, using concepts such as linkage disequilibrium, described in this chapter. In this final section of the chapter, we will examine four other areas in which the principles of population genetics are being to applied to issues affecting modern societies.

Conservation genetics

Conservation biologists attempting to save endangered wild species, and zookeepers attempting to maintain small populations of captive animals, often perform population genetic analyses. Above, we discussed how a genetic bottleneck caused a loss of genetic variation in the California condor and an increase in the frequency of a lethal form of dwarfism. Bottlenecks may also increase the level of inbreeding in a population, perhaps leading to a decline in fitness through inbreeding depression. The issue is complex, however, because inbreeding is not always associated with a decline in fitness. Inbreeding can sometimes help purge deleterious recessive alleles from a population. Purifying selection is more effective at eliminating deleterious recessive alleles since the homozygous recessive class becomes more frequent in inbred populations. Thus, conservation biologists have debated whether they should attempt to maximize genetic diversity and minimize inbreeding or deliberately subject zoo populations to inbreeding with the goal of purging deleterious alleles.

To help address this question, researchers looked for evidence of successful purging among zoo populations. Let's define inbreeding depression as delta (δ)

$$\delta = 1 - \frac{w_f}{w_0}$$

where w_f is the fitness of inbred individuals and w_0 the fitness of non-inbred individuals. The value of δ will be positive when there is a decline in fitness with inbreeding but negative when fitness improves with inbreeding. Researchers calculated δ for 119 zoo populations, including 88 species, and they found evidence that purging had improved fitness (negative values for δ) in 14 populations. Still, it is not clear that deliberate inbreeding of zoo animals is advisable. For one thing, although 14 of the 119 populations improved, the majority of the populations declined in fitness when inbred. Thus, if one starts with a small zoo population and purposely inbreeds the animals, a decline in fitness is the most likely outcome.

Calculating disease risks

In Chapter 2, we saw how alleles for genetic disorders could be traced in pedigrees and we discussed how to calculate the risk that a couple will have a child who inherits such a disorder. Population genetic principles allow us to extend this type of analysis. We will consider two examples.

The disease allele for cystic fibrosis (CF) occurs at a frequency of about 0.025 in Caucasians. In the pedigree for a Caucasian family below, individual II-2 has a first cousin (II-1) with cystic fibrosis. II-2 is married to an unrelated Caucasian (II-3), and they are planning to have a child. What is the chance that the child (III-1) will have cystic fibrosis?

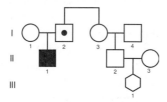

One of II-2's maternal grandparents must have been a carrier. We begin by calculating the probability that III-1 will inherit this cystic fibrosis allele from this grandparent through his father, II-2, using methods already familiar from Chapter 2. The probability that this grandparent transmitted the disease allele to I-3 is 1/2. The probability that I-3 transmitted it to II-2 and that II-2 will transmit it to III-1 are also both 1/2. So the probability that III-1 inherits the CF allele from II-2 is $\left(\frac{1}{2}\right)^3$, or 1/8. We now extend the calculation to determine the probability that III-1 could inherit the cystic fibrosis allele from his mother, II-3. Individual II-3 does not have CF, but we are not sure whether or not she is a carrier. If the frequency (q) of the disease allele in the population is 0.025, then the probability that an unaffected individual such as II-3 is a carrier is $2pq/(1 - q^2) = 0.049$. If II-3 is a carrier, then there is a $\frac{1}{2}$ chance she will transmit the disease allele to III-1. These are all independent probabilities, so we can use the product rule. The probability that III-1 will have cystic fibrosis is

$$\frac{1}{8} \times \frac{1}{2} \times 0.049 = 0.003$$

The frequency of cystic fibrosis among Caucasians is $p^2 = (0.025)^2 = 0.000625$. These calculations tell us that individuals who have a first cousin with cystic fibrosis have a $0.003 \div 0.000625 = 4.9$-fold higher risk of having a child with the disease than members of the general population.

Here is another application of population genetics to assessing disease risk. Sickle-cell anemia, a recessive disease, has a frequency of about 0.25 percent, or 1 in 400, among African Americans (see Chapter 6). Applying the Hardy–Weinberg law, we estimate the frequency of the disease allele (HbS) as 0.05. What would be the expected frequency of this disease among the offspring of African Americans who are first cousins? Using the method described in Box 18-2, we calculate that the inbreeding coefficient (F) for the offspring of first-cousin marriages is 1/16. In the section on inbreeding above, we saw that the frequency of the homozygotes when there is inbreeding is increased, as shown by this equation

$$f_{a/a} = q^2 + pqF$$

Using this equation, we obtain

$$f(Hb^S/Hb^S) = (0.05)^2 + (0.05 \times 0.95)\frac{1}{16} = 0.0055$$

This represents a 2.2-fold increase in the risk of having a child with the disease for first-cousin marriages compared to that in a marriage between unrelated individuals.

DNA forensics

Criminals can leave DNA evidence at the scene of a crime in the form of blood, semen, hairs, or even buccal cells from saliva on a cigarette butt. The polymerase chain reaction (PCR) enables forensic scientists to amplify very tiny amounts of DNA and determine the genotype of the individual who left the specimen. If the DNA found at the crime scene matches that of the suspect, then they "may be" the same individual. The key phrase here is "may be," and this is where population genetics comes into play. Let's see how this works.

Consider two microsatellite loci, each with multiple alleles: $A_1, A_2, \ldots A_n$ and $B_1, B_2, \ldots B_n$. Forensic scientists determine that a DNA specimen from a crime scene and the suspect are both A_3/A_8 B_1/B_7. They have determined that there is a "match" between the evidence and the suspect. Does the match prove that the DNA evidence came from the suspect? Does it prove that that the suspect was at the crime scene?

What population geneticists do with this type of evidence is to test a specific hypothesis: *The evidence came from someone other than the suspect.* This is what statisticians call the "null hypothesis," or the hypothesis that is considered true unless the evidence shows that it is very unlikely (see Chapter 4). To perform the test, we calculate the probability of observing a match between the evidence and the suspect, given that the suspect and the person who left the evidence are different individuals. Symbolically, we write

Prob(match | different individuals)

where "|" means "given." If this probability is very small, then we can reject the null hypothesis and argue in favor of an alternative hypothesis: *The evidence was left by the suspect*. We never formally prove the suspect left the evidence since there could be alternative hypotheses such as *The evidence was left by the suspect's identical twin*.

To calculate the probability of observing a match between the evidence and the suspect if the evidence is from a different individual, we need to know the frequencies of the microsatellite alleles in the population.

$$A_4 \quad 0.03$$
$$A_6 \quad 0.05$$
$$B_1 \quad 0.01$$
$$B_7 \quad 0.12$$

Prob(match | different individuals) is the same as the probability that the evidence came from a randomly chosen individual. We can calculate this probability using the allele frequencies above. First, we will assume that the Hardy-Weinberg law applies and calculate the probability of being A_4/A_6 at the first locus and B_1/B_7 at the second:

$$\text{Prob}(A_4/A_6) = 2pq = 2 \times 0.03 \times 0.05 = 0.003$$
$$\text{Prob}(B_1/B_7) = 2 \times 0.01 \times 0.12 = 0.0024$$

To combine these two probabilities, we need to make one more assumption. We need to assume that the two loci are *independent;* that is, that the loci are at *linkage equilibrium*. By making this assumption, we can apply the *product rule* for independent events (see Chapter 2) and determine that

$$\text{Prob}(\text{match} | \text{different individuals}) = \text{Prob}(A_4/A_6) \times \text{Prob}(B_1/B_7) = 7.2 \times 10^{-6}$$

Thus, the probability under the null hypothesis that the evidence came from someone other than the suspect is 7.2×10^{-6}, or about 7 in a million. That's a small probability, and so the null hypothesis seems unlikely in this case. However, if Prob(match | different individuals) were 0.1, then 10 percent of the population would be a match and could have left the evidence. In that case, we would not want to reject the null hypothesis.

Two microsatellites do not provide very much power to discriminate, so the FBI in the United States uses a set of 13 microsatellites. Microsatellite loci typically have large numbers of alleles (10 to 20 or more); therefore the number of possible genotypes based on 13 microsatellites is astronomically large. With 10 alleles per locus, there are 55 possible genotypes at each locus and 55^{13}, or 4.2×10^{22}, possible multilocus genotypes for 13 loci. The FBI has also assembled a database called CODIS (Combined DNA Index System) that contains the frequencies of different alleles at these loci in the population, including data specific to different ethnic groups and regions of the country.

Googling your DNA mates

In this chapter, we have reviewed the basic principles of population genetics and discussed many applications to human genetics. Basic population genetic theory has been around for nearly 100 years, but only in the last decade has the development of high-throughput DNA-based technologies for genotyping individuals brought the complex patterns of variation among and within human populations into sharp focus. Not only has it been possible to unravel many of the details about how and when humans populated the globe from our African homeland, but geneticists have gained a deep understanding of how forces such as natural selection and genetic drift have shaped who we are.

What does the future hold? Soon, the sequencing of a human genome may cost little more than a new bicycle. A college student might swab the inside of her mouth with a Q-tip and deposit the sample in a kiosk while out to hear a band. Weeks later, she would be able to view her genome sequence at a World

Wide Web site, compare it to those of her friends and relatives, and learn about her ancestry. As we'll see in the next chapter, our ability to predict a person's disease risks, talents, and other traits from his or her genotype is improving. To the extent that someone's taste in music or love of extreme sports has genetic underpinnings, a person could in theory "Google" DNA mates likely to share similar interests. The DNA technology and population genetics theory are already in place; however, there are social and ethical questions to be addressed. Can and how will the information be kept private? Are there any limits on what a person should know about his or her own sequence? Should the government sequence everyone's genome when he or she is born? Could medical insurance providers require that their clients submit their genome sequences? An understanding of the science can aid in determining how these questions are answered.

Summary

Population genetics seeks to understand the laws that govern and forces that influence the amount of genetic variation within populations and changes in genetic variation over time. The concept of the gene pool provides a model for thinking about the transmission of genetic variation from one generation to the next for an entire population. Basic population genetic theory starts with an idealized population that is infinite in size and in which mating is random. In such a population, the Hardy–Weinberg law defines the relationship between allele frequencies in the gene pool and genotype frequencies in the population.

Real populations usually deviate to a small or large degree from the Hardy–Weinberg model. One source of deviation comes in the form of nonrandom or assortative mating. If individuals preferentially mate with others who share a similar phenotype, then there will be an excess of homozygotes at genes controlling that phenotype compared to Hardy–Weinberg expectations. When individuals mate more frequently with relatives than expected by chance, then there will be an excess of homozygous genotypes throughout the entire genome and the population becomes inbred. Even when local populations of a species conform to Hardy–Weinberg expectations, those populations are apt to be isolated from other populations at distant locations. Thus, a species often consists of a series of genetically distinct subpopulations; that is, species show population genetic structure.

Several forces can add new variation to a population or remove existing variation from it. Mutation is the ultimate source of all genetic variation. Population geneticists have determined reasonably precise estimates of the rate at which new mutations arise in populations. Migration can also bring new variation into a population. Migration results in some individuals who are genetically admixed, having ancestry from multiple populations. Genetic recombination can also add variation to populations by recombining alleles into new haplotypes.

Two forces control the fate of genetic variation in populations. First, genetic drift is a random force that can lead to the loss or fixation of an allele as a result of sampling error in finite populations. Drift is a strong force in small populations and a weak force in large ones. Second, natural selection drives changes in allele frequencies in populations over time. Alleles that enhance the fitness of the individuals that carry them will rise in frequency and can become fixed, while deleterious alleles that reduce fitness will be purged from the population.

The fundamental goal of population genetics is to understand the relative contributions made by mating systems, mutation, migration, recombination, drift, and natural selection to the amount and distribution of genetic variation in populations. In this chapter, we have seen how research in population genetics has both developed the basic theory and collected a vast amount of data to achieve this goal. Our understanding of the population genetics of our own species is remarkably detailed.

Finally, the methods and results of population genetics both inform us about the evolutionary process and have practical applications to issues facing modern societies. Population genetic theory and analyses play important roles in the management of endangered species, the identification of perpetrators of crimes, plant and animal breeding, and assessing the risks that a couple will have a child with a disease condition.

KEY TERMS

absolute fitness (p. 667)
adaptation (p. 666)
allele frequency (p. 644)

artificial selection (p. 671)
balancing selection (p. 670)
bottleneck (p. 666)

common SNP (p. 639)
Darwinian fitness (p. 667)
directional selection (p. 669)

SOLVED PROBLEMS

SOLVED PROBLEM 1. About 70 percent of all Caucasians can taste the chemical phenylthiocarbamide, and the remainder cannot. The ability to taste this chemical is determined by the dominant allele *T*, and the inability to taste is determined by the recessive allele *t*. If the population is assumed to be in Hardy-Weinberg equilibrium, what are the genotype and allele frequencies in this population?

Solution

Because 70 percent are tasters (*T/T* and *T/t*), 30 percent must be nontasters (*t/t*). This homozygous recessive frequency is equal to q^2; so, to obtain *q*, we simply take the square root of 0.30:

$$q = \sqrt{0.30} = 0.55$$

Because $p + q = 1$, we can write $p = 1 - q = 1 - 0.55 = 0.45$.

Now we can calculate

$$p^2 = (0.45)^2 = 0.20, \quad \text{the frequency of } T/T$$
$$2pq = 2 \times 0.45 \times 0.55 = 0.50, \quad \text{the frequency of } T/t$$
$$q^2 = 0.3, \quad \text{the frequency of } t/t$$

SOLVED PROBLEM 2. In a large experimental *Drosophila* population, the relative fitness of a recessive phenotype is calculated to be 0.90, and the mutation rate to the recessive allele is 5×10^{-5}. If the population is allowed to come to equilibrium, what allele frequencies can be predicted?

Solution

Here, mutation and selection are working in opposite directions, and so an equilibrium is predicted. Such an equilibrium is described by the formula

$$\hat{q} = \sqrt{\frac{\mu}{s}}$$

In the present question,

$$\mu = 5 \times 10^{-5} \quad \text{and} \quad s = 1 - w = 1 - 0.9 = 0.1$$

Hence

$$\hat{q} = \sqrt{\frac{5 \times 10^{-5}}{0.1}} = 0.022$$
$$\hat{p} = 1 - 0.022 = 0.978$$

SOLVED PROBLEM 3. A colony of 50 horned puffins (*Fratercula corniculata*) is established at a zoo and maintained there for 30 generations.

a. If the inbreeding coefficient of the founding members was zero ($F = 0.0$), what is the expected inbreeding coefficient for this population at present?

b. For a deleterious disease allele with a frequency of 0.001 in the wild, what is the predicted frequency of homozygous affected birds in the wild and in the zoo population at present?

Solution

a. In Box 18-3, we saw that inbreeding will increase as a function of population size (*N*) over time (*t*) as measured in generations according to the following equation:

$$F_{30} = 1 - \left(1 - \frac{1}{2N}\right)^t (1 - F_0)$$

Substituting in $N = 50$, $t = 30$, and $F_0 = 0$, we obtain

$$F_t = 1 - \left(1 - \frac{1}{2 \times 50}\right)^{30} (1 - 0) = 0.26$$

b. If the frequency of a recessive disease allele (*q*) in the wild is 0.001, then by applying the Hardy–Weinberg law we predict that the frequency of homozygous affected individuals in the wild will be $q^2 = 10^{-6}$. For the zoo population, the frequency of homozygotes will be higher because of inbreeding according to the following equation:

$$f_{a/a} = q^2 + pqF$$

Substituting in $q = 0.001$, $p = 0.999$, and $F = 0.26$, we obtain

$$f_{a/a} = 10^{-6} + (0.001 \times 0.999 \times 0.26) = 2.61 \times 10^{-4}$$

The ratio of 2.61×10^{-4} to 10^{-6} shows us that there is 261-fold increase in the expected frequency of affected individuals in the current zoo population compared to the ancestral wild population.

SOLVED PROBLEM 4. At a criminal trial, the prosecutor presents genotypes for three microsatellite loci from the FBI CODIS set. He reports that a DNA sample from the crime scene and one from the suspect both have the genotype FGA_1/FGA_4, $TPOX_1/TPOX_3$, VWA_2/VWA_7 at these three microsatellites. He also presents the allelic frequencies for the general population to which the suspect belongs (see the table that follows). What is the probability that the genotype of the DNA evidence would match that of the suspect given that the person who committed the crime and the suspect are different individuals? What assumptions do you make when calculating this probability?

Allele	Frequency
FGA_1	0.30
FGA_4	0.26
$TPOX_1$	0.32
$TPOX_3$	0.65
VWA_2	0.23
VWA_7	0.59

Solution

The probability that the genotype of the DNA evidence matches that of the suspect given that the person who committed the crime and the suspect are different individuals is the same as the probability that a randomly chosen member of the population would have the same genotype as the DNA evidence. The probability of a randomly chosen person being $FGA_1/FGA_4 = 2pq = 2(0.30)$ $(0.26) = 0.156$ and, similarly, the probability of a random person being $TPOX_1/TPOX_3 = 0.416$ and $VWA_2/VWA_7 = 0.2714$. Applying the multiplicative rule, the probability of a random member of the population being FGA_1/FGA_4, $TPOX_1/TPOX_3$, $VWA_2/VWA_7 = 0.156 \times 0.416 \times 0.2714 = 0.0176$. In calculating this probability, we have assumed that the population is at Hardy–Weinberg equilibrium and that the three loci in question are at linkage equilibrium with one another.

PROBLEMS

Most of the problems are also available for review/grading through the GENETICSPORTAL www.yourgeneticsportal.com.

WORKING WITH THE FIGURES

1. Which individual in Figure 18-3 has the most heterozygous loci and which individual has the fewest?

2. Suppose that the seven haplotypes in Figure 18-4a represent a random sample of haplotypes from a population.

   a. Calculate gene diversity (GD) separately for the indel, the microsatellite locus, and the SNP at position 3.

   b. If the sequence was shortened so that you had data only for positions 1 through 24, how many haplotypes would there be?

   c. Calculate the linkage disequilibrium parameter (D) between the SNPs at positions 29 and 33.

3. Looking at Figure 18-6, can you count how many mitochondrial haplotypes were carried from Asia into the Americas? How many of these made it all the way to South America?

4. In Figure 18-13, the "general" column for Japan is higher than the "general" column for Europe. What does this tell you?

5. In Figure 18-14, some individuals have unique SNP alleles—for example, the T allele at SNP4 occurs only in individual 12. Can you identify two individuals each of whom have unique alleles at two SNPs?

6. Looking at Figure 18-20, do people of the Middle East tend to have higher or lower levels of heterozygosity compared to the people of East Asia? Why might this be the case?

BASIC PROBLEMS

7. What are the forces that can change the frequency of an allele in a population?

8. In a population of mice, there are two alleles of the A locus (A_1 and A_2). Tests showed that, in this population, there are 384 mice of genotype A_1/A_1, 210 of A_1/A_2, and 260 of A_2/A_2. What are the frequencies of the two alleles in the population?

9. In a randomly mating laboratory population of *Drosophila*, 4 percent of the flies have black bodies (encoded by the autosomal recessive *b*), and 96 percent have brown bodies (the wild type, encoded by *B*). If this population

is assumed to be in Hardy–Weinberg equilibrium, what are the allele frequencies of B and b and the genotypic frequencies of B/B and B/b?

10. In a population of a beetle species, you notice that there is a $3:1$ ratio of shiny to dull wing covers. Does this ratio prove that the *shiny* allele is dominant? (Assume that the two states are caused by two alleles of one gene.) If not, what does it prove? How would you elucidate the situation?

11. The relative fitnesses of three genotypes are $w_{A/A} = 1.0$, $w_{A/a} = 1.0$, and $w_{a/a} = 0.7$.

    a. If the population starts at the allele frequency $p = 0.5$, what is the value of p in the next generation?

    b. What is the predicted equilibrium allele frequency if the rate of mutation from $A \rightarrow a$ is 2×10^{-5}?

12. A/A and A/a individuals are equally fertile. If 0.1 percent of the population is a/a, what selection pressure exists against a/a if the $A \rightarrow a$ mutation rate is 10^{-5}? Assume that the frequencies of the alleles are at their equilibrium values.

13. If the recessive allele for an X-linked recessive disease in humans has a frequency of 0.02 in the population, what proportion of individuals in the population will have the disease? Assume that the population is $50:50$ male:female.

14. It seems clear that most new mutations are deleterious. Can you explain why?

15. In a population of 50,000 diploid individuals, what is the probability that a new neutral mutation will ultimately reach fixation? What is the probability that it will ultimately be lost from the population?

CHALLENGING PROBLEMS

16. Figure 18-14 presents haplotype data for the *G6PD* gene in a worldwide sample of people.

    a. Draw a haplotype network for these haplotypes. Label the branches on which each SNP occurs.

    b. Which of the haplotypes has the most connections to other haplotypes?

    c. On what continents is this haplotype found?

    d. Counting the number of SNPs along the branches of your network, how many differences are there between haplotypes 1 and 12?

17. Figure 18-12 shows a pedigree for the offspring of a half-sib mating.

    a. If the inbreeding coefficient for the common ancestor (A) in Figure 18-12 is 1/2, what is the inbreeding coefficient of I?

    b. If the inbreeding coefficient of individual I in Figure 18-12 is 1/8, what is the inbreeding coefficient of the common ancestor, A?

18. Consider 10 populations that have the genotypes shown in the following table:

Population	A/A	A/a	a/a
1	1.0	0.0	0.0
2	0.0	1.0	0.0
3	0.0	0.0	1.0
4	0.50	0.25	0.25
5	0.25	0.25	0.50
6	0.25	0.50	0.25
7	0.33	0.33	0.33
8	0.04	0.32	0.64
9	0.64	0.32	0.04
10	0.986049	0.013902	0.000049

    a. Which of the populations are in Hardy–Weinberg equilibrium?

    b. What are p and q in each population?

    c. In population 10, the $A \rightarrow a$ mutation rate is discovered to be 5×10^{-6}. What must be the fitness of the a/a phenotype if the population is at equilibrium?

    d. In population 6, the a allele is deleterious; furthermore, the A allele is incompletely dominant; so A/A is perfectly fit, A/a has a fitness of 0.8, and a/a has a fitness of 0.6. If there is no mutation, what will p and q be in the next generation?

19. The hemoglobin B gene (*Hb*) has a common allele (*A*) of an SNP (*rs334*) that encodes the Hb^A form of (adult) hemoglobin and a rare allele (*T*) that encodes the sickling form of hemoglobin, Hb^S. Among 57 members of the Yoruba tribe of Nigeria, 44 were A/A and 13 were A/T. No T/T individuals were observed. Use the χ^2 test to determine whether these observed genotypic frequencies fit Hardy–Weinberg expectations.

20. A population has the following gametic frequencies at two loci: $AB = 0.40$ $Ab = 0.1$, $aB = 0.1$, and $ab = 0.4$. If the population is allowed to mate at random until linkage equilibrium is achieved, what will be the expected frequency of individuals that are heterozygous at both loci?

21. Two species of palm trees differ by 50 bp in a 5000-bp stretch of DNA that is thought to be neutral. The mutation rate for these species is 2×10^{-8} substitutions per site per generation. The generation time for these species is five years. Estimate the time since these species had a common ancestor.

22. Color blindness in humans is caused by an X-linked recessive allele. Ten percent of the males of a large and randomly mating population are color-blind. A representative group of 1000 people from this population migrates

to a South Pacific island, where there are already 1000 inhabitants and where 30 percent of the males are color-blind. Assuming that Hardy–Weinberg equilibrium applies throughout (in the two original populations before the migration and in the mixed population immediately after the migration), what fraction of males and females can be expected to be color-blind in the generation immediately after the arrival of the migrants?

23. Using pedigree diagrams, calculate the inbreeding coefficient (F) for the offspring of (a) parent–offspring matings; (b) first-cousin matings; (c) aunt–nephew or uncle–niece matings; (d) self-fertilization of a hermaphrodite.

24. A group of 50 men and 50 women establish a colony on a remote island. After 50 generations of random mating, how frequent would a recessive trait be if it were at a frequency of 1/500 back on the mainland? The population remains the same size over the 50 generations, and the trait has no effect on fitness.

25. Figure 18-22 shows 10 haplotypes from a population before a selective sweep and another 10 haplotypes many generations later after a selective sweep has occurred for this chromosomal region. There are 11 loci defining each haplotype, including one with a red allele that was the target of selection. In the figure, two loci are designated as A and B. These loci each have two alleles: one black and the other gray. Calculate the linkage disequilibrium parameter (D) between A and B, both before and after the selective sweep. What effect has the selective sweep had on the level of linkage disequilibrium?

26. The recombination fraction (r) between linked loci A and B is 0.25. In a population, we observe the following haplotypic frequencies:

$$AB \quad 0.49$$
$$aB \quad 0.49$$
$$Ab \quad 0.00$$
$$ab \quad 0.02$$

   a. What is the level of linkage disequilibrium as measured by D in the present generation?

   b. What will D be in the next generation?

   c. What is the expected frequency of the Ab haplotype in the next generation?

27. Allele B is a deleterious autosomal dominant. The frequency of affected individuals is 4.0×10^{-6}. The reproductive capacity of these individuals is about 30 percent that of normal individuals. Estimate μ, the rate at which b mutates to its deleterious allele B. Assume that the frequencies of the alleles are at their equilibrium values.

28. What is the equilibrium heterozygosity for an SNP in a population of 50,000 when the mutation rate is 3×10^{-8}?

29. Of 31 children born of father–daughter matings, 6 died in infancy, 12 were very abnormal and died in child-

hood, and 13 were normal. From this information, calculate roughly how many recessive lethal genes we have, on average, in our human genomes. (**Hint:** If the answer were 1, then a daughter would stand a 50 percent chance of carrying the lethal allele, and the probability of the union's producing a lethal combination would be $1/2 \times 1/4 = 1/8$. So 1 is not the answer.) Consider also the possibility of undetected fatalities in utero in such matings. How would they affect your result?

30. The sd gene causes a lethal disease of infancy in humans when homozygous. One in 100,000 newborns die each year of this disease. The mutation rate from Sd to sd is 2×10^{-4}. What must the fitness of the heterozygote be to explain the observed gene frequency in view of the mutation rate? Assign a relative fitness of 1.0 to Sd/Sd homozygotes. Assume that the population is at equilibrium with respect to the frequency of sd.

31. If we define the *total selection cost* to a population of deleterious recessive genes as the loss of fitness per individual affected (s) multiplied by the frequency of affected individuals (q^2), then selection cost $= sq^2$.

   a. Suppose that a population is at equilibrium between mutation and selection for a deleterious recessive allele, where $s = 0.5$ and $\mu = 10^{-5}$. What is the equilibrium frequency of the allele? What is the selection cost?

   b. Suppose that we start irradiating individual members of the population so that the mutation rate doubles. What is the new equilibrium frequency of the allele? What is the new selection cost?

   c. If we do not change the mutation rate but we lower the selection coefficient to 0.3 instead, what happens to the equilibrium frequency and the selection cost?

32. Balancing selection acts to maintain genetic diversity at a locus since the heterozygous class has a greater fitness than the homozygous classes. Under this form of selection, the allele frequencies in the population approach an equilibrium point somewhere between 0 and 1. Consider a locus with two alleles A and a with frequencies p and q, respectively. The relative genotypic fitnesses are shown below, where s and g are the selective disadvantages of the two homozygous classes.

Genotype	A/A	A/a	a/a
Relative fitness	$1 - s$	1	$1 - g$

   a. At equilibrium, the mean fitness of the A alleles (w_A) will be equal to the mean fitness of the a alleles (w_a) (see Box 18-7). Set the mean fitness of the A alleles (w_A) equal to the mean fitness of the a alleles (w_a). Solve the resulting equation for the frequency of the A allele. This is the expression for the equilibrium frequency of A ($\hat{p}$).

   b. Using the expression that you just derived, find $\hat{p}$ when $s = 0.2$ and $g = 0.8$.

The Inheritance of Complex Traits

19

Former basketball star Wilt Chamberlain (7 feet, 1 inch tall) and former renowned jockey Willie Shoemaker (4 feet, 11 inches tall) show some of the extremes in human height—a quantitative trait. [© *Annie Leibovitz/(Contact Press Images)*.]

KEY QUESTIONS

- For any particular character, how much of the variation in a population is due to genetic factors and how much to environmental factors?

- How can one use a knowledge of parental phenotypes to predict the phenotype of the parents' offspring?

- How many genes contribute to the genetic variation for a trait?

- What are the specific genes that contribute to variation in quantitative traits in populations?

Look at almost any large group of men or women and you'll notice a considerable range in their heights—some are short, some tall, and some about average. Wilt Chamberlain, a star basketball center of the 1960s, was a towering 7 feet, 1 inch tall, whereas Willie Shoemaker, a renowned jockey who won the Kentucky Derby four times, was a mere 4 feet, 11 inches. You might also have noticed that in some families, the parents and their adult children are all on the tall side, whereas in other families, the parents and adult children are all fairly short. Such observations suggest that genes play a role in determining our heights. Still, people do not segregate cleanly into tall and short categories as we saw for Mendel's pea plants. At first inspection, continuous traits, such as height, do not appear to follow Mendel's laws despite the fact that they are heritable.

Traits such as height that show a continuous range of variation and do not behave in a simple Mendelian fashion are known as **quantitative** or **complex traits.** The term *complex trait* is often preferred because variation for such traits is governed by a "complex" of genetic and environmental factors. How tall you are is partly explained by the genes you inherited from your parents and

OUTLINE

19.1 Measuring quantitative variation

19.2 A simple genetic model for quantitative variation

19.3 Broad-sense heritability: nature versus nurture

19.4 Narrow-sense heritability: predicting phenotypes

19.5 Mapping QTL in populations with known pedigrees

19.6 Association mapping in random-mating populations

partly by environmental factors such as how well you were nourished as a child. Teasing apart the genetic and environmental contributions to an individual phenotype is a substantial challenge, but geneticists have a powerful set of tools to meet it.

In the early 1900s, when Mendel's laws were rediscovered, controversy arose about whether these laws were applicable to continuous traits. A group known as the biometricians discovered that there are correlations between relatives for continuous traits such that tall parents tend to have tall children. However, the biometricians saw no evidence that such traits followed Mendel's laws. Some biometricians concluded that Mendelian loci do not control continuous traits. On the other hand, some adherents of Mendelism thought continuous variation was unimportant and could be ignored when studying inheritance. By 1920, this controversy was resolved with the formulation of the **multifactorial hypothesis.** This hypothesis proposed that continuous traits are governed by a combination of multiple Mendelian loci, each with a small effect on the trait, and environmental factors. The multifactorial hypothesis brought quantitative traits into the realm of Mendelian genetics.

Although the multifactorial hypothesis provided a sensible explanation for continuous variation, classic Mendelian analysis is inadequate for the study of complex traits. If progeny cannot be sorted into categories with expected ratios, then the Mendelian approach has little utility for the analysis of complex traits. In response to this problem, geneticists developed a set of mathematical models and statistical methods for the analysis of complex traits. Through the application of these analytical methods, geneticists have made great strides in answering the key questions posed above. The subfield of genetics that develops and applies these methods to understand the inheritance of complex traits is called **quantitative genetics.**

At the heart of the field of quantitative genetics is the goal of defining the **genetic architecture** of complex traits. Genetic architecture is a description of all of the genetic and environmental factors that influence a trait. It includes the number of genes affecting the trait and the relative contribution of each gene. Some genes may have a large effect on the trait, while others have only a small effect. As we will see in this chapter, genetic architecture is the property of a specific population and can vary among populations of a species. For example, the genetic architecture of a trait such as systolic blood pressure in humans differs among different populations. Because different alleles segregate in different populations and because different populations experience different environments, they are apt to have different architectures for many traits.

Understanding the inheritance of complex traits is one of the most important challenges facing geneticists in the twenty-first century. Complex traits are of paramount importance in medical and agricultural genetics. For humans, blood pressure, body weight, susceptibility to depression, serum cholesterol levels, and the risk of developing cancer or other disorders are all complex traits. For crop plants, yield, resistance to pathogens, ability to tolerate drought stress, efficiency of fertilizer uptake, and even flavor are all complex traits. For livestock, milk production in dairy cows, muscle mass in beef cattle, litter size in pigs, and egg production in chickens are all complex traits. Despite the importance of such traits, we know far less about their inheritance than we do about the inheritance of simply inherited traits such as cystic fibrosis or sickle-cell anemia.

In this chapter, we will explore the inheritance of complex traits. We will begin with a review of some basic statistical concepts. Next, we will develop the mathematical model used to connect the action of genes inside the cell with the phenotypes we observe at the level of the whole organism. Using this model, we will then show how quantitative geneticists partition the phenotypic variation in a population into the parts that are due to genetic and environmental factors. We will review the methods used by plant and animal breeders to predict the phenotype of offspring from the phenotype of their parents. Finally, we will see how a combination of the statistical analysis and molecular markers can be used to identify the specific genes that control quantitative traits.

19.1 Measuring Quantitative Variation

To study the inheritance of quantitative traits, we need some basic statistical tools. In this section, we will introduce the mean (or average), which can be used to describe differences between groups, and the variance, which can be used to quantify the amount of variation that exists within a group. We will also discuss the normal distribution, which is central to understanding quantitative variation in populations. But before discussing the statistical tools, let's define the different types of complex trait variation that can occur in a population.

Types of traits and inheritance

A **continuous trait** is one that can take on a potentially infinite number of states over a continuous range. Height in humans is a good example. People can range from about 140 cm to 230 cm in height. If we measured height precisely, then the number of possible heights is infinite. For example, a person might be 170 cm tall or 170.2 or 170.0002 cm. The possibilities are endless. Continuous traits typically have **complex inheritance** involving multiple genes plus environmental factors.

For some traits, the individuals in a population can be sorted into discrete groups or categories. Such traits are known as **categorical traits.** Examples include purple versus white flowers or tall versus short stems for Mendel's pea plants, as seen in Chapter 2. Categorical traits often exhibit **simple inheritance** such that the progeny of crosses segregate into standard Mendelian ratios such as 3 : 1 for a single gene or 15 : 1 for two genes. The inheritance is simple because only one or two genes are involved and the environment has little or no effect on the phenotype.

Some categorical traits do not show simple inheritance. These include many disease conditions in humans. In medical genetics, individuals can be classified into the categories "affected" or "not affected" by a disease. For example, an individual may or may not have type 2 diabetes. However, type 2 diabetes does not follow simple Mendelian rules or produce Mendelian ratios in pedigrees. Rather, there are multiple genetic and environmental factors that place someone at risk of developing this disease. Individuals who have a certain number of risk factors will exceed a threshold and develop the disease. Type 2 diabetes is a form of a categorical trait called a **threshold trait.** Type 2 diabetes has complex inheritance.

Another type of trait is a **meristic trait,** or counting trait, which takes on a range of discrete values. An example would be the number of offspring that an individual has. You can have 1, 2, 3, or more children, but you can not have 2.49 children. Meristic traits are quantitative, but they are restricted to certain discrete values. They do not take on a continuous range of values. Meristic traits usually have complex inheritance.

Quantitative geneticists seek to understand the inheritance of traits that show complex inheritance resulting from a mix of genetic and environmental factors. They may investigate traits that are categorical, meristic, or continuous. The emphasis is on the type of inheritance—complex. For this reason, the term **complex trait** is often preferred to continuous or quantitative trait because it includes all the types of traits with which quantitative genetics is concerned. Any biological phenomenon for which variation exists may show complex inheritance and can be studied as a complex trait. Thus, size and shape of structures, enzyme kinetics, mRNA levels, circadian rhythms, and bird songs can all be treated as complex traits.

The mean

When quantitative geneticists study the inheritance of a trait, they work with a particular group of individuals, or **population.** For example, we might be interested in the inheritance of height for the population of adult men in Shanghai, China. Here, we are using "population" to denote a group that shares certain features in common such as age, sex, ethnicity, or geographic origin. Since there are more than 5 million adult men in Shanghai, determining each of their heights would be a herculean task. Therefore, quantitative geneticists typically study just a subset or **sample** of the full population. The sample should be *randomly* chosen such that each of the 5 million men has an equal chance of being included in the sample. If the sample meets this criterion, then we can use measurements made on the sample to make inferences about the entire population.

Using the example of height for men from Shanghai, we can describe the population using the **mean** or average value for the trait. We select a random sample of 100 men from the population and measure their heights. Some of the men might be 166 cm tall, others 172 cm tall, and so forth. To calculate the mean, we simply sum all the individual measurements and divide the sum by the size of the sample (n), which in this case is 100. For the data in Table 19-1, the result would be 170 cm, or 5 feet, 7 inches. Since we have a random sample, we can infer that the average height in the entire population is 170 cm.

Height is a *random variable* since it can take on different values, and when we select someone at random from the population, the value we observe is governed by an element of chance. Random variables are usually represented by the letter X in statistics. We have measurements for $X_1, X_2, X_3, \ldots X_{100}$ for the $n = 100$ men in the sample. Symbolically, we can express the mean as

$$\bar{X} = \frac{1}{n} \sum_{i=1}^{n} X_i$$

where $\bar{X}$ represents the sample mean. The uppercase Greek letter sigma (Σ) is the summation sign, indicating that we add all n observed values of X for $i = 1, 2$, through n. (Often, the n above Σ and the $i = 1$ below Σ are omitted to simplify the appearance of equations.)

There is a distinction made between the mean of a sample ($\bar{X}$) and the true mean of the population. To learn the true mean for the height of men in Shanghai, we would need to determine the height of each and every man. The true mean is symbolized by the Greek letter μ, so that we have different symbols for the sample and population means.

Here is another way to calculate the mean, which is often quite useful. We can add the products of each class of values of X in the data set times the frequency of that class in the data set. This operation is symbolized as

$$\bar{X} = \sum_{i=1}^{k} f_i X_i$$

where f_i is the frequency of the ith class of observations, X_i is the value of the ith class, and there are a total of k classes. For the data in Table 19-1, one man of the 100 ($f = 0.01$) is 156 cm tall, two men ($f = 0.02$) are 157 cm tall, and so forth, so we can calculate the sample mean as

$$\bar{X} = (0.01 \times 156) + (0.02 \times 157) + \cdots + (0.02 \times 184) = 170$$

Table 19-1   Simulated Data for the Heights of 100 Men from Shanghai, China

Height (cm)	Count	Frequency × Height
156	1	1.56
157	2	3.14
158	1	1.58
159	2	3.18
160	1	1.60
161	1	1.61
162	2	3.24
164	7	11.48
165	7	11.55
166	1	1.66
167	6	10.02
168	9	15.12
169	7	11.83
170	9	15.30
171	5	8.55
172	5	8.60
173	6	10.38
174	5	8.70
175	6	10.50
176	3	5.28
177	4	7.08
178	2	3.56
179	2	3.58
180	2	3.60
181	2	3.62
184	2	3.68
Sum	100	170.00

The mean is useful both for describing populations and comparing differences between populations. For example, men in urban areas of China are on average 170 cm tall, while men in rural areas of China are 166 cm tall. These values were calculated using samples drawn from each region. One question that a quantitative geneticist might ask about the observed difference in height between rural and urban Chinese men is the following: Is the difference due to genetic factors, or is it due to differences in nutrition, health care, or other environmental factors? Later in the chapter, we'll see how quantitative geneticists tease apart genetic versus environmental contributions to a trait.

Lastly, here is another helpful notation from statistics that can be used to define the mean. The mean of a random variable, X, is the *expectation* or *expected value* of that random variable. The expected value is the average of all the values we would observe if we measured X many times. The expectation is symbolized by E and we write $E(X)$ to signify "the expected value of X." Symbolically, we write

$$E(X) = \bar{X}$$

We will use the notation of expectation in several places in this chapter.

The variance

Besides the mean, we also need a measure of how much variation exists in populations. We can create a visual representation of the variation by plotting the count or frequency of each height class. Figure 19-1 shows such a plot for our simulated height data for 100 men from Shanghai. The x-axis shows different height classes, and the y-axis shows the count or frequency of each class. In this figure, the men were binned into 4-cm groups; for example, between 155 and 158 cm. This type of graph is called a **frequency histogram.** If the values are clustered tightly around the mean, then there is less variation, and if the values are spread out along the x-axis, there is greater variation.

We can quantify the amount of variation in a population using a statistical measure called the **variance.** The variance measures the extent to which individuals in the population deviate from the population mean. If all 100 men in our sample had heights very close to the mean, then the variance would be small. If their heights deviated greatly from the mean, the variance would be large.

Since the variance is a measure of **deviation** from the mean, let's define deviation mathematically. Knowing the mean value for the random variable X, we can calculate the deviation of each individual from the mean by subtracting $\bar{X}$ from the individual observations. We will represent the deviations by a lower case x.

$$x = X - \bar{X}$$

Some individuals will have X values above the mean, and they will have a positive deviation. Others will have X values below the mean, and they will have a negative deviation. For the population overall, the expected value of x is 0, or $E(x) = 0$.

To measure the amount of variation for X in the population, we use the variance, which is the mean of the squared deviations. First, we calculate the sum of the squared deviations (or *sum of squares* for short) as

$$\text{sum of squares} = \sum_i (X_i - \bar{X})^2$$

$$= \sum_i (x_i)^2$$

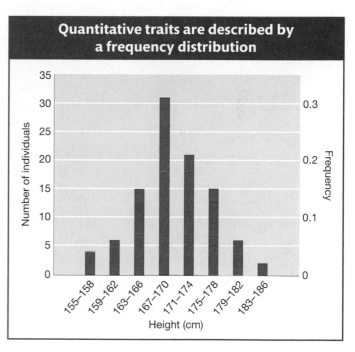

Figure 19-1 Frequency histogram of simulated data for the height of adult men from Shanghai, China.

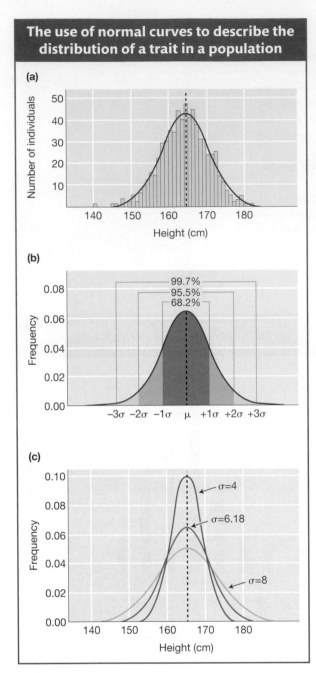

The use of normal curves to describe the distribution of a trait in a population

(a)

(b)

(c)

Since deviations with negative values form positive squares, both negative and positive deviations will contribute positively to the sum of squares. The variance is the mean of the squared deviations (or the sum of squares divided by n). Symbolically, we express the population variance as

$$V_X = \frac{1}{n}\sum_i (X_i - \bar{X})^2$$
$$= \frac{1}{n}\sum_i (x_i)^2$$

where V_X denotes the variance of X. The population variance is sometimes symbolized using the lowercase Greek letter sigma squared (σ^2). In statistics, there is also a distinction made between the population variance (σ^2) and the sample variance (s^2). The latter is calculated by dividing the sums of squares by $n - 1$ rather than n to correct a bias caused by small sample size. For simplicity, we will use the population variance and the formula above throughout this chapter.

There are several points to understand about the variance. First, the variance provides a measure of dispersion about the mean. When the variance is high, the individual values are spread farther apart from the mean; when it is low, then the individual values cluster closer to the mean. Second, the variance is measured in squared units such that if we measure human height in cm, then the variance would be in cm². Third, the variance can range from 0.0 to infinity. Fourth, the variance is equal to the expected value of the squared deviation (x^2) or $E(x^2)$.

The variance of quantitative traits is measured in squared units. These squared units have desirable mathematical properties as we will see below; however, they do not make intuitive sense. If we measure weight in kilograms, then the variance would be in kilograms², which has no clear meaning. Therefore, another statistic used to quantify the extent of deviation from the mean in a population is the **standard deviation** (σ), which is the square root of the variance:

$$\sigma = \sqrt{\sigma^2}$$

The standard deviation is expressed in the same units as the trait itself, so its meaning is more intuitive. We will use the standard deviation in the description of traits below.

Figure 19-2 (a) Frequency histogram of actual data for the height of adult women from the United States. The red line represents the normal curve fit to these data with a mean of 164.4 cm and standard deviation of 6.18 cm. (b) Normal curve for the height of U.S. women showing the predicted percentages of women who will fall within different numbers of standard deviations from the mean. (c) Normal curves with the same mean (164.4 cm) but different standard deviations, showing the effect of the standard deviation on the shape of the curve.

The normal distribution

Even if you have never taken a statistics course, you likely have heard of the **normal distribution,** also known as the "bell curve" in popular culture. The normal distribution is remarkably useful in biology in general and quantitative genetics in particular because the frequency distribution for many biological traits approximates a normal curve. For this reason, geneticists can take advantage of several features of the normal distribution to describe quantitative traits and dissect the underlying genetics.

The normal distribution is a continuous frequency distribution similar to the frequency histogram shown in Figure 19-1. The normal distribution applies to continuous traits. As mentioned above, continuous traits can take on an infinite number of values. A person might be 170 cm tall or 170.2 or 170.002 and so forth. For such traits, the expected frequency of different trait values is better represented by a curve than by a frequency histogram. For the normal distribution, the shape of the curve is determined by two factors—the mean and the standard deviation.

Here is an example using height data for 660 women from the United States collected by the Centers for Disease Control and Prevention. The frequency histogram shows the classic "bell curve" shape with the peak near the mean value of 164.4 cm and the off-mean values distributed symmetrically around the mean (Figure 19-2a). We can fit a normal curve to this distribution using just two pieces of information—the mean and the standard deviation. The shape of the curve is defined by an equation called the normal probability density function, into which the mean and the standard deviation are plugged. The normal distribution allows us to predict the percentage of the observations that will fall within a certain distance from the mean (Figure 19-2b). If we measure distance along the x-axis in standard deviations, then 68 percent of the observations are expected to fall within 1 standard deviation (σ) of the mean and 95.5 percent within 2 standard deviations. For the height data for U.S. women, 71 percent (449 women) fall within 1 standard deviation of the mean and 96 percent (633 women) within 2 standard deviations. These values are very close to the predictions of 68.2 percent and 95.5 percent based on the normal curve.

If we know just the mean and the standard deviation for a trait, we can predict the shape of the distribution of the trait in the population, and we can predict how likely we are to observe certain values when sampling the population. For example, if the mean height for U.S. women is 164.4 cm (5 feet, 5 inches) and the standard deviation is 6.18 cm, we can predict that only 2 percent of women will be more than 177 cm tall, or five feet, 10 inches. As shown in Figure 19-2c, if the standard deviation is greater (for example, 8), then the curve would be flatter and a greater percentage would fall above 177 cm. However, it would still be true that only 2 percent would be more than 2σ above the mean, or 180.4 cm [$(164.4 + (2 \times 8)$].

> **Message** A complex trait is any trait that does not show simple Mendelian inheritance. A complex trait can be either a discontinuous trait such as the presence or absence of a disease condition or a continuously variable trait such as height in humans. The field of quantitative genetics studies the inheritance of complex traits using some basic statistical tools, including the mean, variance, and normal distribution.

Figure 19-3 Basketball star center Yao Ming, who stands 229 cm, or 7 feet, 6 inches, tall, passing a group of power walkers. [*AFP/Getty Images.*]

19.2 A Simple Genetic Model for Quantitative Traits

Genetic and environmental deviations

In this section, we will examine how phenotypes can be decomposed into their genetic and environmental contributions using as an example the height of Yao Ming, the center for the Houston Rockets basketball team. Yao Ming stands out at 229 cm, or 7 feet, 6 inches (Figure 19-3). That's right: Yao Ming is nearly two feet taller than the average man from Shanghai, which happens to be Yao Ming's hometown. As for all of us, Yao Ming's height is the combined result of his genotype and the environment in which he was raised. Let's do an imaginary experiment and see how we can tease apart the genetic and environmental contributions to Yao Ming's exceptional height.

The exceptional height of Yao Ming

First, we will define a simple mathematical model that can be applied to any quantitative trait. The value for an individual for a trait (X) can be expressed in terms of the population mean ($\bar{X}$) and deviations from the mean due to genetic (g) and environmental (e) factors.

$$X = \bar{X} + g + e$$

We are using lowercase g and e for the genetic and environmental deviations, just as we used a lowercase x for the deviation of X from the mean. Thus, in Yao Ming's case, his height can be expressed as the mean value for men from Shanghai (170 cm) plus his specific genetic and environmental deviations ($g + e = 59$ cm). We can simplify the equation above by subtracting $\bar{X}$ from both sides to obtain

$$x = g + e$$

where x represents the individual's phenotypic deviation. For Yao Ming, $x = g + e = 59$.

How can we determine the values of g and e for Yao Ming? One way would be if we had clones of Yao Ming (clones are genetically identical individuals). Let's imagine that we cloned Yao Ming and distributed these clones (as newborns) to a set of randomly chosen households in Shanghai. Twenty-one years later, we locate this army of Yao Ming clones, measure their heights, and determine that their average height is 212 cm. The expectation of e over the many environments

---

Box 19-1	Inbred Lines

An **inbred line** is a specific strain of a plant or animal species that has been self-fertilized or sib-mated for multiple generations such that it becomes homozygous (or inbred) over most of its genome. Self-fertilization can be used in hermaphroditic species such as most plants. In this process, only one seed is used to form each subsequent generation. In maize, for example, a single individual plant is chosen and self-pollinated. Then, in the next generation, a single one of its offspring is chosen and self-pollinated. In the third generation, a single one of its offspring is chosen and self-pollinated, and so forth. Suppose that the original plant is a heterozygote (A/a); then selfing will produce offspring that are $\frac{1}{2}$ heterozygotes and $\frac{1}{2}$ homozygotes ($\frac{1}{4} A/A$ plus $\frac{1}{4} a/a$). Of the ensemble of all heterozygous loci in the genome, then, after one generation of selfing, only 1/2 will still be heterozygous; after two generations, 1/4; after three, 1/8, and so forth. In the nth generation,

$$\text{Het}_n = \frac{1}{2^n} \text{Het}_0$$

where Het_n is the proportion of heterozygous loci in the nth generation and Het_0 is the proportion in the 0 generation.

When selfing is not possible, brother–sister mating will accomplish the same end, although more slowly. The table shows the amount of heterozygosity remaining after n generations of selfing and brother–sister mating.

Remaining heterozygosity		
Generation	Selfing	Brother–sister mating
0	1.000	1.000
1	0.500	0.750
2	0.250	0.625
3	0.125	0.500
4	0.0625	0.406
5	0.03125	0.338
10	0.000977	0.114
20	0.95×10^{-6}	0.014
n	$\text{Het}_n = \frac{1}{2}\text{Het}_{n-1}$	$\text{Het}_n = \frac{1}{2}\text{Het}_{n-1} + \frac{1}{4}\text{Het}_{n-2}$

Inbred lines are enormously important not only in quantitative genetics but in genetics in general. Geneticists have developed many inbred strains for different model organisms, including *Drosophila*, mice, *C. elegans*, yeast, *Arabidopsis*, and maize. If one uses an inbred strain for an experiment, then one knows that individuals receiving different treatments are genetically identical. Therefore, any differences observed between treatments cannot be attributed to genetic differences among the individuals used in the experiment.

Table 19-2 Simulated Data for Days to Pollen Shed for 10 Inbred Lines of Maize Grown in Two Experiments

Experiment I											
Inbred lines	A	B	C	D	E	F	G	H	I	J	Mean
Environment 1	62	64	66	66	68	68	70	70	72	74	68
Environment 2	64	66	68	68	70	70	72	72	74	76	70
Environment 3	66	68	70	70	72	72	74	74	76	78	72
Mean	64	66	68	68	70	70	72	72	74	76	70

Experiment II											
Inbred lines	A	B	C	D	E	F	G	H	I	J	Mean
Environment 4	58	60	62	62	64	64	66	66	68	70	64
Environment 5	64	66	68	68	70	70	72	72	74	76	70
Environment 6	70	72	74	74	76	76	78	78	80	82	76
Mean	64	66	68	68	70	70	72	72	74	76	70

in which the Yao Ming clones were reared is 0.0. In some households, the clones get a positive environment $(+e)$ and in others a negative environment $(-e)$. Overall, $E(e) = 0$. Thus, the mean for the clones minus the population mean equals Yao Ming's genotypic deviation, or $g = (212 - 170) = 42$ cm. The remaining 17 cm of his remarkable 59-cm phenotypic deviation is e for the specific environment in which the real Yao Ming was raised. Plugging these values into the equation above, we obtain

$$229 = 170 + 42 + 17$$

We conclude that Yao Ming's exceptional height is mostly due to exceptional genetics, but he also experienced an environment that boosted his height.

Although our imaginary experiment of cloning Yao Ming is far-fetched, many plant species and some animal species can be clonally propagated with ease. For example, one can use "cuttings" of an individual plant to produce multiple genetically identical individuals. Another way of creating genetically identical individuals is by producing **inbred lines** or **strains** (Box 19-1). All individuals in such strains are genetically identical because they are fully inbred from a common parent or parents. By using clones or inbred lines, geneticists can estimate the genetic and environmental contributions to a trait by rearing the clones in randomly assigned environments. Here is an example.

Table 19-2 (Experiment I) shows simulated data for 10 inbred strains of maize that were grown in three different environments and scored for the number of days between planting and the time that the plants first shed pollen. The overall mean is 70 days. Let's consider line A when grown in environment 1. The mean for all lines in Environment 1 is 68, or 2 less than the overall mean, so e for environment 1 is -2. The mean line A over all three environments is 64, or 6 less than the overall mean, so g for line A is -6. Putting these two values together, we decompose the phenotype of line A when grown in environment 1 as

$$62 = 70 + (-6) + (-2)$$

We could do the same calculations for the other nine inbred lines, and then we'd have a complete description of all the phenotypes in each environment in terms of the extent to which their deviation from the overall mean is due to genetic and environmental factors.

Genetic and environmental variances

We can use the simple model $x = g + e$ to think further about the variance of quantitative traits. Recall that the variance is a way to measure how much

Box 19-2	Genetic and Environmental Variances

To better understand the basic equation $V_X = V_g + V_e$, we need to introduce a new concept from statistics—**covariance.** The covariance provides a measure of association between traits. For two random variables X and Y, their covariance is

$$COV_{X,Y} = \frac{1}{n}\sum_i (X_i - \bar{X})(Y_i - \bar{Y})$$
$$= \frac{1}{n}\sum_i (x_i y_i)$$

where x and y are the deviations of X and Y from their respective means as described in the main text. The term $(X_i - \bar{X})(Y_i - \bar{Y})$ or $(x_i y_i)$ is referred to as the *cross product*. The covariance is obtained by summing all the cross products together and dividing by n. The covariance is the average or expected value, $E(xy)$, of the cross products. The covariance can vary from negative infinity to positive infinity. If large values of X are associated with large values of Y, then the covariance will be positive. If large values of X are associated with small values of Y, then the covariance will be negative. If there is no association between X and Y, then the covariance will be zero. For independent traits, the covariance will be zero.

In the main text, we saw that the variance is the expected value of the squared deviations:

$$V_X = E(x^2)$$

Since the phenotypic deviation (x) is the sum of the genotypic (g) and environmental (e) deviations, we can substitute $(g + e)$ for x and obtain

$$V_X = E[(g + e)^2]$$
$$= E[g^2 + e^2 + 2ge]$$
$$= E(g^2) + E(e^2) + E(2ge)$$

The first term $[E(g^2)]$ is the **genetic variance,** the middle term $[E(e^2)]$ is the **environmental variance,** and the last term is twice the covariance between genotype and environment.

In controlled experiments, different genotypes are placed into different environments at random. In other words, genotype and environment are independent. If genotype and environment are independent, then the covariance between genotype and environment $E(ge) = 0$, and the equation reduces to

$$V_X = E(g^2) + E(e^2)$$
$$= V_g + V_e$$

Thus, the phenotypic variance is the sum of the variance due to the different genotypes in the population and the variance due to the different environments within which the organisms are reared.

individuals deviate from the population mean. Under this model, the trait variance can be partitioned into the genetic and the environmental variances.

$$V_X = V_g + V_e$$

This simple equation tells us that the trait or phenotypic variation (V_X) is the sum of two components—the genetic (V_g) variance and the environmental (V_e) variance. As noted in Box 19-2, there is an important assumption behind this equation; namely, that genotype and environment are not correlated—that is, they are independent. If the best genotypes are placed in the best environments and the worst genotypes in the worst environments, then this equation gives inaccurate results. We will discuss this important assumption later in the chapter.

We can use the data in Table 19-2 (Experiment I) to explore the equation for variances. First, let's use all 30 phenotypic values for the 10 lines in the three environments to calculate the variance. The result is $V_X = 14.67$ days^2. Now, to estimate V_g, we calculate the variance of the means among the 10 inbred lines. The result is $V_g = 12.0$ days^2. Finally, to estimate V_e, we calculate the variance of the means among the three environments. The result is $V_e = 2.67$ days^2. Thus, the phenotypic variance (14.67) is equal to the genetic variance (12.0) plus the environmental variance (2.67). The equation works for these data because genotype and environment are not correlated. All genotypes experience the same range of environments.

If we calculate the standard deviations for the data in Table 19-2 (Experiment I), we'll observe that the phenotypic standard deviation (3.83) is not the sum of the genetic (3.46) and environmental (1.63) standard deviations. Variances can be decomposed into difference sources. Standard deviations cannot be decomposed in this manner. Below, we will see how this property of the variance is helpful for quantifying the extent to which trait variation is heritable versus environmental.

Finally, let's look at what would happen to the variances if genotype and environment are correlated. To do this, imagine that we knew the genetic deviations (g) for nine Thoroughbred horses for the time it takes them to run the Kentucky Derby. We also know the environmental deviations (e) that their trainers contribute to the time it takes each horse to run this race. We will suppose that besides training, there are no other sources of environmental variation. The population mean for this set of Thoroughbreds is 123 seconds to run the Derby. We assign the best horses to the best trainers and the worst horses to the worst trainers. By doing this, we have created a nonrandom relationship or correlation between horses (genotypes) and trainers (environments).

Table 19-3 shows the data for this imaginary experiment. You'll notice that V_X (6.67) is not equal to the sum of V_g (2.22) and V_e (1.33). Because genotype and environment are correlated, we violated the assumption of the equation that states $V_X = V_g + V_e$. The equation only works when genotype and environment are uncorrelated.

Table 19-3 Simulated Data for Time in Seconds (X) that Horses Run the Kentucky Derby Decomposed into the Genetic (g) and Environmental (e) Deviations from the Population Mean

Horse	Population mean	g	Trainer	e	x	X
Secretariat	123	−2	Lucien	−2	−4	119
Decidedly	123	−2	Horatio	−1	−3	120
Barbaro	123	−1	Mike	−1	−2	121
Unbridled	123	−1	Carl	0	−1	122
Ferdinand	123	0	Charlie	0	0	123
Cavalcade	123	1	Bob	0	1	124
Meridian	123	1	Albert	1	2	125
Whiskery	123	2	Fred	1	3	126
Gallant Fox	123	2	Jim	2	4	127
Mean (sec)	123	0		0	0	123
Variance (sec²)		2.22		1.33	6.67	6.67

Correlation between variables

If genotype and environment are correlated, then the $V_X = V_g + V_e$ equation does not apply. Rather, for this equation to be appropriate, genotype and environment must be uncorrelated, or independent. Let's look a little more closely at the concept of **correlation,** the existence of a relationship between two variables. This is a critical concept to quantitative genetics, as we will see throughout this chapter.

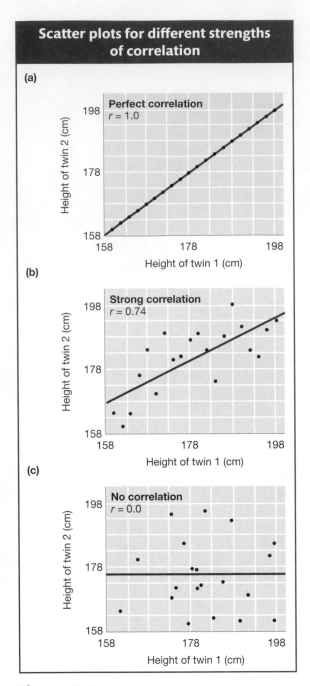

Scatter plots for different strengths of correlation

(a)
Perfect correlation
$r = 1.0$

Height of twin 2 (cm)
198
178
158
158   178   198
Height of twin 1 (cm)

(b)
Strong correlation
$r = 0.74$

Height of twin 2 (cm)
198
178
158
158   178   198
Height of twin 1 (cm)

(c)
No correlation
$r = 0.0$

Height of twin 2 (cm)
198
178
158
158   178   198
Height of twin 1 (cm)

Figure 19-4 Scatter plots for the case of a perfect correlation (a), strong correlation (b), and no correlation (c). Red lines have a slope that is equal to the correlation coefficient.

To visualize the degree of correlation between two variables, we can construct scatter plots, or scatter diagrams. Figure 19-4 shows the scatter plots that we would see under several different strengths of correlation between two variables. These plots use simulated data for the heights of imaginary sets of identical adult male twins. The top panel of the figure shows a perfect correlation, which is what we would see if the height of one twin was exactly the same as that of the other twin for all sets of twins. The middle panel shows a strong but not perfect correlation. Here, when one twin is short, the other also tends to be short, and when one is tall, the other tends to be tall. The bottom panel shows the relationship we would see if the height of one twin was uncorrelated with that of the other twin of the set. Here, the height of one twin of each set is random with respect to the other twin of the set. In the next section, we will see that the data for real twins would look something like the middle panel.

In statistics, there is a specific measure of correlation called the **correlation coefficient,** which is symbolized by a lowercase r. It is a measure of association between two variables. The correlation coefficient is related to the covariance, which was introduced in Box 19-2; however, it is scaled to vary between -1 and $+1$. If we symbolize one random variable by X and the other by Y, then the correlation coefficient between X and Y is

$$r_{X,Y} = \frac{COV_{X,Y}}{\sqrt{V_X V_Y}}$$

The term $\sqrt{V_X V_Y}$ is used to scale the covariance to vary between -1 and $+1$. The expanded equation for the correlation coefficient is

$$r_{X,Y} = \frac{\Sigma(X_i - \bar{X})(Y_i - \bar{Y})}{\sqrt{\Sigma(X_i - \bar{X})^2 \Sigma(Y_i - \bar{Y})^2}}$$

The equation is cumbersome, and in practice, the calculation of correlation coefficients is done with the aid of computers. For two variables that are perfectly correlated, $r = +1.0$ if as one variable gets larger, the other gets larger, or $r = -1.0$ if as one gets larger, the other gets smaller. For completely independent variables, $r = 0.0$.

In Figure 19-4, the correlation coefficient is shown on each panel. It is 1.0 in the top panel for a perfect positive correlation, 0.74 in the middle panel for a strong correlation, and 0.0 in the bottom panel for no correlation (independence of X and Y). The slope of the red line on each panel is equal to the correlation coefficient and provides a visual indicator of the strength of the correlation.

As an exercise, use the data in Table 19-3 to construct a scatter diagram and calculate the correlation coefficient. This would best be done with a computer and spreadsheet software. Use the genetic deviations (g) for the x-axis and the environmental deviations (e) for the y-axis. Then calculate the correlation coefficient between g and e. The scatter diagram will be similar to the one in Figure 19-4 (middle panel), and the correlation coefficient will be 0.90. Thus, when the best horses are placed with the best trainers, genetics and environment are correlated and the $V_X = V_g + V_e$ model cannot be used.

> **Message** An individual's phenotype for a trait can be expressed in terms of its deviation from the population mean. The phenotypic deviation (x) of an individual is composed of two parts—its genetic deviation (g) and its environmental deviation (e). Experiments with clones or inbred lines can be used to decompose an individual's phenotype into its genetic and environmental components.
>
> The phenotypic variation in a population for a trait (V_X) can be decomposed into the genetic (V_g) and the environmental (V_e) variances. This decomposition assumes that the genotypes and environments are uncorrelated.

19.3 Broad-Sense Heritability: Nature Versus Nurture

One of the key questions posed above is, How much of the variation in a population is due to genetic factors and how much to environmental factors? In the popular press, this question is often phrased in terms of *nature* versus *nurture*—that is, what is the influence of innate (genetic) factors compared to external (environmental) factors? For certain controversial topics such as IQ or antisocial behaviors such as alcoholism and criminal behavior, the "nature-versus-nurture" debate is often heated. Some participants in the debate view humans as essentially genetically predetermined for these traits, and others contend that environmental influences can adequately explain the variation in human populations.

Answers to some nature-versus-nurture questions are of practical importance. If high blood pressure is primarily due to lifestyle choices (environment), then changes in diet or exercise habits would be most appropriate. However, if high blood pressure is largely predetermined by our genes, then drug therapy may be recommended.

Quantitative geneticists have developed the statistical tools needed to estimate, with reasonable precision, the extent to which variation in complex traits is due to genes versus the environment. Below, we will describe these tools. At the end of this section, we will discuss the assumptions underlying these estimates and the limits to their utility.

Let's begin by defining **broad-sense heritability** (H^2) as the part of the phenotypic variance that is due to genetic differences among individuals in a population. Mathematically, we write this as the ratio of the genetic variance to the total variance in the population:

$$H^2 = \frac{V_g}{V_X}$$

The H is squared because it is the ratio of two variances, which are measured in squared units. H^2 can vary from 0 to 1.0. When all of the variation in a population is due to environmental sources and there is no genetic variation, then H^2 is 0. When all of the variation in a population is due to genetic sources, then V_g equals V_X and H^2 is 1.0. H^2 is called "broad sense" since it encompasses several different ways by which genes contribute to variation. For example, some of the variation will be due to the contributions of individual genes. Additional genetic variation can be contributed by the way genes work together, the interactions between genes, or epistasis. Later in the chapter, we will explore the different components of the genetic variance.

In Section 19.2, we showed how we can calculate the genetic and environmental variances when we have inbred lines or clones. For the imaginary example of days to pollen shed for maize inbred lines in Table 19-2 (Experiment I), we saw that V_g is 12.0 and V_X is 14.67. Using these values, the heritability of the trait is 12.0/14.67 equals 0.82, or 82 percent. This estimate of H^2 tells that genes contribute most of the variation and environmental factors contribute a more modest share of the variation. Thus, we might conclude that days to pollen shed is a highly heritable trait in maize.

Let's look at the data for Experiment II in Table 19-2. The genotypes are exactly the same as in Experiment I; these are the genotypes of the inbred lines A through J. In this case, however, the lines are reared in more extreme environments. If we calculate the variance for the means of the inbred line in Experiment II, V_g will be 12.0 days^2 as in Experiment I. Since the genotypes are the same in both experiments, the genetic variance is the same. If we calculate the variance for the means of the different environments (V_e) in Experiment II, we will obtain 24.0 days^2, which is much larger than the value for V_e in Experiment I (2.67). Since the environments are more extreme, the environmental variance is larger. Finally, if we calculate H^2 for Experiment II, we obtain

$$H^2 = \frac{V_g}{V_g + V_e} = \frac{12}{12 + 24} = 0.33$$

The estimate of H^2 for Experiment II is on the small side—closer to 0 than to 1. Thus, we might conclude that days to pollen shed is *not* a highly heritable trait in maize.

The contrast between the estimates of the heritability for the same set of maize inbred lines reared in different environments highlights the point that *heritability is the proportion of the phenotypic variance (V_X) due to genetics*. Since $V_X = V_g + V_e$, as V_e increases, then V_g will represent a smaller part of V_X and H^2 will go down. Similarly, if the environmental variance is kept to a minimum, then V_g will represent a larger part of V_X and H^2 will go up. H^2 is a moving target, and results from one study may not apply to another.

Measuring heritability in humans using twin studies

How can we measure heritability in humans? Although we don't have inbred lines for humans, we do have genetically identical individuals—monozygotic or identical twins (Figure 19-5). In most cases, identical twins are raised in the same household and so experience a similar environment. When individuals with the same genotypes are reared in the same environments, we have violated the assumption of our genetic model that genes and environment are independent. So, to estimate heritability in humans, we need to use sets of identical twins who were separated shortly after birth and reared apart by unrelated adoptive parents.

The equation for estimating H^2 in studies of identical twins who are reared apart is relatively simple. It makes use of the statistical measure called the covariance, which was introduced in Box 19-2. As explained in Box 19-3, the covariance between identical twins who are reared apart is equal to the genetic variance (V_g). Thus, we can estimate H^2 in humans by using this covariance as the numerator and the trait variance (V_X) as the denominator:

$$H^2 = \frac{COV_{X', X''}}{V_X}$$

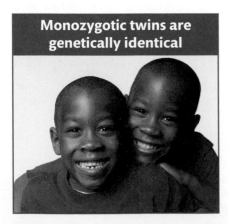

Monozygotic twins are genetically identical

Figure 19-5 A set of identical twins.
[*Barbara Penoyar/Photodisc/Getty Images.*]

Here's how it's done. For each set of twins, let's designate the trait value for one twin as X' and the other as X''. If we have n sets of twins, then the trait values for the n sets could be designated $X_1'\ X_1''$, $X_2'\ X_2''$, ... $X_n'\ X_n''$.

Suppose we had IQ measurements for five sets of twins as follows:

	Twin	
	X'	X''
1	100	110
2	125	118
3	97	90
4	92	104
5	86	89

Using these data and the formula for the covariance from Box 19-2, we calculate that $COV_{x',x''}$ is 119.2 points^2. Using the formula for trait variance, we would calculate that the value of V_X is 154.3 points^2. Thus, we obtain

$$H^2 = \frac{119.2 \text{ points}^2}{154.3 \text{ points}^2} = 0.77$$

The points^2 in the numerator and denominator cancel out, and we are left with a unitless measure that is the proportion of the total variance that is due to genetics.

Box 19-3 provides some additional details about estimating H^2 from twin data, including the derivation of the formula we just used. It also discusses the relationship between the ratio $COV_{x',x''}/V_X$ and the correlation coefficient. Quantitative geneticists have developed several means for estimating heritability using the correlation among relatives. Identical twins share 100 percent of their genes, while brothers, sisters, and dizygotic twins share 50 percent of their genes. The strength of the correlation between different types of relatives can be scaled for the proportion of their genes that they share and the results used to estimate the genetic and environmental contributions to trait variation.

| **Box 19-3** | **Estimating Heritability from Human Twin Studies** |

If we had many sets of identical twins who were reared apart, how could we use them to measure H^2? Let's arbitrarily represent the trait value for one member of each pair of twins as X' and the trait value for the other as X''. We have many (n) sets of twins: $X_1'\ X_1''$, $X_2'\ X_2''$… $X_n'\ X_n''$. We can express the phenotypic deviations for one set of twins as the sum of their genetic and environmental deviations,

$$x' = g + e' \quad \text{and} \quad x'' = g + e''$$

using x' as the deviation for one twin and x'' for the other twin. Notice that g is the same because the twins are genetically identical, but e' and e'' are different because the twins were reared in separate households. Next, we develop an expression for the covariance between the twins. In Box 19-2, we saw that the covariance is the average or expected value of the cross products $E(xy)$. Using our notation for twins, x' and x'', in place of x and y, we get

$$COV_{X',X''} = E(x'x'')$$

We can substitute $(g + e')$ for x' and $(g + e'')$ for x'', giving us

$$COV_{X',X''} = E[(g \times e')(g \times e'')]$$

$$= E(g^2 + ge' + ge'' + e'e'')$$

$$= E(g^2) + E(ge') + E(ge'') + E(e'e'')$$

Let's consider the last three terms of this expression. Under our model, the twins are assigned randomly to households, and thus there should be no correlation between the environments to which the X' and X'' twin of each pair are assigned. Accordingly, the covariance between the environments $[E(e'e'')]$ will be 0.0. Similarly, because the assignment of twins to households is random, we expect no correlation between the genetic deviation of twins (g) and the household to which they are assigned, so $E(ge')$ and $E(ge'')$ will be 0.0. Therefore, the equation for the covariance among twins reduces to

$$COV_{X',X''} = E(g^2) = V_g$$

In other words, the covariance among identical twins reared apart is equal to the genetic variance. If we have a large set of identical twins who were reared apart, we can use the covariance between the twins to estimate the amount of genetic variation for a trait in the general population. If we divide this covariance by the phenotypic variance, then we have an estimate of H^2:

$$H^2 = \frac{COV_{X',X''}}{V_X}$$

This equation is essentially the correlation coefficient between the twins. The variance for the twin of each set designated X' and that for the twin designated X'' are expected to be the same over a large sample. Thus, we can rewrite the denominator of the equation as follows:

$$r_{X',X''} = \frac{COV_{X',X''}}{\sqrt{V_{X'}V_{X''}}} = H^2$$

and we will see that H^2 is equivalent to the correlation between twins.

Table 19-4   Broad-Sense Heritability for Some Traits in Humans as Determined by Twin Studies

Trait	H^2
Physical attributes	
Height	0.88
Chest circumference	0.61
Waist circumference	0.25
Fingerprint ridge count	0.97
Systolic blood pressure	0.64
Heart rate	0.49
Mental attributes	
IQ	0.69
Speed of spatial processing	0.36
Speed of information acquisition	0.20
Speed of information processing	0.56
Personality attributes	
Extraversion	0.54
Conscientiousness	0.49
Neuroticism	0.48
Positive emotionality	0.50
Antisocial behavior in adults	0.41
Psychiatric disorders	
Autism	0.90
Schizophrenia	0.80
Major depression	0.37
Anxiety disorder	0.30
Alcoholism	0.50–0.60
Beliefs and political attitudes	
Religiosity among adults	0.30–0.45
Conservatism among adults	0.45–0.65
Views of school prayer	0.41
Views on pacifism	0.38

Sources: J. R. Alford et al., *American Political Science Review* 99, 2005, 1–15; T. Bouchard et al., *Science* 250, 1990, 223–228; T. Bouchard, *Curr. Dir. Psych. Sci.* 13, 2004, 148-151; P. J. Clark, *Am. J. Hum. Genet.* 7, 1956, 49–54; C. M. Freitag, *Mol. Psychiatry* 12, 2007, 2–22.

Over the last 100 years, there have been extensive genetic studies of twins and other sets of relatives. A great deal has been learned about heritable variation in humans from these studies. Table 19-4 lists some results from twin studies. It may or may not be surprising to you, but there is a genetic contribution to the variance for many different traits, including physique, physiology, personality attributes, psychiatric disorders, and even our social attitudes and political beliefs. We readily observe that traits such as hair and eye color run in families, and we know these traits are the manifestation of genetically controlled biochemical, developmental processes. In this context, it is not so surprising that other aspects of who we are as people also have a genetic influence.

Twin studies and the estimates of heritability that they provide can easily be over- or misinterpreted. Here are a few important points to keep in mind. First, H^2 is a property of a particular population and environment. For this reason, estimates of H^2 can differ widely among different populations and environments. We saw this phenomenon above in the case of the days to pollen shed for maize inbred lines. Second, the twin sets used in many studies were separated at birth and placed into adoptive homes. Adoption agencies do not assign babies randomly to the full range of households in a society; rather, they place babies in economically, socially, and emotionally stable households. As a result, V_e is smaller than in the general population, and the estimate of H^2 will be inflated. Accordingly, the published estimates likely lead us to underestimate the importance of environment and overestimate the importance of genetics. Third, for twins, prenatal effects could cause a positive correlation between genotype and environment. As we saw in the case of Thoroughbreds and jockeys above, such a correlation violates our model and will bias H^2 upward.

Finally, heritability is not useful for interpreting differences between groups. Table 19-4 shows that the heritability for height in humans can be very high: 0.88. However, this high value for heritability does not tell us anything about whether groups with different heights differ because of genetics or the environment. For example, men in the Netherlands today average 184 cm in height, while around 1800, men in the Netherlands were about 168 cm tall on average, a 16-cm difference. The gene pool of the Netherlands has probably not changed appreciably over that time, so genetics cannot explain the huge difference in height between the current population and the one of 200 years ago. Rather, improvements in health and nutrition are the likely cause. Thus, even though height is highly heritable and the past and present Dutch populations differ greatly in height, the difference has an environmental basis.

Message Broad-sense heritability (H^2) is the ratio of the genetic (V_g) to the phenotypic (V_X) variance. H^2 provides a measure of the extent to which differences among individuals within a population are due to genetic versus environmental factors. Estimates of H^2 apply only to the population and environment in which they were made. H^2 is not useful for interpreting differences in trait means among populations.

19.4 Narrow-Sense Heritability: Predicting Phenotypes

Broad-sense heritability tells us the proportion of the variance in a population *within a single generation* that is due to genetic factors. Broad-sense heritability expresses the degree to which the differences in the phenotypes among individuals in a population are determined by differences in their genotypes. However, even when there is genetic variation in a population as measured by broad-sense heritability, it may not be transmissible to the next generation in a predictable way. In this section, we will explore how genetic variation comes in two forms—additive and dominance (nonadditive) variation. Whereas additive variation is predictably transmitted from parent to offspring, dominance variation is not. We will also define another form of heritability called **narrow-sense heritability,** which is the ratio of the additive variance to the phenotypic variance. Narrow-sense heritability provides a measure of the degree to which the genetic constitution of individuals determines the phenotypes of their offspring.

The different modes of **gene action** (interaction among alleles at a locus) are at the heart of understanding narrow-sense heritability, so we will briefly review them. Consider a locus, B, that controls the number of flowers on a plant. The locus has two alleles B_1 and B_2 and three genotypes—B_1/B_1, B_1/B_2, and B_2/B_2. As diagrammed in Figure 19-6a, plants with the B_1/B_1 genotype have 1 flower, B_1/B_2 plants have 2 flowers, and B_2/B_2 plants have 3 flowers. In a case like this, when the heterozygote's trait value is midway between those of the two homozygous classes, gene action is defined as **additive.** In Figure 19-6b, the heterozygote has 3 flowers, the same as the B_2/B_2 homozygote. Here, the B_2 allele is dominant to the B_1 allele. In this case, the gene action is defined as **dominant.** (We could also define this gene action as recessive with the B_1 allele being recessive to the B_2 allele.) Gene action need not be purely additive or dominant but can show **partial dominance.** For example, if B_1/B_2 heterozygotes had 2.5 flowers on average, then we would say that the B_2 allele shows partial dominance.

Gene action and the transmission of genetic variation

Let's work through a simple example to show how the mode of gene action influences heritability. Suppose a plant breeder wants to create an improved plant population with more flowers per plant. Flower number is controlled by the B locus, which has two alleles, B_1 and B_2, as diagrammed in Figure 19-6a. The frequencies of the B_1 and B_2 alleles are both 0.5, and the frequencies of the B_1/B_1, B_1/B_2, and B_2/B_2 genotypes are 0.25, 0.50, and 0.25, respectively. Plants with the B_1/B_1 genotype have 1 flower, B_1/B_2 plants have 2 flowers, and B_2/B_2 plants have 3 flowers. The mean number of flowers per plant in the population is 2.0. (Remember that we can calculate the mean as the sum of the products of frequency of each class times the value for that class.)

Genotype	Frequency	Trait value (no. of flowers)	Contribution to the mean (frequency × value)
B_1/B_1	0.25	1	0.25
B_1/B_2	0.50	2	1.0
B_2/B_2	0.25	3	0.75
			Mean = 2.0

Since the heterozygote has a phenotype that is midway between the two homozygous classes, gene action is additive. There are no environmental effects, and the genotype alone determines the number of flowers, so H^2 is 1.0. If the plant breeder selects 3-flowered plants (B_2/B_2), intermates them, and grows the

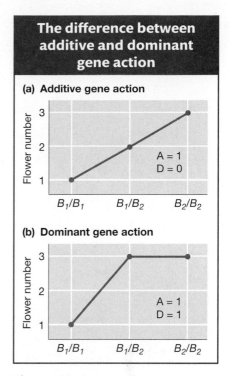

The difference between additive and dominant gene action

(a) Additive gene action

A = 1
D = 0

(b) Dominant gene action

A = 1
D = 1

Figure 19-6 Plot of genotype (*x*-axis) by phenotype (*y*-axis) for a hypothetical locus, *B,* that regulates number of flowers per plant. (a) Additive gene action. (b) Dominant gene action.

offspring, then all the offspring will be B_2B_2 and the mean number of flowers per plant among the offspring will be 3.0. When gene action is completely additive and there are no environmental effects, the phenotype is fully heritable. Selection as practiced by the plant breeder works perfectly.

Now let's consider the case diagrammed in Figure 19-6b, in which the B_2 allele is dominant to the B_1. In this case, the B_1B_2 heterozygote is 3-flowered. The frequency of the B_1 and B_2 alleles are both 0.5, and the frequencies of the B_1/B_1, B_1/B_2, and B_2/B_2 genotypes are 0.25, 0.50, and 0.25, respectively. Again, there is no environmental contribution to the differences among individuals, so H^2 is 1.0. The mean number of flowers per plant in the starting population is 2.5.

Genotype	Frequency	Phenotype	Contribution to the mean (frequency × value)
B_1/B_1	0.25	1	0.25
B_1/B_2	0.50	3	1.5
B_2/B_2	0.25	3	0.75
			Mean = 2.5

If the plant breeder selects a group of 3-flowered plants, 2/3 will be B_1/B_2 and 1/3 B_2/B_2. When the breeder intermates the selected plants, 0.44 (2/3 × 2/3) of the crosses would be between heterozygotes, and 1/4 of the offspring from these crosses would be B_1/B_1 and thus 1-flowered. The remainder of the offspring would be either B_1/B_2 or B_2/B_2 and thus 3-flowered. The overall mean for the offspring would be 2.78, although the mean of their parents was 3.0. When there is dominance, the phenotype is not fully heritable. Selection as practiced by the plant breeder worked but not perfectly because some of the differences among individuals are due to dominance.

In conclusion, when there is dominance, we cannot strictly predict the offspring's phenotypes from the parents' phenotypes. Some of the differences (variation) among the individuals in the parental generation are due to the dominance interactions between alleles. Since parents transmit their genes but not their genotypes to their offspring, these dominance interactions are not transmitted to the offspring.

The additive and dominance effects

In this section, we will show how quantitative geneticists quantify additivity and dominance. Let's again consider the B locus that controls the number of flowers on a plant (see Figure 19-6). The **additive effect (A)** provides a measure of the degree of change in the phenotype that occurs with the substitution of one B_2 allele for one B_1 allele. The additive effect is calculated as the difference between the two homozygous classes divided by 2. For example, as shown in Figure 19-6a, if the trait value of the B_1/B_1 genotype is 1 and the trait value of the B_2/B_2 genotype is 3, then

$$A = \frac{X_{B_2B_2} - X_{B_1B_1}}{2} = \frac{3-1}{2} = 1$$

The **dominance effect** (D) is the deviation of the heterozygote (B_1/B_2) from the midpoint of the two homozygous classes. As shown in Figure 19-6b, if the trait value of the B_1/B_1 genotype is 1, of the B_1/B_2 genotype, 3, and of the B_2/B_2 genotype, 3, then

$$D = X_{B_1B_2} - \left(\frac{X_{B_2B_2} + X_{B_1B_1}}{2}\right) = 3 - 2 = 1$$

If you calculate D for the situation depicted in Figure 19-6a, you'll find D = 0; that is, no dominance.

The ratio of D/A provides a measure of the degree of dominance. For Figure 19-6a, D/A = 0.0, indicating pure additivity or no dominance. For Figure 19-6b, D/A = 1.0, indicating complete dominance. A D/A ratio of −1 would indicate a complete recessive. (The distinction between dominance and recessivity depends on how the phenotypes are coded and is in this sense arbitrary.) Values that are greater than 0 and less than 1 represent partial dominance, and values that are less than 0 and greater than −1 represent partial recessivity.

Here is an example of calculating additive and dominance effects at a single locus. Three-spined sticklebacks (*Gasterosteus aculeatus*) have marine populations with long pelvic spines and populations that live near the bottoms of freshwater lakes with highly reduced pelvic spines (Figure 19-7a). The spines are thought to play a role in defense against predation. The bottom-dwelling freshwater populations are derived from the ancestral marine populations. A change in predation between the marine and freshwater environments may explain the loss of spines in the freshwater environments (see Chapter 20).

Pitx1 is one of several genes that contributes to pelvic-spine length in sticklebacks. This gene encodes a transcription factor that regulates the development of the pelvis in vertebrates, including the growth of pelvic spines in sticklebacks. Michael Shapiro and his colleagues at Stanford University measured the pelvic-spine length in an F_2 population that segregated for the marine or long (*l*) allele and freshwater or short (*s*) allele of *Pitx1*. They recorded the following

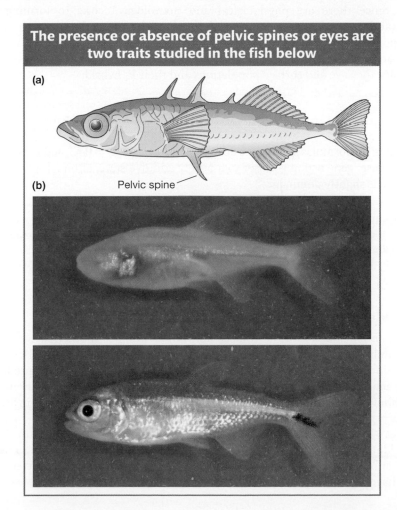

Figure 19-7 (a) Three-spined stickleback (*Gasterosteus aculeatus*). (b) Blind cave fish (*Astyanax mexicanus*) (*top*) and its sighted, surface relative (*bottom*). [*Masato Yoshizawa and William Jeffery, University of Maryland.*]

mean values (in units of proportion of body length) for pelvic-spine length for the three genotypic classes:

s/s	s/l	l/l
0.068	0.132	0.148

Using these values and the formulas above, we can calculate the additive and dominance effects. The additive effect (A) is

$$(0.148 - 0.068)/2 = 0.04$$

or 4 percent of body length. The dominance effect (D) is

$$0.132 - [(0.148 + 0.068)/2] = 0.024$$

The dominance/additivity ratio is

$$0.024/0.04 = 0.6$$

The 0.6 value for the ratio indicates that the long (l) allele of $Pitx1$ is partially dominant to the short (s) allele.

One can also calculate additive and dominance effects averaged over all the genes in the genome that affect the trait. Here is an example using cave fish (*Astyanax mexicanus*) and their surface relatives (Figure 19-7b). The cave populations have highly reduced (small-diameter) eyes compared to the surface populations. Populations colonizing lightless caves do not benefit from having eyes. Since there are physiological and neurological costs to forming and maintaining eyes, evolution may have favored a reduction in the size of the eye in cave populations.

Horst Wilkins at the University of Hamburg measured mean eye diameter (in mm) for the cave and surface populations and their F_1 hybrid:

Cave	F_1	Surface
2.10	5.09	7.05

Using the formulas above, we calculate that A = 2.48, D = 0.52, and D/A = 0.21. In this case, gene action is closer to a purely additive state, although the surface genome is slightly dominant.

Message When the trait value for the heterozygous class is midway between the two homozygous classes, gene action is called additive. Any deviation of the heterozygote from the midpoint between the two homozygous classes indicates a degree of dominance of one allele. The additive (A) and dominance (D) effects and their ratio (D/A) provide metrics for quantifying the mode of gene action.

A model with additivity and dominance

The example above with the B locus and flower number shows that we cannot accurately predict offspring phenotypes from parental phenotypes when there is dominance, although we can do so in cases of pure additivity. When predicting the phenotypes of offspring, we need to separate the additive and dominance contributions. To do this, we need to modify the simple model introduced in Section 19.2, $x = g + e$.

Let's begin by looking more closely at the situation depicted in Figure 19-6b. Individuals with the B_1/B_2 and B_2/B_2 genotypes have the same phenotype, 3 flowers. If we subtract the population mean (2.5) from their trait value (3), we see that they have the same genotypic deviation (g):

$$g_{B_1B_2} = g_{B_2B_2} = 0.5$$

Now let's calculate the mean phenotypes of their offspring. If we self-pollinate a B_1/B_2 individual, the offspring will be $\frac{1}{4}$ B_1/B_1, $\frac{1}{2}$ B_1/B_2, and $\frac{1}{4}$ B_2/B_2, and the mean trait value of these offspring would be 2.75. However, if we self-pollinate a B_2/B_2 individual, the offspring will all be B_2/B_2, and the mean trait value of these offspring would be 3.0. Even though the B_1/B_2 and B_2/B_2 individuals have the same trait value and the same value for their genotypic deviation (g), they do not produce the equivalent offspring because the underlying basis of their phenotypes is different. The phenotype of the B_1/B_2 individual depends on the dominance effect (D), while that of the B_2/B_2 individual does not involve dominance.

We can expand the simple model ($x = g + e$) to incorporate the additive and dominance contributions. The genotypic deviation (g) is the sum of two components—a the additive deviation, which is transmitted to offspring, and d the dominance deviation, which is not transmitted to offspring. We can rewrite the simple model and separate out these two components as follows:

$$x = g + e$$

$$x = a + d + e$$

The additive deviation is transmitted from parent to offspring in a predictable way. The dominance deviation is not transmitted from parent to offspring since new genotypes and thus new interactions between alleles are created each generation.

Let's look at how the genetic deviation is decomposed into the additive and dominance deviations for the case shown in Figure 19-6b.

	B_1B_1	B_1B_2	B_2B_2
Trait value	1	3	3
Genetic deviation (g)	−1.5	0.5	0.5
Additive deviation (a)	−1	0	1
Dominance deviation (d)	−0.5	0.5	−0.5

The genotypic deviations (g) are simply calculated by subtracting the population mean (2.5) from the trait value for each genotype. Each genotypic deviation is then decomposed into the additive (a) and dominance (d) deviations using formulas that are beyond the scope of this book. These formulas include the additive (A) and dominance (D) effects as well as the frequencies of the B_1 and B_2 allele in the population. You'll notice that $a + d$ sum to g. The additive (a) and dominance (d) deviations are dependent on the allele frequencies because the phenotype of an offspring receiving a B_1 allele from one parent will depend on whether that allele combines with a B_1 or B_2 allele from the other parent, and that outcome depends on the frequencies of the alleles in the population.

The additive deviation (a) has an important meaning in plant and animal breeding. It is the **breeding value,** or the part of an individual's deviation from the population mean that is due to additive effects. This is the part that is transmitted to its progeny. Thus, if we wanted to increase the number of flowers per plant in the population, the B_2B_2 individuals have the highest breeding value. Breeding values can also be calculated for the genome overall for an individual. Animal breeders estimate the genomic breeding values of individual animals, and these estimates can determine the economic value of the animal.

We have partitioned the genetic deviation (g) into the additive (a) and dominance (d) deviations. Using algebra similar to that described in Box 19-2, we can also partition the genetic variance into the additive and dominance variances as follows:

$$V_g = V_a + V_d$$

where V_a is the **additive variance** and V_d is the **dominance variance.** V_a is the variance of the additive deviations or the variance of the breeding values. It is the part of the genetic variation that is transmitted from parents to their offspring. V_d is the variance of the dominance deviations. Finally, we can substitute these terms in the equation for the phenotypic variance presented earlier in the chapter:

$$V_X = V_g + V_e$$

$$V_X = V_a + V_d + V_e$$

where V_e is the environmental variance. This equation assumes that the additive and dominance components are not correlated with the environmental effects. This assumption will be true in experiments in which individuals are randomly assigned to environments.

Thus far, we have described models with genetic, environmental, additive, and dominance deviations and variances. In quantitative genetics, the models can get even more complex. In particular, the models can be expanded to include interaction between factors. If one factor alters the effect of another factor, then there is an interaction. For example, if one gene alters the effect of another gene, then there is an interaction. Box 19-4 briefly reviews how interactions are factored into quantitative genetic models.

> **Message** The genetic deviation (g) of an individual from the population mean is composed of two parts—its additive deviation (a) and its dominance deviation (d). The additive deviation is known as the breeding value, and it represents the component of an individual's phenotype that is transmitted to its offspring.
>
> The genetic variation for a trait in a population (V_g) can be decomposed into the additive (V_a) and the dominance (V_e) variances. The additive variance is the fraction of the genetic variation that is transmitted from parent to offspring.

Narrow-sense heritability

We can now define **narrow-sense heritability,** which is symbolized by a lowercase h squared (h^2), as the ratio of the additive variance to the total phenotypic variance:

$$h^2 = \frac{V_a}{V_X} = \frac{V_a}{V_a + V_d + V_e}$$

This form of heritability measures the extent to which variation among individuals in a population is predictably transmitted to their offspring.

To estimate h^2, we need to measure V_a, but how can this be accomplished? Using algebra and logic similar to that we used to show that V_g can be estimated using the covariance between monozygotic twins reared separately (see Box 19-3), it can also be shown that the covariance between a parent and its offspring is equal to one-half the additive variance:

$$COV_{P,O} = \tfrac{1}{2}V_a$$

The parent–offspring covariance is one-half of V_a because the offspring inherits only one-half of its genes from the parent. Combining this formula with the one for h^2, we get

$$h^2 = \frac{V_a}{V_X} = \frac{2COV_{P,O}}{V_X}$$

Box 19-4	Interaction Effects

The simple model for decomposing traits into genetic and environmental deviations, $x = g + e$, assumes that there is no genotype–environment interaction. By this statement, we mean that the differences between genotypes do not change across environments. In other words, a *genotype–environment interaction* occurs when the performance of different genotypes is unequally affected by a change in the environment. Here's an example. Consider two inbred lines, IL1 and IL2, that have different genotypes. We rear both of these inbred lines in two environments, E1 or E2. We can visualize the performance of these two lines in the two environments using a graph (below). This type of graph, which shows the pattern of trait values of different genotypes across two or more environments is called a *reaction norm*.

If there is no interaction, then the difference in trait value between the inbred lines will be the same in both environments, as shown by the graph on the left.

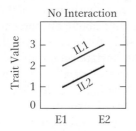
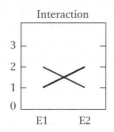

With no interaction, the difference between the two inbreds is 1.0 in both environments, and so the difference between the lines averaged over the two environments is 1.0.

$$\text{Environment 1: IL1} - \text{IL2} = 2 - 1 = 1.0$$

$$\text{Environment 2: IL1} - \text{IL2} = 3 - 2 = 1.0$$

The difference in the overall mean shows that the lines are genetically different. The mean over both environments is 2.5 for IL1 and 1.5 for IL2.

The graph on the right shows a case of an interaction between genotype and environment. IL1 does well in Environment 1 but poorly in Environment 2. The opposite is true for IL2. The difference in the trait value between the two lines is +1.0 in Environment 1 but −1.0 in Environment 2.

$$\text{Environment 1: IL1} - \text{IL2} = 2 - 1 = +1.0$$

$$\text{Environment 2: IL1} - \text{IL2} = 1 - 2 = -1.0$$

The difference between the lines averaged over the two environments is 0.0, so we might incorrectly conclude that these inbreds are genetically equivalent if we looked just at the overall mean.

The simple model can be expanded to include a genotype–environment interaction term ($g{\times}e$):

$$x = g + e + g{\times}e$$

and

$$V_X = V_g + V_e + V_{g{\times}e}$$

where $V_{g{\times}e}$ is the variance of the genotype–environment interaction. If the interaction term is not included in the model, then there is an implicit assumption that there are no genotype–environment interactions.

Interactions can also occur between the alleles at separate genes. This type of interaction is called epistasis. Let's look at how epistatic interactions affect variation in quantitative traits.

Consider two genes, A with alleles A_1 and A_2 and B with allele B_1 and B_2. The left side of the table below shows the case of no interaction between these genes. Starting with the A_1/A_1; B_1/B_1 genotype, whenever you substitute an A_2 allele for an A_1 allele, the trait value goes up by 1 regardless of the genotype at the B locus. The same is true when substituting alleles at the B locus. The effects of alleles at the A locus are independent of those at the B locus and vice versa. There is no interaction or epistasis.

	No interaction				Interaction		
	B_1/B_1	B_1/B_2	B_2/B_2		B_1/B_1	B_1/B_2	B_2/B_2
A_1/A_1	0	1	2	A_1/A_1	0	1	2
A_1/A_2	1	2	3	A_1/A_2	0	1	3
A_2/A_2	2	3	4	A_2/A_2	0	1	4

Now look at the right side of the table. Starting with the A_1/A_1; B_1/B_1 genotype, substituting an A_2 allele for an A_1 allele only has an effect on the trait value when the genotype at the B locus is B_2/B_2. The effects of alleles at the A locus are *dependent* of those at the B locus. There is an interaction or epistasis between the genes.

The genetic model can be expanded to include an epistatic or interaction term (i):

$$x = a + d + i + e$$

and

$$V_X = V_a + V_d + V_i + V_e$$

where V_i is the interaction or epistatic variance.

If the interaction term is not included in the model, then there is an implicit assumption that the genes work independently; that is, there is no epistasis. The interaction variance (V_i), like the dominance variance, is not transmitted from parents to their offspring since new genotypes and thus new epistatic relationships are formed with each generation.

To estimate V_a using the covariance between parents and offspring requires controlling environmental factors in experiments. This can be a challenge because parents and offspring are necessarily reared at different times. V_a can also be estimated using the covariance between half-sibs, in which case all individuals in the experiment can be reared at the same time in the same environment. Half-sibs share one-fourth of their genes, so V_a equals $4 \times$ the covariance between half-sibs.

If you compare the equation for h^2 to the one for H^2 (see Box 19-3), you will see that both involve the ratio of a covariance to a variance. The correlation coefficient introduced earlier in the chapter is also the ratio of a covariance to a variance. We are using the degree of correlation among relatives to infer the extent to which traits are heritable.

Here is an exercise that your class can try. Have each student submit his or her height and the height of their same-sex parent. Using these data and spreadsheet computer software, calculate the covariance between parents and their offspring (the students). Then estimate h^2 as two times the covariance divided by the phenotypic variance. For the total phenotypic variance (V_X) in the denominator of the equation, you can use the variance among the parents. Data for male and female students should be analyzed separately.

Typically, values for narrow-sense heritability of height in humans are about 0.8, meaning that about 80 percent of the variance is additive, or transmissible from parent to offspring. The results for your class could deviate from this value for several reasons. First, if your class is small, sampling error can affect the accuracy of your estimate of h^2. Second, you will not be conducting a randomized experiment. If parents re-create in their households the growth-promoting (or growth-limiting) environments that they experienced as children, then there will be a correlation between the environments of the parents and their offspring. This correlation of environments violates an assumption of the analysis. Third, the population of students in your class may not be representative of the population in which the 0.8 value was obtained.

Figure 19-8 is a scatter plot with the height data for male and female students and their parents. There is a clear correlation between the heights of the students and their same-sex parent. These data give estimates of narrow-sense heritability of 0.86 for mother–daughter and 0.82 for father–son. The results are close to the value of h^2 equals 0.8 obtained from studies in which the children were separated at birth from their parents and reared in adoptive households.

Here are a few more points about narrow-sense heritability. First, when $h^2 = 1.0$ ($V_a = V_X$), offspring's phenotypes will exactly equal the mid–parent value. All the variation in the population is additive and heritable in the narrow sense. Second, when $h^2 = 0.0$ ($V_a = 0$), the expected value of any offspring's phenotype will be the population mean. All the variation in the population is due either to dominance or to environmental factors, and thus it is not transmissible to offspring. Finally, as with broad-sense heritability (H^2), narrow-sense heritability is the property of the specific environment and population in which it was measured. An estimate from one population and environment may not be meaningful for another population or environment.

Narrow-sense heritability is an important concept both in plant and animal breeding and in evolution. For a breeder, h^2 indicates which traits can be improved

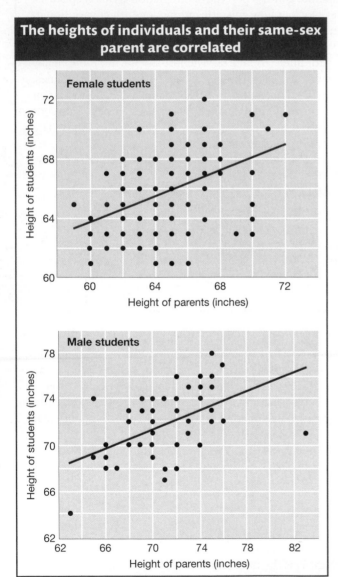

The heights of individuals and their same-sex parent are correlated

Figure 19-8 Scatter diagrams for height in inches of female (*top*) and male (*bottom*) students and their same-sex parent. The plots show positive correlations between the heights of the students and their parents. The slope of the diagonal line is equal to the correlation coefficient.

by artificial selection. For an evolutionary biologist, h^2 is critical to understanding how populations will change in response to natural selection imposed by a changing environment. Table 19-5 lists estimates of narrow-sense heritability for some traits and organisms.

Table 19-5   Narrow-Sense Heritability for Some Traits in Several Different Species

Trait	h^2 (%)
Agronomic species	
Body weight in cattle	65
Milk yield in cattle	35
Back-fat thickness in pig	70
Litter size in pig	5
Body weight in chicken	55
Egg weight in chicken	50
Natural species	
Bill length in Darwin's finch	65
Flight duration in milkweed bug	20
Plant height in jewelweed	8
Fecundity in red deer	46
Life span in collared flycatchers	15

Source: D. F. Falconer and T. F. C. Mackay, *Introduction to Quantitative Genetics,* Longman, 1996; J. C. Conner and D. L. Hartl, *A Primer in Ecological Genetics,* Sinauer, 2004.

Predicting offspring phenotypes

In order to efficiently improve crops and livestock for traits of agronomic importance, the breeder must be able to predict an offspring's phenotype from its parents' phenotypes. Such predictions are made using the breeder's knowledge of narrow-sense heritability. An individual's phenotypic deviation (x) from the population mean is the sum of the additive, dominance, and environmental deviations:

$$x = a + d + e$$

The additive part is the heritable part that is transmitted to the offspring. Let's look at a set of parents with phenotypic deviations x' for the mother and x'' for the father:

$$\text{Mom} \qquad\qquad \text{Dad}$$
$$x' = a' + d' + e' \qquad\qquad x'' = a'' + d'' + e''$$

$$\text{Offspring}$$

$$x_{off} = \frac{a' + a''}{2} = \bar{a}_{par}$$

The parents' dominance deviations (d' and d'') are not transmitted to their offspring since new genotypes and new dominance interactions are created with each generation. Similarly, the parents do not transmit their environmental deviations (e' and e'') to their offspring. Thus, the only factors that parents transmit to their offspring are their additive deviations (a' and a''). Accordingly, we can estimate the offspring's phenotypic deviation (x_{off}) as the mean of the additive deviations of its parents.

So to predict the offspring's phenotype, we need to know its parents' additive deviations. We cannot directly observe the parents' additive deviations, but we can estimate them. The additive deviation of an individual is the heritable part of its phenotypic deviation; that is,

$$\hat{a} = h^2 x$$

where $\hat{a}$ signifies an estimate of the additive deviation or breeding value. Thus, we can estimate the mean of the parents' additive deviations as the product of the h^2 times the mean of their phenotypic deviation and this product will be the phenotypic deviation of the offspring ($\hat{x}_o$):

$$\hat{x}_o = h^2 \left(\frac{x' + x''}{2} \right)$$

or

$$\hat{x}_o = h^2 \bar{x}_p$$

The offspring will have its own dominance and environmental deviations. However, these cannot be predicted. Since they are deviations, they will be zero on average over a large number of offspring.

Here is an example. Icelandic sheep are prized for the quality of their fleece. The average adult sheep in a particular population produces 6 lb of fleece per year. A sire that produces 6.5 lb per year is mated with a dam that produces 7.0 lb per year. The narrow-sense heritability of fleece production in this population is 0.4. What is the predicted fleece production for offspring of this mating? First, calculate the phenotypic deviations for the parents by subtracting the population mean from their phenotypic values:

Sire	$6.5 - 6.0 = 0.5$
Dam	$7.0 - 6.0 = 1.0$
Parent mean ($\bar{x}_p$)	$(0.5 + 1.0)/2 = 0.75$

Now multiply h^2 times $\bar{x}_p$ to determine $\hat{x}_o$, the estimated phenotypic deviation of the offspring:

$$0.4 \times 0.75 = 0.3$$

Finally, add the population mean (6.0) to the predicted phenotypic deviation of the offspring (0.3) and obtain the result that the predicted phenotype of the offspring is 6.3 lb of fleece per year.

It may seem surprising that the offspring are predicted to produce less fleece than either parent. However, this outcome is expected for a trait with a modest heritability of 0.4. Most (60 percent) of the superior performance of the parents is due to dominance and environmental factors that are not transmitted to the offspring. If the heritability was 1.0, then the predicted value for the offspring would be midway between the parents'. If the heritability was 0.0, then the predicted value for the offspring would be at the population mean since all the variation would be due to nonheritable factors.

Selection on complex traits

Our final topic regarding narrow-sense heritability is the application of selection over the long term to improve the performance of a population for a complex trait. By applying selection, plant breeders over the past 10,000 years transformed a host of wild plant species into the remarkable array of fruit, vegetable, cereal, and spice crops that we enjoy today. Similarly, animal breeders applied selection to domesticate many wild species, transforming wolves into dogs, jungle fowl into chickens, and wild boar into pigs.

Selection is a process by which only individuals with certain features contribute to the gene pool that forms the next generation (see Chapters 18 and 20). Selection applied by humans to improve a crop or livestock population is termed *artificial selection* to distinguish it from natural selection. Let's look at an example of how artificial selection works.

Provitamin A is a precursor in the biosynthesis of vitamin A, an important nutrient for healthy eyes and a well-functioning immune system. Plant products are an important source of provitamin A for humans; however, people in many areas of the globe have too little provitamin A in their diets. To solve this problem, a plant breeder seeks to increase the provitamin A content of a maize population used in parts of Latin America where vitamin A deficiency is common. At present, this population produces 1.25 µg of provitamin A per gram of kernels. The variance for the population is 0.06 µg² (Figure 19-9). To improve the population, the breeder selects a group of plants that produce 1.5 µg or more of provitamin A per gram of kernels. The mean for the selected group is 1.63 µg. The breeder randomly intermates the selected plants and grows the offspring to produce the next generation, which has a mean of 1.44 µg per gram of kernels.

If the narrow-sense heritability of a trait is not known before performing an artificial selection experiment, one can use the results of such experiments to estimate it. Here's an example using the case of provitamin A in maize. Let's start with the equation from above:

$$\hat{x}_o = h^2 \overline{x}_p$$

and rewrite it as

$$h^2 = \frac{\overline{x}_o}{\overline{x}_p}$$

$\overline{x}_p$ is the mean deviation of the parents (the selected plants) from the population mean. This is known as the **selection differential (S),** the difference between the mean of the selected group and that of the base population. For our example,

$$\overline{x}_p = 1.63 - 1.25 = 0.38$$

$\overline{x}_o$ is the mean deviation of the offspring from the population mean. This is known as the **selection response (R),** the difference between the mean of the offspring and that of the base population. For our example,

$$\overline{x}_o = 1.44 - 1.25 = 0.19$$

Now we can calculate the narrow-sense heritability for this trait in this population as

$$h^2 = \frac{R}{S} = \frac{\overline{x}_o}{\overline{x}_p} = \frac{0.19}{0.38} = 0.5$$

The underlying logic of this calculation is that the response represents the heritable or additive part of the selection differential.

Over the last century, quantitative geneticists have conducted a large number of selection experiments like this. Typically, these experiments are performed over many generations and are referred to as long-term selection studies. Each

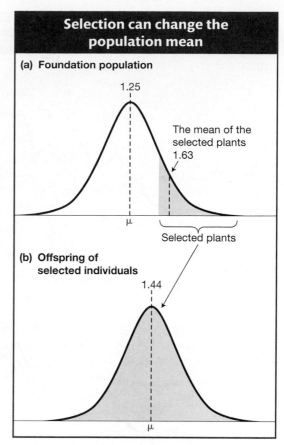

Selection can change the population mean

(a) Foundation population

The mean of the selected plants 1.63

Selected plants

(b) Offspring of selected individuals

Figure 19-9 Distribution of trait values for provitamin A in maize kernels in a starting population (a) and offspring population (b) after one generation of selection. The starting population had a mean of 1.25 µg/g, the selected individuals a mean of 1.63 µg/g, and the offspring population a mean of 1.44 µg/g.

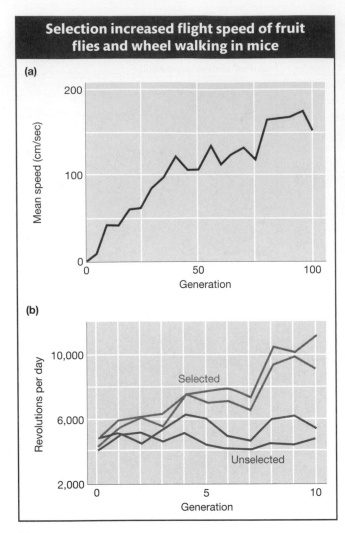

Selection increased flight speed of fruit flies and wheel walking in mice

(a)

(b)

Figure 19-10 Results of long-term selection experiments. (a) Selection for an increase in flight speed of fruit flies. Speed was tested in a wind tunnel in which flies flew against the wind to reach a light source. (b) Selection for an increase in the amount of voluntary wheel walking done by mice. [(a) From K. E. Weber, Genetics 144, 1996, 205–213, (b) J. G. Swallow et al., Behav. Genet. 28, 1998, 227–237.]

generation, the best individuals are selected to produce the subsequent generation. Such studies have been performed in economically important species such as crop plants and livestock and in many model organisms such as *Drosophila*, mice, and nematodes. This work has shown that virtually any species will respond to selection for virtually any trait. Populations contain deep pools of additive genetic variation.

Here are two examples of long-term selection experiments. In the first experiment, fruit flies were selected for increased flight speed over a period of 100 generations (Figure 19-10a). Each generation, the speediest flies were selected and bred to form the next generation. Over the 100 generations, the average flight speed of the flies in the population increased from 2 to 170 cm/sec, and neither the flies nor the gains made by selection showed any signs of slowing down after 100 generations. In the second experiment, mice were selected over 10 generations for the amount of "wheel running" they did per day (Figure 19-10b). There was a 75 percent increase over just 10 generations. These studies and many more like them demonstrate the tremendous power of artificial selection and deep pools of additive genetic variation in species.

Message Narrow-sense heritability (h^2) is the proportion of the phenotypic variance that is attributable to additive effects. This form of heritability measures the extent to which variation among individuals in a population is predictably transmitted to their offspring. The value of h^2 can be estimated in two ways: (1) using the correlation between parents and offspring and (2) the ratio of the selection response to the selection differential. The value of h^2 is an important quantity in plant and animal breeding since it provides a measure of how well a trait will respond to selective breeding.

19.5 Mapping QTL in Populations with Known Pedigrees

The genes that control variation in quantitative (or complex traits) are known as **quantitative trait loci,** or **QTL** for short. As we will see below, QTL are genes just like any others that you have learned about in this book. They may encode metabolic enzymes, cell-surface proteins, DNA-repair enzymes, transcription factors, or any of many other classes of genes. What is of interest here is that QTL have allelic variants that typically make relatively small, quantitative contributions to the phenotype.

We can visualize the contributions of the alleles at a QTL to the trait value by looking at the frequency distributions associated with each genotype at a QTL as shown in Figure 19-11. The QTL locus is *B* and the genotypic classes are *B/B*, *B/b*, and *b/b*. The *B/B* individuals tend to have higher trait values, *B/b* intermediate values, and *b/b* small values. However, their distributions overlap, and we cannot determine genotype simply by looking at an individual's phenotype as we can for genes that segregate in Mendelian ratios. In Figure 19-11, an individual with an intermediate trait value could be either *B/B*, *B/b*, or *b/b*.

Because of this property of QTL, we need special tools to determine their location in the genome and characterize their effects on trait variation. In this section, we will review a powerful form of analysis for accomplishing the first of these goals. This form of analysis is called **QTL mapping.** Over the past two decades, QTL mapping has revolutionized our understanding of the inheritance of quantitative traits. Pioneering work in QTL mapping was performed with crop plants such as tomato and corn. However, it has been broadly applied in model organisms such as mouse, *Drosophila*, and *Arabidopsis*. More recently, evolutionary biologists have employed QTL mapping to investigate the inheritance of quantitative traits in natural populations.

The fundamental idea behind QTL mapping is that one can identify the location of QTL in the genome using marker loci linked to a QTL. Here is how the method works. Suppose you make a cross between two inbred strains—parent one (P_1) with a high trait value and parent two (P_2) with a low trait value. The F_1 can be backcrossed to P_1 to create a BC_1 population in which the alleles at all the genes in the two parental genomes will segregate. Marker loci such as SNPs or microsatellites can be scored unambiguously as homozygous P_1 or heterozygous for each BC_1 individual. If there is a QTL linked to the marker locus, then the mean trait value for individuals that are homozygous P_1 at the marker locus will be different from the mean trait value for the heterozygous individuals. Based on such evidence, one can infer that a QTL is located near the marker locus. Let's look in more detail at how this works.

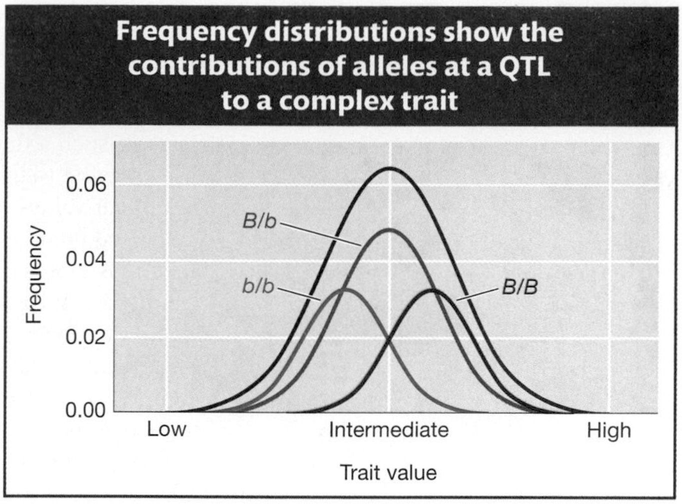

The basic method

There are a variety of experimental designs that can be used in QTL mapping experiments. We will begin by describing a simple design. Let's say we have two inbred lines of tomato that differ in fruit weight—Beefmaster with fruits of 230 grams in weight and Sungold with fruits of 10 grams in weight (Figure 19-12). We cross the two lines to produce an F_1 hybrid and then backcross the F_1 to the Beefmaster line to produce a BC_1 generation. We grow several hundred BC_1 plants to maturity and measure the weight of the fruit on each. We also extract DNA from each of the BC_1 plants. We use these DNA samples to determine the genotype of

Frequency distributions show the contributions of alleles at a QTL to a complex trait

Figure 19-11 Frequency distributions showing how the distributions for the different genotypic classes at QTL locus *B* relate to the overall distribution for the population (black line).

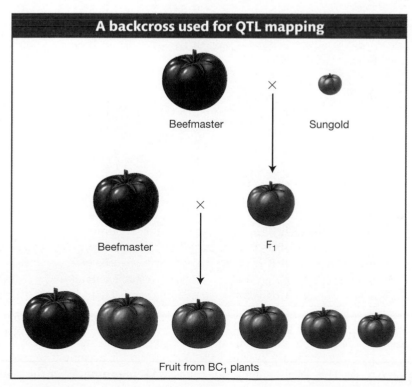

A backcross used for QTL mapping

Fruit from BC_1 plants

Figure 19-12 Breeding scheme for a backcross population between Beefmaster and Sungold tomatoes. In the BC_1 generation, there is a continuous range of fruit sizes.

each plant at a set of marker loci (SNPs or SSRs) that are distributed across all of the chromosomes such that we have a marker locus every 5 to 10 centimorgans.

From this process, we would assemble a data set for several hundred plants and 100 or more marker loci distributed around the genome. Table 19-6 shows part of such a data set for just 20 plants and 5 marker loci. For each BC_1 plant, we have the weight of its fruit and the genotypes at the marker loci. You'll notice that trait values for the BC_1 plants are intermediate between the two parents as expected but closer to the Beefmaster value because this is a BC_1 population and Beefmaster was the backcross parent. Also, since this is a backcross population, the genotypes at each marker locus are either homozygous for the Beefmaster allele (B/B) or heterozygous (B/S). In Table 19-6, you can see the positions of crossovers between the marker loci that occurred during meiosis in the F_1 parent of the BC_1 generation. For example, plant BC_1-001 has a recombinant chromosome with a crossover between marker loci $M3$ and $M4$.

The overall mean fruit weight for the BC_1 population is 175.7. We can also calculate the mean for the two genotypic classes at each marker locus as shown in Table 19-6. For marker $M1$, the means for the B/B (176.3) and B/S (175.3) genotypic classes are very close to the overall mean (175.7). This is the expectation if there is no QTL affecting fruit weight near $M1$. For marker $M3$, the means for the B/B (180.7) and B/S (169.6) genotypic classes are quite different from the overall mean (175.7) and from each other. This is the expectation if

Table 19-6 Simulated Fruit Weight and Marker-Locus Data for a Backcross Population between Two Tomato Inbred Lines—Beefmaster and Sungold

Plant	Fruit wt. (g)	*M1*	*M2*	*M3*	*M4*	*M5*
Beefmaster	230	B/B	B/B	B/B	B/B	B/B
Sungold	10	S/S	S/S	S/S	S/S	S/S
BC_1-001	183	B/B	B/B	B/B	B/S	B/S
BC_1-002	176	B/S	B/S	B/B	B/B	B/B
BC_1-003	170	B/B	B/S	B/S	B/S	B/S
BC_1-004	185	B/B	B/B	B/B	B/S	B/S
BC_1-005	182	B/B	B/B	B/B	B/B	B/B
BC_1-006	170	B/S	B/S	B/S	B/S	B/B
BC_1-007	170	B/B	B/S	B/S	B/S	B/S
BC_1-008	174	B/S	B/S	B/S	B/S	B/S
BC_1-009	171	B/S	B/S	B/S	B/B	B/B
BC_1-010	180	B/S	B/S	B/B	B/B	B/B
BC_1-011	185	B/S	B/B	B/B	B/S	B/S
BC_1-012	169	B/S	B/S	B/S	B/S	B/S
BC_1-013	165	B/B	B/B	B/S	B/S	B/S
BC_1-014	181	B/S	B/S	B/B	B/B	B/S
BC_1-015	169	B/S	B/S	B/S	B/B	B/B
BC_1-016	182	B/B	B/B	B/B	B/S	B/S
BC_1-017	179	B/S	B/S	B/B	B/B	B/B
BC_1-018	182	B/S	B/B	B/B	B/B	B/B
BC_1-019	168	B/S	B/S	B/S	B/B	B/B
BC_1-020	173	B/B	B/B	B/B	B/B	B/B
Mean of B/B	—	176.3	179.6	180.7	176.1	175.0
Mean of B/S	—	175.3	173.1	169.6	175.3	176.4
Overall mean	175.7					

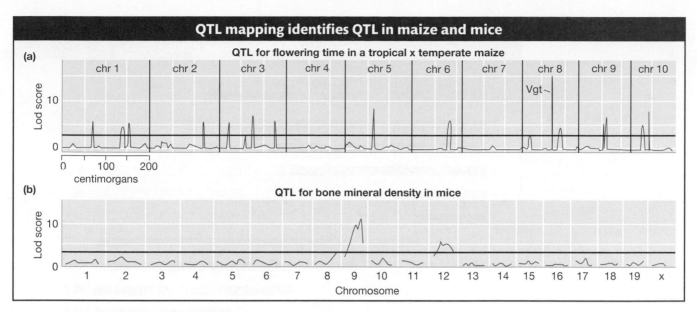

Figure 19-15 Plot of Lod scores from genomic scans for QTL. (a) Results from a scan for flowering time QTL in maize. (b) Results from a scan for bone-mineral-density QTL in mice. [(a) From E. S. Buckler et al., *Science* 523, 2009, 714–718; (b) N. Ishimori et al., *J. Bone Min. Res.* 23, 2008, 1529–1537.]

Much has been learned about genetic architecture from QTL-mapping studies in diverse organisms. Here are two examples. First, flowering time in maize is a classic quantitative or continuous trait. Flowering time is a trait of critical importance in maize breeding since the plants must flower and mature before the end of the growing season. Maize from Canada is adapted to flower within 45 days after planting, while maize from Mexico can require 120 days or longer. QTL mapping has shown that the genetic architecture for flowering time in maize involves more than 50 genes. Results from one experiment are shown in Figure 19-15a; these results show evidence for 15 QTL. QTL for maize flowering time generally have a small effect, such that substituting one allele for another at a QTL alters flowering time by only one day or less. Thus, the difference in flowering time between tropical and temperate maize involves many QTL.

Second, mice have been used to map QTL for many disease-susceptible traits. What one learns about disease-susceptibility genes in mice is often true in humans as well. Figure 19-15b shows the results of a genomic scan in mice for QTL for bone mineral density (BMD), the trait underlying osteoporosis. This scan identified two QTL, one on chromosome 9 and one on chromosome 12. From studies such as this, researchers have indentified over 80 QTL in mice that may contribute to susceptibility to osteoporosis. Similar studies have been done on dozens of other disease conditions.

From QTL to gene

QTL mapping does not typically reveal the identity of the gene(s) at the QTL. At its best, the resolution of QTL mapping is on the order of 1 to 10 cM, the size of a region that can contain 100 or more genes. To go from QTL to a single gene requires additional experiments to **fine-map** a QTL. To do this, the researcher creates a set of genetic homozygous stocks (also called lines), each with a crossover near the QTL. These stocks or lines differ from one another near the QTL, but they are identical to one another **(isogenic)** throughout the rest of their genomes. Lines that are identical throughout their genomes except for a small region of interest are called **congenic** or **nearly isogenic lines.** The isolation of QTL in an

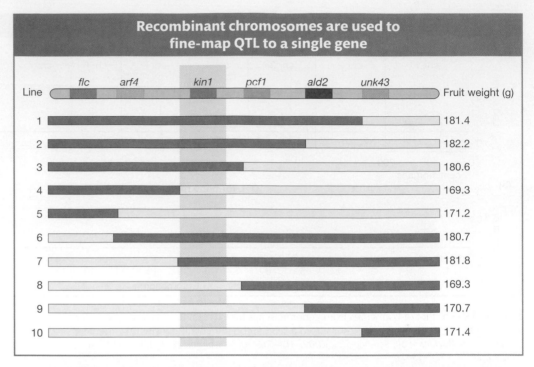

Figure 19-16 A tomato chromosomal segment for a set of 10 congenic lines that have crossovers near a QTL for fruit weight. Red chromosomal segments are derived from the Beefmaster line and yellow segments from the Sungold line. Differences in fruit weight among the lines make it possible to identify the *kin1* gene as the gene underlying this QTL.

isogenic background is critical because only the single QTL region differs between the congenic lines. Thus, the use of congenic lines eliminates the complications caused by having multiple QTL segregate at the same time.

Using the tomato fruit weight example from above, the chromosome region for a set of such congenic lines is shown in Figure 19-16. The genes (*flc, arf4,...*) are shown at the top and the break point for each crossover is indicated by the switch in color from red (Beefmaster genotype) to yellow (Sungold genotype). The mean fruit weight for the congenic lines carrying these recombinant chromosomes is indicated on the right. By inspection of Figure 19-16, you'll notice that all lines with the Beefmaster allele of *kin1* (a kinase gene) have fruit of ~180 g, while those with the Sungold allele of *kin1* have fruit of about ~170 g. None of the other genes are associated with fruit weight in this way. If confirmed by appropriate statistical tests, this result allows us to identify *kin1* as the gene underlying this QTL.

Table 19-7 lists a small sample of the hundreds of genes or QTL affecting quantitative variation from different species that have been identified. The list includes the gene for maize flowering time, *Vgt*, that underlies one of the Lod peaks in Figure 19-15a. One notable aspect of this list is the diversity of gene functions. There does not appear to be a rule that only particular types of genes can be a QTL. Most, if not all, genes in the genomes of organisms are likely to contribute to quantitative variation in populations.

Message Quantitative trait locus (QTL) mapping is a procedure for identifying the genomic locations of the genes (QTL) that control variation for quantitative or complex traits. QTL mapping evaluates the progeny of controlled crosses for their genotypes at molecular markers and for their trait values. If the different genotypes at a marker locus have different mean values for the trait, then there is evidence for a QTL near the marker. Once a region of the genome containing a QTL has been identified, QTL can be mapped to single genes using congenic lines.

Table 19-7   Some Genes Contributing to Quantitative Variation that Were First Identified Using QTL Mapping

Organism	Trait	Gene	Gene function
Yeast	High-temperature growth	*RHO2*	GTPase
Arabidopsis	Flowering time	*CRY2*	Cryptochrome
Maize	Branching	*Tb1*	Transcription factor
Maize	Flowering time	*Vgt*	Transcription factor
Rice	Photoperiod sensitivity	*Hd1*	Transcription factor
Rice	Photoperiod sensitivity	*CK2α*	Casein kinase α subunit
Tomato	Fruit-sugar content	*Brix9-2-5*	Invertase
Tomato	Fruit weight	*Fw2.2*	Cell-cell signaling
Drosophila	Bristle number	*Scabrous*	Secreted glycoprotein
Cattle	Milk yield	*DGAT1*	Diacylglycerol acyltransferase
Mice	Colon cancer	*Mom1*	Modifier of a tumor-suppressor gene
Mice	Type 1 diabetes	*I-Aβ*	Histocompatibility antigen
Humans	Asthma	*ADAM33*	Metalloproteinase-domain-containing protein
Humans	Alzheimer's disease	*ApoE*	Apolipoprotein
Humans	Type 1 diabetes	*HLA-DQA*	MHC class II surface glycoprotein

Source: A. M. Glazier et al., *Science* 298, 2002, 2345–2349.

19.6 Association Mapping in Random-Mating Populations

If you have read a news report recently announcing that researchers have identified a susceptibility gene for autism, diabetes, hypertension, or some other disorder, there is an excellent chance that the gene was discovered using the technique we are about to review, which is called **association mapping.** Association mapping is a method for finding QTL in the genome based on naturally occurring linkage disequilibrium (see Chapter 18) between a marker locus and the QTL in a random-mating population. Because it uses linkage disequilibrium, the method is also called linkage-disequilibrium mapping. As we will see, this method often allows researchers to directly identify the specific genes that control the differences in phenotype among members of a population.

The basic idea behind association mapping has been around and used for decades. Here is an example from the 1990s for the *ApoE* gene in humans, a gene involved in lipoprotein (lipid-protein-complex) metabolism. Because of its role in lipoprotein metabolism, *ApoE* was considered a **candidate gene** for a causative role in cardiovascular disease, the accumulation of fatty (lipid) deposits in the arteries. Researchers looked for statistical associations between the alleles of *ApoE* that people carry and whether they had cardiovascular disease. They found an association between the *e4* allele of this gene and the disease—people carrying the *e4* allele were more likely to have the disease than those who carried other alleles. Although this type of study was successful, it required that a candidate gene suspected to affect the trait be known in advance.

Over the past decade, advances in genomic technologies have catalyzed the broad-scale application of association mapping. In particular, association mapping has been revolutionized by the development of genome-wide SNP maps and high-throughput genotyping technologies that allow scoring of hundreds of thousands of SNPs in tens of thousands of individuals (see Chapter 18). Association mapping is now routinely used to scan the entire genome for genes contributing to quantitative variation. This type of study is known as a **genome-wide association (GWA)** study. A major advantage of GWA studies is that candidate genes are not required since one is scanning every gene in the genome.

Association mapping offers several advantages over QTL mapping. First, since it is performed with random-mating populations, there is no need to make controlled crosses or work with human families with known parent–offspring relationships. Second, it tests many alleles at a locus at once. In QTL-mapping studies, there are two parents (Beefmaster and Sungold tomatoes in the example above) and so only two alleles are being compared. With association mapping, all the alleles in the population are being assayed at the same time. Finally, association mapping can lead to the direct identification of the genes at the QTL without the need for subsequent fine-mapping studies. This is possible because the SNPs in any gene that influences the trait will show stronger associations with the trait than SNPs in other genes. Let's take a look at how it works.

The basic method

Let's begin by looking at how genetic variation is patterned across the genome in a population. In Chapter 18, we discussed linkage disequilibrium (LD), or the

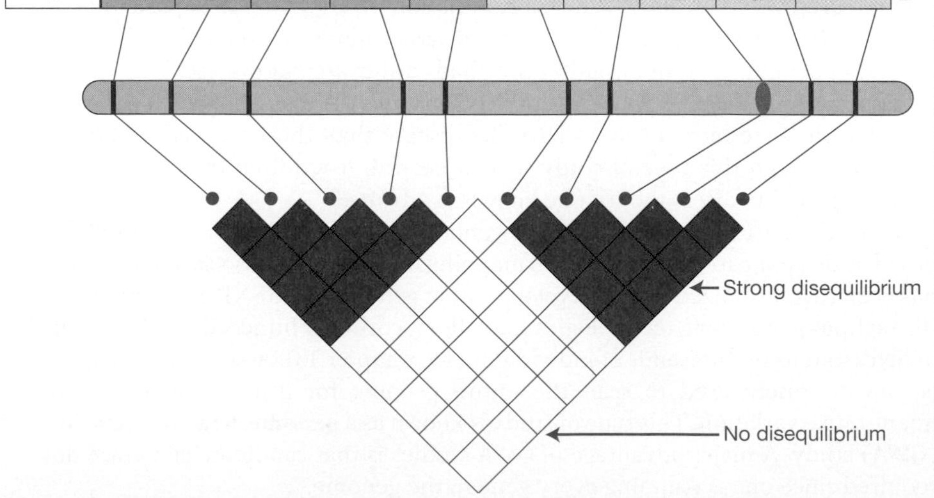

Figure 19-17 (*top*) Diagram of the distribution of SNPs and haplotypes for a chromosomal segment. Haplotypes often occur in blocks (regions of lower recombination) separated from one another by recombination hotspots. (The column of S's and D's at the right are for Problem 19-4.) (*bottom*) Read whether two SNPs show disequilibrium by noting the color of the square where the rows for the markers intersect. Within a haplotype block, SNPs show strong disequilibrium. SNPs in different haplotype blocks show weak or no disequilibrium. [*Adapted from David Altshuler et al., Science 322, 2008, 881–888.*]

nonrandom association of alleles at two loci. Figure 19-17 shows how LD could appear among a sample of chromosomes. SNPs (or other polymorphisms) that are close to each other tend to be in strong disequilibrium, while those that are farther apart are in weak or no disequilibrium. Genomes also tend to have recombination hotspots, points where crossing over occurs at a high frequency. Hotspots disrupt linkage disequilibrium such that SNPs on either side of the hotspot are in equilibrium with each other. SNPs that are not separated by a hotspot form a haplotype block of strongly correlated SNPs.

Suppose SNP8 in Figure 19-17 is an SNP in a gene that causes a difference in phenotype such that individuals with the A/A genotype have a different phenotype than those with either A/G or G/G. SNP8 could affect phenotype by causing an amino acid change or affecting gene expression. SNP8 or any SNPs that directly affect a phenotype are called functional SNPs. Since SNP8 is in strong disequilibrium with other SNPs in the block (SNPs 6, 7, 9, and 10), any of these other SNPs can serve as a proxy for the functional SNP8. Individuals who are T/T at SNP7 will have the same phenotype as those who are A/A at SNP8 because SNP7 and SNP8 are in LD. When the SNP genotypes are correlated (in disequilibrium), then the trait values will be correlated. For this reason, GWA studies do not need to survey the actual functional SNPs, but they do need to have SNPs in every haplotype block.

To conduct a GWA study for a disease condition in humans, we might survey 2000 individuals with a disorder such as adult-onset, or type 2, diabetes. We would also select another 2000 control individuals who do not have this disorder. Each of the 4000 participants would donate blood from which their DNA would be extracted. The DNA samples would be genotyped for a set of 300,000 SNPs that are distributed across the entire genome. We want a sufficient number of SNPs so that each of the haplotype blocks in the genome is marked by one or more SNPs (see Figure 19-17). The resulting data set would be enormous—consisting of 300,000 genotypes in 4000 individuals—a total of 1.2 billion data points. A small part of such a data set is shown in Table 19-8.

Once the data are assembled, the researcher performs a statistical test on each SNP to determine whether one of its alleles is more frequently associated with diabetes than expected by chance. In the case of a categorical trait such as being "affected" or "not affected" by diabetes, statistical tests similar to the χ^2 test (see Chapter 3) can be used. A statistical test is performed separately on each SNP and the P values plotted along the chromosome. The null hypothesis is that the SNP is not associated with the trait. If the P value for an SNP falls below 0.05, then the evidence for the null hypothesis is weak and we will favor the alternative hypothesis that the different genotypes at the SNP are associated with different phenotypes for the trait. Association mapping does not actually prove that a gene or an SNP within a gene affects a trait. It only provides statistical evidence for an

Table 19-8 Part of a Simulated Data Set for an Association-Mapping Experiment

Individual	SNP1	SNP2	SNP3	Type 2 diabetes	Height (cm)
1	C/C	A/G	T/T	yes	173
2	C/C	A/A	C/C	yes	170
3	C/G	G/G	T/T	no	183
4	C/G	G/G	C/T	no	180
5	C/C	G/G	C/T	no	173
6	G/G	A/G	C/T	yes	178
7	G/G	A/G	C/T	no	163
8	C/G	G/G	C/T	no	168
9	C/G	A/G	C/T	yes	165
10	G/G	A/A	C/C	yes	157

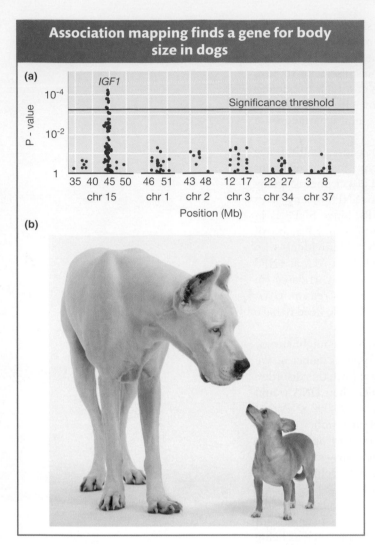

Association mapping finds a gene for body size in dogs

(a)

(b)

Figure 19-18 (a) Results from an association-mapping experiment for body size in dogs. Each dot in the plot represents the P value for a test of association between an SNP and body size. Dots above the "threshold line" show evidence for a statistically significant association. (b) Examples of a small and large breed of dog. [*Tetra Images/Corbis.*]

association between the SNP and the trait. Formal proof requires molecular characterization of the gene and its different alleles.

Figure 19-18a shows the results of an association-mapping study for body size in dogs. Each dot plotted along the chromosomes (*x*-axis) represents the P value (*y*-axis) for a test of association between body size and an SNP. The P values are plotted using an inverse scale such that the higher up the *y*-axis, the smaller the value. On chromosome 15, there is a cluster of SNPs above the threshold line, indicating that the null hypothesis of no association can be rejected for these SNPs in favor of the alternative hypothesis that a gene affecting body size in dogs is located at this position. The strong peak on chromosome 15 involves SNPs in the *insulin-like growth factor-1* (*IGF1*) gene, a gene that encodes a hormone involved in juvenile growth in mammals. This gene is the major contributor to the difference in size between small and large breeds of dogs (Figure 19-18b).

GWA, genes, disease, and heritability

Over the past 10 years, a large number of GWA studies have been performed, and much has been learned from them about heritable variation in humans and other species. Let's look at one of the largest studies, which was a search for disease-risk genes in a group of 17,000 people using 500,000 SNPs. Figure 19-19 shows plots of the P values for associations between SNPs and several common diseases. Green dots are the statistically significant associations. Notice the spike of green dots on chromosome 6 for rheumatoid arthritis and type 1 (juvenile) diabetes. These are two autoimmune diseases, and this spike is positioned over a human leukocyte antigen (*HLA*) gene of the major histocompatibility complex (MHC) of genes that regulates immune response in humans and other vertebrates. Thus, genes active in the normal immune response are implicated as a cause of autoimmune diseases. The gene *PTPN22* is also associated with risk for type 1 diabetes. *PTPN22* encodes the protein tyrosine phosphatase, which is expressed in lymphoid cells of the immune system. For coronary artery disease, there is a significant association with the *ApoE* gene, confirming an earlier study mentioned above.

GWA studies have identified over 300 risk genes for some 70 diseases, and the numbers are growing. These data are ushering in a new era of **personal genomics,** in which an individual can have his or her genome scanned to determine their genotype at genes known to increase disease risk. Although this science is in its infancy, it is possible to identify individuals who have a 10-fold higher risk for certain diseases than other members of the population. Such information can be used to initiate preventative measures and changes in lifestyle (environment) that contribute to disease risk. Some companies are proposing to offer for purchase at your local drugstore "genetic test kits" for specific diseases such as Alzheimer's disease. Bioethicists have expressed concern that consumers are not prepared to evaluate the results appropriately without the counsel of medical professionals.

Since height in humans is a classic quantitative trait, quantitative geneticists had great interest in performing GWA studies for this trait. GWA studies have identified over 180 genes affecting height. Each of these genes has a small additive effect (∼1 to 4 mm), as expected for a trait governed by many genes. However, a perplexing result was that the 180 genes accounted for only 10 percent of the genetic variance in height. This falls far short of the 80 to 90 percent value for broad-sense heritability for height. The difference between 10 percent and 80 percent has been dubbed the missing heritability. For disease risk, there is also much missing heritability. For example, GWA studies have succeeded in explaining only 10 percent of the genetic variation for Crohn's disease and only 5 percent of the genetic variation for type 2 diabetes.

It has come as a surprise to many geneticists that GWA studies with hundreds of thousands of SNPs blanketing the genome and samples of over 10,000 individuals should be able to account for only a tiny fraction of the heritable variation. Currently, it is unknown why this is the case. Researchers had expected that common diseases such as type 2 diabetes would be caused by common alleles; that is, alleles with frequencies between 5 and 50 percent. GWA studies are designed to detect the effects of common alleles, but they are not designed to detect the effects of rare alleles. Thus, one hypothesis is that

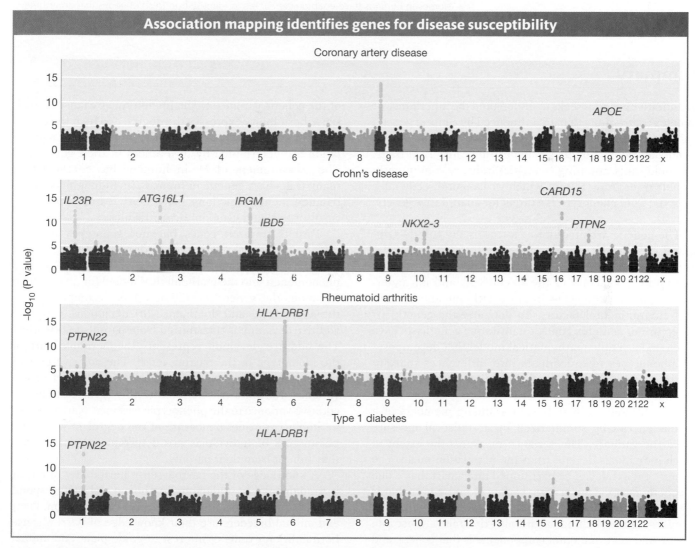

Figure 19-19 Results from a genome-wide association study of common diseases in humans. The 23 human chromosomes are arrayed from left to right. The *y*-axis shows the P value for the statistical test of an association between the disease and each SNP. Significant test results are shown as green dots. The names of some genes identified by this analysis are shown in red. [*The Wellcome Trust Case Control Consortium, Nature 447, 2007, 661–678.*]

susceptibility for many common diseases (or height variation) is caused by a large number of rare alleles. In other words, the disease-susceptibility alleles that segregate in one family are different from those in another, unrelated family.

Despite the inability of GWA studies to explain all the heritable variation for traits, this approach has provided a major advance in understanding quantitative genetic variation. Hundreds of new genes contributing to quantitative variation for disease risk have been identified. These genes are now the targets for the development of new therapies. Beyond humans, GWA studies have advanced our understanding of the inheritance of quantitative traits in *Arabidopsis, Drosophila,* yeast, and maize.

> **Message** Association mapping is a method for identifying statistical associations between molecular markers and phenotypic variation for complex traits. Linkage disequilibrium in a population between the marker locus and a functional variant in a gene can cause the association. If molecular markers over the entire genome are available, then a genome-wide association (GWA) study can be accomplished. GWA studies in humans have enabled geneticists to identify hundreds of genes that contribute to the risks of developing many common diseases.

Summary

Quantitative genetics seeks to understand the inheritance of complex traits—traits that are influenced by a mix of genetic and environmental factors and do not segregate in simple Mendelian ratios. Complex traits can be categorical traits, threshold traits, counting (meristic) traits, or continuously variable traits. Any trait for which we cannot directly infer genotype from phenotype is a target for quantitative genetic analysis.

The genetic architecture of a trait is the full description of the number of genes affecting the trait, their relative contributions to the phenotype, the contribution of environmental factors to the phenotype, and an understanding of how the genes interact with one another and with environmental factors. To decipher the genetic architecture of complex traits, quantitative geneticists have developed a simple mathematical model that decomposes the phenotypes of individuals into differences that are due to genetic factors (g) and those that are due to environmental factors (e).

The differences in trait values among members of a population can be summarized by a statistical measure called the variance. The variance measures the extent to which individuals deviate from the population mean. The variance for a trait can be partitioned into a part that is due to genetic factors (the genetic variance) and a part that is due to environmental factors (the environmental variance). A key assumption behind partitioning the trait variance into genetic and environmental components is that genetic and environmental factors are uncorrelated or independent.

The degree to which variation for a trait in a population is explained by genetic factors is measured by the broad-sense heritability (H^2) of the trait. H^2 is the ratio of the genetic variance to the phenotypic variance. Broad-sense heritability expresses the degree to which the differences in the phenotypes among the individuals in a population are determined by differences in their genotypes. The measurement of H^2 in humans has revealed that many traits have genetic influences, including physical attributes, mental functions, personality features, psychiatric disorders, and even political attitudes.

Parents transmit genes but not genotypes to their offspring. Each generation, new dominance interactions between the alleles at a locus are created. To incorporate this phenomenon into the mathematical model for quantitative variation, the genetic deviation (g) is decomposed into the additive (a) and dominance (d) deviations. Only the additive deviation is transmitted from parents to offspring. The additive deviation represents the heritable part of the phenotype in the narrow sense. The additive part of the variance in a population is the heritable part of the variance. Narrow-sense heritability (h^2) is the ratio of the additive variance to the phenotypic variance. Narrow-sense heritability provides a measure of the degree to which the phenotypes of individuals are determined by the genes they inherit from their parents.

A knowledge of the narrow-sense heritability of a trait is fundamental to understanding how a trait will respond to selective breeding or the force of natural selection. Plant and animal breeders use their knowledge of narrow-sense heritability for traits of interest to guide plant and animal improvement programs. Narrow-sense heritability is used to predict the phenotypes of offspring and estimate the breeding value of individual members of the breeding population.

The genetic loci underlying variation in complex traits are known as quantitative trait loci, or QTL for short. There are two experimental methods for charactering QTL and determining their locations in the genome. First, QTL mapping looks for statistical correlations between the genotypes at marker loci and trait values in populations with known pedigrees such as a BC_1 population. QTL mapping provides estimates of the number of genes controlling a trait, whether the alleles at the QTL exhibit additivity or dominance, and whether each QTL has a small or large effect on the trait. Second, association mapping looks for statistical correlations between the genotypes at marker loci and trait values in random-mating populations. Association mapping can allow researchers to identify the genes that underlie the QTL. Genome-wide association (GWA) studies use markers blanketing the entire genome.

Most traits of importance in medicine, agriculture, and evolutionary biology show complex inheritance. Examples include disease risk in humans, yield in soybeans, milk production in dairy cows, and the full spectrum of phenotypes that differentiate all the species of plants, animals, and microbes on earth. Quantitative genetic analyses are at the forefront of understanding the genetic basis of these critical traits.

KEY TERMS

additive effect (A) (p. 700)
additive gene action (p. 699)
additive variation (p. 704)
association mapping (p. 717)
breeding value (p. 703)
broad-sense heritability (H^2) (p. 695)
candidate gene (p. 717)
categorical trait (p. 685)
complex inheritance (p. 685)
complex trait (p. 683, 685)
congenic line (p. 715)
continuous trait (p. 685)
correlation (p. 693)
correlation coefficient (p. 694)
covariance (p. 692)
deviation (p. 687)
dominance effect (p. 700)

dominance variation (p. 704)
dominant gene action (p. 699)
environmental variance (p. 692)
fine-map (p. 715)
frequency histogram (p. 687)
gene action (p. 699)
genetic architecture (p. 684)
genetic variance (p. 692)
genome-wide association (GWA) (p. 717)
inbred line or strain (p. 691)
isogenic (p. 715)
mean (p. 686)
meristic trait (p. 685)
multifactorial hypothcsis (p. 684)
narrow-sense heritability (h^2) (p. 699)

nearly isogenic line (p. 715)
normal distribution (p. 688)
partial dominance (p. 699)
personal genomics (p. 720)
population (p. 686)
QTL mapping (p. 710)
quantitative genetics (p. 684)
quantitative trait (p. 683)
quantitative trait loci (QTL) (p. 710)
recessive gene action (p. 000)
sample (p. 686)
selection differential (S) (p. 709)
selection response (R) (p. 709)
simple inheritance (p. 685)
standard deviation (p. 688)
threshold trait (p. 685)
variance (p. 687)

SOLVED PROBLEMS

SOLVED PROBLEM 1. In a flock of 100 broiler chickens, the mean weight is 700 g and the standard deviation is 100 g. Assume the trait values follow the normal distribution.

a. How many of the chickens are expected to weigh more than 700 g?

b. How many of the chickens are expected to weigh more than 900 g?

c. If H^2 is 1.0, what is the genetic variance for this population?

Solution

a. Since the normal distribution is symmetrical about the mean, 50 percent of the population will have a trait value above the mean and the other 50 percent will have a trait value below the mean. In this case, 50 of the 100 chickens are expected to weigh more than 700 g.

b. The value of 900 g is 2 standard deviations greater than the mean. Under the normal distribution, 95.5 percent of the population will fall within 2 standard deviations of the mean and the remaining 4.5 percent will lie more than 2 standard deviations from the mean. Of this 4.5 percent, one-half (2.25 percent) will be more than 2 standard deviations less than the mean, and the other half (2.25 percent) will be more than 2 standard deviations greater than the mean. Thus, we expect about 2.25 percent of the 100 chickens (or roughly 2 chickens) to weigh more than 900 g.

c. When H^2 is 1.0, then all of the variance is genetic. We know that the standard deviation is 100, and the variance is the square of the standard deviation

$$\text{Variance} = \sigma^2$$

Thus, the genetic variance would be $(100)^2 = 10,000 \text{ g}^2$.

SOLVED PROBLEM 2. Two inbred lines of beans are intercrossed. In the F_1, the variance in bean weight is measured at 15 g^2. The F_1 is selfed; in the F_2, the variance in bean weight is 61 g^2. Estimate the broad heritability of bean weight in the F_2 population of this experiment.

Solution

The key here is to recognize that all the variance in the F_1 population must be environmental because all individuals have the same genotype. Furthermore, the F_2 variance must be a combination of environmental and genetic components, because all the genes that are heterozygous in the F_1 will segregate in the F_2 to give an array of different genotypes that relate to bean weight. Hence, we can estimate

$$V_e = 15 \text{ g}^2$$

$$V_g + V_e = 61 \text{ g}^2$$

Therefore,

$$V_g = 61 - 15 = 46 \text{ g}^2$$

and broad heritability is

$$H^2 = \frac{46}{61} = 0.75 \ (75\%)$$

SOLVED PROBLEM 3. In an experimental population of *Tribolium* (flour beetles), body length shows a continuous distribution with a mean of 6 mm. A group of males and females with a mean body length of 9 mm are removed and interbred. The body lengths of their offspring average 7.2 mm. From these data, calculate the heritability in the narrow sense for body length in this population.

Solution

The selection differential (S) is $9 - 6 = 3$ mm, and the selection response (R) is $7.2 - 6 = 1.2$ mm. Therefore, the heritability in the narrow sense is

$$h^2 = \frac{R}{S} = \frac{1.2}{3.0} = 0.4 \ (40\%)$$

SOLVED PROBLEM 4. One research team reports that the broad-sense heritability for height in humans is 0.5 based on a study of identical twins reared apart in Iceland. Another team reports that the narrow-sense heritability for human height is 0.8 based on a study of parent–offspring correlation in the United States. What seems unexpected about these results? How could the unexpected results be explained?

Solution

Broad-sense heritability is the ratio of the total genetic variance (V_g) to the phenotypic variance (V_X). The total genetic variance includes both the additive (V_a) and the dominance (V_d) variance

$$h^2 = \frac{V_g}{V_X} = \frac{V_a + V_d}{V_X}$$

Narrow-sense heritability is the ratio of the additive variance (V_a) to the phenotypic variance (V_X).

$$H^2 = \frac{V_a}{V_X}$$

Thus, all other variables being equal, H^2 should be greater than or equal to h^2. It will be equal to h^2 when V_d is 0.0. It is unexpected that h^2 should be greater than H^2. However, the two research teams studied different populations—in Iceland and in the United States. Estimates of heritability apply only to the population and environment in which they were measured. Estimates made in one population can be different from those made in another population because the two populations may segregate for different alleles at numerous genes and the two populations experience different environments.

PROBLEMS

Most of the problems are also available for review/grading through the GENETICSPORTAL www.yourgeneticsportal.com.

WORKING WITH THE FIGURES

1. Figure 19-9 shows the trait distributions before and after a cycle of artificial selection. Does the variance of the trait appear to have changed as a result of selection? Explain.

2. Figure 19-11 shows the expected distributions for the three genotypic classes if the *B* locus is a QTL affecting the trait value.

   a. As drawn, what is the dominance/additive (D/A) ratio?

   b. How would you redraw this figure if the *B* locus had no effect on the trait value?

   c. How would the positions along the *x*-axis of the curves for the different genotypic classes of the *B* locus change if D/A = 1.0?

3. Figure 19-16 shows the results of a QTL fine-mapping experiment. Which gene would be implicated as controlling fruit weight if the mean fruit weight for each line was as follows?

Line	Fruit weight (g)
1	181.4
2	169.3
3	170.7
4	171.2
5	171.4
6	182.2
7	180.6
8	180.7
9	181.8
10	169.3

4. Figure 19-17 shows a set of haplotypes. Suppose these are haplotypes for a chromosomal segment from 18 haploid yeast strains. On the right edge of the figure, the S and D indicate whether the strain survives (S) or dies (D) at high temperature (40°C). Using the χ^2 test (see Chapter 3) and Table 3-1, does either SNP1 or SNP6 show evidence for an association with the growth phenotype? Explain.

5. Figure 19-18a shows a plot of P values (represented by the dots) along the chromosomes of the dog genome. Each P value is the result of a statistical test of association between an SNP and body size. Other than the cluster of small P values near *IGF1*, do you see any chromosomal regions with evidence for a significant association between an SNP and body size? Explain.

6. Figure 19-19 shows plots of P values (represented by the dots) along the chromosomes of the human genome. Each P value is the result of a statistical test of association between an SNP and a disease condition. There is a cluster, or spike, of statistically significant P values (green dots) at the gene *HLA-DRB1* for two diseases. Why might this particular gene contribute to susceptibility for the autoimmune diseases rheumatoid arthritis and type 1 diabetes?

BASIC PROBLEMS

7. Distinguish between continuous and discontinuous variation in a population, and give some examples of each.

8. The table below shows a distribution of bristle number in a *Drosophila* population. Calculate the mean, variance, and standard deviation for these data.

Bristle number	Number of individuals
1	1
2	4
3	7
4	31
5	56
6	17
7	4

9. Suppose that the mean IQ in the United States is roughly 100 and the standard deviation is 15 points. People with IQs of 145 or higher are considered "geniuses" on some scales of measurement. What percentage of the population is expected to have an IQ of 145 or higher? In a country with 300 million people, how many geniuses are there expected to be?

10. A bean breeder is working with a population in which the mean number of pods per plant is 50 and the variance is 10 pods². The broad-sense heritability is known to be 0.8. Given this information, can the breeder be assured that the population will respond to selection for an increase in the number of pods per plant in the next generation?

CHALLENGING PROBLEMS

11. In a large herd of cattle, three different characters showing continuous distribution are measured, and the variances in the following table are calculated:

	Characters		
Variance	Shank length	Neck length	Fat content
Phenotypic	310.2	730.4	106.0
Environmental	248.1	292.2	53.0
Additive genetic	46.5	73.0	42.4
Dominance genetic	15.6	365.2	10.6

a. Calculate the broad- *and* narrow-sense heritabilities for each character.

b. In the population of animals studied, which character would respond best to selection? Why?

c. A project is undertaken to decrease mean fat content in the herd. The mean fat content is currently 10.5 percent. Animals with a mean of 6.5 percent fat content are interbred as parents of the next generation. What mean fat content can be expected in the descendants of these animals?

12. Suppose that two triple heterozygotes *A/a* ; *B/b* ; *C/c* are crossed. Assume that the three loci are on different chromosomes.

a. What proportions of the offspring are homozygous at one, two, and three loci, respectively?

b. What proportions of the offspring carry 0, 1, 2, 3, 4, 5, and 6 alleles (represented by capital letters), respectively?

13. In Problem 12, suppose that the average phenotypic effect of the three genotypes at the *A* locus is *A/A* = 3, *A/a* = 2, and *a/a* = 1 and that similar effects exist for the *B* and *C* loci. Moreover, suppose that the effects of loci add to one another. Calculate and graph the distribution of phenotypes in the population (assuming no environmental variance).

14. In a species of the Darwin's finches (*Geospiza fortis*), the narrow-sense heritability of bill depth has been estimated to be 0.79. Bill depth is correlated with the ability of the finches to eat large seeds. The mean bill depth for the population is 9.6 mm. A male with a bill depth of 10.8 mm is mated with a female with a bill depth of 9.8 mm. What is the expected value for bill depth for the offspring of this mating pair?

15. The table below contains measurements of total serum cholesterol (mg/dl) for 10 sets of monozygotic twins who were reared apart. Calculate the following: overall mean, overall variance, covariance between the twins, broad-sense heritability (*H*²).

X′	X″
228	222
186	152
204	220
142	185
226	210
217	190
207	226
185	213
179	159
170	129

16. Population A consists of 100 hens that are fully isogenic and that are reared in a uniform environment. The average weight of the eggs they lay is 52 g, and the variance is 3.5 g². Population B consists of 100 genetically variable hens that produce eggs with a mean weight of 52 g and a variance of 21.0 g². Population B is raised in an environment that is equivalent to that of Population A. What is the environmental variance (*V*ₑ) for egg weight? What is the genetic variance in Population B? What is the broad-sense heritability in Population B?

17. Maize plants in a population are on average 180 cm tall. Narrow-sense heritability for plant height in this population is 0.5. A breeder selects plants that are 10 cm taller on average than the population mean to produce the next generation, and the breeder continues applying

this level of selection for eight generations. What will the average height of the plants be after eight generations of selection? Assume that *h*² remains 0.5 and *V*ₑ does not change over the course of the experiment.

18. GWA studies reveal statistical correlations between the genotypes at marker loci in genes and complex traits. Do GWA studies prove that allelic variation in a gene actually causes the variation in the trait? If not, what experiments could prove that allelic variants in a gene in a population are responsible for variation in a trait?

19. The *ocular albinism-2* (*OCA2*) gene and the *melanocortin-1-receptor* (*MC1R*) gene are both involved in melanin metabolism in skin cells in humans. To test whether variation at these genes contributes to sun sensitivity and the associated risk of being afflicted with skin cancer, you perform association analyses. A sample of 1000 people from Iceland were asked to classify themselves as having tanning or burning (nontanning) skin when exposed to the sun. The individuals were also genotyped for an SNP in each gene (rs7495174 and rs1805007). The table shows the number of individuals in each class.

	OCA2 (rs7495174)			MC1R (rs1805007)		
	A/A	A/G	G/G	C/C	C/T	T/T
Burning	245	56	1	192	89	21
Tanning	555	134	9	448	231	19

a. What are the frequencies of tanning and burning phenotypes in Iceland?

b. What are the allelic frequencies at each locus (SNP)?

c. Using the χ² test (see Chapter 3) and Table 3-1, test the null hypothesis that there is no association between these SNPs and sun-sensitive skin. Does either SNP show evidence for an association?

d. If you find evidence for an association between the gene and the trait, what is the mode of gene action?

e. If the P value is greater than 0.05, does that prove that the gene does not contribute to variation for sun sensitivity? Why?

Evolution of Genes and Traits

20

The theory of evolution by natural selection was developed independently by two intrepid British naturalists, Charles Darwin (1809–1872) and Alfred Russel Wallace (1823–1913), in the course of their respective long voyages. [*Darwin, at left: The Gallery Collection/Corbis; Wallace, at right: Hutton Archive/Getty Images.*]

Charles Darwin (1809–1882) arrived in the Galápagos Islands in 1835, well into the fourth year of what was supposed to be a two-year voyage. One might think that these islands, now inextricably linked with Darwin's name, were the young naturalist's paradise. Far from it. Darwin found the islands hellishly hot, their broken black volcanic rock scorching under the hot sun. In his diary he observed that "the stunted trees show little signs of life . . . the plants also smell unpleasantly. . . . The black lava rocks on the beach are frequented by large (2–3 ft.) most disgusting clumsy lizards. . . . They assuredly well become the land they inhabit." Other than the lizards and the tortoises, the animal life on the islands was scant and unimpressive. He could not wait to leave the place. The 26-year-old explorer did not know that his five weeks in the Galápagos would inspire a series of radical ideas that, some 24 years later with the publication of his *On the Origin of Species* (1859), would change our perception of the world and our place in it.

Several months after leaving the islands, on the last leg of the voyage home to England, Darwin had his first flash of insight. He had begun to organize his copious field notes from his nearly five years of exploration and collecting. His plan was for experts back in England to lead the study of his collections of fossils, plants, animals, and rocks. Turning to his observations on the birds of the Galápagos, he recalled that he had found slightly different forms of mockingbirds on three different islands. Now, there was a puzzle. The prevailing view of the origin of species in 1835, held by most of Darwin's teachers and much of the scientific establishment, was that species were specially created by God in their present form, unchangeable, and placed in the habitat to which they were best suited. Why, then, would there be slightly different birds on such similar islands? Darwin jotted in his ornithology notebook:

> When I see these Islands in sight of each other and possessed of but a scanty stock of animals, tenanted by these birds but slightly differing in structure filling the same place in Nature, I must suspect they are only varieties. . . . If there is the slightest foundation for these remarks, the zoology of Archipelagoes will be well worth examining; *for such facts would undermine the stability of species* [emphasis added].

Darwin's insight was that species might change. This was not what he had learned at Cambridge University. This was heresy. Although Darwin decided to keep such dangerous thoughts to himself, he was gripped by the idea. After arriving home in England, he filled a series of notebooks with thoughts about species changing. Within a year he had convinced himself that species arise naturally from preexisting species, as naturally as children are born from parents and parents from grandparents. He then pondered *how* species change and adapt to their particular circumstances. In 1838, just two years after the conclusion of his voyage and before he had yet turned 30, he conceived his answer— **natural selection.** In this competitive process, individuals bearing some relative advantage over others live longer and produce more offspring, which in turn inherit the advantage.

Darwin knew that to convince others of these two ideas—the descent of species from ancestors and natural selection—he would need more evidence. He spent the next two decades marshaling all of the facts he could from botany, zoology, embryology, and the fossil record.

He received crucial information from experts who helped to sort out and characterize his collections. Ornithologist John Gould pointed out to Darwin that what the young naturalist thought were blackbirds, grossbeaks, and finches from the Galápagos were actually 12 (now recognized as 13) new and distinct species of ground finches (Figure 20-1). The Galápagos species, though clearly finches, exhibit an immense variation in feeding behavior and in the bill shape that corresponds to their food sources. For example, the vegetarian tree finch uses its heavy bill to eat fruits and leaves, the insectivorous finch has a bill with a biting tip for eating large insects, and, most remarkable of all, the woodpecker finch grasps a twig in its bill and uses it to obtain insect prey by probing holes in trees.

This diversity of species, Darwin deduced, must have arisen from an original population of finch that arrived in the Galápagos from the mainland of South America and populated the islands. The descendants of the original colonizers spread to the different islands and formed local populations that diverged from one another and eventually formed different species.

The finches illustrate the process of **adaptation,** in which the characteristics of a species become modified to suit the environments in which they live. Darwin provided one level of explanation for the process, natural selection, but he could not explain how traits varied or how they changed with time because he did not understand the mechanisms of inheritance. Understanding the genetic basis of adaptation has been one of the long-standing goals of evolutionary biology.

A first step toward this goal was taken when Mendel's work pointing to the existence of genes was rediscovered two decades after Darwin died. Another key emerged a half century later, when the molecular basis of inheritance and the genetic code were deciphered. For many decades since, biologists have known

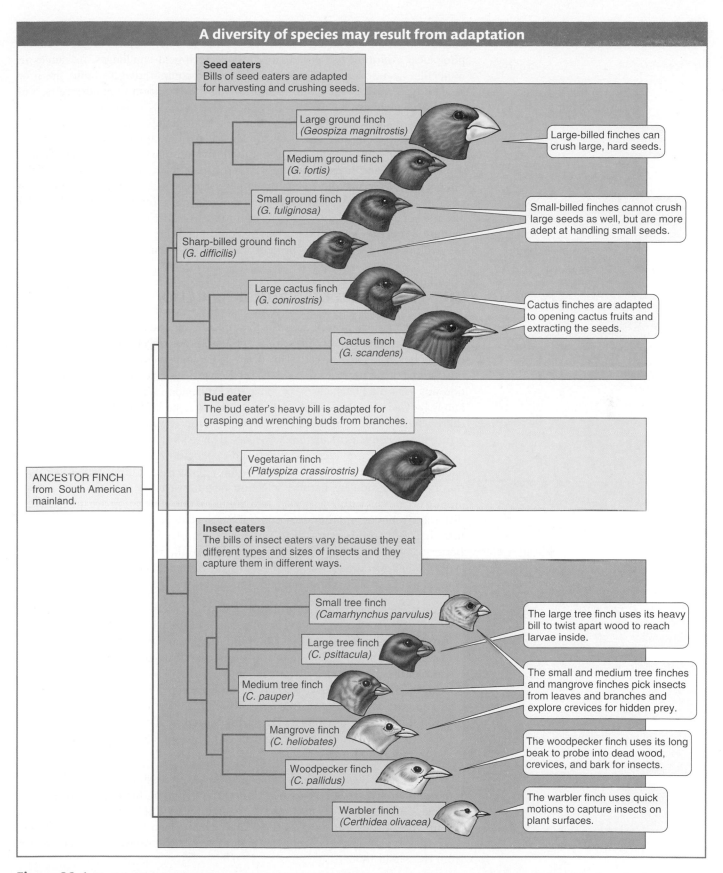

A diversity of species may result from adaptation

Seed eaters
Bills of seed eaters are adapted for harvesting and crushing seeds.

Large ground finch
(*Geospiza magnitrostis*)

Medium ground finch
(*G. fortis*)

Small ground finch
(*G. fuliginosa*)

Sharp-billed ground finch
(*G. difficilis*)

Large cactus finch
(*G. conirostris*)

Cactus finch
(*G. scandens*)

Large-billed finches can crush large, hard seeds.

Small-billed finches cannot crush large seeds as well, but are more adept at handling small seeds.

Cactus finches are adapted to opening cactus fruits and extracting the seeds.

Bud eater
The bud eater's heavy bill is adapted for grasping and wrenching buds from branches.

Vegetarian finch
(*Platyspiza crassirostris*)

ANCESTOR FINCH
from South American mainland.

Insect eaters
The bills of insect eaters vary because they eat different types and sizes of insects and they capture them in different ways.

Small tree finch
(*Camarhynchus parvulus*)

Large tree finch
(*C. psittacula*)

Medium tree finch
(*C. pauper*)

Mangrove finch
(*C. heliobates*)

Woodpecker finch
(*C. pallidus*)

Warbler finch
(*Certhidea olivacea*)

The large tree finch uses its heavy bill to twist apart wood to reach larvae inside.

The small and medium tree finches and mangrove finches pick insects from leaves and branches and explore crevices for hidden prey.

The woodpecker finch uses its long beak to probe into dead wood, crevices, and bark for insects.

The warbler finch uses quick motions to capture insects on plant surfaces.

Figure 20-1 The 13 species of finches found in the Galápagos Islands. [*After W. K. Purves, G. H. Orians, and H. C. Heller, Life: The Science of Biology, 4th ed. Sinauer Associates/W. H. Freeman and Company, 1995, Fig. 20.3, p. 450.*]

that species and traits evolve through changes in DNA sequence. However, the elucidation of specific changes in DNA sequence underlying physiological or morphological evolution has posed considerable technical challenges. Advances in molecular genetics, developmental genetics, and comparative genomics are now revealing the diverse mechanisms underlying the evolution of genes, traits, and organismal diversity.

In this chapter, we will examine the molecular genetic mechanisms underlying the variation in and evolution of traits and the adaptation of organisms to their environments. We will first examine the evolutionary process in general and then focus on specific examples for which the genetic and molecular bases of the phenotypic differences between populations or species have been pinpointed. All of the examples will focus on the evolution of relatively simple traits controlled by a single gene. These relatively simple examples are sufficient to illustrate the fundamental process of evolution at the DNA level and the variety of ways in which the evolution of genes affects the gain, loss, and modification of traits.

20.1 Evolution by Natural Selection

The modern theory of evolution is so completely identified with Darwin's name, many people think Darwin himself first proposed the concept that organisms have evolved, but that is not the case. The idea that life changed over time was circulating in scientific circles for many decades before Darwin's historic voyage. The great question was, *How* did life change? For some, the explanation was a series of special creations by God. To others, such as Jean-Baptiste Lamarck (1744–1829), change was caused by the environment acting directly on the organism, and those changes acquired in an organism's lifetime were passed on to its offspring.

What Darwin provided was a detailed explanation of the mechanism of the evolutionary process that correctly incorporated the role of inheritance. Darwin's theory of evolution by natural selection begins with the variation that exists among organisms within a species. Individuals of one generation are qualitatively different from one another. Evolution of the species as a whole results from the fact that the various types differ in their rates of survival and reproduction. Better-adapted types leave more offspring, and so the relative frequencies of the types change over time. Thus, the three critical ingredients to evolutionary change Darwin put forth were variation, selection, and time:

> Can it, then, be thought improbable. . . that variations useful in some way to each being in the great and complex battle of life, should sometimes occur in the course of thousands of generations? . . . Can we doubt (remembering that many more individuals are born than can possibly survive) that individuals having any advantage, however slight, over others, would have the best chance of surviving and of procreating their kind? On the other hand, we may feel sure that any variation in the least degree injurious would be rigidly destroyed. This preservation of favorable variations and the rejection of injurious variations I call Natural Selection. (*On the Origin of Species*, Chapter IV)

Darwin's writings and ideas are well known, and justifiably so, but it is very important to note that he was not alone in arriving at this concept of natural selection. Alfred Russel Wallace (1823–1913), a fellow Englishman who explored the jungles of the Amazon and the Malay Archipelago for a total of 12 years, reached a very similar conclusion in a paper that was co-published with an excerpt from Darwin in 1858:

> The life of wild animals is a struggle for existence. . . . Perhaps all the variations from the typical form of a species must have some definite effect, however

slight, on the habits or capacities of the individuals. . . . It is also evident that most changes would affect, either favourably or adversely, the powers of prolonging existence. . . . If, on the other hand, any species should produce a variety having slightly increased powers of preserving existence, that variety must inevitably in time acquire a superiority in numbers." (*On the Tendency of Varieties to Depart Indefinitely from the Original Type,* 1858)

While today Darwin's name tends to be exclusively linked to evolution by natural selection, in their day, the theory was recognized as the Darwin-Wallace theory. Perhaps the current perception is at least in part due to Wallace himself, who was always deferential to Darwin and referred to the emergent theory of evolution as "Darwinism."

> **Message** Darwin and Wallace proposed a new explanation to account for the phenomenon of evolution. They understood that the population of a given species at a given time includes individuals of varying characteristics. They realized that the population of succeeding generations will contain a higher frequency of those types that most successfully survive and reproduce under the existing environmental conditions. Thus, the frequencies of various types within the species will change over time.

There is an obvious similarity between the process of evolution as Darwin and Wallace described it and the process by which the plant or animal breeder improves a domestic stock. The plant breeder selects the highest-yielding plants from the current population and uses them as the parents of the next generation. If the characteristics causing the higher yield are heritable, then the next generation should produce a higher yield. It was no accident that Darwin chose the term *natural selection* to describe his model of evolution through differences in the rates of reproduction shown by different variants in the population. As a model for this evolutionary process in the wild, he had in mind the selection that breeders exercise on successive generations of domestic plants and animals.

We can summarize the theory of evolution by natural selection in three principles:

1. *Principle of variation.* Among individuals within any population, there is variation in morphology, physiology, and behavior.

2. *Principle of heredity.* Offspring resemble their parents more than they resemble unrelated individuals.

3. *Principle of selection.* Some forms are more successful at surviving and reproducing than other forms in a given environment.

A selective process can produce change in the population composition only if there are some variations among which to select. If all individuals are identical, no differences in the reproductive rates of individuals, no matter how extreme, will alter the composition of the population. Furthermore, the variation must be in some part heritable if these differences in reproductive rates are to alter the population's genetic composition. If large animals within a population have more offspring than do small ones but their offspring are no larger on average than those of small animals, then there will be no change in population composition from one generation to another. Finally, if all variant types leave, on average, the same number of offspring, then we can expect the population to remain unchanged.

> **Message** The principles of variation, heredity, and selection must all apply for evolution to take place through a variational mechanism.

Heritable variation provides the raw material for successive changes within a species and for the multiplication of new species. The basic mechanisms of those

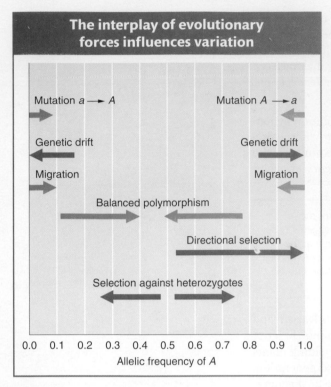

The interplay of evolutionary forces influences variation

Mutation $a \rightarrow A$ Mutation $A \rightarrow a$

Genetic drift   Genetic drift

Migration   Migration

Balanced polymorphism

Directional selection

Selection against heterozygotes

0.0 0.1 0.2 0.3 0.4 0.5 0.6 0.7 0.8 0.9 1.0
Allelic frequency of A

Figure 20-2 The effects on allele frequency of various forces of evolution. The blue arrows show a tendency toward increased variation within the population; the red arrows, decreased variation.

changes (as discussed in Chapter 18) are the origin of new genetic variation by mutation, the change in frequency of alleles within populations by selective and random processes, the divergence of different populations because the selective forces are different or because of random drift, and the reduction of variation between populations by migration (Figure 20-2). From those basic mechanisms, a set of principles governing changes in the genetic composition of populations can be derived. The application of these principles of population genetics provides a genetic theory of evolution. Generally, as Table 20-1 shows, forces that increase or maintain variation within populations prevent populations from diverging from one another, whereas forces that make each population less variable (e.g., homozygous) cause populations to diverge.

> **Message** Evolution, the change in populations or species over time, is the conversion of heritable variation between individuals within populations into heritable differences between populations in time and in space by population genetic mechanisms.

We will now turn from this general genetic description of evolutionary change to the examination of evolution at the molecular level.

20.2 Molecular Evolution: The Neutral Theory

Darwin and Wallace conceived of evolution largely as "changes in organisms brought about by natural selection." Indeed, this is what most people think of as the meaning of "evolution." However, a century after Darwin's theory, as molecular biologists began to confront evolution at the level of proteins and DNA molecules, they encountered and identified another dimension of the evolutionary process, neutral molecular evolution, which did not involve natural selection. An understanding of neutral molecular evolution is crucial to grasping how genes change over time.

The development of the neutral theory

By the 1940s, three branches of evolutionary biology—population genetics, systematics (the definition and classification of species), and paleontology—had become incorporated in what was dubbed the "Modern Synthesis" of evolutionary theory. This synthesis reflected general agreement that the principles operating at the level of populations, as illuminated by population genetics, were sufficient to account for the larger-scale evolution of species

Table 20-1 How the Forces of Evolution Increase (+) or Decrease (−) Variation Within and Between Populations

Force	Variation within populations	Variation between populations
Inbreeding or genetic drift	−	+
Mutation	+	−
Migration	+	−
Directional selection	−	+/−
Balancing	+	−
Incompatible	−	+

and higher taxa as documented, for example, in the fossil record. However, at this time there was no understanding of the molecular basis of either heredity (DNA) or evolutionary change.

In the 1950s and early 1960s, methods were developed that enabled biologists to determine the amino acid sequences of proteins. This new capability raised the prospect that the fundamental basis of evolutionary change was finally at hand. However, as the sequences of proteins from a variety of species were deciphered, a paradox emerged. The sequences of globins and cytochrome c, for example, typically differ between any two species at a number of amino acids, and that number increases with the time elapsed since their divergence from a common ancestor (Figure 20-3). Yet, the function of these proteins is the same in different species—to carry and deliver oxygen to tissues in the case of hemoglobin and to shuttle electrons during cellular respiration in the case of cytochrome c.

The puzzle then was whether the amino acid replacements between species reflected changes in protein function and adaptations to selective conditions. Biochemists Linus Pauling and Emile Zuckerkandl did not think so. They observed that many substitutions were of one amino acid for another with similar properties. They concluded that most amino acid substitutions were "neutral" or "nearly neutral" and did not change the function of a protein whatsoever.

This line of reasoning was rejected at first by many evolutionary biologists, who at the time viewed all evolutionary changes as the result of natural selection and adaptation. Paleontologist George Gaylord Simpson, an architect of the Modern Synthesis, argued that "there is a strong consensus that completely neutral genes or alleles must be rare if they exist at all. To an evolutionary biologist it therefore seems highly improbable that proteins . . . should change in a regular but non-adaptive way."

Zuckerkandl and Pauling asserted that the similarity or differences among organisms need not be reflected at the level of protein—that molecular change and visible change were not necessarily linked or proportional.

The debate was resolved by an onslaught of empirical data and the deciphering of the genetic code. Because multiple codons encode the same amino acid, a mutation that changes, say, CAG to CAC does not change the amino acid encoded. Therefore, variation can exist at the DNA level that has no effect on protein sequences, and thus neutral alleles do exist. But even more important for population genetics was the development of the "neutral theory of molecular evolution" by Motoo Kimura, Jack L. King, and Thomas Jukes. These authors proposed that most, but not all, mutations that are fixed are neutral or nearly neutral and any differences between species at such sites in DNA evolve by random genetic drift.

The "neutral theory" marked a profound conceptual shift away from a view of evolution as always guided by natural selection. Moreover, it provided a baseline assumption of how DNA should change over time if no other agent such as natural selection intervened.

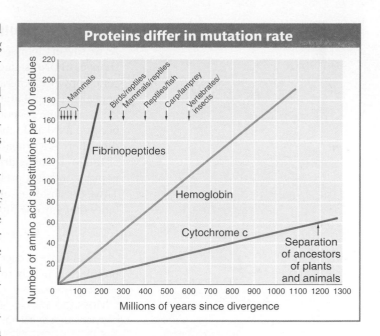

Figure 20-3 Number of amino acid substitutions in the evolution of the vertebrates as a function of time since divergence. The three proteins—fibrinopeptides, hemoglobin, and cytochrome c—differ in substitution rate because different proportions of their amino acid substitutions are selectively neutral.

Message The neutral theory of molecular evolution proposed that most mutations in DNA or amino acid replacements between species are functionally neutral or nearly neutral and fixed by random genetic drift. The assumption of neutrality offers a baseline expectation of how DNA should change over time when natural selection is absent.

The rate of neutral substitutions

As we saw in Chapter 18 (see Box 18-5), we can calculate the expected rate of neutral changes in DNA sequences over time. If μ is the rate of new mutations at a locus per gene copy per generation, then the absolute number of new mutations that will appear in a population of N diploid individuals is $2N\mu$. The new mutations are subject to random genetic drift: most will be lost from the population, while a few will become fixed and replace the previous allele. If a newly arisen mutation is neutral, then there is a probability of $1/(2N)$ that it will replace the previous allele because of random genetic drift. Each one of the $2N\mu$ new mutations that will appear in a population has a probability of $1/(2N)$ of eventually taking over that population. Thus, the absolute substitution rate k is the mutation rate multiplied by the probability that any one mutation will eventually take over by drift:

$$K = \text{rate of neutral substitution} = 2N\mu \times 1/(2N) = \mu$$

That is, we expect that, in every generation, there will be μ substitutions in the population, purely from the genetic drift of neutral mutations.

> **Message** The rate of substitutions in DNA in evolution resulting from the random genetic drift of neutral mutations is equal to the mutation rate to such alleles, μ.

The signature of purifying selection on DNA

When measurements of molecular change deviate from what is expected for neutral changes, that is an important signal—a signal that selection has intervened. That signal may reveal that selection has favored some specific change or that it has rejected others. We will examine the latter case first, as the most pervasive influence of natural selection on DNA is to conserve gene function and sequence. The case of positive natural selection will be covered in Section 20.4.

All classes of DNA sequences, including exons, introns, regulatory sequences, and sequences in between genes, show nucleotide diversity among individuals within populations and between species. Of course, not all mutations in DNA are neutral with respect to gene function or organismal fitness. Mutations of DNA may also be deleterious, reducing the probability of the survival and reproduction of their carriers. The laboratory mutants used by experimental geneticists usually have mutations with some deleterious effect on fitness. A third possibility is that mutations may increase fitness by increasing efficiency, by expanding the range of environmental conditions in which the species can make a living, or by enabling the organism to adjust to changes in the environment.

The constant rate of neutral substitutions predicts that, if the number of nucleotide differences between two species were plotted against the time since their divergence from a common ancestor, the result should be a straight line with slope equal to μ. That is, evolution should proceed according to a **molecular clock** that is ticking at the rate μ. Figure 20-4 shows such a plot for the β-globin gene. The results are quite consistent with the claim that nucleotide substitutions have been neutral in the past 500 million years. Two sorts of neutral nucleotide substitutions are plotted: **synonymous substitutions,** which are from one alternative codon to another, making no change in the amino acid, and **nonsynonymous substitutions,** which result in an amino acid change. Figure 20-4 shows a much lower slope for nonsynonymous substitutions than for synonymous changes, which means that the

Figure 20-4 The amount of nucleotide divergence at synonymous sites is greater than the amount of divergence at nonsynonymous sites of the β-globin gene.

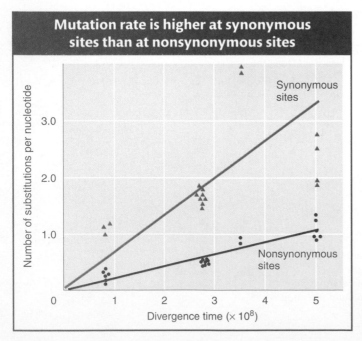

Mutation rate is higher at synonymous sites than at nonsynonymous sites

Synonymous sites

Nonsynonymous sites

Number of substitutions per nucleotide

3.0

2.0

1.0

0   1   2   3   4   5

Divergence time ($\times 10^8$)

mutation rate to neutral nonsynonymous substitutions is much lower than that to synonymous neutral substitutions.

This outcome is precisely what we expect under natural selection. Mutations that cause an amino acid substitution should have a deleterious effect more often than synonymous substitutions, which do not change the protein. Such deleterious variants will be removed from populations by *purifying selection* (see Chapter 18). A lower-than-expected ratio of nonsynonymous to synonymous changes is a signature of purifying selection. It is important to note that these observations do not show that synonymous substitutions have no selective constraints on them; rather, they show that these constraints are, on the average, not as strong as those for mutations that change amino acids. So, a synonymous change, although it has no effect on the amino acid sequence, does change the mRNA for that sequence and thus may affect mRNA stability or efficiency at which the message is translated.

Purifying selection is the most widespread, but often overlooked, facet of natural selection. The "rejection of injurious variations," as Darwin termed it, is pervasive. Purifying selection explains why we find many protein sequences that are unchanged or nearly unchanged over vast spans of evolutionary time. For example, there are several dozen genes that exist in all domains of life—Archaea, bacteria, fungi, plants, and animals—and encode proteins whose sequences have been largely conserved over 3 billion years of evolution. To preserve such sequences, variants that have arisen at random in billions of individuals in tens of millions of species have been rejected by selection over and over again.

> **Message** Purifying selection is a pervasive aspect of natural selection that reduces genetic variation and preserves DNA and protein sequences over aeons of time.

Another prediction of the theory of neutral evolution is that different proteins will have different clock rates, because the metabolic functions of some proteins will be much more sensitive to changes in their amino acid sequences. Proteins in which every amino acid makes a difference will have a lower rate of neutral mutation because a smaller proportion of their mutations will be neutral compared with proteins that are more tolerant of substitution. Figure 20-3 shows a comparison of the clocks for fibrinopeptides, hemoglobin, and cytochrome c. That fibrinopeptides have a much higher proportion of neutral mutations is reasonable because these peptides are merely a nonmetabolic safety catch, cut out of fibrinogen to activate the blood-clotting reaction. It is less obvious why hemoglobins are less sensitive to amino acid changes than is cytochrome c.

> **Message** The rate of neutral evolution for the amino acid sequence of a protein depends on the sensitivity of the protein's function to amino acid changes.

The conservation of gene sequences by purifying selection and the neutral evolution of gene sequences are two crucial dimensions of the evolutionary process, but neither of them account for the origin of adaptations. In the next three sections of the chapter, we will illustrate several examples of the ways in which genetic changes are linked to organismal diversity.

20.3 Natural Selection in Action: An Exemplary Case

For nearly a century after the publication of *On the Origin of Species* there was not one example of natural selection that had been fully elucidated, that is, where the agent of natural selection was known, the effect on different

Figure 20-5 A colorized electron micrograph showing sickle cells among normal red blood cells. [*Eye of Science/ Science Source.*]

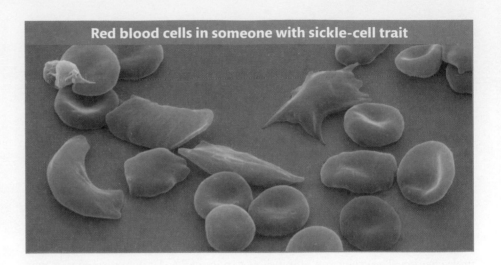

Red blood cells in someone with sickle-cell trait

genotypes could be measured, the genetic and molecular basis of variation was identified, and the physiological role of the gene or protein involved was well understood.

The first such "integrated" example of natural selection on a molecular variant was elucidated in the 1950s, before the genetic code was even deciphered. Remarkably, this trailblazing work revealed natural selection operating on humans. It still stands today as one of the most detailed and important examples of evolution by natural selection in any species.

The story began when Tony Allison, a Kenyan-born, Oxford medical student undertook a field of study of blood types among Kenyan tribes. One of the blood tests he ran was for sickle cells, red blood cells that form a sickle shape on exposure to the reducing agent sodium betasulfite or after standing for a few days (Figure 20-5). The deformed cells are a hallmark of sickle-cell anemia, a disease first described in 1910. These cells cause pathological complications by occluding blood vessels and lead to early mortality.

In 1949, the very year Allison went into the field, Linus Pauling's research group demonstrated that patients with sickle-cell anemia had a hemoglobin protein with an abnormal charge (Hemoglobin S, or HbS) in their blood, compared with the hemoglobin of unaffected individuals (Hemoglobin A, or HbA). This was the first demonstration of a molecular abnormality linked to a complex disease. It was generally understood at the time that carriers of sickle cell were heterozygous and thus had a mixture of HbA and HbS (denoted *AS*), whereas afflicted individuals were homozygous for the *HbS* allele (denoted *SS*).

Allison collected blood specimens from members of the Kikuyo, Masai, Luo, and other tribes across the very diverse geography of Kenya. While he did not see any particularly striking association between ABO or MN blood types among the tribes, he measured remarkably different frequencies of *HbS*. In tribes living in arid central Kenya or in the highlands, the frequency of *HbS* was less than 1 percent; however, in tribes living on the coast or near Lake Victoria, the frequency of *HbS* often exceeded 10 percent and approached 40 percent in some locations (Table 20-2).

Table 20-2   Frequency of *HbS* in Particular Kenyan Tribes

Tribe	Ethnic affinity	District/region	% *HbS*
Luo	Nilotic	Kisumu (Lake Victoria)	25.7
Suba	Bantu	Rusingo Island	27.7
Kikuyu	Bantu	Nairobi	0.4

The allele frequencies were surprising for two reasons. First, since sickle-cell anemia was usually lethal, why were the frequencies of the *HbS* allele so high? And second, given the relatively short distances between regions, why was the *HbS* frequency high in some places and not others?

Allison's familiarity with the terrain, tribes, and tropical diseases of Kenya led him to the crucial explanation. Allison realized that the *HbS* allele was at high frequency in low-lying humid regions with very high levels of malaria and nearly absent at high altitudes such as around Nairobi. Carried by mosquitoes, the intracellular parasite *Plasmodium falciparum*, which causes malaria, multiplies inside red blood cells (Figure 20-6). Mosquitoes and the disease are prevalent throughout sub-Saharan Africa in humid, low-lying regions near bodies of water where the mosquitoes reproduce. Allison surmised that the *HbS* allele might, by altering red blood cells, confer some degree of resistance to malarial infection.

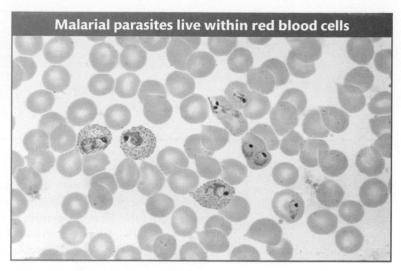

Malarial parasites live within red blood cells

Figure 20-6 A blood smear of an individual infected with malarial parasites. A red blood cell sample was treated with Giemsa stain to reveal parasites within cells (red dots). [*Photo courtesy of Dr. Mae Melvin, CDC Public Health Image Library.*]

The selective advantage of *HbS*

In order to test this idea, Allison carried out a much larger survey of *HbS* frequencies across eastern Africa, including Uganda, Tanzania, and Kenya. He examined about 5000 individuals representing more than 30 different tribes. Again he found *HbS* frequencies of up to 40 percent in malarial areas and frequencies as low as 0 percent where malaria was absent.

The link suggested that the *HbS* allele might affect parasite levels, so Allison also undertook a study of the level of parasites in the blood of heterozygous *AS* children versus wild-type *AA* children. In a study of nearly 300 children, he found the incidence of malarial parasites was indeed lower in *AS* children (27.9 percent) than in *AA* children (45.7 percent) and that parasite density was also lower in *AS* children. The results indicated that AS children had a lower incidence and severity of malarial infection and would thus have a selective advantage in areas where malaria was prevalent.

The advantage to *AS* heterozygotes was especially striking in light of the disease suffered by *SS* homozygotes. Allison noted:

> The proportion of individuals with sickle cells in any population, then, will be the result of a balance between two factors: the severity of malaria, which will tend to increase the frequency of the gene, and the rate of elimination of the sickle-cell genes in individuals dying of sickle-cell anaemia. . . . Genetically speaking, this is a *balanced polymorphism* [emphasis added], where the heterozygote has an advantage over either homozygote.

In other words, the sickle-cell mutation was under *balancing selection* (see Chapter 18) in areas where malaria was present. Positive selection operating on AS individuals is balanced by natural selection operating against *AA* individuals susceptible to malaria and *SS* individuals who would succumb to sickle-cell anemia.

How much of an advantage do *AS* individuals experience? This can be calculated by measuring the frequency of the *HbS* allele in populations and examining how these frequencies differ from the frequencies expected under the assumptions of the Hardy–Weinberg equation (see Chapter 18). A large-scale survey of 12,387 West Africans has revealed an *HbS* allele frequency (q) of 0.123. The frequencies calculated from the Hardy–Weinberg equation are lower for the

Table 20-3   The Fitness Advantage of Sickle-Cell Heterozygotes

Genotype	Observed phenotype frequency	Expected phenotype frequency	Ratio of observed/ expected	W (relative fitness)	Selective advantage
SS	29	187.4	0.155	0.155/1.12 = 0.14	
AS	2993	2672.4	1.12	1.12/1.12 = 1.00	**1.0/0.88 = 1.136**
AA	9365	9527.2	0.983	0.983/1.12 = 0.88	
Total	12,387	12,387			

homozygous phenotypes and higher for the heterozygous phenotype (Table 20-3). If it is assumed that the *AS* heterozygote has a fitness of 1.0, then the relative fitness of the other genotypes can be estimated from these differences (see Table 20-3). The relative fitness of the heterozygous *AS* genotype is 1.0/0.88 = 1.136, which corresponds to a selective advantage of approximately 14 percent.

This selective advantage has been well documented by long-term survival studies of *AA*, *AS*, and *SS* children in Kenya. These studies have found that *AS* individuals have a pronounced survival advantage over *AA* and *SS* individuals in the first few years of life (Figure 20-7).

> **Message** The sickle-cell hemoglobin allele, *HbS,* is under balancing selection in malarial zones and conveys a large survival advantage in heterozygotes over the first few years of life.

The molecular origins of *HbS*

After Allison's discovery, there was keen interest in determining the molecular basis of the difference(s) between *HbS* and *HbA*. Protein sequencing determined

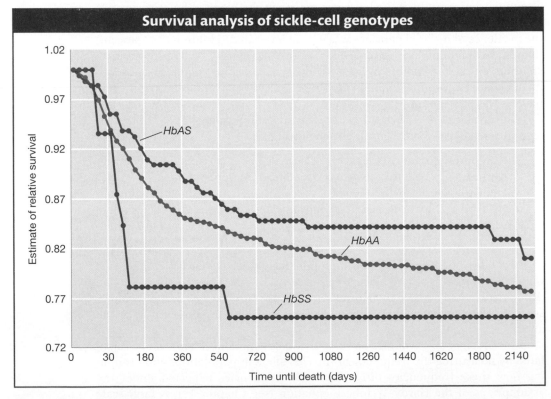

Figure 20-7 The relative survival of approximately 1000 children from Kisumu is plotted from birth through the first few years of life. Sickle-cell heterozygotes experienced a significant advantage in overall survival from ages 2 to 16 months. [*From M. Aidoo et al., The Lancet 359, 2002, 1311–1312.*]

that *HbS* differs from *HbA* by just one amino acid, a valine in the place of a glutamic acid residue. This single amino acid change alters the charge of hemoglobins and causes it to aggregate into long rodlike structures within red blood cells. Once the genetic code was deciphered and methods for sequencing DNA were developed, *HbS* was determined to be caused by a single point mutation (GAG → GTG) in the glutamic acid codon encoding the sixth amino acid of the β-globin subunit within the hemoglobin protein.

Interestingly, Allison also noted a high incidence of *HbS* outside of Africa, including in Italy, Greece, and India. Other blood-type markers did not indicate strong genetic relationships among these populations. Rather, Allison observed that these were also areas with a high incidence of malaria. The correlation between *HbS* frequency and the incidence of malaria held across not only East Africa, but the African continent, southern Europe, and the Indian subcontinent. Allison composed maps showing these striking correlations (Figure 20-8) and inferred that the *HbS* alleles in different regions arose independently, rather than through spreading by migration. Indeed, with the advent of tools for DNA genotyping, it is clear that the *HbS* mutation has arisen independently in five different haplotypes and then increased to high frequency in particular regions. Based on the limited genetic diversity of malarial populations, it is believed that *HbS* mutations arose in just the past several thousand years, once populations began living around bodies of water with the advent of agriculture.

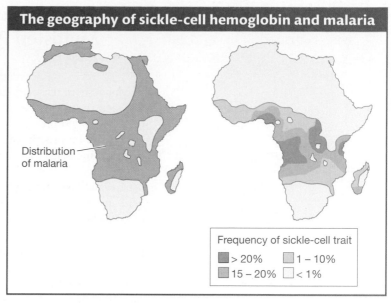

The geography of sickle-cell hemoglobin and malaria

Distribution of malaria

Frequency of sickle-cell trait
> 20%   1 – 10%
15 – 20%   < 1%

Figure 20-8 These maps show the close correspondence between the distribution of malaria (*left*) and the frequency of the sickle-cell trait (*right*) across Africa. [*Based on A. C. Allison, Genetics 66, 2004, 1591; redrawn by Leanne Olds.*]

Message The role of sickle-cell hemoglobin *S* mutation in conferring resistance to malaria was the first example of natural selection to be elucidated where the agent of selection was demonstrated, the relative fitness of different genotypes could be measured, and the genetic and molecular basis of functional variation was pinpointed.

The role of *HbS* in conferring resistance to malaria illustrated three important facets of the evolutionary process:

1. *Evolution can and does repeat itself.* The multiple independent origins and expansions of the *HbS* mutation demonstrate that given sufficient population size and time, the same mutations can arise and spread repeatedly. Many other examples are now known of the precise, independent repetition of the evolution of adaptive mutations, and we will encounter several more in this chapter.

2. *Fitness is a very relative, conditional status.* Whether a mutation is advantageous or disadvantageous, or neither, depends very much on environmental conditions. In the absence of malaria, *HbS* is very rare and disfavored. Where malaria is present, it can reach high frequencies despite the disadvantages imparted to *SS* homozygotes. In African Americans, the frequency of *HbS* is on the decline because of selection against the allele in the absence of malaria in North America.

3. *Natural selection acts on whatever variation is available, and not necessarily by the best means imaginable.* The *HbS* mutation, while protective against malaria, also causes a life-threatening condition. In areas where malaria is prevalent, which includes over 40 percent of the world's population, the imperative of combating malaria counterbalances the deleterious effect of the sickle-cell mutation.

20.4 Cumulative Selection and Multistep Paths to Functional Change

Because so much sequence evolution is neutral, there is no simple relation between the amount of change in a gene's DNA and the amount of change, if any, in the encoded protein's function. At one extreme, almost the entire amino acid sequence of a protein can be replaced while maintaining the original function if those amino acids that are substituted maintain the enzyme's three-dimensional structure.

In contrast, the function of an enzyme can be changed by a single amino acid substitution. The sheep blowfly, *Lucilia cuprina,* has evolved resistance to organophosphate insecticides used widely to control it. Richard Newcombe, Peter Campbell, and their colleagues showed that this resistance is the consequence of a single substitution of an aspartic acid for a glycine residue in the active site of an enzyme that is ordinarily a carboxylesterase (splits a carboxyl ester, R–COO–R, into an alcohol and a carboxylate). The mutation causes complete loss of the carboxylesterase activity and its replacement by esterase activity (splits any ester, R–O–R, into an acid and an alcohol). Three-dimensional modeling of the molecule indicates that the substituted protein gains the ability to bind a water molecule close to the site of attachment of the organophosphate. The water molecule then reacts with the organophosphate, splitting it in two.

> **Message** There is no proportionate relation between how much DNA change takes place in evolution and how much change in function results.

Selection clearly plays a role in the evolution of insect carboxylesterase and insecticide resistance. In many cases, however, the amino acid replacements that alter the function of the protein are more numerous and accumulate through repeated rounds of mutation and selection, what is referred to as **cumulative selection.** The power of cumulative selection to drive greater changes in a molecule's function is one of the least appreciated facets of evolution by natural selection. One reason is that the role of selection in each of the multiple replacements is more difficult to ascertain.

> **Message** Cumulative selection can drive the fixation of many changes in evolving molecules.

In order to understand the role of selection in cases of multiple substitutions, two major approaches are taken: empirical experimental analysis and statistical methods. We will illustrate the former first.

Multistep pathways in evolution

When mutations arise at multiple sites in the evolution from one phenotypic state to another, there are multiple possible orders in which these mutations can appear, each representing a different pathway through the genetic space that evolution might take. Such multistep pathways of evolutionary change are referred to as **adaptive walks.**

Suppose that the difference between the original phenotype and the evolved form is a consequence of mutations at five sites, A, B, C, D, and E. There are many different orders in which these mutations could have occurred over evolutionary time. First, site A may have been fixed in the population, then D, then C, then E, and, finally, B. On the other hand, the order of fixation might have been E, D, A, B, C. For five sites there are $5 \times 4 \times 3 \times 2 \times 1 = 120$ possible orders. Two important questions in understanding evolution are, How many of these alternative evolutionary pathways are possible? and, What are the probabilities of the different possible pathways relative to one another?

Daniel Weinreich and his colleagues have characterized in detail such a set of adaptive walks through genetic space in their study of the evolution of antibiotic resistance in the bacterium *E. coli*. Resistance to the antibiotic cefotaxime is acquired through the accumulation of five mutations at different sites in the bacterial β-lactamase gene. Four of the mutations lead to amino acid changes, and the fifth is a noncoding mutation. When all five mutations are present, the minimum concentration of antibiotic required to inhibit bacterial growth increases by a factor of 100,000. The experimenters first measured the resistance conferred by a mutation at a given site in the presence of all $2^4 = 16$ possible combinations of mutants and nonmutants at the other four sites. In most combinations, but not all, a mutant at one site was more resistant, irrespective of the state of the other four sites. For example, a mutant at site G238S showed significant resistance, irrespective of the mutant or nonmutant state of the other four sites (Table 20-4). On the other hand, the mutation at the noncoding site g4205a conferred significant resistance in eight combinations, negligible change in resistance in six cases, and a *decrease* in resistance in two combinations (see Table 20-4). This dependency of the fitness advantage or disadvantage of a new mutation on the mutations that have previously been fixed is what the experimenters call **sign epistasis.**

Weinrich and colleagues measured the resistance at every stage in the temporal sequence of adding mutations one site after another. If a mutation in one of the 120 possible orderings did not confer a higher resistance, then presumably that evolutionary path would terminate, because there would be no selection either in favor of the mutation or even against it. They found that, of the 120 possible pathways through the mutational history, only 18 provided increased resistance at each mutational step. Thus, $102/120 = 85$ percent of the possible mutational pathways to maximum resistance were not accessible to evolution by natural selection. Finally, we assume that, in a population evolving resistance, the likelihood that a particular accessible pathway will actually be followed is proportional to the magnitude of the increased resistance at each step. Under that assumption, only 10 of the 18 accessible pathways will account

Table 20-4 **The Dependence of the Fitness Effects of Mutations on Prior Mutations in *E. coli***

Mutation*	Number of alleles on which mean mutational effect is			Mean proportional increase
	Positive	Negative	Negligible	
g4205a	8	2	6	1.4
A42G	12	0	4	5.9
E104K	15	1	0	9.7
M182T	8	3	5	2.8
G238S	16	0	0	1.0×10^3

*The mutations leading to antibiotic resistance are designated by their nucleotide or amino acid position. Of the 16 possible allelic combinations of the four other sites, the positive, negative, or neutral effects of the mutation are indicated along with the mean proportional increase in fitness for mutations at the indicated site.

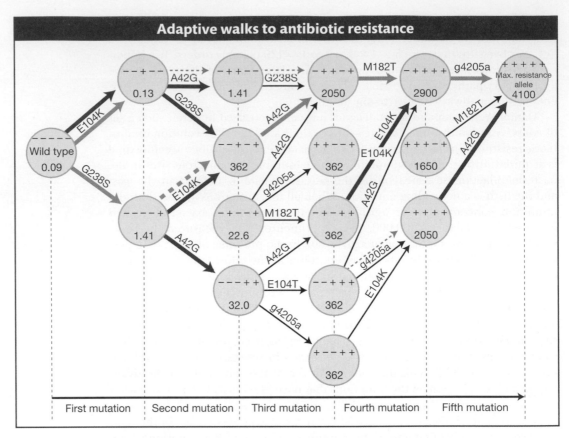

Figure 20-9 The mutational steps for the 10 most probable trajectories from wild-type susceptibility to the antibiotic cefotaxime to maximal resistance. Each circle represents an allele whose identity is denoted by a string of five + or − symbols corresponding (*left to right*) to the presence or absence of mutations g4205a, A42G, E104K, M182T, and G238S, respectively. Numbers indicate degree of cefotaxime resistance in micrograms per milliliter. The relative probability of each beneficial mutation is represented by the color and width of arrows: green—wide, highest; purple—medium, moderate; blue—narrow, low; and red—very narrow, lowest. [*From D. Weinrich et al., Science 312, 2006, 111–114.*]

for 90 percent of the cases of evolution of bacterial resistance to the antibiotic (Figure 20-9).

> **Message** The order in which mutations occur is of critical importance in determining the path of evolution and whether evolution by natural selection will or will not actually reach the most advantageous state. Because the order of occurrence of mutations is random, many advantageous phenotypes may never be achieved even though the individual mutations occur.

A key factor, then, in determining the evolutionary path a population may follow is the randomness of the mutational process. After the initial genetic variation is exhausted by the selective and random fixation of alleles, new variation arising from mutation can be the source of yet further evolutionary change. The particular direction of this further evolution depends on the particular mutations that occur and the temporal order in which they arise.

A very clear illustration of this historical contingency of adaptive walks is a selection experiment carried out by Holly Wichman and her colleagues. They forced the bacteriophage φX174 to reproduce at high temperatures and on the host *Salmonella typhimurium* instead of its normal host, *Escherichia coli*. Two independent lines of viruses were established, labeled TX and ID, and kept separate, although both were exposed to the same conditions. Both evolved the ability to reproduce at high temperatures in the new host. In one of the two lines, the ability to reproduce on *E. coli* still existed, but, in the other line, the ability was lost. The bacteriophage has only 11 genes, and so the experimenters were able to record the successive changes in the DNA for all these genes and in the proteins encoded by them during the selection process. There were 15 DNA changes in strain TX, located in six different genes; in strain ID, there were 14 changes located in four different genes. In seven cases, the changes to the two strains were identical, including a large deletion, but even these identical changes appeared in each line in a different order (Table 20-5). So, for example, the change at DNA site 1533, causing a substitution of isoleucine for threonine, was the third change in the ID strain but the 14th change in the TX strain (see Table 20-5).

Thus, the course of evolution followed by the initially identical viruses depended on the mutations available at any given time in the cumulative selection process. Contrast this situation with the repeated origin of the sickle-cell allele *HbS*: in this case, the same mutation arose and spread five times. Clearly, in some cases there are many molecular "solutions" to selective conditions and in others just one or very few.

> **Message** Under identical conditions of natural selection, two populations may arrive at identical or two different genetic compositions as a direct result of natural selection.

The experimental dissection of evolutionary pathways is very time consuming and expensive. In addition, it is often not practical for experimenters to engineer every possible genotype in an adaptive walk in populations or to attempt to measure relative fitness of many organisms in the wild. The antibiotic-resistance and viral-host examples are cases where both genetic engineering and fitness measurements are readily executed in bacteria and their viruses in the laboratory. In other situations, statistical methods have been devised to uncover a signature indicating that selection has acted on DNA and protein sequences.

Table 20-5 Molecular Substitutions in Two ϕ-X174 Bacteriophages, TX and ID, During Adaptation

Order*	TX site	Amino acid change	ID site	Amino acid change
1	782	E72, T → 1	2167	F388, H → Q
2	1727	F242, L→ F	1613	F204, T → S
3	2085	F361, A→ V	**1533⁶**	F177, T → I
4	319	C63, V→ F	1460	F153, Q → E
5	2973	H15, G → S	1300	F99, silent
6	323	C64, D → G	**1305³**	F101, G → D
7	**4110³**	A44, H → Y	1308	F102, Y → C
8	1025	F8, E → K	**4110¹**	A44, H → Y
9	**3166⁷**	H79, A → V	4637	A219, silent
10	5185	A402, T → M	**965-91⁴**	deletion
11	**1305²**	F101, G → D	**5365⁵**	A462, M → T
12	**965-91⁴**	deletion	**4168⁷**	A63, Q → R
13	**5365⁵**	A462, M → T	**3166²**	H79, A → V
14	**1533¹**	F177, T → I	1809	F269, K → R
15	**4168⁶**	A63, Q → R		

*Changes are listed in the order in which they appeared in each of the two bacteriophage selection lines. The nucleotide position is listed, followed by the protein affected, A–H, with the number of the amino acid residue and the nature of the amino acid substitution. Parallel changes are shown in boldface, and a superscript indicates the order of those changes in the other virus selection line.
Source: H. A. Wichman et al., *Science* 285, 1999, 422–424.

Table 20-6 Synonymous and Nonsynonymous Polymorphisms and Species Differences for Alcohol Dehydrogenase in Three Species of *Drosophila*

Organism	Species differences	Polymorphisms
Nonsynonymous	7	2
Synonymous	17	42
Ratio	0.29 : 0.71	0.05 : 0.95

Source: J. McDonald and M. Kreitman, "Adaptive Protein Evolution at the *Adh* locus in *Drosophila*," *Nature* 351, 1991, 652–654.

The signature of positive selection on DNA sequences

The demonstration of the molecular clock argues that most nucleotide substitutions that have occurred in evolution were neutral, but it does not tell us how much of molecular evolution has been adaptive change driven by positive selection. One way of detecting the adaptive evolution of a protein is by comparing the synonymous and nonsynonymous nucleotide polymorphisms within species with the synonymous and nonsynonymous nucleotide changes between species. Under the operation of neutral evolution by random genetic drift, polymorphism within a species is simply a stage in the eventual fixation of a new allele. So, if all mutations are neutral, the ratio of nonsynonymous to synonymous nucleotide polymorphisms within a species should be the same as the ratio of nonsynonymous to synonymous nucleotide substitutions between species. On the other hand, if the amino acid changes between species have been driven by positive selection, there ought to be an excess of nonsynonymous changes between species. Table 20-6 shows an application of this principle by John MacDonald and Martin Kreitman to the alcohol dehydrogenase gene in three closely related species of *Drosophila*. Clearly, there is an excess of amino acid replacements between species over what is expected. Therefore, we conclude that some of the amino acid replacements that helped form the enzyme were adaptive changes driven by natural selection.

20.5 Morphological Evolution

One of the most obvious and interesting categories of evolving traits is that of organism morphology. Among animals, for example, there is great diversity in the number, kind, size, shape, and color of body parts. Since adult form is the product of embryonic development, changes in form must be the result of changes in what happens during development. Recent advances in understanding the genetic

Figure 20-10 Lava flows in the Pinacate desert have produced outcrops of black-colored rock adjacent to sandy-colored substrates. [*Photograph courtesy of Michael Nachman, University of Arizona.*]

control of development (see Chapter 13) have enabled researchers to investigate the genetic and molecular bases of the evolution of animal form. We will see that some dramatic changes in animal form have a relatively simple genetic and molecular basis, while the evolution of traits governed by many "toolkit" genes involves molecular mechanisms that are distinct from those we have examined thus far. We will examine cases in which coding substitutions, gene inactivation, and regulatory sequence evolution, respectively, underlie morphological divergence.

Adaptive changes in a pigment-regulating protein

Some of the most striking and best-understood examples of morphological divergence are found in animal body-color patterns. Mammalian-coat, bird–plumage, fish-scale, and insect-wing color schemes are wonderfully diverse. Investigators have made much progress in understanding the genetic control of color formation and its role in the evolution of color differences within and between species.

In the Pinacate region of southwestern Arizona, dark rocky outcrops are surrounded by lighter-colored sandy granite (Figure 20-10). The rock pocket mouse, *Chaetodipus intermedius*, inhabits the Pinacate as well as other rocky areas of the Southwest. The mice found on the lava outcrops are typically dark in color, whereas those found in surrounding areas of sandy-colored granite or on the desert floor are usually light colored (Figure 20-11). Field studies suggest that color matching of coat color and environment protects mice against being seen by predators.

The rock pocket mice give an example of *melanism*—the occurrence of a dark form within a population or species. Melanism is one of the most common types of phenotypic variation in animals. The dark color of the fur is due to heavy deposition of the pigment *melanin,* the most widespread pigment in the animal kingdom. In mammals, two types of melanin are produced in melanocytes (the pigment cells of the epidermis and hair follicles): eumelanin, which forms black or brown pigments, and phaeomelanin, which forms yellow or red pigments. The relative amounts of eumelanin and phaeomelanin are controlled by the products of several genes. Two key proteins are the melanocortin 1 receptor (MC1R) and the agouti protein. During the hair-growth cycle, the α-melanocyte-stimulating

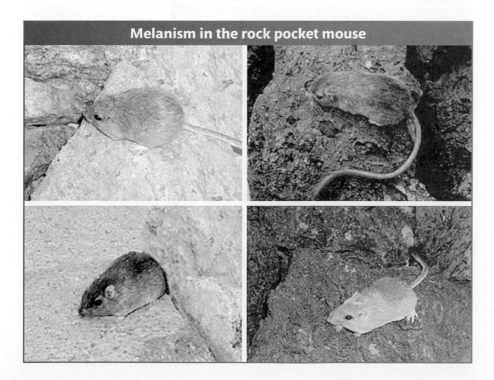

Melanism in the rock pocket mouse

Figure 20-11 Light- and dark-colored *Chaetodipus intermedius* from the Pinacate region of Arizona are shown on sandy-colored and dark lava-rock backgrounds. [*Photographs courtesy of Michael Nachman, from M. W. Nachman, H. E. Hoekstra, and S. L. D'Agostino, "The Genetic Basis of Adaptive Melanism in Pocket Mice," Proc. Natl. Acad. Sci. USA 100, 2003, 5268–5273.*]

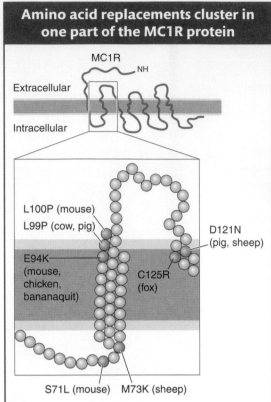

Amino acid replacements cluster in one part of the MC1R protein

Figure 20-12 Amino acid replacements (orange circles) associated with melanism vary slightly in location in different species but are located in the same part of the MC1R protein. The upper part of the figure shows the general topology of the MC1R protein. The region in which replacements are located is enlarged in the lower part of the figure. [*After E. Eizirik et al., "Molecular Genetics and Evolution in the Cat Family," Curr. Biol. 13, 2003, 448–453; reprinted with permission from Elsevier.*]

hormone (α-MSH) binds to the MC1R protein, which triggers the induction of pigment-producing enzymes. The agouti protein blocks MC1R activation and inhibits the production of eumelanin.

Michael Nachman and his colleagues examined the DNA sequences of the *mc1r* genes of light- and dark-colored pocket mice. They found the presence of four mutations in the *mc1r* gene in dark mice that cause the MC1R protein to differ at four amino acid residues from the corresponding protein in light mice. Findings from biochemical studies suggest that such mutations cause the MC1R protein to be constitutively active (active at all times), bypassing the regulation of receptor activity by the agouti protein. Indeed, mutations in *mc1r* are associated with melanism in all sorts of wild and domesticated vertebrates. Many of these mutations alter residues in the same part of the MC1R protein, and the same mutations have occurred independently in some species (Figure 20-12).

In many ways, we can think of these dark mice as analogs of Darwin's finches and the lava outcrops as new "island" habitats produced by the same volcanic activity that produced the Galápagos Islands. The sandy-colored form of the mouse appears to be the ancestral type, akin to the continental ancestral finch that colonized the Galápagos. The advantage of being less visible to predators resulted in natural selection for coat color, and the invasion of the lava-rock islands by the mice led to the spread of an allele that was favored on the black-rock background and selected against on the sandy-colored background. New mutations in the *mc1r* gene were essential to this adaptation to the changing landscape.

The evolution of melanism in the pocket mice illustrates how fitness depends on the conditions in which an organism lives. The new black mutation was favored on the lava outcrops but disfavored in the ancestral population living on sandy-colored terrain.

> **Message** The relative fitness of a new variant depends on the immediate selective conditions. A mutation that may be beneficial in one population may be deleterious in another.

Gene inactivation

It has long been noted that cave-dwelling animals are often blind and uncolored. Darwin noted in *The Origin of Species* that "several animals belonging to the most different classes, which inhabit the caves of Carniola [in Slovenia] and Kentucky, are blind. As it is difficult to imagine that eyes, though useless, could be in any way injurious to animals living in darkness, their loss may be attributed to disuse."

Many species of fish that live in caves have lost their eyes and body color. Because these species belong to many different families that include surface-dwelling, eye-bearing species, the loss of eyes and pigmentation has clearly occurred repeatedly. For example, the Mexican blind cave fish (*Astyonax mexicanus*) belongs to the same order as the piranha and the colorful neon tetra. About 30 cave populations of fish in Mexico have lost the body color of their surface-dwelling relatives (Figure 20-13).

Genetic studies have indicated that albinism in the Pachón cave fish population is due to a single recessive mutation. Furthermore, a cross between a Molino cave individual and a Pachón cave individual produced only albino offspring, suggesting that the albinism in the two populations is due to the same genetic locus. To identify the gene responsible for albinism in the fish, researchers investigated the genotypes of fish at several pigmentation loci known to cause albinism in mice or humans. They found that one of these genes, *Oca2,* mapped to the albino locus.

They also found that there was a perfect association between the genotype of the *Oca2* locus and the phenotype of albinism in F$_2$ offspring that were a backcross between Molino and Molino/surface F$_1$ progeny or Pachón and Pachón/surface F$_1$ progeny.

Further inspection of the *Oca2* gene revealed that the Pachón population was homozygous for a deletion that extended from an intron through most of an exon and that the Molino population was homozygous for the deletion of a different exon. Functional analyses proved that each deletion in the *Oca2* gene caused loss of *Oca2* function.

The identification of different lesions in the *Oca2* gene of the two cave populations indicates that albinism evolved separately in the two cave populations. There is also evidence that a third cave population carries yet a third, distinct *Oca2* mutation. It is known from other vertebrates that albinism can evolve through mutations in other genes. What might account for the repeated inactivation of the *Oca2* gene? There are two likely explanations. First, *Oca2* mutations appear to cause no serious collateral defects other than loss of pigmentation and vision. Some other pigmentation genes, when mutated in fish, cause dramatic reductions in viability. The effects of *Oca2* mutations appear, then, to be less *pleiotropic* and have effects on overall fitness that are less harmful than those of mutations in other fish pigmentation genes. Second, the *Oca2* locus is very large, spanning some 345 kb in humans and containing 24 exons. It presents a very large target for random mutations that would disrupt gene function; *Oca2* mutations are therefore more likely to arise than are mutations at smaller loci.

The loss of gene function is not what we usually think about when we think about evolution. But gene inactivation is certainly what we should predict to happen when selective conditions change or when populations or species shift their habitats or lifestyles and certain gene functions are no longer necessary.

> **Message** Gene-inactivating mutations may occur and rise to high frequency when habitat or lifestyle changes relax natural selection on traits and underlying gene functions.

Regulatory-sequence evolution

As discussed above, a major constraint on gene evolution is the potential for harmful side effects caused by mutations in coding regions that alter protein function. These effects can be circumvented by mutations in regulatory sequences, which play a major role in the evolution of gene regulation and body form.

The examples of body-coloration evolution we have looked at thus far have the coat or scale pattern changing over the entire body. The evolution of solid black or entirely unpigmented body coloration can arise through mutations in pigmentation genes. However, many color schemes are often made up of two or more colors in some spatial pattern. In such cases, the expression of pigmentation genes must differ in areas of the body that will be of different colors. In different populations or species, the regulation of pigmentation genes must evolve by some mechanism that does not disrupt the function of pigmentation proteins.

The species of the fruit-fly genus *Drosophila* display extensive diversity of body and wing markings. A common pattern is the presence of a black spot near the tip of the wing in males (Figure 20-14). The production of the black spots requires enzymes that synthesize melanin, the same pigment made in pocket mice. Many genes controlling the melanin synthesis pathway have been well studied in the model organism *Drosophila melanogaster*. One gene is named

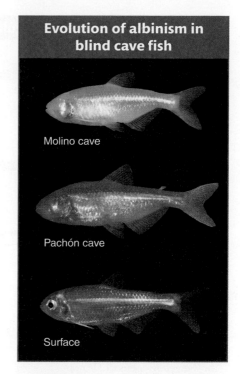

Evolution of albinism in blind cave fish

Molino cave

Pachón cave

Surface

Figure 20-13 Surface forms of the fish *Astyanu mexicanus* appear normal, but cave populations, such as those from the Molino and Pachón caves in Mexico, have repeatedly evolved blindness and albinism. [*Photographs courtesy of Meredith Protas and Cliff Tabin, Harvard Medical School.*]

Wing spots on fruit flies

Figure 20-14 *Drosophila melanogaster* males lack wing spots (*top*), whereas *Drosophila biarmipes* males (*bottom*) have dark wing spots that are displayed in a courtship ritual. This simple morphological difference is due to differences in the regulation of pigmentation genes. [*Photographs by Nicolas Gompel.*]

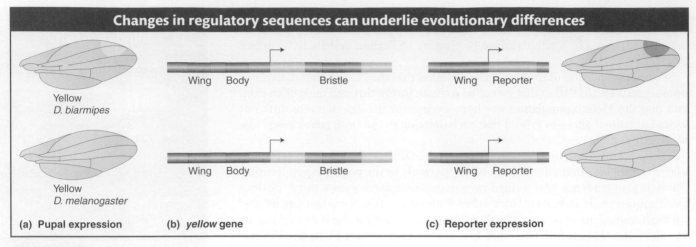

Figure 20-15 The evolution of gene regulation and morphology in the case shown is due to evolution in cis-acting regulatory sequences. (a) In spotted fruit flies, the Yellow pigmentation protein is expressed at high levels in cells that will produce large amounts of melanin. (b) The *yellow* locus of *Drosophila* species contains several discrete cis-acting regulatory elements (red) that govern *yellow* transcription in different body parts. Exons are shown in gold. Arrows indicate the point of the start and direction of transcription of the gene. (c) The "wing" regulatory element from *D. biarmipes* drives reporter-gene expression in a spot pattern in the developing wing, whereas the homologous element from the unspotted *D. melanogaster* does not drive a spot pattern of reporter expression. This difference in wing cis-acting-regulatory-element activities demonstrates that changes in the cis-acting-regulatory-element function underlie differences in Yellow expression and pigmentation between the two species.

yellow because mutations in the gene cause darkly pigmented areas of the body to appear yellowish or tan. The *yellow* gene plays a central role in the development of divergent melanin patterns. In species with spots, the Yellow protein is expressed at high levels in wing cells that will produce the black spot, whereas in species without spots, Yellow is expressed at a low level throughout the wing blade (Figure 20-15a).

The difference in Yellow expression between spotted and unspotted species could be due to differences in how the *yellow* gene is regulated in the two species. Either or both of two possible mechanisms could be at play: the species could differ in the spatial deployment of transcription factors that regulate *yellow* (that is, changes in trans-acting sequences to the *yellow* gene) or they could differ in cis-acting regulatory sequences that govern how the *yellow* gene is regulated. To examine which mechanisms are involved, investigators examined the activity of *yellow* cis-acting regulatory sequences from different species by placing them upstream of a reporter gene and introducing them into *D. melanogaster*.

The *yellow* gene is regulated by an array of separate cis-acting regulatory sequences that govern gene transcription in different tissues and cell types and at different times in development (Figure 20-15b). These regulatory sequences include those controlling transcription in the larval mouthparts, the pupal thorax and abdomen, and the developing wing blade. It was discovered that, whereas the wing-blade cis-acting regulatory element from unspotted species drives low-level expression of a reporter gene across the wing blade, the corresponding element from a spotted species, such as *D. biarmipes* or *D. elegans*, drives a high level of reporter expression in a spot near the tip of the wing (Figure 20-15c). These observations show that changes in sequence and function of a cis-acting regulatory element are responsible for the change in *yellow* regulation and contribute to the origin of the wing spot. The cis-acting regulatory element of the spotted species has apparently acquired binding sites for transcription factors that now drive high levels of gene transcription in a spot pattern in the developing wing.

Thus, evolutionary changes in cis-acting regulatory sequences play a critical role in the evolution of body form. The location of change in the regulatory sequence rather than the gene itself can be best explained in light of the many different effects that can appear as the result of a coding mutation in a "toolkit" gene. In this instance, the *yellow* gene is highly pleiotropic: it is required for the pigmentation of many structures and for functions in the nervous system as well. A coding mutation that alters Yellow protein activity would alter Yellow activity in *all* tissues, which might have a negative consequence for fitness. However, because individual cis-acting regulatory sequences usually affect only one aspect of gene expression, mutations in these sequences provide a mechanism for changing one aspect of gene expression while preserving the role of protein products in other developmental processes.

> **Message** Evolutionary changes in cis-acting regulatory sequences play a critical role in the evolution of gene expression. They circumvent the pleiotropic effects of mutations in the coding sequences of genes that have multiple roles in development.

Loss of characters through regulatory-sequence evolution

Morphological characters may be lost as well as gained as the result of adaptive changes in cis-acting regulatory sequences. If there is no selective pressure to maintain a character, it can be lost over time. But some losses are beneficial because they facilitate some change in lifestyle. Hind limbs, for example, have been lost many times in vertebrates—in snakes, lizards, whales, and manatees—as these organisms adapted to different habitats and means of locomotion. Evolutionary changes in cis-acting regulatory sequences are also linked to these dramatic changes.

The evolutionary forerunners of the hind limbs of four-legged vertebrates are the pelvic fins of fish. Dramatic differences in pelvic-fin anatomy have evolved in closely related fish populations. The three-spine stickleback fish occurs in two forms in many lakes in North America—an open-water form that has a full spiny pelvis and a shallow-water, bottom-dwelling form with a dramatically reduced pelvis and spines. In open water, the long spines help protect the fish from being swallowed by larger predators. But on the lake bottom, those spines are a liability because they can be grasped by dragonfly larvae that feed on the young fish (Figure 20-16, top).

Figure 20-16 Deletions within a *Pitx1* cis-regulatory element underlie the adaptive evolution of the pelvic skeleton of stickleback fish. (a) One form of the three-spine stickleback fish inhabits shallow water, and a different form inhabits open water. (b) The shallow-water form has a reduced pelvic skeleton (*left, black arrow*) relative to the open-water form (*right*). (c) This reduction is due to the selective loss of expression of the *Pitx1* gene (orange) from the pelvic fin bud during development of the stickleback larvae (black arrows). (d) The loss of *Pitx1* expression in turn is due to the mutation of an enhancer of the *Pitx1* gene specific to the pelvic fin (X marks the mutated enhancer). Other enhancers of the *Pitx1* gene, which control expression of the gene elsewhere in the developing body, are unaffected and function similarly in both forms of the fish.

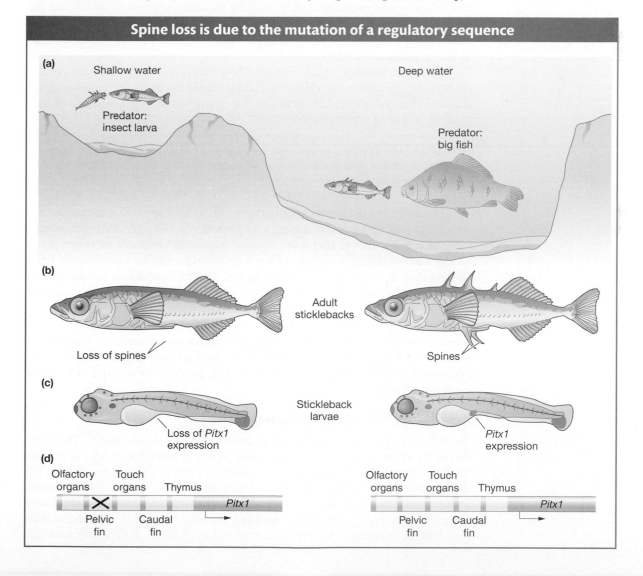

Spine loss is due to the mutation of a regulatory sequence

The differences in pelvic morphology have evolved repeatedly in just the past 10,000 years, since the recession of the glaciers of the last ice age. Many separate lakes were colonized by long-spined oceanic sticklebacks, and forms with reduced pelvic spines evolved independently several times. Because the fish are so closely related and interbreed in the laboratory, geneticists can map the genes involved in the reduction of the pelvis. David Kingsley's group at Stanford University along with Dolph Schluter's group at the University of British Columbia mapped one major factor involved in pelvic differences to the *Pitx1* gene, which encodes a transcription factor. Like most other developmental toolkit genes, the *Pitx1* gene has several distinct functions in fish development. However, in the form of the stickleback with a reduced pelvis, its expression is lost from the area of the developing fish embryo that will give rise to the pelvic-fin bud and spines (see Figure 20-16).

The fact that the difference in pelvic morphology between the two forms mapped to the *Pitx1* locus and was associated with the loss of gene expression suggested that changes in *Pitx1* regulatory sequences were responsible for the difference in phenotypes. Like most pleiotropic toolkit genes, the expression of the *Pitx1* gene in different parts of the developing fish is controlled by separate cis-acting regulatory elements. Frank Chan and colleagues demonstrated that the regulatory element that controls *Pitx1* expression in the developing pelvis has been inactivated by large deletion mutations in multiple, independent populations of pelvic-reduced fish (Figure 20-16, bottom). Furthermore, it was observed that heterozygosity was reduced around the cis-acting sequences controlling pelvic expression relative to other nearby sequences. This observation is consistent with the deletion allele being favored by natural selection acting on the bottom-dwelling, pelvic-reduced form.

Thus, these findings further illustrate how mutations in regulatory sequences circumvent the pleiotropic effects of coding mutations in toolkit genes and that adaptive changes in morphology can be due to the loss as well as the gain of gene expression during development.

> **Message** Adaptive changes in morphology can result from inactivation of regulatory sequences and loss of gene expression as well as the modification of regulatory sequences and the gain of gene expression.

Circumventing the potentially harmful side effects of coding mutations is a very important factor in explaining why evolution acts by generating new roles for transcription factors that may regulate dozens to hundreds of target genes. Changes in the coding sequences of a transcription factor—for example, the DNA-binding domain—may affect all target genes, with catastrophic consequences for the animal. The constraint on the coding sequences of highly pleiotropic proteins, with many functions, explains the extraordinary conservation of the DNA-binding domains of Hox proteins (see Figure 13-8) and many other transcription factors over vast expanses of evolutionary time. But, although the proteins' biochemical functions are constrained, their regulation does diverge. The evolution of the expression patterns of *Hox* and other toolkit genes plays a major role in the evolution of body form.

Regulatory evolution in humans

Regulatory evolution is not limited to genes affecting development. The level, timing, or spatial pattern of the expression of any gene may vary within populations or diverge between species. For example, as noted earlier (see Chapter 18), the frequencies of alleles at the *Duffy* blood-group locus vary widely in human populations. The *Duffy* locus (denoted *Fy*) encodes a glycoprotein that serves as a receptor for multiple intercellular signaling proteins. In sub-Saharan Africa, most members of indigenous populations carry the Fy^{null} allele. Individuals with this allele do not express any of the Duffy glycoprotein on red blood cells, although the protein is still being produced in other cell types. How and why is the Duffy glycoprotein lacking on these individuals' red blood cells?

The molecular explanation for the lack of Duffy glycoprotein expression on red blood cells is the presence of a point mutation in the promoter region of the *Duffy* gene at position -46. This mutation lies in a binding site for a transcription factor specific to red blood cells called GATA1 (Figure 20-17). Mutation of this site abolishes the activity of a *Duffy* gene enhancer in reporter-gene assays.

An evolutionary explanation suggests that the lack of Duffy glycoprotein expression on red blood cells among Africans is the result of natural selection favoring resistance to malarial infection. The malarial parasite *Plasmodium vivax* is the second most prevalent form of malarial parasite in most tropical and subtropical regions of the world but at present is absent from sub-Saharan Africa. The parasite gains entry to red blood cells and red-blood-cell precursors by binding to the Duffy glycoprotein (see Figure 20-17). The very high frequency of Fy^{null} homozygotes in Africa prevents *P. vivax* from being common there. Moreover, if we suppose that *P. vivax* was common in Africa in the past, then the Fy^{null} allele would have been selected for.

The complete absence of Duffy protein on the red blood cells of a large subpopulation raises the question of whether the Duffy protein has any necessary function, because it is apparently dispensable. But it is not the case that these individuals lack Duffy protein expression altogether. The protein is expressed on endothelial cells of the vascular system and the Purkinje cells of the cerebellum. As with the evolution of Yellow expression in wing-spotted fruit flies and of Pitx expression in stickleback fish, the regulatory mutation at the *Fy* locus allows one aspect of gene expression (in red blood cells) to change without disrupting others (see Figure 20-17).

Modifications to coding and regulatory sequences are common means to evolutionary change. They illustrate how diversity can arise without the number of genes in a species changing. However, larger-scale mutational changes can and do happen in DNA that result in the expansion of gene number, and this expansion provides raw material for evolutionary innovation.

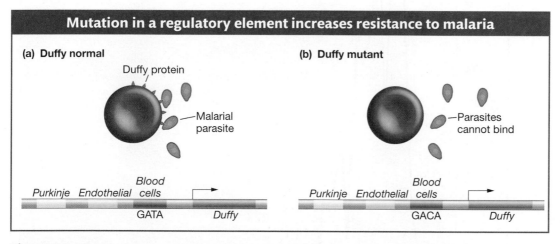

Figure 20-17 A regulatory mutation in a human *Duffy* gene enhancer is associated with resistance to malaria. (a) The Duffy protein (dark blue) is typically expressed on blood cells as well as on Purkinje cells in the brain and endothelial cells. (b) A high proportion of West Africans lack Duffy expression on their red blood cells due to a mutation in a blood-cell enhancer (the GATA sequence is mutated to GACA). Since the Duffy protein is part of the receptor for the *P. vivax* malarial parasite (orange), individuals with the regulatory mutation are resistant to infection but have normal Duffy expression elsewhere in the body.

20.6 The Origin of New Genes and Protein Functions

Evolution consists of more than the substitution of one allele for another at loci of defined function. A large fraction of protein-coding and RNA-encoding genes belong to **gene families,** groups of genes that are related in sequence and typically in biochemical function as well. For example, there are over 1000 genes encoding structurally related olfactory receptors in a mouse and three structurally related opsin genes that encode proteins necessary for color vision in humans. Within families such as these, new functions have evolved that have made possible new capabilities. These new functions may be expansions of existing capabilities. In the examples above, new receptors appeared in mice with the ability to detect new chemicals in the environment, or in the case of humans and their Old World primate relatives, new opsin proteins appeared that can detect wavelengths of light that other mammals cannot. In other cases, the evolution of new gene families may lead to entirely novel functions that open up new ways of living, such as the acquisition of antifreeze proteins in polar fish. Here, we will ask, Where does the DNA for new genes come from? What are the fates of new genes? And how do new protein functions evolve?

Expanding gene number

There are several genetic mechanisms that can expand the number of genes or parts of genes. One large-scale process for the expansion of gene number is the formation of *polyploids,* individuals with more than two chromosome sets. Polyploids

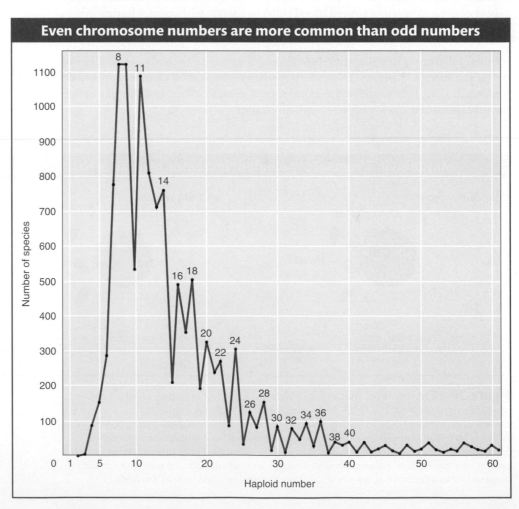

Figure 20-18 Frequency distribution of haploid chromosome numbers in dicotyledonous plants. [*After Verne Grant, The Origin of Adaptations. Columbia University Press, 1963.*

result from the duplication of the entire genome. Much more common in plants than in animals (see Chapter 17), the formation of polyploids has played a major role in the evolution of plant species. Consider the frequency distribution of haploid chromosome numbers among the dicotyledonous plant species shown in Figure 20-18. Above a chromosome number of about 12, even numbers are much more common than odd numbers—a consequence of frequent polyploidy.

A second mechanism that can increase gene number is **gene duplication.** Misreplication of DNA during meiosis can cause segments of DNA to be duplicated. The lengths of the segments duplicated can range from just one or two nucleotides up to substantial segments of chromosomes containing scores or even hundreds of genes. Detailed analyses of human-genome variation has revealed that individual humans commonly carry small duplications that result in variation in gene-copy number.

A third mechanism that can generate gene duplications is *transposition.* Sometimes, when a transposable element is transposed to another part of the genome, it may carry along additional host genetic material and insert a copy of some part of the genome into another location (see Chapter 15).

A fourth mechanism that can expand gene number is **retrotransposition.** Many animal genomes harbor retroviral-like genetic elements (see Chapter 15) that encode reverse-transcriptase activity. Retrotransposons themselves make up approximately 40 percent of the human genome. Occasionally, host genome mRNA transcripts are reverse transcribed into cDNA and inserted back in the genome, producing an intronless gene duplicate.

The fate of duplicated genes

It was once thought that because the ancestral function is provided by the original gene, duplicate genes are essentially spare genetic elements that are free to evolve new functions (termed **neofunctionalization**), and that would be a common fate. However, the detailed analysis of genomes and population-genetic considerations has led to a better understanding of the alternative fates of new gene duplicates, with the evolution of new function being just one pathway.

For simplicity's sake, let's consider a duplication event that results in the duplication of the entire coding and regulatory region of a gene (Figure 20-19a). Many different outcomes can unfold from such a duplication. The simplest result is that the allele bearing the duplicate is lost from the population before it rises to

Figure 20-19 The alternative fates of duplicated genes. (a) The duplication of a gene. The orange, yellow, and pink boxes denote cis-regulatory elements; the beige box denotes the coding region. After duplication, several alternative fates of the duplicates are possible: (b) any inactivating mutation in a coding region will render that duplicate into a pseudogene, and purifying selection will then operate on the remaining paralog; (c) mutations may arise that alter the function of a protein and may be favored by positive selection (neofunctionalization); (d) mutations may affect a subfunction of either duplicate, and so long as the two paralogs together provide the ancestral functions, different subfunctions may be retained, resulting in the evolution of two complementary loci (subfunctionalization). [*Diagram modified from V. E. Prince and F. B. Pickett, Nat. Rev. Genet. 3, 2002, 827–837.*]

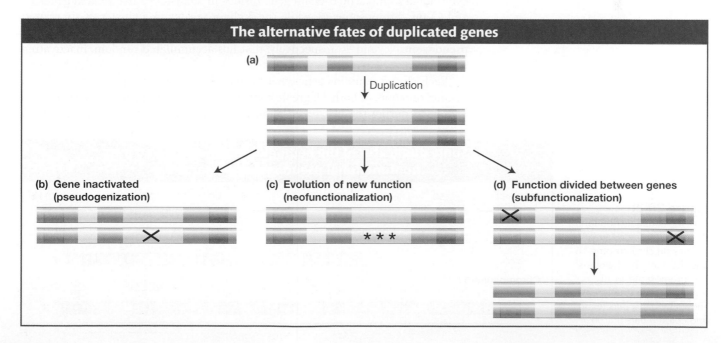

The alternative fates of duplicated genes

(a)

↓ Duplication

(b) Gene inactivated (pseudogenization)

(c) Evolution of new function (neofunctionalization)

* * *

(d) Function divided between genes (subfunctionalization)

any significant frequency, as is the fate of many new mutations (see Chapter 18). But let's consider next the more interesting scenarios: suppose the duplication survives and new mutations begin to occur within the duplicate gene pair. Keeping in mind that the original and duplicated genes are initially exact copies and therefore redundant, once new mutations arise, there are several possible fates:

1. An inactivating mutation may occur in the coding region of either duplicate. The inactivated paralog is called a **pseudogene** and will generally be invisible to natural selection. Thus, it will accumulate more mutations and evolve by random genetic drift, while natural selection will maintain the functional paralog (Figure 20-19b).

2. Mutations may occur that alter the regulation of one duplicate or the activity of one encoded protein. These alleles may then become subject to positive selection and acquire a new function (neofunctionalization) (Figure 20-19c).

3. In cases where the ancestral gene has more than one function and more than one regulatory element, as for most toolkit genes, a third possible outcome is that initial mutations inactivate or alter one regulatory element in each duplicate. The original gene function is now divided between the duplicates, which complement each other. In order to preserve the ancestral function, natural selection will maintain the integrity of both gene-coding regions. Loci that follow this path of duplication and mutation that produce complementary paralogs are said to be **subfunctionalized** (Figure 20-19d).

Some of these alternative fates of gene duplicates are illustrated in the history of the evolution of human globin genes. The evolution of our lineage, from fish ancestors to terrestrial amniotes that laid eggs to placental mammals, has required a series of innovations in tissue oxygenation. These include the evolution of additional globin genes with novel patterns of regulation and the evolution of hemoglobin proteins with distinct oxygen-binding properties.

Adult hemoglobin is a tetramer consisting of two α polypeptide chains and two β chains, each with its bound heme molecule. The gene encoding the adult α chain is on chromosome 16, and the gene encoding the β chain is on chromosome 11. The two chains are about 49 percent identical in their amino acid sequences; this similarity reflects their common origin from an ancestral globin gene deep in evolutionary time. The α-chain gene resides in a cluster of five related genes (α and ζ) on chromosome 16, while the β chain resides in a cluster of six related genes on chromosome 11 (ε, β, δ, and γ) (Figure 20-20). Each cluster contains a pseudogene, Ψ_α and Ψ_β respectively, that has accumulated random, inactivating mutations.

Each cluster contains genes that have evolved distinct expression profiles, a distinct function, or both. Of greatest interest are the two γ genes. These genes are expressed during the last seven months of fetal development to produce fetal

Figure 20-20
Chromosomal distribution of the genes for the α family of globins on chromosome 16 and the β family of globins on chromosome 11 in humans. Gene structure is shown by black bars (exons) and colored bars (introns).

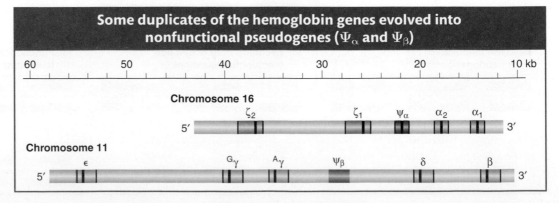

hemoglobin (also known as hemoglobin F), which is composed of two α chains and two γ chains. Fetal hemoglobin has a greater affinity for oxygen than does adult hemoglobin, which allows the fetus to extract oxygen from the mother's circulation via the placenta. At birth, up to 95 percent of hemoglobin is the fetal type, then expression of the adult β form replaces γ and a small amount of δ globin is also produced. The order of appearance of globin chains during development is orchestrated by a complex set of cis-acting regulatory sequences and, remarkably, follows the order of genes on each chromosome.

The γ genes are restricted to placental mammals. Their distinct developmental regulation and protein products mean that these duplicates have evolved differences in function that have contributed to the evolution of the placental lifestyle. Interestingly, regulatory variants of these genes are known that cause expression of the fetal hemoglobin to persist into childhood and adulthood. These naturally occurring variants appear to moderate the severity of sickle-cell anemia by suppressing the levels of *HbS* produced. One widespread strategy for the treatment of sickle-cell anemia is to administer drugs that stimulate the reactivation of fetal-hemoglobin expression.

Summary

The theory of evolution by natural selection explains the changes that take place in populations of organisms as being the result of changes in the relative frequencies of different variants in the population. If there is no variation within a species for some trait, there can be no evolution. Moreover, that variation must be influenced by genetic differences. If differences are not heritable, they cannot evolve, because the reproductive advantage of a variant will not carry across generational lines. It is crucial to understand that the mutational processes that generate variation within the genome act at random, but that the selective process that sorts out the advantageous and disadvantageous variants is not random.

The ability to study evolution at the level of DNA and proteins has transformed our understanding of the evolutionary process. Before we had the ability to study evolution at the molecular level, there was no inkling that much of evolution was in fact a result of genetic drift and not natural selection. A great deal of molecular evolution seems to be the replacement of one protein sequence by another one of equivalent function. Among the evidence for the prevalence of neutral evolution is that the number of amino acid differences between two different species in some molecule—for example, hemoglobin—is directly proportional to the number of generations since their divergence from a common ancestor in the evolutionary past. We would not expect such a "molecular clock" with a constant rate of change to exist if the selection of differences were dependent on particular changes in the environment.

So much sequence evolution is neutral that there is no simple relation between the amount of change in a gene's DNA sequence and the amount of change, if any, in the encoded protein's function. Some protein functions can change through a single amino acid substitution, whereas others require a suite of substitutions brought about through cumulative selection. Such multistep, adaptive walks may follow different paths even when the conditions of natural selection are the same. This is because the paths available to any population at any given moment depend on the chance occurrence of mutations that may not arise in the same order in different populations. Furthermore, the previous steps taken may affect whether a new mutation is favored, disfavored, or neutral.

Before the advent of molecular genetics, it was not possible to know whether independent evolutionary events might have given rise to the same adaptation multiple times. By pinpointing the genes and exact mutations involved in changes in function, we now appreciate that evolution can and does repeat itself by acting on the same genes to produce similar results in independent cases. For example, changes to the same genes are responsible for independently arising cases of melanism and albinism in some vertebrates or for the loss of pelvic spines in different stickleback-fish populations. Evolution may repeat itself by altering the very same nucleotide in the case of independently arising sickle-cell mutations that lead to adaptive resistance to malaria.

An important constraint on the evolution of coding sequences is the potentially harmful side effects of mutations. If a protein serves multiple functions in different tissues, as is the case for many genes involved in the regulation of developmental processes, mutations in coding sequences may affect all functions and decrease fitness. The potential pleiotropic effects of coding mutations can be circumvented by mutations in noncoding regulatory sequences. Mutations in these sequences may selectively change gene expression in only one tissue or body part and not others. The evolution of cis-acting regulatory sequences is central to the evolution of morphological traits and the expression of toolkit genes that control development.

New protein functions often arise through the duplication of genes and subsequent mutation. New DNA may arise by duplication of the entire genome (polyploidy), a frequent occurrence in plants, or by various mechanisms that produce duplicates of individual genes or sets of genes. The fate of duplicate genes depends a great deal on the nature of mutations acquired after duplication. Possible fates are the inactivation of one duplicate, the splitting of function between two duplicates, or the gain of new functions.

Overall, genetic evolution is subject to historical contingency and chance, but it is constrained by the necessity of organisms to survive and reproduce in a constantly changing world. "The fittest" is a conditional status, subject to change as the planet and habitats change.

KEY TERMS

adaptation (p. 728)

adaptive walk (p. 740)

cumulative selection (p. 740)

gene duplication (p. 753)

gene family (p. 752)

molecular clock (p. 734)

natural selection (p. 728)

neofunctionalization (p. 753)

nonsynonymous substitution (p. 734)

pseudogene (p. 754)

retrotransposition (p. 753)

sign epistasis (p. 741)

subfunctionalization (p. 754)

synonymous substitution (p. 734)

SOLVED PROBLEMS

SOLVED PROBLEM 1. Two closely related species of bacteria are found to be fixed for two different electrophoretically detected alleles at a locus encoding an enzyme involved in breaking down a nutrient. How could you test statistically whether that divergence is more likely to be a result of natural selection or of neutral evolution?

Solution

a.   Obtain DNA sequences of the gene from a number of separate individuals or strains from each of the two species. Ten or more sequences from each species would be desirable. The polymorphic sites within populations are then classified into nonsynonymous (a) and synonymous (b) sites. The fixed nucleotide differences between species are classified into nonsynonymous (c) and synonymous (d) differences.

b.   Tabulate the nucleotide differences among individuals within each species (polymorphisms), and classify these differences as either those that result in amino acid changes (replacement polymorphisms; a in the table below) or those that do not change the amino acid (synonymous polymorphisms; b in the accompanying table).

c.   Make the same tabulation of replacement (c) and synonymous (d) changes for the fixed differences between the species, counting only those differences that completely differentiate the species. That is, do not count a polymorphism in one species that includes a variant that is seen in the other species.

d.   If the divergence between the species is purely the result of random genetic drift, then we expect a/b to be equal to c/d. If, on the other hand, there has been selective divergence, there should be an excess of fixed nonsynonymous differences, and so a/b should be less than c/d.

e.   Test the statistical significance of the observed ratio by a 2×2 χ^2 test of the following table:

		Polymorphisms	
		Replacement	Synonymous
Species	Replacement	a	b
differences	Synonymous	c	d

$$\chi^2 = \frac{(a+b+c+d)(ad-bc)^2}{(a+c)(b+d)(a+b)(c+d)}$$

SOLVED PROBLEM 2. In the above example, how could you test experimentally whether the divergence in enzyme sequences may have caused differences in function and fitness?

Solution

In order to test whether the enzymes have different functional properties, one could devise both in vitro and in vivo experiments. If the substrates and properties of the enzyme are known, one could purify the enzyme from each species and measure directly whether there are functional differences. Alternatively, an indirect test would be whether each species grew as well on the particular nutrient that the enzyme broke down.

Ideally, in order to measure fitness differences, one would replace the enzyme-coding region of one species with the enzyme-coding region from the second species and vice versa. Then the growth of each wild-type and transgenic strain could be compared on the same nutrient-containing media, with growth being an indicator of fitness. If there are differences in the relative fitness of the transgenic and wild-type strains, then it is possible that the enzymes have diverged under natural selection. If not, then it is likely that the enzymes have evolved neutrally or that the effect of selection is too small to measure experimentally.

PROBLEMS

Most of the problems are also available for review/grading through the GENETICSPORTAL www.yourgeneticsportal.com.

WORKING WITH THE FIGURES

1. Examining Figure 20-4, explain why the rate of evolution at nonsynonymous sites is lower. Do you expect this to be true only of globin genes or of most genes?

2. In Figure 20-7, note that the difference in survival rates between AS and AA genotypes declines as children get older. Offer one possible explanation for this observation.

3. From Table 20-4, would you expect the noncoding mutation g4205a to be fixed before or after the coding mutation G238S in a population of bacteria evolving resistance to the antibiotic cefotaxime? Give at least two reasons for your answer.

4. Examining Table 20-5, what do you think would be the order of mutations fixed during selection in a third evolving virus line? Would the mutations become fixed in the same order as the TX or ID virus?

5. Using Figure 20-17, explain how the mutation in the GATA sequence of the *Duffy* gene imparts resistance to *P. vivax* infection.

6. In Figure 20-18, what is the evidence that polyploid formation has been important in plant evolution?

BASIC PROBLEMS

7. Compare Darwin's description of natural selection as quoted on page 730 with Wallace's description of the tendency of varieties to depart from the original type quoted on page 731. What ideas do they have in common?

8. What are the three principles of the theory of evolution by natural selection?

9. Why was the neutral theory of molecular evolution a revolutionary idea?

10. What would you predict to be the relative rate of synonymous and nonsynonymous substitutions in a globin pseudogene?

11. Are *AS* heterozygotes completely resistant to malarial infection? Explain the evidence for your answer.

CHALLENGING PROBLEMS

GENETICSPORTAL **Unpacking the Problem** **12.** If the mutation rate to a new allele is 10^{-5}, how large must isolated populations be to prevent chance differentiation among them in the frequency of this allele?

13. Glucose-6-phosphate dehydrogenase (G6PD) is a critical enzyme involved in the metabolism of glucose, especially in red blood cells. Deficiencies in the enzyme are the most common human enzyme defect and occur at a high frequency in Tanzanian children.

    a. Offer one hypothesis for the high incidence of G6PD mutations in Tanzanian children.

    b. How would you test your hypothesis further?

    c. Scores of different G6PD mutations affecting enzyme function have been found in human populations. Offer one explanation for the abundance of different G6PD mutations.

14. Large differences in *HbS* frequencies among Kenyan and Ugandan tribes had been noted in surveys conducted by researchers other than Tony Allison. These researchers offered alternative explanations different from the malarial linkage proposed by Allison. Offer one counterargument to, or experimental test for, the following alternative hypotheses:

    a. The mutation rate is higher in certain tribes.

    b. There is a low degree of genetic mixing among tribes, so the allele rose to high frequency through inbreeding in certain tribes.

15. How many potential evolutionary paths are there for an allele to evolve six different mutations? Seven different mutations? Ten different mutations?

16. The *MC1R* gene affects skin and hair color in humans. There are at least 13 polymorphisms of the gene in European and Asian populations, 10 of which are nonsynonymous. In Africans, there are at least 5 polymorphisms of the gene, none of which are nonsynonymous. What might be one explanation for the differences in *MC1R* variation between Africans and non-Africans?

17. Opsin proteins detect light in photoreceptor cells of the eye and are required for color vision. The nocturnal owl monkey, the nocturnal bush baby, and the subterranean blind mole rat have different mutations in an opsin gene that render it nonfunctional. Explain why all three species can tolerate mutations in this gene that operates in most other mammals.

18. Full or partial limblessness has evolved many times in vertebrates (snakes, lizards, manatees, whales). Do you expect the mutations that occurred in the evolution of limblessness to be in the coding or noncoding sequences of toolkit genes? Why?

19. Several *Drosophila* species with unspotted wings are descended from a spotted ancestor. Would you predict the loss of spot formation to entail coding or noncoding changes in pigmentation genes? How would you test which is the case?

20. It has been claimed here that "evolution repeats itself." What is the evidence for this claim from

   a. the analysis of *HbS* alleles?

   b. the analysis of antibiotic resistance in bacteria?

   c. the analysis of experimentally selected bacteriophage ϕX174?

   d. the analysis of *Oca2* mutations in cave fish?

   e. the analysis of stickleback *Pitx1* loci?

21. What is the molecular evidence that natural selection includes the "rejection of injurious change"?

22. What are three alternative fates of a new gene duplicate?

23. What is the evidence that gene duplication has been the source of the α and β gene families for human hemoglobin?

24. DNA-sequencing studies for a gene in two closely related species produce the following numbers of sites that vary:

Synonymous polymorphisms	50
Nonsynonymous species differences	2
Synonymous species differences	18
Nonsynonymous polymorphisms	20

Does this result support neutral evolution of the gene? Does it support an adaptive replacement of amino acids? What explanation would you offer for the observations?

25. In humans, two genes encoding the opsin visual pigments that are sensitive to green and red wavelengths of light are found adjacent to one another on the X chromosome. They encode proteins that are 96 percent identical. Nonprimate mammals possess just one gene encoding an opsin sensitive to the red/green wavelength.

   a. Offer one explanation for the presence of the two opsin genes on the human X chromosome.

   b. How would you test your explanation further and pinpoint when in evolutionary history the second gene arose?

26. About 9 percent of Caucasian males are color-blind and cannot distinguish red-colored from green-colored objects.

   a. Offer one genetic model for color blindness.

   b. Explain why and how color blindness has reached a frequency of 9 percent in this population.

A Brief Guide to Model Organisms

Escherichia coli • Saccharomyces cerevisiae • Neurospora crassa • Arabidopsis thaliana
Caenorhabditis elegans • Drosophila melanogaster • Mus musculus

This brief guide collects in one place the main features of model organisms as they relate to genetics. Each of seven model organisms is given its own two-page spread; the format is consistent, allowing readers to compare and contrast the features of model organisms. Each treatment focuses on the special features of the organism that have made it useful as a model; the special techniques that have been developed for studying the organism; and the main contributions that studies of the organism have made to our understanding of genetics. Although many differences will be apparent, the general approaches of genetic analysis are similar but have to be tailored to take account of the individual life cycle, ploidy level, size and shape, and genomic properties, such as the presence of natural plasmids and transposons.

Model organisms have always been at the forefront of genetics. Initially, in the historical development of a model organism, a researcher selects the organism because of some feature that lends itself particularly well to the study of a genetic process in which the researcher is interested. The advice of the past hundred years has been, "Choose your organism well." For example, the ascomycete fungi, such as *Saccharomyces cerevisiae* and *Neurospora crassa,* are well suited to the study of meiotic processes, such as crossing over, because their unique feature, the ascus, holds together the products of a single meiosis.

Different species tend to show remarkably similar processes, even across the members of large groups, such as the eukaryotes. Hence, we can reasonably expect that what is learned in one species can be at least partly applied to others. In particular, geneticists have kept an eye open for new research findings that may apply to our own species. Compared with other species, humans are relatively difficult to study at the genetic level, and so advances in human genetics owe a great deal to more than a century of work on model organisms.

All model organisms have far more than one useful feature for genetic or other biological study. Hence, after a model organism has been developed by a few people with specific interests, it then acts as a nucleus for the development of a research community—a group of researchers with an interest in various features of one particular model organism. There are organized research communities for all the model organisms mentioned in this summary. The people in these communities are in touch with one another regularly, share their mutant strains, and often meet at least annually at conferences that may attract thousands of people. Such a community makes possible the provision of important services, such as databases of research information, techniques, genetic stocks, clones, DNA libraries, and genomic sequences.

Another advantage to an individual researcher in belonging to such a community is that he or she may develop "a feeling for the organism" (a phrase of maize geneticist and Nobel laureate Barbara McClintock). This idea is difficult to convey, but it implies an understanding of the general ways of an organism. No living process takes place in isolation, and so knowing the general ways of an organism is often beneficial in trying to understand one process and to interpret it in its proper context.

As the database for each model organism expands (which it currently is doing at a great pace thanks to genomics), geneticists are more and more able to take a holistic view, encompassing the integrated workings of all parts of the organism's makeup. In this way, model organisms become not only models for isolated processes but also models of integrated life processes. The term *systems biology* is used to describe this holistic approach.

Escherichia coli

Key organism for studying:

• Transcription, translation, replication, recombination
• Mutation
• Gene regulation
• Recombinant DNA technology

The unicellular bacterium *Escherichia coli* is widely known as a disease-causing pathogen, a source of food poisoning and intestinal disease. However, this negative reputation is undeserved. Although some strains of *E. coli* are harmful, others are natural and essential residents of the human gut. As model organisms, strains of *E. coli* play an indispensable role in genetic analyses. In the 1940s, several groups began investigating the genetics of *E. coli*. The need was for a simple organism that could be cultured inexpensively to produce large numbers of individual bacteria to be able to find and analyze rare genetic events. Because *E. coli* can be obtained from the human gut and is small and easy to culture, it was a natural choice. Work on *E. coli* defined the beginning of "black box" reasoning in genetics: through the selection and analysis of mutants, the workings of cellular processes could be deduced even though an individual cell was too small to be seen.

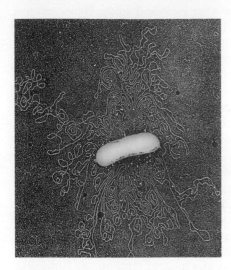

***E. coli* genome.** Electron micrograph of the genome of the bacterium *E. coli*, released from the cell by osmotic shock. [*Dr. Gopal Murti/Science Photo Library/Photo Researchers.*]

Special features

Much of *E. coli*'s success as a model organism can be attributed to two statistics: its 1-μm cell size and a 20-minute generation time. (Replication of the chromosome takes 40 minutes, but multiple replication forks allow the cell to divide in 20 minutes.) Consequently, this prokaryote can be grown in staggering numbers—a feature that allows geneticists to identify mutations and other rare genetic events such as intragenic recombinants. *E. coli* is also remarkably easy to culture. When cells are spread on plates of nutrient medium, each cell divides in situ and forms a visible colony. Alternatively, batches of cells can be grown in liquid shake culture. Phenotypes such as colony size, drug resistance, ability to obtain energy from particular carbon sources, and colored dye production take the place of the morphological phenotypes of eukaryotic genetics.

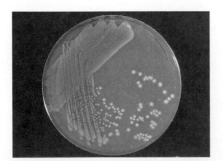

Bacterial colonies. [*Biophoto Associates/ Science Source/Photo Researchers.*]

Life Cycle

Escherichia coli reproduces asexually by simple cell fission; its haploid genome replicates and partitions with the dividing cell. In the 1940s, Joshua Lederberg and Edward Tatum discovered that *E. coli* also has a type of sexual cycle in which cells of genetically differentiated "sexes" fuse and exchange some or all of their genomes, sometimes leading to recombination (see Chapter 5). "Males" can convert "females" into males by the transmission of a particular plasmid. This circular extragenomic 100-kb DNA plasmid, called F, determines a type of "maleness." F⁺ cells acting as male "donors" transmit a copy of the F plasmid to a recipient cell. The F plasmid can integrate into the chromosome to form an Hfr cell type, which transmits the chromosome linearly into F⁻ recipients. Other plasmids are found in *E. coli* in nature. Some carry genes whose functions equip the cell for life in specific environments; R plasmids that carry drug-resistance genes are examples.

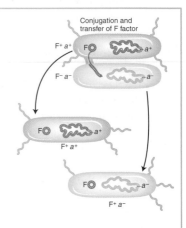

Length of life cycle: 20 minutes

Geneticists have also taken advantage of some unique genetic elements associated with *E. coli*. Bacterial plasmids and phages are used as vectors to clone the genes of other organisms within *E. coli*. Transposable elements from *E. coli* are harnessed to disrupt genes in cloned eukaryotic DNA. Such bacterial elements are key players in recombinant DNA technology.

Genetic analysis

Spontaneous *E. coli* mutants show a variety of DNA changes, ranging from simple base substitutions to the insertion of transposable elements. The study of rare spontaneous mutations in *E. coli* is feasible because large populations can be screened. However, mutagens also are used to increase mutation frequencies.

To obtain specific mutant phenotypes that might represent defects in a process under study, screens or selections must be designed. For example, nutritional mutations and mutations conferring resistance to drugs or phages can be obtained on plates supplemented with specific chemicals, drugs, or phages. Null mutations of any essential gene will result in no growth; these mutations can be selected by adding penicillin (an antibacterial drug isolated from a fungus), which kills dividing cells but not the nongrowing mutants. For conditional lethal mutations, replica plating can be used: mutated colonies on a master plate are transferred by a felt pad to other plates that are then subjected to some toxic environment. Mutations affecting the expression of a specific gene of interest can be screened by fusing it to a reporter gene such as the *lacZ* gene, whose protein product can make a blue dye, or the *GFP* gene, whose product fluoresces when exposed to light of a particular wavelength.

After a set of mutants affecting the process of interest have been obtained, the mutations are sorted into their genes by recombination and complementation. These genes are cloned and sequenced to obtain clues to function. Targeted mutagenesis can be used to tailor mutational changes at specific protein positions (see page 517).

In *E. coli*, crosses are used to map mutations and to produce specific cell genotypes (see Chapter 5). Recombinants are made by mixing Hfr cells (having an integrated F plasmid) and F⁻ cells. Generally an Hfr donor transmits part of the bacterial genome, forming a temporary merozygote in which recombination takes place. Hfr crosses can be used to perform mapping by time-of-marker entry or by recombinant frequency. By transfer of F′ derivatives carrying donor genes to F⁻, it is possible to make stable partial diploids to study gene interaction or dominance.

Techniques of Genetic Modification

Standard mutagenesis:

Chemicals and radiation	Random somatic mutations
Transposons	Random somatic insertions

Transgenesis:

On plasmid vector	Free or integrated
On phage vector	Free or integrated
Transformation	Integrated

Targeted gene knockouts:

Null allele on vector	Gene replacement by recombination
Engineered allele on vector	Site-directed mutagenesis by gene replacement

Genetic engineering

Transgenesis. *E. coli* plays a key role in introducing transgenes to other organisms (see Chapter 10). It is the standard organism used for cloning genes of any organism. *E. coli* plasmids or bacteriophages are used as vectors, carrying the DNA sequence to be cloned. These vectors are introduced into a bacterial cell by transformation, if a plasmid, or by transduction, if a phage, where they replicate in the cytoplasm. Vectors are specially modified to include unique cloning sites that can be cut by a variety of restriction enzymes. Other "shuttle" vectors are designed to move DNA fragments from yeast ("the eukaryotic *E. coli*") into *E. coli*, for its greater ease of genetic manipulation, and then back into yeast for phenotypic assessment.

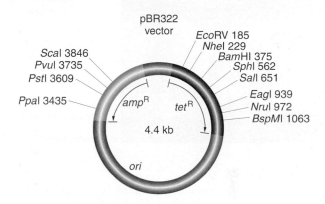

A plasmid designed as a vector for DNA cloning. Successful insertion of a foreign gene into the plasmid is detected by inactivation of either drug-resistance gene (*tet*ᴿ or *amp*ᴿ). Restriction sites are identified.

Targeted gene knockouts. A complete set of gene knockouts is being accumulated. In one procedure, a kanamycin-resistance transposon is introduced into a cloned gene in vitro (by using a transposase). The construct is transformed in, and resistant colonies are knockouts produced by homologous recombination.

Main contributions

Pioneering studies for genetics as a whole were carried out in *E. coli*. Perhaps the greatest triumph was the elucidation of the universal 64-codon genetic code, but this achievement is far from alone on the list of accomplishments attributable to this organism. Other fundamentals of genetics that were first demonstrated in *E. coli* include the spontaneous nature of mutation (the fluctuation test, page 558), the various types of base changes that cause mutations, and the semiconservative replication of DNA (the Meselson and Stahl experiment, page 262). This bacterium helped open up whole new areas of genetics, such as gene regulation (the *lac* operon, pages 385ff.) and DNA transposition (IS elements, page 529). Last but not least, recombinant DNA technology was invented in *E. coli*, and the organism still plays a central role in this technology today.

Other areas of contribution

- Cell metabolism
- Nonsense suppressors
- Colinearity of gene and polypeptide
- The operon
- Plasmid-based drug resistance
- Active transport

Saccharomyces cerevisiae

Genetic "Vital Statistics"

Genome size:	12 Mb
Chromosomes:	$n = 16$
Number of genes:	6000
Percentage with human homologs:	25%
Average gene size:	1.5 kb, 0.03 intron/gene
Transposons:	Small proportion of DNA
Genome sequenced in:	1996

Key organism for studying:

- Genomics
- Systems biology
- Genetic control of cell cycle
- Signal transduction
- Recombination
- Mating type
- Mitochondrial inheritance
- Gene interaction; two-hybrid

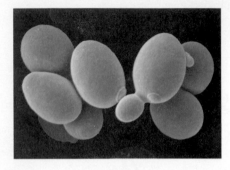

The ascomycete *S. cerevisiae*, alias "baker's yeast," "budding yeast," or simply "yeast," has been the basis of the baking and brewing industries since antiquity. In nature, it probably grows on the surfaces of plants, using exudates as nutrients, although its precise niche is still a mystery. Although laboratory strains are mostly haploid, cells in nature can be diploid or polyploid. In approximately 70 years of genetic research, yeast has become "the *E. coli* of the eukaryotes." Because it is haploid, unicellular, and forms compact colonies on plates, it can be treated in much the same way as a bacterium. However, it has eukaryotic meiosis, cell cycle, and mitochondria, and these features have been at the center of the yeast success story.

Yeast cells, *Saccharomyces cerevisiae*.

Special features

As a model organism, yeast combines the best of two worlds: it has much of the convenience of a bacterium, but with the key features of a eukaryote. Yeast cells are small (10 μm) and complete their cell cycle in just 90 minutes, allowing them to be produced in huge numbers in a short time. Like bacteria, yeast can be grown in large batches in a liquid medium that is continuously shaken. And, like bacteria, yeast produces visible colonies when plated on agar medium, can be screened for mutations, and can be replica plated. In typical eukaryotic manner, yeast has a mitotic cell-division cycle, undergoes meiosis, and contains mitochondria housing a small unique genome. Yeast cells can respire anaerobically by using the fermentation cycle and hence can do without mitochondria, allowing mitochondrial mutants to be viable.

Genetic analysis

Performing crosses in yeast is quite straightforward. Strains of opposite mating type are simply mixed on an appropriate medium. The resulting a/α diploids are induced to undergo meiosis by using a special sporulation medium. Investigators can isolate ascospores from a single tetrad by using a machine called a micromanipulator. They also have the option of synthesizing a/a or α/α diploids for special purposes or creating partial diploids by using specially engineered plasmids.

Because a huge array of yeast mutants and DNA constructs are available within the research community, special-purpose strains for screens and selections can be built by crossing various yeast types. Additionally, new mutant alleles can be mapped by crossing with strains containing an array of phenotypic or DNA markers of known map position.

The availability of both haploid and diploid cells provides flexibility for mutational studies. Haploid cells are convenient for large-scale selections or screens because mutant phenotypes are expressed directly. Diploid cells are convenient for obtaining dominant mutations, sheltering lethal mutations, performing complementation tests, and exploring gene interaction.

Life Cycle

Yeast is a unicellular species with a very simple life cycle consisting of sexual and asexual phases. The asexual phase can be haploid or diploid. A cell divides asexually by budding: a mother cell throws off a bud into which is passed one of the nuclei that result from mitosis. For sexual reproduction, there are two mating types, determined by the alleles *MATα* and *MATa*. When haploid cells of different mating type unite, they form a diploid cell, which can divide mitotically or undergo meiotic division. The products of meiosis are a nonlinear tetrad of four ascospores.

Total length of life cycle: 90 minutes to complete cell cycle

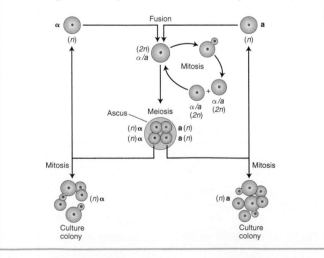

Techniques of Genetic Manipulation

Standard mutagenesis:

Chemicals and radiation	Random somatic mutations
Transposons	Random somatic insertions

Transgenesis:

Integrative plasmid	Inserts by homologous recombination
Replicative plasmid	Can replicate autonomously (2μ or ARS origin of replication)
Yeast artificial chromosome	Replicates and segregates as a chromosome
Shuttle vector	Can replicate in yeast or *E. coli*

Targeted gene knockouts:

Gene replacement	Homologous recombination replaces wild-type allele with null copy

Genetic engineering

Transgenesis. Budding yeast provides more opportunities for genetic manipulation than any other eukaryote (see Chapter 10). Exogenous DNA is taken up easily by cells whose cell walls have been partly removed by enzyme digestion or abrasion. Various types of vectors are available. For a plasmid to replicate free of the chromosomes, it must contain a normal yeast replication origin (ARS) or a replication origin from a 2-μm plasmid found in certain yeast isolates. The most elaborate vector, the yeast artificial chromosome (YAC), consists of an ARS, a yeast centromere, and two telomeres. A YAC can carry large transgenic inserts, which are then inherited in the same way as Mendelian chromosomes. YACs have been important vectors in cloning and sequencing large genomes such as the human genome.

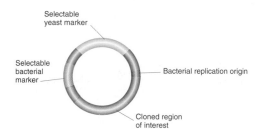

A simple yeast vector. This type of vector is called a yeast integrative plasmid (YIp).

Targeted knockouts. Transposon mutagenesis (transposon tagging) can be accomplished by introducing yeast DNA into *E. coli* on a shuttle vector; the bacterial transposons integrate into the yeast DNA, knocking out gene function. The shuttle vector is then transferred back into yeast, and the tagged mutants replace wild-type copies by homologous recombination. Gene knockouts can also be accomplished by replacing wild-type alleles with an engineered null copy through homologous recombination. By using these techniques, researchers have systematically constructed a complete set of yeast knockout strains (each carrying a different knockout) to assess null function of each gene at the phenotypic level.

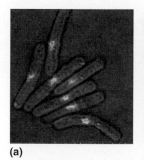
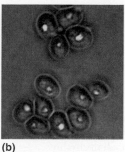

(a)   **(b)**

Cell-cycle mutants. (a) Mutants that elongate without dividing. (b) Mutants that arrest without budding.

Main contributions

Thanks to a combination of good genetics and good biochemistry, yeast studies have made substantial contributions to our understanding of the genetic control of cell processes.

Cell cycle. The identification of cell-division genes through their temperature-sensitive mutants (*cdc* mutants) has led to a powerful model for the genetic control of cell division. The different Cdc phenotypes reveal the components of the machinery required to execute specific steps in the progression of the cell cycle. This work has been useful for understanding the abnormal cell-division controls that can lead to human cancer.

Recombination. Many of the key ideas for the current molecular models of crossing over (such as the double-strand-break model) are based on tetrad analysis of gene conversion in yeast (see page 150). Gene conversion (aberrant allele ratios such as 3:1) is quite common in yeast genes, providing an appropriately large data set for quantifying the key features of this process.

Gene interactions. Yeast has led the way in the study of gene interactions. The techniques of traditional genetics have been used to reveal patterns of epistasis and suppression, which suggest gene interactions (see Chapter 6). The two-hybrid plasmid system for finding protein interactions was developed in yeast and has generated complex interaction maps that represent the beginnings of systems biology (see page 514). Synthetic lethals—lethal double mutants created by intercrossing two viable single mutants—also are used to plot networks of interaction (see page 233).

Mitochondrial genetics. Mutants with defective mitochondria are recognizable as very small colonies called "petites." The availability of these petites and other mitochondrial mutants enabled the first detailed analysis of mitochondrial genome structure and function in any organism.

Genetics of mating type. Yeast *MAT* alleles were the first mating-type genes to be characterized at the molecular level. Interestingly, yeast undergoes spontaneous switching from one mating type to the other. A silent "spare" copy of the opposite *MAT* allele, residing elsewhere in the genome, enters into the mating-type locus, replacing the resident allele by homologous recombination. Yeast has provided one of the central models for signal transduction during detection and response to mating hormones from the opposite mating type.

Other areas of contribution

- Genetics of switching between yeastlike and filamentous growth
- Genetics of senescence

Neurospora crassa

Key organism for studying:
- Genetics of metabolism and uptake
- Genetics of crossing over and meiosis
- Fungal cytogenetics
- Polar growth
- Circadian rhythms
- Interactions between nucleus and mitochondria

Genetic "Vital Statistics"	
Genome size:	43 Mb
Chromosomes:	7 autosomes ($n = 7$)
Number of genes:	10,000
Percentage with human homologs:	6%
Average gene size:	1.7 kb, 1.7 introns/gene
Transposons:	rare
Genome sequenced in:	2003

Neurospora crassa, the orange bread mold, was one of the first eukaryotic microbes to be adopted by geneticists as a model organism. Like yeast, it was originally chosen because of its haploidy, its simple and rapid life cycle, and the ease with which it can be cultured. Of particular significance was the fact that it will grow on a medium with a defined set of nutrients, making it possible to study the genetic control of cellular chemistry. In nature, it is found in many parts of the world growing on dead vegetation. Because fire activates its dormant ascospores, it is most easily collected after burns—for example, under the bark of burnt trees and in fields of crops such as sugar cane that are routinely burned before harvesting.

Neurospora crassa growing on sugarcane.

Special features

Neurospora holds the speed record for fungi because each hypha grows more than 10 cm per day. This rapid growth, combined with its haploid life cycle and ability to grow on defined medium, has made it an organism of choice for studying biochemical genetics of nutrition and nutrient uptake.

Another unique feature of *Neurospora* (and related fungi) allows geneticists to trace the steps of single meioses. The four haploid products of one meiosis stay together in a sac called an ascus. Each of the four products of meiosis undergoes a further mitotic division, resulting in a linear octad of eight ascospores (see pages 96–97). This feature makes *Neurospora* an ideal system in which to study crossing over, gene conversion, chromosomal rearrangements, meiotic nondisjunction, and the genetic control of meiosis itself. Chromosomes, although small, are easily visible, and so meiotic processes can be studied at both the genetic and the chromosomal levels. Hence, in *Neurospora*, fundamental studies have been carried out on the mechanisms underlying these processes (see page 127).

Genetic analysis

Genetic analysis is straightforward (see page 96). Stock centers provide a wide range of mutants affecting all aspects of the biology of the fungus. *Neurospora* genes can be mapped easily by crossing them with a bank of strains with known mutant loci or known RFLP alleles. Strains of opposite mating type are crossed simply by growing them together. A geneticist with a handheld needle can pick out a single ascospore for study. Hence, analyses in which

Life Cycle

N. crassa has a haploid eukaryotic life cycle. A haploid asexual spore (called a conidium) germinates to produce a germ tube that extends at its tip. Progressive tip growth and branching produce a mass of branched threads (called *hyphae*), which forms a compact colony on growth medium. Because hyphae have no cross walls, a colony is essentially one cell containing many haploid nuclei. The colony buds off millions of asexual spores, which can disperse in air and repeat the asexual cycle.

In *N. crassa*'s sexual cycle, there are two identical-looking mating types MAT-*A* and MAT-*a*, which can be viewed as simple "sexes." As in yeast, the two mating types are determined by two alleles of one gene. When colonies of different mating type come into contact, their cell walls

Sexual spores grow to adults

Asci

Meiosis

Synchronous division and fusion to form diploid meiocytes

Cross-fertilization

Maternal nucleus Mating type *A*

Maternal nucleus Mating type *a*

and nuclei fuse. Many transient diploid nuclei arise, each of which undergoes meiosis, producing an octad of ascopores. The ascospores germinate and produce colonies exactly like those produced by asexual spores.

Length of life cycle: 4 weeks for sexual cycle

Wild-type (*left*) and mutant (*right*) *Neurospora* grown in a petri dish.

either complete asci or random ascospores are used are rapid and straightforward.

Because *Neurospora* is haploid, newly obtained mutant phenotypes are easily detected with the use of various types of screens and selections. A favorite system for study of the mechanism of mutation is the *ad-3* gene, because *ad-3* mutants are purple and easily detected.

Although vegetative diploids of *Neurospora* are not readily obtainable, geneticists are able to create a "mimic diploid," useful for complementation tests and other analyses requiring the presence of two copies of a gene (see page 225). Namely, the fusion of two different strains produces a heterokaryon, an individual containing two different nuclear types in a common cytoplasm. Heterokaryons also enable the use of a version of the specific-locus test, a way to recover mutations in a specific recessive allele. (Cells from a $+/m$ heterokaryon are plated and m/m colonies are sought.)

Techniques of Genetic Manipulation

Standard mutagenesis:

Chemicals and radiation	Random somatic mutations
Transposon mutagenesis	Not available

Transgenesis:

Plasmid-mediated transformation	Random insertion

Targeted gene knockouts:

RIP	GC → AT mutations in transgenic duplicate segments before a cross
Quelling	Somatic posttranscriptional inactivation of transgenes

---

Genetic engineering

Transgenesis. The first eukaryotic transformation was accomplished in *Neurospora*. Today, *Neurospora* is easily transformed with the use of bacterial plasmids carrying the desired transgene, plus a selectable marker such as hygromycin resistance to show that the plasmid has entered. No plasmids replicate in *Neurospora*, and so a transgene is inherited only if it integrates into a chromosome.

Targeted knockouts. In special strains of *Neurospora*, transgenes frequently integrate by homologous recombination. Hence, a transgenic strain normally has the resident gene plus the homologous

transgene, inserted at a random ectopic location. Because of this duplication of material, if the strain is crossed, it is subject to RIP, a genetic process that is unique to *Neurospora*. RIP is a premeiotic mechanism that introduces many GC-to-AT transitions into both duplicate copies, effectively disrupting the gene. RIP can therefore be harnessed as a convenient way of deliberately knocking out a specific gene.

Main contributions

George Beadle and Edward Tatum used *Neurospora* as the model organism in their pioneering studies on gene–enzyme relations, in which they were able to determine the enzymatic steps in the synthesis of arginine (see page 219). Their work with *Neurospora* established the beginning of molecular genetics. Many comparable studies on the genetics of cell metabolism with the use of *Neurospora* followed.

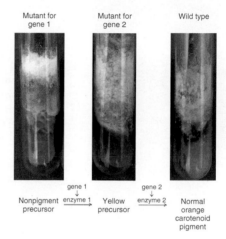

Pathway synthesizing orange caroteinoid pigment in *Neurospora*.

Pioneering work has been done on the genetics of meiotic processes, such as crossing over and disjunction, and on conidiation rhythms. Continuously growing cultures show a daily rhythm of conidiospore formation. The results of pioneering studies using mutations that alter this rhythm have contributed to a general model for the genetics of circadian rhythms.

Neurospora serves as a model for the multitude of pathogenic filamentous fungi affecting crops and humans because these fungi are often difficult to culture and manipulate genetically. It is even used as a simple eukaryotic test system for mutagenic and carcinogenic chemicals in the human environment.

Because crosses can be made by using one parent as female, the cycle is convenient for the study of mitochondrial genetics and nucleus–mitochondria interaction. A wide range of linear and circular mitochondrial plasmids have been discovered in natural isolates. Some of them are retroelements that are thought to be intermediates in the evolution of viruses.

Other areas of contribution

- Fungal diversity and adaptation
- Cytogenetics (chromosomal basis of genetics)
- Mating-type genes
- Heterokaryon-compatibility genes (a model for the genetics of self and nonself recognition)

Arabidopsis thaliana

Genetic "Vital Statistics"

Genome size:	125 Mb
Chromosomes:	diploid, 5 autosomes ($2n = 10$)
Number of genes:	25,000
Percentage with human homologs:	18%
Average gene size:	2 kb, 4 introns/gene
Transposons:	10% of the genome
Genome sequenced in:	2000

Key organism for studying:

- Development
- Gene expression and regulation
- Plant genomics

Arabidopsis thaliana, a member of the Brassicaceae (cabbage) family of plants, is a relatively late arrival as a genetic model organism. Most work has been done in the past 20 years. It has no economic significance: it grows prolifically as a weed in many temperate parts of the world. However, because of its small size, short life cycle, and small genome, it has overtaken the more traditional genetic plant models such as corn and wheat and has become the dominant model for plant molecular genetics.

***Arabidopsis thaliana* growing in the wild.** The versions grown in the laboratory are smaller. [*FloralImages/Alamy.*]

Special features

In comparison with other plants, *Arabidopsis* is small in regard to both its physical size and its genome size—features that are advantageous for a model organism. *Arabidopsis* grows to a height of less than 10 cm under appropriate conditions; hence, it can be grown in large numbers, permitting large-scale mutant screens and progeny analyses. Its total genome size of 125 Mb made the genome relatively easy to sequence compared with other plant model organism genomes, such as the maize genome (2500 Mb) and the wheat genome (16,000 Mb).

Genetic analysis

The analysis of *Arabidopsis* mutations through crossing relies on tried and true methods—essentially those used by Mendel. Plant stocks carrying useful mutations relevant to the experiment in hand are obtained from public stock centers. Lines can be manually crossed with each other or self-fertilized. Although the flowers are small, cross-pollination is easily accomplished by removing undehisced anthers (which are sometimes eaten by the experimenter as a convenient means of disposal). Each pollinated flower then produces a long pod containing a large number of seeds. This abundant production of offspring (thousands of seeds per plant) is a boon to geneticists searching for rare mutants or other rare events. If a plant carries a new recessive mutation in the germ line, selfing allows progeny homozygous for the recessive mutation to be recovered in the plant's immediate descendants.

Life Cycle

Arabidopsis has the familiar plant life cycle, with a dominant diploid stage. A plant bears several flowers, each of which produces many seeds. Like many annual weeds, its life cycle is rapid: it takes only about 6 weeks for a planted seed to produce a new crop of seeds.

Total length of life cycle: 6 weeks

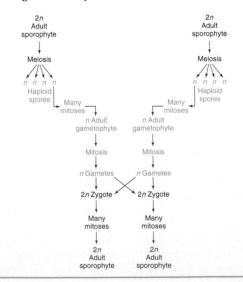

Arabidopsis **mutants.** (*Left*) Wild-type flower of *Arabidopsis*. (*Middle*) The *agamous* mutation (*ag*), which results in flowers with only petals and sepals (no reproductive structures). (*Right*) A double-mutant *ap1, cal,* which makes a flower that looks like a cauliflower. (Similar mutations in cabbage are probably the cause of real cauliflowers.) [*Photos from George Haughn.*]

Techniques of Genetic Modification

Standard mutagenesis:

Chemicals and radiation	Random germ-line or somatic mutations
T-DNA itself or transposons	Random tagged insertions

Transgenesis:

T-DNA carries the transgene	Random insertion

Targeted gene knockouts:

T-DNA or transposon-mediated mutagenesis	Random insertion; knockouts selected with PCR
RNAi	Mimics targeted knockout

Genetic engineering

Transgenesis. *Agrobacterium* T-DNA is a convenient vector for introducing transgenes (see Chapter 10). The vector–transgene construct inserts randomly throughout the genome. Transgenesis offers an effective way to study gene regulation. The transgene is spliced to a reporter gene such as GUS, which produces a blue dye at whatever positions in the plant the gene is active.

Targeted knockouts. Because homologous recombination is rare in *Arabidopsis*, specific genes cannot be easily knocked out by homologous replacement with a transgene. Hence, in *Arabidopsis*, genes are knocked out by the random insertion of a T-DNA vector or transposon (maize transposons such as *Ac-Ds* are used), and then specific gene knockouts are selected by applying PCR analysis to DNA from large pools of plants. The PCR uses a sequence in the T-DNA or in the transposon as one primer and a sequence in the gene of interest as the other primer. Thus, PCR amplifies only copies of the gene of interest that carry an insertion. Subdividing the pool and repeating the process lead to the specific plant carrying the knockout. Alternatively, RNAi may be used to inactivate a specific gene.

Large collections of T-DNA insertion mutants are available; they have the flanking plant sequences listed in public databases; so, if you are interested in a specific gene, you can see if the collection contains a plant that has an insertion in that gene. A convenient feature of knockout populations in plants is that they can be easily and inexpensively maintained as collections of seeds for many years, perhaps even decades. This feature is not possible for most populations of animal models. The worm *Caenorhabditis elegans* can be preserved as a frozen animal, but fruit flies (*Drosophila melanogaster*) cannot be frozen and revived. Thus, lines of fruit-fly mutants must be maintained as living organisms.

Main contributions

As the first plant genome to be sequenced, *Arabidopsis* has provided an important model for plant genome architecture and evolution. In addition, studies of *Arabidopsis* have made key contributions to our understanding of the genetic control of plant development. Geneticists have isolated homeotic mutations affecting flower development, for example. In such mutants, one type of floral part is replaced by another. Integration of the action of these mutants has led to an elegant model of flower-whorl determination based on overlapping patterns of regulatory-gene expression in the flower meristem. *Arabidopsis* has also contributed broadly to the genetic basis of plant physiology, gene regulation, and the interaction of plants and the environment (including the genetics of disease resistance). Because *Arabidopsis* is a natural plant of worldwide distribution, it has great potential for the study of evolutionary diversification and adaptation.

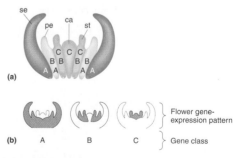

The establishment of whorl fate. (a) Patterns of gene expression corresponding to the different whorl fates. From outermost to innermost, the fates are sepal (se), petal (pe), stamen (st), and carpel (ca). (b) The shaded regions of the cross-sectional diagrams of the developing flower indicate the gene-expression patterns for the genes of the A, B, and C classes.

Other areas of contribution

- Environmental-stress response
- Hormone control systems

Caenorhabditis elegans

Key organism for studying:
- Development
- Behavior
- Nerves and muscles
- Aging

Caenorhabditis elegans may not look like much under a microscope, and, indeed, this 1-mm-long soil-dwelling roundworm (a nematode) is relatively simple as animals go. But that simplicity is part of what makes *C. elegans* a good model organism. Its small size, rapid growth, ability to self, transparency, and low number of body cells have made it an ideal choice for the study of the genetics of eukaryotic development.

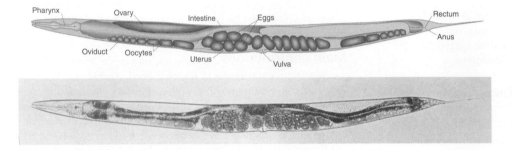

Photomicrograph and drawing of an adult *Caenorhabditis elegans*.

Special features

Geneticists can see right through *C. elegans*. Unlike other multicellular model organisms, such as fruit flies or *Arabidopsis*, this tiny worm is transparent, making it efficient to screen large populations for interesting mutations affecting virtually any aspect of anatomy or behavior. Transparency also lends itself well to studies of development: researchers can directly observe all stages of development simply by watching the worms under a light microscope. The results of such studies have shown that *C. elegans*'s development is tightly programmed and that each worm has a surprisingly small and consistent number of cells (959 in hermaphrodites and 1031 in males). In fact, biologists have tracked the fates of specific cells as the worm develops and have determined the exact pattern of cell division leading to each adult organ. This effort has yielded a lineage pedigree for every adult cell (see page 476).

Genetic analysis

Because the worms are small and reproduce quickly and prolifically (selfing produces about 300 progeny and crossing yields

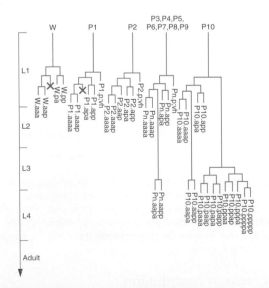

A symbolic representation of the lineages of 11 cells. A cell that undergoes programmed cell death is indicated by a blue X at the end of a branch of a lineage.

Life Cycle

C. elegans is unique among the major model animals in that one of the two sexes is hermaphrodite (XX). The other is male (XO). The two sexes can be distinguished by the greater size of the hermaphrodites and by differences in their sex organs. Hermaphrodites produce both eggs and sperm, and so they can be selfed. The progeny of a selfed hermaphrodite also are hermaphrodites, except when a rare nondisjunction leads to an XO male. If hermaphrodites and males are mixed, the sexes copulate, and many of the resulting zygotes will have been fertilized by the males' amoeboid sperm. Fertilization and embryo production take place within the hermaphrodite, which then lays the eggs. The eggs finish their development externally.

Total length of life cycle: $3\frac{1}{2}$ days

about 1000), they produce large populations of progeny that can be screened for rare genetic events. Moreover, because hermaphroditism in *C. elegans* makes selfing possible, individual worms with homozygous recessive mutations can be recovered quickly by selfing the progeny of treated individual worms. In contrast, other animal models, such as fruit flies or mice, require matings between siblings and take more generations to recover recessive mutations.

Techniques of Genetic Modification

Standard mutagenesis:

Chemical (EMS) and radiation	Random germ-line mutations
Transposons	Random germ-line insertions

Transgenesis:

Transgene injection of gonad	Unintegrated transgene array; occasional integration

Targeted gene knockouts:

Transposon-mediated mutagenesis	Knockouts selected with PCR
RNAi	Mimics targeted knockout
Laser ablation	Knockout of one cell

Genetic engineering

Transgenesis. The introduction of transgenes into the germ line is made possible by a special property of *C. elegans* gonads. The gonads of the worm are syncitial, meaning that there are many nuclei in a common cytoplasm. The nuclei do not become incorporated into cells until meiosis, when formation of the individual egg or sperm begins. Thus, a solution of DNA containing the transgene injected into the gonad of a hermaphrodite exposes more than 100 germ-cell precursor nuclei to the transgene. By chance, a few of these nuclei will incorporate the DNA (see Chapter 10).

Transgenes recombine to form multicopy tandem arrays. In an egg, the arrays do not integrate into a chromosome, but transgenes from the arrays are still expressed. Hence, the gene carried on a wild-type DNA clone can be identified by introducing it into a specific recessive recipient strain (functional complementation). In some but not all cases, the transgenic arrays are passed on to progeny. To increase the chance of inheritance, worms are exposed to ionizing radiation, which can induce the integration of an array into an ectopic chromosomal position, and, in this site, the array is reliably transmitted to progeny.

Targeted knockouts. In strains with active transposons, the transposons themselves become agents of mutation by inserting into random locations in the genome, knocking out the interrupted genes. If we can identify organisms with insertions into a specific gene of interest, we can isolate a targeted gene knockout. Inserts into specific genes can be detected by using PCR if one PCR primer is based on the transposon sequence and another one is based on the sequence of the gene of interest. Alternatively, RNAi can be used to nullify the function of specific genes. As an alternative to mutation, individual cells can be killed by a laser beam to observe the effect on worm function or development (laser ablation).

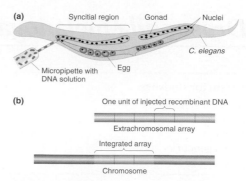

Creation of *C. elegans* transgenes. (a) Method of injection. (b) Extrachromosomal and integrated arrays.

Main contributions

C. elegans has become a favorite model organism for the study of various aspects of development because of its small and invariant number of cells. One example is programmed cell death, a crucial aspect of normal development. Some cells are genetically programmed to die in the course of development (a process called apoptosis). The results of studies of *C. elegans* have contributed a useful general model for apoptosis, which is also known to be a feature of human development.

Another model system is the development of the vulva, the opening to the outside of the reproductive tract. Hermaphrodites with defective vulvas still produce progeny, which in screens are easily visible clustered within the body. The results of studies of hermaphrodites with no vulva or with too many have revealed how cells that start off completely equivalent can become differentiated into different cell types (see page 476).

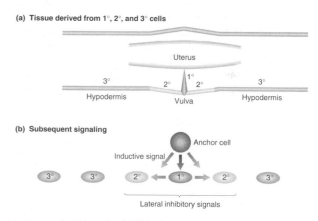

Production of the *C. elegans* vulva. (a) The final differentiated tissue. (b) Method of differentiation. The cells begin completely equivalent. An anchor cell behind the equivalent cells sends a signal to the nearest cells, which become the vulva. The primary vulva cell then sends a lateral signal to its neighbors, preventing them from becoming primary cells, even though they, too, have received the signal from the anchor cell.

Behavior also has been the subject of genetic dissection. *C. elegans* offers an advantage in that worms with defective behavior can often still live and reproduce. The worm's nerve and muscle systems have been genetically dissected, allowing behaviors to be linked to specific genes.

Other area of contribution

• Cell-to-cell signaling

Drosophila melanogaster

Genetic "Vital Statistics"

Genome size:	180 Mb
Chromosomes:	Diploid, 3 autosomes, X and Y
	(2*n* = 8)
Number of genes:	13,000
Percentage with human homologs:	~ 50%
Average gene size:	3 kb, 4 exons/gene
Transposons:	*P* elements, among others
Genome sequenced in:	2000

Key organism for studying:

- Transmission genetics
- Cytogenetics
- Development
- Population genetics
- Evolution

Polytene chromosomes.

The fruit fly *Drosophila melanogaster* (loosely translated as "dusky syrup-lover") was one of the first model organisms to be used in genetics. It was chosen in part because it is readily available from ripe fruit, has a short life cycle of the diploid type, and is simple to culture and cross in jars or vials containing a layer of food. Early genetic analysis showed that its inheritance mechanisms have strong similarities to those of other eukaryotes, underlining its role as a model organism. Its popularity as a model organism went into decline during the years when *E. coli*, yeast, and other microorganisms were being developed as molecular tools. However, *Drosophila* has experienced a renaissance because it lends itself so well to the study of the genetic basis of development, one of the central questions of biology. *Drosophila*'s importance as a model for human genetics is demonstrated by the discovery that approximately 60 percent of known disease-causing genes in humans, as well as 70 percent of cancer genes, have counterparts in *Drosophila*.

Special features

Drosophila came into vogue as an experimental organism in the early twentieth century because of features common to most model organisms. It is small (3 mm long), simple to raise (originally, in milk bottles), quick to reproduce (only 12 days from egg to adult), and easy to obtain (just leave out some rotting fruit). It proved easy to amass a large range of interesting mutant alleles that were used to lay the ground rules of transmission genetics. Early researchers also took advantage of a feature unique to the fruit fly: polytene chromosomes (see page 609). In salivary glands and certain other tissues, these "giant chromosomes" are produced by multiple rounds of DNA replication without chromosomal segregation. Each polytene chromosome displays a unique banding pattern, providing geneticists with landmarks that could be used to correlate recombination-based maps with actual chromosomes. The momentum provided by these early advances, along with the large amount of accumulated knowledge about the organism, made *Drosophila* an attractive genetic model.

Genetic analysis

Crosses in *Drosophila* can be performed quite easily. The parents may be wild or mutant stocks obtained from stock centers or as new mutant lines.

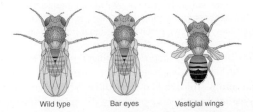

Wild type   Bar eyes   Vestigial wings

Two morphological mutants of *Drosophila*, with the wild type for comparison.

Life Cycle

Drosophila has a short diploid life cycle that lends itself well to genetic analysis. After hatching from an egg, the fly develops through several larval stages and a pupal stage before emerging as an adult, which soon becomes sexually mature. Sex is determined by X and Y sex chromosomes (XX is female, XY is male), although, in contrast with humans, the number of X's in relation to the number of autosomes determines sex (see page 50).

Total length of life cycle: 12 days from egg to adult

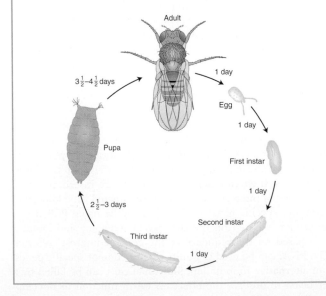

To perform a cross, males and females are placed together in a jar, and the females lay eggs in semisolid food covering the jar's bottom. After emergence from the pupae, offspring can be anesthetized to permit counting members of phenotypic classes and to distinguish males and females (by their different abdominal stripe patterns). However, because female progeny stay virgin for only a few hours after emergence from the pupae, they must immediately be isolated if they are to be used to make controlled crosses. Crosses designed to build specific gene combinations must be carefully planned, because crossing over does not take place in *Drosophila* males. Hence, in the male, linked alleles will not recombine to help create new combinations.

For obtaining new recessive mutations, special breeding programs (of which the prototype is Muller's ClB test) provide convenient screening systems. In these tests, mutagenized flies are crossed with a stock having a balancer chromosome (see page 617). Recessive mutations are eventually brought to homozygosity by inbreeding for one or two generations, starting with single F_1 flies.

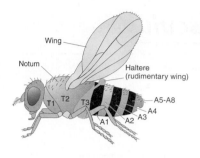

The normal thoracic and abdominal segments of *Drosophila*.

Techniques of Genetic Modification

Standard mutagenesis:

Chemical (EMS) and radiation	Random germ-line and somatic mutations

Transgenesis:

P element mediated	Random insertion

Targeted gene knockouts:

Induced replacement	Null ectopic allele exits and recombines with wild-type allele
RNAi	Mimics targeted knockout

Genetic engineering

Transgenesis. Building transgenic flies requires the help of a *Drosophila* transposon called the *P* element. Geneticists construct a vector that carries a transgene flanked by *P*-element repeats. The transgene vector is then injected into a fertilized egg along with a helper plasmid containing a transposase. The transposase allows the transgene to jump randomly into the genome in germinal cells of the embryo (see Chapter 15).

Targeted knockouts. Targeted gene knockouts can be accomplished by, first, introducing a null allele transgenically into an ectopic position and, second, inducing special enzymes that cause excision of the null allele. The excised fragment (which is linear) then finds and replaces the endogenous copy by homologous crossing over. However, functional knockouts can be produced more efficiently by RNAi.

Main contributions

Much of the early development of the chromosome theory of heredity was based on the results of *Drosophila* studies. Geneticists working with *Drosophila* made key advances in developing techniques for gene mapping, in understanding the origin and nature of gene mutation, and in documenting the nature and behavior of chromosomal rearrangements (see pages 123 and 129).

Their discoveries opened the door to other pioneering studies:

- Early studies on the kinetics of mutation induction and the measurement of mutation rates were performed with the use of *Drosophila*. Muller's ClB test and similar tests provided convenient screening methods for recessive mutations.

- Chromosomal rearrangements that move genes adjacent to heterochromatin were used to discover and study position-effect variegation.

- In the last part of the twentieth century, after the identification of certain key mutational classes such as homeotic and maternal-effect mutations, *Drosophila* assumed a central role in the genetics of development, a role that continues today (see Chapter 13). Maternal-effect mutations that affect the development of embryos, for example, have been crucial in the elucidation of the genetic determination of the *Drosophila* body plan; these mutations are identified by screening for abnormal developmental phenotypes in the embryos from a specific female. Techniques such as enhancer trap screens have enabled the discovery of new regulatory regions in the genome that affect development. Through these methods and others, *Drosophila* biologists have made important advances in understanding the determination of segmentation and of the body axes. Some of the key genes discovered, such as the homeotic genes, have widespread relevance in animals generally.

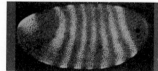

(a) Bicoid protein   **(b) Hairy pair-rule protein**

Photomicrographs showing gradients of body plan determinants. (a) Antibodies to the Bicoid protein are shown localized to the anterior (left-hand) tip of the embryo. (b) Antibodies to the Hairy pair-rule protein are shown localized in segments. The distribution of these and other proteins determines the body axis (a) and segmentation (b). [*(a) Ruth Lehmann; (b) James Langeland.*]

Other areas of contribution

- Population genetics
- Evolutionary genetics
- Behavioral genetics

Mus musculus

Key organism for studying:

- Human disease
- Mutation
- Development
- Coat color
- Immunology

Genetic "Vital Statistics"	
Genome size:	2600 Mb
Chromosomes:	19 autosomes, X and Y ($2n = 40$)
Number of genes:	30,000
Percentage with human homologs:	99%
Average gene size:	40 kb, 8.3 exons/gene
Transposons:	Source of 38% of genome
Genome sequenced in:	2002

Because humans and most domesticated animals are mammals, the genetics of mammals is of great interest to us. However, mammals are not ideal for genetics: they are relatively large in size compared with other model organisms, thereby taking up large and expensive facilities, their life cycles are long, and their genomes are large and complex. Compared with other mammals, however, mice (*Mus musculus*) are relatively small, have short life cycles, and are easily obtained, making them an excellent choice for a mammal model. In addition, mice had a head start in genetics because mouse "fanciers" had already developed many different interesting lines of mice that provided a source of variants for genetic analysis. Research on the Mendelian genetics of mice began early in the twentieth century.

An adult mouse and its litter.

Special features

Mice are not exactly small, furry humans, but their genetic makeup is remarkably similar to ours. Among model organisms, the mouse is the one whose genome most closely resembles the human genome. The mouse genome is about 14 percent smaller than that of humans (the human genome is 3000 Mb), but it has approximately the same number of genes (current estimates are just under 30,000). A surprising 99 percent of mouse genes seem to have homologs in humans. Furthermore, a large proportion of the genome is syntenic with that of humans; that is, there are large blocks containing the same genes in the same relative positions (see page 507). Such genetic similarities are the key to

the mouse's success as a model organism; these similarities allow mice to be treated as "stand-ins" for their human counterparts in many ways. Potential mutagens and carcinogens that we suspect of causing damage to humans, for example, are tested on mice, and mouse models are essential in studying a wide array of human genetic diseases.

Genetic analysis

Mutant and "wild type" (though not actually from the wild) mice are easy to come by: they can be ordered from large stock centers that provide mice suitable for crosses and various other types of experiments. Many of these lines are derived from mice bred in past centuries by mouse fanciers. Controlled crosses can be performed simply by pairing a male with a nonpregnant female. In most cases, the parental genotypes can be provided by male or female.

Human chromosomes

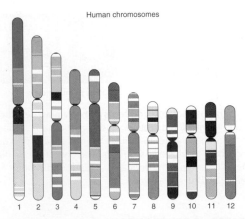

A mouse–human synteny map of 12 chromosomes from the human genome. Color coding is used to depict the regional matches of each block of the human genome to the corresponding sections of the mouse genome. Each color represents a different mouse chromosome.

Life Cycle

Mice have a familiar diploid life cycle, with an XY sex-determination system similar to that of humans. Litters are from 5 to 10 pups; however, the fecundity of females declines after about 9 months, and so they rarely have more than five litters.

Total length of life cycle: 10 weeks from birth to giving birth, in most laboratory strains

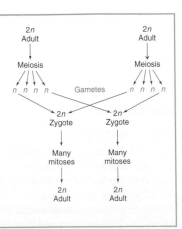

Most of the standard estimates of mammalian mutation rates (including those of humans) are based on measurements in mice. Indeed, mice provide the final test of agents suspected of causing mutations in humans. Mutation rates in the germ line are measured with the use of the specific-locus test: mutagenize $+/+$ gonads, cross to m/m (m is a known recessive mutation at the locus under study), and look for m^*/m progeny (m^* is a new mutation). The procedure is repeated for seven sample loci. The measurement of somatic mutation rates uses a similar setup, but the mutagen is injected into the fetus. Mice have been used extensively to study the type of somatic mutation that gives rise to cancer.

Techniques of Genetic Modification

Standard mutagenesis:

Chemicals and radiation	Germ-line and somatic mutations

Transgenesis:

Transgene injection into zygote	Random and homologous insertion
Transgene uptake by stem cells	Random and homologous insertion

Targeted gene knockouts:

Null transgene uptake by stem cells	Targeted knockouts selected

---

Genetic engineering

Transgenesis. The creation of transgenic mice is straightforward but requires the careful manipulation of a fertilized egg (see Chapter 10). First, mouse genomic DNA is cloned in *E. coli* with the use of bacterial or phage vectors. The DNA is then injected into a fertilized egg, where it integrates at ectopic (random) locations in the genome or, less commonly, at the normal locus. The activity of the transgene's protein can be monitored by fusing the transgene with a reporter gene such as *GFP* before the gene is injected. With the use of a similar method, the somatic cells of mice also can be modified by transgene insertion: specific fragments of DNA are inserted into individual somatic cells and these cells are, in turn, inserted into mouse embryos.

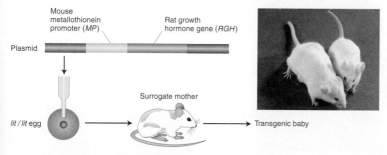

Producing a transgenic mouse. The transgene, a rat growth-hormone gene joined to a mouse promoter, is injected into a mouse egg homozygous for dwarfism (*lit/lit*).

Targeted knockouts. Knockouts of specific genes for genetic dissection can be accomplished by introducing a transgene con-

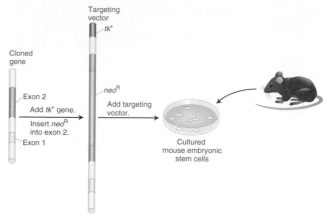

Production of ES cells with a gene knockout

Producing a gene knockout. A drug-resistance gene (*neo*^R) is inserted into the transgene, both to serve as a marker and to disrupt the gene, producing a knockout. (The *tk* gene is a second marker.) The transgene construct is then injected into mouse embryo cells.

taining a defective allele and two drug-resistance markers into a wild-type embryonic stem cell (see Chapter 10). The markers are used to select those specific transformant cells in which the defective allele has replaced the homologous wild-type allele. The transgenic cells are then introduced into mouse embryos. A similar method can be used to replace wild-type alleles with a functional transgene (gene therapy).

Main contributions

Early in the mouse's career as a model organism, geneticists used mice to elucidate the genes that control coat color and pattern, providing a model for all fur-bearing mammals, including cats, dogs, horses, and cattle (see page 216). More recently, studies of mouse genetics have made an array of contributions with direct bearing on human health:

- A large proportion of human genetic diseases have a mouse counterpart—called a "mouse model"—useful for experimental study.
- Mice serve as models for the mechanisms of mammalian mutation.
- Studies on the genetic mechanisms of cancer are performed on mice.
- Many potential carcinogens are tested on mice.
- Mice have been important models for the study of mammalian developmental genetics (see page 459). For example, they provide a model system for the study of genes affecting cleft lip and cleft palate, a common human developmental disorder.
- Cell lines that are fusion hybrids of mouse and human genomes played an important role in the assignment of human genes to specific human chromosomes. There is a tendency for human chromosomes to be lost from such hybrids, and so loss of specific chromosomes can be correlated with loss of specific human alleles.

Other areas of contribution

- Behavioral genetics
- Quantitative genetics
- The genes of the immune system

Appendix A

Genetic Nomenclature

There is no universally accepted set of rules for naming genes, alleles, protein products, and associated phenotypes. At first, individual geneticists developed their own symbols for recording their work. Later, groups of people working on any given organism met and decided on a set of conventions that all would use. Because *Drosophila* was one of the first organisms to be used extensively by geneticists, most of the current systems are variants of the *Drosophila* system. However, there has been considerable divergence. Some scientists now advocate a standardization of this symbolism, but standardization has not been achieved. Indeed, the situation has been made more complex by the advent of DNA technology. Whereas most genes previously had been named for the phenotypes produced by mutations within them, the new technology has shown the precise nature of the products of many of these genes. Hence, it seems more appropriate to refer to them by their cellular function. However, the old names are still in the literature, so many genes have two parallel sets of nomenclature.

The following examples by no means cover all the organisms used in genetics, but most of the nomenclature systems follow one of these types.

Drosophila melanogaster (insect)

ry	A gene that when mutated causes rosy eyes
ry^{502}	A specific recessive mutant allele producing rosy eyes in homozygotes
ry$^+$	The wild-type allele of *rosy*
ry	The rosy mutant phenotype
ry$^+$	The wild-type phenotype (red eyes)
RY	The protein product of the *rosy* gene
XDH	Xanthine dehydrogenase, an alternative description of the protein product of the *rosy* gene; named for the enzyme that it encodes
D	*Dichaete;* a gene that when mutated causes a loss of certain bristles and wings to be held out laterally in heterozygotes and causes lethality in homozygotes
D^3	A specific mutant allele of the *Dichaete* gene
D$^+$	The wild-type allele of *Dichaete*
D	The Dichaete mutant phenotype
D$^+$	The wild-type phenotype
D	(Depending on context) the protein product of the *Dichaete* gene (a DNA-binding protein)

Neurospora crassa (fungus)

arg	A gene that when mutated causes arginine requirement
arg-1	One specific *arg* gene
arg-1	An unspecified mutant allele of the *arg* gene
arg-1 (1)	A specific mutant allele of the *arg-1* gene
arg-1$^+$	The wild-type allele
arg-1	The protein product of the *arg-1*$^+$ gene
Arg$^+$	A strain not requiring arginine
Arg$^-$	A strain requiring arginine

Saccharomyces cerevisiae (fungus)

ARG	A gene that when mutated causes arginine requirement
ARG1	One specific *ARG* gene
arg1	An unspecified mutant allele of the *ARG* gene
urg1-1	A specific mutant allele of the *ARG1* gene
ARG1$^+$	The wild-type allele
ARG1p	The protein product of the *ARG1*$^+$ gene
Arg$^+$	A strain not requiring arginine
Arg$^-$	A strain requiring arginine

Homo sapiens (mammal)

ACH	A gene that when mutated causes achondroplasia
ACH1	A mutant allele (dominance not specified)
ACH	Protein product of *ACH* gene; nature unknown
FGFR3	Recent name for gene for achondroplasia
FGFR3^1 or *FGFR3**1 or *FGFR3* <1>	Mutant allele of *FGFR3* (unspecified dominance)
FGFR3 protein	Fibroblast growth factor receptor 3

Mus musculus (mammal)

Tyrc	A gene for tyrosinase
+^{Tyrc}	The wild-type allele of this gene
Tyrc^{ch} or *Tyrc-ch*	A mutant allele causing chinchilla color
Tyrc	The protein product of this gene
+TYRC	The wild-type phenotype
TYRCch	The chinchilla phenotype

Escherichia coli (bacterium)

lacZ	A gene for utilizing lactose
lacZ$^+$	The wild-type allele
lacZ1	A mutant allele
LacZ	The protein product of that gene
Lac$^+$	A strain able to use lactose (phenotype)
Lac$^-$	A strain unable to use lactose (phenotype)

Arabidopsis thaliana (plant)

YGR	A gene that when mutant produces yellow-green leaves
YGR1	A specific *YGR* gene
YGR1	The wild-type allele
ygr1-1	A specific recessive mutant allele of *YGR1*
ygr1-2D	A specific dominant (D) mutant allele of *YGR1*
YGR1	The protein product of *YGR1*
Ygr$^-$	Yellow-green phenotype
Ygr$^+$	Wild-type phenotype

Appendix B

Bioinformatic Resources for Genetics and Genomics

"You certainly usually find something, if you look, but it is not always quite the something you were after." —*The Hobbit*, J. R. R. Tolkien

The field of bioinformatics encompasses the use of computational tools to distill complex data sets. Genetic and genomic data are so diverse that it has become a considerable challenge to identify the authoritative site(s) for a specific type of information. Furthermore, the landscape of Web-accessible software for analyzing this information is constantly changing as new and more powerful tools are developed. This appendix is intended to provide *some* valuable starting points for exploring the rapidly expanding universe of online resources for genetics and genomics.

1. Finding Genetic and Genomic Web Sites

Here are listed several central resources that contain large lists of relevant Web sites:

- The scientific journal called *Nucleic Acids Research* (*NAR*) publishes a special issue every January listing a wide variety of online database resources at http://nar.oupjournals.org/.
- The Virtual Library has Model Organisms and Genetics subdivisions with rich arrays of Internet resources at http://ceolas.org/VL/mo/ and http://www.ornl.gov/TechResources/Human_Genome/organisms.html.
- The National Human Genome Research Institute (NHGRI) maintains a list of genome Web sites at http://www.nhgri.nih.gov/10000375/.
- The Department of Energy (DOE) maintains a Human Genome Project site at the Oak Ridge National Laboratory at http://public.ornl.gov/hgmis/.
- SwissProt maintains Amos' WWW links page at http://www.expasy.ch/alinks.html.

2. General Databases

Nucleic Acid and Protein Sequence Databases By international agreement, three groups collaborate to house the primary DNA and mRNA sequences of all species: the National Center for Biotechnology Information (NCBI) houses GenBank, the European Bioinformatics Institute (EBI) houses the European Molecular Biology Laboratory (EMBL) Data Library, and the National Institute of Genetics in Japan houses the DNA DataBase of Japan (DDBJ).

Primary DNA sequence records, called accessions, are submitted by individual research groups. In addition to providing access to these DNA sequence records, these sites provide many other data sets. For example, NCBI also houses RefSeq, a summary synthesis of information on the DNA sequences of fully sequenced genomes and the gene products that are encoded by these sequences.

Many other important features can be found at the NCBI, EBI, and DDBJ sites. Home pages and some other key Web sites are

- **NCBI** http://www.ncbi.nlm.nih.gov/
- **NCBI-Genomes** http://www.ncbi.nlm.nih.gov/Genomes/index.html
- **NCBI-RefSeq** http://www.ncbi.nlm.nih.gov/LocusLink/refseq.html
- **The UCSC Genome Bioinformatics Site** http://genome.ucsc.edu/

This outstanding site contains the reference sequence and working draft assemblies for a large collection of genomes and a number of tools for exploring those genomes. The Genome Browser zooms and scrolls over chromosomes, showing the work of annotators worldwide. The Gene Sorter shows expression, homology, and other information on groups of genes that can be related in many ways. Blat quickly maps sequences to the genome. The Table Browser provides access to the underlying database.

- **EBI** http://www.ncbi.nlm.nih.gov/
- **DDBJ** http://www.nig.ac.jp/

The harsh reality is that, with so much biological information, the goal of making these online resources "transparent" to the user is not fully achieved. Thus, exploration of these sites will entail familiarizing yourself with the contents of each site and exploring some of the ways the site helps you to focus your queries so you get the right answer(s) to your queries. For one example of the power of these sites, consider a search for a nucleotide sequence at NCBI. Databases typically store information in separate bins called "fields." By using queries that limit the search to the appropriate field, a more directed question can be asked. Using the "Limits" option, a query phrase can be used to identify or locate a specific species, type of sequence (genomic or mRNA), gene symbol, or any of several other data fields. Query engines usually support the ability to join multiple query statements together. For example: retrieve all DNA sequence records that are from the species *Caenorhabditis elegans* **AND** that were published after January 1, 2000. Using the "History" option, the results of multiple queries can be joined together, so that only those hits common to multiple queries will be retrieved. By proper use of the available query options on a site, a great many false positives can be computationally eliminated while not discarding any of the relevant hits.

Because protein sequence predictions are a natural part of the analysis of DNA and mRNA sequences, these same sites provide access to a variety of protein databases. One important protein database is SwissProt/TrEMBL. TrEMBL sequences are automatically predicted from DNA and/or mRNA sequences. SwissProt sequences are curated, meaning that an expert scientist reviews the output of computational analysis and makes expert decisions about which results to accept or reject. In addition to the primary protein sequence records, SwissProt also offers databases of protein domains and protein signatures (amino acid sequence strings that are characteristic of proteins of a particular type). The SwissProt home page is http://www.ebi.ac.uk/swissprot/.

Protein Domain Databases The functional units within proteins are thought to be local folding regions called domains. Prediction of domains within newly discovered proteins is one way to guess at their function. Numerous protein domain databases have emerged that predict domains in somewhat different ways. Some of the individual domain databases are Pfam, PROSITE, PRINTS, SMART, ProDom, TIGRFAMs, BLOCKS, and CDD. InterPro allows querying multiple protein domain databases simultaneously and presents the combined results. Web sites for some domain databases are

- **InterPro** http://www.ebi.ac.uk/interpro/
- **Pfam** http://www.sanger.ac.uk/Software/Pfam/index.shtml
- **PROSITE** http://www.expasy.ch/prosite/
- **PRINTS** http://www.bioinf.man.ac.uk/dbbrowser/PRINTS/
- **SMART** http://smart.embl-heidelberg.de/
- **ProDom** http://prodes.toulouse.inra.fr/prodom/doc/prodom.html
- **TIGRFAMs** http://www.tigr.org/TIGRFAMs/
- **BLOCKS** http://blocks.fhcrc.org/
- **CDD** http://www.ncbi.nlm.nih.gov/Structure/cdd/cdd.shtml

centromere A specialized region of DNA on each eukaryotic chromosome that acts as a site for the binding of kinetochore proteins.

character An attribute of individual members of a species for which various heritable differences can be defined.

charged tRNA A transfer RNA molecule with an amino acid attached to its 3′ end. Also called aminoacyl-tRNA.

ChIP *See* **chromatin immunoprecipitation.**

chi-square (χ²) test A statistical test used to determine the probability of obtaining observed proportions by chance, under a specific hypothesis.

chloroplast DNA (cpDNA) The small genomic component found in the chloroplasts of plants, concerned with photosynthesis and other functions taking place within that organelle.

chromatid One of the two side-by-side replicas produced by chromosome division.

chromatin The substance of chromosomes; now known to include DNA and chromosomal proteins.

chromatin immunoprecipitation (ChIP) The use of antibodies to isolate specific regions of chromatin and to identify the regions of DNA to which regulatory proteins are bound.

chromatin remodeling Changes in nucleosome position along DNA.

chromosome A linear end-to-end arrangement of genes and other DNA, sometimes with associated protein and RNA.

chromosome map A representation of all chromosomes in the genome as lines, marked with the positions of genes known from their mutant phenotypes, plus molecular markers. Based on analysis of recombinant frequency.

chromosome mutation Any type of change in chromosome structure or number.

chromosome walk A method for the dissection of large segments of DNA, in which a cloned segment of DNA, usually eukaryotic, is used to screen recombinant DNA clones from the same genome bank for other clones containing neighboring sequences.

cis-acting element A site on a DNA (or RNA) molecule that functions as a binding site for a sequence-specific DNA- (or RNA-) binding protein. The term *cis-acting* indicates that protein binding to this site affects only nearby DNA (or RNA) sequences on the same molecule.

cis conformation In a heterozygote having two mutant sites within a gene or within a gene cluster, the arrangement A_1A_2/a_1a_2.

class 1 element A transposable element that moves through an RNA intermediate. Also called an RNA or retro element.

class 2 element A transposable element that moves directly from one site in the genome to another. Also called a DNA element.

cloning vector In cloning, the plasmid or phage chromosome used to carry the cloned DNA segment.

CNV *See* **copy number variation.**

co-activator A special class of eukaryotic regulatory complex that serves as a bridge to bring together regulatory proteins and RNA polymerase II.

c.o.c. *See* **coefficient of coincidence.**

Cockayne syndrome A genetic disorder caused by defects in the nucleotide-excision-repair system and leading to symptoms of premature aging. Individuals with Cockayne syndrome have a mutation in one of two proteins thought to recognize transcription complexes that are stalled owing to DNA damage.

coding strand The nontemplate strand of a DNA molecule having the same sequence as that in the RNA transcript.

codominance A situation in which a heterozygote shows the phenotypic effects of both alleles equally.

codon A section of RNA (three nucleotides in length) that encodes a single amino acid.

coefficient of coincidence (c.o.c.) The ratio of the observed number of double recombinants to the expected number.

cointegrate The product of the fusion of two circular transposable elements to form a single, larger circle in replicative transposition.

colony A visible clone of cells.

common SNP A single nucleotide polymorphism (SNP) for which the less common allele occurs at a frequency of about 5 percent or greater.

comparative genomics Analysis of the relations of the genome sequences of two or more species.

complementary (base pairs) Refers to specific pairing between adenine and thymine and between guanine and cytosine.

complementary DNA (cDNA) DNA transcribed from a messenger RNA template through the action of the enzyme reverse transcriptase.

complementation test A test for determining whether two mutations are in different genes (they complement) or the same gene (they do not complement).

complete dominance Describes an allele that expresses itself the same in single copy (heterozygote) as in double copy (homozygote).

complex inheritance The type of inheritance exhibited by traits affected by a mix of genetic and environmental factors. Continuous traits, such as height, typically have complex inheritance.

complex trait A trait exhibiting complex inheritance.

composite transposon A type of bacterial transposable element containing a variety of genes that reside between two nearly identical insertion sequence (IS) elements.

congenic lines Strains or stocks of a species that are identical throughout their genomes except for a small region of interest.

conjugation The union of two bacterial cells during which chromosomal material is transferred from the donor to the recipient cell.

consensus sequence The nucleotide sequence of a segment of DNA that is derived by aligning similar sequences (either from the same or different organisms) and determining the most common nucleotide at each position.

consensus sequence The nucleotide sequence of a segment of DNA that is in agreement with most sequence reads of the same segment from different individuals.

conservative replication A disproved model of DNA synthesis suggesting that one-half of the daughter DNA molecules should have both strands composed of newly polymerized nucleotides.

conservative substitution Nucleotide-pair substitution within a protein-coding region that leads to the replacement of an amino acid by one of similar chemical properties.

conservative transposition A mechanism of transposition that moves a mobile element to a new location in the genome as it removes it from its previous location.

constitutive expression Refers to genes that are expressed continuously regardless of biological conditions.

constitutive heterochromatin Chromosomal regions of permanently condensed chromatin usually around the telomeres and centromeres.

constitutive mutation A change in a DNA sequence that causes a gene that is repressed at times to be expressed continuously, or "constitutively."

continuous trait A trait that can take on a potentially infinite number of states over a continuous range, such as height in humans.

coordinately controlled genes Genes whose products are simultaneously activated or repressed in parallel.

copia-**like element** A transposable element (retrotransposon) of *Drosophila* that is flanked by long terminal repeats and typically encodes a reverse transcriptase.

"copy and paste" A descriptive term for a transposition mechanism in which a class 1 retrotransposon is copied from the donor site and a double stranded DNA copy is inserted (pasted) into a new target site. *See also* **cut and paste.**

copy number variation (CNV) The number of copies of parts of individual genes, entire genes, or sets of genes that differ in humans, including repeats and duplications that increase copy number and deletions that reduce it.

corepressor A repressor that facilitates gene repression but is not itself a DNA-binding repressor.

correlation The tendency of one variable to vary in proportion to another variable, either positively or negatively.

correlation coefficient A statistical measure of association that signifies the extent to which two variables vary together.

cosuppression An epigenetic phenomenon whereby a transgene becomes reversibly inactivated along with the gene copy in the chromosome.

cotranscriptional processing The simultaneous transcription and processing of eukaryotic pre-mRNA.

cotransductants Two donor alleles that simultaneously transduce a bacterial cell; their frequency is used as a measure of closeness of the donor genes on the chromosome map.

covariance A statistical measure of the extent to which two variables change together. It is used in computing the correlation coefficient between two variables.

cpDNA Chloroplast DNA.

CpG island Unmethylated CG dinucleotides found in clusters near gene promoters.

cross The deliberate mating of two parental types of organisms in genetic analysis.

crossing over The exchange of corresponding chromosome parts between homologs by breakage and reunion.

crossover products Meiotic product cells with chromosomes that have engaged in a crossover.

CTD *See* **carboxyl tail domain.**

cumulative selection The situation when natural selection promotes multiple substitutions that alter the function of a protein or regulatory element through repeated rounds of mutation and selection.

"cut and paste" A descriptive term for a transposition mechanism in which a class 2 (DNA) transposon is excised (cut) from the donor site and inserted (pasted) into a new target site. *See also* **copy and paste.**

C-**value** The DNA content of a haploid genome.

C-**value paradox** The discrepancy (or lack of correlation) between the DNA content of an organism and its biological complexity.

cyclic adenosine monophosphate (cAMP) A molecule containing a diester bond between the 3′ and 5′ carbon atoms of the ribose part of the nucleotide. This modified nucleotide cannot be incorporated into DNA or RNA. It plays a key role as an intracellular signal in the regulation of various processes.

cytidine (C) A nucleoside containing cytosine as its base.

cytoplasmic segregation Segregation in which genetically different daughter cells arise from a progenitor that is a cytohet.

cytosine (C) A pyrimidine base that pairs with guanine.

Darwinian fitness The relative probability of survival and reproduction for a genotype.

daughter molecule One of the two products of DNA replication composed of one template strand and one newly synthesized strand.

decoding center The region in the small ribosomal subunit where the decision is made whether an aminoacyl-tRNA can bind in the A site. This decision is based on complementarity between the anticodon of the tRNA and the codon of the mRNA.

degenerate code A genetic code in which some amino acids may be encoded by more than one codon each.

deletion The removal of a chromosomal segment from a chromosome set.

deletion loop The loop formed at meiosis by the pairing of a normal chromosome and a deletion-containing chromosome.

deletion mapping The use of a set of known deletions to map new recessive mutations by pseudodominance.

deoxyribonucleic acid *See* **DNA.**

deoxyribose The pentose sugar in the DNA backbone.

deviation Difference of an individual trait value from the mean trait value for the population.

dicentric bridge In a dicentric chromosome, the segment between centromeres being drawn to opposite poles at nuclear division.

dicentric chromosome A chromosome with two centromeres.

Dicer A protein complex that recognizes long double-stranded RNA molecules and cleaves them into double-stranded siRNAs. Dicer plays a key role in RNA interference.

dideoxy (Sanger) sequencing The most popular method of DNA sequencing. It uses dideoxynucleotide triphosphates mixed with standard nucleotide triphosphates to produce a ladder of DNA strands whose synthesis is blocked at different lengths. This method has been incorporated into automated DNA-synthesis machines. Also called Sanger sequencing after its inventor, Frederick Sanger.

dihybrid A double heterozygote such as $A/a \cdot B/b$.

dihybrid cross A cross between two individuals identically heterozygous at two loci—for example, $A \, B/a \, b \times A \, B/a \, b$.

dimorphism A polymorphism with only two forms.

dioecious species A plant species in which male and female organs are on separate plants.

diploid A cell having two chromosome sets or an individual organism having two chromosome sets in each of its cells.

directional selection Selection that changes the frequency of an allele in a constant direction, either toward or away from fixation for that allele.

disassortive mating *See* **negative assortative mating.**

discovery panel A group of individuals used to detect variable nucleotide sites by comparing the partial genome sequences of these individuals with one another.

disomic An abnormal haploid carrying two copies of one chromosome.

dispersive replication A disproved model of DNA synthesis suggesting more or less random interspersion of parental and new segments in daughter DNA molecules.

***Dissociation* (*Ds*) element** A nonautonomous transposable element named by Barbara McClintock for its ability to break chromosome 9 of maize but only in the presence of another element called *Activator* (*Ac*).

distributive enzyme An enzyme that can add only a limited number of nucleotides before falling off the DNA template.

DNA (deoxyribonucleic acid) A chain of linked nucleotides (having deoxyribose as their sugars). Two such chains in double helical form are the fundamental substance of which genes are composed.

DNA-binding domain The site on a DNA-binding protein that directly interacts with specific DNA sequences.

DNA clone A section of DNA that has been inserted into a vector molecule, such as a plasmid or a phage, and then replicated to produce many copies.

DNA element A class 2 transposable element found in both prokaryotes and eukaryotes and so named because the DNA element participates directly in transposition.

DNA fingerprint The autoradiographic banding pattern produced when DNA is digested with a restriction enzyme that cuts outside a family of VNTRs (variable number of tandem repeats) and a Southern blot of the electrophoretic gel is probed with a VNTR-specific probe. Unlike true fingerprints, these patterns are not unique to each individual organism.

DNA ligase An important enzyme in DNA replication and repair that seals the DNA backbone by catalyzing the formation of phosphodiester bonds.

DNA methylation The addition of methyl groups to DNA residues after replication.

DNA palindrome A segment of DNA in which both strands have the same nucleotide sequence but in antiparallel orientation.

DNA polymerase III holoenzyme (DNA pol III holoenzyme) In *E. coli*, the large multisubunit complex at the replication fork consisting of two catalytic cores and many accessory proteins.

DNA technology The collective techniques for obtaining, amplifying, and manipulating specific DNA fragments.

DNA template library A group of single-stranded DNA molecules that can be amplified into many copies.

DNA transposon *See* **DNA element.**

domain A region of a protein associated with a particular function. Some proteins contain more than one domain.

dominance effect The difference between the trait value for the heterozygous class at a QTL and the midpoint between the trait values of the two homozygous classes.

dominant The phenotype shown by a heterozygote.

dominant allele An allele that expresses its phenotypic effect even when heterozygous with a recessive allele; thus, if *A* is dominant over *a*, then *A*/*A* and *A*/*a* have the same phenotype.

dominant gene action The situation when the trait value for the heterozygous class at a QTL is equal to the trait value for one of the two homozygous classes.

dominant negative mutation A mutant allele that in single dose (a heterozygote) wipes out gene function by a spoiler effect on the protein.

donor Bacterial cell used in studies of unidirectional DNA transmission to other cells; examples are Hfr in conjugation and phage source in transduction.

donor DNA Any DNA to be used in cloning or in DNA-mediated transformation.

dosage compensation The process in organisms using a chromosomal sex-determination mechanism (such as XX versus XY) that allows standard structural genes on the sex chromosome to be expressed at the same levels in females and males, regardless of the number of sex chromosomes. In mammals, dosage compensation operates by maintaining only a single active X chromosome in each cell; in *Drosophila*, it operates by hyperactivating the male X chromosome.

double helix The structure of DNA first proposed by James Watson and Francis Crick, with two interlocking helices joined by hydrogen bonds between paired bases.

double (mixed) infection Infection of a bacterium with two genetically different phages.

double mutant Genotype with mutant alleles of two different genes.

double-strand break A DNA break cleaving the sugar–phosphate backbones of both strands of the DNA double helix.

double-stranded RNA (dsRNA) An RNA molecule comprised of two complementary strands.

double transformation Simultaneous transformation by two different donor markers.

downstream A way to describe the relative location of a site in a DNA or RNA molecule. A downstream site is located closer to the $3'$ end of a transcription unit.

Down syndrome An abnormal human phenotype, including mental retardation, due to a trisomy of chromosome 21; more common in babies born to older mothers.

drift *See* **random genetic drift.**

dsRNA *See* **double-stranded RNA.**

duplication More than one copy of a particular chromosomal segment in a chromosome set.

dyad A pair of sister chromatids joined at the centromere, as in the first division of meiosis.

ectopic integration In a transgenic organism, the insertion of an introduced gene at a site other than its usual locus.

elongation The stage of transcription that follows initiation and precedes termination.

embryoid A small dividing mass of monoploid cells, produced from a cell destined to become a pollen cell by exposing it to cold.

endogenote *See* **merozygote.**

endogenous gene A gene that is normally present in an organism, in contrast with a foreign gene from a different organism that might be introduced by transgenic techniques.

enhanceosome The macromolecular assembly responsible for interaction between enhancer elements and the promoter regions of genes.

enhancer A set of regulatory proteins consisting of transcription factors that bind to *cis-acting* regulatory sequences in the DNA.

enhancer-blocking insulators Regulatory elements positioned between a promoter and an enhancer. Their presence prevents the promoter from being activated by the enhancer.

enhancer element A cis-acting regulatory sequence that can elevate levels of transcription from an adjacent promoter. Many tissue-specific enhancers can determine spatial patterns of gene expression in higher eukaryotes. Enhancers can act on promoters over many tens of kilobases of DNA and can be 5′ or 3′ of the promoters that they regulate.

enol form *See* **tautomeric shift.**

environmental variance The part of the phenotypic variation among individuals in a population that is due to the different environments the individuals have experienced.

epigenetic Nongenetic chemical changes in histones or DNA that alter gene function without altering the DNA sequence.

epigenetic inheritance Heritable modifications in gene function not due to changes in the base sequence of the DNA of the organism. Examples of epigenetic inheritance are paramutation, X-chromosome inactivation, and parental imprinting.

epigenetic mark A heritable alteration, such as DNA methylation or a histone modification, that leaves the DNA sequence unchanged.

epigenetic silencing The repression of the expression of a gene by virtue of its position in the chromosome rather than by a mutation in its DNA sequence. Epigenetic silencing can be inherited from one cell generation to the next.

epistasis A situation in which the differential phenotypic expression of a genotype at one locus depends on the genotype at another locus; a mutation that exerts its expression while canceling the expression of the alleles of another gene.

E site *See* **exit site.**

EST *See* **expressed sequence tag.**

euchromatin A less-condensed chromosomal region, thought to contain most of the normally functioning genes.

euploid A cell having any number of complete chromosome sets or an individual organism composed of such cells.

excise Describes what a transposable element does when it leaves a chromosomal location. Also called transpose.

exconjugant A female bacterial cell that has just been in conjugation with a male and contains a fragment of male DNA.

exit (E) site The site on the ribosome where the deacylated tRNA can be found. 9

exogenote *See* **merozygote.**

exon Any nonintron section of the coding sequence of a gene; together, the exons correspond to the mRNA that is translated into protein.

expressed sequence tag (EST) A cDNA clone for which only the 5′ or the 3′ ends or both have been sequenced; used to identify transcript ends in genomic analysis.

expressivity The degree to which a particular genotype is expressed in the phenotype.

extranuclear Refers to a small specialized fraction of eukaryotic genomes found in mitochondria or chloroplasts.

F⁻ cell In *E. coli,* a cell having no fertility factor; a female cell.

F⁺ cell In *E. coli,* a cell having a free fertility factor; a male cell.

F factor *See* **fertility factor.**

F′ factor A fertility factor into which a part of the bacterial chromosome has been incorporated.

F′ plasmid *See* **F′ factor.**

F₁ generation The first filial generation, produced by crossing two parental lines.

F₂ generation The second filial generation, produced by selfing or intercrossing the F₁ generation.

fertility factor (F factor) A bacterial episome whose presence confers donor ability (maleness).

fibrous protein A protein with a linear shape such as the components of hair and muscle.

fine mapping Finding the genomic location of a gene of interest (or a functional region within a gene) with marker loci that are very tightly linked to it.

first-division segregation pattern (M₁ pattern) A linear pattern of spore phenotypes within an ascus for a particular allele pair, produced when the alleles go into separate nuclei at the first meiotic division, showing that no crossover has taken place between the allele pair and the centromere.

first filial generation (F₁) The progeny individuals arising from a cross of two homozygous diploid lines.

5′ untranslated region (5′ UTR) The region of the RNA transcript at the 5′ end upstream of the translation start site.

fixed allele An allele for which all members of the population under study are homozygous, and so no other alleles for this locus exist in the population.

fluctuation test A test used in microbes to establish the random nature of mutation or to measure mutation rates.

forward genetics The classical approach to genetic analysis, in which genes are first identified by mutant alleles and mutant phenotypes and later cloned and subjected to molecular analysis.

fosmid A vector that can carry a 35- to 45-kb insert of foreign DNA.

founder effect A random difference in the frequency of an allele or a genotype in a new colony as compared to the parental population that results from a small number of founders.

frameshift mutation The insertion or deletion of a nucleotide pair or pairs, causing a disruption of the translational reading frame.

frequency histogram A "step curve" in which the frequencies of various arbitrarily bounded classes are graphed.

full dominance *See* **complete dominance.**

functional complementation (mutant rescue) The use of a cloned fragment of wild-type DNA to transform a mutant into wild type; used in identifying a clone containing one specific gene.

functional genomics The study of the patterns of transcript and protein expression and of molecular interactions at a genome-wide level.

functional RNA An RNA type that plays a role without being translated.

G *See* **guanine; guanosine.**

gap gene In *Drosophila,* a class of cardinal genes that are activated in the zygote in response to the anterior–posterior gradients of positional information.

GD *See* **gene diversity.**

gel electrophoresis A method of molecular separation in which DNA, RNA, or proteins are separated in a gel matrix according to molecular size, with the use of an electrical field to draw the molecules through the gel in a predetermined direction.

gene The fundamental physical and functional unit of heredity, which carries information from one generation to the next; a segment of DNA composed of a transcribed region and a regulatory sequence that makes transcription possible.

gene action Interaction among alleles at a locus.

gene balance The idea that a normal phenotype requires a 1:1 relative proportion of genes in the genome.

gene complex A group of adjacent functionally and structurally related genes that typically arise by gene duplication in the course of evolution.

gene discovery The process whereby geneticists find a set of genes affecting some biological process of interest by the single-gene inheritance patterns of their mutant alleles or by genomic analysis.

gene diversity (*GD*) The probability that two alleles drawn at random from the gene pool will be different.

gene-dosage effect (1) Proportionality of the expression of some biological function to the number of copies of an allele present in the cell. (2) A change in phenotype caused by an abnormal number of wild-type alleles (observed in chromosomal mutations).

gene duplication The duplication of genes or segments of DNA through misreplication of DNA.

gene family A set of genes in one genome, all descended from the same ancestral gene.

gene flow *See* **migration.**

gene knockout The inactivation of a gene by either a naturally occurring mutation or through the integration of a specially engineered introduced DNA fragment. In some systems, such inactivation is random, with the use of transgenic constructs that insert at many different locations in the genome. In other systems, it can be carried out in a directed fashion. *See also* **targeted gene knockout.**

gene locus The specific place on a chromosome where a gene is located.

gene pair The two copies of a particular type of gene present in a diploid cell (one in each chromosome set).

gene pool The sum total of all alleles in the breeding members of a population at a given time.

generalized transduction The ability of certain phages to transduce any gene in the bacterial chromosome.

general transcription factor (GTF) A eukaryotic protein complex that does not take part in RNA synthesis but binds to the promoter region to attract and correctly position RNA polymerase II for transcription initiation.

gene replacement The insertion of a genetically engineered transgene in place of a resident gene; often achieved by a double crossover.

gene silencing A gene that is not expressed owing to epigenetic regulation. Unlike genes that are mutant due to DNA sequence alterations, genes inactivated by silencing can be reactivated.

gene therapy The correction of a genetic deficiency in a cell by the addition of new DNA and its insertion into the genome. Different techniques have the potential to carry out gene therapy only in somatic tissues or, alternatively, to correct the genetic deficiency in the zygote, thereby correcting the germ line as well.

genetic admixture The mix of genes that results when individuals have ancestry from more than one subpopulation.

genetically modified organism (GMO) A popular term for a transgenic organism, especially applied to transgenic agricultural organisms.

genetic architecture All of the genetic and environmental factors that influence a trait.

genetic code A set of correspondences between nucleotide triplets in RNA and amino acids in protein.

genetic dissection The use of recombination and mutation to piece together the various components of a given biological function.

genetic drift The change in the frequency of an allele in a population resulting from chance differences in the actual numbers of offspring of different genotypes produced by different individual members.

genetic engineering The process of producing modified DNA in a test tube and reintroducing that DNA into host organisms.

genetic load The total set of deleterious alleles in an individual genotype.

genetic map unit (m.u.) A distance on the chromosome map corresponding to 1 percent recombinant frequency.

genetic marker An allele used as an experimental probe to keep track of an individual organism, a tissue, a cell, a nucleus, a chromosome, or a gene.

genetics (1) The study of genes. (2) The study of inheritance.

genetic switch A segment of regulatory DNA and the regulatory protein(s) that binds to it that govern the transcriptional state of a gene or set of genes.

genetic variance The part of the phenotypic variation among individuals in a population that is due to the genetic differences among the individuals.

genome The entire complement of genetic material in a chromosome set.

genome project A large-scale, often multilaboratory effort required to sequence a complex genome.

genome-wide association (GWA) Association mapping that uses marker loci throughout the entire genome.

genomic imprinting A phenomenon in which a gene inherited from one of the parents is not expressed, even though both gene copies are functional. Imprinted genes are methylated and inactivated in the formation of male or female gametes.

genomic library A library encompassing an entire genome.

genomics The cloning and molecular characterization of entire genomes.

genotype The allelic composition of an individual or of a cell—either of the entire genome or, more commonly, of a certain gene or a set of genes.

genotype frequency The proportion of individuals in a population having a particular genotype.

GGR *See* **global genomic repair.**

global genomic repair (GGR) A type of nucleotide-excision repair that takes place at nontranscribed sequences.

globular protein A protein with a compact structure, such as an enzyme or an antibody.

GMO *See* **genetically modified organism.**

GU-AG rule So named because the GU and AG dinucleotides are almost always at the 5′ and 3′ ends, respectively, of introns, where they are recognized by components of the spliceosome.

guanine (G) A purine base that pairs with cytosine.

GWA *See* **genome-wide association.**

H *See* **heterozygosity.**

haploid A cell having one chromosome set or an organism composed of such cells.

haploid number The number of chromosomes in the basic genomic set of a species.

haplosufficient Describes a gene that, in a diploid cell, can promote wild-type function in only one copy (dose).

haplotype The type (or form) of a haploid segment of a chromosome as defined by the alleles present at the loci within that segment.

haplotype network A network that shows relationships among haplotypes and the positions of the mutations defining the haplotypes on the branches.

HapMap A genome-wide haplotype map.

Hardy–Weinberg equilibrium The stable frequency distribution of genotypes A/A, A/a, and a/a, in the proportions of p^2, $2pq$, and q^2, respectively (where p and q are the frequencies of the alleles A and a), that is a consequence of random mating in the absence of mutation, migration, natural selection, or random drift.

Hardy–Weinberg law An equation used to describe the relationship between allelic and genotypic frequencies in a random-mating population.

helicase An enzyme that breaks hydrogen bonds in DNA and unwinds the DNA during movement of the replication fork.

hemimethylated DNA DNA sequence with one methylated strand and one unmethylated strand.

hemizygous gene A gene present in only one copy in a diploid organism—for example, an X-linked gene in a male mammal.

heterochromatin Densely staining condensed chromosomal regions, believed to be for the most part genetically inert.

heterochromatin protein-1 (HP-1) A protein necessary for the maintenance of heterochromatin.

heteroduplex DNA DNA in which there is one or more mismatched nucleotide pairs in a gene under study.

heterogametic sex The sex that has heteromorphic sex chromosomes (e.g., XY) and hence produces two different kinds of gametes with respect to the sex chromosomes.

heterogeneous genetic disorder A disorder caused by mutations in any one of several genes encoding a particular process.

heterokaryon A culture of cells composed of two different nuclear types in a common cytoplasm.

heteroplex DNA DNA in which there is a mismatched nucleotide pair in a gene under study.

heterozygote An individual organism having a heterozygous gene pair.

heterozygosity A measure of the genetic variation in a population; with respect to one locus, stated as the frequency of heterozygotes for that locus.

heterozygous gene pair A gene pair having different alleles in the two chromosome sets of the diploid individual—for example, A/a or A^1/A^2.

hexaploid A cell having six chromosome sets or an organism composed of such cells.

Hfr *See* **high frequency of recombination cell.**

high frequency of recombination (Hfr) cell In *E. coli*, a cell having its fertility factor integrated into the bacterial chromosome; a donor (male) cell.

histone A type of basic protein that forms the unit around which DNA is coiled in the nucleosomes of eukaryotic chromosomes.

histone code Refers to the pattern of modification (e.g., acetylation, methylation, phosphorylation) of the histone tails that may carry information required for correct chromatin assembly.

histone deacetylase The enzymatic activity that removes an acetyl group from a histone tail, which promotes the repression of gene transcription.

histone tail The end of a histone protein protruding from the core nucleosome and subjected to posttranslational modification. *See also* **histone code.**

homeobox (homeotic box) A family of quite similar 180-bp DNA sequences that encode a polypeptide sequence called a homeodomain, a sequence-specific DNA-binding sequence. Although the homeobox was first discovered in all homeotic genes, it is now known to encode a much more widespread DNA-binding motif.

homeodomain A highly conserved family of sequences, 60 amino acids in length and found within a large number of transcription factors, that can form helix-turn-helix structures and bind DNA in a sequence-specific manner.

homeologous chromosomes Partly homologous chromosomes, usually indicating some original ancestral homology.

homogametic sex The sex with homologous sex chromosomes (e.g., XX).

homolog A member of a pair of homologous chromosomes.

homologous chromosomes Chromosomes that pair with each other at meiosis or chromosomes in different species that have retained most of the same genes during their evolution from a common ancestor.

homology-dependent repair system A mechanism of DNA repair that depends on the complementarity or homology of the template strand to the strand being repaired.

homozygote An individual organism that is homozygous.

homozygous Refers to the state of carrying a pair of identical alleles at one locus.

homozygous dominant Refers to a genotype such as A/A.

homozygous recessive Refers to a genotype such as a/a.

housekeeping gene An informal term for a gene whose product is required in all cells and carries out a basic physiological function.

***Hox* genes** Members of this gene class are the clustered homeobox-containing, homeotic genes that govern the identity of body parts along the anterior–posterior axis of most bilateral animals.

hybrid dysgenesis A syndrome of effects including sterility, mutation, chromosome breakage, and male recombination in the hybrid progeny of crosses between certain laboratory and natural isolates of *Drosophila*.

hybridize (1) To form a hybrid by performing a cross. (2) To anneal complementary nucleic acid strands from different sources.

hybrid vigor A situation in which an F_1 is larger or healthier than its two different pure parental lines.

hyperacetylation An overabundance of acetyl groups attached to certain amino acids of the histone tails. Transcriptionally active chromatin is usually hyperacetylated.

hypoacetylation An underabundance of acetyl groups on certain amino acids of the histone tails. Transcriptionally inactive chromatin is usually hypoacetylated.

IBD *See* **identical by descent.**

identical by descent (IBD) When two copies of a gene in an individual trace back to the same copy in an ancestor.

imino form *See* **tautomeric shift.**

inbred line A stock consisting of genetically identical individuals that were fully inbred from a common parent(s).

inbreeding Mating between relatives.

inbreeding coefficient (F) The probability that the two alleles at a locus in an individual are identical by descent.

inbreeding depression A reduction in vigor and reproductive success from inbreeding.

incomplete dominance A situation in which a heterozygote shows a phenotype quantitatively (but not exactly) intermediate between the corresponding homozygote phenotypes. (Exact intermediacy means no dominance.)

indel mutation A mutation in which one or more nucleotide pairs is added or deleted.

independent assortment *See* **Mendel's second law.**

induced mutation A mutation that arises through the action of an agent that increases the rate at which mutations occur.

inducer An environmental agent that triggers transcription from an operon.

induction (1) The relief of repression of a gene or set of genes under negative control. (2) An interaction between two or more cells or tissues that is required for one of those cells or tissues to change its developmental fate.

initiation The first stage of transcription or translation. Its main function in transcription is to correctly position RNA polymerase before the elongation stage, and in translation it is to correctly position the first aminoacyl-tRNA in the P site.

initiation factor A protein required for the correct initiation of translation.

initiator A special tRNA that inserts the first amino acid of a polypeptide chain into the ribosomal P site at the start of translation. The amino acid carried by the initiator in bacteria is *N*-formylmethionine.

insertional duplication A duplication in which the extra copy is not adjacent to the normal one.

insertional mutagenesis The situation when a mutation arises by the interruption of a gene by foreign DNA, such as from a transgenic construct or a transposable element.

insertion sequence (IS) element A mobile piece of bacterial DNA (several hundred nucleotide pairs in length) capable of inactivating a gene into which it inserts.

interactome The entire set of molecular interactions within cells, including in particular protein–protein interactions.

intercalating agent A mutagen that can insert itself between the stacked bases at the center of the DNA double helix, causing an elevated rate of *indel mutations*.

interference A measure of the independence of crossovers from each other, calculated by subtracting the coefficient of coincidence from 1.

interrupted mating A technique used to map bacterial genes by determining the sequence in which donor genes enter recipient cells.

intervening sequence An intron; a segment of largely unknown function within a gene. This segment is initially transcribed, but the transcript is not found in the functional mRNA.

intragenic deletion A deletion within a gene.

intron *See* **intervening sequence.**

inversion A chromosomal mutation consisting of the removal of a chromosome segment, its rotation through 180°, and its reinsertion in the same location.

inversion heterozygote A diploid with a normal and an inverted homolog.

inversion loop A loop formed by meiotic pairing of homologs in an inversion heterozygote.

inverted repeat (IR) sequence A sequence found in identical (but inverted) form—for example, at the opposite ends of a DNA transposon.

IR sequence *See* **inverted repeat sequence.**

IS element *See* **insertion sequence element.**

isoforms Related by different proteins. They can be generated by alternative splicing of a gene.

isolation by distance A bias in mate choice that arises from the amount of geographic distance between individuals, causing individuals to be more apt to mate with a neighbor than another member of their species farther away.

keto form *See* **tautomeric shift.**

Klinefelter syndrome An abnormal human male phenotype due to an extra X chromosome (XXY).

lagging strand In DNA replication, the strand that is synthesized apparently in the 3′-to-5′ direction by the ligation of short fragments synthesized individually in the 5′-to-3′ direction.

λ (lambda) attachment site Where the λ prophage inserts in the *E. coli* chromosome.

law of equal segregation The production of equal numbers (50 percent) of each allele in the meiotic products (e.g., gametes) of a heterozygous meiocyte.

LD *See* **linkage disequilibrium.**

leader sequence The sequence at the 5′ end of an mRNA that is not translated into protein.

leading strand In DNA replication, the strand that is made in the 5′-to-3′ direction by continuous polymerization at the 3′ growing tip.

leaky mutation A mutation that confers a mutant phenotype but still retains a low but detectable level of wild-type function.

lethal allele An allele whose expression results in the death of the individual organism expressing it.

LINE *See* **long interspersed element.**

linkage disequilibrium (LD) Deviation in the frequencies of different haplotypes in a population from the frequencies expected if the alleles at the loci defining the haplotypes are associated at random.

linkage equilibrium A perfect fit of haplotype frequencies in a population to the frequencies expected if the alleles at the loci defining the haplotypes are associated at random.

linkage map A chromosome map; an abstract map of chromosomal loci that is based on recombinant frequencies.

linked The situation in which two genes are on the same chromosome as deduced by recombinant frequencies less than 50 percent.

lncRNA *See* **long noncoding RNA.**

locus (plural, **loci**) *See* **gene locus.**

long interspersed element (LINE) A type of class 1 transposable element that encodes a reverse transcriptase. LINEs are also called non-LTR retrotransposons.

long noncoding RNA (lncRNA) Nonprotein-coding transcripts that are over approximately 200 nucleotides in length.

long terminal repeat (LTR) A direct repeat of DNA sequence at the 5′ and 3′ ends of retroviruses and retrotransposons.

LTR *See* **long terminal repeat.**

LTR-retrotransposon A type of class 1 transposable element that terminates in long terminal repeats and encodes several proteins including reverse transcriptase.

lysate Population of phage progeny.

lysis The rupture and death of a bacterial cell on the release of phage progeny.

lysogen *See* **lysogenic bacterium.**

lysogenic bacterium A bacterial cell containing an inert prophage integrated into, and that is replicated with, the host chromosome.

lysogenic cycle The life cycle of a normal bacterium when it is infected by a wild-type λ phage and the phage genome is integrated into the bacterial chromosome as an inert prophage.

lytic cycle The bacteriophage life cycle that leads to lysis of the host cell.

M_I pattern *See* **first-division segregation pattern.**

M_II pattern *See* **second-division segregation pattern.**

major groove The larger of the two grooves in the DNA double helix.

mapping function A formula expressing the relation between distance in a linkage map and recombinant frequency.

map unit (m.u.) The "distance" between two linked gene pairs where 1 percent of the products of meiosis are recombinant; a unit of distance in a linkage map.

maternal-effect gene A gene that produces an effect only when present in the mother.

maternal imprinting The expression of a gene only when inherited from the father, because the copy of the gene inherited from the mother is inactive due to methylation in the course of gamete formation.

maternal inheritance A type of uniparental inheritance in which all progeny have the genotype and phenotype of the parent acting as the female.

M cytotype Laboratory stocks of *Drosophila melanogaster* that completely lack the *P* element transposon, which is found in stocks from the wild (P cytotype).

mean The arithmetic average.

mean fitness The mean of the fitness of all individual members of a population.

mediator complex A protein complex that acts as an adaptor that interacts with transcription factors bound to regulatory sites and with general initiation factors for RNA polymerase II–mediated transcription.

meiocyte A cell in which meiosis takes place.

meiosis Two successive nuclear divisions (with corresponding cell divisions) that produce gametes (in animals) or sexual spores (in plants and fungi) that have one-half of the genetic material of the original cell.

meiotic recombination Recombination from assortment or crossing over at meiosis.

Mendel's first law The two members of a gene pair segregate from each other in meiosis; each gamete has an equal probability of obtaining either member of the gene pair.

Mendel's second law The law of independent assortment; unlinked or distantly linked segregating gene pairs assort independently at meiosis.

meristic trait A counting trait, taking on a range of discrete values.

merozygote A partly diploid *E. coli* cell formed from a complete chromosome (the endogenote) plus a fragment (the exogenote).

messenger RNA *See* **mRNA.**

microarray A set of DNAs containing all or most genes in a genome deposited on a small glass chip.

microRNA (miRNA) A class of functional RNA that regulates the amount of protein produced by a eukaryotic gene.

microsatellite A locus composed of several to many copies (repeats) of a short (about 2 to 6 bp) sequence motif. Difference alleles have different numbers of repeats.

microsatellite marker A difference in DNA at the same locus in two genomes that is due to different repeat lengths of a microsatellite.

migration The movement of individuals (or gametes) between populations.

miniature inverted repeat transposable element (MITE) A type of nonautonomous DNA transposon that can form by deletion of the transposase gene from an autonomous element and attain very high copy numbers.

minimal medium Medium containing only inorganic salts, a carbon source, and water.

minisatellite marker Heterozygous locus representing a variable number of tandem repeats of a unit 15 to 100 nucleotides long.

minor groove The smaller of the two grooves in the DNA double helix.

miRNA *See* **microRNA.**

mismatch-repair system A system for repairing damage to DNA that has already been replicated.

missense mutation Nucleotide-pair substitution within a protein-coding region that leads to the replacement of one amino acid by another amino acid.

MITE *See* **miniature inverted repeat transposable element.**

mitochondrial DNA (mtDNA) The subset of the genome found in the mitochondrion, specializing in providing some of the organelle's functions.

mitosis A type of nuclear division (occurring at cell division) that produces two daughter nuclei identical with the parent nucleus.

mixed (double) infection The infection of a bacterial culture with two different phage genotypes.

modifier A mutation at a second locus that changes the degree of expression of a mutated gene at a first locus.

molecular clock The constant rate of substitution of amino acids in proteins or nucleotides in nucleic acids over long evolutionary time.

molecular genetics The study of the molecular processes underlying gene structure and function.

molecular marker A site of DNA heterozygosity, not necessarily associated with phenotypic variation, used as a tag for a particular chromosomal locus.

monohybrid A single-locus heterozygote of the type A/a.

monohybrid cross A cross between two individuals identically heterozygous at one gene pair—for example, $A/a \times A/a$.

monoploid A cell having only one chromosome set (usually as an aberration) or an organism composed of such cells.

monosomic A cell or individual organism that is basically diploid but has only one copy of one particular chromosome type and thus has chromosome number $2n + 1$.

morph One form of a genetic polymorphism; the morph can be either a phenotype or a molecular sequence.

mRNA (messenger RNA) An RNA molecule transcribed from the DNA of a gene; a protein is translated from this RNA molecule by the action of ribosomes.

mtDNA Mitochondrial DNA.

m.u. *See* **map unit.**

multifactorial hypothesis A hypothesis that explains quantitative variation by proposing that traits are controlled by a large number of genes, each with a small effect on the trait.

multigenic deletion A deletion of several adjacent genes.

multiple alleles The set of forms of one gene, differing in their DNA sequence or expression or both.

mutagen An agent capable of increasing the mutation rate.

mutagenesis An experiment in which experimental organisms are treated with a mutagen and their progeny are examined for specific mutant phenotypes.

mutant An organism or cell carrying a mutation.

mutant rescue *See* **functional complementation.**

mutation (1) The process that produces a gene or a chromosome set differing from that of the wild type. (2) The gene or chromosome set that results from such a process.

mutation rate The probability that a copy of an allele changes to some other allelic form in one generation.

NAHR *See* **nonallelic homologous recombination.**

narrow-sense heritability (h^2) The proportion of phenotypic variance that can be attributed to additive genetic variance.

natural selection The differential rate of reproduction of different types in a population as the result of different physiological, anatomical, or behavioral characteristics of the types.

ncRNA *See* **non-protein-coding RNA.**

nearly isogenic line *See* **congenic line.**

negative assortative mating Preferential mating between phenotypically unlike partners.

negative control Regulation mediated by factors that block or turn off transcription.

negative selection The elimination of a deleterious trait from a population by natural selection.

neofunctionalization The evolution of a new function by a gene.

NER *See* **nucleotide-excision repair system.**

neutral allele An allele that has no effect on the fitness of individuals that possess it.

neutral evolution Nonadaptive evolutionary changes due to random genetic drift.

NH *See* **number of haplotypes.**

NHEJ *See* **nonhomologous end joining.**

NLS *See* **nuclear localization sequence.**

nonallelic homologous recombination (NAHR) Crossing over between short homologous units found at different chromosomal loci.

nonautonomous element A transposable element that relies on the protein products of autonomous elements for its mobility. *Dissociation* (*Ds*) is an example of a nonautonomous transposable element.

nonconservative substitution Nucleotide-pair substitution within a protein-coding region that leads to the replacement of an amino acid by one having different chemical properties.

nondisjunction The failure of homologs (at meiosis) or sister chromatids (at mitosis) to separate properly to opposite poles.

nonhomologous end joining (NHEJ) A mechanism used by eukaryotes to repair double-strand breaks.

non-protein-coding RNA (ncRNA) RNA that is not translated into protein.

nonsense mutation Nucleotide-pair substitution within a protein-coding region that changes a codon for an amino acid into a termination (nonsense) codon.

nonsynonymous substitution Mutational replacement of an amino acid with one having different chemical properties.

normal distribution A continuous distribution defined by the normal density function with a specified mean and standard deviation showing the expected frequencies for different values of a random variable (the "bell curve").

Northern blotting The transfer of electrophoretically separated RNA molecules from a gel onto an absorbent sheet, which is then immersed in a labeled probe that will bind to the RNA of interest.

nuclear localization sequence (NLS) Part of a protein required for its transport from the cytoplasm to the nucleus.

nucleosome The basic unit of eukaryotic chromosome structure; a ball of eight histone molecules that is wrapped by two coils of DNA.

nucleotide A molecule composed of a nitrogen base, a sugar, and a phosphate group; the basic building block of nucleic acids.

nucleotide diversity Heterozygosity or gene diversity averaged over all the nucleotide sites in a gene or any other stretch of DNA.

nucleotide-excision-repair (NER) system An excision-repair pathway that breaks the phosphodiester bonds on either side of a damaged base, removing that base and several on either side followed by repair replication.

null allele An allele whose effect is the absence either of normal gene product at the molecular level or of normal function at the phenotypic level.

null hypothesis A hypothesis that proposes no difference between two or more data sets.

nullisomic Refers to a cell or individual organism with one chromosomal type missing, with a chromosome number such as $n - 1$ or $2n - 2$.

null mutation A mutation that results in complete absence of function for the gene.

number of haplotypes (NH) A simple count of the number of haplotypes at a locus in a population.

O *See* **origin of replication.**

octad An ascus containing eight ascospores, produced in species in which the tetrad normally undergoes a postmeiotic mitotic division.

Okazaki fragment A small segment of single-stranded DNA synthesized as part of the lagging strand in DNA replication.

oncogene A gain-of-function mutation that contributes to the production of a cancer.

oncoprotein The protein product of an oncogene mutation.

one-gene–one-polypeptide hypothesis A mid-twentieth-century hypothesis that originally proposed that each gene (nucleotide sequence) encodes a polypeptide sequence; generally true, with the exception of untranslated functional RNA.

open reading frame (ORF) A gene-sized section of a sequenced piece of DNA that begins with a start codon and ends with a stop codon; it is presumed to be the coding sequence of a gene.

operator A DNA region at one end of an operon that acts as the binding site for a repressor protein.

operon A set of adjacent structural genes whose mRNA is synthesized in one piece, plus the adjacent regulatory signals that affect transcription of the structural genes.

ORF *See* **open reading frame.**

origin (O) *See* **origin of replication.**

origin of replication (O) The point of a specific sequence at which DNA replication is initiated.

orthologs Genes in different species that evolved from a common ancestral gene by speciation.

outgroup Taxa outside of a group of organisms among which evolutionary relationships are being determined.

paired-end reads In whole-genome shotgun sequence assembly, the DNA sequences corresponding to both ends of a genomic DNA insert in a recombinant clone.

pair-rule gene In *Drosophila*, a member of a class of zygotically expressed genes that act at an intermediary stage in the process of establishing the correct numbers of body segments. Pair-rule mutations have half the normal number of segments, owing to the loss of every other segment.

paracentric inversion An inversion not including the centromere.

paralogs Genes that are related by gene duplication in a genome.

parental generation The two strains or individual organisms that constitute the start of a genetic breeding experiment; their progeny constitute the F_1 generation.

parsimony To favor the simplest explanation involving the smallest number of evolutionary changes.

parthenogenesis The production of offspring by a female with no genetic contribution from a male.

partial diploid *See* **merozygote.**

partial dominance Gene action under which the phenotype of heterozygotes is intermediate between the two homozygotes but more similar to that of one homozygote than the other.

paternal imprinting The expression of a gene only when inherited from the mother, because the allele of the gene inherited from the father is inactive due to methylation in the course of gamete formation.

PCNA *See* **proliferating cell nuclear antigen.**

PCR *See* **polymerase chain reaction.**

P cytotype Natural stocks of *Drosophila melanogaster* that contain 20 to 50 copies of the *P* element. Laboratory stocks have none. *See* **M cytotype.**

pedigree analysis Deducing single-gene inheritance of human phenotypes by a study of the progeny of matings within a family, often stretching back several generations.

P element A DNA transposable element in *Drosophila* that has been used as a tool for insertional mutagenesis and for germ-line transformation.

penetrance The proportion of individuals with a specific genotype that manifest that genotype at the phenotype level.

pentaploid An individual organism with five sets of chromosomes.

peptidyl (P) site The site in the ribosome to which a tRNA with the growing polypeptide chain is bound.

peptidyltransferase center The site in the large ribosomal subunit at which the joining of two amino acids is catalyzed.

pericentric inversion An inversion that includes the centromere.

permissive temperature The temperature at which a temperature-sensitive mutant allele is expressed the same as the wild-type allele.

personal genomics The analysis of the genome of an individual to better understand his or her ancestry or the genetic basis of phenotypic traits such as his or her risk of developing a disease.

PEV *See* **position-effect variegation.**

phage *See* **bacteriophage.**

phage recombination The production of recombinant phage genotypes as a result of doubly infecting a bacterial cell with different "parental" phage genotypes.

phenotype (1) The form taken by some character (or group of characters) in a specific individual. (2) The detectable outward manifestations of a specific genotype.

phosphate An ion formed of four oxygen atoms attached to a phosphorus atom or the chemical group formed by the attachment of a phosphate ion to another chemical species by an ester bond.

phylogenetic inference Determining the state of a character or the direction of change in a character based on the distribution of that character within a phylogeny of organisms.

phylogeny The evolutionary history of a group.

physical map The ordered and oriented map of cloned DNA fragments on the genome.

PIC *See* **preinitiation complex.**

piRNA *See* **piwi-interacting RNA.**

piwi-interacting RNA (piRNA) An RNA that helps to protect the integrity of plant and animal genomes and to prevent the spread of transposable elements to other chromosomal loci. piRNAs restrain transposable elements in animals.

P site *See* **peptidyl site.**

plaque A clear area on a bacterial lawn, left by lysis of the bacteria through progressive infections by a phage and its descendants.

plasmid An autonomously replicating extrachromosomal DNA molecule.

plating Spreading the cells of a microorganism (bacteria, fungi) on a dish of nutritive medium to allow each cell to form a visible colony.

pleiotropic allele An allele that affects several different properties of an organism.

point mutation A small lesion, usually the insertion or deletion of a single base pair.

Poisson distribution A mathematical distribution giving the probability of observing various numbers of a particular event in a sample when the mean probability of an event on any one trial is very small.

pol III holoenzyme *See* **DNA polymerase III holoenzyme.**

poly(A) tail A string of adenine nucleotides added to mRNA after transcription.

polygene (quantitative trait locus) A gene whose alleles are capable of interacting additively with alleles at other loci to affect a phenotype (trait) showing continuous distribution.

polymerase chain reaction (PCR) An in vitro method for amplifying a specific DNA segment that uses two primers that hybridize to opposite ends of the segment in opposite polarity and, over successive cycles, prime exponential replication of that segment only.

polymerase III holoenzyme *See* **DNA polymerase III holoenzyme.**

polymorphism The occurrence in a population (or among populations) of several phenotypic forms associated with alleles of one gene or homologs of one chromosome.

polypeptide A chain of linked amino acids; a protein.

polyploid A cell having three or more chromosome sets or an organism composed of such cells.

polytene chromosome A giant chromosome in specific tissues of some insects, produced by an endomitotic process in which the multiple DNA sets remain bound in a haploid number of chromosomes.

population (1) A group of individuals that mate with one another to produce the next generation. (2) A group of individuals from which a sample is drawn.

population genetics The study of genetic variation in populations and changes over time in the amount or patterning of that variation resulting from mutation, migration, recombination, random genetic drift, natural selection, and mating systems.

population structure The division of a species or population into multiple genetically distinct subpopulations.

positional cloning The identification of the DNA sequences encoding a gene of interest based on knowledge of its genetic or cytogenetic map location.

positional information The process by which chemical cues that establish cell fate along a geographic axis are established in a developing embryo or tissue primordium.

position effect Describes a situation in which the phenotypic influence of a gene is altered by changes in the position of the gene within the genome.

position-effect variegation (PEV) Variegation caused by the inactivation of a gene in some cells through its abnormal juxtaposition with heterochromatin.

positive assortative mating A situation in which like phenotypes mate more commonly than expected by chance.

positive control Regulation mediated by a protein that is required for the activation of a transcription unit.

positive selection The process by which a favorable allele is brought to a higher frequency in a population because individuals carrying that allele have more viable offspring than other individuals.

post-transcriptional gene silencing Occurs when the mRNA of a particular gene is destroyed or its translation blocked. The mechanism of silencing usually involves RNAi or miRNA.

post-transcriptional processing Modifications of amino acid side groups after a protein has been translated.

preinitiation complex (PIC) A very large eukaryotic protein complex comprising RNA polymerase II and the six general transcription factors (GTFs), each of which is a multiprotein complex.

pre-mRNA *See* **primary transcript.**

primary structure of a protein The sequence of amino acids in the polypeptide chain.

primary transcript (pre-mRNA) Eukaryotic RNA before it has been processed.

primase An enzyme that makes RNA primers in DNA replication.

primer An RNA or DNA oligonucleotide that can serve as a template for DNA synthesis by DNA polymerase when annealed to a longer DNA molecule.

primosome A protein complex at the replication fork whose central component is primase.

probe Labeled nucleic acid segment that can be used to identify specific DNA molecules bearing the complementary sequence, usually through autoradiography or fluorescence.

processed pseudogene A pseudogene that arose by the reverse transcription of a mature mRNA and its integration into the genome.

processive enzyme As used in Chapter 7, describes the behavior of DNA polymerase III, which can perform thousands of rounds of catalysis without dissociating from its substrate (the template DNA strand).

product of meiosis One of the (usually four) cells formed by the two meiotic divisions.

product rule The probability of two independent events occurring simultaneously is the product of the individual probabilities.

prokaryote An organism composed of a prokaryotic cell, such as a bacterium or a blue-green alga.

proliferating cell nuclear antigen (PCNA) Part of the replisome, PCNA is the eukaryotic version of the prokaryotic sliding clamp protein.

promoter A regulator region that is a short distance from the 5′ end of a gene and acts as the binding site for RNA polymerase.

promoter-proximal element The series of transcription-factor binding sites located near the core promoter.

property A characteristic feature of an organism, such as size, color, shape, or enzyme activity.

prophage A phage "chromosome" inserted as part of the linear structure of the DNA chromosome of a bacterium.

propositus In a human pedigree, the person who first came to the attention of the geneticist.

proteome The complete set of protein-coding genes in a genome.

proto-oncogene The normal cellular counterpart of a gene that can be mutated to become a dominant oncogene.

prototroph A strain of organisms that will proliferate on minimal medium (*compare* **auxotroph**).

provirus The chromosomally inserted DNA genome of a retrovirus.

pseudoautosomal regions 1 and 2 Small regions at the ends of the X and Y sex chromosomes; they are homologous and undergo pairing and crossing over at meiosis.

pseudodominance The sudden appearance of a recessive phenotype in a pedigree, due to the deletion of a masking dominant gene.

pseudogene A mutationally inactive gene for which no functional counterpart exists in wild-type populations.

pseudolinkage The appearance of linkage of two genes on translocated chromosomes.

pulse–chase experiment An experiment in which cells are grown in radioactive medium for a brief period (the pulse) and then transferred to nonradioactive medium for a longer period (the chase).

pure line A population of individuals all bearing the identical fully homozygous genotype.

purifying selection Natural selection that removes deleterious variants of a DNA or protein sequence, thus reducing genetic diversity.

purine A type of nitrogen base; the purine bases in DNA are adenine and guanine.

pyrimidine A type of nitrogen base; the pyrimidine bases in DNA are cytosine and thymine.

pyrosequencing DNA sequencing technology that is based on the generation and detection of a pyrophosphate group liberated from a nucleotide triphosphate.

QTL *See* **quantitative trait locus.**

quantitative genetics The subfield of genetics that studies the inheritance of complex or quantitative traits.

quantitative trait Any trait exhibiting complex inheritance because it is controlled by a mix of genetic and/or environmental factors.

quantitative trait locus (QTL) A gene contributing to the phenotypic variation in a trait that show complex inheritance, such as height and weight.

quantitative trait locus mapping A method for locating QTL in the genome and characterizing the effects of QTL on trait variation.

quaternary structure of a protein The multimeric constitution of a protein.

random genetic drift Changes in allele frequency that result because the genes appearing in offspring are not a perfectly representative sampling of the parental genes.

rare SNP A single nucleotide polymorphism (SNP) for which the less common allele occurs at a frequency below 5 percent.

rearrangement The production of abnormal chromosomes by the breakage and incorrect rejoining of chromosomal segments; examples are inversions, deletions, and translocations.

recessive allele An allele whose phenotypic effect is not expressed in a heterozygote.

recipient The bacterial cell that receives DNA in a unilateral transfer between cells; examples are F⁻ in a conjugation or the transduced cell in a phage-mediated transduction.

recombinant Refers to an individual organism or cell having a genotype produced by recombination.

recombinant DNA A novel DNA sequence formed by the combination of two nonhomologous DNA molecules.

recombinant frequency (RF) The proportion (or percentage) of recombinant cells or individuals.

recombination (1) In general, any process in a diploid or partly diploid cell that generates new gene or chromosomal combinations not previously found in that cell or in its progenitors. (2) At meiosis, the process that generates a haploid product of meiosis whose genotype is different from either of the two haploid genotypes that constituted the meiotic diploid.

recombination map A chromosome map in which the positions of loci shown are based on recombinant frequencies.

regulon Genes that are transcribed in a manner that is coordinated by the same regulatory protein (e.g., sigma factor).

relative fitness A measure of the fitness of an individual or genotype relative to some other individual or genotype, usually the most fit individual or genotype in the population.

release factor (RF) A protein that binds to the A site of the ribosome when a stop codon is in the mRNA.

replica plating In microbial genetics, a way of screening colonies arrayed on a master plate to see if they are mutant under other environments; a felt pad is used to transfer the colonies to new plates.

replication fork The point at which the two strands of DNA are separated to allow the replication of each strand.

replicative transposition A mechanism of transposition that generates a new insertion element integrated elsewhere in the genome while leaving the original element at its original site of insertion.

replisome The molecular machine at the replication fork that coordinates the numerous reactions necessary for the rapid and accurate replication of DNA.

reporter gene A gene whose phenotypic expression is easy to monitor; used to study tissue-specific promoter and enhancer activities in transgenes.

repressor A protein that binds to a cis-acting element such as an operator or a silencer, thereby preventing transcription from an adjacent promoter.

resistant mutant A mutant that can grow in a normally toxic environment.

restriction enzyme An endonuclease that will recognize specific target nucleotide sequences in DNA and break the DNA chain at those points; a variety of these enzymes are known, and they are extensively used in genetic engineering.

restriction fragment A DNA fragment resulting from cutting DNA with a restriction enzyme.

restriction fragment length polymorphism (RFLP) A difference in DNA sequence between individuals or haplotypes that is recognized as different restriction fragment lengths. For example, a nucleotide-pair substitution can cause a restriction-enzyme-recognition site to be present in one allele of a gene and absent in another. Consequently, a probe for this DNA region will hybridize to different-sized fragments within restriction digests of DNAs from these two alleles.

restrictive temperature The temperature at which a temperature-sensitive mutation expresses the mutant phenotype.

retrotransposition A mechanism of transposition characterized by the reverse flow of information from RNA to DNA.

retrotransposon A transposable element that uses reverse transcriptase to transpose through an RNA intermediate. *See* **class 1 element.**

retrovirus An RNA virus that replicates by first being converted into double-stranded DNA.

reverse genetics An experimental procedure that begins with a cloned segment of DNA or a protein sequence and uses it (through directed mutagenesis) to introduce programmed mutations back into the genome to investigate function.

reverse transcriptase An enzyme that catalyzes the synthesis of a DNA strand from an RNA template.

revertant An allele with wild-type function arising by the mutation of a mutant allele; caused either by a complete reversal of the original event or by a compensatory second-site mutation.

RF *See* **recombinant frequency; release factor.**

R factors Plasmids carrying genes that encode resistance to several antibiotics.

RFLP *See* **restriction fragment length polymorphism.**

ribonucleic acid *See* **RNA.**

ribose The pentose sugar of RNA.

ribosomal RNA *See* **rRNA.**

ribosome A complex organelle that catalyzes the translation of messenger RNA into an amino acid sequence; composed of proteins plus rRNA.

ribozyme An RNA with enzymatic activity—for instance, the self-splicing RNA molecules in *Tetrahymena*.

RISC (RNA-induced silencing complex) A multisubunit protein complex that associates with siRNAs and is guided to a target mRNA by base complementarity. The target mRNA is cleaved by RISC activity.

RNA (ribonucleic acid) A single-stranded nucleic acid similar to DNA but having ribose sugar rather than deoxyribose sugar and uracil rather than thymine as one of the bases.

RNA blotting *See* **Northern blotting.**

RNAi *See* **RNA interference.**

RNA interference (RNAi) A system in eukaryotes to control the expression of genes through the action of siRNAs and miRNAs. *See* **gene silencing.**

RNA polymerase An enzyme that catalyzes the synthesis of an RNA strand from a DNA template. Eukaryotes possess several classes of RNA polymerase; structural genes encoding proteins are transcribed by RNA polymerase II.

RNA polymerase holoenzyme The bacterial multisubunit complex composed of the four subunits of the core enzyme plus the σ factor.

RNA processing The collective term for the modifications to eukaryotic RNA, including capping and splicing, that are necessary before the RNA can be transported into the cytoplasm for translation.

RNA splicing A reaction found largely in eukaryotes that removes introns and joins together exons in RNA.

RNA world The name of a popular theory that RNA must have been the genetic material in the first cells because only RNA is known to both encode genetic information and catalyze biological reactions.

rolling circle replication A mode of replication used by some circular DNA molecules in bacteria (such as plasmids) in which the circle seems to rotate as it reels out one continuous leading strand.

R plasmid A plasmid containing one or several transposons that bear resistance genes.

rRNA (ribosomal RNA) A class of RNA molecules, encoded in the nucleolar organizer, that have an integral (but poorly understood) role in ribosome structure and function.

s *See* **selection coefficient.**

S *See* **segregating site** or **selection differential.**

safe haven A site in the genome where the insertion of a transposable element is unlikely to cause a mutation, thus preventing harm to the host.

sample A small group of individual members or observations meant to be representative of a larger population from which the group has been taken.

Sanger sequencing *See* **dideoxy sequencing.**

scaffold (1) The central framework of a chromosome to which the DNA solenoid is attached as loops; composed largely of topoisomerase. (2) In genome projects, an ordered set of contigs in which there may be unsequenced gaps connected by paired-end sequence reads.

screen A mutagenesis procedure in which essentially all mutagenized progeny are recovered and are individually evaluated for mutant phenotype; often the desired phenotype is marked in some way to enable its detection.

SDSA *See* **synthesis-dependent strand annealing.**

secondary structure of a protein A spiral or zigzag arrangement of the polypeptide chain.

second-division segregation pattern (M_{II} pattern) A pattern of ascospore genotypes for a gene pair showing that the two alleles separate into different nuclei only at the second meiotic division, as a result of a crossover between that gene pair and its centromere; can be detected only in a linear ascus.

second filial generation (F_2) The progeny of a cross between two individuals from the F_1 generation.

segmental duplication Presence of two or more large nontandem repeats.

segment-polarity gene In *Drosophila*, a member of a class of genes that contribute to the final aspects of establishing the correct number of segments. Segment-polarity mutations cause a loss of or change in a comparable part of each of the body segments.

segregating site (*S*) The number of variable or polymorphic nucleotide sites in a set of homologous DNA sequences.

selection (1) An experimental procedure in which only a specific type of mutant can survive. (2) The production of different average numbers of offspring by different genotypes in a population as a result of the different phenotypic properties of those genotypes.

selection coefficient (*s*) The loss of fitness in (or selective disadvantage of) one genotype relative to another genotype.

selection differential (*S*) The difference between the mean of a population and the mean of the individual members selected to be parents of the next generation.

selection response (*R*) The amount of change in the average value of some phenotypic character between the parental generation and the offspring generation as a result of the selection of parents.

selective system A mutational selection technique that enriches the frequency of specific (usually rare) genotypes by establishing environmental conditions that prevent the growth or survival of other genotypes.

self To fertilize eggs with sperms from the same individual.

self-splicing intron The first example of catalytic RNA; in this case, an intron that can be removed from a transcript without the aid of a protein enzyme.

semiconservative replication The established model of DNA replication in which each double-stranded molecule is composed of one parental strand and one newly polymerized strand.

semisterility (half-sterility) The phenotype of an organism heterozygotic for certain types of chromosome aberration; expressed as a reduced number of viable gametes and hence reduced fertility.

sequence assembly The compilation of thousands or millions of independent DNA sequence reads into a set of contigs and scaffolds.

sequence contig A group of overlapping cloned segments.

serially reiterated structures Body parts that are members of repeated series, such as digits, ribs, teeth, limbs, and segments.

sex chromosome A chromosome whose presence or absence is correlated with the sex of the bearer; a chromosome that plays a role in sex determination.

sex linkage The location of a gene on a sex chromosome.

Shine–Dalgarno sequence A short sequence in bacterial RNA that precedes the initiation AUG codon and serves to correctly position this codon in the P site of the ribosome by pairing (through base complementarity) with the 3′ end of the 16S RNA in the 30S ribosomal subunit.

short interspersed element (SINE) A type of class 1 transposable element that does not encode reverse transcriptase but is thought to use the reverse transcriptase encoded by LINEs. *See also* **Alu.**

sigma (σ) factor A bacterial protein that, as part of the RNA polymerase holoenzyme, recognizes the –10 and –35 regions of bacterial promoters, thus positioning the holoenzyme to initiate transcription correctly at the start site. The *σ* factor dissociates from the holoenzyme before RNA synthesis.

signal sequence The amino-terminal sequence of a secreted protein; it is required for the transport of the protein through the cell membrane.

sign epistasis The dependency of the fitness advantage or disadvantage of a new mutation on the mutations that have been previously fixed.

simple inheritance A form of inheritance in which only one (or a few) genes are involved and the environment has little or no effect on the phenotype; categorical traits often exhibit simple inheritance.

simple sequence length polymorphism (SSLP) The existence in the population of individuals showing different numbers of copies of a short simple DNA sequence at one chromosomal locus.

simple transposon A type of bacterial transposable element containing a variety of genes that reside between short inverted repeat sequences.

SINE *See* **short interspersed element.**

single nucleotide polymorphism (SNP) (snip) A nucleotide-pair difference at a given location in the genomes of two or more naturally occurring individuals.

single-strand-binding (SSB) protein A protein that binds to DNA single strands and prevents the duplex from re-forming before replication.

siRNA *See* **small interfering RNA.**

small interfering RNA (siRNA) Short double-stranded RNAs produced by the cleavage of long double-stranded RNAs by Dicer.

small nuclear RNA (snRNA) Any of several short RNAs found in the eukaryotic nucleus, where they assist in RNA processing events.

SNP *See* **single nucleotide polymorphism.**

snRNA *See* **small nuclear RNA.**

solo LTR A single copy of an *LTR.*

SOS (repair) system An error-prone process whereby a bypass polymerase replicates past DNA damage at a stalled replicating fork by inserting nonspecific bases.

Southern blot The transfer of electrophoretically separated fragments of DNA from a gel to an absorbent sheet such as paper; this sheet is then immersed in a solution containing a labeled probe that will bind to a fragment of interest.

specialized transduction The situation in which a particular phage will transduce only specific regions of the bacterial chromosome.

spliceosome The ribonucleoprotein processing complex that removes introns from eukaryotic mRNAs.

splicing A reaction that removes introns and joins together exons in RNA.

spontaneous lesion DNA damage occurring in the absence of exposure to mutagens; due primarily to the mutagenic action of the by-products of cellular metabolism.

spontaneous mutation A mutation occurring in the absence of exposure to mutagens.

SRY gene The maleness gene, residing on the Y chromosome.

SSB *See* **single-strand-binding protein.**

SSLP *See* **short-sequence-length polymorphism.**

standard deviation The square root of the variance.

structure-based drug design The use of basic information about cellular processes and machinery to develop drugs.

subfunctionalization A path of gene duplication and mutation that produces *paralogs* with complementary functions.

subunit As used in Chapter 9, a single polypeptide in a protein containing multiple polypeptides.

sum rule The probability that one or the other of two mutually exclusive events will occur is the sum of their individual probabilities.

supercontig *See* **scaffold** (2).

suppressor A secondary mutation that can cancel the effect of a primary mutation, resulting in wild-type phenotype.

synergistic effect A feature of eukaryotic regulatory proteins for which the transcriptional activation mediated by the interaction of several proteins is greater than the sum of the effects of the proteins taken individually.

synonymous mutation A mutation that changes one codon for an amino acid into another codon for that same amino acid. Also called *silent mutation.*

synonymous substitution *See* **synonymous mutation.**

synteny A situation in which genes are arranged in similar blocks in different species.

synthesis-dependent strand annealing (SDSA) An error-free mechanism for correcting double-strand breaks that occur after the replication of a chromosomal region in a dividing cell.

synthetic lethal Refers to a double mutant that is lethal, whereas the component single mutations are not.

systems biology An attempt to interpret a genome as a holistic interacting system.

tandem duplication Adjacent identical chromosome segments.

targeted gene knockout The introduction of a null mutation into a gene by a designed alteration in a cloned DNA sequence that is then introduced into the genome through homologous recombination and replacement of the normal allele.

targeting A feature of certain transposable elements that facilitates their insertion into regions of the genome where they are not likely to insert into a gene causing a mutation.

target-site duplication A short direct-repeat DNA sequence (typically from 2 to 10 bp in length) adjacent to the ends of a transposable element that was generated during the element's integration into the host chromosome.

TATA-binding protein (TBP) A general transcription factor that binds to the TATA box and assists in attracting other general transcription factors and RNA polymerase II to eukaryotic promoters.

TATA box A DNA sequence found in many eukaryotic genes that is located about 30 bp upstream of the transcription start site.

tautomeric shift The spontaneous isomerization of a nitrogen base from its normal keto form to an alternative hydrogen-bonding enol (or imino) form.

tautomerization *See* **tautometic shift.**

TBP *See* **TATA-binding protein.**

TC-NER *See* **transcription-coupled nucleotide-excision repair.**

telomerase An enzyme that, with the use of a special small RNA as a template, adds repetitive units to the ends of linear chromosomes to prevent shortening after replication.

telomere The tip, or end, of a chromosome.

temperate phage A phage that can become a prophage.

temperature-sensitive mutation A conditional mutation that produces the mutant phenotype in one temperature range and the wild-type phenotype in another temperature range.

template A molecular "mold" that shapes the structure or sequence of another molecule; for example, the nucleotide sequence of DNA acts as a template to control the nucleotide sequence of RNA during transcription.

termination The last stage of transcription; it results in the release of the RNA and RNA polymerase from the DNA template.

terminus The end represented by the last added monomer in the unidirectional synthesis of a polymer such as RNA or a polypeptide.

tertiary structure of a protein The folding or coiling of the secondary structure to form a globular molecule.

testcross A cross of an individual organism of unknown genotype or a heterozygote (or a multiple heterozygote) with a tester.

tester An individual organism homozygous for one or more recessive alleles; used in a testcross.

tetrad (1) Four homologous chromatids in a bundle in the first meiotic prophase and metaphase. (2) The four haploid product cells from a single meiosis.

tetraploid A cell having four chromosome sets; an organism composed of such cells.

theta (θ) structure An intermediate structure in the replication of a circular bacterial chromosome.

three-point testcross (three-factor testcross) A testcross in which one parent has three heterozygous gene pairs.

3′ untranslated region (3′ UTR) The region of the RNA transcript at the 3′ end downstream of the site of translation termination.

threshold trait A categorical trait for which the expression of the different phenotypic states depends on a combination of multiple genetic and/or environmental factors that place an individual above or below a critical value for trait expression.

thymine (T) A pyrimidine base that pairs with adenine.

Ti plasmid A circular plasmid of *Agrobacterium tumifaciens* that enables the bacterium to infect plant cells and produce a tumor (crown gall tumor).

Tn *See* **transposon.**

topoisomerase An enzyme that can cut and re-form polynucleotide backbones in DNA to allow it to assume a more relaxed configuration.

trait More or less synonymous with phenotype.

trans-acting factor A diffusible regulatory molecule (almost always a protein) that binds to a specific cis-acting element.

trans conformation In a heterozygote with two mutant sites within a gene or gene cluster, the arrangement $a_1 +/+ a_2$.

transcript The RNA molecule copied from the DNA template strand by RNA polymerase.

transcription The synthesis of RNA from a DNA template.

transcriptional gene silencing Occurs when a gene cannot be transcribed because it is located in heterochromatin.

transcription bubble The site at which the double helix is unwound so that RNA polymerase can use one of the DNA strands as a template for RNA synthesis.

transcription-coupled nucleotide-excision repair (TC-NER) A form of nucleotide-excision repair that is activated by stalled transcription complexes and corrects DNA damage in transcribed regions of the genome.

transduction The movement of genes from a bacterial donor to a bacterial recipient with a phage as the vector.

transfer RNA *See* **tRNA.**

transformation The directed modification of a genome by the external application of DNA from a cell of different genotype.

transgene A gene that has been modified by externally applied recombinant DNA techniques and reintroduced into the genome by germ-line transformation.

transgene silencing Refers to the presence of a foreign gene in a transgenic organism that does not produce an mRNA or protein product owing to epigenetic modifications.

transgenic organism An organism whose genome has been modified by externally applied new DNA.

transition A type of nucleotide-pair substitution in which a purine replaces another purine or in which a pyrimidine replaces another pyrimidine—for example, G–C to A–T.

translation The ribosome- and tRNA-mediated production of a polypeptide whose amino acid sequence is derived from the codon sequence of an mRNA molecule.

translesion DNA synthesis A damage-tolerance mechanism in eukaryotes that uses *bypass polymerases* to replicate DNA past a site of damage.

translesion polymerases A family of DNA polymerases that can continue to replicate DNA past a site of damage that would halt replication by the normal replicative polymerase. Also known as *bypass polymerases.*

translocation The relocation of a chromosomal segment to a different position in the genome.

transposase An enzyme encoded by transposable elements that undergo conservative transposition.

transpose To move from one location in the genome to another; said of a mobile genetic element.

transposition A process by which mobile genetic elements move from one location in the genome to another.

transposon (Tn) A mobile piece of DNA that is flanked by terminal repeat sequences and typically bears genes encoding transposition functions. Bacterial transposons can be simple or composite.

transposon tagging A method used to identify and isolate a host gene through the insertion of a cloned transposable element in the gene.

transversion A type of nucleotide-pair substitution in which a pyrimidine replaces a purine or vice versa—for example, G–C to T–A.

trinucleotide repeat *See* **triplet expansion.**

triplet Three nucleotide pairs that compose a codon.

triplet expansion The expansion of a 3-bp repeat from a relatively low number of copies to a high number of copies that is responsible for a number of genetic diseases, such as fragile X syndrome and Huntington disease.

triploid A cell having three chromosome sets or an organism composed of such cells.

trisomic Basically a diploid with an extra chromosome of one type, producing a chromosome number of the form $2n + 1$.

trivalent Refers to the meiotic pairing arrangement of three homologs in a triploid or trisomic.

tRNA (transfer RNA) A class of small RNA molecules that bear specific amino acids to the ribosome in the course of translation; an amino acid is inserted into the growing polypeptide chain when the anticodon of the corresponding tRNA pairs with a codon on the mRNA being translated.

tumor-suppressor gene A gene encoding a protein that suppresses tumor formation. The wild-type alleles of tumor-suppressor genes are thought to function as negative regulators of cell proliferation.

Turner syndrome An abnormal human female phenotype produced by the presence of only one X chromosome (XO).

two-hybrid test A method for detecting protein–protein interactions, typically performed in yeast.

Ty element A yeast LTR retrotransposon; the first isolated from any organism.

U *See* **uracil; uridine.**

UAS *See* **upstream activation sequence.**

ubiquitin A protein that, when attached as a multicopy chain to another protein, targets that protein for degradation by a protease called the 26S proteasome. The addition of single ubiquitin residues to a protein can change protein–protein interactions, as in the case of PCNA and bypass polymerases.

ubiquitinization The process of adding ubiquitin chains to a protein targeted for degradation.

unbalanced rearrangement A rearrangement in which chromosomal material is gained or lost in one chromosome set.

uniparental inheritance Inheritance pattern in which the progeny have the genotype and phenotype of one parent only, for example, inheritance of mitochondrial genomes.

univalent A single unpaired meiotic chromosome, as is often found in trisomics and triploids.

unselected marker In a bacterial recombination experiment, an allele scored in progeny for the frequency of its cosegregation with a linked selected allele.

unstable phenotype A phenotype characterized by frequent reversion either somatically or germinally or both due to the interaction of transposable elements with a host gene.

upstream Refers to a DNA or RNA sequence located on the 5′ side of a point of reference.

upstream activation sequence (UAS) A DNA sequence of yeast located 5′ of the gene promoter; a transcription factor binds to the UAS to positively regulate gene expression.

uracil (U) A pyrimidine base in RNA in place of the thymine found in DNA.

uridine (U) A nucleoside having uracil as its base.

UTR *See* **3′ untranslated region; 5′ untranslated region.**

variable number tandem repeat (VNTR) A chromosomal locus at which a particular repetitive sequence is present in different numbers in different individuals or in the two different homologs in one diploid individual.

variance A statistical measure used to quantify the degree to which the trait values of individuals deviate from the population mean.

vector *See* **cloning vector.**

virulent phage A phage that cannot become a prophage; infection by such a phage always leads to lysis of the host cell.

virus A particle consisting of nucleic acid and protein that must infect a living cell to replicate and reproduce.

VNTR *See* **variable number tandem repeat.**

wild type The genotype or phenotype that is found in nature or in the standard laboratory stock for a given organism.

wobble The ability of certain bases at the third position of an anticodon in tRNA to form hydrogen bonds in various ways, causing alignment with several different possible codons.

X chromosome One of a pair of sex chromosomes, distinguished from the Y chromosome.

xeroderma pigmentosum (XP) A disorder caused by mutations in the transcription-coupled nucleotide-excision-repair system that leads to the frequent development of skin cancers.

X linkage The inheritance pattern of genes found on the X chromosome but not on the Y chromosome.

XP *See* **xeroderma pigmentosum.**

Y chromosome One of a pair of sex chromosomes, distinguished from the X chromosome.

Y linkage The inheritance pattern of genes found on the Y chromosome but not on the X chromosome (rare).

zygote A cell formed by the fusion of an egg and a sperm; the unique diploid cell that will divide mitotically to create a differentiated diploid organism.

zygotic induction The sudden release of a lysogenic phage from an Hfr chromosome when the prophage enters the F⁻ cell followed by the subsequent lysis of the recipient cell.

Answers to Selected Problems

This section includes selected answers to Basic Problems and Challenging Problems from all chapters except Chapter 1. Answers to Chapter 1 problems are not included here because they are discussion questions. Answers to Working with the Figures may be found in the GeneticsPortal. Answers to all the problems are available in the Solutions Manual.

Chapter 2

16. PFGE separates DNA molecules by size. When DNA is carefully isolated from *Neurospora* (which has seven different chromosomes), seven bands should be produced with the use of this technique. Similarly, the pea has seven different chromosomes and will produce seven bands (homologous chromosomes will comigrate as a single band).

19. The key function of mitosis is to generate two daughter cells genetically identical with the original parent cell.

23. As cells divide mitotically, each chromosome consists of identical sister chromatids that are separated to form genetically identical daughter cells. Although the second division of meiosis appears to be a similar process, the "sister" chromatids are likely to be different from each other. Recombination in earlier meiotic stages will have swapped regions of DNA between sister and nonsister chromosomes such that the two daughter cells of this division are typically not genetically identical.

27. Yes. Half of our genetic makeup is derived from each parent, half of each parent's genetic makeup is derived from half of each of their parents', etc.

31. (5) Synapsis (chromosome pairing)

36. The progeny ratio is approximately 3 : 1, indicating classic heterozygous-by-heterozygous mating. Because Black (*B*) is dominant over white (*b*),

Parents: $B/b \times B/b$
Progeny: 3 black : 1 white (1 B/B : 2 B/b : 1 b/b)

40. The fact that about half of the F_1 progeny are mutant suggests that the mutation that results in three cotyledons is dominant and the original mutant was heterozygous. If C — the mutant allele and c — the wild-type allele, the cross is as follows:

P $C/c \times c/c$
F_1 C/c three cotyledons
   c/c two cotyledons

45. p (child has galactosemia) = p (John is G/g) $\times$ p (Martha is G/g) $\times$ p (both parents passed g to the child) = $(2/3)(1/4)(1/4) = 2/48 = 1/24$

51. a. The disorder appears to be dominant because all affected individuals have an affected parent. If the trait were recessive, then I-1, II-2, III-1, and III-8 would all have to be carriers (heterozygous for the rare allele).

b. With the assumption of dominance, the genotypes are

I: d/d, D/d
II: D/d, d/d, D/d, d/d
III: d/d, D/d, d/d, D/d, d/d, d/d, D/d, d/d
IV: D/d, d/d, D/d, d/d, d/d, d/d, d/d, D/d, d/d

c. The probability of an affected child (D/d) equals 1/2, and the probability of an unaffected child (d/d) equals 1/2. Therefore, the chance of having four unaffected children (since each is an independent event) is $(1/2) \times (1/2) \times (1/2) \times (1/2) = 1/16$.

57. a. Sons inherit the X chromosome from their mothers. The mother has free earlobes; the son has attached earlobes. If the allele for free earlobes is dominant and the allele for attached earlobes is recessive, then the mother could be heterozygous for this trait and the gene could be X linked.

b. It is not possible from the data given to decide which allele is dominant. If attached earlobes is dominant, then the father would be heterozygous and the son would have a 50% chance of inheriting the dominant attached earlobes" allele. If attached earlobes is recessive, then the trait could be autosomal or X linked, but, in either case, the mother would be heterozygous.

61. Let H = hypophosphatemia and h = normal. The cross is $H/Y \times h/h$, yielding H/h (females) and h/Y (males). The answer is 0%.

66. a. X^C/X^c, X^c/X^c
   b. p (color-blind) $\times$ p (male) = $(1/2)(1/2) = 1/4$
   c. The girls will be 1 normal (X^C/X^c) : 1 color-blind (X^c/X^c).
   d. The cross is $X^C/X^c \times X^c/Y$, yielding 1 normal : 1 color-blind for both sexes.

74. a. The pedigree suggests that the allele causing red hair is recessive because most red-haired individuals are from parents without this trait.

b. Observation of those around us makes the allele appear to be somewhat rare.

78. Note that only males are affected and that, in all but one case, the trait can be traced through the female side. However, there is one example of an affected male having affected sons. If the trait is X linked, this male's wife must be a carrier. Depending on how rare this trait is in the general population, that could be unlikely, suggesting that the disorder is caused by an autosomal dominant allele with expression limited to males.

Chapter 3

13. The genotype of the daughter cells will be identical with that of the original cell: (f) A/a ; B/b.

18. Mitosis produces daughter cells having the same genotype as that of the original cell: A/a ; B/b ; C/c.

21. His children will have to inherit the satellite-containing 4 (probability = 1/2), the abnormally staining 7 (probability = 1/2), and the Y chromosome (probability = 1/2). To inherit all three, the probability is $(1/2)(1/2)(1/2) = 1/8$.

26. With the assumption of independent assortment and simple dominant–recessive relations of all genes, the number of genotypic classes expected from selfing a plant heterozygous for n gene pairs is $3n$ and the number of phenotypic classes expected is $2n$.

29. a. and **b.** Cross 2 indicates that purple (*G*) is dominant over green (*g*), and cross 1 indicates that cut (*P*) is dominant over potato (*p*).

Cross 1: G/g ; $P/p \times g/g$; P/p	There are 3 cut : 1 potato, and 1 purple : 1 green.
Cross 2: G/g ; $P/p \times G/g$; p/p	There are 3 purple : 1 green, and 1 cut : 1 potato.
Cross 3: G/G ; $P/p \times g/g$; P/p	There are no green, and there are 3 cut : 1 potato.
Cross 4: G/g ; $P/P \times g/g$; p/p	There are no potato, and there are 1 purple : 1 green.
Cross 5: G/g ; $p/p \times g/g$; P/p	There are 1 cut : 1 potato, and there are 1 purple : 1 green.

34. The crosses are

Cross 1: stop-start female × wild-type male →
                                    all stop-start progeny
Cross 2: wild-type female × stop-start male →
                                    all wild-type progeny

mtDNA is inherited only from the "female" in *Neurospora*

40. a. There should be nine classes corresponding to 0, 1, 2, 3, 4, 5, 6, 7, 8 "doses."

   b. There should be 13 classes corresponding to 0, 1, 2, 3, 4, 5, 6, 7, 8, 9, 10, 11, 12 "doses."

45. Progeny plants inherited only normal cpDNA (lane 1), only mutant cpDNA (lane 2), or both (lane 3). To obtain homoplasmic cpDNA (all chloroplasts containing the same DNA), seen in lanes 1 and 2, chloroplasts had to have segregated.

50. a. and **b.** Begin with any two of the three lines and cross them. If, for example, you began with a/a ; B/B ; $C/C \times A/A$; b/b ; C/C, all the progeny would be A/a ; B/b ; C/C. Crossing two of them would yield

   9    $A/-$; $B/-$; C/C

   3   a/a ; $B/-$; C/C

   3    $A/-$; b/b ; C/C

   1   a/a ; b/b ; C/C

The a/a ; b/b ; C/C genotype has two of the genes in a homozygous recessive state and is found in 1/16 of the offspring. If that genotype were crossed with A/A ; B/B ; c/c, all the progeny would be A/a ; B/b ; C/c. Crossing two of them (or "selfing") would lead to a $27 : 9 : 9 : 9 : 3 : 3 : 3 : 1$ ratio, and 1/64 of the progeny would be the desired a/a ; b/b ; c/c.

There are several different routes to obtaining a/a ; b/b ; c/c, but the one just outlined requires only four crosses.

55. a. In a diploid cell, expect two chromosomes (a pair of homologs) to each have a single locus of radioactivity.

   b. Expect many regions of radioactivity scattered throughout the chromosomes. The exact number and pattern would depend on the specific sequence in question and on where and how often it is present within the genome.

   c. The multiple copies of the genes for ribosomal RNA are organized into large tandem arrays called nucleolar organizers. Therefore, expect broader areas of radioactivity compared with that in part *a*. The number of these regions would equal the number of nuclear organizers present in the organism.

   d. Expect each chromosome end to be labeled by telomeric DNA.

   e. The multiple repeats of this heterochromatic DNA are organized into large tandem arrays. Therefore, expect broader areas of radioactivity compared with that in part *a*. There also may be more than one area in the genome of the same simple repeat.

59. a. Let B = brachydactylous, b = normal, T = taster, and t = nontaster. The genotypes of the couple are B/b ; T/t for the male and b/b ; T/t for the female.

   b. For all four children to be brachydactylous, $p = (1/2)^4 = 1/16$.

   c. For none of the four children to be brachydactylous, $p = (1/2)^4 = 1/16$.

   d. For all to be tasters, $p = (3/4)^4 = 81/256$.

   e. For all to be nontasters, $p = (1/4)^4 = 1/256$.

   f. For all to be brachydactylous tasters, $p = (1/2 \times 3/4)^4 = 81/4096$.

   g. The probability of not being a brachydactylous taster is $1 -$ (the probability of being a brachydactylous taster), or $1 - (1/2 \times 3/4) = 5/8$. The probability that all four children are not brachydactylous tasters is $(5/8)^4 = 625/4096$.

   h. The probability that at least one is a brachydactylous taster is $1 -$ (the probability of none being a brachydactylous taster), or $1 - (5/8)^4$.

Chapter 4

13. P $A\,d/A\,d \times a\,D/a\,D$

   F_1 $A\,d/a\,D$

   F_2 $1\ A\,d/A\,d$ phenotype: $A\ d$

          $2\ A\,d/a\,D$ phenotype: $A\ D$

          $1\ a\,D/a\,D$ phenotype: $a\ D$

16. Because only parental types are recovered, the two genes must be tightly linked and recombination must be very rare. Knowing how

many progeny were looked at would give an indication of how close the genes are.

21. a. The three genes are linked.

   b. A comparison of the parentals (most frequent) with the double crossovers (least frequent) reveals that the gene order is *v p b*. There were 2200 recombinants between *v* and *p*, and 1500 between *p* and *b*. The general formula for map units is

m.u. = 100%(number of recombinants)/total number of progeny

Therefore, the map units between *v* and $p = 100\%(2200)/10,000 = 22$ m.u., and the map units between *p* and $b = 100\%(1500)/10,000 = 15$ m.u. The map is

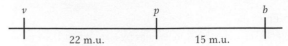

   c. $I = 1 -$ observed double crossovers/expected double crossovers

      $= 1 - 132/(0.22)(0.15)(10,000)$

      $= 1 - 0.4 = 0.6$

27. a.

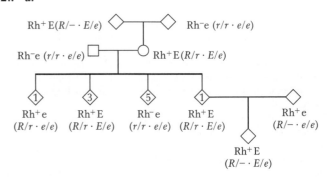

   b. Yes.

   c. Dominant.

   d. As drawn, the pedigree hints at linkage. If unlinked, expect that the phenotypes of the 10 children should be in a $1 : 1 : 1 : 1$ ratio of Rh^+ E, Rh^+ e, Rh^- E, and Rh^- e. There are actually five Rh^- e, four Rh^+ E, and one Rh^+ e. If linked, this last phenotype would represent a recombinant, and the distance between the two genes would be $100\%(1/10) = 10$ m.u. However, there is just not enough data to strongly support that conclusion.

33. a. If the genes are unlinked, the cross is

   P hyg/hyg ; $her/her \times hyg^+/hyg^+$; her^+/her^+

   F_1 hyg^+/hyg ; $her^+/her \times hyg^+/hyg$; her^+/her

   F_2 9/16 $hyg^+/-$; $her^+/-$

          3/16 $hyg^+/-$; her/her

          3/16 hyg/hyg ; $her^+/-$

          1/16 hyg/hyg ; her/her

So only 1/16 (or 6.25%) of the seeds are expected to germinate.

   b. and **c.** No. More than twice the expected seeds germinated; so assume that the genes are linked. The cross then is

   P $hyg\ her\ /hyg\ her \times hyg^+\ her^+/hyg^+her^+$

   F_1 $hyg^+\ her^+/hyg\ her \times hyg^+\ her^+/hyg\ her$

   F_2 13% $hyg\ her/hyg\ her$

Because this class represents the combination of two parental chromosomes, it is equal to

$$p(hyg\ her) \times p(hyg\ her) = (\tfrac{1}{2}\ \text{parentals})^2 = 0.13$$

and

$$\text{parentals} = 0.72$$

So

$$\text{recombinants} = 1 - 0.72 = 0.28$$

Therefore, a testcross of *hyg⁺ hyg⁺/hyg her* will give

    36%   *hyg⁺ her⁺/hyg her*

    36%   *hyg her/hyg her*

    14%   *hyg⁺ her/hyg her*

    14%   *hyg her⁺/hyg her*

and 36% of the progeny will grow (the *hyg her/hyg her* class).

37. The formula for this problem is $f(i) = e^{-m}m^i/i!$ where $m = 2$ and $i = 0, 1,$ or 2.

    a. $f(0) = e^{-2}2^0/0! = e^{-2} = 0.135$, or 13.5%.

    b. $f(1) = e^{-2}2^1/1! = e^{-2}(2) = 0.27$, or 27%.

    c. $f(2) = e^{-2}2^2/2! = e^{-2}(2) = 0.27$, or 27%.

43. **a.** The cross was *pro* + × + *his*, which makes the first tetrad class NPD (6 nonparental ditypes), the second tetrad class T (82 tetratypes), and the third tetrad class PD (112 parental ditypes). When PD >> NPD, you know that the two genes are linked.

    b. Map distance can be calculated by using the formula RF = [NPD + (1/2)T]100%. In this case, the frequency of NPD is 6/200, or 3%, and the frequency of T is 82/200, or 41%. Map distance between these two loci is therefore 23.5 cM.

```
    |                                      |
          23.5 cM
   pro                                    his
```

    c. To correct for multiple crossovers, the Perkins formula can be used. Thus, map distance = (T + 6NPD)50%, or (0.41 + 0.18)50% = 29.5 cM.

47.   **a.** The cross is *W e F/W e F × w E f/w E f* and the F₁ are *W e F/w E f.* Progeny that are *ww ee ff* from a testcross of this F₁ must have inherited one of the double-crossover recombinant chromosomes (*w e f*). With the assumption of no interference, the expected percentage of double crossovers is 8% × 24% = 1.92%, half of which is 0.96%.

    b. To obtain a *ww ee ff* progeny from a self cross of this F₁ requires the independent inheritance of two doubly recombinant *w e f* chromosomes. Its chances of happening, based on the answer to part *a* of this problem, are 0.96 × 0.96 = 0.009%.

53. The short answer is that the results tell us little about linkage. Although the number of recombinants (3) is less than the number of parentals (5), one can have no confidence in the fact that the RF is <50%. The main problem is that the sample size is small, so just one individual more or less in a genotypic class can dramatically affect the ratios. Even the chi-square test is unreliable at such small sample sizes. It *is* probably safe to say that there is not tight linkage because several recombinants were found in a relatively small sample. However, one cannot distinguish between more distant linkage and independent assortment. A larger sample size is required.

58. The data given for each of the three-point testcrosses can be used to determine the gene order when one realizes that the rarest recombinant classes are the result of double-crossover events. A comparison of these chromosomes with the "parental" types reveals that the alleles that have switched represent the gene in the middle.

    For example, in data set 1, the most common phenotypes (+ + + and *a b c*) represent the parental-allele combinations. A comparison of these phenotypes with the rarest phenotypes of this data set (+ *b c* and *a* + +) indicates that the *a* gene is recombinant and must be in the middle. The gene order is *b a c.*

    For data set 2, + *b c* and *a* + + (the parentals) should be compared with + + + and *a b c* (the rarest recombinants) to indicate that the *a* gene is in the middle. The gene order is *b a c.*

    For data set 3, compare + *b* + and *a* + *c* with *a b* + and + + *c,* which gives the gene order *b a c.*

    For data set 4, compare + + *c* and *a b* + with + + + and *a b c,* which gives the gene order *a c b.*

    For data set 5, compare + + + and *a b c* with + + *c* and *a b* +, which gives the gene order *a c b.*

64. **a.** Cross 1 reduces to

    P $A/A \cdot B/B \cdot D/D \times a/a \cdot b/b \cdot d/d$

    F₁ $A/a \cdot B/b \cdot D/d \times a/a \cdot b/b \cdot d/d$

The testcross progeny indicate that these three genes are linked (CO = crossover, DCO = double crossover).

Testcross	A B D	316	parental
progeny	a b d	314	parental
	A B d	31	CO B–D
	a b D	39	CO B–D
	A b d	130	CO A–B
	a B D	140	CO A–B
	A b D	17	DCO
	a B d	13	DCO

  A–B: 100%(130 + 140 + 17 + 13)/1000 = 30 m.u.

  B–D: 100%(31 + 39 + 17 + 13)/1000 = 10 m.u.

Cross 2 reduces to

    P $A/A \cdot C/C \cdot E/E \times a/a \cdot c/c \cdot e/e$

    F₁ $A/a \cdot C/c \cdot E/e \times a/a \cdot c/c \cdot e/e$

The testcross progeny indicate that these three genes are linked.

Testcross	A C E	243	parental
progeny	a c e	237	parental
	A c e	62	CO A–C
	a C E	58	CO A–C
	A C e	155	CO C–E
	a c E	165	CO C–E
	a C e	46	DCO
	A c E	34	DCO

  A–B: 100%(62 + 58 + 46 + 34)/1000 = 20 m.u.

  B–D: 100%(155 + 165 + 46 + 34)/1000 = 40 m.u.

The map that accommodates all the data is

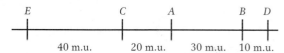

```
E              C        A            B    D
|              |        |            |    |
   40 m.u.       20 m.u.    30 m.u.   10 m.u.
```

  b. Interference (I) = 1 − [(observed DCO)/(expected DCO)]

    For cross 1: I = 1 − {30/[(0.30)(0.10)(1000)]} = 1 − 1 = 0, no interference

    For cross 2: I = 1 − {80/[(0.20)(0.40)(1000)]} = 1 − 1 = 0, no interference

69. **a.** and **b.** The data support the independent assortment of two genes (call them *arg1* and *arg2*). The cross becomes *arg1 ; arg2⁺ × arg1⁺ ; arg2* and the resulting tetrads are

4 : 0 (PD)	3 : 1 (T)	2 : 2 (NPD)
arg1 ; arg2⁺	*arg1 ; arg2⁺*	*arg1 ; arg2*
arg1 ; arg2⁺	*arg1⁺ ; arg2*	*arg1 ; arg2*
arg1⁺ ; arg2	*arg1 ; arg2*	*arg1⁺ ; arg2⁺*
arg1⁺ ; arg2	*arg1⁺ ; arg2⁺*	*arg1⁺ ; arg2⁺*

Because PD = NPD, the genes are unlinked.

Chapter 5

19. An Hfr strain has the fertility factor F integrated into the chromosome. An F⁺ strain has the fertility factor free in the cytoplasm. An F⁻ strain lacks the fertility factor.

23. Although the interrupted-mating experiments will yield the gene order, it will be relative only to fairly distant markers. Thus, the mutation cannot be precisely located with this technique. Generalized transduction will yield information with regard to very close markers, which makes it a poor choice for the initial experiments because of the massive amount of screening that would have to be done. Together, the two techniques allow, first, for localization of the mutant (interrupted mating) and, second, for the precise determination of the location of the mutant (generalized transduction) within the general region.

28. The best explanation is that the integrated F factor of the Hfr looped out of the bacterial chromosome abnormally and is now an F' that contains the pro^+ gene. This F' is rapidly transferred to F^- cells, converting them into pro^+ (and F^+).

33. The expected number of double recombinants is $(0.01)(0.002)(100,000) = 2$. Interference $= 1 -$ (observed double crossover/expected double crossover) $= 1 - 5/2 = -1.5$. By definition, the interference is negative.

37. a. This process appears to be specialized transduction. It is characterized by the transduction of specific markers based on the position of the integration of the prophage. Only those genes near the integration site are possible candidates for misincorporation into phage particles that then deliver this DNA to recipient bacteria.

b. The only media that supported colony growth were those lacking either cysteine or leucine. These media selected for cys^+ or leu^+ transductants and indicate that the prophage is located in the cys–leu region.

42. No. Closely linked loci would be expected to be cotransduced; the greater the contransduction frequency, the closer the loci are. Because only 1 of 858 $metE^+$ was also $pyrD^+$, the genes are not closely linked. The lone $metE^+$ $pyrD^+$ could be the result of cotransduction, it could be a spontaneous mutation of $pyrD$ to $pyrD^+$, or it could be the result of coinfection by two separate transducing phages.

47. a. To determine which genes are close, compare the frequencies of double transformants. Pair-by-pair testing gives low values whenever B is included but fairly high rates when any drug but B is included. This finding suggests that the gene for B resistance is not close to the other three genes.

b. To determine the relative order of genes for resistance to A, C, and D, compare the frequencies of double and triple transformants. The frequency of resistance to AC is approximately the same as that of resistance to ACD, which strongly suggests that D is in the middle. Additionally, the frequency of AD coresistance is higher than AC (suggesting that the gene for A resistance is closer to D than to C) and the frequency of CD is higher than AC (suggesting that C is closer to D than to A).

51. To isolate the specialized transducing particles of phage φ80 that carried lac^+, the researchers would have had to lysogenize the strain with φ80, induce the phage with UV, and then use these lysates to transduce a Lac⁻ strain to Lac⁺. The Lac⁺ colonies would then have been used to make a new lysate, which should have been highly enriched for the lac^+ transducing phage.

Chapter 6

13. With the assumption of homozygosity for the normal gene, the mating is $A/A \cdot b/b \times a/a \cdot B/B$. The children would be normal, $A/a \cdot B/b$.

16. a. Red **b.** Purple

c. 9	$M_1/-$; $M_2/-$	purple
3	m_1/m_1 ; $M_2/-$	blue
3	$M_1/-$; m_2/m_2	red
1	m_1/m_1 ; m_2/m_2	white

d. The mutant alleles do not produce functional enzyme. However, enough functional enzyme must be produced by the single wild-type allele of each gene to synthesize normal levels of pigment.

20. a. The original cross was a dihybrid cross. Both oval and purple must represent an incomplete dominant phenotype.

b. A long, purple × oval, purple cross is as follows:

P L/L ; R/R' × L/L' ; R/R''

F₁ $\frac{1}{2}L/L \times \frac{1}{2}R/R'$ $\frac{1}{4}R/R$ $\frac{1}{8}$ long, red

$\frac{1}{4}R/R'$ $\frac{1}{4}$ long, purple

$\frac{1}{4}R'/R'$ $\frac{1}{8}$ long, white

$\frac{1}{2}L/L' \times \frac{1}{2}R/R'$ $\frac{1}{4}R/R$ $\frac{1}{8}$ oval, red

$\frac{1}{4}R/R'$ $\frac{1}{4}$ oval, purple

$\frac{1}{2}R'/R'$ $\frac{1}{4}$ oval, white

23.

	Parents	Child
a.	AB × O	B
b.	A × O	A
c.	A × AB	AB
d.	O × O	O

27. a. The sex ratio is expected to be 1 : 1.

b. The female parent was heterozygous for an X-linked recessive lethal allele, which would result in 50% fewer males than females.

c. Half of the female progeny should be heterozygous for the lethal allele and half should be homozygous for the nonlethal allele. Individually mate the F₁ females and determine the sex ratio of their progeny.

30. a. The mutations are in two different genes because the heterokaryon is prototrophic (the two mutations complemented each other).

b. $leu1^+$; $leu2^-$ and $leu1^-$; $leu2^+$

c. With independent assortment, expect

$\frac{1}{4}$ $leu1^+$; $leu2^-$

$\frac{1}{4}$ $leu1^-$; $leu2^+$

$\frac{1}{4}$ $leu1^-$; $leu2^-$

$\frac{1}{4}$ $leu1^+$; $leu2^+$

34. a. P A/a (frizzle) × A/a (frizzle)

F₁   1 A/A (normal) : 2 A/a (frizzle) : 1 a/a (woolly)

b. If A/A (normal) is crossed with a/a (woolly), all offspring will be A/a (frizzle).

36. The production of black offspring from two pure-breeding recessive albino parents is possible if albinism results from mutations in two different genes. If the cross is designated

$$A/A ; b/b \times a/a ; B/B$$

all offspring would be

$$A/a ; B/b$$

and they would have a black phenotype because of complementation.

40. The purple parent can be either A/a ; b/b or a/a ; B/b for this answer. Assume that the purple parent is A/a ; b/b. The blue parent must be A/a ; B/b.

43. The cross is gray × yellow, or $A/-$; $R/-$ × $A/-$; r/r. The F₁ progeny are

$\frac{3}{8}$ yellow $\frac{1}{8}$ black $\frac{3}{8}$ gray $\frac{1}{8}$ white

For white progeny, both parents must carry an r and an a allele. Now the cross can be rewritten as A/a ; $R/r \times A/a$; r/r.

46. The original brown dog is w/w ; b/b and the original white dog is W/W ; B/B. The F₁ progeny are W/w ; B/b and the F₂ progeny are

9 $W/-$; $B/-$	white
3 w/w ; $B/-$	black
3 $W/-$; b/b	white
1 w/w ; b/b	brown

50. Pedigrees such as this one are quite common. They indicate lack of penetrance due to epistasis or environmental effects. Individual A must have the dominant autosomal gene.

52. a. Let W^O = oval, W^S = sickle, and W^R = round. The three crosses are

Cross 1: $W^S/W^S \times W^R/Y \rightarrow W^S/W^R$ and W^S/Y

Cross 2: $W^S/W^S \times W^S/Y \rightarrow W^S/W^R$ and W^R/Y

Cross 3: $W^S/W^S \times W^O/Y \rightarrow W^O/W^S$ and W^S/Y

b. $W^O/W^S \times W^R/Y$

$\frac{1}{4} W^O/W^R$ female oval

$\frac{1}{4} W^S/W^R$ female sickle

$\frac{1}{4} W^O/Y$ male oval

$\frac{1}{4} W^S/Y$ male sickle

55. a. The genotypes are

P	$B/B \,;\, i/i \times b/b \,;\, I/I$	
F$_1$	$B/b \,;\, I/i$	hairless
F$_2$	$9\, B/- \,;\, I/-$	hairless
	$3\, B/- \,;\, i/i$	straight
	$3\, b/b \,;\, I/-$	hairless
	$1\, b/b \,;\, i/i$	bent

b. The genotypes are $B/b \,;\, I/i \times B/b \,;\, i/i$.

58. There are a total of 159 progeny that should be distributed in a 9 : 3 : 3 : 1 ratio if the two genes are assorting independently. You can see that

Observed	Expected
88 $P/- \,;\, Q/-$	90
32 $P/- \,;\, q/q$	30
25 $p/p \,;\, Q/-$	30
14 $p/p \,;\, q/q$	10

61. a. Cross-feeding is taking place, whereby a product made by one strain diffuses to another strain and allows growth of the second strain.

b. For cross-feeding to take place, the growing strain must have a block that is present earlier in the metabolic pathway than the block in the strain from which the growing strain is obtaining the product for growth.

c. The data suggest that the metabolic pathway is

$$trpE \rightarrow trpD \rightarrow trpB$$

d. Without some tryptophan, there would be no growth at all, and the cells would not have lived long enough to produce a product that could diffuse

63. a. The best explanation is that Marfan's syndrome is inherited as a dominant autosomal trait.

b. The pedigree shows both pleiotropy (multiple affected traits) and variable expressivity (variable degree of expressed phenotype).

c. Pleiotropy indicates that the gene product is required in a number of different tissues, organs, or processes. When the gene is mutant, all tissues needing the gene product will be affected. Variable expressivity of a phenotype for a given genotype indicates modification by one or more other genes, random noise, or environmental effects.

66. a. This type of gene interaction is called *epistasis*. The phenotype of e/e is epistatic to the phenotypes of $B/-$ or b/b.

b. The inferred genotypes are as follows:

I   1 ($B/b\ E/e$)   2 ($B/b\ E/e$)

II   1 ($b/b\ E/e$)   2 ($B/b\ E/e$)   3 ($-/-\ e/e$)   4 ($b/b\ E/-$)
    5 ($B/b\ E/e$)   6 ($b/b\ E/e$)

III   1 ($B/b\ E/-$)   2 ($-/b\ e/e$)   3 ($b/b,\ E/-$)   4 ($B/b\ E/-$)
    5 ($b/b\ E/-$)   6 ($B/b\ E/-$)   7 ($-/b\ e/e$)

69. a. A multiple-allelic series has been detected: superdouble > single > double.

b. Although the explanation for part *a* does rationalize all the crosses, it does not take into account either the female sterility or the origin of the superdouble plant from a double-flowered variety.

71. a. A trihybrid cross would give a 63 : 1 ratio. Therefore, there are three R loci segregating in this cross.

b.

P	$R_1/R_1 \,;\, R_2/R_2 \,;\, R_3/R_3 \times r_1/r_1 \,;\, r_2/r_2 \,;\, r_3/r_3$	
F$_1$	$R_1/r_1 \,;\, R_2/r_2 \,;\, R_3/r_3$	

F$_2$	27	$R_1/- \,;\, R_2/- \,;\, R_3/-$	red
	9	$R_1/- \,;\, R_2/- \,;\, r_3/r_3$	red
	9	$R_1/- \,;\, r_2/r_2 \,;\, R_3/-$	red
	9	$r_1/r_1 \,;\, R_2/- \,;\, R_3/-$	red
	3	$R_1/- \,;\, r_2/r_2 \,;\, r_3/r_3$	red
	3	$r_1/r_1 \,;\, R_2/- \,;\, r_3/r_3$	red
	3	$r_1/r_1 \,;\, r_2/r_2 \,;\, R_3/-$	red
	1	$r_1/r_1 \,;\, r_2/r_2 \,;\, r_3/r_3$	white

c. (1) To obtain a 1 : 1 ratio, only one of the genes can be heterozygous. A representative cross is $R_1/r_1 \,;\, r_2/r_2 \,;\, r_3/r_3 \times r_1/r_1 \,;\, r_2/r_2 \,;\, r_3/r_3$.

(2) To obtain a 3 red : 1 white ratio, two alleles must be segregating and they cannot be within the same gene. A representative cross is $R_1/r_1 \,;\, R_2/r_2 \,;\, r_3/r_3 \times r_1/r_1 \,;\, r_2/r_2 \,;\, r_3/r_3$.

(3) To obtain a 7 red : 1 white ratio, three alleles must be segregating, and they cannot be within the same gene. The cross is $R_1/r_1 \,;\, R_2/r_2 \,;\, R_3/r_3 \times r_1/r_1 \,;\, r_2/r_2 \,;\, r_3/r_3$.

d. The formula is $1 - (\frac{1}{4})n$, where $n =$ the number of loci that are segregating in the representative crosses in part *c*.

75. a. and **b.** Epistasis is implicated, and the homozygous recessive white genotype seems to block the production of color by a second gene.

Assume the following dominance relations: red > orange > yellow. Let the alleles be designated as follows:

red	A^R
orange	A^O
yellow	A^Y

Crosses 1 through 3 now become

P	$A^O/A^O \times A^Y/A^Y$	$A^R/A^R \times A^O/A^O$	$A^R/A^R \times A^Y/A^Y$
F$_1$	A^O/A^Y	A^R/A^O	A^R/A^Y
F$_2$	$3\, A^O/- : 1\, A^Y/A^Y$	$3\, A^R/- : 1\, A^O/A^O$	$3\, A^R/- : 1\, A^Y/A^Y$

Cross 4: To do this cross, you must add a second gene. You must also rewrite crosses 1 through 3 to include the second gene. Let B allow color expression and b block its expression, producing white. The first three crosses become

P	$A^O/A^O \,;\, B/B \times A^Y/A^Y \,;\, B/B$
	$A^R/A^R \,;\, B/B \times A^O/A^O \,;\, B/B$
	$A^R/A^R \,;\, B/B \times A^Y/A^Y \,;\, B/B$
F$_1$	$A^O/A^Y \,;\, B/B$
	$A^R/A^O \,;\, B/B$
	$A^R/A^Y \,;\, B/B$
F$_2$	$3\, A^O/- \,;\, B/B : 1\, A^Y/A^Y \,;\, B/B$
	$3\, A^R/- \,;\, B/B : 1\, A^O/A^O \,;\, B/B$
	$3\, A^R/- \,;\, B/B : 1\, A^Y/A^Y \,;\, B/B$

The fourth cross is

P	$A^R/A^R \,;\, B/B \times A^R/A^R \,;\, b/b$
F$_1$	$A^R/A^R \,;\, B/b$
F$_2$	$3\, A^R/A^R \,;\, B/- : 1\, A^R/A^R \,;\, b/b$

Cross 5: To do this cross, note that there is no orange. Therefore, the two parents must carry the alleles for red and yellow, and the expression of red must be blocked.

P	$A^Y/A^Y \,;\, B/B \times A^R/A^R \,;\, b/b$	
F$_1$	$A^R/A^Y \,;\, B/b$	
F$_2$	$9\, A^R/- \,;\, B/-$	red
	$3\, A^R/- \,;\, b/b$	white
	$3\, A^Y/A^Y \,;\, B/-$	yellow
	$1\, A^Y/A^Y \,;\, b/b$	white

Cross 6: This cross is identical with cross 5 except that orange replaces yellow.

P A^O/A^O ; B/B × A^R/A^R ; b/b

F_1 A^R/A^O ; B/b

F_2 9 $A^R/-$; $B/-$ red
3 $A^R/-$; b/b white
3 A^O/A^O ; $B/-$ orange
1 A^O/A^O ; b/b white

Cross 7: In this cross, yellow is suppressed by b/b.

P A^R/A^R ; B/B × A^Y/A^Y ; b/b

F_1 A^R/A^Y ; B/b

F_2 9 $A^R/-$; $B/-$ red
3 $A^R/-$; b/b white
3 A^Y/A^Y ; $B/-$ yellow
1 A^Y/A^Y ; b/b white

77. **a.** Intercrossing mutant strains that all have a common recessive phenotype is the basis of the complementation test. This test is designed to identify the number of different genes that can mutate to a particular phenotype. In this problem, if the progeny of a given cross still express the wiggle phenotype, the mutations fail to complement and are considered alleles of the same gene; if the progeny are wild type, the mutations complement and the two strains carry mutant alleles of separate genes.

b. These data identify five complementation groups (genes).

c. mutant 1: $a^1/a^1 \cdot b^+/b^+ \cdot c^+/c^+ \cdot d^+/d^+ \cdot e^+/e^+$ (although only the mutant alleles are usually listed)

mutant 2: $a^+/a^+ \cdot b^2/b^2 \cdot c^+/c^+ \cdot d^+/d^+ \cdot e^+/e^+$

mutant 5: $a^5/a^5 \cdot b^+/b^+ \cdot c^+/c^+ \cdot d^+/d^+ \cdot e^+/e^+$

$\frac{1}{5}$ hybrid: $a^1/a^5 \cdot b^+/b^+ \cdot c^+/c^+ \cdot d^+/d^+ \cdot e^+/e^+$
phenotype: wiggles

Conclusion: 1 and 5 are both mutant for gene A.

(The relevant cross $a^+/a^+ \cdot b^2/b^2 \times a^2/a^2 \cdot b^5/b^5$ gives the following hybrid.)

$\frac{2}{5}$ hybrid: $a^+/a^5 \cdot b^+/b^2 \cdot c^+/c^+ \cdot d^+/d^+ \cdot e^+/e^+$
phenotype: wild type

Conclusion: 2 and 5 are mutant for different genes.

Chapter 7

6. The DNA double helix is held together by two types of bonds: covalent and hydrogen. Covalent bonds are found within each linear strand and strongly bond the bases, sugars, and phosphate groups (both within each component and between components). Hydrogen bonds are found between the two strands; a hydrogen bond forms between a base in one strand and a base in the other strand in complementary pairing. These hydrogen bonds are individually weak but collectively quite strong.

9. Helicases are enzymes that disrupt the hydrogen bonds that hold the two DNA strands together in a double helix. This breakage is required for both RNA and DNA synthesis. Topoisomerases are enzymes that create and relax supercoiling in the DNA double helix. The supercoiling itself is a result of the twisting of the DNA helix when the two strands separate.

11. No. The information of DNA depends on a faithful copying mechanism. The strict rules of complementarity ensure that replication and transcription are reproducible.

14. The chromosome would become hopelessly fragmented.

17. **b.** The RNA would be more likely to contain errors.

21. If the DNA is double stranded, A = T, G = C, and A + T + C + G = 100%. If T = 15%, then C = [100 − 15(2)]/2 = 35%.

22. If the DNA is double stranded, G = C = 24% and A = T = 26%.

26. Yes. DNA replication is also semiconservative in diploid eukaryotes.

28. 5′ . . . .CCTTAAGACTAACTACTTACTGGGATC. . . . 3′

30. Without functional telomerase, the telomeres would shorten at each replication cycle, leading to eventual loss of essential coding information and death. In fact, some current observations indicate that decline or loss of telomerase activity plays a role in the mechanism of aging in humans.

32. Chargaff's rules are that A = T and G = C. Because these equalities are not observed, the most likely interpretation is that the DNA is single stranded. The phage would first have to synthesize a complementary strand before it could begin to make multiple copies of itself.

Chapter 8

9. In prokaryotes, translation is beginning at the 5′ end while the 3′ end is still being synthesized. In eukaryotes, processing (capping, splicing) is taking place at the 5′ end while the 3′ end is still being synthesized.

15. Yes. Both replication and transcription are performed by large, multisubunit molecular machines (the replisome and RNA polymerase II, respectively), and both require helicase activity at the fork of the bubble. However, transcription proceeds in only one direction and only one DNA strand is copied.

17. **a.** The original sequence represents the −35 and −10 consensus sequences (with the correct number of intervening spaces) of a bacterial promoter. The σ factor, as part of the RNA polymerase holoenzyme, recognizes and binds to these sequences.

b. The mutated (transposed) sequences will not be a binding site for the σ factor. The orientation of the two regions with respect to each other is not correct; therefore, they will not be recognized as a promoter.

22. Self-splicing introns are capable of excising themselves from a primary transcript without the need of additional enzymes or energy source. They are one of many examples of RNA molecules that are catalytic, and, for this property, they are also known as ribozymes. With this additional function, RNA is the only known biological molecule to encode genetic information and catalyze biological reactions. In simplest terms, life possibly began with an RNA molecule or group of molecules that evolved the ability to self-replicate.

26. Double-stranded RNA, composed of a sense strand and a complementary antisense strand, can be used in *C. elegans* (and likely all organisms) to selectively prevent the synthesis of the encoded gene product (a discovery for which the 2006 Nobel Prize in Physiology or Medicine was awarded). This process, called gene silencing, blocks the synthesis of the encoded protein from the endogenous gene and is thus equivalent to "knocking out" the gene. To test whether a specific mRNA encodes an essential embryonic protein, inject the double-stranded RNA produced from the mRNA into eggs or very early embryos, thus activating the RNAi pathway. The effects of knocking out the specified gene product can then be followed by observing what happens in these embryos compared with controls. If the encoded protein is essential, embryonic development should be perturbed when your gene is silenced.

Chapter 9

13. **a.** and **b.** 5′ UUG GGA AGC 3′

c. and **d.** With the assumption that the reading frame starts at the first base,

$$NH_3 - Leu - Gly - Ser - COOH$$

For the bottom strand, the mRNA is 5′ GCU UCC CAA 3′ and, with the assumption that the reading frame starts at the first base, the corresponding amino acid chain is

$$NH_3 - Ala - Ser - Gln - COOH$$

17. There are three codons for isoleucine: 5′ AUU 3′, 5′ AUC 3′, and 5′ AUA 3′. Possible anticodons are 3′ UAA 5′ (complementary), 3′ UAG 5′ (complementary), and 3′ UAI 5′ (wobble). Although complementary,

5′ UAU 3′ also would base-pair with 5′ AUG 3′ (methionine) owing to wobble and therefore would not be an acceptable alternative.

22. Quaternary structure is due to the interactions of subunits of a protein. In this example, the enzyme activity being studied may be that of a protein consisting of two different subunits. The polypeptides of the subunits are encoded by separate and unlinked genes.

26. No. The enzyme may require posttranslational modification to be active. Mutations in the enzymes required for these modifications would not map to the isocitrate lyase gene.

29. With the assumption that all three mutations of gene *P* are nonsense mutations, three different possible stop codons (amber, ochre, or opal) might be the cause. A suppressor mutation would be specific to one type of nonsense codon. For example, amber suppressors would suppress amber mutants but not opal or ochre.

33. Single amino acid changes can result in changes in protein folding, protein targeting, or post-translational modifications. Any of these changes could give the results indicated.

40. If the anticodon on a tRNA molecule was altered by mutation to be four bases long, with the fourth base on the 5′ side of the anticodon, it would suppress the insertion. Alterations in the ribosome also can induce frameshifting.

41. f, d, j, e, c, i, b, h, a, g.

Chapter 10

14. Ligase is an essential enzyme within all cells that seals breaks in the sugar–phosphate backbone of DNA. In DNA replication, ligase joins Okazaki fragments to create a continuous strand, and, in cloning, it is used to join the various DNA fragments with the vector. If it were not added, the vector and cloned DNA would simply fall apart.

15. Each cycle takes 5 minutes and doubles the DNA. In 1 hour, there would be 12 cycles; so the DNA would be amplified $2^{12} = 4096$-fold.

18. You could isolate DNA from the suspected transgenic plant and probe for the presence of the transgene by Southern hybridization.

22. a. The transformed phenotype will map to the same locus. If gene replacement was due to double crossing over, the transformed cells will not contain vector DNA. If a single crossing over took place, the entire vector will now be part of the linear *Neurospora* chromosome.

    b. The transformed phenotype will map to a different locus from that of the auxotroph if the transforming gene was inserted ectopically (i.e., at another location). Ecotopic incorporation could also be inferred by reverse PCR.

23. Size, translocations between known chromosomes, and hybridization to probes of known location can all be useful in identifying which band on a pulsed-field gel corresponds to a particular chromosome.

28. The region of DNA that encodes tyrosinase in "normal" mouse genomic DNA contains two *Eco*RI sites. Thus, after *Eco*RI digestion, three different-size fragments hybridize to the cDNA clone. When genomic DNA from certain albino mice is subjected to similar analysis, no DNA fragments contain complementary sequences to the same cDNA. This result indicates that these mice lack the ability to produce tyrosinase because the DNA that encodes the enzyme must have been deleted.

31. The promoter and control regions of the plant gene of interest must be cloned and joined in the correct orientation with the glucuronidase gene, which places the reporter gene under the same transcriptional control as the gene of interest. The text describes the methodology used to create transgenic plants. Transform plant cells with the reporter gene construct, and, as discussed in the text, grow them into transgenic plants. The glucuronidase gene will now be expressed in the same developmental pattern as that of the gene of interest, and its expression can be easily monitored by bathing the plant in an X-Gluc solution and assaying for the blue reaction product.

Chapter 11

8. O^C mutants are changes in the DNA sequence of the operator that impair the binding of the *lac* repressor. Because an operator controls only the genes on the same DNA strand, it is cis (on the same strand).

11. A gene is turned off or inactivated by the "modulator" (usually called a *repressor*) in negative control, and the repressor must be removed for transcription to take place. A gene is turned on by the "modulator" (usually called an *activator*) in positive control, and the activator must be added or converted into an active form for transcription to take place.

16. If an operon were governing both genes, then a frameshift mutation could cause the stop codon separating the two genes to be read as a sense codon. Therefore, the second gene product would be incorrect for almost all amino acids. However, there are no known multigene operons in eukaryotes. The alternative, and better, explanation is that both enzymatic functions are performed by different parts of the same gene product. In this case, a frameshift mutation beyond the first function, carbamyl phosphate synthetase, will result in the second half of the protein molecule being nonfunctional.

19. The *S* mutation is an alteration in *lacI* such that the repressor protein binds to the operator, regardless of whether inducer is present. In other words, it is a mutation that inactivates the allosteric site that binds to inducer but does not affect the ability of the repressor to bind to the operator site. The dominance of the *S* mutation is due to the binding of the mutant repressor, even under circumstances when normal repressor does not bind to DNA (i.e., in the presence of inducer). The constitutive reverse mutations that map to *lacI* are mutational events that inactivate the ability of this repressor to bind to the operator. The constitutive reverse mutations that map to the operator alter the operator DNA sequence such that it will not permit binding to any repressor molecules (wild-type or mutant repressor).

22. Mutations in *cI*, *cII*, and *cIII* would all affect lysogeny: *cI* encodes the repressor, *cII* encodes an activator of P_{RE}, and *cIII* encodes a protein that protects *cII* from degradation. Mutations in *N* (an antiterminator) also would affect lysogeny because its function is required for transcription of the *cII* and *cIII* genes, but it is also necessary for genes having roles in lysis. Mutations in the gene encoding the integrase (*int*) also would affect the ability of a mutant phage to lysogenize.

Chapter 12

12. In general, the ground state of a bacterial gene is "on." Thus, transcription initiation is prevented or reduced if the binding of RNA polymerase is blocked. In contrast, the ground state of eukaryotes is "off." Thus, the transcriptional machinery (including RNA polymerase II and associated general transcription factors) cannot bind to the promoter in the absence of other regulatory proteins.

16. Among the mutations that might prevent a strain of yeast from switching mating type would be mutations in the *HO* and *HMRa* genes. The *HO* gene encodes an endonuclease that cuts the DNA to initiate switching and the *HMRa* locus contains the "cassette" of unexpressed genetic information for the MATa mating type.

19. The term *epigenetic inheritance* is used to describe heritable alterations in which the DNA sequence itself is not changed. It can be defined operationally as the inheritance of chromatin states from one cell generation to next. Genomic imprinting, X-chromosome inactivation, and position-effect variegation are several such examples.

23. The inheritance of chromatin structure is thought to be responsible for the inheritance of epigenetic information. This inheritance is due to the inheritance of the histone code and may also include the inheritance of DNA methylation patterns.

26. a. D through J; the primary transcript will include all exons and introns.

    b. E, G, and I; all introns will be removed.

    c. A, C, and L; the promoter and enhancer regions will bind various transcription factors that may interact with RNA polymerase.

30. A gene not expressed owing to alteration of its DNA sequence will never be expressed and will be inherited from generation to generation. An epigenetically inactivated gene may still be regulated. Chromatin structure can change in the course of the cell cycle; for example, when transcription factors modify the histone code.

32. Chromatin structure greatly affects gene expression. Transgenes inserted into regions of euchromatin would more likely be capable of expression than those inserted into regions of heterochromatin.

Chapter 13

11. The primary pair-rule gene *eve* (*even-skipped*) would be expressed in seven stripes along the A–P axis of the late blastoderm.

15. If you diagram these results, you will see that the deletion of a gene that functions posteriorly allows the next most anterior segments to extend in a posterior direction. Deletion of an anterior gene does not allow extension of the next most posterior segment in an anterior direction. The gap genes activate *Ubx* in both thoracic and abdominal segments, whereas the *abd-A* and *Abd-B* genes are activated only in the middle and posterior abdominal segments. The functioning of the *abd-A* and *Abd-B* genes in those segments somehow prevents *Ubx* expression. However, if the *abd-A* and *Abd-B* genes are deleted, *Ubx* can be expressed in these regions.

18. **a.** A pair-rule gene.

b. Look for expression of the mRNA from the candidate gene in a repeating pattern of seven stripes along the A–P axis of the developing embryo.

c. No. An embryo mutant for the gap gene *Krüppel* would be missing many anterior segments. This effect would be epistatic to the expression of a pair-rule gene.

21. **a.** The homeodomain is a conserved protein domain containing 60 amino acids found in a significant number of transcription factors. Any protein that contains a functional homeodomain is almost certainly a sequence-specific DNA-binding transcription factor.

b. The *eyeless* gene (named for its mutant phenotype) regulates eye development in *Drosophila*. You would expect that it is expressed only in those cells that will give rise to the eyes. To test this prediction, visualization of the location of *eyeless* mRNA expression by in situ hybridization and the Eyeless protein by immunological methods should be performed. Through genetic manipulation, the *eyeless* gene can be expressed in tissues in which it is not ordinarily expressed. For example, when *eyeless* is turned on in cells destined to form legs, eyes form on the legs.

c. Transgenic experiments have shown that the mouse *Small eye* gene and the *Drosophila eyeless* gene are so similar that the mouse gene can substitute for *eyeless* when introduced into *Drosophila*. As in the answer to part *b*, when the mouse *Small eye* gene is expressed in *Drosophila*, even in cells destined to form legs, eyes form on the legs. (However, the "eyes" are not mouse eyes, because *Small eye* and *eyeless* act as master switches that turn on the entire cascade of genes needed to build the eye—in this case, the *Drosophila* set to build a *Drosophila* eye.)

25. GLP-1 protein is localized to the two anterior cells of the four-cell *C. elegans* embryo by repression of its translation in the two posterior cells. The repression of GLP-1 translation requires the 3′ UTR spatial control region (SCR). Deletion of the SCR will allow *glp-1* expression in both anterior and posterior cells. In both heterozygous and homozygous mutants, you would expect GLP-1 protein expression in all cells.

Chapter 14

11. Because bacteria have small genomes (roughly 3-Mb pairs) and essentially no repeating sequences, the whole-genome shotgun approach would be used.

13. A scaffold is also called a supercontig. A contig is a sequence of overlapping reads assembled into a unit, and a scaffold is a collection of joined-together contigs.

18. Yes. The operator is the location at which repressor functionally binds through interactions between the DNA sequence and the repressor protein.

23. You can determine whether the cDNA clone is a monster or not by the alignment of the cDNA sequence against the genomic sequence. (Computer programs for doing such alignments are available.) Is the sequence derived from two different sites? Does the cDNA map within

one (gene-size) region in the genome or to two different regions? Introns may complicate the matter.

27. **a.** Because the triplet code is redundant, changes in the DNA nucleotide sequence (especially at those nucleotides encoding the third position of a codon) can occur without changing its encoded protein.

b. Protein sequences can be expected to evolve and diverge more slowly than do the genes that encode them.

32. The correct assembly of large and nearly identical regions is problematic with either method of genomic sequencing. However, the whole-genome shotgun method is less effective at finding these regions than the clone-based method. This method also has the added advantage of easy access to the suspect clone(s) for further analysis.

35. 15 percent are essential gene functions (such as enzymes required for DNA replication or protein synthesis).

25 percent are auxotrophs (enzymes required for the synthesis of amino acids or the metabolism of sugars, etc.).

60 percent are redundant or pathways not tested (genes for histones, tubulin, ribosomal RNAs, etc., are present in multiple copies; the yeast may require many genes under only unique or special situations or in other ways that are not necessary for life in the laboratory).

Chapter 15

9. Boeke, Fink, and their coworkers demonstrated that transposition of the *Ty* element in yeast is through an RNA intermediate. They constructed a plasmid by using a *Ty* element into which they inserted not only a promoter that can be activated by galactose but also an intron into the *Ty* element's coding region. First, the frequency of transposition was greatly increased by the addition of galactose, indicating that an increase in transcription (and production of RNA) was correlated to rates of transposition. More importantly, after transposition, they found that the newly transposed *Ty* DNA lacked the intron sequence. Because intron splicing takes place only during RNA processing, there must have been an RNA intermediate in the transposition event.

13. Some transposable elements have evolved strategies to insert into safe havens, regions of the genome where they will do minimal harm. Safe havens include duplicate genes (such as tRNA or rRNA genes) and other transposable elements. Safe havens in bacterial genomes might be very specific sequences between genes or the repeated rRNA genes.

16. The staggered cut will lead to a nine-base-pair target-site duplication that flanks the inserted transposon.

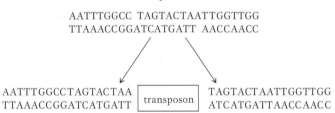

19. It would not be surprising to find a SINE element in an intron of a gene rather than in an exon. Processing of the pre-mRNA would remove the transposable element as part of the intron and translation of the FB enzyme would not be effected.

Chapter 16

8. You need to know the reading frame of the possible message.

11. With the assumption of single-base-pair substitutions, CGG can be changed to CGU, CGA, CGC, or AGG and will still encode arginine.

14. The following list of observations argues "cancer is a genetic disease":

(1) Certain cancers are inherited as highly penetrant simple Mendelian traits.

(2) Most carcinogenic agents are also mutagenic.

(3) Various oncogenes have been isolated from tumor viruses.

(4) A number of genes that lead to the susceptibility to particular types of cancer have been mapped, isolated, and studied.

(5) Dominant oncogenes have been isolated from tumor cells.

(6) Certain cancers are highly correlated to specific chromosomal rearrangements.

17. The mismatched T would be corrected to C and the resulting ACG, after transcription, would be 5′ UGC 3′ and encode cysteine. Or, if the other strand were corrected, ATG would be transcribed to 5′ UAC 3′ and encode tyrosine.

24. Many repair systems are available: direct reversal, excision repair, transcription-coupled repair, and nonhomologous end joining.

25. Yes, it is mutagenic. It will cause CG-to-TA transitions.

30. a. A lack of revertants suggests either a deletion or an inversion within the gene.

b. To understand these data, recall that half the progeny should come from the wild-type parent.

Prototroph A: Because 100 percent of the progeny are prototrophic, a reversion at the original mutant site may have occurred.

Prototroph B: Half the progeny are parental prototrophs, and the remaining prototrophs, 28 percent, are the result of the new mutation. Notice that 28 percent is approximately equal to the 22 percent auxotrophs. The suggestion is that an unlinked suppressor mutation occurred, yielding independent assortment with the *nic-2* mutant.

Prototroph C: There are 496 "revertant" prototrophs (the other 500 are parental prototrophs) and four auxotrophs. This suggests that a suppressor mutation occurred in a site very close to the original mutation and was infrequently separated from the original mutation by recombination [$100\%(4 \times 2)/1000 = 0.8$ m.u.].

33. Xeroderma pigmentosum is a heterogeneous genetic disorder and is caused by mutations in any one of several genes taking part in the process of NER (nucleotide excision repair). As the discovery of yet another protein in the NHEJ pathway through research on cell line 2BN attests, this patient could have a mutation in an as yet unknown gene that encodes a protein necessary for NER.

Chapter 17

21. MM N OO would be classified as $2n - 1$; MM NN OO would be classified as $2n$; and MMM NN PP would be classified as $2n + 1$.

24. There would be one possible quadrivalent.

27. Seven chromosomes.

29. Cells destined to become pollen grains can be induced by cold treatment to grow into embryoids. These embryoids can then be grown on agar to form monoploid plantlets.

31. Yes.

34. No.

36. An acentric fragment cannot be aligned or moved in meiosis (or mitosis) and is consequently lost.

39. Very large deletions tend to be lethal, likely owing to genomic imbalance or the unmasking of recessive lethal genes. Therefore, the observed very large pairing loop is more likely to be from a heterozygous inversion.

41. Williams syndrome is the result of a deletion of the 7q11.23 region of chromosome 7. Cri du chat syndrome is the result of a deletion of a significant part of the short arm of chromosome 5 (specifically bands 5p15.2 and 5p15.3). Both Turner syndrome (XO) and Down syndrome (trisomy 21) result from meiotic nondisjunction. The term *syndrome* is used to describe a set of phenotypes (often complex and varied) that are generally present together.

46. The order is *b a c e d f*.

Allele	Band
b	1
a	2
c	3
e	4
d	5
f	6

47. The data suggest that one or both breakpoints of the inversion are located within an essential gene, causing a recessive lethal mutation.

50. a. When crossed with yellow females, the results would be

X^e/Y^{e+} gray males

X^e/X^e yellow females

b. If the e^+ allele was translocated to an autosome, the progeny would be as follows, where "A" indicates autosome:

P $A^{e+}/A\,;X^e/Y \times A/A\,;X^e/X^e$

F$_1$ $A^{e+}/A\,;X^e/X^e$ gray female

$A^{e+}/A\,;X^e/Y$ gray male

$A/A\,;X^e/X^e$ yellow female

$A/A\,;X^e/Y$ yellow male

52.

Klinefelter syndrome	XXY male
Down syndrome	trisomy 21
Turner syndrome	XO female

56. a. If a hexaploid were crossed with a tetraploid, the result would be pentaploid.

b. Cross A/A with $a/a/a/a$ to obtain $A/a/a$.

c. The easiest way is to expose the A/a^* plant cells to colchicine for one cell division, which will result in a doubling of chromosomes to yield $A/A/a^*/a^*$.

d. Cross a hexaploid ($a/a/a/a/a/a$) with a diploid (A/A) to obtain $A/a/a/a$.

58. a. The ratio of normal-leaved to potato-leaved plants will be 5 : 1.

b. If the gene is not on chromosome 6, there should be a 1 : 1 ratio of normal-leaved to potato-leaved plants.

62. a. The aberrant plant is semisterile, which suggests an inversion. Because the *d–f* and *y–p* frequencies of recombination in the aberrant plant are normal, the inversion must implicate *b* through *x*.

b. To obtain recombinant progeny when there has been an inversion requires the occurrence of either a double crossover within the inverted region or single crossovers between *f* and the inversion, which occurred someplace between *f* and *b*.

64. The original plant is homozygous for a translocation between chromosomes 1 and 5, with break points very close to genes *P* and *S*. Because of the close linkage, a ratio suggesting a monohybrid cross, instead of a dihybrid cross, was observed, both with selfing and with a testcross. All gametes are fertile because of homozygosity.

original plant: *P S/p s*

tester: *p s/p s*

F$_1$ progeny: heterozygous for the translocation:

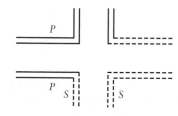

The easiest way to test this hypothesis is to look at the chromosomes of heterozygotes in meiosis I.

70. The original parents must have had the following chromosome constitution:

G. hirsutum	26 large, 26 small
G. thurberi	26 small
G. herbaceum	26 large

G. hirsutum is a polyploid derivative of a cross between the two Old World species, which could easily be checked by looking at the chromosomes.

72. a. Loss of one X in the developing fetus after the two-cell stage.

b. Nondisjunction leading to Klinefelter syndrome (XXY), followed by a nondisjunctive event in one cell for the Y chromosome after the two-cell stage, resulting in XX and XXYY.

c. Nondisjunction of the X at the one-cell stage.

d. Fused XX and XY zygotes (from the separate fertilizations either of two eggs or of an egg and a polar body by one X-bearing and one Y-bearing sperm).

e. Nondisjunction of the X at the two-cell stage or later.

75. a. Each mutant is crossed with wild type, or

$$m \times m^+$$

The resulting tetrads (octads) show 1 : 1 segregation, indicating that each mutant is the result of a mutation in a single gene.

b. The results from crossing the two mutant strains indicate either that both strains are mutant for the same gene:

$$m_1 \times m_2$$

or that they are mutant in different but closely linked genes:

$$m_1\, m_2^+ \times m_1^+\, m_2$$

c. and d. Because phenotypically black offspring can result from nondisjunction (notice that, in cases C and D, black appears in conjunction with aborted spores), mutant 1 and mutant 2 are likely to be mutant in different but closely linked genes. The cross is therefore

$$m_1\, m_2^+ \times m_1^+\, m_2$$

Case A is an NPD tetrad and would be the result of a four-strand double crossover.

$m_1^+\, m_2^+$	black
$m_1^+\, m_2^+$	black
$m_1\, m_2$	fawn
$m_1\, m_2$	fawn

Case B is a tetratype and would be the result of a single crossover between one of the genes and the centromere.

$m_1^+\, m_2^+$	black
$m_1^+\, m_2$	fawn
$m_1\, m_2^+$	fawn
$m_1\, m_2$	fawn

Case C is the result of nondisjunction in meiosis I.

$m_1^+\, m_2^+$; $m_1^+\, m_2^+$	black
$m_1^+\, m_2^+$; $m_1^+\, m_2^+$	black
no chromosome	abort
no chromosome	abort

Case D is the result of recombination between one of the genes and the centromere followed by nondisjunction in meiosis II. For example,

$m_1^+\, m_2$; $m_1\, m_2^+$	black
no chromosome	abort
$m_1\, m_2^+$	fawn
$m_1^+\, m_2$	fawn

Chapter 18

7. The frequency of an allele in a population can be altered by natural selection, mutation, migration, and genetic drift.

9. The frequency of b is $q = \sqrt{0.04} = 0.2$, and the frequency of B is $p = 1 - q = 0.8$. The frequency of B/B is $p^2 = 0.64$, and the frequency of B/b is $2pq = 0.32$.

11. a. $p' = 0.5[(0.5)(1.0) + 0.5(1.0)]/[(0.25)(1.0) + (0.5)(1.0)$
$\qquad\qquad + (0.25)(0.7)] = 0.54$

b. 0.008

13. $\frac{1}{2}(0.02) + \frac{1}{2}(0.02)^2 = 0.0102$

15. probability of fixation $= \dfrac{1}{2N} = \dfrac{1}{100,000}$;

probability of loss $= 1 - \dfrac{1}{2N} = \dfrac{99,999}{100,000}$

17. a. $F_I = (\frac{1}{2})^3 \times (1 + \frac{1}{2}) = 3/16$

b. $1/8 = (\frac{1}{2})^3 \times (1 + F_A)$, so $F_A = 0$

20. $p_A = p_a = p_B = p_b = 0.5$. At equilibrium, the frequency of doubly heterozygous individuals is $2(p_A p_a) \times 2(p_B p_b) = 0.25$

22. Before migration, $q_A = 0.1$ and $q_B = 0.3$ in the two populations. Because the two populations are equal in number, immediately after migration $q_{A+B} = \frac{1}{2}(q_A + q_B) = \frac{1}{2} 0.1 + 0.3) = 0.2$. At the new equilibrium, the frequency of affected males is $q = 0.2$, and then frequency of affected females is $q^2 = (0.2)^2 = 0.04$. (Color blindness is an X-linked trait.)

24. $q^2 = 0.002$. $q = 0.045$. Assuming F in the founders is 0.0, the $F_{50} = 0.222$ (see Box 18-3). $f_{a/a} = q^2 + pqF = 0.012$.

28. $\hat{H} = [4 \times 50,000 \times (3 \times 10^{-8})]/[4 \times 50,000 \times (3 \times 10^{-8}) + 1]$
$\qquad = 5.96 \times 10^{-3}$

31. a. $\hat{q} = \sqrt{1.0 \times 10^{-5}/0.5} = 4.47 \times 10^{-3}$

Genetic cost $= sq^2 = 0.5(4.47 \times 10^{-3})^2 = 10^{-5}$

b. $\hat{q} = 6.32 \times 10^{-3}$

Genetic cost $= sq^2 = 0.5(6.32 \times 10^{-3})^2 = 2 \times 10^{-5}$

c. Genetic cost $= sq^2 = 0.3(5.77 \times 10^{-3})^2 = 10^{-5}$

Chapter 19

7. Many traits vary more or less continuously over a wide range. For example, height, weight, shape, color, reproductive rate, metabolic activity, etc., vary quantitatively rather than qualitatively. Continuous variation can often be represented by a bell-shaped curve, where the "average" phenotype is more common than the extremes. Discontinuous variation describes the easily classifiable, discrete phenotypes of simple Mendelian genetics: seed shape, auxotrophic mutants, sickle-cell anemia, etc. These traits often show a simple relation between genotype and phenotype, although discontinuous traits such as affected versus not-affected for a disease condition can also exhibit complex inheritance.

8. The mean is 4.7 bristles, the variance is 1.11 bristles², and the standard deviation is 1.05 bristles.

10. The breeder cannot be assured that this population will respond to selective breeding even though the broad-sense heritability is high. Broad-sense heritability is ratio of the genetic variance to the phenotypic variance. The genetic variance is the sum of the additive and dominance variances. Only additive variance is transmitted from parent to offspring. Dominance variance is not transmitted from parent to offspring. If all the genetic variance in the population is dominance variance, then selective breeding will not succeed.

14. $\bar{x}_{par} = [(9.8 + 10.8)/2] - 9.6 = 0.7$ mm, $\bar{a}_{par} = 0.79 \times 0.7 = 0.55$ mm, $\hat{x}_{off} = 9.6 + 0.55 = 10.15$ mm

16. $V_e = 3.5$ g², V_g in population B is $21.0 - 3.5 = 17.5$ g², $H^2 = 17.5/21.0 = 0.83$.

Chapter 20

8. The three principles are (1) individuals within any one population vary from one another, (2) offspring resemble their parents more than they resemble unrelated individuals, and (3) some forms are more successful at surviving and reproducing than other forms in a given environment.

10. The relative rate of synonymous and nonsynonymous substitutions would not be higher than expected in a globin pseudogene because a pseudogene is inactive and has no function to be preserved.

12. A population will not differentiate from other populations by local inbreeding if:

$$\mu \geq 1/N$$

and so

$$N \geq 1/\mu$$
$$N \geq 105$$

16. When amino acid changes have been driven by positive adaptive selection, there should be an excess of nonsynonymous changes. The *MC1R* gene (melanocortin 1 receptor) encodes a key protein controlling the amount of melanin in skin and hair. Asian and European populations appear to have experienced positive adaptive selection for more lightly pigmented skin relative to their African counterparts.

18. Noncoding sequences. A major constraint on gene evolution comprises the potential pleiotropic effects of mutations in coding regions. These effects can be circumvented by mutations in regulatory sequences, which play a major role in the evolution of body form. Changes in noncoding sequences provide a mechanism for altering one aspect of gene expression while preserving the role of pleiotropic proteins in other essential developmental processes.

20 a. The HbS mutation has arisen independently in five different haplotypes in different regions and then increased to high frequency.

c. Two independent lines of bacteriophage evolved the ability to reproduce at high temperatures in a new host.

22. A new gene duplicate can (1) evolve a new function, (2) become inactivated, or (3) perform part of the original function, sharing full function with the original gene.

24. For polymorphic sites with a species, let nonsynonymous = a and synonymous = b. For polymorphic sites between the species, let nonsynonymous = c and synonymous = d. If divergence is due to neutral evolution, then

$$a/b = c/d$$

If divergence is due to selection, then

$$a/b < c/d$$

However, in this example, $a/b = 20/50 > c/d = 2/18$, which fits neither expectation. Because the ratio of nonsynonymous to synonymous polymorphisms (a/b) is relatively high, the gene being studied may encode a protein tolerant of relatively fewer species differences. The relatively fewer species differences may suggest that speciation was a recent event so new polymorphisms have been fixed in one species that are not variants in the other.

Index

Note: Page numbers followed by f indicate figures; those followed by t indicate tables. **Boldface** page numbers indicate Key Terms.

Index to Model Organisms

The following table provides page references to discussions of specific model organisms in the text.

	Bacterium (E. coli)	Baker's yeast (S. cerevisiae)	Bread mold (N. crassa)	Mustard weed (A. thaliana)	Roundworm (C. elegans)	Fruit fly (D. melanogaster)	Mouse (M. musculus)
MAIN FEATURES pp.	p. 760	p. 762	p. 764	p. 766	p. 768	p. 770	p. 772
SPOTLIGHTS	p. 173	p. 420	p. 98		p. 476	p. 52, pp. 452–454	p. 218
CHAPTERS							
1. The Genetics Revolution in the Life Sciences	description of organism, p. 22	description of organism, p. 22	description of organism, p. 22	description of organism, p. 22	description of organism, p. 22	description of organism, p. 22	description of organism, p. 23
2. Single-Gene Inheritance		genetic analysis using ascus, p. 39	mycelium development mutants, p. 28	flower development mutants, p. 28		identifying a gene for wing development, pp. 47–48, sex determination, pp. 50–51	
3. Independent Assortment of Genes		meiotic recombination, pp. 99–101	life cycle, p. 96 observation of independent assortment, pp. 94, 98 maternal inheritance (poky mutants), p. 105 cytoplasmic segregation, pp. 105–106			X-linked inheritance (eye color), p. 97	
4. Mapping Eukaryote Chromosomes by Recombination			centromere mapping, pp. 142–143 tetrad analysis of crossovers, pp. 126–127			map of Drosophila chromosome, p. 121 Morgan's experiments, linkage, pp. 123–125 dihybrid linkage analysis, pp. 129–131 three-point testcross, pp. 132–134 interference, pp. 135–136 no crossing over in males, pp. 135–136	
5. The Genetics of Bacteria and Their Viruses	conjugation, p. 175 phage crosses, pp. 190–192 mapping rII gene, pp. 191–192 classical transduction experiments, pp. 192–195 genome map, pp. 197–200						

(Continued)

	(E. coli)	(S. cerevisiae)	(N. crassa)	(A. thaliana)	(C. elegans)	(D. melanogaster)	(M. musculus)
6. Gene Interaction		suppression, pp. 230–231 modifier mutations, pp. 232–233	Beadle-Tatum, experiments, pp. 219–220 complementation, p. 225			eye-color suppression, pp. 231–232	example of haploinsufficient gene, p. 213 coat coloration, pp. 216–217
7. DNA: Structure and Replication	Hershey-Chase experiment, pp. 254–255 Meselson-Stahl experiment, pp. 262–263 DNA polymerases, pp. 264–265 replication speed, p. 268 replisome, pp. 268–270 origin of replication, p. 272	replisome, p. 273 origins of replication, p. 272 cell-cycle control, pp. 273–274					Griffith experiment, pp. 252–253
8. RNA: Transcription and Processing	number of genes, p. 283 Volkin-Astrachan pulse-chase experiment, p. 285 stages of transcription, pp. 291–292 promoter sequences, pp. 290–291 sigma factors, p. 291 termination, p. 292 gene density, p. 293	number of genes, p. 283 origin recognition complex, p. 274 RNA polymerase II, p. 294 frequency of introns, p. 296				number of genes, p. 283 gene density, p. 293	
9. Proteins and Their Synthesis	Crick codon length experiment, p. 315	abundance of RNA transcripts, p. 311 Nirenberg genetic code elucidation, p. 317		number of kinase genes, p. 331	interactome, pp. 331–332		
10. Gene Isolation and Manipulation	restriction enzymes from, pp. 345, 348–350 cloning with bacteriophage vectors, p. 352 cloning with fosmids, p. 353	genetic engineering with yeast vectors, pp. 370–371			genome size and cloning, p. 355 transgenesis in, p. 373		transgenesis in, p. 373 targeted gene knockout, p. 374
11. Regulation of Gene Expression in Bacteria and Their Viruses	Jacob-Monod *lac* operon experiments, pp. 387–388 *trp* operon, pp. 399–401						
12. Regulation of Gene Expression in Eukaryotes		GAL system, pp. 420–424 SWI-SNF mutants, p. 428 histone modification, pp. 430–431 control of mating type, pp. 424–425 gene silencing and mating-type switching, pp. 435–437 signal transduction for mating type switch pp. 424–426				epigenetic silencing, p. 438 position effect variegation, pp. 438–441 dosage compensation, p. 444	genomic imprinting, pp. 441–443